Food Analysis

Fourth Edition

For other titles published in this series, go to
www.springer.com/series/5999

Food Analysis

Fourth Edition

edited by

S. Suzanne Nielsen

Purdue University
West Lafayette, IN, USA

Dr. S. Suzanne Nielsen
Purdue University
Dept. Food Science
745 Agriculture Mall Dr.
West Lafayette IN 47907-2009
USA
nielsens@purdue.edu

ISBN 978-1-4419-1477-4 e-ISBN 978-1-4419-1478-1
DOI 10.1007/978-1-4419-1478-1
Springer New York Dordrecht Heidelberg London

Library of Congress Control Number: 2010924120

Printed on acid-free paper

Springer is part of Springer Science+Business Media (www.springer.com)

Contents

Contributing Authors vii

Preface and Acknowledgments ix

List of Abbreviations xi

Part I. General Information

1. Introduction to Food Analysis 3
S. Suzanne Nielsen

2. United States Government Regulations and International Standards Related to Food Analysis 15
S. Suzanne Nielsen

3. Nutrition Labeling 35
Lloyd E. Metzger

4. Evaluation of Analytical Data 53
J. Scott Smith

5. Sampling and Sample Preparation 69
Rubén O. Morawicki

Part II. Compositional Analysis of Foods

6. Moisture and Total Solids Analysis 85
Robert L. Bradley, Jr.

7. Ash Analysis 105
Maurice R. Marshall

8. Fat Analysis 117
David B. Min and Wayne C. Ellefson

9. Protein Analysis 133
Sam K. C. Chang

10. Carbohydrate Analysis 147
James N. BeMiller

11. Vitamin Analysis 179
Ronald B. Pegg, W.O. Landen, Jr., and Ronald R. Eitenmiller

12. Traditional Methods for Mineral Analysis 201
Robert E. Ward and Charles E. Carpenter

Part III. Chemical Properties and Characteristics of Foods

13. pH and Titratable Acidity 219
George D. Sadler and Patricia A. Murphy

14. Fat Characterization 239
Sean F. O'Keefe and Oscar A. Pike

15. Protein Separation and Characterization Procedures 261
Denise M. Smith

16. Application of Enzymes in Food Analysis 283
Joseph R. Powers

17. Immunoassays 301
Y-H. Peggy Hsieh

18. Analysis of Food Contaminants, Residues, and Chemical Constituents of Concern 317
Baraem Ismail, Bradley L. Reuhs, and S. Suzanne Nielsen

19. Analysis for Extraneous Matter 351
Hulya Dogan, Bhadriraju Subramanyam, and John R. Pedersen

20. Determination of Oxygen Demand 367
Yong D. Hang

Part IV. Spectroscopy

21. Basic Principles of Spectroscopy 375
Michael H. Penner

22. Ultraviolet, Visible, and Fluorescence Spectroscopy 387
Michael H. Penner

23. Infrared Spectroscopy 407
Randy L. Wehling

24. Atomic Absorption Spectroscopy, Atomic Emission Spectroscopy, and Inductively Coupled Plasma-Mass Spectrometry 421
Dennis D. Miller and Michael A. Rutzke

25. Nuclear Magnetic Resonance 443
Bradley L. Reuhs and Senay Simsek

26. Mass Spectrometry 457
J. Scott Smith and Rohan A. Thakur

Part V. Chromatography

27. Basic Principles of Chromatography 473
Baraem Ismail and S. Suzanne Nielsen

28. High-Performance Liquid Chromatography 499
Bradley L. Reuhs and Mary Ann Rounds

29. Gas Chromatography 513
Michael C. Qian, Devin G. Peterson, and Gary A. Reineccius

Part VI. Physical Properties of Foods

30. Rheological Principles for Food Analysis 541
Christopher R. Daubert and E. Allen Foegeding

31. Thermal Analysis 555
Leonard C. Thomas and Shelly J. Schmidt

32. Color Analysis 573
Ronald E. Wrolstad and Daniel E. Smith

Index 587

Contributing Authors

James N. BeMiller
Department of Food Science,
Purdue University,
West Lafayette, IN 47907-1160, USA

Robert L. Bradley, Jr.
Formerly, Department of Food Science,
University of Wisconsin, Madison, WI 53706, USA

Charles E. Carpenter
Department of Nutrition and Food Sciences,
Utah State University, Logan, UT 84322-8700, USA

Sam K.C. Chang
Department of Cereal and Food Sciences,
North Dakota State University,
Fargo, ND 58105, USA

Christopher R. Daubert
Department of Food, Bioprocessing
and Nutritional Sciences,
North Carolina State University,
Raleigh, NC 27695-7624, USA

Hulya Dogan
Department of Grain Science and Industry,
Kansas State University,
Manhattan, KS 66506, USA

Ronald R. Eitenmiller
Department of Food Science and Technology,
The University of Georgia,
Athens, GA 30602-7610, USA

Wayne C. Ellefson
Nutritional Chemistry and Food Safety,
Covance Laboritories, Madison, WI 53714, USA

E. Allen Foegeding
Department of Food, Bioprocessing
and Nutritional Sciences,
North Carolina State University,
Raleigh, NC 27695-7624, USA

Yong D. Hang
Department of Food Science and Technology,
Cornell University, Geneva, NY 14456, USA

Y-H. Peggy Hsieh
Department of Nutrition, Food and Exercise Sciences,
Florida State University,
Tallahassee, FL 32306-1493, USA

Baraem Ismail
Department of Food Science and Nutrition,
University of Minnesota, St. Paul, MN 55108-6099, USA

W.O. Landen, Jr.
Department of Food Science and Technology,
The University of Georgia,
Athens, GA 30602-7610, USA

Maurice R. Marshall
Department of Food Science and Human Nutrition,
University of Florida,
Gainesville, FL 32611-0370, USA

Lloyd E. Metzger
Department of Dairy Science,
University of South Dakota,
Brookings, SD 57007, USA

Dennis D. Miller
Department of Food Science, Cornell University,
Ithaca, NY 14853-7201, USA

David B. Min
Department of Food Science and Technology,
The Ohio State University,
Columbus, OH 43210, USA

Rubén Morawicki
Department of Food Science,
University of Arkansas,
Fayetteville, AR 72703, USA

Patricia A. Murphy
Department of Food Science
and Human Nutrition,
Iowa State University,
Ames, IA 50011, USA

S. Suzanne Nielsen
Department of Food Science,
Purdue University,
West Lafayette, IN 47907-1160, USA

Sean F. O'Keefe
Department of Food Science and Technology,
Virginia Tech, Blacksburg, VA 24061, USA

John R. Pedersen
Formerly, Department of Grain Science and Industry,
Kansas State University,
Manhattan, KS 66506-2201, USA

Ronald B. Pegg
Department of Food Science and Technology,
The University of Georgia,
Athens, GA 30602-7610, USA

Michael H. Penner
Department of Food Science and Technology,
Oregon State University,
Corvallis, OR 97331-6602, USA

Devin G. Peterson
Department of Food Science and Nutrition,
University of Minnesota,
St. Paul, MN 55108-6099, USA

Oscar A. Pike
Department of Nutrition, Dietetics,
and Food Science,
Brigham Young University, Provo, UT 84602, USA

Joseph R. Powers
School of Food Science,
Washington State University,
Pullman, WA 99164-6376, USA

Michael C. Qian
Department of Food Science and Technology,
Oregon State University,
Corvallis, OR 97331-6602, USA

Gary A. Reineccius
Department of Food Science and Nutrition,
University of Minnesota,
St. Paul, MN 55108-6099, USA

Bradley L. Reuhs
Department of Food Science,
Purdue University,
West Lafayette, IN 47907-2009, USA

Mary Ann Rounds (deceased)
Formerly, Department of Physics, Purdue University,
West Lafayette, IN 47907, USA

Michael A. Rutzke
Department of Food Science,
Cornell University,
Ithaca, NY 14853-7201, USA

George D. Sadler
PROVE IT LLC
204 Deerborne ct.
Geneva, IL 60134, USA

Shelly J. Schmidt
Department of Food Science and Human Nutrition,
University of Illinois at Urbana-Champaign,
Urbana, IL 61801, USA

Senay Simsek
Department of Plant Sciences,
North Dakota State University,
Fargo, ND 58108-6050, USA

Daniel E. Smith
Department of Food Science and Technology,
Oregon State University,
Corvallis, OR 97331-6602, USA

Denise M. Smith
Department of Food Science and Technology,
The Ohio State University,
Columbus, OH 43210, USA

J. Scott Smith
Food Science Institute, Kansas State University,
Manhattan, KS 66506-1600, USA

Bhadrirju Subramanyam
Department of Grain Science and Industry,
Kansas State University, Manhattan, KS 66506, USA

Rohan A. Thakur
Taylor Technology, Princeton, NJ 08540, USA

Leonard C. Thomas
DSC Solutions LLC, Smyrna, DE 19977, USA

Robert E. Ward
Department of Nutrition and Food Sciences,
Utah State University,
Logan, UT 84322-8700, USA

Randy L. Wehling
Department of Food Science and Technology,
University of Nebraska,
Lincoln, NE 68583-0919, USA

Ronald E. Wrolstad
Department of Food Science and Technology,
Oregon State University,
Corvallis, OR 97331-6602, USA

Preface and Acknowledgments

The intent of this book is the same as that described in the Preface to the first three editions – a text primarily for undergraduate students majoring in food science, currently studying the analysis of foods. However, comments from users of the first three editions have convinced me that the book is also a valuable text for persons in the food industry who either do food analysis or interact with analysts.

The big focus of this edition was to do a general update, adding many new methods and topics and deleting outdated/unused methods. The following summarizes changes from the third edition: (1) general updates, including addition and deletion of methods, (2) combined two chapters to create one chapter focused on food contaminants, residues, and chemical constituents of concern, (3) some chapters rewritten by new authors (e.g., Immunoassays, Extraneous Matter Analysis, Color Analysis, Thermal Analysis), (4) reorganized some chapters (e.g., Atomic Absorption and Atomic Emission Spectroscopy; Basic Chromatography), (5) added chapter on nuclear magnetic resonance, (6) added calculations for all practice problems, and (7) added table to some chapters to summarize methods (e.g., Vitamin Analysis, HPLC), and (8) newly drawn figures and photographs.

Regrettably, in an effort to keep the book at a manageable size and cost, especially for students, some suggestions by users to add chapters could not be accommodated. For specialized topics (e.g., phytochemicals) that utilize the methods included in this text book, readers are referred to detailed books on those topics.

As stated for the first three editions, the chapters in this textbook are not intended as detailed references, but as general introductions to the topics and the techniques. Course instructors may wish to provide more details on a particular topic to students. The chapters focus on principles and applications of techniques. Procedures given are meant to help explain the principles and give some examples, but are not meant to be presented in the detail adequate to actually conduct a specific analysis. As in the first three editions, all chapters have summaries and study questions, and key words or phrases are in italics type, to help students focus their studies. As done for the third edition, the chapters are organized into the following sections: I. Introduction, II. Compositional Analysis of Foods, III. Chemical Properties and Characteristics of Foods, IV. Spectroscopy, V. Chromatography, and VI. Physical Properties of Foods. Instructors are encouraged to cover the topics from this text in whatever order is most suitable for their course. Also, instructors are invited to contact me for additional teaching materials related to this text book.

Starting with the third edition, the new competency requirements established by the Institute of Food Technologists were considered. Those requirements relevant to food analysis are as follows: (1) understanding the principles behind analytical techniques associated with food, (2) being able to select the appropriate analytical technique when presented with a practical problem, and (3) demonstrating practical proficiency in food analysis laboratory. This textbook should enable instructors to meet the requirements and develop learning objectives relevant to the first two of these requirements. The laboratory manual, now in its second edition, should be a useful resource to help students meet the third requirement.

I am grateful to all chapter authors for agreeing to be a part of this project. Many authors have drawn on their experience of teaching students and/or experience with these analyses to give chapters the appropriate content, relevance, and ease of use. I wish to thank the authors of articles and books, and well as the publishers and industrial companies, for their permission to reproduce materials used here. Special thanks are extended to the following persons: Baraem (Pam) Ismail and Brad Reuhs for valuable discussions about the content of the book and assistance with editing; Jonathan DeVries for input that helped determine content; Brooke Sadler for her graphic art work in draw/redrawing many figures; Gwen Shoemaker for keeping track of all the figures and help on equations; and Kirsti Nielsen (my daughter) for word processing assistance.

S. Suzanne Nielsen

List of Abbreviations

AACC	American Association of Cereal Chemists
AAS	Atomic absorption spectroscopy
ADI	Acceptable daily intake
AE-HPLC	Anion exchange high performance liquid chromatography
AES	Atomic emission spectroscopy
AMS	Accelerator mass spectrometer
AMS	Agricultural Marketing Service
AOAC	Association of Official Analytical Chemists
AOCS	American Oil Chemists' Society
AOM	Active oxygen method
APCI	Atmospheric pressure chemical ionization
APE	Atmosphere-pressure ionization
APHA	American Public Health Association
APPI	Atmospheric pressure photo-ionization
ASE	Accelerated solvent extraction
ASTM	American Society for Testing Materials
ATCC	American Type Culture Collection
ATP	Adenosine-5′-triphosphate
ATR	Attenuated total reflectance
a_w	Water activity
B_0	External magnetic field
BAW	Base and acid washed
BCA	Bicinchoninic acid
BCR	Community Bureau of Reference
Bé	Baumé modulus
BHA	Butylated hydroxyanisole
BHT	Butylated hydroxytoluene
BOD	Biochemical oxygen demand
BPA	Bisphenol A
BSA	Bovine serum albumin
BSDA	*Bacillus stearothermophilis* disk assay
Bt	*Bacillus thuringiensis*
CAST	Calf antibiotic and sulfa test
CCD	Charge-coupled device
CDC	Centers for Disease Control
CFR	Code of Federal Regulations
CFSAN	Center for Food Safety and Applied Nutrition
cGMP	Current Good Manufacturing Practices
CI	Chemical ionization
CI	Confidence interval
CID	Charge injection device
CID	Collision-induced dissociation
CID	Commercial Item Description
CIE	Commission Internationale d'Eclairage
CLA	Conjugated linoleic acid
CLND	Chemiluminescent nitrogen detector
COA	Certificate of analysis
COD	Chemical oxygen demand
C-PER	Protein efficiency ratio calculation method
CPG	Compliance policy guidance
CP-MAS	Cross polarization magic angle spinning
CQC	2,6-Dichloroquinonechloroimide
CRC	Collision reaction cells
CSLM	Confocal scanning laser microscopy
CT	Computed technology
CT	Computed tomography
CV	Coefficient of variation
CVM	Center for Veterinary Medicine
DAL	Defect action level
DDT	Dichlorodiphenyltrichloroethane
DE	Degree of esterification
dE*	Total color difference
DF	Dilution factor
DFE	Dietary folate equivalent
DHHS	Department of Health and Human Services
DMA	Dynamic mechanical analysis
DMD	D-Malate dehydrogenase
DMSO	Dimethyl sulfoxide
DNA	Deoxyribronucleic acid
DNFB	1-Fluoro-2,4-dinitrobenzene
dNTPs	Deoxynucleoside triphosphates
DON	Deoxynivalenol
DRI	Dietary references intake
DRIFTS	Diffuse relectrance Fourier-transform spectroscopy
DRV	Daily Reference Value
DSC	Differential scanning calorimetry
DSHEA	Dietary Supplement Health and Education Act
DSPE	Dispersive solid-phase extraction

DTGS	Deuterated triglycine sulfate
DV	Daily value
DVB	Divinylbenzene
dwb	Dry weight basis
E_a	Activation energy
EAAI	Essential amino acid index
EBT	Eriochrome black T
ECD	Electron capture detector
EDL	Electrode-less discharge lamp
EDS	Energy dispersive spectroscopy
EDTA	Ethylenediaminetetraacetic acid
EEC	European Economic Community
EFSA	European Food Safety Authority
EI	Electron impact
EIE	Easily ionized elements
ELCD	Electrolytic conductivity detector
ELISA	Enzyme linked immunosorbent assay
EPA	Environmental Protection Agency
EPSPS	5-Enolpyruvyl-shikimate-3-phsophate synthase
Eq	Equivalents
ERH	Equilibrium relative humidity
ESI	Electrospray interface
ESI	Electrospray ionization
ETO	Ethylene oxide
EU	European Union
Fab	Fragment antigen binding
FAME	Fatty acid methyl esters
FAO/WHO	Food and Agricultural Organization/World Health Organization
FAS	Ferrous ammonium sulfate
FBs	Fumonisins
Fc	Fragment crystallizable
FCC	Food Chemicals Codex
FDA	Food and Drug Administration
FDAMA	Foods and Drug Administration Modernization Act
FD&C	Food, Drug and Cosmetic
FDNB	1-Fluoro-2,4-dinitrobenzene
FFA	Free fatty acid
FID	Flame ionization detector
FID	Free induction decay
FIFRA	Federal Insecticide, Fungicide, and Rodenticide Act
FNB/NAS	Food and Nutrition Board of the National Academy of Sciences
FOS	Fructooligosaccharide
FPD	Flame photometric detector
FPIA	Fluorescence polarization immunoassay
FSIS	Food Safety and Inspection Service
FT	Fourier transform
FTC	Federal Trade Commission
FT-ICR	Fourier transform – ion cyclotrons
FTIR	Fourier transform infrared
G6PDH	Glucose-6-phosphate dehydrogenase
GATT	General Agreement on Tariffs and Trade
GC	Gas chromatography
GC-AED	Gas chromatography – atomic emission detector
GC-FTIR	Gas chromatography – Fourier transform infrared
GC-MS	Gas chromatography – mass spectrometry
GFC	Gel-filtration chromatography
GIPSA	Grain Inspection, Packers and Stockyard Administration
GLC	Gas–liquid chromatography
GMA	Grocery Manufacturers of America
GMO	Genetically modified organism
GMP	Good Manufacturing Practices (also Current Good Manufacturing Practice in Manufacturing, Packing, or Holding Human Food)
GOPOD	Glucose oxidase/peroxidase
GPC	Gel-permeation chromatography
GRAS	Generally recognized as safe
HACCP	Hazard Analysis Critical Control Point
HCL	Hollow cathode lamp
HETP	Height equivalent to a theoretical plate
HFS	High fructose syrup
HILIC	Hydrophilic interaction liquid chromatography
HK	Hexokinase
H-MAS	High-resolution magic angle spinning
HMDS	Hexamethyldisilazane
HPLC	High performance liquid chromatography
HPTLC	High performance thin-layer chromatography
HRGC	High resolution gas chromatography
HS	Headspace
HVP	Hydrolyzed vegetable protein
IC	Ion chromatography
IC_{50}	Median inhibition concentration
ICP	Inductively coupled plasma
ICP-AES	Inductively coupled plasma – atomic emission spectroscopy
ICP-MS	Inductively coupled plasma – mass spectrometer
ID	Inner diameter
IDK	Insect damaged kernels
Ig	Immunoglobulin
IgE	Immunoglobulin E
IgG	Immunoglobulin G
IMS	Interstate Milk Shippers
InGaAs	Indium–gallium–arsenide
IR	Infrared

IRMM	Institute for Reference Materials and Measurements
ISA	Ionic strength adjustor
ISE	Ion-selective electrode
ISO	International Organization for Standardization
ITD	Ion-trap detector
IT-MS	Ion traps mass spectrometry
IU	International Units
IUPAC	International Union of Pure and Applied Chemistry
JECFA	Joint FAO/WHO Expert Committee on Food Additives
kcal	Kilocalorie
KDa	Kilodalton
KFR	Karl Fischer reagent
KFReq	Karl Fischer reagent water equivalence
KHP	Potassium acid phthalate
LALLS	Low-angle laser light scattering
LC	Liquid chromatography
LC-MS	Liquid chromatography – mass spectroscopy
LFS	Lateral flow strip
LIMS	Laboratory information management system
LOD	Limit of detection
LOQ	Limit of quantitation
LTM	Low thermal mass
LTP	Low-temperature plasma probe
3-MCPD	3-Monochloropropane 1,2-diol
MALDI-TOF	Matrix-assisted laser desorption time-of-flight
MALLS	Multi-angle laser light scattering
MAS	Magic angle spinning
MASE	Microwave-assisted solvent extraction
MCL	Maximum contaminant level
MCT	Mercury:cadmium:telluride
MDGC	Multidimensional gas chromatography
MDL	Method detection limit
MDSC™	Modulated differential scanning calorimeter™
mEq	Milliequivalents
MES-TRIS	2-(*N*-morpholino)ethanesulfonic acid-tris(hydroxymethyl)aminomethane
MLR	Multiple linear regression
MRI	Magnetic resonance imaging
MRL	Maximum residue level
MRM	Multiresidue method
MS	Mass spectrometry (or spectrometer)
MS/MS	Tandem MS
Ms^n	Multiple stages of mass spectrometry
MW	Molecular weight
m/z	Mass-to-charge ratio
NAD	Nicotinamide-adenine dinucleotide
NADP	Nicotinamide-adenine dinucleotide phosphate
NADPH	Reduced NADP
NCM	*N*-methyl carbamate
NCWM	National Conference on Weights and Measures
NIR	Near-infrared
NIRS	Near-infrared spectroscopy
NIST	National Institute of Standards and Technology
NLEA	Nutrition Labeling and Education Act
NMFS	National Marine Fisheries Service
NMR	Nuclear magnetic resonance
NOAA	National Oceanic and Atmospheric Administration
NOAEL	No observed adverse effect level
NPD	Nitrogen phosphorus detector or thermionic detector
NSSP	National Shellfish Sanitation Program
NVOC	Non-volatile organic compounds
OC	Organochlorine
OD	Outer diameter
ODS	Octadecylsilyl
OES	Optical emission spectroscopy
OMA	Official Methods of Analysis
OP	Organophosphate/organophosphorus
OPA	*O*-phthalaldehyde
OSI	Oil stability index
OT	Orbitrap
OTA	Ochratoxin A
PAD	Pulsed-amperometric detector
PAGE	Polyacrylamide gel electrophoresis
PAM I	Pesticide Analytical Manual, Volume I
PAM II	Pesticide Analytical Manual, Volume II
P_c	Critical pressure
PCBs	Polychlorinated biphenyls
PCR	Polymerase chain reaction
PCR	Principal components regression
PDCAAS	Protein digestibility – corrected amino acid score
PDMS	Polydimethylsiloxane
PEEK	Polyether ether ketone
PER	Protein efficiency ratio
PFPD	Pulsed flame photometric detector
pI	Isoelectric point
PID	Photoionization detector
PLE	Pressurized liquid extraction
PLOT	Porous-layer open tabular
PLS	Partial least squares
PMO	Pasteurized Milk Ordinance
PMT	Photomultiplier tube
ppb	Parts per billion
PPD	Purchase Product Description

ppm Parts per million
ppt Parts per trillion
PUFA Polyunsaturated fatty acids
PVPP Polyvinylpolypyrrolidone
qMS Quadrupole mass spectrometry
QqQ Triple quadrupole
Q-trap Quadruple-ion trap
QuEChERS Quick, Easy, Cheap, Effective, Rugged and Safe
RAC Raw agricultural commodity
RAE Retinol activity equivalents
RASFF Rapid Alert System for Food and Feed
RDA Recommended Daily Allowance
RDI Reference Daily Intake
RE Retinol equivalent
RF Radiofrequency
R_f Relative mobility
RF Response factor
RI Refractive index
RIA Radioimmunoassay
ROSA Rapid One Step Assay
RPAR Rebuttable Presumption Against Registration
RVA RapidViscoAnalyser
SASO Saudi Arabian Standards Organization
SBSE Stir bar sorptive extraction
SD Standard deviation
SDS Sodium dodecyl sulfate
SDS-PAGE Sodium dodecyl sulfate – polyacrylamide gel electrophoresis
SEC Size-exclusion chromatography
SEM Scanning electron microscopy
SFC Solid fat content
SFC Supercritical-fluid chromatography
SFC-MS Supercritical-fluid chromatography mass spectrometry
SFE Supercritical fluid extraction
SFE-GC Supercritical fluid extraction – gas chromatography
SFI Solid fat index
SI International Scientific
SKCS Single kernel characteristics system
SMEDP Standard Methods for the Examination of Dairy Products
SO Sulfite oxidase
SPDE Solid-phase dynamic extraction
SPE Solid-phase extraction
SPME Solid-phase microextraction
SRF Sample response factor
SRM Single residue method
SSD Solid state detector
STOP Swab test on premises
SVOC Semi-volatile organic compounds
TBA Thiobarbituric acid
TBARS TBA reactive substances
TCD Thermal conductivity detector
TDA Total daily intake
TDF Total dietary fiber
T-DNA Transfer of DNA
TEM Transmission electron microscopies
TEMED Tetramethylethylenediamine
Tg Glass transition temperature
TGA Thermogravimetric analysis
Ti Tumor-inducing
TIC Total ion current
TLC Thin-layer chromatography
TMA Thermomechanical analysis
TMCS Trimethylchlorosilane
TMS Trimethylsilyl
TOF Time-of-flight
TOF-MS Time-of-flight mass spectrometry
TPA Texture profile analysis
TS Total solids
TSQ Triple stage quadrupole
TSS Total soluble solids
TSUSA Tariff Schedules of the United States of America
TTB Alcohol and Tobacco Tax and Trade Bureau
TWI Total weekly intake
UHPC Ultra-high pressure chromatography
UHPLC Ultra-high performance liquid chromatography
US United States
USA United States of America
US RDA United States Recommended Dietary Allowance
USCS United States Customs Service
USDA United States Department of Agriculture
USDC United States Department of Commerce
USP United States Pharmacopeia
UV Ultraviolet
UV–Vis Ultraviolet–visible
Vis Visible
VOC Volatile organic compounds
wt Weight
wwb Wet weight basis
XMT X-ray microtomography
ZEA Zearalenone

General Information

Introduction to Food Analysis

S. Suzanne Nielsen
Department of Food Science, Purdue University, West Lafayette, IN 47907-2009, USA
nielsens@purdue.edu

1.1 Introduction 5
1.2 Trends and Demands 5
  1.2.1 Consumers 5
  1.2.2 Food Industry 5
  1.2.3 Government Regulations and International Standards and Policies 6
1.3 Types of Samples Analyzed 6
1.4 Steps in Analysis 6
  1.4.1 Select and Prepare Sample 6
  1.4.2 Perform the Assay 7
  1.4.3 Calculate and Interpret the Results 7
1.5 Choice and Validity of Method 7
  1.5.1 Characteristics of the Method 7
  1.5.2 Objective of the Assay 7
  1.5.3 Consideration of Food Composition and Characteristics 8
  1.5.4 Validity of the Method 9
1.6 Official Methods 10
  1.6.1 AOAC International 10
  1.6.2 Other Endorsed Methods 11
1.7 Summary 12
1.8 Study Questions 12
1.9 Ackowledgments 13
1.10 References 13
1.11 Relevant Internet Addresses 13

S.S. Nielsen, *Food Analysis*, Food Science Texts Series, DOI 10.1007/978-1-4419-1478-1_1,

1.1 INTRODUCTION

Investigations in food science and technology, whether by the food industry, governmental agencies, or universities, often require determination of food composition and characteristics. Trends and demands of consumers, the food industry, and national and international regulations challenge food scientists as they work to monitor food composition and to ensure the quality and safety of the food supply. All food products require analysis as part of a **quality management** program throughout the development process (including raw ingredients), through production, and after a product is in the market. In addition, analysis is done of problem samples and competitor products. The characteristics of foods (i.e., chemical composition, physical properties, sensory properties) are used to answer specific questions for regulatory purposes and typical quality control. The nature of the sample and the specific reason for the analysis commonly dictate the choice of analytical methods. **Speed, precision, accuracy**, and **ruggedness** often are key factors in this choice. **Validation** of the method for the specific **food matrix** being analyzed is necessary to ensure usefulness of the method. Making an appropriate choice of the analytical technique for a specific application requires a good knowledge of the various techniques (Fig. 1-1). For example, your choice of method to determine the salt content of potato chips would be different if it is for nutrition labeling than for quality control. The success of any analytical method relies on the proper selection and preparation of the food sample, carefully performing the analysis, and doing the appropriate calculations and interpretation of the data. Methods of analysis developed and endorsed by several nonprofit scientific organizations allow for standardized comparisons of results between different laboratories and for evaluation of less standard procedures. Such **official methods** are critical in the analysis of foods, to ensure that they meet the legal requirements established by governmental agencies. **Government regulations** and **international standards** most relevant to the analysis of foods are mentioned here but covered in more detail in Chap. 2, and nutrition labeling regulations in the USA are covered in Chap. 3. Internet addresses for many of the organizations and government agencies discussed are given at the end of this chapter.

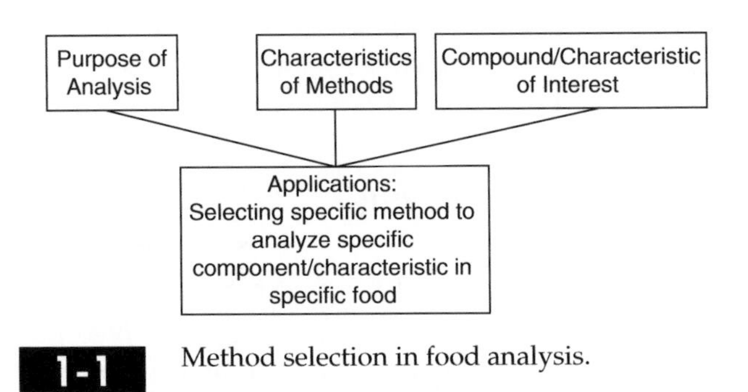

1-1 figure Method selection in food analysis.

1.2 TRENDS AND DEMANDS

1.2.1 Consumers

Consumers have many choices regarding their food supply, so they can be very selective about the products they purchase. They demand a wide variety of products that are of high quality, nutritious, and offer a good value. Also, consumers are concerned about the safety of foods, which has increased the testing of foods for allergens, pesticide residues, and products from genetic modification of food materials. Many consumers are interested in the relationship between diet and health, so they utilize nutrient content and health claim information from food labels to make purchase choices. These factors create a challenge for the food industry and for its employees. For example, the demand for foods with lower fat content has challenged food scientists to develop food products that contain fat content claims (e.g., free, low, reduced) and certain health claims (e.g., the link between dietary fat and cancer; dietary saturated fat and cholesterol and risk of coronary heart disease). Analytical methods to determine and characterize fat content provide the data necessary to justify these statements and claims. Use of fat substitutes in product formulations makes possible many of the lower fat foods, but these fat substitutes can create challenges in the accurate measurement of fat content (1). Likewise, there has been growing interest in functional foods that may provide health benefits beyond basic nutrition. However, such foods present some unique challenges regarding analytical techniques and in some cases questions of how these components affect the measurement of other nutrients in the food (2).

1.2.2 Food Industry

To compete in the marketplace, food companies must produce foods that meet the demands of consumers as described previously. Management of product quality by the food industry is of increasing importance, beginning with the raw ingredients and extending to the final product eaten by the consumer. Analytical methods must be applied across the entire food supply chain to achieve the desired final product quality. Downsizing in response to increasing competition in the food industry often has pushed the responsibility for ingredient quality to the suppliers. Companies increasingly rely on others to supply high-quality and safe raw ingredients and packaging materials. Many companies have **select suppliers**, on whom they rely to perform the analytical tests to ensure compliance with detailed specifications for ingredients/raw materials. These specifications, and the associated tests, target various chemical, physical, and microbiological properties. Results of these analytical tests related to the predetermined specifications are delivered as a **Certificate of Analysis** (COA) with the ingredient/raw material. Companies must have in place a means to maintain control of these COAs and react to them. With careful control over the quality of raw ingredients/materials, less testing is required during processing and on the final product.

In some cases, the cost of goods is linked directly to the composition as determined by analytical tests. For example, in the dairy field, butterfat content of bulk tank raw milk determines how much money the milk producer is paid for the milk. For flour, the protein content can determine the price and food application for the flour. These examples point to the importance for accurate results from analytical testing.

Traditional quality control and quality assurance concepts are only a portion of a comprehensive quality management system. Food industry employees responsible for quality management work together in teams with other individuals in the company responsible for product development, production, engineering, maintenance, purchasing, marketing, and regulatory and consumer affairs.

Analytical information must be obtained, assessed, and integrated with other relevant information about the food system to address quality-related problems. Making appropriate decisions depends on having knowledge of the analytical methods and equipment utilized to obtain the data related to the quality characteristics. To design experiments in product and process development, and to assess results, one must know the operating principles and capabilities of the analytical methods. Upon completion of these experiments, one must critically evaluate the analytical data collected to determine whether product reformulation is needed or what parts of the process need to be modified for future tests. The situation is similar in the research laboratory, where knowledge of analytical techniques is necessary to design experiments, and the evaluation of data obtained determines the next set of experiments to be conducted.

1.2.3 Government Regulations and International Standards and Policies

To market safe, high-quality foods effectively in a national and global marketplace, food companies must pay increasing attention to government regulations and guidelines and to the policies and standards of international organizations. Food scientists must be aware of these regulations, guidelines, and policies related to food safety and quality and must know the implications for food analysis. Government regulations and guidelines in the USA relevant to food analysis include **nutrition labeling regulations** (Chap. 3), **mandatory and voluntary standards** (Chap. 2), **Good Manufacturing Practice** (GMP) regulations (now called Current Good Manufacturing Practice in Manufacturing Packing, or Holding Human Food) (Chap. 2), and **Hazard Analysis Critical Control Point** (HACCP) systems (Chap. 2). The HACCP system is highly demanded of food companies by auditing firms and customers. The HACCP concept has been adopted not only by the US **Food and Drug Administration** (FDA) and other federal agencies in the USA, but also by the **Codex Alimentarius Commission**, an international organization that has become a major force in world food trade. Codex is described in Chap. 2, along with other organizations active in developing international standards and safety practices relevant to food analysis that affect the import and export of raw agricultural commodities and processed food products.

1.3 TYPES OF SAMPLES ANALYZED

Chemical analysis of foods is an important part of a quality assurance program in food processing, from ingredients and raw materials, through processing, to the finished products (3–7). Chemical analysis also is important in formulating and developing new products, evaluating new processes for making food products, and identifying the source of problems with unacceptable products (Table 1-1). For each type of product to be analyzed, it may be necessary to determine either just one or many components. The nature of the sample and the way in which the information obtained will be used may dictate the specific method of analysis. For example, process control samples are usually analyzed by rapid methods, whereas nutritive value information for **nutrition labeling** generally requires the use of more time-consuming methods of analysis endorsed by scientific organizations. Critical questions, including those listed in Table 1-1, can be answered by analyzing various types of samples in a food processing system.

1.4 STEPS IN ANALYSIS

1.4.1 Select and Prepare Sample

In analyzing food samples of the types described previously, all results depend on obtaining a representative sample and converting the sample to a form

Types of Samples Analyzed in a Quality Assurance Program for Food Products

Sample Type	*Critical Questions*
Raw materials	Do they meet your specifications? Do they meet required legal specifications? Are they safe and authentic? Will a processing parameter have to be modified because of any change in the composition of raw materials? Are the quality and composition the same as for previous deliveries? How does the material from a potential new supplier compare to that from the current supplier?
Process control samples	Did a specific processing step result in a product of acceptable composition or characteristics? Does a further processing step need to be modified to obtain a final product of acceptable quality?
Finished product	Does it meet the legal requirements? What is the nutritive value, so that label information can be developed? Or is the nutritive value as specified on an existing label? Does it meet product claim requirements (e.g., "low fat")? Will it be acceptable to the consumer? Will it have the appropriate shelf life? If unacceptable and cannot be salvaged, how do you handle it (trash? rework? seconds?)
Competitor's sample	What are its composition and characteristics? How can we use this information to develop new products?
Complaint sample	How do the composition and characteristics of a complaint sample submitted by a customer differ from a sample with no problems?

Adapted and updated from (8, 9).

that can be analyzed. Neither of these is as easy as it sounds! **Sampling** and **sample preparation** are covered in detail in Chap. 5.

Sampling is the initial point for sample identification. Analytical laboratories must keep track of incoming samples and be able to store the analytical data from the analyses. This analytical information often is stored on a **laboratory information management system**, or LIMS, which is a computer database program.

1.4.2 Perform the Assay

Performing the assay is unique for each component or characteristic to be analyzed and may be unique to a specific type of food product. Single chapters in this book address sampling and sample preparation (Chap. 5) and data handling (Chap. 4), while the remainder of the book addresses the step of actually performing the assay. The descriptions of the various specific procedures are meant to be overviews of the methods. For guidance in actually performing the assays, details regarding chemicals, reagents, apparatus, and step-by-step instructions are found in the books and articles referenced in each chapter. Numerous chapters in this book, and other recent books devoted to food analysis (10–14), make the point that for food analysis we increasingly rely on expensive equipment, some of which requires considerable expertise. Also, it should be noted that numerous analytical methods utilize automated instrumentation, including autosamplers and robotics to speed the analyses.

1.4.3 Calculate and Interpret the Results

To make decisions and take action based on the results obtained from performing the assay that determined the composition or characteristics of a food product, one must make the appropriate calculations to interpret the data correctly. **Data handling**, covered in Chap. 4, includes important statistical principles.

1.5 CHOICE AND VALIDITY OF METHOD

1.5.1 Characteristics of the Method

Numerous methods often are available to assay food samples for a specific characteristic or component. To select or modify methods used to determine the chemical composition and characteristics of foods, one must be familiar with the principles underlying the procedures and the critical steps. Certain properties of methods and criteria described in Table 1-2 are useful to evaluate the appropriateness of a method in current use or a new method being considered.

1.5.2 Objective of the Assay

Selection of a method depends largely on the objective of the measurement. For example, methods used for rapid online processing measurements may be less accurate than official methods (see Sect. 1.6) used for nutritional labeling purposes. Methods referred to as reference, definitive, official, or primary are most applicable in a well-equipped and staffed analytical lab. The more rapid secondary or field methods may be more applicable on the manufacturing floor in a food processing facility. For example, refractive index may be used as a rapid, secondary method for sugar

table 1-2 Criteria for Choice of Food Analysis Methods

Characteristic	*Critical Questions*
Inherent properties	
Specificity/selectivity	Is the property being measured the same as that claimed to be measured, and is it the only property being measured? Are there interferences? What steps are being taken to ensure a high degree of specificity?
Precision	What is the precision of the method? Is there within-batch, batch-to-batch, or day-to-day variation? What step in the procedure contributes the greatest variability?
Accuracy	How does the new method compare in accuracy to the old or a standard method? What is the percent recovery?
Applicability of method to laboratory	
Sample size	How much sample is needed? Is it too large or too small to fit your needs? Does it fit your equipment and/or glassware? Can you obtain representative sample?[a]
Reagents	Can you properly prepare them? What equipment is needed? Are they stable? For how long and under what conditions?
Equipment	Is the method very sensitive to slight or moderate changes in the reagents? Do you have the appropriate equipment? Are personnel competent to operate equipment?
Cost	What is the cost in terms of equipment, reagents, and personnel?
Usefulness	
Time required	How fast is it? How fast does it need to be?
Reliability	How reliable is it from the standpoints of precision and stability?
Need	Does it meet a need or better meet a need?
Personnel	Is any change in method worth the trouble of the change?
Safety	Are special precautions necessary?
Procedures	Who will prepare the written description of the procedures and reagents? Who will do any required calculations?

[a]In-process samples may not accurately represent finished product; Must understand what variation can and should be present.

analysis (see Chaps. 6 and 10), with results correlated to those of the primary method, **high-performance liquid chromatography** (HPLC) (see Chaps. 10 and 28). Moisture content data for a product being developed in the pilot plant may be obtained quickly with a moisture balance unit that has been calibrated using a more time-consuming hot air oven method (see Chap. 6). Many companies commonly use unofficial, rapid methods, but validate them against official methods.

1.5.3 Consideration of Food Composition and Characteristics

Proximate analysis of foods refers to determining the major components of moisture (Chap. 6), ash (total minerals) (Chap. 7), lipids (Chap. 8), protein (Chap. 9), and carbohydrates (Chap. 10). The performance of many analytical methods is affected by the **food matrix** (i.e., its major chemical components, especially lipid, protein, and carbohydrate). In food analysis, it is usually the food matrix that presents the greatest challenge to the analyst (15). For example, high-fat or high-sugar foods can cause different types of interferences than low-fat or low-sugar foods. Digestion procedures and extraction steps necessary for accurate analytical results can be very dependent on the food matrix. The complexity of various food systems often requires having not just one technique available for a specific food component, but multiple techniques and procedures, as well as the knowledge about which to apply to a specific food matrix.

A task force of **AOAC International**, formerly known as the **Association of Official Analytical Chemists** (AOAC), suggested a "triangle scheme" for dividing foods into matrix categories (16–20) (Fig. 1-2). The apexes of the triangle contain food groups that were either 100% fat, 100% protein, or 100% carbohydrate. Foods were rated as "high," "low," or "medium" based on levels of fat, carbohydrate, and proteins, which are the three nutrients expected to

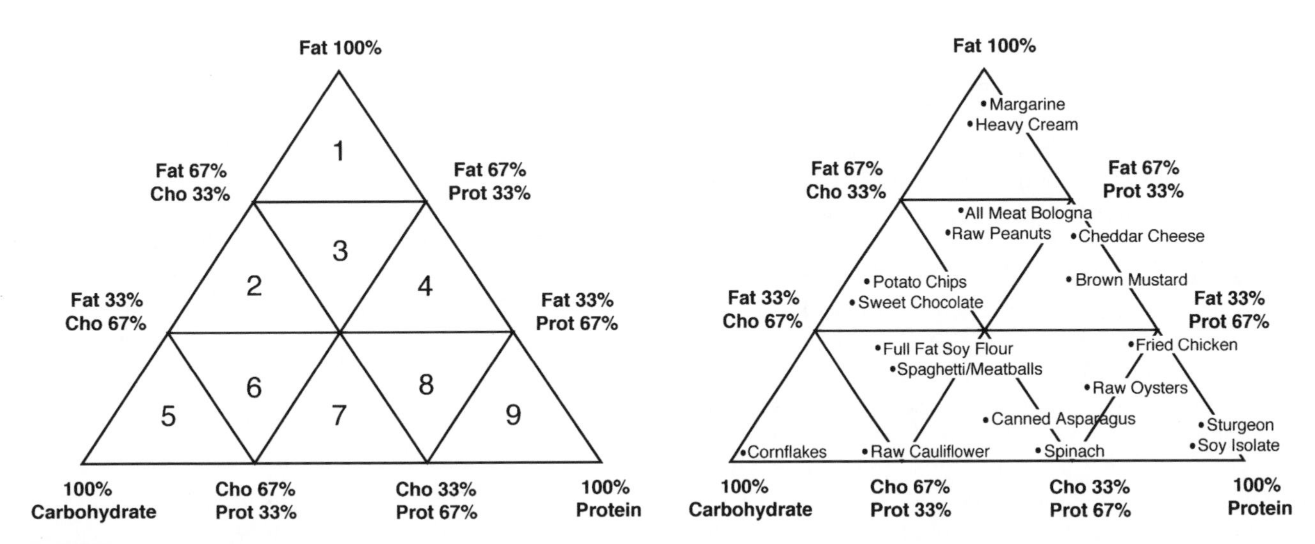

1-2 figure Schematic layout of food matrixes based on protein, fat, and carbohydrate content, excluding moisture and ash. [Reprinted with permission from (20), *Inside Laboratory Management*, September 1997, p. 33. Copyright 1997, by AOAC International.]

have the strongest effect on analytical method performance. This created nine possible combinations of high, medium, and low levels of fat, carbohydrate, and protein. Complex foods were positioned spatially in the triangle according to their content of fat, carbohydrate, and protein, on a normalized basis (i.e., fat, carbohydrate, and protein normalized to total 100%). General analytical methods ideally would be geared to handle each of the nine combinations, replacing more numerous matrix-dependent methods developed for specific foods. For example, using matrix-dependent methods, one method might be applied to potato chips and chocolates, both of which are low-protein, medium-fat, medium-carbohydrate foods, but another might be required for a high-protein, low-fat, high-carbohydrate food such as nonfat dry milk (17). In contrast, a robust general method could be used for all of the food types. The **AACC International**, formerly known as the **American Association of Cereal Chemists** (AACC), has approved a method studied using this approach (18).

1.5.4 Validity of the Method

Numerous factors affect the usefulness and validity of the data obtained using a specific analytical method. One must consider certain characteristics of any method, such as specificity, precision, accuracy, and sensitivity (see Table 1-2 and Chap. 4). However, one also must consider how the variability of data from the method for a specific characteristic compares to differences detectable and acceptable to a consumer, and the variability of the specific characteristic inherent in processing of the food. One must consider the nature of the samples collected for the analysis, how representative the samples were of the whole, and the number of samples analyzed (Chap. 5). One must ask whether details of the analytical procedure were followed adequately, such that the results are accurate, repeatable, and comparable to data collected previously. For data to be valid, equipment to conduct the analysis must be standardized and appropriately used, and the performance limitations of the equipment must be recognized.

A major consideration for determining method validity is the analysis of materials used as controls, often referred to as **standard reference materials** or **check samples** (21). Analyzing check samples concurrently with test samples is an important part of quality control (22). Standard reference materials can be obtained in the USA from the National Institute of Standards and Technology (NIST) and from US Pharmacopeia, in Canada from the Center for Land and Biological Resource Research, in Europe from the Institute for Reference Materials and Measurements (IRMM), and in Belgium from the Community Bureau of Reference (BCR). Besides government-related groups, numerous organizations offer check sample services that provide test samples to evaluate the reliability of a method (21). For example, AACC International has a check sample service in which a subscribing laboratory receives specifically prepared test samples from AACC International. The subscribing laboratory performs the specified analyses on the samples and returns the results to AACC International. The AACC International then provides a statistical evaluation of the analytical results and compares the subscribing laboratory's data with those of other laboratories to inform the subscribing laboratory of its degree of accuracy. The AACC International offers check samples such as flours, semolina, and other cereal-based samples, for analyses such as moisture, ash, protein, vitamins, minerals, sugars, sodium, total dietary fiber, soluble and insoluble dietary fiber, and β-glucan. Samples also are available for testing physical properties and for microbiological and sanitation analyses.

The **American Oil Chemists' Society** (AOCS) has a reference sample program for oilseeds, oilseed meals, marine oils, aflatoxin, cholesterol, trace metals, specialty oils suitable for determination of *trans* fatty acids, and formulations for nutritional labeling. Laboratories from many countries participate in the program to check the accuracy of their work, their reagents, and their laboratory apparatus against the statistical norm derived from the group data.

Standard reference materials are important tools to ensure reliable data. However, such materials need not necessarily be obtained from outside organizations. Control samples internal to the laboratory can be prepared by carefully selecting an appropriate type of sample, gathering a large quantity of the material, mixing and preparing to ensure homogeneity, packaging the sample in small quantities, storing the samples appropriately, and routinely analyzing the control sample when test samples are analyzed. Whatever the standard reference materials used, these should match closely the matrix of the samples to be analyzed by a specific method. AOAC International has begun a peer-review program of matching reference materials with respective official methods of analysis.

1.6 OFFICIAL METHODS

The choice of method for a specific characteristic or component of a food sample is often made easier by the availability of **official methods**. Several nonprofit scientific organizations have compiled and published these methods of analysis for food products, which have been carefully developed and standardized. They allow for comparability of results between different laboratories that follow the same procedure and for evaluating results obtained using new or more rapid procedures.

1.6.1 AOAC International

AOAC International is an organization begun in 1884 to serve the analytical methods needs of government regulatory and research agencies. The goal of AOAC International is to provide methods that will be fit for their intended purpose (i.e., will perform with the necessary accuracy and precision under usual laboratory conditions).

This volunteer organization functions as follows:

1. Methods of analysis from published literature are selected or new methods are developed by AOAC International volunteers.
2. Methods are collaboratively tested using multilaboratory studies in volunteers' laboratories.
3. Methods are given a multilevel peer review by expert scientists, and if found acceptable, they are adopted as official methods of analysis.
4. Adopted methods are published in the *Official Methods of Analysis*, which covers a wide variety of assays related to foods, drugs, cosmetics, agriculture, forensic science, and products affecting public health and welfare.
5. AOAC International publishes manuals, methods compilations in specific areas of analysis, monographs, and the monthly magazine *Inside Laboratory Management*.
6. AOAC International conducts training courses of interest to analytical scientists and other laboratory personnel.

Methods validated and adopted by AOAC International and the data supporting the method validation are published in the *Journal of AOAC International*. Such methods must be successfully validated in a formal interlaboratory collaborative study before being accepted as an official first action method by AOAC International. Details of the validation program (e.g., number of laboratories involved, samples per level of analyte, controls, control samples, and the review process) are given in the front matter of the AOAC International's *Official Methods of Analysis*. First action methods are subject to scrutiny and general testing by other scientists and analysts for some time period before final action adoption. Adopted first action and final action methods for many years were compiled in books published and updated every 4–5 years as the *Official Methods of Analysis* (23) of AOAC International. In 2007, AOAC International created an online version of the book as a "continuous edition," with new and revised methods posted as soon as they are approved. The *Official Methods of Analysis* of AOAC International includes methods appropriate for a wide variety of products and other materials (Table 1-3). These methods often are specified by the FDA with regard to legal requirements for food products. They are generally the methods followed by the FDA and the **Food Safety and Inspection Service** (FSIS) of the **United States Department of Agriculture** (USDA) to check the nutritional labeling information on foods and to check foods for the presence or absence of undesirable residues or residue levels.

Table of Contents of 2007 Official Methods of Analysis of AOAC International (23)

Chapter	Title
1	Agriculture liming materials
2	Fertilizers
3	Plants
4	Animal feed
5	Drugs in feeds
6	Disinfectants
7	Pesticide formulations
8	Hazardous substances
9	Metals and other elements at trace levels in foods
10	Pesticide and industrial chemical residues
11	Waters; and salt
12	Microchemical methods
13	Radioactivity
14	Veterinary analytical toxicology
15	Cosmetics
16	Extraneous materials: isolation
17	Microbiological methods
18	Drugs: Part I
19	Drugs: Part II
20	Drugs: Part III
21	Drugs: Part IV
22	Drugs: Part V
23	Drugs and feed additives in animal tissues
24	Forensic sciences
25	Baking powders and baking chemicals
26	Distilled liquors
27	Malt beverages and brewing materials
28	Wines
29	Nonalcoholic beverages and concentrates
30	Coffee and tea
31	Cacao bean and its products
32	Cereal foods
33	Dairy products
34	Eggs and egg products
35	Fish and other marine products
36	Flavors
37	Fruits and fruit products
38	Gelatin, dessert preparations, and mixes
39	Meat and meat products
40	Nuts and nut products
41	Oils and fats
42	Vegetable products, processed
43	Spices and other condiments
44	Sugars and sugar products
45	Vitamins and other nutrients
46	Color additives
47	Food additives: Direct
48	Food additives: Indirect
49	Natural toxins
50	Infant formulas, baby foods, and enteral products
51	Dietary supplements

1.6.2 Other Endorsed Methods

The AACC International publishes a set of approved laboratory methods, applicable mostly to cereal products (e.g., baking quality, gluten, physical dough tests, staleness/texture). The AACC International process of adopting the *Approved Methods of Analysis* (24) is consistent with the process used by the AOAC International and AOCS. Approved methods of the AACC International are continuously reviewed, critiqued, and updated (Table 1-4), and are now available online.

The AOCS publishes a set of official methods and recommended practices, applicable mostly to fat and oil analysis (e.g., vegetable oils, glycerin, lecithin) (25) (Table 1-5). AOCS is a widely used methodology source on the subjects of edible fats and oils, oilseeds

1-4 table **Table of Contents of 2010 Approved Methods of AACC International (24)**

Chapter	Title
2	Acidity
4	Acids
6	Admixture of flours
7	Amino acids
8	Total ash
10	Baking quality
11	Biotechnology
12	Carbon dioxide
14	Color and pigments
20	Ingredients
22	Enzymes
26	Experimental milling
28	Extraneous matter
30	Crude fat
32	Fiber
33	Sensory analysis
38	Gluten
39	Infrared analysis
40	Inorganic constituents
42	Microorganisms
44	Moisture
45	Mycotoxins
46	Nitrogen
48	Oxidizing, bleaching, and maturing agents
54	Physical dough tests
55	Physical tests
56	Physicochemical tests
58	Special properties of fats, oils, and shortenings
61	Rice
62	Preparation of sample
64	Sampling
66	Semolina, pasta, and noodle quality
68	Solutions
74	Staleness/texture
76	Starch
78	Statistical principles
80	Sugars
82	Tables
86	Vitamins
89	Yeast

Table 1-5 Table of Contents of 2009 Official Methods and Recommended Practices of the American Oil Chemists' Society (25)

Section	Title
A	Vegetable oil source materials
B	Oilseed by-products
C	Commerical fats and oils
D	Soap and synthetic detergents
E	Glycerin
F	Sulfonated and sulfated oils
G	Soapstocks
H	Specifications for reagents, and solvents and apparatus
J	Lecithin
M	Evaluation and design of test methods
S	Official listings
T	Recommended practices for testing industrial oils and derivatives

Table 1-6 Contents of Chap. 15 on Chemical and Physical Methods in Standard Methods for the Examination of Dairy Products (26)

15.010	Introduction
15.020	Acidity tests
15.030	Adulterant tests
15.040	Ash tests
15.050	Chloride tests
15.060	Contaminant tests
15.070	Extraneous material tests
15.080	Fat determination methods
15.090	Lactose and galactose tests
15.100	Minerals and food additives
15.110	Moisture and solids tests
15.120	Multicomponent tests
15.130	Protein/nitrogen tests
15.140	Rancidity tests
15.150	Sanitizer tests
15.160	Vitamins A, D2, and D3 in milk products, HPLC method
15.170	Functional tests
15.180	Cited references

and oilseed proteins, soaps and synthetic detergents, industrial fats and oils, fatty acids, oleochemicals, glycerin, and lecithin.

Standard Methods for the Examination of Dairy Products (26), published by the American Public Health Association, includes methods for the chemical analysis of milk and dairy products (e.g., acidity, fat, lactose, moisture/solids, added water) (Table 1-6). *Standard Methods for the Examination of Water and Wastewater* (27) is published jointly by the American Public Health Association, American Water Works Association, and the Water Environment Federation. *Food Chemicals Codex* (28), published by US Pharmacopeia, contains methods for the analysis of certain food additives. Some trade associations publish standard methods for the analysis of their specific products.

1.7 SUMMARY

Food scientists and technologists determine the chemical composition and physical characteristics of foods routinely as part of their quality management, product development, or research activities. For example, the types of samples analyzed in a quality management program of a food company can include raw materials, process control samples, finished products, competitors' samples, and consumer complaint samples. Consumer, food industry, and government concern for food quality and safety has increased the importance of analyses that determine composition and critical product characteristics.

To successfully base decisions on results of any analysis, one must correctly conduct all three major steps in the analysis: (1) select and prepare samples, (2) perform the assay, and (3) calculate and interpret the results. The choice of analysis method is usually based on the objective of the analysis, characteristics of the method itself (e.g., specificity, accuracy, precision, speed, cost of equipment, and training of personnel), and the food matrix involved. Validation of the method is important, as is the use of standard reference materials to ensure quality results. Rapid methods used for quality assessment in a production facility may be less accurate but much faster than official methods used for nutritional labeling. Endorsed methods for the chemical analyses of foods have been compiled and published by AOAC International, AACC International, AOCS, and certain other nonprofit scientific organizations. These methods allow for comparison of results between different laboratories and for evaluation of new or more rapid procedures.

1.8 STUDY QUESTIONS

1. Identify six reasons you might need to determine certain chemical characteristics of a food product as part of a quality management program.
2. You are considering the use of a new method to measure Compound X in your food product. List six factors you will consider before adopting this new method in your quality assurance laboratory.
3. In your work at a food company, you mentioned to a coworker something about the *Official Methods of Analysis* published by AOAC International. The coworker asks you what AOAC International does, and what the *Official Methods of Analysis* is. Answer your coworker's questions.

4. For each type of product listed below, identify a publication in which you can find standard methods of analysis appropriate for the product:
   (a) Ice cream
   (b) Enriched flour
   (c) Wastewater (from food processing plant)
   (d) Margarine

1.9 ACKOWLEDGMENTS

The author thanks the numerous former students, working in quality assurance in the food industry, who reviewed this chapter and contributed ideas for its revision.

1.10 REFERENCES

1. Flickinger B (1997) Challenges and solutions in compositional analysis. Food Quality 3(19):21–26
2. Spence JT (2006) Challenges related to the composition of functional foods. J Food Compost Anal 19 Suppl 1: S4–S6
3. Alli I (2003) Food quality assurance: principles and practices. CRC, Boca Raton, FL
4. Vasconcellos JA (2004) Quality assurances for the food industry: a practical approach. CRC, Boca Raton, FL
5. Multon J-L (1995) Analysis and control methods for foods and agricultural products, vol 1: quality control for foods and agricultural products. Wiley, New York
6. Linden G, Hurst WJ (1996) Analysis and control methods for foods and agricultural products, vol 2: analytical techniques for foods and agricultural products. Wiley, New York
7. Multon J-L, Stadleman WJ, Watkins BA (1997) Analysis and control methods for foods and agricultural products, vol 4: analysis of food constituents. Wiley, New York
8. Pearson D (1973) Introduction – some basic principles of quality control, Ch. 1. In: Laboratory techniques in food analysis. Wiley, New York, pp 1–26
9. Pomeranz Y, Meloan CE (1994) Food analysis: theory and practice, 3rd edn. Chapman & Hall, New York
10. Jones L (2005) Chemical analysis of food: an introduction. Campden & Chorleywood Food Research Association, Gloucestershire, UK
11. Tothill IE (2003) Rapid and on-line instrumentation for food quality assurance. Woodhead, CRC, Boca Raton, FL
12. Nollett LML (2004) Handbook of food analysis, 2nd edn, vol 1: physical characterization and nutrient analysis, vol 2: residues and other food component analysis. CRC, Boca Raton, FL
13. Otles S (2005) Methods of analysis of food components and additives. Woodhead, Cambridge, England
14. Otles S (2008) Handbook of food analysis instruments. CRC, Boca Raton, FL
15. Wetzel DLB, Charalambous G (eds) (1998) Instrumental methods in food and beverage analysis. Elsevier Science, Amsterdam, The Netherlands
16. AOAC International (1993) A food matrix organizational system applied to collaborative studies. Referee 17(7): 1, 6, 7
17. Lovett RA (1997) U.S. food label law pushes fringes of analytical chemistry. Inside Lab Manage 1(4):27–28
18. DeVries JW, Silvera KR (2001) AACC collaborative study of a method for determining vitamins A and E in foods by HPLC (AACC Method 86–06). Cereal Foods World 46(5):211–215
19. Sharpless KE, Greenberg RR, Schantz MM, Welch MJ, Wise SA, Ihnat M (2004) Filling the AOAC triangle with food-matrix standard reference materials. Anal Bioanal Chem 378:1161–1167
20. Ellis C, Hite D, van Egmond H (1997) Development of methods to test all food matrixes unrealistic, says OMB. Inside Lab Manage 1(8):33–35
21. Latimer GW Jr (1997) Check sample programs keep laboratories in sync. Inside Lab Manage 1(4):18–20
22. Ambrus A (2008) Quality assurance, Ch. 5. In: Tadeo JL (ed) Analysis of pesticides in food and environmental samples. CRC, New York, p 145
23. AOAC International (2007) Official methods of analysis, 18th edn., 2005; current through revision 2, 2007 (Online). AOAC International, Gaithersburg, MD
24. AACC International (2010) Approved methods of analysis, 11th edn (online). AACC International, St. Paul, MN
25. AOCS (2009) Official methods and recommended practices, 6th edn. American Oil Chemists' Society, Champaign, IL
26. Wehr HM, Frank JF (eds) (2004) Standard methods for the examination of dairy products, 17th edn. American Public Health Association, Washington, DC
27. Eaton AD, Clesceri LS, Rice EW, Greenberg AE (eds) (2005) Standard methods for the examination of water and wastewater, 21st edn. American Public Health Association, Washington, DC
28. U.S. Pharmacopeia (USP) (2008) Food chemicals codex, 6th edn. United Book, Baltimore, MD

1.11 RELEVANT INTERNET ADDRESSES

American Association of Cereal Chemists – http://www.aaccnet.org/
American Oil Chemists' Society – http://www.aocs.org/
American Public Health Association – http://www.apha.org/
AOAC International – http://www.aoac.org
Code of Federal Regulations – http://www.gpoaccess.gov/cfr/index.html
Codex Alimentarius Commission – http://www.codexalimentarius.net/web/index_en.jsp

Food Chemicals Codex –
http://www.usp.org/fcc/
Food and Drug Administration –
http://www.fda.gov
Center for Food Safety & Applied Nutrition –
http://www.cfsan.fda.gov/
Current Good Manufacturing Practices –
http://www.cfsan.fda.gov/~dms/cgmps.html
Food Labeling and Nutrition –
http://vm.cfsan.fda.gov/label.html
Hazard Analysis Critical Control Point –
http://www.cfsan.fda.gov/~lrd/haccp.html

National Institute of Standards and Technology –
http://www.nist.gov/
U.S. Department of Agriculture –
http://www.usda.gov/wps/portal/usdahome
Food Safety and Inspection Service –
http://www.fsis.usda.gov
HACCP/Pathogen Reduction –
http://www.fsis.usda.gov/Science/Hazard_Analysis_&_Pathogen_Reduction/index.asp

United States Government Regulations and International Standards Related to Food Analysis

S. Suzanne Nielsen
Department of Food Science, Purdue University, West Lafayette, IN 47907-2009, USA
nielsens@purdue.edu

2.1 Introduction 17
2.2 US Federal Regulations Affecting Food Composition 17
2.2.1 US Food and Drug Administration 17
2.2.1.1 Legislative History 17
2.2.1.1.1 Federal Food, Drug, and Cosmetic Act of 1938 17
2.2.1.1.2 Amendments and Additions to the 1938 FD&C Act 17

S.S. Nielsen, *Food Analysis*, Food Science Texts Series, DOI 10.1007/978-1-4419-1478-1_2,

2.2.1.1.3 Other FDA Regulations 18
2.2.1.2 Food Definitions and Standards 18
2.2.1.3 Inspection and Enforcement 19
2.2.2 US Department of Agriculture 21
2.2.2.1 Standards of Identity for Meat Products 21
2.2.2.2 Grade Standards 21
2.2.2.3 Inspection Programs 22
2.2.3 US Department of Commerce 22
2.2.3.1 Seafood Inspection Service 22
2.2.3.2 Interaction with FDA and EPA 22
2.2.4 US Bureau of Alcohol, Tobacco, Firearms and Explosives 22
2.2.4.1 Regulatory Responsibility for Alcoholic Beverages 22
2.2.4.2 Standards and Composition of Beer, Wine, and Distilled Beverage Spirits 22
2.2.5 US Environmental Protection Agency 23
2.2.5.1 Pesticide Registration and Tolerance Levels 23
2.2.5.2 Drinking Water Standards and Contaminants 24
2.2.5.3 Effluent Composition from Food Processing Plants 24
2.2.6 US Customs Service 25
2.2.6.1 Harmonized Tariff Schedule of the US 25
2.2.6.2 Food Composition and the TSUSA 25
2.2.7 US Federal Trade Commission 25
2.2.7.1 Enforcement Authority 25
2.2.7.2 Food Labels, Food Composition, and Deceptive Advertising 26
2.3 Regulations and Recommendations for Milk 26
2.3.1 FDA Responsibilities 26
2.3.2 USDA Responsibilities 27
2.3.3 State Responsibilities 27
2.4 Regulations and Recommendations for Shellfish 27
2.4.1 State and Federal Shellfish Sanitation Programs 27
2.4.2 Natural and Environmental Toxic Substances in Shellfish 28
2.5 Voluntary Federal Recommendations Affecting Food Composition 28
2.5.1 Food Specifications, Food Purchase, and Government Agencies 28
2.5.2 National Conference on Weights and Measures: State Food Packaging Regulations 28
2.6 International Standards and Policies 29
2.6.1 Codex Alimentarius 29
2.6.2 ISO Standards 30
2.6.3 Other Standards 30
2.7 Summary 30
2.8 Study Questions 30
2.9 Acknowledgments 31
2.10 References 31
2.11 Relevant Internet Addresses 33

2.1 INTRODUCTION

Knowledge of government regulations relevant to the chemical analysis of foods is extremely important to persons working in the food industry. Federal laws and regulations reinforce the efforts of the food industry to provide wholesome foods, to inform consumers about the nutritional composition of foods, and to eliminate economic frauds. In some cases, they dictate what ingredients a food must contain, what must be tested, and the procedures used to analyze foods for safety factors and quality attributes. This chapter describes the US federal regulations related to the composition of foods. The reader is referred to references (1–4) for comprehensive coverage of US food laws and regulations. Many of the regulations referred to in this chapter are published in the various titles of the ***Code of Federal Regulations*** (CFR) (5). This chapter also includes information about food standards and safety practices established by international organizations. Internet addresses are given at the end of this chapter for many of the government agencies, organizations, and documents discussed.

2.2 US FEDERAL REGULATIONS AFFECTING FOOD COMPOSITION

2.2.1 US Food and Drug Administration

The **Food and Drug Administration** (FDA) is a government agency within the **Department of Health and Human Services** (DHHS). The FDA is responsible for regulating, among other things, the safety of foods, cosmetics, drugs, medical devices, biologicals, and radiological products. It acts under laws passed by the US Congress to monitor the affected industries and ensure the consumer of the safety of such products. A comprehensive collection of federal laws, guidelines, and regulations relevant to foods and drugs has been published by the Food and Drug Law Institute (1,2).

2.2.1.1 Legislative History

2.2.1.1.1 Federal Food, Drug, and Cosmetic Act of 1938 The **Federal Food, Drug, and Cosmetic** (FD&C) **Act of 1938** was intended to assure consumers that foods are safe and wholesome, produced under sanitary conditions, and packaged and labeled truthfully. This law, which broadened the scope of the Food and Drug Act of 1906, further defined and set regulations on adulterated and misbranded foods. The FDA was given power to seize illegal products and to imprison and fine violators. An important part of the 1938 Act relevant to food analysis is the section that authorizes food definitions and standards of identity, as further described below.

2.2.1.1.2 Amendments and Additions to the 1938 FD&C Act The 1938 FD&C Act has been amended several times to increase its power. The Miller Pesticide Amendment was added in 1954 to specify the acceptable amount of pesticide residues on fresh fruits, vegetables, and other raw agricultural products when they enter the marketplace. This Amendment, then under the authority of the FDA, is now administered by the **Environmental Protection Agency** (EPA).

The **Food Additives Amendment** enacted in 1958 was designed to protect the health of consumers by requiring a food additive to be proven safe before addition to a food and to permit the food industry to use food additives that are safe at the intended level of use. The highly controversial **Delaney Clause**, attached as a rider to this amendment, prohibits the FDA from setting any tolerance level as a food additive for substances known to be carcinogenic.

The **Color Additives Amendment** of 1960 defines color additives, sets rules for both certified and uncertified colors, provides for the approval of color additives that must be certified or are exempt from certification, and empowers the FDA to list color additives for specific uses and set quantity limitations. Similar to the Food Additives Amendment, the Color Additives Amendment contains a Delaney Clause.

The **Nutrition Labeling and Education Act** of 1990 (NLEA), described further in Chap. 3, made nutrition labeling mandatory on most food products under FDA jurisdiction and established definitions for health and nutrient claims. The NLEA emphasized the relationship between diet and health and provided consumers a means to choose foods based on complete and truthful label information.

The **Dietary Supplement Health and Education Act** (1994) (DSHEA) changed the definition and regulations for dietary supplements from those in the FD&C Act and in acts relevant to dietary supplements passed prior to 1994. The DSHEA defined supplements as "dietary ingredients" (defined in specific but broad terms), set criteria to regulate claims and labeling, and established government agencies to handle regulation. Classified now as "dietary ingredients" rather than by the previously used term "food additives," dietary supplements are not subject to the Delaney Clause of the FD&C Act. Regulations for dietary supplements permit claims not allowed for traditional foods. Control and regulation of dietary supplements have been separated from those for traditional foods.

The **Food Quality Protection Act** (1996) amended both the FD&C Act and the Federal Insecticide, Fungicide and Rodenticide Act (FIFRA), as further described in Sect. 2.2.5.1.

2.2.1.1.3 Other FDA Regulations The FDA has developed many administrative rules, guidelines, and action levels, in addition to the regulations described above, to implement the FD&C Act of 1938. Most of them are published in Title 21 of the CFR. They include the **Current Good Manufacturing Practice in Manufacturing, Packing, or Holding Human Food** (GMP) regulations (21 CFR 110), regulations regarding **food labeling** (21 CFR 101), **recall policy** (21 CFR 7.40), and **nutritional quality guidelines** (21 CFR 104). The food labeling regulations include nutritional labeling requirements and guidelines and specific requirements for nutrient content, health claims, and descriptive claims (discussed in Chap. 3).

There has been increased responsibility placed on the food industry and different regulatory agencies to better ensure the safe handling of foods eaten by consumers. Both FDA and USDA have placed considerable emphasis on GMP regulations and on **Hazard Analysis Critical Control Point** (HACCP) systems in an effort to improve food safety and quality programs. HACCP is an important component of an interagency initiative to reduce the incidence of food-borne illness, and it includes the FDA, USDA, EPA, **Centers for Disease Control** (CDC), and state/local regulatory agencies (6).

The **GMP regulations**, legally based on the FD&C Act, but not established as a proposed rule until 1969, are designed to prevent adulterated food in the marketplace (7, 8). The GMP regulations define requirements for acceptable sanitary operation in food plants and include the following relevant to food processing:

1. General provisions that define and interpret the detailed regulations
2. Requirements and expectations for maintaining grounds, buildings, and facilities
3. Requirements and expectations for design, construction, and maintenance of equipment
4. Requirements for production and process controls
5. Defect action levels (DALs) for natural and unavoidable defects

In addition to general GMPs (21 CFR 110), specific GMPs exist for thermally processed low-acid canned foods (21 CFR 113), acidified foods (21 CFR 114), and bottled drinking water (21 CFR 129).

HACCP is an internationally recognized systematic approach that is used to prevent and/or control microbial, chemical, and physical hazards within the food supply. The "farm to the fork" approach was originally designed to be used by the food processing industry to produce zero defect (no hazard) food for astronauts to consume on space flights (9). Within the past few decades, the overall approach has expanded to be used as the most effective method of hazard risk reduction and control in all areas of the food flow, including production, agriculture, distribution, manufacturing, and retail food establishments. Both FDA and USDA have adopted the HACCP concept as part of the overall inspection programs. The HACCP approach begins with a description of the food being produced. A flow diagram of the process is developed to further describe the process used. For each step in the process flow, the HACCP program approach is based on seven principles identified below:

1. Determine potential microbial, chemical, and physical hazards in each step of the process flow.
2. Identify critical control points in the process.
3. Establish control limits for each critical control point.
4. Establish procedures to monitor control points.
5. Establish corrective actions when limits of control point are exceeded.
6. Establish appropriate system of record keeping.
7. Establish program to verify and validate efficacy of program.

While HACCP-based principles are used for many different foods and many different food processes, from a regulatory standpoint, HACCP is only required for the meat and poultry industry (9 CFR Part 417), the seafood industry (21 CFR Part 123), and juice industry (21 CFR Part 120).

2.2.1.2 Food Definitions and Standards

The food definitions and standards established by the FDA are published in 21 CFR 100-169 and include standards of identity, quality, and fill. The **standards of identity**, which have been set for a wide variety of food products, are most relevant to the chemical analysis of foods because they specifically establish which ingredients a food must contain. They limit the amount of water permitted in certain products. The minimum levels for expensive ingredients are often set, and maximum levels for inexpensive ingredients are sometimes set. The kind and amount of certain vitamins and minerals that must be present in foods labeled "enriched" are specified. The standards of identity for some foods include a list of optional ingredients. The standard of identity for sour cream (21 CFR 131.160) is given in Fig. 2-1. Table 2-1 summarizes

§131.160 Sour cream.

(a) **Description.** Sour cream results from the souring, by lactic acid producing bacteria, of pasteurized cream. Sour cream contains not less than 18 percent milkfat; except that when the food is characterized by the addition of nutritive sweeteners or bulky flavoring ingredients, the weight of the milkfat is not less than 18 percent of the remainder obtained by subtracting the weight of such optional ingredients from the weight of the food; but in no case does the food contain less than 14.4 percent milkfat. Sour cream has a titratable acidity of not less than 0.5 percent, calculated as lactic acid.

(b) **Optional ingredients.**

(1) Safe and suitable ingredients that improve texture, prevent syneresis, or extend the shelf life of the product.
(2) Sodium citrate in an amount not more than 0.1 percent may be added prior to culturing as a flavor precursor.
(3) Rennet.
(4) Safe and suitable nutritive sweeteners.
(5) Salt.
(6) Flavoring ingredients, with or without safe and suitable coloring, as follows:
(i) Fruit and fruit juice (including concentrated fruit and fruit juice).
(ii) Safe and suitable natural and artificial food flavoring.

(c) **Methods of analysis.** Referenced methods in paragraph (c) (1) and (2) of this section are from "Official Methods of Analysis of the Association of Official Analytical Chemists," 13th Ed. (1980), which is incorporated by reference. Copies may be obtained from the AOAC INTERNATIONAL, 481 North Frederick Ave., suite 500, Gaithersburg, MD 20877, or may be examined at the National Archives and Records Administration (NARA). For information on the availability of this material at NARA, call 202-741-6030, or go to: http://www.archives.gov/federal_register/code_of_federal_regulations/ibr_location.html
(1) Milkfat content - "Fat-Official Final Action," section 16.172.
(2) Titratable acidity - "Acidity-Official Final Action," section 16.023.

(d) **Nomenclature.** The name of the food is "Sour cream" or alternatively "Cultured sour cream". The full name of the food shall appear on the principal display panel of the label in type of uniform size, style, and color. The name of the food shall be accompanied by a declaration indicating the presence of any flavoring that characterizes the product, as specified in §101.22 of this chapter. If nutritive sweetener in an amount sufficient to characterize the food is added without addition of characterizing flavoring, the name of the food shall be preceded by the word "sweetened".

(e) **Label declaration.** Each of the ingredients used in the food shall be declared on the label as required by the applicable sections of parts 101 and 130 of this chapter. [42 FR 14360, Mar. 15, 1977, as amended at 47 FR 11824, Mar. 19, 1982; 49 FR 10092, Mar. 19, 1984; 54 FR 24893, June 12, 1989; 58 FR 2891, Jan. 6, 1993]

2-1 figure Standard of identity for sour cream. [From 21 CFR 131.160 (2009).]

the standards of identity relevant to food analysis for a number of other foods. Note that the standard of identity often includes the recommended analytical method for determining chemical composition.

Although standards of quality and fill are less related to the chemical analysis of foods than are standards of identity, they are important for economic and quality control considerations. **Standards of quality**, established by the FDA for some canned fruits and vegetables, set minimum standards and specifications for factors such as color, tenderness, weight of units in the container, and freedom from defects. The **standards of fill** established for some canned fruits and vegetables, tomato products, and seafood state how full a container must be to avoid consumer deception.

2.2.1.3 Inspection and Enforcement

The FDA has the broadest regulatory authority over most foods (generally, all foods other than meat, poultry, eggs; water supplies; imported foods). However, the FDA shares responsibilities with other regulatory agencies for certain foods, as described in later sections of this chapter. The FDA has responsibility for enforcing the 1938 FD&C Act as amended, which prohibits adulteration and misbranding of food products. Relevant to food analysis and for FDA-regulated foods, the FDA inspects food processing facilities for compliance with GMP regulations and for any mandatory HACCP inspection programs. The FDA monitors appropriate foods for composition and characteristics relevant to the standards of identity, standards of quality, standards of fill, nutrition labeling, and other labeling regulations. It regulates color additives and the use of food additives for all foods.

The FDA, together with other federal agencies described in this chapter and with state and local governments, works to help ensure the quality and safety of food in the USA. Some specific examples of how FDA interacts with other agencies regarding the safety and analysis of foods follow. Working with the National Marine Fisheries Service to ensure seafood safety, the FDA sets and enforces allowable levels of contaminants and pathogenic microorganisms in seafood. The FDA has jurisdiction over some alcoholic beverages and cooking wines and handles questions

Selected Chemical Composition Requirements of Some Foods with Standards of Identity

Section in 21 CFR[a]	Food Product	Requirement	AOAC Method[b] Number in 13th Edn.	Number in 18th Edn.	Name/Description
131.110	Milk	Milk solids nonfat $\geq 8^1/_4$%	16.032	990.19	Total solids, by forced air oven after steam table
		Milkfat $\geq 3^1/_4$%	16.059	905.02	Roese-Gottlieb
		Vitamin A (if added) $\geq$ 2,000 IU[c]/qt[d]			
		Vitamin D (if added) – 400 IU[c]/qt[d]	43.195–43.208	936.14	Bioassay line test with rats
133.113	Cheddar cheese	Milkfat $\geq$ 50% by wt. of solids	16.255 and calculation	933.05	Digest with HCl, Roese-Gottlieb
		Moisture $\leq$ 39% by wt.	16.233	926.08	Vacuum oven
		Phosphatase level $\leq$ 3 µg phenol equivalent/0.25 g[e]	16.275–16.277	946.03–946.03C	Residual phosphatase
137.165	Enriched flour	Moisture $\leq$ 15%	14.002, 14.003	925.09, 925.09B	Vacuum oven
		Ascorbic acid $\leq$ 200 ppm (if added as dough conditioner)			
		Ash[f] $\leq 0.35 + (1/20$ of the percent of protein, calculated on dwb[g])	14.006	923.03	Dry ashing
		(Protein)	2.057	955.04C	Kjeldahl, for nitrate-free samples
		Thiamine, 2.9 mg/lb			
		Riboflavin, 1.8 mg/lb			
		Niacin, 24 mg/lb			
		Iron, 20 mg/lb			
		Calcium (if added), 960 mg/lb			
		Folic acid 0.7 mg/lb			
146.185	Pineapple juice	Soluble solids $\geq$ 10.5°Brix[h]	31.009	932.14A	Hydrometer
		Total acidity $\leq$ 1.35 g/100 ml (as anhydrous citric acid)			Titration with NaOH[i]
		Brix/acid ratio $\geq$ 12			Calculated[j]
		Insoluble solids $\geq$ 5% and $\leq$ 30%			Calculated from volume of sediment[k]
163.113	Cocoa	Cocoa fat $\leq$ 22% and $\geq$ 10%		963.15	Petroleum ether extraction with Soxhlet unit
164.150	Peanut butter	Fat $\leq$ 55%	27.006(a)	948.22	Ether extraction with Soxhlet unit

[a] *CFR*, Code of Federal Regulations (2009).
[b] *Official Methods of Analysis* of AOAC International.
[c] IU, International units.
[d] Within limits of good manufacturing practice.
[e] If pasteurized dairy ingredients are used.
[f] Excluding ash resulting from any added iron or salts of iron or calcium or wheat germ.
[g] dwb, moisture-free or dry weight basis.
[h] Exclusive of added sugars, without added water. As determined by refractometer at 20°C uncorrected for acidity and read as °Brix on International Sucrose Scales. Exception stated for juice from concentrate.
[i] Detailed titration method given in 21 CFR, 145.180 (b)(2)(ix).
[j] Calculated from °Brix and total acidity values, as described in 21 CFR 146.185 (b)(2)(ii).
[k] Detailed method given in 21 CFR 146.185 (b)(2)(iv).

of deleterious substances in alcoholic beverages. Regulations on tolerance levels of pesticide residues in foods and agricultural commodities set by the EPA are enforced by the FDA. Imported food products regulated by the FDA are subject to inspection upon entry through US Customs, and products must comply with US laws and regulations. The FDA works with individual states and the United States Department of Agriculture (USDA) to ensure the safety and wholesomeness of dairy products. Also, the FDA has regulatory power over shellfish sanitation for products shipped interstate.

When violations of the FD&C Act are discovered by the FDA through periodic inspections of facilities and products and through analysis of samples, the FDA can use warning letters, seizures, injunctions, or recalls, depending on the circumstances. The FDA cannot file criminal charges, but rather recommends to the Justice Department that court action be taken that might result in fines or imprisonment of offenders. Details of these enforcement activities of the FDA are given in references (1–4).

2.2.2 US Department of Agriculture

The USDA administers several federal statutes relevant to food standards, composition, and analysis. These include standards of identity for meat products, grade standards, and inspection programs. Some programs for fresh and processed food commodities are mandatory and others are voluntary.

2.2.2.1 Standards of Identity for Meat Products

Standards of identity have been established by the **Food Safety Inspection Service** (FSIS) of the USDA for many meat products (9 CFR 319). These commonly specify percentages of meat, fat, and water. Analyses are to be conducted using an AOAC method, if available.

2.2.2.2 Grade Standards

Grade standards developed for many foods by the USDA classify products in a range from substandard to excellent in quality. While most grade standards are not mandatory requirements (but they are mandatory for certain grains), they are widely used, voluntarily, by food processors and distributors as an aid in wholesale trading, because the quality of a product affects its price. Such grade standards often are used as quality control tools. Consumers are familiar generally with grade standards for beef, butter, and eggs, but buyers for the retail market utilize grade standards for a wide variety of foods. Major users of standards include institutions such as schools, hospitals, restaurants, prisons, and the Department of Defense (see also Sect. 2.5).

The USDA has issued grade standards for more than 300 food products under authority of the Agricultural Marketing Act of 1946 and related statutes. Standards for grades are not required to be stated on the label, but if they are stated, the product must comply with the specifications of the declared grade. Official USDA grading services are provided, for a fee, to pickers, processors, distributors, and others who seek official certification of the grades of their products.

While complete information regarding the standards was published previously in the CFR, currently only some standards are published in the CFR because they are USDA Agricultural Marketing Service (AMS) Administrative Orders. All grade standards are available as pamphlets from USDA and also are accessible on the Internet.

Grade standards, issued by the AMS of the USDA for agricultural products and by the Department of Commerce for fishery products, must not be confused with standards of quality set by the FDA or standards of identity set by the FDA or FSIS of the USDA, as discussed previously. Grade standards exist for many types of meats, poultry, dairy products, fruits, vegetables, and grains, along with eggs, domestic rabbits, certain preserves, dry beans, rice, and peas. Additional information about grade standards for dairy products is given in Sect. 2.3, but examples of grade standards for several other types of foods follow here.

Standards for grades of processed fruits and vegetables often include factors such as color, texture or consistency, defects, size and shape, tenderness, maturity, flavor, and a variety of chemical characteristics. Sampling procedures and methods of analysis are commonly given. As an example, the quality and analytical factors that determine the grade standards of frozen concentrated orange juice (10) are given in Table 2-2.

2-2 table **USDA Standards for Grades Determinants of Frozen Concentrated Orange Juice**

Quality	*Analytical*
Appearance	Concentrate
Reconstitution	Brix
Color	Brix/acid ratio
Defects	Reconstituted Juice
Flavor	Brix
	Soluble orange solids
	Recoverable oil

Grades for various grains (e.g., wheat, corn, soybeans, oats) are determined by factors such as test weight per bushel and percentages of heat-damaged kernels, broken kernels, and foreign material. Also, a grade limit is set commonly for moisture content. Grade standards for rice, beans, peas, and lentils are determined commonly by factors such as defects, presence of foreign material, and insect infestation, and sometimes moisture content is specified.

2.2.2.3 Inspection Programs

The USDA administers some programs on inspection and certification that are mandatory, and some inspection programs are voluntary. Comprehensive inspection manuals specific to various types of foods have been developed to assist inspectors and industry personnel in interpreting and utilizing the regulations. Under the Federal Meat Inspection Act, the Poultry Products Inspection Act, and the Egg Products Inspection Act, the FSIS of the USDA inspects all meat, poultry, and egg products in interstate commerce (9 CFR 200-End). This includes a review of foreign inspection systems and packing plants that export meat and poultry to the USA. Imported products are reinspected at ports of entry. HACCP is a major component of FSIS rules for all slaughter and processing plants, to improve safety of meat and poultry. A program within the **Grain Inspection, Packers and Stockyard Administration** (GIPSA) of the USDA administers the mandatory requirements of the US Grain Standards Act (7 CFR 800). Regulations to enforce this act provide for a national inspection system for grain and mandatory official grade standards of numerous types of grain. Another program of the USDA standardizes, grades, and inspects fruits and vegetables under various voluntary programs. The inspection programs rely heavily on the HACCP concept.

2.2.3 US Department of Commerce

2.2.3.1 Seafood Inspection Service

The **National Oceanic and Atmospheric Administration's** (NOAA) **National Marine Fisheries Service** (NMFS), a division of the **United States Department of Commerce** (USDC), provides a seafood inspection service. The USDC Seafood Inspection Program ensures the safety and quality of seafoods consumed in the USA and certified for export through **voluntary grading, standardization, and inspection programs**, as described in 50 CFR 260. The inspection programs rely heavily on the HACCP concept. USDC Handbook 25 is a comprehensive manual on these subjects entitled *Fishery Products Inspection Manual* (11). The US Standards for Grades of Fishery Products (50 CFR 261) are intended to help the fishing industry maintain and improve quality and to thereby increase consumer confidence in seafoods. Standards are based on attributes such as color, size, texture, flavor, odor, workmanship defects, and consistency.

2.2.3.2 Interaction with FDA and EPA

The FDA and the EPA work with the NMFS for the assurance of seafood safety. The FDA, under the FD&C Act, is responsible for ensuring that seafood shipped or received in interstate commerce is safe, wholesome, and not misbranded or deceptively packaged. The FDA has primary authority in setting and enforcing allowable levels of contaminants and pathogenic microorganisms in seafood. The EPA assists the FDA in identifying the range of chemical contaminants that pose a human health risk and are most likely to accumulate in seafood. A tolerance of 2.0 parts per million (ppm) for total polychlorinated biphenyls (PCBs) (21 CFR 109.30) is the only formal tolerance specified by the FDA to mitigate human health impacts in seafood. However, the EPA has established tolerances for certain pesticide residues, and the FDA has established guidance levels for the toxic elements arsenic, cadmium, chromium, lead, nickel, and methyl mercury (12).

2.2.4 US Bureau of Alcohol, Tobacco, Firearms and Explosives

2.2.4.1 Regulatory Responsibility for Alcoholic Beverages

Beer, wines, liquors, and other alcoholic beverages are termed "food" according to the FD&C Act of 1938. However, regulatory control over their quality, standards, manufacture, and other related aspects is specified by the **Federal Alcohol Administration Act**, which is enforced by the **Alcohol and Tobacco Tax and Trade Bureau** (TTB) of the **US Department of Treasury**. Issues regarding the composition and labeling of most alcoholic beverages are handled by the Bureau. However, the FDA has jurisdiction over certain other alcoholic beverages and cooking wines. The FDA also deals with questions of sanitation, filth, and the presence of deleterious substances in alcoholic beverages.

2.2.4.2 Standards and Composition of Beer, Wine, and Distilled Beverage Spirits

Information related to **definitions, standards of identity**, and certain **labeling requirements** for beer, wine, and distilled beverage spirits is given in 27 CFR 1-30. Standards of identity for these types of beverages stipulate the need for analyses such as percent alcohol by volume, total solids content, volatile acidity, and calculated acidity. For example, the fruit juice used for the production of wine is often specified by its °Brix and total solids content. The maximum volatile acidity (calculated as acetic acid and exclusive of sulfur dioxide) for grape wine must not be more than 0.14 g/100 ml (20°C) for natural red wine and 0.12 g/100 ml for other grape wines (27 CFR 4.21). The percent alcohol by volume is often used as a criterion for class or type designation of alcoholic beverages. For example, dessert wine is grape wine with an alcoholic content in excess of 14% but not in excess of 24% by volume, while table wines have an alcoholic content not in excess of 14% alcohol by volume (27 CFR 4.21). No product with less than 0.5% alcohol by volume is permitted to be labeled "beer," "lager beer," "lager," "ale," "porter," "stout," or any other class or type designation normally used for malt beverages with higher alcoholic content (27 CFR 7.24).

2.2.5 US Environmental Protection Agency

The EPA was established as an independent agency in 1970 through a reorganization plan to consolidate certain federal government environmental activities. The EPA regulatory activities most relevant to this book are control of pesticide residues in foods, drinking water safety, and the composition of effluent from food processing plants.

2.2.5.1 Pesticide Registration and Tolerance Levels

Pesticides are chemicals intended to protect our food supply by controlling harmful insects, diseases, rodents, weeds, bacteria, and other pests. However, most pesticide chemicals can have harmful effects on people, animals, and the environment if they are improperly used. The three federal laws relevant to protection of food from pesticide residues are certain provisions of the Federal FD&C Act, the Federal Insecticide, Fungicide, and Rodenticide Act (FIFRA), as amended, and the **Food Quality Protection Act** of 1996. FIFRA, supplemented by the FD&C Act, authorizes a comprehensive program to regulate the manufacturing, distribution, and use of pesticides, along with a research effort to determine the effects of pesticides.

The Food Quality Protection Act amends both the FD&C Act and FIFRA, to take pesticides out of the section of the FD&C Act that includes the Delaney Clause. This was done by changing the definition of a "food additive" to exclude pesticides. This redefinition leaves the Delaney Clause greatly reduced in scope and less relevant.

The EPA registers approved pesticides and sets tolerances for pesticide residues (see also Chap. 18, Sect. 18.3). The EPA is authorized to establish an **allowable limit** or **tolerance** for any detectable pesticide residues that might remain in or on a harvested food or feed crop. The **tolerance level** is often many times less than the level expected to produce undesirable health effects in humans or animals. Tolerances are established based on factors that include registration data, consumption pattern, age groups, mode of action, chemistry of the compound, toxicological data, plant and animal physiology, efficacy data, and risk assessment. While the EPA establishes the tolerance levels, the FDA enforces the regulations by collecting and analyzing food samples, mostly agricultural commodities. Livestock and poultry samples are collected and analyzed by the USDA. Pesticide residue levels that exceed the established tolerances are considered in violation of the FD&C Act.

The Food Quality Protection Act of 1996 requires an explicit determination that tolerances are safe for children and have consideration of children's special sensitivity and exposure to pesticide chemicals. It includes an additional safety factor of up to tenfold, if necessary, to account for uncertainty in data relative to children. The 1996 law requires that all existing tolerances be reviewed within 10 years to make sure they meet the requirements of new health-based safety standards established by law.

Regulations regarding pesticide tolerances in foods are given in 40 CFR 180, which specifies general categories of products and specific commodities with tolerances or exemptions, and in some cases which part of the agricultural product is to be examined. Products covered include a wide variety of both plants (e.g., fruits, vegetables, grains, legumes, nuts) and animals (e.g., poultry, cattle, hogs, goats, sheep, horses, eggs, milk). Unless otherwise noted, the specific tolerances established for the pesticide chemical apply to residues resulting from their application prior to harvest or slaughter. Tolerances are expressed in terms of parts by weight of the pesticide chemical per one million parts by weight of the product (i.e., ppm). Tolerance levels for selected pesticides and insecticides permitted in foods as food additives are given in Table 2-3.

The analytical methods to be used for determining whether pesticide residues are in compliance with the tolerance established are identified among the methods contained or referenced in the *Pesticide Analytical Manual* (13) maintained by and available from the FDA. The methods must be sensitive and reliable at and above the tolerance level. Pesticides are generally detected and quantitated by gas chromatographic or high-performance liquid chromatographic methods (see Chaps. 18, 28, and 29).

2-3 table **Tolerance for Selected Insecticides (I), Fungicides (F), and Herbicides (H) Classified as Food Additives Permitted in Foods for Human Consumption**

Section	*Food Additive*	*Chemical Classification*	*Food*	*Tolerance*[a]
180.294	Benomyl (F)	Carbamate	Apples	7
			Cattle, meat	0.1
			Milk	0.1
			Grapes	10
			Raisins	50
			Strawberry	5
			Tomato products, conc.	50
180.342	Chloropyrifos[b](I)	Organophosphate	Apples	0.01
			Cattle, meat	0.05
			Corn oil	0.25
			Strawberry	0.2
180.435	Deltamethrin (I)	Pyrethroid	Cattle, meat	0.02
			Tomatoes	0.02
			Tomato products, conc.	1.0
180.292	Picloram (H)	Chloropyridine–carboxylic acid	Cattle, meat	0.02
			Milk	0.05
			Corn oil	2.5
			Wheat, grain	0.5

Adapted from 40 CFR 180 (2009).
[a]Parts per million.
[b]Also known as Dursban™ and Lorsban™.

2.2.5.2 Drinking Water Standards and Contaminants

The EPA administers the **Safe Drinking Water Act** of 1974, which is to provide for the safety of drinking water supplies in the USA and to enforce national drinking water standards. The EPA has identified potential contaminants of concern and established their maximum acceptable levels in drinking water. The EPA has primary responsibility to establish the standards, while the states enforce them and otherwise supervise public water supply systems and sources of drinking water. **Primary and secondary drinking water regulations** (40 CFR 141 and 143, respectively) have been established. Recently, concerns have been expressed regarding the special standardization of water used in the manufacturing of foods and beverages.

Maximum contaminant levels (MCL) for primary drinking water are set for certain **inorganic** and **organic chemicals**, **turbidity**, **certain types of radioactivity**, and **microorganisms**. Sampling procedures and analytical methods for the analysis of chemical contaminants are specified, with common reference to *Standard Methods for the Examination of Water and Wastewater* (14) published by the American Public Health Association; *Methods of Chemical Analysis of Water and Wastes* (15), published by the EPA; and *Annual Book of ASTM Standards* (16), published by the American Society for Testing Materials (ASTM). Methods commonly specified for the analysis of inorganic contaminants in water include atomic absorption (direct aspiration or furnace technique), inductively coupled plasma (see Chap. 24), ion chromatography (see Chap. 28), and ion selective electrode (see Chap. 12).

2.2.5.3 Effluent Composition from Food Processing Plants

In administering the **Federal Water Pollution and Control Act**, the EPA has developed effluent guidelines and standards that cover various types of food processing plants. Regulations prescribe effluent limitation guidelines for existing sources, standards of performance for new sources, and pretreatment standards for new and existing sources. Point sources of discharge of pollution are required to comply with these regulations, where applicable. Regulations are prescribed for specific foods under the appropriate point source category: dairy products processing (40 CFR 405), grain mills (40 CFR 406), canned and preserved fruits and vegetables processing (40 CFR 407), canned and preserved seafood processing (40 CFR 408), sugar processing (40 CFR 409), and meat and poultry products (40 CFR 432). **Effluent characteristics** commonly prescribed for food processing plants are **biochemical oxygen demand** (BOD) (see Chap. 20), **total soluble solids** (TSS) (see Chap. 6), and **pH** (see Chap. 13), as

Effluent Limitations for Plants Processing Natural and Processed Cheese

	Effluent Characteristics					
	Metric Units[a]			English Units[b]		
Effluent Limitations	*BOD 5*[c]	*TSS*[d]	*pH*	*BOD 5*	*TSS*	*pH*
Processing more than 100,000 lb/day of milk equivalent						
Maximum for any 1 day	0.716	1.088	([e])	0.073	0.109	([e])
Average of daily values for 30 consecutive days shall not exceed	0.290	0.435	([e])	0.029	0.044	([e])
Processing less than 100,000 lb/day of milk equivalent						
Maximum for any 1 day	0.976	1.462	([e])	0.098	0.146	([e])
Average of daily values for 30 consecutive days shall not exceed	0.488	0.731	([e])	0.049	0.073	([e])

Adapted from 40 CFR 405.62 (2009).
[a]Kilograms per 1000 kg of BOD 5 input.
[b]Pounds per 100 lbs of BOD 5 input.
[c]BOD 5 refers to biochemical oxygen demand measurement after 5 days of incubation.
[d]TSS refers to total soluble solids.
[e]Within the range 6.0–9.0.

shown in Table 2-4 for effluent from a plant that makes natural and processed cheese. The test procedures for measurement of effluent characteristics are prescribed in 40 CFR 136.

2.2.6 US Customs Service

Over 100 countries export food, beverages, and related edible products to the USA. The **United States Customs Service** (USCS) assumes the central role in ensuring that imported products are taxed properly, safe for human consumption, and not economically deceptive. The USCS receives assistance from the FDA and USDA as it assumes these responsibilities. The major regulations promulgated by the USCS are given in Title 19 of the CFR.

2.2.6.1 Harmonized Tariff Schedule of the US

All goods imported into the United States are subject to duty or duty-free entry according to their classification under applicable items in the **Harmonized Tariff Schedule of the United States** (TSUSA). The US tariff system has official tariff schedules for over 400 edible items exported into the USA (17). The TSUSA specifies the food product in detail and gives the general rate of duty applicable to that product coming from most countries and any special higher or lower rates of duty for certain other countries.

2.2.6.2 Food Composition and the TSUSA

The **rate of duty** for certain food products is determined by their chemical composition. For example, the rate of duty on some dairy products is determined in part by the fat content. The tariff for some syrups is determined by the fructose content, for some chocolate products by the sugar or butterfat content, for butter substitutes by the butterfat content, and for some wines by their alcohol content (percent by volume).

2.2.7 US Federal Trade Commission

The **Federal Trade Commission** (FTC) is the most influential of the federal agencies that have authority over various aspects of advertising and sales promotion practices for foods in the USA. The major role of the FTC is to keep business and trade competition free and fair.

2.2.7.1 Enforcement Authority

The **Federal Trade Commission Act** of 1914 authorizes the FTC to protect both the consumer and the business person from anticompetitive behavior and unfair or deceptive business and trade practices. The FTC periodically issues industry guides and trade regulations and rules that tell businesses what they can and cannot do. These issuances are supplemented with advisory opinions given to corporations and individuals upon request. The FTC not only has guidance and preventive functions but is also authorized to issue complaints or shutdown orders and sue for civil penalties for violation of trade regulation rules. The **Bureau of Consumer Protection** is one of the FTC bureaus that enforce and develop trade regulation rules.

2.2.7.2 Food Labels, Food Composition, and Deceptive Advertising

While the **Fair Packaging and Labeling Act** of 1966 is administered by the FTC, that agency does not have specific authority over the packaging and labeling of foods. The FTC and FDA have agreed upon responsibilities: The FTC has primary authority over advertising of foods and the FDA has primary authority over labeling of foods.

Grading, standards of identity, and labeling of foods regulated by several federal agencies as described previously have eliminated many potential problems in the advertising of foods. Such federal regulations and voluntary programs have reduced the scope of advertising and other forms of product differentiation. Misleading, deceptive advertising is less likely to be an issue and is more easily controlled. For example, foods such as ice cream, mayonnaise, and peanut butter have standards of identity that set minimum ingredient standards. If these standards are not met, the food must be given a different generic designation (e.g., salad dressing instead of mayonnaise) or be labeled "imitation." Grading, standards, and labeling of food aid consumers in making price–quality comparisons. Once again, analyses of chemical composition play an important role in developing and setting these grades, standards, and labels. In many cases in which the FTC intervenes, data from a chemical analysis become central evidence for all parties involved.

2.3 REGULATIONS AND RECOMMENDATIONS FOR MILK

The safety and quality of milk and dairy products in the USA are the responsibility of both federal (FDA and USDA) and state agencies. The FDA has regulatory authority over the dairy industry interstate commerce, while the USDA involvement with the dairy industry is voluntary and service oriented. Each state has its own regulatory office for the dairy industry within that state. The various regulations for milk involve several types of chemical analyses.

2.3.1 FDA Responsibilities

The FDA has responsibility under the FD&C Act, the Public Health Service Act, and the Import Milk Act to assure consumers that the US milk supply and imported dairy products are safe, wholesome, and not economically deceptive. Processors of both Grade A and Grade B milk are required under FDA regulations to take remedial action when conditions exist that could jeopardize the safety and wholesomeness of milk and dairy products being handled. As described in Sect. 2.2.1.2, the FDA also promulgates standards of identity and labeling, quality, and fill-of-container requirements for milk and dairy products moving in interstate commerce.

For **Grade A milk and dairy products**, each state shares with the FDA the responsibility of ensuring safety, wholesomeness, and economic integrity. This is done through a **Memorandum of Understanding with the National Conference on Interstate Milk Shipments**, which comprises all 50 states. In cooperation with the states and the dairy industry, the FDA has also developed for state adoption model regulations regarding sanitation and quality aspects of producing and handling Grade A milk. These regulations are contained in the ***Grade A Pasteurized Milk Ordinance*** (PMO) (18), which all states have adopted as minimum requirements.

The standards for Grade A pasteurized milk and milk products and bulk-shipped heat-treated milk products under the PMO are given in Table 2-5. The PMO specifies that "all sampling procedures, including the use of approved in-line samples, and required laboratory examinations shall be in substantial compliance with the most current edition of *Standard Methods for the Examination of Dairy Products* (SMEDP) of the American Public Health Association, and the most current edition of *Official Methods of Analysis* of the *AOAC INTERNATIONAL (OMA)*" (18–20).

The FDA monitors state programs for compliance with the PMO and trains state inspectors. To facilitate movement of Grade A milk in interstate commerce,

Pasteurized Milk Ordinance Standards for Grade A Pasteurized Milk and Milk Products and Bulk-Shipped Heat-Treated Milk Products

Criteria	*Requirement*
Temperature	Cooled to 7°C (45°F) or less and maintained thereat
Bacterial limits[a]	20,000 per ml
Coliform[b]	Not to exceed 10 per ml. Provided, that in the case of bulk milk transport tank shipments, shall not exceed 100 per ml
Phosphatase[b]	Less than 350 milliunits/L for fluid products and other milk products by the Fluorometer or Charm ALP or equivalent
Drugs[c]	No positive results on drug residue detection methods

Adapted from (18).

[a]Not applicable to acidified or cultured products.

[b]Not applicable to bulk-shipped heat-treated milk products.

[c]Reference to specific laboratory techniques.

a federal-state certification program exists: the **Interstate Milk Shippers** (IMS) **Program**. This program is maintained by the **National Conference on Interstate Milk Shipments**, which is a voluntary organization that includes representatives from each state, the FDA, the USDA, and the dairy industry. In this program, the producers of Grade A pasteurized milk are required to pass inspections and be rated by cooperating state agencies, based on PMO sanitary standards, requirements, and procedures. The ratings appear in the *IMS List* (21), which is published by the FDA, and made available to state authorities and milk buyers to ensure the safety of milk shipped from other states.

2.3.2 USDA Responsibilities

Under authority of the Agricultural Marketing Act of 1946, the **Dairy Quality Program** of the USDA offers **voluntary grading services** for manufactured or processed dairy products (7 CFR 58). If USDA inspection of a dairy manufacturing plant shows that good sanitation practices are being followed to meet the requirements in the *General Specifications for Dairy Plants Approved for USDA Inspection and Grading Service* (22), the plant qualifies for the USDA services of grading, sampling, testing, and certification of its products. A product such as nonfat dry milk is graded based on flavor, physical appearance, and various laboratory analyses (Table 2-6).

As with the USDA voluntary grading programs for other foods described in Sect. 2.2.2.2, the USDA has no regulatory authority regarding dairy plant inspections and cannot require changes in plant operations. The USDA, under an arrangement with the FDA, assists states in establishing safety and quality regulations for manufacturing-grade milk. Much as described previously for the FDA with Grade A milk, the USDA has developed model regulations for state adoption regarding the quality and sanitation aspects of producing and handling manufacturing-grade milk. These regulations are given in the *Milk for Manufacturing Purposes and Its Production and Processing, Recommended Requirements* (23). The states that have **Grade B milk** have essentially adopted these model regulations.

2-6 table **US Standards for Grades of Nonfat Dry Milk (Spray Process)**

Laboratory Tests[a]	*US Extra Grade*	*US Standard Grade*
Bacterial estimate, standard plate count per gram	10,000	75,000
Milkfat content (%)	1.25	1.50
Moisture content (%)	4.0	5.0
Scorched particle content (mg)	15.0	22.5
Solubility index (ml)	1.2	2.0
US High-heat	2.0	2.5
Titratable acidity (lactic acid) (%)	0.15	0.17

US Standards: http://www.ams.usda.gov/AMSv1.0/getfile?dDocName_STELDEV3004466.

[a]All values are maximum allowed.

2.3.3 State Responsibilities

As described previously, individual states have enacted safety and quality regulations for Grade A and manufacturing-grade milk that are essentially identical to those in the PMO and the USDA Recommended Requirements, respectively. The department of health or agriculture in each state normally is responsible for enforcing these regulations. The states also establish their own standards of identity and labeling requirements for milk and dairy products, which are generally similar to the federal requirements.

2.4 REGULATIONS AND RECOMMENDATIONS FOR SHELLFISH

Shellfish include fresh or frozen oysters, clams, and mussels. They may transmit intestinal diseases such as typhoid fever or act as carriers of natural or chemical toxins. This makes it very important that they be obtained from unpolluted waters and handled and processed in a sanitary manner.

2.4.1 State and Federal Shellfish Sanitation Programs

The growing, handling, and processing of shellfish must comply not only with the general requirements of the FD&C Act but also with the requirements of state health agencies cooperating in the **National Shellfish Sanitation Program** (NSSP), a federal, state, industry voluntary cooperative program, administered by the FDA (24). The FDA has no regulatory power over shellfish sanitation unless the product is shipped interstate. However, the Public Health Service Act authorizes the FDA to make recommendations and to cooperate with state and local authorities to ensure the safety and wholesomeness of shellfish. Under special agreement, certain other countries are in the NSSP and are subject to the same sanitary controls as required in the USA. Through the NSSP, state health personnel continually inspect and survey bacteriological conditions in shellfish-growing areas. Any contaminated location is supervised or patrolled so that shellfish cannot be harvested from the area. State inspectors check harvesting boats and shucking plants before issuing approval **certificates**, which serve as operating licenses. The certification number of the approved plant is placed on each shellfish package shipped.

2.4.2 Natural and Environmental Toxic Substances in Shellfish

A major concern is the ability of shellfish to concentrate radioactive material, insecticides, and other chemicals from their environment. Thus, one aspect of the NSSP is to ensure that shellfish-growing areas are free from sewage pollution and toxic industrial waste. **Pesticide residues** in shellfish are usually quantitated by gas chromatographic techniques, and **heavy metals** such as mercury are commonly quantitated by atomic absorption spectroscopy (e.g., AOAC Method 977.15). Another safety problem with regard to shellfish is the control of **natural toxins**, which is a separate issue from sanitation. The naturally occurring toxins are produced by planktonic organisms, and testing is conducted using a variety of assays. Control of this toxicity is achieved by a careful survey followed by prohibition of harvesting from locations inhabited by toxic shellfish.

2.5 VOLUNTARY FEDERAL RECOMMENDATIONS AFFECTING FOOD COMPOSITION

2.5.1 Food Specifications, Food Purchase, and Government Agencies

Large amounts of food products are purchased by federal agencies for use in domestic (e.g., school lunch) and foreign programs, prisons, veterans' hospitals, the armed forces, and other organizations. Specifications or descriptions developed for many food products are used by federal agencies in procurement of foods, to ensure the safety and quality of the product specified. Such specifications or descriptions often include information that requires assurance of chemical composition. These specifications include the following:

1. Federal Specifications
2. Commercial Item Descriptions (CIDs)
3. Purchase Product Description (PPD)
4. USDA Specifications
5. Commodity Specifications
6. Department of Defense Specifications

Various CIDs, PPDs, Federal Specifications, or USDA Specifications are used by the USDA to purchase meat products for programs such as school lunches. For example, the CID for canned tuna (25) specifies salt/sodium levels and the required method of analysis. The Institutional Meat Purchase Specification (a USDA specification) for frozen ground pork (26) and frozen ground beef products (27) states maximum allowable fat contents.

Commodity Specifications for various poultry and dairy products have been issued by the USDA. For example, the Commodity Specification for dried egg mix (28) specifies that the vegetable oil in the product must meet specifications for the following, as determined by American Oil Chemists' Society (AOCS) test methods: free fatty acid value, peroxide value, linolenic acid, moisture and volatile matter, iodine value, and Lovibond color values (see Chap. 14 for some of these tests). The Commodity Specifications issued by USDA for various dairy products also include compositional requirements. For example, the milk fat content, pH, and moisture and fat contents of pasteurized process American cheese (29) and mozzarella cheese (30) are specified.

The Defense Personnel Support Center of the Defense Logistics Agency, Department of Defense, utilizes a variety of specifications, standards, and notes in the purchase of food for the military. For example, they use CIDs for syrup (specifies Brix, ash content, and color) (31), instant tea (specifies moisture and sugar contents, and titratable acidity) (32), and peanut butter (specifies salt and aflatoxin contents) (33).

2.5.2 National Conference on Weights and Measures: State Food Packaging Regulations

Consumers assume that the weighing scale for a food product is accurate and that a package of flour, sugar, meat, or ice cream contains the amount claimed on the label. While this assumption is usually correct, city or county offices responsible for weights and measures need to police any unfair practices. Leadership in this area is provided by the **National Conference on Weights and Measures** (NCWM), which was established by the **National Institute of Standards and Technology** (NIST) (formerly the National Bureau of Standards) (part of the US Department of Commerce). The NCWM has no regulatory power, but it develops many technical, legal, and general recommendations in the field of weights and measures administration and technology. The NCWM is a membership organization comprising state and local weights and measures regulatory officers, other officials of federal, state, and local governments, and representatives of manufacturers, industry, business, and consumer organizations.

The NIST Handbook 133, *Checking the Net Contents of Packaged Goods* (34), gives model state **packaging and labeling regulations** that have been adopted by a majority of states. The Handbook specifies that the average quantity of contents of packages must at least equal the labeling quantity, with the variation between the individual package net contents and the labeled quantity not too "unreasonably large." Variations are permitted within the bounds of GMPs and are due to gain or loss of moisture (within the bounds of good distribution practice). For certain products (e.g., flour, pasta, rice), this requires careful monitoring of moisture content and control of storage conditions by the manufacturer.

2.6 INTERNATIONAL STANDARDS AND POLICIES

With the need to compete in the worldwide market, employees of food companies must be aware that allowed food ingredients, names of food ingredients, required and allowed label information, and standards for foods and food ingredients differ between countries (35). For example, colorings and preservatives allowed in foods differ widely between countries, and nutritional labeling is not universally required. To develop foods for, and market foods in, a global economy, one must seek such information from international organizations and from organizations in specific regions and countries.

2.6.1 Codex Alimentarius

The **Codex Alimentarius Commission** (Codex Alimentarius is Latin for "code concerned with nourishment") was established in 1962 by two United Nations organizations, the Food and Agriculture Organization (FAO) and the World Health Organization (WHO), to develop international standards and safety practices for foods and agricultural products (35,36). The standards, published in the *Codex Alimentarius*, are intended to protect consumers' health, ensure fair business practices in food trade, and facilitate international trade of foods (36).

The *Codex Alimentarius* is published in 13 volumes: one on general requirements (includes labeling, food additives, contaminants, irradiated foods, import/export inspection, and food hygiene), nine on standards and codes of practice compiled on a commodity basis, two on residues of pesticides and veterinary drugs in foods, and one on methods of analysis and sampling (Table 2-7). Codex has efforts to validate and harmonize methods of food safety analysis among countries and regions, to help maintain the smooth flow of international commerce, and ensure appropriate decisions on food exports and imports. Codex has adopted the HACCP concept as the preferred means to ensure the safety of perishable foods and is determining how HACCP will be implemented in *Codex Alimentarius*.

2-7 table Content of the Codex Alimentarius (38)

Volume	*Subject*
1A	General requirements
1B	General requirements (food hygiene)
2A	Pesticide residues in foods (general text)
2B	Pesticide residues in foods (maximum residue limits)
3	Residues in veterinary drugs in foods
4	Foods for special dietary uses
5A	Processed and quick-frozen fruits and vegetables
5B	Fresh fruits and vegetables
6	Fruit juices
7	Cereals, pulses (legumes) and derived products and vegetable proteins
8	Fats and oils and related products
9	Fish and fishery products
10	Meat and meat products, soups, and broths
11	Sugars, cocoa products and chocolate, and miscellaneous Products
12	Milk and milk products
13	Methods of analysis and sampling

Codex has strengthened its commitment to base food standards on strong science, rather than on social or cultural factors, economics, or trade policies. The setting of international standards on food quality by Codex has been a high priority in world trade to minimize "nontariff" trade barriers. International trade of food and raw agricultural products has increased due to reduced economic trade restrictions and tariffs imposed, but food standards set in the past by some countries created nontariff trade barriers. Food standards developed by Codex are intended to overcome the misuse of standards by a country, when the standards do more to protect products in a country from the competition of imports than to protect the health of consumers.

Decisions at the 1994 Uruguay Round of the **General Agreement on Tariffs and Trade** (GATT) strengthened the role of Codex as the principal standard-setting group internationally for the quality and safety of foods. The USA is among the 156 countries that are members of Codex. The USA recognizes treaty obligations related to Codex that have arisen from GATT. As a result, representatives of the FDA, USDA, and EPA (the three US federal agencies that participate in Codex) developed in 1995 a strategic plan for Codex that included greater US acceptance of Codex standards. In the USA, there is increased participation of nongovernmental organizations [e.g., Grocery Manufacturers Association (GMA)] in the Codex process, with many food companies working through these organizations.

2.6.2 ISO Standards

In addition to food standards and policies established by the Codex Alimentarius Commission, the **International Organization for Standardization** (ISO) has the 9000 series of standards on quality management and quality performance (37–39). The intent of the quality management standards is to establish a quality system, maintain product integrity, and satisfy customers. ISO 9001:2000 focuses on a process approach to quality management. Companies can elect to become registered only in the relevant parts of the ISO standards. Some manufacturers and retailers require food industry suppliers to be ISO certified. Relevant to food analysis, ISO standards include sampling procedures and food standards.

2.6.3 Other Standards

Other international, regional, and country-specific organizations publish standards relevant to food composition and analysis. For example, the **Saudi Arabian Standards Organization** (SASO) publishes standards documents (e.g., labeling, testing methods) important in the Middle East (except Israel), and the **European Commission** sets standards for foods and food additives for countries in the European Economic Community (EEC). In the USA, the Food Ingredients Expert Committee, which operates as part of the US Pharmacopeia, sets standards for the identification and purity of food additives and chemicals, published as the **Food Chemicals Codex** (FCC) (40). For example, a company may specify in the purchase of a specific food ingredient that it be "FCC grade." Countries other than the USA adopt FCC standards (e.g., Australia, Canada). At an international level, the **Joint FAO/WHO Expert Committee on Food Additives** (JECFA) sets standards for purity of food additives (41). The Codex Alimentarius Commission is encouraged to utilize the standards established by JECFA. Standards established by FCC and JECFA are used by many countries as they develop their own standards.

2.7 SUMMARY

Various kinds of standards set for certain food products by federal agencies make it possible to get essentially the same food product whenever and wherever purchased in the USA. The standards of identity set by the FDA and USDA define what certain food products must consist of. The USDA and NMFS of the Department of Commerce have specified grade standards to define attributes for certain foods. Grading programs are voluntary, while inspection programs may be either voluntary or mandatory, depending on the specific food product.

While the FDA has broadest regulatory authority over most foods, responsibility is shared with other regulatory agencies for certain foods. The USDA has significant responsibilities for meat and poultry, the NOAA and the NMFS for seafood, and the ATF for alcoholic beverages. The FDA, the USDA, state agencies, and the dairy industry work together to ensure the safety, quality, and economic integrity of milk and milk products. The FDA, the EPA, and state agencies work together in the NSSP to ensure the safety and wholesomeness of shellfish. The EPA shares responsibility with the FDA for control of pesticide residues in foods and has responsibility for drinking water safety and the composition of effluent from food processing plants. The Customs Service receives assistance from the FDA and USDA in its role to ensure the safety and economic integrity of imported foods. The FTC works with the FDA to prevent deceptive advertising of food products, as affected by food composition and labels. The NCWM, under the NIST within the Department of Commerce, has developed model packaging and labeling regulations related to weights and measures of food packages.

The chemical composition of foods is often an important factor in determining the quality, grade, and price of a food. Government agencies that purchase foods for special programs often rely on detailed specifications that include information on food composition.

International organizations have developed food standards and safety practices to protect consumers, ensure fair business practices, and facilitate international trade. The Codex Alimentarius Commission is the major international standard-setting group for food safety and quality. The International Organization for Standardization has a series of standards that focus on documentation of procedures, with some relevant to food analysis. Certain regional and country-specific organizations also publish standards related to food composition and analysis.

2.8 STUDY QUESTIONS

1. Define the abbreviations FDA, USDA, and EPA, and give two examples for each of what they do or regulate relevant to food analysis.
2. Differentiate "standards of identity," "standards of quality," and "grade standards" with regard to what they are and which federal agency establishes and regulates them.
3. Government regulations regarding the composition of foods often state the official or standard method by which the food is to be analyzed. Give the full name of three

organizations that publish commonly referenced sources of such methods.

4. For each type of product listed below, identify the governmental agency (or agencies) that has regulatory or other responsibility for quality assurance. Specify the general nature of that responsibility and, if given, the specific types of analyses that would be associated with that responsibility.
   (a) Frozen fish sticks
   (b) Contaminants in drinking water
   (c) Dessert wine
   (d) Grade A milk
   (e) Frozen oysters
   (f) Imported chocolate products
   (g) Residual pesticide on wheat grain
   (h) Corned beef
5. Food products purchased by federal agencies often have specifications that include requirements for chemical composition. Give the names of four such specifications.
6. You are developing a food product that will be marketed in another country. What factors will you consider as you decide what ingredients to use and what information to include on the food label? What resources should you use as you make these decisions?
7. Upon completing your college degree, you are employed by a major US food company that processes fruits and vegetables.
   (a) Where, specifically, would you look to find if a standard of identity exists for each of your processed products? What kind of information does such a standard include?
   (b) What US governmental agency sets the standards of identity for such products?
   (c) What are the minimum standards called that are set for some fruit and vegetable products?
   (d) What governmental agency sets the grade standards that you may want to use as a quality control tool and in marketing your products?
   (e) You are concerned about pesticide tolerances for the fruits and vegetables you process. What governmental agency sets those tolerances?
   (f) What governmental agency enforces the pesticide tolerances?
   (g) For nutrition labeling purposes for your products, you want to check on official methods of analysis. Where, specifically, should you look?
   (h) You want to check the detailed rules on nutrition labeling that would apply to your products. Where, specifically, would you look to find those rules?
   (i) You are considering marketing some of your products internationally. What resource could you check to determine if there are international standards and safety practices specified for those products?
   (j) You understand that a quality assurance inspection service, based on "HACCP," is being offered to the fruit and vegetable industry. What does "HACCP" stand for, and what is its intent?

2.9 ACKNOWLEDGMENTS

The author thanks numerous employees of the various agencies and organizations who contributed information and reviewed sections of this chapter.

2.10 REFERENCES

1. Food and Drug Law Institute (2009) Food and drug law and regulation. Food and Drug Law Institute, Washington, DC
2. Fortin ND (2009) Food regulation: law, science, policy, and practice. Wiley-Blackwell, San Francisco, CA
3. Piña KR, Pines WL (2008) A practical guide to food and drug law and regulation, 3rd edn. Food and Drug Law Institute, Washington, DC
4. Curtis PA (2005) Guide to food laws and regulations. Wiley-Blackwell, San Francisco, CA
5. Anonymous (2009) Code of federal regulations. Titles 7, 9, 21, 27, 40, 50. US Government Printing Office, Washington, DC
6. Anonymous (1997) Food safety from farm to table. A national food-safety initiative. A report to the president. Environmental Protection Agency, Department of Health and Human Services, US Department of Agriculture, Washington, DC
7. Cramer MM (2006) Food plant sanitation: design, maintenance, and good manufacturing practices. CRC, Boca Raton, FL
8. Hui YH, Bruinsma BL, Gorham JR, Tong WKS, Ventresca P (eds) (2002) Food plant sanitation. Marcel Dekker, New York
9. Pierson MD, Corlett DA Jr (1992) HACCP principles and applications. Van Nostrand Reinhold, New York
10. USDA (1983) U.S. standards for grades of orange juice. 10 Jan 1983. Processed Products Branch, Fruit and Vegetable Division, Agricultural Marketing Service, US Department of Agriculture. http://www.ams.usda.gov/AMSv1.0/getfile?dDocName=STELDEV3019705
11. National Marine Fisheries Service (NMFS) Fishery products inspection manual (updated continuously). National Seafood Inspection Laboratory, Pascagoula, MS
12. FDA (2001) Fish and fisheries products hazards and controls guide, 3rd edn. Center for Food Safety and Applied Nutrition, Office of Seafood, Food and Drug Administration, Washington, DC
13. FDA (1994) Pesticide analytical manual, vol 1 (PAMI) (updated Oct 1999) (Methods which detect multiple residues) and vol 2 (PAMII) (updated Jan 2002) (Methods for individual pesticide residues), 3rd edn. National Technical Information Service, Springfield, VA. Also available on Internet
14. Eaton AD, Clesceri LS, Rice EW, Greenberg AE (eds) (2005) Standard methods for the examination of water and wastewater, 21st edn. American Public Health Association, Washington, DC

15. EPA (1983) Methods of chemical analysis of water and wastes. EPA-600/4-79-020, March 1979. Reprinted in 1983. EPA Environmental Monitoring and Support Laboratory, Cincinnati, OH
16. American Society for Testing Materials (ASTM) (2009) 6 Annual book of ASTM standards, section 11, water and environmental technology, vol 11.02, water (II). ASTM, West Conshohocken, Philadelphia, PA
17. US International Trade Commission (USITC) (2009) Revision 1 to harmonized tariff schedule of the United States. http://www.usitc.gov/tata/hts/bychapter/index.htm
18. US Department of Health and Human Services, Public Health Service, Food and Drug Administration (2003) Grade A pasteurized milk ordinance. http://www.cfsan.fda.gov/~ear/pmo03toc.html
19. Wehr HM, Frank JF (eds) (2004) Standard methods for the examination of dairy products, 17th edn. American Public Health Association, Washington, DC
20. AOAC International (2007) Official methods of analysis, 18th edn, 2005; current through revision 2, 2007 (On-line). AOAC International, Gaithersburg, MD
21. US Department of Health and Human Services, Public Health Service, Food and Drug Administration. IMS list. Sanitation compliance and enforcement ratings of interstate milk shippers. US Food and Drug Administration, Center for Food Safety and Applied Nutrition. http://www.cfsan.fda.gov/~ear/ims-toc.html
22. USDA (2002) General specifications for dairy plants approved for USDA inspection and grading service. Dairy Program, Agricultural Marketing Service, US Department of Agriculture, Washington, DC. http://www.ams.usda.gov/AMSv1.0/getfile?dDocName=STELDEV3004788
23. USDA (2005) Milk for manufacturing purposes and its production and processing, recommended requirements. Dairy Program, Agricultural Marketing Service, US Department of Agriculture, Washington, DC. http://www.ams.usda.gov/AMSv1.0/getfile?dDocName=STELDEV3004791
24. FDA (1995) National shellfish sanitation program. Manuals of operations. Shellfish Program Implementation Branch, Center for Food Safety and Applied Nutrition, Food and Drug Administration, Washington, DC. http://vm.cfsan.fda.gov/ ear/nsspman.html
25. USDA (2004) Tuna, canned or in flexible pouches. A-A-20155C. 8 Oct 2004. Livestock and Seed Division, Agriculture Marketing Service, US Department of Agriculture, Washington, DC. http://www.ams.usda.gov/AMSv1.0/getfile?dDocName=STELDEV3003155
26. USDA (1997) Institutional meat purchase specifications for fresh pork products. Series 400. July 1997. Livestock and Seed Division, Agricultural Marketing Service, US Department of Agriculture, Washington, DC. http://www.ams.usda.gov/AMSv1.0/getfile?dDocName=STELDEV3003285
27. USDA (1996) Institutional meat purchase specifications for fresh beef products. Series 100. June 1996. Livestock and Seed Division, Agricultural Marketing Service, US Department of Agriculture, Washington, DC. http://www.ams.usda.gov/AMSv1.0/getfile?dDocName=STELDEV3003281
28. USDA (2006) Commodity specification of all-purpose egg mix. September 2006. Poultry Division, Agricultural Marketing Service, US Department of Agriculture, Washington, DC. http://www.ams.usda.gov/AMSv1.0/getfile?dDocName=STELPRDC5048747
29. USDA (2007) USDA Commodity requirements, pasteurized process American cheese for use in domestic programs. PCD-6. 5 Nov 2007. Kansas City Commodity Office, Commodity Credit Corporation, US Department of Agriculture, Kansas City, MO. http://www.fsa.usda.gov/Internet/FSA_File/pcd6.pdf
30. USDA (2007) USDA Commodity requirements, Mozzarella cheese for use in domestic programs. Announcement MCD-4. 15 Oct 2007. Kansas City Commodity Office, Commodity Credit Corporation, US Department of Agriculture, Kansas City, MO. http://www.fsa.usda.gov/Internet/FSA_File/mcd4.pdf
31. USDA (2008) Commercial item description. Syrup. A-A-20124D. 17 April 2008. General Services Administration, Specifications Section, Washington, DC. http://www.dscp.dla.mil/subs/support/specs/cids/20124.pdf
32. USDA (2008) Commercial item description. Tea, instant. A-A-220183C. 21 Oct 2008. General Services Administration, Specifications Unit, Washington, DC. http://www.dscp.dla.mil/subs/support/specs/cids/20183.pdf
33. USDA (2006) Commercial item description. Peanut butter. A-A-20328A. 29 Aug 2006. General Service Administration Specification Unit, Washington, DC. http://www.dscp.dla.mil/subs/support/specs/cids/20328.pdf
34. US Department of Commerce, National Institute of Standards and Technology (2008) Checking the net contents of packaged goods. NIST handbook 133, 4th edn. National Institute of Standards and Technology, Rockville, MD
35. Kellam J, Guarino ET (2000) International food law. Prospect Media, St Leonards, NSW
36. FAO/WHO (2005) Codex Alimentarius. Joint FAO/WHO food standards programme. Codex Alimentarius Commission, Food and Agriculture Organization of the United Nations/World Health Organization, Rome, Italy (available on CD in 2006)
37. International Organization for Standardization (2007) ISO 9000 international standards for quality management, compendium, 11th edn. International Organization for Standardization, New York
38. Hoyle D (2009) ISO 9000 quality systems handbook-updated for the ISO 9001: 2008 standard. Elsevier, New York
39. Early R (2008) ISO 9000 in the food industry. Blackwell, Ames, IA
40. US Pharmacopeia (USP) (2008) Food chemicals codex, 6th edn. United Book, Baltimore, MD
41. JECFA (2006) Monograph 1: combined compendium of food additive specifications, vol 4. Joint FAO/WHO Committee on Food Additives (JECFA). 1956–2005. FAO, Rome, Italy. Available at ftp://ftp.fao.org/docrep/fao/009/a0691e/a0691e00a.pdf

2.11 RELEVANT INTERNET ADDRESSES

American Public Health Association – http://www.apha.org/
American Society of Testing Materials – http://www.astm.org/
AOAC International – http://www.aoac.org/
Bureau of Alcohol, Tobacco, Firearms, and Explosives – http://www.atf.gov/
Centers for Disease Control and Prevention – http://www.cdc.gov/
Code of Federal Regulations – http://www.access.gpo.gov/nara/cfr/cfr-table-search.html
Codex Alimentarius Commission – http://www.codexalimentarius.net/web/index_en.jsp
Department of Commerce – http://www.doc.gov/
  National Institute of Standards and Technology – http://www.nist.gov/
    National Conference on Weights and Measures – http://ts.nist.gov/WeightsAndMeasures/
  National Oceanic and Atmospheric Administration – http://www.noaa.gov/
  National Marine Fisheries Service – http://www.nmfs.noaa.gov/
Environmental Protection Agency – http://www.epa.gov/
Federal Trade Commission – http://www.ftc.gov/
Food Chemicals Codex – http://www.usp.org/fcc/
Food and Drug Administration – http://www.fda.gov/
  Center for Food Safety and Applied Nutrition – http://vm.cfsan.fda.gov/list.html
    Food Labeling and Nutrition – http://vm.cfsan.fda.gov/label.html
    Food Safety Team – http://www.fda.gov/opacom/backgrounders/foodteam.html
    Hazard Analysis Critical Control Point – http://vm.cfsan.fda.gov/~lrd/haccp.html
    Milk Safety References – http://vm.cfsan.fda.gov/~ear/prime.html
    Pesticides, Metals, Chemical Contaminants and Natural Toxins – http://www.cfsan.fda.gov/~lrd/pestadd.html
    Seafood Information and Resources – http://vm.cfsan.fda.gov/seafood1.html
International Organization for Standardization – http://www.iso.ch/
National Shellfish Sanitation Program – http://vm.cfsan.fda.gov/~ear/nsspman.html
US Customs and Border Protection – http://www.customs.gov/; http://www.usitc.gov/tata/hts/bychapter/index.htm
US Department of Agriculture – http://www.usda.gov/
  Agricultural Marketing Service – http://www.ams.usda.gov/
    Quality Standards – http://www.ams.usda.gov/standards/
    Laboratory Testing Program – http://www.ams.usda.gov/labserv.htm
  Food Safety and Inspection Service – http://www.fsis.usda.gov/
    HACCP/Pathogen Reduction – http://www.fsis.usda.gov/oa/haccp.imhaccp.htm
  Grain Inspection, Packers, and Stockyards Administration – http://www.usda.gov/gipsa
  Nutrient Database for Standard Reference – http://www.ars.usda.gov/ba/bhnrc/ndl

Nutrition Labeling

Lloyd E. Metzger

Department of Dairy Science, South Dakota State University, Brookings, SD 57007, USA
lloyd.metzger@sdstate.edu

3.1 Introduction 37
3.1.1 1973 Regulations on Nutrition Labeling 37
3.1.2 Nutrition Labeling and Education Act of 1990 37
3.2 Food Labeling Regulations 38
3.2.1 Mandatory Nutrition Labeling 38
3.2.1.1 Basic Format 38
3.2.1.2 Daily Values and Serving Size 39
3.2.1.3 Simplified Format 40
3.2.1.4 Exemptions 40
3.2.1.5 Rounding Rules 40
3.2.1.6 Caloric Content 40
3.2.1.7 Protein Quality 42
3.2.2 Compliance 42
3.2.2.1 Sample Collection 42
3.2.2.2 Methods of Analysis 42
3.2.2.3 Levels for Compliance 42
3.2.3 Nutrient Content Claims 43
3.2.4 Health Claims 43
3.2.5 Designation of Ingredients 49
3.2.6 National Uniformity and Preemptions Authorized by NLEA 49
3.2.7 Other Provisions of NLEA 49
3.3 Summary 49
3.4 Study Questions 50
3.5 References 50
3.6 Relevant Internet Addresses 51

S.S. Nielsen, *Food Analysis*, Food Science Texts Series, DOI 10.1007/978-1-4419-1478-1_3,

3.1 INTRODUCTION

Nutrition labeling regulations differ in countries around the world. The focus of this chapter is on nutrition labeling regulations in the USA, as specified by the **Food and Drug Administration** (FDA) and the **Food Safety and Inspection Service** (FSIS) of the **United States Department of Agriculture** (USDA). A major reason for analyzing the chemical components of foods in the USA is nutrition labeling regulations. Nutrition label information is not only legally required in many countries, but also is of increasing importance to consumers as they focus more on health and wellness.

The FDA was authorized under the 1906 Federal Food and Drug Act and the 1938 **Federal Food, Drug, and Cosmetic** (FD&C) **Act** to require certain types of food labeling (1, 2). This labeling information includes the amount of food in a package, its common or usual name, and its ingredients. The **1990 Nutrition Labeling and Education Act** (NLEA) (2,3) modified the 1938 FD&C Act to regulate nutrition labeling. Additionally, the 1997 **Food and Drug Administration Modernization Act** (FDAMA) (4) also amended the FD&C act and included provisions that speed up the process for approving health and nutrient content claims.

The FDA and FSIS of the USDA (referred to throughout the chapter simply as FSIS) have coordinated their regulations for nutrition labeling. The regulations of both agencies strive to follow the intent of the NLEA, although only the FDA is bound by the legislation. The differences that exist in the regulations are due principally to the inherent differences in the food products regulated by the FDA and USDA (USDA regulates meat, poultry, and egg products only). The two agencies maintain close harmony regarding interpretation of the regulations and changes made in regulations.

Complete details of the current nutrition labeling regulations are available in the **Federal Register** and the **Code of Federal Regulations** (CFR) (5–8). The 1973 regulations on voluntary nutrition labeling and the 1990 NLEA are described briefly below, followed by select aspects of current FDA and FSIS nutrition labeling regulations. In developing a nutrition label for a food product, refer to details of the regulations in the CFR and other references cited. A reference manual that explains nutritional labeling regulations (with continual updating) can be purchased from the National Food Processors Association (a nonprofit organization) (9) and several commercial publishers (10). The FDA has available on the Internet *A Food Labeling Guide* (11), which includes *Food Labeling – Questions and Answers*, available for the industry. Since Internet addresses are frequently updated, the home page for the FDA is recommended as a starting point. From the FDA homepage the search function can be utilized to conduct a search based on the document title provided in the references section. The FDA homepage and other relevant Internet addresses are given at the end of this chapter.

Interpretation of nutrition labeling regulations can be difficult. Additionally, during the product development process the effect of formulation changes on the nutritional label may be important. As an example, a small change in the amount of an ingredient may determine if a product can be labeled low fat. As a result, the ability to immediately approximate how a formulation change will impact the nutritional label can be valuable. The use of nutrient databases and computer programs designed for preparing and analyzing nutritional labels can be valuable and can simplify the process of preparing a nutritional label. The use of computer programs to prepare nutritional labels is beyond the scope of this chapter. However, an example computer program (TechWizard^{TM}) and a description of how this program can be used to prepare a nutrition label are found in the laboratory manual that accompanies the text.

3.1.1 1973 Regulations on Nutrition Labeling

The FDA promulgated regulations in 1973 that permitted, and in some cases required, foods to be labeled with regard to their nutritional value (1, 2). Nutrition labeling was required only if a food contained an added nutrient or if a nutrition claim was made for the food on the label or in advertising. The nutrition label included the following: serving size; number of servings per container; Calories per serving; grams of protein, carbohydrate, and fat per serving; and percentage of **US Recommended Dietary Allowance** (USRDA) per serving of protein, vitamins A and C, thiamine, riboflavin, niacin, calcium, and iron. In 1984, the FDA adopted regulations to include sodium content on the nutritional label (effective from 1985).

3.1.2 Nutrition Labeling and Education Act of 1990

Since the nutrition label was established in 1973, dietary recommendations for better health have focused more on the role of Calories and macronutrients (e.g., total fat) in chronic diseases and less on the role of micronutrients (minerals and vitamins) in deficiency diseases. Therefore, in the early 1990s the FDA revised the content of the nutritional label to make it more consistent with current dietary concerns (see Table 3-1 and Fig. 3-1, which are discussed more in Sect. 3.2.1). The list of specific nutrients to be included on the nutrition label was only one aspect of the NLEA

Mandatory (Bold) and Voluntary Components for Food Label Under Nutrition Labeling and Education Act of 1990

Total Calories
Calories from fat
Calories from saturated fat
Total fat
  Saturated fat
  ***Trans* fat**
  Polyunsaturated fat
  Monounsaturated fat
Cholesterol
Sodium
Potassium
Total carbohydrate
  Dietary fiber
  Soluble fiber
  Insoluble fiber
  Sugars
  Sugar alcohols (e.g., sugar substitutes xylitol, mannitol, and sorbitol)
Other carbohydrates (the difference between total carbohydrate and the sum of dietary fiber, sugars, and sugar alcohols, if declared)
Protein
Vitamin A
% of Vitamin A present as beta-carotene
Vitamin C
Calcium
Iron
Other essential vitamins and minerals

From (7), updated.
Nutrition panel will have the heading "Nutrition Facts." Only components listed are allowed on the nutrition panel, and they must be in the order listed. Components are to be expressed as amount and/or as percent of an established "Daily Value."

Nutrition Facts
Serving Size 1/2 cup (about 82g)
Servings Per Container 8

Amount Per Serving

Calories 200	Calories from Fat 130
	% Daily Value*
Total Fat 14g	**22%**
Saturated Fat 9g	**45%**
Trans Fat 0g	
Cholesterol 55mg	**18%**
Sodium 40mg	**2%**
Total Carbohydrate 17g	**6%**
Dietary Fiber 1g	**4%**
Sugars 14g	
Protein 3g	

Vitamin A 10% • Vitamin C 0%
Calcium 10% • Iron 6%

*Percent Daily Values are based on a 2,000 calorie diet. Your daily values may be higher or lower depending on your calorie needs:

	Calories:	2,000	2,500
Total Fat	Less than	65g	80g
Saturated Fat	Less than	20g	25g
Cholesterol	Less than	300mg	300 mg
Sodium	Less than	2,400mg	2,400mg
Total Carbohydrate		300g	375g
Dietary Fiber		25g	30g

Calories per gram:
Fat 9 • Carbohydrate 4 • Protein 4

An example of the nutrition label, Nutrition Labeling and Education Act of 1990. (Courtesy of the Food and Drug Administration, Washington, DC.)

of 1990 (NLEA) (2, 3), which amended the FD&C Act with regard to five primary changes:

1. Mandatory nutrition labeling for almost all food products
2. Federal regulation of nutrient content claims and health claims
3. Authority for states to enforce certain provisions of FD&C Act
4. Federal preemption over state laws for misbranding provisions
5. Declaration of ingredients

In 2003 the FDA also published a final rule in the Federal Register that amended food labeling regulations to require *trans*-fatty acid declaration (68 FR 41434) (effective date of rule: January 1, 2006). This modification in nutritional labeling resulted from reports that intake of trans fat and other cholesterol-raising fat should be limited (12).

3.2 FOOD LABELING REGULATIONS

For each aspect of nutrition labeling regulations described below, general or FDA labeling requirements are covered, followed by, if applicable, certain FSIS regulations that differ from the FDA requirements. While the focus here is on **mandatory nutrition labeling**, it should be noted that the FDA has guidelines for **voluntary nutrition labeling** of raw fruit, vegetables, and fish (21 CFR 101.45), and FSIS has guidelines for voluntary nutrition labeling of single-ingredient raw meat and poultry products (9 CFR 317.445, 381.445). These FDA and FSIS guidelines for voluntary nutrition labeling differ in issues such as source of nutrient databases used, compliance checks, and use of claims on product labels.

3.2.1 Mandatory Nutrition Labeling

3.2.1.1 Basic Format

The FDA regulations implementing the 1990 NLEA require nutrition labeling for most foods offered for sale and regulated by the FDA (21 CFR 101.9), and FSIS regulations require nutrition labeling of most meat or meat products (9 CFR 317.300 to 317.400) and poultry products (9 CFR 381.400 to 381.500). Certain nutrient information is required on the label, and other information is voluntary (Table 3-1). In addition, while FSIS allows voluntary declaration of stearic acid content on the label, FDA does not, but has been petitioned to do so.

The standard format for nutrition information on food labels [21 CFR 101.9 (d)] is given in Fig. 3-1 and consists of the following:

1. Serving size and servings per container
2. Quantitative amount per serving of each nutrient or dietary component except vitamins and minerals
3. Amount of each nutrient, except sugars and protein, as a percent of the **Daily Value** (i.e., the new label reference values) for a 2000-Calorie (Cal) diet
4. Footnote with Daily Values for selected nutrients based on 2000-Cal and 2500-Cal diets.

3.2.1.2 Daily Values and Serving Size

Daily Value (DV) is a generic term used to describe two separate terms that are (1) **Reference Daily Intake** (RDI) and (2) **Daily Reference Value** (DRV). The term RDI is used for essential vitamins and minerals, and the values are shown in Table 3-2. The term DRV is used for food components (total fat, saturated fat, cholesterol, total carbohydrate, dietary fiber, sodium, potassium, and protein), and the values are shown in Table 3-3. A DRV for sugar and *trans* fat has not been established. The DRVs are based on a 2000 or 2500 reference Calorie intake. Nutrient content values and percent Daily Value calculations for the nutrition label are based on serving size. Serving size regulations of the FDA and FSIS differ in issues such as product categories, reference amounts, and serving size for units or pieces [21 CFR 101.12 (b), 101.9 (b); 9 CFR 317.312 (b), 381.412 (b), 317.309 (b), 381.409 (b)]. The serving size regulations for the FDA are described in more detail below.

For the FDA, use of the term serving or serving size is defined in CFR 101.9(b)(1) as "an amount of food customarily consumed per eating occasion by persons 4 years of age or older which is expressed in a common household measure that is appropriate to the food." Additionally, if the food is targeted for infants or toddlers, the term serving or serving size is defined as "an amount of food customarily consumed per eating occasion by infants up to 12 months of age or by children 1 to 3 years of age respectively." The FDA has defined the "reference amount customarily consumed per eating occasion" in CFR 101.12(b) Table 3-1 and Table 3-2. These data are based on national food consumption surveys as well as the serving size used in dietary guidance recommendations, serving sizes

table 3-2 Reference Daily Intakes (RDIs) for Vitamins and Minerals Essential in Human Nutrition

Nutrient	*RDI*
Vitamin A	5000 IU
Vitamin C	60 mg
Calcium	1000 mg
Iron	18 mg
Vitamin D	400 IU
Vitamin E	30 IU
Vitamin K	80 μg
Thiamin	1.5 mg
Riboflavin	1.7 mg
Niacin	20 mg
Vitamin B_6	2 mg
Folate	400 μg
Vitamin B_{12}	6 μg
Biotin	300 μg
Pantothenic acid	10 mg
Phosphorus	1,000 mg
Iodine	150 μg
Magnesium	400 mg
Zinc	15 mg
Selenium	70 μg
Copper	2 mg
Manganese	2 mg
Chromium	120 μg
Molybdenum	75 μg
Chloride	3400 mg

From 21 CFR 101.9 (c) (8) (iv) (2009).
Values are for adults and children 4 or more years of age. RDI values have also been established for infants, children under 4 years of age, and pregnant and lactating women. RDI values listed by the Food Safety and Inspection Service [9 CFR 317.309 (c) (8) (iv); 9 CFR 381.409 (c) (8) (iv)] are as above but do not include values for chloride, chromium, manganese, vitamin K, molybdenum, and selenium.

table 3-3 Daily Reference Values (DRVs) of Food Components Based on the Reference Calorie Intake of 2000 Calories

Food Component	*DRV*
Fat	65 g
Saturated fatty acids	20 g
Cholesterol	300 mg
Total carbohydrate	300 g
Fiber	25 g
Sodium	2400 mg
Potassium	3500 mg
Protein	50 g

From 21 CFR 101.9 (c) (9) (2009). Same as in 9 CFR 317.309 (c) (9) and 9 CFR 381.409 (c) (9).

recommended in comments, serving size used by manufactures and grocers, or serving sizes used by other countries. Adjustments in the reference amounts can be initiated by the FDA or in response to a petition. The labeled serving size and reference amount are important since the use of nutrient content claims is dependent on the serving size and the reference amount. The use of nutrient content claims is outlined in Sect. 3.2.3.

3.2.1.3 Simplified Format

A simplified format for nutrition information on FDA-regulated foods may be used if seven or more of the 13 required nutrients are present in only insignificant amounts (but does not include Calories from fat) (e.g., soft drinks) [21 CFR 101.9 (f)]. For such foods, information on five core nutrients (Calories, total fat, total carbohydrate, protein, and sodium) must be given. However, if other mandatory nutrients are present in more than insignificant amounts they must be listed. "Insignificant" is defined generally as the amount that allows a declaration of zero on the nutrition label. However, in the cases of protein, total carbohydrate, and dietary fiber, insignificant is the amount that allows a statement of "less than 1 gram." The footnotes required with the basic format are not required for the simplified format label, except that the statement "Percent Daily Values are based on a 2000 Calorie diet" must be included. The statement "Not a significant source of _____" is optional on the simplified format label of an FDA-regulated product, unless a nutrient claim is made on the label or optional nutrients (e.g., potassium) are voluntarily listed on the nutrition label, or if any vitamins or minerals are required to be added as a nutrient supplement to foods for which a standard of identity exists.

For USDA-regulated foods, a simplified nutrition label format may be used when any required nutrient other than a core nutrient (Calories, total fat, sodium, carbohydrate, or protein) is present in an insignificant amount [9 CFR 317.309 (f) (1) and (4), 381.409 (f) (1) and (4)]. Any required nutrient, other than a core nutrient, that is present in an insignificant amount may be omitted from the tabular listing if it is listed in a footnote, "Not a significant source of _____." This option also exists for FDA-regulated foods, but it is known as a "shortened" format [21 CFR 101.9 (c); see listing for each noncore nutrient].

3.2.1.4 Exemptions

Certain foods are exempt from FDA mandatory nutrition labeling requirements [21 CFR 101.9 (j)] (Table 3-4), unless a nutrient content claim or health claim is made or any other nutrition information is provided. Special labeling provisions apply to certain other foods as specified in 21 CFR 101.9(j) (e.g., foods in small packages; foods for young children; game meats, shell eggs; foods sold from bulk containers; unit containers in multiunit packages; foods in gift packs). Infant formula must be labeled in accordance with 21 CFR 107, and raw fruits, vegetables, and fish according to 21 CFR 101.45. Dietary supplements must be labeled in accordance with 21 CFR 101.36.

Table 3-4 Foods Exempt from Mandatory Nutrition Labeling Requirements by the FDA

Food offered for sale by small business
Food sold in restaurants or other establishments in which food is served for immediate human consumption
Foods similar to restaurant foods that are ready to eat but are not for immediate consumption are primarily prepared on site and are not offered for sale outside that location
Foods that contain insignificant amounts of all nutrients subject to this rule, e.g., coffee and tea
Dietary supplements
Infant formula
Medical foods
Foods shipped or sold in bulk form and not for sale to consumers
Raw fruits, vegetables, and fish
Packaged single-ingredient products of fish or game meat
Game meats
Food in small packages
Shell eggs packaged in a carton
Unit containers in a multiunit retail food package that bears a nutrition label
Food products sold from bulk container

Summarized from 21 CFR 101.9 (j) (2009).
See details in regulations for foods exempt from mandatory nutrition labeling requirements.

Exemptions from mandatory nutrition labeling for USDA-regulated foods ([9 CFR 317.400, 381.500]) differ somewhat from those for FDA-regulated foods regarding issues such as definitions of a small business, small package, and retail product.

3.2.1.5 Rounding Rules

Increments for the numerical expression of quantity per serving are specified for all nutrients (Table 3-5, as summarized by FDA) [21 CFR 101.9 (c); 9 CFR 317.309 (c), 381.409 (c)]. For example, Calories are to be reported to the nearest 5 Cal up to and including 50 Cal and to the nearest 10 Cal above 50 Cal. Calories can be reported as zero if there are less than 5 Cal per serving.

3.2.1.6 Caloric Content

Caloric conversion information on the label for fat, carbohydrate, and protein is optional. Calories can be

3-5 table **Rounding Rules for Declaring Nutrients on Nutrition Label**

Nutrient/Serving	*Increment Rounding*[a,b]	*Insignificant Amount*
Calories, Calories from fat, Calories from saturated fat	<5 Cal – express as zero ≤50 Cal – express to nearest 5 Cal increment > 50 Cal – express to nearest 10 Cal increment	<5 Cal
Total fat, *trans* fat, polyunsaturated fat, monounsaturated, saturated fat	<0.5 g – express as zero <5 g – express to nearest 0.5 g increment ≥5 g – express to nearest 1 g increment	<0.5 g
Cholesterol	<2 mg – express as zero 2–5 mg – express as "less than 5 mg" >5 mg – express to nearest 5 mg increment	<2 mg
Sodium, potassium	<5 mg – express as zero 5–140 mg – express to nearest 5 mg increment >140 mg – express to nearest 10 mg increment	<5 mg
Total carbohydrate, sugars, sugar alcohols, other carbohydrates, dietary fiber, soluble fiber, insoluble fiber, protein	<0.5 g – express as zero <1 g – express as "Contains less than 1 g" OR "less than 1 g" ≥1 g – express to nearest 1 g increment	<1 g
Vitamins and minerals	<2% of RDI – may be expressed as: 1. 2% If actual amount is 1.0% or more 2. Zero 3. An asterisk that refers to statement "Contains less than 2% of the Daily Value of this (these) nutrient (nutrients)" 4. For Vitamins A and C, calcium, iron: statement "Not a significant source of _____ (listing the vitamins or minerals omitted)" ≤10% of RDI – express to nearest 2% increment >10% to ≤50% of RDI – express to nearest 5% increment >50% of RDI – express to nearest 10% increment	<2% RDI
Beta-carotene	≤10% of Vitamin A – express to nearest 2% increment >10% to ≤50% of Vitamin A – express to nearest 5% increment >50% of Vitamin A – express to nearest 10% increment	

Summarized from Food Labeling Guide – Appendix H: Rounding the values according to FDA rounding rules (2009) Center for Food Safety and Applied Nutrition, Food and Drug Administration, Washington, DC.

[a]To express to the nearest 1g increment, amounts exactly halfway between two whole numbers or higher (e.g., 2.50–2.99 g) round up (e.g., 3 g), and amounts less than halfway between two whole numbers (e.g., 2.01–2.49 g) round down (e.g., 2 g).

[b]Notes for rounding % Daily Value (DV) for total fat, saturated fat, cholesterol, sodium, total carbohydrate, fiber, and protein:

(1) To calculate %DV, divide either the actual (unrounded) quantitative amount or the declared (rounded) amount by the appropriate RDI or DRV. Use whichever amount will provide the greatest consistency on the food label and prevent unnecessary consumer confusion (21 CFR 101.9 (d)(7)(iii).

(2) When %DV values fall between two whole numbers, rounding shall be as follows:

- For values exactly halfway between two whole numbers or higher (e.g., 2.50–2.99) the values shall round up (e.g., 3%).
- For values less than halfway between whole numbers (e.g., 2.01–2.49) the values shall round down (e.g., 2%).

expressed in numerous ways. A "**calorie**," which is the standard for measurement of the energy value of substances and to express the body's energy requirement, is the amount of heat required to raise the temperature of 1 g of water 1°C (1 calorie = 4.184 joules). The unit used in nutritional work is "**Calorie**" or "kilo-calorie" (kcal), which equals 1000 calories. In this chapter, the term Calorie is used to express caloric content. The FDA regulations specify five methods by which caloric content may be calculated, one of which uses bomb calorimetry [21 CFR 101.9 (c) (1)]:

1. Specific Atwater factors for Calories per gram of protein, total carbohydrate, and total fat
2. The general factors of 4, 4, and 9 Cal/g of protein, total carbohydrate, and total fat, respectively

3. The general factors of 4, 4, and 9 Cal/g of protein, total carbohydrate, less the amount of insoluble dietary fiber, and total fat, respectively
4. Data for specific food factors for particular foods or ingredients approved by the FDA
5. Bomb calorimetry data subtracting 1.25 Cal/g protein to correct for incomplete digestibility

FSIS allows only the calculation procedures 1–4 above and not the use of bomb calorimetry for caloric content [9 CFR 317.309 (c) (1) (i), 381.409 (c) (l) (i)].

3.2.1.7 Protein Quality

For both FDA-regulated and USDA-regulated foods, reporting the amount of protein as a percent of its Daily Value is optional, except if a protein claim is made for the product, or if the product is represented or purported to be used by infants or children under 4 years of age, in which case the statement is required [21 CFR 101.9 (c) (7); 9 CFR 317.309 (c) (7) (i), 381.409 (c) (7) (i)]. For infant foods, the corrected amount of protein per serving is calculated by multiplying the actual amount of protein (g) per serving by the relative protein quality value. This relative quality value is the **protein efficiency ratio** (PER) value of the subject food product divided by the PER value for casein. For foods represented or purported for adults and children 1 year or older, the corrected amount of protein per serving is equal to the actual amount of protein (g) per serving multiplied by the **protein digestibility-corrected amino acid score** (PDCAAS). Both the PER and PDCAAS methods to assess protein quality are described in Chap. 15. The FDA and FSIS allow use of the general factor 6.25 and food-specific factors for this calculation (described in Chap. 9).

3.2.2 Compliance

Compliance procedures of the FDA and FSIS for nutrition labeling differ somewhat in sample collection, specified methods of analysis, and levels required for compliance [21 CFR 101.9 (g); 9 CFR 317.309 (h), 381.409 (h)].

3.2.2.1 Sample Collection

Random sampling techniques are used by the FDA to collect samples to be analyzed for compliance with nutrition labeling regulations. A "lot" is the basis for sample collection by the FDA, defined as "a collection of primary containers or units of the same size, type, and style produced under conditions as nearly uniform as possible, and designated by a common container code or marking, or in the absence of any common container code or marking, a day's production." The sample used by the FDA for nutrient analysis consists of a "composite of 12 subsamples (consumer units), taken 1 from each of 12 different randomly chosen shipping cases, to be representative of a lot" [21 CFR 101.9 (g)].

The FSIS defines a "lot" similar to that of the FDA. However, the sample used by FSIS for compliance analysis is a composite of a minimum of six consumer units: (1) each from a production lot, or (2) each chosen randomly to be representative of a production lot [9 CRF 317.309 (h), 381.409 (h)].

3.2.2.2 Methods of Analysis

The FDA states that unless a particular method of analysis is specified in 21 CFR 101.9(c), appropriate methods of AOAC International published in the *Official Methods of Analysis* (13) are to be used. Other reliable and appropriate methods can be used if no AOAC method is available or appropriate. If scientific knowledge or reliable databases have established that a nutrient is not present in a specific product (e.g., dietary fiber in seafood, cholesterol in vegetables), the FDA does not require analyses for the nutrients. FSIS specifies for nutritional analysis the methods of the *USDA Analytical Chemistry Laboratory Guidebook* (14). If no USDA method is available and appropriate for the nutrient, methods in the *Official Methods of Analysis* of AOAC International (13) are to be used. If no USDA, AOAC International, or specified method is available and appropriate, FSIS specifies the use of other reliable and appropriate analytical procedures as determined by the Agency.

3.2.2.3 Levels for Compliance

The FDA and FSIS both monitor accuracy of nutrient content information for compliance based on two classes of nutrients and an unnamed third group, as described in Table 3-6. For example, a product fortified with iron would be considered misbranded if it contained less than 100% of the label declaration. A product that naturally contains dietary fiber would be considered misbranded if it contained less than 80% of the label declaration. A product would be considered misbranded if it had a caloric content greater than 20% in excess of the label declaration. Reasonable excesses over labeled amounts (of a vitamin, mineral, protein, total carbohydrate, polyunsaturated or monounsaturated fat, or potassium) or deficiencies below label amounts (of Calories, sugars, total fat, saturated fat, cholesterol, or sodium) are acceptable within current

3-6 table **Basis for Compliance of Nutrition Labeling Regulation by Food and Drug Administration and Food Safety and Inspection Service of the US Department of Agriculture**

Class of Nutrients	*Purposes of Compliance*	*Nutrients Regulated*	*% Required*[a]
I	Added nutrients in fortified or fabricated foods	Vitamin, mineral, protein, dietary fiber, potassium	≥100%
II	Naturally occurring (indigenous) nutrients	Vitamin, mineral, protein, total carbohydrate, dietary fiber, other carbohydrate, polyunsaturated or monounsaturated fat, potassium	≥80%
*b		Calories, sugars, total fat, saturated fat, cholesterol, sodium	≤120%

Summarized from 21 CFR 101.9 (g), 9 CFR 317.309 (h), and 9 CFR 381.409(h) (2009).
[a]Amount of nutrient required in food sample as a percentage of the label declaration, or else product is considered misbranded.
[b]* Indicates unnamed class.

Good Manufacturing Practices (cGMP). Noncompliance with regard to a nutrition label can result in warning letters, recalls, seizures, and prosecution (21 CFR 1.21). Compliance Policy Guides and the Regulatory Procedures Manual are available from the FDA Office of Regulatory Affairs (15).

For FDA-regulated foods, compliance with the regulations described above can be obtained by use of FDA-approved databases (16) [21 CFR 101.9 (g) (8)] that have been computed using FDA guidelines, and foods have been handled under cGMP conditions to prevent nutritional losses. For USDA-regulated foods, compliance enforcement described previously is not applicable to single-ingredient, raw meat products (including those frozen previously), when nutrition labeling is based on database values in USDA's National Nutrient Data Bank or in USDA Handbook No. 8 (most recent version on the Internet) (17). The USDA does not preapprove databases, but provides a manual for guidance in using them (18).

3.2.3 Nutrient Content Claims

The FDA and FSIS have defined **nutrient content claims** that characterize the level of a nutrient. Nutrient content claims are based on definitions related to certain nutrient levels. The terms include "free," "low," "lean," "light," "reduced," "less," "fewer," "added," "extra," "plus," "fortified," "enriched," "good source," "contains," "provides," "more," "high," "rich in," "excellent source of," and "high potency" (21 CFR 101.13, 101.54-101.67; 9 CFR 317.313, 317.354-317.363, 381.413, 381.454-381.463) (Tables 3-7 and 3-8 are FDA regulations). The term "healthy" or its derivatives may be used on the label or in labeling of foods under conditions defined by the FDA and FSIS (Table 3-9 is FDA summary). The FDA requirements on nutrient content claims do not apply to infant formulas and medical foods.

Only nutrient descriptors defined by the FDA or FSIS may be used. The terms "less" (or "fewer"), "more," "reduced," "added" (or "extra," "plus," "fortified," and "enriched"), and "light" are relative terms and require label information about the food product that is the basis of the comparison. The percentage difference between the original food (reference food) and the food product being labeled must be listed on the label for comparison. The criteria for selecting an appropriate reference food are dependent on the claim term used. A summary of the criteria is in Table 3-8, and a complete description is found in 21 CFR 101.13(j)(1).

FSIS regulations on nutrient content claims differ from those of the FDA in that the terms "enriched" and "fortified" are not defined in the FSIS regulations. The terms "lean" and "extra lean" are defined and approved for use on all USDA-regulated products, but only for the FDA-regulated products of seafood, game meat, and meal products. The term "____% Lean" is approved for USDA-regulated but not FDA-regulated products. The term "___% Fat Free" is approved for use by both organizations as long as the food meets the definition for "low fat." Conditions for use of the term "light" as part of a brand name to describe sodium content differ somewhat between FDA and USDA regulations [21 CFR 101.56 (b) (4); 9 CFR 356 (d) (3); 9 CFR 381.456 (d) (3)]. FDA and FSIS regulations also differ in the definition of "meal product" as it relates to nutrient content claims [21 CFR 101.13 (l) and (m); 9 CFR 317.313 (l), 318.413 (l)].

3.2.4 Health Claims

The FDA has defined and will allow claims for certain relationships between a nutrient or a food and the risk of a disease or health-related condition (21 CFR 101.14). An overview of the claims that can be made for conventional foods and dietary supplements is available on the Internet (19). The FDA utilizes three different types of oversight to determine which health

Food and Drug Administration: Definitions of Nutrient Content Claims

Nutrients	*Free*[a]	*Low*[b]	*Reduced/Less*[c]
Calories § 101.60(b)	Less than 5 Calories per reference amount and per labeled serving	40 Calories or less per reference amount (and per 50 g if reference amount is small) Meals and main dishes: 120 cal or less per 100 g	At least 25% fewer Calories per reference amount than an appropriate reference food Reference food may not be "Low Calorie" Uses term "Fewer" rather than "Less"
	Comments "Light" or "Lite": If 50% or more of the Calories are from fat, fat must be reduced by at least 50% per reference amount. If less than 50% of Calories are from fat, fat must be reduced at least 50% or Calories reduced at least 1/3 per reference amount "Light" or "Lite" meal or main dish product meets definition for "Low Calorie" or "Low Fat" meal and is labeled to indicate which definition is met For dietary supplements: Calorie claims can only be made when the reference product is greater than 40 Calories per serving		
Total fat § 101.62 (b)	Less than 0.5 g per reference amount and per labeled serving (or for meals and main dishes, less than 0.5 g per labeled serving) No ingredient that is fat or understood to contain fat except as noted below*	3 g or less per reference amount (and per 50 g if reference amount is small) Meals and main dishes: 3 g or less per 100 g and not more than 30% of Calories from fat	At least 25% less fat per reference amount than an appropriate reference food Reference food may not be "Low Fat"
	Comments "_____% Fat Free": OK if food meets the requirements for "Low Fat" 100% Fat Free: Food must be "Fat Free" "Light": See above For dietary supplements: Fat claims cannot be made for products that are 40 Calories or less per serving		
Saturated Fat § 101.62(c)	Less than 0.5 g saturated fat and less than 0.5 g *trans* fatty acids per reference amount and per labeled serving (or for meals and main dishes, less than 0.5 g saturated fat and less than 0.5 g *trans* fatty acid per labeled serving) No ingredient that is understood to contain saturated fat except as noted below*	1 g or less per reference amount and 15% or fewer Calories from saturated fat Meals and main dishes: 1 g or less per 100 g and less than 10% of Calories from saturated fat	At least 25% less saturated fat per reference amount than an appropriate reference food Reference food may not be "Low Saturated Fat"
	Comments Next to all saturated fat claims, must declare the amount of cholesterol if 2 mg or more per reference amount; and the amount of total fat if more than 3 g per reference amount (or 0.5 g or more of total fat for "Saturated Fat Free").[c] For dietary supplements: Saturated fat claims cannot be made for products that are 40 Calories or less per serving		
Cholesterol § 101.62(d)	Less than 2 mg per reference amount and per labeled serving (or for meals and main dishes, less than 2 mg per labeled serving)	20 mg or less per reference amount (and per 50 g of food if reference amount is small) If qualifies by special processing and total fat exceeds 13 g per reference amount and labeled	At least 25% less cholesterol per reference amount than an appropriate reference food Reference food may not be "Low Cholesterol"

(continued)

Food and Drug Administration: Definitions of Nutrient Content Claims (continued)

Nutrients	*Free*[a]	*Low*[b]	*Reduced/Less*[c]
	No ingredient that contains cholesterol except as noted below* If less than 2 mg per reference amount by special processing and total fat exceeds 13 g per reference amount and labeled serving, the amount of cholesterol must be "Substantially Less" (25%) than in a reference food with significant market share (5% of market)	serving, the amount of cholesterol must be "Substantially Less" (25%) than in a reference food with significant market share (5% of market) Meals and main dishes: 20 mg or less per 100 g	
	Comments Cholesterol claims allowed only when food contains 2 g or less saturated fat per reference amount, or for meals and main dish products, per labeled serving size for "free" claims and per 100 g for "low" and "reduced/less" claims Must declare the amount of total fat next to cholesterol claim when fat exceeds 13 g per reference amount of labeled serving (or per 50 g of food if reference amount is small), or when the fat exceeds 19.5 g per labeled serving for main dishes or 26 g for meal products For dietary supplements: cholesterol claims cannot be made for products that are 40 Cal or less per serving		
Sodium § 101.61	Less than 5 mg per reference amount and per labeled serving (or for meals and main dishes, less than 5 mg per labeled serving) No ingredient that is sodium chloride or generally understood to contain sodium except as noted below	140 mg or less per reference amount (and per 50 g if reference amount is small) Meals and main dishes: 140 mg or less per 100 g	At least 25% less sodium per reference amount than an appropriate reference food Reference food may not be "Low Sodium"
	Comments "Light" (for sodium reduced products): If food is "Low Calorie" and "Low Fat" and sodium is reduced by at least 50% "Light in Sodium": If sodium is reduced by at least 50% per reference amount. Entire term "Light in Sodium" must be used in the same type size, color & prominence. "Light in Sodium" for meals = "Low in Sodium" "Very Low Sodium": 35 mg or less per reference amount (and per 50 g if reference amount is small). For meals and main dishes: 35 mg or less per 100 g "Salt Free" must meet criterion for "Sodium Free" "No Salt Added" and "Unsalted" must meet conditions of use and must declare "This is Not A Sodium Free Food" on information panel if food is not "Sodium Free" "Lightly Salted": 50% less sodium than normally added to reference food and if not "Low Sodium" so labeled on information panel		
Sugars § 101.60(c)	"Sugar Free": Less than 0.5 g sugars per reference amount and per labeled serving (or for meals and main dishes, less than 0.5 g per labeled serving) No ingredient that is a sugar or generally understood to contain sugars except as noted below* Disclose Calorie profile (e.g., "Low Calorie")	Not defined. No basis for a recommended intake	At least 25% less sugars per reference amount than an appropriate reference food May not use this claim on dietary supplements of vitamins and minerals

(continued)

3-7 table **Food and Drug Administration: Definitions of Nutrient Content Claims (continued)**

Nutrients	*Free*[a]	*Low*[b]	*Reduced/Less*[c]
	Comments "No Added Sugars" and "Without Added Sugars" are allowed if no sugar or sugar containing ingredient is added during processing. State if food is not "Low" or "Reduced Calorie" The terms "Unsweetened" and "No Added Sweeteners" remain as factual statements Claims about reducing dental caries are implied health claims Does not include sugar alcohols		

From (11), pp. 95–98. updated.
[a]Synonyms for "Free": "Zero," "No," "Without," "Trivial Source of," "Negligible Source of," "Dietarily Insignificant Source of," Definitions for "Free" for meals and main dishes are the stated values per labeled serving.
[b]Synonyms for "Low": "Little" ("Few" for Calories), "Contains a Small Amount of," "Low Source of".
[c]Synonyms for "Reduced/Less": "Lower" ("Fewer" for Calories), "Modified" may be used in statement of identity, Definitions for meals and main dishes are same as for individual foods on a per 100 g basis.
Comments. For "Free," "Very Low," or "Low": must indicate if food meets a definition without benefit of special processing, alteration, formulation, or reformulation; e.g., "broccoli, a fat free food," or "celery, a low Calorie food".
Notes. *Except if the ingredient listed in the ingredient statement has an asterisk that refers to the footnote (e.g. "adds a trivial amount of fat.")
♦ "Reference Amounts" = reference amount customarily consumed.
♦ "Small Reference Amount" = reference amount of 30 g or less or 2 tablespoons or less (for dehydrated foods that are typically consumed when rehydrated with water or a diluent containing an insignificant amount, as defined in 21 CFR 101, p(f)(1) of all nutrients per reference amount, the per 50 g criterion refers to the prepared form of the food).
♦ Statement "See _______ panel for nutrition information" must accompany all content claims. When levels exceed: 13 Fat, 4 g Saturated Fat, 60 mg Cholesterol, or 480 mg Sodium per reference amount, per labeled serving or, for foods with small reference amounts, per 50 g. Disclosure statement is required as part of claim (e.g. "See side panel for ____ content" with the blank filled with nutrient(s) that exceed the prescribed levels).

claims may be used in labeling of food or dietary supplements: 1990 NLEA, 1997 FDAMA, and 2003 FDA Consumer Health Information for Better Nutrition Initiative. These three types of FDA oversight have resulted in three categories of health claims: NLEA-authorized health claims, health claims based on authoritative statements, and qualified health claims based on enforcement discretion.

The **NLEA-authorized health claims** characterize a relationship between a food, food component, dietary ingredient, or dietary supplement and risk of a disease. These health claims are authorized based on an extensive review of scientific literature using the significant scientific agreement standard. An industry guidance document is available that includes the FDA process for evaluating the scientific evidence for a health claim and the meaning of the significant scientific agreement standard (20). The NLEA-authorized claims include the following:

1. Calcium and osteoporosis (21 CFR 101.72)
2. Dietary fat and cancer (21 CFR 101.73)
3. Sodium and hypertension (21 CFR 101.74)
4. Dietary saturated fat and cholesterol and risk of coronary heart disease (21 CFR 101.75)
5. Fiber-containing grain products, fruits, and vegetables and cancer (21 CFR 101.76)
6. Fruits, vegetables, and grain products that contain fiber, particularly soluble fiber, and risk of coronary heart disease (21 CFR 101.77)
7. Fruits and vegetables and cancer (21 CFR 101.78)
8. Folate and neural tube defects (21 CFR 101.79)
9. Dietary noncariogenic carbohydrate sweeteners and dental caries (21 CFR 101.80)
10. Soluble fiber from certain foods and risk of coronary heart disease (21 CFR 101.81)
11. Soy protein and risk of coronary heart disease (21 CFR 101.82)
12. Plant sterol/stanol esters and risk of coronary heart disease (21 CFR 101.83)

The 1997 FDAMA also authorized health claims based on "successful notification to FDA of a health claim based on an authoritative statement from a scientific body of the U.S. Government or the National Academy of Sciences" (19). The **health claims based on authoritative statements** by scientific bodies (such as the National Academy of Sciences – Institute of Medicine) include the following:

1. Whole grain foods (21)
2. Potassium-containing foods (22)
3. Choline-containing foods (23)

3-8 table **Food and Drug Administration: Nutrient Content Claims**

	Accompanying Information for Relative (or Comparative) Claims
• For all relative claims, percent (or fraction) of change and identity of reference food must be declared in immediate proximity to the most prominent claim. Quantitative comparison of the amount of the nutrient in the product per labeled serving with that in the reference food must be declared on the information panel. • For "Light" claims: % reduction for both fat and Calories must be stated but % reduction need not be specified if product is low in the nutrient. Quantitative comparisons must be stated for both fat and Calories.	
	Reference Foods
"Light"	(1) A food representative of the type of food bearing the claim, for example, average value of top three brands for representative value from valid database; (2) Similar food (e.g., potato chips for potato chips); and (3) Not low Calorie *and* low fat (except "light" or sodium reduced foods which *must* be low Calorie and low fat).
"Reduced" and "Added" (or "Extra," "Plus," "Fortified" and "Enriched")	(1) An established regular product or average representative product, and (2) Similar food
"More" and "Less" (or "Fewer")	(1) An established regular product or average representative product, and (2) A dissimilar food in the same product category which may be generally substituted for labeled food (e.g., potato chips for pretzels) or a similar food.
	Other Nutrient Content Claims
"High," "Rich in," or "Excellent Source of"[a]	Contains 20% or more of the Daily Value (DV) per RACC. May be used on meals or main dishes to indicate that product contains a food that meets the definition.
"Good Source," "Contains," or "Provides"	Contains 10–19% of Daily Value (DV) per RACC. These terms may be used on meals or main dishes to indicate that product contains a food that meets the definition.
"More," "Fortified," "Enriched," "Added," "Extra," or "Plus"[a]	10% or more of the DV per reference amount. May be used only for vitamins, minerals, protein, dietary fiber, and potassium.
"Lean"	On seafood or game meat that contains <10 g total fat, 4.5 g or less saturated fat and <95 mg cholesterol per reference amount and per 100 g (for meals and main dishes, meets criteria per 100 g and per label serving). On mixed dishes not measurable with a cup (as defined in 21 CFR 101.12 (b) in Table 2) that contain less than 8 g fat, 3.5 g or less saturated fat and less than 80 mg cholesterol per RACC.
"Extra Lean"	On seafood or game meat products that contains <5 g total fat, <2 g saturated fat and <95 mg cholesterol per reference amount and per 100 g (for meals and main dishes, meets criteria per 100 g and per label serving).
"High Potency"	May be used on foods to describe individual vitamins or minerals that are present at 100% or more of the RDI per RACC or on a multi-ingredient food product that contains 100% or more of the RDI for at least 2/3 of the vitamins and minerals with RDIs and that are present in the product at 2% or more of the RDI (e.g., "High potency multivitamin, multimineral dietary supplement tablets").
"Modified"	May be used in statement of identity that bears a relative claim, (e.g., "Modified Fat Cheese Cake, Contains 35% Less Fat Than Other Regular Cheese Cake.")
"Fiber" Claim	If a fiber claim is made and the food is not low in fat, then the label must disclose the level of total fat per labeled serving. (21 CFR 101.54(d)(1)
Claims using the term "antioxidant"	For claims characterizing the level of antioxidant nutrient in food: (1) An RDI must be established for each of the nutrients that are the subject of the claim; (2) each nutrient must have existing scientific evidence of antioxidant activity; (3) the level of each nutrient must be sufficient to meet the definition for "high," "good source," or "more"; (4) Beta-carotene may be the subject of an antioxidant claim when the level of vitamin A present as beta-carotene in the food is sufficient to qualify for the claim.

(continued)

Food and Drug Administration: Nutrient Content Claims (continued)

Implied Claims

- Claims about a food or ingredient that suggest that the nutrient or ingredient is absent or present in a certain amount or claims about a food that suggest a food may be useful in maintaining healthy dietary practices and which are made with an explicit claim (e.g., "healthy, contains 3 grams of fat") are implied claims and are prohibited unless provided for in a regulation by FDA. In addition, the Agency has devised a petition system whereby specific additional claims may be considered.
- Claims that a food contains or is made with an ingredient that is known to contain a particular nutrient may be made if the product is "Low" in or a "Good Source" of the nutrient associated with the claim (e.g., "good source of oat bran").
- Equivalence claims: "Contains as much [nutrient] as a [food]" may be made if both reference food and labeled food are a "Good Source" of the nutrient on a per serving basis (e.g., "Contains as much vitamin C as an 8 ounce glass of orange juice").
- The following label statements are generally not considered implied claims unless they are made in a nutrition context: (1) avoidance claims for religious, food intolerance, or other non-nutrition related reasons [e.g., "100% milk free"]; (2) statements about non-nutritive substances [e.g., "no artificial color"]; (3) added value statements [e.g., "made with real butter"]; (4) statements of identity [e.g., "corn oil" or "corn oil margarine"]; and (5) special dietary statements made in compliance with a specific Part 105 provision.

Claims on Food for Infants and Children Less than 2 Years of Age

Nutrient content claims are not permitted on foods intended specifically for infants and children less than 2 years of age except:

1. Claims describing the percentage of vitamins and minerals in a food in relation to a Daily Value.
2. Claims on infant formulas provided for in Part 107.
3. The terms "Unsweetened" and "Unsalted" as taste claims.
4. "Sugar Free" and "No Added Sugar" claims on dietary supplements only.

Terms Covered That Are Not Nutrient Content Claims[b]

"Fresh"	A raw food that has not been frozen, thermally processed, or otherwise preserved.[c]
"Fresh Frozen" and Frozen Fresh	Food that was quickly frozen while still fresh.

From (11), pp. 100–103, updated.
[a]Dietary supplements cannot use these claims to describe any nutrient or ingredient (e.g., fiber, protein, psyllium, bran) other than vitamins or minerals.
[b]21 CFR 101.95
[c]Except as provided in 21 CFR 101.95(c).

4. Whole grain foods with moderate fat content (24)
5. Fluoridated water and reduced risk of dental carries (25)
6. Saturated fat, cholesterol, and trans fat, and reduced risk of heart disease (26)
7. Substitution of saturated fat with unsaturated fatty acids and risk of heart disease (27)

The use of **qualified health claims** resulted from the FDA 2003 Consumer Health Information for Better Nutrition Initiative. This initiative was undertaken to "encourage makers of conventional foods and dietary supplements to make accurate, up-to-date, science-based claims about the health benefits of their products, and to help eliminate bogus labeling claims by pursuing marketers of human dietary supplements and other who make false or misleading claims about the health benefits or other effects of their products" (28). Qualified health claims are utilized when there is "emerging evidence for a relationship between a food, food component or dietary supplement and reduced risk of a disease or health related condition" (19). For qualified health claims, the evidence to support the claim does not meet the "significant scientific agreement standard" criteria, and qualifying language is included in the claim. Unlike the health claims described above, which are authorized through rule making, the qualified health claims are used under FDA's enforcement discretion and are not considered authorized claims. A summary of the qualified health claims authorized by FDA is available on the Internet (29).

In addition to the previously described health claims, dietary guidance statements and structure/function claims also may be used (19). **Dietary guidance statements** must be truthful and non-misleading and address the role of dietary patterns or food categories in health. **Structure/function claims** identify the roles of a specific substance in maintaining normal healthy structures or functions of the body. An industry guidance document for structure/function claims is available on the Internet (30). The FDA

3-9 table **Conditions for Use of the Term "Healthy" in Labeling of Foods**

	Individual Food	*Seafood or Game Meat*[a]	*Meal or Main Dish*
Total fat	Low fat	<5 g fat per RACC[b] and per 100 g	Low fat
Saturated fat	Low saturated fat	<2 g fat per RACC and per 100 g	Low saturated fat
Sodium	≤480 mg per RACC[b], per LS and per 50 g if RACC[b] small	≤480 mg per RACC and per LS and per 50 g if RACC small[b]	≤600 mg/LS
Cholesterol	≤Disclosure level	<95 mg/RACC and per 100 g	≤90 mg/LS
Beneficial nutrients	Contains at least 10% of DV[b] per RACC of vitamins A, C, calcium, iron, protein, or fiber[c]	Contains at least 10% of DV/RACC for vitamins A, C, calcium, iron, protein, or fiber	10% DV per LS of 2 nutrients (for a main dish product), or of 3 nutrients (for a meal product) of vitamins A, C, calcium, iron, protein, or fiber
Fortification	Per 21 CFR 104.20	Per 21 CFR 104.20	Per 21 CFR 104.20
Other claims	Food complies with established definition and declaration requirements for any specific nutrient content claim		

From (11), p. 102, updated.
[a]Raw, single-ingredient seafood or game meat once processed becomes an individual food, meal, or main dish.
[b]*RACC*, reference amount customarily consumed; *LS*, labeled serving; *DV*, Daily Value; *RACC, small* 30 g or less or 2 tablespoons or less.
[c]Except: raw fruits and vegetables; frozen or canned single ingredient fruits and vegetables (may include ingredients whose addition does not change the nutrient profile of the fruits or vegetables); enriched cereal-grain products that conform to a standard of identity in 21 CFR 136, 137, or 139.

does not require conventional food manufacturers to notify FDA about their structure/function claim, and disclaimers are not required for conventional foods.

The Labeling and Consumer Protection Staff, USDA's point of contact for developing and answering questions concerning meat and poultry labeling, considers health claims on products as it implements USDA's mandated prior label approval system. Health claims are allowed on meat and poultry products, as long as they are in compliance with FDA regulations.

3.2.5 Designation of Ingredients

The ingredients used to produce a food must be declared on the label as defined in 21 CFR 101.4. The "common or usual" name is used, and the ingredients are listed in descending order by weight in the product. If the ingredient is present at less than 2%, the statement "contains 2 percent or less of _____" can be used. However, certain ingredients defined as "incidental additives" (i.e., present in the food at insignificant levels and having no technical or functional effect) are exempt from labeling requirements as outlined in 21 CFR 101.100(a)(3).

3.2.6 National Uniformity and Preemptions Authorized by NLEA

To provide for national uniformity, the 1990 NLEA authorizes federal preemption of certain state and local labeling requirements that are not identical to federal requirements. This pertains to requirements for food standards, nutrition labeling, claims of nutrient content, health claims, and ingredient declaration. States may petition the FDA for exemption of state requirements from federal preemption.

3.2.7 Other Provisions of NLEA

The 1990 NLEA amends the FD&C Act to allow a state to bring, in its own name in state court, an action to enforce the food labeling provisions of the FD&C Act that are the subject of national uniformity. Criteria are defined for a state to exercise this enforcement power. The rule-making procedure for standards of identity was modified by the 1990 Act. The FD&C Act is also amended to impose several new requirements concerning ingredient labeling intended to make this aspect of labeling more useful to consumers.

3.3 SUMMARY

A major reason for analyzing the chemical components of food in the USA (and many other countries) is nutrition labeling regulation. The FDA and FSIS of the USDA have coordinated their regulations on nutrition labeling. Regulations that implement the

NLEA of 1990 require nutrition labeling for most foods regulated by the FDA, and FSIS requires the same label on most meat and poultry products. The regulations define the format for the nutrition information and give the rules and methods to report specific information. The FDA and FSIS have described the sample collection procedures, the method of analysis to be used, and the nutrient levels required to ensure compliance with nutrition labeling regulations. The FDA and FSIS allow specific nutrient content claims and specific health claims on the nutrition label. The NLEA provides for national uniformity in nutrition labeling, by preempting any existing state regulations, and authorizes to states certain enforcement powers. The goal of current nutrition labeling regulations is to provide consumers in all states with nutrition information on food in their diets consistent with their dietary concerns.

3.4 STUDY QUESTIONS

1. Utilize the data in the table below that you obtained on the nutrient content of your cereal product (actual amount per serving) to help develop a nutrition label that meets FDA requirements under the NLEA. Use appropriate rounding rules to complete the blank columns. Can you make a "low fat" claim? Explain your answer. Could you use the term "healthy" on the label? If you wanted to report the protein content as a percent of the Daily Value, what would you need to do?

	Actual Amount per Serving[a]	*Amount per Serving Reported on Label*	*% Daily Value*[b] *Reported on Label*
Calories	192		–
Calories from fat	11		–
Total fat	1.1 g		
Saturated fat	0 g		
Trans fat	0 g		
Cholesterol	0 mg		
Sodium	268 mg		
Potassium	217 mg		
Total carbohydrate	44.3 g		
Dietary fiber	3.8 g		
Sugars	20.2 g		–
Protein	3.7 g		–
Vitamin A	475 IU	–	
Vitamin C	0 mg	–	
Calcium	210 mg	–	
Iron	18.3 mg	–	

[a]Serving size is 1 cup (55 g).
[b]Percent Daily Value based on a 2000-Cal diet.

2. The FDA and FSIS of the USDA have very similar regulations for nutrition labeling, with differences due primarily to the inherent difference in the food products they regulate. Identify the differences in regulations between FDA and FSIS regarding nutrient content claims that support this generalization.
3. You want to develop a macaroni and cheese product that can be labeled "reduced sodium." The first step is to identify a reference food that you will use for comparison. What are some potential reference foods you could use and how would you select one?
4. You want to label 1% fat milk as "99% fat free". Can you do this?
5. The ingredient list for a cooking spray is canola oil, grain alcohol, and lecithin. The label states that the product is for fat-free cooking. How is this possible when the first ingredient is oil?

3.5 REFERENCES

Note: For FDA references on the Internet go to the FDA home page at http://www.fda.gov/default.htm and utilize the search function to conduct a search based on the document title.

1. Food and Drug Law Institute (2009) Food and drug law and regulation. Food and Drug Law Institute, Washington, DC
2. Piña KR, Pines WL (2008) A practical guide to food and drug law and regulation, 3rd edn. Food and Drug Law Institute, Washington, DC
3. US Congress (1990) US public law 101–535. Nutrition labeling and education act of 1990. 8 Nov 1990. US Congress, Washington, DC
4. US Congress (1997) US public law 105–115. Food and Drug Administration modernization act of 1997. 21 Nov 1997. US Congress, Washington, DC
5. Federal Register (1993) Department of Agriculture. Food Safety and Inspection Service. Part II. 9 CFR Parts 317, 320, and 381. Nutrition labeling of meat and poultry products; final rule. January 6, 1993. 58(3): 631–685. Part III. 9 CFR Parts 317 and 381. Nutrition labeling: use of "healthy" and similar terms on meat and poultry products labeling; proposed rule. January 6, 1993. 58(3): 687–691. Superintendent of documents. US Government Printing Office, Washington, DC
6. Federal Register (1993) 21 CFR Part 1, et al. Food labeling; general provisions; nutrition labeling; label format; nutrient content claims; health claims; ingredient labeling; state and local requirements; and exemptions; final rules. January 6, 1993. 58(3):#2066–2941. Superintendent of documents. US Government Printing Office, Washington, DC
7. Code of Federal Regulations (2009) Nutrition labeling of foods. 21 CFR 101.9-101.108. US Government Printing Office, Washington, DC
8. Code of Federal Regulations (2009) (animal and animal products). 9 CFR 317 subpart B 317. 300–317.400; 9 CFR 381 subpart Y 381.400–381.500. US Government Printing Office, Washington, DC

9. Hildwine R (ed) (2004) Food labeling manual: complying with FDA requirements for the labeling of processed foods. National Food Processors Association, Washington, DC
10. Hutt PB (1991) (updated continuously) Guide to U.S. food labeling law, vols I and II. Thompson, New York
11. FDA (2008) Food labeling guide. Center for Food Safety and Applied Nutrition, Food and Drug Administration, US Department of Health and Human Services, Public Health Administration, Washington, DC
12. FDA (2003) Food labeling: trans fatty acids in nutritional labeling, nutrient content claims, and health claims – small entity compliance guide. Center for Food Safety and Applied Nutrition, Food and Drug Administration, Washington, DC
13. AOAC International (2007) Official methods of analysis, 18th edn., 2005; current through revision 2, 2007 (Online). AOAC International, Gaithersburg, MD
14. USDA (1991–1996) Analytical chemistry laboratory guidebook – residue chemistry (Winter 1991), food chemistry (Spring 1993), residue chemistry supplement (Sept 1995). Science and technology. Nutritional analysis methods. Quality Assurance Branch, Chemistry and Toxicology Division, Food Safety and Inspection Service, US Department of Agriculture, Washington, DC
15. FDA/Office of Regulatory Affairs (2000) FDA/ORA compliance policy guides manual. FDA/Office of Regulatory Affairs, Washington, DC
16. FDA (1998) FDA nutrition labeling manual: a guide for developing and using databases. Office of Food Labeling, Food and Drug Administration (HFS-150), Washington, DC
17. US Department of Agriculture, Agriculture Research Service (2009) USDA national nutrient database for standard reference. Nutrient data laboratory home page. http://www.ars.usda.gov/ba/bhnrc/ndl
18. USDA (1993) Food safety inspection service (FSIS) manual on use of data bases for nutrition labeling. FSIS guidelines for effective use of data bases to develop nutrient declarations for nutrition labeling of meat and poultry products. Food Safety and Inspection Service, US Department of Agriculture, Washington, DC
19. FDA (2003) Claims that can be made for conventional foods and dietary supplements. Center for Food Safety and Applied Nutrition, Food and Drug Administration, Washington, DC
20. FDA (2009) Evidence-based review system for the scientific evaluation of health claims. Center for Food Safety and Applied Nutrition, Food and Drug Administration, Washington, DC
21. FDA (1999) Health claim notification for whole grain foods. Office of Food Labeling, Food and Drug Administration, Dockets Management Branch (Docket No. 99P-2209)
22. FDA (2000) Health claim notification for potassium containing foods. Office of Food Labeling, Food and Drug Administration, Dockets Management Branch (Docket No. 00Q-1582)
23. FDA (2001) Health claim notification for choline containing foods. Office of Food Labeling, Food and Drug Administration, Dockets Management Branch (Docket No. 01Q-0352)
24. FDA (2003) Health claim notification for whole grain foods with moderate fat content. Office of Food Labeling, Food and Drug Administration, Dockets Management Branch (Docket No. 03Q-0547)
25. FDA (2006) Health claim notification for fluoridated water and reduced risk of dental carries. Office of Food Labeling, Food and Drug Administration, Dockets Management Branch (Docket No. 2006Q-0418)
26. FDA (2006) Health claim notification for saturated fat, cholesterol, and trans fat, and reduced risk of heart disease. Office of Food Labeling, Food and Drug Administration, Dockets Management Branch (Docket No. 2006Q-0458)
27. FDA (2007) Health claim notification for substitution of saturated fat in the diet with unsaturated fatty acids and reduced risk of heart disease. Office of Food Labeling, Food and Drug Administration, Dockets Management Branch (Docket No. 2007Q-0192)
28. FDA (2003) Consumer health information for better nutrition initiative. Task force final report. Center for Food Safety and Applied Nutrition, Food and Drug Administration, Washington, DC
29. FDA Summary of qualified health claims subject to enforcement discretion. Center for Food Safety and Applied Nutrition, Food and Drug Administration, Washington, DC
30. FDA Guidance for industry: structure/function claims, small entity compliance guide. Center for Food Safety and Applied Nutrition, Food and Drug Administration, Washington, DC

3.6 RELEVANT INTERNET ADDRESSES

Code of Federal Regulations –
  http://www.access.gpo.gov/nara/cfr/cfr-table-search.html

Food and Drug Administration –
  http://www.fda.gov
  Center for Food Safety and Applied Nutrition –
    http://vm.cfsan.fda.gov/list.html

US Department of Agriculture –
  http://www.usda.gov
  Food Safety and Inspection Service –
    http://www.fsis.usda.gov
  Labeling and Consumer Protection Staff –
    http://www.fsis.usda.gov/About_FSIS/labeling_&_consumer_protection/index.asp
  Nutrient Data Laboratory (USDA Handbook No. 8) –
    http://www.ars.usda.gov/ba/bhnrc/ndl

Evaluation of Analytical Data

J. Scott Smith
Food Science Institute, Kansas State University,
Manhattan, KS 66506-1600, USA
jsschem@ksu.edu

4.1 Introduction 55
4.2 Measures of Central Tendency 55
4.3 Reliability of Analysis 55
4.3.1 Accuracy and Precision 55
4.3.2 Sources of Errors 59
4.3.3 Specificity 59
4.3.4 Sensitivity and Limit of Detection 59
4.4 Curve Fitting: Regression Analysis 60
4.4.1 Linear Regression 60
4.4.2 Correlation Coefficient 61
4.4.3 Errors in Regression Lines 62
4.5 Reporting Results 63
4.5.1 Significant Figures 64
4.5.2 Rejecting Data 65
4.6 Summary 65
4.7 Study Questions 65
4.8 Practice Problems 66
4.9 Resource Materials 67

S.S. Nielsen, *Food Analysis*, Food Science Texts Series, DOI 10.1007/978-1-4419-1478-1_4,

4.1 INTRODUCTION

The field of food analysis, or any type of analysis, involves a considerable amount of time learning principles, methods, and instrument operations and perfecting various techniques. Although these areas are extremely important, much of our effort would be for naught if there were not some way for us to evaluate the data obtained from the various analytical assays. Several mathematical treatments are available that provide an idea of how well a particular assay was performed or how well we can reproduce an experiment. Fortunately, the statistics are not too involved and apply to most analytical determinations.

The focus of this chapter is primarily on how to evaluate replicate analyses of the same sample for accuracy and precision. In addition, considerable attention is given to the determination of best line fits for standard curve data. Keep in mind as you read and work through this chapter that there is a vast array of computer software to perform most types of data evaluation and calculations/plots.

Proper sampling and sample size are not covered in this chapter. Readers should refer to Chap. 5 (especially Sect. 5.3.4) for sampling, in general, and statistical approaches to determine the appropriate sample size and to Chap. 18, Sect. 18.4, for mycotoxin sampling.

4.2 MEASURES OF CENTRAL TENDENCY

To increase accuracy and precision, as well as to evaluate these parameters, the analysis of a sample is usually performed (repeated) several times. At least three assays are typically performed, though often the number can be much higher. Because we are not sure which value is closest to the true value, we determine the mean (or average) using all the values obtained and report the results of the **mean**. The mean is designated by the symbol $\bar{x}$ and calculated according to the equation below.

$$\bar{x} = \frac{x_1 + x_2 + x_3 + \cdots + x_n}{n} = \frac{\sum x_i}{n} \quad [1]$$

where:

$\bar{x}$ = mean

x_1, x_2, etc. = individually measured values (x_i)

n = number of measurements

For example, suppose we measured a sample of uncooked hamburger for percent moisture content four times and obtained the following results: 64.53%, 64.45%, 65.10%, and 64.78%.

$$\bar{x} = \frac{64.53 + 64.45 + 65.10 + 64.78}{4} = 64.72\% \quad [2]$$

Thus, the result would be reported as 64.72% moisture. When we report the mean value, we are indicating that this is the best experimental estimate of the value. We are not saying anything about how accurate or true this value is. Some of the individual values may be closer to the true value, but there is no way to make that determination, so we report only the mean.

Another determination that can be used is the **median**, which is the midpoint or middle number within a group of numbers. Basically, half of the experimental values will be less than the median and half will be greater. The median is not used often, because the mean is such a superior experimental estimator.

4.3 RELIABILITY OF ANALYSIS

Returning to our previous example, recall that we obtained a mean value for moisture. However, we did not have any indication of how repeatable the tests were or how close our results were to the true value. The next several sections will deal with these questions and some of the relatively simple ways to calculate the answers.

4.3.1 Accuracy and Precision

One of the most confusing aspects of data analysis for students is grasping the concepts of accuracy and precision. These terms are commonly used interchangeably in society, which only adds to this confusion. If we consider the purpose of the analysis, then these terms become much clearer. If we look at our experiments, we know that the first data obtained are the individual results and a mean value ($\bar{x}$). The next questions should be: "How close were our individual measurements?" and "How close were they to the true value?" Both questions involve accuracy and precision. Now, let us turn our attention to these terms.

Accuracy refers to how close a particular measure is to the true or correct value. In the moisture analysis for hamburger, recall that we obtained a mean of 64.72%. Let us say the true moisture value was actually 65.05%. By comparing these two numbers, you could probably make a guess that your results were fairly accurate because they were close to the correct value. (The calculations of accuracy will be discussed later.)

The problem in determining accuracy is that most of the time we are not sure what the true value is. For certain types of materials, we can purchase known samples from, for example, the National Institute of Standards and Technology and check our assays against these samples. Only then can we have an indication of the accuracy of the testing procedures. Another approach is to compare our results with those of other labs to determine how well they agree, assuming the other labs are accurate.

A term that is much easier to deal with and determine is **precision**. This parameter is a measure of how reproducible or how close replicate measurements become. If repetitive testing yields similar results, then we would say that the precision of that test was good. From a true statistical view, the precision often is called **error**, when we are actually looking at experimental **variation**. So, the concepts of precision, error, and variation are closely related.

The difference between precision and accuracy can be illustrated best with Fig. 4-1. Imagine shooting a rifle at a target that represents experimental values. The bull's eye would be the true value and where the bullets hit would represent the individual experimental values. As you can see in Fig. 4-1a, the values can be tightly spaced (good precision) and close to the bull's eye (good accuracy), or, in some cases, there can be situations with good precision but poor accuracy (Fig. 4-1b). The worst situation, as illustrated in Fig. 4-1d, is when both the accuracy and precision are poor. In this case, because of errors or variation in the determination, interpretation of the results becomes very difficult. Later, the practical aspects of the various types of error will be discussed.

When evaluating data, several tests are commonly used to give some appreciation of how much the experimental values would vary if we were to repeat the test (indicators of precision). An easy way to look at the variation or scattering is to report the range of the experimental values. The **range** is simply the difference between the largest and smallest observation. This measurement is not too useful and thus is seldom used in evaluating data.

Probably the best and most commonly used statistical evaluation of the precision of analytical data is the standard deviation. The **standard deviation** measures the spread of the experimental values and gives a good indication of how close the values are to each other. When evaluating the standard deviation, one has to remember that we are never able to analyze the entire food product. That would be difficult, if not impossible, and very time consuming. Thus, the calculations we use are only estimates of the unknown true value.

If we have many samples, then the standard deviation is designated by the Greek letter sigma (σ). It is calculated according to Equation [3], assuming all of the food products were evaluated (which would be an infinite amount of assays).

$$\sigma = \sqrt{\frac{\sum (x_i - \mu)^2}{n}} \qquad [3]$$

where:

σ = standard deviation
x_i = individual sample values
μ = true mean
n = total population of samples

Because we do not know the value for the true mean, the equation becomes somewhat simplified so that we can use it with real data. In this case, we now call the σ term the standard deviation of the sample and designate it by SD or σ. It is determined according to the calculation in Equation [4], where $\bar{x}$ replaces the true mean term μ and n represents the number of samples.

$$\text{SD} = \sqrt{\frac{\sum (x_i - \bar{x})^2}{n}} \qquad [4]$$

If the number of replicate determinations is small (about 30 or less), which is common with most assays, the n is replaced by the $n-1$ term, and Equation [5] is used. Unless you know otherwise, Equation [5] is always used in calculating the standard deviation of a group of assays.

$$\text{SD} = \sqrt{\frac{\sum (x_i - \bar{x})^2}{n-1}} \qquad [5]$$

Depending on which of the equations above is used, the standard deviation may be reported as SD_n or σ_n and SD_{n-1} or σ_{n-1}. (Different brands of software and scientific calculators sometimes use different labels for the keys, so one must be careful.) Table 4-1 shows an example of the determination of standard deviation. The sample results would be reported to average 64.72% moisture with a standard deviation of 0.293.

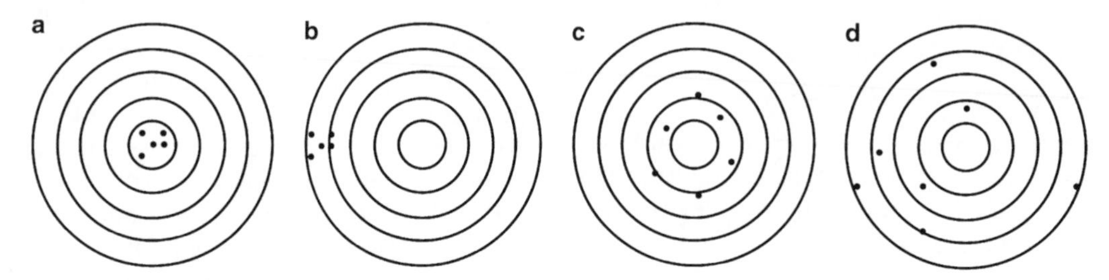

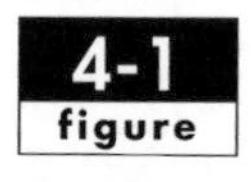

Comparison of accuracy and precision: (**a**) Good accuracy and good precision, (**b**) Good precision and poor accuracy, (**c**) Good accuracy and poor precision, and (**d**) Poor accuracy and poor precision.

4-1 table **Determination of the Standard Deviation of Percent Moisture in Uncooked Hamburger**

Measurement	*Observed % Moisture*	*Deviation from the Mean* $(x_i - \bar{x})$	$(x_i - \bar{x})^2$
1	64.53	−0.19	0.0361
2	64.45	−0.27	0.0729
3	65.10	+0.38	0.1444
4	64.78	+0.06	0.0036
	$\sum x_i = 258.86$		$\sum (x_i - \bar{x})^2 = 0.257$

$$\bar{x} = \frac{\sum x_i}{n} = \frac{258.86}{4} = 64.72$$

$$\mathrm{SD} = \sqrt{\frac{\sum (x_i - \bar{x})^2}{n-1}} = \sqrt{\frac{0.257}{3}} = 0.2927$$

Once we have a mean and standard deviation, we next must determine how to interpret these numbers. One easy way to get a feel for the standard deviation is to calculate what is called the **coefficient of variation** (CV), also known as the **relative standard deviation**. This calculation is shown below for our example of the moisture determination of uncooked hamburger.

$$\text{Coefficient of variation (CV)} = \frac{\mathrm{SD}}{\bar{x}} \times 100\% \qquad [6]$$

$$\mathrm{CV} = \frac{0.293}{64.72} \times 100\% = 0.453\% \qquad [7]$$

The CV tells us that our standard deviation is only 0.453% as large as the mean. For our example, that number is small, which indicates a high level of precision or reproducibility of the replicates. As a rule, a CV below 5% is considered acceptable, although it depends on the type of analysis.

Another way to evaluate the meaning of the standard deviation is to examine its origin in statistical theory. Many populations (in our case, sample values or means) that exist in nature are said to have a normal distribution. If we were to measure an infinite number of samples, we would get a distribution similar to that represented by Fig. 4-2. In a population with a **normal distribution**, 68% of those values would be within ±1 standard deviation from the mean, 95% would be within ±2 standard deviations, and 99.7% would be within ±3 standard deviations. In other words, there is a probability of less than 1% that a sample in a population would fall outside ±3 standard deviations from the mean value.

Another way of understanding the normal distribution curve is to realize that the probability of finding the true mean is within certain confidence intervals as defined by the standard deviation. For large numbers of samples, we can determine the **confidence limit** or

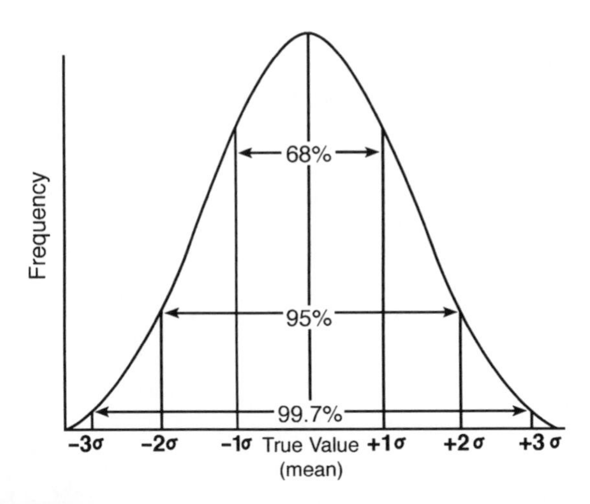

4-2 figure A normal distribution curve for a population or a group of analyses.

4-2 table **Values for Z for Checking Both Upper and Lower Levels**

Degree of Certainty (Confidence) (%)	*Z value*
80	1.29
90	1.64
95	1.96
99	2.58
99.9	3.29

interval around the mean using the statistical parameter called the **Z value**. We do this calculation by first looking up the Z value from statistical tables once we have decided the desired degree of certainty. Some Z values are listed in Table 4-2.

The confidence limit (or interval) for our moisture data, assuming a 95% probability, is calculated according to Equation [8]. Since this calculation is not valid for small numbers, assume we ran 25 samples instead of four.

$$\text{Confidence interval (CI)} = \bar{x} \pm Z\,\text{value} \times \frac{\text{standard deviation (SD)}}{\sqrt{n}} \qquad [8]$$

$$\mathrm{CI}\,(\text{at } 95\%) = 64.72 \pm 1.96 \times \frac{0.2927}{\sqrt{25}} = 64.72 \pm 0.115\% \qquad [9]$$

Because our example had only four values for the moisture levels, the confidence interval should be calculated using statistical t tables. In this case, we have to look up the t value from Table 4-3 based on the **degrees of freedom**, which is the sample size minus one $(n-1)$, and the desired level of confidence.

table 4-3 Values of t for Various Levels of Probability[a]

Degrees of Freedom (n − 1)	Levels of Certainty 95%	99%	99.9%
1	12.7	63.7	636
2	4.30	9.93	31.60
3	3.18	5.84	12.90
4	2.78	4.60	8.61
5	2.57	4.03	6.86
6	2.45	3.71	5.96
7	2.36	3.50	5.40
8	2.31	3.56	5.04
9	2.26	3.25	4.78
10	2.23	3.17	4.59

[a]More extensive t-tables can be found in statistics books.

The calculation for our moisture example with four samples (n) and three degrees of freedom ($n-1$) is given below:

$$\mathrm{CI} = \bar{x} \pm t\,\text{value} \times \frac{\text{standard deviation (SD)}}{\sqrt{n}} \quad [10]$$

$$\mathrm{CI}\,(\text{at } 95\%) = 64.72 \pm 3.18 \times \frac{0.2927}{\sqrt{4}}$$

$$= 64.72 \pm 0.465\% \quad [11]$$

To interpret this number, we can say that, with 95% confidence, the true mean for our moisture will fall within $64.72 \pm 0.465\%$ or between 65.185 and 64.255%.

The expression $\mathrm{SD}/\sqrt{n}$ is often reported as the **standard error of the mean**. It is then left to the reader to calculate the confidence interval based on the desired level of certainty.

Other quick tests of precision used are the relative deviation from the mean and the relative average deviation from the mean. The **relative deviation from the mean** is useful when only two replicates have been performed. It is calculated according to Equation [12], with values below 2% considered acceptable.

$$\text{Relative deviation from the mean} = \frac{x_i - \bar{x}}{\bar{x}} \times 100 \quad [12]$$

where:

x_i = individual sample value

$\bar{x}$ = mean

If there are several experimental values, then the **relative average deviation from the mean** becomes a useful indicator of precision. It is calculated similarly to the relative deviation from the mean, except the average deviation is used instead of the individual deviation. It is calculated according to Equation [13].

Relative average deviation from the mean

$$= \frac{\frac{\sum |x_i - \bar{x}|}{n}}{\bar{x}} \times 1000 = \text{parts per thousand} \quad [13]$$

Using the moisture values discussed in Table 4-1, the $x_i - \bar{x}$ terms for each determination are −0.19, −0.27, +0.38, +0.06. Thus, the calculation becomes:

$$\text{Rel. avg. dev.} = \frac{\frac{0.19 + 0.27 + 0.38 + 0.06}{4}}{64.72} \times 1000$$

$$= \frac{0.225}{64.72} \times 1000$$

$$= 3.47 \text{ parts per thousand} \quad [14]$$

Up to now, our discussions of calculations have involved ways to evaluate precision. If the true value is not known, we can calculate only precision. A low degree of precision would make it difficult to predict a realistic value for the sample.

However, we may occasionally have a sample for which we know the true value and can compare our results with the known value. In this case, we can calculate the error for our test, compare it to the known value, and determine the accuracy. One term that can be calculated is the **absolute error**, which is simply the difference between the experimental value and the true value.

$$\text{Absolute error} = E_{abs} = x - T \quad [15]$$

where:

x = experimentally determined value

T = true value

The absolute error term can have either a positive or a negative value. If the experimentally determined value is from several replicates, then the mean (0) would be substituted for the x term. This is not a good test for error, because the value is not related to the magnitude of the true value. A more useful measurement of error is **relative error**.

$$\text{Relative error} = E_{rel} = \frac{E_{abs}}{T} = \frac{x - T}{T} \quad [16]$$

The results are reported as a negative or positive value, which represents a fraction of the true value.

If desired, the relative error can be expressed as percent relative error by multiplying by 100%. Then the relationship becomes the following, where x can be either an individual determination or the mean (0) of several determinations.

$$\%E_{rel} = \frac{E_{abs}}{T} \times 100\% = \frac{x - T}{T} \times 100\% \quad [17]$$

Using the data for the percent moisture of uncooked hamburger, suppose the true value of the sample is 65.05%. The percent relative error is calculated using our mean value of 64.72% and Equation [17].

$$\%E_{\mathrm{rel}} = \frac{\bar{x} - T}{T} \times 100\% = \frac{64.72 - 65.05}{65.05} \times 100\% = -0.507\% \quad [18]$$

Note that we keep the negative value, which indicates the direction of our error, that is, our results were 0.507% lower than the true value.

4.3.2 Sources of Errors

As you may recall from our discussions of accuracy and precision, error (variation) can be quite important in analytical determinations. Although we strive to obtain correct results, it is unreasonable to expect an analytical technique to be entirely free of error. The best we can hope for is that the variation is small and, if possible, at least consistent. As long as we know about the error, the analytical method often will be satisfactory. There are several sources of error, which can be classified as: systematic error (determinate), random error (indeterminate), and gross error or blunders. Again, note that error and variation are used interchangeably in this section and essentially have the same meaning for these discussions.

Systematic or **determinate error** produces results that consistently deviate from the expected value in one direction or the other. As illustrated in Fig. 4-1b, the results are spaced closely together, but they are consistently off the target. Identifying the source of this serious type of error can be difficult and time consuming, because it often involves inaccurate instruments or measuring devices. For example, a pipette that consistently delivers the wrong volume of reagent will produce a high degree of precision yet inaccurate results. Sometimes impure chemicals or the analytical method itself are the cause. Generally, we can overcome systematic errors by proper calibration of instruments, running blank determinations, or using a different analytical method.

Random or **indeterminate errors** are always present in any analytical measurement. This type of error is due to our natural limitations in measuring a particular system. These errors fluctuate in a random fashion and are essentially unavoidable. For example, reading an analytical balance, judging the endpoint change in a titration, and using a pipette all contribute to random error. Background instrument noise, which is always present to some extent, is a factor in random error. Both positive and negative errors are equally possible. Although this type of error is difficult to avoid, fortunately it is usually small.

Blunders are easy to eliminate, since they are so obvious. The experimental data are usually scattered, and the results are not close to an expected value. This type of error is a result of using the wrong reagent or instrument or of sloppy technique. Some people have called this type of error the "Monday morning syndrome" error. Fortunately, blunders are easily identified and corrected.

4.3.3 Specificity

Specificity of a particular analytical method means that it detects only the component of interest. Analytical methods can be very specific for a certain food component or, in many cases, can analyze a broad spectrum of components. Quite often, it is desirable for the method to be somewhat broad in its detection. For example, the determination of food lipid (fat) is actually the crude analysis of any compound that is soluble in an organic solvent. Some of these compounds are glycerides, phospholipids, carotenes, and free fatty acids. Since we are not concerned about each individual compound when considering the crude fat content of food, it is desirable that the method be broad in scope. On the other hand, determining the lactose content of ice cream would require a specific method. Because ice cream contains other types of simple sugars, without a specific method we would overestimate the amount of lactose present.

There are no hard rules for what specificity is required. Each situation is different and depends on the desired results and type of assay used. However, it is something to keep in mind as the various analytical techniques are discussed.

4.3.4 Sensitivity and Limit of Detection

Although often used interchangeably, the terms sensitivity and limit of detection should not be confused. They have different meanings, yet are closely related. **Sensitivity** relates to the magnitude of change of a measuring device (instrument) with changes in compound concentration. It is an indicator of how little change can be made in the unknown material before we notice a difference on a needle gauge or a digital readout. We are all familiar with the process of tuning in a radio station on our stereo and know how, at some point, once the station is tuned in, we can move the dial without disturbing the reception. This is sensitivity. In many situations, we can adjust the sensitivity of an assay to fit our needs, that is, whether we desire more or less sensitivity. We even may desire a lower sensitivity so that samples with widely varying concentration can be analyzed at the same time.

Limit of detection (LOD), in contrast to sensitivity, is the lowest possible increment that we can detect with some degree of confidence (or statistical significance). With every assay, there is a lower limit at which point we are not sure if something is present or not. Obviously, the best choice would be to concentrate the sample so we are not working close to the detection limit. However, this may not be possible, and we may need to know the LOD so we can work away from that limit.

There are several ways to measure the LOD, depending on the apparatus that is used. If we are using something like a spectrophotometer, gas chromatograph, or high-performance liquid chromatograph (HPLC), the LOD often is reached when the signal to noise ratio is 3 or greater. In other words, when the sample gives a value that is three times the magnitude of the noise detection, the instrument is at the lowest limit possible. Noise is the random signal fluctuation that occurs with any instrument.

A more general way to define the LOD is to approach the problem from a statistical viewpoint, in which the variation between samples is considered. A common mathematical definition of LOD is given below.

$$X_{LD} = X_{Blk} + 3 \times SD_{Blk} \qquad [19]$$

where:

X_{LD} = minimum detectable concentration
X_{Blk} = signal of a blank
SD_{Blk} = standard deviation of the blank readings

In this equation, the variation of the blank values (or noise, if we are talking about instruments) determines the detection limit. High variability in the blank values decreases the LOD.

Another method that encompasses the entire assay method is the **method detection limit** (MDL). According to the US Environmental Protection Agency (EPA), the MDL is defined as "the minimum concentration of a substance that can be measured and reported with 99% confidence that the analyte concentration is greater than zero and is determined from analysis of a sample in a given matrix containing the analyte." What differentiates the MDL from the LOD is that it includes the entire assay and various sample types thus correcting for variability throughout. The MDL is calculated based on values of samples within the assay matrix and thus is considered a more rigorous performance test. The procedures on how to set up the MDL are explained in Appendix B of Part 136 (40 CFR, Vol. 22) of the EPA regulations on environmental testing.

Though the LOD or MDL are often sufficient to characterize an assay, a further evaluation to check is the **limit of quantitation** (LOQ). In this determination, data are collected similar to the LOD except the value is determined as $X_{Blk} + 10 \times SD_{Blk}$ (instead of $X_{Blk} + 3 \times SD_{Blk}$).

4.4 CURVE FITTING: REGRESSION ANALYSIS

Curve fitting is a generic term used to describe the relationship and evaluation between two variables. Most scientific fields use curve fitting procedures to evaluate the relationship of two variables. Thus, curve fitting or curvilinear analysis of data is a vast area as evidenced by the volumes of material describing these procedures. In analytical determinations, we are usually concerned with only a small segment of curvilinear analysis, the standard curve or regression line.

A **standard curve** or **calibration curve** is used to determine unknown concentrations based on a method that gives some type of measurable response that is proportional to a known amount of standard. It typically involves making a group of known standards in increasing concentration and then recording the particular measured analytical parameter (e.g., absorbance, area of a chromatography peak, etc.). What results when we graph the paired x and y values is a scatterplot of points that can be joined together to form a straight line relating concentration to observed response. Once we know how the observed values change with concentration, it is fairly easy to estimate the concentration of an unknown by interpolation from the standard curve.

As you read through the next three sections, keep in mind that not all correlations of observed values to standard concentrations are linear (but most are). There are many examples of nonlinear curves, such as antibody binding, toxicity evaluations, and exponential growth and decay. Fortunately, with the vast array of computer software available today, it is relatively easy to analyze any group of data.

4.4.1 Linear Regression

So how do we set up a standard curve once the data have been collected? First, a decision must be made regarding onto which axis to plot the paired sets of data. Traditionally, the concentration of the standards is represented on the x-axis and the observed readings are on the y-axis. However, this protocol is used for reasons other than convention. The x-axis data are called the **independent variable** and are assumed to be essentially free of error, while the y-axis data (the **dependent variable**) may have error associated with them. This assumption may not be true because error could be incorporated as the standards are made. With modern day instruments, the error can be very small. Although arguments can be made for making the y-axis data concentration, for all practical purposes the end result is essentially the same. Unless there are some unusual data, the *concentration should be associated with the x-axis and the measured values with the y-axis.*

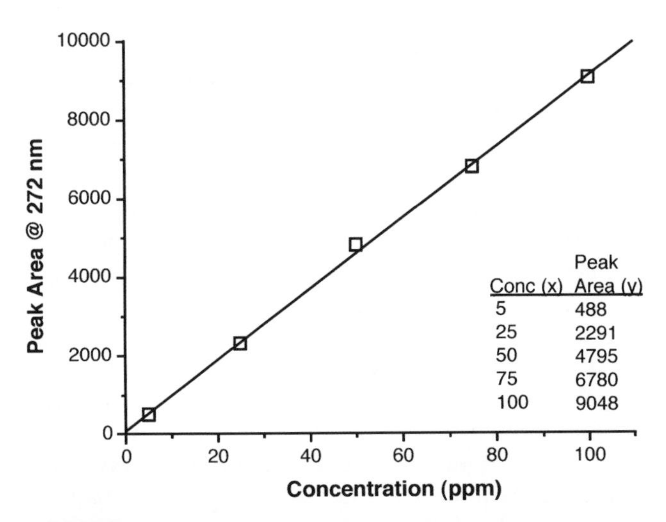

4-3 figure A typical standard curve plot showing the data points and the generated best fit line. The data used to plot the curve are presented on the graph.

Figure 4-3 illustrates a typical standard curve used in the determination of caffeine in various foods. Caffeine is analyzed readily in foods by using HPLC coupled with an ultraviolet detector set at 272 nm. The area under the caffeine peak at 272 nm is directly proportional to the concentration. When an unknown sample (e.g., coffee) is run on the HPLC, a peak area is obtained that can be related back to the sample using the standard curve.

The plot in Fig. 4-3 shows all the data points and a straight line that appears to pass through most of the points. The line almost passes through the origin, which makes sense because zero concentration should produce no signal at 272 nm. However, the line is not perfectly straight (and never is) and does not quite pass through the origin.

To determine the caffeine concentration in a sample that gave an area of say 4000, we could extrapolate to the line and then draw a line down to the x-axis. Following a line to the x-axis (concentration), we can estimate the solution to be at about 42–43 ppm of caffeine.

We can mathematically determine the best fit of the line by using **linear regression**. Keep in mind the equation for a straight line, which is $y = ax + b$, where a is the slope and b is the y-intercept. To determine the slope and y-intercept, the regression equations shown below are used. We determine a and b and thus, for any value of y (measured), we can determine the concentration (x).

$$\text{Slope } a = \frac{\sum (x_i - \bar{x})(y_i - \bar{y})}{\sum (x_i - \bar{x})^2} \qquad [20]$$

$$y\text{-intercept } b = \bar{y} - a\bar{x} \qquad [21]$$

where:

x_i and y_i = individual values

$\bar{x}$ and $\bar{y}$ = means of the individual values

Low-cost calculators and computer spreadsheet software can readily calculate regression equations so no attempt is made to go through the mathematics in the formulas.

The formulas give what is known as the line of regression of y on x, which assumes that the error occurs in the y direction. The regression line represents the average relationship between all the data points and thus is a balanced line. These equations also assume that the straight line fit does not have to go through the origin, which at first does not make much sense. However, there are often background interferences so that even at zero concentration, a weak signal may be observed. In most situations, calculating the origin as going through zero will yield the same results.

Using the data from Fig. 4-3, calculate the concentration of caffeine in the unknown and compare with the graphing method. As you recall, the unknown had an area at 272 m of 4000. Linear regression analysis of the standard curve data gave the y-intercept (b) as 90.727 and the slope (a) as 89.994 ($r^2 = 0.9989$).

$$y = ax + b \qquad [22]$$

or

$$x = \frac{y - b}{a} \qquad [23]$$

$$x\,(\text{conc}) = \frac{4000 - 90.727}{89.994} = 43.4393\,\text{ppm caffeine} \qquad [24]$$

The agreement is fairly close when comparing the calculated value to that estimated from the graph. Using high-quality graph paper with many lines could give us a line very close to the calculated one. However, as we will see in the next section, additional information can be obtained about the nature of the line when using computer software or calculators.

4.4.2 Correlation Coefficient

In observing any type of correlation, including linear ones, questions always surface concerning how to draw the line through the data points and how well the data fit to the straight line. The first thing that should be done with any group of data is to plot it to see if the points fit a straight line. By just looking at the plotted data, it is fairly easy to make a judgment on the linearity of the line. We also can pick out regions on the line where a linear relationship does not exist. The figures below illustrate differences in standard curves; Fig. 4-4a shows a good correlation of the data and

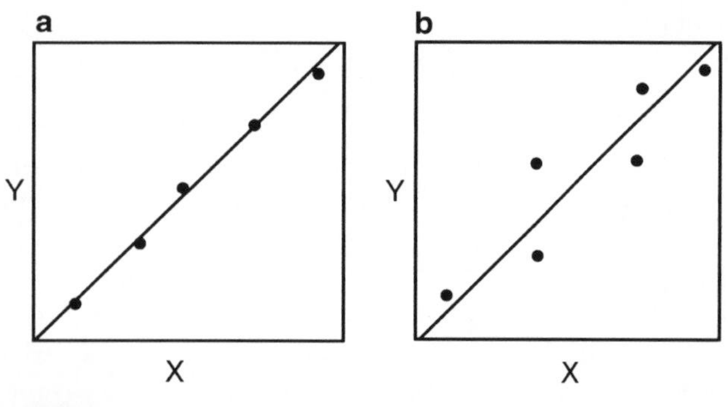

4-4 figure Examples of standard curves showing the relationship between the x and y variables when there is (**a**) a high amount of correlation and (**b**) a lower amount of correlation. Both lines have the same equation.

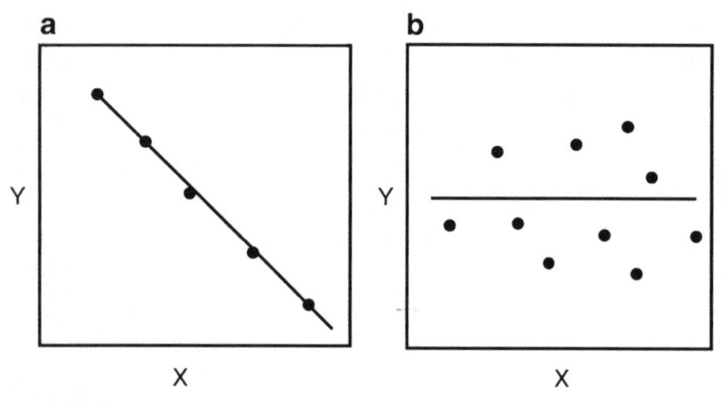

4-5 figure Examples of standard curves showing the relationship between the x and y variables when there is (**a**) a high amount of negative correlation and (**b**) no correlation between x and y values.

Fig. 4-4b shows a poor correlation. In both cases, we can draw a straight line through the data points. Both curves yield the same straight line, but the precision is poorer for the latter.

There are other possibilities when working with standard curves. Figure 4-5a shows a good correlation between x and y, but in the negative direction, and Fig. 4-5b illustrates data that have no correlation at all.

The **correlation coefficient** defines how well the data fit to a straight line. For a standard curve, the ideal situation would be that all data points lie perfectly on a straight line. However, this is never the case, because errors are introduced in making standards and measuring the physical values (observations).

The correlation coefficient and coefficient of determination are defined below. Essentially all spreadsheet and plotting software will calculate the values automatically.

$$\text{Correlation coefficient} = r = \frac{\sum (x_i - \bar{x})(y_i - \bar{y})}{\sqrt{[\sum (x_i - \bar{x})^2]\,[\sum (y_i - \bar{y})^2]}} \qquad [25]$$

For our example of the caffeine standard curve from Fig. 4-3, $r = 0.99943$ (values are usually reported to at least four significant figures).

For standard curves, we want the value of r as close to $+1.0000$ or -1.000 as possible, because this value is a perfect correlation (perfect straight line). Generally, in analytical work, the r should be 0.9970 or better. (This does not apply to biological studies).

The **coefficient of determination** (r^2) is used often because it gives a better perception of the straight line even though it does not indicate the direction of the correlation. The r^2 for the example presented above is 0.99886, which represents the proportion of the variance of absorbance (y) that can be attributed to its linear regression on concentration (x). This means that about 0.114% of the straight line variation ($1.0000 - 0.99886 = 0.00114 \times 100\% = 0.114\%$) does not vary with changes in x and y and thus is due to indeterminate variation. A small amount of variation is expected normally.

4.4.3 Errors in Regression Lines

While the correlation coefficient tells us something about the error or variation in linear curve fits, it does not always give the complete picture. Also, neither linear regression nor correlation coefficient will indicate that a particular set of data have a linear relationship. They only provide an estimate of the fit assuming the line is a linear one. As indicated before, plotting the data is critical when looking at how the data fit on the curve (actually, a line). One parameter that is used often is the ***y*-residuals**, which are simply the differences between the observed values and the calculated or computed values (from the regression line). Advanced computer graphics software can actually plot the residuals for each data point as a function of concentration. However, plotting the residuals is usually not necessary because data that do not fit on the line are usually quite obvious. If the residuals are large for the entire curve, then the entire method needs to be evaluated carefully. However, the presence of one point that is obviously off the line while the rest of the points fit very well probably indicate an improperly made standard.

One way to reduce the amount of error is to include more replicates of the data such as repeating the observations with a new set of standards. The replicate x and y values can be entered into the calculator or spreadsheet as separate points for the regression and coefficient determinations. Another, probably more desirable, option is to expand the concentrations at which the readings are taken. Collecting observations at more data points (concentrations) will produce

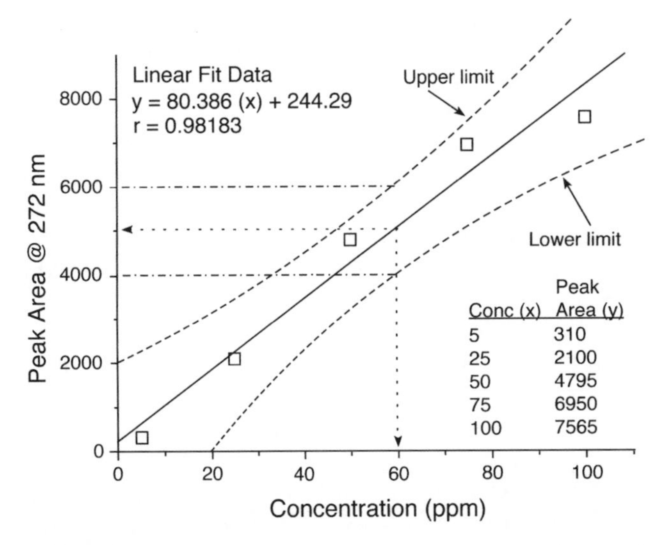

4-6 figure A standard curve graph showing the confidence bands. The data used to plot the graph are presented on the graph as are the equation of the line and the correlation coefficient.

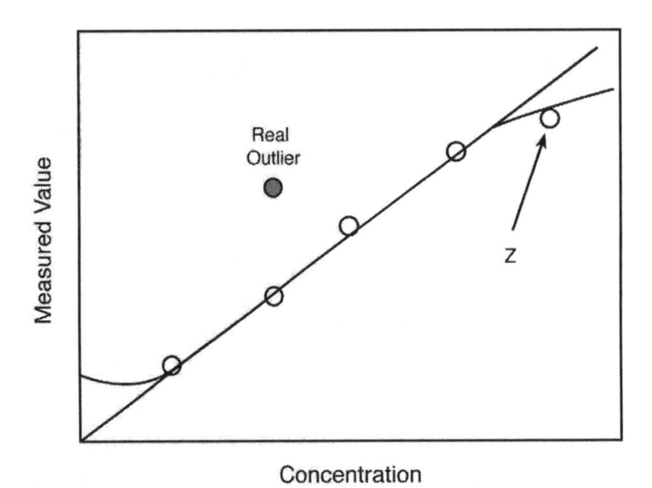

4-7 figure A standard curve plot showing possible deviations in the curve in the upper and lower limits.

a better standard curve. However, increasing the data beyond seven or eight points usually is not beneficial.

Plotting **confidence intervals**, or **bands** or **limits**, on the standard curve along with the regression line is another way to gain insight into the reliability of the standard curve. Confidence bands define the statistical uncertainty of the regression line at a chosen probability (such as 95%) using the t-statistic and the calculated standard deviation of the fit. In some aspects, the confidence bands on the standard curve are similar to the confidence interval discussed in Sect. 4.3.1. However, in this case we are looking at a line rather than a confidence interval around a mean. Figure 4-6 shows the caffeine data from the standard curve presented before, except some of the numbers have been modified to enhance the confidence bands. The confidence bands (dashed lines) consist of both an upper limit and a lower limit that define the variation of the y-axis value. The upper and lower bands are narrowest at the center of the curve and get wider as the curve moves to the higher or lower standard concentrations.

Looking at Fig. 4-6 again, note that the confidence bands show what amount of variation we expect in a peak area at a particular concentration. At 60 ppm concentration, by going up from the x-axis to the bands and extrapolating to the y-axis, we see that with our data the 95% confidence interval of the observed peak area will be 4000–6000. In this case, the variation is large and would not be acceptable as a standard curve and is presented here only for illustration purposes.

Error bars also can be used to show the variation of y at each data point. Several types of error or variation statistics can be used such as standard error, standard deviation, or percentage of data (i.e., 5%). Any of these methods give a visual indication of experimental variation.

Even with good standard curve data, problems can arise if the standard curve is not used properly. One common mistake is to extrapolate beyond the data points used to construct the curve. Figure 4-7 illustrates some of the possible problems that might occur when extrapolation is used. As shown in Fig. 4-7, the curve or line may not be linear outside the area where the data were collected. This can occur in the region close to the origin or especially at the higher concentration level.

Usually a standard curve will go through the origin, but in some situations it may actually tail off as zero concentration is approached. At the other end of the curve, at higher concentrations, it is fairly common for a plateau to be reached where the measured parameter does not change much with an increase in concentration. Care must be used at the upper limit of the curve to ensure that data for unknowns are not collected outside of the curve standards. Point Z in Fig. 4-7 should be evaluated carefully to determine if the point is an outlier or if the curve is actually tailing off. Collecting several sets of data at even higher concentrations should clarify this. Regardless, the unknowns should be measured only in the region of the curve that is linear.

4.5 REPORTING RESULTS

In dealing with experimental results, we are always confronted with reporting data in a way that indicates the sensitivity and precision of the assay. Ideally, we do not want to overstate or understate the sensitivity of the assay, and thus we strive to report a meaningful value, be it a mean, standard deviation, or some other number. The next three sections discuss how we can evaluate experimental values so as to be precise when reporting results.

4.5.1 Significant Figures

The term **significant figure** is used rather loosely to describe some judgment of the number of reportable digits in a result. Often, the judgment is not soundly based, and meaningful digits are lost or meaningless digits are retained. Exact rules are provided below to help determine the number of significant figures to report. However, it is important to keep some flexibility when working with significant figures.

Proper use of significant figures is meant to give an indication of the sensitivity and reliability of the analytical method. Thus, reported values should contain only significant figures. A value is made up of significant figures when it contains all digits known to be true and one last digit that is in doubt. For example, a value reported as 64.72 contains four significant figures, of which three digits are certain (64.7) and the last digit is uncertain. Thus, the 2 is somewhat uncertain and could be either 1 or 3. As a rule, numbers that are presented in a value represent the significant figures, regardless of the position of any decimal points. This also is true for values containing zeros, provided they are bounded on either side by a number. For example, 64.72, 6.472, 0.6472, and 6.407 all contain four significant figures. Note that the zero to the left of the decimal point is used only to indicate that there are no numbers above 1. We could have reported the value as 0.6472, but using the zero is better, since we know that a number was not inadvertently left off our value.

Special considerations are necessary for zeros that may or may not be significant.

1. Zeros after a decimal point are always significant figures. For example, 64.720 and 64.700 both contain five significant figures.
2. Zeros before a decimal point with no other preceding digits are not significant. As indicated before, 0.6472 contains four significant figures.
3. Zeros after a decimal point are not significant if there are no digits before the decimal point. For example, 0.0072 has no digits before the decimal point; thus, this value contains two significant figures. In contrast, the value 1.0072 contains five significant figures.
4. Final zeros in a number are not significant unless indicated otherwise. Thus, the value 7,000 contains only one significant figure. However, adding a decimal point and another zero gives the number 7000.0, which has five significant figures.

A good way to measure the significance of zeros, if the above rules become confusing, is to convert the number to the exponential form. If the zeros can be omitted, then they are not significant. For example, 7000 expressed in exponential form is 7×10^3 and contains one significant figure. With 7000.0, the zeros are retained and the number becomes 7.0000×10^3. If we were to convert 0.007 to exponent form, the value is 7×10^{-3} and only one significant figure is indicated. As a rule, determining significant figures in arithmetic operations is dictated by the value having the least number of significant figures. The easiest way to avoid any confusion is to perform all the calculations and then round off the final answer to the appropriate digits. For example, $36.54 \times 238 \times 1.1 = 9566.172$, and because 1.1 contains only two significant figures, the answer would be reported as 9600 (remember, the two zeros are not significant). This method works fine for most calculations, except when adding or subtracting numbers containing decimals. In those cases, the number of significant figures in the final value is determined by the numbers that follow the decimal point. Thus, when adding $7.45 + 8.725 = 16.175$, the sum is rounded to 16.18 because 7.45 has only two numbers after the decimal point. Likewise, $433.8 - 32.66$ gives 401.14, which rounds off to 401.1.

A word of caution is warranted when using the simple rule stated above, for there is a tendency to underestimate the significant figures in the final answer. For example, take the situation in which we determined the caffeine in an unknown solution to be 43.5 ppm (see Equation [24]). We had to dilute the sample 50-fold using a volumetric flask in order to fit the unknown within the range of our method. To calculate the caffeine in the original sample, we multiply our result by 50 or $43.5\,\mu g/ml \times 50 = 2,175\,\mu g/ml$ in the unknown. Based on our rule above, we then would round the number to one significant figure (because 50 contains one significant figure) and report the value as 2000. However, doing this actually underestimates the sensitivity of our procedure, because we ignore the accuracy of the volumetric flask used for the dilution. A Class-A volumetric flask has a tolerance of 0.05 ml; thus, a more reasonable way to express the dilution factor would be 50.0 instead of 50. We now have increased the significant figures in the answer by two, and the value becomes 2180 mg/ml.

As you can see, an awareness of significant figures and how they are adopted requires close inspection. The guidelines can be helpful, but they do not always work unless each individual value or number is closely inspected.

4.5.2 Rejecting Data

Inevitably, during the course of working with experimental data we will come across a value that does not match the others. Can you reject that value and thus not use it in calculating the final reported results?

The answer is "sometimes," but only after careful consideration. If you are routinely rejecting data to help make your assay look better, then you are misrepresenting the results and the precision of the assay. If the bad value resulted from an identifiable mistake in that particular test, then it is probably safe to drop the value. Again, caution is advised because you may be rejecting a value that is closer to the true value than some of the other values.

Consistently poor accuracy or precision indicates that an improper technique or incorrect reagent was used or that the test was not very good. It is best to make changes in the procedure or change methods rather than try to figure out ways to eliminate undesirable values.

There are several tests for rejecting an aberrant value. One of these tests, the **Q-Test,** is commonly used. In this test, a **Q-value** is calculated as shown below and compared to values in a table. If the calculated value is larger than the table value, then the questionable measurement can be rejected at the 90% confidence level.

$$Q\text{-value} = \frac{x_2 - x_1}{W} \quad [26]$$

where:

x_1 = questionable value

x_2 = next closest value to x_1

W = total spread of all values, obtained by subtracting the lowest value from the highest value

Table 4-4 provides the rejection Q-values for a 90% confidence level.

Q-Values for the Rejection of Results

Number of Observations	Q of Rejection (90% level)
3	0.94
4	0.76
5	0.64
6	0.56
7	0.51
8	0.47
9	0.44
10	0.41

Reprinted with permission from Dean RB, Dixon WJ (1951) Simplified statistics for small numbers of observations. *Anal. Chem.* 23:636–638. Copyright 1951, American Chemical Society.

The example below shows how the test is used for the moisture level of uncooked hamburger for which four replicates were performed giving values of 64.53, 64.45, 64.78, and 55.31. The 55.31 value looks as if it is too low compared to the other results. Can that value be rejected? For our example, x_1 is the questionable value (55.31) and x_2 is the closest neighbor to x_1 (which is 64.45). The spread (W) is the high value minus the low measurement, which is 64.78 − 55.31.

$$Q\text{-value} = \frac{64.45 - 55.31}{64.78 - 55.31} = \frac{9.14}{9.47} = 0.97 \quad [27]$$

From Table 4-4, we see that the calculated Q-value must be greater than 0.76 to reject the data. Thus, we make the decision to reject the 55.31% moisture value and do not use it in calculating the mean.

4.6 SUMMARY

This chapter focuses on statistical methods to measure data variability, precision, etc. and on basic mathematical treatment that can be used in evaluating a group of data. For example, it should be almost second nature to determine a mean, standard deviation, and CV when evaluating replicate analyses of an individual sample. In evaluating linear standard curves, best line fits should always be determined along with the indicators of the degree of linearity (correlation coefficient or coefficient of determination). Fortunately, most computer spreadsheet and graphics software will readily perform the calculations for you. Guidelines are available to enable one to report analytical results in a way that tells something about the sensitivity and confidence of a particular test. A section is included which describes sensitivity and LOD as related to various analytical methods and regulatory agency policies. Additional information includes the proper use of significant figures, rules for rounding off numbers and use of the Q-test to reject grossly aberrant individual values.

4.7 STUDY QUESTIONS

1. Method A to quantitate a particular food component was reported to be more specific and accurate than method B, but method A had lower precision. Explain what this means.
2. You are considering adopting a new analytical method in your lab to measure moisture content of cereal products. How would you determine the precision of the new method and compare it to the old method? Include any equations to be used for any needed calculations.
3. A sample known to contain 20 g/L glucose is analyzed by two methods. Ten determinations were made for each method and the following results were obtained:

Method A	*Method B*
Mean = 19.6	Mean = 20.2
Std. Dev. = 0.055	Std. Dev. = 0.134

(a) Precision and accuracy:

- Which method is more precise? Why do you say this?
- Which method is more accurate? Why do you say this?

(b) In the equation to determine the standard deviation, $n-1$ was used rather than just n. Would the standard deviation have been smaller or larger for each of those values above if simply n had been used?

(c) You have determined that values obtained using Method B should not be accepted if outside the range of two standard deviations from the mean. What range of values will be acceptable?

(d) Do the data above tell you anything about the specificity of the method? Describe what "specificity" of the method means as you explain your answer.

4. Differentiate "standard deviation" from "coefficient of variation," "standard error of the mean," and "confidence interval."
5. Differentiate the terms "absolute error" vs. "relative error." Which is more useful? Why?
6. For each of the errors described below in performing an analytical procedure, classify the error as random error, systematic error, or blunder, and describe a way to overcome the error.

   (a) Automatic pipettor consistently delivered 0.96 ml rather than 1.00 ml.

   (b) Substrate was not added to one tube in an enzyme assay.

7. Differentiate the terms "sensitivity" and "limit of detection."
8. The correlation coefficient for standard curve A is reported as 0.9970. The coefficient of determination for standard curve B is reported as 0.9950. In which case do the data better fit a straight line?

4.8 PRACTICE PROBLEMS

1. How many significant figures are in the following numbers: 0.0025, 4.50, 5.607?
2. What is the correct answer for the following calculation expressed in the proper amount of significant figures?

$$\frac{2.43 \times 0.01672}{1.83215} =$$

3. Given the following data on dry matter (88.62, 88.74, 89.20, 82.20), determine the mean, standard deviation, and CV. Is the precision for this set of data acceptable? Can you reject the value 82.20 since it seems to be different than the others? What is the 95% confidence level you would expect your values to fall within if the test were repeated? If the true value for dry matter is 89.40, what is the percent relative error?
4. Compare the two groups of standard curve data below for sodium determination by atomic emission spectroscopy. Draw the standard curves using graph paper or a computer software program. Which group of data provides a better standard curve? Note that the absorbance of the emitted radiation at 589 nm increases proportionally to sodium concentration. Calculate the amount of sodium in a sample with a value of 0.555 for emission at 589 nm. Use both standard curve groups and compare the results.

Sodium concentration (μg/ml)	*Emission at 589 nm*
Group A – Sodium Standard Curve	
1.00	0.050
3.00	0.140
5.00	0.242
10.0	0.521
20.0	0.998
Group B – Sodium Standard Curve	
1.00	0.060
3.00	0.113
5.00	0.221
10.0	0.592
20.0	0.917

Answers

1. 2, 3, 4
2. 0.0222
3. Mean = 87.19, $SD_{n-1} = 3.34$, and

$$CV = \frac{3.34}{87.18} \times 100\% = 3.83\%$$

thus the precision is acceptable because it is less than 5%.

$$Q\text{-calc value} = \frac{88.62 - 82.20}{89.20 - 82.20} = \frac{6.42}{7.00} = 0.917$$

$Q_{calc} = 0.92$, therefore the value 82.20 can be rejected because it is more than 0.76 from Table 4-4, using 4 as number of observations.

$$\text{CI (at 95\%)} = 87.19 \pm 3.18 \times \frac{3.34}{\sqrt{4}} = 87.19 \pm 5.31$$

Relative error = $\%E_{rel}$ where mean is 87.19 and true value is 89.40

$$\%E_{rel} = \frac{\bar{x} - T}{T} \times 100\% = \frac{87.19 - 89.40}{89.40} \times 100\%$$
$$= -2.47\%$$

4. Using linear regression we get

   Group A: $y = 0.0504x - 0.0029$, $r^2 = 0.9990$
   Group B: $y = 0.0473x + 0.0115$, $r^2 = 0.9708$

   The Group A r^2 is closer to 1.000 and is more linear and thus the better standard curve.
   Sodium in the sample using group A standard curve is:

   $$0.555 = 0.0504x - 0.0029, \; x = 11.1 \mu g/ml$$

   Sodium in the sample using group B standard curve is:

   $$0.555 = 0.0473x + 0.0115, \; x = 11.5 \mu g/ml$$

4.9 RESOURCE MATERIALS

1. Garfield FM, Klestra E, Hirsch J (2000) Quality assurance principles for analytical laboratories, AOAC International, Gaithersburg, MD. This book covers mostly quality assurance issues but has a good chapter (Chap. 3) on the basics of statistical applications to data.
2. Meier PC, Zund RE (2000) Statistical methods in analytical chemistry, 2nd edn. Wiley, New York. This is another excellent text for beginner and advanced analytical chemists alike. It contains a fair amount of detail, sufficient for most analytical statistics, yet works through the material starting at a basic introductory level. The assay evaluation parameters LOD, MDL, and LOQ are thoroughly discussed. The authors also discuss the *Q*-test used for rejecting data. A big plus of this book is the realistic approach taken throughout.
3. Miller JN, Miller JC (2005) Statistics and chemometrics for analytical chemistry, 5th edn. Pearson Prentice Hall, Upper Saddle River, NJ. This is another excellent introductory text for beginner analytical chemists at the undergraduate and graduate level. It contains a fair amount of detail, sufficient for most analytical statistics, yet works through the material starting at a basic introductory level. The authors also discuss the *Q*-test used for rejecting data.
4. Skoog DA, West DM, Holler JF, Crouch SR (2000) Analytical chemistry: an introduction, 7th edn. Brooks/Cole, Pacific Grove, CA. Section I, Chaps. 5–7 do an excellent job of covering most of the statistics needed by an analytical chemist in an easy-to-read style.
5. Code of Federal Regulations (2010) Environmental Protection Agency. 40 CFR, Vol. 22. Chapter I part 136 – Guidelines Establishing Test Procedures for the Analysis of Pollutants. Appendix B to Part 136 – Definition and Procedure for the Determination of the Method Detection Limit – Revision 1.11.

Sampling and Sample Preparation

Rubén O. Morawicki
Department of Food Science, University of Arkansas, Fayetteville, AR 72704, USA
rmorawic@uark.edu

5.1 Introduction 71
5.2 Selection of Sampling Procedures 71
5.2.1 General Information 71
5.2.2 Sampling Plan 71
5.2.3 Factors Affecting the Choice of Sampling Plans 72
5.2.4 Sampling by Attributes and Sampling by Variables 72
5.2.5 Acceptance Sampling 73
5.2.6 Risks Associated with Sampling 74
5.3 Sampling Procedures 74
5.3.1 Introduction and Examples 74
5.3.2 Homogeneous vs. Heterogeneous Populations 74
5.3.3 Manual vs. Continuous Sampling 74
5.3.4 Statistical Considerations 75
5.3.4.1 Probability Sampling 75
5.3.4.2 Nonprobability Sampling 76
5.3.4.3 Mixed Sampling 76
5.3.4.4 Estimating the Sample Size 76
5.3.5 Problems in Sampling 77
5.4 Preparation of Samples 77
5.4.1 General Size Reduction Considerations 77

S.S. Nielsen, *Food Analysis*, Food Science Texts Series, DOI 10.1007/978-1-4419-1478-1_5,

5.4.2 Grinding 78
5.4.2.1 Introduction 78
5.4.2.2 Applications for Grinding Equipment 78
5.4.2.3 Determination of Particle Size 78
5.4.3 Enzymatic Inactivation 79
5.4.4 Lipid Oxidation Protection 80
5.4.5 Microbial Growth and Contamination 80
5.5 Summary 80
5.6 Study Questions 80
5.7 Acknowledgments 81
5.8 References 81

5.1 INTRODUCTION

Quality attributes in food products, raw materials, or ingredients are measurable characteristics that need monitoring to ensure that specifications are met. Some quality attributes can be measured online by using specially designed sensors and results obtained in real time (e.g., color of vegetable oil in an oil extraction plant). However, in most cases quality attributes are measured on small portions of material that are taken periodically from continuous processes or on a certain number of small portions taken from a lot. The small portions taken for analysis are referred to as **samples**, and the entire lot or the entire production for a certain period of time, in the case of continuous processes, is called a **population**. The process of taking samples from a population is called **sampling**. If the procedure is done correctly, the measurable characteristics obtained for the samples become a very accurate estimation of the population.

By sampling only a fraction of the population, a quality estimate can be obtained accurately, quickly, and with less expense and personnel time than if the total population were measured. Moreover, in the case of food products, analyzing a whole population would be practically impossible because of the destructive nature of most analytical methods. Paradoxically, estimated parameters using representative samples (discussed in Sects. 5.2 and 5.3) are normally more accurate than the same estimations done on the whole population (census).

A **laboratory sample** for analysis can be of any size or quantity (1). Factors affecting the sample size and associated problems are discussed in Sects. 5.3 and 5.4, while preparation of laboratory samples for testing is described in Sect. 5.4.

As you read each section of the chapter, consider application of the information to some specific examples of sampling needs in the food industry: sampling for nutrition labeling (see Study Question 7 in this chapter), pesticide analysis (see also Chap. 18, Sect. 18.3), mycotoxin analysis (see also Chap. 18, Sect. 18.4), extraneous matter (see also Chap. 19), or rheological properties (see also Chap. 30). To consider sample collection and preparation for these and other applications subject to government regulations, you are referred also to the sample collection section of compliance procedures established by the Food and Drug Administration (FDA) and Food Safety and Inspection Service (FSIS) of the United States Department of Agriculture (USDA) (see Chap. 3, Sect. 3.2.2.1).

It should be noted that sampling terminology and procedures used may vary between companies and between specific applications. However, the principles described in this chapter are intended to provide a basis for understanding, developing, and evaluating sampling plans and sample handling procedures for specific applications encountered.

5.2 SELECTION OF SAMPLING PROCEDURES

5.2.1 General Information

The first step in any sampling procedure is to clearly define the population that is going to be sampled. The population may vary in size from a production lot, a day's production, to the contents of a warehouse. Information obtained from a sample of a particular production lot in a warehouse must be used strictly to make inferences about that particular lot, but conclusions cannot be extended to other lots in the warehouse.

Once sampling is conducted, a series of stepwise procedures – from sample preparation, laboratory analysis, data processing, and interpretation – is needed to obtain data from the samples. In each step, there is a potential for error that would compromise the certainty, or reliability, of the final result. This final result depends on the cumulative errors at each stage that are usually described by the variance (2, 3). **Variance** is an estimate of the uncertainty. The total variance of the whole testing procedure is equal to the sum of the variances associated with each step of the sampling procedure and represents the **precision** of the process. Precision is a measure of the reproducibility of the data. In contrast, **accuracy** is a measure of how close the data are to the true value. The most efficient way to improve accuracy is to improve the reliability of the step with the greatest variance that is frequently the initial sampling step. The reliability of sampling is dependent more on the sample size than on the population size (4). The larger the sample size the more reliable the sampling. However, sample size is limited by time, cost, sampling methods, and the logistics of sample handling, analysis, and data processing.

5.2.2 Sampling Plan

Most sampling is done for a specific purpose, and the purpose may dictate the nature of the sampling approach. The two primary objectives of sampling are often to estimate the average value of a characteristic and determine if the average value meets the specifications defined in the sampling plan. Sampling purposes vary widely among different food industries; however, the most important categories include the following:

1. Nutritional labeling
2. Detection of contaminants and foreign matter

3. Statistical process control (Quality Assurance)
4. Acceptance of raw materials, ingredients, or products (Acceptance Sampling)
5. Release of lots of finished product
6. Detection of adulterations
7. Microbiological safety
8. Authenticity of food ingredients, etc.

The **International Union of Pure and Applied Chemistry** (IUPAC) defines a sampling plan as: "A predetermined procedure for the selection, withdrawal, preservation, transportation, and preparation of the portions to be removed from a lot as samples" (5). A sampling plan should be a well-organized document that establishes the goals of the sampling plan, the factors to be measured, sampling point, sampling procedure, frequency, size, personnel, preservation of the samples, etc. The primary aim of sampling is to obtain a sample, subject to constraints of size that will satisfy the sampling plan specifications. A sampling plan should be selected on the basis of the sampling objective, the study population, the statistical unit, the sample selection criteria, and the analysis procedures. Depending on the purpose of the sampling plan, samples are taken at different points of the food production system, and the sampling plan may vary significantly for each point.

Factors that Affect the Choice of Sampling Plans

Factors to Be Considered	*Questions*
Purpose of the inspection	Is it to accept or reject the lot? Is it to measure the average quality of the lot? Is it to determine the variability of the product?
Nature of the product	Is it homogeneous or heterogeneous? What is the unit size? How consistently have past populations met specifications? What is the cost of the material being sampled?
Nature of the test method	Is the test critical or minor? Will someone become sick or die if the population fails to pass the test? Is the test destructive or nondestructive? How much does the test cost to complete?
Nature of the population being investigated	Is the lot large but uniform? Does the lot consist of smaller, easily identifiable sublots? What is the distribution of the units within the population?

Adapted from (1).

5.2.3 Factors Affecting the Choice of Sampling Plans

Each factor affecting the choice of sampling plans (Table 5-1) must be considered in the selection of a plan. Once the purpose of the inspection, the nature of the product, the test method, and the nature of the population to be sampled are determined, then a sampling plan that will provide the desired information can be developed.

The choice of a sampling plan is an important consideration, especially when monitoring food safety by measurement of fungal toxins, named mycotoxins, in food systems. Mycotoxins are distributed broadly and randomly within a population and a normal distribution cannot be assumed (1). Such distribution requires a combination of many randomly selected portions to obtain a reasonable estimate of mycotoxin levels. Methods of analysis that are extremely precise are not needed when determining mycotoxin levels, when sampling error is many times greater than analytical error (1). In this case, sampling and good comminution and mixing prior to particle size reduction are more important than the chemical analysis itself. Additional information on sampling for mycotoxin analysis is provided in Chap. 18, Sect. 18.4.

5.2.4 Sampling by Attributes and Sampling by Variables

Sampling plans are designed for examination of either attributes or variables (4). In **attribute sampling**, sampling is performed to decide on the acceptability of a population based on whether the sample possesses a certain characteristic or not. The result has a binary outcome of either conforming or nonconforming. Sampling plans by attributes are based on the hypergeometric, binomial, or Poisson statistical distributions. In the event of a binomial distribution (e.g., presence of *Clostridium botulinum*), the probability of a single occurrence of the event is directly proportional to the size of the sample, which should be at least ten times smaller than the population size. Computing binomial probabilities will allow the investigator to make inferences on the whole lot.

In **variable sampling**, sampling is performed to estimate quantitatively the amount of a substance (e.g., protein content, moisture content, etc.) or a characteristic (e.g., color) on a continuous scale. The estimate obtained from the sample is compared with an acceptable value (normally specified by the label, regulatory agencies, or the customer) and the deviation measured. This type of sampling usually produces data that have a *normal distribution* such as in the percent fill of a container and total solids of a food sample. In general, variable sampling requires smaller sample size than attribute sampling (4), and each characteristic should be sampled for separately when possible. However, when the FDA and the USDA's FSIS perform sampling for compliance of nutrition labeling, a composite of 12 and of at least six subsamples, respectively, is obtained and used for all nutrients to be analyzed.

5.2.5 Acceptance Sampling

Acceptance sampling is a procedure that serves a very specific role: to determine if a shipment of products or ingredients has enough quality to be accepted. **Acceptance sampling** can be performed by the food processor before receiving a lot of materials from a supplier, or by a buyer who is evaluating the processor's output (6). Acceptance sampling is a very broad topic that can be applied to any field; more specific literature can be consulted if needed.

Lot acceptance sampling plans that may be used for evaluation of attributes or variables, or a combination of both, fall into the following categories: **single, double, multiple, sequential, and skip plans**. In **single sampling** plans, the decision of accepting or rejecting a lot is based just on one sample of items taken at random. These plans are usually denoted as (n,c) plans for a sample size n, where the lot is rejected if there are more than c defective samples (7). If results are inconclusive, a second sample is taken and the decision of accepting or rejecting is made based on the combined outcome of both samples. Figure 5-1 shows an example of a **double sampling** plan (i.e., two samples are taken). **Multiple sampling** plans are extensions of double sampling plans for which more than two samples are drawn to reach a conclusion.

The ultimate extension of multiple sampling is **sequential sampling**. Under this plan, a sample is taken, and after analysis a decision of accepting, rejecting, or taking another sample is made. Therefore, the number of total samples to be taken depends exclusively on the sampling process. A sequential sampling graph is presented in Fig. 5-2. In this chart the cumulative observed number of defective samples is plotted against the number of samples taken. Two lines – the rejection and acceptance lines – are drawn, thus dividing the plot in three different regions: accept, reject, and continue sampling. An initial sample is taken and the results are plotted in the graph. If the plotted point falls within the parallel lines, then a second sample is taken and the process is repeated until the reject or accept zones are reached (7). Details about the construction of this plot are beyond the scope of this book, and particulars can be found in more specialized literature.

In **skip lot sampling** only a fraction of the submitted lots is inspected. It is a money-saving sampling procedure, but it can be implemented only when there is enough proof that the quality of the lots is consistent.

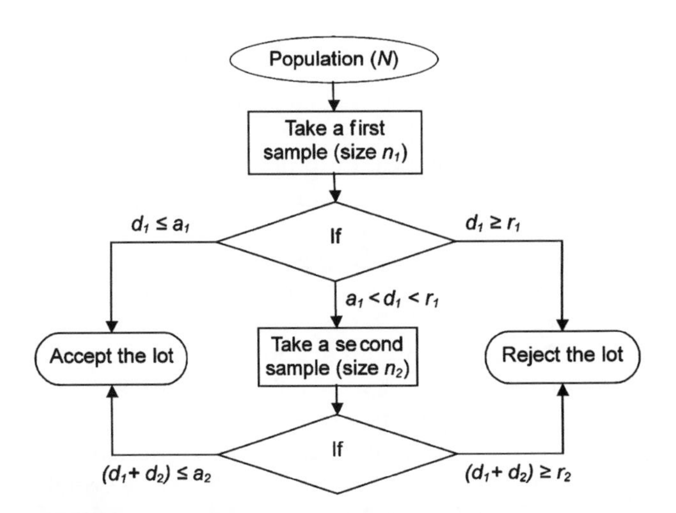

5-1 figure Example of a double sampling plan with two points where the decision of acceptance or rejection can be made [Adapted from (7)]. N, population size; n_1 and n_2, sample size; a_1 and a_2, acceptance numbers; r_1 and r_2, rejection numbers; d_1 and d_2, number of nonconformities. *Subindices* 1 and 2 represent samples 1 and 2, respectively.

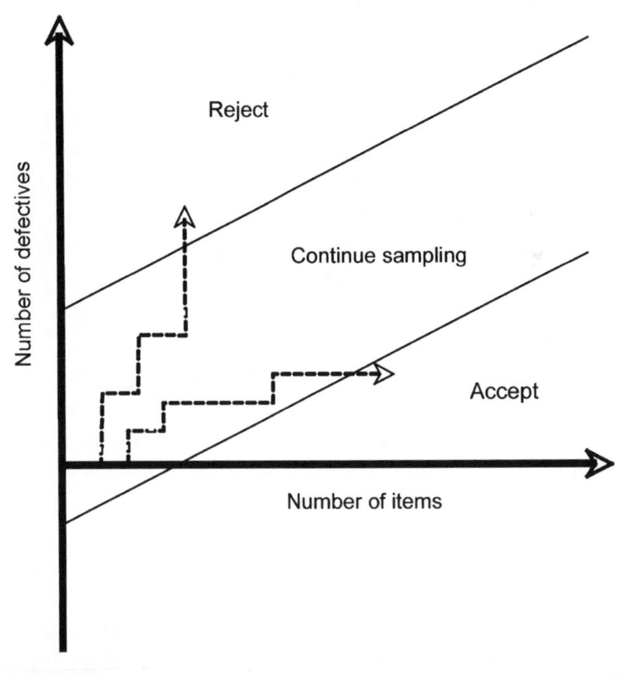

5-2 figure Sequential sampling plan. [Adapted from (7).]

5.2.6 Risks Associated with Sampling

There are two types of risks associated with sampling: the producer's and consumer's risks (6). The **consumer's risk** describes the probability of accepting a poor quality population. This should happen rarely (<5% of the lots), but the actual acceptable probability (β) of a consumer risk depends on the consequences associated with accepting an unacceptable lot. These may vary from major health hazards and subsequent fatalities to a lot being of slightly lower quality than standard lots. Obviously, the former demands a low or no probability of occurring whereas the latter would be allowed to occur more frequently. The **producer risk** is the probability of rejecting (α) an acceptable product. As with consumer's risk, the consequences of an error determine the acceptable probability of the risk. An acceptable probability of producer's risk is usually 5–10%. Further discussion of sampling plans can be found in the following section.

5.3 SAMPLING PROCEDURES

5.3.1 Introduction and Examples

The reliability of analytical data is compromised if sampling is not done properly. As shown in Table 5-1, the use of the data to be obtained will determine the sampling procedure. Details for the sampling of specific food products are described in the *Official Methods of Analysis* of AOAC International (8) and in the *Code of Federal Regulations* (CFR) (9). Two such examples for specific foods follow.

The AOAC Method 925.08 (8) describes the method for sampling flour from sacks. The number of sacks to be sampled is determined by the square root of the number of sacks in the lot. The sacks to be sampled are chosen according to their exposure. The samples that are more frequently exposed are sampled more often than samples that are exposed less. Sampling is done by drawing a core from a corner at the top of the sack diagonally to the center. The sampling instrument is a cylindrical, polished trier with a pointed end. It is 13 mm in diameter with a slit at least one third of the circumference of the trier. A second sample is taken from the opposite corner in a similar manner. The cores are stored for analysis in a clean, dry, airtight container that has been opened near the lot to be sampled. The container should be sealed immediately after the sample is added. A separate container is used for each sack. Additional details regarding the container and the procedure also are described below.

Title 21 CFR specifies the sampling procedures required to ensure that specific foods conform to the standard of identity. In the case of canned fruits, 21 CFR 145.3 defines a sample unit as "container, a portion of the contents of the container, or a composite mixture of product from small containers that is sufficient for the testing of a single unit" (9). Furthermore, a sampling plan is specified for containers of specific net weights. The container size is determined by the size of the lot. A specific number of containers must be filled for sampling of each lot size. The lot is rejected if the number of defective units exceeds the acceptable limit. For example, out of a lot containing 48,001–84,000 units, each weighing 1 kg or less, 48 samples should be selected. If six or more of these units fail to conform to the attribute of interest the lot will be rejected. Based on statistical confidence intervals, this sampling plan will reject 95% of the defective lots examined, that is, 5% consumer risk (9).

The discussion below describes general considerations to take into account when obtaining a sample for analysis.

5.3.2 Homogeneous vs. Heterogeneous Populations

The ideal population would be uniform throughout and identical at all locations. Such a population would be **homogeneous**. Sampling from such a population is simple, as a sample can be taken from any location and the analytical data obtained will be representative of the whole. However, this occurs rarely, as even in an apparently uniform product, such as sugar syrup, suspended particles and sediments in a few places may render the population heterogeneous. In fact, most populations that are sampled are **heterogeneous**. Therefore, the location within a population where a sample is taken will affect the subsequent data obtained. However, sampling plans (Sect. 5.2.2) and sample preparation (Sect. 5.4) can make the sample representative of the population or take heterogeneity into account in some other way.

5.3.3 Manual vs. Continuous Sampling

To obtain a **manual sample** the person taking the sample must attempt to take a "random sample" to avoid human bias in the sampling method. Thus, the sample must be taken from a number of locations within the population to ensure that it is representative of the whole population. For liquids in small containers, this can be done by shaking prior to sampling. When sampling from a large volume of liquid, such as that stored in silos, aeration ensures a homogeneous unit. Liquids may be sampled by pipetting, pumping, or dipping. However, when sampling grain from a rail car, mixing is impossible and samples are obtained by probing from several points at random within the rail car. Such manual sampling of granular or powdered

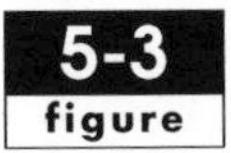

An automatic liquid sampling device that uses air under high pressure to collect multiple 1.5-ml samples. The control box (*left*) regulates the sampling frequency. (Courtesy of Liquid Sampling Systems Inc., Cedar Rapids, IA.)

material is usually achieved with triers or probes that are inserted into the population at several locations. Errors may occur in sampling (10), as rounded particles may flow into the sampling compartments more easily than angular ones. Similarly, hygroscopic materials flow more readily into the sampling devices than do nonhygroscopic materials. Horizontal core samples have been found to contain a larger proportion of small-sized particles than vertical ones (10).

Continuous sampling is performed mechanically. Figure 5-3 shows an automatic sampling device that is used to take liquid samples from a continuous production line. Continuous sampling should be less prone to human bias than manual sampling.

5.3.4 Statistical Considerations

5.3.4.1 Probability Sampling

Probability sampling plans prescribe the selection of a sample from a population based on chance. It provides a statistically sound basis for obtaining representative samples with elimination of human bias (1). The probability of including any item in the sample is known, and sampling error can be calculated. Several probability sampling methods are available to the researcher, and the most common ones are described in the next few paragraphs.

Simple random sampling requires that the number of units in the population be known and each unit is assigned an identification number. Then using a random selection process, a certain number of identification numbers are selected according to the sample size. The sample size is determined according to the lot size and the potential impact of a consumer or vendor error. The random selection of the individuals units is done by using random number tables or computer-generated random numbers. Units selected randomly (sample) are analyzed, and the results can be considered an unbiased estimate of the population.

Systematic sampling is used when a complete list of sample units is not available, but when samples are distributed evenly over time or space, such as on a production line. The first unit is selected at random (random start) and then units are taken every *n*th unit (sampling interval) after that.

Stratified sampling involves dividing the population (size N) into a certain number of mutually exclusive homogeneous subgroups (size N_1, N_2, N_3, etc.) and then applying random or another sampling technique to each subgroup. Stratified sampling is used when subpopulations of similar characteristics can be observed within the whole population. An example of stratified sampling would be a company that produces tomato juice in different plants. If we need to study the residual activity of polygalacturonase in tomato juice we can stratify on production plants and take samples on each plant.

Cluster sampling entails dividing the population into subgroups, or clusters, and then selecting randomly only a certain number of clusters for analysis. The main difference between cluster sampling and stratified sampling is that in the latter samples are taken from every single subgroup, while in cluster sampling only some randomly selected clusters are sampled. The clusters selected for sampling may be either totally inspected or subsampled for analysis. This sampling method is more efficient and less expensive than simple random sampling, if populations can be divided into clusters. Going back to the tomato juice example, when using cluster sampling we would consider all processing plants, but we would select randomly just a few for the purpose of the study.

Composite sampling is used to obtain samples from bagged products such as flour, seeds, and larger items in bulk. Small aliquots are taken from different bags, or containers, and combined in a simple sample (the composite sample) that is used for analysis. Composite sampling also can be used when a representative sample of a whole production day in a continuous process is needed. In this case, a systematic approach is used to take equal aliquots at different times, and then a representative sample is obtained by mixing the individual aliquots. A typical example of composite sampling is the sampling plan mandated by the FDA and FSIS for nutritional labeling. They require a composite of 12 samples with at least six subsamples taken and analyzed for compliance with nutrition labeling regulations (11).

5.3.4.2 Nonprobability Sampling

Randomization is always desired. However, it is not always feasible, or even practical, to take samples based on probability methods. Examples include preliminary studies to generate hypothesis, the estimation of the standard deviation so that a more accurate sampling plan can be designed, or cases for which the bulkiness of the material makes inaccessible the removal of samples. In these cases, nonprobability sampling plans may be more economical and practical than probability sampling. Moreover, in certain cases of adulteration such as rodent contamination, the objective of the sampling plan may be to highlight the adulteration rather than collect a representative sample of the population.

Nonprobability sampling can be done in many ways, but in each case the probability of including any specific portion of the population is not equal because the investigator selects the samples deliberately. Without the use of a methodology that gives every element of the population the same chance to be selected, it is not possible to estimate the sampling variability and possible bias.

Judgment sampling is solely at the discretion of the sampler and therefore is highly dependent on the person taking the sample. This method is used when it is the only practical way of obtaining the sample. It may result in a better estimate of the population than random sampling if sampling is done by an experienced individual and the limitations of extrapolation from the results are understood (1). **Convenience sampling** is performed when ease of sampling is the key factor. The first pallet in a lot or the sample that is most accessible is selected. This is also called "chunk sampling" or "grab sampling." Although this sampling requires little effort, the sample obtained will not be representative of the population, and therefore is not recommended. **Restricted sampling** may be unavoidable when the entire population is not accessible. This is the case if sampling from a loaded boxcar, but the sample will not be representative of the population. **Quota sampling** is the division of a lot into groups representing various categories, and samples are then taken from each group. This sampling method is less expensive than random sampling but also is less reliable.

5.3.4.3 Mixed Sampling

When the sampling plan is a mixture of two or more basic sampling methods that can be random or nonrandom, then the sampling plan is called **mixed sampling**.

5.3.4.4 Estimating the Sample Size

Sample size determination can be based on either **precision analysis** or **power analysis**. Precision and power analysis are done by controlling the confidence level (type I error) or the power (type II error). For the purpose of this section, the precision analysis will be used, and it will be based on the confidence interval approach and the assumptions that the population is normal.

The confidence interval for a sample mean is described by the following equation:

$$\bar{x} \pm z_{\alpha/2}\frac{\mathrm{SD}}{\sqrt{n}} \quad [1]$$

where:

$\bar{x}$ = sample mean

$z_{\alpha/2}$ = z-value corresponding to the level of confidence desired

SD = known, or estimated, standard deviation of the population

n = sample size

In Equation [1],

$$z_{\alpha/2}\frac{\mathrm{SD}}{\sqrt{n}}$$

represents the maximum error (E) that is acceptable for a desired level of confidence. Therefore, we can set the equation

$$E = z_{\alpha/2}\frac{\mathrm{SD}}{\sqrt{n}}$$

and solve for n:

$$n = \left(\frac{z_{\alpha/2}\,\mathrm{SD}}{E}\right)^2 \quad [2]$$

The maximum error, E, in Equation [2] can be expressed in terms of the accuracy (γ) as: $E = \gamma \times \bar{x}$. Then Equation [2] can be rearranged as follows:

$$n = \left(\frac{z_{\alpha/2}\,\mathrm{SD}}{\gamma \times \bar{x}}\right)^2 \quad [3]$$

Now, we have an equation to calculate the sample size, but the equation is dependant on an unknown parameter: the standard deviation. To solve this problem we can follow different approaches. One way is to take few samples using a nonstatistical plan and use the data to estimate the mean and standard deviation. A second approach is using data from the past or data from a similar study. A third method is to estimate the standard deviation as 1/6 of the range of data values (6). A fourth method is to use typical coefficients of variation (defined as 100× [standard deviation/population mean]), assuming we have an estimation of the population mean.

If the estimated sample is smaller than 30, then the Student's t distribution needs to be used instead of the normal distribution by replacing the $z_{\alpha/2}$ with the parameter t, with $n-1$ degrees of freedom. However, the use of the Student's t-test distribution comes with the additional cost or introducing another uncertainty into Equation [3]: the **degrees of freedom**. For the estimation of the t-score, we need to start somewhere by assuming the degrees of freedom, or assuming a t-score, and then calculating the number of samples, recalculating the t-score with $n-1$ degrees of freedom, and calculating the number of samples again. For a level of uncertainty of 95%, a conservative place to start would be assuming a t-score of 2.0 and then calculating the initial sample size. If we use a preliminary experiment to estimate the standard deviation, then we can use the sample size of the preliminary experiment minus 1 to calculate the t-score.

Example: We want to test the concentration of sodium in a lot of a ready-to-eat food product with a level of confidence of 95%. Some preliminary testing showed an average content of 1000 mg of sodium per tray with an estimated standard deviation of 500. Determine the sample size with an accuracy of 10%.

Data: Confidence level $= 95\% \Rightarrow \alpha = 0.05 \Rightarrow z = 1.96; \gamma = 0.1; \bar{x} = 1000; \mathrm{SD} = 500$

$$n = \left(\frac{z_{\alpha/2}\,\mathrm{SD}}{\gamma \times \bar{x}}\right)^2 = \left(\frac{1.96 \times 500}{0.1 \times 1000}\right)^2 = 96\,\text{trays}$$

5.3.5 Problems in Sampling

No mater how reliable our analytical technique is, our ability to make inferences on a population will always depend on the adequacy of sampling techniques. **Sampling bias**, due to nonstatistically viable convenience, may compromise reliability. Errors also may be introduced by not understanding the **population distribution** and subsequent selection of an inappropriate sampling plan.

Unreliable data also can be obtained by nonstatistical factors such as poor **sample storage** resulting in sample degradation. Samples should be stored in a container that protects the sample from moisture and other environmental factors that may affect the sample (e.g., heat, light, air). To protect against changes in moisture content, samples should be stored in an airtight container. Light-sensitive samples should be stored in containers made of opaque glass or in containers wrapped in aluminum foil. Oxygen-sensitive samples should be stored under nitrogen or an inert gas. Refrigeration or freezing may be necessary to protect chemically unstable samples. However, freezing should be avoided when storing unstable emulsions. Preservatives (e.g., mercuric chloride, potassium dichromate, and chloroform) (4) can be used to stabilize certain food substances during storage.

Mislabeling of samples causes mistaken sample identification. Samples should be clearly identified by markings on the sample container in a manner such that markings will not be removed or damaged during storage and transport. For example, plastic bags that are to be stored in ice water should be marked with water-insoluble ink.

If the sample is an official or **legal sample** the container must be sealed to protect against tampering and the seal mark easily identified. Official samples also must include the date of sampling with the name and signature of the sampling agent. The chain of custody of such samples must be identified clearly.

5.4 PREPARATION OF SAMPLES

5.4.1 General Size Reduction Considerations

If the particle size or mass of the sample is too large for analysis, it must be reduced in bulk or particle size (4). To obtain a smaller quantity for analysis the sample can be spread on a clean surface and divided into quarters. The two opposite quarters are combined. If the mass is still too large for analysis, the process is repeated until an appropriate amount is obtained. This method can be modified for homogeneous liquids by pouring into four containers, and it can be automated (Fig. 5-4). The samples are thus homogenized to ensure negligible differences between each portion (1).

AOAC International (8) provides details on the preparation of specific food samples for analysis, which depends on the nature of the food and the analysis to be performed. For example, in the case of meat and meat products (8), it is specified in Method 983.18 that small samples should be avoided, as this results in significant moisture loss during preparation and subsequent handling. Ground meat samples should be stored in glass or similar containers, with air and watertight lids. Fresh, dried, cured, and smoked meats are to be bone free and passed three times through a food chopper with plate openings no more than 3 mm wide. The sample then should be mixed thoroughly and analyzed immediately. If immediate analysis is not possible, samples should be chilled or dried for short-term and long-term storage, respectively.

A further example of size reduction is the preparation of solid sugar products for analysis as described in AOAC Method 920.175 (8). The method prescribes that the sugar should be ground, if necessary, and

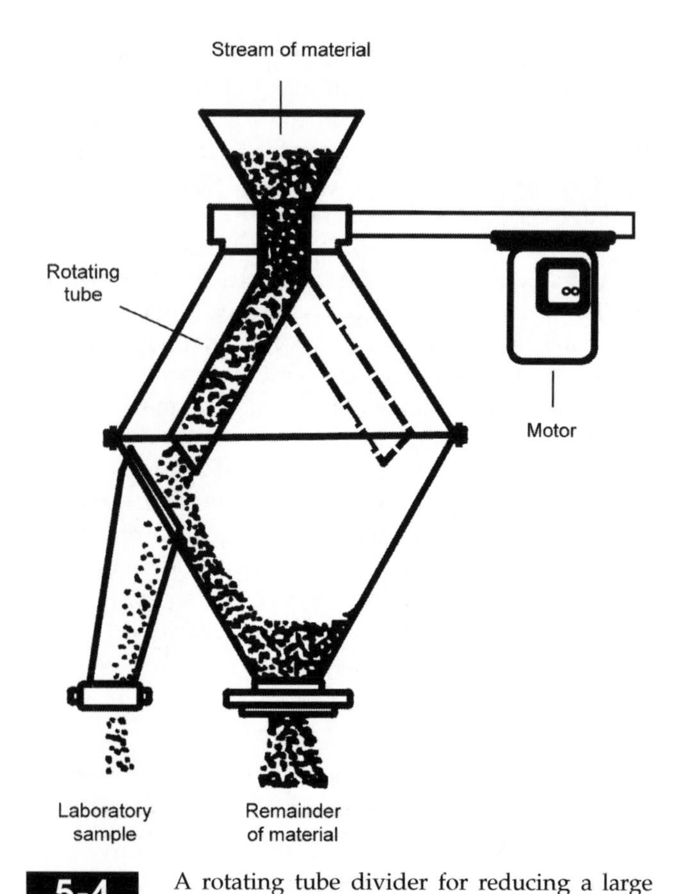

5-4 figure A rotating tube divider for reducing a large sample (ca. 880 kg) of dry, free-flowing material to a laboratory size sample (ca. 0.2 kg). (Courtesy of Glen Mills, Inc., Clifton, NJ.)

mixed to uniformity. Raw sugars should be mixed thoroughly and rapidly with a spatula. Lumps are to be broken by a mortar and pestle or by crushing with a glass or iron rolling pin on a glass plate.

5.4.2 Grinding

5.4.2.1 Introduction

Grinding is important both for sample preparation prior to analysis and for food ingredient processing. Various mills are available for reducing particle size to achieve sample homogenization (11). To homogenize moist samples, bowl cutters, meat mincers, tissue grinders, mortars and pestles, or blenders are used, while mortars and pestles and mills are best for dry samples. Some foods are more easily ground after drying in a desiccator or vacuum oven. Grinding wet samples may cause significant losses of moisture and chemical changes. In contrast, grinding frozen samples reduces undesirable changes. The grinding process should not heat the sample, and therefore the grinder should not be overloaded because heat will be produced through friction. Contact of food with bare metal surfaces should be avoided if trace metal analysis is to be performed (12).

To break up moist tissues, a number of slicing devices are available: bowl cutters can be used for fleshy tubers and leafy vegetables, while meat mincers may be better suited for fruit, root, and meat (13). Addition of sand as an abrasive can provide further subdivision of moist foods. Blenders are effective in grinding soft and flexible foods and suspensions. Rotating knives (25,000 rpm) will disintegrate a sample in suspension. In colloidal mills, a dilute suspension is flown under pressure through a gap between slightly serrated or smooth surfaced blades until they are disintegrated by shear. Sonic and supersonic vibrations disperse foods in suspension and in aqueous and pressurized gas solution. The Mickle disintegrator sonically shakes suspensions with glass particles, and the sample is homogenized and centrifuged at the same time (13). Alternatively, a low-shear continuous tissue homogenizer is fast and handles large volumes of sample.

5.4.2.2 Applications for Grinding Equipment

Mills differ according to their mode of action, being classified as a **burr, hammer, impeller, cyclone, impact, centrifugal, or roller mill** (13). Methods for grinding dry materials range from a simple pestle and mortar to power-driven hammer mills. Hammer mills wear well, and reliably and effectively grind cereals and dry foods, while small samples can be finely ground by ball mills. A ball mill grinds by rotating the sample in a container that is half filled with ceramic balls. This impact grinding can take hours or days to complete. A chilled ball mill can be used to grind frozen foods without predrying and also reduces the likelihood of undesirable heat-initiated chemical reactions occurring during milling (13). Alternatively, dry materials can be ground using an ultracentrifugal mill by beating, impact, and shearing. The food is fed from an inlet to a grinding chamber and is reduced in size by rotors. When the desired particle size is obtained, the particles are delivered by centrifugal force into a collection pan (13). Large quantities can be ground continuously with a cyclone mill.

5.4.2.3 Determination of Particle Size

Particle size is controlled in certain mills by adjusting the distance between burrs or blades or by screen mesh size/number. The **mesh number** is the number of square screen openings per linear inch of mesh. The final particles of dried foods should be 20 mesh for moisture, total protein, or mineral determinations.

Particles of 40-mesh size are used for extraction assays such as lipid and carbohydrate estimation.

In addition to reducing particle size for analysis of samples, it is also important to reduce the particle size of many food ingredients for use in specific food products. For example, rolled oats for a grain-based snack bar may have a specified granulation size described as 15% of the oats maximum passes through a #7 US standard sieve. A higher granulation (i.e., more smaller particles) would mean more fines and less whole oats in the finished bar. This would result in higher incidences of snack bar breakage.

A **Ro-Tap® Sieve Shaker** is an instrument that can achieve separation by size of small particles, to then determine the size distribution of the sample. This is a horizontal rotating hollow cylinder into which materials of varying particle sizes are introduced. The internal surface of the cylinder has many indentations of a specific size that hold particles of a specific size or smaller. In this way the rotating cylinder carries smaller particles away from the mass of material that remains at the bottom of the cylinder.

There are a variety of methods for measuring particle size, each suited for different materials. The simplest way to measure particle sizes of dry materials of less than 50 μm in diameter is by passing the sample through a series of vertically stacked sieves with increasing mesh number. As the mesh number increases, the apertures between the mesh are smaller and only finer and finer particles pass through subsequent sieves (see Table 5-2). Sieve sizes have been specified for salt, sugar, wheat flour, corn meal, semolina, and cocoa. The sieve method is inexpensive and fast, but it is not suitable for emulsions or very fine powders (14).

To obtain more accurate size data for smaller particles (<50 μm), characteristics that correlate to size are measured, and thus size is measured indirectly (15). **Surface area** and **zeta potential** (electrical charge on a particle) are characteristics that are commonly used. Zeta potential is measured by an electroacoustic method whereby particles are oscillated in a high-frequency electrical field and generate a sound wave whose amplitude is proportional to the zeta potential. Optical and electron microscopes are routinely used to measure particle size. Optical microscopes are interfaced with video outputs and video-imaging software to estimate size and shape. The advantage of the visual approach is that a three-dimensional size and detailed particle structure can be observed. The most widely used technique for particle analysis is **dynamic light scattering**. These instruments determine particle size and even the molecular weight of large molecules in solution. This is achieved by measurement of frequency shifts of light scattered by particles due to Brownian motion. This method can be used for particles as small as a few nanometers in diameter.

table 5-2 US Standard Mesh with Equivalents in Inches and Millimeters

US Standard Mesh	Sieve opening	
	Inches	*Millimeters*
4	0.1870	4.760
6	0.1320	3.360
7	0.1110	2.830
8	0.0937	2.380
10	0.0787	2.000
12	0.0661	1.680
14	0.0555	1.410
16	0.0469	1.190
18	0.0394	1.000
20	0.0331	0.841
30	0.0232	0.595
40	0.0165	0.400
50	0.0117	0.297
60	0.0098	0.250
70	0.0083	0.210
80	0.0070	0.177
100	0.0059	0.149
120	0.0049	0.125
140	0.0041	0.105
170	0.0035	0.088
200	0.0029	0.074
230	0.0024	0.063
270	0.0021	0.053
325	0.0017	0.044
400	0.0015	0.037

Understanding of the principles of the instrument used to obtain size data is vital to appreciate the limitations of each method. For example, data obtained from the same sample using sieves and light scattering will differ (15). Sieves separate particles using square holes, and therefore they distinguish size in the smallest dimension, independent of shape. However, light scattering techniques assume that the particle is spherical, and data are derived from the average of all dimensions. Particle size measurement is useful to maintain sample quality, but care must be taken in choosing an appropriate method and interpreting the data.

5.4.3 Enzymatic Inactivation

Food materials often contain enzymes that may degrade the food components being analyzed. Enzyme activity therefore must be eliminated or controlled using methods that depend on the nature of the food. Heat denaturation to inactivate enzymes and freezer storage (−20°C to −30°C) for limiting enzyme activity are common methods. However, some enzymes are more effectively controlled by changing the pH, or by salting out (13). Oxidative enzymes may be controlled by adding reducing agents.

5.4.4 Lipid Oxidation Protection

Lipids present particular problems in sample preparation. High-fat foods are difficult to grind and may need to be ground while frozen. Unsaturated lipids are sensitive to oxidative degradation and should be protected by storing under nitrogen or vacuum. Antioxidants may stabilize lipids and may be used if they do not interfere with the analysis. Light-initiated photooxidation of unsaturated lipids can be avoided by controlling storage conditions. In practice, lipids are more stable when frozen in intact tissues rather than as extracts (13). Therefore, ideally, unsaturated lipids should be extracted just prior to analysis. Low-temperature storage is generally recommended to protect most foods.

5.4.5 Microbial Growth and Contamination

Microorganisms are present in almost all foods and can alter the sample composition. Likewise, microorganisms are present on all but sterilized surfaces, so sample cross-contamination can occur if samples are not handled carefully. The former is always a problem, and the latter is particularly important in samples for microbiological examination. Freezing, drying, and chemical preservatives are effective controls and often a combination of these is used. The preservation methods used are determined by the probability of contamination, the storage conditions, storage time, and the analysis to be performed (13).

5.5 SUMMARY

Food quality is monitored at various processing stages but 100% inspection is rarely possible, or even desirable. To ensure that a representative sample of the population is obtained for analysis, sampling and sample reduction methods must be developed and implemented. The selection of the sampling procedure is determined by the purpose of the inspection, the food product, the test method, and the characteristics of the population. Increasing the sample size will generally increase the reliability of the analytical results, and using *t*-test techniques will optimize the sample size necessary to obtain reliable data. Multiple sampling techniques also can be used to minimize the number of samples to be analyzed. Sampling is a vital process, as it is often the most variable step in the entire analytical procedure.

Sampling may be for attributes or variables. Attributes are monitored for their presence or absence, whereas variables are quantified on a continuous scale. Sampling plans are developed for either attributes or variables and may be single, double, or multiple. Multiple sampling plans reduce costs by rejecting low-quality lots or accepting high-quality lots quickly, while intermediate quality lots require further sampling. There is no sampling plan that is risk free. The consumer risk is the probability of accepting a poor-quality product, while the vendor risk is the probability of rejecting an acceptable product. An acceptable probability of risk depends on the seriousness of a negative consequence.

Sampling plans are determined by whether the population is homogeneous or heterogeneous. Although sampling from a homogeneous population is simple, it rarely is found in practical industrial situations. Sampling from heterogeneous populations is most common, and suitable sampling plans must be used to obtain a representative sample. Sampling methods may be manual or continuous. Ideally, the sampling method should be statistically sound. However, nonprobability sampling is sometimes unavoidable, even though there is not an equal probability that each member of the population will be selected due to the bias of the person sampling. Probability sampling is preferred because it ensures random sampling and is a statistically sound method that allows calculation of sampling error and the probability of any item of the population being included in the sample.

Each sample must be clearly marked for identification and preserved during storage until completion of the analysis. Official and legal samples must be sealed and a chain of custody maintained and identified. Often, only a portion of the sample is used for analysis, and sample size reduction must ensure that the portion analyzed is representative of both the sample and population. Sample preparation and storage should account for factors that may cause sample changes. Samples can be preserved by limiting enzyme activity, preventing lipid oxidation, and inhibiting microbial growth/contamination.

5.6 STUDY QUESTIONS

1. As part of your job as supervisor in a quality assurance laboratory, you need to give a new employee instruction regarding choosing a sampling plan. Which general factors would you discuss with the new employee? Distinguish between sampling for attributes vs. sampling for variables. Differentiate the three basic sampling plans and the risks associated with selecting a plan.
2. Your supervisor wants you to develop and implement a multiple sampling plan. What would you take into account to define the acceptance and rejection lines? Why?
3. Distinguish probability sampling from nonprobability sampling. Which is preferable and why?
4. (a) Identify a piece of equipment that would be useful in *collecting* a representative sample for analysis. Describe precautions to be taken to ensure that a representative

sample is taken and a suitable food product that could be sampled with this device. (b) Identify a piece of equipment that would be useful for *preparing* a sample for analysis. What precautions should be taken to ensure that the sample composition is not changed during preparation?

5. For each of the problems identified below that can be associated with collection and preparation of samples for analysis, state one solution for how the problem can be overcome:
   (a) Sample bias
   (b) Change in composition during storage of sample prior to analysis
   (c) Metal contamination in grinding
   (d) Microbial growth during storage of product prior to analysis
6. The instructions you are following for cereal protein analysis specify grinding a cereal sample to 10 mesh before you remove protein by a series of solvent extractions.
   (a) What does 10 mesh mean?
   (b) Would you question the use of a 10-mesh screen for this analysis? Provide reasons for your answer.
7. You are to collect and prepare a sample of cereal produced by your company for the analyses required to create a standard nutritional label. Your product is considered "low fat" and "high fiber" (see regulations for nutrient claims, and FDA compliance procedures in Chap. 3). What kind of sampling plan will you use? Will you do attribute or variable sampling? What are the risks associated with sampling in your specific case? Would you use probability or nonprobability sampling, and which specific type would you choose? What specific problems would you anticipate in sample collection and in preparation of the sample? How would you avoid or minimize each of these problems?

5.7 ACKNOWLEDGMENTS

The author of this chapter wishes to acknowledge Drs. Andrew Proctor and Jean-François Meullenet, who wrote this chapter for the third edition of the book and offered the use of their chapter content for this fourth edition.

5.8 REFERENCES

1. Puri SC, Ennis D, Mullen K (1979) Statistical quality control for food and agricultural scientists. G.K. Hall and Co., Boston, MA
2. Harris DC (1999) Quantitative chemical analysis, 5th edn. W.H. Freeman and Co., New York
3. Miller JC (1988) Basic statistical methods for analytical chemistry. I. Statistics of repeated measurements. A review. Analyst 113:1351–1355
4. Horwitz W (1988) Sampling and preparation of samples for chemical examination. J Assoc Off Anal Chem 71:241–245
5. IUPAC (1997) Compendium of chemical terminology, 2nd edn (the "Gold Book"). Compiled by McNaught AD, Wilkinson A. Blackwell Scientific, Oxford. XML on-line corrected version: http://goldbook.iupac.org (2006) created by Nic M, Jirat J, Kosata B; updates compiled by Jenkins A. ISBN 0–9678550–9–8. doi:10.1351/goldbook
6. Weiers RM (2007) Introduction to business statistics, 6th edn. South-Western College, Cincinnati, OH
7. NIST/SEMATECH (2009) e-Handbook of statistical methods, chapter 6: process or product monitoring and control. http://www.itl.nist.gov/div898/handbook/
8. AOAC International (2007) Official methods of analysis, 18th edn, 2005; Current through revision 2, 2007 (On-line). AOAC International, Gaithersburg, MD
9. Anonymous (2009) Code of federal regulations. Title 21. US Government Printing Office, Washington, DC
10. Baker WL, Gehrke CW, Krause GF (1967) Mechanism of sampler bias. J Assoc Off Anal Chem 50:407–413
11. Anonymous (2009) Code of federal regulations. 21 CFR 101.9 (g), 9 CFR 317.309 (h), 9 CFR 381.409 (h). US Government Printing Office, Washington, DC
12. Pomeranz Y, Meloan CE (1994) Food analysis: theory and practice, 3rd edn. Chapman & Hall, New York
13. Cubadda F, Baldini M, Carcea M, Pasqui LA, Raggi A, Stacchini P (2001) Influence of laboratory homogenization procedures on trace element content of food samples: an ICP-MS study on soft and duram wheat. Food Addit Contam 18:778–787
14. Kenkel JV (2003) Analytical chemistry for technicians, 3rd edn. CRC, Boca Raton, FL
15. Jordan JR (1999) Particle size analysis. Inside Lab Manage 3(7):25–28

Compositional Analysis of Foods

Moisture and Total Solids Analysis

Robert L. Bradley, Jr.

Department of Food Science, University of Wisconsin, Madison, WI 53706, USA

rbradley@wisc.edu

6.1 Introduction 87
- 6.1.1 Importance of Moisture Assay 87
- 6.1.2 Moisture Content of Foods 87
- 6.1.3 Forms of Water in Foods 87
- 6.1.4 Sample Collection and Handling 87

6.2 Oven Drying Methods 88
- 6.2.1 General Information 88
  - 6.2.1.1 Removal of Moisture 88
  - 6.2.1.2 Decomposition of Other Food Constituents 89
  - 6.2.1.3 Temperature Control 89
  - 6.2.1.4 Types of Pans for Oven Drying Methods 90
  - 6.2.1.5 Handling and Preparation of Pans 90
  - 6.2.1.6 Control of Surface Crust Formation (Sand Pan Technique) 90
  - 6.2.1.7 Calculations 91
- 6.2.2 Forced Draft Oven 91
- 6.2.3 Vacuum Oven 91
- 6.2.4 Microwave Analyzer 92
- 6.2.5 Infrared Drying 93

S.S. Nielsen, *Food Analysis*, Food Science Texts Series, DOI 10.1007/978-1-4419-1478-1_6,

6.2.6 Rapid Moisture Analyzer Technology 93
6.3 Distillation Procedures 93
6.3.1 Overview 93
6.3.2 Reflux Distillation with Immiscible Solvent 93
6.4 Chemical Method: Karl Fischer Titration 94
6.5 Physical Methods 96
6.5.1 Dielectric Method 96
6.5.2 Hydrometry 96
6.5.2.1 Hydrometer 97
6.5.2.2 Pycnometer 97
6.5.3 Refractometry 98
6.5.4 Infrared Analysis 99
6.5.5 Freezing Point 100
6.6 Water Activity 101
6.7 Comparison of Methods 101
6.7.1 Principles 101
6.7.2 Nature of Sample 101
6.7.3 Intended Purposes 102
6.8 Summary 102
6.9 Study Questions 102
6.10 Practice Problems 103
6.11 References 104

6.1 INTRODUCTION

Moisture assays can be one of the most important analyses performed on a food product and yet one of the most difficult from which to obtain accurate and precise data. This chapter describes various methods for moisture analysis – their principles, procedures, applications, cautions, advantages, and disadvantages. Water activity measurement also is described, since it parallels the measurement of total moisture as an important stability and quality factor. With an understanding of techniques described, one can apply appropriate moisture analyses to a wide variety of food products.

6.1.1 Importance of Moisture Assay

One of the most fundamental and important analytical procedures that can be performed on a food product is an assay for the amount of moisture (1–3). The dry matter that remains after moisture removal is commonly referred to as **total solids**. This analytical value is of great economic importance to a food manufacturer because water is an inexpensive filler. The following listing gives some examples in which moisture content is important to the food processor.

1. Moisture is a quality factor in the preservation of some products and affects stability in
   (a) Dehydrated vegetables and fruits
   (b) Dried milks
   (c) Powdered eggs
   (d) Dehydrated potatoes
   (e) Spices and herbs
2. Moisture is used as a quality factor for
   (a) Jams and jellies to prevent sugar crystallization
   (b) Sugar syrups
   (c) Prepared cereals – conventional, 4–8%; puffed, 7–8%
3. Reduced moisture is used for convenience in packaging or shipping of
   (a) Concentrated milks
   (b) Liquid cane sugar (67% solids) and liquid corn sweetener (80% solids)
   (c) Dehydrated products (these are difficult to package if too high in moisture)
   (d) Concentrated fruit juices
4. Moisture (or solids) content is often specified in compositional standards (i.e., Standards of Identity)
   (a) Cheddar cheese must be ≤39% moisture.
   (b) Enriched flour must be ≤15% moisture.
   (c) Pineapple juice must have soluble solids of ≥10.5°Brix (conditions specified).
   (d) Glucose syrup must have ≥70% total solids.
   (e) The percentage of added water in processed meats is commonly specified.
5. Computations of the nutritional value of foods require that you know the moisture content.
6. Moisture data are used to express results of other analytical determinations on a uniform basis [i.e., dry weight basis (dwb), rather than wet weight basis (wwb)].

6.1.2 Moisture Content of Foods

The moisture content of foods varies greatly as shown in Table 6-1 (4). Water is a major constituent of most food products. The approximate, expected moisture content of a food can affect the choice of the method of measurement. It can also guide the analyst in determining the practical level of accuracy required when measuring moisture content, relative to other food constituents.

6.1.3 Forms of Water in Foods

The ease of water removal from foods depends on how it exists in the food product. The three states of water in food products are:

1. **Free water:** This water retains its physical properties and thus acts as the dispersing agent for colloids and the solvent for salts.
2. **Adsorbed water:** This water is held tightly or is occluded in cell walls or protoplasm and is held tightly to proteins.
3. **Water of hydration:** This water is bound chemically, for example, lactose monohydrate; also some salts such as $Na_2SO_4 \cdot 10H_2O$.

Depending on the form of the water present in a food, the method used for determining moisture may measure more or less of the moisture present. This is the reason for official methods with stated procedures (5–7). However, several official methods may exist for a particular product. For example, the AOAC International methods for cheese include: Method 926.08, vacuum oven; 948.12, forced draft oven; 977.11, microwave oven; 969.19, distillation (5). Usually, the first method listed by AOAC International is preferred over others in any section.

6.1.4 Sample Collection and Handling

General procedures for sampling, sample handling and storage, and sample preparation are given in Chap. 5. These procedures are perhaps the greatest potential source of error in any analysis. Precautions must be taken to minimize inadvertent **moisture losses or gains** that occur during these steps.

Moisture Content of Selected Foods

Food Item	Approximate Percent Moisture (Wet Weight Basis)
Cereals, bread, and pasta	
Wheat flour, whole-grain	10.3
White bread, enriched (wheat flour)	13.4
Corn flakes cereal	3.5
Crackers saltines	4.0
Macaroni, dry, enriched	9.9
Dairy products	
Milk, reduced fat, fluid, 2%	89.3
Yogurt, plain, low fat	85.1
Cottage cheese, low fat or 2% milk fat	80.7
Cheddar cheese	36.8
Ice cream, vanilla	61.0
Fats and oils	
Margarine, regular, hard, corn, hydrogenated	15.7
Butter, with salt	15.9
Oil-soybean, salad, or cooking	0
Fruits and vegetables	
Watermelon, raw	91.5
Oranges, raw, California navels	86.3
Apples, raw, with skin	85.6
Grapes, American type, raw	81.3
Raisins	15.3
Cucumbers, with peel, raw	95.2
Potatoes, microwaved, cooked in skin, flesh and skin	72.4
Snap beans, green, raw	90.3
Meat, poultry, and fish	
Beef, ground, raw, 95% lean	73.3
Chicken, broilers and fryers, light meat, meat and skin, raw	68.6
Finfish, flatfish (flounder and sole species), raw	79.1
Egg, whole, raw, fresh	75.8
Nuts	
Walnuts, black, dried	4.6
Peanuts, all types, dry roasted with salt	1.6
Peanut butter, smooth style, with salt	1.8
Sweeteners	
Sugar, granulated	0
Sugar, brown	1.3
Honey, strained or extracted	17.1

From US Department of Agriculture, Agricultural Research Service (2009) USDA National Nutrient Database for Standard Reference. Release 22. Nutrient Data Laboratory Home Page, http://www.ars.usda.gov/ba/bhnrc/ndl

Obviously, any exposure of a sample to the open atmosphere should be as short as possible. Any heating of a sample by friction during grinding should be minimized. Headspace in the sample storage container should be minimal because moisture is lost from the sample to equilibrate the container environment against the sample. It is critical to control temperature fluctuations since moisture will migrate in a sample to the colder part. To control this potential error, remove the entire sample from the container, reblend quickly, and then remove a test portion (8,9).

To illustrate the need for optimum efficiency and speed in weighing samples for analysis, Bradley and Vanderwarn (10) showed, using shredded Cheddar cheese (2–3 g in a 5.5-cm aluminum foil pan), that moisture loss within an analytical balance was a straight line function. The rate of loss was related to the relative humidity. At 50% relative humidity, it required only 5 s to lose 0.01% moisture. This time doubled at 70% humidity or 0.01% moisture loss in 10 s. While one might expect a curvilinear loss, the moisture loss was actually linear over a 5-min study interval. These data demonstrate the necessity of absolute control during collection of samples through weighing, before drying.

6.2 OVEN DRYING METHODS

In **oven drying methods**, the sample is heated under specified conditions, and the loss of weight is used to calculate the moisture content of the sample. The amount of moisture determined is highly dependent on the type of oven used, conditions within the oven, and the time and temperature of drying. Various oven methods are approved by AOAC International for determining the amount of moisture in many food products. The methods are simple, and many ovens allow for simultaneous analysis of large numbers of samples. The time required may be from a few minutes to over 24 h.

6.2.1 General Information

6.2.1.1 Removal of Moisture

Any oven method used to evaporate moisture has as its foundation the fact that the boiling point of water is 100°C; however, this considers only pure water at sea level. Free water is the easiest of the three forms of water to remove. However, if 1 molecular weight (1 mol) of a solute is dissolved in 1.0 L of water, the boiling point would be raised by 0.512°C. This boiling point elevation continues throughout the moisture removal process as more and more concentration occurs.

Moisture removal is sometimes best achieved in a two-stage process. Liquid products (e.g., juices, milk) are commonly predried over a **steam bath** before drying in an oven. Products such as bread and field-dried grain are often air dried, then ground and oven dried, with the moisture content calculated from moisture loss at both air and oven drying steps. Particle size, particle size distribution, sample sizes, and surface area during drying influence the rate and efficiency of moisture removal.

6.2.1.2 Decomposition of Other Food Constituents

Moisture loss from a sample during analysis is a function of time and temperature. Decomposition enters the picture when time is extended too much or temperature is too high. Thus, most methods for food moisture analysis involve a compromise between time and a particular temperature at which limited decomposition might be a factor. One major problem exists in that the physical process must separate all the moisture without decomposing any of the constituents that could release water. For example, carbohydrates decompose at 100°C according to the following reaction:

$$C_6H_{12}O_6 \rightarrow 6C + 6H_2O \quad [1]$$

The moisture generated in carbohydrate decomposition is not the moisture that we want to measure. Certain other chemical reactions (e.g., sucrose hydrolysis) can result in utilization of moisture, which would reduce the moisture for measurement. A less serious problem, but one that would be a consistent error, is the loss of **volatile constituents**, such as acetic, propionic, and butyric acids; and alcohols, esters, and aldehydes among flavor compounds. While weight changes in oven drying methods are assumed to be due to moisture loss, weight gains also can occur due to oxidation of unsaturated fatty acids and certain other compounds.

Nelson and Hulett (11) determined that moisture was retained in biological products to at least 365°C, which is coincidentally near the critical temperature for water. Their data indicate that among the decomposition products at elevated temperatures were CO, CO_2, CH_4, and H_2O. These were not given off at any one particular temperature but at all temperatures and at different rates at the respective temperature in question.

By plotting moisture liberated against temperature, curves were obtained that show the amount of moisture liberated at each temperature (Fig. 6-1). Distinct breaks were shown that indicated the temperature at which decomposition became measurable. None of these curves showed any break before 184°C. Generally, proteins decompose at temperatures somewhat lower than required for starches and celluloses. Extrapolation of the flat portion of each curve to 250°C gave a true moisture content based on the assumption that there was no adsorbed water present at the temperature in question.

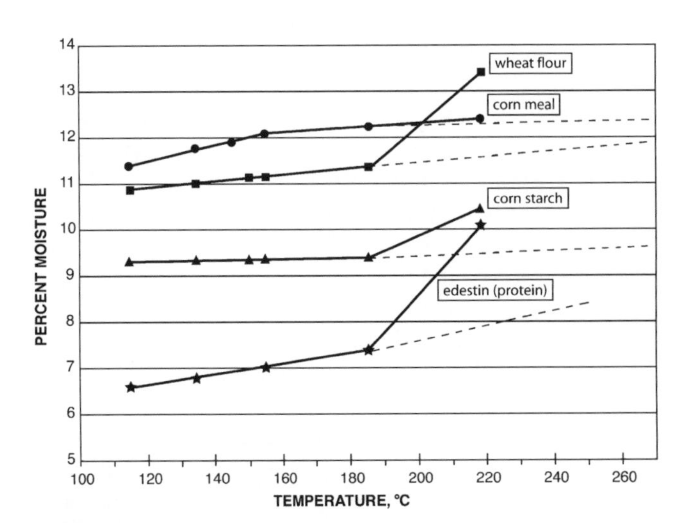

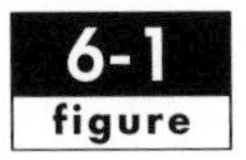
6-1 figure

Moisture content of several foods held at various temperatures in an oven. The hyphenated line extrapolates data to 275°C, the true moisture content. [Reprinted with permission from (11) Nelson OA and Hulett GA. 1920. The moisture content of cereals. *J. Industrial Eng. Chem.* 12:40–45. Copyright 1920, American Chemical Society.]

6.2.1.3 Temperature Control

Drying methods utilize specified drying temperatures and times, which must be carefully controlled. Moreover, there may be considerable variability of temperature, depending on the type of oven used for moisture analysis. One should determine the extent of variation within an oven before relying on data collected from its use.

Consider the temperature variation in three types of ovens: **convection (atmospheric)**, **forced draft**, and **vacuum**. The greatest temperature variation exists in a convection oven. This is because hot air slowly circulates without the aid of a fan. Air movement is obstructed further by pans placed in the oven. When the oven door is closed, the rate of temperature recovery is generally slow. This is dependent also upon the load placed in the oven and upon the ambient temperature. A 10°C temperature differential across a convection oven is not unusual. This must be considered in view of anticipated analytical accuracy and precision. A convection oven should not be used when precise and accurate measurements are needed.

Forced draft ovens have the least temperature differential across the interior of all ovens, usually not greater than 1°C. Air is circulated by a fan that forces air movement throughout the oven cavity. Forced draft ovens with air distribution manifolds appear to have added benefit where air movement is horizontal across shelving. Thus, no matter whether the oven shelves are filled completely with moisture pans or only half filled, the result would be the same for a particular sample. This has been demonstrated using a Lab-Line oven (Melrose Park, IL) in which three stacking configurations for the pans were used (10). In one configuration, the oven shelves were filled with as many pans holding 2–3 g of Cheddar cheese as the forced draft oven could hold. In the two others, one-half of the full load of pans with cheese was used with the pans (1) in orderly vertical rows with the width of one pan between rows, or (2) staggered such that pans on every other shelf were in vertical alignment. The results after drying showed no difference in the mean value or the standard deviation.

Two features of some **vacuum ovens** contribute to a wider temperature spread across the oven. One feature is a glass panel in the door. Although from an educational point of view, it may be fascinating to observe some samples in the drying mode; the glass is a heat sink. The second feature is the way by which air is bled into the oven. If the air inlet and discharge are on opposite sides, conduct of air is virtually straight across the oven. Some newer models have air inlet and discharge manifolds mounted top and bottom. Air movement in this style of vacuum oven is upward from the front and then backward to the discharge in a broad sweep. The effect is to minimize cold spots as well as to exhaust moisture in the interior air.

6.2.1.4 Types of Pans for Oven Drying Methods

Pans used for moisture determinations are varied in shape and may or may not have a cover. The AOAC International (5) moisture pan is about 5.5 cm in diameter with an insert cover. Other pans have covers that slip over the outside edge of the pan. These pans, while reusable, are expensive, in terms of labor costs to clean appropriately to allow reuse.

Pan covers are necessary to control loss of sample by spattering during the heating process. If the cover is metal, it must be slipped to one side during drying to allow for moisture evaporation. However, this slipping of the cover also creates an area where spattering will result in product loss. Examine the interior of most moisture ovens and you will detect odor and deposits of burned-on residue, which, although undetected at the time of occurrence, produce erroneous results and large standard deviations (10).

Consider the use of **disposable pans** whenever possible; then purchase **glass fiber discs** for covers. At 5.5 cm in diameter, these covers fit perfectly inside disposable aluminum foil pans and prevent spattering while allowing the surface to breathe. Paper filter discs foul with fat and thus do not breathe effectively. Drying studies done on cheese using various pans and covers have shown that fat does spatter from pans with slipped covers, and fiberglass is the most satisfactory cover.

6.2.1.5 Handling and Preparation of Pans

The preparation and handling of pans before use requires consideration. Use only **tongs** to handle any pan. Even fingerprints have weight. All pans must be oven treated to prepare them for use. This is a factor of major importance unless disproved by the technologist doing moisture determinations with a particular type of pan. Disposable aluminum pans must be vacuum oven dried for 3 h before use. At 3 and 15 h in either a vacuum or forced draft oven at 100°C, pans varied in their weight within the error of the balance or 0.0001 g (10). Store dried moisture pans in a functioning **desiccator**. The glass fiber covers should be dried for 1 h before use.

6.2.1.6 Control of Surface Crust Formation (Sand Pan Technique)

Some food materials tend to form a semipermeable crust or lump together during drying, which will contribute to erratic and erroneous results. To control this problem, analysts use the **sand pan technique**. Clean, dry sand and a short glass stirring rod are preweighed into a moisture pan. Subsequently, after weighing in a sample, the sand and sample are admixed with the stirring rod left in the pan. The remainder of the procedure follows a standardized method if available; otherwise the sample is dried to constant weight. The purpose of the sand is twofold: to prevent **surface crust** from forming and to disperse the sample so evaporation of moisture is less impeded. The amount of sand used is a function of sample size. Consider 20–30 g sand/3 g sample to obtain desired distribution in the pan. Similar to the procedure, applications, and advantages of using sand, other heat-stable inert materials such as diatomaceous earth can be used in moisture determinations, especially for sticky fruits.

The inert matrices such as sand and **diatomaceous earth** function to disperse the food constituents and minimize the retention of moisture in the food products. However, the analyst must ascertain that the inert matrix used does not give erroneous results for the assay because of decomposition or entrapped moisture loss. Test the sand or other inert matrix for weight loss before using in any method. Add approximately 25 g of sand into a moisture pan and heat at 100°C for 2 h and weigh to 0.1 mg. Add 5 ml of water and mix with the matrix using a glass rod. Heat dish, matrix, cover, and glass rod for at least 4 h at 100°C, reweigh. The difference between weighing must be less than 0.5 mg for any suitable matrix (12).

6.2.1.7 Calculations

Moisture and total solids contents of foods can be calculated as follows using oven drying procedures:

$$\%\text{Moisture (wt/wt)} = \frac{\text{wt } H_2O \text{ in sample}}{\text{wt of wet sample}} \times 100 \quad [2]$$

$$\%\text{Moisture (wt/wt)} = \frac{\text{wt of wet sample} - \text{wt of dry sample}}{\text{wt of wet sample}} \times 100 \quad [3]$$

$$\%\text{Total solids (wt/wt)} = \frac{\text{wt of dry sample}}{\text{wt of wet sample}} \times 100 \quad [4]$$

6.2.2 Forced Draft Oven

When using a forced draft oven, the sample is rapidly weighed into a predried moisture pan covered and placed in the oven for an arbitrarily selected time if no standardized method exists. Drying time periods for this method are 0.75–24 h (Table 6-2), depending on the food sample and its pretreatment; some liquid samples are dried initially on a steam bath at 100°C to minimize spattering. In these cases, drying times are shortened to 0.75–3 h. A forced draft oven is used with or without a steam table predrying treatment to determine the solids content of fluid milks (AOAC Method 990.19, 990.20).

An alternative to selecting a time period for drying is to weigh and reweigh the dried sample and pan until two successive weighings taken 30 min apart agree within a specified limit, for example, 0.1–0.2 mg for a 5-g sample. The user of this second method must be aware of sample transformation, such as browning which suggests moisture loss of the wrong form. Lipid oxidation and a resulting sample weight gain can occur at high temperatures in a forced draft oven. Samples high in carbohydrates should not be dried in a forced draft oven but rather in a vacuum oven at a temperature no higher than 70°C.

6.2.3 Vacuum Oven

By drying under reduced pressure (25–100 mm Hg), one is able to obtain a more complete removal of water and volatiles without decomposition within a 3–6-h drying time. Vacuum ovens need a dry air purge in addition to temperature and vacuum controls to operate within method definition. In older methods, a vacuum flask is used, partially filled with concentrated sulfuric acid as the desiccant. One or two air bubbles per second are passed through the acid. Recent changes now stipulate an air trap that is filled with calcium sulfate containing an indicator to show moisture saturation. Between the trap and the vacuum oven is an appropriately sized rotameter to measure air flow (100–120 ml/min) into the oven.

The following are important points in the use of a vacuum drying oven:

6-2 table **Forced Draft Oven Temperature and Times for Selected Foods**

Product	*Dry on Steam Bath*	*Oven Temperature (°C ± 2)*	*Time in Oven (h)*
Buttermilk, liquid	X[a]	100	3
Cheese, natural type only		100	16.5 ± 0.5
Chocolate and cocoa		100	3
Cottage cheese		100	3
Cream, liquid and frozen	X	100	3
Egg albumin, liquid	X	130	0.75
Egg albumin, dried	X	100	0.75
Ice cream and frozen desserts	X	100	3.5
Milk	X	100	3
Whole, low fat, and skim		100	3
Condensed skim		100	3
Nuts: almonds, peanuts, walnuts		130	3

From (6) p. 492, with permission. Copyright 2004 by the American Public Health Association, Washington, DC.

[a]*X* = samples must be partially dried on steam bath before being placed in oven.

1. **Temperature** used depends on the product, such as 70°C for fruits and other high-sugar products. Even with reduced temperature, there can be some decomposition.
2. If the product to be assayed has a high concentration of **volatiles**, you should consider the use of a correction factor to compensate for the loss.
3. Analysts should remember that in a **vacuum**, heat is not conducted well. Thus pans must be placed directly on the metal shelves to conduct heat.
4. **Evaporation** is an endothermic process; thus, a pronounced cooling is observed. Because of the cooling effect of evaporation, when several samples are placed in an oven of this type, you will note that the temperature will drop. Do not attempt to compensate for the cooling effect by increasing the temperature, otherwise samples during the last stages of drying will be overheated.
5. The **drying time** is a function of the total moisture present, nature of the food, surface area per unit weight of sample, whether sand is used as a dispersant, and the relative concentration of sugars and other substances capable of retaining moisture or decomposing. The drying interval is determined experimentally to give reproducible results.

6.2.4 Microwave Analyzer

Determination of moisture in food products has traditionally been done using a standard oven, which, though accurate, can take many hours to dry a sample. Other methods have been developed over the years including infrared and various types of instruments that utilize halogen lamps or ceramic heating elements. They were often used for "spot checking" because of their speed, but they lacked the accuracy of the standard oven method. The introduction of microwave moisture/solids analyzers in the late 1970s gave laboratories the accuracy they needed and the speed they wanted. **Microwave moisture analysis**, often called **microwave drying**, was the first precise and rapid technique that allowed some segments of the food industry to make in-process adjustment of the moisture content in food products before final packaging. For example, processed cheese could be analyzed and the composition adjusted before the blend was dumped from the cooker. The ability to adjust the composition of a product in-process helps food manufacturers reduce production costs, meet regulatory requirements, and ensure product consistency. Such control could effectively pay for the microwave analyzer within a few months.

A particular microwave moisture/solids analyzer (CEM Corporation, Matthews, NC), or equivalent, is specified in the AOAC International procedures for total solids analysis of processed tomato products (AOAC Method 985.26) and moisture analysis of meat and poultry products (AOAC Method 985.14).

The general procedure for use of a microwave moisture/solids analyzer has been to set the microprocessor controller to a percentage of full power to control the microwave output. Power settings are dependent upon the type of sample and the recommendations of the manufacturer of the microwave moisture analyzer. Next, the internal balance is tared with two sample pads on the balance. As rapidly as possible, a sample is placed between the two pads, then pads are centered on the pedestal, and weighed against the tare weight. Time for the drying operation is set by the operator and "start" is activated. The microprocessor controls the drying procedure, with percentage moisture indicated in the controller window. Some newer models of microwave moisture analyzers have a temperature control feature to precisely control the drying process, removing the need to guess appropriate time and power settings for specific applications. These new models also have a smaller cavity that allows the microwave energy to be focused directly on the sample.

There are some considerations when using a microwave analyzer for moisture determination: (1) the sample must be of a uniform, appropriate size to provide for complete drying under the conditions specified; (2) the sample must be centrally located and evenly distributed, so some portions are not burned and other areas are underprocessed; and (3) the amount of time used to place an appropriate sample weight between the pads must be minimized to prevent moisture loss or gain before weight determination. Sample pads also should be considered. There are several different types, including fiberglass and quartz fiber pads. For optimum results, the pads should not absorb microwave energy, as this can cause the sample to burn, nor should they fray easily, as this causes them to lose weight and can affect the analysis. In addition, they should absorb liquids well.

Another style of microwave oven that includes a vacuum system is used in some food plants. This vacuum microwave oven will accommodate one sample in triplicate or three different samples at one time. In 10 min, the results are reported to be similar to 5 h in a vacuum oven at 100°C. The vacuum microwave oven is not nearly as widely used as conventional microwave analyzers, but can be beneficial in some applications.

Microwave drying provides a fast, accurate method to analyze many foods for moisture content. The method is sufficiently accurate for routine assay.

The distinct advantage of rapid analysis far outweighs its limitation of testing only single samples (13).

6.2.5 Infrared Drying

Infrared drying involves penetration of heat into the sample being dried, as compared with heat conductivity and convection with conventional ovens. Such heat penetration to evaporate moisture from the sample can significantly shorten the required drying time to 10–25 min. The infrared lamp used to supply heat to the sample results in a filament temperature of 2000–2500 K (degrees Kelvin). Factors that must be controlled include distance of the infrared source from the dried material and thickness of the sample. The analyst must be careful that the sample does not burn or case harden while drying. Infrared drying ovens may be equipped with forced ventilation to remove moisture air and an analytical balance to read moisture content directly. No infrared drying moisture analysis techniques are approved by AOAC International currently. However, because of the speed of analysis, this technique is suited for qualitative in-process use.

6.2.6 Rapid Moisture Analyzer Technology

Many rapid moisture/solids analyzers are available to the food industry. In addition to those based on infrared and microwave drying as described previously, compact instruments that depend on high heat are available, such as analyzers that detect moisture levels from 50 ppm to 100% using sample weights of 150 mg to 40 g (e.g., Computrac®, Arizona Instrument LLC, Chandler, AZ). Using a digital balance, the test sample is placed on an aluminum pan or filter paper and the heat control program (with a heating range of 25–275°C) elevates the test sample to a constant temperature. As the moisture is driven from the sample, the instrument automatically weighs and calculates the percentage moisture or solids. This technology is utilized to cover a wide range of applications within the food industry and offers quick and accurate results within minutes. These analyzers are utilized for both production and laboratory use with results comparable to reference methods.

6.3 DISTILLATION PROCEDURES

6.3.1 Overview

Distillation techniques involve codistilling the moisture in a food sample with a high boiling point solvent that is immiscible in water, collecting the mixture that distills off, and then measuring the volume of water. Two distillation procedures are in use today: **direct** and **reflux distillations**, with a variety of solvents. For example, in direct distillation with immiscible solvents of higher boiling point than water, the sample is heated in mineral oil or liquid with a flash point well above the boiling point for water. Other immiscible liquids with boiling point only slightly above water can be used (e.g., toluene, xylene, and benzene). However, reflux distillation with the immiscible solvent toluene is the most widely used method.

Distillation techniques were originally developed as rapid methods for quality control work, but they are not adaptable to routine testing. The distillation method is an AOAC-approved technique for moisture analysis of spices (AOAC Method 986.21), cheese (AOAC Method 969.19), and animal feeds (AOAC Method 925.04). It also can give good accuracy and precision for nuts, oils, soaps, and waxes.

Distillation methods cause less thermal decomposition of some foods than oven drying at high temperatures. Adverse chemical reactions are not eliminated but can be minimized by using a solvent with a lower boiling point. This, however, will increase distillation times. Water is measured directly in the distillation procedure (rather than by weight loss), but reading the volume of water in a receiving tube may be less accurate than using a weight measurement.

6.3.2 Reflux Distillation with Immiscible Solvent

Reflux distillation uses either a solvent less dense than water (e.g., toluene, with a boiling point of 110.6°C; or xylene, with a boiling range of 137–140°C) or a solvent more dense than water (e.g., tetrachlorethylene, with a boiling point of 121°C). The advantage of using this last solvent is that material to be dried floats; therefore it will not char or burn. In addition, there is no fire hazard with this solvent.

A **Bidwell–Sterling moisture trap** (Fig. 6-2) is commonly used as part of the apparatus for reflux distillation with a solvent less dense than water. The distillation procedure using such a trap is described in Fig. 6-3, with emphasis placed on dislodging adhering water drops, thereby minimizing error. When the toluene in the distillation just starts to boil, the analyst will observe a hazy cloud rising in the distillation flask. This is a vaporous emulsion of water in toluene. Condensation occurs as the vapors rise, heating the vessel, the Bidwell–Sterling trap, and the bottom of the condenser. It is also hazy at the cold surface of the condenser, where water droplets are visible. The emulsion inverts and becomes toluene dispersed in water. This turbidity clears very slowly on cooling.

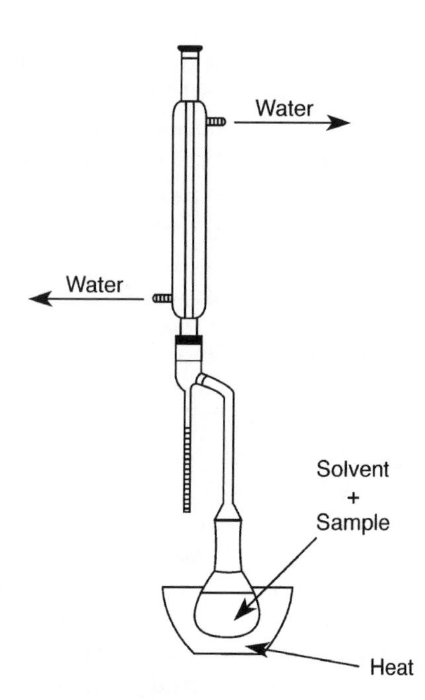

Apparatus for reflux distillation of moisture from a food. Key to this setup is the Bidwell–Sterling moisture trap. This style can be used only where the solvent is less dense than water.

Three potential sources of error with distillation should be eliminated if observed:

1. Formation of emulsions that will not break. Usually this can be controlled by allowing the apparatus to cool after distillation is completed and before reading the amount of moisture in the trap.
2. Clinging of water droplets to dirty apparatus. Clean glassware is essential, but water seems to cling even with the best cleaning effort. A burette brush, with the handle end flattened so it will pass down the condenser, is needed to dislodge moisture droplets.
3. Decomposition of the sample with production of water. This is principally due to carbohydrate decomposition to generate water ($C_6H_{12}O_6 \rightarrow 6H_2O + 6C$). If this is a measurable problem, discontinue method use and find an alternative procedure.

6.4 CHEMICAL METHOD: KARL FISCHER TITRATION

The **Karl Fischer titration** is particularly adaptable to food products that show erratic results when heated or submitted to a vacuum. This is the method

REFLUX DISTILLATION

Place sample in distillation flask and cover completely with solvent.

⇓

Fill the receiving tube (e.g., Bidwell-Sterling Trap) with solvent, by pouring it through the top of the condenser.

⇓

Bring to a boil and distill slowly at first then at increased rate.

⇓

After the distillation has proceeded for approximately 1 hr, use an adapted buret brush to dislodge moisture droplets from the condenser and top part of the Bidwell-Sterling trap.

⇓

Slide the brush up the condenser to a point above the vapor condensing area.

⇓

Rinse the brush and wire with a small amount of toluene to dislodge adhering water drops.

⇓

If water has adhered to the walls of the calibrated tube, invert the brush and use the straight wire to dislodge this water so it collects in the bottom of the tube.

⇓

Return the wire to a point above the condensation point, and rinse with another small amount of toluene.

⇓

After no more water has distilled from the sample, repeat the brush and wire routine to dislodge adhering water droplets.

⇓

Rinse the brush and wire with toluene before removing from the condenser.

⇓

Allow the apparatus to cool to ambient temperatures before measuring the volume of water in the trap.

⇓

Volume of water x 2 (for a 50 g sample) = % moisture

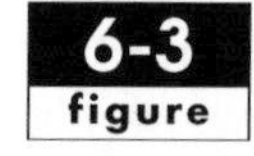

Procedures for reflux distillation with toluene using a Bidwell–Sterling trap. Steps to dislodge adhering moisture drops are given.

of choice for determination of water in many low-moisture foods such as dried fruits and vegetables (AOAC Method 967.19 E-G), candies, chocolate (AOAC Method 977.10), roasted coffee, oils and fats (AOAC Method 984.20), or any low-moisture food high in sugar or protein. The method is quite rapid, is accurate, and uses no heat. This method is based on the fundamental reaction described by Bunsen in 1853 (14) involving the reduction of iodine by SO_2 in the presence of water:

$$2H_2O + SO_2 + I_2 \rightarrow H_2SO_4 + 2HI \quad [5]$$

This was modified to include methanol and pyridine in a four-component system to dissolve the iodine and SO_2:

$$C_5H_5N \cdot I_2 + C_5H_5N \cdot SO_2 + C_5H_5N + H_2O \rightarrow 2C_5H_5N \cdot HI + C_5H_5N \cdot SO_3 \quad [6]$$

$$C_5H_5N \cdot SO_3 + CH_3OH \rightarrow C_5H_5N(H)SO_4 \cdot CH_3 \quad [7]$$

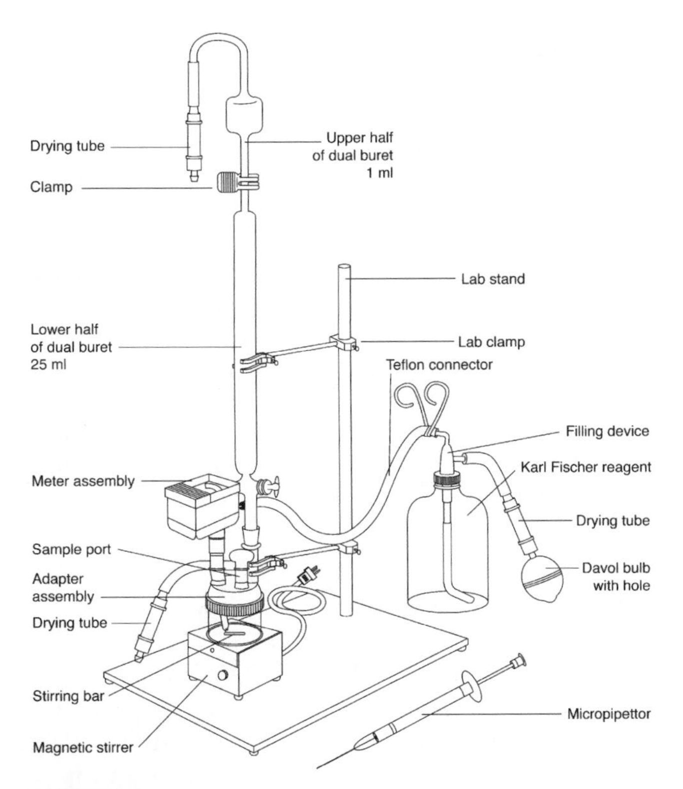

6-4 figure Manual Karl Fischer titration unit. (Courtesy of Lab Industries, Inc., Berkeley, CA.)

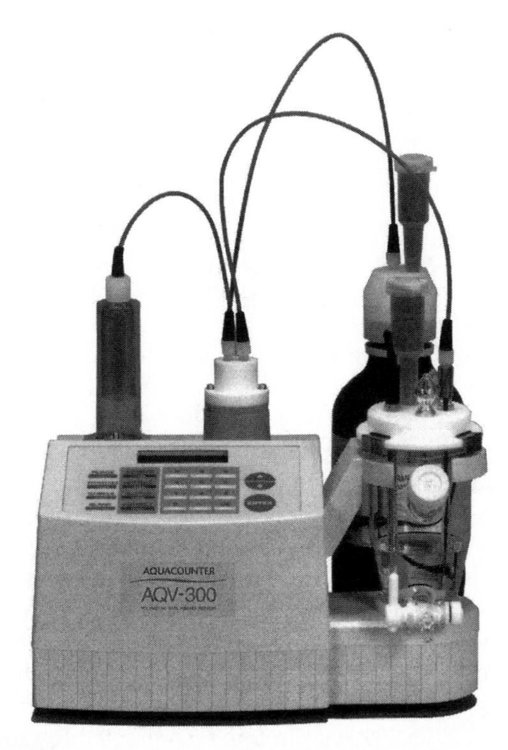

6-5 figure Automated Karl Fischer volumetric titration unit. (Courtesy of Mettler-Toledo, Columbus, OH.)

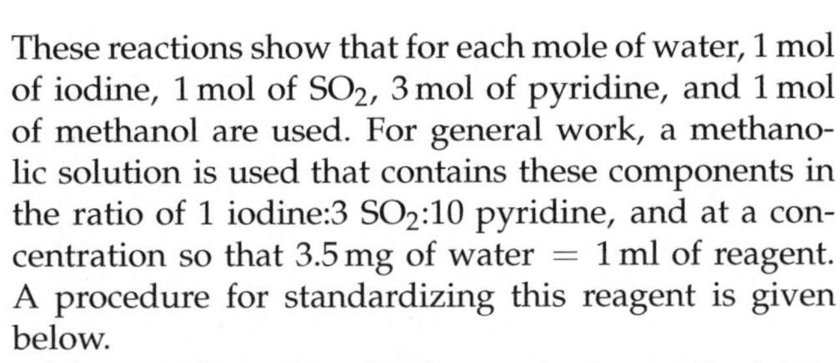

These reactions show that for each mole of water, 1 mol of iodine, 1 mol of SO_2, 3 mol of pyridine, and 1 mol of methanol are used. For general work, a methanolic solution is used that contains these components in the ratio of 1 iodine:3 SO_2:10 pyridine, and at a concentration so that 3.5 mg of water = 1 ml of reagent. A procedure for standardizing this reagent is given below.

In a **volumetric titration** procedure (Fig. 6-4 is manual titration unit; Fig. 6-5 is example of automated titration unit), iodine and SO_2 in the appropriate form are added to the sample in a closed chamber protected from atmospheric moisture. The excess of I_2 that cannot react with the water can be determined **visually**. The endpoint color is dark red-brown. Some instrumental systems are improved by the inclusion of a potentiometer (i.e., **conductometric method**) to electronically determine the endpoint, which increases the sensitivity and accuracy. The volumetric titration can be done manually (Fig. 6-4) or with an automated unit (Fig. 6-5 is one example instrument). The automated volumetric titration units (used for 100 ppm water to very high concentrations) use a pump for mechanical addition of titrant and use the conductometric method for endpoint determination (i.e., detection of excess iodine is by applying a current and measuring the potential).

The volumetric titration procedure described above is appropriate for samples with a moisture content greater than ~0.03%. A second type of titration, referred to as **coulometric titration**, is ideal for products with very low levels of moisture, from 0.03% down to parts per million (ppm) levels. In this method, iodine is electrolytically generated ($2I \rightarrow I_2 + 2e^-$) to titrate the moisture. The amount of iodine required to titrate the moisture is determined by the current needed to generate the iodine. Just like for volumetric titration, automated coulometric titration units are available commercially.

In a Karl Fischer volumetric titration, the **Karl Fischer reagent** (KFR) is added directly as the titrant if the moisture in the sample is accessible. However, if moisture in a solid sample is inaccessible to the reagent, the moisture is extracted from the food with an appropriate solvent (e.g., methanol). (Particle size affects efficiency of extraction directly.) Then the methanol extract is titrated with KFR.

The obnoxious odor of pyridine makes it an undesirable reagent. Therefore, researchers have experimented with other amines capable of dissolving iodine and sulfur dioxide. Some aliphatic amines and several other heterocyclic compounds were found suitable. On the basis of these new amines, **one-component reagents** (solvent and titrant components together) and **two-component reagents** (solvent and titrant components separate) have been prepared. The one-component reagent may be more convenient to use, but the two-component reagent has greater storage stability.

Before the amount of water found in a food sample can be determined, a **KFR water (moisture) equivalence** (KFReq) must be determined. The KFReq value represents the equivalent amount of moisture that reacts with 1 ml of KFR. Standardization must be checked before each use because the KFReq will change with time.

The KFReq can be established with **pure water**, a **water-in-methanol standard**, or **sodium tartrate dihydrate**. Pure water is a difficult standard to use because of inaccuracy in measuring the small amounts required. The water-in-methanol standard is premixed by the manufacturer and generally contains 1 mg of water/ml of solution. This standard can change over prolonged storage periods by absorbing atmospheric moisture. Sodium tartrate dihydrate ($Na_2C_4H_4O_6{\cdot}2H_2O$) is a primary standard for determining KFReq. This compound is very stable, contains 15.66% water under all conditions expected in the laboratory, and is the material of choice to use.

The KFReq is calculated as follows using sodium tartrate dihydrate:

$$\text{KFReq}\,(\text{mg H}_2\text{O/ml}) = \frac{36\,\text{g H}_2\text{O/mol Na}_2\text{C}_4\text{H}_4\text{O}_6 \cdot 2\text{H}_2\text{O} \times S \times 1000}{230.08\,\text{g/mol} \times A} \quad [8]$$

where:

KFReq = Karl Fischer reagent moisture equivalence
S = weight of sodium tartrate dihydrate (g)
A = ml of KFR required for titration of sodium tartrate dehydrate

Once the KFReq is known, the moisture content of the sample is determined as follows:

$$\%\text{H}_2\text{O} = \frac{\text{KFReq} \times \text{Ks}}{S} \times 100 \quad [9]$$

where:

KFReq = Karl Fischer reagent water (moisture) equivalence
Ks = ml of KFR used to titrate sample
S = weight of sample (mg)

The major difficulties and sources of error in the Karl Fischer titration methods are as follows:

1. **Incomplete moisture extraction**. For this reason, fineness of grind (i.e., particle size) is important in preparation of cereal grains and some foods.
2. **Atmospheric moisture**. External air must not be allowed to infiltrate the reaction chamber.
3. **Moisture adhering** to walls of unit. All glassware and utensils must be carefully dried.
4. **Interferences** from certain food constituents. **Ascorbic acid** is oxidized by KFR to dehydroascorbic acid to overestimate moisture content; **carbonyl compounds** react with methanol to form acetals and release water to overestimate moisture content (this reaction also may result in fading endpoints); **unsaturated fatty acids** will react with iodine, so moisture content will be overestimated.

6.5 PHYSICAL METHODS

6.5.1 Dielectric Method

The electrical properties of water are used in the **dielectric method** to determine the moisture content of certain foods, by measuring the change in **capacitance** or **resistance to an electric current** passed through a sample. These instruments require calibration against samples of known moisture content as determined by standard methods. Sample density or weight/volume relationships and sample temperature are important factors to control in making reliable and repeatable measurements by dielectric methods. These techniques can be very useful for process control measurement applications, where continuous measurement is required. These methods are limited to food systems that contain no more than 30–35% moisture.

The moisture determination in dielectric-type meters is based on the fact that the dielectric constant of water (80.37 at 20°C) is higher than that of most solvents. The **dielectric constant** is measured as an index of capacitance. As an example, the dielectric method is used widely for cereal grains. Its use is based on the fact that water has a dielectric constant of 80.37, whereas starches and proteins found in cereals have dielectric constants of 10. By determining this properly on samples in standard metal condensers, dial readings may be obtained and the percentage of moisture determined from a previously constructed standard curve for a particular cereal grain.

6.5.2 Hydrometry

Hydrometry is the science of measuring **specific gravity** or **density**, which can be done using several different principles and instruments. While hydrometry is considered archaic in some analytical circles, it is still widely used and, with proper technique, is highly accurate. Specific gravity measurements with various types of **hydrometers** or with a **pycnometer** are commonly used for routine testing of moisture (or solids) content of numerous food products. These include beverages, salt brines, and sugar solutions. Specific gravity measurements are best applied to the analysis of solutions consisting of only one solute in a medium of water.

6.5.2.1 Hydrometer

A second approach to measuring specific gravity is based on **Archimedes' principle**, which states that a solid suspended in a liquid will be buoyed by a force equal to the weight of the liquid displaced. The weight per unit volume of a liquid is determined by measuring the volume displaced by an object of standard weight. A hydrometer is a standard weight on the end of a spindle, and it displaces a weight of liquid equal to its own weight (Fig. 6-6). For example, in a liquid of low density, the hydrometer will sink to a greater depth, whereas in a liquid of high density, the hydrometer will not sink as far. Hydrometers are available in narrow and wide ranges of specific gravity. The spindle of the hydrometer is calibrated to read specific gravity directly at 15.5 or 20°C. A **hydrometer** is not as accurate as a pycnometer, but the speed with which you can do an analysis is a decisive factor. The accuracy of specific gravity measurements can be improved by using a hydrometer calibrated in the desired range of specific gravities.

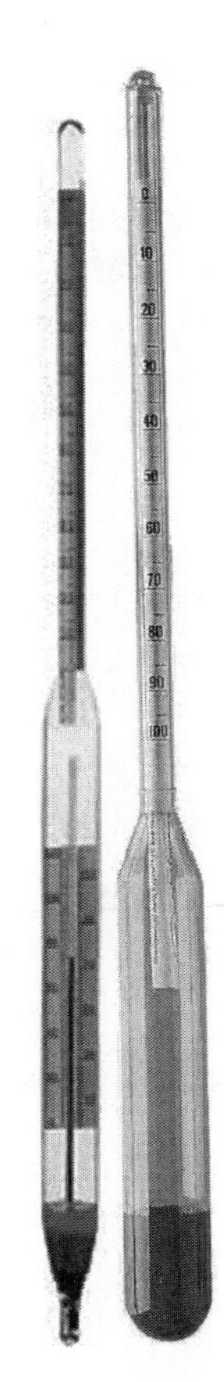

6-6 figure Hydrometers. (Courtesy of Cole-Parmer Instrument Company, Vernon Hills, IL.)

The rudimentary but surprisingly accurate hydrometer comes equipped with various modifications depending on the fluid to be measured:

1. The Quevenne and New York Board of Health **lactometer** is used to determine the density of milk. The Quevenne lactometer reads from 15 to 40 lactometer units and corresponds to 1.015 to 1.040 specific gravity. For every degree above 60°F, 0.1 lactometer unit is added to the reading, and 0.1 lactometer unit is subtracted for every degree below 60°F.
2. The **Baumé hydrometer** was used originally to determine the density of salt solutions (originally 10% salt), but it has come into much wider use. From the value obtained in the Baumé scale, you can convert to specific gravity of liquids heavier than water. For example, it is used to determine the specific gravity of milk being condensed in a vacuum pan.
3. The **Brix hydrometer** is a type of **saccharometer** used for sugar solutions such as fruit juices and syrups and one usually reads directly the percentage of sucrose at 20°C. **Balling saccharometers** are graduated to indicate percentage of sugar by weight at 60°F. The terms **Brix** and **Balling** are interpreted as the weight percentage of pure sucrose.
4. **Alcoholometers** are used to estimate the alcohol content of beverages. Such hydrometers are calibrated in 0.1 or 0.2° proof to determine the percentage of alcohol in distilled liquors (AOAC Method 957.03).
5. The **Twaddell hydrometer** is only for liquids heavier than water.

6.5.2.2 Pycnometer

One approach to measuring specific gravity is a comparison of the weights of equal volumes of a liquid and water in standardized glassware, a **pycnometer** (Fig. 6-7). This will yield density of the liquid compared to water. In some texts and reference books, 20/20 is given after the specific gravity number. This indicates that the temperature of both fluids was 20°C when the weights were measured. Using a clean, dry pycnometer at 20°C, the analyst weighs it empty, fills it to the full point with distilled water at 20°C, inserts the thermometer to seal the fill opening, and then touches off the last drops of water and puts on the cap for the overflow tube. The pycnometer is wiped dry in case of

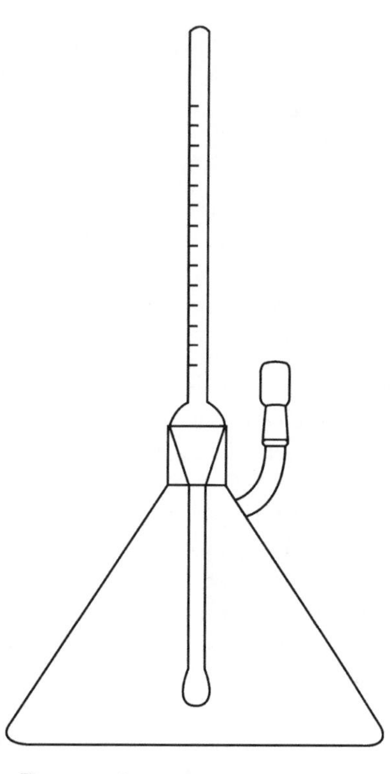

6-7 figure Pycnometer.

any spillage from filling and is reweighed. The density of the sample is calculated as follows:

$$\frac{\text{weight of sample-filled pycnometer} \atop -\text{weight of empty pycnometer}}{\text{weight of water-filled pycnometer} \atop -\text{weight of empty pycnometer}}$$

$$= \text{density of sample} \qquad [10]$$

This method is used for determining alcohol content in alcoholic beverages (e.g., distilled liquor, AOAC Method 930.17), solids in sugar syrups (AOAC Method 932.14B), and solids in milk (AOAC Method 925.22).

6.5.3 Refractometry

Moisture in liquid sugar products and condensed milks can be determined using a Baumé hydrometer (solids), a Brix hydrometer (sugar content), gravimetric means, or a **refractometer**. If it is performed correctly and no crystalline solids are evident, the refractometer procedure is rapid and surprisingly accurate (AOAC Method 9.32.14C, for solids in syrups). The refractometer has been valuable in determining the soluble solids in fruits and fruit products (AOAC Method 932.12; 976.20; 983.17).

The **refractive index** (RI) of an oil, syrup, or other liquid is a dimensionless constant that can be used to describe the nature of the food. While some refractometers are designed only to provide results as refractive indices, others, particularly hand-held, quick-to-use units, are equipped with scales calibrated to read the percentage of solids, percentage of sugars, and the like, depending on the products for which they are intended. Tables are provided with the instruments to convert values and adjust for temperature differences. Refractometers are used not just on the laboratory bench or as hand-held units. Refractometers can be installed in a liquid processing line to monitor the °Brix of products such as carbonated soft drinks, dissolved solids in orange juice, and the percentage of solids in milk (15).

When a beam of light is passed from one medium to another and the density of the two differs, then the beam of light is bent or refracted. Bending of the light beam is a function of the media and the sines of the angles of incidence and refraction at any given temperature and pressure and is thus a constant (Fig. 6-8). The (RI) (η) is a ratio of the sines of the angles:

$$\eta = \frac{\text{sine incident ray angle}}{\text{sine refracted ray angle}} \qquad [11]$$

All chemical compounds have an index of refraction. Therefore, this measurement can be used for the qualitative identification of an unknown compound by comparing its RI with literature values. RI varies with **concentration** of the compound, **temperature**, and **wavelength of light**. Instruments are designed to give a reading by passing a light beam of a specific wavelength through a glass prism into a liquid, the sample. Bench-top or hand-held units use **Amici prisms** to obtain the **D line of the sodium spectrum** or

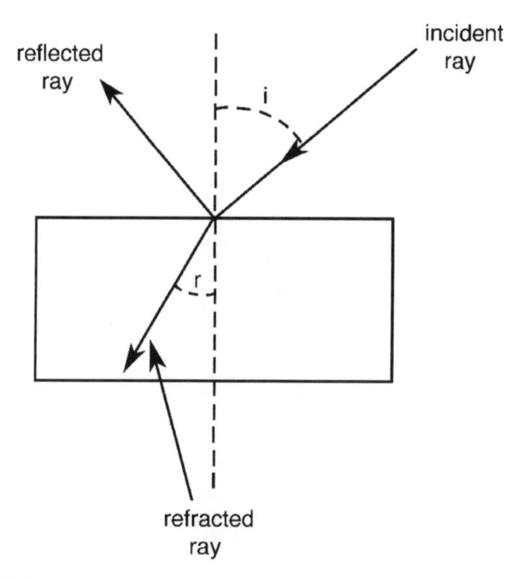

6-8 figure Reflection and refraction concepts of refractometry.

589 nm from white light. Whenever refractive indices of standard fluids are given, these are prefaced with $\eta_D^{20} = a$ value from 1.3000 to 1.7000. The Greek letter η is the symbol for RI; the 20 refers to temperature in °C; and D is the wavelength of the light beam, the D line of the sodium spectrum.

Bench-top instruments are more accurate compared with hand-held units mainly because of temperature control (Fig. 6-9). These former units have provisions for water circulation through the head where the prism and sample meet. **Abbe refractometers** are the most popular for laboratory use. Care must be taken when cleaning the prism surface following use. Wipe the contact surface clean with lens paper and rinse with distilled water and then ethanol. Close the prism chamber and cover the instrument with a bag when not in use to protect the delicate prism surface from dust or other debris that might lead to scratches and inaccuracy.

The fact that the RI of a solution increases with concentration has been exploited in the analysis of total soluble solids of carbohydrate-based foods such as sugar syrups, fruit products, and tomato products. Because of this use, these refractometers are calibrated in °Brix (g of sucrose/100 g of sample), which is equivalent to percentage sucrose on a wt/wt basis. Refractive index measurements are used widely to approximate sugar concentration in foods, even though values are accurate only for pure sucrose solutions.

6-9 figure Rhino Brix hand-held refractometer, R^2 mini digital hand-held refractometer, and Mark III Abbe refractometer. (Courtesy of Reichert Analytical Instrument, Depew, NY.)

6.5.4 Infrared Analysis

Infrared spectroscopy (see Chap. 23) has attained a primary position in monitoring the composition of food products before, during, and following processing (16). It has a wide range of food applications and has proven successful in the laboratory, at-line, and on-line. Infrared spectroscopy measures the absorption of radiation (near- or mid-infrared) by molecules in foods. Different frequencies of infrared radiation are absorbed by different functional groups characteristic of the molecules in food. Similar to the use of ultraviolet (UV) or visible (Vis) light in UV–Vis spectroscopy, a sample is irradiated with a wavelength of infrared light specific for the constituent to be measured. The concentration of that constituent is determined by measuring the energy that is reflected or transmitted by the sample, which is inversely proportional to the energy absorbed. Infrared spectrometers must be calibrated for each analyte to be measured and the analyte must be uniformly distributed in the sample.

For water, near-infrared (NIR) bands (1400–1450; 1920–1950 nm) are characteristic of the –OH stretch of the water molecule and can be used to determine the moisture content of a food. NIR has been applied to moisture analysis of a wide variety of food commodities.

The use of mid-infrared milk analyzers to determine fat, protein, lactose, and total solids in milk (AOAC Method 972.16) is covered in Chap. 23. The midrange spectroscopic method does not yield moisture or solids results except by computer calculation because these instruments do not monitor at wavelengths where water absorbs. The instrument must be calibrated using a minimum of eight milk samples that were previously analyzed for fat (F), protein (P), lactose (L), and total solids (TS) by standard methods. Then, a mean difference value, a, is calculated for all samples used in calibration:

$$a = \Sigma(\mathrm{TS} - F - P - L)/n \qquad [12]$$

where:

a = solids not measurable by the F, P, and L methods
n = number of samples
F = fat percentage
P = protein percentage
L = lactose percentage
TS = total solids percentage

Total solids then can be determined from any infrared milk analyzer results by using the formula

$$\mathrm{TS} = a + F + P + L \qquad [13]$$

The a value is thus a standard value mathematically derived. Newer instruments have the algorithm in their computer software to ascertain this value automatically. Moreover, Fourier transform infrared spectroscopy (FTIR) is the latest development that allows greater flexibility in infrared assays.

6.5.5 Freezing Point

When water is added to a food product, many of the physical constants are altered. Some properties of solutions depend on the number of solute particles as ions or molecules present. These properties are vapor pressure, freezing point, boiling point, and osmotic pressure. Measurement of any of these properties can be used to determine the concentration of solutes in a solution. However, the most commonly practiced assay for milk is the change of the freezing point value. It has economic importance with regard to both raw and pasteurized milk. The **freezing point** of milk is its most constant physical property. The secretory process of the mammary gland is such that the osmotic pressure is kept in equilibrium with blood and milk. Thus, with any decrease in the synthesis of lactose, there is a compensating increase in the concentrations of Na^+ and Cl^-. While termed a physical constant, the freezing point varies within narrow limits, and the vast majority of samples from individual cows fall between −0.503°C and −0.541°C (−0.525 and −0.565°H, temperature in °H or **Hortvet**, the surname of the inventor of the first freezing point apparatus). The average is very close to −0.517°C (−0.540°H). Herd or bulk milk will exhibit a narrower range unless the supply was watered intentionally or accidentally or if the milk is from an area where severe drought has existed. All values today are given in °C by agreement. The following is used to convert °H to °C, or °C to °H (5, 6):

$$°\mathrm{C} = 0.9623°\mathrm{H} - 0.0024 \qquad [14]$$

$$°\mathrm{H} = 1.03916°\mathrm{C} + 0.0025 \qquad [15]$$

The principal utility of freezing point is to measure for **added water**. However, the freezing point of milk can be altered by mastitis infection in cows and souring of milk. In special cases, nutrition and environment of the cow, stage of lactation, and processing operations for the milk can affect the freezing point. If the solute remains constant in weight and composition, the change of the freezing point varies inversely with the amount of solvent present. Therefore, we can calculate the percent H_2O added:

$$\%\mathrm{H_2O\ added} = \frac{0.517 - T}{0.517} \times 100 \qquad [16]$$

where:

0.517 = freezing point in °C of all milk entering a plant
T = freezing point in °C of a sample

The AOAC Method 961.07 for water added to milk uses a **cryoscope** to test for freezing points, and assumes a freezing point for normal milk of −0.527°C (−0.550°H). The Food and Drug Administration will reject all milk with freezing points above −0.503°C (−0.525°H). Since the difference between the freezing points of milk and water is slight and since the freezing point can be used to calculate the amount of water added, it is essential that the method be as precise as possible. The thermister used can sense temperature change to 0.001°C (0.001°H). The general technique is to supercool the solution and then induce crystallization by a vibrating reed. The temperature will rise rapidly to the freezing point or eutectic temperature as the water freezes. In the case of pure water, the temperature remains constant until all the water is frozen. In the case of milk, the temperature is read when there is no further temperature rise.

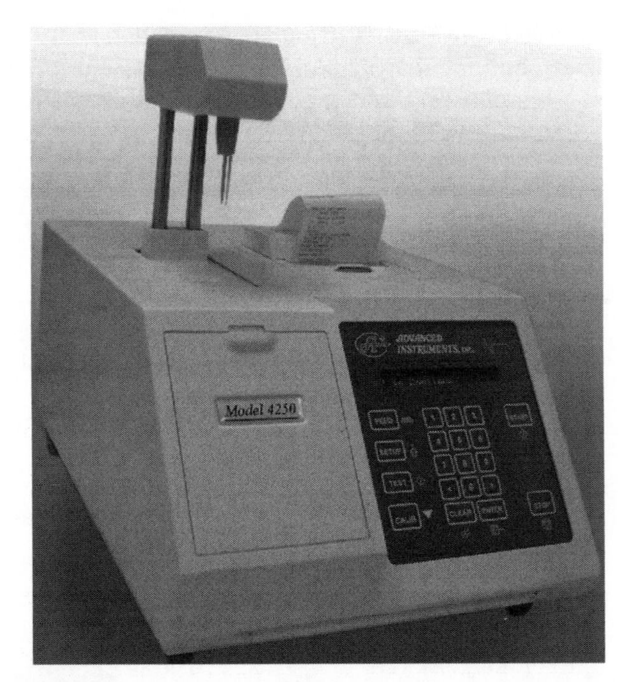

6-10 figure A model 4D3 advanced instruments cryoscope for freezing point determination in milk. (Courtesy of Advanced Instruments, Inc., Norwood, MA.)

Instrumentation available is manufactured by Advanced Instruments (Fig. 6-10). Time required for the automated instruments is 1–2 min per sample using a prechilled sample.

6.6 WATER ACTIVITY

Water content alone is not a reliable indicator of food stability, since foods with the same water content differ in their perishability (17). This is at least partly due to differences in the way that water associates with other constituents in a food. Water tightly associated with other food constituents is less available for microbial growth and chemical reactions to cause decomposition. **Water activity** (a_W) is a better indication of food perishability than is water content. Water activity is defined as follows:

$$a_W = \frac{P}{P_o} \quad [17]$$

$$a_W = \frac{\text{ERH}}{100} \quad [18]$$

where:

a_W = water activity

P = partial pressure of water above the sample

P_o = vapor pressure of pure water at the same temperature (specified)

ERH = equilibrium relative humidity surrounding the product

There are various techniques to measure a_W. A commonly used approach relies on measuring the amount of moisture in the equilibrated headspace above a sample of the food product, which correlates directly with sample a_W. A sample for such analysis is placed in a small closed chamber at constant temperature, and a relative humidity sensor is used to measure the ERH of the sample atmosphere after equilibration. A simple and accurate variation of this approach is the chilled mirror technique in which the water vapor in the headspace condenses on the surface of a mirror that is cooled in a controlled manner. The dew point is determined by the temperature at which condensation takes place, and this determines the relative humidity in the headspace. Two other general approaches to measuring a_W are (1) using the sample freezing point depression and moisture content to calculate a_W, and (2) equilibrating a sample in a chamber held at constant relative humidity (by means of a saturated salt solution) and then using the water content of the sample to calculate a_W (17).

6.7 COMPARISON OF METHODS

6.7.1 Principles

Oven drying methods involve the removal of moisture from the sample and then a weight determination of the solids remaining to calculate the moisture content. Nonwater volatiles can be lost during drying, but their loss is generally a negligible percentage of the amount of water lost. Distillation procedures also involve a separation of the moisture from the solids, and the moisture is quantitated directly by volume. Karl Fischer titration is based on chemical reactions of the moisture present, reflected as the amount of titrant used.

The dielectric method is based on electrical properties of water. Hydrometric methods are based on the relationship between specific gravity and moisture content. The refractive index method is based on how water in a sample affects the refraction of light. Near-infrared analysis of water in foods is based on measuring the absorption at wavelengths characteristic of the molecular vibration in water. Freezing point is a physical property of milk that is changed by a change in solute concentration.

6.7.2 Nature of Sample

While most foods will tolerate oven drying at high temperatures, some foods contain volatiles that are lost at such temperatures. Some foods have constituents that undergo chemical reactions at high temperatures to generate or utilize moisture or other compounds, to affect the calculated moisture content. Vacuum oven drying at reduced temperatures may overcome such problems for some foods. However, a distillation technique is necessary for some food to minimize volatilization and decomposition. For foods very low in moisture or high in fats and sugars, Karl Fischer titration is often the method of choice. The use of a pycnometer, hydrometer, and refractometer requires liquid samples, ideally with limited constituents.

6.7.3 Intended Purposes

Moisture analysis data may be needed quickly for quality control purposes, in which high accuracy may not be necessary. Of the oven drying methods, microwave drying, infrared drying, and the moisture analyzer technique are fastest. Some forced draft oven procedures require less than 1 h drying, but most forced draft oven and vacuum oven procedures require a much longer time. The electrical, hydrometric, and refractive index methods are very rapid but often require correlation to less empirical methods. Oven drying procedures are official methods for a variety of food products. Reflux distillation is an AOAC method for chocolate, dried vegetables, dried milk, and oils and fats. Such official methods are used for regulatory and nutrition labeling purposes.

6.8 SUMMARY

The moisture content of foods is important to food processors and consumers for a variety of reasons. While moisture determination may seem simplistic, it is often one of the most difficult assays in obtaining accurate and precise results. The free water present in food is generally more easily quantitated as compared to the adsorbed moisture and the water of hydration. Some moisture analysis methods involve a separation of moisture in the sample from the solids and then quantitation by weight or volume. Other methods do not involve such a separation but instead are based on some physical or chemical property of the water in the sample. A major difficulty with many methods is attempting to remove or otherwise quantitate all water present. This often is complicated by decomposition or interference by other food constituents. For each moisture analysis method, there are factors that must be controlled or precautions that must be taken to ensure accurate and precise results. Careful sample collection and handling procedures are extremely important and cannot be overemphasized. The choice of moisture analysis method is often determined by the expected moisture content, nature of other food constituents (e.g., highly volatile, heat sensitive), equipment available, speed necessary, accuracy and precision required, and intended purpose (e.g., regulatory or in-plant quality control).

6.9 STUDY QUESTIONS

1. Identify five factors that one would need to consider when choosing a moisture analysis method for a specific food product.
2. Why is standardized methodology needed for moisture determinations?
3. What are the potential advantages of using a vacuum oven rather than a forced draft oven for moisture content determination?
4. In each case specified below, would you likely overestimate or underestimate the moisture content of a food product being tested? Explain your answer.
   (a) Forced draft oven:
   - Particle size too large
   - High concentration of volatile flavor compounds present
   - Lipid oxidation
   - Sample very hygroscopic
   - Alteration of carbohydrates (e.g., Maillard browning)
   - Sucrose hydrolysis
   - Surface crust formation
   - Splattering
   - Desiccator with dried sample not sealed properly

   (b) Toluene distillation:
   - Emulsion between water in sample and solvent not broken
   - Water clinging to condenser

   (c) Karl Fischer:
   - Very humid day when weighing original samples
   - Glassware not dry
   - Sample ground coarsely
   - Food high in vitamin C
   - Food high in unsaturated fatty acids
5. The procedure for an analysis for moisture in a liquid food product requires the addition of 1–2 ml of deionized water to the weighed sample in the moisture pan. Why should you add water to an analysis in which moisture is being determined?
6. A new instrument based on infrared principles has been received in your laboratory to be used in moisture analysis. Briefly describe the way you would ascertain if the new instrument would meet your satisfaction and company standards.
7. A technician you supervise is to determine the moisture content of a food product by the Karl Fischer method. Your technician wants to know what is this "Karl Fischer reagent water equivalence" that is used in the equation to calculate percentage of moisture in the sample, why it is necessary, and how it is determined. Give the technician your answer.

8. To explain and contrast the principles (not procedures) in determining the moisture content of food products by the following method, complete the table below. (Assume that sample selection and handling has been done appropriately.)

	What is actually measured?	How is water removed/ reacted/ identified?	What assumptions are made in trusting the data obtained (or precautions taken to ensure accurate data)?
Microwave oven			
NIR			
Karl Fischer			
Toluene distillation			

9. You are fortunate to have available in your laboratory the equipment for doing moisture analysis by essentially all methods – both official and rapid quality control methods. For each of the food products listed below (with the purpose specified as rapid quality control or official), indicate (a) the name of the method you would use, (b) the principle (not procedure) for the method, (c) a justification for use of that method (as compared to using a hot air drying oven), and (d) two cautions in use of the method to ensure accurate results.
   (a) Ice cream mix (liquid) – quality control
   (b) Milk chocolate – official
   (c) Spices – official
   (d) Syrup for canned peaches – quality control
   (e) Oat flour – quality control
10. You are a manufacturer of processed cheese. The maximum allowed moisture content for your product is 40%. Your current product has a mean moisture content of 38%, with a standard deviation of 0.7. It would be possible to increase your mean moisture content to 39.5% if you could reduce your standard deviation to 0.25. This would result in a saving of $3.4 million per year. You can accomplish this by rapidly analyzing the moisture content of the cheese blend prior to the cooking step of manufacture. The cheese blend is prepared in a batch process, and you have 10 min to adjust the moisture content of each batch.
    (a) Describe the rapid moisture analysis method you would use. Include your rationale for selecting the method.
    (b) How would you ensure the accuracy and precision of this method (you need to be sure your standard deviation is below 0.25)?
11. You work in a milk drying plant. As part of the production process, you need to rapidly analyze the moisture content of condensed milk.
    (a) What rapid secondary method would you use, and what primary method would you use to calibrate the secondary method? Additionally, how would you ensure the accuracy and precision of your secondary method?
    (b) Your results with the secondary method are consistently high (about 1%), based on the secondary method you chose. What are some potential problems and how would you correct them?
12. During a 12-h period, 1000 blocks (40 lbs each) from ten different vats (100 blocks per vat) of Cheddar cheese were produced. It was later realized that the cooking temperature was too low during cheesemaking. You are concerned that this might increase the moisture content of the cheese above the legal requirement. Describe the sampling plan and method of analysis you would use to determine the moisture content of the cheese. You want the results within 48 h so you can determine what to do with the cheese.

6.10 PRACTICE PROBLEMS

1. As an analyst, you are given a sample of condensed soup to analyze to determine if it is reduced to the correct concentration. By gravimetric means, you find that the concentration is 26.54% solids. The company standard reads 28.63%. If the starting volume were 1000 gallons at 8.67% solids and the weight is 8.5 pounds per gallon, how much more water must be removed?
2. Your laboratory just received several sample containers of peas to analyze for moisture content. There is a visible condensate on the inside of the container. What is your procedure to obtain a result?
3. You have the following gravimetric results: weight of dried pan and glass disc is 1.0376 g, weight of pan and liquid sample is 4.6274 g, and weight of the pan and dried sample is 1.7321 g. What was the moisture content of the sample and what is the percent solids?

Answers

1. The weight of the soup initially is superfluous information. By condensing the soup to 26.54% solids from 8.67% solids, the volume is reduced to 326.7 gal [(8.67%/26.54%) × 1000 gal]. You need to reduce the volume further to obtain 28.63% solids [(8.67%/28.63%) × 1000 gal] or 302.8 gal. The difference in the gallons obtained is 23.9 gal (326.7 gal − 302.8 gal), or the volume of water that must be removed from the partially condensed soup to comply with company standards.
2. This problem focuses on a real issue in the food processing industry – when do you analyze a sample and when don't you? It would appear that the peas have lost moisture that should be within the vegetable for correct results. You will need to grind the peas in a food mill or blender. If the peas are in a Mason jar or one that fits a blender head, no transfer is needed. Blend the peas to a creamy texture. If a container transfer was made, then

put the blended peas back into the original container. Mix with the residual moisture to a uniform blend. Collect a sample for moisture analysis. You should note on the report form containing the results of the analysis that the pea samples had free moisture on container walls when they arrived.

3. Note Equations [2]–[4]. To use any of the equations, you must subtract the weight of the dried pan and glass disc. Then you obtain 3.5898 g of original sample and 0.6945 g when dried. By subtracting these results, you have removed water (2.8953 g). Then $(0.6945\,\mathrm{g}/3.5898\,\mathrm{g}) \times 100 = 19.35\%$ solids and $(2.8953\,\mathrm{g}/3.5898\,\mathrm{g}) \times 100 = 80.65\%$ water.

6.11 REFERENCES

1. Pomeranz Y, Meloan C (1994) Food analysis: theory and practice, 3rd edn. Chapman & Hall, New York
2. Aurand LW, Woods AE, Wells MR (1987) Food composition and analysis. Van Nostrand Reinhold, New York
3. Josyln MA (1970) Methods in food analysis, 2nd edn. Academic, New York
4. USDA (2008) USDA Nutrient Database for Standard Reference. Release 21, 2008. http://www.nal.usda.gov/fnic/cgi-bin/nut_search.pl
5. AOAC International (2007) Official methods of analysis, 18th edn, 2005; Current through revision 2, 2007 (On-line). AOAC International, Gaithersburg, MD
6. Wehr HM, Frank JF (eds) (2004) Standard methods for the examination of dairy products, 17th edn. American Public Health Association, Washington, DC
7. AACC International (2010) Approved methods of analysis, 11th edn (On-line). AACC International, St. Paul, MN
8. Emmons DB, Bradley RL Jr, Sauvé JP, Campbell C, Lacroix C, Jimenez-Marquez SA (2001) Variations of moisture measurements in cheese. J AOAC Int 84: 593–604
9. Emmons DB, Bradley RL Jr, Campbell C, Sauve JP (2001) Movement of moisture in refrigeration cheese samples transferred to room temperature. J AOAC Int 84:620–622
10. Bradley RL Jr, Vanderwarn MA (2001) Determination of moisture in cheese and cheese products. J AOAC Int 84:570–592
11. Nelson OA, Hulett GA (1920) The moisture content of cereals. J Ind Eng Chem 12:40–45
12. International Dairy Federation. Provisional Standard 4A (1982) Cheese and processed cheese: determination of the total solids content, Brussels, Belgium
13. Bouraoui M, Richard P, Fichtali J (1993) A review of moisture content determination in foods using microwave oven drying. Food Res Inst Int 26:49–57
14. Mitchell J Jr, Smith DM (1948) Aquametry. Wiley, New York
15. Giese J (1993) In-line sensors for food processing. Food Technol 47(5):87–95
16. Wilson RH, Kemsley EK (1992) On-line process monitoring using infrared techniques. In: Food processing automation II. Proceedings of the American society of agricultural engineers, ASAE Publication 02-92. American Society of Agricultural Engineers, St. Joseph, MI
17. Damodaran S, Parkin KL, Fennema OR (2007) Water and ice (Chapter 2). In: Fennema OR (ed) Fennema's food chemistry, 4rd edn. CRC, Boca Raton, FL

Ash Analysis

Maurice R. Marshall
Department of Food Science and Human Nutrition, University of Florida, Gainesville, FL 32611-0370, USA
martym@ufl.edu

7.1 Introduction 107
7.1.1 Definitions 107
7.1.2 Importance of Ash in Food Analysis 107
7.1.3 Ash Contents in Foods 107
7.2 Methods 108
7.2.1 Sample Preparation 108
7.2.1.1 Plant Materials 108
7.2.1.2 Fat and Sugar Products 108
7.2.2 Dry Ashing 108
7.2.2.1 Principles and Instrumentation 108
7.2.2.2 Procedures 109
7.2.2.3 Special Applications 109
7.2.3 Wet Ashing 109
7.2.3.1 Principle, Materials, and Applications 109
7.2.3.2 Procedures 110
7.2.4 Microwave Ashing 110
7.2.4.1 Microwave Wet Ashing 111
7.2.4.2 Microwave Dry Ashing 112
7.2.5 Other Ash Measurements 112
7.3 Comparison of Methods 112
7.4 Summary 113
7.5 Study Questions 113
7.6 Practice Problems 113
7.7 Acknowledgments 114
7.8 References 114

S.S. Nielsen, *Food Analysis*, Food Science Texts Series, DOI 10.1007/978-1-4419-1478-1_7,

7.1 INTRODUCTION

Ash refers to the inorganic residue remaining after either ignition or complete oxidation of organic matter in a foodstuff. A basic knowledge of the characteristics of various ashing procedures and types of equipment is essential to ensure reliable results. Two major types of ashing are used: dry ashing, primarily for proximate composition and for some types of specific mineral analyses; wet ashing (oxidation), as a preparation for the analysis of certain minerals. Microwave systems now are available for both dry and wet ashing, to speed the processes. Most dry samples (i.e., whole grain, cereals, dried vegetables) need no preparation, while fresh vegetables need to be dried prior to ashing. High-fat products such as meats may need to be dried and fat extracted before ashing. The ash content of foods can be expressed on either a wet weight (as is) or on a dry weight basis. For general and food-specific information on measuring ash content, see references (1–11).

7.1.1 Definitions

Dry ashing refers to the use of a muffle furnace capable of maintaining temperatures of 500–600°C. Water and volatiles are vaporized, and organic substances are burned in the presence of oxygen in air to CO_2 and oxides of N_2. Most minerals are converted to oxides, sulfates, phosphates, chlorides, and silicates. Elements such as Fe, Se, Pb, and Hg may partially volatilize with this procedure, so other methods must be used if ashing is a preliminary step for specific elemental analysis.

Wet ashing is a procedure for oxidizing organic substances by using acids and oxidizing agents or their combinations. Minerals are solubilized without volatilization. Wet ashing often is preferable to dry ashing as a preparation for specific elemental analysis. Wet ashing often uses a combination of acids and requires a special perchloric acid hood if that acid is used.

7.1.2 Importance of Ash in Food Analysis

Ash content represents the total mineral content in foods. Determining the ash content may be important for several reasons. It is a part of proximate analysis for nutritional evaluation. Ashing is the first step in preparing a food sample for specific elemental analysis. Because certain foods are high in particular minerals, ash content becomes important. One can usually expect a constant elemental content from the ash of animal products, but that from plant sources is variable.

7.1.3 Ash Contents in Foods

The average ash content for various food groups is given in Table 7-1. The ash content of most fresh foods rarely is greater than 5%. Pure oils and fats generally contain little or no ash; products such as cured bacon may contain 6% ash, and dried beef may be as high as 11.6% (wet weight basis).

Fats, oils, and shortenings vary from 0.0 to 4.1% ash, while dairy products vary from 0.5 to 5.1%. Fruits, fruit juice, and melons contain 0.2–0.6% ash, while dried fruits are higher (2.4–3.5%). Flours and meals vary from 0.3 to 1.4% ash. Pure starch contains 0.3% and wheat germ 4.3% ash. It would be expected that

Ash Content of Selected Foods

Food Item	Percent Ash (Wet Weight Basis)
Cereals, bread, and pasta	
Rice, brown, long-grain, raw	1.5
Corn meal, whole-grain, yellow	1.1
Hominy, canned, white	0.9
White rice, long-grain, regular, raw, enriched	0.6
Wheat flour, whole-grain	1.6
Macaroni, dry, enriched	0.9
Rye bread	2.5
Dairy products	
Milk, reduced fat, fluid, 2%	0.7
Evaporated milk, canned, with added vitamin A	1.6
Butter, with salt	2.1
Cream, fluid, half-and-half	0.7
Margarine, hard, regular, soybean	2.0
Yogurt, plain, low fat	1.1
Fruits and vegetables	
Apples, raw, with skin	0.2
Bananas, raw	0.8
Cherries, sweet, raw	0.5
Raisins	1.9
Potatoes, raw, skin	1.6
Tomatoes, red, ripe, raw	0.5
Meat, poultry, and fish	
Eggs, whole, raw, fresh	0.9
Fish fillet, battered or breaded, and fried	2.5
Pork, fresh, leg (ham), whole, raw	0.9
Hamburger, regular, single patty, plain	1.9
Chicken, broilers or fryers, breast meat only, raw	1.0
Beef, chuck, arm pot roast, raw	1.1

From US Department of Agriculture, Agricultural Research Service (2009) USDA National Nutrient Database for Standard Reference. Release 22. Nutrient Data Laboratory Home Page: http://www.ars.usda.gov/ba/bhnrc/ndl

grain and grain products with bran would tend to be higher in ash content than such products without bran. Nuts and nut products contain 0.8–3.4% ash, while meat, poultry, and seafoods contain 0.7–1.3% ash.

7.2 METHODS

Principles, materials, instrumentation, general procedures, and applications are described below for various ash determination methods. Refer to methods cited for detailed instructions of the procedures.

7.2.1 Sample Preparation

It cannot be overemphasized that the small sample used for ash, or other determinations, needs to be very carefully chosen so that it represents the original materials. A 2–10-g sample generally is used for ash determination. For that purpose, milling, grinding, and the like probably will not alter the ash content much; however, if this ash is a preparatory step for specific mineral analyses, contamination by microelements is of potential concern. Remember, most grinders and mincers are of steel construction. Repeated use of glassware can be a source of contaminants as well. The water source used in dilutions also may contain contaminants of some microelements. Distilled-deionized water always should be used.

7.2.1.1 Plant Materials

Plant materials are generally dried by routine methods prior to grinding. The temperature of drying is of little consequence for ashing. However, the sample may be used for multiple determinations – protein, fiber, and so on – which require consideration of temperature for drying. Fresh stem and leaf tissue probably should be dried in two stages (i.e., first at a lower temperature of 55°C, then a higher temperature) especially to prevent artifact lignin. Plant material with 15% or less moisture may be ashed without prior drying.

7.2.1.2 Fat and Sugar Products

Animal products, syrups, and spices require treatments prior to ashing because of high fat, moisture (spattering, swelling), or high sugar content (foaming) that may result in loss of sample. Meats, sugars, and syrups need to be evaporated to dryness on a steam bath or with an infrared (IR) lamp. One or two drops of olive oil (which contains no ash) are added to allow steam to escape as a crust is formed on the product.

Smoking and burning may occur upon ashing for some products (e.g., cheese, seafood, spices). Allow this smoking and burning to finish slowly by keeping the muffle door open prior to the normal procedure. A sample may be ashed after drying and fat extraction. In most cases, mineral loss is minimal during drying and fat extraction. Under no circumstances should fat-extracted samples be heated until all the ether has been evaporated.

7.2.2 Dry Ashing

7.2.2.1 Principles and Instrumentation

Dry ashing is incineration at high temperature (525°C or higher). Incineration is accomplished with a muffle furnace. Several models of muffle furnaces are available, ranging from large-capacity units requiring either 208 or 240 V supplies to small benchtop units utilizing 110-V outlets.

Crucible selection becomes critical in ashing because the type depends upon the specific use. **Quartz crucibles** are resistant to acids and halogens, but not alkali, at high temperatures. **Vycor® brand crucibles** are stable to 900°C, but **Pyrex® Gooch crucibles** are limited to 500°C. Ashing at a lower temperature of 500–525°C may result in slightly higher ash values because of less decomposition of carbonates and loss of volatile salts. **Porcelain crucibles** resemble quartz crucibles in their properties, but will crack with rapid temperature changes. Porcelain crucibles are relatively inexpensive and usually the crucible of choice. **Steel crucibles** are resistant to both acids and alkalies and are inexpensive, but they are composed of chromium and nickel, which are possible sources of contamination. **Platinum crucibles** are very inert and are probably the best crucibles, but they are currently far too expensive for routine use for large numbers of samples. **Quartz fiber crucibles** are disposable, unbreakable, and can withstand temperatures up to 1000°C. They are porous, allowing air to circulate around the sample and speed combustion. This reduces ashing times significantly and makes them ideal for solids and viscous liquids. Quartz fiber also cools in seconds, virtually eliminating the risk of burns.

All crucibles should be marked for identification. Marks on crucibles with a felt-tip marking pen will disappear during ashing in a muffle furnace. Laboratory inks scribed with a steel pin are available commercially. Crucibles also may be etched with a diamond point and marked with a 0.5 *M* solution of $FeCl_3$, in 20% HCl. An iron nail dissolved in concentrated HC1 forms brown goo that is a satisfactory marker. The crucibles should be fired and cleaned prior to use.

The *advantages* of conventional dry ashing are that it is a safe method, it requires no added reagents or blank subtraction, and little attention is needed once ignition begins. Usually a large number of crucibles can be handled at once, and the resultant ash can be used additionally in other analyses for most individual elements, acid-insoluble ash, and water-soluble and insoluble ash. The *disadvantages* are the length of time required (12–18 h or overnight) and expensive equipment. There will be a loss of the volatile elements and interactions between mineral components and crucibles. Volatile elements at risk of being lost include As, B, Cd, Cr, Cu, Fe, Pb, Hg, Ni, P, V, and Zn.

7.2.2.2 Procedures

AOAC International has several dry ashing procedures (e.g., AOAC Methods 900.02 A or B, 920.117, 923.03) for certain individual foodstuffs. The general procedure includes the following steps:

1. Weigh a 5–10-g sample into a tared crucible. Predry if the sample is very moist.
2. Place crucibles in a cool muffle furnace. Use tongs, gloves, and protective eyewear if the muffle furnace is warm.
3. Ignite 12–18 h (or overnight) at about 550°C.
4. Turn off muffle furnace and wait to open it until the temperature has dropped to at least 250°C, preferably lower. Open door carefully to avoid losing ash that may be fluffy.
5. Using safety tongs, quickly transfer crucibles to a desiccator with a porcelain plate and desiccant. Cover crucibles, close desiccator, and allow crucibles to cool prior to weighing.

Note. Warm crucibles will heat air within the desiccator. With hot samples, a cover may bump to allow air to escape. A vacuum may form on cooling. At the end of the cooling period, the desiccator cover should be removed gradually by sliding to one side to prevent a sudden inrush of air. Covers with a ground glass sleeve or fitted for a rubber stopper allow for slow release of a vacuum.

The ash content is calculated as follows:

$$\%\ \text{ash (dry basis)} = \frac{\text{wt after ashing} - \text{tare wt of crucible}}{\text{original sample wt} \times \text{dry matter coefficient}} \times 100 \quad [1]$$

where:

$$\text{dry matter coefficient} = \%\ \text{solids}/100$$

For example, if corn meal is 87% dry matter, the dry matter coefficient would be 0.87. If ash is calculated on an as-received or wet weight basis (includes moisture), delete the dry matter coefficient from the denominator. If moisture was determined in the same crucible prior to ashing, the denominator becomes (dry sample wt - tared crucible wt).

7.2.2.3 Special Applications

Some of the AOAC procedures recommend steps in addition to those listed previously. If carbon is still present following the initial incineration, several drops of water or nitric acid should be added; then the sample should be re-ashed. If the carbon persists, such as with high-sugar samples, follow this procedure:

1. Suspend the ash in water.
2. Filter through ashless filter paper because this residue tends to form a glaze.
3. Dry the filtrate.
4. Place paper and dried filtrate in muffle furnace and re-ash.

Other suggestions that may be helpful and accelerate incineration:

1. High-fat samples should be extracted either by using the crude fat determination procedure or by burning off prior to closing the muffle furnace. Pork fat, for example, can form a combustible mixture inside the furnace and burn with the admission of oxygen if the door is opened.
2. Glycerin, alcohol, and hydrogen will accelerate ashing.
3. Samples such as jellies will spatter and can be mixed with cotton wool.
4. Salt-rich foods may require a separate ashing of water-insoluble components and salt-rich water extract. Use a crucible cover to prevent spattering.
5. An alcoholic solution of magnesium acetate can be added to accelerate ashing of cereals. An appropriate blank determination is necessary.

7.2.3 Wet Ashing

7.2.3.1 Principle, Materials, and Applications

Wet ashing is sometimes called **wet oxidation** or **wet digestion**. Its primary use is preparation for specific mineral analysis and metallic poisons. Often, analytical testing laboratories use only wet ashing in preparing samples for certain mineral analyses (e.g., Fe, Cu, Zn, P), because losses would occur by volatilization during dry ashing.

There are several *advantages* to using the wet ashing procedure. Minerals will usually stay in solution, and there is little or no loss from volatilization because of the lower temperature. The oxidation time is short and requires a hood, hot plate, and long tongs, plus safety equipment.

The *disadvantages* of wet ashing are that it takes virtually constant operator attention, corrosive reagents are necessary, and only small numbers of samples can be handled at any one time. If the wet digestion utilizes perchloric acid, all work needs to be carried out in an expensive special fume hood called a **perchloric acid hood**.

Unfortunately, a single acid used in wet ashing does not give complete and rapid oxidation of organic material, so a mixture of acids often is used. Combinations of the following acid solutions are used most often: (1) **nitric acid**, (2) **sulfuric acid-hydrogen peroxide**, and (3) **perchloric acid**. Different combinations are recommended for different types of samples. The nitric–perchloric combination is generally faster than the sulfuric–nitric procedure. While wet digestion with perchloric acid is an AOAC procedure (e.g., AOAC Method 975.03), many analytical laboratories avoid if possible the use of perchloric acid in wet ashing and instead use a combination of nitric acid with either sulfuric acid, hydrogen peroxide, or hydrochloric acid.

Wet oxidation with perchloric acid is *extremely* dangerous since the perchloric acid has a tendency to explode. The perchloric acid hood that must be used has wash-down capabilities and does not contain plastic or glycerol-base caulking compounds. Precautions for use of perchloric acid are found in the AOAC methods under "Safe Handling of Special Chemical Hazards." Cautions must be taken when fatty foods are wet ashed using perchloric acid. While perchloric acid does not interfere with atomic absorption spectroscopy, it does interfere in the traditional colorimetric assay for iron by reacting with iron in the sample to form ferrous perchlorate, which forms an insoluble complex with the *o*-phenanthrolene in the procedure.

7.2.3.2 Procedures

The following is a wet ash procedure using concentrated nitric and sulfuric acids (*to be performed in a fume hood*) (John Budin, Silliker Laboratories, Chicago, IL, personal communication):

1. Accurately weigh a dried, ground 1-g sample in a 125-ml Erlenmeyer flask (previously acid washed and dried).
2. Prepare a blank of 3 ml of H_2SO_4 and 5 ml of HNO_3, to be treated like the samples. (Blank is to be run with every set of samples.)
3. Add 3 ml of H_2SO_4 followed by 5 ml of HNO_3 to the sample in the flask.
4. Heat the sample on a hot plate at ca. 200°C (boiling). Brown-yellow fumes will be observed.
5. Once the brown-yellow fumes cease and white fumes from decomposing H_2SO_4 are observed, the sample will become darker. Remove the flask from the hot plate. Do not allow the flask to cool to room temperature.
6. *Slowly* add 3–5 ml of HNO_3.
7. Put the flask back on the hot plate and allow the HNO_3 to boil off. Proceed to the next step when all the HNO_3 is removed and the color is clear to straw yellow. If the solution is still dark in color, add another 3–5 ml of HNO_3 and boil. Repeat the process until the solution is clear to straw yellow.
8. While on the hot plate, reduce the volume appropriately to allow for ease of final transfer. Allow the sample to cool to room temperature, then quantitatively transfer the sample to an appropriately sized volumetric flask.
9. Dilute the sample to volume with ultrapure water, and mix well. Dilute further, as appropriate, for the specific type of mineral being analyzed.

The following procedure for a modified dry–wet ash sample destruction may be used. It is listed under "Minerals in Infant Formula, Enteral Products, and Pet Foods" (AOAC Method 985.35).

1. Evaporate moist samples (25–50 ml) in an appropriate dish at 100°C overnight or in a microwave drying oven until dry.
2. Heat on a hot plate until smoking ceases.
3. Ash in a 525°C furnace for 3–8 h.
4. Remove dish from furnace and allow to cool. Ash should be grayish white to white and free from carbon.
5. Cool and wet with deionized distilled water plus 0.5–3.0 ml of HNO_3.
6. Dry on a hot plate or steam bath and then return to a 525°C furnace for 1–2 h.
7. Repeat steps 5 and 6 if carbon persists. (*Caution*: Some K may be lost with repeated ashing.)
8. Dissolve the ash in 5 ml of 1 *M* HNO_3 by warming on a hot plate for 2–3 min to aid solution. Transfer to an appropriate size volumetric flask (i.e., 50 ml), then repeat with two additional portions of 1 *M* HNO_3.

7.2.4 Microwave Ashing

Both **wet ashing** and **dry ashing** can be done using microwave instrumentation, rather than the conventional dry ashing in a muffle furnace and wet ashing in a flask or beaker on a hot plate. The CEM Corporation (Matthews, NC) has developed a series of instruments for dry and wet ashing, as well as other laboratory systems for microwave-assisted chemistry. While the ashing procedures by conventional means can take many hours, the use of microwave instrumentation can reduce sample preparation time to minutes, allowing laboratories to increase their sample throughput significantly. This advantage has led to widespread use of microwave ashing, especially for wet ashing, both within analytical laboratories and quality control laboratories within food companies.

7.2.4.1 Microwave Wet Ashing

Microwave wet ashing (acid digestion) may be performed safely in either an open- or closed-vessel microwave system. Choice of the system depends on the amount of sample and the temperatures required for digesting. Because of the ability of the closed vessels to contain higher pressures (some vessels can handle up to 1500 psi), acids may be heated past their boiling points. This ensures a more complete dissolution of hard-to-digest substances. It also allows the chemist to use nitric acid with samples that might normally require a harsher acid, such as sulfuric or perchloric. In closed vessels specifically designed for high-temperatures/high-pressure reactions, nitric acid can reach a temperature of 240°C. Thus, **nitric acid** is often the acid of choice, though hydrochloric, hydrofluoric, and sulfuric acids also are used, depending on the sample and the subsequent analysis being performed. **Closed-vessel microwave digestion systems** (Fig. 7-1) can process up to 40 samples at a time, with vessel liners available in Teflon®, TFM™ Fluoropolymer, and quartz. These systems allow the input of time, temperature, and pressure parameters in a step-by-step format (ramping). In addition, some instruments enable the user to adjust the power and offer "change-on-the-fly" software, which allows the method to be changed while the reaction is running.

Typically, in a closed-vessel microwave system, sample is placed in vessels with the appropriate amount of acid. The vessels are sealed and set on a carousel where the temperature and pressure sensors are connected to a control vessel. The carousel then is placed in the microwave cavity, and the sensors are connected to the instrument. Time, temperature, pressure, and power parameters are chosen and the unit is started. Digestions normally take less than 30 min. Because of the pressure generated by raising the temperature of a reaction, the vessels must be allowed to cool before being opened. The ability to process multiple samples simultaneously provides the chemist with greater throughput than traditional methods. (Note that some closed-vessel microwave digestion systems may also be used for acid concentration, solvent extraction, protein hydrolysis, and synthesis with the proper accessories.)

Open-vessel digestion systems (Fig. 7-2) are used often for larger sample sizes (up to 10 g) and for samples that generate substantial amounts of gas as they are digested. Open-vessel systems can process up to six samples, each according to its own parameters in a sequential or simultaneous format. Teflon®, quartz, or Pyrex® vessels are used, and condensers are added for refluxing. Acid (reagent) is automatically added according to the programmed parameters. Sulfuric and nitric acids are used most often with open-vessel systems, as they process reactions under atmospheric conditions; however, hydrochloric and hydrofluoric acids, as well as hydrogen peroxide, can be used.

7-1 figure Microwave closed-vessel digestion system. (Courtesy of CEM Corporation, Matthews, NC.)

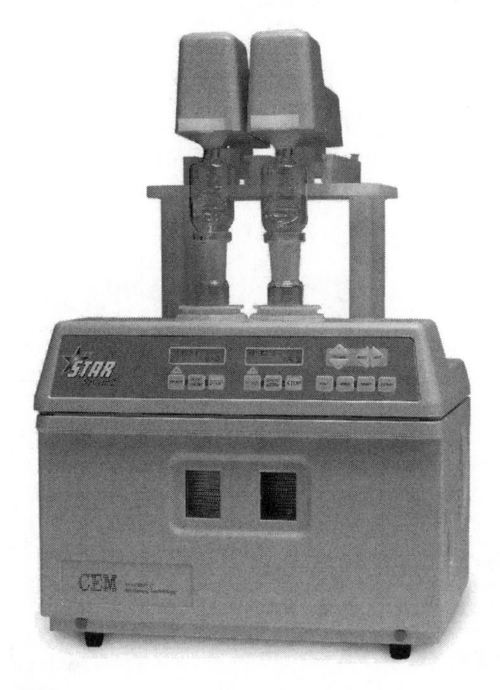

7-2 figure Microwave open-vessel system. (Courtesy of CEM Corporation, Matthews, NC.)

These instruments do not require the use of a fume hood, because a vapor containment system contains and neutralizes harmful fumes.

Generally, in an open-vessel microwave system, the sample is placed in a vessel and the vessel is set in a slot in the microwave system. Time, temperature, and reagent addition parameters are then chosen. The unit is started, the acid is added, and the vapor containment system neutralizes the fumes from the reaction. Samples are typically processed much faster and more reproducibly than on a conventional hot plate. (Note that some open-vessel systems may be used for evaporation and acid concentration as well.)

7.2.4.2 Microwave Dry Ashing

Compared with conventional dry ashing in a muffle furnace that often takes many hours, **microwave muffle furnaces** (Fig. 7-3) can ash samples in minutes, decreasing analysis time by as much as 97%. Microwave muffle furnaces can reach temperatures of up to 1200°C. These systems may be programmed with various methods and to automatically warm up and cool down. In addition, they are equipped with exhaust systems that circulate the air in the cavity to help decrease ashing times. Some also have scrubber systems to neutralize any fumes. Any crucible that may be used in a conventional muffle furnace may be used in a microwave furnace, including those made of porcelain, platinum, quartz, and quartz fiber. Quartz fiber crucibles cool in seconds and are not breakable. Some systems can process up to 15 (25 ml) crucibles at a time.

Typically, in microwave dry ashing, a desiccated crucible is weighed and then sample is added and it is weighed again. The crucible then is placed in the microwave furnace, and the time and temperature parameters are set. A step-by-step (ramping) format may be used when programming the method. The system is started and the program is run to completion. The crucible then is carefully removed with tongs and reweighed. The sample then may be further analyzed, if necessary. Some tests call for acid to be added to a dry ashed sample, which is then digested for further analysis.

7-3 figure Microwave muffle furnace. (Courtesy of CEM Corporation, Matthews, NC.)

A comparative study (9) showed that dry ashing various plants for 40 min using a microwave system (CEM Corporation, Matthews NC) was similar to the 4-h time in a conventional muffle furnace. Twenty minutes was shown to be adequate for the plant material used except for Cu determinations, which needed 40 min to obtain similar results. Other comparative examples include dried egg yolks, which can be ashed in 20 min in a microwave system, but require 4 h in a conventional muffle furnace. It takes 16 h to ash lactose in a conventional muffle furnace, but only 35 min in a microwave furnace. Though microwave furnaces may not hold as many samples as a conventional furnace, their speed actually allows significantly more samples to be processed in the same amount of time. Also, microwave furnaces do not require fume hood space.

7.2.5 Other Ash Measurements

The following are several special ash measurements and their applications:

1. **Soluble and insoluble ash** (e.g., AOAC Method 900.02) – Applied to fruits.
2. **Ash insoluble in acid** – A measure of the surface contamination of fruits and vegetables and wheat and rice coatings; contaminants are generally silicates and remain insoluble in acid, except HBr.
3. **Alkalinity of ash** (e.g., AOAC Method 900.02, 940.26) – Ash of fruits and vegetable is alkaline; ash of meats and some cereals is acid.
4. **Sulfated ash** (AOAC Method 900.02, 950.77) – Applied to sugars, syrups, and color additives.

7.3 COMPARISON OF METHODS

Ash determination by dry ashing requires expensive equipment, especially if many samples are analyzed. The muffle furnace may have to be placed in a heat room along with drying ovens and it requires a 220-V outlet. It is important to make sure that large furnaces of that type are equipped with a double-pole, single-throw switch. Heating coils are generally exposed, and care must be taken when taking samples in and out with metal tongs. Desktop furnaces (110 V) are available for fewer samples. Wet ashing requires a hood (a special hood if perchloric acid is used), corrosive reagents, and constant operator attention. While wet oxidation causes little volatilization, dry ashing will result in the loss of volatile elements. The type of further elemental analyses will dictate the equipment. Some micro- and most volatile elements will require special equipment and procedures. Refer to Chaps. 12 and 24 for specific preparation procedures for elemental analyses. Both dry and wet ashing can be done using microwave systems that utilize relatively expensive instrumentation, but they greatly reduce the time for ashing and do not require use of a fume hood.

7.4 SUMMARY

The two major types of ashing, dry ashing and wet oxidation (ashing), can be done by conventional means or using microwave systems. The procedure of choice depends upon the use of ash following its determination, and limitations based on cost, time, and sample numbers. Conventional dry ashing is based upon incineration at high temperatures in a muffle furnace. Except for certain elements, the residue may be used for further specific mineral analyses. Wet ashing (oxidation) often is used as a preparation for specific elemental analysis by simultaneously dissolving minerals and oxidizing all organic material. Wet ashing conserves volatile element, but requires more operator time than dry ashing and is limited to a smaller number of samples. Dry and wet ashing using microwave technology reduces the time for analyses and requires little additional equipment (special fume hood) or space (heat room).

7.5 STUDY QUESTIONS

1. Identify four potential sources of error in the preparation of samples for ash analysis and describe a way to overcome each.
2. You are determining the total ash content of a product using the conventional dry ashing method. Your boss asks you to switch to a conventional wet ashing method because he/she has heard it takes less time than dry ashing.
   (a) Do you agree or disagree with your boss concerning the time issue, and why?
   (b) Not considering the time issues, why might you want to continue using dry ashing, *and* why might you change to wet ashing?
3. Your lab technician was to determine the ash content of buttermilk by conventional dry ashing. The technician weighed 5 g of buttermilk into one weighed platinum crucible, immediately put the crucible into the muffle furnace using a pair of all stainless steel tongs, and ashed the sample for 48 h at 800°C. The crucible was removed from the muffle furnace and set on a rack in the open until it was cool enough to reweigh. Itemize the instructions you should have given your technician before beginning, so there would not have been the mistakes made as described above.
4. How would you recommend to your technician to overcome the following problems that could arise in conventional dry ashing of various foods?
   (a) You seem to be getting volatilization of phosphorus, when you want to later determine the phosphorus content.
   (b) You are getting incomplete combustion of a product high in sugar after a typical dry ashing procedure (i.e., the ash is dark colored, not white or pale gray).
   (c) The typical procedure takes too long for your purpose. You need to speed up the procedure, but you do not want to use the standard wet ashing procedure.
   (d) You have reason to believe the compound you want to measure after dry ashing may be reacting with the porcelain crucibles being used.
   (e) You want to determine the iron content of some foods but cannot seem to get the iron solubilized after the dry ashing procedure.
5. Identify an advantage and disadvantage of using microwave wet digesters or microwave muffle furnaces compared with conventional units.

7.6 PRACTICE PROBLEMS

1. A grain was found to contain 11.5% moisture. A 5.2146-g sample was placed into a crucible (28.5053 g tare). The ashed crucible weighed 28.5939 g. Calculate the percentage ash on (a) an as-received (wet weight) basis and (b) a dry matter basis.
2. A vegetable (23.5000 g) was found to have 0.0940-g acid-insoluble ash. What is the percentage of acid-insoluble ash?
3. You wish to have at least 100 mg of ash from a cereal grain. Assuming 2.5% ash on average, how many grams of the grain should be weighed for ashing?
4. You wish to have a coefficient of variation (CV) below 5% with your ash analyses. The following ash data are obtained: 2.15%, 2.12%, 2.07%. Are these data acceptable, and what is the CV?
5. The following data were obtained on a sample of hamburger: sample wt, 2.034 g; wt after drying, 1.0781 g; wt after ether extraction, 0.4679 g; and wt of ash, 0.0233 g. What is the percentage ash on (a) a wet weight basis and (b) a fat-free basis?

Answers

1. (a) 1.70%, (b) 1.92%
   Calculate ash from sample:

Crucible + ash:	28.5939 g
Tared crucible:	28.5053 g
Ash:	0.0886 g

(a) Calculate for ash on a wet weight basis (a):

$$\frac{0.0886\,\text{g ash}}{5.2146\,\text{g sample}} \times 100\% = 1.70\%\ \text{or}\ 1.7\%$$

(b) Calculate for ash on a dry weight basis (b):

$$0.0886\,\text{g ash} \div \left[5.2146\,\text{g sample} \times \left(\frac{100\% - 11.5\%}{100\%}\ \text{dry matter coeff}\right)\right] \times 100\%$$

$$= 1.92\%$$

or

$$5.214\,\text{g sample} \times \frac{11.5\,\text{g water}}{100\,\text{g sample}} = 0.5997\,\text{g water}$$

$$5.214\,\text{g sample} - 0.5997\,\text{g water} = 4.6149\,\text{g sample dry wt}$$

$$\frac{0.0886\,\text{g ash}}{4.6149\,\text{g dry wt sample}} \times 100\% = 1.92\%$$

2. 0.4%

Calculate % insoluble ash:

$$\frac{0.0940\,\text{g acid insoluble ash}}{23.5\,\text{g sample}} \times 100\% = 0.4\%$$

3. 4 g

100 mg = 0.1 g ash
2.5% = 2.5 g ash/100 g sample

$$\frac{2.5\,\text{g ash}}{100\,\text{g sample}} = \frac{0.1\,\text{g ash}}{x}$$

$$2.5x = 10$$

$$x = 4\,\text{g sample}$$

4. Yes, 1.9%

Calculate the mean:

$$\frac{2.15 + 2.12\% + 2.07\%}{3} = 2.11\%$$

Calculation of mean and standard deviation was done using Excel:

1. 2.15%
2. 2.12%
3. 2.07%

Average = 2.11%
Std. deviation = 0.0404

$$\text{Coefficient of variation (CV)} = \frac{\text{SD}}{x} \times 100\%$$

$$\text{CV} = \frac{0.0404}{2.11} \times 100\% = 1.91\%$$

It is within the 5% level for CV? YES.

5. (a) 1.1%, (b) 1.64%

Sample wet wt: 2.034 g
Sample dry wt: 1.0781 g
Wt after extraction: 0.4679 g
Wt of ash: 0.0233 g

(a) Calculate for wet weight basis:

$$\frac{0.0233\,\text{g ash}}{2.034\,\text{g sample}} \times 100\% = 1.15\%$$

(b) Calculate for fat-free basis:

$$2.034\ \text{g wet sample} - 1.0781\ \text{g solids} = 0.9559\ \text{g water this is 47\% moisture)}$$

$$1.0781\ \text{g solids dry wt} - 0.4679\ \text{g solids after extraction} = 0.6102\ \text{g fat}$$

$$2.034\ \text{g wet sample} - 0.6102\ \text{g fat} = 1.4238\ \text{g wet sample wt without fat}$$

$$\frac{0.0233\ \text{g ash}}{(1.4238\ \text{g wet sample wt without fat})} \times 100\%$$

$$= 1.64\%\ \text{ash, fat-free basis}$$

7.7 ACKNOWLEDGMENTS

The author of this chapter wishes to acknowledge the contributions of Dr. Leniel H. Harbers (Emeritus Professor, Kansas State University) for previous editions of this chapter. Also acknowledged in the preparation of this chapter is the assistance of Dr. John Budin (Silliker Laboratories, Chicago Heights, IL) and Ms. Michelle Horn (CEM Corporation, Matthews, NC).

7.8 REFERENCES

1. Analytical Methods Committee (1960) Methods for the destruction of organic matter. Analyst 85:643–656. This report gives a number of methods for wet and dry combustion and their applications, advantages, disadvantages, and hazards
2. AOAC International (2007) Official methods of analysis, 18th edn., 2005; Current through revision 2, 2007 (On-line). AOAC International, Gaithersburg, MD. This contains the official methods for many specific food ingredients. It may be difficult for the beginning student to follow
3. Aurand LW, Woods AE, Wells MR (1987) Food composition and analysis. Van Nostrand Reinhold, New York. The chapters that deal with ash are divided by foodstuffs. General dry procedures are discussed under each major heading

4. Bakkali K, Martos NR, Souhail B, Ballesteros E (2009) Characterization of trace metals in vegetables by graphite furnace atomic absorption spectrometry after closed vessel microwave digestion. Food Chem 116(2):590–594
5. Mesko MF, De Moraes DP, Barin JS, Dressler VL, Knappet G (2006) Digestion of biological materials using the microwave-assisted sample combustion technique. Microchem J 82:183–188
6. Neggers YH, Lane RH (1995) Minerals, ch. 8. In: Jeon IJ, Ikins WG (eds) Analyzing food for nutrition labeling and hazardous contaminants. Marcel Dekker, New York. This chapter compares wet and dry ashing and summarizes in tables the following: losses of specific elements during dry ashing; acids used in wet oxidation related to applications; AOAC methods for specific elements related to food applications
7. Pomeranz Y, Meloan C (1994) Food analysis: theory and practice, 3rd edn. Chapman & Hall, New York. Chapter 35 on ash and minerals gives an excellent narrative on ashing methods and is easy reading for a student in food chemistry. A good reference list of specific mineral losses is given at the end of the chapter. No stepwise procedures are given
8. Smith GF (1953) The wet ashing of organic matter employing hot concentrated perchloric acid. The liquid fire reaction. Anal Chim Acta 8:397–421. The treatise gives an in-depth review of wet ashing with perchloric acid. Tables on reaction times with foodstuffs and color reactions are informative. It is easy for the food scientist to understand
9. Wehr HM, Frank JF (eds) (2004) Standard methods for the examination of dairy products, 17th edn. American Public Health Association, Washington, DC. This text gives detailed analytical procedures for ashing dairy products
10. Wooster HA (1956) Nutritional data, 3rd edn. H.J. Heinz, Pittsburgh, PA
11. Zhang H, Dotson P (1994) Use of microwave muffle furnace for dry ashing plant tissue samples. Commun Soil Sci Plant Anal 25(9/10):1321–1327

Fat Analysis

David B. Min
Department of Food Science and Technology, The Ohio State University, Columbus, OH 43210, USA
min2@osu.edu

and

Wayne C. Ellefson
Nutritional Chemistry and Food Safety, Covance Laboratories, 3301 Kinsman Boulevard, Madison, WI 53714, USA
Wayne.Ellefson@covance.com

8.1 Introduction 119
8.1.1 Definitions 119
8.1.2 General Classification 119
8.1.2.1 Simple Lipids 119
8.1.2.2 Compound Lipids 119
8.1.2.3 Derived Lipids 119
8.1.3 Content of Lipids in Foods 119
8.1.4 Importance of Analysis 120
8.2 General Considerations 120
8.3 Solvent Extraction Methods 120
8.3.1 Sample Preparation 121
8.3.1.1 Predrying Sample 121

S.S. Nielsen, *Food Analysis*, Food Science Texts Series, DOI 10.1007/978-1-4419-1478-1_8,

8.3.1.2 Particle Size Reduction 121
8.3.1.3 Acid Hydrolysis 121
8.3.2 Solvent Selection 122
8.3.3 Continuous Solvent Extraction Method: Goldfish Method 122
8.3.3.1 Principle and Characteristics 122
8.3.3.2 Procedure 122
8.3.3.3 Calculations 122
8.3.4 Semicontinuous Solvent Extraction Method: Soxhlet Method 123
8.3.4.1 Principle and Characteristics 123
8.3.4.2 Preparation of Sample 123
8.3.4.3 Procedure 123
8.3.4.4 Calculation 123
8.3.5 Discontinuous Solvent Extraction Methods 123
8.3.5.1 Mojonnier Method 123
8.3.5.1.1 Principle and Characteristics 123
8.3.5.1.2 Procedure: Milk Fat Method (AOAC Method 989.05) 124
8.3.5.2 Chloroform–Methanol Procedure 124
8.3.5.2.1 Principle and Characteristics 124
8.3.5.2.2 Procedure (Modified Folch Extraction) (9, 11) 125
8.3.6 Total Fat by GC for Nutrition Labeling (AOAC Method 996.06) 125
8.3.6.1 Principle 125
8.3.6.2 Sample Preparation Procedures 125
8.3.6.3 Chromatographic Conditions 127
8.3.6.4 Calculations 127
8.4 Nonsolvent Wet Extraction Methods 127
8.4.1 Babcock Method for Milk Fat (AOAC Method 989.04 and 989.10) 127
8.4.1.1 Principle 127
8.4.1.2 Procedure 127
8.4.1.3 Applications 130
8.4.2 Gerber Method for Milk Fat 130
8.4.2.1 Principle 130
8.4.2.2 Procedure 130
8.4.2.3 Applications 130
8.5 Instrumental Methods 130
8.5.1 Infrared Method 130
8.5.2 Specific Gravity (Foss-Let Method) 130
8.5.3 Nuclear Magnetic Resonance 131
8.6 Comparison of Methods 131
8.7 Summary 131
8.8 Study Questions 131
8.9 Practice Problems 132
8.10 References 132

8.1 INTRODUCTION

8.1.1 Definitions

Lipids, proteins, and carbohydrates constitute the principal structural components of foods. Lipids are a group of substances that, in general, are soluble in ether, chloroform, or other organic solvents but are sparingly soluble in water. However, there exists no clear scientific definition of a lipid, primarily due to the water solubility of certain molecules that fall within one of the variable categories of food lipids (1). Some lipids, such as triacylglycerols, are very hydrophobic. Other lipids, such as di- and monoacylglycerols, have both hydrophobic and hydrophilic moieties in their molecules and are soluble in relatively polar solvents (2). Short-chain fatty acids such as C1–C4 are completely miscible in water and insoluble in nonpolar solvents (1). The most widely accepted definition is based on solubility as previously stated. While most macromolecules are characterized by common structural features, the designation of "lipid" being defined by solubility characteristics is unique to lipids (2). Lipids comprise a broad group of substances that have some common properties and compositional similarities (3). Triacylglycerols are fats and oils that represent the most prevalent category of the group of compounds known as lipids. The terms lipids, fats, and oils are often used interchangeably. The term "lipid" commonly refers to the broad, total collection of food molecules that meet the definition previously stated. Fats generally refer to those lipids that are solid at room temperature and oils generally refer to those lipids that are liquid at room temperature. While there may not be an exact scientific definition, the US Food and Drug Administration (FDA) has established a regulatory definition for nutrition labeling purposes. The FDA has defined total fat as the sum of fatty acids from C4 to C24, calculated as triglycerides. This definition provides a clear path for resolution of any nutrition labeling disputes.

8.1.2 General Classification

The general classification of lipids that follows is useful to differentiate lipids in foods (3).

8.1.2.1 Simple Lipids

Ester of fatty acids with alcohol:

- **Fats:** Esters of fatty acids with glycerol – triacylglycerols
- **Waxes:** Esters of fatty acids with long-chain alcohols other than glycerols (e.g., myricyl palmitate, cetyl palmitate, vitamin A esters, and vitamin D esters)

8.1.2.2 Compound Lipids

Compounds containing groups in addition to an ester of a fatty acid with an alcohol:

- **Phospholipids:** Glycerol esters of fatty acids, phosphoric acids, and other groups containing nitrogen (e.g., phosphatidyl choline, phosphatidyl serine, phosphatidyl ethanolamine, and phosphatidyl inositol)
- **Cerebrosides:** Compounds containing fatty acids, a carbohydrate, and a nitrogen moiety (e.g., galactocerebroside and glucocerebroside)
- **Sphingolipids:** Compounds containing fatty acids, a nitrogen moiety, and phosphoryl group (e.g., sphingomyelins)

8.1.2.3 Derived Lipids

Derived lipids are substances derived from neutral lipids or compound lipids. They have the general properties of lipids – examples are fatty acids, long-chain alcohols, sterols, fat-soluble vitamins, and hydrocarbons.

8.1.3 Content of Lipids in Foods

Foods may contain any or all types of the lipid compounds previously mentioned. The lipid content in bovine milk (Table 8-1) illustrates the complexity and variability of lipids in a food system, having lipids that differ in polarity and concentrations.

Foods contain many types of lipids, but those which tend to be of greatest importance are the triacylglycerols and the phospholipids. **Liquid triacylglycerols** at room temperature are referred to as **oils,**

8-1 table Lipids of Bovine Milk

Kinds of Lipids	*Percent of Total Lipids*
Triacylglycerols	97–99
Diacylglycerols	0.28–0.59
Monoacylglycerols	0.016–0.038
Phospholipids	0.2–1.0
Sterols	0.25–0.40
Squalene	Trace
Free fatty acids	0.10–0.44
Waxes	Trace
Vitamin A	(7–8.5 μg/g)
Carotenoids	(8–10 μg/g)
Vitamin D	Trace
Vitamin E	(2–5 μg/g)
Vitamin K	Trace

Adapted from (4) with permission of S. Patton and (5) *Principles of Dairy Chemistry.* Jenness R. and Patton S. Copyright ©1959, John Wiley & Sons, Inc with permission.

such as soybean oil and olive oil, and are generally of plant origin. **Solid triacylglycerols** at room temperature are termed as **fats**. Lard and tallow are examples of fats, which are generally from animals. The term *fat* is applicable to all triacylglycerols whether they are normally solid or liquid at ambient temperatures. Table 8-2 shows the wide range of lipid content in different foods.

8.1.4 Importance of Analysis

An accurate and precise quantitative and qualitative analysis of lipids in foods is important for accurate nutritional labeling, determination of whether the food meets the standard of identity, and to ensure that the product meets manufacturing specifications. Inaccuracies in analysis may prove costly for manufacturers and could result in a product of undesirable quality and functionality.

8.2 GENERAL CONSIDERATIONS

By definition, lipids are soluble in organic solvents and insoluble in water. Therefore, water insolubility is the essential analytical property used as the basis for the separation of lipids from proteins, water, and carbohydrates in foods. Glycolipids are soluble in alcohols and have a low solubility in hexane. In contrast, triacylglycerols are soluble in hexane and petroleum ether, which are nonpolar solvents. The wide range of relative hydrophobicity of different lipids makes the selection of a single universal solvent impossible for lipid extraction of foods. Some lipids in foods are components of complex lipoproteins and liposaccharides; therefore, successful extraction requires that bonds between lipids and proteins or carbohydrates be broken so that the lipids can be freed and solubilized in the extracting organic solvents.

8.3 SOLVENT EXTRACTION METHODS

The total lipid content of a food is commonly determined by organic solvent extraction methods or by alkaline or acid hydrolysis followed by Mojonnier extraction. For multicomponent food products, acid hydrolysis is often the method of choice. Both acid hydrolysis and alkaline hydrolysis methods can be performed using Mojonnier extraction equipment. The use of acid hydrolysis eliminates some of the matrix effects that may be exhibited by simple solvent extraction methods. The accuracy of direct solvent extraction methods (i.e., without prior acid or alkaline hydrolysis) greatly depends on the solubility of the lipids in the solvent used and the ability to separate the lipids from complexes with other macromolecules. The lipid content of a food determined by extraction with one solvent may be quite different from the

Fat Content of Selected Foods

Food Item	*Percent Fat (Wet Weight Basis)*
Cereals, bread, and pasta	
Rice, white, long-grain, regular, raw, enriched	0.7
Sorghum	3.3
Wheat, soft white	2.0
Rye	2.5
Wheat germ, crude	9.7
Rye bread	3.3
Cracked-wheat bread	3.9
Macaroni, dry, enriched	1.5
Dairy products	
Milk, reduced fat, fluid, 2%	2.0
Skim milk, fluid	0.2
Cheddar cheese	33.1
Yogurt, plain, whole milk	3.2
Fats and oils	
Lard, shortening, oils	100.0
Butter, with salt	81.1
Margarine, regular, hard, soybean	80.5
Salad dressing	
Italian, commercial, regular	28.3
Thousand Island, commercial, regular	35.1
French, commercial, regular	44.8
Mayonnaise, soybean oil, with salt	79.4
Fruits and vegetables	
Apples, raw, with skin	0.2
Oranges, raw, all commercial varieties	0.1
Blackberries, raw	0.5
Avocados, raw, all commercial varieties	14.7
Asparagus, raw	0.1
Lima beans, immature seeds, raw	0.9
Sweet corn, yellow, raw	1.2
Legumes	
Soybeans, mature seeds, raw	19.9
Black beans, mature seed, raw	1.4
Meat, poultry, and fish	
Beef, flank, separable lean and fat	5.0
Chicken, broilers or fryers, breast meat only	1.2
Bacon, pork, cured, raw	45.0
Pork, fresh, loin, whole, raw	12.6
Finfish, halibut, Atlantic and Pacific, raw	2.3
Finfish, cod, Atlantic, raw	0.7
Nuts	
Coconut meat, raw	33.5
Almonds, dried, unblanched, dry roasted	52.8
Walnuts, black, dried	56.6
Egg, whole, raw, fresh	10.0

From US Department of Agriculture, Agricultural Research Service (2009) USDA National Nutrient Database for Standard Reference. Release 22. Nutrient Data Laboratory Home Page, http://www.ars.usda.gov/ba/bhnrc/ndl

content determined with another solvent of different polarity. In addition to solvent extraction methods, there are nonsolvent wet extraction methods and several instrumental methods that utilize the physical and chemical properties of lipids in foods for fat content determination. For nutrition labeling purposes, total fat is most commonly determined by gas chromatography (GC) analysis.

Many of the methods cited in this chapter are official methods of AOAC International. Refer to these methods and other original references cited for detailed instructions of procedures. There are many methods available for the determination of lipid content. This chapter will focus on some of the primary methods in common use.

8.3.1 Sample Preparation

The validity of the fat analysis of a food depends on proper sampling and preservation of the sample before the analysis (see also Chap. 5). An ideal sample should be as close as possible in all of its intrinsic properties to the material from which it is taken. However, a sample is considered satisfactory if the properties under investigation correspond to those of the bulk material within the limits of the test (7).

The sample preparation for lipid analysis depends on the type of food and the type and nature of lipids in the food (8). The extraction method for lipids in liquid milk is generally different from that for lipids in solid soybeans. To analyze the lipids in foods effectively, knowledge of the structure, the chemistry, and the occurrence of the principal lipid classes and their constituents is necessary. Therefore, there is no single standard method for the extraction of all kinds of lipids in different foods. For the best results, sample preparation should be carried out under an inert atmosphere of nitrogen at low temperature to minimize chemical reactions such as lipid oxidation.

Several preparatory steps are common in lipid analysis. These act to aid in extraction by removal of water, reduction of particle size, or separation of the lipid from bound proteins and/or carbohydrates.

8.3.1.1 Predrying Sample

Lipids cannot be effectively extracted with ethyl ether from moist food because the solvent cannot easily penetrate the moist food tissues due to the hydrophobicity of the solvents used or the hydroscopic nature of the solvents. The ether, which is hygroscopic, becomes saturated with water and inefficient for lipid extraction. Drying the sample at elevated temperatures is undesirable because some lipids become bound to proteins and carbohydrates, and bound lipids are not easily extracted with organic solvents. Vacuum oven drying at low temperature or lyophilization increases the surface area of the sample for better lipid extraction. Predrying makes the sample easier to grind for better extraction, breaks fat–water emulsions to make fats dissolve easily in the organic solvent, and helps to free fat from the tissues of foods (7).

8.3.1.2 Particle Size Reduction

The extraction efficiency of lipids from dried foods depends on particle size; therefore, adequate grinding is very important. The classical method of determining fat in oilseeds involves the extraction of the ground seeds with selected solvent after repeated grinding at low temperature to minimize lipid oxidation. For better extraction, the sample and solvent are mixed in a high-speed comminuting device such as a blender. It can be difficult to extract lipids from whole soybeans because of the limited porosity of the soybean hull and its sensitivity to dehydrating agents. The lipid extraction from soybeans is easily accomplished if the beans are broken mechanically by grinding. Extraction of fat from finished products can be a challenge, based on the ingredients (e.g., energy bars with nuts, caramel, protein, granola, soybean oil). Such products may best be ground after freezing with liquid nitrogen.

8.3.1.3 Acid Hydrolysis

A significant portion of the lipids in foods such as dairy, bread, flour, and animal products is bound to proteins and carbohydrates, and direct extraction with nonpolar solvents is inefficient. Such foods must be prepared for lipid extraction by acid hydrolysis. This includes a significant percentage of finished food products. Table 8-3 shows the inaccuracy that can occur if samples are not prepared by acid hydrolysis. Acid hydrolysis can break both covalently and ionically bound lipids into easily extractable lipid forms. The sample can be predigested by refluxing for 1 h

8-3 table **Effects of Acid Digestion on Fat Extraction from Foods**

	Percent Fat	
	Acid Hydrolysis	*No Acid Hydrolysis*
Dried egg	42.39	36.74
Yeast	6.35	3.74
Flour	1.73	1.20
Noodles	3.77–4.84	2.1–3.91
Semolina	1.86–1.93	1.1–1.37

Adapted from (6), p. 154, with permission.

with 3 N hydrochloric acid. Ethanol and solid hexametaphosphate may be added to facilitate separation of lipids from other components before food lipids are extracted with solvents (6, 7). For example, the acid hydrolysis of two eggs requires 10 ml of HCl and heating in a water bath at 65°C for 15–25 min or until the solution is clear (6).

8.3.2 Solvent Selection

Ideal solvents for fat extraction should have a high solvent power for lipids and low or no solvent power for proteins, amino acids, and carbohydrates. They should evaporate readily and leave no residue, have a relatively low boiling point, and be nonflammable and nontoxic in both liquid and vapor states. The ideal solvent should penetrate sample particles readily, be in single component form to avoid fractionation, and be inexpensive and nonhygroscopic (6,7). It is difficult to find an ideal fat solvent to meet all of these requirements. Ethyl ether and petroleum ether are the most commonly used solvents, but pentane and hexane are used to extract oil from soybeans.

Ethyl ether has a boiling point of 34.6°C and is a better solvent for fat than petroleum ether. It is generally expensive compared to other solvents, has a greater danger of explosion and fire hazards, is hygroscopic, and forms peroxides (6). **Petroleum ether** is the low boiling point fraction of petroleum and is composed mainly of pentane and hexane. It has a boiling point of 35–38°C and is more hydrophobic than ethyl ether. It is selective for more hydrophobic lipids, cheaper, less hygroscopic, and less flammable than ethyl ether. The detailed properties of petroleum ether for fat extraction are described in AOAC Method 945.16 (8).

A combination of two or three solvents is frequently used. The solvents should be purified and peroxide free and the proper solvent-to-solute ratio must be used to obtain the best extraction of lipids from foods (7).

8.3.3 Continuous Solvent Extraction Method: Goldfish Method

8.3.3.1 Principle and Characteristics

For continuous solvent extraction, solvent from a boiling flask continuously flows over the sample held in a ceramic thimble. Fat content is measured by weight loss of the sample or by weight of the fat removed.

The continuous methods give faster and more efficient extraction than semicontinuous extraction methods. However, they may cause channeling which results in incomplete extraction. The Goldfish (as well as the Wiley and Underwriters) tests are examples of continuous lipid extraction methods (6,7).

8.3.3.2 Procedure (See Fig. 8-1)

1. Weigh predried porous ceramic extraction thimble. Place vacuum oven dried sample in thimble and weigh again. (Sample could instead be combined with sand in thimble and then dried.)
2. Weigh predried extraction beaker.
3. Place ceramic extraction thimble into glass holding tube and then up into condenser of apparatus.
4. Place anhydrous ethyl ether (or petroleum ether) in extraction beaker and put beaker on heater of apparatus.
5. Extract for 4 h.
6. Lower the heater and let sample cool.
7. Remove the extraction beaker and let air dry overnight, then at 100°C for 30 min. Cool beaker in desiccator and weigh.

8.3.3.3 Calculations

$$\text{Weight of fat in sample} = (\text{beaker} + \text{fat}) - \text{beaker} \quad [1]$$

$$\%\ \text{Fat on dry weight basis} = (\text{g of fat in sample/g of dried sample}) \times 100 \quad [2]$$

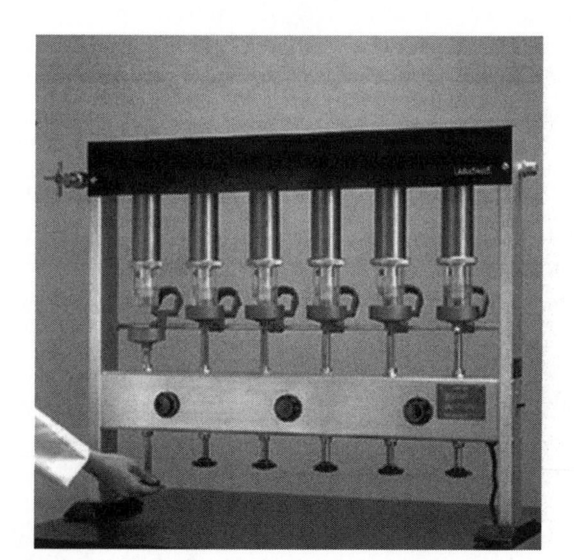

8-1 figure Goldfish fat extractor. (Courtesy of Labconco Corp., Kansas City, MO.) (http://www.labconco.com/_scripts/EditItem.asp?ItemID=487)

8.3.4 Semicontinuous Solvent Extraction Method: Soxhlet Method

The Soxhlet method (AOAC Method 920.39C for Cereal Fat; AOAC Method 960.39 for Meat Fat) (8) is an example of the semicontinuous extraction method and is described below.

8.3.4.1 Principle and Characteristics

For semicontinuous solvent extraction, the solvent builds up in the extraction chamber for 5–10 min and completely surrounds the sample and then siphons back to the boiling flask. Fat content is measured by weight loss of the sample or by weight of the fat removed.

This method provides a soaking effect of the sample and does not cause channeling. However, this method requires more time than the continuous method. Instrumentation for a more rapid and automated version of the Soxhlet method is available (e.g., Soxtec™, FOSS in North America, Eden Prairie, MN) and is used for some quality control applications.

8.3.4.2 Preparation of Sample

If the sample contains more than 10% H_2O, dry the sample to constant weight at 95–100°C under pressure $\leq$ 100 mm Hg for about 5 h (AOAC Method 934.01).

8.3.4.3 Procedure (See Fig. 8-2)

1. Weigh, to the nearest mg, about 2 g of predried sample into a predried extraction thimble, with porosity permitting a rapid flow of ethyl ether. Cover sample in thimble with glass wool.
2. Weigh predried boiling flask.
3. Put anhydrous ether in boiling flask. *Note*: The anhydrous ether is prepared by washing commercial ethyl ether with two or three portions of H_2O, adding NaOH or KOH, and letting stand until most of H_2O is absorbed from the ether. Add small pieces of metallic Na and let hydrogen evolution cease (AOAC Method 920.39B). Petroleum ether may be used instead of anhydrous ether (AOAC Method 960.39).
4. Assemble boiling flask, Soxhlet flask, and condenser.
5. Extract in a Soxhlet extractor at a rate of five or six drops per second by condensation for about 4 h, or for 16 h at a rate of two or three drops per second by heating solvent in boiling flask.
6. Dry boiling flask with extracted fat in an air oven at 100°C for 30 min, cool in desiccator, and weigh.

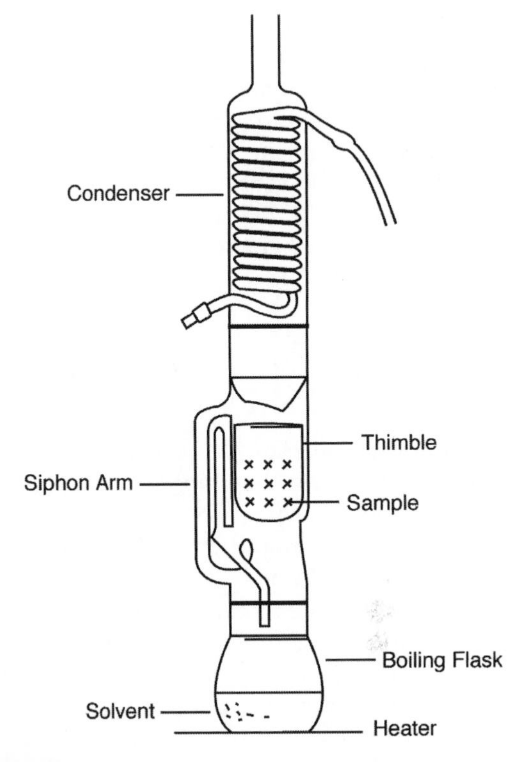

8-2 figure Soxhlet extraction apparatus.

8.3.4.4 Calculation

% Fat on dry weight basis

$$= (\text{g of fat in sample}/\text{g of dried sample}) \times 100 \quad [3]$$

8.3.5 Discontinuous Solvent Extraction Methods

8.3.5.1 Mojonnier Method

8.3.5.1.1 Principle and Characteristics Fat is extracted with a mixture of ethyl ether and petroleum ether in a Mojonnier flask, and the extracted fat is dried to a constant weight and expressed as percent fat by weight.

The Mojonnier test is an example of the discontinuous solvent extraction method and does not require removal of moisture from the sample. It can be applied to both liquid and solid samples. If petroleum ether is used to purify the extracted fat, this method is very similar to the **Roese-Gottlieb Method** (AOAC Method 905.02) in both principle and practice. The Mojonnier flasks (Fig. 8-3) are used not only for the Mojonnier and Roese-Gottlieb methods, but also to do the hydrolysis (acid, alkaline, or combination) prior to fat extraction and GC analysis to determine fat content and fatty acid profile (Sect. 8.3.6). (Note that sometimes the terms Mojonnier, Roese Gottlieb, and alkaline hydrolysis are used interchangeably.)

The Mojonnier method was developed for and is applied primarily to dairy foods (procedure as described below for milk fat), but is applicable to other foods. Specifically, methods for fat in flour (AOAC Method 922.06) and fat in pet food (AOAC Method 954.02) both involve an acid hydrolysis with HCl, followed by extraction with a combination of ethyl ether and petroleum ether as described in AOAC Method 989.05 below for milk fat.

8.3.5.1.2 Procedure: Milk Fat Method (AOAC Method 989.05)

1. **Preparation of Sample**. Bring the sample to about 20°C; mix to prepare a homogeneous sample by pouring back and forth between clean beakers. Promptly weigh or measure the test portion. If lumps of cream do not disperse, warm the sample in a water bath to about 38°C and keep mixing until it is homogeneous, using a "rubber policeman" if necessary to reincorporate the cream adhering to the container or stopper. When it can be done without interfering with dispersal of the fat, cool warmed samples to about 20°C before transferring the test portion.
2. **Procedure**
   (a) Weigh, to the nearest 0.1 mg, 10 g of milk into a Mojonnier fat extraction flask (Fig. 8-3).
   (b) Add 1.5 ml of NH_4OH and shake vigorously. Add 2 ml if the sample is sour. NH_4OH neutralizes the acidic sample and dissolves protein.
   (c) Add 10 ml of 95% ethanol and shake for 90 s. The alcohol prevents possible gel formation.
   (d) Add 25 ml of ethyl ether and shake for 90 s. The ether dissolves the lipid.
   (e) Cool if necessary, and add 25 ml of petroleum ether and shake for 90 s. The petroleum ether removes moisture from the ethyl ether extract and dissolves more nonpolar lipid.
   (f) Centrifuge for 30 s at 600 rpm.
   (g) Decant ether solution from the Mojonnier flask into the previously weighed Mojonnier fat dish.
   (h) Perform second and third extractions in the same manner as for the first extraction described previously (ethanol, ethyl ether, petroleum ether, centrifugation, decant).
   (i) Evaporate the solvent in the dish on the electric hot plate at ≤100°C in a hood.
   (j) Dry the dish and fat to a constant weight in a forced air oven at 100°C ± 1°C.
   (k) Cool the dish to room temperature and weigh.
3. **Calculations**

$$\% \text{ Fat} = 100 \times \{[(\text{wt dish} + \text{fat}) - (\text{wt dish})] - (\text{avg wt blank residue})\}/\text{wt sample} \quad [4]$$

A pair of reagent blanks must be prepared every day. For reagent blank determination, use 10 ml of distilled water instead of milk sample. The reagent blank should be <0.002 g. Duplicate analyses should be <0.03% fat.

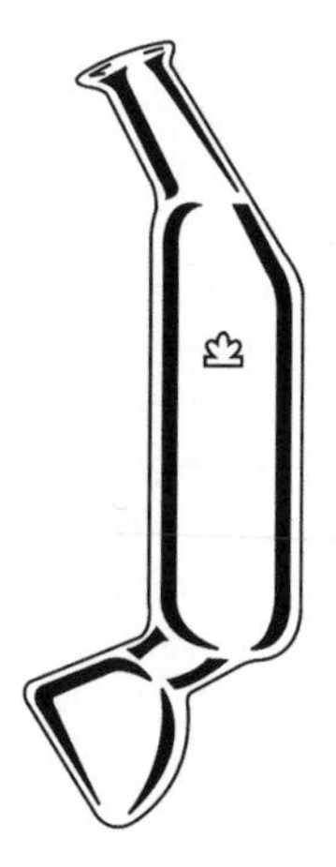

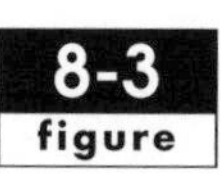

8-3 figure Mojonnier fat extraction flask. (Courtesy of Kontes Glass Co., Vineland, NJ.)

8.3.5.2 Chloroform–Methanol Procedure

8.3.5.2.1 Principle and Characteristics The combination of chloroform and methanol has been used commonly to extract lipids. The "Folch extraction" (9) applied to small samples, and the "Bligh and Dyer extraction" (10) applied to large samples of high moisture content, both utilize this combination of solvents to recover lipids from foods. These methods have been reviewed and procedures modified by Christie (11) and others. The Bligh and Dyer procedure (10) is a modification of the Folch extraction (9), designed for more efficient solvent usage for low-fat samples. The Christie modification (11) of these former methods replaced water with 0.88% potassium chloride aqueous solution to create two phases.

In both the modified Folch extraction and Bligh and Dyer procedure, food samples are mixed/homogenized in a chloroform–methanol solution, and the homogenized mixture is filtered into a collection tube. A 0.88% potassium chloride aqueous solution is added to the chloroform–methanol mixture containing the extracted fats. This causes the solution to break into two phases: the aqueous phase (top) and the chloroform phase containing the lipid (bottom). The phases are further separated in a separatory funnel or by centrifugation. After evaporation of the chloroform, the fat can be quantitated by weight.

The various methanol–chloroform extraction procedures are rapid, well suited to low-fat samples, and can be used to generate lipid samples for subsequent fatty acid compositional analysis. The procedure has been more applied to basic commodities, rather than to finished product samples. For consistent results, the procedures must be followed carefully, including the ratio of chloroform and methanol. A cautionary note is that chloroform and methanol are highly toxic, so the extraction procedure must be done in well-ventilated areas.

8.3.5.2.2 Procedure (Modified Folch Extraction) (9, 11)

1. Weigh 1 g accurately and homogenize 1 min in 10 ml of methanol.
2. Add 20 ml of chloroform and homogenize for 2 min.
3. Filter on a Buchner funnel and reextract cake and filter paper with 20 ml of chloroform plus 10 ml methanol.
4. Filter and combine two filtrates. Measure volume. Calculate 25% of the total volume and add this amount of 0.88% KCl. Let separate in a separatory funnel.
5. Remove the lower lipid-containing layer and measure the volume. Calculate 25% of the volume and add this volume of water to wash the sample.
6. Remove water from the sample by adding sodium sulfate. Filter to separate sample from hydrated sodium sulfate. Use chloroform to wash the sodium sulfate on the filter paper.
7. Place the chloroform–lipid solution in a weighed round bottom flask and remove chloroform from the lipid using a rotary evaporator with water temperature below 50°C. (If the extract is cloudy, add 20 ml of acetone and 20 ml of chloroform and reevaporate.) Weigh the flask to determine the % lipid in the sample.

8.3.6 Total Fat by GC for Nutrition Labeling (AOAC Method 996.06)

8.3.6.1 Principle

AOAC International (8) gives an excellent description of the principle of AOAC Method 996.06: "Fat and fatty acids are extracted from food by hydrolytic methods (acidic hydrolysis for most products, alkaline hydrolysis for dairy products, and combination for cheese). Pyrogallic acid is added to minimize oxidative degradation of fatty acids during analysis. Triglyceride, triundecanoin ($C_{11:0}$), is added as internal standard. Fat is extracted into ether, then methylated to *fatty acid methyl esters* (*FAMEs*) using BF_3 (boron trifluoride) in methanol. FAMEs are quantitatively measured by capillary gas chromatography (GC) against $C_{11:0}$ internal standard. Total fat is calculated as sum of individual fatty acids expressed as triglyceride equivalents. Saturated and monounsaturated fats are calculated as sum of respective fatty acids. Monounsaturated fat includes only *cis* form." Trans fat can be quantified utilizing this method in conjunction with identification criteria established by the American Association of Oil Chemists (AOCS Method Ce 1h-05) (12) and Golay et al. (13).

8.3.6.2 Sample Preparation Procedures

Samples are methylated by either a standard or alternative method (Fig. 8-4) to form the FAMEs prior to GC analysis. In both methods, the sample is combined with a specified amount of the **internal standard** triglyceride, triundecanoin.

Most foods and 100% fat products can be subjected to the standard sample preparation procedure. Before using the standard method, unless the sample is 100% fat, it is subjected to the appropriate fat extraction procedure (including, but not limited to, acid hydrolysis, alkaline hydrolysis, and Soxhlet extraction). Before any samples are subjected to acid or alkaline hydrolysis, 1.0 ml of triundecanoin internal standard solution is added to the sample. In the case of a sample extracted by the Soxhlet method, or if the sample is 100% fat, 1.0 ml of the internal standard is added directly to the flask used as the sample is methylated. Extracted samples can be stored refrigerated (5 ± 3°C), with hexane added, for up to 2 weeks if methylation will not be done immediately. (The hexane is dried off on a steam bath with a gentle stream of nitrogen before the sample is weighed and methylation occurs.)

The alternative sample preparation method is efficient for infant formulas and other similar matrices containing short chain fatty acids (e.g., butyric acid) and long chain fatty acids (e.g., eicosapentaenoic or docosahexaenoic acids) and for products containing microencapsulated fatty acids. This alternative procedure provides for more thorough recovery of short chain fatty acids, such as butyric acid, and eliminates the preliminary fat extraction step, thereby saving time.

Standard Sample Preparation Procedure:

1. For oil samples (100% lipid) or samples that were extracted by the Soxhlet method, accurately weigh a portion of the lipid (usually 0.15–0.2 g) into a 125-ml, or 50-ml, round bottom flask, containing the internal standard. For low lipid amounts, the lipid may be transferred into the flask using hexane. The weight of lipid is the difference in the weight of the empty flask and the weight of the flask after the hexane is evaporated on a steam both under a gentle stream of nitrogen.
2. Add 4–5 ml of 0.5 *N* sodium hydroxide (NaOH) in methanol (MeOH), then a few boiling chips.
3. Attach a water-cooled condenser and reflux on a boiling steam bath for ~15 min.
4. Add 5 ml of boron trifluoride (BF3): methanol (MeOH) (14%, w/v) through the top of the condenser.
5. Reflux for ~5 min.
6. Add an appropriate amount of heptane, usually 10 ml.
7. Reflux for ~1 min.
8. Raise the round bottom flask still attached to the condenser above the steam bath and let it cool for ~15 min. Disconnect the condensers.
9. Add 2–3 ml of saturated sodium chloride (NaCl) solution to the flask and stopper the flask.
10. Let stand until it reaches room temperature (~15 min).
11. Add saturated NaCl solution so that the heptane layer will float in the neck of the flask.
12. Place a small scoop of sodium sulfate in a vial.
13. Transfer the heptane layer containing the methyl esters into the vial and place the screw cap on the vial. (Sample is stable for 2 wk)
14. Inject an appropriate amount of the heptane layer into the GC system.

Alternative Sample Preparation Procedure:

Note: Samples with butryric and caproic acids (typically found in matrices containing dairy products) are treated differently after Step 8 than those without. Step 9 provides direction concerning this.

1. For a final volume of 5 ml (usually samples with less than 10% fat, or known to have low concentrations of the fatty acids of interest), add 1.0 ml of the tridecanoic tritridecanoin internal standard to a 50-ml round-bottom flask. If a final volume of 10 ml is used (usually samples known to contain more than 10% fat), add 2.0 ml of the tritridecanoic tritridecanoin internal standard. Evaporate the solvent on a steambath under a gentle stream of nitrogen until dry. Let the flask cool to room temperature.
2. Accurately weigh an appropriate amount of sample (depending on the amount of lipid in the sample) into the round-bottom flask.
3. Add approximately 6 ml of the 0.5 *N* NaOH in MeOH solution and a magnetic stir bar to the round-bottom flask.
4. Attach the flask to a condenser and reflux the solution for 1–2 hr on a heating stirplate.
5. Dry moist samples or powder samples under a gentle nitrogen stream without heat to avoid bumping. Liquid samples may be dried on a steam bath with a gentle nitrogen stream.
6. Place the flask in an oven set to maintain 100°C to dry completely for (usually 20–30 min) (cover samples that splatter in oven). Cool to room temperature. At this point, the samples may be covered and left at room temperature overnight (6–12 hr).
7. Add approximately 5 ml MeOH and a few boiling chips to the round-bottom flask. Connect the flask to condensers over a steambath.
8. Bring the samples to a boil and add approximately 5 ml of BF_3 in MeOH (14% w/v) through the condenser.
9. For samples not containing butyric and caproic acids skip to Step 11; samples containing butyric and caproic acid continue to Step 10.
10. Methylation of fatty acids including butyric and caproic acids.
    - 10.1. Continue to reflux the sample for ~1 min.
    - 10.2. Raise the sample above the steam bath and add an appropriate amount of methylene chloride [either 5.0 or 10.0 ml depending on the amount (1 or 2 ml, respectively) of tridecanoic tritridecanoin internal standard added in Step 1].
    - 10.3. Place the flask in a cup containing ice chips to quickly cool the sample, while still leaving it attached to the condenser and let sample cool.
    - 10.4. Remove flask from condenser and add 5–20 ml saturated NaCl solution. Stopper and shake.
    - 10.5. Transfer contents of flask to a 50-ml glass centrifuge tube and centrifuge to achieve separation of the methylene chloride.
    - 10.6. Continue as specified in Steps 11–14 of the Standard Sample Preparation section, replacing heptane with methylene chloride. The methylene chloride layer will be the lower layer.
11. Methylation of fatty acids excluding butyric and caproic acids.
    - 11.1. Continue to reflux the sample for ~2–5 min.
    - 11.2. Add an appropriate amount of heptane through the condenser [either 5 or 10 ml, depending on the amount (1 or 2 ml, respectively) of tridecanoic tritridecnoin internal standard added in Step 1]. Continue to reflux the solution for ~1 min.
    - 11.3. Continue as specified in Steps 9–14 of the Standard Sample Preparation section.

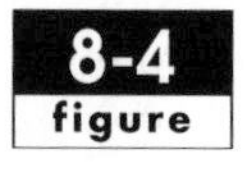

Standard and alternative procedures to methylate samples prior to gas chromatographic analysis for fatty acid composition.

8.3.6.3 Chromatographic Conditions

The following are two sets of conditions that might be used, but conditions may be modified as needed to optimize separation. Electronic pressure control of gases may be used.

Column	Any 0.25 mm ID × 100 m long, 0.20 μm film thickness 100% biscyanopropanol (nonbonded) capillary column such as SP-2560, HP-88, CP-Sil 88, etc.
Column temperature	170°C, hold 11.10 min; increase 7.9°C/min to 200°C, hold 1.27 min; increase 7.9°C/min to 210°C, hold 16 min
Injection port temperature	250°C
Detector temperature	300°C
Carrier gas (hydrogen) flow	1.2 ml/min (26 cm/s)
Split ratio	100:1
Hydrogen flow	30 ml/min
Air flow	300 ml/min
Makeup gas flow	30 ml/min

OR

Column	Any 0.25 mm ID × 100 m long, 0.20 μm film thickness 100% cyanopropanol capillary column such as SP-2560, HP-88, CP-Sil 88, etc.
Column temperature	170°C, hold 5.0 min; increase 2.0°C/min to 190°C, hold 5.0 min; increase 10.0°C/min to 210°C, hold 5.0 min; increase 10.0°C/min to 230°C, hold 9 min

8.3.6.4 Calculations

Calculations can be made by peak height or by integrated area (Fig. 8-5; Table 8-4). Fatty acids may be calculated as triglycerides, methyl esters, ethyl esters, or acids. Use a current validated computer software (e.g., Waters Empower Chromatography Manager, Waters Corporation, Milford, MA) to perform calculations, entering data for sample weight (mg) and final volume (ml).

$$\begin{aligned}&\%\text{ Fatty acid on a lipid basis}\\&\quad= \text{mg/mg of fatty acid} \times 100\end{aligned} \qquad [5]$$

or

$$\begin{aligned}&\%\text{ Fatty acids on a sample basis}\\&\quad= \text{mg/mg of fatty acid} \times \%\text{ lipid}\end{aligned} \qquad [6]$$

Alternative calculations can be made using peak areas obtained from current validated computer software. Calculate the response factor (RF) for each standard (by dividing the peak area by the internal standard area) and the standard factor (F) (by dividing the response factor by the internal standard concentration, mg/ml). Measure the peak area of each respective fatty acid and the internal standard, then calculate the sample response factor (SRF) by dividing the sample fatty acid peak area by the internal standard peak area.

$$\%\text{ Fatty acid} = \frac{\text{SRF}}{F} \times \frac{\text{final sample volume (ml)}}{\text{sample weight (mg)}} \times 100 \qquad [7]$$

where:

SRF = sample response factor

F = standard factor

8.4 NONSOLVENT WET EXTRACTION METHODS

8.4.1 Babcock Method for Milk Fat (AOAC Method 989.04 and 989.10)

8.4.1.1 Principle

In the Babcock method, H_2SO_4 is added to a known amount of milk in the Babcock bottle. The sulfuric acid digests protein, generates heat, and releases the fat. Centrifugation and hot water addition isolate fat for quantification in the graduated portion of the test bottle. The fat is measured volumetrically, but the result is expressed as percent fat by weight.

8.4.1.2 Procedure

1. Accurately pipette the milk sample (17.6 ml) into a Babcock test bottle (Fig. 8-6).
2. Add reagent grade (1.82 specific gravity) sulfuric acid (17.5 ml) to the bottle, allowing the acid to flow gently down the neck of the bottle as it is being slowly rotated. The acid digests proteins to liberate the fat.
3. Centrifuge the mixture for 5 min and liquid fat will rise into the calibrated bottle neck. The centrifuge must be kept at 55–60°C during centrifugation.
4. Add hot water to bring liquid fat up into the graduated neck of the Babcock bottle.
5. The direct percentage of fat by weight is read to the nearest 0.05% from the graduation mark of the bottle.

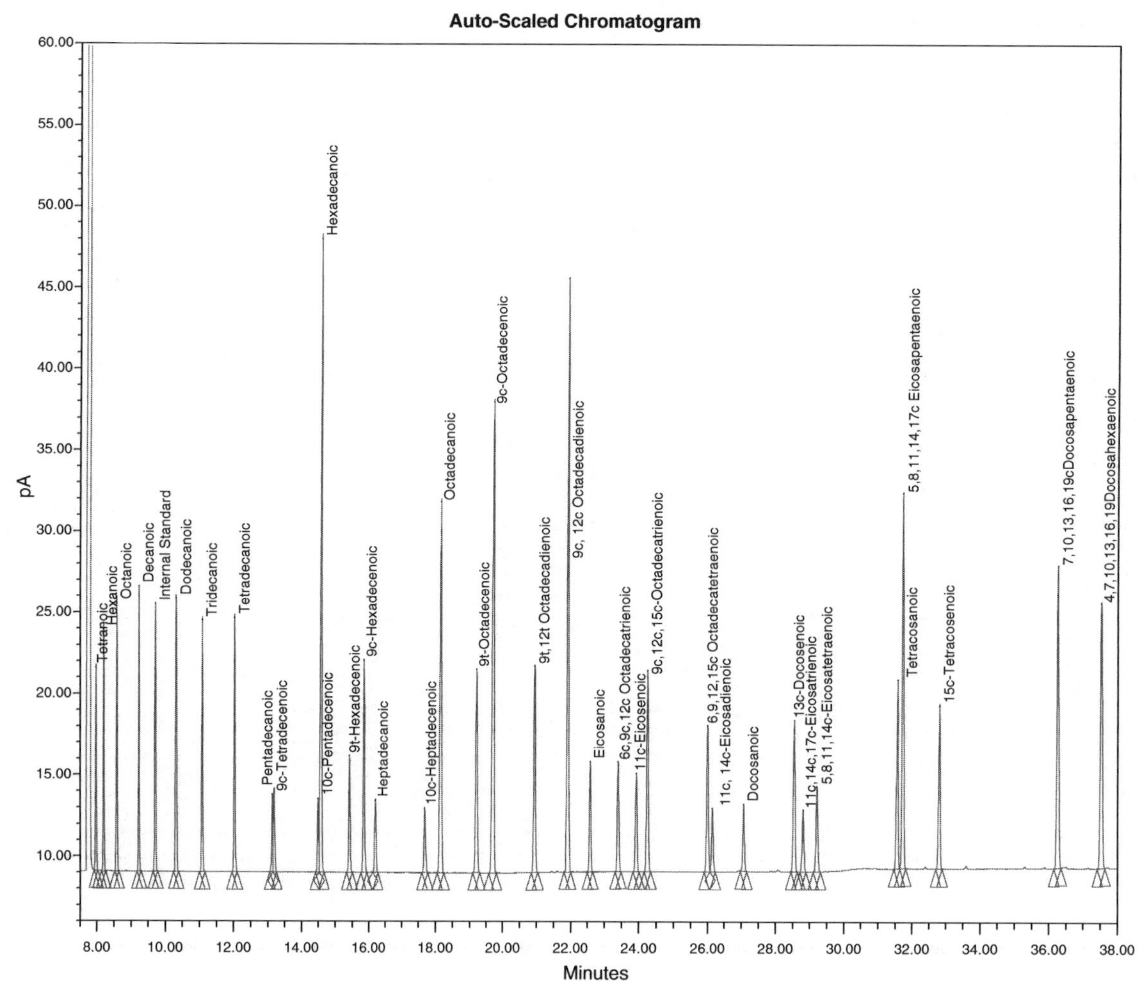

8-5 figure Example of chromatogram from gas chromatography analysis.

8-4 table **Fatty Acid Conversion Table**

	Molecular Weight				Conversion Factors[b]		
Fatty Acid[a]	*Acid*	*Methyl Ester*	*Triglyceride*	*1/3 Triglyceride*	*Triglyceride/ Methyl Ester*	*Acid/Methyl Ester*	*Acid/ Triglyceride*
4:0 Butyric Tetranoic	88.11	102.14	302.38	100.79	0.9868	0.8627	0.8742
5:0 Valeric Pentanoic	102.40	116.43	345.25	115.08	0.9885	0.8795	0.8898
6:0 Caproic Hexanoic	116.16	130.19	386.53	128.84	0.9897	0.8923	0.9016
7:0 Heptanoic	130.19	144.22	428.62	142.87	0.9907	0.9027	0.9112
8:0 Caprylic Octanoic	144.21	158.24	470.69	156.90	0.9915	0.9114	0.9192
10:0 Capric Decanoic	172.27	186.30	554.85	184.95	0.9928	0.9247	0.9314
12:0 Lauric Dodecanoic	200.35	214.38	639.02	213.01	0.9937	0.9346	0.9405
13:0 Tridecanoic	214.35	228.38	681.10	227.03	0.9941	0.9386	0.9441

(continued)

Fatty Acid Conversion Table (continued)

Fatty Acid[a]	Molecular Weight				Conversion Factors[b]		
	Acid	Methyl Ester	Triglyceride	1/3 Triglyceride	Triglyceride/ Methyl Ester	Acid/Methyl Ester	Acid/ Triglyceride
14:0 Myristic Tetradecanoic	228.38	242.41	723.18	241.06	0.9945	0.9421	0.9474
14:1 Myristoleic 9-Tetradecenoic	226.38	240.41	717.18	239.06	0.9944	0.9417	0.9469
15:0 Pentadecanoic	242.41	256.44	765.26	255.09	0.9948	0.9453	0.9503
15:1 Pentadecenoic 10-Pentadecenoic	240.40	254.43	759.26	253.09	0.9947	0.9449	0.9499
16:0 Palmitic Hexadecanoic	256.43	270.46	807.34	269.11	0.9950	0.9481	0.9529
16:1 Palmitoleic 9-Hexadecenoic	254.43	268.46	801.46	267.11	0.9950	0.9477	0.9525
17:0 Heptadecanoic	270.48	284.51	849.42	283.14	0.9953	0.9507	0.9552
17:1 Heptadecenoic 10-Heptadecenoic	268.48	282.51	843.42	281.14	0.9952	0.9503	0.9549
18:0 Stearic Octadecanoic	284.48	298.51	891.50	297.17	0.9955	0.9530	0.9573
18:1 Oleic 9-Octadecenoic	282.48	296.51	885.50	295.17	0.9955	0.9527	0.9570
18:2 Linoleic 9-12 Octadecadienoic	280.48	294.51	879.50	293.17	0.9954	0.9524	0.9567
18:3 Gamma Linolenic 6-9-12 Octadecatrienoic	278.48	292.51	873.50	291.17	0.9954	0.9520	0.9564
18:3 Linolenic 9-12-15 Octadecatrienoic	278.48	292.51	873.50	291.17	0.9954	0.9520	0.9564
18:4 Octadecatetraenoic 6-9-12-15 Octadecatetraenoic	276.48	290.51	867.50	289.17	0.9954	0.9517	0.9561
20:0 Arachidic Eicosanoic	312.54	326.57	975.66	325.22	0.9959	0.9570	0.9610
20:1 Eicosenoic 11-Eicosenoic	310.54	324.57	969.66	323.22	0.9959	0.9568	0.9608
20:2 Eicosadienoic 11-14 Eicosadienoic	308.53	322.56	963.66	321.22	0.9958	0.9565	0.9605
20:3 Eicosatrienoic 11-14-17 Eicosatrienoic	306.53	320.56	957.66	319.22	0.9958	0.9562	0.9603
20:4 Arachidonic 5-8-11-14 Eicosatetraenoic	304.52	318.55	951.66	317.22	0.9958	0.9560	0.9600
20:5 Eicosapentaenoic 5-8-11-14-17 Eicosapenatenoic	302.52	316.55	945.66	315.22	0.9958	0.9557	0.9598
22:0 Behenic Docosanoic	340.59	354.62	1,059.82	353.27	0.9962	0.9604	0.9641
22:1 Erucic 13-Docosenoic	338.59	352.63	1,053.82	351.27	0.9962	0.9602	0.9639
22:5 Docosapentaenoic 7-10-13-1619 Docosapentaenoic	330.50	344.53	1,029.55	343.18	0.9961	0.9593	0.9630
22:6 Docosahexaenoic 4-7-10-13-16-19 Docosahexaenoic	328.57	342.60	1,023.82	341.27	0.9961	0.9591	0.9628
24:0 Lignoceric Tetracosanoic	368.64	382.67	1,143.98	381.33	0.9965	0.9633	0.9667
24:1 Nervonic 15-Tetracosenoic	366.63	380.66	1,137.98	379.33	0.9965	0.9632	0.9666

[a]Top number indicates carbon chain length and degree of unsaturation. Bottom number(s) indicate position of double bond(s) on the carbon chain.

[b]Conversion factors: To convert methyl ester to triglyceride: multiply by T/M ratio; To convert methyl ester to acid: multiply by A/M ratio; To convert triglyceride to acid: multiply by A/T ratio; To convert tridecanoic to methyl ester: multiply by 1.065 or divide by A/M ratio.

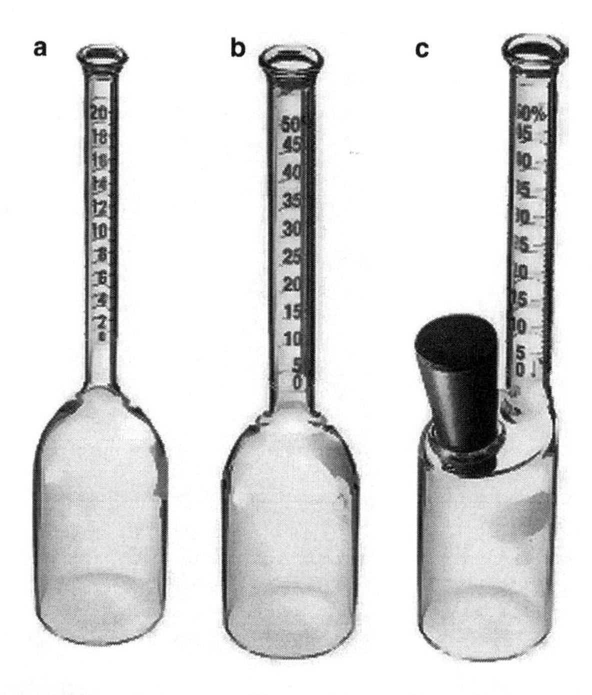

8-6 figure Babcock milk test bottles for milk (**a**), cream (**b**), and cheese (Paley bottle) (**c**) testing. (Courtesy of Kimble Glass Co., Vineland, NJ.)

8.4.1.3 Applications

The Babcock method, which is a common official method for the determination of fat in milk, takes about 45 min and duplicate tests should agree within 0.1%. The Babcock method does not determine the phospholipids in the milk products. It is not applicable to products containing chocolate or added sugar without modification because of charring of chocolate and sugars by sulfuric acid. A modified Babcock method is used to determine essential oil in flavor extracts (AOAC Method 932.11) and fat in seafood (AOAC Method 964.12).

8.4.2 Gerber Method for Milk Fat

8.4.2.1 Principle

The principle of the Gerber method is similar to that of the Babcock method, but it uses sulfuric acid and amyl alcohol. The sulfuric acid digests proteins and carbohydrates, releases fat, and maintains the fat in a liquid state by generating heat.

8.4.2.2 Procedure

1. Transfer 10 ml of H_2SO_4 at 15–21°C into a Gerber milk bottle.
2. Accurately measure milk sample (11 ml) into the Gerber bottle, using a Gerber pipette.
3. Add 1 ml of isoamyl alcohol to the bottle.
4. Tighten the stopper and mix by shaking the bottle.
5. Centrifuge the bottle for 4 min.
6. Place the bottle in a water bath at 60–63°C for 5 min and then read the fat content from the graduations on the bottle neck.

8.4.2.3 Applications

The Gerber method is comparable to the Babcock method but is simpler and faster and has wider application to a variety of dairy products (14). The isoamyl alcohol generally prevents the charring of sugar found with the regular Babcock method. This test is more popular in Europe than in America.

8.5 INSTRUMENTAL METHODS

Instrumental methods offer numerous attractive features compared with the previously described extraction methods. In general, they are rapid, nondestructive, and require minimal sample preparation and chemical consumption. However, the equipment can be expensive and measurements often require the establishment of calibration curves specific to various compositions. Despite these drawbacks, several of the following instrumental methods are very widely used in quality control as well as research and product development applications. The following section describes several of these instrumental methods.

8.5.1 Infrared Method

The infrared (IR) method is based on absorption of IR energy by fat at a wavelength of 5.73 μm. The more the energy absorption at 5.73 μm, the higher is the fat content of the sample (15). Mid-IR spectroscopy is used in Infrared Milk Analyzers to determine milk fat content (AOAC Method 972.16). Near-infrared (NIR) spectroscopy has been used to measure the fat content of commodities such as meats, cereals, and oilseeds in the laboratory and is being adapted for on-line measurement. See Chap. 23 for a discussion of IR spectroscopy.

8.5.2 Specific Gravity (Foss-Let Method)

Fat content by the Foss-Let method (Foss North America, Eden Prairie, MN) is determined as a function of the specific gravity of a sample solvent extract. A sample of known weight is extracted for 1.5–2 min in a vibration-reaction chamber with perchloroethylene. The extract is filtered, and using a thermostatically controlled device with digital readout, its specific gravity is determined. The reading can then be converted to oil or fat percentage using a conversion chart.

8.5.3 Nuclear Magnetic Resonance

Nuclear magnetic resonance (NMR) can be used to measure lipids in food materials in a nondestructive way. It is one of the most popular methods for use in determining lipid melting curves to measure solid fat content (see Chap. 14), and with more affordable instruments is becoming more popular for measuring total fat content. Total fat content can be measured using low-resolution pulsed NMR. The principles and applications of NMR are described in Chap. 25. NMR analysis is a very rapid and accurate method, and while the principles of NMR are relatively complex, the use of NMR can be quite simple, especially due to the high degree of automation and computer control.

8.6 COMPARISON OF METHODS

Soxhlet extraction or its modified method is a common crude fat determination method in many food commodities. However, this method requires a dried sample for the hydroscopic ethyl ether extraction. If the samples are moist or liquid foods, the Mojonnier method is generally applicable to determination of the fat content. Acid hydrolysis or alkaline hydrolysis is widely used on many finished food products. The instrumental methods such as IR and NMR are very simple, reproducible, and fast, but are available only for fat determination for specific foods. The application of instrumental methods for fat determination generally requires a standard curve between the signal of the instrument analysis and the fat content obtained by a standard solvent extraction method. However, a rapid instrumental method could be used as a quality control method for fat determination of a specific food.

Major uses of the Goldfish, Soxhlet, and Mojonnier (or Roese-Gottlieb) methods include the following: (1) extract fat prior to GC analysis, (2) quality control of formulated products, (3) determine fat content during product development, (4) verify when fat content is <0.5 g per serving (so nutrient content claim can be made), and (5) defat samples prior to fiber analysis. Compared with GC analysis of fat content by AOAC Method 996.06, these three methods are faster and cheaper, but give a higher fat content (which must be recognized when using these methods for product development).

8.7 SUMMARY

Lipids are generally defined by their solubility characteristics rather than by some common structural feature. Lipids in foods can be classified as simple, compound, or derived lipids. The lipid content of foods varies widely, but quantitation is important because of regulatory requirements, nutritive value, and functional properties. To analyze food for the fat content accurately and precisely, it is essential to have a comprehensive knowledge of the general compositions of the lipids in the foods, the physical and chemical properties of the lipids as well as the foods, and the principles of fat determination. There is no single standard method for the determination of fats in different foods. The validity of any fat analysis depends on proper sampling and preservation of the sample prior to analysis. Predrying of the sample, particle size reduction, and acid hydrolysis prior to analysis also may be necessary. The total lipid content of foods is commonly determined by organic solvent extraction methods, which can be classified as continuous (e.g., Goldfish), semicontinuous (e.g., Soxhlet), discontinuous (e.g., Mojonnier, Folch), or by GC analysis for nutrition labeling. Nonsolvent wet extraction methods, such as the Babcock or Gerber, are commonly used for certain types of food products. Instrumental methods, such as NMR, infrared, and Foss-Let, are also available for fat determination of specific foods. These methods are rapid and so may be useful for quality control but generally require correlation to a standard solvent extraction method.

8.8 STUDY QUESTIONS

1. What are some important considerations when selecting solvents to be used in continuous and noncontinuous solvent extraction methods?
2. To extract the fat from a food sample, you have the choice of using ethyl ether or petroleum ether as the solvent, and you can use either a Soxhlet or a Goldfish apparatus. What combination of solvent and extraction would you choose? Give all the reasons for your choice.
3. Itemize the procedures that may be required to prepare a food sample for accurate fat determination by a solvent extraction method (e.g., Soxhlet method). Explain why each of these procedures may be necessary.
4. You performed fat analysis on a new superenergy shake (high carbohydrate and protein) using standard Soxhlet extraction. The value obtained for fat content was much lower than that expected. What could have caused the measured fat content to be low and how would you modify the standard procedure to correct the problem?

5. What is the purpose of the following chemicals used in the Mojonnier method?
   (a) Ammonium hydroxide
   (b) Ethanol
   (c) Ethyl ether
   (d) Petroleum ether
6. What is a key application of the GC method and what does it specifically quantify?
7. What is the purpose of the following procedures used in Babcock method?
   (a) Sulfuric acid addition
   (b) Centrifugation and addition of hot water
8. Which of the following methods are volumetric and which are gravimetric determinations of lipid content: Babcock, Soxhlet, Mojonnier, Gerber?
9. Explain and contrast the principles (not procedures) involved in determining the fat content of a food product by the following methods. Indicate for each method the type of sample and application that would be appropriate for analysis.
   (a) Soxhlet
   (b) Babcock
   (c) Mojonnier
   (d) GC analysis

8.9 PRACTICE PROBLEMS

1. To determine the fat content of a semimoist food by the Soxhlet method, the food was first vacuum oven dried. The moisture content of the product was 25%. The fat in the dried food was determined by the Soxhlet method. The fat content of the dried food was 13.5%. Calculate the fat content of the original semimoist product.
2. The fat content of 10 g of commercial ice cream was determined by the Mojonnier method. The weights of extracted fat after the second extraction and the third extraction were 1.21 g and 1.24 g, respectively. How much of fat, as a percentage of the total, was extracted during the third extraction?

Answers

1. If the sample weight of a semimoist food is 10 g and the moisture content is 25%, the dried weight of the original food is 7.5 g ($10\,g \times 75\% = 7.5\,g$). If the fat content of the dried food is 13.5%, the 7.5 g of dried sample has 1.0125 g fat (7.5 g dried food $\times$13.5% fat = 1.0125 g fat). The 10 g of semimoist food contains the same amount of fat, i.e., 1.0125 g. Therefore, the fat content of the semimoist food is 10.125% (1.0125 g fat/10 g semimoist food).
2. $[(1.24\,g–1.21\,g)/10\,g] \times 100 = 0.3\%$

8.10 REFERENCES

1. O'Keefe SF (1998) Nomenclature and classification of lipids, Chap. 1. In: Akoh CC, Min DB (eds) Food lipids: chemistry, nutrition and biotechnology. Marcel Dekker, New York, pp 1–36
2. Belitz HD, Grosch W (1987) Food chemistry. Springer, Berlin
3. Nawar WW (1996) Lipids, Chap. 5. In: Fennema OR (ed) Food chemistry, 3rd edn. Marcel Dekker, New York, pp 225–319
4. Patton S, Jensen RG (1976) Biomedical aspects of lactation. Pergamon, Oxford, p 78
5. Jenness R, Patton S (1959) Principles of dairy chemistry. Wiley, New York
6. Joslyn MA (1970) Methods in food analysis, 2nd edn. Academic, New York
7. Pomeranz Y, Meloan CF (1994) Food analysis: theory and practice, 3rd edn. Van Nostrand Reinhold, New York
8. AOAC International (2007) Official methods of analysis, 18th edn, 2005; Current through revision 2, 2007 (Online). AOAC International, Gaithersburg, MD
9. Folch J, Lees M, Stanley GHS (1957) A simple method for the isolation and purification of total lipids from animal tissues. J Biol Chem 226:297–509
10. Bligh EG, Dyer WJ (1959) A rapid method of total lipid extraction and purification. Can J Physiol 37:911–917
11. Christie WW (1982) Lipid analysis. Isolation, separation, identification, and structural analysis of lipids, 2nd edn. Pergamon, Oxford
12. AOCS (2009) Official methods and recommended practices, 6th edn. American Oil Chemists' Society, Champaign, IL
13. Golay P-A, Dionisi F, Hug B, Giuffrida F, Destaillats F (2006) Direct quantification of fatty acids in dairy powders, with special emphasis on *trans* fatty acid content. Food Chem 106:115–1120
14. Wehr HM, Frank JF (eds) (2004) Standard methods for the examination of dairy products, 17th edn. American Public Health Association, Washington, DC
15. Cronin DA, McKenzie K (1990) A rapid method for the determination of fat in foodstuffs by infrared spectrometry. Food Chem 35:39–49

Protein Analysis

Sam K. C. Chang

Department of Cereal and Food Sciences, North Dakota State University, Fargo, ND 58105, USA
kow.chang@ndsu.edu

9.1 Introduction 135
9.1.1 Classification and General Considerations 135
9.1.2 Importance of Analysis 135
9.1.3 Content in Foods 135
9.2 Methods 135
9.2.1 Kjeldahl Method 136
9.2.1.1 Principle 136
9.2.1.2 Historical Background 136
9.2.1.2.1 Original Method 136
9.2.1.2.2 Improvements 136
9.2.1.3 General Procedures and Reactions 137
9.2.1.3.1 Sample Preparation 137
9.2.1.3.2 Digestion 137
9.2.1.3.3 Neutralization and Distillation 137
9.2.1.3.4 Titration 137
9.2.1.3.5 Calculations 137
9.2.1.3.6 Alternate Procedures 137
9.2.1.4 Applications 138

S.S. Nielsen, *Food Analysis*, Food Science Texts Series, DOI 10.1007/978-1-4419-1478-1_9,

9.2.2 Dumas (Nitrogen Combustion) Method 138
9.2.2.1 Principle 138
9.2.2.2 Procedure 138
9.2.2.3 Applications 138
9.2.3 Infrared Spectroscopy 138
9.2.3.1 Principle 138
9.2.3.2 Procedure 138
9.2.3.3 Applications 138
9.2.4 Biuret Method 139
9.2.4.1 Principle 139
9.2.4.2 Procedure 139
9.2.4.3 Applications 139
9.2.5 Lowry Method 139
9.2.5.1 Principle 139
9.2.5.2 Procedure 140
9.2.5.3 Applications 140
9.2.6 Dye-Binding Methods 140
9.2.6.1 Anionic Dye-Binding Method 140
9.2.6.1.1 Principle 140
9.2.6.1.2 Procedure 140
9.2.6.1.3 Applications 140
9.2.6.2 Bradford Dye-Binding Method 141
9.2.6.2.1 Principle 141
9.2.6.2.2 Procedure 141
9.2.6.2.3 Applications 141
9.2.7 Bicinchoninic Acid Method 141
9.2.7.1 Principle 141
9.2.7.2 Procedure 142
9.2.7.3 Applications 142
9.2.8 Ultraviolet 280 nm Absorption Method 142
9.2.8.1 Principle 142
9.2.8.2 Procedure 142
9.2.8.3 Applications 142
9.3 Comparison of Methods 143
9.4 Special Considerations 143
9.5 Summary 144
9.6 Study Questions 144
9.7 Practice Problems 145
9.8 References 146

9.1 INTRODUCTION

9.1.1 Classification and General Considerations

Proteins are an abundant component in all cells, and almost all except storage proteins are important for biological functions and cell structure. Food proteins are very complex. Many have been purified and characterized. Proteins vary in molecular mass, ranging from approximately 5000 to more than a million Daltons. They are composed of elements including hydrogen, carbon, nitrogen, oxygen, and sulfur. Twenty α-amino acids are the building blocks of proteins; the amino acid residues in a protein are linked by peptide bonds. Nitrogen is the most distinguishing element present in proteins. However, nitrogen content in various food proteins ranges from 13.4 to 19.1% (1) due to the variation in the specific amino acid composition of proteins. Generally, proteins rich in basic amino acids contain more nitrogen.

Proteins can be classified by their composition, structure, biological function, or solubility properties. For example, simple proteins contain only amino acids upon hydrolysis, but conjugated proteins also contain non-amino-acid components.

Proteins have unique conformations that could be altered by denaturants such as heat, acid, alkali, $8\,M$ urea, $6\,M$ guanidine-HCl, organic solvents, and detergents. The solubility as well as functional properties of proteins could be altered by denaturants.

The analysis of proteins is complicated by the fact that some food components possess similar physicochemical properties. Nonprotein nitrogen could come from free amino acids, small peptides, nucleic acids, phospholipids, amino sugars, porphyrin, and some vitamins, alkaloids, uric acid, urea, and ammonium ions. Therefore, the total organic nitrogen in foods would represent nitrogen primarily from proteins and to a lesser extent from all organic nitrogen-containing nonprotein substances. Depending upon methodology, other major food components, including lipids and carbohydrates, may interfere physically with analysis of food proteins.

Numerous methods have been developed to measure protein content. The basic principles of these methods include the determinations of nitrogen, peptide bonds, aromatic amino acids, dye-binding capacity, ultraviolet absorptivity of proteins, and light scattering properties. In addition to factors such as sensitivity, accuracy, precision, speed, and cost of analysis, what is actually being measured must be considered in the selection of an appropriate method for a particular application.

9.1.2 Importance of Analysis

Protein analysis is important for:

1. **Nutrition labeling**
2. **Pricing:** The cost of certain commodities is based on the protein content as measured by nitrogen content (e.g., cereal grains; milk for making certain dairy products, e.g., cheese).
3. **Functional property investigation:** Proteins in various types of food have unique food functional properties: for example, gliadin and glutenins in wheat flour for breadmaking, casein in milk for coagulation into cheese products, and egg albumen for foaming (see Chap. 15).
4. **Biological activity determination:** Some proteins, including enzymes or enzyme inhibitors, are relevant to food science and nutrition: for instance, the proteolytic enzymes in the tenderization of meats, pectinases in the ripening of fruits, and trypsin inhibitors in legume seeds are proteins. To compare between samples, enzymes activity often is expressed in terms of specific activity, meaning units of enzyme activity per mg of protein.

Protein analysis is required when you want to know:

1. Total protein content
2. Content of a particular protein in a mixture
3. Protein content during isolation and purification of a protein
4. Nonprotein nitrogen
5. Amino acid composition (see Chap. 15)
6. Nutritive value of a protein (see Chap. 15)

9.1.3 Content in Foods

Protein content in food varies widely. Foods of animal origin and legumes are excellent sources of proteins. The protein contents of selected food items are listed in Table 9-1.

9.2 METHODS

Principles, general procedures, and applications are described below for various protein determination methods. Refer to the referenced methods for detailed instructions of the procedures. The Kjeldahl, Dumas (N combustion), and infrared spectroscopy methods cited are from the *Official Methods of Analysis* of AOAC International (3) and are used commonly in nutrition labeling and quality control. The other methods

Protein Content of Selected Foods (2)

Food Item	*Percent Protein (Wet Weight Basis)*
Cereals and pasta	
Rice, brown, long-grain raw	7.9
Rice, white, long-grain, regular, raw, enriched	7.1
Wheat flour, whole-grain	13.7
Corn flour, whole-grain, yellow	6.9
Spaghetti, dry, enriched	13.0
Cornstarch	0.3
Dairy products	
Milk, reduced fat, fluid, 2%	3.2
Milk, nonfat, dry, regular, with added vitamin A	36.2
Cheese, cheddar	24.9
Yogurt, plain, low fat	5.3
Fruits and vegetables	
Apple, raw, with skin	0.3
Asparagus, raw	2.2
Strawberries, raw	0.7
Lettuce, iceberg, raw	0.9
Potato, whole, flesh and skin	2.0
Legumes	
Soybeans, mature seeds, raw	36.5
Beans, kidney, all types, mature seeds, raw	23.6
Tofu, raw, firm	15.8
Tofu, raw, regular	8.1
Meats, poultry, fish	
Beef, chuck, arm pot roast	21.4
Beef, cured, dried beef	31.1
Chicken, broilers or fryers, breast meat only, raw	23.1
Ham, sliced, regular	16.6
Egg, raw, whole, fresh	12.6
Finfish, cod, Pacific, raw	17.9
Finfish, tuna, white, canned in oil, drained solids	26.5

From US Department of Agriculture, Agricultural Research Service (2009). USDA National Nutrient Database for Standard Reference. Release 22. Nutrient Data Laboratory Home Page, http://www.ars.usda.gov/ba/bhnrc/ndl

described are used commonly in research laboratories working on proteins. Many of the methods covered in this chapter are described in somewhat more detail in recent books on food proteins (4–6).

9.2.1 Kjeldahl Method

9.2.1.1 Principle

In the Kjeldahl procedure, proteins and other organic food components in a sample are digested with sulfuric acid in the presence of catalysts. The **total organic nitrogen** is converted to ammonium sulfate. The digest is neutralized with alkali and distilled into a boric acid solution. The borate anions formed are titrated with standardized acid, which is converted to nitrogen in the sample. The result of the analysis represents the crude protein content of the food since nitrogen also comes from nonprotein components (note that the Kjeldahl method also measures nitrogen in any ammonia and ammonium sulfate).

9.2.1.2 Historical Background

9.2.1.2.1 Original Method In 1883, Johann Kjeldahl developed the basic process of today's Kjeldahl method to analyze organic nitrogen. General steps in the original method include the following:

1. **Digestion** with sulfuric acid, with the addition of powdered potassium permanganate to complete oxidation and conversion of nitrogen to ammonium sulfate.
2. **Neutralization** of the diluted digest, followed by **distillation** into a known volume of standard acid, which contains potassium iodide and iodate.
3. **Titration** of the liberated iodine with standard sodium thiosulfate.

9.2.1.2.2 Improvements Several important modifications have improved the original Kjeldahl process:

1. Metallic catalysts such as mercury, copper, and selenium are added to sulfuric acid for complete digestion. Mercury has been found to be the most satisfactory. Selenium dioxide and copper sulfate in the ratio of 3:1 have been reported to be effective for digestion. Copper and titanium dioxide also have been used as a mixed catalyst for digestion (AOAC Method 988.05) (3). The use of titanium dioxide and copper poses less **safety concern** than mercury in the postanalysis disposal of the waste.
2. Potassium sulfate is used to increase the boiling point of the sulfuric acid to accelerate digestion.
3. Sulfide or sodium thiosulfate is added to the diluted digest to help release nitrogen from mercury, which tends to bind ammonium.
4. The ammonia is distilled directly into a boric acid solution, followed by titration with standard acid.
5. Colorimetry Nesslerization, or ion chromatography to measure ammonia, is used to determine nitrogen content after digestion.

An excellent book to review the Kjeldahl method for total organic nitrogen was written by Bradstreet (7). The basic AOAC Kjeldahl procedure is Method 955.04.

Semiautomation, automation, and modification for microgram nitrogen determination (micro Kjeldahl method) have been established by AOAC in Methods 976.06, 976.05, and 960.52, respectively.

9.2.1.3 General Procedures and Reactions

9.2.1.3.1 Sample Preparation Solid foods are ground to pass a 20-mesh screen. Samples for analysis should be homogeneous. No other special preparations are required.

9.2.1.3.2 Digestion Place sample (accurately weighed) in a Kjeldahl flask. Add acid and catalyst; digest until clear to get complete breakdown of all organic matter. Nonvolatile ammonium sulfate is formed from the reaction of nitrogen and sulfuric acid.

$$\text{Protein} \xrightarrow[\text{Heat, catalyst}]{\text{Sulfuric acid}} (NH_4)_2SO_2 \quad [1]$$

During digestion, protein nitrogen is liberated to form ammonium ions; sulfuric acid oxidizes organic matter and combines with ammonium formed; carbon and hydrogen elements are converted to carbon dioxide and water.

9.2.1.3.3 Neutralization and Distillation The digest is diluted with water. Alkali-containing sodium thiosulfate is added to neutralize the sulfuric acid. The ammonia formed is distilled into a boric acid solution containing the indicators methylene blue and methyl red (AOAC Method 991.20).

$$(NH_4)_2SO_4 + 2NaOH \rightarrow 2NH_3 + Na_2SO_4 + 2H_2O \quad [2]$$

$$NH_3 + H_3BO_3\ (\text{boric acid}) \rightarrow NH_4 + \underset{(\text{borate ion})}{H_2BO_3^-} \quad [3]$$

9.2.1.3.4 Titration Borate anion (proportional to the amount of nitrogen) is titrated with standardized HCl.

$$H_2BO_3^- + H^+ \rightarrow H_3BO_3 \quad [4]$$

9.2.1.3.5 Calculations

$$\text{Moles of HCl} = \text{moles of } NH_3 = \text{moles of N in the sample} \quad [5]$$

A reagent blank should be run to subtract reagent nitrogen from the sample nitrogen.

$$\%N = N\,HCl \times \frac{\text{Corrected acid volume}}{\text{g of sample}} \times \frac{14\,\text{g N}}{\text{mol}} \times \frac{100}{1000} \quad [6]$$

table 9-2 Nitrogen to Protein Conversion Factors for Various Foods

	Percent N in Protein	Factor
Egg or meat	16.0	6.25
Milk	15.7	6.38
Wheat	18.76	5.33
Corn	17.70	5.65
Oat	18.66	5.36
Soybean	18.12	5.52
Rice	19.34	5.17

Data from (1, 8).

where:

NHCl = normality of HCl, in mol/1000 ml

Corrected acid vol. = (ml std. acid for sample) – (ml std. acid for blank)

14 = atomic weight of nitrogen

A factor is used to convert percent N to percent crude protein. Most proteins contain 16% N, so the conversion factor is 6.25 (100/16 = 6.25).

$$\%N/0.16 = \%\text{protein} \quad [7]$$

or

$$\%N \times 6.25 = \%\text{protein}$$

Conversion factors for various foods are given in Table 9-2.

9.2.1.3.6 Alternate Procedures In place of distillation and titration with acid, ammonia or nitrogen can be quantitated by:

1. Nesslerization

$$\underset{\text{mercuric iodide}}{4NH_4OH + 2HgI_2 + 4KI + 3KOH} \rightarrow \underset{\text{ammonium dimercuric iodide, red-orange, 440 nm}}{NH_2Hg_2IO} + 7KI + 2H_2O \quad [8]$$

   This method is rapid and sensitive, but the ammonium dimercuric iodide is colloidal and color is not stable.
2. $NH_3 + \text{phenol} + \text{hypochloride} \xrightarrow[OH^-]{} \text{indophenol (blue, 630 nm)}$ [9]
3. pH measurement after distillation into known volume of boric acid
4. Direct measurement of ammonia, using ion chromatographic method

9.2.1.4 Applications

Advantages:

1. Applicable to all types of foods
2. Inexpensive (if not using an automated system)
3. Accurate; an official method for crude protein content
4. Has been modified (micro Kjeldahl method) to measure microgram quantities of proteins

Disadvantages:

1. Measures total organic nitrogen, not just protein nitrogen
2. Time consuming (at least 2 h to complete)
3. Poorer precision than the biuret method
4. Corrosive reagent

9.2.2 Dumas (Nitrogen Combustion) Method

9.2.2.1 Principle

The combustion method was introduced in 1831 by Jean-Baptiste Dumas. It has been modified and automated to improve accuracy since that time. Samples are combusted at high temperatures (700–1000°C) with a flow of pure oxygen. All carbon in the sample is converted to carbon dioxide during the flash combustion. Nitrogen-containing components produced include N_2 and nitrogen oxides. The nitrogen oxides are reduced to nitrogen in a copper reduction column at a high temperature (600°C). The total nitrogen (including inorganic fraction, i.e., including nitrate and nitrite) released is carried by pure helium and quantitated by **gas chromatography** using a **thermal conductivity detector** (TCD) (9). Ultra-high purity acetanilide and EDTA (ethylenediamine tetraacetate) may be used as the standards for the calibration of the nitrogen analyzer. The nitrogen determined is converted to protein content in the sample using a protein conversion factor.

9.2.2.2 Procedure

Samples (approximately 100–500 mg) are weighed into a tin capsule and introduced to a combustion reactor in automated equipment. The nitrogen released is measured by a built-in gas chromatograph. Figure 9-1 shows the flow diagram of the components of a Dumas nitrogen analyzer.

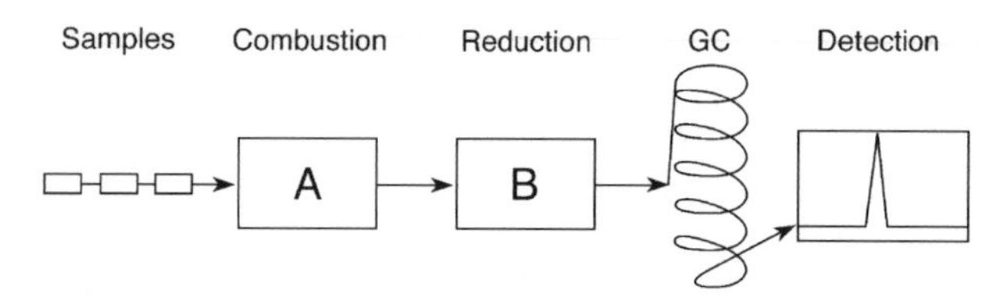

9-1 figure General components of a Dumas nitrogen analyzer. *A*, the incinerator; *B*, copper reduction unit for converting nitrogen oxides to nitrogen; and *GC*, gas chromatography column.

9.2.2.3 Applications

The combustion method is an alternative to the Kjeldahl method (10) and is suitable for all types of foods. AOAC Method 992.15 and Method 992.23 are for meat and cereal grains, respectively.

Advantages:

1. Requires no hazardous chemicals.
2. Can be accomplished in 3 min.
3. Recent automated instruments can analyze up to 150 samples without attention.

Disadvantages:

1. Expensive equipment is required.
2. Measures total organic nitrogen, not just protein nitrogen.

9.2.3 Infrared Spectroscopy

9.2.3.1 Principle

Infrared spectroscopy measures the **absorption of radiation** (near- or mid-infrared regions) by molecules in food or other substances. Different functional groups in a food absorb different frequencies of radiation. For proteins and peptides, various **mid-infrared** bands (6.47 μm) and **near-infrared** (NIR) bands (e.g., 3300–3500 nm; 2080–2220 nm; 1560–1670 nm) characteristic of the **peptide bond** can be used to estimate the protein content of a food. By irradiating a sample with a wavelength of infrared light specific for the constituent to be measured, it is possible to predict the concentration of that constituent by measuring the energy that is reflected or transmitted by the sample (which is inversely proportional to the energy absorbed) (11).

9.2.3.2 Procedure

See Chap. 23 for a detailed description of instrumentation, sample handling, and calibration and quantitation methodology.

9.2.3.3 Applications

Mid-infrared spectroscopy is used in Infrared Milk Analyzers to determine milk protein content, while near-infrared spectroscopy is applicable to a wide range of food products (e.g., grains; cereal, meat, and dairy products) (3,12,13) (AOAC Method 997.06). Instruments are expensive and they must be calibrated properly. However, samples can be analyzed rapidly (30 s to 2 min) by analysts with minimal training.

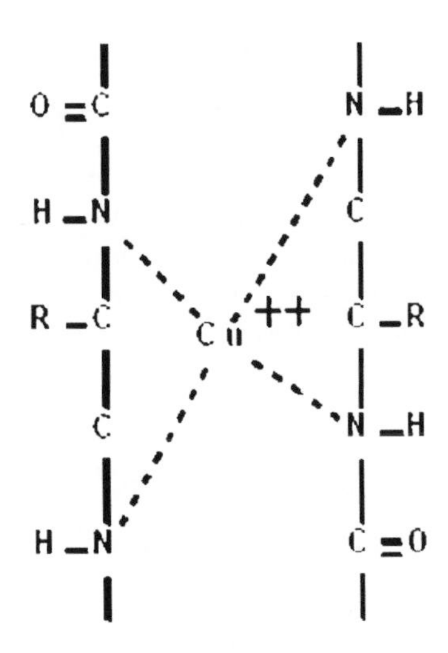

9-2 figure Reaction of peptide bonds with cupric ions.

9.2.4 Biuret Method

9.2.4.1 Principle

A violet-purplish color is produced when **cupric ions** are complexed with **peptide bonds** (substances containing at least two peptide bonds, i.e., biuret, large peptides, and all proteins) under **alkaline conditions** (Fig. 9-2). The absorbance of the color produced is read at 540 nm. The color intensity (absorbance) is proportional to the protein content of the sample (14).

9.2.4.2 Procedure

1. A 5-ml biuret reagent is mixed with a 1-ml portion of protein solution (1–10 mg protein/ml). The reagent includes copper sulfate, NaOH, and potassium sodium tartrate, which is used to stabilize the cupric ion in the alkaline solution.
2. After the reaction mix is allowed to stand at room temperature for 15 or 30 min, the absorbance is read at 540 nm against a reagent blank.
3. Filtration or centrifugation before reading absorbance is required if the reaction mixture is not clear.
4. A standard curve of concentration versus absorbance is constructed using **bovine serum albumin** (BSA).

9.2.4.3 Applications

The biuret method has been used to determine proteins in cereal (15, 16), meat (17), soybean proteins (18), and as a qualitative test for animal feed [AOAC Method 935.11 (refers to Methods 22.012–22.013, AOAC, 10th edn, 1965)] (19). The biuret method also can be used to measure the protein content of isolated proteins.

Advantages:

1. Less expensive than the Kjeldahl method; rapid (can be completed in less than 30 min); simplest method for analysis of proteins.
2. Color deviations are encountered less frequently than with Lowry, ultraviolet (UV) absorption, or turbidimetric methods (described below).
3. Very few substances other than proteins in foods interfere with the biuret reaction.
4. Does not detect nitrogen from nonpeptide or nonprotein sources.

Disadvantages:

1. Not very sensitive as compared to the Lowry method; requires at least 2–4 mg protein for assay.
2. Absorbance could be contributed from bile pigments if present.
3. High concentration of ammonium salts interfere with the reaction.
4. Color varies with different proteins; gelatin gives a pinkish-purple color.
5. Opalescence could occur in the final solution if high levels of lipid or carbohydrate are present.
6. Not an absolute method: color must be standardized against known protein (e.g., BSA) or against the Kjeldahl nitrogen method.

9.2.5 Lowry Method

9.2.5.1 Principle

The Lowry method (20, 21) combines the **biuret reaction** with the reduction of the **Folin–Ciocalteau phenol reagent** (phosphomolybdic-phosphotungstic acid) by **tyrosine** and **tryptophan** residues in the proteins (Fig. 9-3). The bluish color developed is read at 750 nm (high sensitivity for low protein concentration) or 500 nm (low sensitivity for high protein concentration). The original procedure has been modified by Miller (22) and Hartree (23) to improve the linearity of the color response to protein concentration.

9-3 figure Side chains of amino acids tyrosine between *C* and H_2 (**a**) and tryptophan between *C* and H_2, and between *N* and *H* (**b**).

9.2.5.2 Procedure

The following procedure is based on the modified procedure of Hartree (23):

1. Proteins to be analyzed are diluted to an appropriate range (20–100 μg).
2. K Na Tartrate-Na_2CO_3 solution is added after cooling and incubated at room temperature for 10 min.
3. $CuSO_4$-K Na Tartrate-NaOH solution is added after cooling and incubated at room temperature for 10 min.
4. Freshly prepared Folin reagent is added and then the reaction mixture is mixed and incubated at 50°C for 10 min.
5. Absorbance is read at 650 nm.
6. A standard curve of BSA is carefully constructed for estimating protein concentration of the unknown.

9.2.5.3 Applications

Because of its simplicity and sensitivity, the Lowry method has been widely used in protein biochemistry. However, it has not been widely used to determine proteins in food systems without first extracting the proteins from the food mixture.

Advantages:

1. Very sensitive
   (a) 50–100 times more sensitive than biuret method
   (b) 10–20 times more sensitive than 280-nm UV absorption method (described below)
   (c) Similar sensitivity as Nesslerization; however, more convenient than Nesslerization
2. Less affected by turbidity of the sample.
3. More specific than most other methods.
4. Relatively simple; can be done in 1–1.5 h.

Disadvantages:

For the following reasons, the Lowry procedure requires careful standardization for particular applications:

1. Color varies with different proteins to a greater extent than the biuret method.
2. Color is not strictly proportional to protein concentration.
3. The reaction is interfered with to varying degrees by sucrose, lipids, phosphate buffers, monosaccharides, and hexoamines.
4. High concentrations of reducing sugars, ammonium sulfate, and sulfhydryl compounds interfere with the reaction.

9.2.6 Dye-Binding Methods

9.2.6.1 Anionic Dye-Binding Method

9.2.6.1.1 Principle The protein-containing sample is mixed with a known excess amount of **anionic dye** in a buffered solution. Proteins bind the dye to form an insoluble complex. The unbound soluble dye is measured after equilibration of the reaction and the removal of insoluble complex by centrifugation or filtration.

$$\text{Protein} + \text{excess dye} \rightarrow \text{Protein} - \text{dye insoluble complex} + \text{unbound soluble dye} \quad [10]$$

The anionic sulfonic acid dye, including acid orange 12, orange G, and Amido Black 10B, binds cationic groups of the **basic amino acid residues** (imidazole of histidine, guanidine of arginine, and ϵ-amino group of lysine) and the **free amino terminal group** of the protein (24). The amount of the unbound dye is inversely related to the protein content of the sample (24).

9.2.6.1.2 Procedure

1. The sample is finely ground (60 mesh or smaller sizes) and added to an excess dye solution with known concentration.
2. The content is vigorously shaken to equilibrate the dye binding reactions and filtered or centrifuged to remove insoluble substances.
3. Absorbance of the unbound dye solution in the filtrate or supernatant is measured and dye concentration is estimated from a dye standard curve.
4. A straight calibration curve can be obtained by plotting the unbound dye concentration against total nitrogen (as determined by Kjeldhal method) of a given food covering a wide range of protein content.
5. Protein content of the unknown sample of the same food type can be estimated from the calibration curve or from a regression equation calculated by the least squares method.

9.2.6.1.3 Applications Anionic dye binding has been used to estimate proteins in milk (25, 26), wheat flour (27), soy products (18), and meats (17). The AOAC approved methods include two dye-binding methods [Method 967.12 using Acid Orange 12 and Method 975.17 using Amido Black (10B) for analyzing proteins in milk]. AACC Method 46–14.02 uses Acid Orange 12 binding for measuring proteins in wheat flour and soy samples (28). An automated Sprint Rapid Protein

Analyzer has been developed by the CEM Company (Matthews, NC) based on the anionic dye-binding method. This automated method requires calibration for each type of food protein determined using other official methods.

Advantages:

1. Rapid (15 min or less), inexpensive, and relatively accurate for analyzing protein content in food commodities.
2. May be used to estimate the changes in available lysine content of cereal products during processing since the dye does not bind altered, unavailable lysine. Since lysine is the limiting amino acid in cereal products, the available lysine content represents protein nutritive value of the cereal products (29).
3. No corrosive reagents.
4. Does not measure nonprotein nitrogen.
5. More precise than the Kjeldahl method.

Disadvantages:

1. Not sensitive; milligram quantities of protein are required.
2. Proteins differ in basic amino acid content and so differ in dye-binding capacity. Therefore, a calibration curve for a given food commodity is required.
3. Not suitable for hydrolyzed proteins due to binding to N-terminal amino acids.
4. Some nonprotein components bind dye (i.e., starch) or protein (calcium or phosphate) and cause errors in final results. The problem with calcium and heavy metal ions can be eliminated using properly buffered reagent that contains oxalic acid.

9.2.6.2 Bradford Dye-Binding Method

9.2.6.2.1 Principle When **Coomassie Brilliant Blue G-250** binds to protein, the **dye changes color** from reddish to bluish, and the absorption maximum of the dye is shifted from 465 to 595 nm. The change in the absorbance at 595 nm is proportional to the protein concentration of the sample (30). Like other dye-binding methods, the Bradford relies on the **amphoteric nature of proteins**. When the protein-containing solution is acidified to a pH less than the isoelectric point of the protein(s) of interest, the dye added binds electrostatically. Binding efficiency is enhanced by hydrophobic interaction of the dye molecule with the polypeptide backbone adjoining positively charged residues in the protein (4). In the case of the Bradford method, the dye bound to protein has a change in absorbance spectrum relative to the unbound dye.

9.2.6.2.2 Procedure

1. Coomassie Brilliant Blue G-250 is dissolved in 95% ethanol and acidified with 85% phosphoric acid.
2. Samples containing proteins (1–100 μg/ml) and standard BSA solutions are mixed with the Bradford reagent.
3. Absorbance at 595 nm is read against a reagent blank.
4. Protein concentration in the sample is estimated from the BSA standard curve.

9.2.6.2.3 Applications The Bradford method has been used successfully to determine protein content in worts and beer products (31) and in potato tubers (32). This procedure has been improved to measure microgram quantities of proteins (33). Due to its rapidity, sensitivity, and fewer interferences than the Lowry method, the Bradford method has been used widely for the analysis of low concentrations of proteins and enzymes in their purification and characterizations.

Advantages:

1. Rapid; reaction can be completed in 2 min
2. Reproducible
3. Sensitive; several fold more sensitive than the Lowry method
4. No interference from ammonium sulfate, polyphenols, carbohydrates such as sucrose, or cations such as K^+, Na^+, and Mg^{+2}
5. Measures protein or peptides with molecular mass approximately equal to or greater than 4000 Da

Disadvantages:

1. Interfered with by both nonionic and ionic detergents, such as Triton X-100 and sodium dodecyl sulfate. However, errors due to small amounts (0.1%) of these detergents can be corrected using proper controls.
2. The protein–dye complex can bind to quartz cuvettes. The analyst must use glass or plastic cuvettes.
3. Color varies with different types of proteins. The standard protein must be selected carefully.

9.2.7 Bicinchoninic Acid Method

9.2.7.1 Principle

Proteins and peptides (as short as dipeptides) reduce **cupric ions** to **cuprous ions** under **alkaline conditions** (34), which is similar in principle to that of the biuret reaction. The cuprous ion then reacts with the apple-greenish **bicinchoninic acid** (BCA) **reagent** to

^{-}OOC N N COO^{-} Cu^{+1} ^{-}OOC N N COO^{-}

9-4 figure Protein reaction with cupric ions under alkaline conditions to form cuprous ions, which react with bicinchoninic acid (BCA) to form purple color, measured at 562 nm. (Figure Courtesy of Pierce Biotechnology Technical Library, Thermo Fisher Scientific, Inc., Rockford, IL.)

form a purplish complex (one cuprous ion is chelated by two BCA molecules) (Fig. 9-4). The color measured at 562 nm is near linearly proportional to protein concentration over a wide range of concentration from micrograms up to 2 mg/ml. Peptide bonds and four amino acids (cysteine, cystine, tryptophan, and tyrosine) contribute to the color formation with BCA.

9.2.7.2 Procedure

1. Mix (one step) the protein solution with the BCA reagent, which contains BCA sodium salt, sodium carbonate, NaOH, and copper sulfate, pH 11.25.
2. Incubate at 37°C for 30 min, or room temperature for 2 h, or 60°C for 30 min. The selection of the temperature depends upon sensitivity desired. A higher temperature gives a greater color response.
3. Read the solution at 562 nm against a reagent blank.
4. Construct a standard curve using BSA.

9.2.7.3 Applications

The BCA method has been used in protein isolation and purification. The suitability of this procedure for measuring protein in complex food systems has not been reported.

Advantages:

1. Sensitivity is comparable to that of the Lowry method; sensitivity of the micro-BCA method (0.5–10 μg) is better than that of the Lowry method.
2. One-step mixing is easier than in the Lowry method.
3. The reagent is more stable than for the Lowry reagent.
4. Nonionic detergent and buffer salts do not interfere with the reaction.
5. Medium concentrations of denaturing reagents (4 *M* guanidine-HCl or 3 *M* urea) do not interfere.

Disadvantages:

1. Color is not stable with time. The analyst needs to carefully control the time for reading absorbance.
2. Any compound capable of reducing Cu^{+2} to Cu^{+} will lead to color formation.
3. Reducing sugars interfere to a greater extent than in the Lowry method. High concentrations of ammonium sulfate also interfere.
4. Color variations among proteins are similar to those in the Lowry method.

9.2.8 Ultraviolet 280 nm Absorption Method

9.2.8.1 Principle

Proteins show strong absorption in the region at **ultraviolet** (UV) **280 nm**, primarily due to **tryptophan** and **tyrosine** residues in the proteins. Because the content of tryptophan and tyrosine in proteins from each food source is fairly constant, the absorbance at 280 nm could be used to estimate the concentration of proteins, using **Beer's law**. Since each protein has a unique aromatic amino acid composition, the extinction coefficient (E_{280}) or molar absorptivity (E_m) must be determined for individual proteins for protein content estimation.

9.2.8.2 Procedure

1. Proteins are solubilized in buffer or alkali.
2. Absorbance of protein solution is read at 280 nm against a reagent blank.
3. Protein concentration is calculated according to the equation

$$A = abc \quad [11]$$

where:

A = absorbance
a = absorptivity
b = cell or cuvette path length
c = concentration

9.2.8.3 Applications

The UV 280-nm method has been used to determine the protein contents of milk (35) and meat products (36). It has not been used widely in food systems. This technique is better applied in a purified protein system or to proteins that have been extracted in alkali or denaturing agents such as 8 *M* urea. Although peptide bonds in proteins absorb more strongly at 190–220 nm than at 280 nm, the low UV region is more difficult to measure.

Advantages:

1. Rapid and relatively sensitive; At 280 nm, 100 μg or more protein is required; several times more sensitive than the biuret method.
2. No interference from ammonium sulfate and other buffer salts.
3. Nondestructive; samples can be used for other analyses after protein determination; used very widely in postcolumn detection of proteins.

Disadvantages:

1. Nucleic acids also absorb at 280 nm. The absorption 280 nm/260 nm ratios for pure protein and nucleic acids are 1.75 and 0.5, respectively. One can correct the absorption of nucleic acids at 280 nm if the ratio of the absorption of 280 nm/260 nm is known. Nucleic acids also can be corrected using a method based on the absorption difference between 235 and 280 nm (37).
2. Aromatic amino acid contents in the proteins from various food sources differ considerably.
3. The solution must be clear and colorless. Turbidity due to particulates in the solution will increase absorbance falsely.
4. A relatively pure system is required to use this method.

9.3 COMPARISON OF METHODS

- **Sample preparation:** The Kjeldahl, Dumas, and infrared spectroscopy methods require little preparation. Sample particle size of 20 mesh or smaller generally is satisfactory for these methods. Some of the newer NIR instruments can make measurements directly on whole grains and other coarsely granulated products without grinding or other sample preparation. Other methods described in this chapter require fine particles for extraction of proteins from the complex food systems.
- **Principle:** The Dumas and Kjeldahl methods measure directly the nitrogen content of foods. However, the Kjeldahl method measures only organic nitrogen plus ammonia, while Dumas measures total nitrogen, including the inorganic fraction. (Therefore, Dumas gives a higher value for products that contain nitrates/nitrites.) Other methods of analysis measure the various properties of proteins. For instance, the biuret method measures peptide bonds, and the Lowry method measures a combination of peptide bonds and the amino acids tryptophan and tyrosine. Infrared spectroscopy is an indirect method to estimate protein content, based on the energy absorbed when a sample is subjected to a wavelength of infrared radiation specific for the peptide bond.
- **Sensitivity:** Kjeldahl, Dumas, and biuret methods are less sensitive than Lowry, Bradford, BCA, or UV methods.
- **Speed:** After the instrument has been properly calibrated, infrared spectroscopy is likely the most rapid of the methods discussed. In most other methods involving spectrophotometric (colorimetric) measurements, one must separate proteins from the interfering insoluble materials before mixing with the color reagents or must remove the insoluble materials from the colored protein–reagent complex after mixing. However, the speed of determination in the colorimetric methods and in the Dumas method is faster than with the Kjeldahl method.
- **Applications:** Although both Kjeldahl and Dumas methods can be used to measure N content in all types of foods, in recent years the Dumas method has largely replaced the Kjeldahl method for nutrition labeling (since Dumas method is faster, has a lower detection limit, and is safer). However, the Kjeldahl method is the preferred method for high-fat samples/products since fat may cause an instrument fire during the incineration procedure in the Dumas method. Also, the Kjeldahl method is specified to correct for protein content in an official method to measure the fiber content of foods (see Chap. 10, Sect. 10.5). Melamine, a toxic nitrogen adulterant, is included in the total nitrogen content if measured by the Kjeldahl or Dumas methods.

9.4 SPECIAL CONSIDERATIONS

1. To select a particular method for a specific application, sensitivity, accuracy, and reproducibility as well as physicochemical properties of food materials must be considered. The data should be interpreted carefully to reflect what actually is being measured.
2. Food processing methods, such as heating, may reduce the extractability of proteins for analysis and cause an underestimation of the protein content measured by methods involving an extraction step (9).
3. Except for the Dumas and Kjeldahl methods, and the UV method for purified proteins, all

methods require the use of a standard or reference protein or a calibration with the Kjeldahl method. In the methods using a standard protein, proteins in the samples are assumed to have similar composition and behavior compared with the standard protein. The selection of an appropriate standard for a specific type of food is important.

4. **Nonprotein nitrogen** is present in practically all foods. To determine **protein nitrogen**, the samples usually are extracted under alkaline conditions then precipitated with trichloroacetic acid or sulfosalicylic acid. The concentration of the acid used affects the precipitation yield. Therefore, nonprotein nitrogen content may vary with the type and concentration of the reagent used. Heating could be used to aid protein precipitation by acid, alcohol, or other organic solvents. In addition to acid precipitation methods used for nonprotein nitrogen determination, less empirical methods such as dialysis and ultrafiltration and column chromatography could be used to separate proteins from small nonprotein substances.
5. In the determination of the nutritive value of food proteins, including **protein digestibility** and **protein efficiency ratio** (PER), the Kjeldahl method with a 6.25 conversion factor usually is used to determine crude protein content. The PER could be underestimated if a substantial amount of nonprotein nitrogen is present in foods. A food sample with a higher nonprotein nitrogen content (particularly if the nonprotein nitrogen does not have many amino acids or small peptides) may have a lower PER than a food sample containing similar protein structure/composition and yet with a lower amount of nonprotein nitrogen.

9.5 SUMMARY

Methods based on the unique characteristics of proteins and amino acids have been described to determine the protein content of foods. The Kjeldahl and Dumas methods measure nitrogen. Infrared spectroscopy is based on absorption of a wavelength of infrared radiation specific for the peptide bond. Copper–peptide bond interactions contribute to the analysis by the biuret and Lowry methods. Amino acids are involved in the Lowry, dye-binding, and UV 280 nm methods. The BCA method utilizes the reducing power of proteins in an alkaline solution. The various methods differ in their speed and sensitivity.

In addition to the commonly used methods discussed, there are other methods available for protein quantification. Because of the complex nature of various food systems, problems may be encountered to different degrees in protein analysis by available methods. Rapid methods may be suitable for quality control purposes, while a sensitive method is required for work with a minute amount of protein. Indirect colorimetric methods usually require the use of a carefully selected protein standard or a calibration with an official method (e.g., Kjeldahl).

9.6 STUDY QUESTIONS

1. What factors should one consider when choosing a method for protein determination?
2. The Kjeldahl method of protein analysis consists of three major steps. List these steps in the order they are done and describe in words what occurs in each step. Make it clear why milliliters of HCl can be used as an indirect measure of the protein content of a sample.
3. Why is the conversion factor from Kjeldahl nitrogen to protein different for various foods, and how is the factor of 6.25 obtained?
4. How can Nesslerization or the procedure that uses phenol and hypochlorite be used as part of the Kjeldahl procedure, and why might they be best for the analysis?
5. Differentiate and explain the chemical basis of the following techniques that can be used to quantitate proteins in quality control/research:
   (a) Kjeldahl method
   (b) Dumas method (N combustion)
   (c) Infrared spectroscopy
   (d) Biuret method
   (e) Lowry method
   (f) Bradford method
   (g) Bicinchoninic acid method
   (h) Absorbance at 280 nm
   (i) Absorbance at 220 nm
6. Differentiate the principles of protein determination by dye binding with an anionic dye such as Amido Black vs. with the Bradford method, which uses the dye Coomassie Blue G-250.
7. With the anionic dye-binding method, would a sample with a higher protein content have a higher or a lower absorbance reading than a sample with a low protein content? Explain your answer.
8. For each of the situations described below, identify a protein assay method most appropriate for use, and indicate the chemical basis of the method (i.e., what does it really measure?)
   (a) Nutrition labeling
   (b) Intact protein eluting from a chromatography column; qualitative or semiquantitative method
   (c) Intact protein eluting from a chromatography column; colorimetric, quantitative method
   (d) Rapid, quality control method for protein content of cereal grains
9. The FDA found melamine (see structure below) in pet food linked to deaths of pets in the United States. The

FDA also found evidence of melamine in wheat gluten imported from China used as one of the ingredients in the production of the pet food. Melamine is a nitrogen-rich chemical used to make plastic and sometimes used as a fertilizer.

NH_2 / N N / H_2N N NH_2

(a) Knowing that each ingredient is tested and analyzed when imported, explain how melamine in wheat gluten could have escaped detection.
(b) How can the adulteration of wheat gluten be detected (not necessarily detecting melamine specifically), using a combination of protein analysis methods? Explain your answer.

9.7 PRACTICE PROBLEMS

1. A dehydrated precooked pinto bean was analyzed for crude protein content in duplicate using the Kjeldahl method. The following data were recorded:
   - Moisture content = 8.00%
   - Wt of Sample 1 = 1.015 g
   - Wt of Sample 2 = 1.025 g
   - Normality of HCl used for titration = 0.1142 *N*
   - HCl used for Sample 1 = 22.0 ml
   - HCl used for Sample 2 = 22.5 ml
   - HCl used for reagent blank = 0.2 ml

   Calculate crude protein content on both wet and dry weight basis of the pinto bean, assuming pinto bean protein contains 17.5% nitrogen.
2. A 20 ml protein fraction recovered from a column chromatography was analyzed for protein using the BCA method. The following data were the means of a duplicate analysis using BSA as a standard:

BSA (mg/ml)	*Mean Absorbance at 562 nm*
0.2	0.25
0.4	0.53
0.6	0.74
0.8	0.95
1.0	1.15

The average absorbance of a 1-ml sample was 0.44. Calculate protein concentration (mg/ml) and total protein quantity of this column fraction.

Answers

1. Protein content = 19.75% on a wet weight basis; 21.47% on a dry weight basis.

Calculations:

$$\%\,N = N\,HCl \times \frac{\text{Corrected acid volume}}{\text{g of sample}} \times \frac{14\,\text{g N}}{\text{mol}} \times 100 \quad [6]$$

where:

*N*HCl = normality of HCl, in mol/1000 ml
Corrected acid vol. = (ml std. acid for sample) – (ml std. acid for blank)
14 = atomic weight of nitrogen

Corrected acid volume for Sample 1
= 22.0 ml – 0.2 ml = 21.8 ml
Corrected acid volume for Sample 2
= 22.5 ml – 0.2 ml = 22.3 ml

%N for Sample 1

$$= \frac{0.1142\,\text{mol}}{1000\,\text{ml}} \times \frac{21.8\,\text{ml}}{1.015\,\text{g}} \times \frac{14\,\text{g N}}{\text{mol}} \times 100\% = 3.433\%$$

%N for Sample 2

$$= \frac{0.1142\,\text{mol}}{1000\,\text{ml}} \times \frac{22.3\,\text{ml}}{1.025\,\text{g}} \times \frac{14\,\text{g N}}{\text{mol}} \times 100\% = 3.478\%$$

Protein conversion factor = 100%/17.5% N = 5.71
Crude protein content for Sample 1
= 3.433% × 5.71 = 19.6%
Crude protein content for Sample 2
= 3.478% × 5.71 = 19.9%

The average for the duplicate data = (19.6% + 19.9%)/2 = 19.75% = ~19.8% wet weight basis.

To calculate protein content on a dry weight basis: Sample contain 8% moisture, therefore, the sample contains 92% dry solids, or 0.92 g out of 1-g sample. Therefore, protein on a dry weight basis can be calculated as follows = 19.75%/0.92 g dry solids = 21.47% = ~21.5% dry weight basis.

2. Protein content = 0.68 mg/ml. Total protein quantity = 6.96 mg

Calculations:
Plot absorbance (*y*-axis, absorbance at 562 nm) vs. BSA protein concentration (*x*-axis, mg/ml)) using the data above. Determine the equation of the line ($y = 1.11x + 0.058$), then use this equation and the given absorbance ($y = 0.44$) to calculate the concentration ($x = 0.344$ mg/ml). Since 1 ml of sample gives a concentration of 0.344 mg/ml and we have a total of 20 ml collected from column chromatography, we will have a total of (0.344 mg/ml × 20 ml) = 6.88 mg protein in this collected column fraction.

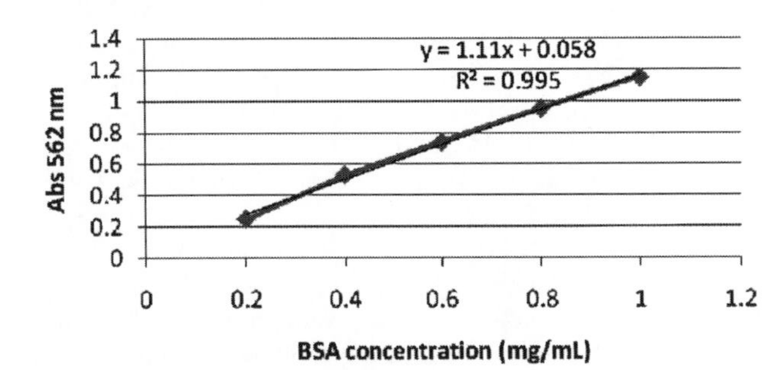

9.8 REFERENCES

1. Jones DB (1931) Factors for converting percentages of nitrogen in foods and feeds into percentages of proteins. US Dept. Agric. Circular No. 183, August. USDA, Washington, DC
2. US Department of Agriculture, Agricultural Research Service (2008) USDA National Nutrient Database for Standard References, Release 21. Nutrient Data Laboratory Home Page. http://www.ars.usda.gov/ba/bhnrc/ndl
3. AOAC International (2007) Official Methods of Analysis, 18th edn, 2005; Current through revision 2, 2007 (On-line). AOAC International, Gaithersburg, MD
4. Yada RY, Jackman RL, Smith JL, Marangoni AG (1996) Analysis: quantitation and physical characterization (Chapter 7). In: Nakai S, Modler HW (eds) Food proteins. Properties and characterization. VCH, New York, pp 333–403
5. Kolakowski E (2001) Protein determination and analysis in food system (Chapter 4). In: Sikorski ZE (ed) Chemical and functional properties of food protein. Technomic Publishing, Lancaster, PA, pp 57–112
6. Owusu-Apenten RK (2002) Food protein analysis. Quantitative effects on processing. Marcel Dekker, New York
7. Bradstreet RB (1965) The Kjeldahl method for organic nitrogen. Academic, New York
8. Mossé J (1990) Nitrogen to protein conversion factor for ten cereals and six legumes or oilseeds. A reappraisal of its definition and determination. Variation according to species and to seed protein content. J Agric Food Chem 38:18–24
9. Wilson PR (1990) A new instument concept for nitrogen/protein analysis. A challenge to the Kjeldahl method. Aspects Appl Biol 25:443–446
10. Wiles PG, Gray I, Kissling RC (1998) Routine analysis of proteins by Kjeldahl and Dumas methods: review and interlaboratory study using dairy products. J AOAC Int 81:620–632
11. O'Sullivan A, O'Connor B, Kelly A, McGrath MJ (1999) The use of chemical and infrared methods for analysis of milk and dairy products. Int J Dairy Technol 52:139–148
12. Luinge HJ, Hop E, Lutz ETG, van Hemert JA, de Jong EAM (1993) Determination of the fat, protein and lactose content of milk using Fourier transform infrared spectrometry. Anal Chim Acta 284:419–433
13. Krishnan PG, Park WJ, Kephart KD, Reeves DL, Yarrow GL (1994) Measurement of protein and oil content of oat cultivars using near-infrared reflectance. Cereal Foods World 39(2):105–108
14. Robinson HW, Hodgen CG (1940) The biuret reaction in the determination of serum protein. 1. A study of the conditions necessary for the production of the stable color which bears a quantitative relationship to the protein concentration. J Biol Chem 135:707–725
15. Jennings AC (1961) Determination of the nitrogen content of cereal grain by colorimetric methods. Cereal Chem 38:467–479
16. Pinckney AJ (1961) The biuret test as applied to the estimation of wheat protein. Cereal Chem 38:501–506
17. Torten J, Whitaker JR (1964) Evaluation of the biuret and dye-binding methods for protein determination in meats. J Food Sci 29:168–174
18. Pomeranz Y (1965) Evaluation of factors affecting the determination of nitrogen in soya products by the biuret and orange-G dye-binding methods. J Food Sci 30:307–311
19. AOAC (1965) Official methods of analysis, 10th edn. Association of Official Analytical Chemists, Washington, DC
20. Lowry OH, Rosebrough NJ, Farr AL, Randall RJ (1951) Protein measurement with the Folin phenol reagent. J Biol Chem 193:265–275
21. Peterson GL (1979) Review of the Folin phenol protein quantitation method of Lowry, Rosebrough, Farr, and Randall. Anal Biochem 100:201–220
22. Miller GL (1959) Protein determination for large numbers of samples. Anal Chem 31:964
23. Hartree EF (1972) Determination of protein: a modification of the Lowry method that gives a linear photometric response. Anal Biochem 48:422–427
24. Fraenkel-Conrat H, Cooper M (1944) The use of dye for the determination of acid and basic groups in proteins. J Biol Chem 154:239–246
25. Udy DC (1956) A rapid method for estimating total protein in milk. Nature 178:314–315
26. Tarassuk NP, Abe N, Moats WA (1966) The dye binding of milk proteins. Technical bulletin no. 1369. USDA Agricultural Research Service in cooperation with California Agricultural Experiment Station. Washington, DC
27. Udy DC (1954) Dye-binding capacities of wheat flour protein fractions. Cereal Chem 31:389–395
28. AACC International (2010) Approved methods of American association of cereal chemists international, 11th edn (On-line). American Association of Cereal Chemists, St. Paul, MN
29. Hurrel RF, Lerman P, Carpenter KJ (1979) Reactive lysine in foodstuffs as measured by a rapid dye-binding procedure. J Food Sci 44:1221–1227
30. Bradford M (1976) A rapid and sensitive method for the quantitation of microgram quantities of protein utilizing the principle of protein–dye binding. Anal Biochem 72:248–254
31. Lewis MJ, Krumland SC, Muhleman DJ (1980) Dye-binding method for measurement of protein in wort and beer. J Am Soc Brew Chem 38:37–41
32. Snyder J, Desborou S (1978) Rapid estimation of potato tuber total protein content with Coomassie Brilliant Blue G-250. Theor Appl Genet 52:135–139
33. Bearden JC Jr (1978) Quantitation of submicrogram quantities of protein by an improved protein–dye binding assay. Biochim Biophys Acta 533:525–529
34. Smith PK, Krohn RL, Hermanson GT, Mallia AK, Gartner FH, Provensano MD, Fujimoto EK, Goeke NM, Olson BJ, Klenk DC (1985) Measurement of protein using bicinchoninic acid. Anal Biochem 150:76–85
35. Nakai S, Wilson HK, Herreid EO (1964) Spectrophotometric determination of protein in milk. J Dairy Sci 47:356–358
36. Gabor E (1979) Determination of the protein content of certain meat products by ultraviolet absorption spectrophotometry. Acta Alimentaria 8(2):157–167
37. Whitaker JR, Granum PE (1980) An absolute method for protein determination based on difference in absorbance at 235 and 280 nm. Anal Biochem 109:156–159

Carbohydrate Analysis

James N. BeMiller
Department of Food Science, Purdue University, West Lafayette, IN 47907–2009, USA
bemiller@purdue.edu

10.1 Introduction 149
10.2 Sample Preparation 151
10.3 Mono- and Oligosaccharides 152
10.3.1 Extraction 152
10.3.2 Total Carbohydrate: Phenol-Sulfuric Acid Method 152
10.3.2.1 Principle and Characteristics 152
10.3.2.2 Outline of Procedure 153
10.3.3 Total Reducing Sugar 153
10.3.3.1 Somogyi–Nelson Method 153
10.3.3.1.1 Principle 153
10.3.3.1.2 Outline of Procedure 154
10.3.3.2 Other Methods (3) 154
10.3.4 Specific Analysis of Mono- and Oligosaccharides 155
10.3.4.1 High-performance Liquid Chromatography 155
10.3.4.1.1 Stationary Phases 155
10.3.4.1.2 Detectors 156
10.3.4.2 Gas Chromatography 157

S.S. Nielsen, *Food Analysis*, Food Science Texts Series, DOI 10.1007/978-1-4419-1478-1_10,

10.3.4.2.1 Neutral Sugars: Outline of Procedure (38) 157
10.3.4.2.2 Hydrolyzates of Polysaccharides Containing Uronic Acids: Outline of Procedure (41) 158
10.3.4.3 Enzymic Methods 159
10.3.4.3.1 Overview 159
10.3.4.3.2 Sample Preparation 159
10.3.4.3.3 Enzymic Determination of D-Glucose 159
10.3.4.4 Mass Spectrometry 159
10.3.4.5 Thin-layer Chromatography 160
10.3.4.6 Capillary Electrophoresis (46,47) 160
10.4 Polysaccharides 160
10.4.1 Starch 160
10.4.1.1 Total Starch 160
10.4.1.1.1 Principle 160
10.4.1.1.2 Potential Problems 160
10.4.1.1.3 Outline of Procedure 161
10.4.1.2 Degree of Gelatinization of Starch 162
10.4.1.3 Degree of Retrogradation of Starch 162
10.4.2 Nonstarch Polysaccharides (Hydrocolloids/Food Gums) 162
10.4.2.1 Overview 162
10.4.2.2 Hydrocolloid/Food Gum Content Determination 163
10.4.2.3 Pectin 165
10.4.2.3.1 Nature of Pectin 165
10.4.2.3.2 Pectin Content Determination 165
10.4.2.3.3 Degree of Esterification 165
10.5 Dietary Fiber 165
10.5.1 Introduction 165
10.5.1.1 Importance of Dietary Fiber 166
10.5.1.2 Definition 166
10.5.2 Major Components of Dietary Fiber 167
10.5.2.1 Cell-Wall Polysaccharides of Land Plants 167
10.5.2.1.1 Cellulose 167
10.5.2.1.2 Hemicelluloses 167
10.5.2.1.3 Pectins 167
10.5.2.2 Hydrocolloids/Food Gums as Dietary Fiber 167
10.5.2.3 Resistant Starch 167
10.5.2.4 Lignin 168
10.5.3 General Considerations 168
10.5.4 Methods 168
10.5.4.1 Overview 168
10.5.4.2 Sample Preparation 168
10.5.4.3 Methods 169
10.5.4.3.1 Overview 169
10.5.4.3.2 AOAC Method 991.43 (AACC Method 32-07.01) 169
10.6 Physical Methods 171
10.6.1 Microscopy 171
10.6.2 Mass and NIR Transmittance Spectrometry 171
10.6.3 Specific Gravity 172
10.6.4 Refractive Index 172
10.7 Summary 172
10.8 Study Questions 173
10.9 Practice Problems 173
10.10 References 175

10.1 INTRODUCTION

Carbohydrates are important in foods as a major source of energy, to impart crucial textural properties, and as dietary fiber which influences physiological processes. Digestible carbohydrates, which are converted into monosaccharides, which are absorbed, provide metabolic energy. Worldwide, carbohydrates account for more than 70% of the caloric value of the human diet. It is recommended that all persons should limit calories from fat (the other significant source) to not more than 30% and that most of the carbohydrate calories should come from starch. Nondigestible polysaccharides (all those other than starch) comprise the major portion of dietary fiber (Sect. 10.5). Carbohydrates also contribute other attributes, including bulk, body, viscosity, stability to emulsions and foams, water-holding capacity, freeze-thaw stability, browning, flavors, aromas, and a range of desirable textures (from crispness to smooth, soft gels). They also provide satiety. Basic carbohydrate structures, chemistry, and terminology can be found in references (1,2).

Major occurrences of major carbohydrates in foods are presented in Table 10-1. Ingested carbohydrates are almost exclusively of plant origin, with milk lactose being the major exception. Of the **monosaccharides** (sometimes called **simple sugars**), only D-glucose and D-fructose are found in other than minor amounts. Monosaccharides are the only carbohydrates that can be absorbed from the small intestine. Higher saccharides (**oligo-** and **polysaccharides**) must first be digested (i.e., hydrolyzed to monosaccharides) before absorption and utilization can occur. (Note: There is no official definition of an oligosaccharide. Most sources consider an oligosaccharide to be a carbohydrate composed of from 2 to 10 sugar (saccharide) units. A polysaccharide usually contains from 30 to at least 60,000 monosaccharide units.) Humans can digest only sucrose, lactose, maltooligosaccharides/maltodextrins, and starch. All are digested with enzymes found in the small intestine.

At least 90% of the carbohydrate in nature is in the form of polysaccharides. As stated above, starch polymers are the only polysaccharides that humans can digest and use as a source of calories and carbon. All other polysaccharides are nondigestible. **Nondigestible polysaccharides** can be divided into **soluble** and **insoluble** classes. Along with lignin and other nondigestible, nonabsorbed substances, they make up **dietary fiber** (Sect. 10.5). As dietary fiber, they regulate normal bowel function, reduce the postprandial hyperglycemic response, and may lower serum cholesterol, among other effects. However, nondigestible polysaccharides most often are added to processed foods because of the functional properties they impart, rather than for a physiological effect. Nondigestible oligosaccharides serve as prebiotics and are, therefore, increasingly used as ingredients in functional foods and neutraceuticals. The foods in which dietary fiber components can be used, and particularly the amounts that can be incorporated, are limited because addition above a certain level usually changes the characteristics of the food product. Indeed, as already stated, they are used often as ingredients because of their ability to impart important functional properties at a low level of usage.

Carbohydrate analysis is important from several perspectives. Qualitative and quantitative analysis is used to determine compositions of foods, beverages, and their ingredients. **Qualitative analysis** ensures that ingredient labels present accurate compositional information. **Quantitative analysis** ensures that added components are listed in the proper order on ingredient labels. Quantitative analysis also ensures that amounts of specific components of consumer interest, for example, **β-glucan**, are proper and that caloric content can be calculated. Both qualitative and quantitative analysis can be used to authenticate (i.e., to detect adulteration of) food ingredients and products.

In this chapter, the most commonly used methods of carbohydrate determination are presented. [A thorough description of the analytical chemistry of carbohydrates was published in 1998 (3).] However, methods often must be made specific to a particular food product because of the nature of the product and the presence of other constituents. Approved methods are referenced, but method approval has not kept pace with methods development; so where better methods are available, they are also presented. Methods that have been in long-time use, although not giving as much or as precise information as newer methods, nevertheless may be useful for quality assurance and product standardization in some cases.

In general, evolution of analytical methods for carbohydrates has followed the succession: qualitative color tests, adaptation of the color test for reducing sugars based on reduction of Cu(II) to Cu(I) (Fehling test) to quantitation of reducing sugars, qualitative paper chromatography, quantitative paper chromatography, gas chromatography (GC) of derivatized sugars, qualitative and quantitative thin-layer chromatography, enzymic methods, and high-performance liquid chromatography (HPLC). Multiple official methods for the analysis of mono- and disaccharides in foods are currently approved by AOAC International (4, 5); some are outdated, but still used. Methods continue to be developed and refined. Methods employing nuclear magnetic resonance, near-infrared (NIR) spectrometry (Sect. 10.6.2 and Chap. 23), antibodies (Immunoassays; Chap. 17),

Occurrences of Some Major Carbohydrates in Foods

Carbohydrate	*Source*	*Constituent(s)*
Monosaccharides[a]		
D-Glucose (Dextrose)	Naturally occurring in honey, fruits, and fruit juices. Added as a component of corn (glucose) syrups and high-fructose syrups. Produced during processing by hydrolysis (inversion) of sucrose.	
D-Fructose	Naturally occurring in honey, fruits, and fruit juices. Added as a component of high-fructose syrups. Produced during processing by hydrolysis (inversion) of sucrose.	
Sugar alcohol[a]		
Sorbitol (D-Glucitol)	Added to food products, primarily as a humectant	
Disaccharides[a]		
Sucrose	Widely distributed in fruit and vegetable tissues and juices in varying amounts. Added to food and beverage products	D-Fructose D-Glucose
Lactose	In milk and products derived from milk	D-Galactose D-Glucose
Maltose	In malt. In varying amounts in various corn (glucose) syrups and maltodextrins	D-Glucose
Higher oligosaccharides[a]		
Maltooligosaccharides	Maltodextrins. In varying amounts in various glucose (corn) syrups	D-Glucose
Raffinose	Small amounts in beans	D-Glucose D-Fructose D-Galactose
Stachyose	Small amounts in beans	D-Glucose D-Fructose D-Galactose
Polysaccharides		
Starch[b]	Widespread in cereal grains and tubers. Added to processed foods.	D-Glucose
Food gums/hydrocolloids[c]		
Algins Carboxymethylcelluloses Carrageenans Curdlan Gellan Guar gum Gum arabic Hydroxypropylmethyl-celluloses Inulin Konjac glucomannan Locust bean gum Methylcelluloses Pectins Xanthan	Added as ingredients	d
Cell-wall polysaccharides[c]		
Pectin (native) Cellulose Hemicelluloses Beta-glucan	Naturally occurring	

[a]For analysis, see Sect. 10.3.4.
[b]For analysis, see Sect. 10.4.1.1.
[c]For analysis, see Sect. 10.4.2.
[d]For compositions, characteristics, and applications, see reference (2) and Table 10-2.

fluorescence spectrometry (Chap. 22), capillary electrophoresis (Sect. 10.3.4.6), and mass spectrometry (Sect. 10.3.4.4) have been published, but are not yet in general use for carbohydrate analysis.

It should be noted that, according to the nutrition labeling regulations of the US Food and Drug Administration, the "**total carbohydrate**" content of a food (Table 10-2), which is declared in relation to a serving, which is defined as the amount of food customarily consumed per eating occasion by persons 4-years of age or older [(6), paragraph (b)(1)], must be calculated by subtraction of the sums of the weights of crude protein, total fat, moisture, and ash in a serving from the total weight of the food in a serving [(6), paragraph (c)(6)] (i.e., carbohydrate is determined by difference). The grams of dietary fiber (Sect. 10.5) in a serving also must be stated on the label [(6), paragraph (c)(6)(i)]. The content of "**other carbohydrate**" (formerly called "**complex carbohydrate**") is obtained by calculating the difference between the amount of "total carbohydrate" and the sum of the amounts of dietary fiber and sugars (Table 10-1). For labeling purposes, **sugars** are defined as the sum of all free monosaccharides (viz., D-glucose and -fructose) and disaccharides [viz., sucrose, lactose, and maltose (if a maltodextrin or glucose/corn syrup has been added)] [(6), paragraph (c)(6)(ii)] (Table 10-1). Other carbohydrates are likely to be **sugar alcohols** (alditols, polyhydroxy alcohols, polyols), such as sorbitol and xylitol, the specific declaring of which is also voluntary [(6), paragraph (c)(6)(iii)].

10-2 table **Total Carbohydrate Contents of Selected Foods[a]**

Food	*Approximate Percent Carbohydrate (Wet Weight Basis)*
Cereals, bread, and pasta	
Corn flakes	80.4
Macaroni, dry, enriched	74.7
Bread, white, commercially prepared	50.6
Dairy products	
Ice cream, chocolate	28.2
Yogurt, plain, low fat (12 g protein/8 oz)	7.0
Milk, reduced fat, fluid, 2%	4.7
Milk, chocolate, commercial, whole	10.3
Fruits and vegetables	
Apple sauce, canned, sweetened, with salt	19.9
Grapes, raw	17.2
Apples, raw, with skin	13.8
Potatoes, raw, with skin	12.4
Orange juice, raw	10.4
Carrots, raw	9.6
Broccoli, raw	6.6
Tomato, tomato juice, canned, with salt	4.2
Meat, poultry, and fish	
Fish fillets, battered or breaded, fried	17.0
Bologna, beef	4.0
Chicken, broilers or fryers, breast meat	0
Other	
Honey	82.4
Salad dressing, Italian, fat free	11.0
Salad dressing, Italian, regular	10.4
Carbonated beverage, cola, contains caffeine	9.6
Cream of mushroom soup, from condensed and canned	6.7
Light beer	1.6

[a]In part from US Department of Agriculture, Agricultural Research Service (2009). USDA National Nutrient Database for Standard Reference. Release 22. Nutrient Data Laboratory Home Page, http://www.ars.usda.gov/ba/bhnrc/ndl

10.2 SAMPLE PREPARATION

Sample preparation is related to the specific raw material, ingredient, or food product being analyzed and the specific carbohydrate being determined, because carbohydrates have such a wide range of solubilities. However, some generalities can be presented (Fig. 10-1).

For most foods, the first step is drying, which also can be used to determine moisture content. For other than beverages, drying is done by placing a weighed amount of material in a vacuum oven and drying to constant weight at 55°C and 1 mm Hg pressure. Then, the material is ground to a fine powder, and lipids are extracted using 19:1 vol/vol chloroform–methanol

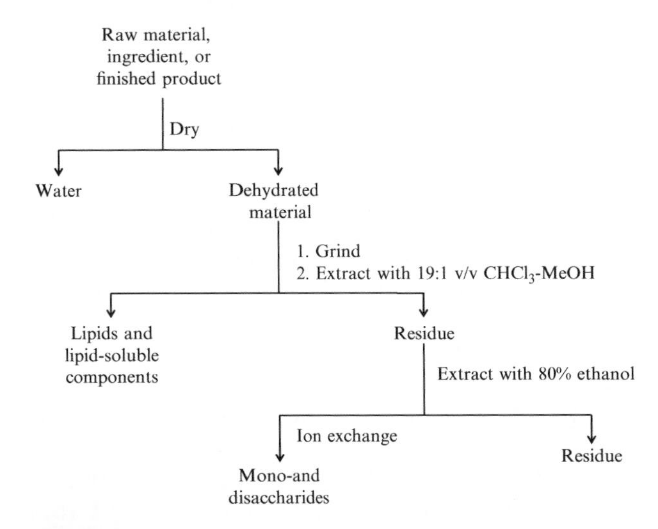

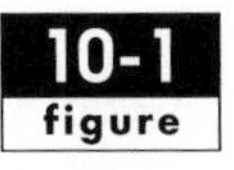
10-1 figure Flow diagram for sample preparation and extraction of mono- and disaccharides.

in a Soxhlet extractor (Chap. 8). (Note: Chloroform–methanol forms an azeotrope boiling at 54°C with a mole ratio of 0.642:0.358 or a vol/vol ratio of 3.5:1 in the vapor.) Prior extraction of lipids makes extraction of carbohydrates easier and more complete.

However, other sample preparation schemes may be required. For example, the AOAC International method (3) for presweetened, ready-to-eat breakfast cereals calls for removal of fats by extraction with petroleum ether (hexane) rather than the method described above and extraction of sugars with 50% ethanol (AOAC Method 982.14), rather than the method described below.

10.3 MONO- AND OLIGOSACCHARIDES

10.3.1 Extraction

Food raw materials and products and some ingredients are complex, heterogeneous, biological materials. Thus, it is quite likely that they may contain substances that interfere with measurement of the mono- and oligosaccharides present, especially if a spectrophotometric method is used. Interference may arise either from compounds that absorb light of the same wavelength used for the carbohydrate analysis or from insoluble, colloidal material that scatters light, since light scattering will be measured as absorbance. Also, the aldehydo or keto group of the sugar can react with other components, especially amino groups of proteins, a reaction (the **nonenzymatic browning** or **Maillard reaction**) that simultaneously produces color and destroys the sugar. Even if chromatographic methods, such as HPLC (Sect. 10.3.4.1), are used for analysis, the mono- and oligosaccharides must usually be separated from other components of the food before chromatography. Thus, for determination of any mono- (glucose, fructose), di- (sucrose, lactose, maltose), tri- (raffinose), tetra- (stachyose), or other oligo- (maltodextrins) saccharides present, the dried, lipid-free sample is extracted with **hot 80% ethanol** (final concentration) in the presence of precipitated calcium carbonate to neutralize any acidity (AOAC Method 922.02, 925.05). Higher oligosaccharides from added malto- or fructooligosaccharides also may be extracted. Carbohydrates are soluble in polar solvents. However, much of the composition of a food (other than water) is in the form of polymers, and almost all polysaccharides and proteins are insoluble in hot 80% ethanol. Thus, this extraction is rather specific. Extraction is done by a batch process. Refluxing for 1 h, cooling, and filtering is standard practice. (A Soxhlet apparatus cannot be used because aqueous ethanol undergoes azeotropic distillation as 95% ethanol.) Extraction should be done at least twice to check for and ensure completeness of extraction. If the foodstuff or food product is particularly acidic, for example a low-pH fruit, neutralization before extraction may be necessary to prevent hydrolysis of sucrose, which is particularly acid labile; thus, precipitated calcium carbonate is routinely added.

The 80% ethanol extract will contain components other than carbohydrates, in particular ash, pigments, organic acids, and perhaps free amino acids and low-molecular-weight peptides. Because the mono- and oligosaccharides are neutral and the contaminants are charged, the contaminants can be removed by **ion-exchange** techniques (Chap. 27). Because reducing sugars can be adsorbed onto and be isomerized by strong anion-exchange resins in the hydroxide form, a weak anion-exchange resin in the carbonate (CO_3^{2-}) or hydrogencarbonate (HCO_3^-) form is used. [**Reducing sugars** are those mono- and oligosaccharides that contain a free carbonyl (aldehydo or keto) group and, therefore, can act as reducing agents; see Sect. 10.3.3.] Because sucrose and sucrose-related oligosaccharides are very susceptible to acid-catalyzed hydrolysis, the anion-exchange resin should be used before the cation-exchange resin. However, because the anion-exchange resin is in a carbonate or hydrogencarbonate form, the cation-exchange resin (in H^+ form) cannot be used in a column because of CO_2 generation. Mixed-bed columns are not recommended for the same reason. AOAC Method 931.02C reads basically as follows for cleanup of ethanol extracts: Place a 50-ml aliquot of the ethanol extract in a 250-ml Erlenmeyer flask. Add 3 g of anion-exchange resin (hydroxide form) and 2 g of cation-exchange resin (acid form). Let it stand for 2 h with occasional swirling.

The aqueous alcohol of the ethanol extract is removed under reduced pressure using a **rotary evaporator** (Fig. 10-2) and a temperature of 45–50°C. The residue is dissolved in a known, measured amount of water. Filtration should not be required, but should be used if necessary. Some methods employ a final passage through a hydrophobic column such as a Sep-Pak C18 cartridge (Waters Associates, Milford, MA) as a final cleanup step to remove any residual lipids, proteins, and/or pigments, but this should not be necessary if the lipids and lipid-soluble components were properly removed prior to extraction. (Extracts may contain minor carbohydrates, such as cyclitols and naturally occurring or added sugar alcohols. These are not considered in Sects. 10.3.2 or 10.3.3.)

10.3.2 Total Carbohydrate: Phenol-Sulfuric Acid Method

10.3.2.1 Principle and Characteristics

Carbohydrates are destroyed by strong acids and/or high temperatures. Under these conditions, a series of

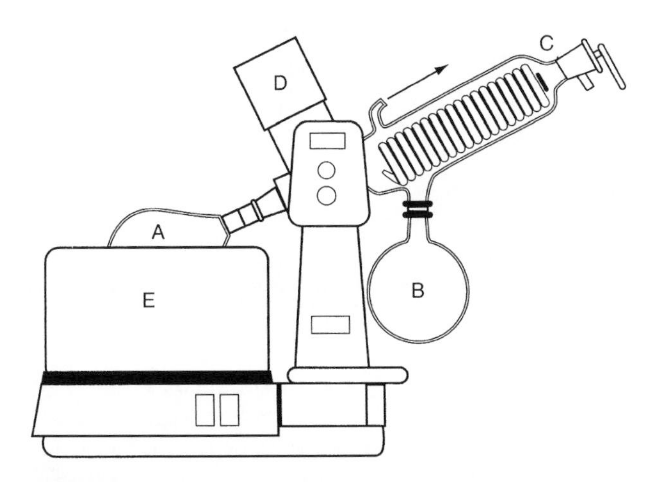

10-2 figure Diagram of a rotary evaporator. The solution to be concentrated is placed in the round-bottom Flask A in a water bath (E) at a controlled temperature. The system is evacuated by means of a water aspirator or pump; connecting tubing is attached at the arrow. Flask (A) turns (generally slowly). Evaporation is relatively rapid from a thin film on the inside walls of flask (A) produced by its rotation because of the reduced pressure, the large surface area, and the elevated temperature. C is a condenser. D is the motor. Condensate collects in flask B. The stopcock at the top of the condenser is for releasing the vacuum.

complex reactions takes place, beginning with a simple dehydration reaction as shown in Equation [1].

$$-\underset{\text{OH}}{\overset{\text{H}}{\text{C}}}-\underset{\text{OH}}{\overset{\text{H}}{\text{C}}}- \xrightarrow{-H_2O} -\overset{\text{H}}{\text{C}}=\underset{\text{OH}}{\text{C}}- \rightleftharpoons -CH_2-\underset{\text{O}}{\overset{}{\text{C}}}(=\text{O})- \quad [1]$$

Continued heating in the presence of acid produces various furan derivatives (Fig. 10-3). These products then condense with themselves and other products to produce brown and black substances. They will also condense with various phenolic compounds, such as phenol, resorcinol, orcinol, α-naphthol, and napthoresorcinol, and with various aromatic amines, such as aniline and *o*-toluidine, to produce colored compounds that are useful for carbohydrate analysis (3, 6).

The most often used condensation is with phenol itself (3, 7–10) (AOAC Method 44.1.30). This method is simple, rapid, sensitive, accurate, specific for carbohydrates, and widely applied. The reagents are inexpensive, readily available, and stable. Virtually all classes of sugars, including sugar derivatives and oligo- and polysaccharides, can be determined with the phenol-sulfuric acid method. (Oligo- and polysaccharides react because they undergo hydrolysis in the presence of the hot, strong acid, releasing monosaccharides.) A stable color is produced, and results are reproducible. Under proper conditions, the phenol-sulfuric method is accurate to ±2%.

Neither this method nor those for measuring reducing sugar content (Sect. 10.3.3) involves stoichiometric reactions. The extent of reaction is, in part, a function of the structure of the sugar. Therefore, a standard curve must be used. Ideally, the standard curve will be prepared using mixtures of the same sugars present in the same ratio as they are found in the unknown. If this is not possible, for example, if a pure preparation of the sugar being measured is not available, or if more than one sugar is present either as free sugars in unknown proportions or as constituent units of oligo- or polysaccharides or mixtures of them, D-glucose is used to prepare the standard curve. In these cases, accuracy is determined by conformity of the standard curve made with D-glucose to the curve that would be produced from the exact mixture of carbohydrates being determined. In any analysis, the concentrations used to construct the standard curve must span the sample concentrations and beyond (i.e., all sample concentrations must fall within the limits of the standard concentrations), and both must be within the limits reported for sensitivity of the method. If any concentrations are greater than the upper limit of the sensitivity range, dilutions should be used.

The phenol-sulfuric acid procedure is often used as a qualitative test for the presence of carbohydrate. Neither sorbitol nor any other alditol (polyol, polyhydroxyalchol) gives a positive test.

10.3.2.2 Outline of Procedure

1. A clear, aqueous solution of carbohydrate(s) is transferred using a pipette into a small tube. A blank of water also is prepared.
2. An aqueous solution of phenol is added, and the contents are mixed.
3. Concentrated sulfuric acid is added rapidly to the tube so that the stream produces good mixing. The tube is agitated. (Adding the sulfuric acid to the water produces considerable heat.) A yellow-orange color results.
4. Absorbance is measured at 490 nm.
5. The average absorbance of the blanks is subtracted, and the amount of sugar is determined by reference to a standard curve.

10.3.3 Total Reducing Sugar

10.3.3.1 Somogyi–Nelson Method

10.3.3.1.1 Principle Oxidation is a loss of electrons; reduction is a gain of electrons. Reducing sugars are those sugars that have an aldehydo group (aldoses) that can give up electrons (i.e., act as a reducing agent)

Furaldehyde (Furfaral)

Hydroxymethylfuraldehyde (HMF)

10-3 figure Furan products that could arise from, in order, pentoses and hexuronic acids, hexoses, 6-deoxyhexoses, and ketohexoses (1).

to an oxidizing agent, which is reduced by receiving the electrons. Oxidation of the aldehydo group produces a carboxylic acid group. Under alkaline conditions, ketoses behave as weak reducing sugars because they will partially isomerize to aldoses.

The most often used method to determine amounts of reducing sugars is the Somogyi–Nelson method (7, 11–14), also at times referred to as the Nelson–Somogyi method. This and other reducing sugar methods (Sect. 10.3.3.2) can be used in combination with enzymic methods (Sect. 10.3.4.3) for determination of oligo- and polysaccharides. In enzymic methods, specific hydrolases are used to convert the oligo- or polysaccharide into its constituent monosaccharide or repeating oligosaccharide units, which are measured using a reducing sugar method.

$$R{-}\overset{O}{\overset{\|}{C}}{-}H + 2Cu(OH+)_2\ NaOH \longrightarrow R{-}\overset{O}{\overset{\|}{C}}{-}O^{-}Na^{+} + Cu_2O + 3H_2O \quad [2]$$

The Somogyi–Nelson method is based on the **reduction of Cu(II) ions to Cu(I) ions by reducing sugars**. The Cu(I) ions then reduce an arsenomolybdate complex, prepared by reacting ammonium molybdate $[(NH_4)_6Mo_7O_{24}]$ and sodium arsenate (Na_2HAsO_7) in sulfuric acid. Reduction of the arsenomolybdate complex produces an intense, stable blue color that is measured spectrophotometrically. This reaction is not stoichiometric and must be used with a standard curve of the sugar(s) being determined or D-glucose.

10.3.3.1.2 Outline of Procedure

1. A solution of copper(II) sulfate and an alkaline buffer are added by pipettes to a solution of reducing sugars(s) and a water blank.
2. The resulting solution is heated in a boiling water bath.
3. A reagent prepared by mixing solutions of acidic ammonium molybdate and sodium arsenate is added.
4. After mixing, dilution, and remixing, absorbance is measured at 520 nm.
5. After subtracting the absorbance of the reagent blank, the A_{250} is converted into glucose equivalents using a standard plot of micrograms of glucose vs. absorbance.

10.3.3.2 Other Methods (3)

The **dinitrosalicylic acid method** (15) will measure reducing sugars naturally occurring in foods or released by enzymes, but is not much used. In this reaction, 3,5-dinitrosalicylate is reduced to the reddish monoamine derivative.

There are other methods that, like the Somogyi–Nelson method, are based on the **reduction of Cu(II) ions in alkaline solution to Cu(I) ions** that precipitate as the brick-red oxide Cu_2O. Tartrate or citrate ions are added to keep the Cu(II) ions in solution under the alkaline conditions. The **Munson–Walker method** (AOAC Method 906.03) has various forms. The precipitate of cuprous oxide can be determined **gravimetrically** (AOAC Method 31.039), by **titration** with sodium thiosulfate (AOAC Method 31.040), by titration with potassium permanganate (AOAC Method 31.042), by titration in the presence of methylene blue (the **Lane–Eynon method**; AOAC Method 923.09, 920.183b), and **electrolytically** (AOAC Method 31.044). These methods also must be used with standard curves because each reducing sugar reacts differently. Because assay conditions affect the outcome, they generally also must be done by trained, experienced analysts so that they always are done in exactly the same way. They are still used where specified.

A keto group cannot be oxidized to a carboxylic acid group, and thus ketoses are not reducing sugars. However, under the alkaline conditions employed, ketoses are isomerized to aldoses (1) and, therefore, are measured as reducing sugars. The response is less with ketoses, so a standard curve made with D-fructose as one of the sugars in the mixture of sugars should be used if it is present.

Methods that both identify individual carbohydrates present and determine their amounts are preferred over general reducing sugar methods and are described next.

10.3.4 Specific Analysis of Mono- and Oligosaccharides

10.3.4.1 High-performance Liquid Chromatography

HPLC (Chap. 28) is the method of choice for analysis of mono- and oligosaccharides and can be used for analysis of polysaccharides after hydrolysis (Sect. 10.4.2). HPLC gives both qualitative analysis (identification of the carbohydrate) and, with peak integration, quantitative analysis. HPLC analysis is rapid, can tolerate a wide range of sample concentrations, and provides a high degree of precision and accuracy. HPLC requires no prior derivatization of carbohydrates, unlike GC of sugars (Sect.10.3.4.2), but does require micron-filter filtration prior to injection. Complex mixtures of mono- and oligosaccharides can be analyzed. The basic principles and important parameters of HPLC (the stationary phase, the mobile phase, and the detector) are presented and discussed in Chap. 28. Some details related to carbohydrate analysis are discussed here. Use of HPLC to determine soluble food and other carbohydrates has been reviewed many times. Selected reviews can be found in references (16–21). Specific details of methods of analyses of specific food ingredients or products should be obtained from the literature. The USDA Food Safety and Inspection Service recommends the use of HPLC for determination of sugars and sugar alcohols in meat and poultry products (22). Results of interlaboratory collaborative studies done to validate the reproducibility of HPLC methods are available (20,23,24).

10.3.4.1.1 Stationary Phases Stationary phases are presented in probable order of use for carbohydrate analysis.

1. **Anion-exchange chromatography** (*AE-HPLC*). Carbohydrates have pK_a values in the pH range 12–14 and are, therefore, very weak acids. In a solution of high pH, some carbohydrate hydroxyl groups are ionized, allowing sugars to be separated on columns of anion-exchange resins. Special column packings have been developed for this purpose. The general elution sequence is sugar alcohols (alditols), monosaccharides, disaccharides, and higher oligosaccharides.

AE-HPLC is most often used in conjunction with electrochemical detection (see Chap. 27 and Section "Detectors") (18–21, 23–27). AE-HPLC has been used to examine the complex oligosaccharide patterns of many food components and products. The method has the advantage of being applicable to baseline separation within each class of carbohydrates [see Fig. 10-4 for separation of some monosaccharides, disaccharides, alditols (sugar alcohols), and raffinose] and of providing separation of homologous series of oligosaccharides into their components (27,28).

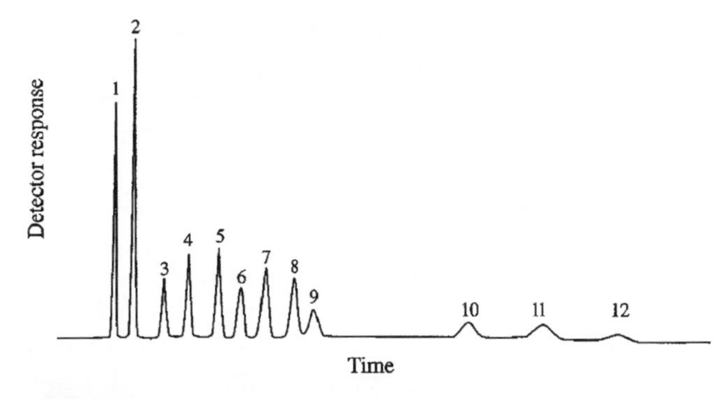

10-4 figure High-performance liquid chromatogram of some common monosaccharides, disaccharides, alditols, and the trisaccharide raffinose at equal wt/vol concentrations separated by anion-exchange chromatography and detected by pulsed amperometric detection (see Sect. 10.3.4.1.2). Peak *1*, glycerol; *2*, erythritol; *3*, L-rhammose; *4*, D-glucitol (sorbitol); *5*, mannitol; *6*, L-arabinose; *7*, D-glucose; *8*, D-galactose; *9*, lactose; *10*, sucrose; *11*, raffinose; *12*, maltose.

2. **Normal-phase chromatography** (29). Normal-phase chromatography is a widely used HPLC method for carbohydrate analysis. In normal-phase chromatography, the stationary phase is polar and elution is accomplished by employing a mobile phase of increasing polarity. Silica gel that has been derivatized with one or more of several reagents to incorporate amino groups is often used. These so-called amine-bonded stationary phases that are generally used with acetonitrile–water (50–85% acetonitrile) as the eluent are effective in carbohydrate separations. The elution order is monosaccharides and sugar alcohols, disaccharides, and higher oligosaccharides. Amine-bonded silica gel columns have been used successfully to analyze the low-molecular-weight carbohydrate content of foods (17).

A severe disadvantage of amine-bonded silica gel is the tendency for reducing sugars to react with the amino groups of the stationary phase, which results in a deterioration of column performance over time and loss of some of the carbohydrate being measured. This situation can be partially alleviated through the use of amine-modified silica gel columns. To prepare amine-modified silica gel columns, small amounts of modifiers, which are soluble amine compounds, are added to the mobile phase to modify the packing in situ. The modifier must have at least two amino groups, for one is needed to adsorb to the silica gel and the other must be free to interact with the carbohydrate. Because the modifier is in the eluent, the column is continuously regenerated.

3. **Cation-exchange chromatography**. Microparticulate spheres of sulfonated resin are used for cation-exchange stationary phases. The resin is loaded with one of a variety of metal counter ions, depending on the type of separation desired. Usually Ca^{2+}, Pb^{2+}, or Ag^{+} is used as the counter ion. The mobile phase used with these columns is water plus varying amounts (typically <40%) of an organic solvent such as acetonitrile and/or methanol. These columns normally are operated at elevated temperatures (>80°C) to increase column efficiency by increasing the mass transfer rate between the stationary and mobile phases which effects peak narrowing and improved resolution (30).

Carbohydrate elution from cation-exchange resins takes place in the order of decreasing molecular weight. Oligosaccharides with a degree of polymerization (DP) greater than 3 elute first, followed by trisaccharides, disaccharides, monosaccharides, and alditols. There is some resolution of disaccharides, but the real strength of this stationary phase is in the separation of individual monosaccharides.

4. **Reversed-phase chromatography**. In reversed-phase chromatography, the stationary phase is hydrophobic, and the mobile phase is largely water. The hydrophobic stationary phase is made by reacting silica gel with a reagent that adds alkyl chains, such as an 18-carbon-atom alkyl chain (a C18 column) or a phenyl group (a phenyl column). Reversed-phase chromatography has been used for separation of mono-, di-, and trisaccharides by groups (31, 32) (Fig. 10-5).

A major disadvantage of this stationary phase is the short retention times of monosaccharides, which result in elution as a single unresolved peak. The addition of salts (such as sodium chloride) can increase retention on the stationary phase and the utility of this method for monosaccharide analysis. Reversed-phase chromatography is complicated by peak doubling and/or peak broadening due to the presence of anomers. This problem can be alleviated by the addition of an amine to the mobile phase to accelerate mutarotation (anomerization), but separation may be negatively affected by the shorter retention times that usually result.

A wide variety of stationary phases is available, including phases not included in one of the four groups given above, and new improved phases continue to be developed. Both normal- and reversed-phase columns have long lives, have good stability over a wide range of solvent compositions and pH values (from pH 2 to pH 10), are suitable for the separation of a range of carbohydrates, and are of relatively low cost. All silica-based stationary phases share the disadvantage that silica dissolves to a small extent in water-rich eluents.

10.3.4.1.2 Detectors Detectors and their limitations and detector limits have been reviewed (19).

1. **Refractive index detection. The refractive index** (*RI*) detector is commonly employed for carbohydrate analysis (33) (Sect. 10.6.4). RI measurements are linear over a wide range of carbohydrate concentrations and can be universally applied to all carbohydrates, but the RI detector has its drawbacks. RI is a bulk physical property that is sensitive to changes

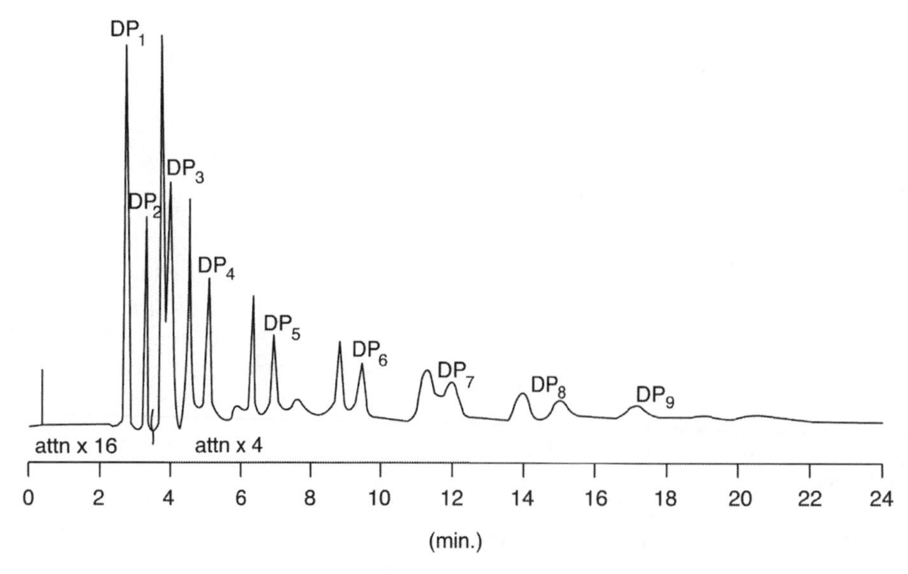

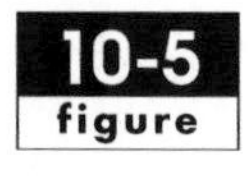

High-performance, reversed-phase liquid chromatogram of maltodextrins (DP 1–9). [From (32), used with permission.]

in flow, pressure, and temperature; but with modern HPLC equipment and a temperature-controlled detector, problems arising from these changes can be minimized. The most significant limiting factor with RI detection is that gradient elution cannot be used. The other is that, since an RI detector measures mass, it is not sensitive to low concentrations.

2. **Electrochemical detection**. The triple-pulsed electrochemical detector, called a **pulsed-amperometric detector** (PAD), which relies on oxidation of carbohydrate hydroxyl and aldehydo groups, is universally used with AE-HPLC (18–21, 23–27, 34). It requires a high pH. Gradient and graded elutions can be used with the PAD. The solvents employed are simple and inexpensive (sodium hydroxide solution, with or without sodium acetate). (Water may be used, but when it is, postcolumn addition of a sodium hydroxide solution is required.) The detector is suitable for both reducing and nonreducing carbohydrates. Limits are approximately 1.5 ng for monosaccharides and 5 ng for di-, tri-, and tetrasaccharides.

3. **Postcolumn derivatization** (35). The purpose of pre- and postcolumn derivatization is to increase detection sensitivity by addition of a substituent whose concentration can be measured using an ultraviolet (UV) or fluorescence detector. However, with the development of the PAD detector, neither pre- nor postcolumn derivatization is much used. Postcolumn derivatization involves addition of reagents that will provide compounds whose concentration can be measured using absorbance (visible) or fluorescence detection. It is straightforward; requires only one or two additional pumps, a mixing coil, and a thermostatted bath; and provides greater sensitivity than does an RI detector.

4. **Precolumn derivatization** (35). Precolumn derivatization reactions must be stoichiometric. Oligosaccharides derivatized with aromatic groups are often separated with higher resolution in normal-phase HPLC.

10.3.4.2 Gas Chromatography

GC (**gas-liquid chromatography**, GLC), like HPLC, provides both qualitative and quantitative analysis of carbohydrates. For GC, sugars must be converted into volatile derivatives. The most commonly used derivatives are the alditol peracetates (and aldonic acid pertrimethylsilyl ethers from uronic acids) (36–39). These derivatives are prepared as illustrated in Fig. 10-6 for D-galactose and D-galacturonic acid. Conversion of sugars into peracetylated aldononitrile (aldoses) and peracetylated ketooxime (ketoses) derivatives for GC has also been done (40), although this procedure is not used nearly as much as the preparation of peracetylated aldoses and aldonic acids. A **flame ionization detector** is the detector of choice for peracetylated carbohydrate derivatives.

The most serious problem with GC for carbohydrate analysis is that two preparation steps are involved: reduction of aldehyde groups to primary alcohol groups and conversion of the reduced sugar into a volatile peracetate ester or pertrimethylsilyl ether derivative. Of course, for the analysis to be successful, each of these steps must be 100% complete (i.e., stoichiometric). The basic principles and important parameters of GC (the stationary phase, temperature programming, and detection) are presented and discussed in Chap. 29.

10.3.4.2.1 Neutral Sugars: Outline of Procedure (38)

1. **Reduction to alditols**. Neutral sugars from the 80% ethanol extract (Sect. 10.3.1) or from hydrolysis of a polysaccharide (see Sects. 10.4.2.2 and "Overview") are reduced with an excess of sodium or potassium borohydride dissolved in dilute ammonium hydroxide solution. After reaction at 40°C, glacial acetic acid is added dropwise until no more hydrogen is evolved. This treatment destroys excess borohydride. The acidified solution is evaporated to dryness. Borate ions may be removed as methyl borate by successive additions and evaporation of methanol, but this step is not necessary.

A potential problem is that, if fructose is present, either as a naturally occurring sugar, from the hydrolysis of inulin, or as an additive [from high fructose syrup (HFS), invert sugar, or honey], it will be reduced to a mixture of D-glucitol (sorbitol) and D-mannitol (Fig. 10-7).

2. **Acetylation of alditols**. Acetic anhydride and 1-methylimidazole (as a catalyst) are added. After 10 min at room temperature, water and dichloromethane are added. The dichloromethane layer is washed with water and evaporated to dryness. The residue of alditol peracetates is dissolved in a polar organic solvent (usually acetone) for chromatography.

3. **GC of alditol peracetates** (38, 39). Alditol acetates may be chromatographed isothermally and identified by their retention times relative to that of inositol hexaacetate, inositol being added as an internal standard prior to acetylation. It is wise to run standards of the additol peracetates of the sugars being determined with inositol hexaacetate as an internal standard.

$NaBH_4$ → Ac_2O →

D-Galactose

D-Galactitol

D-Galactitol pentaacetate

Ac = —C(=O)—CH_3

$NaBH_4$ → ≡ Me_3Si-NH-$SiMe_3$ →

D-Galacturonic Acid, Sodium salt

L-Galactonic Acid, Sodium salt

Trimethylsilyl L-Galactonate Penta(trimethylsilyl) Ether

TMS = — $Si(CH_3)_3$

10-6 figure Modification of D-galactose and D-galacturonic acid in preparation for gas chromatography.

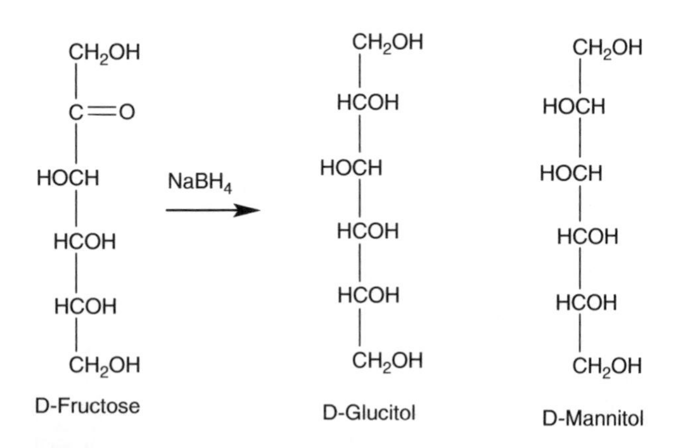

10-7 figure Reduction of D-fructose to a mixture of alditols.

10.3.4.2.2 Hydrolyzates of Polysaccharides Containing Uronic Acids: Outline of Procedure (41) A method different from that used for neutral sugars (Sect. 10.3.4.2.1) is required when uronic acids are present.

1. **Reduction**. As with hydrolyzates containing only neutral sugars, the hydrolyzate is evaporated to dryness. The residue is dissolved in sodium carbonate solution and treated with an excess of sodium borohydride. Excess borohydride is decomposed by addition of glacial acetic acid; borate may be removed by addition and evaporation of methanol (Sect. 10.3.4.2.1). This procedure reduces uronic acids to aldonic acids and aldoses to alditols (Fig. 6).

2. **Preparation and chromatography of trimethylsilyl (TMS) derivatives**. The aldonic acids are converted into per-TMS ethers rather than peracetate esters (Fig. 10-6). Trimethylsilyation of free aldonic acids gives derivatives of lactones (predominately the 1,4-lactone), while trimethylsilyation of the sodium salt produces the ester. Several procedures and packaged reagents have been developed for this etherification. The reaction mixture is injected directly into the chromatograph. Temperature programming is required. Components are identified by their retention times.

10.3.4.3 Enzymic Methods

10.3.4.3.1 Overview The method of choice for the determination of starch employs a combination of enzymes in sequential **enzyme-catalyzed reactions** and is specific for starch, as long as purified enzyme preparations are used (Sect. 10.4.1.1).

Other enzymic methods for the determination of carbohydrates have been developed (Table 10-3) [see also Equation (3) and Chap. 16]. They are often, but not always, specific for the substance being measured. Kits for several enzymic methods have been developed and marketed. The kits contain specific enzymes, other required reagents, buffer salts, and detailed instructions that must be followed because enzyme concentration, substrate concentration, concentration of other required reagents, pH, and temperature all affect reaction rates and results. A good description of a method will point out any interferences and other limitations.

Limits of detection by methods involving enzyme- or coupled enzyme-catalyzed reactions are generally low. In addition, enzymic methods are usually quite specific for a specific carbohydrate, although not always 100% specific. However, it is not often that determination of a single component is desired, the notable exception being the determination of starch (Sect. 10.4.1.1). Other exceptions are the identification and quantitative determination of β-glucan and inulin. Thus, chromatographic methods (Sects. 10.3.4.1 and 10.3.4.2) that give values for each of the sugars present are preferred.

10.3.4.3.2 Sample Preparation It sometimes is recommended that the **Carrez treatment** (7), which breaks emulsions, precipitates proteins, and absorbs some colors, be applied to food products prior to determination of carbohydrates by enzymic methods. The Carrez treatment involves addition of a solution of potassium hexacyanoferrate ($K_4[Fe(CN)_6]$, potassium ferrocyanide), followed by addition of a solution of zinc sulfate ($ZnSO_4$), followed by addition of a solution of sodium hydroxide. The suspension is filtered, and the clear filtrate is used directly in enzyme-catalyzed assays.

10.3.4.3.3 Enzymic Determination of D-Glucose The enzyme **glucose oxidase** oxidizes D-glucose quantitatively to D-glucono-1,5-lactone (glucono-delta-lactone), the other product being hydrogen peroxide (Fig. 10-8). To measure the amount of D-glucose present, **peroxidase** is added along with a colorless compound that can be oxidized to a colored compound. In a second enzyme-catalyzed reaction, the leuco dye is oxidized to a colored compound which is measured spectrophotometrically. Various dyes are used in commercial kits. The method using this combination of two enzymes and an oxidizable colorless compound is known as the **GOPOD (glucose oxidase-peroxidase) method**.

Table 10-3 Selected Enzymic Methods of Carbohydrate Analysis

Carbohydrate	*Reference*	*Kit Form*[a]
Monosaccharides		
Pentoses		
L-Arabinose	(42, 43)	
D-Xylose	(42, 43)	
Hexoses		
D-Fructose	(42, 43)	x
D-Galactose	(42, 43)	x
D-Galacturonic acid	(42)	
D-Glucose		
Using glucose oxidase	(43), Sect. 10.3.4.3.3	x
Using glucose dehydrogenase	(42, 43)	
Using glucokinase (hexokinase)	(42, 43)	x
D-Mannose	(42, 43)	
Monosaccharide derivatives		
D-Gluconate/D-glucono-δ-lactone	(42, 43)	x
D-Glucitol/sorbitol	(42, 43)	x
D-Mannitol	(42, 43)	
Xylitol	(42, 43)	x
Oligosaccharides		
Lactose	(42, 43)	x
Maltose	(42, 43)	x
Sucrose	(42, 43)	x
Raffinose, stachyose, verbascose	(42, 43)	x
Polysaccharides		
Amylose, amylopectin (contents and ratio)		x
Cellulose	(42, 43)	
Galactomannans (guar and locust bean gums)	(42)	
β-Glucan (mixed-linkage)	(42)	x
Glycogen	(42, 43)	
Hemicellulose	(42, 43)	
Inulin	(42, 43)	x
Pectin/poly(D-galacturonic acid)	(42, 43)	
Starch	Sect. 10.4.1.1 (42, 43)	x

[a]Available in kit form from companies such as R-Biopharm, Megazyme, and Sigma-Aldrich.

10.3.4.4 Mass Spectrometry

There are many different variations of mass spectrometry (MS) (Chap. 26). With carbohydrates most of the techniques are used for structural analysis; MS has

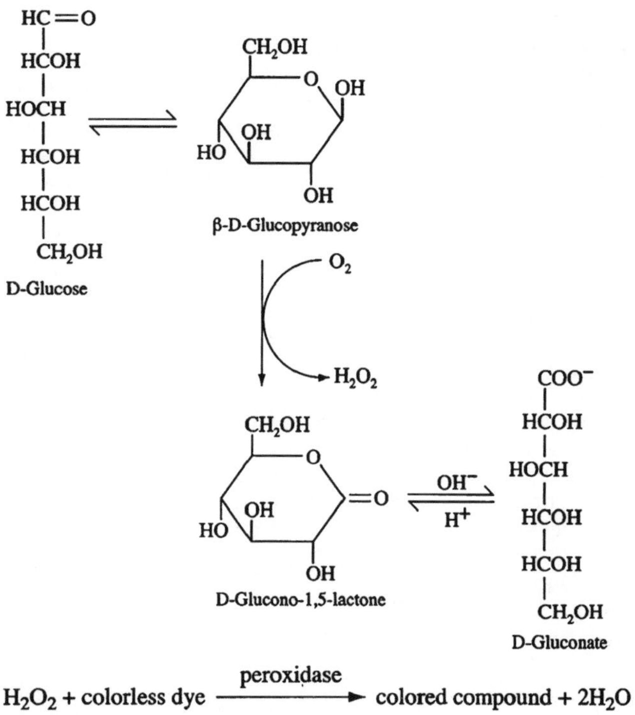

$$H_2O_2 + \text{colorless dye} \xrightarrow{\text{peroxidase}} \text{colored compound} + 2H_2O$$

10-8 figure Coupled enzyme-catalyzed reactions for the determination of D-glucose.

been used for analysis of carbohydrates, but not in a routine manner (44). Particularly useful is the **matrix-assisted laser desorption time-of-flight** (MALDI-TOF) technique for analysis of a homologous series of oligosaccharides (Fig. 10-9). A comparison was made between anion-exchange HPLC (Sect. 10.3.4.1) (the most used carbohydrate analysis technique today), capillary electrophoresis (Sect. 10.3.4.6), and MALDI-TOF mass spectrometry for the analysis of maltooligosaccharides, with the conclusion that the latter technique gave the best results (28).

10.3.4.5 Thin-layer Chromatography

Thin-layer chromatography has been used for identification and quantitation of the sugars present in the molasses from sugar beet and cane processing (45). It is particularly useful for rapid screening of several samples simultaneously.

10.3.4.6 Capillary Electrophoresis (46, 47)

Capillary zone electrophoresis (Chap. 15) has also been used to separate and measure carbohydrates, but because carbohydrates lack chromophores, pre-column derivatization and detection with a UV or fluorescence detector is required (35). Generally, this method provides no advantage over HPLC methods for carbohydrate analysis.

10.4 POLYSACCHARIDES

10.4.1 Starch

Starch is second only to water as the most abundant component of food. Starch is found in all parts of plants (leaves, stems, roots, tubers, seeds). A variety of commercial starches are available worldwide as food additives. These include corn (maize), waxy maize, high-amylose corn (amylomaize), potato, wheat, rice, tapioca (cassava), arrowroot, and sago starches. In addition, starch is the main component of wheat, rye, barley, oat, rice, corn, mung bean, and pea flours and certain roots and tubers such as potatoes, sweet potatoes, and yams.

10.4.1.1 Total Starch

10.4.1.1.1 Principle The only reliable method for determination of total starch is based on complete conversion of the starch into D-glucose by purified enzymes specific for starch and determination of the D-glucose released by an enzyme specific for it (Fig. 10-8) (see also Chap. 16).

10.4.1.1.2 Potential Problems Starch-hydrolyzing enzymes (amylases) must be purified to eliminate any other enzymic activity that would release D-glucose (e.g., cellulases, invertase or sucrase, β-glucanase) and catalase, which would destroy the hydrogen peroxide on which the enzymic determination of D-glucose depends (Sect. 10.3.4.3.3). The former contamination would give false high values and the latter, false low values. Even with purified enzymes, problems can be encountered with this method. It may not be quantitative for high-amylose or another starch at least partially resistant to enzyme-catalyzed hydrolysis. **Resistant starch** (RS), by definition, is composed of starch and starch-degradation products that escape digestion in the small intestine (48). There are generally considered to be four starch sources that are resistant to digestion or so slowly digested that they pass through the small intestine:

1. Starch that is physically inaccessible to amylases because it is trapped within a food matrix (RS1),
2. Starch that resists enzyme-catalyzed hydrolysis because of the nature of the starch granule (uncooked starch) (RS2),
3. Retrograded starch (i.e., starch polymers that have recrystallized after gelatinization of the granules, e.g., cooled cooked potatoes contain resistant starch) (Sect. 10.4.1.3) (RS3), and
4. Starch that has been modified structurally in such a way as to make it less susceptible to digestion (RS4).

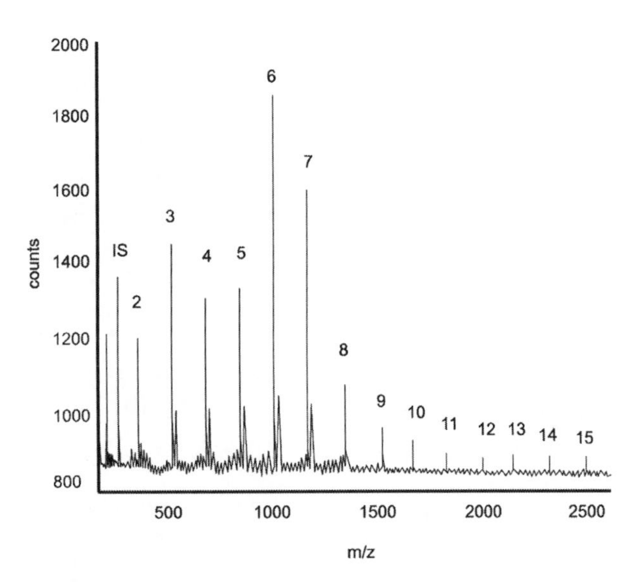

10-9 figure MALDI-TOF mass spectrum of maltooligosaccharides produced by hydrolysis of starch. *Numbers* indicate DP. *IS*, internal standard. [From (28), used with permission, Copyright Springer-Verlag, 1998.]

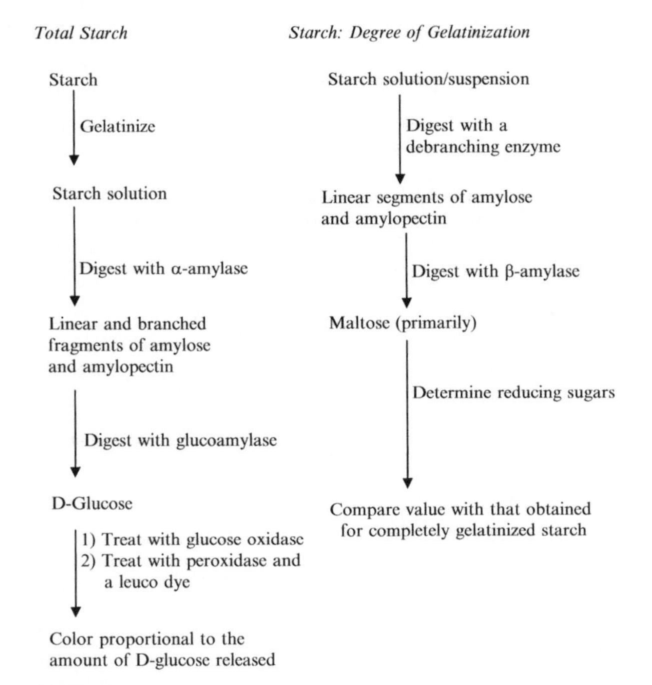

10-10 figure Flow diagrams for determination of total starch (Sect. 10.4.1.1) and determination of the degree of starch gelatinization (Sect. 10.4.1.2).

RS is at best only partially converted into D-glucose by the method described below to measure starch; rather most of it is usually included in the analysis for dietary fiber (Sect. 10.5.4.1).

One method of starch analysis purports to overcome at least the first three of these problems (49). In it, the starch is dispersed in dimethyl sulfoxide (DMSO) and then is converted quantitatively to D-glucose by treatment with a thermostable α-amylase to effect depolymerization and solubilization of the starch (Fig. 10-10). Glucoamylase (amyloglucosidase) effects quantitative conversion of the fragments produced by the action of α-amylase into D-glucose. D-glucose is determined with a glucose oxidase-peroxidase (GOPOD) reagent (Sect. 10.3.4.3.3) (AOAC Method 969.39; AACC Method 76-13). This reagent contains a colorless (leuco) dye that is oxidized to a colored compound by the hydrogen peroxide (produced by the glucose oxidase-catalyzed oxidation of glucose, Fig. 10-8) in a reaction catalyzed by peroxidase. The method determines total starch. It does not reveal the botanical source of the starch or whether it is native starch or modified food starch. The botanical source of the starch may be determined microscopically (Sect. 10.6.1) if the material being analyzed has not been cooked. Some information about modification also may be determined with a microscope.

10.4.1.1.3 Outline of Procedure

1. A sample of finely milled material is placed in a glass test tube and wetted with 80% vol/vol ethanol. DMSO is added to the ethanol-wetted sample, and the contents of the tube are mixed vigorously. The tube is then heated in a boiling water bath.
2. A buffered solution of a thermostable α-amylase is added. Tube contents are vortex mixed, and the tube is returned to the boiling water bath.
3. After 5 min, the tube is brought to 50°C. Sodium acetate buffer, pH 4.5, and glucoamylase (amyloglucosidase) solution is added, and the contents are mixed. The tube then is incubated at 50°C.
4. The tube contents are transferred quantitatively to a volumetric flask using distilled water to

wash the tube and to adjust the contents to volume.

5. After thorough mixing of the flask, aliquots are removed, treated with GOPOD reagent, and incubated at 50°C. Absorbance of the test sample and a reagent blank is measured at the wavelength required by the GOPOD reagent being used.

Glucose and a starch low in protein and lipid content (such as potato starch) are used as standards after determination of their moisture contents. Addition of DMSO can be omitted, and diluted thermostable α-amylase solution can be added directly to the ethanol-wetted sample if it is known from experience that no starch resistant to the α-amylase under the conditions used is present in the samples being analyzed.

10.4.1.2 Degree of Gelatinization of Starch

When starch granules are heated in water to a temperature specific for the starch being cooked, they swell, lose their crystallinity and birefringence, and become much more susceptible to enzyme-catalyzed hydrolysis. Heating starch in water produces phenomena that result from two processes: **gelatinization** and **pasting**, often together referred to simply as gelatinization, which are very important in determining the texture and digestibility of foods containing starch.

Several methods have been developed that make use of the fact that certain enzymes act much more rapidly on cooked starch than they do on native starch. A particularly sensitive method employs a combination of pullulanase and β-amylase, neither of which is able to act on uncooked starch granules (50). With gelatinized or pasted starch, the enzyme **pullulanase** debranches amylopectin and any branched amylose molecules, giving a mixture of linear segments of various sizes. (Another debranching enzyme, **isoamylase**, may also be used.) ***β*-Amylase** then acts on the linear chains, releasing the disaccharide maltose, starting at the nonreducing ends (Fig. 10-10) and a small amount of maltotriose (from chains containing an odd number of glucosyl units). The **degree of gelatinization** is determined by measuring the amount of reducing sugar formed (Sect. 10.3.3).

10.4.1.3 Degree of Retrogradation of Starch

Upon storage of a product containing cooked starch, the two starch polymers, **amylose** and **amylopectin**, associate with themselves and with each other, forming polycrystalline arrays. This process of reordering is called **retrogradation**. (Retrogradation is a contributing factor to the staling of bread and other bakery products, for example.) Retrograded starch, like native starch, is acted on very slowly by the combination of pullulanase plus β-amylase. Therefore, the basic method described in Sect. 10.4.1.2 can be used to determine retrogradation. The decrease in reducing power (from maltose released by action of the enzyme combination) after storage is a measure of the amount of retrograded starch at the time of analysis and/or the degree of retrogradation.

10.4.2 Nonstarch Polysaccharides (Hydrocolloids/Food Gums)

10.4.2.1 Overview

A starch (or starches) may be used as ingredients in a food product, either as isolated starch or as a component of a flour, or may occur naturally in a fruit or vegetable tissue. Other polysaccharides are almost always added as ingredients, although there are exceptions. These added polysaccharides, along with the protein gelatin, comprise the group of ingredients known as **food gums** or **hydrocolloids**. Their use is widespread and extensive. They are used in everything from processed meat products to chocolate products, from ice cream to salad dressings.

Analytical methods are required for these polysaccharides to enable both suppliers and food processors to determine the purity of a gum product, to ensure that label declarations of processors are correct, and to monitor that hydrocolloids have not been added to standardized products in which they are not allowed. It also may be desirable to determine such things as the ***β*-glucan** content of oat or barley flour or a breakfast cereal for a label claim or the **arabinoxylan** content of wheat flour to set processing parameters. Another processor may want to determine other polysaccharides not declared on the ingredient label, such as those introduced by microorganisms during fermentation in making yogurt and yogurt-based products.

Food gum analysis is problematic because polysaccharides present a variety of chemical structures, solubilities, and molecular weights. Plant polysaccharides do not have uniform, repeating-unit structures; rather the structure of a specific polysaccharide such as κ-carrageenan varies from molecule to molecule. In addition, the average structure can vary with the source and the conditions under which the plant is grown. Some polysaccharides are neutral; some are anionic. Some are linear; some are branched. Some of the branched polysaccharides are still effectively linear; some are bushlike. Some contain ether, ester, and/or cyclic acetal groups in addition to sugar units, either naturally or as a result of chemical modification. Some are soluble only in hot water; some are soluble only in room temperature or colder water; some are soluble in both hot and cold water, and some require aqueous solutions of acids, bases, or metal ion-chelating compounds to dissolve them. And all polysaccharide preparations are composed of a mixture of molecules with a range of molecular weights. All this structural diversity complicates qualitative analysis of food gums when their nature is unknown or when more than one is present, and structural heterogeneity complicates quantitative analysis of a specific gum.

Current methods depend on extraction of the gum(s), followed by fractionation of the extract. Fractionation invariably results in some loss of material. Most often, an isolated gum is identified by identifying and quantitating its constituent sugars after acid-catalyzed hydrolysis. However, sugars are released from polysaccharides by hydrolysis at different rates and are destroyed by hot acids at different rates, so the exact monosaccharide composition of a polysaccharide may be difficult to determine. Problems associated with the determination of gums in foods and various procedures that have been used for their measurement have been reviewed (51,52).

Qualitative identification tests, specifications, and analytical methods for many food-approved gums/hydrocolloids, including modified starches, have been established for the United States (53) and Europe (54). None of the qualitative methods is conclusive. AOAC International has established methods for analysis of some specific food products. But not all gums approved for food use are included; not all methods that determine total gums can be used if starch is present; and not all methods can be used to determine all gums. Hydrocolloid/gum suppliers and food processors usually have their own specifications of purity and properties.

10.4.2.2 Hydrocolloid/Food Gum Content Determination

Several schemes, some published, some unpublished, have been developed for analysis of food products for food gums. Most are targeted to a specific group of food products, as it is difficult, perhaps impossible, to develop a universal scheme. A general scheme that is reported to work successfully (41) is presented here. Figure 10-11 presents the scheme for isolation and purification of nonstarch, water-soluble polysaccharides. Letters in the parentheses below refer to the same letters in Fig. 10-11. Many of the steps in the method utilize principles previously described.

(a) It is usually difficult to extract polysaccharides quantitatively when fats, oils, waxes, and proteins are present. Therefore, lipid-soluble substances are removed first. Before this can be effected, the sample must be dried. Freeze drying is recommended. If the dried material contains lumps, it must be ground to a fine powder. A known weight of dry sample is placed in a Soxhlet apparatus, and the lipid-soluble substances are removed with 19:1 vol/vol chloroform–methanol (see note in Sect. 10.2). (*n*-Hexane has also been used.) Solvent is removed from the sample by air drying in a hood, then by placing the sample in a desiccator, which is then evacuated.
(b) Although not in the published scheme, soluble sugars, other low-molecular-weight compounds, and ash can be removed at this point using hot 80% ethanol as described in Sect. 10.3.1. (Hot 80% methanol has also been used.)
(c) Protein is removed by enzyme-catalyzed hydrolysis. The cited procedure (41) uses papain as the protease. Bacterial alkaline proteases are recommended by some because carbohydrases have acidic pH optima. However, one must always be aware of the fact that commercial enzyme preparations, especially those from bacteria or fungi, almost always have carbohydrase activities in addition to proteolytic activity. In this procedure, proteins are denatured for easier digestion by dispersion of the sample in sodium acetate buffer, pH 6.5, containing sodium chloride and heating the mixture. Papain [activated by dispersing it in sodium acetate buffer, pH 6.5, containing cysteine and ethylenediaminetetraacetic acid (EDTA)] is added to the sample, and the mixture is incubated.
(d) Any solubilized polysaccharides are precipitated by addition of sodium chloride to the cooled dispersion, followed by the addition of four volumes of absolute ethanol (to give an ethanol concentration of 75%). The mixture is centrifuged.
(e) The pellet is suspended in acetate buffer, usually pH 4.5. To this suspension is added a freshly prepared solution of glucoamylase (amyloglucosidase) in the same buffer. This suspension is then incubated. Just as in the analysis of starch, highly purified enzyme must be used to minimize hydrolytic breakdown of other polysaccharides (Sect. 10.4.1.1.2). This step may be omitted if it is known that no starch is present. Centrifugation after removal of starch isolates and removes insoluble fiber (cellulose, some hemicelluloses, lignin) (Sect. 10.5).

The presence of starch can be tested for by adding a solution of iodine and potassium iodide and observing the color. A color change to blue or brownish-red indicates the presence of starch. A microscope may be used to look for stained intact or swollen granules or granule

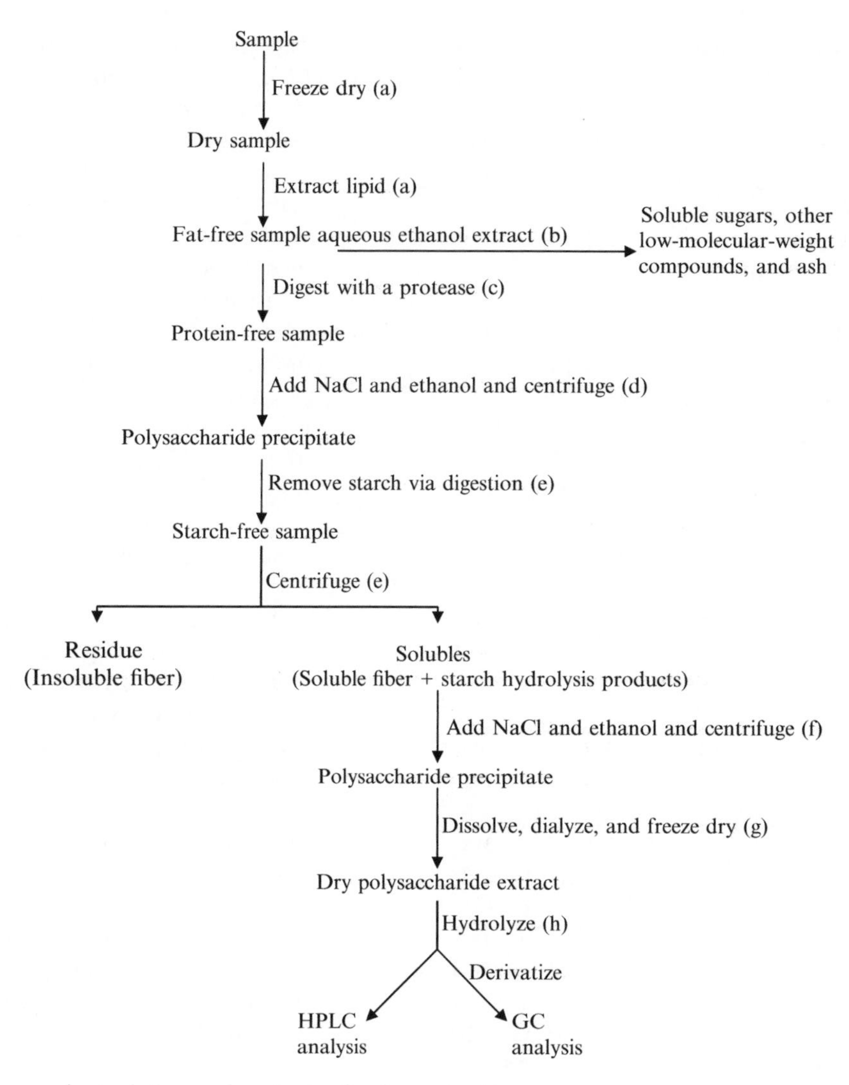

10-11 figure Flow diagram for isolation and analysis of polysaccharides.

fragments (Sect. 10.6.1). However, unless a definite blue color appears, the test may be inconclusive. A better check is to analyze the ethanol-soluble fraction from step (f) for the presence of glucose (Sect. 10.3.4). If no glucose is found, the starch digestion part of step (e) may be omitted in future analyses of the same product.

(f) Solubilized polysaccharides are reprecipitated by addition of sodium chloride to the cooled dispersion, followed by the addition of four volumes of absolute ethanol (to give an ethanol concentration of 75%). The mixture is centrifuged. The precipitate (pellet) of water-soluble polysaccharides (often added hydrocolloids/food gums) is soluble dietary fiber (Sect. 10.5).

(g) The pellet is suspended in deionized water, transferred to dialysis tubing, and dialyzed against frequent changes of sodium azide solution (used to prevent microbial growth). Finally, dialysis against deionized water is done to remove the sodium azide. The retentate is recovered from the dialysis tubing and freeze dried.

(h) Polysaccharide identification relies on hydrolysis to constituent monosaccharides and identification of these sugars (Sect. 10.3.4). For hydrolysis, polysaccharide material is added to a Teflon-lined, screw-capped vial. Trifluoroacetic acid solution is added (usually 2 *M*), and the vial is tightly capped and heated (usually for 2 h at 120°C). After cooling, the contents are evaporated to dryness in a hood with a stream of air or nitrogen. Then, sugars are determined by HPLC (Sect. 10.3.4.1) or GC (Sect. 10.3.4.2). If GC is used, inositol is added as an internal standard. Qualitative and quantitative analysis of the polysaccharides present can be determined by sugar analysis. For example, guaran, the polysaccharide component of guar gum, yields D-mannose and D-galactose in an approximate molar ratio of 1.00:0.56.

The described acid-catalyzed hydrolysis procedure does not release uronic acids quantitatively. The presence of **uronic acids** can be indicated by either the modified **carbazole assay** (55, 56), the ***m*-hydroxydiphenyl assay** (11, 57, 58), or the **3,5-dimethylphenol assay** (58). All three methods are based on the same principle as the phenol-sulfuric acid assay (Sect. 10.3.2) (i.e., condensation of dehydration products with a phenolic compound to produce colored compounds that can be measured quantitatively by means of spectrophotometry).

10.4.2.3 Pectin

10.4.2.3.1 Nature of Pectin Even though pectin is a very important food polysaccharide, no official methods for its determination have been established. What few methods have been published basically involve its precipitation (by addition of ethanol) from jams, jellies, etc. in which it is the only polysaccharide present.

Even the definition of pectin is somewhat ambiguous. What may be called "**pectin**" in a native fruit or vegetable is a complex mixture of polysaccharides whose structures depend on the source, including the stage of development (degree of ripeness) of the particular fruit or vegetable. Generally, much of this native material can be described as a main chain of α-D-galactopyranosyluronic acid units (some of which are in the methyl ester form) interrupted by L-rhamnopyranosyl units (1, 2). Many of the rhamnosyl units have arabinan, galactan, or arabinogalactan chains attached to them. Other sugars, such as D-apiose, also are present. In the manufacture of commercial pectin, much of the neutral sugar part is removed. Commercial pectin is, therefore, primarily poly(α-D-galacturonic acid methyl ester) with various degrees of esterification and sometimes amidation.

Enzyme action during development/ripening or during processing can partially deesterify and/or depolymerize native pectin. These enzyme-catalyzed reactions are important determinants of the stability of fruit juices, tomato sauce, tomato paste, apple butter, etc. in which some of the texture/body is supplied by pectin and its interaction with calcium ions. It is probable that the fact that pectin is not a single substance has precluded development of methods for its determination (see also Sect. 10.5.2.1.3).

10.4.2.3.2 Pectin Content Determination The constant in pectins is **D-galacturonic acid** as the principal component (often at least 80%). However, glycosidic linkages of uronic acids are difficult to hydrolyze without decomposition, so methods involving acid-catalyzed hydrolysis to release D-galacturonic acid and chromatography are generally not applicable.

One method employed for pectin uses saponification in sodium hydroxide solution, followed by acidification, and addition of Ca^{2+} to precipitate the pectin. **Calcium pectate** is collected, washed, dried, and measured gravimetrically. Precipitation with the quaternary ammonium salt cetylpyridinium bromide has been used successfully because there is a much lower critical electrolyte concentration for its salt formation with pectin than with other acidic polysaccharides (60), and because pectin and other acidic polysaccharides are not likely to be found together. For a review of methods for determination of pectin, see references (61, 62).

Because of the dominance of D-galacturonic acid in its structure, pectins are often determined using the **carbazole** or ***m*-hydroxydiphenyl methods** (Sect. 10.4.2.2). Isolation of crude pectin usually precedes analysis.

10.4.2.3.3 Degree of Esterification The **degree of esterification** (DE) is a most important parameter in both natural products and added pectin. DE may be measured directly by titration before and after saponification. First, the isolated pectin (Sect. 10.4.2.2) is washed with acidified alcohol to convert carboxylate groups into free carboxylic acid groups and then washed free of excess acid. Then, a dispersion of the pectinic acid in water is titrated with dilute base, such as standardized sodium hydroxide solution, to determine the percentage of nonesterified carboxyl ester groups. Excess base is added to saponify the methyl ester groups. Back-titration with standardized acid to determine excess base following saponification gives the DE. Also, DE can be determined by measuring methanol released by saponification via GC (63) and by nuclear magnetic resonance (NMR) (see Chap. 25) (64, 65).

10.5 DIETARY FIBER

10.5.1 Introduction

Although there is an ongoing discussion about what constitutes dietary fiber within both domestic and international organizations (66), **dietary fiber** is essentially the sum of the nondigestible components of a foodstuff or food product. Most, but not all, dietary fiber is plant cell-wall material (cellulose, hemicelluloses, lignin) and thus is composed primarily of polysaccharide molecules (see Sect. 10.5.1.2 for definitions of dietary fiber). Because only the amylose and amylopectin molecules in cooked starch are digestible (Sect. 10.4.1.2), all other polysaccharides are also components of dietary fiber. Some are components of insoluble fiber; some make up soluble fiber. **Insoluble dietary fiber** components are cellulose, microcrystalline cellulose added as a food ingredient, lignin, hemicelluloses entrapped in a lignocellulosic matrix, and resistant starch (Sect. 10.4.1.1.2). Other polysaccharides, including many, but not all, hemicelluloses not entrapped in a lignocellulosic matrix, much of the native pectin, and the majority of hydrocolloids/food gums (Sect. 10.4.2), are classified as **soluble dietary fiber**. Often, their determination is important in terms of making food label claims and is described in Sect. 10.4.2. Determination of the β-glucan content of products made with oat or barley flours is an example. (Nondigestible protein is not considered to be a significant contributor to dietary fiber.)

Since the scheme presented in Fig. 10-11 is designed to separate nonstarch, water-soluble polysaccharides from other components for quantitative and/or qualitative analysis, the pellet from the centrifugation step (e) is insoluble fiber, and those components precipitated from the supernatant with alcohol [step (f)] constitute soluble fiber; but specific fiber determination methods have been established and are presented in Sect. 10.5.4.3.

Measurement of insoluble fiber is important not only in its own right, but also for calculating the caloric content of a food. According to nutrition labeling regulations, one method allowed to calculate calories involves subtracting the amount of insoluble dietary fiber from the value for total carbohydrate, before calculating the calories based on protein, fat, and carbohydrate content (approximately 4, 9, and 4 Calories per gram, respectively) (Chap. 3). This method ignores the fact that soluble fiber, like insoluble fiber, is also essentially noncaloric. [Fiber components can contribute calories via absorption of products of fermentation (mostly short-chain fatty acids) from the colon].

10.5.1.1 Importance of Dietary Fiber

In 1962, it was postulated that the prevalence of heart disease and certain cancers in Western societies was related to inadequate consumption of dietary fiber (67). Much research has been done since then to test the fiber hypothesis. While the research has not always produced consistent results, it is clear that adequate consumption of dietary fiber is important for optimum health.

Adequate consumption of dietary fiber from a variety of foods will help protect against colon cancer and also help to keep blood lipids within the normal range, thereby reducing the risk of obesity, hypertension, and cardiovascular disease in general. Certain types of fiber can slow D-glucose absorption and reduce insulin secretion, which is of great importance for diabetics and probably contributes to the well-being of nondiabetics as well. Fiber helps prevent constipation and diverticular disease. However, dietary fiber is not a magic potion that will correct or prevent all diseases. Rather, dietary fiber is an essential component of a well-balanced diet that will help minimize some common health problems. References (68–73) provide an extensive compilation of articles related to the physiological action of dietary fiber.

The Dietary Reference Intake (DRI) value for dietary fiber to promote optimal health has been set at 25 g per 2000 kcal per day. However, dietary fiber includes a variety of materials that in turn produce a variety of physiological actions (68–73). For example, the pentosan fraction of dietary fiber seems to be most beneficial in preventing colon cancer and reducing cardiovascular disease. Pectin and the hydrocolloids are most beneficial in slowing glucose absorption and in lowering insulin secretion. A mixture of hemicellulose and cellulose will help prevent diverticulosis and constipation.

Recognition of the importance of dietary fiber and of the fact that certain physiologic effects can be related to specific fiber components has led to the emergence of a number of methodologies for determining dietary fiber.

10.5.1.2 Definition

Because labeling of food products for dietary fiber content is required, an official analytical method(s) for its determination is required. The first step in adopting a method must be agreement on what constitutes dietary fiber. Then, there must be a method that measures what is included in the definition. A definition and a method related to it are also needed: (a) to determine the dietary fiber content of any new ingredient such as new resistant or slowly digesting starch products, and (b) to ensure that scientific studies of the physiological effects of dietary fiber are based on the same measure of dietary fiber content.

Following extensive international consultation, the American Association of Cereal Chemists (now AACC International) adopted the following definition in 2001 (74–78). "**Dietary fiber** is the edible parts of plants or analogous carbohydrates that are resistant to digestion and absorption in the human small intestine with complete or partial fermentation in the large intestine. Dietary fiber includes polysaccharides, oligosaccharides, lignin, and associated plant substances. Dietary fiber promotes beneficial physiological effects, such as laxation, and/or blood cholesterol attenuation, and/or blood glucose attenuation."

No polysaccharide other than starch is digested in the human small intestine; so all polysaccharides other than nonresistant starch are included in this definition of fiber. Of the oligosaccharides, only sucrose, lactose, and those derived from starch (maltooligosaccharides/maltodextrins) are digested. The term **analogous carbohydrates** is defined as those carbohydrate-based food ingredients that are nondigestible and nonabsorbable, but are not natural plant components. Wax (suberin and cutin) is included within associated substances. The definition also includes some of the health benefits known to be associated with ingestion of dietary fiber.

Since adoption of the definition, modified versions have been adopted by both governmental and nongovernmental organizations around the world. However, there is yet no consensus of national and international organizations as to a definition (66). One reason that formulating a definition acceptable to all is so difficult is that dietary fiber materials from different sources are often different mixtures of nondigestible and nonabsorbable carbohydrates and other substances with different effects on human physiology. However, there is general agreement that dietary fiber consists of oligo- and polysaccharides, lignin, and other substances not digested by digestive enzymes in the human stomach or small intestine.

10.5.2 Major Components of Dietary Fiber

The major components of natural dietary fiber are **cellulose, hemicelluloses, lignin,** and **other nonstarch plant polysaccharides** such as pectin. In a food product, added hydrocolloids/food gums, resistant starch, and certain oligosaccharides such as those derived from inulin are included because they are also nondigestible and provide certain of the physiological benefits of dietary fiber. An example is polydextrose, which is often, but not always, used in product formulations specifically because it is considered to be soluble dietary fiber.

10.5.2.1 Cell-Wall Polysaccharides of Land Plants

10.5.2.1.1 Cellulose Cellulose is a linear polymer of β-D-glucopyranosyl units (1). Some molecules may contain 10,000 or more glucosyl units. Hydrogen bonding between parallel polymers forms strong microfibrils. Cellulose microfibrils provide the strength and rigidity required in primary and secondary plant cell walls.

10.5.2.1.2 Hemicelluloses Hemicelluloses are a heterogeneous group of polysaccharides, the only similarity between them being their association with cellulose in plant cell walls (1). Units of D-xylose, D-mannose, and D-galactose frequently form the main-chain structures of hemicelluloses; units of L-arabinose, D-galactose, and uronic acids are often present as branch units or in side chains. Hemicelluloses may be soluble or insoluble in water. Molecular sizes and degrees of branching vary widely.

10.5.2.1.3 Pectins What food scientists generally call pectin (Sect. 10.4.2.3) is (like the hemicelluloses) a family of polysaccharides, although in this case there is structural similarity. The main feature of all commercial pectins is a linear chain of 1,4-linked α-D-galactopyranosyluronic acid units. Interspersed segments of neutral sugar units may be branched, sometimes with other polysaccharides. The carboxylic acid groups of the D-galacturonic acid units are often in the methyl ester form. When present primarily in a calcium and/or magnesium salt form, they are generally water insoluble and extractable only with dilute solutions of acid, chelators such as EDTA, or ammonium oxalate. These molecules are present in the middle lamella of plant tissues.

10.5.2.2 Hydrocolloids/Food Gums as Dietary Fiber

As mentioned in Sect. 10.5.1, all polysaccharides other than those in cooked starch are nondigestible and, therefore, classified as dietary fiber. Therefore, those polysaccharides classified as food gums or hydrocolloids (Sect. 10.4.2) fall within the definition of dietary fiber. Those obtained from marine algae (alginates and carrageenans) and certain of those from higher land plants (cellulose, the hemicelluloses, and the pectic polysaccharides) are cell-wall or middle lamella structural components. Others are either nonstructural plant polysaccharides (guar gum, locust bean gum, and inulin) or bacterial polysaccharides (xanthan and gellan), but neither do we humans have small intestinal enzymes that can digest them.

10.5.2.3 Resistant Starch

See Sect. 10.4.1.1.2. Resistant starch content in a food or food ingredient can be determined using AOAC Method 2002.02 (AACC Method 32-40.01).

10.5.2.4 Lignin

Lignin is a noncarbohydrate, three-dimensional, water-insoluble polymer and a major component of the cell walls of higher land plants (79). Lignin may be covalently linked to hemicellulose.

10.5.3 General Considerations

Fiber components or subfractions of them are usually not distinct entities, rather their compositions are methodology dependent. Although considerable progress has been made in relating fiber composition to physiological effects, much remains to be learned, and improving the nutritional value of foods by adding fiber or modifying resistant starch content remains a challenge for the food scientist.

10.5.4 Methods

10.5.4.1 Overview

Dietary fiber is often determined **gravimetrically**. In such a procedure, digestible carbohydrates, lipids, and proteins are selectively solubilized by chemicals or removed by enzyme-catalyzed hydrolysis. Then, nonsolubilized and/or nondigested materials are collected by filtration, and the fiber residue is recovered, dried, and weighed.

The food component that may be most problematic in fiber analysis is **starch**. In any method for determination of dietary fiber, it is essential that all digestible starch be removed, for incomplete removal of digestible starch increases the residue weight and inflates the estimate of fiber. [Resistant starch (Sect. 10.4.1.1.2) is a component of dietary fiber.]

Alpha-amylase, debranching enzymes, and glucoamylase (amyloglucosidase) are enzymes used in starch analysis (76). **α-Amylase** catalyzes hydrolysis of unbranched segments of 1,4-linked α-D-glucopyranosyl units forming primarily maltooligosaccharides composed of 3–6 units. **Debranching enzymes** (both pullulanase and isoamylase are used) catalyze hydrolysis of the 1,6 linkages that constitute the branch points and thereby produce short linear molecules. **Glucoamylase** (amyloglucosidase) starts at the nonreducing ends of starch chains and releases D-glucose, one unit at a time; it will catalyze hydrolysis of both 1, 4 and 1, 6 α-D-glucosyl linkages.

All fiber methods include a heating step (95–100°C for 35 min) to **gelatinize starch granules** and make them susceptible to hydrolysis. Resistant starch molecules (Sect. 10.4.1.1.2) remain unhydrolyzed and, therefore, are usually measured as dietary fiber, but not all nondigestible products made from starch may be determined as dietary fiber by the approved methods.

Nondigestible oligosaccharides such as those derived from inulin and certain specially prepared maltodextrins also are problematic in an analytical sense since they are in the soluble portion that is not precipitated with ethanol.

It is essential either that all digestible materials be removed from the sample so that only nondigestible polysaccharides remain or that the nondigestible residue be corrected for remaining digestible contaminants. **Lipids** are removed easily from the sample with organic solvents (Sect. 10.5.4.2) and generally do not pose analytical problems for the fiber analyst. **Protein** and **minerals** that are not removed from the sample during the solubilization steps should be corrected for by Kjeldahl nitrogen analysis (Chap. 9) and by ashing (Chap. 7) portions of the fiber residue.

Because labeling of dietary fiber content is required, because dietary fiber is a complex heterogeneous material containing several substances with different solubilities and other properties, and because of its physiological importance, methods for fiber determination continue to be researched and refined (76,77).

10.5.4.2 Sample Preparation

Measures of fiber are most consistent when the samples are low in fat (less than 10% lipid), dry, and finely ground. If necessary, the sample is ground to pass through a 0.3–0.5-mm mesh screen. If the sample contains more than 10% lipid, the lipid is removed by extraction with 25 parts (vol/wt) of petroleum ether or hexane in an ultrasonic water bath. The mixture is then centrifuged and the organic solvent decanted. This extraction is repeated. The sample is air dried to remove the organic solvent. It may then be dried overnight in a vacuum oven at 70°C if a measure of lipid and moisture content is required. Loss of weight due to fat and moisture removal is recorded, and the necessary correction is made in the calculation of the percentage dietary fiber value determined in the analysis.

If samples contain large amounts of soluble sugars (mono-, di-, and trisaccharides), they should be extracted three times with 80% aqueous ethanol in an ultrasonic water bath at room temperature for 15 min. The supernatant liquid is discarded and the residue is dried at 40°C.

Nonsolid samples with less than 10% fiber are best analyzed after freeze drying. Nonsolid samples with greater than 10% fiber can be analyzed without drying if the sample is homogeneous and low in fat and if particle size is sufficiently small to allow efficient removal of digestible carbohydrate and protein.

10.5.4.3 Methods

10.5.4.3.1 Overview A variety of methods have been developed and used at different times for different products. *AOAC International Official Methods of Analysis* in reference (5) and *AACC International Approved Methods* in reference (80) are listed in Table 10-4. It is obvious from the list that methods are generally specific for the type of fiber or the fiber component desired to be measured. For example, when inulin (a fructan) or its breakdown products (fructooligosaccharides, FOS) are added to food products, not all of the inulin and perhaps none of the FOS are precipitated by addition of four volumes of alcohol (because of their low molecular weights) and measured as soluble dietary fiber, although both inulin and FOS undergo fermentation in the colon and are, therefore, components of dietary fiber. As a result special methods have been designed for them. The same is true of polydextrose and resistant maltodextrins. In other cases, determination of a specific component of dietary fiber, such as β-glucan and resistant starch, may be desired. The most widely used general method for total, soluble, and insoluble dietary fiber (AOAC Method 991.43, AACC Method 32-07.01) is outlined below. Table 10-5 gives the fiber content of select foods analyzed by this method.

10.5.4.3.2 AOAC Method 991.43 (AACC Method 32-07.01) This method determines soluble, insoluble, and total dietary fiber in cereal products, fruits and vegetables, processed foods, and processed food ingredients.

1. **Principle**. Starch and protein are removed from a sample by treating the sample sequentially with a thermostable α-amylase, a protease, and glucoamylase

table 10-4 Official Methods of Analysis for Dietary Fiber

AOAC Method No. (5)	*AACC Method No. (80)*	*Description of Method and Measured Substance*
994.13	32-25.01	Total dietary fiber determined as neutral sugar and uronic acid monomer units and Klason lignin by a gas chromatographic–spectrophotometric–gravimetric method
993.21		Nonenzymic-gravimetric method for total dietary fiber applicable to determination of >10% TDF in foods and food products with <2% starch
985.29	32-05.01	Enzymic-gravimetric method for total dietary fiber in cereal grains and cereal grain-based products
	32-06.01	A rapid gravimetric method for total dietary fiber
991.42		Enzymic-gravimetric method for insoluble dietary fiber in vegetables, fruits, and cereal grains
993.19		Enzymic-gravimetric method for soluble dietary fiber
991.43	32-07.01	Enzymic-gravimetric method for total, soluble, and insoluble dietary fiber in grain and cereal products, processed foods, fruits, and vegetables
2002.02	32-40.01	Enzymic method for RS2 and RS3 in products and plant materials
	32-21.01	Enzymic-gravimetric method for insoluble and soluble dietary fiber in oats and oat products
	32-32.01	Enzymic-spectrophotometric method for total fructan (inulin and fructooligosaccharides) in foods
993.03		Enzymic-spectrophotometric method for fructan (inulin) in foods (not applicable to fructooligosaccharides)
997.08	32-31.01	Anion-exchange chromatographic method for fructan in foods and food products applicable to the determination of added inulin in processed foods
2000.11	32-28.01	Anion-exchange chromatographic method for polydextrose in foods
	32-22.01	Enzymic method for β-glucan in oat fractions and unsweetened oat cereals
	32-23.01	Rapid enzymic procedure for β-glucan content of barley and oats
2001.03	32-41.01	Enzymic-gravimetric and liquid chromatographic method for dietary fiber containing added resistant maltodextrin
2001.02	32-33.01	Anion-exchange chromatographic method for *trans*-galactooligosaccharides (TGOS) applicable to added TGOS in selected food products

table 10-5 Total, Soluble, and Insoluble Dietary Fiber in Foods as Determined by AOAC Method 991.43[a]

Food	*Soluble*[b]	*Insoluble*[b]	*Total*[b]
Barley	5.02	7.05	12.14
High-fiber cereal	2.78	30.52	33.30
Oat bran	7.17	9.73	16.90
Soy bran	6.90	60.53	67.56
Apricots	0.53	0.59	1.12
Prunes	5.07	4.17	9.37
Raisins	0.73	2.37	3.03
Carrots	1.10	2.81	3.92
Green beans	1.02	2.01	3.03
Parsley	0.64	2.37	3.01

[a]Adapted from *Official Methods of Analysis*, 18th edn. Copyright 2005 by AOAC International.
[b]Grams of fiber per 100 g of food on a fresh weight basis.

(amyloglucosidase). The insoluble residue is recovered and washed (**insoluble dietary fiber**). Ethanol is added to the soluble portion to precipitate soluble polysaccharides (**soluble dietary fiber**). To obtain **total dietary fiber** (TDF), the alcohol is added after digestion with the glucoamylase, and the soluble and insoluble dietary fiber fractions are collected together, dried, weighed, and ashed.

2. **Outline of procedure**. A flow diagram outlining the general procedure for the method is given in Fig. 10-12. Letters in the parentheses refer to the same letters in Fig. 10-12. If necessary, lipids are removed by extraction (Sect. 10.5.4.2).

(a) To samples devoid of significant lipid solvent-soluble substances is added 2-(*N*-morpholino)ethanesulfonic acid-tris(hydroxymethyl)aminomethane (MES-TRIS) buffer (0.05 *M* each, pH 8.2) and a thermostable α-amylase. The mixture is heated 35 min at 95–100°C to gelatinize any starch so that the α-amylase can break it down.
(b) After cooling to 60°C, a protease is added, and the mixture is incubated at 60°C for 35 min to break down the protein.
(c) The pH is adjusted to 4.1–4.8, glucoamylase is added, and the mixture is incubated at 60°C for 30 min to complete the digestion of any starch.
(d) To determine TDF, four volumes of 95% ethanol are added. The residue plus precipitate is collected by filtration, washed with 78% ethanol, 95% ethanol, and acetone in that order, dried, and weighed (see below). Protein and ash are determined on duplicate samples and the weight is corrected for them. Alternatively, TDF can be calculated as the sum of the insoluble and soluble dietary fiber determined in the remainder of the procedure.

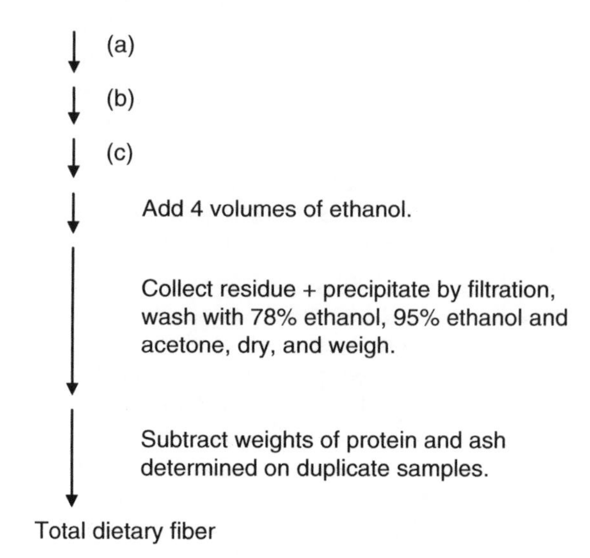

(e) The mixture obtained after step (c) is filtered through a crucible containing fritted glass disk and preashed Celite (a siliceous filter aid).
(f) The residue is washed with water, 95% ethanol, and acetone in that order, dried, and weighed.
(g) The dried residue is analyzed for protein using the Kjeldahl method (Chap. 9). A duplicate residue is analyzed for ash (Chap. 7). The weights of protein and ash are subtracted from the residue weight obtained in step (f) to determine insoluble dietary fiber.
(h) To determine soluble dietary fiber, to the filtrate and washings from steps (e) and (f) at 60°C are added four volumes of 95% ethanol (to give an ethanol concentration of 76%). The precipitate is collected by filtration through a crucible containing a fritted glass disk and preashed Celite. The residue is washed with 78% ethanol, 95% ethanol, and acetone. The crucible is dried at 103°C and weighed.
(i) Protein and ash are determined as in step (f) and the weights of protein and ash are subtracted from the residue weight obtained in step (h) to determine soluble dietary fiber. Total dietary fiber may be determined as described in (d) or obtained by adding the values for insoluble (g) and soluble (i) dietary fiber.

Duplicate reagent blanks must be run through the entire procedure for each type of fiber determination. Table 10-6 shows a sample and blank sheet

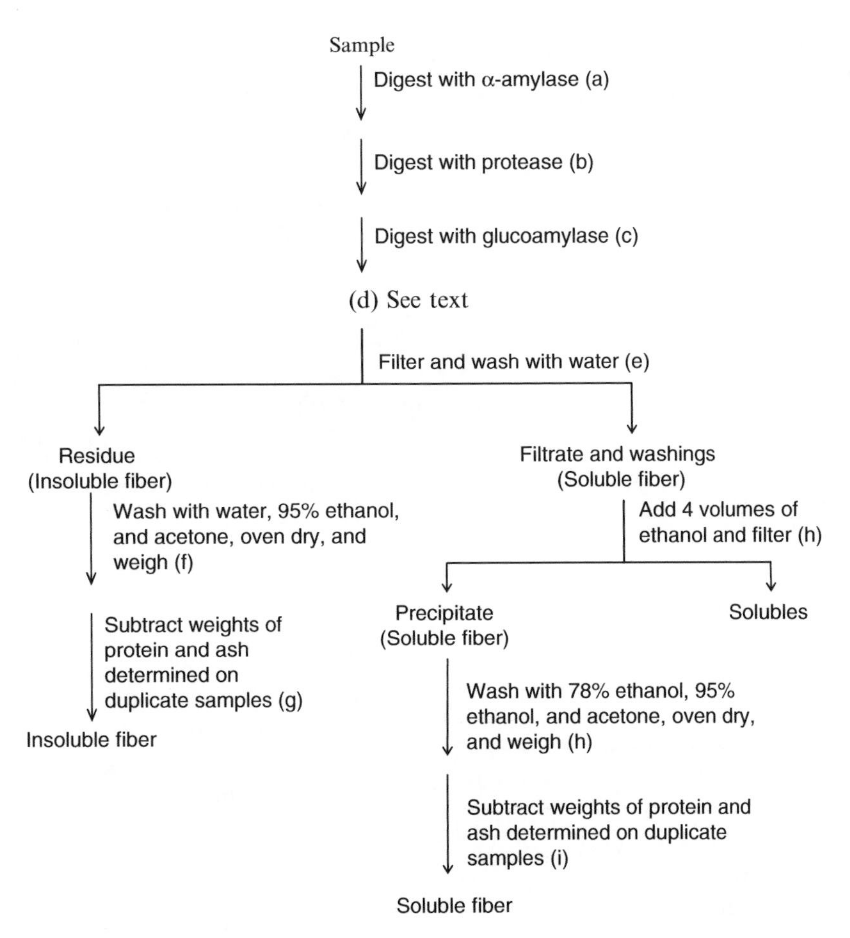

Flow diagram of AOAC Method 991.43 (AACC International Method 32-07.01) for determining soluble, insoluble, and total dietary fiber.

used to calculate fiber percentages. Using the equations shown, percent dietary fiber is expressed on a dry weight basis if the sample weights are for dried samples. If it is believed that resistant starch is present, it can be determined separately using AOAC Method 2002.02 (AACC Method 32-40.01).

10.6 PHYSICAL METHODS

10.6.1 Microscopy

Microscopy can be a valuable tool in food analysis. Various kinds of microscopy [light, fluorescence, confocal scanning laser (CSLM), Fourier transform infrared (FTIR), scanning electron (SEM), and transmission electron (TEM) microscopies] have been used to study the organization of food products and the stability of emulsions and foams and to identify extraneous matter and its amount (Chap. 19). Microscopy is particularly useful in examinations of starchy foods. Granule size, shape, and form, the birefringence endpoint temperature determined using a polarizing microscope with a hot stage, and, in some cases, iodine-staining characteristics can be used to identify the starch source (81). In cooked starch products, the extent of retrogradation (82) and the effects of storage on microstructure have been evaluated by iodine staining and light microscopy (83–89). The degree that starch has been damaged mechanically during dry milling (90), the extent of digestion by enzymes, and whether the starch-based product has been overcooked, undercooked, or correctly cooked also can be determined microscopically. Quantitative microscopy has been employed for analysis of the nonstarch polysaccharides of cereal grains (91).

10.6.2 Mass and NIR Transmittance Spectrometry

Mass and NIR transmittance spectrometry have been used to determine sugar content (92). NIR spectrometry is described in Chap. 23. Mass spectrometry is mentioned in Sect. 10.3.4.4.

10-6 table Dietary Fiber Data Sheet[a]

	Sample					Blank			
	Insoluble Fiber		Soluble Fiber			Insoluble Fiber		Soluble Fiber	
Sample wt (mg)	m_1	m_2	■	■		■	■	■	■
Crucible + Celite wt (mg)									
Crucible + Celite + residue wt (mg)									
Residue wt (mg)	R_1	R_2	R_1	R_2		R_1	R_2	R_1	R_2
Protein (mg) P		■		■			■		■
Crucible + Celite + ash wt (mg)	■		■			■		■	
Ash wt (mg) A	■		■			■		■	
Blank wt (mg) B[b]	■	■	■	■					
Fiber (%)[c]						■	■	■	■

[a]Adapted with permission from J AOAC Int (1988), 71:1019.

[b]$\text{Blank(mg)} = \frac{R_1 + R_2}{2} - P - A$

[c]$\text{Fiber(\%)} = \frac{\frac{R_1+R_2}{2} - P - A - B}{\frac{m_1+m_2}{2}} \times 100$

10.6.3 Specific Gravity

Specific gravity is defined as the ratio of the density of a substance to the density of a reference substance (usually water), both at a specified temperature. The concentration of a carbohydrate solution can be determined by measuring the specific gravity of the solution, then referring to appropriate specific gravity tables (11).

Measurement of specific gravity as a means of determining sugar concentration is accurate only for pure sucrose or other solutions of a single pure substance (AOAC Method 932.14), but it can be, and is, used for obtaining approximate values for liquid products (Chap. 6). Two basic means of determining specific gravity are used. By far the most common is use of a hydrometer calibrated either in °**Brix**, which corresponds to sucrose concentrations by weight, or in **Baumé Modulus** (Bé). The values obtained are converted into concentrations by use of tables constructed for the substance in the pure solution, e.g., sucrose or glucose syrups.

10.6.4 Refractive Index

When electromagnetic radiation passes from one medium to another, it changes direction (i.e., is bent or refracted). The ratio of the sine of the angle of incidence to the sine of the angle of refraction is termed the **refractive index** (RI). The RI varies with the nature of the compound, the temperature, the wavelength of light, and the concentration of the compound. By holding the first three variables constant, the concentration of the compound can be determined by measuring the RI. Thus, measurement of refractive index is another way to determine total solids in solution (Chap. 6). Like determination of specific gravity, use of RI to determine concentrations is accurate only for pure sucrose or other solutions of a single pure substance, and also like the determination of specific gravity, it is used for obtaining approximate sugar concentrations in liquid products (11). In this case, the solution must be clear. Refractometers that read directly in sucrose units are available.

10.7 SUMMARY

For determination of low-molecular-weight carbohydrates, older colorimetric methods for total carbohydrate, various reducing sugar methods, and physical measurements have largely been replaced by chromatographic methods. The older chemical methods suffer from the fact that they are not stoichiometric and, therefore, require standard curves. This makes them particularly problematic when a mixture of sugars is being determined. Physical measurements are not specific for carbohydrates. Chromatographic methods (HPLC and GC) separate mixtures into the component sugars, identify each component by retention time, and provide a measurement of the mass of each component. Enzymic methods are specific and sensitive, but seldom, except in the case of starch, is determination of only a single component desired. HPLC is widely used for identification and measurement of mono- and oligosaccharides.

Polysaccharides are important components of many food products. Yet there is no universal procedure for their analysis. Generally, isolation must precede measurement. Isolation introduces errors because no extraction or separation technique is stochiometric. Identification and measurement are done by hydrolysis to constituent monosaccharides and their determination. An exception is starch, which can be digested to glucose using specific enzymes (amylases), followed by measurement of the glucose released.

Insoluble dietary fiber, soluble dietary fiber, and total dietary fiber are each composed primarily of nonstarch polysaccharides. The method for the determination of starch is based on its complete conversion to, and determination of, D-glucose. Methods for the determination of total dietary fiber and its components rely on removal of the digestible starch in the same way and often on removal of digestible protein with a protease, leaving nondigestible components.

10.8 STUDY QUESTIONS

1. Give three reasons why carbohydrate analysis is important.
2. "Proximate composition" refers to analysis for moisture, ash, fat, protein, and carbohydrate. Identify which of these components of "proximate composition" are actually required on a nutrition label. Also, explain why it is important to measure the nonrequired components quantitatively if one is developing a nutrition label.
3. Distinguish chemically between monosaccharides, oligosaccharides, and polysaccharides, and explain how solubility characteristics can be used in an extraction procedure to separate monosaccharides and oligosaccharides from polysaccharides.
4. Discuss why mono- and oligosaccharides are extracted with 80% ethanol rather than with water. What is the principle involved?
5. Define reducing sugar. Classify each of the following as a reducing or nonreducing carbohydrate: D-glucose, D-fructose (Conditions must be described. Why?), sorbitol, sucrose, maltose, raffinose, maltotriose, cellulose, amylopectin, κ-carrageenan.
6. Briefly describe a method that could be used for each of the following:
   (a) To prevent hydrolysis of sucrose when sugars are extracted from fruits via a hot alcohol extraction
   (b) To remove proteins from solution for an enzymic analysis
   (c) To measure total carbohydrate
   (d) To measure total reducing sugars
   (e) To measure the sucrose concentration in a pure sucrose solution by a physical method
   (f) To measure glucose enzymically
   (g) To measure simultaneously the concentrations of individual free sugars
7. What are the principles behind total carbohydrate determination using the phenol-sulfuric acid method? Give an example of another assay procedure based on the same principle.
8. What is the principle behind determination of total reducing sugars using the Somogyi–Nelson and similar methods?
9. The Munson–Walker, Lane–Eynon, and Somogyi–Nelson methods can be used to measure reducing sugars. Explain the similarities and differences among these methods with regard to the principles involved and the procedures used.
10. Describe the principle behind AE-HPLC of carbohydrates.
11. Describe the general procedure for preparation of sugars for GC. What is required for this method to be successful?
12. What difference is there between the preparation of an extract of reducing sugars for GC and the preparation of polysaccharide hydrolyzates containing uronic acids for GC? What two differences are there in the final derivatives?
13. Why has HPLC largely replaced GC for analysis of carbohydrates?
14. Compare and contrast RI and PAD detectors.
15. What is the advantage of an enzymic method? What is the limitation (potential problem)?
16. Describe the principles behind the enzymic determination of starch. What are the advantages of this method? What are potential problems?
17. Describe the principle behind each step in Fig. 10-11. What is the reason for each step?
18. Describe the principles behind separation and analysis of water-soluble gums and starch.
19. Describe two methods for determination of pectin.
20. Describe the principles behind and the limitations of determining sugar (sucrose) concentrations by (a) specific gravity determination and (b) RI measurement.
21. Define dietary fiber.
22. List the major constituents of dietary fiber.
23. Explain how measurement of dietary fiber relates to calculating the caloric content of a food product.
24. Explain the purpose(s) of each of the steps in the AOAC Method 994.13 for total dietary fiber listed below as applied to determination of the dietary fiber content of a high-fiber snack food.
    (a) Heating sample and treating with α-amylase
    (b) Treating sample with glucoamylase
    (c) Treating sample with protease
    (d) Adding four volumes of 95% ethanol to sample after treatment with glucoamylase and protease
    (e) After drying and weighing the filtered and washed residue, heating one duplicate final product to 525°C in a muffle furnace and analyzing the other duplicate sample for protein.
25. What is the physiological definition and the chemical nature of resistant starch? What types of foods have relatively high levels of resistant starch?

10.9 PRACTICE PROBLEMS

1. The following data were obtained when an extruded breakfast cereal was analyzed for total fiber by AOAC Method 991.43 (AACC Method 32-07).

Sample wt (mg)	1002.8
Residue wt (mg)	151.9
Protein wt (mg)	13.1
Ash wt (mg)	21.1
Blank wt (mg)	6.1
Resistant starch (mg)	35.9

What is percent total fiber (a) without and (b) with correction for resistant starch, determined to the appropriate number of significant figures?

2. The following tabular data were obtained when a high-fiber cookie was analyzed for fiber content by AOAC Method 991.43 (AACC Method 32-07).

	Sample			
	Insoluble		*Soluble*	
Sample wt (mg)	1002.1	1005.3		
Crucible + Celite wt (mg)	31,637.2	32,173.9	32,377.5	33,216.4
Crucible + Celite + residue wt (mg)	31,723.5	32,271.2	32,421.6	33,255.3
Protein (mg)	6.5		3.9	
Crucible + Celite + ash wt (mg)		32,195.2		33,231.0

	Blank			
	Insoluble		*Soluble*	
Crucible + Celite wt (mg)	31,563.6	32,198.7	33,019.6	31,981.2
Crucible + Celite + residue wt (mg)	31,578.2	32,213.2	33,033.4	33,995.6
Protein (mg)	3.2		3.3	
Crucible + Celite + ash wt (mg)		32,206.8		31,989.1

What is the (a) insoluble, (b) soluble, and (c) total fiber content of the cookie determined to the appropriate number of significant figures?

Answers

1. Number of significant figures = 2 (6.1 mg)

$$\text{(a)}\ \frac{151.9 - 13.1 - 21.1 - 6.0}{1002.8} \times 100 = 11\%$$

$$\text{(b)}\ \frac{151.9 - 13.1 - 21.1 - 6.1 - 35.9}{1002.8} \times 100 = 7.5\%$$

2. (a) 6.1%, (b) 2.0%, (c) 8.1%
(Calculations are done a little differently than those at the bottom of Table 10-6.)

a. *Insoluble dietary fiber*

Number of significant figures = 2 (6.5 mg, 3.2 mg)

$$\text{Blank residue} = 31{,}578.2\,\text{mg} - 31{,}563.6\,\text{mg} = 14.6\,\text{mg}$$
$$32{,}231.2\,\text{mg} - 32{,}198.7\,\text{mg} = 14.5\,\text{mg}$$
$$\text{Average} = 14.6\,\text{mg}$$

$$\text{Blank ash} = 32{,}206.8\,\text{mg} - 32{,}198.7\,\text{mg} = 8.1\,\text{mg}$$

First sample residue:

$$= 31{,}723.5\,\text{mg} - 31{,}637.2\,\text{mg} = 86.3\,\text{mg}$$
$$\text{Ash} = 32{,}195.2\,\text{mg} - 32{,}173.9\,\text{mg} = 21.3\,\text{mg}$$

86.3 mg (residue weight)
– 14.6 mg (blank)
– 3.3 mg (protein, 6.5 – 3.2 [blank])
– 13.2 mg (ash, 21.3 – 8.1 [blank])
= 55.2 mg

(55.2 mg ÷ 1,002.1 [sample wt.]) × 100 = 5.5%

Second sample residue :

$$= 32{,}271.2\,\text{mg} - 32{,}173.9\,\text{mg} = 97.3\,\text{mg}$$
$$97.3 - 14.5 - 3.3 - 13.2 = 66.3\,\text{mg}$$

(66.3 mg ÷ 1005.3 mg [sample wt.]) × 100 = 6.6%

Average of 5.5% and 6.6% = 6.1%

b. *Soluble dietary fiber*

Number of significant figures = 2 (3.9 mg, 3.3 mg)

$$\text{Blank residue} = 33{,}033.4\,\text{mg} - 33{,}019.6\,\text{mg} = 13.8\,\text{mg}$$
$$33{,}995.6\,\text{mg} - 31{,}981.2\,\text{mg} = 14.4\,\text{mg}$$
$$\text{Average} = 14.1\,\text{mg}$$

$$\text{Blank ash} = 31{,}989.1\,\text{mg} - 31{,}981.2\,\text{mg} = 7.9\,\text{mg}$$

First sample residue:

$$= 32{,}421.6\,\text{mg} - 32{,}377.5\,\text{mg} = 44.1\,\text{mg}$$
$$\text{Ash} = 33{,}231.0\,\text{mg} - 33{,}216.4\,\text{mg} = 14.6\,\text{mg}$$

44.1 mg (residue weight)
– 14.1 mg (blank)
– 0.6 mg (protein, 3.9 – 3.3 [blank])
– 6.7 mg (ash, 14.6 – 7.9 [blank])
= 22.7 mg

(22.7 mg ÷ 1,002.1 [sample wt.]) × 100 = 2.3%

Second sample residue:

$$= 33{,}255.3\,\text{mg} - 33{,}216.4\,\text{mg} = 38.9\,\text{mg}$$
$$38.9 - 14.1 - 0.6 - 6.7 = 17.5\,\text{mg}$$

(17.5 mg ÷ 1005.3 [sample wt.]) × 100 = 1.7%

Average of 2.3% and 1.7% = 2.0%

c. *Total dietary fiber* (TDF)

$$TDF = 6.1\% \text{ (insolublefiber)} + 2.0\% \text{ (solublefiber)} = 8.1\%$$

10.10 REFERENCES

1. BeMiller JN (2007) Carbohydrate chemistry for food scientists, 2nd edn. AACC International, St. Paul, MN
2. BeMiller JN, Huber K (2008) Carbohydrates (Chapter 4). In: Damodaran S, Parkin KL, Fennema OR (eds) Food chemistry, 4th edn. Marcel Dekker, New York
3. Scherz H, Bonn G (1998) Analytical chemistry of carbohydrates. Georg Thieme Verlag, Stuttgart
4. AOAC International (1995) Official methods of analysis, 16th edn. AOAC International, Gaithersburg, MD
5. AOAC International (2007) Official methods of analysis, 18th edn., 2005; Current through revision 2, 2007 (On-line). AOAC International, Gaithersburg, MD
6. Anonymous (2009) Code of federal regulations, Title 21, Part 101.9 – Food nutrition labeling of food. US Government Printing Office, Washington, DC
7. Southgate DAT (1976) Determination of food carbohydrates. Applied Science Publishers, London
8. Chaplin MF, Kennedy JF (eds) (1994) Carbohydrate analysis. A practical approach, 2nd edn. IRL Press, Oxford
9. Dubois M, Gilles KA, Hamilton JK, Rebers PA, Smith F (1956) Colorimetric method for determination of sugars and related substances. Anal Chem 28:350
10. Sakano Y, Kobayashi T (1994) Enzymic preparation of panose and isopanose from pullulan. Methods Carbohydr Chem 10:249
11. Hodge JE, Hofreiter BT (1962) Determination of reducing sugars and carbohydrates. Methods Carbohydr Chem 1:380
12. Nelson N (1944) A photometric adaptation of the Somogyi method for the determination of glucose. J Biol Chem 153:376
13. Somogyi M (1952) Notes on sugar determination. J Biol Chem 195:19
14. Wood TM (1994) Enzymic conversion of cellulose into D-glucose. Methods Carbohydr Chem 10:219
15. Miller GL, Blum R, Glennon WEG, Burton AL (1960) Measurement of carboxymethylcellulase activity. Anal Biochem 1:127
16. Hicks KB (1988) High-performance liquid chromatography of carbohydrates. Adv Carbohydr Chem Biochem 46:17
17. Ball GFM (1990) The application of HPLC to the determination of low molecular weight sugars and polyhydric alcohols in foods: a review. Food Chem 35:117
18. Corradini C (1994) Recent advances in carbohydrate analysis by high-performance anion-exchange chromatography coupled with pulsed amperometric detection (HPAEC-PAD). Ann Chim-Rome 84:385
19. El Rassi Z (ed) (1995) Carbohydrate analysis (Journal of Chromatography Library, vol 58). Elsevier, Amsterdam
20. Cserháti T, Forgács E (1999) Chromatography in food science and technology. Technomic Publishing, Lancaster, PA
21. Cataldi TRI, Campa C, DeBenedetto GE (2000) Carbohydrate analysis by high-performance anion-exchange chromatography with pulsed amperometric detection: the potential is still growing. Fresen J Anal Chem 368:739
22. Jeon IJ (1995) Carbohydrates and sugars. Food Sci Technol (NY) 65:87
23. Bugner E, Feinberg M (1992) Determination of mono- and disaccharides in foods by interlaboratory study: quantitation of bias components for liquid chromatography. J AOAC Int 75:443
24. Huber CG, Bonn GK (1995) HPLC of carbohydrates with cation- and anion-exchange silica and resin-based stationary phases (Chapter 4 in reference 19)
25. Andersen R, Sørensen A (2000) Separation and determination of alditols and sugars by high-pH anion-exchange chromatography with pulsed amperometric detection. J Chromatogr A 897:195
26. Hanko VP, Rohrer JS (2000) Determination of carbohydrates, sugar alcohols, and glycols in cell cultures and fermentation broths using high-performance anion-exchange chromatography with pulsed amperometric detection. Anal Biochem 283:192
27. Ammeraal RN, Delgado GA, Tenbarge FL, Friedman RB (1991) High-performance anion-exchange chromatography with pulsed amperometric detection of linear and branched glucose oligosaccharides. Carbohydr Res 215:179
28. Kazmaier T, Roth S, Zapp J, Harding M, Kuhn R (1998) Quantitative analysis of maltooligosaccharides by MALDI-TOF mass spectrometry, capillary electrophoresis, and anion exchange chromatography. Fresen J Anal Chem 361:473; 362:552
29. Churms SC (1995) High performance hydrophilic interaction chromatography of carbohydrates with polar sorbents (Chapter 3 in reference 19)
30. Verzele M, Simoens G, Van Damme F (1987) A critical review of some liquid chromatography systems for the separation of sugars. Chromatographia 23:292
31. El Rassi Z (1995) Reversed-phase and hydrophobic interaction chromatography of carbohydrates and glycoconjugates (Chapter 2 in reference 19)
32. Rajakyla E (1986) Use of reversed-phase chromatography in carbohydrate analysis. J Chromatogr 353:1
33. Bruno AE, Krattiger B (1995) On-column refractive index detection of carbohydrates separated by HPLC and CE (Chapter 11 in reference 19)
34. Johnson DC, Lacourse WR (1995) Pulsed electrochemical detection of carbohydrates at gold electrodes following liquid chromatographic separations (Chapter 10 in reference 19)
35. Hase S (1995) Pre- and post-column detection-oriented derivatization techniques in HPLC of carbohydrates (Chapter 15 in reference 19)
36. Biermann CJ, McGinnis GD (eds) (1989) Analysis of carbohydrates by GLC and MS. CRC, Boca Raton, FL
37. Biermann CJ (1989) Introduction to analysis of carbohydrates by gas-liquid chromatography (Chapter 1 in reference 35)
38. Sloneker JH (1972) Gas–liquid chromatography of alditol acetates. Methods Carbohydr Chem 6:20

39. Fox A, Morgan SL, Gilbart J (1989) Preparation of alditol acetates and their analysis by gas chromatography (GC) and mass spectrometry (MS) (Chapter 5 in reference 35)
40. Seymour FR (1993) Identification and characterization of saccharides by GLC separation and MS analysis of their peracetylated aldononitrile (PAAN) and ketooxime (PAKO) derivatives. Methods Carbohydr Chem 9:59
41. Harris P, Morrison A, Dacombe C (1995) A practical approach to polysaccharide analysis (Chapter 18). In: Stephen AM (ed) Food polysaccharides and their applications. Marcel Dekker, New York
42. BeMiller JN (ed) (1994) Methods in carbohydrate chemistry, vol 10, Enzymic methods. Wiley, New York
43. Bergmeyer HU (ed) (1984) Methods of enzymatic analysis, vol 6, Metabolites 1: carbohydrates, 3rd edn. Verlag Chemie, Weinheim, Germany
44. Setttineri CA, Burlingame AL (1995) Mass spectrometry of carbohydrates and glycoconjugates (Chapter 12 in reference 19)
45. Vaccari G, Lodi G, Tamburini E, Bernardi T, Tosi S (2001) Detection of oligosaccharides in sugar products using planar chromatography. Food Chem 74:99
46. Klockow A, Paulus A, Figueiredo V, Amado R, Widmer HM (1994) Determination of carbohydrates in fruit juices by capillary electrophoresis and high-performance liquid chromatography. J Chromatogr A 680:187
47. El Rassi Z, Nashabeth W (1995) High performance capillary electrophoresis of carbohydrates and glycoconjugates (Chapter 8 in reference 19)
48. Asp NG, Bjoerck I (1992) Resistant starch. Trends Food Sci Technol 3:111
49. McCleary BV, Gibson TS, Mugford DC (1997) Collaborative evaluation of a simplified assay for total starch in cereal products (AACC Method 76-13). Cereal Foods World 42:476
50. Kainuma K (1994) Determination of the degree of gelatinization and retrogradation of starch. Methods Carbohydr Chem 10:137
51. Baird JK (1993) Analysis of gums in foods (Chapter 23). In: Whistler RL, BeMiller JN (eds) Industrial gums, 3rd edn. Academic, San Diego, CA
52. BeMiller JN (1996) Gums/hydrocolloids: analytical aspects (Chapter 6). In: Eliasson AC (ed) Carbohydrates in food. Marcel Dekker, New York
53. National Academy of Sciences (1996) Food chemicals codex, 4th edn. Food and Nutrition Board, National Research Council, National Academy Press, Washington, DC
54. Joint FAO/WHO Expert Committee on Food Additives (JECFA) (1992) Compendium of food additive specifications, vols 1 and 2. FAO Food and Nutrition Paper 52/1, Food and Agriculture Organization of the United Nations, Rome, Italy
55. Bitter T, Muir HM (1962) A modified uronic acid carbazole reaction. Anal Biochem 4:330
56. Chandrasekaran EV, BeMiller JN (1980) Constituent analysis of glycosaminoglycans. Methods Carbohydr Chem 8:89
57. Blumenkrantz N, Absoe-Hansen G (1973) New method for quantitative determination of uronic acids. Anal Biochem 54:484
58. Kintner PK III, Van Buren JP (1982) Carbohydrate interference and its correction in pectin analysis using the *m*-hydroxydiphenyl method. J Food Sci 47:756
59. Scott RW (1979) Colorimetric determination of hexuronic acids in plant materials. Anal Chem 51:936
60. Scott JE (1965) Fractionation by precipitation with quaternary ammonium salts. Methods Carbohydr Chem 5:38
61. Baker RA (1997) Reassessment of some fruit and vegetable pectin levels. J Food Sci 62:225
62. Walter RH (1991) Analytical and graphical methods for pectin (Chapter 10). In: Walter RH (ed) The chemistry and technology of pectin. Academic, San Diego, CA
63. Walter RH, Sherman RM, Lee CY (1983) A comparison of methods for polyuronide methoxyl determination. J Food Sci 48:1006
64. Grasdalen H, Bakøy OE, Larsen B (1988) Determination of the degree of esterification and the distribution of methylated and free carboxyl groups in pectins by ^{1}H-N.M.R. spectroscopy. Carbohydr Res 184:183
65. Westerlund E, Åman P, Andersson RE, Andersson R (1991) Investigation of the distribution of methyl ester groups in pectin by high-field ^{13}C NMR. Carbohydr Polym 14:179
66. Gordon DT (2007) Dietary fiber definitions at risk. Cereal Foods World 52:112
67. Burkitt SP, Trowell MC (1975) Refined carbohydrate foods and disease. Academic, London
68. Leeds AR, Avenell A (eds) (1985) Dietary fibre perspectives: reviews and bibliography 1. John Libby and Company, London
69. Leeds AR, Burley V (eds) (1990) Dietary fibre perspectives: reviews and bibliography 2. John Libby and Company, London
70. Muir JC, Young OP, O'Dea K, Cameron-Smith S, Brown IL, Collier GR (1993) Resistant starch the neglected "dietary fiber"? Implications for health. Diet Fiber Bibliogr Rev 1:33
71. Valnouny CV, Kritchevsky D (eds) (1986) Dietary fiber: basic and clinical aspects. Plenum, New York
72. Salovaara H, Gates F, Tenkanen M (eds) (2007) Dietary fibre components and functions. Wageningen Academic, Wageningen, The Netherlands
73. Gordon DT, Goda T (eds) (2008) Dietary fiber. AACC International, St. Paul, MN
74. McCleary BV, Prosky L (eds) (2001) Advanced dietary fibre technology. Blackwell Science, London
75. Prosky L (2001) What is dietary fibre? A new look at the definition (Chapter 6 in reference 69)
76. Asp N-G (2001) Development of dietary fibre methodology (Chapter 7 in reference 69)
77. McCleary BV (2001) Measurement of dietary fibre components: the importance of enzyme purity, activity and specificity (Chapter 8 in reference 69)
78. Anon (2001) The definition of dietary fiber. Cereal Foods World 46:112
79. Lewis NG, Davin LB, Sarkanen S (1999) The nature and function of lignins (Chapter 18). In: Pinto BM (ed) Comprehensive natural products chemistry. Elsevier, Oxford
80. AACC International (2010) Approved methods, 11th edn (On-line). AACC International, St. Paul, MN

81. Fitt LE, Snyder EM (1984) Photomicrographs of starches (Chapter 23). In: Whistler RL, BeMiller JN, Paschall EF (eds) Starch: chemistry and technology, 2nd edn. Academic, Orlando, FL
82. Jacobson MR, Obanni M, BeMiller JN (1997) Retrogradation of starches from different botanical sources. Cereal Chem 74:511
83. Langton M, Hermansson A-M (1989) Microstructural changes in wheat starch dispersions during heating and cooling. Food Microstruct 8:29
84. Autio K (1990) Rheological and microstructural changes of oat and barley starches during heating and cooling. Food Struct 9:297
85. Svegmark K, Hermansson A-M (1991) Distribution of amylose and amylopectin in potato starch pastes: effect of heating and shearing. Food Struct 10:117
86. Autio K, Poutanen K, Suortti T, Pessa E (1992) Heat-induced structural changes in acid-modified barley starch dispersions. Food Struct 11:315
87. Svegmark K, Hermansson A-M (1992) Microstructure and viscoelastic behavior of potato starch pastes. In: Phillips GO, Wedlock DJ, Williams PA (eds) Gums and stabilizers for the food industry, vol 6. Oxford University Press, Oxford, p 93
88. Svegmark K, Hermansson A-M (1993) Microstructure and rheological properties of composites of potato starch granules and amylose; a comparison of observed and predicted structures. Food Struct 12:181
89. Virtanen T, Autio K, Suortti T, Poutanen K (1993) Heat-induced changes in native and acid-modified oat starch pastes. J Cereal Sci 17:137
90. Sandstedt RM, Schroeder H (1960) A photomicrographic study of mechanically damaged wheat starch. Food Technol 14:257
91. Fulcher RG, Faubion JM, Ruan R, Miller SS (1994) Quantitative microscopy in carbohydrate analysis. Carbohydr Polym 25:285
92. Mehrübeoglu M, Coté GL (1997) Determination of total reducing sugars in potato samples using near-infrared spectroscopy. Cereal Foods World 42:409 and references therein

Vitamin Analysis

Ronald B. Pegg, W.O. Landen, Jr., and Ronald R. Eitenmiller*

Department of Food Science and Technology, The University of Georgia, 100 Cedar Street, Athens, GA 30602–7610, USA

rpegg@uga.edu
woland@uga.edu
eiten@uga.edu

11.1 Introduction 181
- 11.1.1 Definition and Importance 181
- 11.1.2 Importance of Analysis 181
- 11.1.3 Vitamin Units 181

11.2 Methods 182
- 11.2.1 Overview 182
- 11.2.2 Extraction Methods 184
- 11.2.3 Bioassay Methods 184
- 11.2.4 Microbiological Assays 185
  - 11.2.4.1 Applications 185
  - 11.2.4.2 Principle 185
  - 11.2.4.3 Niacin 185
  - 11.2.4.4 Folate 185
    - 11.2.4.4.1 Principle 187
    - 11.2.4.4.2 Critical Points 187
    - 11.2.4.4.3 Procedure 187

S.S. Nielsen, *Food Analysis*, Food Science Texts Series, DOI 10.1007/978-1-4419-1478-1_11,

11.2.4.4.4 Calculations 187
11.2.5 Chemical Methods 188
11.2.5.1 Vitamin A 188
11.2.5.1.1 Principle 188
11.2.5.1.2 Critical Points 188
11.2.5.1.3 Procedure 188
11.2.5.1.4 Calculations 188
11.2.5.2 Vitamin E (Tocopherols and Tocotrienols) 189
11.2.5.2.1 Vitamin E Compounds 189
11.2.5.2.2 Principle 189
11.2.5.2.3 Critical Points 189
11.2.5.2.4 Procedure 189
11.2.5.2.5 Calculation 190
11.2.5.3 Vitamin C 190
11.2.5.3.1 2,6-Dichloroindophenol Titrimetric Method (AOAC Method 967.21, 45.1.14) (2,9) 190
11.2.5.3.2 Microfluorometric Method (AOAC Method 967.22, 45.1.15) (2,19) 191
11.2.5.4 Thiamin (Vitamin B_1) in Foods, Thiochrome Fluorometric Procedure (AOAC Method 942.23) (2) 191
11.2.5.4.1 Principle 191
11.2.5.4.2 Critical Points 191
11.2.5.4.3 Procedure 192
11.2.5.4.4 Calculations 192
11.2.5.5 Riboflavin (Vitamin B_2) in Foods and Vitamin Preparations, Fluorometric Method (AOAC Method 970.65, 45.1.08) (2) 193
11.2.5.5.1 Principle 193
11.2.5.5.2 Critical Points 193
11.2.5.5.3 Procedure 193
11.2.5.5.4 Calculations 193
11.3 Comparison of Methods 194
11.4 Summary 195
11.5 Study Questions 195
11.6 Practice Problems 195
11.7 References 199

11.1 INTRODUCTION

11.1.1 Definition and Importance

Vitamins are defined as relatively low-molecular-weight compounds which humans, and for that matter, any living organism that depends on organic matter as a source of nutrients, require small quantities for normal metabolism. With few exceptions, humans cannot synthesize most vitamins and therefore need to obtain them from food and supplements. Insufficient levels of vitamins result in deficiency diseases [e.g., scurvy and pellagra, which are due to the lack of ascorbic acid (vitamin C) and niacin, respectively].

11.1.2 Importance of Analysis

Vitamin analysis of food and other biological samples has played a critical role in determining animal and human nutritional requirements. Furthermore, accurate food composition information is required to determine dietary intakes to assess diet adequacy and improve human nutrition worldwide. From the consumer and industry points of view, reliable assay methods are required to ensure accuracy of food labeling. This chapter provides an overview of techniques for analysis of the vitamin content of food and some of the problems associated with these techniques. Please note that the sections below on bioassay, microbiological, and chemical methods are not comprehensive, but rather just give examples of each type of analysis.

11.1.3 Vitamin Units

When vitamins are expressed in units of mg or μg per tablet or food serving, it is very easy to grasp how much is present. Vitamins can also be expressed as **international units** (IU), **United States Pharmacopeia** (USP) **units**, and % **Daily Value** (DV). To many, these definitions are unclear. When analysis of a foodstuff or dietary supplement is required for its content of vitamins, as might be the case for labeling and quality control purposes, being able to report the findings on different bases becomes important.

The IU is a unit of measurement for the amount of a substance, based on measured biological activity or effect. It is used for vitamins, hormones, vaccines, and similar biologically active substances. The precise definition of 1 IU differs from substance to substance, but has been established by international agreement for each substance. There is no equivalence among different substances; that is, 1 IU or USP unit of vitamin E does not contain the same number of micrograms as 1 IU or USP unit of vitamin A. Although IUs are still employed in food fortification and for nutrition labeling in the US (e.g., dietary supplements), many regulators feel that their use should be abandoned.

Concerning vitamin E, the USP discontinued the use of the IU in the US after 1980 and replaced it with USP units derived from the same biological activity values as the IU. Thus, 1 USP unit is defined as the activity of 1 mg of all-*rac*-α-tocopheryl acetate on the basis of biological activity measured by the rat fetal resorption assay. This equals the activity of 0.67 mg of *RRR*-α-tocopherol or 0.74 mg of *RRR*-α-tocopheryl acetate. Biological activities relative to *RRR*-α-tocopherol have been a convenient way to compare the different forms of vitamin E on the basis of IU or USP units, and were used to calculate milligram α-tocopherol equivalent (mg α-TE) values for reporting vitamin E contents. As vitamin E is available in different forms, conversion factors have been established (1) (Table 11-1).

Some other IU definitions for vitamins include the following:

- 1 IU of vitamin A is the biological equivalent of 0.3 μg retinol, 0.6 μg β-carotene, and 1.2 μg of other provitamin A active carotenoids (e.g., α-carotene and β-cryptoxanthin). One retinol equivalent (RE) is defined as 1 μg of all-*trans*-retinol. Varying dietary sources of vitamin A have different potencies. For calculation of RE values in foods, 100% efficiency of absorption of all-*trans*-retinol is assumed; however, incomplete absorption and conversion of β-carotene as well as other provitamin A active carotenoids must be taken into account. The conversion factors of 1 RE equals 6 μg and 12 μg for β-carotene and other provitamin A active carotenoids, respectively, are applied. A more recent international standard of measure of vitamin A established by the Institute of Medicine of the National Academies is to report μg retinol activity equivalents (RAE). For example, 2 μg of β-carotene in oil provided as a supplement can be converted by the body to 1 μg of retinol giving it an RAE ratio of 2:1, whereas 12 μg of all-*trans*-β-carotene from foods are required to provide the body with 1 μg of retinol giving dietary β-carotene an RAE ratio of 12:1. Other provitamin A carotenoids in foods are less easily absorbed than β-carotene resulting in RAE ratios of 24:1. So in food, unlike a dietary supplement, there is no direct comparison between an IU and μg RE or RAE. As a guide to convert IUs of vitamin A to μg RE, multiply the number of IUs by 0.1 if the food is of plant origin and by 0.2 if it is of animal origin. The result will be the approximate number of μg RE in the food.

11-1 table **Conversion Factors to Calculate α-Tocopherol from International Units or United States Pharmacopeia Units to Meet Dietary Reference Intakes for Vitamin E**

	USP unit (IU) mg^{-1}	mg USP $unit^{-1}$ (IU^{-1})	μmol USP $unit^{-1}$ (IU^{-1})	αT mg USP $unit^{-1}$(IU^{-1})
Natural vitamin E				
RRR-α-Tocopherol	1.49	0.67	1.56	0.67
RRR-α-Tocopheryl acetate	1.36	0.74	1.56	0.67
RRR-α-Tocopheryl acid succinate	1.21	0.83	1.56	0.67
Synthetic vitamin E				
all-*rac*-α-Tocopherol	1.10	0.91	2.12	0.45
all-*rac*-α-Tocopheryl acetate	1.00	1.00	2.12	0.45
all-*rac*-α-Tocopheryl acid succinate	0.89	1.12	2.12	0.45

USP, United States Pharmacopeia; *IU*, international unit; *αT*, α-tocopherol.
From: Reference (1), used with permission of Taylor & Francis Group, CRC Press, Boca Raton, FL.

- 1 IU of vitamin C is the biological equivalent of 50 μg L-ascorbic acid.
- 1 IU of vitamin D is the biological equivalent of 0.025 μg cholecalciferol/ergocalciferol.

The % Daily Value (DV) is a newer dietary reference value designed to help consumers to use label information to plan a healthy overall diet (see also Chap. 3). The DVs are reference numbers based on Recommended Dietary Allowances (RDAs) established by the Food and Nutrition Board of the Institute of Medicine. On food labels, the numbers tell you the % DV that one serving of this food provided as a percentage of established standards. In fact, DVs actually comprise two sets of reference values for nutrients: Daily Reference Values, or DRVs, and Reference Daily Intakes, or RDIs. The % DV is based on a 2000-Calorie diet for adults older than 18.

11.2 METHODS

11.2.1 Overview

Vitamin assays can be classified as follows:

1. **Bioassays** involving humans and animals.
2. **Microbiological assays** making use of protozoan organisms, bacteria, and yeast.
3. **Physicochemical assays** that include spectrophotometric, fluorometric, chromatographic, enzymatic, immunological, and radiometric methods.

In terms of ease of performance, but not necessarily with regard to accuracy and precision, the three systems follow the reverse order. It is for this reason that bioassays, on a routine basis at least, are limited in their use to those instances in which no satisfactory alternative method is available.

The selection criteria for a particular assay depend on a number of factors, including accuracy and precision, but also economic factors and the sample load to be handled. Applicability of certain methods for a particular matrix also must be considered. It is important to bear in mind that many official methods presented by regulatory agencies are limited in their applicability to certain matrices, such as vitamin concentrates, milk, or cereals, and thus cannot be applied to other matrices without some procedural modifications, if at all.

On account of the sensitivity of certain vitamins to adverse conditions such as light, oxygen, pH, and heat, proper precautions need to be taken to prevent any deterioration throughout the analytical process, regardless of the type of assay employed. Such precautionary steps need to be followed with the test material in bioassays throughout the feeding period. They are required with microbiological and physicochemical methods during extraction as well as during the analytical procedure.

Just as with any type of analysis, proper sampling and subsampling as well as the preparation of a homogeneous sample are critical aspects of vitamin analysis. General guidelines regarding this matter are provided in Chap. 5 of this book.

The principles, critical points, procedures, and calculations for various vitamin analysis methods are described in this chapter. Many of the methods cited are official methods of AOAC International (2), the European Committee for Standardization (3–10), or the US Pharmacopeial Convention (11). Refer to these methods and other original references cited for detailed instructions on procedures. A summary of commonly used regulatory methods is provided in Table 11-2. The sections below on bioassay, microbiological, and chemical methods are not comprehensive, but rather just give examples of each type of analysis.

Commonly Used Regulatory Methods for Vitamin Analysis

Vitamin	*Method Designation*	*Application*	*Approach*
Fat-Soluble Vitamins			
Vitamin A (and precursors)			
Retinol	AOAC Method 992.04 (2)	Vitamin A in milk-based infant formula	LC[a] 340 nm
Retinol	AOAC Method 2001.13 (2)	Vitamin A in foods	LC 328 or 313 nm
all-*trans*-retinol 13-*cis*-retinol	EN 1283-1 (3)	All foods	LC 325 nm or Fluorometric[b] $E_x\ \lambda = 325$ nm $E_m\ \lambda = 475$ nm
β-Carotene	AOAC Method 2005.07 (2)	β-Carotene in supplements and raw materials	LC 445 or 444 nm
β-Carotene	EN 1283–2 (3)	All foods	LC 450 nm
Vitamin D			
Cholecalciferol Ergocalciferol	AOAC Method 936.14 (3)	Vitamin D in foods	Bioassay
Cholecalciferol Ergocalciferol	AOAC Method 995.05 (3)	Vitamin D in infant formula and enteral products	LC 265 nm
Cholecalciferol Ergocalciferol	EN 1282172 (5)	Vitamin D in foods	LC 265 nm
Vitamin E			
R, *R*, *R* – tocopherols	EN 12822 (6)	Vitamin E in foods	LC Fluorescence $E_x\ \lambda = 295$ nm $E_m\ \lambda = 330$ nm
Vitamin K			
Phylloquinone	AOAC Method 999.15 (2)	Vitamin K in milk and infant formulas	LC postcolumn reduction Fluorescence $E_x\ \lambda = 243$ nm $E_m\ \lambda = 430$ nm
Phylloquinone	EN 14148 (7)	Vitamin K in foods	LC postcolumn reduction Fluorescence $E_x\ \lambda = 243$ nm $E_m\ \lambda = 430$ nm
Water-Soluble Vitamins			
Ascorbic acid (Vitamin C)			
Ascorbic acid	AOAC Method 967.21 (2)	Vitamin C in juices and vitamin preparations	2,6-Dichloroindophenol titration
Ascorbic acid	AOAC Method 967.22 (2)	Vitamin C in vitamin preparations	Fluorescence $E_x\ \lambda = 350$ nm $E_m\ \lambda = 430$ nm
Ascorbic acid	EN 14130 (8)	Vitamin C in foods	LC 265 nm
Thiamin (Vitamin B_1)			
Thiamin Thiamin·HCl	AOAC Method 942.23 (2)	Thiamin in foods	Thiochrome Fluorescence $E_x\ \lambda = 365$ nm $E_m\ \lambda = 435$ nm
Thiamin	EN 14122 (9)	Thiamin in foods	LC Thiochrome Fluorescence $E_x\ \lambda = 366$ nm $E_m\ \lambda = 420$ nm

(continued)

Commonly Used Regulatory Methods for Vitamin Analysis

Vitamin	*Method Designation*	*Application*	*Approach*
Riboflavin (Vitamin B_2)			
Riboflavin	AOAC Method 970.65 (2)	Riboflavin in foods and vitamin preparations	Fluorescence $E_x\ \lambda = 440$ nm $E_m\ \lambda = 565$ nm
Riboflavin	EN 14152 (10)	Riboflavin in foods	LC Fluorescence $E_x\ \lambda = 468$ nm $E_m\ \lambda = 520$ nm
Niacin			
Nicotinic acid Nicotinamide	AOAC Method 944.13 (2)	Niacin and niacinamide in vitamin preparations	Microbiological
Vitamin B_6			
Pyridoxine Pyridoxal Pyridoxamine	AOAC Method 2004.07 (1,2)	Total Vitamin B_6 in infant formula	LC Fluorescence $E_x\ \lambda = 290$ nm $E_m\ \lambda = 395$ nm
Folic Acid, Folate			
Total folates	AOAC Method 2004.05 (2)	Total folates in cereals and cereal products – Trienzyme procedure	Microbiological
Vitamin B_{12}			
Cyanocobalamin	AOAC Method 986.23 (2)	Cobalamin (Vitamin B_{12}) in milk-based infant formula	Microbiological
Biotin			
Biotin	USP29/NF24, Dietary supplements official monograph (11)	Biotin in dietary supplements	LC 200 nm or Microbiological
Pantothenic acid			
Ca pantothenate	AOAC Method 992.07 (2)	Pantothenic acid in milk-based infant formula	Microbiological

[a] LC, liquid chromatography (high-performance liquid chromatography).
[b] Fluorometric test, giving excitation (E_x) and emission (E_m) wavelengths.

11.2.2 Extraction Methods

With the exception of some biological feeding studies, vitamin assays in most instances involve the extraction of a vitamin from its biological matrix prior to analysis. This generally includes one or several of the following treatments: **heat**, **acid**, **alkali**, **solvents**, and **enzymes**.

In general, extraction procedures are specific for each vitamin and designed to stabilize the vitamin. In some instances, some procedures are applicable to the combined extraction of more than one vitamin, for example, for thiamin and riboflavin as well as some of the fat-soluble vitamins (1, 2, 13). Typical extraction procedures are as follows:

- *Ascorbic acid*: Cold extraction with metaphosphoric acid/acetic acid.
- *Vitamin B_1 and B_2*: Boiling or autoclaving in acid plus enzyme treatment.
- *Niacin*: Autoclaving in acid (noncereal products) or alkali (cereal products).
- *Folate*: Enzyme extraction with α-amylase, protease and γ-glutamyl hydrolase(conjugase)
- *Vitamins A, E, or D*: Organic solvent extraction, saponification, and re-extraction with organic solvents. For unstable vitamins such as these, antioxidants are routinely added to inhibit oxidation.

Analysis of fat-soluble vitamins may require **saponification**, generally either overnight at room temperature or by refluxing at 70°C. In the latter case, an air-cooled reflux vessel as depicted in Fig. 11-1 provides excellent control of conditions conducive to oxidation.

11.2.3 Bioassay Methods

Outside of vitamin bioavailability studies, bioassays at the present are used only for the analysis of **vitamins B_{12}** and **D**. For the latter, it is the reference standard method of analysis of food materials (AOAC Method

Reflux vessel useful for saponification.

936.14), known as the **line test** (Fig. 11-2), based on bone calcification. Because the determination of vitamin D involves deficiency studies as well as sacrificing the test organisms, it is limited to animals rather than humans as test organisms.

11.2.4 Microbiological Assays

11.2.4.1 Applications

Microbiological assays are limited to the analysis of water-soluble vitamins. The methods are very sensitive and specific for each vitamin. The methods are somewhat time consuming, and strict adherence to the analytical protocol is critical for accurate results. All microbiological assays can use microtiter plates (96-well) in place of test tubes. Microplate usage results in significant savings in media and glassware, as well as labor.

11.2.4.2 Principle

The growth of microorganisms is proportional to their requirement for a specific vitamin. Thus, in microbiological assays the growth of a certain microorganism in an extract of a vitamin-containing sample is compared against the growth of this microorganism in the presence of known quantities of that vitamin. Bacteria, yeast, or protozoans are used as test organisms. **Growth** can be measured in terms of **turbidity**, **acid production**, **gravimetry**, or by **respiration**. With bacteria and yeast, turbidimetry is the most commonly employed system. If turbidity measurements are involved, clear sample and standard extracts vs. turbid ones, are essential. With regard to incubation time, turbidity measurement is also a less time-consuming method. The microorganisms are specified by ATCC^{TM} numbers and are available from the *American Type Culture Collection* (ATCC^{TM}) (12301 Parkway Drive, Rockville, MD 20852).

11.2.4.3 Niacin

The procedural sequence for the microbiological analysis of niacin is outlined in Fig. 11-3 (AOAC Method 944.13, 45.2.04) (2, 14). *Lactobacillus plantarum* ATCC^{TM} 8014 is the test organism. A stock culture needs to be prepared and maintained by inoculating the freeze-dried culture on Bacto Lactobacilli agar followed by incubation at 37°C for 24 h prior to sample and standard inoculation. A second transfer may be advisable in the case of poor growth of the inoculum culture.

In general, growth is measured by turbidity. If lactobacilli are employed as the test organism, acidimetric measurements can be used as well. The latter may be necessary if a clear sample extract cannot be obtained prior to inoculation, and incubation (which is a prerequisite for turbidimetry) cannot be obtained. In making a choice between the two methods of measurement, one needs to bear in mind that a prolonged incubation period of 72 h is required for acidimetry.

11.2.4.4 Folate

Folate is the general term including folic acid (pteroylglutamate, PteGln) and poly-γ-glutamyl conjugates with the biological activity of folic acid. Folates present a diverse array of compounds that vary by oxidation state of the pteridine ring structure, one-carbon moieties carried by the specific folate, and the number of conjugated glutamate residues on the folate. Folates are labile to oxidation, light, thermal losses, and leaching when foods are processed. Because of the presence of multiple forms in food products and its instability, folate presents a rather difficult analytical problem. To account for differences in biological availability of synthetic folic acid used for food fortification and food folate, the Institute of Medicine Panel on Folate, Other B Vitamins and Choline established the dietary folate equivalent (DFE) value (5). Based on research showing that folic acid is 85% bioavailable whereas food folate

VITAMIN D BIOASSAY PROCEDURE

Sample Preparation

AOAC International provides specific instructions for preparation of various matrices for the bioassay. In some cases, saponification is used.

Depletion Period

Rats are suitable for depletion at age $\leq$30 days with body weight of $\geq$44 g but $\leq$60 g. A rachitogenic diet is fed for 18–25 days.

Assay Period

The assay period is the interval of life of the rat between the last day of the depletion period and the eighth or eleventh day thereafter. Feeding protocols are specified. During the assay, depleted rats are fed known and unknown amounts of vitamin D from standards and samples, respectively.

Potency of Sample

Vitamin D in the sample is determined by the line test from staining of the proximal end of the tibia or distal end of the radius or ulna.

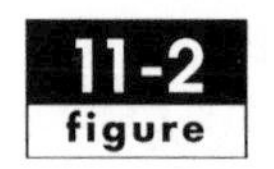

The bioassay of vitamin D by the line test, AOAC Method 936.14, 45.3.01 (2).

NIACIN MICROBIOLOGICAL ASSAY PROCEDURE

Test Sample Preparation

Weigh out a sufficient amount of sample to contain $\leq$5.0 mg niacin/ml, add volume of 1 *N* H_2SO_4 equal in ml to $\geq$10X dry weight of test portion in g, macerate, autoclave 30 min at 121–123°C, and cool. If dissolved protein is not present, adjust mixture to pH 6.8 with NaOH solution, dilute with deionized H_2O to volume (*ca.* 0.1–0.4 μg niacin/ml), mix, and filter.

Assay Tube Preparation

In at least duplicate use 0.0, 0.5, 1.0, 2.0, 3.0, 4.0, and 5.0 ml test sample filtrate and make up the difference to 5.0 ml with deionized H_2O, then add 5.0 ml of Difco™ Niacin Assay Medium to each tube, autoclave 10 min at 121–123°C, and cool.

Niacin Standard Preparation

Prepare assay tubes in at least duplicate using 0.0, 0.5, 1.0, 1.5, 2.0, 2.5, 3.0, 4.0, and 5.0 ml standard working solution (0.1–0.4 μg niacin/ml), make up difference to 5.0 ml with deionized H_2O, then add 5.0 ml of Difco™ Niacin Assay Medium and treat identically as the sample tubes.

Inoculation and Incubation

Prepare inoculum using *Lactobacillus plantarum* ATCC™ 8014 in Difco™ Lactobacilli Broth AOAC. Add one drop of inoculum to each tube, cover tubes, and then incubate at 37°C for 16–24 hr; that is, until maximum turbidity is obtained as demonstrated by lack of significant change during a 2-hr additional incubation period in tubes containing the highest concentration of niacin.

Determination

Measure %T at any specific wavelength between 540 and 660 nm. Set transmittance to 100% with the inoculated blank sample. Prepare a standard concentration-response curve by plotting %T readings for each level of standard solution used against amount of reference standard contained in respective tubes. Determine the amount of niacin for each level of the test solution by interpolation from the standard curve.

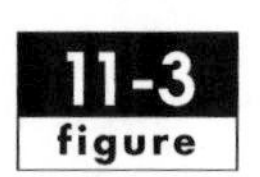

The microbiological assay of niacin, AOAC Method 944.13, 45.2.04 (2).

is only 50%, it can be stated that folic acid in fortified products is 85/50 or 1.7 times more bioavailable than food folate. Therefore, the μg of DFEs provided equals the μg of food folate plus (1.7 × μg folic acid). Calculation of the μg DFE for any food requires quantitation of folic acid as a separate entity from food folate. Currently, quite sophisticated liquid chromatography methods are necessary for accurate quantitation of folic acid and the multiple forms of folates in foods. A collaborated microbiological procedure based on the trienzyme extraction quantifies only total folate and cannot differentiate between added folic acid and food folate. The microbiological assay for total food folate with *Lactobacillus casei* (spp. *rhamnosus*) ATCC™ 7469 and trienzyme digestion follows AOAC International (2,15).

11.2.4.4.1 Principle Folate in the sample is extracted with a buffer at 100°C (boiling water bath). The extract is then digested with α-amylase and protease (i.e., to free macromolecularly bound folates) and conjugase (i.e., to cleave poly-γ-glutamyl folates to $PteGln_3$ or lower.) Growth response of the assay microorganism is measured by percent transmittance. Transmittance depends on folate concentration.

11.2.4.4.2 Critical Points Care must be exercised to protect labile folates from oxidation and photochemical degradation. Reducing agents including ascorbic acid, β-mercaptoethanol, and dithiothreitol are effective in preventing oxidation. Strict adherence to microbiological assay techniques is necessary to assay folate with accuracy and precision.

11.2.4.4.3 Procedure Analysis of food folate by *Lactobacillus casei* (spp. *rhamnosus*) ATCC™ 7469 and a trienzyme extraction procedure (Fig. 11-4) is provided by AOAC International (2). The analytical protocol has also been easily adapted using 96-well microtiter plates and a reader (16).

11.2.4.4.4 Calculations Results are calculated manually or from the regression line of the standard curve responses using 4th degree polynomial plots and a computer program written to conform to the AOAC microbiological analysis protocol. Software provided for microplate readers is suitable for calculating results from analyses using 96-well microplates. Results are reported as micrograms of vitamin per 100 g or per serving.

FOLATE MICROBIOLOGICAL ASSAY PROCEDURE

Sample Preparation

To 1.2–2.0 g of sample, add 50 ml of specified buffer, homogenize, and proceed to digestion step. (Note: High fat samples should be extracted with hexane, and all samples should be protected from light and air.)

Trienzyme Digestion

Boil samples for 5 min and cool to room temperature. Digest each sample with specified α-amylase, protease, and conjugase. Deactivate enzymes by boiling for 5 min. Cool tubes, filter, and dilute an appropriate aliquot to a final concentration of *ca.* 0.15 ng/ml.

Preparation of Standard Curve and Blank Tubes

Construct an 8-point standard curve using a working standard solution of folate. Add 5 ml of *Lactobacillus casei* (spp. *rhamnosus*) ATCC™ 7469 assay medium to each tube. Prepare an uninoculated blank and an inoculated blank to zero the spectrophotometer, and an enzyme blank to determine the contribution of the enzymes to microbial growth.

Assay

Folic acid is assayed by the growth of *Lactobacillus casei* (spp. *rhamnosus*) ATCC™ 7469 according to AOAC International (2). Prepared tubes of samples, standard curve, inoculated and uninoculated blanks, and enzyme blank are autoclaved at 121–123°C for 5 min and then inoculated with one drop of the prepared inoculum per tube. After tubes have been incubated at 37°C for 20–24 hr, the growth response is measured by percent transmittance at $\lambda = 550$ nm.

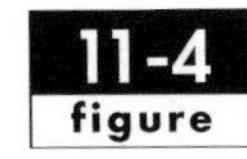

Analysis of folate in cereals and cereal products or other foods using *Lactobacillus casei* (spp. *rhamnosus*) ATCC™ 7469 and a trienzyme extraction procedure (2).

11.2.5 Chemical Methods

11.2.5.1 Vitamin A

Vitamin A is sensitive to ultraviolet (UV) light, air (and any prooxidants, for that matter), high temperatures, and moisture. Therefore, steps must be taken to avoid any adverse changes in this vitamin due to such effects. Steps include using low actinic glassware, nitrogen, and/or vacuum, as well as avoiding excessively high temperatures. The addition of an antioxidant at the onset of the procedure is highly recommended. **High-performance liquid chromatographic** (HPLC) methods are considered the only acceptable methods to provide accurate food measurements of vitamin A activity.

Details follow for the HPLC method of vitamin A (i.e., retinol isomers) in milk and milk-based infant formula (AOAC Method 992.04, 50.1.02) (2):

11.2.5.1.1 Principle The test sample is saponified with ethanolic KOH, vitamin A (retinol) is extracted into organic solvent and then concentrated. Vitamin A isomers – all-*trans*-retinol and 13-*cis*-retinol – levels are determined by HPLC on a silica column.

11.2.5.1.2 Critical Points All work must be performed in subdued artificial light. Care must be taken to avoid oxidation of the retinol throughout the entire procedure. Solvent evaporation should be completed under nitrogen, and hexadecane is added to prevent destruction during and after solvent evaporation.

11.2.5.1.3 Procedure Figure 11-5 outlines the procedural steps of the assay. Pyrogallol is added prior to saponification as an antioxidant.

11.2.5.1.4 Calculations

$$\text{all-}\textit{trans}\text{-retinol(ng/ml milk or diluted formula)} = (A_t/A_{st}) \times W_t \times C_t \times \text{DF} \quad [1]$$

where:

A_t = peak area, all-*trans*-retinol in test sample
A_{st} = peak area, all-*trans*-retinol in standard
W_t = weight, mg, oil solution used to prepare working standard solution
C_t = concentration, ng/ml, all-*trans*-retinol in oil solution
DF = dilution factor $= 1/50 \times 25/15 \times 100/3 \times 1/2 \times 1/40 = 5/360$

VITAMIN A HPLC ANALYSIS PROCEDURE

Test Sample Saponification

Transfer 40 ml of ready-to-use formula or fluid milk to a 100-ml digestion flask containing a stirring bar. For saponification, add 10 ml of ethanolic pyrogallol solution (*i.e.*, 2% (w/v) pyrogallol in 95% ethanol) and 40-ml ethanolic KOH (*i.e.*, 10% (w/v) KOH in 90% ethanol). Wrap the flask in aluminum foil and stir at room temperature for 18 hr, or at 70°C using the reflux vessel as depicted in Fig. 11-1. Dilute to volume with ethanolic pyrogallol solution.

Extraction of Digest

Pipet 3 ml of digestate into a 15-ml centrifuge tube and add 2 ml of deionized H_2O. Extract Vitamin A with 7-ml of hexane:diethyl ether (85:15, v/v). Repeat extraction 2X with 7-ml portions of extractant. After extractions, transfer the organic layer to a 25-ml volumetric flask. Add 1 ml of hexadecane solution (*i.e.*, 1-ml hexadecane in 100-ml hexane) and dilute to volume with hexane. Pipette 15 ml of diluted extract into a test tube and evaporate under nitrogen. Dissolve the residue in 0.5 ml of heptane.

Chromatography Parameters

Column	4.6 mm x 150 mm packed with 3-µm silica (Apex 3-µm silica)
Mobile Phase	Isocratic elution; heptane containing 2-propanol (1-5%, v/v)
Injection Volume	100 µl
Detection	UV, 340 nm
Flow Rate	1-2 ml/min

Inject 100-µl standard working solutions (see AOAC Method 992.04, 50.1.02 for details) into the HPLC. Inject 100-µl test extract. Measure peak areas for all-*trans*-retinol and 13-*cis*-retinol.
Note: The exact mobile phase composition and flow rate are determined by system suitability test to give retention times of 4.5 and 5.5 min for 13-*cis*-retinol and all-*trans*-retinol, respectively.

11-5 figure The HPLC analysis of vitamin A in milk and milk-based infant formula, AOAC Method 992.04, 50.1.02 (2).

$$13\text{-}cis\text{-retinol(ng/ml milk or diluted formula)} = (A_c/A_{sc}) \times W_c \times C_c \times DF \quad [2]$$

where:

A_c = peak area, 13-*cis*-retinol in test sample
A_{sc} = peak area, 13-*cis*-retinol in standard
W_c = weight, mg, oil solution used to prepare working standard solution
C_c = concentration, ng/ml, 13-*cis*-retinol in oil solution
DF = dilution factor = $1/50 \times 25/15 \times 100/3 \times 1/2 \times 1/40 = 5/360$

11.2.5.2 Vitamin E (Tocopherols and Tocotrienols)

11.2.5.2.1 Vitamin E Compounds Vitamin E is present in foods as eight different compounds: all are 6-hydroxychromans. The vitamin E family is comprised of α-, β-, γ-, and δ-tocopherol, characterized by a saturated side chain of three isoprenoid units and the corresponding unsaturated tocotrienols (α-, β-, γ-, and δ-). All homologs in nature are (*R, R, R*)-isomers. Recently, the Institute of Medicine Panel on Dietary Antioxidants and Related Compounds recommended that human requirements for vitamin E include only the 2R-stereoisomeric forms of α-tocopherol for establishment of recommended intakes (17). For the past two decades human requirements have been stated in terms of α-tocopherol equivalents.

Details follow of vitamin E analysis in food products using HPLC (18):

11.2.5.2.2 Principle

1. General food products. The sample is saponified under reflux (see Fig. 11-1), extracted with hexane, and injected onto a normal phase HPLC column connected to a fluorescence detector, $E_x\ \lambda = 290\,\text{nm}$, $E_m\ \lambda = 330\,\text{nm}$ (E_x, excitation; E_m, emission; see Chap. 22, Sect. 22.3).
2. Margarine and vegetable oil spreads. The sample is dissolved in hexane, anhydrous $MgSO_4$ is added to remove water, and the filtered extracts are assayed by HPLC.
3. Oils. Oil is dissolved in hexane and injected directly onto the HPLC column.

11.2.5.2.3 Critical Points Vitamin E is subject to oxidation. Therefore, saponification is completed under reflux, in the presence of the antioxidant, pyrogallol, with the reaction vessel protected from light.

11.2.5.2.4 Procedure The vitamin E assay is detailed in Fig. 11-6 and an example chromatogram is depicted in Fig. 11-7.

VITAMIN E HPLC ANALYSIS PROCEDURE

Sample Preparation

a. ***General food products***: Add 10 ml of 6% (w/v) pyrogallol in 95% ethanol to sample, mix, and flush with N_2. Heat at 70°C for 10 min with sonication. Add 2 ml of 60% (w/v) KOH solution, mix, and flush with N_2. Digest for 30 min at 70°C. Sonicate 5 min. Cool to room temperature, and add sodium chloride and deionized H_2O. Extract 3X with hexane (containing 0.1% BHT). Combine hexane extracts. Add 0.5 g of anhydrous $MgSO_4$ and mix. Filter through a Millipore filtration apparatus (0.45 μm). Dilute to volume with hexane. Inject sample into HPLC.
b. ***Margarine and vegetable oil spreads***: Add 40 ml of hexane (containing 0.1% BHT) to a 10-g sample and mix. Add 3 g of anhydrous $MgSO_4$, mix, let stand ≥2 hr. Filter and dilute combined filtrate to volume with hexane (0.1% BHT). Inject sample into HPLC.

Chromatography Parameters

Column	Hibar® LiChrosorb Si 60 (4 mm × 250 mm, 5-μm particle size) and LiChromCART® 4-4 guard column packed with LiChrospher® Si 60 (5 μm)
Mobile Phase	Isocratic, 0.85% (v/v) 2-propanol in hexane
Injection Volume	20 μl
Flow	1 ml/min
Detector	Fluorescence, $E_x\lambda = 290\,\text{nm}$, $E_m\lambda = 330\,\text{nm}$

(Note: Determine recovery for each food product.)

Analysis of vitamin E in food products using HPLC. [Adapted from (18).] Refer to (18) for details on applications.

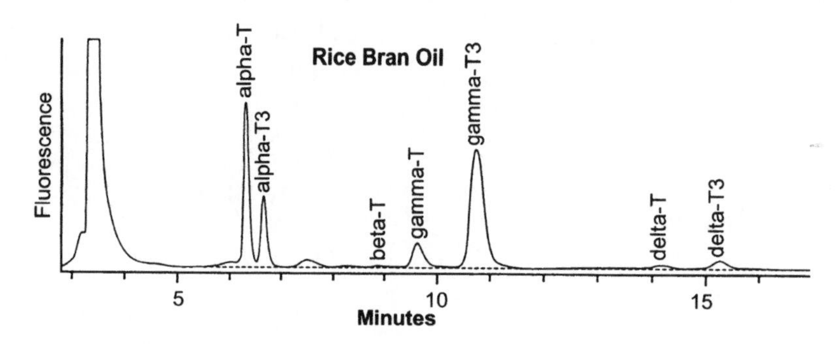

11-7 figure Chromatogram of rice bran oil showing tocopherols and tocotrienols.

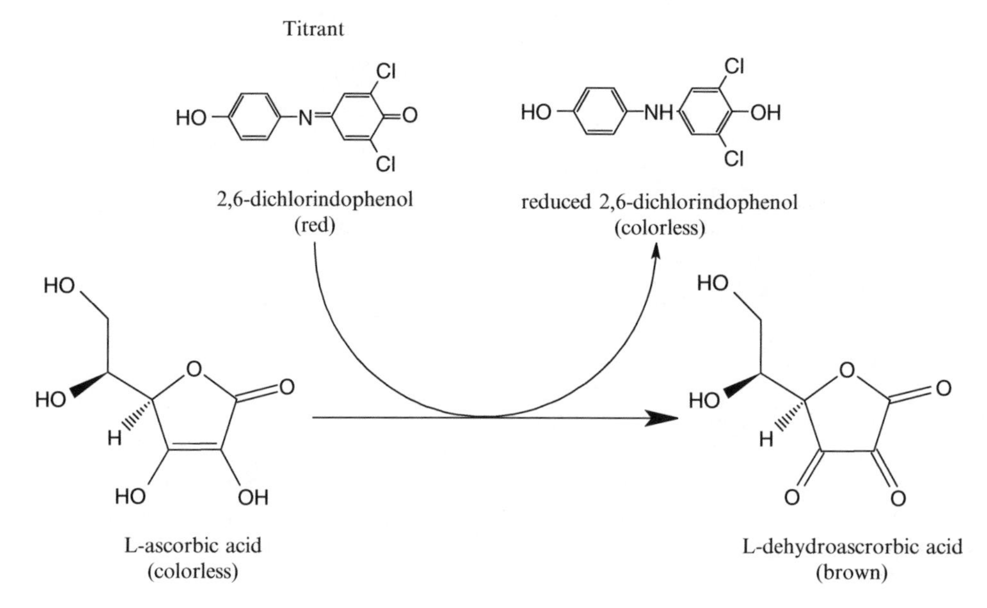

11-8 figure Chemical reaction between L-ascorbic acid and the indicator dye, 2,6-dichloroindophenol.

11.2.5.2.5 Calculation Vitamin E is quantitated by external standards from peak area by linear regression.

11.2.5.3 Vitamin C

The vitamin (**L-ascorbic acid** and **L-dehydroascorbic acid**) is very susceptible to oxidative deterioration, which is enhanced by high pH and the presence of ferric and cupric ions. For these reasons, the entire analytical procedure needs to be performed at low pH and, if necessary, in the presence of a chelating agent.

Mild oxidation of ascorbic acid results in the formation of dehydroascorbic acid, which is also biologically active and is reconvertible to ascorbic acid by treatment with **reducing agents** such as β-mercaptoethanol and dithiothreitol.

11.2.5.3.1 2,6-Dichloroindophenol Titrimetric Method (AOAC Method 967.21, 45.1.14) (2, 9)

1. **Principle**. L-ascorbic acid is oxidized to L-dehydroascorbic acid by the oxidation–reduction indicator dye, 2,6-dichloroindophenol. At the endpoint, excess unreduced dye appears rose-pink in acid solution (see Fig. 11-8).
2. **Procedure**. Figure 11-9 outlines the protocol followed for this method. In the presence of significant amounts of ferrous Fe, cuprous Cu, and stannous Sn ions in the biological matrix to be analyzed, it is advisable to include a **chelating agent** such as ethylenediaminetetraacetic acid (EDTA) with the extraction to avoid overestimation of the ascorbic acid content.

   The light but distinct rose-pink endpoint should last more than 5 s to be valid. With colored samples such as red beets or heavily

VITAMIN C ASSAY PROCEDURE 2,6-DICHLOROINDOPHENOL TITRATION

Sample Preparation

Weigh and extract by homogenizing test sample in metaphosphoric acid-acetic acid solution (*i.e.*, 15 g of HPO_3 and 40 ml of HOAc in 500 ml of deionized H_2O). Filter (and/or centrifuge) sample extract, and dilute appropriately to a final concentration of 10–100 mg of ascorbic acid/100 ml.

Standard Preparation

Weigh 50 mg of USP L-ascorbic acid reference standard and dilute to 50 ml with HPO_3-HOAc extracting solution.

Titration

Titrate three replicates each of the standard (*i.e.*, to determine the concentration of the indophenol solution as mg ascorbic acid equivalents to 1.0 ml of reagent), test sample, and blank with the indophenol reagent (*i.e.*, prepared by dissolving 50 mg of 2,6-dichloroindophenol sodium salt and 42 mg of $NaHCO_3$ to 200 ml with deionized H_2O) to a light but distinctive rose pink endpoint lasting $\geq$5 sec.

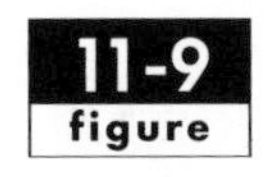

Analysis of vitamin C by the 2,6-dichloroindophenol titration, AOAC Method 967.21, 45.1.14 (2). [Adapted from (19), pp. 334–336.]

browned products, the endpoint is impossible to detect by human eyes. In such cases it, therefore, needs to be determined by observing the change of transmittance using a spectrophotometer with the wavelength set at 545 nm.

3. **Calculations**.

$$\text{mg of ascorbic acid/g or ml of sample} = (X - B) \times (F/E) \times (V/Y) \quad [3]$$

where:

X = average ml for test solution titration
B = average ml for test blank titration
F = mg ascorbic acid equivalents to 1.0-ml indophenol standard solution
E = sample weight (g) or volume (ml)
V = volume of initial test solution
Y = volume of test solution titrated

Note. The (V/Y) term represents the dilution factor employed.

11.2.5.3.2 Microfluorometric Method (AOAC Method 967.22, 45.1.15) (2, 19)

1. **Principle**. This method measures both ascorbic acid and dehydroascorbic acid. Ascorbic acid, following oxidation to dehydroascorbic acid, is reacted with **o-phenylenediamine** to produce a **fluorescent quinoxaline compound**.
2. **Procedure**. The procedural sequences for this method are outlined in Fig. 11-10. To compensate for the presence of interfering extraneous material, blanks need to be run using boric acid prior to the addition of the *o*-phenylenediamine solution.
3. **Calculations**.

$$\text{mg of ascorbic acid/g or ml} = [(X - D)/(C - B)] \times S \times (\text{DF}/E) \quad [4]$$

where:

X and C = average fluorescence of sample and standard, respectively
D and B = average fluorescence of sample blank and standard blank, respectively
S = concentration of standard in mg/ml
DF = dilution factor
E = sample weight, g, or sample volume, ml

11.2.5.4 Thiamin (Vitamin B_1) in Foods, Thiochrome Fluorometric Procedure (AOAC Method 942.23) (2)

11.2.5.4.1 Principle Following extraction, enzymatic hydrolysis of thiamin's phosphate esters and chromatographic cleanup (i.e., purification), this method is based on the fluorescence measurement of the oxidized form of thiamin, **thiochrome**, in the test solution compared to that from an oxidized thiamin standard solution.

11.2.5.4.2 Critical Points Thiochrome is light sensitive. Therefore, the analytical steps following the oxidation must be performed under subdued light. Thiamin is sensitive to heat, especially at alkaline

VITAMIN C MICROFLUOROMETRIC ASSAY PROCEDURE

Sample Preparation

Prepare sample extract as outlined in Fig. 11-8. To 100 ml each of the ascorbic acid standard and test sample solutions add 2 g of acid-washed Norit® Neutral, shake vigorously, and filter, discarding the first few ml.

Ascorbic Acid Standard and Test Sample Blanks

Transfer 5 ml of each filtrate to separate 100-ml volumetric flasks containing 5 ml of H_3BO_3-NaOAc solution. Let stand 15 min, swirling occasionally. Designate as standard or test sample blank. Development of the fluorescent quinoxaline is prevented by the formation of a H_3BO_3-dehydroascorbic acid complex prior to the addition of the *o*-phenylenediamine reagent. At the appropriate time, dilute blank solutions to volume with deionized H_2O. Transfer 2 ml of these solutions to each of three fluorescence tubes.

Ascorbic Acid Standard and Test Samples

Transfer 5 ml of each standard and test sample filtrate to separate 100-ml volumetric flasks containing 5 ml of 50% (w/v) NaOAc trihydrate and *ca.* 75 ml of H_2O, swirl contents, and then dilute to volume with deionized H_2O. Transfer 2 ml of these solutions to each of three fluorescence tubes.

Formation of Quinoxaline

To all sample and blank tubes, add 5 ml of 0.02% (w/v) aqueous *o*-phenylenediamine reagent, swirl contents using a Vortex mixer, and allow to stand at room temperature with protection from light for 35 min.

Determination

Measure fluorescence of sample and blank tubes at E_x $\lambda = 350$ nm, E_m $\lambda = 430$ nm.

Analysis of vitamin C by the microfluorometric method, AOAC Method 967.22, 45.1.15 (2). [Adapted from (20), pp. 338–341.]

pH. The analytical steps beginning with the oxidation of thiamin through to the fluorescence measurement (Fig. 11-11) must be carried out rapidly and precisely according to the instructions.

11.2.5.4.3 Procedure Figure 11-11 outlines the procedural sequence of the thiamin analysis. The enzymatic treatment and subsequent chromatographic cleanup may not be necessary with certain matrices, such as vitamin concentrates that contain nonphosphorylated thiamin and no significant amounts of substances that could interfere with the determination.

11.2.5.4.4 Calculations

$$\mu\text{g of thiamin in 5 ml of test solution} = [(I - b/S - d)] \quad [5]$$

where:

I and b = fluorescence of extract from oxidized test sample and sample blank, respectively

S and d = fluorescence of extract from oxidized standard and standard blank, respectively

$$\mu\text{g of thiamin/g} = [(I - b/S - d)] \times C/A \times 25/V_p \times V_o/\text{WT} \quad [6]$$

where:

I and b = fluorescence of extract from oxidized test sample and sample blank, respectively

S and d = fluorescence of extract from oxidized standard and standard blank, respectively

C = concentration of thiamin · HCl standard, μg/ml

A = aliquot taken, ml

25 = final volume of column eluate, ml

V_p = volume passed through the chromatographic column, ml

V_o = dilution volume of original sample, ml

WT = sample weight, g

THIAMIN ANALYSIS BY THIOCHROME PROCEDURE

Sample Preparation

Weigh out sample containing thiamin, add volume of 0.1 *M* HCl equal in ml to ≥10X dry weight of test portion in g, mix, autoclave for 30 min at 121–123°C, then cool. Dilute with 0.1 *M* HCl to measured volume containing *ca.* 0.2–5 μg thiamin/ml.

Enzyme Hydrolysis

Take aliquot containing *ca.* 10–25 μg thiamin, dilute to *ca.* 65 ml with 0.1 *M* HCl and adjust pH to 4.0–4.5 with *ca.* 5-ml 2 *M* CH_3COONa. Add 5 ml of enzyme solution, mix, incubate for 3 hr at 45–50°C. Cool, adjust to *ca.* pH 3.5, dilute to 100 ml with deionized H_2O, and filter.

Sample Extract Cleanup

Apply an aliquot of the test sample extract containing *ca.* 5 μg thiamin to a specified ion-exchange resin column, and wash column with 3X 5-ml portions of almost boiling water. Then elute thiamin from the resin with 5X 4.0 to 4.5-ml portions of almost boiling acid-KCl solution. Collect the eluate in a 25-ml volumetric flask and dilute to volume with acid-KCl solution. Treat standards identically.

Oxidation of Thiamin to Thiochrome

To a test tube, add 1.5 g of NaCl and 5 ml of the thiamin · HCl standard solution (1 μg/ml). Add 3 ml of oxidizing reagent [*i.e.*, basic $K_3Fe(CN)_6$], swirl contents, then add 13 ml of isobutanol, shake vigorously, and centrifuge. Repeat steps for the standard blank but instead replace the oxidizing reagent with 3 ml of 15% (w/v) NaOH. Decant the isobutanol extracts (*i.e.*, the standard and blank) into fluorescence reading tubes, and measure at E_x $\lambda = 365$ nm and E_m $\lambda = 435$ nm. Treat the test solution identically, and record the fluorescence intensity of the test sample and blank.

Analysis of thiamin (vitamin B_1) by the thiochrome fluorometric procedure, AOAC Method 942.2.3, 45.1.05 (2). Refer to (2) for more details on procedure.

11.2.5.5 Riboflavin (Vitamin B_2) in Foods and Vitamin Preparations, Fluorometric Method (AOAC Method 970.65, 45.1.08) (2)

11.2.5.5.1 Principle Following extraction, cleanup, and compensation for the presence of interfering substances, riboflavin is determined fluorometrically.

11.2.5.5.2 Critical Points Due to the extreme sensitivity of the vitamin to UV radiation, all operations need to be conducted under subdued light. The analyst also needs to be aware that exact adherence to the permanganate oxidation process is essential for reliable results.

11.2.5.5.3 Procedure An outline of the procedural protocol for this analysis is shown in Fig. 11-12. In spite of the fact that riboflavin is classified as a water-soluble vitamin, it does not readily dissolve in water. When preparing the standard solution, the analyst must pay special attention and ensure that the riboflavin is completely dissolved.

11.2.5.5.4 Calculations

$$\text{mg of riboflavin/ml final test solution} = [(B - C)/(X - B)] \times 0.10 \times 0.001 \quad [7]$$

where:

B and C = fluorescence of test sample containing water and sodium dithionite, respectively

X = fluorescence of test sample containing riboflavin standard

Note. Value of $[(B - C)/(X - B)]$ must be ≥0.66 and ≤1.5)

$$\text{mg of riboflavin/g of sample} = [(B - C)/(X - B)]\ x\ (CS/V)\ x\ (DF/WT) \quad [8]$$

where:

B and C = fluorescence of sample containing water and sodium hydrosulfite, respectively

X = fluorescence of sample containing riboflavin standard

CS = concentration of standard expressed as mg/ml

RIBOFLAVIN ASSAY PROCEDURE BY FLUORESCENCE

Sample Preparation

Weigh out homogenized sample, add volume of 0.1 *M* HCl equal in ml to $\geq$10X dry weight of test portion in g; the resulting solution must contain $\leq$0.1 mg riboflavin/ml. Mix contents, autoclave for 30 min at 121–123°C and then cool. Precipitate interfering substances by adjusting pH to 6.0–6.5 with dilute NaOH immediately followed by a pH readjustment to 4.5 with dilute HCl. Dilute with deionized H_2O to *ca.* 0.1 μg of riboflavin/ml, and filter.

Oxidation of Interfering Materials

Oxidize as follows: Transfer 10 ml of test filtrate to each of four tubes. To two of these tubes, add 1.0 ml of deionized H_2O, and to the remaining ones add 1.0 ml of a standard solution (*i.e.*, 1 μg/ml of riboflavin). Then to each tube, one at a time, add 1.0 ml of glacial HOAc followed by 0.5 ml of 4% (w/v) $KMnO_4$. Allow the mixture to stand for 2 min, and then add 0.5 ml of 3% (v/v) H_2O_2. Shake vigorously until excess O_2 is expelled.

Measurement of Fluorescence

Measure fluorescence at E_x $\lambda = 440$ nm and E_m $\lambda = 565$ nm. First read test samples containing 1 ml of added standard riboflavin solution, and then samples containing 1 ml of deionized H_2O. Add, with mixing, 20 mg of $Na_2S_2O_4$ to two of the tubes, and measure the minimum fluorescence within 5 sec.

Analysis of riboflavin (vitamin B_2) by fluorescence, AOAC Method 970.65, 45.1.08 (2).

V = volume of sample for fluorescence measurement, ml
DF = dilution factor
WT = weight of sample, g

11.3 COMPARISON OF METHODS

Each type of method has its advantages and disadvantages. In selecting a certain method of analysis for a particular vitamin or vitamins, a number of factors need to be considered, some of which are listed below:

1. Method accuracy and precision.
2. The need for bioavailability information.
3. Time and instrumentation requirements.
4. Personnel requirements.
5. The type of biological matrix to be analyzed.
6. The number of samples to be analyzed.
7. Regulatory requirements – Must official AOAC International methods be used?

Bioassays are extremely time consuming. Their employment is generally limited to those instances in which no suitable alternate method is available, or for cases in which bioavailability of the analyte is desired, especially if other methods have not been demonstrated to provide this information. Bioassays have the advantage that they sometimes do not require the preparation of an extract, thus eliminating the potential of undesirable changes of the analyte during the extract preparation. On the other hand, in the case of deficiency development requirements prior to analysis, bioassays are limited to animals rather than humans.

Both microbiological and physicochemical methods require vitamin extraction (i.e., solubilization prior to analysis). In general, the results obtained through these methods represent the total content of a particular vitamin in a certain biological matrix, such as food, and not necessarily its bioavailability to humans.

The applicability of microbiological assays is limited to water-soluble vitamins, and most commonly applied to niacin, B_{12}, and pantothenic acid. Though somewhat time consuming, they generally can be used for the analysis of a relatively wide array of biological matrices without major modifications. Furthermore, less sample preparation is often required compared to physicochemical assays.

Because of their relative simplicity, accuracy, and precision, the physicochemical methods, in particular the chromatographic methods using HPLC, are preferred. For example, standard HPLC is commonly employed as an official method of analysis for vitamins A, E, and D, and as a quality control method for vitamin C. While HPLC involves a high capital outlay, it is applicable to most vitamins and lends itself in some instances to simultaneous analysis of several vitamins and/or vitamers (i.e., isomers of vitamins). Implementation of multianalyte procedures for the analysis of water-soluble vitamins can result in assay efficiency with savings in time and materials. To be useful, a simultaneous assay must not lead to loss of sensitivity, accuracy, and precision when compared to single analyte methods. In general terms, multi-analyte methods for water-soluble vitamin assay of high concentration products including pharmaceuticals, supplements, and vitamin premixes are quite easily developed. Though the applicability of HPLC has been demonstrated to a wide variety of biological matrices with no or only minor modifications in some cases, one must always bear in mind that all chromatographic techniques, including HPLC, are separation and not identification methods. Therefore, during adaptation of an existing HPLC method to a new matrix, establishing evidence of peak identity and purity is an essential step of the method adaptation or development.

Over the past decade, liquid chromatography in combination with mass spectrometry (MS) (see Chap. 26) has added a new dimension to vitamin analysis. In general, LC–MS methods are now available for each fat- and water-soluble vitamin. Detection by MS leads to increased sensitivity as well as unequivocal identification and characterization of the vitamin. The LC–MS assays are rapidly becoming a mainstay of accurate, cost-effective vitamin analyses. For example, LC–MS is commonly employed for verification of vitamin D content of products with difficult matrices (i.e., comparing results to those with standard LC analysis), and LC–MS/MS for folate (vs. the microbiological method). The reader is referred to reference (12) for applications of LC–MS to specific vitamins.

When selecting a system for analysis, at least initially, it is wise to consider the use of official methods that have been tested through interlaboratory studies and that are published by such organizations as AOAC International (2), the European Committee for Standardization (3–10), the US Pharmacopeial Convention (11), or the AACC International (21). Again, one must realize that these methods are limited to certain biological matrices.

11.4 SUMMARY

The three most used types of methods for the analysis of vitamins – bioassays and microbiological and physicochemical assays – have been outlined in this chapter. They are, in general, applicable to the analysis of more than one vitamin and several food matrices. However, the analytical procedures must be properly tailored to the analyte in question and the biological matrix to be analyzed; issues concerning sample preparation, extraction, and quantitative measurements are also involved. It is essential to validate any new application appropriately by assessing its accuracy and precision. Method validation is especially important with chromatographic methods such as HPLC, because these methods basically accent separations rather than identification of compounds. For this reason, it is essential to ensure not only identity of these compounds but also, just as important, their purity.

11.5 STUDY QUESTIONS

1. What factors should be considered in selecting the assay for a particular vitamin?
2. To be quantitated by most methods, vitamins must be extracted from foods. What treatments are commonly used to extract the vitamins? For one fat-soluble vitamin and one water-soluble vitamin, give an appropriate extraction procedure.
3. What two vitamins must be listed on the standard nutritional label?
4. The standard by which all chemical methods to measure vitamin D content are compared is a bioassay method. Describe this bioassay method.
5. Explain why it is possible to use microorganisms to quantitate a particular vitamin in a food product, and describe such a procedure.
6. Niacin and folate both can be quantitated by microbiological methods. What extra procedures and precautions are necessary in the folate assay compared to the niacin assay, and why?
7. There are two commonly used AOAC methods to measure the vitamin C content of foods. Identify these two methods; then compare and contrast them with regard to the principles involved.
8. Would the vitamin C content as determined by the 2,6-dichloroindophenol method be underestimated or overestimated in the case of heat processed juice samples? Explain your answer.
9. What are the advantages and disadvantages of using HPLC for vitamin analysis?
10. Vitamin contents can be presented as units of mg or μg, as International Units (IU), or as % DV. Discuss the differences between these approaches for reporting the result.

11.6 PRACTICE PROBLEMS

1. A 3.21-g tuna sample (packed in water and drained before sampling) was analyzed for its niacin content. The sample was digested in 50 ml of 1 *N* H_2SO_4. After dissolved protein was removed by precipitation according to the AOAC method, a 20-ml aliquot was diluted to 100 ml, and then a 25-ml aliquot of the intermediate solution was taken and diluted to 250 ml. The concentration of niacin in the working solution was determined to be 0.168 μg/ml. (a) How much niacin is present in the tuna sample, and (b) how closely does this value compare with

that provided in the USDA Nutrient Database for Standard Reference (i.e., fish, tuna, light, canned in water, drained solids)?

2. Vitamin C in a nutraceutical formulation was assayed using the 2,6-dichloroindophenol titrimetric method. Determine the concentration of vitamin C (mg/g) in the nutraceutical based on the data given below from the assay.
   - Sample weight. 101.7 g, diluted to 500 ml with HPO_3/HOAc solution and filtered
   - Volume of sample filtrate titrated: 25 ml
   - Volume of dye used for the test solution titration: 9.2 ml
   - Volume of dye used for the test blank titration: 0.1 ml
   - mg ascorbic acid equivalents to 1.0 ml of indophenol standard solution: 0.175 mg/ml
3. Thiamin in a pet food sample was analyzed using the AOAC fluorometric method. Based on the assay conditions described below, determine the concentration of thiamin (μg/g) in the original pet food sample.
   - Sample weight. 2.0050 g
   - Dilutions. Diluted sample to 100 ml, applied 25 ml onto the Bio-Rex 70 ion-exchange column, then diluted the eluate to 25 ml and used 5 ml for fluorometry
   - Concentration of thiamin·HCl standard working solution: 0.1 μg/ml
   - Fluorometry reading ratio: 0.850
4. Riboflavin in raw almonds was analyzed using the AOAC fluorometric method. Based on the assay conditions described below, (a) determine the concentration of riboflavin (mg/g) in the almonds and (b) how closely this value compares with that provided in the USDA Nutrient Database for Standard Reference.
   - Sample weight: 1.0050 g
   - Dilutions: to 50 ml; used 10 ml for fluorometry
   - Fluorometry readings: $B_{60}/X_{85}/C_{10}$
   - Concentration or riboflavin standard solution: 1 μg/ml
5. 1.7 g of a braised, loin pork chop was analyzed for thiamin. The sample was digested with 20 ml of 0.1 *M* HCl and then diluted to 50 ml. A 40-ml aliquot of the digest was treated with the enzyme preparation and eventually diluted to 100 ml. A 45-ml aliquot of the enzyme-treated filtrate was purified using a Bio-Rex 70 ion-exchange column. The analyte from the 45-ml aliquot was applied to the column and recovered with five 4.0-ml portions of hot acid-KCl solution. The portions were pooled in a 25-ml volumetric flask and diluted to mark. A 5-ml aliquot as well as appropriate blanks and standard (concentration = 1 μg/ml) were converted to thiochrome and measured spectrofluorometrically. The following results were found:
   - Fluorescent intensity of the oxidized test sample and blank were 62.8 and 7.3, respectively
   - Fluorescent intensity of the thiamin·HCl standard and blank were 60.4 and 5.2, respectively

   Determine (a) how many μg of thiamin are in 5 ml of the test solution; (b) how many μg of thiamin/g braised loin pork chop; (c) how the answer from (b) compares with that reported in the USDA Nutrient Database for Standard Reference.
6. A graduate student is analyzing the antioxidant activity of phenolic compounds in apple juice samples and purchases some apple cider from a Farmer's market. The student worries that the product might have been fortified with ascorbic acid, and if so, will mess up his antioxidant assay. So, a vitamin C analysis is conducted by the 2,6-dichloroindophenol titrimetric method. 120 ml of the apple cider was mixed with 120 ml of the HPO_3/HOAc solution. In triplicate, a 10-ml aliquot of the resultant solution was taken and titrated against the indophenol standard solution. The following results were found:
   - An average of 13.3 and 0.1 ml of the indophenol standard was consumed during titration of the test sample and blank, respectively
   - From preliminary work it was determined that 0.1518 mg of ascorbic acid was equivalent to 1.0 ml of indophenol standard solution

   Determine (a) how many mg of ascorbic acid/ml of apple cider; (b) how do these results compare to tinned or bottled fresh apple juice according to the USDA Nutrient Database for Standard Reference; (c) has the apple cider been fortified with vitamin C; and (d) can this apple cider sample be used for the intended antioxidant activity study?
7. In the 2,6-dichloroindophenol titrimetric method, the indophenol reagent is prepared by dissolving 50.0 mg of 2,6-dichloroindophenol sodium salt plus some sodium bicarbonate in deionized water to 200 ml. The ascorbic acid standard solution is prepared by dissolving 50.0 mg of USP ascorbic acid reference standard in 50 ml of deionized water. If a 4.5-ml aliquot of the ascorbic acid reference standard is treated with 5.0 ml of the HPO_3/HOAc solution and then titrated against the indophenol reagent, how many milliliters should be consumed? Note: Remember the stoichiometry for the reaction.
   - FW of 2,6-dichloroindophenol sodium salt is 290.08 g/mol
   - FW of ascorbic acid is 176.12 g/mol
8. A new infant formula was developed for delivery to a third world country. Unfortunately, the dried infant formula was stored outdoors for 3 months in the sunlight before use and there is fear that the vitamin A has degraded. The reported vitamin A content (expressed as all-*trans* retinol equivalents) in the formula at the point of shipping was 3.0 μg/g dry formula. The following test was performed: 140 g of dry formula (i.e., after 3 months of storage outdoors in the sunlight) were dissolved in water and made up to 1 L. The AOAC method was followed with the following observations:
   - Peak area for the all-*trans*-retinol in the standard was 934
   - Weight of oil solution used to prepare the working retinol standard was 53 mg
   - Concentration of all-*trans*-retinol in oil standard solution was 1996 ng/ml
   - Peak area for the test solution after outdoor storage was 89

Determine (a) the concentration (ng/ml) of all-*trans*-retinol in the rehydrated infant formula; (b) by what percentage has the vitamin A content, expressed as all-*trans*-retinol, in the dried infant formula degraded?

Answers

1. (a) 420 µg in the 3.2-g test portion; (b) there are 13.08 mg niacin/100 g tuna in the test sample. The USDA Nutrient Database for Standard Reference lists a value of 13.280 ± 0.711 mg niacin/100 g tuna; so, the values compare well.

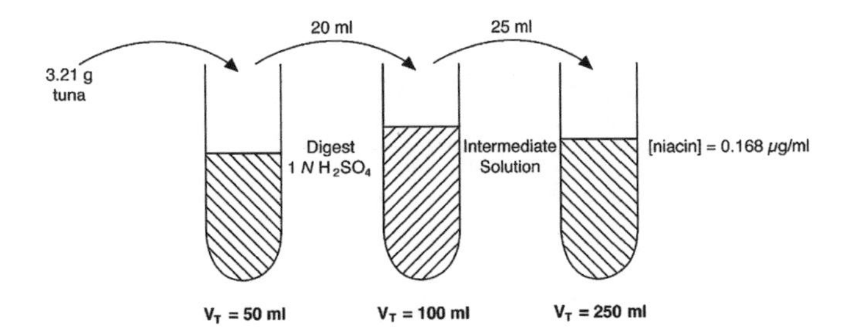

Calculations:

(a) Niacin present in tuna sample is . . .

$$0.168\,\frac{\mu g}{ml} \times \frac{250\,ml}{25\,ml} \times \frac{100\,ml}{20\,ml} \times 50\,ml$$

$$= 420\,\mu g = 0.420\,mg\ niacin$$

So,

$$\frac{0.420\,mg\ niacin}{3.21\,g}$$

$$= 0.1308\,mg/g\ tuna\ or\ 13.08\,mg\ niacin/100\,g\ tuna$$

(b) The USDA Nutrient Database for Standard Reference lists a niacin content of 13.28 mg/100 g tuna for "fish, tuna, light, canned in water, drained solids." Thus, the 13.08 mg niacin/100 g tuna value is very close to the 13.28 mg value reported in the database.

2. 0.313 mg ascorbic acid/g nutraceutical

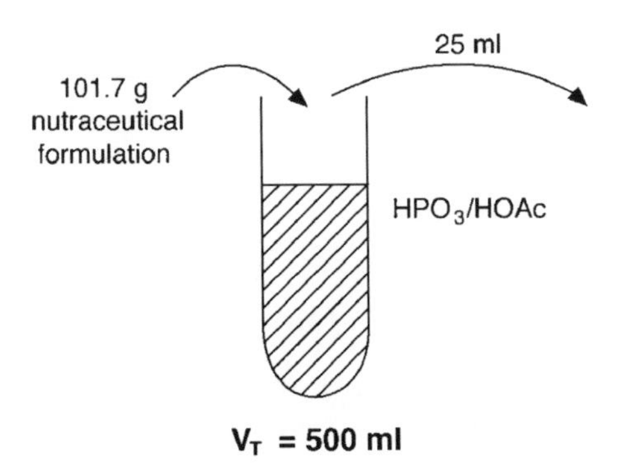

Calculations:

$F = 0.175$ mg AA eq./ml indophenol standard
$X = 9.2$ ml of dye for test solution titration
$B = 0.1$ ml of dye for test blank titration

mg ascorbic acid/g sample

$$= (X - B) \times (F/E) \times (V/Y)$$

$$= \frac{(9.2\,ml - 0.1\,ml) \times 0.175\,mg/ml}{101.7\,g} \times \frac{500\,ml}{25\,ml}$$

$$= 0.313\ mg\ of\ ascorbic\ acid/g\ nutraceutical\ formulation$$

3. 0.8479 µg thiamin/g pet food

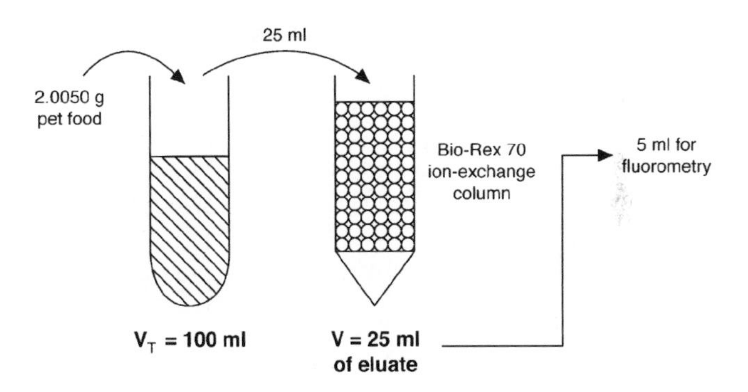

Calculations:

Concentration of thiamin·HCl standard $= 0.1\,\mu g/ml$

$$\text{Fluorometry reading ratio} = 0.850\left[\frac{(I-b)}{(S-d)}\right]$$

$$\mu g\ thiamin/g = \left[\frac{(I-b)}{(S-d)}\right] \times \frac{C}{A} \times \frac{25}{V_P} \times \frac{V_O}{WT}$$

$$= 0.850 \times \frac{0.1\,\mu g/ml}{5\,ml} \times \frac{25}{25\,ml} \times \frac{100\,ml}{2.0050\,g}$$

$$= 0.8479\,\mu g\ thiamin/g\ pet\ food$$

4. (a) 9.95×10^{-3} mg riboflavin/g raw almonds; (b) there is 0.995 mg riboflavin/100 g raw almonds in the test sample. The USDA Nutrient Database for Standard Reference lists a value of 1.014 ± 0.025 mg riboflavin/100 g raw almonds; so, the values compare well.

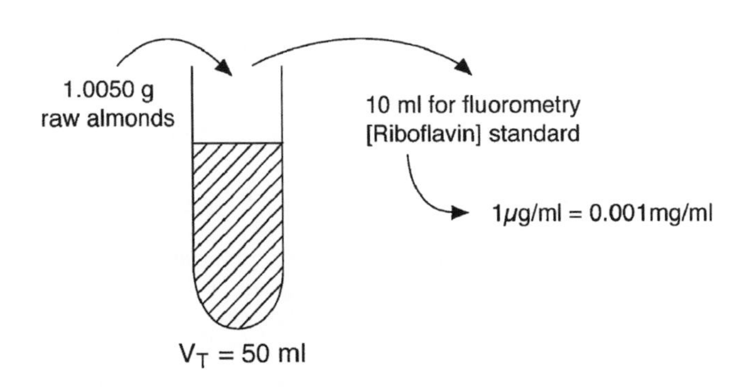

Calculations:
(a) mg of riboflavin/g raw almonds

$$= \left[\frac{(B-C)}{(X-B)}\right] \times \frac{CS}{V} \times \frac{DF}{WT}$$

$$= \left[\frac{(60-10)}{(85-60)}\right] \times \frac{0.001\,\text{mg/ml}}{10\,\text{ml}} \times \frac{50\,\text{ml}}{1.0050\,\text{g}}$$

$$= 9.95 \times 10^{-3}\,\text{mg of riboflavin/g raw almonds}$$

(b) The USDA Nutrient Database for Standard Reference lists a riboflavin content of 1.014 ± 0.025 mg/100 g for "nuts, almonds." Thus, the 9.95×10^{-3} mg of riboflavin/g raw almonds is very close to the value reported in the database.

5. (a) 1.005 μg in 5 ml; (b) 8.214 μg/g; (c) very close, as each 100 g of braised, pork loin chop contains 0.822 mg of thiamin.

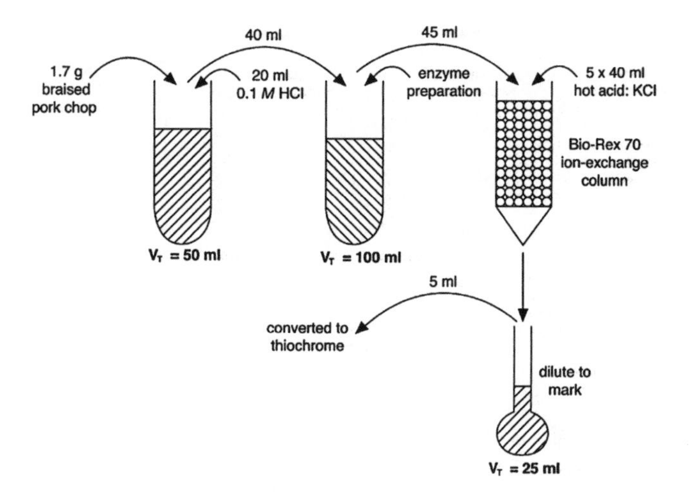

Calculations:
(a) μg of thiamin in 5 ml of test solution

$$= \left[\frac{(I-b)}{(S-d)}\right] = \frac{(62.8-7.3)}{(60.4-5.2)} = \frac{55.5}{55.2}$$

$$= 1.005\,\mu\text{g thiamin/5 ml}$$

(b) μg of thiamin/g braised pork chops

$$= \left[\frac{(I-b)}{(S-d)}\right] \times \frac{C}{A} \times \frac{25}{V_p} \times \frac{V_o}{WT}$$

$$= \left[\frac{(62.8-7.3)}{(60.4-5.2)}\right] \times \frac{1\,\mu\text{g/ml}}{5\,\text{ml}} \times \frac{25\,\text{ml}}{45\,\text{ml}}$$

$$\times 50\,\text{ml} \times \frac{100\,\text{ml}}{40\,\text{ml}} / 1.7\,\text{g}$$

$$= 8.214\,\mu\text{g/g braised pork chop}$$

(c) The USDA Nutrient Database for Standard Reference lists a thiamin content of 0.822 mg/100 g for "pork, fresh loin, center loin (chops), bone-in, separable lean only, cooked, braised." Thus, the 8.214 μg thiamin/g braised pork chops is very close to the value reported in the database.

6. (a) 0.401 mg ascorbic acid/ml; (b) tinned or bottled juice contains 38.5 mg/100 g juice, so the results are similar; (c) Yes. The cider has been fortified with vitamin C; and (d) No. The high content of vitamin C will likely sacrifice itself as the antioxidant in the antioxidant activity assay before any endogenous phenolic compounds in the cider do so.

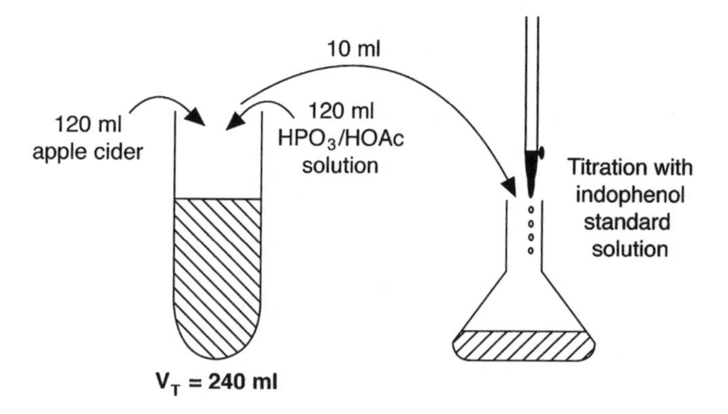

Calculations:
(a) mg of ascorbic acid (AA)/ml apple cider

$$= (X-B) \times \frac{F}{E} \times \frac{V}{Y}$$

$$= (13.3\,\text{ml} - 0.1\,\text{ml})$$

$$\times \frac{0.1518\,\text{mg AA eq./1.0 ml indophenol std. solution}}{120\,\text{ml}}$$

$$\times \frac{240\,\text{ml}}{10\,\text{ml}}$$

$$= 0.401\,\text{mgAA/ml apple cider}$$

(b) According to the USDA Nutrient Database for Standard Reference, "apple juice, canned or bottled, unsweetened, with added ascorbic acid" contains 38.5 mg ascorbic acid/100 g apple juice; so, the results are similar.
(c) Yes. The cider has been fortified with vitamin C.
(d) No. The high content of vitamin C will likely sacrifice itself as the antioxidant in the antioxidant assay before any endogenous phenolic compounds in the cider do so.

7. 29.6 ml
Calculations:
Formula Weight: 2,6-dichloroindophenol sodium salt = 290.08 g/mol
Formula Weight: ascorbic acid = 176.12 g/mol

$$\text{moles (mol)} = \frac{\text{mass (g)}}{\text{formula weight (g/mol)}}$$

$$\text{mol} = \frac{0.25\,\text{g}}{290.08\,\text{g/mol}}$$

So, concentration of 2,6-dichloroindophenol sodium salt solution is $= 8.618 \times 10^{-4}$ mol/L

$$\text{mol} = \frac{\text{g}}{\text{g/mol}};\ \text{mol} = \frac{1\,\text{g}}{176.12\,\text{g/mol}}$$

So, concentration of ascorbic acid stock solution $= 5.678 \times 10^{-3}$ mol/L

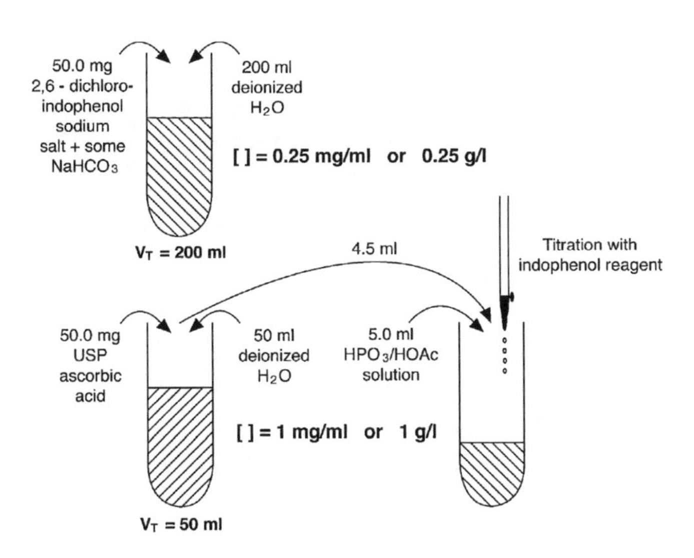

In the reaction vessel, there is:

$$5.678 \times 10^{-3}\,\text{mmol/ml} \times 4.5\,\text{ml}$$
$$= 2.555 \times 10^{-2}\,\text{mmol ascorbic acid}$$

The reaction stoichiometry is 1:1. See Fig. 11-8 for the chemical reaction. So, 2.555×10^{-2} mmol of 2,6-dichloroindophenol reagent need to be consumed.

$$8.618 \times 10^{-4}\text{mmol/ml} \times X\,\text{ml} = 2.555 \times 10^{-2}\text{mmol}$$
$$x = 29.6\,\text{ml}$$

8. (a) 140 ng all-*trans*-retinol/ml; and (b) 66.7%, or 2/3.

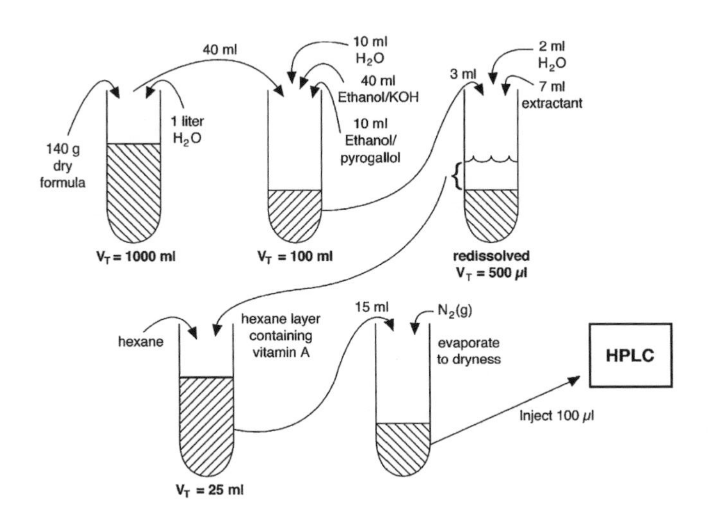

Calculations:

(a) all-*trans*-retinol (ng/ml rehydrated formula)

$$= \frac{A_t}{As_t} \times W_t \times C_t \times DF$$
$$= \frac{89}{934} \times 53\,\text{mg} \times 1996\,\text{ng/ml} \times \frac{5}{360}$$
$$= 140\,\text{ng all-}trans\text{-retinol/ml rehydrated milk}$$

(b) Originally there was 3 µg all-*trans* retinol/g dry formula

So, in 140 g of formula → 420 µg all-*trans*-retinol → in 1 L → 420 ng/ml

$$420\,\text{ng/ml} \times x = 140\,\text{ng/ml}$$
$$X = 0.333 \text{ or } {}^{1}/_{3}$$

Therefore, the vitamin A content, expressed as all-*trans*-retinol, had degraded by 66.7% or 2/3.

11.7 REFERENCES

1. Eitenmiller RR, Ye L, Landen WO, Jr (2008) Vitamin analysis for the health and food sciences, 2nd edn., p. 135. Taylor & Francis Group, CRC Press, Boca Raton, FL
2. AOAC International (2007) Official methods of analysis, 18th ed., 2005; Current through revision 2, 2007 (On-line). AOAC International, Gaithersburg, MD
3. European Committee for Standardization, Technical Committee CEN/TC 275 (2000) Foodstuffs – determination of Vitamin A by high performance liquid chromatography – Part 1: measurement of all-*trans*-Retinol and 13-*cis*-Retinol, EN 12823-1
4. European Committee for Standardization, Technical Committee CEN/TC 275 (2000) Foodstuffs – determination of Vitamin A by high performance liquid chromatography – Part 2: measurement of β-Carotene, EN 12823-2
5. European Committee for Standardization, Technical Committee CEN/TC 275 (2000) Foodstuffs – determination of Vitamin D by high performance liquid chromatography – measurement of cholecalciferol (D_3) and ergocalciferol (D_2), EN 12821
6. European Committee for Standardization, Technical Committee CEN/TC 275 (2000) Foodstuffs – determination of Vitamin E by high performance liquid chromatography – measurement of α-, β-, γ- and δ-Tocopherols, EN 12822
7. European Committee for Standardization, Technical Committee CEN/TC 275 (2003) Foodstuffs – determination of Vitamin K by HPLC, EN 14148
8. European Committee for Standardization, Technical Committee CEN/TC 275 (2003) Foodstuffs – determination of Vitamin C by HPLC, EN 14130
9. European Committee for Standardization, Technical Committee CEN/TC 275 (2003) Foodstuffs – determination of Vitamin B_1 by HPLC, EN 14122

10. European Committee for Standardization, Technical Committee CEN/TC 275 (2003) Foodstuffs – determination of Vitamin B_2 by HPLC, EN 14152
11. United States Pharmacopeial Convention (2005) US Pharmacopoeia National Formulary, USP29/NF24, Nutritional Supplements, Official Monographs, United States Pharmacopoeial Convention, Rockville, MD
12. Ball GFM (2006) Vitamins in foods. Analysis, bioavailability, and stability. Taylor & Francis Group, CRC Press, Boca Raton, FL.
13. Blake CJ (2007) Analytical procedures for water-soluble vitamins in foods and dietary supplements: a review. Anal Bioanal Chem 389:63–76
14. Eitenmiller RR, De Souza S (1985) Niacin. In: Augustin J, Klein BP, Becker DA, Venugopal PB (eds) Methods of vitamin assay, 4th edn. Wiley, New York, pp 685–397.
15. De Souza S, Eitenmiller RR (1990) Effects of different enzyme treatments on extraction of total folate from various foods prior to microbiological assay and radioassay. J Micronutrient Anal 7:37–57.
16. Tamura T (1990) Microbiological assay of folates. In: Picciano MF, Stokstad ELR, Gregory JF III (eds) Folic acid metabolism in health and disease, contemporary issues in clinical nutrition, Vol. 13. Wiley-Liss, New York, pp 121–137
17. Institute of Medicine (2000) Dietary reference intakes for vitamin c, vitamin e, selenium and carotenoids. National Academy Press, Washington, DC
18. Lee J, Ye L, Landen WO Jr, Eitenmiller RR (2000) Optimization of an extraction procedure for the quantification of Vitamin E in tomato and broccoli using response surface methodology. J Food Compos Anal 13:45–57
19. Pelletier O (1985) Vitamin C (L-ascorbic and dehydro-L-ascorbic acid). In: Augustin J, Klein BP, Becker DA, Venugopal PB (eds) Methods of vitamin assay, 4th edn. Wiley, New York, pp 335–336
20. Pelletier O (1985) Vitamin C (L-ascorbic and dehydro-L-ascorbic acid). In: Augustin J, Klein BP, Becker DA, Venugopal PB (eds) Methods of vitamin assay, 4th edn. Wiley, New York, pp 338–341
21. AACC International (2010) Approved methods of analysis, 11th edn. (On-line), AACC International, St. Paul, MN

Traditional Methods for Mineral Analysis

Robert E. Ward and Charles E. Carpenter*

Department of Nutrition and Food Sciences, Utah State University, Logan, UT 84322-8700, USA
robert.ward@usu.edu
chuck.carpenter@usu.edu

12.1 Introduction 203
12.1.1 Importance of Minerals in the Diet 203
12.1.2 Minerals in Food Processing 203
12.2 Basic Considerations 204
12.2.1 Nature of Analyses 204
12.2.2 Sample Preparation 204
12.2.3 Interferences 205
12.3 Methods 205
12.3.1 EDTA Complexometric Titration 205
12.3.1.1 Principles 205
12.3.1.2 Procedure: Hardness of Water Using EDTA Titration 205
12.3.1.3 Applications 206

S.S. Nielsen, *Food Analysis*, Food Science Texts Series, DOI 10.1007/978-1-4419-1478-1_12,

12.3.2 Precipitation Titration 206
12.3.2.1 Principles 206
12.3.2.2 Procedures 207
12.3.2.2.1 Mohr Titration of Salt in Butter (AOAC Method 960.29) 207
12.3.2.2.2 Volhard Titration of Chloride in Plant Material (AOAC Method 915.01) 207
12.3.2.3 Applications 207
12.3.3 Colorimetric Methods 208
12.3.3.1 Principles 208
12.3.3.2 Procedures: Colorimetric Determination of Iron in Meat 209
12.3.3.3 Applications 209
12.3.4 Ion-Selective Electrodes 209
12.3.4.1 Principles 209
12.3.4.2 General Methodology 210
12.3.4.3 Electrode Calibration and Determination of Concentration 211
12.3.4.4 Applications 211
12.4 Comparison of Methods 212
12.5 Summary 212
12.6 Study Questions 212
12.7 Practice Problems 213
12.8 References 214

12.1 INTRODUCTION

This chapter describes traditional methods for analysis of minerals involving titrimetric and colorimetric procedures, and the use of ion selective electrodes. Other traditional methods of mineral analysis include gravimetric titration (i.e., insoluble forms of minerals are precipitated, rinse, dried, and weighed) and redox reactions (i.e., mineral is part of an oxidation–reduction reaction, and product is quantitated). However, these latter two methods will not be covered because they currently are used little in the food industry. The traditional methods that will be described have maintained widespread usage in the food industry despite the development of more modern instrumentation such as atomic absorption spectroscopy and inductively coupled plasma-atomic emission spectroscopy (Chap. 24). Traditional methods generally require chemicals and equipment that are routinely available in an analytical laboratory and are within the experience of most laboratory technicians. Additionally, traditional methods often form the basis for rapid analysis kits (e.g., Quantab® for salt determination) that are increasingly in demand. Procedures for analysis of minerals of major nutritional or food processing concern are used for illustrative purposes. For additional examples of traditional methods refer to references (1–6). Slight modifications of these traditional methods are often needed for specific foodstuffs to minimize interferences or to be in the range of analytical performance. For analytical requirements for specific foods see the *Official Methods of Analysis* of AOAC International (5) and related official methods (6).

12.1.1 Importance of Minerals in the Diet

Calcium, phosphorus, sodium, potassium, magnesium, chlorine, and sulfur make up the dietary macro minerals, those minerals required at more than 100 mg/day by the adult (7–9). An additional ten minerals are required in milli- or microgram quantities per day and are referred to as **trace minerals**. These include iron, iodine, zinc, copper, chromium, manganese, molybdenum, fluoride, selenium, and silica. There is also a group of minerals called **ultra trace minerals**, including vanadium, tin, nickel, arsenic, and boron, that are being investigated for possible biological function, but that currently do not have clearly defined biochemical roles. Some mineral elements have been documented to be **toxic** to the body and should, therefore, be avoided in the diet. These include lead, mercury, cadmium, and aluminum. Essential minerals such as fluoride and selenium also are known to be harmful if consumed in excessive quantities, even though they do have beneficial biochemical functions at proper dietary levels.

The Nutrition Labeling and Education Act of 1990 (NLEA) mandated labeling of **sodium**, **iron**, and **calcium** contents largely because of their important roles in controlling hypertension, preventing anemia, and impeding the development of osteoporosis, respectively (see Fig. 3-1, Chap. 3). The content of these minerals in several foods is shown in Table 12-1. The content of other minerals may be included on the label at the producer's option, although this becomes mandatory if the mineral is the subject of a nutrient claim on the label. Implementation of the NLEA has led to an increased need for more rapid and accurate analysis of minerals and other food components.

12.1.2 Minerals in Food Processing

Minerals are of **nutritional** and **functional** importance, and for that reason their levels need to be known and/or controlled. Some minerals are contained at high levels in natural foodstuffs. For example, milk is a good source of calcium, containing about 300 mg of calcium per 8-ounce cup. However, direct acid cottage cheese is very low in calcium because of the action of the acid causing the calcium bound to the casein to be freed and consequently lost in the whey fraction. Similarly, a large portion of the phosphorus, zinc, manganese, chromium, and copper found in a grain kernel is lost when the bran layer is removed in processing. The enrichment law for flour requires that iron be replaced in white flour to the level at which it occurred naturally in the wheat kernel before removal of the bran.

Fortification of some foods has allowed addition of minerals above levels ever expected naturally. Prepared breakfast cereals often are fortified with minerals such as calcium, iron, and zinc, formerly thought to be limited in the diet. Fortification of salt with iodine has almost eliminated goiter in the USA. In other cases, minerals may be added for functionality. Salt is added for flavor, to modify ionic strength that effects solubilization of protein and other food components, and as a preservative. This increases significantly the sodium content of products such as processed meats, pickles, and processed cheese. Phosphorus may be added as phosphates to increase water-holding capacity. Calcium may be added to promote gelation of proteins and gums.

Water is an integral part of food processing, and **water quality** is a major factor to be considered in the food processing industry. Water is used for washing, rinsing, blanching, cooling, and as an ingredient in formulations. Microbiological safety of water used in food processing is very important. Also important, but generally not appreciated by the consuming public, is the mineral content of water used in food processing.

Mineral Content of Selected Foods

Food Item	mg/g (Wet Weight Basis)		
	Calcium	Iron	Sodium
Cereals, bread, and pasta			
Rice, brown, long-grain, raw	23	2	7
Corn flakes, plain	3	19	950
White rice, long-grain, regular, real, enriched	28	4	5
Wheat flour, whole-grain	34	4	5
Wheat flour, white, all-purpose, unenriched	15	1	2
Macaroni, dry, enriched	21	3	6
Rye bread	73	3	660
Dairy products			
Milk, whole, fluid, 3.3% fat	110	<1	40
Evaporated milk, whole	260	<1	110
Butter, with salt	24	<1	580
Cream, fluid, half and half	110	<1	41
Cheese, cottage, low fat, 2% milk fat	91	<1	330
Yogurt, plain, low fat	200	<1	77
Fruits and vegetables			
Apples, raw, with skin	6	<1	1
Bananas, raw	5	<1	1
Cherries, sweet, raw	13	<1	<1
Raisins, seedless	50	2	11
Potatoes, raw, skin	30	3	10
Tomatoes, red, ripe, raw	10	<1	5
Meats, poultry, and fish			
Eggs, whole, raw, fresh	53	2	40
Fish fillet, battered or breaded, and fried	18	2	530
Pork, fresh, leg (ham), whole, raw	5	<1	47
Bologna, chicken, pork, beef	92	1	1,100
Chicken, broilers or fryers, breast meat only, raw	11	<1	65
Beef, chuck, arm pot roast, raw	16	2	74

From US Department of Agriculture, Agricultural Research Service (2009) USDA National Nutrient Database for Standard Reference. Release 22. Nutrient Data Laboratory Home Page, http://www.ars.usda.gov/ba/bhnrc/ndl

Waters that contain excessive minerals can result in clouding of beverages. Textural properties of fruits and vegetables can be influenced by the "**hardness**" or "**softness**" of the water used during processing.

12.2 BASIC CONSIDERATIONS

12.2.1 Nature of Analyses

Mineral analysis is a valuable model for understanding the basic structure of analysis procedures to separate and measure. Separation of minerals from the food matrix is often specific, such as **complexometric titrations** (Sect. 12.3.1) or **precipitation titrations** (Sect. 12.3.2). In these cases of specific separation, nonspecific measurements such as volume of titrant are made and are later converted to mass of mineral based on fundamental stoichiometric relationships. In other cases, separation of mineral involves nonspecific procedures such as **ashing** or **acid extraction**. These nonspecific separations require that a specific measurement be made as provided by **colorimetry** (Sect. 12.3.3), **ion-selective electrodes** (ISE) (Sect. 12.3.4), **atomic absorption spectroscopy**, or **inductively coupled plasma-atomic emission spectroscopy** (Chap. 24).

Because determination of mass of mineral is the final objective of analysis, measures other than mass are considered to be surrogate, or stand-in, measures. **Surrogate measures** are converted into mass of mineral via fundamental stoichiometric and physiochemical relationships or by empirical relationships. Empirical relationships are those associations that need to be established by experimentation because they do not follow any well-established physiochemical relationship. An example of a surrogate measurement is the absorbance of a chromogen–mineral complex (Sect. 12.3.3). It may be possible to convert absorbance into mass of mineral using the fundamental relationships defined by the molar absorptivity and stoichiometry of the chromogen–mineral complex. However, it is more commonly required that the absorbance: concentration relationship be empirically developed using a series of standards (i.e., a standard curve).

12.2.2 Sample Preparation

Some sample preparation is generally required for traditional methods of mineral analysis to ensure a well-mixed and representative sample and to make the sample ready for the procedure to follow. A major concern in mineral analysis is **contamination** during sample preparation. **Comminution** (e.g., grinding or chopping) and mixing using metallic instruments can add significant mineral to samples and, whenever possible, should be performed using nonmetallic instruments or instruments not composed of the sample mineral. For example, it is standard practice in our laboratories to use an aluminum grinder for comminution of meat samples undergoing iron analysis. **Glassware** used in sample preparation and analysis should be scrupulously cleaned using acid washes and triple rinsed in the purest water. The latter may necessitate installation of an **ultrapure water system** in the laboratory to further purify the general supply of distilled water.

Solvents, including water, can contain significant quantities of minerals. Therefore, all procedures involving mineral analysis require the use of the purest reagents available. In some cases, the cost of ultrapure reagents may be prohibitive. When this is the case, the alternative is to always work with a reagent blank. A **reagent blank** is a sample of reagents used in the sample analysis, quantitatively the same as that used in the sample but without any of the material being analyzed. This reagent blank, representing the sum of the mineral contamination in the reagents, is then subtracted from the sample values to more accurately quantify the mineral.

A method such as near-infrared spectroscopy (Chap. 23) allows for mineral estimation without destruction of the carbon matrix of carbohydrates, fats, protein, and vitamins that make up foods. However, traditional methods generally require that the minerals be freed from this organic matrix in some manner. Chapter 7 describes the various methods used to ash foods in preparation for determination of specific mineral components of the food. In water samples, minerals may be determined without further preparation.

12.2.3 Interferences

Factors such as **pH**, **sample matrix**, **temperature**, and other **analytical conditions** and **reagents** can interfere with the ability of an analytical method to quantify a mineral. Often there are specific interfering substances that must be removed or suppressed for accurate analysis. Two of the more common approaches are to isolate the sample mineral, or remove interfering minerals, using selective precipitations or ion exchange resins. Water may need to be boiled to remove carbonates that interfere with several traditional methods of mineral analysis.

If other interferences are suspected, it is a common practice to develop the standard curve using sample mineral dissolved in a background matrix containing interfering elements known to be in the food sample. For example, if a food sample is to be analyzed for calcium content, a **background matrix solution** of the known levels of sodium, potassium, magnesium, and phosphorus should be used to prepare the calcium standards for developing the standard curve. In this manner, the standard curve more closely represents the analysis response to the sample mineral when analyzing a food sample. Alternatively, the standard curve can be developed using a series of sample mineral spikes added to the food sample. A **spike** is a small volume of a concentrated standard that is added to the sample. The volume is small enough so as to not appreciably change the overall composition of the sample, except for the mineral of interest. Thus, measurements of both the standards and the sample are made in the presence of the same background. If the spikes are added before implementation of the analysis protocol, possible effects of incomplete extractions, sample mineral degradation, and other losses are integrated into the standard curve.

12.3 METHODS

12.3.1 EDTA Complexometric Titration

12.3.1.1 Principles

The hexadentate ligand **ethylenediaminetetraacetate** (EDTA) forms stable 1:1 complexes with numerous mineral ions. This gives complexometric titration using EDTA broad application in mineral analysis. Stability of mineral–EDTA complexes generally increases with valence of the ion, although there is significant variation among ions of similar valence due to their coordination chemistry. The complexation equilibrium is strongly pH dependent. With decreasing pH the chelating sites of EDTA become protonated, thereby decreasing its effective concentration. Endpoints are detected using mineral chelators that have coordination constants lower than EDTA (i.e., less affinity for mineral ions) and that produce different colors in each of their complexed and free states. **Calmagite** and **Eriochrome Black T** (EBT) are such indicators that change from blue to pink when they complex with calcium or magnesium. The endpoint of a complexometric EDTA titration using either Calmagite or EBT as the indicator is detected as the color changes from pink to blue.

The pH affects a complexometric EDTA titration in several ways and must be controlled for best performance. The pH must be 10 or more for calcium or magnesium to form stable complexes with EDTA. Also, the sharpness of the endpoint increases with increasing pH. However, magnesium and calcium precipitate as their hydroxides at pH 12, and titration pH should probably be no more than 11 to ensure their solubility. Considering all factors, EDTA complexometric titration of calcium and magnesium is specified at pH 10 ± 0.1 using an ammonia buffer (10).

12.3.1.2 Procedure: Hardness of Water Using EDTA Titration

Water hardness is determined by EDTA complexometric titration of the total of calcium and magnesium, in the presence of Calmagite, and expressed as the equivalents of calcium carbonate (mg/L) (*Standard Methods for the Examination of Water and Wastewater*, Method 2340, Hardness) (10) (Fig. 12-1). The calcium–Calmagite complex is not stable, and calcium alone cannot be titrated using the Calmagite indicator. However, Calmagite becomes an effective indicator for

WATER HARDNESS-EDTA TITRATION

Titration of Water Sample

Dilute 25 ml sample (or such volume to require <15 ml titrant) to 50 ml in a flask.

⇓

Bring pH to 10±0.1 by adding 1–2 ml buffer solution (NH_4 in NH_4OH, combined with Na_2EDTA and $MgSO_4$ or $MgCl_2$) and 1–2 drops Calmagite indicator solution.

⇓

Titrate with a standard solution of ca. 0.01 *M* EDTA to a blue endpoint.

Standardization of EDTA

Weigh 1.000 mg $CaCO_3$ into a 500-ml Erlenmeyer flask and add HCl (1 : 1 dilution with water) until dissolved. Add 200 ml H_2O and boil a few minutes to expel CO_2. Let cool.

⇓

Add a few drops of methyl red indicator and adjust to intermediate orange color with 3 *N* NH_4OH or HCl (1 : 1) as required. Transfer to 1 L flask and dilute to volume.

⇓

Titrate calcium standard solution with EDTA solution, to Calmagite endpoint.

⇓

Determine $CaCO_3$ equivalents as mg $CaCO_3$/ml EDTA solution.

Calculations

Hardness (EDTA) as mg $CaCO_3$/L = (mg $CaCO_3$/ml EDTA x ml EDTA)/L sample

Procedure for determination of water hardness by EDTA titration. *Standard Methods for the Examination of Water and Wastewater*, Method 2340, Hardness. [Adapted from (10).]

calcium titration if we include in the buffer solution a small amount of neutral magnesium salt and enough EDTA to bind all magnesium. Upon mixing sample into the buffer solution, calcium in the sample replaces the magnesium bound to EDTA. The free magnesium binds to Calmagite, and the pink magnesium–Calmagite complex persists until all calcium in the sample has been titrated with EDTA. The first excess of EDTA removes magnesium from Calmagite and produces a blue endpoint.

12.3.1.3 Applications

The major application of EDTA complexometric titration is testing calcium plus magnesium as an indicator of water hardness (10). However, EDTA complexometric titration is suitable for determining calcium in the ash of fruits and vegetables (AOAC Method 968.31) (5) and other foods that have calcium without appreciable magnesium or phosphorus. The water hardness application of the EDTA complexometric titration is made easy using test strips impregnated with Calmagite and EDTA (e.g., **AquaChek**, Environmental Test Systems, Inc., a HACH Company, Elkhart, IN). The strips are dipped into the water to test for total hardness caused by calcium and magnesium. The calcium displaces the magnesium bound to EDTA, and the released magnesium binds to Calmagite, causing the test strip to change color.

12.3.2 Precipitation Titration

12.3.2.1 Principles

When at least one product of a titration reaction is an insoluble precipitate, it is referred to as **precipitation titrimetry**. Few of the many gravimetric methods, however, can be adapted to yield accurate volumetric methods. Some of the major factors blocking the adaptation are long times necessary for complete precipitation, failure of the reaction to yield a single product of definite composition, and lack of an endpoint indicator for the reaction.

Nonetheless, precipitation titration has resulted in at least two methods that are used widely in the food industry today. The **Mohr method** for chloride determination is a direct or **forward titration** method, based on the formation of an orange-colored solid, silver chromate, after silver from silver nitrate has complexed with all the available chloride.

$$Ag^+ + Cl^- \rightarrow AgCl \text{ (until all } Cl^- \text{ is complexed)} \quad [1]$$

$$2Ag^+ + CrO_4^{2-} \rightarrow Ag_2CrO_4 \text{ (orange only after } Cl^- \text{ is all complexed)} \quad [2]$$

The **Volhard method** is an indirect or **back-titration** method in which an excess of a standard solution of silver nitrate is added to a chloride-containing sample solution. The excess silver is then back-titrated using a standardized solution of potassium or ammonium thiocyanate with ferric ion as an indicator. The amount of silver that is precipitated with chloride in the sample solution is calculated by subtracting the excess silver from the original silver content.

$$Ag^+ + Cl^- \rightarrow AgCl \text{ (until all } Cl^- \text{ is complexed)} \quad [3]$$

$$Ag^+ + SCN^- \rightarrow AgSCN \text{ (to quantitate silver not complexed with chloride)} \quad [4]$$

$$SCN^- + Fe^{3+} - [FeSCN]^{2+} \text{ (red when there is any } SCN^- \text{ not complexed to } Ag^+) \quad [5]$$

12.3.2.2 Procedures

12.3.2.2.1 Mohr Titration of Salt in Butter (AOAC Method 960.29) Salt in foods may be estimated by titrating the chloride ion with silver (Fig. 12-2). The orange endpoint in this reaction occurs only when all chloride ion is complexed, resulting in an excess of silver to form the colored silver chromate. The endpoint of this reaction is therefore at the first hint of an orange color. When preparing reagents for this assay, use boiled water to avoid interferences from carbonates in the water.

12.3.2.2.2 Volhard Titration of Chloride in Plant Material (AOAC Method 915.01) In the Volhard method (Fig. 12-3), water must be boiled to minimize errors due to interfering carbonates, because the solubility product of silver carbonate is much less than the solubility product of silver chloride. Once chloride is determined by titration, the chloride weight is multiplied by 1.648 to obtain salt weight, if salt content is desired.

12.3.2.3 Applications

Precipitation titration methods are well suited for any foods that may be high in chlorides. Because of added salt in processed cheeses and meats, these products should certainly be considered for using this method to detect chloride; then salt content is estimated by calculation. Precipitation titrations are easily automated, thus ensuring that these traditional methods will see continued use in the analytical food laboratory. For example, the automatic titration system commonly used to rapidly measure the salt content of potato

SALT — MOHR TITRATION

Titration of Butter Sample

Weigh about 5 g of butter into 250-ml Erlenmeyer flask and add 100 ml of boiling H_2O.

⇓

Let stand 5–10 min with occasional swirling.

⇓

Add 2 ml of a 5% solution of K_2CrO_4 in d H_2O.

⇓

Titrate with 0.1 *N* $AgNO_3$ standardized as below until an orange-brown color persists for 30 sec.

Standardization of 0.1 *N* $AgNO_3$

Accurately weigh 300 mg of recrystallized dried KCl and transfer to a 250-ml Erlenmeyer flask with 40 ml of water.

⇓

Add 1 ml of K_2CrO_4 solution and titrate with $AgNO_3$ solution until first perceptible pale red-brown appears.

⇓

From the titration volume subtract the milliliters of the $AgNO_3$ solution required to produce the endpoint color in 75 ml of water containing 1 ml of K_2CaO_4.

⇓

From the next volume of $AgNO_3$ calculate normality of the $AgNO_3$ as:

$$\text{Normality } AgNO_3 = \frac{\text{mg KCl}}{\text{ml } AgNO_3 \times 74.555 \text{ g KCl/mole}}$$

Calculating Salt in Butter

$$\text{Percent salt} = \frac{\text{ml } 0.1\,N\ AgNO_3 \times 0.585}{\text{g of sample}}$$

$$[0.585 = (58.5 \text{ g NaCl/mol})/100]$$

12-2 figure Procedure of Mohr titration of salt in butter. AOAC Method 960.29 [Adapted from (5)].

SALT — VOLHARD TITRATION

Titration of Sample

Moisten 5 g of sample in crucible with 20 ml of 5% $Na_2 CO_3$ in water.

⇓

Evaporate to dryness.

⇓

Char on a hot plate under a hood until smoking stops.

⇓

Combust at 500°C for 24 hr.

⇓

Dissolve residue in 10 ml of 5 *N* HNO_3.

⇓

Dilute to 25 ml with d H_2O.

⇓

Titrate with standardized $AgNO_3$ solution (from the Mohr method) until white AgCl stops precipitating and then add a slight excess.

⇓

Stir well, filter through a retentive filter paper, and wash AgCl thoroughly.

⇓

Add 5 ml of a saturated solution of $FeNH_4(SO_4)2 \bullet 12H_2O$ to the combined titrate and washings.

⇓

Add 3 ml of 12 *N* HNO_3 and titrate excess silver with 0.1 *N* potassium thiocyanate.

Standardization of Potassium Thiocyanate Standard Solution

Determine working titer of the 0.1 *N* potassium thiocyanate standard solution by accurately measuring 40–50 ml of the standard $AgNO_3$ and adding it to 2 ml of $FeNH_4(SO_4)2 \cdot 12H_2O$ indicator solution and 5 ml of 9 *N* HNO_3.

⇓

Titrate with thiocyanate solution until solution appears pale rose after vigorous shaking.

Calculating Cl Concentration

Net volume of the $AgNO_3$ = Total volume $AgNO_3$ added – Volume titrated with thiocyanate

1 ml of 0.1 *M* $AgNO_3$ = 3.506 mg chloride

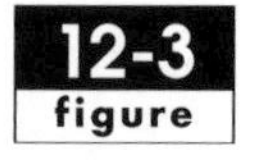

Procedure for Volhard titration of chloride in plant material. AOAC Method 915.01. [Adapted from (5).]

chips is simply doing a Mohr titration. Also, the **Quantab® chloride titration** used in AOAC Method 971.19 is an adaptation of the principles involved in the Mohr titration method. This test strip adaptation allows for very rapid quantitation of salt in food products and is accurate to ±10% over a range of 0.3–10% NaCl in food products.

12.3.3 Colorimetric Methods

12.3.3.1 Principles

Chromogens are chemicals that, upon reaction with the compound of interest, form a colored product. Chromogens are available that selectively react with a wide variety of minerals. Each chromogen reacts with its corresponding mineral to produce a soluble colored product that can be quantified by absorption of light at a specified wavelength. The relationship between concentration and absorbance is given by **Beer's law** as detailed in Chap. 22. Generally, concentration of mineral in a sample is determined from a standard curve developed during the analysis, although in some cases it is possible to directly calculate concentration based on molar absorptivity of the chromogen–mineral complex.

Samples generally must be ashed or treated in some other manner to isolate and/or release the minerals from organic complexes that would otherwise inhibit their reactivity with the chromogen. The mineral of interest must be solubilized from a dry ash and subsequently handled in a manner that prevents its precipitation. The soluble mineral may need to be treated (e.g., reduced or oxidized) to ensure that all mineral is in a form that reacts with the chromogen (2). Ideally, the chromogen reacts rapidly to produce a stable product. This is not always the case in practice, and time constraints may be established for color development and reading of absorbance. As with all mineral analysis of food, special efforts must be put in place to avoid contamination during sampling and analysis.

12.3.3.2 Procedures: Colorimetric Determination of Iron in Meat

The total iron content of foods can be quantified spectrophotometrically as shown in Fig. 12-4. In this method, the absorption of light at 562 nm is converted to iron concentration in the sample via a regression equation generated from a standard curve developed during the analysis using a standard solution. In meat systems, this method has been coupled with a method specific for heme iron to determine the ratio of heme iron to total iron, which is important nutritionally as the former is more bioavailable (11). Another interesting aspect of this method of interest to the food scientist is the fact that the ferrozine reagent only reacts with ferrous iron, and not ferric. The addition of ascorbic acid in the second to last step is necessary to convert all ionic iron to the detectable ferrous form. Repeating the procedure with and without ascorbic acid allows determination of total and ferrous ionic iron, respectively. Ferric iron is calculated by difference.

12.3.3.3 Applications

Colorimetry is used for the detection and quantification of a wide variety of minerals in food, and it is often a viable alternative to atomic absorption spectroscopy and other mineral detection methods. Colorimetric methods generally are very specific and usually can be performed in the presence of other minerals, thereby avoiding extensive separation to isolate the mineral of interest. They are particularly robust and often immune to matrix effects that can limit the usefulness of other methods for mineral analysis. With minimal effort and expense, many colorimetric methods will perform with precision and accuracy similar to that obtained by experienced personnel using atomic absorption spectroscopy (11).

12.3.4 Ion-Selective Electrodes

12.3.4.1 Principles

Many electrodes have been developed for the selective measurement of various cations and anions, such as bromide, calcium, chloride, fluoride, potassium, sodium, and sulfide (12, 13). The pH electrode described in Chap. 13 is a specific example of an **ISE**. For any ISE, an **ion-selective sensor** is placed such that it acts as a "bridging electrode" between two reference electrodes carefully designed to produce a constant and reproducible **potential**. The sensor can take on many forms (e.g., glass, single crystal, precipitate based, solvent polymer), although each provides an ion-selective electronic coupling that allows a potential to develop across the sensor. The exact mechanism(s) of charge transport across the sensor is not completely understood, but it is brought about by ion-selective species incorporated within the sensor itself and has been described by analogy to the response of billiard balls to an impact. If the ion-selective species within the sensor are imagined to be a row of billiard balls, it can be envisioned how the impact of sample ions on one sensor surface is translocated to the other surface. In this manner, the potential within the

IRON DETERMINATION OF MEAT—COLORIMETRIC ASSAY

Preparation of Standards

Prepare solutions of 10, 8, 6, 4, 2 μg iron/ml from a stock solution of 10 μg iron/ml.
Make dilutions using 0.1 *N* HCl.

Analysis of Sample

Place ~5 g sample into crucible and accurately weigh.

⇓

Heat on hot plate until well charred and sample has stopped smoking.

⇓

Ash in furnace at ca 550°C until ash is white.

⇓

Dissolve ash in small amount 1 *N* HCl and dilute to 50 ml volume with 0.1 *N* HCl

⇓

Transfer 0.500 ml of diluted sample and standards into 10 ml test tubes.

⇓

Add 1.250 ml ascorbic acid (0.02% in 0.2 *N* HCl, made fresh daily). Vortex and let set 10 min.

⇓

Add 2.000 ml 30% ammonium acetate. Vortex. (pH needs to be >3 for color development)

⇓

Measure absorbance at 562 nm. Determine iron concentration in sample digest (μg iron/ml) from standard curve.

Procedure for determination of iron in meat by colorimetry. [Adapted from (11).]

sensor remains constant, while potentials develop at the sensor surfaces according to the **Nernst equation** (Sect. 13.3.2.2), dependent on sample **ion activity** in the solutions contacting each surface.

Typically, the inside surface of the ion-selective sensor is in contact with the **negative reference electrode** (anode) via a filling solution having a constant concentration of sample ion, while the outside surface of the ion-selective sensor is in contact with the **positive reference electrode** (cathode) via sample solutions having varying concentration of sample ion. Because sample ion activity of the internal solution is fixed, the potential varies across the sensor depending solely on the activity of sample ion in the sample solution. As described by the Nernst equation, the potential of the outside sensor surface increases by $0.059/n$V (where n is the number of electrons involved in the half reaction of the sample ion) for each tenfold increase in activity of a mineral cation. Conversely, the potential decreases by $0.059/n$V for each tenfold increase in activity of mineral anion. These changes have a direct effect on the overall ISE potential because the outside sensor surface is orientated toward the positive reference electrode. **Ion concentration** is generally substituted for ion activity, which is a reasonable approximation at low concentrations and controlled ionic strength environments. Indeed, this is observed within limitations set by electrode and instrumental capabilities (Fig. 12-5).

12.3.4.2 General Methodology

For ISE analysis, one simply attaches an ISE for the sample ion to a **pH meter** set on the **mV scale** and follows instructions for determination. However, the performance of an ISE must be considered when first selecting an electrode and later when designing sampling and analysis protocols. Detailed information regarding the performance of specific ISEs is available from vendor catalogs. Typical ISEs likely to be employed for analysis of foods operate in the range of $1–10^{-6}$ M, although the electrode response may be distinctly nonlinear at the lower concentrations.

Electrode performance is affected by the presence of **interfering ions**, often with the strongest interference from those ions having size and charge similar to the ion of interest. Relative response of ISE to interfering ions may be expressed as selectivity coefficients or as concentration of interfering ion that leads to 10% error. If the selectivity coefficient relative to an interfering ion is 1000 (i.e., the ISE is 1000-fold more responsive to the sample ion than the interfering ion), 10% or greater error can be expected when measuring μM levels of the sample ion with interfering ion present at mM levels. Most ISEs operate over a broad pH range, although pH may need to be controlled for best performance. Minimum response times for ISEs fall in the range of 20 s to 1 min.

Despite inherent limitations of electrode design and construction, the analyst can adjust the sample and control measurement conditions to minimize many practical problems that otherwise limit the specificity and precision of ISEs. Because the ISE responds to ionic activity, it is important that the **activity coefficient** be kept constant in samples and calibration standards. The activity coefficient (γ) is used to relate ion activity (A) to ion concentration (C) ($A = \gamma C$). Activity coefficient is a function of ionic strength, so **ionic strength adjustment** (ISA) **buffers** are used to adjust the samples and standards to the same ionic strength. These ISA buffers are commercially available. The use of ISA buffers also adjusts the pH, which may be necessary if H^+ or OH^- activities affect the ion-specific sensor or if they interact with the analyte. In the case of metals having insoluble hydroxides, it is necessary to work at a pH that precludes their precipitation. Depending on the selectivity of the ISE, it may be necessary to remove interfering ions from the sample by selective precipitation or complexation.

In view of temperature effects on standard potentials and slopes of electrodes (see Nernst Equation [8] in Chap. 13), it is important to keep the electrode and solutions at a constant temperature. This may involve working in a room that is thermostatically controlled to 25 °C (one of the internationally accepted temperatures for electrochemical measurements), and allowing sufficient time for all samples and standards to equilibrate to this temperature. Solutions should be gently stirred during the measurement to attain rapid equilibrium and to minimize concentration gradients at the sensor surface. Finally, it is important to allow sufficient time for the electrode to stabilize

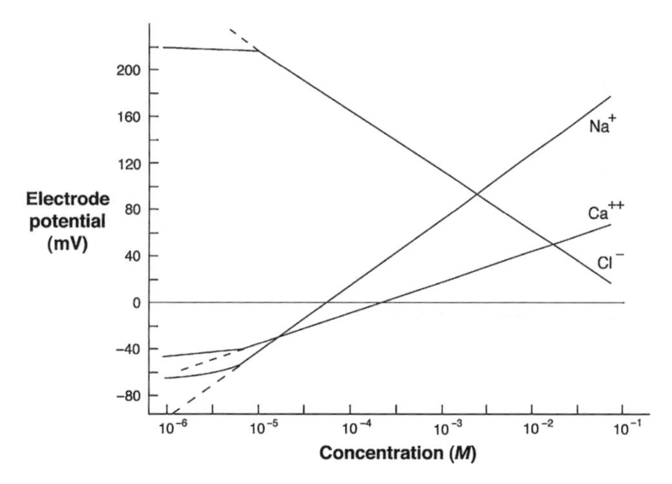

12-5 figure Examples of ion-selective electrode calibration curves for ions important in foods. (Courtesy of Van London pHoenix Co., Houston, TX.)

before taking a reading. ISEs may not completely stabilize within a practical timeframe, so a decision needs to be made of when to take the reading. The reading may be taken when the rate of change has fallen below some predetermined value or at a fixed time after the electrode was placed in solution. A problem with the latter is that many ISEs respond more rapidly, as samples are changed, to an increase in concentration of sample ion as compared to a decrease in concentration of sample ion.

12.3.4.3 Electrode Calibration and Determination of Concentration

In using an ISE, ion concentration can be determined using either a calibration curve, standard addition, or endpoint titration. It is common practice to develop a **calibration curve** when working with an ISE because it allows a large number of samples to be measured rapidly. The electrode potential (volts) is developed in a series of solutions of known concentration and plotted on **semilog paper** against the standard concentrations. Examples of calibration curves for various ions are given in Fig. 12-5. Upon analysis of a test sample, the observed electrode potential is used to determine ion concentration by referring to the calibration curve. Note the nonlinear region of the curve at the lowest concentrations. Total ionic strength and the concentration of interfering ions are especially important factors limiting selective detection of low levels of ions.

The **standard addition method** is of great value when only a few samples are to be measured and time does not permit the development of a calibration curve. This method also eliminates complex and unknown background effects that cannot be replicated when developing a calibration curve using standards. The ISE is immersed in the sample and the resulting voltage is recorded (E_{sample}). An aliquot, or **spike**, containing a known amount of the measured species is added to the sample, and a second measurement of electrode potential is determined (E_{spike}). Concentration of active species in the original sample is determined from the absolute difference in the voltage readings ($\Delta E = |E_{spike} - E_{sample}|$) according to the following relationship algebraically derived from the Nernst equations.

$$C_O = \frac{C_\Delta}{(10^{\Delta E/S} - 1)} \quad [6]$$

where:

C_O = original concentration of sample ion (mol/L)

C_Δ = change in sample ion concentration when spike was added (mol/L)

E = difference in potential between the two readings (V)

S = 0.059/number of electrons in the 1/2 reaction of the sample ion

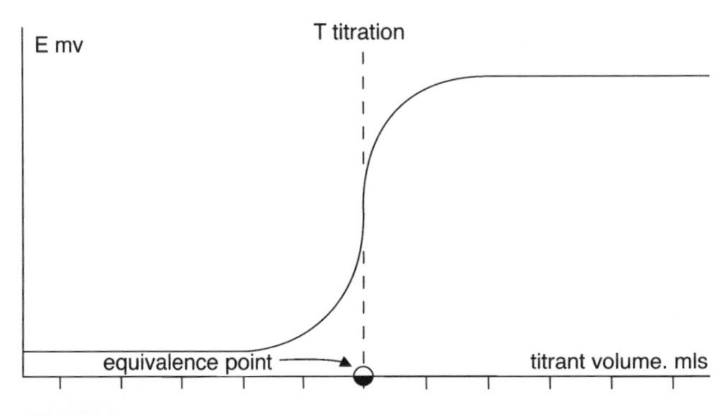

12-6 figure A typical T-type titration. [From (2), used with permission.]

Finally, ISEs can be used to detect the **endpoints of titrations** using species that form a precipitate or strong complex with the sample ion. If an ISE is selected that detects titrant species, a T-type titration curve results from the large increase in titrant activity detected at the equivalence point (see Fig. 12-6 for a cation titrant). If an ISE is selected that detects the sample ion, an S-type titration curve results from the removal of sample ion activity at the equivalence point. In either case, sample concentration is calculated from titrant volume to reach the equivalence point and the stoichiometric relationship between titrant species and sample ion.

12.3.4.4 Applications

Some examples of applications of ISEs are salt and nitrate in processed meats, salt content of butter and cheese, calcium in milk, sodium in low-sodium ice cream, carbon dioxide in soft drinks, potassium and sodium levels in wine, and nitrate in canned vegetables. An ISE method applicable to foods containing $<$100 mg sodium/100 g is an official method of AOAC International (Method 976.25). This method employs a sodium combination ISE, pH meter, magnetic stirrer, and a semilog graph paper for plotting a standard curve. Obviously, there are many other applications, but the above serve to demonstrate the versatility of this valuable measuring tool.

A major *advantage* of ISEs is their ability to measure many anions and cations directly. Such measurements are relatively simple compared to most other analytical techniques, particularly because a pH meter may be used as the voltmeter. Analyses are independent of sample volume when making direct measurements, while turbidity, color, and viscosity are all of no concern.

A major *disadvantage* in the use of ISEs is their inability to measure below 2–3 ppm, although there are some electrodes that are sensitive down to 1 part per billion. At low levels of measurement (below 10^{-4} M), the electrode response time is slow. Finally, some electrodes have had a high rate of premature failure or a short operating life and possible excessive noise characteristics.

12.4 COMPARISON OF METHODS

For labeling, processing, and even practical nutrition, we are concerned only with a few minerals, which generally can be analyzed by traditional methods. The traditional methods available for mineral analysis are varied, and only a very limited number of examples have been given in this chapter. Choice of methods for mineral analysis must be made considering method performance regarding accuracy, sensitivity, detection limit, specificity, and interferences. Information on method performance is available from the collaborative studies referenced with official methods (Chap. 1). Other factors to be considered include cost per analysis completed, equipment availability, and analytical time compared to analytical volume.

Generally, for a small laboratory with skilled analytical personnel, the traditional methods can be carried out rapidly, with accuracy, and at minimal costs. If a large number of samples of a specific element are to be run, there is certainly a time factor in favor of using atomic absorption spectroscopy or emission spectroscopy, depending on the mineral being analyzed. The graphite furnace on the atomic absorption spectrophotometer is capable of sensitivity in the parts per billion range. This is beyond the limits of the traditional methods. However, for most minerals of practical concern in the food industry, this degree of sensitivity is not required.

Modern instrumentation has made it possible to quantify an entire spectrum of minerals in one process, some into the parts per billion range. Instrumentation capable of such analysis is expensive and beyond the financial resources of many quality assurance laboratories. Large numbers of samples to be analyzed may justify the automation of some routine analyses and perhaps the expense of some of the modern pieces of equipment. However, the requirements for only occasional samples to be analyzed for a specific mineral will not justify the initial costs of much instrumentation. This leaves the options of sending samples out to certified laboratories for analysis or utilizing one of the more traditional methods for analysis.

12.5 SUMMARY

The mineral content of water and foodstuffs is important because of their nutritional value, toxicological potential, and interactive effects with processing and texture of some foods. Traditional methods for mineral analysis include titrimetric and colorimetric procedures. The basic principles of these methods are described in this chapter, along with discussion of ISE methodology that has general application for mineral analysis.

Procedures are described in this chapter that illustrate use of these traditional methods to quantify minerals of concern in the food industry. These procedures generally require chemicals and equipment routinely available in an analytical laboratory and do not require expensive instrumentation. These methods may be suited to a small laboratory with skilled analytical personnel and a limited number of samples to be analyzed. The traditional procedures will often perform similarly to procedures requiring more instrumentation and may be more robust in actual practice.

Foods are typically ashed prior to traditional analyses because the methods generally require that the minerals be freed from the organic matrix of the foods. Sample preparation and analysis must include steps necessary to prevent contamination or loss of volatile elements and must deal with a variety of potential interferences. Various approaches are described to account for these possible errors including use of reagent blanks, addition of spikes, and development of standard curves using appropriate mineral matrix background.

Traditional methods for mineral analysis are often automated or adapted to test kits for rapid analysis. Tests for water hardness and the Quantab® for salt determination are examples currently being used. The basic principles involved in traditional methods will continue to be utilized to develop inexpensive rapid methods for screening mineral content of foods and beverages. Familiarity with the traditional principles will allow the food analyst to obtain the best possible performance with the kits and adapt to problems that may be encountered.

12.6 STUDY QUESTIONS

1. What is the major concern in sample preparation for specific mineral analysis? How can this concern be addressed?
2. If the ammonia buffer is pH 11.5 rather than pH 10 in the EDTA complexometric titration to determine the hardness of water, would you expect to overestimate or underestimate the hardness? Explain your answer.

3. This chapter includes descriptions of the EDTA complexometric titration method and ISE methodology for quantifying calcium. Differentiate these techniques with regard to the principles involved, and discuss primary advantages and disadvantages of these two techniques.
4. The Mohr and Volhard titration methods often are used to determine the NaCl content of foods. Compare and contrast these two methods, as you explain the principles involved.
5. In a back-titration procedure, would overshooting the endpoint in the titration cause an over- or underestimation of the compound being quantified? Explain your answer.
6. Describe how and why to employ standards in background matrix, spikes, and reagent blanks.
7. Explain the principles of using an ISE to measure the concentration of a particular inorganic element in food. List the factors to control, consider, or eliminate for an accurate measure of concentration by the ISE method.
8. You have decided to purchase an ISE to monitor the sodium content of foods produced by your plant. List the advantages this would have over the Mohr/Volhard titration method. List the problems and disadvantages of ISE that you should anticipate.
9. What factors should be considered in selecting a specific method for mineral analysis for a food product?

12.7 PRACTICE PROBLEMS

1. If a given sample of food yields 0.750 g of silver chloride in a gravimetric analysis, what weight of chloride is present?
2. A 10-g food sample was dried, then ashed, and analyzed for salt (NaCl) content by the Mohr titration method ($AgNO_3 + Cl \rightarrow AgCl$). The weight of the dried sample was 2 g, and the ashed sample weight was 0.5 g. The entire ashed sample was titrated using a standardized $AgNO_3$ solution. It took 6.5 ml of the $AgNO_3$ solution to reach the endpoint, as indicated by the red color of Ag_2CO_4 when K_2CrO_4 was used as an indicator. The $AgNO_3$ solution was standardized using 300 mg of dried KCl as described in Fig. 12-2. The corrected volume of $AgNO_3$ solution used in the titration was 40.9 ml. Calculate the salt (NaCl) content of the original food sample as percent NaCl (wt/wt).
3. A 25-g food sample was dried, then ashed, and finally analyzed for salt (NaCl) content by the Volhard titration method. The weight of the dried sample was 5 g, and the ashed sample weighed 1 g. Then 30 ml of 0.1 *N* $AgNO_3$ was added to the ashed sample, the resultant precipitate was filtered out, and a small amount of ferric ammonium sulfate was added to the filtrate. The filtrate was then titrated with 3 ml of 0.1 *N* KSCN to a red endpoint.
   (a) What was the moisture content of the sample, expressed as percent H_2O (wt/wt)?
   (b) What was the ash content of the sample, expressed as percent ash (wt/wt) on a dry weight basis?
   (c) What was the salt content of the original sample in terms of percent (wt/wt) NaCl? (molecular weight Na = 23; molecular weight Cl = 35.5)
4. Compound X in a food sample was quantified by a colorimetric assay. Use the following information and Beer's law to calculate the content of Compound X in the food sample, in terms of mg Compound X/100 g sample:
   (a) A 4-g sample was ashed.
   (b) Ashed sample was dissolved with 1 ml of acid and the volume brought to 250 ml.
   (c) A 0.75-ml aliquot was used in a reaction in which the total volume of the sample to be read in the spectrophotometer was 50 ml.
   (d) Absorbance at 595 nm for the sample was 0.543.
   (e) The absorptivity constant for the reaction (i.e., extinction coefficient) was known to be 1574 L/M cm.
   (f) Inside diameter of cuvette for spectrophotometer was 1 cm.
5. Colorimetric analysis
   (a) You are using a colorimetric method to determine the concentration of Compound A in your liquid food sample. This method allows a sample volume of 5 ml. This volume must be held constant but can comprise diluted standard solution and water. For this standard curve, you need standards that contain 0, 0.25, 0.50, 0.75, and 1.0 mg of Compound A. Your stock standard solution contains 5 g/L of Compound A.

   Devise a dilution scheme(s) for preparing the samples for this standard curve that could be followed by a lab technician. Be specific. In preparing the dilution scheme, use no volumes less than 0.5 ml.
   (b) You obtain the following absorbance values for your standard curve:

Sample (mg)	*Absorbance (500 nm)*
0.00	0.00
0.25	0.20
0.50	0.40
0.75	0.60
1.00	0.80

   Construct a standard curve and determine the equation of the line.
   (c) A 5-ml sample is diluted to 500 ml, and 3 ml of this solution is analyzed as per the standard samples; the absorbance of 0.50 units at 500 nm. Use the equation of the line calculated in part (b) and information about the dilutions to calculate what the concentration is of Compound A in your original sample in terms of g/L.
6. What is the original concentration of copper in a 100-ml sample that shows a potential change of 6 mV after the addition of 1 ml of 0.1 *M* $Cu(NO_3)_2$?

Answers

1.

$$\frac{x \text{ g Cl}}{0.750 \text{ g AgCl}} = \frac{35.45 \text{ g/mol}}{143.3 \text{ g/mol}}$$

$$x = 0.186 \text{ g Cl}$$

2.

$$NAgNO_3 = \frac{0.300\,\text{g KCl}}{\text{ml AgNO}_3 \times 74.555\,\text{g KCl/mol}}$$

$$0.0984\,N = \frac{0.300\,\text{g}}{40.9\,\text{ml} \times 74.555}$$

Percent salt

$$= \left(\frac{0.0065\,\text{L} \times 0.0984\,N\,\text{AgNO}_3 \times 58.5\,\text{g/mol}}{10\,\text{g}}\right) \times 100$$

Percent salt = 0.37%

3.

(a) $\frac{25\,\text{g wet sample} - 5\,\text{g dry sample}}{25\,\text{g wet sample}} \times 100 = 80\%$

(b) $\frac{1\,\text{g ash}}{5\,\text{g dry sample}} \times 100 = 20\%$

(c) mol Ag added = mol Cl^- in sample +mol SCN^- added

$$\text{mol Ag} = (0.1\,\text{mol/L}) \times (0.03\,\text{L}) = 0.003\,\text{mol}$$

$$\text{mol SCN}^- = (0.1\,\text{mol/L}) \times (0.003\,\text{L}) - 0.0003\,\text{mol}$$

$$0.003\,\text{mol Ag} = \text{mol Cl}^- + 0.0003\,\text{mol SCN}^-$$

$$0.0027\,\text{mol} = \text{mol Cl}^-$$

$$(0.0027\,\text{mol Cl}^-) \times \frac{58.5\,\text{g NaCl}}{\text{mol}} = 0.1580\,\text{g NaCl}$$

$$\frac{0.1580\,\text{g NaCl}}{25\,\text{g wet sample}} = \frac{0.00632\,\text{g NaCl}}{\text{g wet sample}} \times 100$$

$$= 0.63\%\ \text{NaCl (w/w)}$$

4.

$$A = abc$$

$$0.543 = (1574\,\text{L g}^{-1}\,\text{cm}^{-1})(1\,\text{cm})\,c$$

$$c = 3.4498 \times 10^{-4}\,\text{g/L}$$

$$c = 3.4498 \times 10^{-4}\,\text{mg/ml}$$

$$\frac{3.4498 \times 10^{-4}\,\text{mg}}{\text{ml}} \times 50\,\text{ml} = 1.725 \times 10^{-2}\,\text{mg}$$

$$\frac{1.725 \times 10^{-2}\,\text{mg}}{0.75\,\text{ml}} \times \frac{250\,\text{ml}}{4\,\text{g}} = 1.437\,\text{mg/g}$$

$$= 143.7\,\text{mg/100 g}$$

5.

(a) Lowest dilution volume for 1 ml of stock:

$$\frac{0.25\,\text{mg}}{0.5\,\text{ml}} = \frac{1\,\text{ml}}{x\,\text{ml}} \times \frac{5\,\text{mg}}{\text{ml}}$$

$$x = 10\,\text{ml}$$

Therefore, use a volumetric pipette to add 1 ml of stock to a 10-ml volumetric flask. Bring to volume with ddH_20 to give a diluted stock solution of 0.25 mg/0.5 ml. Use this to make up standards according to the following table.

mg A/5 ml	*ml Diluted stock solution*	*ml H_2O*
0	0	5.0
0.25	0.5	4.5
0.50	1.0	4.0
0.75	1.5	3.5
1.0	2.0	3.0

(b)

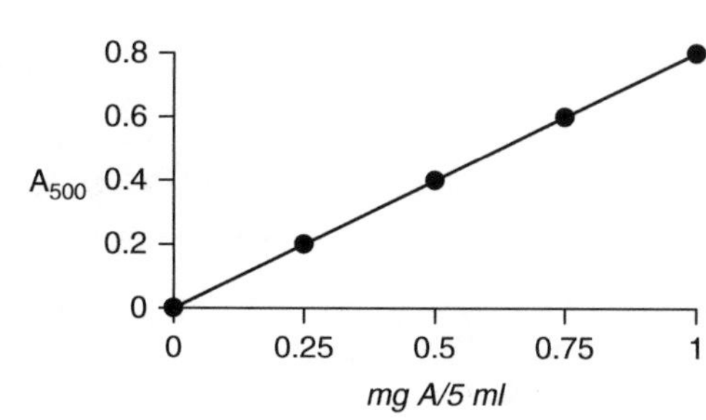

Equation of the line: $y = 0.8x + 0$

(c)

$$A_{500} = 0.50 = y$$

$$0.50 = 0.8x + 0$$

$$x = 0.625$$

$$\frac{0.625\,\text{mg}}{5\,\text{ml}} \times \frac{5\,\text{ml}}{3\,\text{ml}} \times \frac{500\,\text{ml}}{5\,\text{ml}} = 20.8\,\text{mg/ml}$$

$$= 20.8\,\text{g/L}$$

6.

$$C_O = \frac{0.001\,\text{L} \times \frac{0.001\,\text{moles}}{\text{L}} \times \frac{1}{0.100\,\text{L}}}{10^{0.006/0.0285} - 1} = 1.6\,\text{m}M$$

12.8 REFERENCES

1. Schwendt G (1997) The essential guide to analytical chemistry. Wiley, New York
2. Kirk RS, Sawyer R (1991) Pearson's composition and analysis of foods, 9th edn. Longman Scientific and Technical, Essex, England
3. Skoog DA, West DM, Holler JF, Crouch SR (2000) Analytical chemistry: an introduction, 7th edn. Brooks/Cole, Pacific Grove, CA
4. Harris DC (1999) Quantitative chemical analysis, 5th edn. W.H. Freeman and Co., New York
5. AOAC International (2007) Official methods of analysis, 18th edn., 2005; Current through revision 2, 2007 (On-line). AOAC International, Gaithersburg, MD

6. Sullivan DM, Carpenter DE (eds) (1993) Methods of analysis for nutritional labeling. AOAC International, Arlington, VA
7. Food and Nutrition Board, Institute of Medicine (1997) Dietary reference intakes for calcium, phosphorus, magnesium, vitamin D, and fluoride. National Academy Press, Washington, DC
8. Food and Nutrition Board, Institute of Medicine (2000) Dietary reference intakes for vitamin C, vitamin E, selenium, and carotenoids. National Academy Press, Washington, DC
9. Food and Nutrition Board, Institute of Medicine (2002) Dietary reference intakes for vitamin A, vitamin K, arsenic, boron, chromium, copper, iodine, iron, manganese, molybdenum, nickel, silicon, vanadium, and zinc. National Academy Press, Washington, DC
10. Eaton AD, Clesceri LS, Rice EW, Greenburg AE (eds) (2005) Standard methods for the examination of water and wastewater, 21st edn. Method 2340, hardness. American Public Health Association, American Water Works Association, Water Environment Federation, Washington, DC, pp 2–37 to 2–39
11. Carpenter C, Clark E (1995) Evaluation of iron methods used in meat iron analysis and iron content of raw and cooked meats. J Agric Food Chem 43:1824–1827
12. Covington AK (ed) (1980) Ion selective electrode methodology. CRC, Boca Raton, FL
13. Wang J (2000) Analytical electrochemistry, 2nd edn. Wiley, New York

Chemical Properties and Characteristics of Foods

pH and Titratable Acidity

George D. Sadler
PROVE IT, LLC,
Geneva, IL 60134, USA
gsadler@proveitllc.com

and

Patricia A. Murphy
Department of Food Science and Human Nutrition, Iowa State University,
Ames, IA 50011, USA
pmurphy@iastate.edu

13.1 Introduction 221
13.2 Calculation and Conversion for Neutralization Reactions 221
13.2.1 Concentration Units 221
13.2.2 Equation for Neutralization and Dilution 223
13.3 pH 223
13.3.1 Acid–Base Equilibria 223

S.S. Nielsen, *Food Analysis*, Food Science Texts Series, DOI 10.1007/978-1-4419-1478-1_13,

13.3.2 pH Meter 224
13.3.2.1 Activity vs. Concentration 224
13.3.2.2 General Principles 224
13.3.2.3 Reference Electrode 226
13.3.2.4 Indicator Electrode 226
13.3.2.5 Combination Electrodes 227
13.3.2.6 Guidelines for Use of pH Meter 227
13.4 Titratable Acidity 227
13.4.1 Overview and Principle 227
13.4.2 General Considerations 228
13.4.2.1 Buffering 228
13.4.2.2 Potentiometric Titration 229
13.4.2.3 Indicators 229
13.4.3 Preparation of Reagents 230
13.4.3.1 Standard Alkali 230
13.4.3.2 Standard Acid 231
13.4.4 Sample Analysis 231
13.4.5 Calculation of Titratable Acidity 232
13.4.6 Acid Content in Food 232
13.4.7 Volatile Acidity 233
13.4.8 Other Methods 233
13.5 Summary 233
13.6 Study Questions 234
13.7 Practice Problems 234
13.8 References 237

13.1 INTRODUCTION

There are two interrelated concepts in food analysis that deal with acidity: **pH** and **titratable acidity**. Each of these quantities is analytically determined in separate ways and each has its own particular impact on food quality. **Titratable acidity** deals with measurement of the **total acid concentration** contained within a food (also called **total acidity**). This quantity is determined by exhaustive titration of intrinsic acids with a standard base. Titratable acidity is a better predictor of acid's impact on flavor than pH.

Total acidity does not tell the full story, however. Foods establish elaborate buffering systems that dictate how **hydrogen ions** (H^+), the fundamental unit of acidity, are expressed. Even in the absence of buffering, less than 3% of any food acid is ionized into H^+ and its anionic parent species (its **conjugate base**). This percentage is further suppressed by buffering. In aqueous solution, hydrogen ions combine with water to form **hydronium ions,** H_3O^+. The ability of a microorganism to grow in a specific food is an important example of a process that is more dependent on hydronium ion concentration than on titratable acidity. The need to quantify only the free H_3O^+ concentration leads to the second major concept of acidity, that of **pH** (also called **active acidity**). In nature, the H_3O^+ concentration can span a range of 14 orders of magnitudes. The term pH is a mathematical shorthand for expressing this broad continuum of H_3O^+ concentration in a concise and convenient notation. In contemporary food analysis, pH is usually determined instrumentally with a pH meter; however, chemical pH indicators also exist.

For general and food-specific information on measuring pH and titratable acidity, see references (1–15). For the actual pH and titratable acidity of select foods, see reference (8).

13.2 CALCULATION AND CONVERSION FOR NEUTRALIZATION REACTIONS

13.2.1 Concentration Units

This chapter deals with the theory and practical application of titratable acidity calculation and pH determination. To quantitatively measure components of foods, solutions must be prepared to accurate concentrations and diluted into the desired working range.

The terms used for concentration in food analysis should be reviewed. The most common concentration terms are given in Table 13-1. Molarity and normality are the most common SI (International Scientific) terms used in food analysis, but solutions also can be expressed as percentages. It is important that the analyst be able to convert between both systems.

Molarity (M) is a concentration unit representing the number of moles of the solute per liter of solution. **Normality** (N) is a concentration unit representing the number of equivalents (Eq) of a solute per liter of solution. In acid and base solutions, the normality represents the concentration or moles of H^+ or OH^- per liter that will be exchanged in a neutralization reaction when taken to completion. For oxidation–reduction reagents, the normality represents the concentration or moles of electrons per liter to be exchanged when the reaction is taken to completion. The following are some examples of molarity vs. normality (equivalents):

Acid–Base Reactions

$1\,M\,H_2SO_4 = 2\,N\,H_2SO_4$
2 equivalents of H^+ per mol of acid
$1\,M\,NaOH = 1\,N\,NaOH$
1 equivalent of OH^- per mol of base
$1\,M\,CH_3COOH = 1\,N$ acetic acid
1 equivalent of H^+ per mol of acid
$1\,M\,H_2C_4H_4O_5 = 2\,N$ malic acid
2 equivalents of H^+ per mol of acid

Oxidation–Reduction Reactions
For example, $HSO_3^- + I_2 + H_2O \rightleftarrows SO_4^2 + 2I^- + 3H^+$

$1\,M\,I_2 = 2\,N$ iodine
2 equivalents of electrons gained per mol of I_2
$1\,M\,HSO_3^- = 2\,N$ bisulfite
2 equivalents of electrons lost per mol of bisulfate

Many analytical determinations in food analysis use the concept of equivalents to measure the amount of an unknown. Perhaps the most familiar of these are **acid–base reactions** in which hydrogen ions are exchanged and can be quantified through stoichiometric neutralization with a standard base. Acid–base reactions are used to determine nitrogen in the Kjeldahl protein determination (see Chap. 9), benzoic acid in sodas, and in determining percent titratable acidity. The concept of equivalents also is used in oxidation–reduction problems to quantify unknown analytes that are capable of direct electron transfer.

Equivalent weight can be defined as the molecular weight divided by the number of equivalents in the reactions. For example, the molecular weight of H_2SO_4 is 98.08 g. Since there are 2 equivalents per mole of H_2SO_4, the equivalent weight of H_2SO_4 is 49.04 g. Table 13-2 provides a list of molecular and equivalent weights for acids important in food analysis. In working with normality and milliliters, the term **milliequivalents** (mEq) is usually preferred. Milliequivalent weight is the equivalent weight divided by 1000.

Percentage concentrations are the mass amount of solute or analyte per 100 ml or 100 g of material. Percentage can be expressed for solutions or for solids

Concentration Expressions Terms

Unit	Symbol	Definition	Relationship
Molarity	*M*	Number of moles of solute per liter of solution	$M = \frac{\text{moles}}{\text{liter}}$
Normality	*N*	Number of equivalents of solute per liter of solution	$N = \frac{\text{equivalents}}{\text{liter}}$
Percent by weight (parts per hundred)	wt%	Ratio of weight of solute to weight of solute plus weight of solvent × 100	$\text{wt} = \frac{\text{wt solute} \times 100}{\text{total wt}}$
	wt/vol%	Ratio of weight of solute to total vol. × 100	$\text{wt\%} = \frac{\text{wt solute} \times 100}{\text{total volume}}$
Percent by volume	vol%	Ratio of volume of solute to total volume	$\text{vol\%} = \frac{\text{vol solute} \times 100}{\text{total volume}}$
Parts per million	ppm	Ratio of solute (wt or vol) to total wt or vol. × 1,000,000	$\text{ppm} = \frac{\text{mg solute}}{\text{kg solution}} = \frac{\mu\text{g solute}}{\text{g solution}} = \frac{\text{mg solute}}{\text{liter solution}} = \frac{\mu\text{g solute}}{\text{ml solution}}$
Parts per billion	ppb	Ratio of solute (wt or vol) to total wt or vol. × 1,000,000,000	$\text{ppb} = \frac{\mu\text{g solute}}{\text{liter solution}} = \frac{\mu\text{g solute}}{\text{kg}} = \frac{\text{ng solute}}{\text{ml}} = \frac{\text{ng solute}}{\text{g}}$

Molecular and Equivalent Weights of Common Food Acids

Acid	Chemical Formula	Molecular Weight	Equivalents per Mole	Equivalent Weight
Citric (anhydrous)	$H_3C_6H_5O_7$	192.12	3	64.04
Citric (hydrous)	$H_3C_6H_5O_7 \cdot H_2O$	210.14	3	70.05
Acetic	$HC_2H_3O_2$	60.06	1	60.05
Lactic	$HC_3H_5O_3$	90.08	1	90.08
Malic	$H_2C_4H_4O_5$	134.09	2	67.05
Oxalic	$H_2C_2O_4$	90.04	2	45.02
Tartaric	$H_2C_4H_4O_6$	150.09	2	75.05
Ascorbic	$H_2C_6H_6O_6$	176.12	2	88.06
Hydrochloric	HCl	36.47	1	36.47
Sulfuric	H_2SO_4	98.08	2	49.04
Phosphoric	H_3PO_4	98.00	3	32.67
Potassium acid phthalate	$KHC_8H_4O_4$	204.22	1	204.22

and can be on a volume basis or mass basis. When the percentage becomes a number less than 1% **parts per million** (ppm), **parts per billion** (ppb) and even **parts per trillion** (ppt) usually are preferred. If percentage is defined as the mass of the solute or analyte per mass (or volume) of sample × 100, then ppm is simply the same ratio of mass of solute per mass of sample × 1,000,000.

13.2.2 Equation for Neutralization and Dilution

There are some general rules in evaluating equilibrium reactions that are helpful in most situations. At full neutralization the millequivalents (mEq) of one reactant in the neutralization equal the milliequivalents of the other reactant. This can be expressed mathematically as follows:

$$(\text{ml of } X)(N \text{ of } X) = (\text{ml of } Y)(N \text{ of } Y) \qquad [1]$$

Equation [1] also can be used to solve dilution problems where X represents the stock solution and Y represents the working solution. When Equation [1] is used for dilution problems, any value of concentration (grams, moles, ppm, etc.) can be substituted for N. Units should be recorded with each number. Cancellation of units provides a quick check on proper setup of the problem. (See Practice Problems 1–8 at the end of Chap. 13.)

13.3 pH

13.3.1 Acid–Base Equilibria

The Brønsted-Lowry theory of neutralization is based upon the following definitions for acid and base:

Acid: A substance capable of donating protons. In food systems the only significant proton donor is the hydrogen ion.

Base: A substance capable of accepting protons.

Neutralization is the reaction of an acid with a base to form a salt as shown below:

$$HCl + NaOH \rightleftarrows NaC1 + H_2O \qquad [2]$$

Acids form hydrated protons called **hydronium ions** (H_3O^+) and bases form **hydroxide ions** (OH^-) in aqueous solutions:

$$H_3O^+ + OH^- \rightleftarrows 2H_2O \qquad [3]$$

At any temperature, the product of the molar concentrations (moles/liter) of H_3O^+ and OH^- is a constant referred to as the **ion product constant for water** (K_w):

$$[H_3O^+][OH^-] = K_w \qquad [4]$$

K_w varies with the temperature. For example, at 25 °C, $K_w = 1.04 \times 10^{-14}$ but at 100 °C, $K_w = 58.2 \times 10^{-14}$.

The above concept of K_w leads to the question of what the concentrations of $[H_3O^+]$ and $[OH^-]$ are in pure water. Experimentation has revealed that the concentration of $[H_3O^+]$ is approximately 1.0×10^{-7} M, as is that of the $[OH^-]$ at 25 °C. Because the concentrations of these ions are equal, pure water is referred to as being **neutral**.

Suppose that a drop of acid is added to pure water. The $[H_3O^+]$ concentration would increase. However, K_w would remain constant (1.0×10^{-14}), revealing a decrease in the $[OH^-]$ concentration. Conversely, if a drop of base is added to pure water, the $[H_3O^+]$ would decrease while the $[OH^-]$ would increase, maintaining the K_w at 1.0×10^{-14} at 25 °C.

How did the term pH derive from the above considerations? In approaching the answer to this question, one must observe the concentrations of $[H_3O^+]$ and $[OH^-]$ in various foods, as shown in Table 13-3.

13-3 table Concentrations of H_3O^+ and OH^- in Various Foods at 25 °C

Food	*$[H_3O^+]^a$*	*$[OH^-]^a$*	*K_w*
Cola	2.24×10^{-3}	4.66×10^{-12}	1×10^{-14}
Grape juice	5.62×10^{-4}	1.78×10^{-11}	1×10^{-14}
SevenUp	3.55×10^{-4}	2.82×10^{-11}	1×10^{-14}
Schlitz beer	7.95×10^{-5}	1.26×10^{-10}	1×10^{-14}
Pure water	1.00×10^{-7}	1.00×10^{-7}	1×10^{-14}
Tap water	4.78×10^{-9}	2.09×10^{-6}	1×10^{-14}
Milk of magnesia	7.94×10^{-11}	1.26×10^{-4}	1×10^{-14}

From (12), used with permission. Copyright 1971 American Chemical Society.

[a]Moles per liter.

Calculating the pH of the cola:

Step 1. Substitute the [H+] into the pH equation:

$$pH = -\log(H^+)$$

$$pH = -\log(2.24 \times 10^{-3})$$

Step 2. Separate 2.24×10^{-3} into two parts; determine the logarithm of each part:

$$\log 2.24 = 0.350$$

$$\log 10^{-3} = -3$$

Step 3. Add the two logs together since adding logs is equivalent to multiplying the two numbers:

$$0.350 + (-3) = -2.65$$

Step 4. Place the value into the pH equation:

$$pH = -(-2.65)$$

$$pH = 2.65$$

The numerical values found in Table 13-3 for $[H_3O^+]$ and $[OH^-]$ are bulky and led a Swedish chemist, S.L.P. Sørensen, to develop the pH system in 1909.

pH is defined as the logarithm of the reciprocal of the hydrogen ion concentration. It also may be defined as the negative logarithm of the molar concentration of hydrogen ions. Thus, a $[H_3O^+]$ concentration of 1×10^{-6} is expressed simply as pH 6. The $[OH^-]$ concentration is expressed as pOH and would be pOH 8 in this case, as shown in Table 13-4.

While the use of pH notation is simpler from the numerical standpoint, it is a confusing concept in the minds of many students. One must remember that it is a logarithmic value and that a change in one pH unit is actually a tenfold change in the concentration of $[H_3O^+]$. (See Practice Problems 9–12 at the end of Chap. 13.)

It is important to understand that pH and titratable acidity are not the same. Strong acids such as hydrochloric, sulfuric, and nitric acids are almost fully dissociated at pH 1. Only a small percentage of food acid molecules (citric, malic, acetic, tartaric, etc.) dissociate in solution. This point may be illustrated by comparing the pH of 0.1 N solutions of hydrochloric and acetic acids.

$$HCl \rightleftarrows H^+ + Cl^- \quad [5]$$

$$CH_3COOH \rightleftarrows H^+ + CH_3COO^- \quad [6]$$

The HCl fully dissociates in solution to produce a pH of 1.02 at 25 °C. By contrast, only about 1% of CH_3COOH is ionized at 25 °C, producing a significantly higher pH of 2.89. The calculation and significance of partial dissociation on pH is presented in more detail in Sect. 13.4.2.1.

table 13-4 Relationship of $[H^+]$ vs. pH and $[OH^-]$ vs. pOH at 25 °C

$[H^+]$[a]	*pH*	$[OH^-]$[a]	*pOH*
1×10^0	0	1×10^{-14}	14
10^{-1}	1	10^{-13}	13
10^{-2}	2	10^{-12}	12
10^{-3}	3	10^{-11}	11
10^{-4}	4	10^{-10}	10
10^{-5}	5	10^{-9}	9
10^{-6}	6	10^{-8}	8
10^{-7}	7	10^{-7}	7
10^{-8}	8	10^{-6}	6
10^{-9}	9	10^{-5}	5
10^{-10}	10	10^{-4}	4
10^{-11}	11	10^{-3}	3
10^{-12}	12	10^{-2}	2
10^{-13}	13	10^{-1}	1
10^{-14}	14	10^0	0

From (12), used with permission. Copyright 1971 American Chemical Society.

[a]Moles per liter. Note that the product of $[H^+][OH^-]$ is always 1×10^{-14}.

Calculation of $[H^+]$ of a beer with pH 4.30:

Step 1. Substitute numbers into the pH equation:

$$pH = -\log[H^+]$$
$$4.30 = -\log[H^+]$$
$$-4.30 = \log[H^+]$$

Step 2. Divide the −4.30 into two parts so that the first part contains the decimal places and second part the whole number:

$$-4.30 = 0.70 - 5 = \log[H^+]$$

Step 3. Find the antilogs:

$$\text{antilog of } 0.70 = 5.0$$
$$\text{antilog of } -5 = 10^{-5}$$

Step 4. Multiply the two antilogs to get $[H^+]$:

$$5 \times 10^{-5} = [H^+]$$
$$H^+ = 5 \times 10^{-5}\,M$$

13.3.2 pH Meter

13.3.2.1 Activity vs. Concentration

In using pH electrodes, the concept of activity vs. concentration must be considered. **Activity** is a measure of expressed chemical reactivity, while **concentration** is a measure of all forms (free and bound) of ions in solution. Because of the interactions of ions between themselves and with the solvent, the effective concentration or activity is, in general, lower than the actual concentration, although activity and concentration tend to approach each other at infinite dilution. Activity and concentration are related by the following equation:

$$A = \gamma C \quad [7]$$

where:

A = activity
γ = activity coefficient
C = concentration

The **activity coefficient** is a function of ionic strength. Ionic strength is a function of the concentration of, and the charge on, all ions in solution. Activity issues can become significant for hydronium ions below pH 1 and for hydroxyl ions at pH 13 and above.

13.3.2.2 General Principles

The pH meter is a good example of a **potentiometer** (a device that measures voltage at infinitesimal current flow). The basic principle of **potentiometry** (an electrochemical method of voltammetry at zero current) involves the use of an electrolytic cell composed of two electrodes dipped into a test solution. A voltage develops, which is related to the ionic concentration of the solution. Since the presence of current could alter the concentration of surrounding ions or produce irreversible reactions, this voltage is measured under conditions such that infinitesimal current (10^{-12} amperes or less) is drawn.

Four major parts of the pH system are needed: (1) reference electrode, (2) indicator electrode (pH sensitive), (3) voltmeter or amplifier that is capable of measuring small voltage differences in a circuit of very high resistance, and (4) the sample being analyzed (Fig. 13-1).

One notes that there are two electrodes involved in the measurement. Each of these electrodes is designed carefully to produce a constant, reproducible potential. Therefore, in the absence of other ions, the potential difference between the two electrodes is fixed and easily calculated. However, H_3O^+ ions in solution contribute a new potential across an ion-selective glass membrane built into the indicating electrode. This alters the potential difference between the two electrodes in a way that is proportional to H_3O^+ concentration. The new potential resulting from the combination of all individual potentials is called the **electrode potential** and is readily convertible into pH readings.

Hydrogen ion concentration (or more accurately, activity) is determined by the voltage that develops between the two electrodes. The **Nernst equation** relates the electrode response to the activity where:

$$E = E_o + 2.303\frac{RT}{NF}\log A \quad [8]$$

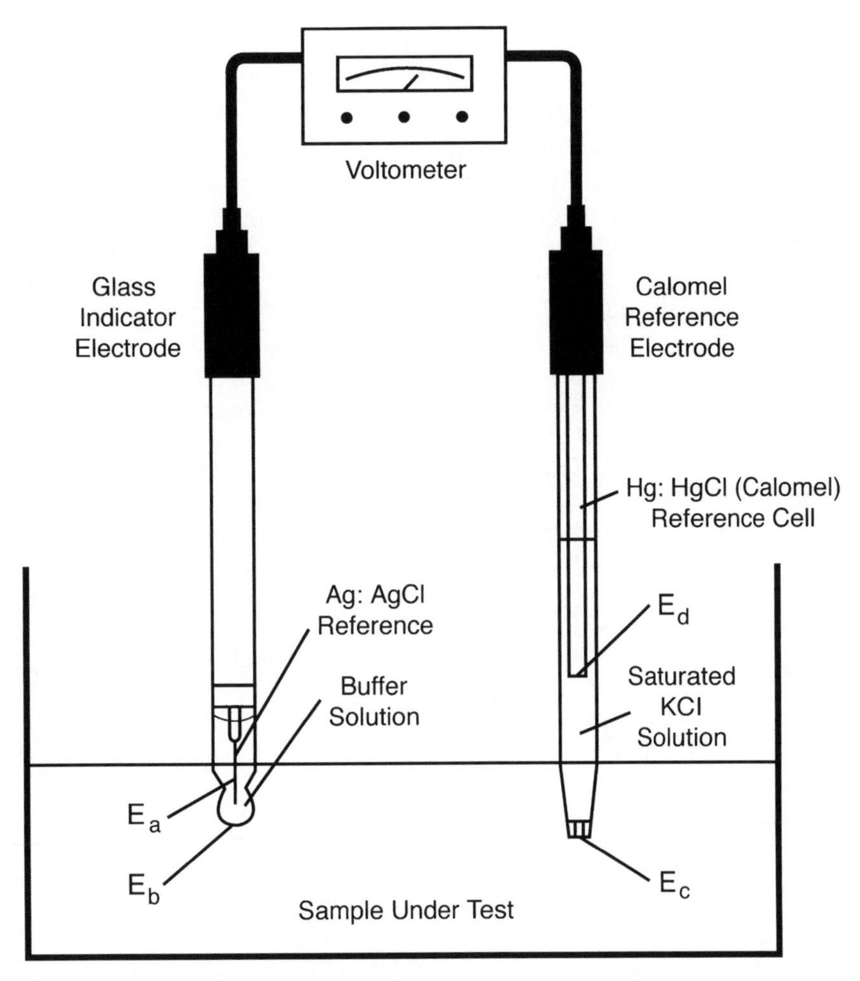

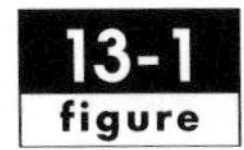

The measuring circuit of the potentiometric system. E_a: contact potential between Ag:AgCl electrode and inner liquid. E_a is independent of pH of the test solution but is temperature dependent. E_b: potential developed at the pH-sensitive glass membrane. E_b varies with the pH of the test solution and also with temperature. In addition to this potential the glass electrode also develops an asymmetry potential, which depends upon the composition and shape of the glass membrane. It also changes as the electrode ages. E_c: diffusion potential between saturated KCl solution and test sample. E_c is essentially independent of the solution under test. E_d: contact potential between calomel portion of electrode and KCl salt bridge. E_d is independent of the solution under test but is temperature dependent. [From (3), used with permission.]

where:

E = measured electrode potential

E_o = standard electrode potential, a constant representing the sum of the individual potentials in the system at a standard temperature, ion concentration, and electrode composition

R = universal gas constant, 8.313 Joules/degree/ g mole wt

F = Faraday constant, 96,490 coulombs/g equiv wt

T = absolute temperature (Kelvin)

N = number of charges on the ion

A = activity of the ion being measured

For monovalent ions (such as the hydronium ion) at 25 °C, the relationship of 2.303 *RT/F* is calculated to be 0.0591, as follows:

$$\frac{2.303 \times 8.316 \times 298}{96,490} = 0.0591 \quad [9]$$

Thus, voltage produced by the electrode system is a linear function of the pH, the electrode potential being essentially +59 mV (0.059 V) for each change of one pH unit. At neutrality (pH 7), the electrode potential is 0 mV. At pH 6, the electrode potential is +60 mV, while at pH 4, the electrode potential is +180 mV. Conversely, at pH 8, the electrode potential is −60 mV.

It must be emphasized that the above relationship between millivolts and pH exists only at 25 °C, and changes in temperature will erroneously alter the pH reading. For example, at 0 °C, the electrode potential is 54 mV, while at 100 °C it is 70 mV. Modern pH meters have a sensitive attenuator (temperature compensator) built into them in order to account for this effect of temperature.

13.3.2.3 Reference Electrode

The **reference electrode** is needed to complete the circuit in the pH system. This half cell is one of the most troublesome parts of the pH meter. Problems in obtaining pH measurements are often traced to a faulty reference electrode.

The **saturated calomel electrode** (Fig. 13-1) is the most common reference electrode. It is based upon the following reversible reaction:

$$Hg_2Cl_2 + 2e^- \rightleftarrows 2Hg + 2Cl^- \quad [10]$$

The $E_{0,25\,°C}$ for the saturated KCl salt bridge is +0.2444 V vs. a standard hydrogen electrode; the Nernst equation for the reaction is as follows:

$$E = E_0 - 0.059/2 \log[Cl^-]^2 \quad [11]$$

Thus, one observes that the potential is dependent upon the chloride ion concentration, which is easily regulated by the use of saturated KCl solution in the electrode.

A calomel reference electrode has three principal parts: (1) a platinum wire covered with a mixture of calomel (Hg_2Cl_2), (2) a filling solution (saturated KCl), and (3) a permeable junction through which the filling solution slowly migrates into the sample being measured. Junctions are made of ceramic or fibrous material. These junctions tend to clog up, causing a slow, unstable response and inaccurate results.

A less widely used reference electrode is the **silver–silver chloride electrode**. Because the calomel electrode is unstable at high temperatures (80°C) or in strongly basic samples (pH > 9), a silver–silver chloride electrode must be used for such application. It is a very reproducible electrode based upon the following reaction:

$$AgCl(s) + e \rightleftarrows Ag(s) + Cl^- \quad [12]$$

The internal element is a silver-coated platinum wire, the surface silver being converted to silver chloride by hydrolysis in hydrochloric acid. The filling solution is a mixture of 4 *M* KCl, saturated with AgCl that is used to prevent the AgCl surface of the internal element from dissolving. The permeable junction is usually of the porous ceramic type. Because of the relative insolubility of AgCl, this electrode tends to clog more readily than the calomel reference electrode. However, it is possible to obtain a double-junction electrode in which a separate inner body holds the Ag/AgCl internal element electrolyte and ceramic junction. An outer body containing a second electrolyte and junction isolates the inner body from the sample.

13.3.2.4 Indicator Electrode

The **indicator electrode** most commonly used in measuring pH today is referred to as the **glass electrode**. Prior to its development, the hydrogen electrode and the quinhydrone electrode were used.

The history of the glass electrode goes back to 1875, when it was suggested by Lord Kelvin that glass was an electrical conductor. Cremer discovered the glass electrode potential 30 years later when he observed that a thin glass membrane placed between two aqueous solutions exhibited an electrical potential sensitive to changes in acidity. Subsequently, the reaction was shown to be dependent upon the hydrogen ion concentration. These observations were of great importance in the development of the pH meter.

What is the design of the glass electrode? This electrode (Fig. 13-1) also has three principal parts: (1) a silver–silver chloride electrode with a mercury connection that is needed as a lead to the potentiometer; (2) a buffer solution consisting of 0.01 *N* HCl, 0.09 *N* KCl, and acetate buffer used to maintain a constant pH (E_a); and (3) a small pH-sensitive glass membrane for which the potential (E_o) varies with the pH of the test solution. In using the glass electrode as an indicator electrode in pH measurements, the measured potential (measured against the calomel electrode) is directly proportional to the pH as discussed earlier, $E = E_0 - 0.059\,\text{pH}$.

Conventional glass electrodes are suitable for measuring pH in the range of pH 1–9. However, this electrode is sensitive to higher pH, especially in the presence of sodium ions. Thus, equipment manufacturers have developed modern glass electrodes that are usable over the entire pH range of 0–14 and feature a very low sodium ion error, such as <0.01 pH at 25 °C.

13.3.2.5 Combination Electrodes

Today, most food analysis laboratories use **combination electrodes** that combine both the pH and reference electrodes along with the temperature sensing probe in a single unit or probe. These combination electrodes are available in many sizes and shapes from very small microprobes to flat surface probes, from all glass to plastic, and from exposed electrode tip to jacketed electrode tips to prevent glass tip breakage. Microprobes may be used to measure pH of very small systems such as inside a cell or a solution on a microscope slide. Flat surface electrode probes can be used to measure pH of semisolid and high-viscosity substances such as meat, cheese, and agar plates and small volumes as low as 10 μl.

13.3.2.6 Guidelines for Use of pH Meter

It is very important that the pH meter be operated and maintained properly. One should always follow the specific instructions provided by the manufacturer. For maximum accuracy, the meter should be standardized using two buffers (**two-point calibration**). Select two buffers of pH values about 3 pH units apart, bracketing that of the anticipated sample pH. The three standardization buffers used most widely in laboratories are a pH 4.0 buffer, a pH 7.0 buffer, and a pH 9.0 buffer (at 25 °C). These are the typical pink, yellow, and blue solutions found adjacent to pH meters in many laboratories.

When standardizing the pH electrode, follow manufacturer's instructions for **one-point calibration**; rinse thoroughly with distilled water and blot dry. Immerse electrode in the second buffer (e.g., pH 4) and perform a second standardization. This time, the pH meter slope control is used to adjust the reading to the correct value of the second buffer. Repeat these two steps, if necessary, until a value within 0.1 pH unit of the correct value of the second buffer is displayed. If this cannot be achieved, the instrument is not in good working condition. Electrodes should be checked, remembering that the reference electrode is more likely in need of attention. One should always follow the electrode manufacturer's specific directions for storage of a pH electrode. In this way, the pH meter is always ready to be used and the life of the electrodes is prolonged. One precaution that should be followed pertains to a calomel reference electrode. The storage solution level always should be at least 2 cm below the saturated KCl solution level in the electrode to prevent diffusion of storage solution into the electrode (Fig. 13-2).

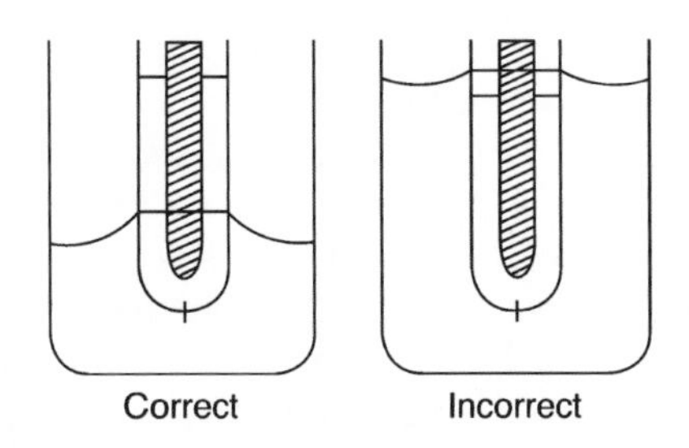

Correct and incorrect depth of calomel electrodes in solutions. [Reprinted with permission from (12). Copyright 1971 American Chemical Society.]

13.4 TITRATABLE ACIDITY

13.4.1 Overview and Principle

The titratable acidity measures the total acid concentration in a food. Food acids are usually organic acids, with citric, malic, lactic, tartaric, and acetic acids being the most common. However, inorganic acids such as phosphoric and carbonic (arising from carbon dioxide in solution) acids often play an important and even predominant role in food acidulation. The organic acids present in foods influence the flavor (i.e., tartness), color (though their impact on anthocyanin and other pH-influenced pigments), microbial stability (via inherent pH-sensitive characteristics of organisms), and keeping quality (arising from varying chemical sensitivities of food components to pH). The titratable acidity of fruits is used, along with sugar content, as an indicator of maturity (Sect. 13.4.6). While organic acids may be naturally present in the food, they also may be formed through fermentation or they may be added as part of a specific food formulation.

Titratable acidity is determined by neutralizing the acid present in a known quantity (weight or volume) of food sample using a standard base. The endpoint for titration is usually either a target pH or the color change of a pH-sensitive dye, typically phenolphthalein. The volume of titrant used, along with the normality of the base and the volume (or weight) of sample, is used to calculate the titratable acidity, expressed in terms of the predominant organic acid.

13.4.2 General Considerations

Many food properties correlate better with pH than with acid concentration. The pH is also used to determine the endpoint of an acid–base titration. The pH determination can be achieved directly with a pH meter, but more commonly using a pH-sensitive dye. In some cases, the way pH changes during titration can lead to subtle problems. Some background in acid theory is necessary to fully understand titration and to appreciate the occasional problems that might arise.

13.4.2.1 Buffering

Although pH can hypothetically range from −1 to 14, pH readings below 1 are difficult to obtain due to incomplete dissociation of hydrogen ions at high acid concentrations. At 0.1 *N*, strong acids are assumed to be fully disassociated. Therefore, for titrations involving strong acids, fully dissociated acid is present at all titrant concentrations; the pH at any point in the titration is equal to the hydrogen ion concentration of the remaining acid (Fig. 13-3).

By contrast, all food acids are weak acids. Less than 3% of their ionizable hydrogens are dissociated from the parent molecule. When free hydrogen ions are removed through titration, new hydrogen ions can arise from other previously undissociated parent molecules. This tends to cushion the solution from abrupt changes in pH. This property of a solution to resist change in pH is termed buffering. **Buffering** occurs in foods whenever a weak acid and its salt are present in the same medium. Because of buffering, a graph of pH vs. titrant concentration is more complex for weak acids than for strong acids. However, this relationship can be estimated by the **Henderson–Hasselbalch equation**.

$$\mathrm{pH} = \mathrm{p}K_\mathrm{a} + \log \frac{[\mathrm{A}^-]}{[\mathrm{HA}]} \qquad [13]$$

[HA] represents the concentration of undissociated acid. $[\mathrm{A}^-]$ represents the concentration of its salt, also known as the **conjugated base**. The conjugated base is equal in concentration to the **conjugated acid** $[\mathrm{H_3O^+}]$.

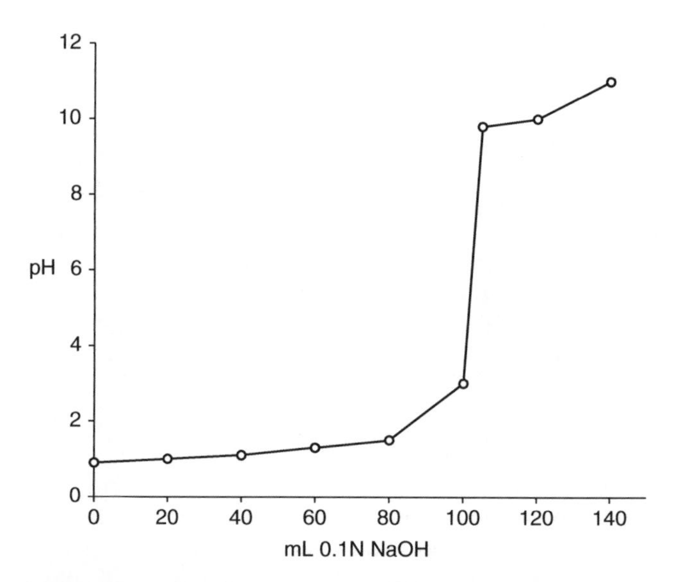

Titration of a strong acid with strong base. The pH at any point in the titration is dictated by the hydrogen ion concentration of the acid remaining after partial neutralization with base.

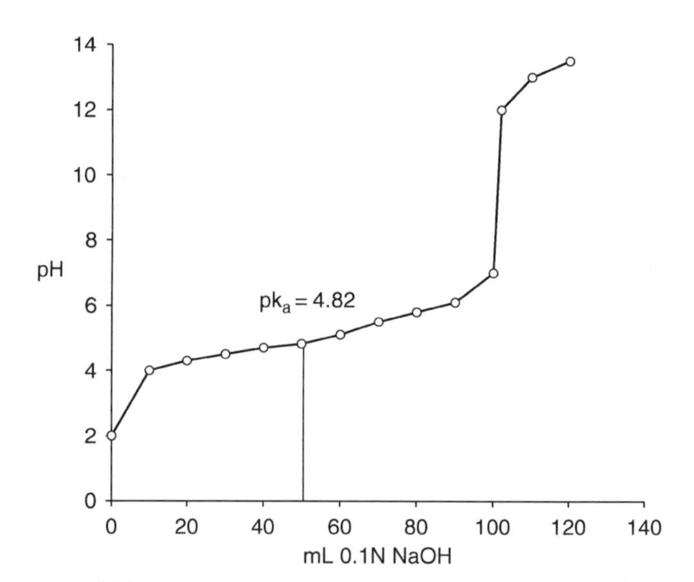

13-4 figure
Titration of a weak monoprotic acid with a strong base. A buffering region is established around the pK_a (4.82). The pH at any point is described by the Henderson–Hasselbalch equation.

The $\mathbf{pK_a}$ is the pH at which equal quantities of undissociated acid and conjugated base are present. The equation indicates that maximum buffering capacity will exist when the pH equals the pK_a. A graph showing the titration of 0.1 *N* acetic acid with 0.1 *N* NaOH illustrates this point (Fig. 13-4).

Di- and triprotic acids will have two and three buffering regions, respectively. A pH vs. titrant graph of citric acid is given in Fig. 13-5. If the pK_a steps

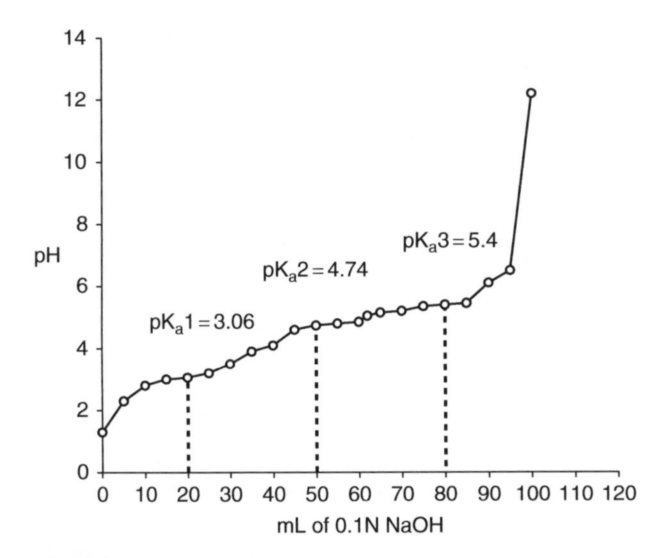

13-5 figure

Titration of a weak polyprotic acid with a strong base. Buffering regions are established around each pK_a. The Henderson–Hasselbalch equation can predict the pH for each pK_a value if pK_a steps are separated by more than three units. However, complex transition mixtures between pK_a steps make simple calculations of transition pH values impossible.

13-5 table pK_a **Values for Some Acids Important in Food Analysis**

Acid	pK_a^1	pK_a^2	pK_a^3
Oxalic	1.19	4.21	–
Phosphoric	2.12	7.21	12.30
Tartaric	3.02	4.54	–
Malic	3.40	5.05	–
Citric	3.06	4.74	5.40
Lactic	3.86	–	–
Ascorbic	4.10	11.79	–
Acetic	4.76	–	–
Potassium acid phthalate	5.40	–	–
Carbonic	6.10	10.25	–

in polyprotic acids differ by three or more pK_a units, then the Henderson–Hasselbalch equation can predict the plateau corresponding to each step. However, the transition region between steps is complicated by the presence of protons and conjugate bases arising from other disassociation state(s). Consequently, the Henderson–Hasselbalch equation breaks down near the equivalence point between two pK_a steps. However, the pH at the equivalence point is easily calculated. The pH is simply $(pK_a1 + pK_a2)/2$. Table 13-5 lists pK_a values of acids important in food analysis.

Precise prediction of pH by the Henderson–Hasselbalch equation requires that all components form ideal solutions. An approximation to ideal solutions occurs at infinite dilution for all active components. However, real solutions may not behave ideally. For such solutions, the Henderson–Hasselbalch equation may only provide a good estimate of pH.

13.4.2.2 Potentiometric Titration

At the **equivalence point** in a titration, the number of acid equivalents exactly equals the number of base equivalents, and total acid neutralization is achieved. As the equivalence point is approached, the denominator [HA] in the Henderson–Hasselbalch equation becomes insignificantly small and the quotient $[A^-]/[HA]$ increases exponentially. As a result, the solution pH rapidly increases and ultimately approaches the pH of the titrant. The exact equivalent point is the halfway mark on this slope of abrupt pH increase. The use of a pH meter to identify the endpoint is called the **potentiometric method** for determining titratable acidity. The advantage of determining the equivalence point potentiometrically is that the precise equivalence point is identified. Since a rapid change in pH (and not some final pH value per se) signals the end of titration, accurate calibration of the pH meter is not even essential. However, in order to identify the equivalence point, a careful record of pH vs. titrant must be kept. This and the physical constraints of pH probes and slow response with some electrodes make the potentiometric approach somewhat cumbersome.

13.4.2.3 Indicators

For simplicity in routine work, an indicator solution is often used to approximate the equivalence point. This approach tends to overshoot the equivalence point by a small amount. When indicators are used, the term **endpoint** or **colorimetric endpoint** is substituted for equivalence point. This emphasizes that the resulting values are approximate and dependent on the specific indicator. Phenolphthalein is the most common indicator for food use. It changes from clear to red in the pH region 8.0–9.6. Significant color change is usually present by pH 8.2. This pH is termed the **phenolphthalein endpoint**.

A review of pK_a values in Table 13-5 indicates that naturally occurring food acids do not buffer in the region of the phenolphthalein endpoint. However, phosphoric acid (used as an acidulant in some soft drinks) and carbonic acid (carbon dioxide in aqueous solution) do buffer at this pH. Consequently, taking the solution from the true equivalence point to the endpoint may require a large amount of titrant when quantifying these acids. Indistinct endpoints and erroneously large titration values may result. When these acids are titrated, potentiometric analysis is usually preferred. Interference by CO_2 can be removed by boiling the sample and titrating the remaining acidity to a phenolphthalein endpoint.

Deeply colored samples also present a problem for endpoint indicators. When colored solutions obscure the endpoint, a potentiometric method is normally used. For routine work, pH vs. titrant data are not collected. Samples are simply titrated to pH 8.2 (the phenolphthalein endpoint). Even though this is a potentiometric method, the resulting value is an endpoint and not the true equivalence point, since it simply reflects the pH value for the phenolphthalein endpoint.

A pH of 7 may seem to be a better target for a potentiometric endpoint than 8.2. This pH, after all, marks the point of true neutrality on the pH scale. However, once all acid has been neutralized, the conjugate base remains. As a result, the pH at the equivalence point is slightly greater than 7. Confusion also might arise if pH 7 was the target for colored samples and pH 8.2 was the target for noncolored samples.

Dilute acid solutions (e.g., vegetable extracts) require dilute solutions of standard base for optimal accuracy in titration. However, a significant volume of dilute alkali may be required to take a titration from the equivalence point to pH 8.2. Bromothymol blue is sometimes used as an alternative indicator in low-acid situations. It changes from yellow to blue in the pH range 6.0–7.6. The endpoint is usually a distinct green. However, endpoint identification is somewhat more subjective than the phenolphthalein endpoint.

Indicator solutions rarely contain over a few tenths percent dye (wt/vol). All indicators are either weak acids or weak bases that tend to buffer in the region of their color change. If added too liberally, they can influence the titration by conferring their own acid/base character to the sample under analysis. Therefore, indicator solutions should be held to the minimum amount necessary to impart effective color. Typically, two to three drops of indicator are added to the solution to be titrated. The lower the indicator concentration, the sharper will be the endpoint.

13.4.3 Preparation of Reagents

13.4.3.1 Standard Alkali

Sodium hydroxide (NaOH) is the most commonly used base in titratable acidity determinations. In some ways, it appears to be a poor candidate for a **standard base**. Reagent grade NaOH is very hygroscopic and often contains significant quantities of insoluble sodium carbonate (Na_2CO_3). Consequently, the normality of working solutions is not precise, so each new batch of NaOH must be standardized against an acid of known normality. However, economy, availability, and long tradition of use for NaOH outweigh these shortcomings. Working solutions are normally made from a stock solution containing 50% sodium hydroxide in water (wt/vol). Sodium carbonate is essentially insoluble in concentrated alkali and gradually precipitates out of solution over the first 10 days of storage.

The NaOH can react with dissolved and atmospheric CO_2 to produce new Na_2CO_3. This reduces alkalinity and sets up a carbonate buffer that can obscure the true endpoint of a titration. Even just CO_2 and water react to form buffering compounds and generate hydrogen ions, as shown in the following equations:

$$H_2O + CO_2 \leftrightarrow H_2CO_3(\text{carbonate}) \quad [14]$$

$$H_2CO_3 \leftrightarrow H^+ + HCO_3^-(\text{bicarbonate}) \quad [15]$$

$$HCO_3^- \leftrightarrow H^+ + CO_3^{-2} \quad [16]$$

Therefore, CO_2 should be removed from water prior to making the stock solution. This can be achieved by purging water with CO_2-free gas for 24 h or by boiling distilled water for 20 min and allowing it to cool before use. During cooling and long-term storage, air (with accompanying CO_2) will be drawn back into the container. Carbon dioxide can be stripped from reentering air with a soda lime (20% NaOH, 65% CaO, 15% H_2O) or **ascarite trap** (NaOH-coated silica base). Air passed through these traps also can be used as purge gas to produce CO_2-free water.

Stock alkali solution of 50% in water is approximately 18 *N*. A working solution is made by diluting stock solution with **CO_2-free water**. There is no ideal container for strong alkali solutions. Glass and plastic are both used, but each has its drawbacks. If a glass container is used it should be closed with a rubber or thick plastic closure. Glass closures should be avoided since, over time, strong alkali dissolves glass, resulting in permanent fusion of the contact surfaces. Reaction with glass also lowers the normality of the alkali. These liabilities also are relevant to long-term storage of alkali in burettes. NaOH has a low surface tension. This predisposes to leakage around the stopcock. Stopcock leakage during titration will produce erroneously high acid values. Slow evaporation of titrating solution from the stopcock valve during long periods of nonuse also creates a localized region of high pH with ensuing opportunities for fusion between the stopcock and burette body. After periods of nonuse, burettes should be emptied, cleaned, and refilled with fresh working solution.

Long-term storage of alkali in plastic containers also requires special vigilance because CO_2 permeates freely through most common plastics. Despite this shortcoming, plastic containers are usually preferred for long-term storage of stock alkali solutions. Whether glass or plastic is used for storage, working solutions should be restandardized weekly to correct for alkalinity losses arising from interactions with glass and CO_2.

13.4.3.2 Standard Acid

The impurities and hygroscopic nature of NaOH make it unsuitable as a primary standard. Therefore, NaOH titrating solutions must be standardized against a **standard acid**. **Potassium acid phthalate** (KHP) is commonly used for this purpose.

KHP's single ionizable hydrogen ($pK_a = 5.4$) provides very little buffering at pH 8.2. It can be manufactured in very pure form, it is relatively non-hygroscopic, and it can be dried at 120 °C without decomposition or volatilization. Its high molecular weight also favors accurate weighing.

KHP should be dried for 2 h at 120 °C and allowed to cool to room temperature in a desiccator immediately prior to use. An accurately measured quantity of KHP solution is titrated with a base of unknown normality. The base is always the titrant. CO_2 is relatively insoluble in acidic solutions. Consequently, stirring an acid sample to assist in mixing will not significantly alter the accuracy of the titration.

13.4.4 Sample Analysis

A number of official methods exist for determining titratable acidity in various foods (1). However, determining titratable acidity on most samples is relatively routine, and various procedures share many common steps. An aliquot of sample (often 10 ml) is titrated with a standard alkali solution (often 0.1 *N* NaOH) to a phenolphthalein endpoint. Potentiometric endpoint determination is used when sample pigment makes use of a color indicator impractical.

Typical titration setups are illustrated in Fig. 13-6 for potentiometric and colorimetric endpoints. Erlenmeyer flasks are usually preferred for samples when endpoint indicators are used. A magnetic stirring bar may be used; but mixing the sample with hand swirling is usually adequate. When hand mixing is used the sample flask is swirled with the right hand. The stopcock is positioned on the right side. Four fingers on the left hand are placed behind the stopcock valve and the thumb is placed on the front of the valve. Titrant is dispensed at a slow, uniform rate until the endpoint is approached and then added dropwise until the endpoint does not fade after standing for some predetermined period of time, usually 5–10 s.

The bulkiness of the pH electrode usually demands that beakers be used instead of Erlenmeyer flasks when samples are analyzed potentiometrically. Mixing is almost always achieved through magnetic stirring, and loss of sample through splashing is more likely with beakers than with Erlenmeyer flasks. Otherwise, titration practices are identical to

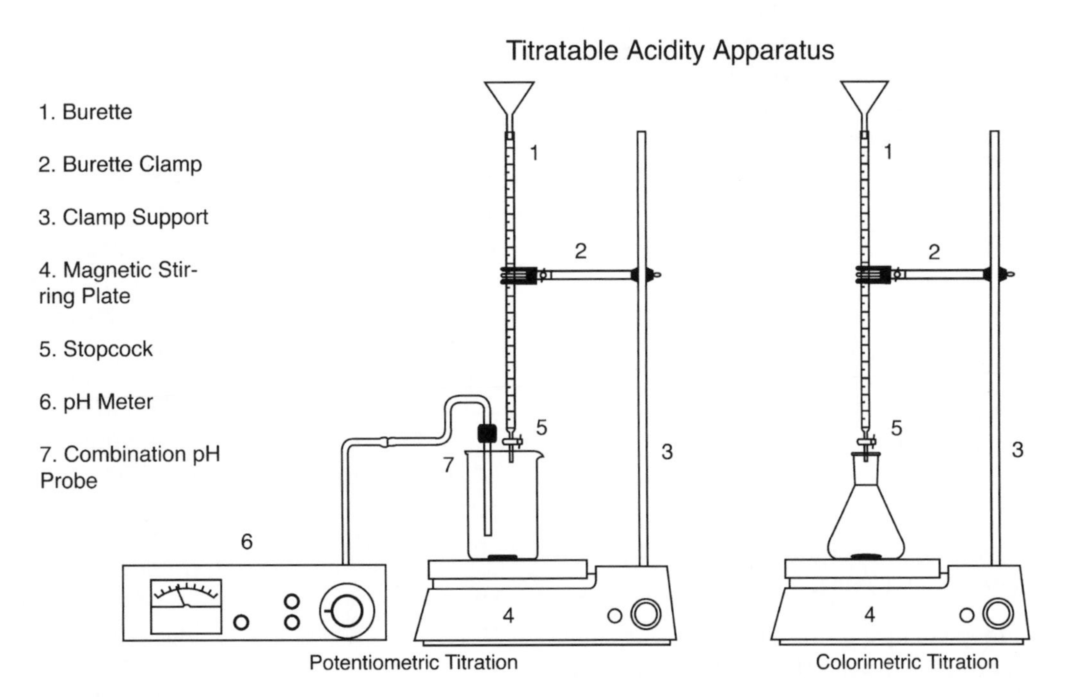

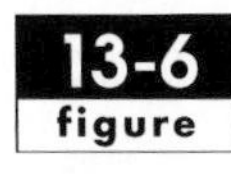
13-6 figure

Titratable acidity apparatus.

those described previously for colorimetric endpoint titrations.

Problems may arise when concentrates, gels, or particulate-containing samples are titrated. These matrices prevent rapid diffusion of acid from densely packed portions of sample material. This slow diffusion process results in a fading endpoint. Concentrates can simply be diluted with CO_2-free water. Titration then is performed, and the original acid content is calculated from dilution data. Starch and similar weak gels often can be mixed with CO_2-free water, stirred vigorously, and titrated in a manner similar to concentrates. However, some pectin and food gum gels require mixing in a blender to adequately disrupt the gel matrix. Thick foams are occasionally formed in mixing. Antifoam or vacuum can be used to break the foams.

Immediately following processing, the pH values of particulate samples often vary from one particulate piece to another. Acid equilibration throughout the entire mass may require several months. As a result, particulate-containing foods should be liquefied in a blender before titrating. The comminuting process may incorporate large quantities of air. Air entrapment makes the accuracy of volumetric measurements questionable. Aliquots often are weighed when air incorporation may be a problem.

13.4.5 Calculation of Titratable Acidity

In general chemistry, acid strength is frequently reported in normality (equivalents per liter) and can be calculated using the equation $N_{\text{titrant}} \times V_{\text{titrant}} = N_{\text{sample}} \times V_{\text{sample}}$, where N is normality and V is volume (often in milliliters). However, food acids are usually reported as percent of total sample weight. Thus, the equation for titratable acidity is as follows:

$$\%\,\text{acid (wt/wt)} = \frac{N \times V \times \text{Eq wt}}{W \times 1000} \times 100 \qquad [17]$$

where:

N = normality of titrant, usually NaOH (mEq/ml)
V = volume of titrant (ml)
Eq. wt. = equivalent weight of predominant acid (mg/mEq)
W = mass of sample (g)
1000 = factor relating mg to grams (mg/g) (1/10 = 100/1000)

Note that the normality of the titrant is expressed in milliequivalents (mEq) per ml, which is a typical way of reporting normality for small volumes. This value is numerically the same as equivalents/liter. Also note that it is easier to report sample mass in grams instead of milligrams, so multiplying sample mass by the factor of 1000 mg/g allows units to cancel.

For routine titration of fruit juices, milliliters can be substituted for sample weight in grams, as shown in Equations [15] and [16]. Depending on the soluble solids content of the juice, the resulting acid values will be high by 1–6%. However, this is common practice.

$$\%\,\text{acid (wt/vol)} = \frac{N \times V_1 \times \text{Eq wt}}{V_2 \times 1000} \times 100 \qquad [18]$$

or

$$\%\,\text{acid (wt/vol)} = \frac{N \times V_1 \times \text{Eq wt}}{V_2 \times 10} \qquad [19]$$

where:

N = normality of titrant, usually NaOH (mEq/ml)
V_1 = volume of titrant (ml)
Eq. wt. = Equivalent weight of predominant acid (mg/mEq)
V_2 = volume of sample (ml)
1000 = factor relating mg to grams (mg/g) (1/10 = 100/1000)

For example, if it takes 17.5 ml of 0.085 N NaOH to titrate a 15-ml sample of a juice, the total titratable acidity of that juice, expressed as percent citric acid (molecular weight = 192; equivalent weight = 64), would be 0.635%, wt/vol, citric acid:

$$\%\,\text{acid (wt/vol)} = \frac{(0.085)(17.5)(64)}{(15)(10)} = 0.635\% \qquad [20]$$

Notice that the equivalent weight of anhydrous (vs. hydrous) citric acid always is used in calculating and reporting the results of titration.

13.4.6 Acid Content in Food

Most foods are as chemically complex as life itself. As such, they contain the full complement of Krebs cycle acids (and their derivatives), fatty acids, and amino acids. Theoretically, all of these contribute to titratable acidity. Routine titration cannot differentiate between individual acids. Therefore, titratable acidity is usually stated in terms of the **predominant acid**. For most foods this is unambiguous. In some cases, two acids are present in large concentrations, and the predominant acid may change with maturity. In grapes, malic acid often predominates prior to maturity while tartaric acid typically predominates in the ripe fruit. A similar phenomenon is observed with malic and citric acids in pears. Fortunately, the equivalent weights of common food acids are fairly similar. Therefore, percent titratable acidity is not substantially affected by mixed predominance or incorrect selection of the predominant acid.

13-6 table Acid Composition and °Brix of Some Commercially Important Fruits

Fruit	*Principal Acid*	*Typical Percent Acid*	*Typical °Brix*
Apples	Malic	0.27–1.02	9.12–13.5
Bananas	Malic/citric (3:1)	0.25	16.5–19.5
Cherries	Malic	0.47–1.86	13.4–18.0
Cranberries	Citric	0.9–1.36	
	Malic	0.70–0.98	12.9–14.2
Grapefruit	Citric	0.64–2.10	7–10
Grapes	Tartaric/malic (3:2)	0.84–1.16	13.3–14.4
Lemons	Citric	4.2–8.33	7.1–11.9
Limes	Citric	4.9–8.3	8.3–14.1
Oranges	Citric	0.68–1.20	9–14
Peaches	Citric	1–2	11.8–12.3
Pears	Malic/citric	0.34–0.45	11–12.3
Pineapples	Citric	0.78–0.84	12.3–16.8
Raspberries	Citric	1.57–2.23	9–11.1
Strawberries	Citric	0.95–1.18	8–10.1
Tomatoes	Citric	0.2–0.6	4

The range of acid concentrations in foods is very broad. Acids can exist at levels below detection limits or they can be the preeminent substance present in the food. The contribution of acids to food flavor and quality is not told by acid content alone. The tartness of acids is reduced by sugars. Consequently, the **Brix/acid ratio** (often simply called ratio) is usually a better predictor of an acid's flavor impact than Brix or acid alone. Acids tend to decrease with the maturity of fruit while sugar content increases. Therefore, the Brix/acid ratio often is often an index of fruit maturity. For mature fruit, this ratio can also be affected by climate, variety, and horticultural practices. Table 13-6 gives typical acid composition and sugar levels for many commercially important fruits at maturity. Citric and malic acids are the most common acids in fruits and most vegetables; however, leafy vegetables also may contain significant quantities of oxalic acid. Lactic acid is the most important acid in dairy foods for which titratable acidity is commonly used to monitor the progress of lactic acid fermentations in cheese and yogurt production (15).

Organic acids contribute to the refractometer reading of soluble solids. When foods are sold on the basis of pound solids, Brix readings are sometimes corrected for acid content. For citric acid, 0.20 °Brix is added for each percent titratable acidity.

13.4.7 Volatile Acidity

In acetic acid fermentations, it is sometimes desirable to know how much acidity comes from the acetic acid and how much is contributed naturally by other acids in the product. This can be achieved by first performing an initial titration to measure **titratable acidity** as an indicator of total acidity. The acetic acid is then boiled off, the solution is allowed to cool, and a second titration is performed to determine the **fixed acidity**. The difference between fixed and total acidity is the **volatile acidity**. A similar practice is used sometimes in the brewing industry to separate acidity due to dissolved CO_2 from fixed acids. Fixed acids are titrated after CO_2 is removed by low heat (40 °C) and gentle agitation.

13.4.8 Other Methods

High-performance liquid chromatography (HPLC) and electrochemistry both have been used to measure acids in food samples. Both methods allow identification of specific acids. HPLC uses refractive index, ultraviolet, or for some acids electrochemical detection. Ascorbic acid has a strong electrochemical signature and significant absorbance at 265 nm. Significant absorbance of other prominent acids does not occur until 200 nm or below.

Many acids can be measured with such electrochemical techniques as voltammetry and polarography. In ideal cases, the sensitivity and selectivity of electrochemical methods are exceptional. However, interfering compounds often reduce the practicality of electrochemical approaches.

Unlike titration, chromatographic and electrochemical techniques do not differentiate between an acid and its conjugate base. Both species inevitably exist side by side as part of the inherent food-buffer system. As a result, acids determined by instrumental methods may be 50% higher than values determined by titration. It follows that Brix/acid ratios can be based only on acid values determined by titration.

13.5 SUMMARY

Organic acids have a pronounced impact on food flavor and quality. Unlike strong acids that are fully dissociated, food acids are only partially ionized. Some properties of foods are affected only by this ionized fraction of acid molecules while other properties are affected by the total acid content. It is impractical to quantify only free hydronium ions in solution by chemical methods. Once the free ions are removed by chemical reaction, others arise from previously undissociated molecules. Indicator dyes, which change color depending on the hydronium ion environment, exist but they only identify when a certain pH threshold has been achieved and do not stoichiometrically quantify free hydronium ions. The best that can be done is to identify the secondary effect of the hydronium ion environment on some property of the system such as the color of the indicator dyes or the electrochemical potential of the medium. The pH meter measures the change in electrochemical potential established by the hydronium ion across a semipermeable glass membrane on an indicator electrode. The shift in the indicator electrode potential is indexed against the potential of a reference electrode. The difference in millivolt reading between the two electrodes can be converted into pH using the Nernst equation. The hydronium ion concentration can be back-calculated from pH using the original definition of pH as the negative log of hydrogen ion concentration. Buffer solutions of any pH can be created using the Henderson–Hasselbalch equation. However, the predictions of all these equations are somewhat approximate unless the activity of acids and conjugate bases is taken into account.

Titratable acidity provides a simple estimate of the total acid content of a food. In most cases, it is only an estimate since foods often contain many acids that cannot be differentiated through titration. Titratable acidity is not a good predictor of pH, since pH is a combined function of titratable acid and conjugate base. Instrumental methods such as HPLC and electrochemical approaches measure acids and their conjugate bases as a single compound and, therefore, tend to produce acid contents that are higher than those determined by titration. Titratable acidity, somewhat curiously, is a better predictor of tartness than the concentration of free hydronium ions as reflected by pH. The perception of tartness is strongly influenced by the presence of sugars. Indicator dyes are used commonly to identify the endpoint of acidity titrations, although pH meters can be used in critical work or when sample color makes indicators impractical.

13.6 STUDY QUESTIONS

1. Explain the theory of potentiometry and the Nernst equation as they relate to being able to use a pH meter to measure H^+ concentration.
2. Explain the difference between a saturated calomel electrode and a silver–silver chloride electrode; describe the construction of a glass electrode and a combination electrode.
3. You return from a 2-week vacation and ask your lab technician about the pH of the apple juice sample you gave him or her before you left. Having forgotten to do it before, the technician calibrates a pH meter with one standard buffer stored next to the meter and then reads the pH of the sample of unpasteurized apple juice immediately after removing it from the refrigerator (40 °F), where it has been stored for 2 weeks. Explain the reasons why this stated procedure could lead to inaccurate or misleading pH values.
4. For each of the food products listed below, what acid should be used to express titratable acidity?
   (a) Orange juice
   (b) Yogurt
   (c) Apple juice
   (d) Grape juice
5. What is a "Brix/acid ratio," and why is it often used as an indicator of flavor quality for certain foods, rather than simply Brix or acid alone?
6. How would you recommend determining the endpoint in the titration of tomato juice to determine the titratable acidity? Why?
7. The titratable acidity was determined by titration to a phenolphthalein endpoint for a boiled and unboiled clear carbonated beverage. Which sample would you expect to have a higher calculated titratable acidity? Why? Would you expect one of the samples to have a fading endpoint? Why?
8. Why and how is an ascarite trap used in the process of determining titratable acidity?
9. Why is volatile acidity useful as a measure of quality for acetic acid fermentation products, and how is it determined?
10. What factors make KHP a good choice as a standard acid for use in standardizing NaOH solutions to determine titratable acidity?
11. Could a sample that is determined to contain 1.5% acetic acid also be described as containing 1.5% citric acid? Why or why not?
12. An instructor was grading lab reports of students who had determined the titratable acidity of grape juice. One student had written that the percent titratable acidity was 7.6% citric acid. Give two reasons why the answer was marked wrong. What would have been a more reasonable answer?

13.7 PRACTICE PROBLEMS

1. How would you prepare 500 ml of 0.1 *M* NaH_2PO_4 starting with the solid salt?
2. Starting with reagent grade sulfuric acid (36 *N*), how would you prepare 1 L of 2 *M* H_2SO_4? How many milliliters of 10 *N* NaOH would be required to neutralize this acid?
3. How would you prepare 250 ml of 2 *N* HCl starting with reagent grade HCl (12 *N*)?
4. How would you prepare 1 L of 0.04 *M* acetic acid starting with reagent grade HOAc (17 *M*)?
5. How would you prepare 150 ml of 10% NaOH?
6. If about 8.7 ml of saturated NaOH is required to prepare 1 L of 0.1 *N* NaOH, how would you prepare 100 ml of 1 *N* NaOH?
7. What is the normality of a (1 + 3) HCl solution?
8. You are performing a titration on duplicate samples and duplicate blanks that require 4 ml of 1 *N* NaOH per titration sample. The lab has 10% NaOH and saturated NaOH. Choose one and describe how you would prepare the needed amount of NaOH solution.
9. Is a 1% HOAc solution the same as a 0.1 *M* solution? Show calculations.

10. Is a 10% NaOH solution the same as a 1 N solution? Show calculations.
11. What is the normality of a 40% NaOH solution?
12. You are performing duplicate titrations on five samples that require 15 ml of 6 N HCl each. How would you prepare the needed solution from reagent grade HCl?
13. What is the pH of a 0.057 M HCl solution?
14. Vinegar has a $[H^+]$ of 1.77×10^{-3} M. What is the pH? What is the major acid found in vinegar, and what is its structure?
15. Orange juice has a $[H^+]$ of 2.09×10^{-4} M. What is the pH? What is the major acid found in orange juice and what is its structure?
16. A sample of vanilla yogurt has a pH of 3.59. What is the $[H^+]$? What is the major acid found in yogurt and what is its structure?
17. An apple pectin gel has a pH of 3.30. What is the $[H^+]$? What is the major acid found in apples, and what is its structure?
18. How would you make 100 ml of a 0.1 N solution of KHP?
19. How would you make 100 ml of a citrate buffer that is 0.1 N in both citric acid (anhydrous) and potassium citrate $KH_2C_6H_5O_7$ (MW 230.22)?
20. What would be the pH of the 0.1 N citrate buffer described in Problem 19?
21. How would you make 1 L of 0.1 N NaOH solution from an 18 N stock solution?
22. A stock base solution assumed to be 18 N was diluted to 0.1 N. KHP standardization indicated that the normality of the working solution was 0.088 N. What was the actual normality of the solution?
23. A 20-ml sample of juice requires 25 ml of 0.1 N NaOH titrant. What would be the percent acid if the juice is (1) apple juice, (2) orange juice, (3) grape juice?
24. A lab analyzes a large number of orange juice samples. All juice samples will be 10 ml. It is decided that 5 ml of titrant should equal 1% citric acid. What base normality should be used?
25. A lab wishes to analyze apple juice. They would like each milliliter of titrant to equal 0.1% malic acid. Sample aliquots will all be 10 ml. What base normality should be used?

Answers

1. The question asks for 500 ml of a 0.1 M NaH_2PO_4 solution. The molecular weight of this salt is 120 g/mol. You can use Equation [1] to solve this problem.

$$\begin{aligned}&(500 \text{ ml of sodium phosphate})\\&\times(\text{molarity of sodium phosphate})\\&= \text{millimoles of sodium phosphate}\end{aligned}$$

$$\frac{(500\,\text{ml})(0.1\,M)(120\,\text{g/mol})}{1000\,\text{ml/L}} = 6\,\text{g}\,NaH_2PO_4$$

2(a) 1000 ml of 2 M H_2SO_4 is required. Reagent grade H_2SO_4 is 36 N and 18 M. Therefore,

$$(18\,\text{M})(x\,\text{ml}) = (2\,\text{M})(1{,}000\text{ml}).$$

$$x\,\text{ml} = 111.1\,\text{ml of conc. acid diluted to 1L.}$$

(When diluting concentrated acids, always add concentrated acid to about one half the final volume of water to dilute and to dissipate the heat generated by mixing. Never add the water to the concentrated acid!)

(b)

$$\begin{aligned}&(1000\,\text{ml}\,H_2SO_4)(2\,M\,H_2SO_4)(2\,N/1\,M)\\&= (x\,\text{ml NaOH})(10\,N\,\text{NaOH})\\&x\,\text{ml} = 400\,\text{ml NaOH}\end{aligned}$$

3. Using Equation [1]:

$$\begin{aligned}(250\,\text{ml})(2\,N\,\text{HCl}) &= (x\,\text{ml})(12\,N\,\text{HCl})\\x\,\text{ml} &= 41.67\,\text{ml of conc. HCl}\\&\quad\text{diluted with water to 250 ml}\end{aligned}$$

4. Using Equation [1]:

$$(0.04\,M\,\text{HOAc})(1\,\text{L})(1000\,\text{ml/L}) = (x\,\text{ml})(17\,M\,\text{HOAc})$$

$$x\,\text{ml} = 2.35\,\text{ml conc. acetic acid that is diluted to 1 L}$$

5. Usually with a solid starting material like NaOH, the percent is a weight-to-volume percent (or percent wt/vol). Therefore, 10% NaOH = 10 g NaOH/100 ml of solution. Thus, 150 ml of 10% NaOH requires 15 g NaOH = 15 g NaOH/150 ml = 10% NaOH.
6. If about 8.7 ml of saturated NaOH diluted to 1 L gives 0.1 N, this equals $(0.1\,N)(1000\,\text{ml}) = 100$ mEq. Since both solutions contain the same number of milliequivalents, they both must require the same volume of saturated NaOH, 8.7 ml.
7. The convention (1 + 3) HCl, as used for some analytical food methods (e.g., AOAC Methods), means 1 part concentrated acid and 3 parts distilled water, or a 1-in-4 dilution. Starting with concentrated HCl at 12 N, a 1-in-4 dilution will yield $(1/4)(12\,N\,\text{HCl}) = 3.00\,N$ HCl.
8. Four titrations of 4 ml each will be performed requiring a total of about 16 ml of 1 N NaOH. For simplicity, 20 ml of 1 N NaOH can be prepared. If a 10% NaOH stock solution is used, then

$$10\,\text{g NaOH/100 ml} = 100\,\text{g NaOH/L} = 2.5\,N\,\text{NaOH}$$

$$\begin{aligned}(20\,\text{ml})(1\,N\,\text{NaOH}) &= (x\,\text{ml})(2.5\,N)\\x\,\text{ml} &= 8\,\text{ml of 10\% diluted to 20 ml}\\&\quad\text{with distilled water}\end{aligned}$$

If saturated NaOH is used, remember from Problem 6 that approximately 8.7 ml of saturated NaOH diluted to 100 ml yields 1.0 N. Therefore, 1.87 ml or 2 ml of saturated NaOH diluted to 20 ml with distilled water will yield about 1 N NaOH

9. 1% HOAc = 1 g HOAc/100 ml = (10 g HOAc/L)/(60.05 g/mol) = 0.17 mole/L = 0.17 M

and

$$0.1\,M\,\text{HOAc} = 0.1\,\text{mol HOAc/L} \times 60.05\,\text{g/mol}$$

$$6.005\,\text{g HOAc/L} = 0.60\,\text{g/100 ml} = 0.60\%\,\text{HOAc}$$

Therefore, the two acetic acid solutions are not the same, differing by a factor of about 2.

10. 10% NaOH = 10 g NaOH/100 ml = 100 g NaOH/L
100 g NaOH/(40 g/mol)/L = 2.5 *M* NaOH

and

$$1\,N\ \text{NaOH} = 1\ \text{mol NaOH/L} = 40\ \text{g NaOH/L}$$
$$4\ \text{g NaOH/100 ml} = 4\%\ \text{NaOH}$$

No, the solutions are not the same.

11. 40% NaOH = 40 g NaOH/100 ml = 400 g NaOH/L
(400 g NaOH/L)/(40 g NaOH/mol) = 10 mol/L = 10 *N*

12. A total of (5 samples)(2 duplicates)(15 ml) = 150 ml of 6 *N* HCl

$$(150\ \text{ml})(6\,N\ \text{HCl}) = (x\ \text{ml})(12\,N\ \text{HCl})$$
$$x\ \text{ml} = 75\ \text{ml concentrated HCl diluted with distilled water to 150 ml}$$

13. Since HCl is a strong acid, it will be completely dissociated. Therefore, the molar concentration of HCl is the molar concentration of H^+ and of Cl^-.

$$\begin{aligned}(H^+) &= 0.057\,N = 5.7 \times 10^{-2}M\\ \text{pH} &= -\log(5.7 \times 10^{-2})M\\ &= (0.76 - 2)\\ &= -(-1.24)\\ &= (1.24)\end{aligned}$$

What is the pH of a 0.025 *N* NaOH solution?

$$\begin{aligned}(OH^-) &= 0.025\,M = 2.5 \times 10^{-2}M\\ \text{pOH} &= -\log(2.5 \times 10^{-2})M\\ &= -(0.40 - 2)\\ &= 1.6\\ \text{pH} &= 14 - 1.6\\ &= 12.40\end{aligned}$$

How many grams of NaOH are required to make 100 ml of 0.5 *N* NaOH?

$$100\ \text{ml NaOH} \times 0.5\,N = 50\ \text{mEq or } 0.050\ \text{Eq}$$

Since NaOH has molecular weight of 40.0 g/mol and one equivalent per mole, the equivalent weight is 40.0 g per equivalent.

14. 2.75; acetic acid;

```
      H
      |
  H — C — COOH
      |
      H
```

(Use the equation in Step 1 of Table 13-4, pH = −log[H^+], to solve Problems 14–17.)

15. 3.68; citric acid;

```
      COOH       COOH       COOH
      |          |          |
  H — C ———————— C ———————— C — H
      |          |          |
      H          OH         H
```

16. 1.1×10^{-4} *M*; lactic acid;

```
           OH
           |
    CH3 —  C — COOH
           |
           H
```

17. 5.0×10^{-4} *M*; malic acid;

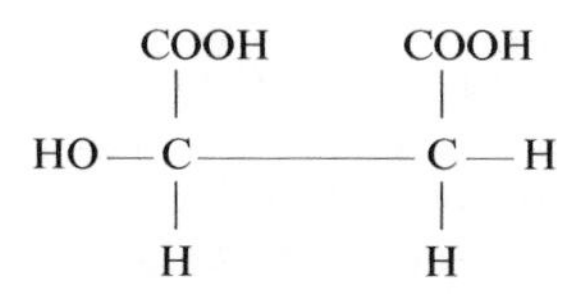

18. From Table 13-2, the equivalent weight of KHP is 204.22 g/Eq. The weight of KHP required can be calculated from the equation.

$$\text{Acid wt.} = \frac{\text{Desired volume (ml)}}{1000\ \text{ml/L}} \times \text{Eq wt. (g/Eq)} \times \text{desired } N\ \text{(Eq/L)}$$

Therefore,

$$\text{KHP wt} = \frac{100\ \text{ml}}{1000\ \text{ml/L}} \times 204\ \text{g/Eq} \times 0.1\ \text{Eq/L} = 2.0422\ \text{g}$$

The solution can be made by weighing exactly 2.0422 g of cool, dry KHP into a 100-ml volumetric flask and diluting to volume.

19. This problem is the same as Problem 18, except that two components are being added to 100 ml of solution. From Table 13-2, the equivalent weight of citric acid (anhydrous) is 64.04 g/Eq. Therefore, the weight of citric acid (CA) would be

$$\text{CA wt} = \frac{100\ \text{ml}}{1000\ \text{ml/L}} \times 64.04\ \text{g/Eq} \times 0.1\ \text{Eq/L} = 0.6404\ \text{g}$$

Potassium citrate (PC) is citric acid with one of its three hydrogen ions removed. Consequently, it has one less equivalent per mole than CA. The equivalent weight of PC would be its molecular weight (230.22) divided by its two remaining hydrogen ions, or 115.11 g per equivalent. Therefore, the weight contribution of PC would be

$$\text{PC wt} = \frac{100\,\text{ml}}{1000\,\text{ml}} \times 115.11\,\text{g/Eq} \times 0.1\,\text{Eq/L} = 1.511\,\text{g}$$

20. The relationship between pH and conjugate acid/base pair concentrations is given by the Henderson–Hasselbalch equation.

$$\text{pH} = \text{p}K_\text{a} + \log\frac{[\text{A}^-]}{[\text{HA}]}$$

When acid and conjugate base concentrations are equal, $[\text{A}^-]/[\text{HA}] = 1$. Since the log of 1 is 0, the pH will equal the pK_a of the acid. Because CA and PC are both 0.1 N, the pH will equal the p$K_{\text{a}1}$ of citric acid given in Table 13-5 (pH = 3.2).

21. Using Equation [1] and solving for volume of concentrate, we get

$$\text{ml concentrated solution} = \frac{\text{final}\,N \times \text{final ml}}{\text{beginning}\,N} = \frac{0.1\,N \times 1000\,\text{ml}}{18\,N} = 5.55\,\text{ml}$$

Consequently, 5.55 ml would be dispensed into a 1-L volumetric flask. The flask would then be filled to volume with distilled CO_2-free water.

The normality of this solution will only be approximate since NaOH is not a primary standard. Standardization against a KHP solution or some other primary standard is essential. It is useful sometimes to back-calculate the true normality of the stock solution. Even under the best circumstances, the normality will decrease with time, but back-calculating will permit a closer approximation of the target normality the next time a working standard is prepared.

22. This answer is a simple ratio.

$$\frac{0.088}{0.100} \times 18 = 15.85\,N$$

23. Table 13-6 indicates that the principal acids in apple, orange, and grape juice are malic, citric, and tartaric acids, respectively. Table 13-2 indicates that the equivalent weight of these acids are malic (67.05), citric (64.04), and tartaric (75.05). The percent acid for each of these juices would be as follows:

Malic acid

$$= \frac{0.1\,\text{mEq/ml NaOH} \times 25\,\text{ml} \times 67.05\,\text{mg/mEq}}{20\,\text{ml}(10)}$$

$$= 0.84\%$$

Citric acid

$$= \frac{0.1\,\text{mEq/ml NaOH} \times 25\,\text{ml} \times 64.04\,\text{mg/mEq}}{20\,\text{ml}(10)}$$

$$= 0.80\%$$

Tartaric acid

$$= \frac{0.1\,\text{mEq/ml NaOH} \times 25\,\text{ml} \times 75.05\,\text{mg/mEq}}{20\,\text{ml}(10)}$$

$$= 0.94\%$$

24. Quality control laboratories often analyze a large number of samples having a specific type of acid. Speed and accuracy are increased if acid concentration can be read directly from the burette. It is possible to adjust the normality of the base to achieve this purpose. The proper base normality can be calculated from the equation:

$$N = \frac{10 \times A}{B \times C}$$

where:

A = weight (or volume) of the sample to be titrated
B = volume (ml) of titrant you want to equal 1% acid
C = equivalent weight of the acid

$$N = \frac{10 \times 10}{5 \times 64.04} = 0.3123\,N$$

In actuality, the standard alkali solution used universally by the Florida citrus industry is 0.3123 N.

25. Since each milliliter will equal 0.1% malic acid, 1% malic acid will equal 10 ml. Therefore,

$$N = \frac{10 \times 10}{5 \times 67.05} = 0.1491\,N$$

13.8 REFERENCES

1. AOAC International (2007) Official methods of analysis, 18th edn., 2005; Current through revision 2, 2007 (On-line). AOAC International, Gaithersburg, MD
2. Beckman Instruments (1995) The Beckman handbook of applied electrochemistry. Bulletin No. BR-7739B. Fullerton, CA
3. Dicker DH (1969) The laboratory pH meter. American Laboratory, February
4. Efiok BJS, Eduok EE (2000) Basic calculations for chemical and biological analysis, 2nd edn. AOAC International, Gaithersburg, MD
5. Fisher Scientific (1996) Fisher electrode handbook, 7th edn. Bulletin No. 120P. Pittsburgh, PA
6. Gardner WH (1996) Food acidulants. Allied Chemical Co., New York
7. Harris DC (2002) Quantitative chemical analysis, 6th edn. Macmillan, New York
8. Joslyn MA (1970) pH and buffer capacity, ch. 12, and acidimetry, ch. 13. In: Methods in food analysis, Academic, New York
9. Kenkel J (1988) Analytical chemistry for technicians. Lewis, Chelsea, MI
10. Mohan C (1995) Buffers. Calbiochem – Novabiochem International, La Jolla, CA

11. Nelson PE, Tressler DK (1980) Fruit and vegetable juice process technology, 3rd edn. AVI, Westport, CT
12. Pecsok RL, Chapman K, Ponder WH (1971) Modern chemical technology, vol 3, revised edn. American Chemical Society, Washington, DC
13. Pomeranz Y, Meloan CE (1994) Food analysis: theory and practice, 3rd edn. Chapman & Hall, New York
14. Skogg DA, West DM, Holler JF, Crouch SR (2000) Analytical chemistry: an introduction, 7th edn. Brooks/Cole, Pacific Grove, CA
15. Wehr HM, Frank JF (eds) (2004) Standard method for examination of dairy products, 17th edn. American Public Health Association, Washington, DC

Fat Characterization

Sean F. O'Keefe
Department of Food Science and Technology, Virginia Tech, Blacksburg, VA 24061, USA
okeefes@vt.edu

and

Oscar A. Pike
Department of Nutrition, Dietetics, and Food Science, Brigham Young University, Provo, UT 84602, USA
oscar_pike@byu.edu

14.1 Introduction 241
14.1.1 Definitions and Classifications 241
14.1.2 Importance of Analyses 243
14.1.3 Lipid Content in Foods and Typical Values 243
14.2 General Considerations 243
14.3 Methods for Bulk Oils and Fats 244
14.3.1 Sample Preparation 244
14.3.2 Refractive Index 244
14.3.2.1 Principle 244

S.S. Nielsen, *Food Analysis*, Food Science Texts Series, DOI 10.1007/978-1-4419-1478-1_14,

14.3.2.2 Procedure 245
14.3.2.3 Applications 245
14.3.3 Melting Point 245
14.3.3.1 Principle 245
14.3.3.2 Applications 245
14.3.4 Smoke, Flash, and Fire Points 245
14.3.4.1 Principle 245
14.3.4.2 Procedure 245
14.3.4.3 Applications 245
14.3.5 Cold Test 246
14.3.5.1 Principle 246
14.3.5.2 Procedure 246
14.3.5.3 Applications 246
14.3.6 Cloud Point 246
14.3.6.1 Principle 246
14.3.6.2 Procedure 246
14.3.7 Color 246
14.3.7.1 Procedure 246
14.3.7.2 Applications 246
14.3.8 Iodine Value 246
14.3.8.1 Principle 246
14.3.8.2 Procedure 246
14.3.8.3 Applications 247
14.3.9 Saponification Value 247
14.3.9.1 Principle 247
14.3.9.2 Procedure 247
14.3.9.3 Applications 248
14.3.10 Free Fatty Acids and Acid Value 248
14.3.10.1 Principle 248
14.3.10.2 Procedure 248
14.3.10.3 Applications 248
14.3.11 Solid Fat Index and Solid Fat Content 248
14.3.11.1 Principle 248
14.3.11.2 Procedure 249
14.3.11.3 Applications 249
14.3.12 Consistency and Spreadability 249
14.3.13 Polar Components in Frying Fats 249
14.3.13.1 Principle 250
14.3.13.2 Procedure 250
14.3.13.3 Applications 250
14.4 Lipid Oxidation: Measuring Present Status 250
14.4.1 Overview 250
14.4.2 Sample Preparation 251
14.4.3 Peroxide Value 251
14.4.3.1 Principle 251
14.4.3.2 Procedure 251
14.4.3.3 Applications 251
14.4.4 *p-Anisidine* Value and Totox Value 251
14.4.4.1 Principle 251
14.4.4.2 Procedure 251
14.4.4.3 Applications 251
14.4.5 Hexanal (Volatile Organic Compounds) 252
14.4.5.1 Principle 252
14.4.5.2 Procedure 252
14.4.5.3 Applications 252
14.4.6 Thiobarbituric Acid Reactive Substances Test 252
14.4.6.1 Principle 252
14.4.6.2 Procedure 252
14.4.6.3 Applications 252
14.4.7 Conjugated Dienes and Trienes 253
14.4.7.1 Principle 253
14.4.7.2 Procedure 253
14.4.7.3 Applications 253
14.5 Lipid Oxidation: Evaluating Oxidative Stability 253
14.5.1 Overview 253
14.5.2 Oven Storage Test 253
14.5.3 Oil Stability Index and Active Oxygen Method 254
14.5.3.1 Principle 254
14.5.3.2 Applications 254
14.5.4 Oxygen Bomb 254
14.5.4.1 Principle 254
14.5.4.2 Procedure 254
14.5.4.3 Applications 254
14.6 Methods for Lipid Components 254
14.6.1 Overview 254
14.6.2 Fatty Acid Composition and Fatty Acid Methyl Esters 255
14.6.2.1 Principle 255
14.6.2.2 Procedure 255
14.6.2.3 Applications 256
14.6.3 *trans* Isomer Fatty Acids Using Infrared Spectroscopy 256
14.6.3.1 Principle 257
14.6.3.2 Procedure 257
14.6.3.3 Applications 257
14.6.4 Mono-, Di-, and Triacylglycerols 257
14.6.5 Cholesterol and Phytosterols 257
14.6.5.1 Principle 257
14.6.5.2 Procedure 257
14.6.5.3 Applications 257
14.6.6 Separation of Lipid Fractions by TLC 258
14.6.6.1 Procedure 258
14.6.6.2 Applications 258
14.7 Summary 258
14.8 Study Questions 258
14.9 Practice Problems 259
14.10 References 260

14.1 INTRODUCTION

Methods for characterizing edible lipids, fats, and oils can be separated into two categories: those developed to analyze bulk oils and fats, and those focusing on analysis of foodstuffs and their lipid extracts. In evaluating foodstuffs, it is usually necessary to extract the lipids prior to analysis. In these cases, if sufficient quantities of lipids are available, methods developed for bulk fats and oils can be utilized.

The methods described in this chapter are divided into four sections. The first are traditional analytical methods for bulk fats and oils, many involving "wet chemistry." Then, two sections discuss methods of measuring lipid oxidation. Some of these methods utilize intact foodstuffs, but most require the lipids to be extracted from foodstuffs. Last addressed are methods for the analysis of lipid fractions, including fatty acids, triacylglycerols, and cholesterol.

Numerous methods exist for the characterization of lipids, fats, and oils (1–11). In this chapter are included those methods required for the nutritional labeling of food and others appropriate for an undergraduate food analysis course. Many traditional "wet chemistry" methods have been supplemented or superseded by instrumental methods such as gas chromatography (GC), high-performance liquid chromatography (HPLC), nuclear magnetic resonance (NMR), and Fourier transform infrared (FTIR) spectroscopy. Nonetheless, an understanding of basic concepts derived from traditional methods is valuable in learning more sophisticated instrumental methods.

Many of the methods cited are official methods of the AOAC International (1), American Oil Chemists' Society (2), or the International Union of Pure and Applied Chemists (3). The principles, general procedures, and applications are described for the methods. Refer to the specific methods cited in Table 14-1 for detailed information on procedures.

14.1.1 Definitions and Classifications

As explained in Chap. 8, the term **lipids** refers to a wide range of compounds soluble in organic solvents but only sparingly soluble in water. Chapter 8 also outlines the general classification scheme for lipids. The majority of lipids present in foodstuffs are of the following types: fatty acids and their glycerides, including mono-, di-, and triacylglycerols; phospholipids; sterols (including cholesterol); waxes; and lipid-soluble pigments and vitamins. The commonly used terms monoglyceride, diglyceride, and triglyceride are synonymous with the proper nomenclature terms monoacylglycerol, diacylglycerol, and triacylglycerol, respectively.

In contrast to lipids, the terms fats and oils often refer to bulk products of commerce, crude or refined, that have already been extracted from animal products or oilseeds and other plants grown for their lipid content. The term **fat** signifies extracted lipids that are solid at room temperature, and **oil** refers to those that are liquid. However, the three terms, lipid, fat, and oil, often are used interchangeably.

The FDA has defined **fat content** for nutritional labeling purposes as the total lipid fatty acids expressed as triglyceride, rather than the extraction and gravimetric procedures used in the past (see also Chap. 8, Sect. 8.3.6.1).

> *"Fat, total" or "Total fat": A statement of the number of grams of total fat in a serving defined as total lipid fatty acids and expressed as triglycerides. Amounts shall be expressed to the nearest 0.5 (1/2) gram increment below 5 grams and to the nearest gram increment above 5 grams. If the serving contains less than 0.5 gram, the content shall be expressed as zero [21 CFR 101.9(c)(2)].*

Fatty acids included in this definition may be derived from triacylglycerols, partial glycerides, phospholipids, glycolipids, sterol esters, or free fatty acids (FFAs), but the concentration will be expressed as grams of triacylglycerols. This change in definition for nutritional labeling purpose requires a methodology change from extraction and gravimetry to GC of **fatty acid methyl esters** (FAMEs). Although this requires more complex and expensive analytical equipment, the new fat definition and analysis provides a better estimation of the fat calories in foods.

Saturated fat is defined for nutritional labeling purposes as the sum (in grams) of all fatty acids without double bonds. Saturated fat is expressed to the closest 0.5 g below 5 g/serving and to the closest gram above 5 g. A food with less than 0.5 g of saturated fat per serving has the content expressed as zero. The optional category of **polyunsaturated fat** (PUFA) is defined as *cis*, *cis*-methylene-interrupted polyunsaturated fatty acids and has the same gram reporting requirements as saturated fats. Another optional category (unless certain label claims are made), **monounsaturated fat**, is defined as *cis*-monounsaturated fatty acids. The requirement that the fatty acids be *cis* prevents including fatty acids that contain *trans* isomers. **Trans fatty acids** must now be included on the nutritional label in the USA. The definition for nutritional labeling requires that the *trans* acids are not conjugated. Most *trans* fatty acids found in foods are monounsaturated, and methods for monounsaturated fatty acid and *trans* fatty acid analysis must distinguish between *cis* and *trans* monoenes, as well as the conjugated *cis–trans* fatty acids such as **conjugated linoleic** acid (CLA). There are several CLA isomers identified in ruminant lipids, although the 9 *cis* 11 *trans* isomer

Correlation of Selected AOCS (2), AOAC (1), and IUPAC (3) Methods

Method	*AOCS*	*AOAC*	*IUPAC*
Bulk fats and oils			
Refractive Index	Cc 7-25	921.08	2.102
Melting			
Capillary tube melting point	Cc 1-25	920.157	
Slip melting point	Cc 3-25		
	Cc 3b-92		
DSC melting properties	Cj 1-94		
Dropping point	Cc 18-80		
Wiley melting point	Cc 2-38[a]	920.156	
Smoke, flash, and fire points	Cc 9a-48		
	Cc 9b-55		
Cold test	Cc 11-53	929.08	
Cloud point	Cc 6-25		
Color			
Lovibond	Cc 13e-92		
	Cc 13j-97		
Spectrophotometric	Cc 13c-50		2.103
Iodine value	Cd 1-25[a]	920.159	2.205
	Cd 1d-92	993.20	
	Cd 1c-85		
Saponification number	Cd 3-25	920.160	2.202
	Cd 3c-91		
	Cd 3a-94		
Free fatty acids (FFAs)	Ca 5a-40	940.28	
Acid value	Cd 3d-63		2.201
Solid fat index (SFI)	Cd 10-57		2.141
Solid fat content (SFC)	Cd 16b-93		2.150
Consistency, penetrometer method	Cc 16-60		
Spreadability	Cj 4-00		
Polar components in frying fats	Cd 20-91	982.27	2.507
Lipid oxidation – present status			
Peroxide value	Cd 8-53	965.33	2.501
	Cd 8b-90		
p-Anisidine value	Cd 18-90		2.504
Hexanal (volatile organic compounds)	Cg 4-94		
Thiobarbituric acid (TBA) test	Cd 19-90		2.531
Conjugated dienes and trienes	Ti 1a-64	957.13	
	Ch 5-91		
Lipid oxidation – oxidative stability			
Oven storage test	Cg 5-97		
Oil stability index (OSI)	Cd 12b-92		
Active oxygen method (AOM)	Cd 12-57[a]		2.506
Oxygen bomb			
Lipid fractions			
Fatty acid composition (including saturated/unsaturated, *cis/trans*)	Ce 1-62	963.22	2.302
		996.06	
	Ce 1b-89		
	Ce 1e-91		
	Ce 1f-96		
Fatty acid methyl esters (FAMEs)	Ce 2-66	969.33	2.301
trans Isomer fatty acids using IR spectroscopy	Cd 14-95		2.207
	Cd 14d-96		
Mono- and diacylglycerols	Cd 11-57	966.18	
	Cd 11b-91		
	Cd 11d-96		
Triacylglycerols		986.19	
	Ce 5b-89		
	Ce 5c-93		
Cholesterol (and other sterols)		976.26	2.403

AOCS, American Oil Chemists' Society; *AOAC*, AOAC International; *IUPAC*, International Union of Pure and Applied Chemists.

[a]Though no longer current, these methods are included for reference because of their previous common use.

appears to be found in greatest abundance (4). CLA isomers are thought to reduce the risk of cancer and other diseases.

Trans fat" or "Trans": A statement of the number of grams of trans fat in a serving, defined as the sum of all unsaturated fatty acids that contain one or more isolated (i.e., nonconjugated) double bonds in a trans configuration, except that label declaration of trans fat content information is not required for products that contain less than 0.5 gram of total fat in a serving if no claims are made about fat, fatty acid or cholesterol content. [21 CFR 101.9(c)(2)(ii)].

14.1.2 Importance of Analyses

Such issues as the effect of dietary fat on health and food labeling requirements necessitate that food scientists be able to not only measure the total lipid content of a foodstuff but also to characterize it (5–10). Health concerns require the measurement of such parameters as cholesterol and phytosterol contents and amounts of *trans*, n-3/ω3, saturated, mono- and polyunsaturated fatty acids. Lipid stability impacts not only the shelf life of food products, but also their safety, since some oxidation products (e.g., malonaldehyde, cholesterol oxides) have toxic properties. Another area of interest is the analysis of oils and fats used in deep-fat frying operations (11). Total polar materials or acid value are used as quality standards in deep fat frying oil. Finally, the development of food ingredients composed of lipids that are not bioavailable (e.g., sucrose polyesters such as Olestra®) or lipids not contributing the normal 9 Cal/g to the diet (e.g., short- and medium-chain triglycerides such as Salatrim® and Caprenin®) accentuates the need to characterize the lipids present in food.

14.1.3 Lipid Content in Foods and Typical Values

Commodities containing significant amounts of fats and oils include butter, cheese, imitation dairy products such as margarine, spreads, shortening, frying fats, cooking and salad oils, emulsified dressings such as mayonnaise, peanut butter, confections, and muscle foods such as meat, poultry, and fish (12, 13). Information is available summarizing the total fat content of foods (see Table 8-2) as well as their constituent fatty acids [e.g., Section I of AOCS Official Methods (2)]. Ongoing studies are refining the quantities of saturated and unsaturated fat, *trans* isomers, cholesterol, cholesterol oxides, phytosterols, and other specific parameters in foods.

Because of their usefulness as food ingredients, it sometimes is important to know the physical and chemical characteristics of bulk fats and oils.

14-2 table **Typical Values of Selected Parameters for Fats and Oils**

Fat/Oil Source	*Refractive Index (40° C)*	*Melting Point (° C)*	*Iodine Value*	*Saponification Value*
Beef tallow	1.454–1.458	40–48	40–48	190–199
Butterfat (bovine)	1.453–1.456	28–35	26–42	210–233
Cocoa butter	1.456–1.458	31–35	32–40	192–200
Coconut oil	1.448–1.450	23–26	6–11	248–265
Corn	1.465–1.468		107–128	187–195
Cottonseed	1.458–1.466		100–115	189–198
Lard	1.459–1.461	33–46	53–77	190–202
Menhaden	1.472		148–160	189–193
Olive			75–94	184–196
Palm	1.449–1.455	33–40	50–55	190–209
Palm kernel	1.452–1.488	24–26	14–21	230–254
Peanut	1.460–1.465		86–107	187–196
Rapeseed	1.465–1.467		94–120	168–181
Safflower	1.467–1.470		136–148	186–198
Soybean	1.466–1.470		124–139	189–195
Sunflower	1.467–1.469		118–145	188–194

Compiled from (2), *Physical and chemical characteristics of oils, fats, and waxes, section 1*, with permission from AOCS Press, copyright 1998.

Definitions and specifications for bulk fats and oils (e.g., soybean oil, corn oil, coconut oil), including values for many of the tests described in this chapter, can be found in Section I of the *AOCS Official Methods* (2), in the *Merck Index* (14), and in *Fats and Oils* (15). Table 14-2 gives typical values for several of the tests for some of the common commercial fats and oils. It must be remembered that bulk fats and oils can vary markedly in such parameters due to differences in source, composition, and susceptibility to deterioration. Foods containing even minor amounts of lipids (e.g., <1%) can have a shelf life limited by lipid oxidation and subsequent rancidity.

14.2 GENERAL CONSIDERATIONS

Various fat extraction solvents and methods are discussed in Chap. 8. For lipid characterization, extraction of fat or oil from foodstuffs can be accomplished by homogenizing with a solvent combination such as hexane–isopropanol (3:2, vol/vol) or chloroform–methanol (2:1, vol/vol). The solvent then can be removed using a rotary evaporator or by evaporation under a stream of nitrogen gas. Lipid oxidation during extraction and testing can be minimized by adding antioxidants [e.g., 10–100 mg of butylated hydroxytoluene (BHT)/L] to solvents and by taking other precautions such as flushing containers with nitrogen and avoiding exposure to heat and light (4).

Sample preparation is hastened through the use of **solid-phase extraction** (SPE), which consists of passing the lipid extract through a commercially available prepackaged absorbent (e.g., silica gel) that separates contaminants or various fractions based on polarity [see Chap. 18, Sect. 18.2.2.2.2 and Chap. 29, Sect. 29.2.2.5]. Constituents present in lipid extractions that may present problems in lipid characterization include phosphatides, gossypol, carotenoids, chlorophyll, sterols, tocopherols, vitamin A, and metals.

Bulk oils such as soybean oil typically undergo the following purification processes: degumming, alkali or physical refining [removal of **free fatty acids** (FFAs)], bleaching, and deodorization after extraction from their parent source. Modifications such as fractionation, winterization, interesterification, and hydrogenation also may be a part of the processing, depending on the commodity being produced. Various methods discussed in this chapter can be used to monitor the refining process.

Changes that lipids undergo during processing and storage include hydrolysis (lipolysis), oxidation, and thermal degradation including polymerization (such as during deep-fat frying operations). These changes are discussed in the following sections on methods.

14.3 METHODS FOR BULK OILS AND FATS

Numerous methods exist to measure the characteristics of fats and oils. Some methods (e.g., titer test) have limited use for edible oils (in contrast to soaps and industrial oils). Other methods may require special apparatuses not commonly available or may have been antiquated by modern instrumental procedures [e.g., volatile acid methods (Reichert–Meissl, Polenske, and Kirschner values) have been replaced largely by determination of fatty acid composition using GC]. Methods to determine impurities, including moisture, unsaponifiable material in refined vegetable oil, and insoluble impurities, also are not covered in this chapter. Defined methods exist for the sensory evaluation of fats and oils (see AOCS Methods Cg 1-83 and Cg 2-83) but are outside the scope of this text.

14.3.1 Sample Preparation

Ensure that samples are visually clear and free of sediment. When required (e.g., iodine value), dry the samples prior to testing (AOAC Method 981.11). Because exposure to heat, light, or air promotes lipid oxidation, avoiding these conditions during sample storage will retard rancidity. Sampling procedures are available for bulk oils and fats (AOCS Method C 1-47; 7).

14.3.2 Refractive Index

14.3.2.1 Principle

The **refractive index** (RI) of an oil is defined as the ratio of the speed of light in air (technically, a vacuum) to the speed of light in the oil. When a ray of light shines obliquely on an interface separating two materials, such as air and oil, the light ray is refracted in a manner defined by Snell's law, as shown in Equation [1].

$$\theta_1 n_1 = \theta_2 n_2 \qquad [1]$$

where:

θ_1 = angle of the incident light
n_1 = refractive index of material 1
θ_2 = angle of the refracted light
n_2 = refractive index of material 2

As can be seen in Fig. 14-1 and from Equation [1], if the angles of incidence and refraction and the refractive index (n) of one of the two materials are known, the refractive index of the other material can be determined. In practice, the θ_1 and n_1 are constant, so n_2 is determined by measuring θ_2.

Because the frequency of light affects its refraction (violet light is refracted more than red light), white light can be dispersed or split after refraction through two materials of different refractive indexes (explaining the color separation of diamonds and rainbows). Refractometers often use monochromatic light (or nearly monochromatic light from the sodium doublet D line, that has 589.0 and 589.6 nm wavelengths or light emitting diodes to provide 589.3 nm) to avoid errors from variable refraction of the different wavelengths of visible light.

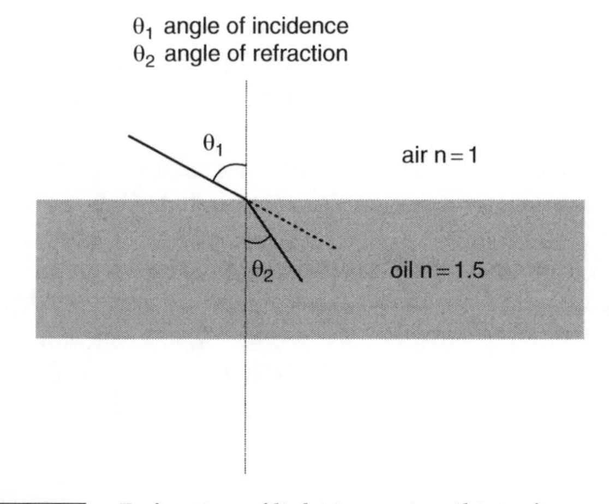

14-1 figure Refraction of light in an air–oil interface.

14.3.2.2 Procedure

Samples are measured with a refractometer at 20°C for oils and at specified higher temperatures for fats, depending on the temperature at which the fat is completely liquid.

14.3.2.3 Applications

RI is related to the amount of saturation in a lipid; the RI decreases linearly as iodine value (a measure of total unsaturation) decreases. RI also is used as a measure of purity and as a means of identification, since each substance has a characteristic RI. However, RI is influenced by such factors as FFA content, oxidation, and heating of the fat or oil. The RI of different lipids is shown in Table 14-2. A relatively saturated lipid such as coconut oil has a different RI ($n = 1.448$–1.450) compared with a relatively unsaturated lipid such as menhaden oil ($n = 1.472$).

14.3.3 Melting Point

14.3.3.1 Principle

Melting point may be defined in various ways, each corresponding to a different residual amount of solid fat. The **capillary tube melting point**, also known as the **complete melting point** or **clear point**, is the temperature at which fat heated at a given rate becomes completely clear and liquid in a one-end closed capillary. The **slip melting point** is performed similarly to the capillary tube method and measures the temperature at which a column of fat moves in an open capillary when heated. The **dropping melting point**, or **dropping point**, is the temperature at which the sample flows through a 0.11-in. hole in a sample cup placed in a specialized furnace. The **Wiley melting point** measures the temperature at which a 1/8 × 3/8-in. disc of fat, suspended in an alcohol–water mixture of similar density, changes into a sphere.

14.3.3.2 Applications

It appears that the predominant method in the USA for measuring melting point is the dropping melting point. The procedure has been automated and therefore is not labor intensive. The capillary tube method is less useful for oils and fats (in comparison to pure compounds) since they lack a sharp melting point due to their array of various components. The slip melting point often is used in Europe whereas the Wiley melting point was preferred previously in the USA; the latter is no longer a current AOCS Method. A disadvantage of the Wiley melting point is the subjective determination as to when the disc is spherical. A disadvantage of the slip melting point is its 16-h stabilization time.

14.3.4 Smoke, Flash, and Fire Points

14.3.4.1 Principle

The **smoke point** is the temperature at which the sample begins to smoke when tested under specified conditions. The **flash point** is the temperature at which a flash appears at any point on the surface of the sample; volatile gaseous products of combustion are produced rapidly enough to permit ignition. The **fire point** is the temperature at which evolution of volatiles (by decomposition of sample) proceeds with enough speed to support continuous combustion.

14.3.4.2 Procedure

A cup is filled with oil or melted fat and heated in a well-lighted container. The smoke point is the temperature at which a thin, continuous stream of bluish smoke is given off. The flash point and fire point are obtained with continued heating, during which a test flame is passed over the sample at 5°C intervals. For fats and oils that flash at temperatures below 149°C, a closed cup is used.

14.3.4.3 Applications

These tests reflect the volatile organic material in oils and fats, especially FFAs (Fig. 14-2) and residual extraction solvents. Frying oils and refined oils should have smoke points above 200°C and 300°C, respectively.

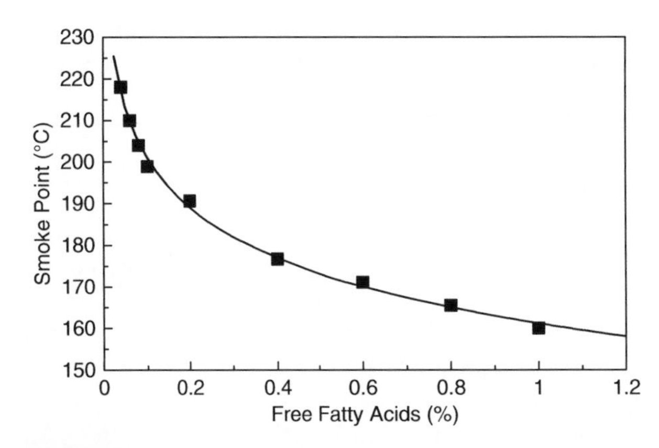

Effect of free fatty acid content on smoke point of olive oil.

14.3.5 Cold Test

14.3.5.1 Principle

The **cold test** is a measure of the resistance of an oil to crystallization. Absence of crystals or turbidity indicates proper winterizing.

14.3.5.2 Procedure

Oil is stored in an ice bath (0°C) for 5.5 h and observed for crystallization.

14.3.5.3 Applications

The cold test is a measure of success of the winterizing process. It ensures that oils remain clear even when stored at refrigerated temperatures. Winterizing subjects an oil to cold temperatures and then separates the crystallized material from the bulk. This results in oil that will not cloud at low temperature. This is useful for ensuring that oils remain clear in products, such as salad dressings, that will be stored at refrigeration temperatures.

14.3.6 Cloud Point

14.3.6.1 Principle

The **cloud point** is the temperature at which a cloud is formed in a liquid fat due to the beginning of crystallization.

14.3.6.2 Procedure

The sample is heated to 130°C and then cooled with agitation. The temperature of first crystallization is taken to be the point at which a thermometer in the fat is no longer visible.

14.3.7 Color

Two methods for measuring the color of fats and oils are the **Lovibond** method and the **spectrophotometric** method.

14.3.7.1 Procedure

In the Lovibond method, oil is placed in a standard-sized glass cell and visually compared with red, yellow, blue, and neutral color standards. Results are expressed in terms of the numbers associated with the color standards. Automated colorimeters are available.

For the spectrophotometric method, the sample is heated to 25–30 °C, placed in a cuvette, and absorbance read at the following wavelengths: 460, 550, 620, and 670 nm. The photometric color index is calculated as shown in Equation [2] AOCS Method Cc 13c-50.

$$\text{Photometric color index} = 1.29\,(A_{460}) + 69.7\,(A_{550}) + 41.2\,(A_{620}) - 56.4\,(A_{670}) \quad [2]$$

14.3.7.2 Applications

The color of fats and oils is most commonly evaluated using the Lovibond method. Oils and fats from different sources vary in color. But if refined oil is darker than expected, it is probably indicative of improper refinement or abuse (13). Though specifically developed for testing the color of cottonseed, soybean, and peanut oils, the spectrophotometric method is probably applicable to other fats and oils as well.

14.3.8 Iodine Value

14.3.8.1 Principle

The iodine value (or iodine number) is a measure of degree of unsaturation, which is the number of carbon–carbon double bonds in relation to the amount of fat or oil. **Iodine value** is defined as the grams of iodine absorbed per 100 g of sample. The higher the amount of unsaturation, the more iodine is absorbed and the higher the iodine value.

A common practice is to determine **calculated iodine value** from the fatty acid composition (see Sect. 14.6.1) using AOCS Recommended Practice Cd 1c-85. The calculated iodine value is not meant to be a rapid method, but instead gives two results (iodine value of triacylglycerols and FFAs) from one analysis (fatty acid composition).

14.3.8.2 Procedure

A quantity of fat or oil dissolved in solvent is reacted, avoiding light, with a measured amount of iodine or some other halogen such as ICl or IBr. Halogen addition to double bonds takes place Equation [3]. A solution of potassium iodide is added to reduce excess ICl to free iodine Equation [4]. The liberated iodine is then titrated with a sodium thiosulfate standard using a starch indicator Equation [5], and the iodine value is calculated Equation [6].

$$\underset{\text{(excess)}}{\text{ICl}} + \text{R}-\text{CH}=\text{CH}-\text{R} \rightarrow \text{R}-\text{CHI}-\text{CHCl}-\text{R} + \underset{\text{(remaining)}}{\text{ICl}} \quad [3]$$

$$ICl + 2KI \rightarrow KCl + KI + I_2 \quad [4]$$

$$I_2 + \underset{\text{(blue)}}{\text{starch}} + 2Na_2S_2O_3 \rightarrow 2NaI + \underset{\text{(colorless)}}{\text{starch}} + Na_2S_4O_6 \quad [5]$$

$$\text{Iodine value} = \frac{(B - S) \times N \times 126.9}{(W \times 1000)} \quad [6]$$

where:

iodine value = g iodine absorbed per 100 g of sample
B = volume of titrant (ml) for blank
S = volume of titrant (ml) for sample
N = normality of $Na_2S_2O_3$ (mol/1000 ml)
126.9 = MW of iodine (g/mol)
W = sample mass (g)

Calculated iodine value is obtained from fatty acid composition using Equation [7] for triacylglycerols. A similar equation allows calculation of the iodine value of FFAs.

$$\begin{aligned}\text{Iodine value (triglycerides)} &= (\%\ \text{hexadecenoic acid} \times 0.950) \\ &+ (\%\ \text{octadecenoic acid} \times 0.860) \\ &+ (\%\ \text{octadecadienoic acid} \times 1.732) \\ &+ (\%\ \text{octadecatrienoic acid} \times 2.616) \\ &+ (\%\ \text{eicosenoic acid} \times 0.785) \\ &+ (\%\ \text{docosenoic acid} \times 0.723) \quad [7]\end{aligned}$$

14.3.8.3 Applications

Iodine value is used to characterize oils, to follow the hydrogenation process in refining, and as an indication of lipid oxidation, since there is a decline in unsaturation during oxidation. The calculated value tends to be low for materials with a low iodine value and for oils with greater than 0.5% unsaponifiable material (e.g., fish oils). The Wijs iodine procedure uses ICl and the Hanus procedure uses IBr. The Wijs procedure may be preferable for highly unsaturated oils as it reacts faster with the double bonds.

14.3.9 Saponification Value

14.3.9.1 Principle

Saponification is the process of breaking down or degrading a neutral fat into glycerol and fatty acids by treatment of the fat with alkali Equation [8].

$$\underset{\text{triacylglycerol}}{\begin{array}{l} H_2C{-}O{-}C(=O){-}R_1 \\ HC{-}O{-}C(=O){-}R_2 \\ H_2C{-}O{-}C(=O){-}R_3 \end{array}} + 3\ KOH \xrightarrow{\text{heat}} \underset{\text{glycerol}}{\begin{array}{l} H_2C{-}OH \\ HC{-}OH \\ H_2C{-}OH \end{array}} + \underset{\text{potassium salt of fatty acids}}{\begin{array}{l} K^{+}\,{}^{-}O{-}C(=O){-}R_1 \\ K^{+}\,{}^{-}O{-}C(=O){-}R_2 \\ K^{+}\,{}^{-}O{-}C(=O){-}R_3 \end{array}} \quad [8]$$

The **saponification value** (or saponification number) is defined as the amount of alkali necessary to saponify a given quantity of fat or oil. It is expressed as the milligrams of KOH required to saponify 1 g of the sample. The saponification value is an index of the mean molecular weight of the triacylglycerols in the sample. The mean molecular weight of the triacylglycerols may be divided by 3 to give an approximate mean molecular weight for the fatty acids present; the smaller the saponification value, the longer the average fatty acid chain length.

In common practice, the **calculated saponification value** is determined from the fatty acid composition (see Sect. 14.6.2) using AOCS Recommended Practice Cd 3a-94.

14.3.9.2 Procedure

Excess alcoholic potassium hydroxide is added to the sample and the solution is heated to saponify the fat Equation [8]. The unreacted potassium hydroxide is back-titrated with standardized HCl using phenolphthalein as the indicator, and the saponification value is calculated Equation [9].

$$\text{Saponification value} = \frac{(B - S) \times N \times 56.1}{W} \quad [9]$$

where:

saponification value = mg KOH per g of sample
B = volume of titrant (ml) for blank
S = volume of titrant (ml) for sample
N = normality of HCl (mmol/ml)
56.1 = MW of KOH (mg/mmol)
W = sample mass (g)

The **calculated saponification value** is obtained from fatty acid composition using Equation [10]. The **fractional molecular weight** of each fatty acid in the sample must be determined first by multiplying the fatty acid percentage (divided by 100) by its molecular weight. The **mean molecular weight** is the sum of the fractional weights of all the fatty acids in the sample.

$$\text{Calculated saponification value} = \frac{3 \times 56.1 \times 1000}{(\text{mean molecular weight} \times 3) + 92.09 - (3 \times 18)} \quad [10]$$

where:

calculated saponification value = mg KOH per g of sample
3 = the number of fatty acids per triacylglycerol
56.1 = MW of KOH (g/mol)
1000 = conversion of units (mg/g)
92.09 = MW of glycerol (g/mol)
18 = MW of water (g/mol)

14.3.9.3 Applications

The calculated saponification value is not applicable to fats and oils containing high amounts of unsaponifiable material, FFAs (>0.1%), or mono- and diacylglycerols (>0.1%).

14.3.10 Free Fatty Acids and Acid Value

14.3.10.1 Principle

Measures of fat acidity normally reflect the amount of fatty acids hydrolyzed from triacylglycerols Equation [11].

$$\begin{array}{l} H_2C{-}O{-}C(=O){-}R_1 \\ \;\;| \\ HC{-}O{-}C(=O){-}R_2 \\ \;\;| \\ H_2C{-}O{-}C(=O){-}R_2 \end{array} \quad + \quad 3H_2O$$

triacylglycerol

$$\rightarrow \begin{array}{l} H_2C{-}OH \\ \;\;| \\ HC{-}OH \\ \;\;| \\ H_2C{-}OH \end{array} \quad + \quad \begin{array}{l} HO{-}C(=O){-}R_1 \\ HO{-}C(=O){-}R_2 \\ HO{-}C(=O){-}R_3 \end{array} \quad [11]$$

glycerol fatty acids

FFA is the percentage by weight of a specified fatty acid (e.g., percent oleic acid). **Acid value** (AV) is defined as the mg of KOH necessary to neutralize the free acids present in 1 g of fat or oil. The AV is often used as a quality indicator in frying oils, where a limit of 2 mg KOH/g oil is sometimes used. In addition to FFAs, acid phosphates and amino acids also can contribute to acidity. In samples containing no acids other than fatty acids, FFA and acid value may be converted from one to the other using a conversion factor Equation [12]. Acid value conversion factors for lauric and palmitic are 2.81 and 2.19, respectively.

$$\%\,\text{FFA (as oleic)} \times 1.99 = \text{acid value} \quad [12]$$

Sometimes the acidity of edible oils and fats is expressed as milliliters of NaOH (of specified normality) required to neutralize the fatty acids in 100 g of fat or oil (9).

14.3.10.2 Procedure

To a liquid fat sample, neutralized 95% ethanol and phenolphthalein indicator are added. The sample then is titrated with NaOH and the percent FFA is calculated Equation [13].

$$\%\,FFA\ (\text{as oleic}) = \frac{V \times N \times 282}{(W \times 1000)} \times 100 \quad [13]$$

where:

% FFA = percent free fatty acid (g/100 g) expressed as oleic acid
V = volume of NaOH titrant (ml)
N = normality of NaOH titrant (mol/1000 ml)
282 = MW of oleic acid (g/mol)
W = sample mass (g)

14.3.10.3 Applications

In crude fat, FFA or acid value estimates the amount of oil that will be lost during refining steps designed to remove fatty acids. In refined fats, a high acidity level means a poorly refined fat or fat breakdown after storage or use. However, if a fat seems to have a high amount of FFAs, it may be attributable to acidic additives (e.g., citric acid added as a metal chelator) since any acid will participate in the reaction (13). If the fatty acids liberated are volatile, FFA or acid value may be a measure of hydrolytic rancidity.

14.3.11 Solid Fat Index and Solid Fat Content

14.3.11.1 Principle

Originally, the amount of solids in a fat was estimated using the **solid fat index** (SFI). SFI is measured using dilatometry, which determines the change

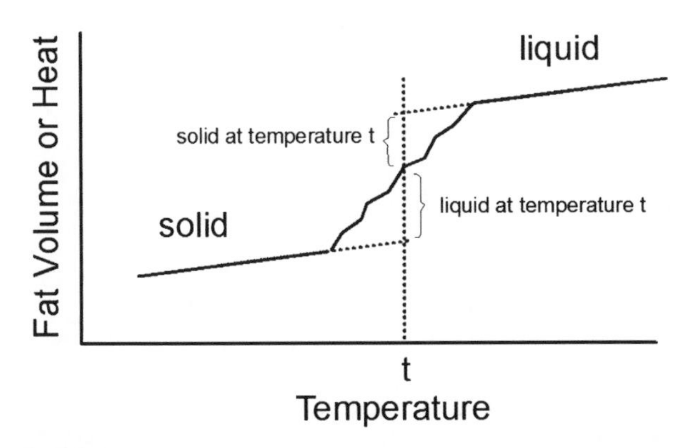

Melting curve of a glyceride mixture. [Adapted from (13), p 249, by Courtesy of Marcel Dekker.]

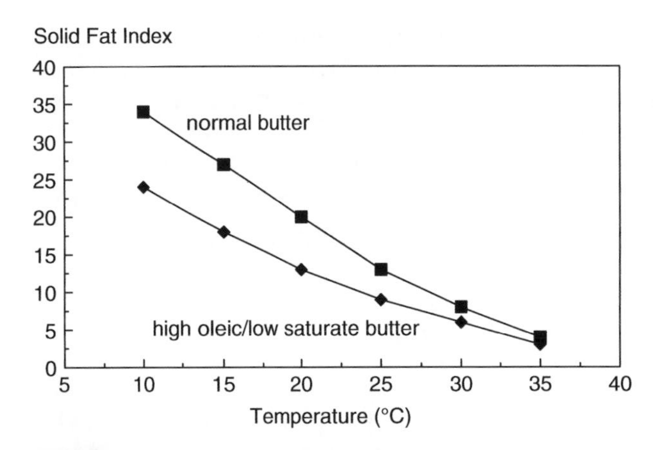

SFI curves of butter with normal and high oleic/low saturated fatty acid compositions.

in volume with change in temperature. As solid fat melts, it increases in volume. Plotting volume against temperature gives a line at which the fat is solid, a line at which it is liquid, and a melting curve in between (Fig. 14-3). However, because the solid fat line is difficult to determine experimentally, a line is placed 0.100 specific volume units (ml/g) below the liquid line with the same slope. The SFI is the volume of solid fat divided by the volume between the upper and lower lines, expressed as a percentage (15).

Preferably, the actual percent solid fat in a sample, termed the **solid fat content** (SFC), can be determined using either continuous wave or pulsed NMR (see Chap. 25).

14.3.11.2 Procedure

For SFI, fat dilatometers consist of a bulb connected to a calibrated capillary tube. As fat in the bulb expands upon heating, it forces a liquid (i.e., colored water or mercury) up the capillary tube.

14.3.11.3 Applications

The amount of solid fat phase present in a plastic fat (e.g., margarine, shortening) depends on the type of fat, its history, and the temperature of measurement. The proportion of solids to liquids in the fat and how quickly the solids melt have an impact on functional properties, such as the mouthfeel of a food. An example of the SFI use is shown in Fig. 14-4; the butter with high oleic and low saturated fatty acid composition has lower solid fat and is softer as it melts over the temperature range 10–35°C, and it is more easily spreadable at refrigeration temperatures.

Chapter 25 explains NMR as it relates to measuring the solid content of fat and other foods. Though the equipment is expensive, SFC is preferred over SFI because it measures the actual fat content, is less subject to error, and takes less time. Comparison between samples must be made using SFC (or SFI) values taken at the same temperature.

14.3.12 Consistency and Spreadability

The textural properties of plasticized fats (e.g., shortenings, margarine, butter) can be measured using such tests as consistency and spreadability. The consistency method described has been used for several decades, whereas the spreadability method is a recently approved method that utilizes modern texture analysis instruments. The **penetrometer method** of determining consistency measures the distance a cone-shaped weight will penetrate a fat in a given time period. The **spreadability** test delineates the parameters for using a Texture Technologies TA-XT2 Texture Analyzer® (or similar instrument) to determine the force needed to compress a sample. See Chap. 30 for general approaches to characterizing the rheological properties of foods; many aspects can be applied to fats and oils.

The penetrometer method is useful for measuring the consistency of plastic fats and solid fat emulsions. Like SFI and SFC, consistency is dependent on the type of fat, its history, and the temperature during measurement. The spreadability method is applicable to lipid-containing solid suspensions, emulsion, and pastes that can maintain their shape at the temperature used for the analysis, including products such as peanut butter and mayonnaise.

14.3.13 Polar Components in Frying Fats

Methods used to monitor the quality of the oil or fat used in deep-fat frying operations are based on the physical and chemical changes that occur, which include an increase in each of the following parameters: viscosity, foaming, FFAs, degree of saturation, hydroxyl and carbonyl group formation, and saponification value. Standard tests used in the evaluation of frying fats include quantitating polar components, conjugated dienoic acids, polymers, and FFAs. In addition, there are several rapid tests useful in day-to-day quality assurance of deep-fat frying operations (11).

14.3.13.1 Principle

Deterioration of used frying oils and fats can be monitored by measuring the polar components, which include monoacylglycerols, diacylglycerols, FFAs, and oxidation products formed during heating of foodstuffs. Nonpolar compounds are primarily unaltered triacylglycerols. The polar compounds in a sample can be separated from nonpolar compounds using chromatographic techniques.

14.3.13.2 Procedure

Polar components are measured by dissolving the fat sample in light petroleum ether–diethyl ether (87:13), then applying the solution to a silica gel column. Polar compounds are adsorbed onto the column. Nonpolar compounds are eluted, the solvent evaporated, the residue weighed, and the total polar components estimated by difference. Quality of the determination can be verified by eluting polar compounds and separating polar and nonpolar components using thin-layer chromatography (TLC).

14.3.13.3 Applications

A suggested limit of 27% polar components in frying oil is a guide for when it should be discarded. A limitation of this method is the sample run time of 3.5 h (14). The acid value is often determined as an alternate indicator of frying oil deterioration; however, rapid procedures based on dielectric constant are becoming more widely used because of their speed.

14.4 LIPID OXIDATION: MEASURING PRESENT STATUS

14.4.1 Overview

The term **rancidity** refers to the off odors and flavors resulting from lipolysis (**hydrolytic rancidity**) or lipid oxidation (**oxidative rancidity**). **Lipolysis** is the hydrolysis of fatty acids from the glyceride molecule. Because of their volatility, hydrolysis of short-chain fatty acids can result in off odors. Fatty acids shorter than C12 (lauric acid) can produce off-odors in foods.

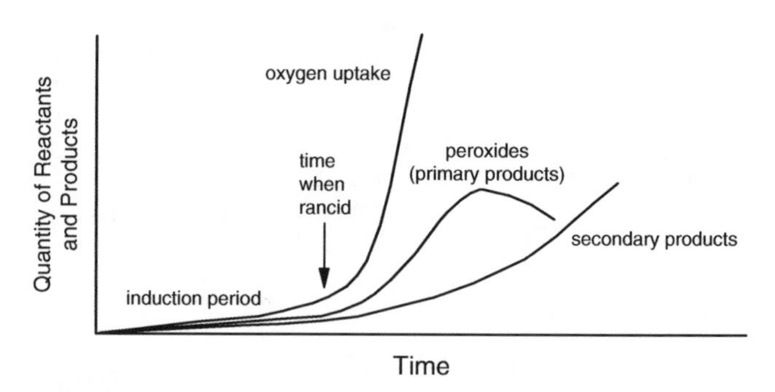

14-5 figure Changes in quantities of lipid oxidation reactants and products over time. [Adapted from (15), with permission from (16), Copyright CRC Press, Boca Raton, FL, ©1971.]

Free C12 is often associated with a soapy taste but no aroma. FFAs longer than C12 do not cause significant impairment in taste or odor.

Lipid oxidation (also called autoxidation) as it occurs in bulk fats and oils proceeds via a self-sustaining free radical mechanism that produces hydroperoxides (initial or primary products) that undergo scission to form various secondary products including aldehydes, ketones, organic acids, and hydrocarbons (final or secondary products) (16) (see Fig. 14-5).

In biological tissues, including foodstuffs, abstraction reactions and rearrangements of alkoxyl and peroxyl radicals may result in the production of endoperoxides and epoxides as secondary products. Many methods have been developed to measure the different compounds as they form or degrade during lipid oxidation. Since the system is dynamic, it is recommended that two or more methods be used to obtain a more complete understanding of lipid oxidation.

Measuring the current quality of a fat or oil in regard to lipid oxidation can be achieved using such procedures as peroxide value, *p*-anisidine value, measurement of volatile organic compounds (VOCs) (e.g., hexanal), and the thiobarbituric reactive substances (TBARS) test. Some of these procedures have been modified (especially with respect to sample size) for use in biological tissue assays (17). Other methods that monitor lipid oxidation (and that vary in usefulness) include the iodine value, acid value, Kreis test, and oxirane test, as well as the measurement of conjugated dienes and trienes, total and volatile carbonyl compounds, polar compounds, and hydrocarbon gases (7,13).

While quantitating lipid oxidation by one or more of the methods listed above is usually adequate, in some cases it may be necessary to *visualize the location* of lipid molecules and lipid oxidation within a food or raw ingredient. **Fluorescence microscopy** with stains specific to lipids can be applied to such a problem. For example, the dye Nile Blue (with the active ingredient Nile Red) can be combined with a lipid-containing sample and the preparation viewed under a fluorescence microscope (18–20). Lipids will appear an intense yellow fluorescence, with the intensity of the fluorescence changed by the nature of the lipids and by lipid oxidation. Examples of applications include localizing oxidized lipids in a cereal product, visualizing interactions between lipids and emulsifiers, and localizing lipids in cheeses, frosting, and chocolates.

14.4.2 Sample Preparation

Most methods require lipid extraction prior to analysis (see Sect. 14.2). However, variations of some methods (e.g., some TBARS tests) may begin with the original foodstuff.

14.4.3 Peroxide Value

14.4.3.1 Principle

Peroxide value is defined as the milliequivalents (mEq) of peroxide per kilogram of sample. It is a redox titrimetric determination. The assumption is made that the compounds reacting under the conditions of the test are peroxides or similar products of lipid oxidation.

14.4.3.2 Procedure

The fat or oil sample is dissolved in glacial acetic acid–isooctane (3:2). Upon addition of excess potassium iodide, which reacts with the peroxides, iodine is produced Equation [14]. The solution then is titrated with standardized sodium thiosulfate using a starch indicator Equation [14]. Peroxide value is calculated as shown in Equation [15].

$$\mathrm{ROOH} + \underset{\text{(excess)}}{\mathrm{K^+I^-}} \xrightarrow{\mathrm{H^+,\ heat}} \mathrm{ROH} + \mathrm{K^+OH^-} + \mathrm{I_2} \quad [14]$$

$$\mathrm{I_2} + \underset{\text{(blue)}}{\text{starch}} + 2\mathrm{Na_2S_2O_3} \rightarrow 2\mathrm{NaI} + \underset{\text{(colorless)}}{\text{starch}} + \mathrm{Na_2S_4O_6} \quad [15]$$

$$\text{Peroxide value} = \frac{(S - B) \times N}{W} \times 1000 \quad [16]$$

where:

peroxide value = mEq peroxide per kg of sample
S = volume of titrant (ml) for sample
B = volume of titrant (ml) for blank
N = normality of $Na_2S_2O_3$ solution (mEq/ml)
1000 = conversion of units (g/kg)
W = sample mass (g)

14.4.3.3 Applications

Peroxide value measures a transient product of oxidation, (i.e., after forming, peroxides and hydroperoxides break down to form other products). A low value may represent either the beginning of oxidation or advanced oxidation (see Fig. 14-5), which can be distinguished by measuring peroxide value over time or by using a procedure that measures secondary products of oxidation. For determination in foodstuffs, a disadvantage of this method is the 5 g fat or oil sample size required; it is difficult to obtain sufficient quantities from foods low in fat. This method is empirical and any modifications may change results. Despite its drawbacks, peroxide value is one of the most common tests of lipid oxidation.

High-quality, freshly deodorized fats and oils will have a peroxide value of zero. Peroxide values >20 correspond to very poor quality fats and oils, which normally would have significant off flavors. For soybean oil, peroxide values of 1–5, 5–10, and >10 correspond to low, medium, and high levels of oxidation, respectively (AOCS Method Cg 3-91).

14.4.4 *p-Anisidine* Value and Totox Value

14.4.4.1 Principle

The ***p*-anisidine value** estimates the amount of α- and β- unsaturated aldehydes (mainly 2-alkenals and 2,4-dienals), which are secondary oxidation products in fats and oils. The aldehydes react with *p*-anisidine to form a chromogen that is measured spectrophotometrically. The **totox value** tends to indicate the total oxidation of a sample using both the peroxide and *p*-anisidine values Equation [16].

$$\text{Totox value} = p\text{-anisidine value} + (2 \times \text{peroxide value}) \quad [17]$$

14.4.4.2 Procedure

The ***p*-anisidine value** by convention is defined as 100 times the absorbance at 350 nm of a solution containing precisely 1 g of test oil diluted to 100 ml using a mixture of solvent (isooctane) and *p*-anisidine.

14.4.4.3 Applications

Since peroxide value measures hydroperoxides (which increase and then decrease) and *p*-anisidine value measures aldehydes (decay products of hydroperoxides, which continually increase), the totox value usually rises continually during the course of lipid oxidation. Fresh soybean oil should have a *p*-anisidine value <2.0 and a totox value <4.0 (AOCS Method Cg 3-91).

Though not common in the USA, *p*-anisidine and totox values are extensively used in Europe (13).

14.4.5 Hexanal (Volatile Organic Compounds)

14.4.5.1 Principle

Volatile organic compounds (VOCs) present in fats and oils are related to flavor, quality, and oxidative stability. These compounds include secondary products of lipid oxidation that can be responsible for the off flavors and odors of oxidized fats and oils (AOCS Method Cg 3-91). The compounds formed will vary depending on the fatty acid composition of the sample and environmental conditions. Commonly measured compounds include pentane, pentanal, hexanal, and 2,4-decadienal. Thus, measurement of **headspace hexanal** is a means of monitoring the extent of lipid oxidation (21). It is commonly performed using **static (equilibrium) headspace analysis**, which entails the chromatographic analysis of a set volume of vapor obtained from the headspace above a sample held in a closed container (see Chap. 29, Sect. 29.2.2.1). The introduction of **solid-phase microextraction** (SPME) has improved the sensitivity of static headspace analysis of VOC because the volatiles are concentrated on the SPME fiber.

14.4.5.2 Procedure

General guidelines regarding GC parameters are given in AOCS Method Cg 4-94, under the heading of static headspace; a review of research literature will indicate current practices used with various commodities. Typically, a small sample of the commodity is placed in a container having a septum cap. An internal standard may be added, for example, 4-heptanone (21). The container is sealed and then heated for a given time. Heating increases the concentration of headspace volatiles (22). Using a gas-tight syringe, an aliquot of the vapors in the container headspace then is removed and injected into a GC equipped with a flame ionization detector or mass selective detector. Automated headspace samplers are available that both heat the sample and ensure that a constant volume is being analyzed. The quantity of hexanal then is calculated from the peak area (see Chaps. 27 and 29).

14.4.5.3 Applications

Hexanal may correlate well with sensory determination of lipid oxidation since it is a major contributor to off flavors in some food commodities. The quantity of other volatile compounds resulting from lipid oxidation can be obtained simultaneously with hexanal measurement and may enhance the characterization of lipid oxidation in various food commodities. An advantage of this method is that lipid extraction is not required (i.e., intact foodstuffs can be analyzed).

14.4.6 Thiobarbituric Acid Reactive Substances Test

14.4.6.1 Principle

The **thiobarbituric acid reactive substances** (TBARS) test, also known as the **thiobarbituric acid** (TBA) test, measures secondary products of lipid oxidation, primarily **malonaldehyde**. It involves reaction of malonaldehyde (or malonaldehyde-type products including unsaturated carbonyls) with TBA to yield a colored compound that is measured spectrophotometrically. The food sample may be reacted directly with TBA, but is often distilled to eliminate interfering substances, and then the distillate is reacted with TBA. Many modifications of the test have been developed over the years.

14.4.6.2 Procedure

In contrast to the direct method for fats and oils (see Table 14-1), a commonly used procedure (23, 24) is outlined here that requires distillation of the food commodity prior to determining TBARS. A weighed sample is combined with distilled water and mixed. The pH is adjusted to 1.2 and the sample is transferred to a distillation flask. After addition of *t*-butylhydroxytoluene (BHT) (optional), antifoam reagent, and boiling beads, the sample is distilled rapidly and the first 50 ml collected. An aliquot of the distillate is combined with TBA reagent and heated in a boiling water bath for 35 min. Absorbance of the solution is determined at 530 nm and, using a standard curve, absorbance readings are typically converted to milligrams of malonaldehyde (or TBARS) per kilogram of sample.

14.4.6.3 Applications

The TBA test correlates better with sensory evaluation of rancidity than does peroxide value but, like peroxide value, it is a measure of a transient product of oxidation (i.e., malonaldehyde and other carbonyls readily react with other compounds). Despite its limited specificity and the large sample sizes possibly required (depending on the method), the TBA test with minor modifications is frequently used to measure lipid oxidation, especially in meat products. An alternative to the spectrophotometric method described is to determine the actual content of malonaldehyde using HPLC analysis of the distillate.

14.4.7 Conjugated Dienes and Trienes

14.4.7.1 Principle

Double bonds in lipids are changed from nonconjugated to conjugated bonds upon oxidation. Fatty acid **conjugated diene hydroperoxides** formed on lipid oxidation absorb UV light at about 232 nm and **conjugated trienes** at about 270 nm. Conjugated products of lipid oxidation should not be confused with CLA isomers, which are produced by incomplete hydrogenation in ruminants.

14.4.7.2 Procedure

A homogeneous lipid sample is weighed into a graduated flask and brought to volume with a suitable solvent (e.g., isooctane, cyclohexane). A 1% solution (e.g., 0.25 g sample in 25 ml) is made, then diluted or concentrated if necessary to obtain an absorbance between 0.1 and 0.8. The diluted solution must be completely clear. Absorbance is measured with an ultraviolet (UV) spectrophotometer, using the pure solvent as a blank.

14.4.7.3 Applications

The conjugated dienes and trienes methods are useful for monitoring the early stages of oxidation. The magnitude of the changes in absorption is not easily related to the extent of oxidation in advanced stages. Olive oil grades are based on ultraviolet absorbance.

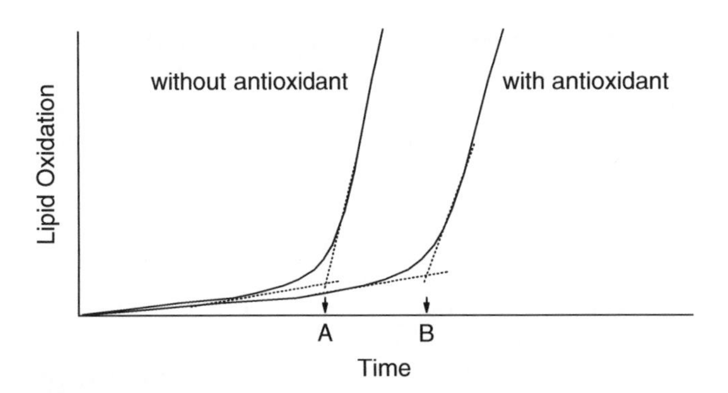

A plot of lipid oxidation over time, showing the effect of an antioxidant on induction period. Time A is induction period of sample without antioxidant and Time B is induction period of sample with antioxidant.

14.5 LIPID OXIDATION: EVALUATING OXIDATIVE STABILITY

14.5.1 Overview

Because of their inherent properties (e.g., the amount of unsaturation and the presence of natural antioxidants) as well as external factors (e.g., added antioxidants, processing and storage conditions), lipids and lipid-containing foodstuffs vary in their susceptibility to rancidity. The resistance of lipids to oxidation is known as **oxidative stability**. Inasmuch as determining oxidative stability using actual shelf life determinations at ambient conditions of storage (usually room temperature) requires months or even years, accelerated tests have been developed to evaluate the oxidative stability of bulk oils and fats, and foodstuffs. **Accelerated tests** artificially hasten lipid oxidation by exposing samples to heat, oxygen, metal catalysts, light, or enzymes. A major problem with accelerated tests is assuming that reactions that occur at elevated temperatures or under other artificial conditions are the same as normal reactions occurring at the actual storage temperature of the product. An additional difficulty is ensuring that the apparatus is clean and completely free of metal contaminants and oxidation products from previous runs. Therefore, assuming lipid oxidation is the factor that limits shelf life, shelf life determinations at ambient conditions should accompany and hopefully validate the results of accelerated tests of oxidative stability.

The **induction period** is defined as the length of time before detectable rancidity, or time before rapid acceleration of lipid oxidation (see Fig. 14-6). Induction period can be determined by such methods as calculating the maximum of the second derivative with respect to time or manually drawing tangents to the lines (Fig. 14-6). Measurement of the induction period allows a comparison of the oxidative stability of samples that contain differing ingredients or of samples held at varying storage conditions, and it provides an indication of the **effectiveness of** various **antioxidants** in preventing lipid oxidation.

14.5.2 Oven Storage Test

As a means of accelerating the determination of oxidative stability, the **oven storage test** is often used. Previously this test was an ill-defined procedure known as the Schaal oven test. It is now a recommended practice of the AOCS (AOCS Method Cg 5-97). This protocol consists of placing a fat or oil of known volume in a forced-draft oven at a temperature above ambient but less than 80°C, with 60°C being recommended. Such temperatures are desirable accelerated storage temperatures since the mechanism of oxidation in this range is the same as oxidation at room temperature. Tests should be conducted in the dark, the initial quality of the sample oil should be high, and the surface to volume ratio of the oil must be kept constant for all samples.

To determine an induction period and thus oxidative stability, the oven storage test must be combined with other methods of detecting rancidity, for example, sensory evaluation and peroxide value. More than one method should be used to determine the extent of oxidation; it is recommended that one method measure primary products of lipid oxidation (e.g., peroxide value, conjugated dienes) and the other method measure secondary products (e.g., VOC, *p*-anisidine value, TBARS, or sensory evaluation). Results of oxidative stability determinations obtained at approximately 60 °C correlate well with actual shelf life determinations (10).

14.5.3 Oil Stability Index and Active Oxygen Method

14.5.3.1 Principle

The **oil stability index** (OSI) determines induction period by bubbling purified air through an oil or fat sample held at an elevated temperature (often 110 °C or 130 °C), then passing the acidic volatiles (primarily formic acid) into a deionized water trap. The conductivity of the water is measured continuously, resulting in data similar to those shown in Fig. 14-6. Results should specify the temperature used as well as induction period time. Two instruments that automate this method are the Rancimat® (Brinkmann Instruments, Inc.) and the Oxidative Stability Instrument® (Omnion, Inc.). The more familiar but outdated and labor-intensive **active oxygen method** (AOM) is similar to the OSI except that induction period is determined by discontinuous measurements of either peroxide value or sensory evaluation of rancid odor.

14.5.3.2 Applications

These methods were designed originally to measure the effectiveness of antioxidants. The OSI is determined much faster than tests performed at oven storage test temperatures, but results from the latter may correlate better with actual shelf life. Extrapolating OSI results to actual shelf life is difficult because the high temperatures used result in the formation of compounds not present in samples held at ambient or slightly elevated temperature conditions (10,25,26).

Specification sheets for fats and oils often report AOM values to accommodate individuals working in this area who are familiar with AOM values. OSI values can be converted to AOM values.

Applicable to all fats and oils, the OSI has also been researched for applicability to certain low-moisture snack foods (e.g., potato chips and corn chips). Because of the continuous exposure to circulating air, samples that contain more than negligible amounts of water tend to dehydrate during the determination and are not likely to give reliable results because water activity can affect oxidation rates.

14.5.4 Oxygen Bomb

14.5.4.1 Principle

Inasmuch as lipid oxidation results in the uptake of oxygen from the surrounding environment (see Fig. 14-2), measuring the time required for the onset of rapid disappearance of oxygen in a closed system provides a means of determining oxidative stability.

14.5.4.2 Procedure

The oxygen bomb consists of a heavy-walled container that has a pressure recorder attached. The sample is placed in the container and oxygen is used to pressurize the container to 100 psi. The container then is placed in a boiling water bath. Induction period is determined by measuring the time until a sharp drop in pressure occurs, which corresponds with the rapid absorption of oxygen by the sample.

14.5.4.3 Applications

Oxygen bomb results may have a better correlation with rancidity shelf life tests than AOM values. Another advantage, compared with the OSI (or AOM), is that the oxygen bomb method may be used with intact foodstuffs instead of extracted lipids (15).

14.6 METHODS FOR LIPID COMPONENTS

14.6.1 Overview

The lipid present in food commodities or bulk fats and oils can be characterized by measuring the amount of its various fractions, which include fatty acids, mono-, di-, and triacylglycerols, phospholipids, sterols (including cholesterol and phytosterols), and lipid-soluble pigments and vitamins. Another means of categorizing lipid fractions is inherent in nutrition labeling, which involves the measurement of not only total fat, but may also require quantification of saturated fat, monounsaturated fat, PUFA, and *trans* isomer fatty acids. In addition, foods may contain lipids that do not contribute the same caloric content as normal lipids, for example, sucrose polyesters (e.g., Olestra®), medium-chain triglycerides, and triglycerides that contain short-chain fatty acids (e.g., Salatrim®, Caprenin®). Many of these fractions are determined readily by evaluating the component fatty acids. From the fatty acid composition, calculations are made to determine such parameters as total fat, saturated fat, calculated iodine value, and calculated saponification value.

In contrast, the measurement of total and saturated fat in foods containing Olestra® requires special consideration. AOAC peer-verified Method PVM 4:1995 outlines the use of lipase on the lipid extract, which yields fatty acids and unreacted Olestra®. The fatty acids are converted to calcium soaps and Olestra® is extracted and discarded. The precipitated soaps are converted back to fatty acids, which are subsequently analyzed via capillary GC (27). Other procedures that have been studied include chromatographic separation using HPLC with an evaporative light scattering (ELSD) detector (28). Olestra®, like most lipids, will not absorb UV or VIS light appreciably, preventing analysis by using the common UV-VIS HPLC detectors. For this reason, detectors such as the refractive index (RID), transport flame ionization, or ELSD are required for analysis of most lipids.

GC (see Chap. 29) is ideal for the analysis of many lipid components. GC can be used for determinations such as total fatty acid composition, distribution and position of fatty acids in lipid, sterols, studies of fat stability and oxidation, assaying heat or irradiation damage to lipids, and detection of adulterants and antioxidants (10). Methods exist that detail the analysis of various lipid fractions using GC (6). GC combined with **mass spectrometry** (MS) (see Chap. 26) is a powerful tool used in identification of compounds. **HPLC** (see Chap. 28) also is useful in lipid analyses, especially for components that are not readily volatilized, such as hydroperoxides and triacylglycerols (26). **Thin-layer chromatography** (TLC) (see Chap. 27) has been used extensively in the past by lipid chemists. Partly due to low cost and ease, TLC is still useful, although many assays may be more quantitative or have better resolution using GC or HPLC.

14.6.2 Fatty Acid Composition and Fatty Acid Methyl Esters

The **fatty acid composition**, or **fatty acid profile**, of a food product is determined by quantifying the kind and amount of fatty acids that are present, usually by extracting the lipids and analyzing them using **capillary GC** (also described in Chap. 8, Sect. 8.3.6.1).

14.6.2.1 Principle

To increase volatility before GC analysis, triacylglycerols are typically esterified to form **FAMEs** (Equation [18]). Acyl lipids are readily transesterified using base such as sodium hydroxide and methanol. Sodium methoxide produced by this combination will create FAMEs from acyl lipids rapidly, but will not react with FFAs. Acidic reagents such as methanolic HCL or boron trifluoride (BF_3) react rapidly with FFAs, but more slowly with acyl lipids. Procedures such as the AOCS Method Ce 1b-89 (this is a joint method with AOAC 991.39) use a two-step methylation, first reacting the lipid with 0.5 *N* NaOH and then with excess BF_3/methanol. This allows a rapid methylation of FFAs, acyl lipids, and phospholipids. The sodium hydroxide step is not a saponification procedure (i.e., hydrolysis of acyl groups); it is a direct transmethylation.

$$\begin{array}{l} H_2C-O-C(=O)-R_1 \\ HC-O-C(=O)-R_2 \\ H_2C-O-C(=O)-R_3 \end{array} + CH_3OH \xrightarrow[\text{NaOH then } BF_3]{} \begin{array}{l} H_2C-OCH_3 \\ HC-OCH_3 \\ H_2C-OCH_3 \end{array} + \begin{array}{l} H_3C-O-C(=O)-R_1 \\ H_3C-O-C(=O)-R_2 \\ H_3C-O-C(=O)-R_3 \end{array} \quad [18]$$

triacylglycerol → FAMEs

14.6.2.2 Procedure

The lipid is extracted from the food, for example, by homogenizing with a suitable solvent such as hexane–isopropanol (3:2, vol/vol) and then evaporating the solvent. The FAMEs are prepared by combining the extracted lipid with sodium hydroxide-methanol and internal standard in isooctane and then heating at 100 °C for 5 min. The sample is cooled and then excess BF_3-methanol is added with further heating (100 °C for 30 min). After addition of saturated aqueous sodium chloride, additional isooctane and mixing, the upper isooctane solution containing the FAMEs is removed and dried with anhydrous Na_2SO_4, then diluted to a concentration of 5–10% for injection onto the GC.

Several methods (see Table 14-1) describe procedures and conditions for using GC to determine fatty acid composition. AOCS Method Ce 1b-89 is specific for marine oils, and AOCS Method Ce 1f-96 is specifically suited for determining *trans* isomer fatty acids.

14.6.2.3 Applications

Determination of the fatty acid composition of a product permits the **calculation** of the following categories of fats that pertain to health issues and food labeling: percent saturated fatty acids, percent unsaturated fatty acids, percent monounsaturated fatty acids, percent polyunsaturated fatty acids, CLAs, and percent *trans* isomer fatty acids. Calculation of fatty acids as a percentage is referred to as **normalization**, that is, the areas of all of the FAMEs are summed and the percent area of each fatty acid is calculated relative to the total area. This is a reasonable procedure because with flame ionization detectors (FID), the weight of fatty acids in a mixture closely parallels the area on the chromatogram. However, this is not absolutely correct. Theoretical correction factors are needed to correct for the FID response, which is different depending on the level of unsaturation in FAMEs (29). A chromatogram showing separation of FAMEs of varying length and unsaturation is shown in Chap. 8, Fig. 8-5. The separation of FAMEs on this SP2560 column is typical of what is seen when using a highly polar (biscyanopropyl polysiloxane) column (30).

The separation of FAMEs on GC columns depends on the polarity of the liquid phase. On nonpolar liquid phases [such as 100% dimethyl polysiloxane (DB-1, HP-1, CPSil5CB) or 95% dimethyl, 5% diphenyl polysiloxane (DB5, HP5, CPSil8 CB)], FAMEs are separated largely based on their boiling points. This results in the elution order $18:3n-3 > 18:3n-6 > 18:1n-9 > 18:0 > 20:0$. On phases of medium polarity [such as 50% cyanopropylphenyl polysiloxane (DB225, HP225, CPSil43CB)], the order of elution is changed because of the interaction of the pi electrons of the double bonds with the liquid phase. The order of elution on these columns would be $18:0 > 18:1n-9 > 18:2n-6 > 18:3n-3 > 20:0$ (first eluted to last). When the polarity of the liquid is increased further with 100% biscyanopropyl polysiloxane columns (SP2560, CPSil88), the greater interaction of the double bonds with the very polar liquid phase results in an elution pattern $18:0 > 18:1n-9 > 18:2n-6 > 20:0 > 18:3n-3$. As the liquid phase polarity increases, the effect of double bonds on retention time increases. Additionally, *trans* fatty acids interact less effectively with the liquid phase than *cis* acids for steric reasons, so *trans* acids will elute before the corresponding *cis* acid; see Fig. 8-5 where $18:1\Delta 9$ *trans* (elaidate) elutes before $18:1\Delta 9$ (oleic acid) and $18:2\Delta 9$ *trans*, $\Delta 12$ *trans* (linoelaidate) elutes before linoleic acid ($18:2n-6$, $18:2\Delta 9$ *cis* $\Delta 12$ *cis*) on this highly polar 100% biscyanopropyl polysiloxane column. It also can be seen that gamma linolenic ($18:3n-6$) elutes before linolenic acid ($18:3n-3$). Because the double bonds are closer to the methyl side of the FAMEs in $18:3n-3$, the double bonds can more effectively interact with the liquid phase, resulting in greater retention by the column.

The complexity of the fatty acids found in various foods will affect the details of the GC analysis that can be used. Analysis of FAMEs of a vegetable oil is quite simple and can easily be accomplished in less than 20 min using a column with a medium polarity liquid phase. The fatty acids present in most vegetable oils range from C14 to C24. Coconut and palm kernel oils also contain shorter chain fatty acids such as C8–C12. Dairy fats contain butyric acid (C4) and other short chain fatty acids whereas peanut oil contains C26 at around 0.4–0.5% of the total FAMEs. Marine lipids contain a much wider range of fatty acids and require care in the separation and identification of FAMEs, many of which have no commercially available standards.

Trans fatty acids in foods originate from three main sources: biohydrogenation in ruminants, incomplete hydrogenation in the conversion of liquid oils to plastic fats, and high-temperature exposure during deodorization. The *trans* fatty acids formed from these three processes are quite different and require careful attention to achieve accurate analysis.

Separation of *trans* FAMEs is facilitated by selection of the most polar column phases available. Currently, Sulepco SP2560 and Chrompak CPSil88 are most often used for analysis of *trans* fatty acids (30). These columns have liquid phases based on 100% biscyanopropyl polysiloxane. Even with optimized temperature programming and column selection, resolution of *trans* isomers from partially hydrogenated vegetable oil mixtures is incomplete and facilitated by use of a Fourier transform infrared (FTIR) detector (Chap. 23) or mass spectrometer (Chap. 26).

14.6.3 *trans* Isomer Fatty Acids Using Infrared Spectroscopy

Most natural fats and oils extracted from plant sources contain only isolated (i.e., methylene interrupted, nonconjugated) *cis* double bonds. Fats and oils extracted from animal sources may contain small amounts of *trans* double bonds. Inasmuch as the *trans* isomer is more thermodynamically stable, additional amounts of *trans* double bonds can be formed in fats and oils that undergo oxidation or during processing treatments such as extraction, heating, and hydrogenation.

Ongoing studies are evaluating the health effects of dietary lipids that contain *trans* fatty acids.

Measurement of *trans* isomer fatty acids is commonly done using GC techniques, such as AOCS Method Ce 1f-96 (see Sect. 14.6.2). However, this section describes the use of infrared (IR) spectroscopy for determining *trans* isomer fatty acids.

14.6.3.1 Principle

The concentration of *trans* fatty acids is measurable in lipids from an absorption peak at 966 cm^{-1} in the IR spectrum.

14.6.3.2 Procedure

AOCS Method Cd 14-95 requires that liquid samples be converted to methyl esters and dissolved in an appropriate solvent that does not absorb in the IR region strongly, due to planar (carbon disulfide) or tetrahedral (carbon tetrachloride) symmetry. The absorbance spectra between 1050 and 900 cm^{-1} are obtained using an infrared spectrometer (see Chap. 23). Methyl elaidate is used as an external standard in calculating the content of *trans* double bonds. Alternately, AOCS Method Cd 14d-96 determines total *trans* fatty acids using attenuated total reflection-Fourier transform infrared (ATR-FTIR) spectroscopy (see Chap. 23).

14.6.3.3 Applications

The methods described will only detect isolated (i.e., nonconjugated) *trans* isomers. This is especially important when oxidized samples are of interest since oxidation results in a conversion from nonconjugated to conjugated double bonds. Also, AOCS Method Cd 14-95 is restricted to samples containing at least 5% *trans* isomers and AOCS Method Cd 14d-96 is limited to samples containing at least 0.8% *trans* isomer fatty acids. For samples containing less than 0.8% *trans* double bonds, a capillary GC method (AOCS Method Ce 1f-96) is recommended.

14.6.4 Mono-, Di-, and Triacylglycerols

Mono-, di-, and triacylglycerols may be determined using various techniques (see Table 14-1). Older methods use titrimetric approaches whereas newer methods utilize chromatographic techniques, including HPLC and GC. Short nonpolar columns and very high temperatures are needed for analysis of intact triacylglycerols by GC. Section 14.6.6 describes the use of TLC to separate lipid classes, including mono-, di-, and triacylglycerols.

14.6.5 Cholesterol and Phytosterols

Many methods exist for the quantification of cholesterol and phytosterols in various matrices. Consulting research literature will give an indication of current practice and methods that may be less laborious or adapted for use with specific foodstuffs.

14.6.5.1 Principle

The lipid extracted from the food is saponified. The saponification process is a hydrolysis, with acyl lipids being converted to water-soluble FFA salts. Other components (called the unsaponifiable or nonsaponifiable matter) do not change in solubility after hydrolysis, and thus they remain soluble in organic solvents. Cholesterol (in the unsaponifiable fraction) is extracted and derivatized to form trimethylsilyl (TMS) ethers or acetate esters. This increases their volatility and reduces problems of peak tailing during chromatography. Quantitation is achieved using capillary GC.

14.6.5.2 Procedure

AOAC Method 976.26 outlined here is representative of the various procedures available for cholesterol determination. Lipids are extracted from the food, saponified, and the unsaponifiable fraction is extracted. This is accomplished by filtering an aliquot of the chloroform layer through anhydrous sodium sulfate and evaporating to dryness in a water bath using a stream of nitrogen gas. Concentrated potassium hydroxide and ethanol are added and the solution is refluxed. Aliquots of benzene and 1 *N* potassium hydroxide are added and then shaken. The aqueous layer is removed, and the process is repeated with 0.5 *N* potassium hydroxide. After several washes with water, the benzene layer is dried with anhydrous sodium sulfate, and an aliquot is evaporated to dryness on a rotary evaporator. The residue is taken up in dimethylformamide. An aliquot of this sample is derivatized by adding hexamethyldisilazane (HMDS) and trimethylchlorosilane (TMCS). Water (to react with and inactivate excess reagent) and an internal standard in heptane are added, then the solution is centrifuged. A portion of the heptane layer is injected into a GC equipped with a nonpolar column. The HMDS and TMCS reagents are rapidly inactivated by water, and thus the reaction conditions must remain anhydrous.

14.6.5.3 Applications

GC quantitation of cholesterol is recommended since many spectrophotometric methods are not specific for cholesterol. In the past, samples such as eggs and shrimp have had their cholesterol contents overestimated by relying on less specific colorimetric procedures. Other GC, HPLC, and enzymatic methods are available. For example, cholesterol methods developed for frozen foods (31) and meat products (32) eliminate the fat extraction step, directly saponifying the sample; compared to the AOAC method outlined previously, they are more rapid and avoid exposure to toxic solvents.

Cholesterol oxidation products as well as phytosterols can be quantified using the GC procedure outlined for cholesterol. A wide range of methods for analysis of sterols exist in the literature; most use TMS ether formation to increase volatility of the hydroxide-containing sterols and to improve chromatographic resolution (reduce peak tailing).

14.6.6 Separation of Lipid Fractions by TLC

14.6.6.1 Procedure

TLC is performed using silica gel G as the adsorbent and hexane–diethyl ether–formic acid (80:20:2 vol/vol/vol) as the eluting solvent system (Fig. 14-7). Plates are sprayed with 2′,7′-dichlorofluorescein in methanol and placed under ultraviolet light to view yellow bands against a dark background (5).

14.6.6.2 Applications

This procedure permits rapid analysis of the presence of the various lipid fractions in a food lipid extract. For small-scale preparative purposes, TLC plates can be scraped to remove various bands for further analysis using GC or other means. Many variations in TLC parameters are available that will separate various lipids. Thin layer plates can be impregnated with silver nitrate to allow separation of FAMEs based on their number of double bonds. FAMEs with six double bonds are highly retained by the silver ions on the plate; FAMEs with no double bonds are only slightly retained. This allows a separation of FAMEs based on number of double bonds, which can be useful when identifying FAMEs in complex mixtures (bands are scraped off, eluted with solvent, and then analyzed by GC).

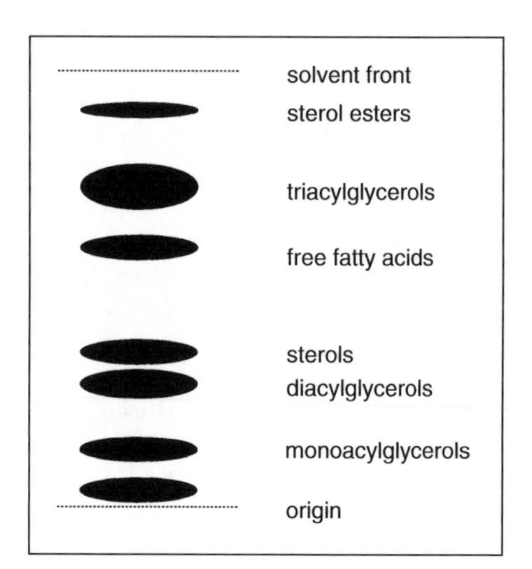

Schematic thin-layer chromatography (TLC) separation of lipid fractions on Silica gel G. [Adapted with permission from (5).]

14.7 SUMMARY

The importance of fat characterization is evident in many aspects of the food industry, including ingredient technology, product development, quality assurance, product shelf life, and regulatory aspects. The effort to reduce the amount of calories consumed as fat in the USA accentuates the significance of understanding the lipid components of food. Lipids are closely associated with health; the cholesterol or phytosterol compositions and amounts of *trans*, saturated, and $n-3/\omega 3$ fatty acids are of great concern to consumers.

The methods described in this chapter help to characterize bulk oils and fats and the lipids in foodstuffs. Methods described for bulk oils and fats can be used to determine characteristics such as melting point; smoke, flash, and fire points; color; degree of unsaturation; average fatty acid chain length; and amount of polar components. The peroxide value, TBA, and hexanal tests can be used to measure the present status of a lipid with regard to oxidation, while the OSI can be used to predict the susceptibility of a lipid to oxidation and the effectiveness of antioxidants. Lipid fractions, including fatty acids, triacylglycerols, phospholipids, and cholesterol, are commonly analyzed by chromatographic techniques such as GC and TLC.

The methods discussed in this chapter represent only a few of the many tests that have been developed to characterize lipid material. Consult the references cited for additional methods or more detailed explanations. Time, funding, availability of equipment and instruments, required accuracy, and purpose all will dictate the choice of method to characterize oils, fats, and foodstuffs containing lipids.

14.8 STUDY QUESTIONS

1. You want to compare several fat/oil samples for the chemical characteristics listed below. For each characteristic, name one test (give full name, not abbreviation) that could be used to obtain the information desired:
   (a) Degree of unsaturation

(b) Predicted susceptibility to oxidative rancidity
(c) Present status with regard to oxidative rancidity
(d) Average fatty acid molecular weight
(e) Amount of solid fat at various temperatures
(f) Hydrolytic rancidity

2. Your analysis of an oil sample gives the following results. What does each of these results tell you about the characteristics of the sample? Briefly describe the principle for each method used:
   (a) Large saponification value
   (b) Low iodine value
   (c) High TBA number
   (d) High FFA content
   (e) High OSI
3. Define solid fat content and explain the usefulness of this measurement.
4. Peroxide value, TBA number, and hexanal content all can be used to help characterize a fat sample.
   (a) What do the results of these tests tell you about a fat sample?
   (b) Differentiate these three tests as to what chemical is being measured.
5. What methods would be useful in determining the effectiveness of various antioxidants added to an oil?
6. You are responsible for writing the specifications for vegetable oil purchased from your supplier for use in deep-fat frying several foods processed by your company. Itemize the tests you should require in your list of specifications (specific values for the tests are not needed). For each test, briefly state what useful information is obtained.
7. The Nutrition Education and Labeling Act of 1990 (see Chap. 3) requires that the nutrition label on food products contains information related to lipid constituents. In addition to the amount of *total fat* (see Chap. 8), the label must state the content of *saturated fat*, *cholesterol*, and *trans fat*.
   (a) For a product such as traditional potato chips, explain an appropriate method for the analysis of each of these lipid constituents.
   (b) Compared with assays on traditional chips, how would the assays for total fat and saturated fat differ for potato chips made with Olestra®?
8. You have developed a new butter containing added fish oil, which is high in PUFA. Before placing the product on the market, you need to determine its shelf life. What method or methods would you use and why?
9. You work in quality control for a company that makes peanut butter. You received information that between July and August several lots of peanut butter may have been improperly stored and you are concerned about potential lipid oxidation in the product. A total of 50 lots of peanut butter are involved.
   (a) What test(s) would you use to measure lipid oxidation? Include your rationale for selecting the method.
   (b) How would you decide what to do with the peanut butter?

14.9 PRACTICE PROBLEMS

1. A 5.00-g sample of oil was saponified with excess KOH. The unreacted KOH was then titrated with 0.500 *N* HCl (standardized). The difference between the blank and the sample was 25.8 ml of titrant. Calculate the saponification value.
2. A sample (5.0 g) of food grade oil was reacted with excess KI to determine peroxide value. The free iodine was titrated with a standardized solution of 0.10 *N* $Na_2S_2O_3$. The amount of titrant required was 0.60 ml (blank corrected). Calculate the peroxide value of the oil.
3. You analyze the saponification value of an unknown. You use 4.0 g oil and titrate this with 44.0 ml 0.5 *N* HCl. The blank titration is 1.0 ml. What is the average fatty acid molecular weight of the oil?
4. You analyze an oil FAME by GC and find the following areas for your identified peaks:

16:0	2,853,369
18:0	1,182,738
$18{:}1n-9$	38,999,438
$18{:}2n-6$	14,344,172
$18{:}3n-3$	2,148,207

Report the fatty acid composition as % and tentatively identify the oil.

Answers

1. Saponification value

$$= \frac{25.8\,\text{ml} \times 0.500\,\text{meq/ml} \times 56.1}{5.00\,\text{g}} = 145$$

2. Peroxide value

$$= \frac{0.60\,\text{ml} \times 0.10\,\text{meq/ml} \times 1000}{5.0\,\text{g}} = 12\,\text{meq/kg}$$

3. Saponification value

$$= \frac{(44\,\text{ml} - 1\,\text{ml}) \times 0.500\,\text{meq/ml} \times 56.1}{4.0\,\text{g}} = 302$$

$$302 = \frac{(3 \times 56.1 \times 1000)}{\{[(\text{Mean mol. wt} \times 3) + 92.09] - (3 \times 18)\}}$$

Mean mol. wt = 172

4. Sum of all areas = 58,416,924

$$\text{Area \% } 16:0 = 100 \times 2,853,369/58,416,924 = 4.9\%$$

so,

16:0	4.9%
18:0	2.0%
$18{:}1n-9$	66.8%
$18{:}2n-6$	22.7%
$18{:}3n-3$	3.7%

Based on the fatty acid composition, the oil is probably canola oil.

14.10 REFERENCES

1. AOAC International (2007) Official methods of analysis, 18th edn., 2005; Current through revision 2, 2007 (online). AOAC International, Gaithersburg, MD
2. AOCS (2009) Official methods and recommended practices of the AOCS, 6th edn. American Oil Chemists' Society, Champaign, IL
3. IUPAC (1987) Standard methods for analysis of oils, fats, and derivatives and supplements, 7th edn. International Union of Pure and Applied Chemistry, Commission on Oils, Fats and Derivatives, Paquot C, Hautfenne A (eds). Blackwell Scientific, Oxford
4. Khanal RC, Dhiman TR (2004) Biosynthesis of conjugated linoleic acid (CLA): a review. Pak J Nutr 3:72–81
5. Christie WW (1982) Lipid analysis. Isolation, separation, identification, and structural analysis of lipids, 2nd edn. Pergamon, Oxford
6. Christie WW (1989) Gas chromatography and lipids. A practical guide. The Oily Press, Ayr, Scotland
7. Gray JI (1978) Measurement of lipid oxidation: a review. J Am Oil Chem Soc 55:539–546
8. Hamilton RJ, Rossell JB (1986) Analysis of oils and fats. Elsevier Applied Science, London
9. Melton SL (1983) Methodology for following lipid oxidation in muscle foods. Food Technol 37(7):105–111, 116
10. Pomeranz Y, Meloan CE (1994) Food analysis: theory and practice, 3rd edn. Chapman & Hall, New York
11. White PJ (1991) Methods for measuring changes in deep-fat frying oils. Food Technol 45(2):75–80
12. Hui YH (ed) (1996) Bailey's industrial oil and fat products, 5th edn. Wiley, New York
13. Nawar WW (1996) Lipids, ch. 5. In: Fennema OR (ed) Food chemistry, 3rd edn. Marcel Dekker, New York
14. Budavari S (ed) (1996) The Merck index. An encyclopedia of chemicals, drugs, and biologicals, 12th edn. Merck, Whitehouse Station, NJ
15. Stauffer CE (1996) Fats and oils. Eagan Press handbook series. American Association of Cereal Chemists, St. Paul, MN
16. Labuza TP (1971) Kinetics of lipid oxidation in foods. CRC Crit Rev Food Technol 2:355–405
17. Buege JA, Aust SD (1978) Microsomal lipid peroxidation. Methods Enzymol 52:302–310
18. Fulcher RG, Irving DW, de Franciso A (1989) Fluorescence microscopy: applications in food analysis, ch. 3. In: Munck L (ed) Fluorescence analysis in foods. Longman Scientific & Technical, copublished in the U.S. with Wiley, New York, pp 59–109
19. Green FJ (1990) The Sigma-Aldrich handbook of stains, dyes and indicators. Aldrich Chemical, Milwaukee, WI
20. Smart MG, Fulcher RG, Pechak DG (1995) Recent developments in the microstructural characterization of foods, ch. 11. In: Gaonkar AG (ed) Characterization of food: emerging methods. Elsevier Science, New York, pp 233–275
21. Fritsch CW, Gale JA (1977) Hexanal as a measure of rancidity in low fat foods. J Am Oil Chem Soc 54:225
22. Dupey HP, Fore SP (1970) Determination of residual solvent in oilseed meals and flours: volatilization procedure. J Am Oil Chem Soc 47:231–233
23. Tarladgis BG, Watts BM, Younathan MT, Dugan LR (1960) A distillation method for the quantitative determination of malonaldehyde in rancid foods. J Am Oil Chem Soc 37:1
24. Rhee KS, Watts BM (1966) Evaluation of lipid oxidation in plant tissues. J Food Sci 31:664–668
25. Frankel EN (1993) In search of better methods to evaluate natural antioxidants and oxidative stability in food lipids. Trends Food Sci Technol 4:220–225
26. Perkins EG (1991) Analyses of fats, oils, and lipoproteins. American Oil Chemists' Society, Champaign, IL
27. Schul D, Tallmadge D, Burress D, Ewald D, Berger B, Henry D (1998) Determination of fat in olestra-containing savory snack products by capillary gas chromatography. J AOAC Int 81:848–868
28. Tallmadge DH, Lin PY (1993) Liquid chromatographic method for determining the percent of olestra in lipid samples. J AOAC Int 76:1396–1400
29. Ackman RG, Sipos JC (1964) Application of specific response factors in the gas chromatographic analysis of methyl esters of fatty acids with flame ionization detectors. J Am Oil Chem Soc 41:377–378
30. Ratnayake WMN, Hansen SL, Kennedy MP (2006) Evaluation of the CP-Sil88 and SP-2560 GC columns used in the recently approved AOCS Official Method Ce 1h-05: determination of *cis*-, trans-, saturated, monounsaturated, and polyunsaturated fatty acids in vegetable or non-ruminant animal oils and fats by capillary GLC method. J Am Oil Chem Soc 83:475–488
31. Al-Hasani SM, Shabany H, Hlavac J (1990) Rapid determination of cholesterol in selected frozen foods. J Assoc Off Anal Chem 73:817–820
32. Adams ML, Sullivan DM, Smith RL, Richter EF (1986) Evaluation of direct saponification method for determination of cholesterol in meats. J Assoc Off Anal Chem 69:844–846

Protein Separation and Characterization Procedures

Denise M. Smith

Department of Food Science and Technology, The Ohio State University, Columbus, OH 43210, USA
smith.5732@osu.edu

15.1 Introduction 263
15.2 Methods of Protein Separation 263
15.2.1 Initial Considerations 263
15.2.2 Separation by Differential Solubility Characteristics 263
15.2.2.1 Principle 263
15.2.2.2 Procedures 263
15.2.2.2.1 Salting Out 263
15.2.2.2.2 Isoelectric Precipitation 263
15.2.2.2.3 Solvent Fractionation 264
15.2.2.2.4 Denaturation of Contaminating Proteins 264
15.2.2.3 Applications 264
15.2.3 Separation by Adsorption 264
15.2.3.1 Principle 264
15.2.3.2 Procedures 264
15.2.3.2.1 Ion-Exchange Chromatography 264
15.2.3.2.2 Affinity Chromatography 265
15.2.3.2.3 High-Performance Liquid Chromatography 265

S.S. Nielsen, *Food Analysis*, Food Science Texts Series, DOI 10.1007/978-1-4419-1478-1_15,

15.2.3.3 Applications 265
15.2.4 Separation by Size 265
15.2.4.1 Principle 265
15.2.4.2 Procedures 265
15.2.4.2.1 Dialysis 265
15.2.4.2.2 Membrane Processes 266
15.2.4.2.3 Size-Exclusion Chromatography 266
15.2.4.3 Applications 267
15.2.5 Separation by Electrophoresis 267
15.2.5.1 Polyacrylamide Gel Electrophoresis 267
15.2.5.1.1 Principle 267
15.2.5.1.2 Procedures 268
15.2.5.1.3 Applications 269
15.2.5.2 Isoelectric Focusing 269
15.2.5.2.1 Principle 269
15.2.5.2.2 Procedure 269
15.2.5.2.3 Applications 270
15.2.5.3 Capillary Electrophoresis 270
15.2.5.3.1 Principle 270
15.2.5.3.2 Procedure 270
15.2.5.3.3 Applications 271
15.3 Protein Characterization Procedures 271
15.3.1 Amino Acid Analysis 271
15.3.1.1 Principle 271
15.3.1.2 Procedures 271
15.3.1.3 Applications 272
15.3.2 Protein Nutritional Quality 272
15.3.2.1 Introduction 272
15.3.2.2 Protein Digestibility-Corrected Amino Acid Score 273
15.3.2.2.1 Principle 273
15.3.2.2.2 Procedure 273
15.3.2.2.3 Applications 273
15.3.2.3 Protein Efficiency Ratio 274
15.3.2.3.1 Principle 274
15.3.2.3.2 Procedure 274
15.3.2.3.3 Application 274
15.3.2.4 Other Protein Nutritional Quality Tests 274
15.3.2.4.1 Essential Amino Acid Index 274
15.3.2.4.2 In Vitro Protein Digestibility 274
15.3.2.4.3 Lysine Availability 275
15.3.3 Assessment of Protein Functional Properties 275
15.3.3.1 Solubility 275
15.3.3.1.1 Principle 275
15.3.3.1.2 Procedures 275
15.3.3.2 Emulsification 275
15.3.3.2.1 Principle 275
15.3.3.2.2 Procedures 276
15.3.3.3 Foaming 276
15.3.3.3.1 Principle 276
15.3.3.3.2 Procedures 276
15.3.3.4 Applications of Testing for Solubility, Emulsification, and Foaming 277
15.3.3.5 Gelation and Dough Formation 277
15.3.3.5.1 Gelation 277
15.3.3.5.2 Dough Formation 277
15.4 Summary 278
15.5 Study Questions 278
15.6 Practice Problems 279
15.7 References 280

15.1 INTRODUCTION

Many protein separation techniques are available to food scientists. Several of the separation techniques described in this chapter are used commercially for the production of food or food ingredients, whereas others are used to purify a protein from a food for further study in the laboratory. In general, separation techniques exploit the biochemical differences in protein solubility, size, charge, adsorption characteristics, and biological affinities for other molecules. These physical characteristics then are used to purify individual proteins from complex mixtures.

The biochemical, nutritional, and functional properties of food proteins can be characterized in a variety of ways. This chapter describes methods of amino acid analysis and several methods for protein nutritional quality analysis. Finally, protein solubility, emulsification, and foaming tests are described, along with gelation and dough formation, for the characterization of protein functional properties.

15.2 METHODS OF PROTEIN SEPARATION

15.2.1 Initial Considerations

Usually, several separation techniques are used in sequence to purify a protein from a food. In general, the more the separation steps used, the higher the purity of the resulting preparation. Food ingredients such as protein concentrates may be prepared using only one separation step because high purity is not necessary. Three or more separation steps are often used in sequence to prepare a pure protein for laboratory study.

Before starting a separation sequence, it is necessary to learn as much as possible about the biochemical properties of a protein, such as molecular mass, isoelectric point (pI), solubility properties, and denaturation temperature, to determine any unusual physical characteristics that will make separation easier. The first separation step should be one that can easily be used with large quantities of material. This is often a technique that utilizes the differential solubility properties of a protein. Each succeeding step in a purification sequence will use a different mode of separation. Some of the most common methods of purification are described in this section and include precipitation, ion-exchange chromatography, affinity chromatography, and size-exclusion chromatography. More detailed information about the various purification techniques can be found in a variety of sources (1,2).

15.2.2 Separation by Differential Solubility Characteristics

15.2.2.1 Principle

Separation by precipitation exploits the differential solubility properties of proteins in solution. Proteins are polyelectrolytes; thus, solubility characteristics are determined by the type and charge of amino acids in the molecule. Proteins can be selectively precipitated or solubilized by changing **buffer pH**, **ionic strength**, **dielectric constant**, or **temperature**. These separation techniques are advantageous when working with large quantities of material, are relatively quick, and are not usually influenced by other food components. Precipitation techniques are used most commonly during early stages of a purification sequence.

15.2.2.2 Procedures

15.2.2.2.1 Salting Out Proteins have unique solubility profiles in neutral salt solutions. Low concentrations of neutral salts usually increase the solubility of proteins; however, proteins are precipitated from solution as ionic strength is increased. This property can be used to precipitate a protein from a complex mixture. **Ammonium sulfate** [$(NH_4)_2SO_4$] is commonly used because it is highly soluble, although other neutral salts such as NaCl or KCl may be used to salt out proteins. Generally a two-step procedure is used to maximize separation efficiency. In the first step, $(NH_4)_2SO_4$ is added at a concentration just below that necessary to precipitate the protein of interest. When the solution is centrifuged, less soluble proteins are precipitated while the protein of interest remains in solution. The second step is performed at an $(NH_4)_2SO_4$ concentration just above that necessary to precipitate the protein of interest. When the solution is centrifuged, the protein is precipitated, while more soluble proteins remain in the supernatant. One disadvantage of this method is that large quantities of salt contaminate the precipitated protein and often must be removed before the protein is resolubilized in buffer. Tables and formulas are available in many biochemistry books (2) and online (type "ammonium sulfate calculator" into your Web browser) to determine the proper amount of $(NH_4)_2SO_4$ to achieve a specific concentration.

15.2.2.2.2 Isoelectric Precipitation The **isoelectric point** (pI) is defined as the pH at which a protein has no net charge in solution. Proteins aggregate and precipitate at their pI because there is no electrostatic repulsion between molecules. Proteins have different pIs; thus, they can be separated from each other by adjusting solution pH. When the pH of a solution is adjusted to the pI of a protein, the protein precipitates while proteins with different pIs remain in solution. The precipitated protein can be resolubilized in another solution of different pH.

15.2.2.2.3 Solvent Fractionation Protein solubility at a fixed pH and ionic strength is a function of the **dielectric constant** of a solution. Thus, proteins can be separated based on solubility differences in **organic solvent–water** mixtures. The addition of water-miscible organic solvents, such as **ethanol** or **acetone**, decreases the dielectric constant of an aqueous solution and decreases the solubility of most proteins. Organic solvents decrease ionization of charged amino acids, resulting in protein aggregation and precipitation. The optimum quantity of organic solvent to precipitate a protein varies from 5 to 60%. Solvent fractionation is usually performed at 0°C or below to prevent protein denaturation caused by temperature increases that occur when organic solvents are mixed with water.

15.2.2.2.4 Denaturation of Contaminating Proteins Many proteins are **denatured** and precipitated from solution when heated above a certain temperature or by adjusting a solution to highly acid or basic pHs. Proteins that are stable at high temperatures or at extremes of pH are most easily separated by this technique because many contaminating proteins can be precipitated while the protein of interest remains in solution.

15.2.2.3 Applications

All of the above techniques are commonly used to fractionate proteins. The differential solubility of selected muscle proteins in $(NH_4)_2SO_4$ and acetone and temperature stability at 55°C are illustrated in Table 15-1. These three techniques can be combined in sequence to prepare muscle proteins of high purity.

One of the best examples of the commercial use of differential solubility to separate proteins is in production of protein concentrates. Soy protein concentrate can be prepared from defatted soybean flakes or flour using several methods. Soy proteins can be precipitated from other soluble constituents in the flakes or flour using a 60–80% aqueous alcohol solution, by isoelectric precipitation at pH 4.5 (which is the pI of many soy proteins) or by denaturation with moist heat. These methods have been used to produce concentrates containing greater than 65% protein. Two or three separation techniques can be combined in sequence to produce soy protein isolates with protein concentrations above 90%.

Table 15-1 Conditions for Fractionating Water-Soluble Muscle Proteins Using Differential Solubility Techniques

	Precipitation Range		
Enzyme	*$(NH_4)_2SO_4$, pH 5.5, 10°C (Percent saturation)*	*Acetone pH 6.5, −5°C (Percent vol/vol)*	*Stability,[a] pH 5.5, 55°C*
Phosphorylase	30–40	18–30	U
Pyruvate kinase	55–65	25–40	S
Aldolase	45–55	30–40	S
Lactate dehydrogenase	50–60	25–35	S
Enolase	60–75	35–45	U
Creatine kinase	60–80	35–45	U
Phosphoglycerate kinase	60–75	45–60	S
Myoglobin	70–90	45–60	U

Adapted from (3) with permission of the University of Wisconsin Press. From Briskey EJ, Cassens RG, Marsh BB. *The Physiology and Biochemistry of Muscle as Food.* Copyright 1970.
[a]U, Unstable; S, Stable at heating temperature.

15.2.3 Separation by Adsorption

15.2.3.1 Principle

Adsorption chromatography is defined as the separation of compounds by adsorption to, or desorption from, the surface of a solid support by an eluting solvent. Separation is based on differential affinity of the protein for the adsorbent or eluting buffer. Ion-exchange chromatography and affinity chromatography are two types of adsorption chromatography that will be described briefly below (see Chap. 27, Sects. 27.4.3 and 27.4.5, respectively, for a more detailed description).

15.2.3.2 Procedures

15.2.3.2.1 Ion-Exchange Chromatography Ion-exchange chromatography is defined as the reversible adsorption between charged molecules and ions in solution and a charged solid support matrix. Ion-exchange chromatography is the most commonly used protein separation technique and results in an average eightfold purification. A positively charged matrix is called an **anion exchanger** because it binds negatively charged ions or molecules in solution. A negatively charged matrix is called a **cation exchanger** because it binds positively charged ions or molecules. The most commonly used exchangers for protein purification are anionic diethylaminoethyl derivatized supports, followed by carboxymethyl and phospho cation-exchangers.

The protein of interest is first adsorbed to the ion exchanger under buffer conditions (ionic strength and pH) that maximize the affinity of the protein for the matrix. Contaminating proteins of different charges pass through the exchanger unabsorbed. Proteins bound to the exchangers are selectively eluted from the column by gradually changing the ionic strength or pH of the eluting solution. As the composition of the eluting buffer changes, the charges of the proteins change and their affinity for the ion-exchange matrix is decreased.

15.2.3.2.2 Affinity Chromatography Affinity chromatography is a type of adsorption chromatography in which a protein is separated in a chromatographic matrix containing a **ligand** covalently bound to a solid support. A ligand is defined as a molecule with a reversible, specific, and unique binding affinity for a protein. Ligands include enzyme inhibitors, enzyme substrates, coenzymes, antibodies, and certain dyes. Many covalently bound ligands and associated buffers are commercially available as kits.

The protein is passed through a column containing the ligand bound to a solid support, under buffer conditions (pH, ionic strength, temperature, and protein concentration) that maximize binding of the protein to the ligand. Contaminating proteins and molecules that do not bind the ligand are eluted. The bound protein is then desorbed or eluted from the column under conditions that decrease the affinity of the protein for the bound ligand, by changing the pH, temperature, or concentration of salt or ligand in the eluting buffer.

Affinity chromatography is a very powerful technique and is a commonly used protein purification procedure. The average purification achieved by affinity chromatography is approximately 100-fold, although 1000-fold increases in purification have been reported. This technique is more powerful than size-exclusion, ion-exchange, and other separation methods that usually achieve less than a 12-fold purification. The development of new affinity chromatography procedures can be very time consuming because many variables must be optimized, which is a major disadvantage to the method. Also, the affinity materials are often more expensive than other separation media.

15.2.3.2.3 High-Performance Liquid Chromatography Many chromatographic methods have been adapted for use with **high-performance liquid chromatography** (HPLC) **systems**. The use of HPLC to separate proteins was made possible by development of macroporous, microparticulate packing materials that withstand high pressures. This technique is discussed in more detail in Chap. 28.

15.2.3.3 Applications

Ion-exchange chromatography is commonly used to separate proteins in the laboratory and can be used for quantification of amino acids in a protein as described in Sect. 15.3.5. Ion exchange chromatography is used to isolate proteins while removing lactose, minerals, and fat from sweet dairy whey. Whey protein isolates (containing greater than 90% protein) and several protein fractions, alpha-lactalbumin, lactoperoxidase, and lactoferrin, are purified from sweet dairy whey using cation exchange chromatography (4). Whey protein isolates are used as supplements in nutrition bars and beverages, since the high-quality protein is soluble and digestible.

Affinity chromatography has many uses in the analytical lab and may be used for commercial preparation of protein reagents by chemical suppliers, but is not generally used for commercial production of food protein ingredients due to the high costs involved. Glycoproteins, commonly purified by affinity chromatography, can be separated from other proteins in a complex mixture by utilization of the high carbohydrate-binding affinity of lectins. Lectins, such as concanavalin A, are carbohydrate-binding proteins that can be bound to a solid support and used to bind the carbohydrate moiety of glycoproteins that are applied to the column. Once the glycoproteins are bound to the column, they can be desorbed using an eluting buffer containing an excess of lectin. The glycoproteins bind preferentially to the free lectins and elute from the column.

15.2.4 Separation by Size

15.2.4.1 Principle

Protein molecular masses range from about 10,000 to over 1,000,000; thus, size is a logical parameter to exploit for separations. Actual separation occurs based on the **Stokes radius** of the protein, not on the molecular mass. Stokes radius is the average radius of the protein in solution and is determined by protein conformation. For example, a globular protein may have an actual radius very similar to its Stokes radius, whereas a fibrous or rod-shaped protein of the same molecular mass may have a Stokes radius that is much larger than that of the globular protein. As a result, the two proteins may separate as if they had different molecular mass.

15.2.4.2 Procedures

15.2.4.2.1 Dialysis Dialysis is used to separate molecules in solution by the use of semipermeable membranes that permit passage of small molecules but

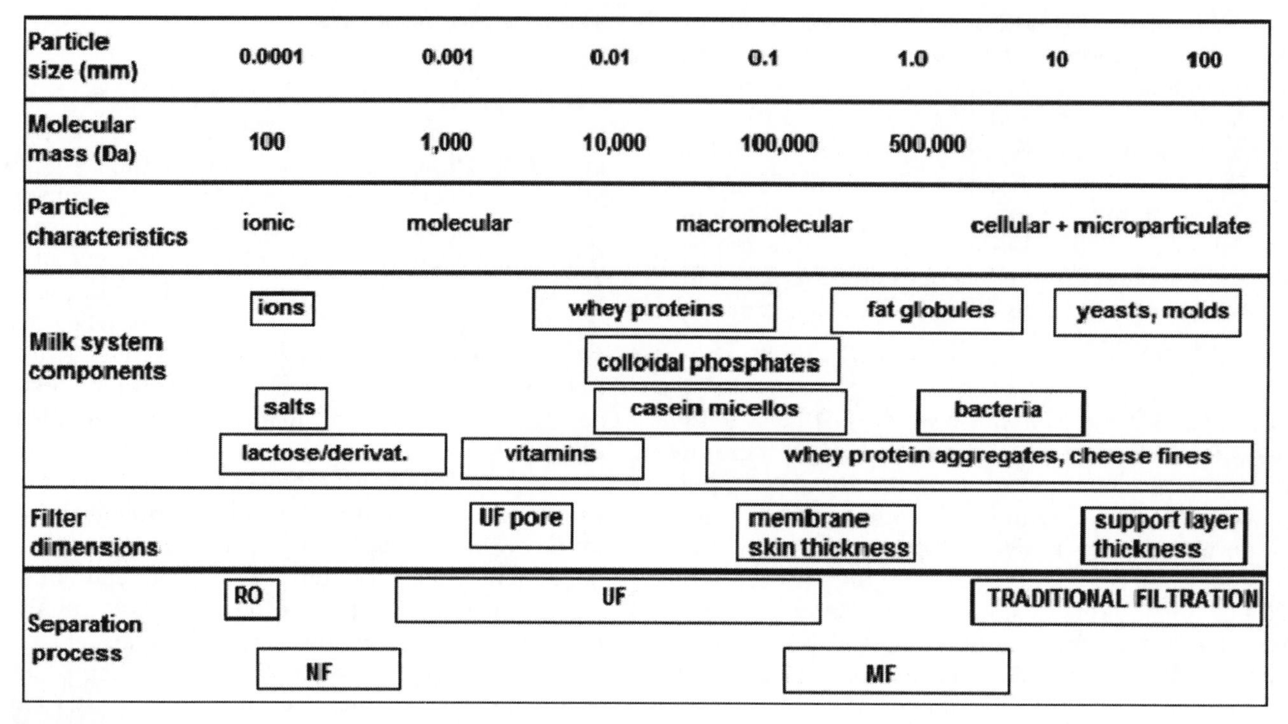

15-1 figure Range of particle sizes used for various membrane filtration techniques compared to particle sizes of milk components and microorganisms. *RO*, reverse osmosis; *NF*, nanofiltration; *UF*, ultrafiltration; *MF*, microfiltration. [Adapted from (5), with permission of the International Dairy Federation.]

not larger molecules. To perform dialysis, a protein solution is placed into dialysis tubing that has been tied or clamped at one end. The other end of the tubing is sealed, and the bag is placed in a large volume of water or buffer (usually 500–1000 times greater than the sample volume inside the dialysis tubing), which is slowly stirred. Solutes of low molecular mass diffuse from the bag, while buffer diffuses into the bag. Dialysis is simple; however, it is a relatively slow method, usually requiring at least 12 h and one change of buffer. The protein solution inside the bag is often diluted during dialysis, due to osmotic strength differences between the solution and dialysis buffer.

15.2.4.2.2 Membrane Processes Microfiltration, ultrafiltration, nanofiltration, and reverse osmosis all are processes that use a semipermeable membrane for the separation of solutes on the basis of size under an applied pressure. These methods are similar to dialysis but are much faster and are applicable to large-scale separations. Molecules larger than the membrane cutoff are retained and become part of the retentate, while smaller molecules pass through the membrane and become part of the filtrate.

These membrane processes differ mainly in the porosity of the membranes and in the operating pressure used. The porosity of the membrane sequentially decreases and the pressure used sequentially increases, respectively, for microfiltration, ultrafiltration, nanofiltration, and reverse osmosis. The approximate pore size of each membrane process relative to the different components present in milk is shown in Fig. 15-1.

Ultrafiltration is used commonly in protein research laboratories, with various commercial units available. A stirred cell ultrafiltration unit is illustrated in Fig. 15-2. The protein solution in the stirred cell is filtered through the semipermeable membrane by gas pressure, leaving a concentrated solution of proteins larger than the membrane cutoff point inside the cell. Disposable ultrafiltration units are available for small sample volumes. Molecules smaller than the membrane pore size are forced through the membrane by overriding air pressure or centrifugation resulting in concentration and purification of the protein in the retentate.

15.2.4.2.3 Size-Exclusion Chromatography Size-exclusion chromatography, also known as gel filtration or gel permeation chromatography, is a column technique that can be used to separate proteins on the basis of size. A protein solution is allowed to flow down a column packed with a solid support of porous beads made of a cross-linked polymeric material such as agarose or dextran. Beads of different average pore sizes that allow for efficient fractionation of proteins

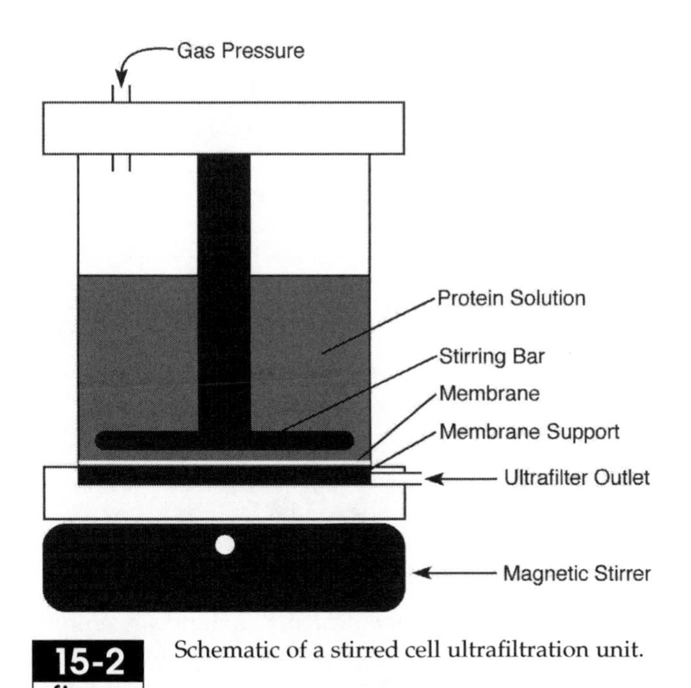

15-2 figure Schematic of a stirred cell ultrafiltration unit.

of different molecular masses are commercially available. Molecules larger than the pores in the beads are excluded, moving quickly through the column and eluting from the column in the shortest times. Small molecules enter the pores of the beads and are retarded, thus moving very slowly through the column. Molecules of intermediate sizes partially interact with the porous beads and elute at intermediate times. Consequently, molecules are eluted from the column in order of decreasing size.

Molecular mass can be calculated by chromatographing the unknown protein and several proteins of known molecular mass. Standards of known molecular mass are commercially available and can be used to prepare a standard curve. A plot of the **elution volume** (V_e) of each protein vs. **log of the molecular mass** yields a straight line. Size-exclusion techniques generally can be used to estimate molecular mass within $\pm 10\%$; however, errors can occur if the Stokes radii of the unknown protein and standards are quite different. More information on size-exclusion chromatography is available in Chap. 27, Sect. 27.4.4.

15.2.4.3 Applications

Microfiltration can be used to remove particles and microorganisms and has been applied to wastewater treatment and to remove the bacteria from milk and beer. Ultrafiltration and nanofiltration are used to concentrate a protein solution, remove salts, exchange buffer, or fractionate proteins on the basis of size. Ultrafiltration is used to preconcentrate milk for cheesemaking and to manufacture whey protein products, whereas nanofiltration has been used to remove monovalent ions from salt whey. Ultrafiltration is used to preconcentrate whole liquid egg and liquid egg white prior to spray drying. Reverse osmosis is often used to remove aqueous salts, metal ions, simple sugars, and other compounds with molecular mass below 2000. The various membrane systems can be used in combination, for example, ultrafiltration and reverse osmosis in sequence are used to concentrate whey proteins, then remove salts and lactose.

Dialysis and size-exclusion chromatography are primarily used in the analytical laboratory in a protein separation sequence. Dialysis may be used to change the buffer to one of the appropriate pH and ionic strength during purification or prior to electrophoresis of a protein sample. Dialysis is usually performed after $(NH_4)_2SO_4$ precipitation of a protein to remove excess salt and other small molecules and to solubilize protein in a new buffer. Size-exclusion chromatography is used to remove salts, change buffers, fractionate proteins, and estimate protein molecular mass.

15.2.5 Separation by Electrophoresis

15.2.5.1 Polyacrylamide Gel Electrophoresis

15.2.5.1.1 Principle Electrophoresis is defined as the migration of charged molecules in a solution through an electrical field. The most common type of electrophoresis performed with proteins is zonal electrophoresis in which proteins are separated from a complex mixture into bands by migration in aqueous buffers through a solid polymer matrix called a gel. **Polyacrylamide gels** are the most common matrix for zonal electrophoresis of proteins, although other matrices such as starch and agarose may be used. Gel matrices can be formed in glass tubes or as slabs between two glass plates.

Separation depends on the friction of the protein within the matrix and the charge of the protein molecule as described by the following equation:

$$\text{Mobility} = \frac{(\text{Applied voltage})(\text{Net charge on molecule})}{\text{Friction of the molecule}} \quad [1]$$

Proteins are positively or negatively charged, depending on solution pH and their pI. A protein is negatively charged if solution pH is above its pI, whereas a protein is positively charged if solution pH is below its pI. The magnitude of the charge and applied voltage will determine how far a protein will migrate in an electrical field. The higher the voltage and stronger the charge on the protein, the greater the migration within the electrical field. Molecular size and shape, which determine the Stokes radius of a protein, also

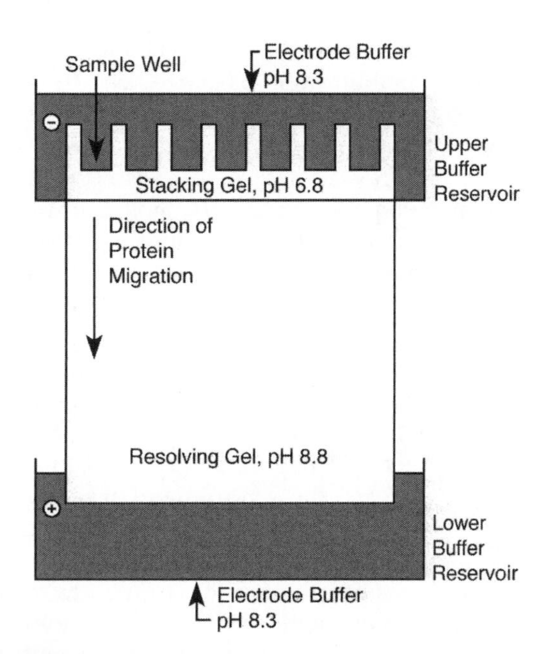

15-3 figure Schematic of a slab gel electrophoresis unit indicating the pHs of the stacking and resolving gels and the electrode buffer in an anionic discontinuous buffer system.

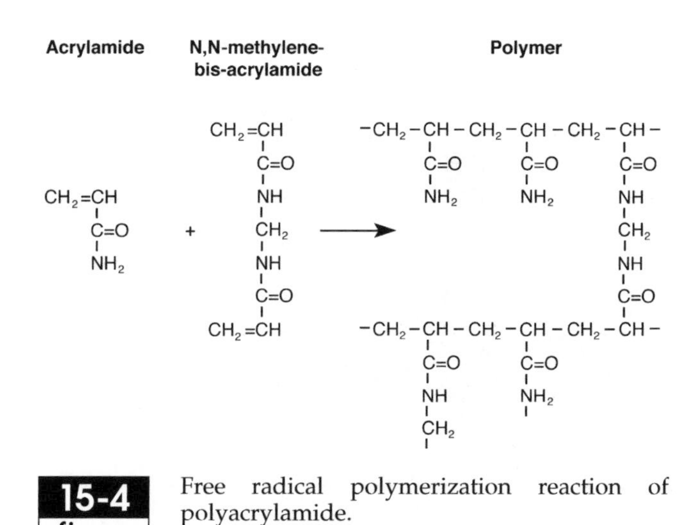

15-4 figure Free radical polymerization reaction of polyacrylamide.

determine migration distance within the gel matrix. Mobility of proteins decreases as molecular friction increases due to an increase in Stokes radius; thus, smaller proteins tend to migrate faster through the gel matrix. Similarly, a decrease in pore size of the gel matrix will decrease mobility.

In nondenaturing or **native electrophoresis**, proteins are separated in their native form based on charge, size, and shape of the molecule. Another form of electrophoresis commonly used for separating proteins is denaturing electrophoresis. **Polyacrylamide gel electrophoresis** (PAGE) with an anionic detergent, **sodium dodecyl sulfate** (SDS), is used to separate protein subunits by size. Proteins are solubilized and dissociated into subunits in a buffer containing SDS and a reducing agent. **Reducing agents**, such as mercaptoethanol or dithiothreitol, are used to reduce disulfide bonds within a protein subunit or between subunits. Proteins bind SDS, become negatively charged, and are separated based on size alone.

15.2.5.1.2 Procedures A power supply and electrophoresis apparatus containing the polyacrylamide gel matrix and two buffer reservoirs are necessary to perform a separation. A representative slab gel and electrophoresis unit is shown in Fig. 15-3. The power supply is used to make the electric field by providing a source of constant current, voltage, or power. The electrode buffer controls the pH to maintain the proper charge on the protein and conducts the current through the polyacrylamide gel. Commonly used buffer systems include an anionic tris-(hydroxymethyl)aminomethane buffer with a resolving gel at pH 8.8 and a cationic acetate buffer at pH 4.3.

The polyacrylamide gel matrix is formed by polymerizing **acrylamide** and a small quantity (usually 5% or less) of the **cross-linking reagent, *N,N′*-methylenebisacrylamide**, in the presence of a **catalyst, tetramethylethylenediamine** (TEMED), and **source of free radicals, ammonium persulfate**, as illustrated in Fig. 15-4. Gels can be made in the laboratory or purchased precast.

A discontinuous gel matrix is usually used to improve resolution of proteins within a complex mixture. The discontinuous matrix consists of a **stacking gel** with a large pore size (usually 3–4% acrylamide) and a **resolving gel** of a smaller pore size. The stacking gel, as its name implies, is used to stack or concentrate the proteins into very narrow bands prior to their entry into the resolving gel. At pH 6.8, a voltage gradient is formed between the chloride (high negative charge) and glycine ions (low negative charge) in the electrode buffer, which serves to stack the proteins into narrow bands between the ions. Migration into the resolving gel of a different pH disrupts this voltage gradient and allows separation of the proteins into discrete bands.

The pore size of the resolving gel is selected based on the molecular mass of the proteins of interest and is varied by altering the concentration of acrylamide in solution. Proteins are usually separated on resolving gels that contain 4–15% acrylamide. Acrylamide concentrations of 15% may be used to separate proteins with molecular mass below 50,000. Proteins greater than 500,000 Da are often separated on gels with acrylamide concentrations below 7%. A **gradient gel** in which the acrylamide concentration increases from the top to the bottom of the gel is often used to separate a mixture of proteins with a large molecular mass range.

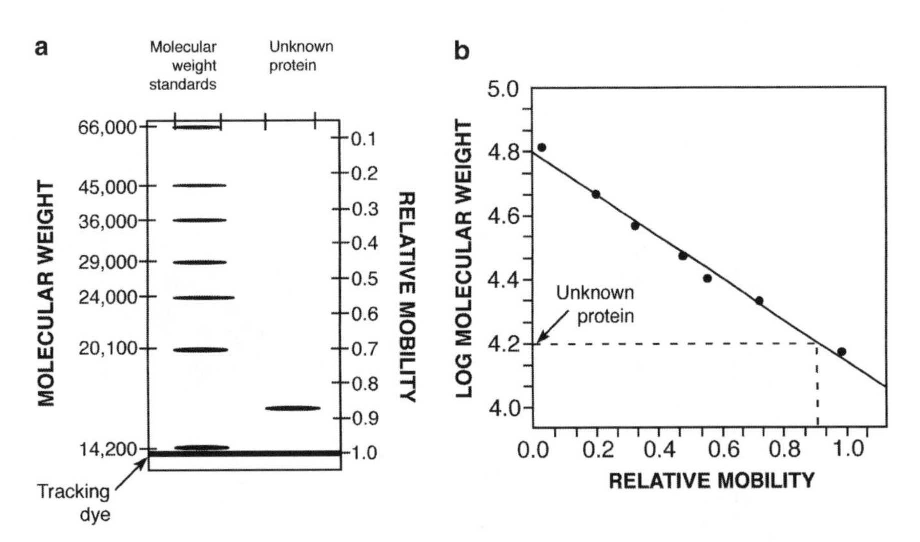

15-5 figure Use of SDS-PAGE to determine the molecular mass of a protein. (**a**) Separation of molecular mass standards and the unknown protein. (**b**) Standard curve for estimating protein molecular mass.

To perform a separation, proteins in a buffer of the appropriate pH are loaded on top of the stacking gel. **Bromophenol blue tracking dye** is added to the protein solution. This dye is a small molecule that migrates ahead of the proteins and is used to monitor the progress of a separation. After an electrophoresis run, the bands on the gels are generally visualized using a **protein stain** such as **Coomassie Brilliant Blue** or **silver stain**. Specific enzyme stains or antibodies can be used to detect a protein.

The electrophoretic or relative mobility (R_m) of each protein band is calculated as

$$R_m = \frac{\text{Distance protein migrated from start of resolving gel}}{\text{Distance between start of running gel and tracking dye}} \quad [2]$$

Additional procedural details can be found in several sources (1,2).

15.2.5.1.3 Applications Electrophoretic techniques can be used as one step in a purification process or to aid in the biochemical characterization of a protein. Commercially available preparative electrophoresis units are used to purify large quantities of protein. Alternatively, small quantities of protein can be eluted and collected from electrophoresis gels using electroelution techniques (2). A discussion of electroelution techniques is beyond the scope of this chapter. Electrophoresis can also be used to determine the purity of a protein extract.

Electrophoresis is often used to determine the protein composition of a food product. For example, differences in the protein composition of soy protein concentrates and whey protein concentrates produced by different separation techniques can be detected.

SDS-PAGE is used in characterization protocols to determine subunit composition of a protein and to estimate subunit molecular mass. Molecular mass can usually be estimated within an error of $\pm 5\%$, although highly charged proteins or glycoproteins may be subject to a larger error. Molecular mass is determined by comparing R_m of the protein subunit with R_m of protein standards of known molecular mass (Fig. 15-5). Commercially prepared protein standards are available in several molecular mass ranges. To prepare a standard curve, logarithms of protein standard molecular mass are plotted against their corresponding R_m values. The molecular mass of the unknown protein is determined from its R_m value using the standard curve.

15.2.5.2 Isoelectric Focusing

15.2.5.2.1 Principle Isoelectric focusing is a modification of electrophoresis, in which proteins are separated by charge in an electric field on a gel matrix in which a pH gradient has been generated using ampholytes. Proteins are focused or migrate to the location in the gradient at which pH equals the pI of the protein. Resolution is among the highest of any protein separation technique and can be used to separate proteins with pIs that vary less than 0.02 of a pH unit.

15.2.5.2.2 Procedure A pH gradient is formed using **ampholytes**, which are small polymers (molecular mass of about 5000 Da) containing both positively and negatively charged groups. An ampholyte mixture is composed of thousands of polymers that exhibit a range of pH values. Ampholytes are added to the

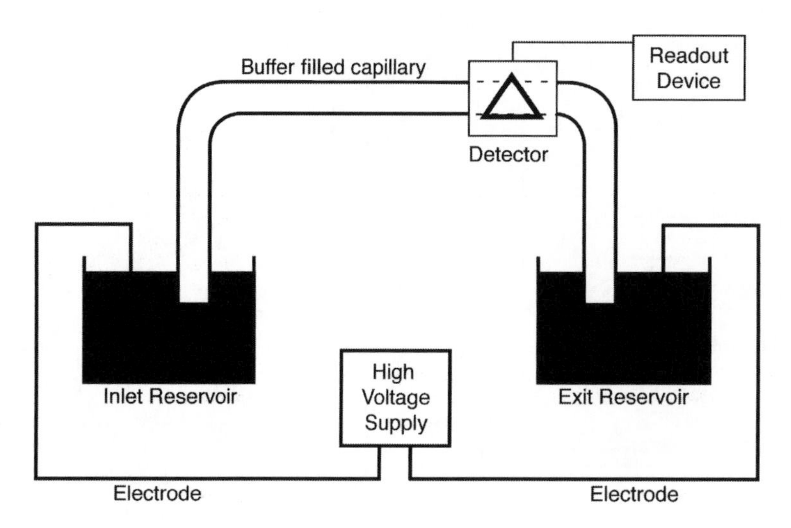

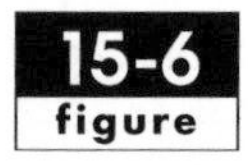
Schematic of a capillary electrophoresis system.

gel solution prior to polymerization. After the gel is formed and a current applied, the ampholytes migrate to produce the pH gradient; negatively charged ampholytes migrate toward the anode while positively charged ampholytes migrate toward the cathode. Ampholyte mixtures are available that cover a narrow pH range (2–3 units) or a broad range (pH 3–10) and should be selected for use based on properties of the proteins to be separated.

15.2.5.2.3 Applications Isoelectric focusing is the method of choice for determining the isoelectric point of a protein and is an excellent method for determining the purity of a protein preparation. For example, isozymes of polyphenol oxidase and other plant and animal proteins are identified using isoelectric focusing. Isoelectric focusing is used to differentiate closely related fish species based on protein patterns.

Isoelectric focusing and SDS-PAGE can be combined to produce a two-dimensional electrophoretogram that is extremely useful for separating very complex mixtures of proteins. This technique is called **two-dimensional electrophoresis**. Proteins are first separated in tube gels by isoelectric focusing. The tube gel containing the separated proteins is then placed on top of an SDS-PAGE slab gel, and proteins are separated. Thus, proteins are separated first on the basis of charge and then according to size and shape. Over 1000 proteins in a complex mixture have been resolved using this technique.

15.2.5.3 Capillary Electrophoresis

15.2.5.3.1 Principle Similar principles apply for the separation of proteins by both capillary and conventional electrophoretic techniques; proteins can be separated on the basis of charge or size in an electric field. The primary difference between capillary electrophoresis and conventional electrophoresis (described previously) is that **capillary tubing** is used in place of acrylamide gels cast in tubes or slabs. **Electroosmotic flow** within the capillary also can influence separation of proteins in capillary electrophoresis and is discussed briefly in Sect. 15.2.5.3.3.

15.2.5.3.2 Procedure A schematic diagram of a capillary electrophoresis system is shown in Fig. 15-6. A capillary electrophoresis system comprises a capillary column, power supply, detector, and two buffer reservoirs. The sample is introduced into the inlet side of the capillary tube by simply replacing the inlet buffer reservoir with the sample solution and applying low pressure or voltage across the capillary until the desired volume of sample has been loaded onto the column. Capillaries are composed of fused silica with internal diameters that commonly range from 25 to 100 μm. Column length varies from a few centimeters to 100 cm. High electric fields (100–500 V/cm) can be used as the narrow columns dissipate heat very effectively, allowing for short run times of 10–30 min.

At the end of a run protein bands are not visualized by staining as in conventional electrophoresis. Instead, protein bands are detected on the column as they migrate past a detector. The detectors are similar to those used in HPLC described in Chap. 28. Ultraviolet (UV)–visible detectors are most common, although fluorescence and conductivity detectors are available. Proteins can be labeled with a fluorescent derivative to increase sensitivity when a fluorescent detector is used. The data obtained from a capillary electrophoresis run look like a typical chromatogram from a high-performance liquid chromatograph or gas chromatograph (see Chaps. 28 and 29).

15.2.5.3.3 Applications Capillary electrophoresis is used primarily in analytical labs, although use for routine quality control purposes is increasing. There are three variations of capillary electrophoresis commonly used for protein separations.

Capillary zone electrophoresis or **free solution electrophoresis** is much like native PAGE, except proteins are separated in free solution inside capillary tubes filled with buffer of the desired pH. Diffusion is prevented within the narrow diameter of the capillaries, so a gel matrix is not required. In capillary zone electrophoresis, electroosmotic flow also influences the separation of proteins within capillary tubes. The negatively charged fused silica capillary wall [containing silanol groups (SiO^-)] attracts positively charged ions (cations) from the buffer to form a double ion layer at the interface between the capillary column wall and the buffer. When the electric field is applied, the cations forming the double layer are attracted toward the cathode and "pull" other molecules (independent of charge) in the same direction. Thus, in free solution capillary electrophoresis, cations, anions, and uncharged molecules can be separated in a single run. Electroosmotic flow can be controlled by changing the pH or ionic strength of the buffer to alter the charge on the capillary wall and change the rate of protein migration. Capillary zone electrophoresis is used for a variety of applications, including the fractionation of milk, cereal, soybean, and muscle proteins (6).

SDS capillary gel electrophoresis techniques can be used to separate proteins by size and to determine molecular mass. In this technique proteins are denatured and dissociated in the presence of SDS and a reducing agent, then fractionation occurs in polyacrylamide gel-filled capillaries of specific pore sizes. Alternatively, linear polymers, such as methyl cellulose, dextrans, or polyethylene glycol, are added to the buffer within the capillary in a technique called dynamic sieving capillary electrophoresis. These entangled polymers act like the pores of the polyacrylamide gel to slow migration of the larger proteins and allow separation by size.

Proteins also can be separated on the basis of their isoelectric points, in a technique called **capillary isoelectric focusing**. Ampholytes (described in Sect. 15.2.5.2.2) are used to form a pH gradient within the capillary tube. A gel matrix is not needed. In this technique, electroosmotic flow is minimized by coating the capillary walls with buffer additives to prevent undesirable effects caused by surface charge.

15.3 PROTEIN CHARACTERIZATION PROCEDURES

15.3.1 Amino Acid Analysis

15.3.1.1 Principle

Amino acid analysis is used to quantitatively determine the amino acid composition of a protein. The protein sample is first hydrolyzed to release the amino acids. Amino acids are then separated using chromatographic techniques and quantified. Ion-exchange chromatography, reversed-phase liquid chromatography, and gas–liquid chromatography are three separation techniques used. This section will describe the use of ion-exchange and reversed-phase liquid chromatography.

15.3.1.2 Procedures

In general, a protein sample is **hydrolyzed** in constant boiling 6 *N* HCl for 24 h to release amino acids prior to chromatography. Accurate quantification of some amino acids is difficult because they react differently during hydrolysis. Consequently, special hydrolysis procedures must be used to prevent errors.

Tryptophan is completely destroyed by acid hydrolysis. Methionine, cysteine, threonine, and serine are progressively destroyed during hydrolysis; thus, the duration of hydrolysis will influence results. Asparagine and glutamine are quantitatively converted to aspartic and glutamic acid, respectively, and cannot be measured. Isoleucine and valine are hydrolyzed more slowly in 6 *N* HCl than other amino acids, while tyrosine may be oxidized.

In general, losses of threonine and serine can be estimated by hydrolysis of samples for three periods of time (i.e., 24, 48, and 72 h) followed by amino acid analysis. Compensation for amino acid destruction may be made by calculation to zero time assuming first-order kinetics. Valine and isoleucine are often estimated from a 72-h hydrolysate. Cysteine and cystine can be converted to the more stable compound, cysteic acid, by hydrolysis in performic acid and then hydrolyzed in 6 *M* HCl and chromatographed. Tryptophan can be separated chromatographically after a basic hydrolysis or analyzed using a method other than amino acid analysis.

In the original method developed by Moore and Stein (7) and later revised by Stein and colleagues (8), amino acids were separated by **cation-exchange chromatography** using a stepwise elution with three buffers of increasing **pH** and **ionic strength**. In a procedure called **postcolumn derivatization**, amino acids eluting from the column were derivatized and quantified by reaction with **ninhydrin** (reacts with primary amino group of amino acids) to produce a colored product that was measured spectrophotometrically. The method was automated in the late 1970s and adapted for use with high-performance liquid chromatographs in the 1980s, as new ion-exchange resins were developed that could withstand high pressures and extremes of pH, ionic strength, and temperature. Amino acids eluting from the column also may be derivatized with *o*-**phthalaldehyde** (OPA) (reacts with primary amino group of amino acids), then measured with a fluorescence detector. (Note: The ninhydrin and OPA methods can be used not only for amino acid analysis but also to monitor **hydrolysis of proteins** and to assay for **protease activity**, since these result in an increase of free primary amino groups.)

Other methods were developed in the 1980s using **precolumn derivatization** of the amino acids followed by reversed-phase HPLC. The hydrolyzed amino acids are derivatized prior to chromatography with **phenylthiocarbamyl** (reacts with primary amino group of amino acids) or other compounds, separated by reversed-phase HPLC, and quantified by UV spectroscopy. Methods using precolumn derivatizations can detect picomole quantities of amino acids. Chromatographic runs usually take 30 min or less. A chromatogram showing the separation of amino acids in an infant formula is shown in Fig. 15-7.

The quantity of each amino acid in a peak is usually determined by spiking the sample with a known quantity of internal standard. The **internal standard** is usually an amino acid, such as norleucine, not commonly found in a food product. Results are usually expressed as mole percent. This quantity is calculated by dividing the mass of each amino acid (determined from the chromatogram) by its molecular mass, summing the values for all amino acids, dividing each by the total moles, and multiplying the result by 100.

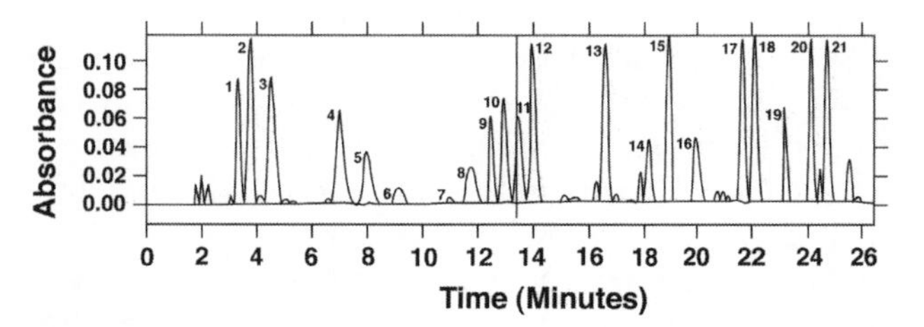

15-7 figure High-performance liquid chromatographic analysis of phenylthiocarbamyl-derived amino acids from infant formula separated on a reversed-phase column. Sample was spiked with taurine (1. Asp, 2. Glu, 3. internal standard, 4. Ser, 5. Gly, 6. His, 7. Tau, 8. Arg, 9. Thr, 10. Ala, 11. NH_3, 12. Pro, 13. internal standard, 14. Tyr, 15. Val, 16. Met, 17. Ile, 18. Leu, 19. Phe, 20. reagent, 21. Lys). [Adapted from (9), with permission from Millipore Corporation, © 1990 by Millipore Corporation.]

15.3.1.3 Applications

Amino acid analysis is used to determine the amino acid composition of a protein, determine quantities of essential amino acids to evaluate protein quality, identify proteins based on the amino acid profile, detect odd amino acids, and corroborate synthetic or recombinant protein structures. Amino acid analysis provides information for estimating the molecular mass of a protein. Proteins used in animal diets, infant formulas, sports nutrition products, and therapeutic human diets are often analyzed for protein quality to ensure adequate quantities of essential amino acids.

15.3.2 Protein Nutritional Quality

15.3.2.1 Introduction

The nutritional quality of a protein is determined by the **amino acid composition** and the **digestibility** of that protein. **Antinutritional factors** can affect the nutritional quality of a protein. However, foods that contain heat-labile antinutritional factors (e.g., trypsin inhibitors) are usually cooked prior to consumption, thereby inactivating the inhibitor that might reduce protein digestibility. Some foods contain heat-stable antinutritional factors (e.g., tannins) that can decrease the nutritive value of a protein.

Many protein quality assessment methods utilize information about the essential amino acid content of a food. **Essential amino acids** are those that cannot be synthesized in the body and must be present in the diet. Although there are some special cases due to age and medical status of an individual, the amino acids generally categorized as essential (or indispensable) include histidine, isoleucine, leucine, lysine, methionine, phenylalanine, threonine, tryptophan, and valine. Requirements for these amino acids have been determined for various age groups of humans (10) (Table 15-2). The **first-limiting amino acid** of a human food is defined as the essential amino acid present in the lowest amount compared to a reference protein or to human requirements.

A food scientist's concerns regarding protein nutritional quality include meeting the requirements of nutrition labeling, formulating products of high protein quality, and testing the effects of food processing on protein digestibility. Protein nutritional quality assays may utilize animals in biological tests (in vivo), chemical or biochemical assays (in vitro), and/or simply calculations (11,12). Because of the time and expense of in vivo methods, in vitro assays and

15-2 table Amino Acid Requirements of Infants, Preschool Children, Adolescents, and Adults (Males and Females Combined)

Age (Years)		*His*	*Ile*	*Leu*	*Lys*	*SAA*	*AAA*	*Thr*	*Trp*	*Val*
				Amino Acid Requirements (mg/kg per day)						
0.5	Infants	22	36	73	64	31	59	34	9.5	49
1–2	Preschool children	15	27	54	45	22	40	23	6.4	36
11–14	Adolescent	12	22	44	35	17	30	18	4.8	29
>18	Adult	10	20	39	30	15	25	15	4.0	26
				Scoring Pattern (mg/g Protein Requirement)						
0.5	Infants	20	32	66	57	28	52	31	8.5	43
1–2	Preschool children	18	31	63	52	26	46	27	7.4	42
11–14	Adolescent	16	30	60	48	23	41	25	6.5	40
>18	Adult	15	30	59	45	22	38	23	6.0	39

His, Histidine; *Ile*, Isoleucine; *Leu*, Leucine; *Lys*, Lysine; *SAA*, Sulfur amino acids; *AAA*, Aromatic amino acids; *Thr*, Threonine; *Trp*, Tryptophan; *Val*, Valine. Adapted from (10).

calculations based on amino acid content alone often are used to estimate protein quality. This section of the chapter covers the tests and calculations required for nutrition labeling and mentions briefly several other protein quality methods for specialized applications.

15.3.2.2 Protein Digestibility-Corrected Amino Acid Score

15.3.2.2.1 Principle The **Protein Digestibility – Corrected Amino Acid Score** (PDCAAS method) (13) estimates protein nutritional quality by combining information from (1) a calculation that compares the amount of the first-limiting amino acid in a protein to the amount of that amino acid in a reference protein, and (2) an in vivo assay measuring the digestibility of the protein by rats.

15.3.2.2.2 Procedure

1. Determine the amino acid composition of the food.
2. Calculate the **amino acid score** for the first-limiting amino acid, using the requirements of preschool age children as a reference pattern.

$$\text{Amino acid score} = \frac{\text{mg of amino acid in 1 g of test protein}}{\text{mg of amino acid in 1 g of reference protein}} \quad [3]$$

3. Feed male weanling rats standardized diets with 10% test protein or with no protein, following the procedure for true protein digestibility (AOAC Method 991.29) (14). **True digestibility** is calculated based on nitrogen ingested and feed intake, corrected for metabolic losses in the feces. If available, published values of true digestibility for the test protein can be used.
4. Calculate PDCAAS:

$$\text{Amino acid score} \times \%\,\text{True digestibility} \quad [4]$$

5. For nutritional labeling: (50 g = Daily Value for protein)

$$\%\,\text{Daily Value} = \frac{100 \times (\text{g protein/serving} \times \text{PDCAAS value})}{50\,\text{g protein}} \quad [5]$$

15.3.2.2.3 Applications The Nutrition Labeling and Education Act (NLEA) requires that the percent Daily Value used on nutrition labels must be determined using the PDCAAS method, except for foods intended for consumption by infants (13) (see also Chap. 3, Sect. 3.2.1.7). Because of the time and cost associated with the PDCAAS method, protein on the nutrition label is usually expressed only as amount and not as a percent of the Daily Value. However, if a food label includes any claim regarding the protein, the nutrition label must include protein expressed as a percent of the Daily Value (15) [21CFR 101.9 (c) (7)].

The PDCAAS method generally is thought to better estimate protein quality for humans than the **Protein Efficiency Ratio** (PER) method, which measures rat growth (11). Rat growth is not comparable to that of adult humans, but it is more comparable to that of human infants. Therefore, the PER method is used to estimate protein quality of only infant foods. The actual digestion of protein by rats is thought to be fairly comparable to that by humans, so the protein digestibility portion of the PDCAAS method utilizes true digestibility as determined with rats.

The PDCAAS method includes information on both amino acid composition and protein digestibility, since these are the factors that determine protein nutritional quality. However, perhaps as a limitation to the PDCAAS method, the amino acid score portion of the PDCAAS method includes only information about the first-limiting amino acid, and not other essential amino acids. There is no differentiation in amino acid score between two proteins limiting to the same extent in one amino acid, but with the one protein only limiting in that amino acid and another protein limiting in many amino acids.

15.3.2.3 Protein Efficiency Ratio

15.3.2.3.1 Principle The PER method (AOAC Method 960.48) (14) estimates protein nutritional quality in an in vivo assay by measuring rat growth as weight gain per gram of protein fed.

15.3.2.3.2 Procedure

1. Determine the nitrogen content of the test protein-containing sample and calculate the protein content.
2. Formulate a standardized test protein diet and a casein control diet to each containing 10% protein.
3. Feed groups of male weanling rats the diet and water *ad libitum* for 28 days.
4. Record the weight of each animal at the beginning of the assay, at least every 7 days during the assay, and at the end of 28 days.
5. Record the food intake of each animal during the 28-day feeding trial.
6. Calculate the PER using the average total weight gain and average total protein intake for each diet group at day 28:

$$\text{PER} = \frac{\text{Total weight gain of test group (g)}}{\text{Total protein consumed (g)}} \quad [6]$$

7. Normalize the PER value for the test protein (i.e., compare the quality of the test protein to that of casein) by assigning casein a PER of 2.5.

$$\text{Adjusted or corrected PER} = \frac{\text{PER of test protein}}{\text{PER of casein control}} \quad [7]$$

15.3.2.3.3 Application The Food and Drug Administration (FDA) requires that the PER method be used when determining protein as a percent of the Daily Value on the nutrition label of foods intended for consumption by infants (13) (see also Chap. 3, Sect. 3.2.1.7). The PER method is limited to this application because the essential amino acid requirements for young rats are similar to those of human infants, but not to those of other age groups. The PER method is time consuming, and it does not give any value to a protein for its ability to simply maintain body weight (i.e., a protein that produces no weight gain in the assay has a PER of zero).

15.3.2.4 Other Protein Nutritional Quality Tests

15.3.2.4.1 Essential Amino Acid Index **Essential amino acid index** (EAAI) estimates protein nutritional quality based on the content of all essential amino acids compared to a reference protein (or human requirements). The EAAI is a rapid method to evaluate and optimize the amino acid content of food formulations. Unlike the amino acid score component of the PDCAAS method that considers only the first-limiting amino acid, the EAAI method accounts for all essential amino acids. However, EAAI does not include any estimate of protein digestibility, which could be affected by processing method. The essential amino acid content of the test protein (determined by amino acid analysis or from literature values) is compared to that of a reference protein (e.g., casein or human requirements) as follows: (Note: Use values for methionine plus cystine, and phenylalanine plus tyrosine, because they can substitute for one another as essential amino acids.)

$$\text{Essential amino acid index} = \sqrt[9]{\left(\frac{\text{mg of lysine in 1 g of test protein}}{\text{mg of lysine in 1 g reference protein}}\right) \times (\text{etc. for other 8 essential amino acids})} \quad [8]$$

15.3.2.4.2 In Vitro Protein Digestibility The **pH shift method** is an in vitro protein digestibility assay used to estimate the digestibility of a protein by measuring the extent of protein hydrolysis upon reaction under standardized conditions with commercial digestive enzymes. The digestion procedure is designed to simulate human digestion of protein by using the enzymes trypsin, chymotrypsin, peptidase, and a bacterial protease. The pH of the protein solution drops when proteases break peptide bonds, release carboxyl groups, and liberate hydrogen ions. The pH at the end of the digestion period is used to calculate protein digestibility.

The pH shift method is the enzyme digestibility part of the **Protein Efficiency Ratio-Calculation Method** (C-PER) (AOAC Method 982.30) (14), which combines a calculation of the essential amino acid composition with a calculation based on the in vitro assay of digestibility. The C-PER assay is intended for routine quality control screening of foods and food ingredients, to estimate the PER as it would be determined with the rat bioassay method.

The in vitro digestibility assay can provide a rapid and inexpensive means to determine and compare the protein digestibility of food products. This assay could be used to determine the effect of processing conditions on the protein digestibility of food products with the same formulation.

15.3.2.4.3 Lysine Availability If the nutritional quality of a heat-processed product is lower than expected from its amino acid composition, one should assess the availability of lysine in the product. The **free ε-amino group** of the essential amino acid lysine can react with many food constituents during processing and storage to form biologically unavailable complexes with reduced nutritional quality. Lysine can readily complex with reducing sugars in the Maillard browning reaction, oxidized polyphenols, and oxidized lipids. Such reactions are accelerated upon heating and under alkaline conditions. **Lysinoalanine**, often found in alkali-treated proteins, decreases both the digestibility of the protein and the availability of lysine as an essential amino acid. The most commonly used method to measure **available lysine** is a spectrophotometric method that utilizes the reagent 1-fluoro-2,3-dinitrobenzene (DNFB, also referred to as FDNB). The DNFB method, detailed in AOAC Method 975.44 (14), can be used to determine if food processing operations have reduced the availability of lysine in a food.

15.3.3 Assessment of Protein Functional Properties

Protein functionality has been defined as the physical and chemical properties of protein molecules that affect their behavior in food products during processing, storage, and consumption. The functional properties of proteins contribute to the quality attributes, organoleptic properties, and processing yields of food. It is often desirable to characterize the functional properties of food proteins to optimize their use in a food product. Three of the most important protein functional properties in foods include **solubility**, **emulsification**, and **foaming**. This chapter will highlight only a few of the many methods available to measure these three functional properties. Two other functional properties of proteins, **gelation** and **dough formation**, are closely related to viscosity. It should be pointed out that there is no single test to measure these functional properties that is applicable to all food systems, so careful test selection is imperative.

15.3.3.1 Solubility

15.3.3.1.1 Principle One of the most popular tests of protein functionality is solubility. Proteins usually need to be soluble under the conditions of use for optimal functionality in food systems. Many other important functional attributes of proteins are influenced by protein solubility, such as thickening (viscosity effects), foaming, emulsification, water binding, and gelation properties.

Solubility is dependent on the balance of hydrophobic and hydrophilic amino acids that make up the protein, especially those amino acids on the surface of the molecule. Protein solubility also is dependent on the thermodynamic interactions between the protein and the solvent. Protein solubility is influenced by solvent polarity, pH, ionic strength, ion composition, and interactions with other food components, such as lipids or carbohydrates. Common food processing operations, such as heating, freezing, drying, and shearing, may all influence the solubility of proteins in a food system. Proteolysis by endogenous proteases also may alter protein solubility.

15.3.3.1.2 Procedures There are many standardized methods to measure protein solubility including those published by the American Oil Chemists' Society (16) and the AACC International (17). Other names synonymous with protein solubility include protein dispersibility index and nitrogen solubility index.

In a typical solubility procedure, a protein is dispersed in water or buffer at a specified pH, and the dispersion is centrifuged using defined conditions. Protein insoluble under the test conditions precipitates while soluble protein remains in the supernatant. Buffer and test conditions must be carefully controlled as these parameters have a large influence on results. Total protein and protein in the supernatant are measured, usually by the Kjeldahl method or a colorimetric procedure, such as the Bradford assay or the bicinchoninic acid (BCA) assay (Chap. 9). Percentage protein solubility is measured by dividing the protein in the supernatant (soluble protein) by the total protein and multiplying by 100.

15.3.3.2 Emulsification

15.3.3.2.1 Principle Food **emulsions** include margarine, butter, milk, cream, infant formulas, mayonnaise, processed cheese, salad dressings, ice cream, and some highly comminuted meat products, such as bologna. Emulsions are mixtures of two or more immiscible liquids, one of which is dispersed as droplets in the other. Oil and water are the two most common immiscible liquids found in food emulsions, although many other food components are usually present. The droplets are collectively called the **discontinuous** or **dispered phase**, whereas the liquid surrounding the droplets is the **continuous phase**. Energy, in the form of homogenization, blending, or shaking, is used to disperse one immiscible liquid into the other.

Proteins are used as emulsifiers to lower the interfacial energy between phases to facilitate emulsion formation and to improve the stability of emulsions. Proteins migrate to the surface of a droplet during emulsion formation to form a protective layer or membrane on the surface, thus reducing interactions between the two immiscible phases. Emulsions are inherently unstable. The quality of an emulsion is dictated by many factors including droplet size, droplet size distribution, density differences between the two phases, viscosity of the two phases, electrostatic and steric interactions between molecules at the interface, and thickness and viscosity of the adsorbed protein layer (18).

15.3.3.2.2 Procedures Food emulsions usually are highly complex systems with multiple ingredients, so industrial scientists often choose to investigate the properties of emulsions in simplified model systems containing only a few of the most important ingredients. For a protein-based emulsion, this model system may contain only water or buffer, oil, and protein. It must be remembered that pH, salt concentration, temperature, type and amount of oil, protein concentration, energy input, and temperature during emulsion formation have a large effect on the properties of the final emulsion. These parameters must be selected prior to establishing a procedure.

The droplet size of the dispersed phase in an emulsion has a large influence on the emulsion quality. Droplet size can influence appearance, stability, and rheological properties of an emulsion, and hence the quality of a food product. Smaller droplets of more uniform size indicate a better emulsion. Droplet size can be determined by turbidimetric techniques, microscopy, laser diffraction, and electrical pulse counting (19).

An efficient emulsifier can prevent the breakdown or phase separation of an emulsion during storage. Emulsions can be stable for a long period of time (months to years), so often test protocols include a destabilization step involving physical or chemical stress. **Emulsion stability** can be tested by centrifugation or agitation of an emulsion at a given speed and time to determine the amount of creaming or oil separation that occurs. This is a fairly rapid test, but may not adequately represent the breakdown of the emulsion during normal storage conditions. Another method involves measuring the change in particle size distribution of the dispersed phase over time (19).

Many other more sophisticated techniques are available to measure the properties of food emulsions and include measures of interfacial properties, measurement of the dispersed phase volume fraction, characterization of emulsion rheology, and investigations into droplet charge (18).

15.3.3.3 Foaming

15.3.3.3.1 Principle Foams are coarse dispersions of gas bubbles in a liquid or semisolid continuous phase. Like emulsions, foams require energy input during formation and are inherently unstable. Whipping, shaking, and sparging (gas injection) are three common methods of foam formation. Proteins or other large macromolecules in the continuous phase lower the surface tension between the two phases during foam formation and impart stability to films formed around the gas bubbles. Foams are found in cakes, breads, marshmallows, whipped cream, meringues, ice cream, souffles, mousses, and beer.

15.3.3.3.2 Procedures Foam volume and foam stability are two important parameters used to evaluate foams. **Foam volume** is dependent on the ability of a protein to lower surface tension between the aqueous phase and gas bubbles during foam formation. The volume of foam generated during a standardized foaming process is recorded and can be compared to other foams made under identical conditions. Often the foam is formed in a blender and then transferred into a graduated cylinder for measurement. Another common approach is to measure foam overrun or foam expansion. This is an indirect measure of the amount of air incorporated into the foam. Percent overrun can be calculated as follows:

$$\text{Overrun}(\%) = \frac{\text{Weight 100 ml liquid} - \text{Weight 100 ml foam}}{\text{Weight 100 ml foam}} \times 100 \quad [9]$$

Foam stability depends on the properties of the protein film formed around the gas droplets. Free liquid is released as a foam breaks down. A more stable foam usually takes a longer time to collapse. Foam stability is often expressed as a half-life. A sample is whipped for a fixed time under standardized conditions. In one of the simplest methods, the foam is placed in a funnel over a graduated cylinder. The time for half of the original weight or volume of the foam to drain away from the foam is recorded. The greater the half-life, the more stable the foam.

Foam volume and stability are influenced by energy input, pH, temperature, heat treatment, and by the type and concentration of ions, sugars, lipids, and proteins in the foam. Hence all of these variables, except the one under test, must be standardized when designing a procedure to measure foam properties. Other methods can be used to better understand the molecular properties of the foam including: (1) determination of interfacial properties such as surface pressure and film thickness, (2) characterization of the viscoelastic properties of the film, and (3) characterization of the bubble size and distribution.

15.3.3.4 Applications of Testing for Solubility, Emulsification, and Foaming

Proteins perform a variety of functions in food systems. The quality of many foods is dependent on the successful manipulation of protein functional properties during processing. The solubility of a protein under a given set of conditions must be known for optimizing the use of that protein in a food product. For example, proteins must be soluble for optimal function in most beverages. Slight formulation modifications may change product pH and subsequently protein solubility.

Many proteins become insoluble and less functional once they are denatured. Thus, solubility often is used as an index of protein denaturation that may occur during processing. In general, proteins must be soluble to migrate to surfaces during foaming and emulsification. Any change in a process or ingredient that affects solubility also may alter the foaming and emulsification properties of that protein. Freezing, heating, shearing, and other processes can influence the functional properties of proteins in food systems. Thus, it may be necessary to reevaluate solubility, emulsion, and foaming properties of proteins if any processing changes are made.

Food product developers often need to understand the effect of ingredient substitutions on emulsion or foaming properties and subsequent product quality. Variations in raw materials can lead to differences in protein functionality and ultimately product quality. Protein functional tests can be used to compare two ingredients from different manufacturers or to verify the quality of protein ingredients in each batch purchased from a manufacturer. For example, the solubility of different commercial whey and soy protein concentrates may vary, leading to differences in other functional responses and, ultimately, product quality. Also, if a product developer is trying to substitute soy protein for egg protein in a formulation, it might be important to know how the functional attributes of each ingredient are affected by environmental conditions, such as pH or salt concentration. It often is desirable to know how long an emulsion or foam will be stable under a certain set of storage conditions. For example, Fligner et al. (20) used emulsion stability tests to evaluate the long-term stability of infant formulas.

15.3.3.5 Gelation and Dough Formation

Two other functional properties of proteins, gelation and dough formation, are closely related to viscosity and are described here only briefly.

15.3.3.5.1 Gelation Protein gels are made by treating a protein solution with heat, enzymes, or divalent cations under appropriate conditions. While most food **protein gels** are made by heating protein solutions, some can be made by limited enzymatic proteolysis (e.g., chymosin action on casein micelle to form cheese curd), and some are made by addition of the divalent cations Ca^{2+} or Mg^{2+} (e.g., tofu from soy proteins). Proteins are transformed from the "soluble" state to a cross-linked "gel-like" state in protein gelation. The continuous, cross-linked network structure formed may involve covalent (i.e., disulfide bond) and/or noncovalent interactions (i.e., hydrogen bonds, hydrophobic interactions, electrostatic interactions). Some protein gel networks made by heating a protein solution are thermally reversible (e.g., gelatin gel, that is formed by heating and then cooling gelatin, but reverts to a liquid when reheated), while some protein gels are thermally irreversible. The stability of a protein gel is affected by a variety of factors, such as the nature and concentration of the protein, temperature, rates of heating and cooling, pH, ionic strength, and the presence of other food constituents. Therefore, these variables must be standardized and controlled to measure and compare gelation properties of proteins. Techniques used to measure rheological properties of foods (Chap. 30) are applied to determine the properties of protein gels.

15.3.3.5.2 Dough Formation Wheat protein is unique in its ability to form a **viscoelastic dough** suitable for making bread and other bakery products. Gluten, the major storage protein of wheat, is a heterogeneous mixture of the proteins gliadins and glutenins. These proteins have a unique amino acid composition that makes possible the formation of a viscoelastic dough, which can entrap carbon dioxide gas during yeast fermentation. Bread-making quality of wheat varieties is tested commonly by measuring dough strength after combining wheat flour with water and then kneading the dough. The effect of other dough ingredients (e.g., whey or soy proteins, phospholipids, other surfactants) on viscoelastic properties can be tested in the same way.

Dough strength is measured commonly under standardized conditions with a mixograph, farinograph, or **RapidViscoAnalyser** (RVA). The

mixograph is used to test the mixing properties of flour, so the dough will have the proper consistency, which is essential for automated manufacture of baked products. A **Farinograph®** is a torque meter to determine the water absorption of flour and the mixing properties of dough (21). The **RVA** is a cooking viscometer to rapidly test starch pasting (22). It incorporates heating and cooling, and variable shear, to test the viscosity of starches, cereals, and other foods.

15.4 SUMMARY

There are a variety of techniques used to separate and characterize proteins. Separation techniques rely on the differences in the solubility, size, charge, and adsorption characteristics of protein molecules. Ion-exchange chromatography is used to separate proteins on the basis of charge. Affinity chromatography utilizes ligands, such as enzyme inhibitors, coenzymes, or antibodies, to specifically bind proteins to a solid support. Proteins can be separated by size using dialysis, ultrafiltration, and size-exclusion chromatography. Electrophoresis can be used to separate proteins from complex mixtures on the basis of size and charge. SDS-PAGE also can be used to determine the molecular mass and subunit composition of a protein. Isoelectric focusing can be used to determine the isoelectric point of a protein. Capillary electrophoresis is an adaptation of conventional electrophoresis in which proteins are separated in capillary tubes. Chromatographic techniques are used in amino acid analysis to determine the amino acid composition of a protein. The nutritional quality of a protein is determined by the amino acid composition and protein digestibility. The PDCAAS is thought to be a better method than the PER method. The functional properties of proteins are used to characterize a protein for a particular application in a food. Common tests of protein functionality include solubility, emulsification, foaming, and gelation. No single test is applicable to all food systems. Microscopy in combination with specific dyes is used to visualize the location of protein within foods.

15.5 STUDY QUESTIONS

1. For each of the techniques listed below, identify the basis by which it can be used to separate proteins within a protein solution (e.g., precipitation, adsorption, size, charge) and give a brief explanation of how/why it works in that way.
   - (a) Dialysis
   - (b) Adjustment of pH to pI
   - (c) Addition of ammonium sulfate
   - (d) Ultrafiltration
   - (e) Heating to high temperature
   - (f) Addition of ethanol
   - (g) Affinity chromatography
   - (h) Size-exclusion chromatography
2. You have a protein system with the following characteristics:

Protein	*Solubility in $(NH_4)_2SO_4$ (%)*	*Solubility in ethanol (%)*	*pI*	*Denaturation temperature (°C)*
1	10–20	5–10	4.6	80
2	70–80	10–20	6.4	40
3	60–75	10–20	4.6	40
4	50–70	5–10	6.4	70

   Describe how you would separate protein 4 from the others.
3. Compare and contrast the principles and procedures of SDS-PAGE vs. isoelectric focusing to separate proteins. Include in your explanation how and why it is possible to separate proteins by each method and what you can learn about the protein by running it on each type of system.
4. Explain how capillary electrophoresis differs from SDS-PAGE.
5. Briefly describe what each of the following tells you about the characteristics of the proteins of interest described in the statement (Note: protein is not the same one in each statement):
   - (a) When subjected to dialysis using tubing with a molecular mass cutoff of 3000 Da, a protein of interest is found in the retentate (i.e., not in the filtrate).
   - (b) When subjected to ultrafiltration using a membrane with a molecular mass cutoff of 10,000 Da, a protein of interest is found in the filtrate (i.e., not in the retentate).
   - (c) When the protein was subjected to ion-exchange chromatography using an anion-exchange column and a buffer of pH 8.0, a protein of interest bound to the column.
   - (d) When a protein of interest was subjected to IEF, the protein migrated to a position of approximately pH 7.2 in the pH gradient of the gel.
   - (e) When a protein of interest was subjected to SDS-PAGE in both the presence and absence of mercaptoethanol, the protein appeared as three bands at molecular mass 42,000, 45,000, and 48,000 Da.
   - (f) When a solution with various proteins was heated to 60°C, the protein of interest was found in the precipitate obtained upon centrifugation of the solution.
6. You are submitting a soy protein sample to a testing laboratory with an amino acid analyzer (ion-exchange chromatography) so that you can obtain the amino acid composition. Explain how (a) the sample will be treated initially and (b) the amino acids will be quantified as they elute from the ion exchange column. Describe the procedures. (Note: You want to quantify all the amino acids.)

(a) How will samples be treated initially?
(b) How will amino acids be quantified?

7. In amino acid analysis, a protein sample hydrolyzed to individual amino acids is applied to a cation-exchange column. The amino acids are eluted by gradually increasing the pH of the mobile phase.
   (a) Describe the principles of ion-exchange chromatography.
   (b) Differentiate anion vs. cation exchangers.
   (c) Explain why changing the pH allows different amino acids to elute from the column at different times.
8. For each of the protein quality assay methods listed below, state what the method measures, and briefly describe an appropriate application of the method (e.g., for what situations/food samples is each required and/or ideally used):

	What is measured	*Application*
PDCAAS		
PER		
In vitro digestibility		
DNFB method for available lysine		

9. Briefly describe the differences between the following assay procedures:
   (a) Amino acid score vs. essential amino acid index
   (b) PDCAAS vs. amino acid score
10. You are helping to develop a new process for making a high-protein snack food from cereal grains and soy. You want to determine the protein quality of the snack food under various processing (toasting and drying) conditions. Considering the number of samples to be tested, you cannot afford an expensive in vivo assay, and you cannot wait more than a few days to get the results.
    (a) What method would you use to compare the protein quality of the snack food made under different processing conditions? Include an explanation of the principles involved.
    (b) You suspect that certain time–temperature combinations lead to overprocessed products. Your testing from (7a) shows that these samples have a lower nutritional quality. What amino acid(s) in the snack food would you suspect to be the most adversely affected by thermal abuse?
    (c) What test(s) could you use to confirm that amino acid(s) have become nutritionally unavailable by the overprocessing? How are these tests conducted?
11. Define "protein functionality" and list three important functional properties of foods.
12. Describe a functional test used to measure the following:
    (a) Protein solubility
    (b) Emulsion stability
    (c) Foam volume
13. You need to reconstitute nonfat dry milk (NFDM) powder for use in a yogurt formula. You have experienced a large amount of variation in the hydration times required for different lots of NFDM. As the quality control manager, you need to develop a test to measure the solubility and hydration characteristics of each lot of NFDM before it is used (in hopes of avoiding future rehydration problems).
    (a) How would you measure solubility and what precautions would you take with the method you develop?
    (b) Assume that you did the following for the test you developed: (1) mix 20 g of NFDM with 200 g water at 60°C, (2) blend for 2 min, (3) transfer 50 ml to a centrifuge tube, (4) centrifuge at $10{,}000 \times g$ for 5 min, and (5) measure the protein content of the supernatant after centrifugation. If the supernatant contains 2.95% protein and the starting NFDM contained 36.4% protein, what is the percent protein solubility?
14. You work for a fluid milk manufacturer that sells milk used in steamed coffee products. With this application, the milk needs to form large amounts of stable foam. Periodically, you have been receiving complaints that the milk does not produce adequate foam when steamed. As a result, you need to design a quality control method to measure foam characteristics of each lot of milk before it is shipped. Describe the test(s) you would perform.

15.6 PRACTICE PROBLEMS

1. Using the data provided in the table below:
   (a) Calculate the EAAI for defatted soy flour.
   (b) Determine the amino acid score for the soy flour.
   (c) Calculate the PDCAAS, using the true digestibility value of 87% for defatted soy flour.

Amino Acid	*Soy*[a] *(mg/g protein)*	*Reference Pattern*[b] *(mg/g protein)*
Histidine	26	18
Isoleucine	46	31
Leucine	78	63
Lysine	64	52
Methionine/cystine	26	26
Phenylalanine/tyrosine	88	46
Threonine	39	27
Tryptophan	14	7.4
Valine	46	42

[a]From (23).
[b]From (10), for preschool-age children.

2. You work for a manufacturer of protein supplements sold to body builders. You need to screen several proteins that may be used in a new protein supplement. You have three samples (A, B, and C) to evaluate. The amino acid

profiles of these three samples and the reference profile (i.e., amino acid requirements of preschool age children) are shown below.

Amino Acid	*Reference profile*	*Sample A*	*Sample B*	*Sample C*
Histidine	18	26	35	24
Isoleucine	31	50	55	35
Leucine	63	65	46	32
Lysine	52	80	92	80
Methionine/ cysteine	26	70	48	50
Phenylalanine/ tyrosine	46	70	90	85
Threonine	27	51	40	39
Tryptophan	7.4	16	22	25
Valine	42	60	64	42

(a) Calculate the PDCAAS for each supplement. (Assume that the true digestiblities of Samples A, B, and C are 87%, 93%, and 64%, respectively.)
(b) Which sample would you use if Sample A costs \$1.25/lb, Sample B costs \$3.25/lb, and Sample C costs \$1.15/lb?

Answers

1.

(a) Essential amino acid index

$$= \sqrt[9]{\begin{array}{c}(26/18)\,(46/31)\,(78/63)\,(64/52)\,(26/26)\,(88/46)\\(39/27)\,(14/7.4)\,(46/42)\end{array}}$$

$$= \sqrt[9]{\begin{array}{c}(1.44)\,(1.48)\,(1.24)\,(1.23)\,(1.00)\,(1.91)\\(1.44)\,(1.89)\,(1.10)\end{array}}$$

$$= \sqrt[9]{18.5866}$$

$$= 1.38$$

(b) Amino acid score = 26/26 = 1.00; lowest ratio represents the limiting amino acid, methionine/cystine.
(c) PDCAAS = amino acid score × true digestibility = 1.00 (0.87) = 0.87

2.

(a) The first limiting amino acid for all three samples is leucine. (Determine by determining ratio of each amino acid compared to reference profile.)
PDCAAS = amino acid score for first limiting amino acid × true digestibility
Sample A = (65/63) × 0.87 = 0.90
Sample B = (46/63) × 0.93 = 0.68
Sample C = (32/63) × 0.64 = 0.33
(b) The cost-to-protein quality ratio for each sample is as follows:
Sample A = (\$1.25/0.90) = \$1.39
Sample B = (\$3.25/0.68) = \$4.78
Sample C = (\$1.15/0.33) = \$3.48
Sample A provides the highest amount of usable protein per dollar, suggesting it would be best to use.

15.7 REFERENCES

1. Bonner PLR (2007) Protein purification. Taylor and Francis Group, New York
2. Coligan JE, Dunn BM, Speicher DW, Wingfield PT (eds) (2007) Current protocols in protein science. Wiley, New York
3. Scopes RK (1970) Characterization and study of sarcoplasmic proteins. Ch. 22. In: Briskey EJ, Cassens RG, Marsh BB (eds) Physiology and biochemistry of muscle as a food, vol 2. University of Wisconsin Press, Madison, WI, pp 471–492
4. Doultani S, Turhan KN, Etzel MR (2004) Fractionation of proteins from whey using cation exchange chromatrography. Process Biochem 39:1737–1743
5. Jelen P (1991) Pressure-driven membrane processes: principles and definitions. In New applications of membrane processes. Document No. 9201. pp 6–41, International Dairy Federation, Brussels, Belgium
6. Dolnik V (2008) Capillary electrophoresis of proteins 2005–2007. Electrophoresis 29:143–156
7. Moore S, Stein WH (1951) Chromatography of amino acids on sulfonated polystyrene resins. J Biol Chem 192:663–681
8. Moore S, Spackman DH, Stein WH (1958) Chromatography of amino acids on sulfonated polystyrene resins: an improved system. Anal Chem 30:1185–1190
9. Millipore Corp (1990) Liquid chromatography analysis of amino acids in feeds and foods using a modification of the Pico-Tag method. Technical Bulletin, Millipore Corp., Milford, MA
10. World Health Organization (2007) Protein and amino acid requirements in human nutrition. Joint WHO/FAO/UNU Expert Consultation. WHO Tech. Report Ser. No. 935. World Health Organization, Geneva, Switzerland
11. Friedman M (1996) Nutrition. Ch. 6 In: Nakai S Modler HW (eds) Food proteins: properties and characterization VCH Publishers, Inc., New York, pp 281–331
12. Owusu-Apenten RK (2002) Food protein analysis. Quantitative effects on processing. Part 5. Protein nutrient value (Ch. 12–14), pp 341–446. Marcel Dekker, Inc., New York
13. Schaafsma G (2005) The protein digestibility-corrected amino acid score (PDCAAS). A concept for describing protein quality in foods and food ingredients: a critical review. J AOAC Int 88(3):988–994
14. AOAC International (2007) Official methods of analysis, 18th edn., 2005; Current through revision 2, 2007 (On-line). AOAC International, Gaithersburg, MD
15. Federal Register (2008) Title 21 code of federal regulations part 101. Food labeling; Superintendent of documents. U.S. Government Printing Office, Washington, DC

16. AOCS (2009) Official methods and recommended practices of the AOCS, 6th edn. American Oil Chemists Society, Champaign, IL
17. AACC International (2000) Approved methods, 10th edn. AACC International, St. Paul, MN
18. McClements DJ (1999) Food emulsions. CRC Press, Boca Raton, Florida
19. Wrolstad RE, Acree TE, Decker EA, Penner MH, Reid DS, Schwartz SJ, Shoemaker CF, Smith D, Sporns P (eds) (2005) Handbook of food analytical chemistry vol. 1. Wiley, Hoboken, NJ
20. Fligner KL, Fligner MA Mangino ME (1991) Accelerated tests for predicting long-term creaming stability of infant formula emulsions Food Hydrocolloids 5:269–280
21. Appolinia BLC, Kunerth WH (1984) The farinograph handbook, 3rd edn., American Association of Cereal Chemists, St. Paul, MN
22. Crosbie GB, Ross AS (2007) RVA handbook. Wiley, Hoboken, NJ
23. Cavins JF, Kwolek DR, Inglett GE, Cowen JC (1972) Amino acid analysis of soybean meal: interlaboratory study. J Assoc Off Chem 55:686–694

Application of Enzymes in Food Analysis

Joseph R. Powers
School of Food Science, Washington State University, Pullman, WA 99164-6376, USA
powersjr@wsu.edu

16.1 Introduction 285
16.2 Principles 285
16.2.1 Enzyme Kinetics 285
16.2.1.1 Overview 285
16.2.1.2 Order of Reactions 287
16.2.1.3 Determination of Michaelis Constant (K_m) and V_m 287
16.2.2 Factors that Affect Enzyme Reaction Rate 288
16.2.2.1 Effect of Enzyme Concentration 288
16.2.2.2 Effect of Substrate Concentration 289
16.2.2.3 Environmental Effects 289
16.2.2.3.1 Effect of Temperature on Enzyme Activity 289
16.2.2.3.2 Effect of pH on Enzyme Activity 290
16.2.2.4 Activators and Inhibitors 291
16.2.2.4.1 Activators 291

S.S. Nielsen, *Food Analysis*, Food Science Texts Series, DOI 10.1007/978-1-4419-1478-1_16,

16.2.2.4.2 Inhibitors 292
16.2.3 Methods of Measurement 292
16.2.3.1 Overview 292
16.2.3.2 Coupled Reactions 293
16.3 Applications 293
16.3.1 Substrate Assays 294
16.3.1.1 Sample Preparation 294
16.3.1.2 Total Change/Endpoint Methods 294
16.3.1.3 Specific Applications 294
16.3.1.3.1 Measurement of Sulfite 294
16.3.1.3.2 Colorimetric Determination of Glucose 295
16.3.1.3.3 Starch/Dextrin Content 295
16.3.1.3.4 Determination of D-Malic Acid in Apple Juice 295
16.3.2 Enzyme Activity Assays 296
16.3.2.1 Peroxidase Activity 296
16.3.2.2 Lipoxygenase 296
16.3.2.3 Phosphatase Assay 296
16.3.2.4 α-Amylase Activity 297
16.3.2.5 Rennet Activity 297
16.3.3 Biosensors/Immobilized Enzymes 297
16.4 Summary 298
16.5 Study Questions 298
16.6 References 299

16.1 INTRODUCTION

Enzymes are protein catalysts that are capable of very great specificity and reactivity under physiological conditions. Enzymatic analysis is the measurement of compounds with the aid of added enzymes or the measurement of endogenous enzyme activity to give an indication of the state of a biological system including foods. The fact that enzyme catalysis can take place under relatively mild conditions allows for measurement of relatively unstable compounds not amenable to some other techniques. In addition, the specificity of enzyme reactions can allow for measurement of components of complex mixtures without the time and expense of complicated chromatographic separation techniques.

There are several uses of enzyme analyses in food science and technology. In several instances, enzyme activity is a useful measure for adequate processing of a food product. The thermal stability of enzymes has been used extensively as a measure of heat treatment; for example, peroxidase activity is used as a measure of adequacy of blanching of vegetable products. Enzyme activity assays are also used by the food technologist to assess potency of enzyme preparations used as processing aids.

The food scientist can also use commercially available enzyme preparations to measure constituents of foods that are enzyme substrates. For example, glucose content can be determined in a complex food matrix containing other monosaccharides by using readily available enzymes. A corollary use of commercially available enzymes is to measure enzyme activity as a function of enzyme inhibitor content in a food. Organophosphate insecticides are potent inhibitors of the enzyme acetylcholinesterase and hence the activity of this enzyme in the presence of a food extract is a measure of organophosphate insecticide concentration in the food. Also of interest is the measurement of enzyme activity associated with food quality. For example, catalase activity is markedly increased in milk from mastitic udders. Catalase activity also parallels the bacterial count in milk. Another use of enzyme assays to determine food quality is estimation of protein nutritive value by monitoring the activity of added proteases on food protein samples (see Chap. 15). Enzymes can be used to measure the appearance of degradation products such as trimethylamine in fish during storage. Enzymes are also used as preparative tools in food analysis. Examples include the use of amylases and proteases in fiber analysis (Chap. 10) and the enzymatic hydrolysis of thiamine phosphate esters in vitamin analysis.

To successfully carry out enzyme analyses in foods, an understanding of certain basic principles of enzymology is necessary. After a brief overview of these principles, examples of the use of enzymatic analyses in food systems are examined.

16.2 PRINCIPLES

16.2.1 Enzyme Kinetics

16.2.1.1 Overview

Enzymes are biological catalysts that are proteins. A catalyst increases the rate (velocity) of a thermodynamically possible reaction. The enzyme does not modify the equilibrium constant of the reaction, and the enzyme catalyst is not consumed in the reaction. Because enzymes affect rates (velocities) of reactions, some knowledge of **enzyme kinetics** (study of rates) is needed for the food scientist to effectively use enzymes in analysis. To measure the rate of an enzyme-catalyzed reaction, typically one mixes the enzyme with the substrate under specified conditions (pH, temperature, ionic strength, etc.) and follows the reaction by measuring the amount of product that appears or by measuring the disappearance of substrate. Consider the following as a simple representation of an enzyme-catalyzed reaction:

$$\mathrm{S} + \mathrm{E} \rightleftharpoons \mathrm{ES} \rightarrow \mathrm{P} + \mathrm{E} \quad [1]$$

where:

S = substrate
E = enzyme
ES = enzyme–substrate complex
P = product

The time course of an enzyme-catalyzed reaction is illustrated in Fig. 16-1. The formation of the enzyme substrate complex is very rapid and is not normally seen in the laboratory. The brief time in which the enzyme–substrate complex is initially formed is on the millisecond scale and is called the **pre-steady-state period**. The slope of the linear portion of the curve following the pre-steady-state period gives us the **initial velocity** (v_0). After the pre-steady-state period, a **steady-state period** exists in which the concentration of the enzyme–substrate complex is constant. A time course needs to be established experimentally by using a series of points or a continuous assay to establish the appropriate time frame for the measurement of the initial velocity.

The rate of an enzyme-catalyzed reaction depends on the concentration of the enzyme and also depends on the substrate concentration. With a fixed enzyme concentration, increasing substrate concentration will result in an increased velocity (see Fig. 16-2). As substrate concentration increases further, the increase in velocity slows until, with a very large concentration of substrate, no further increase in velocity is noted.

The velocity of the reaction at this very large substrate concentration is the **maximum velocity** (V_m) of the reaction under the conditions of that particular assay. The substrate concentration at which one-half V_m is observed is defined as the **Michaelis constant** or K_m. K_m is an important characteristic of an enzyme. It is an indication of the relative binding affinity of the enzyme for a particular substrate. The lower the K_m, the greater the affinity of the enzyme for the substrate. Both K_m and V_m are affected by environmental conditions such as pH and temperature.

If we examine relationships that hold in the steady state period, the Michaelis–Menten equation can be derived for the simplified enzyme-catalyzed reaction:

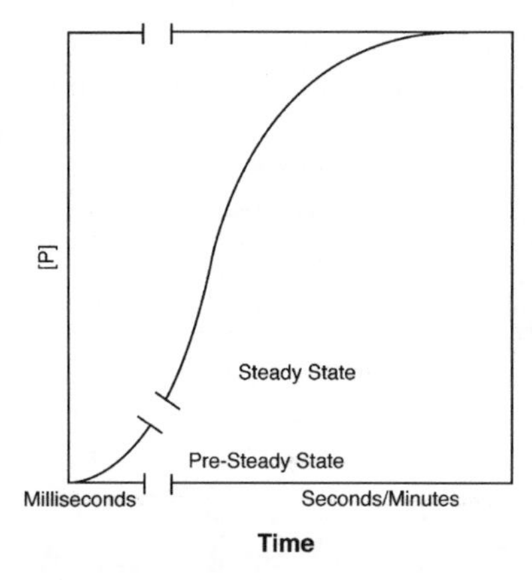

16-1 figure Time course for a typical enzyme-catalyzed reaction showing the pre-steady-state and steady-state periods. [P] = product concentration.

$$E + S \underset{k_{-1}}{\overset{k_1}{\rightleftharpoons}} ES \xrightarrow{k_2} E + P \quad [2]$$

where:

k_1, k_{-1}, k_2 = reaction rate constants for reactions indicated

In the steady state, the rate of change in enzyme–substrate complex concentration is zero: $dES/dt = 0$ and:

Rate of disappearance of ES

$$= k_{-1}[ES] + k_2[ES] \quad [3]$$

Rate of appearance of ES

$$= k_1[E][S] \quad [4]$$

then

$$k_1[E][S] = k_{-1}[ES] + k_2[ES] \quad [5]$$

$$[E_o] = [E] + [ES] \quad [6]$$

where:

E_o = total enzyme

E = free enzyme

ES = enzyme–substrate complex

Substituting

$$[E] = [E_o] - [ES] \quad [7]$$

$$k_1([E_o] - [ES])[S] = k_{-1}[ES] + k_2[ES] = (k_{-1} + k_2)[ES] \quad [8]$$

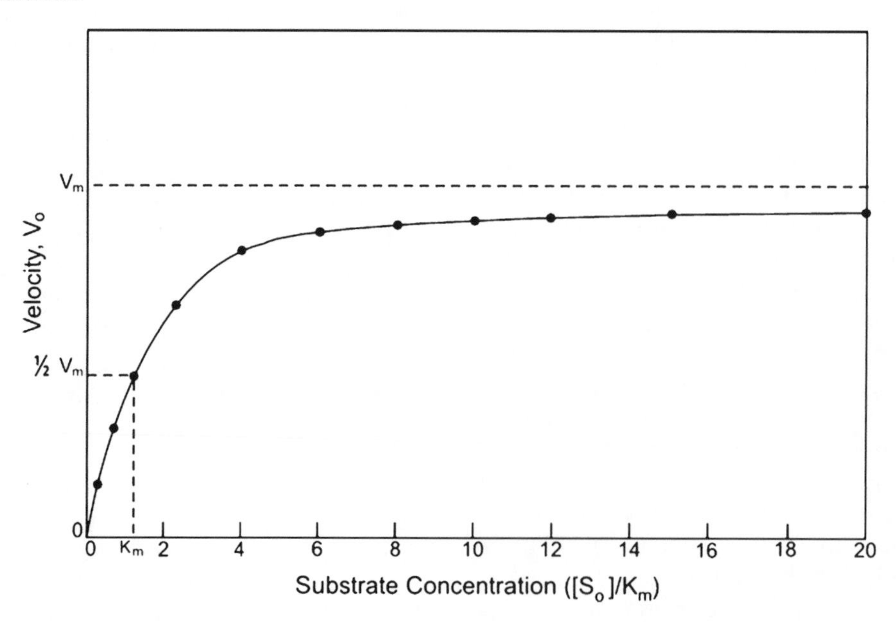

16-2 figure Effect of substrate concentration on the rate of an enzyme-catalyzed reaction. Plotted according to the Michaelis–Menten equation.

Rearranging and solving for [ES]:

$$\mathrm{ES} = \frac{k_1[\mathrm{E_o}][\mathrm{S}]}{k_1[\mathrm{S}] + (k_{-1} + k_2)} = \frac{[\mathrm{E_o}][\mathrm{S}]}{k_{-1} + k_2/k_1} + [\mathrm{S}] \quad [9]$$

If the collection of rate constants in the denominator is defined as the Michaelis constant (K_m):

$$K_m = \frac{k_{-1} + k_2}{k_1} \quad [10]$$

Note that K_m is *not* affected by enzyme *or* substrate concentration.

Then:

$$\mathrm{ES} = \frac{[\mathrm{E_0}][\mathrm{S}]}{K_m + [\mathrm{S}]} \quad [11]$$

If we define the velocity (v_0) of the enzyme-catalyzed reaction as:

$$v_0 = k_2[\mathrm{ES}] \quad [12]$$

then:

$$v_0 = \frac{k_2[\mathrm{E_0}][\mathrm{S}]}{K_m + [\mathrm{S}]} \quad [13]$$

When the enzyme is saturated, i.e., all substrate binding sites on the enzyme are occupied, at the large substrate concentrations in Fig. 16-2, we have maximum velocity, V_m. All of E_o is in the ES form and

$$k_2[\mathrm{ES}] = k_2[\mathrm{E_o}] \text{ at } [\mathrm{S}] \gg K_m \quad [14]$$

and

$$v_0 = \frac{V_m[\mathrm{S}]}{K_m + [\mathrm{S}]} \quad [15]$$

This is the Michaelis–Menten equation, the equation for a right hyperbola; the data plotted in Fig. 16-2 fit such an equation. A convenient way to verify this equation is to simply remember that $v_0 = 1/2\ V_m$ when $[\mathrm{S}] = K_m$. Therefore, by simple substitution

$$1/2\,V_m = \frac{V_m K_m}{K_m + K_m} = \frac{V_m}{2} \quad [16]$$

16.2.1.2 Order of Reactions

The velocity of an enzyme-catalyzed reaction increases as substrate concentration increases (see Fig. 16-2). A **first-order reaction** with respect to substrate concentration is obeyed in the region of the curve where substrate concentration is small ($[\mathrm{S}] \ll K_m$). This means that the velocity of the reaction is directly proportional to the substrate concentration in this region. When the substrate concentration is further increased, the velocity of the reaction no longer increases linearly, and the reaction is **mixed order**. This is seen in the figure as the curvilinear portion of the plot. If substrate concentration is increased further, the velocity asymptotically approaches the maximum velocity (V_m). In this linear, nearly zero slope portion of the plot, the velocity is independent of substrate concentration. However, note that at large substrate concentrations ($[\mathrm{S}] \gg K_m$), the velocity is directly proportional to enzyme concentration ($V_m = k_2\ [\mathrm{E_o}]$). Thus, in this portion of the curve where $[\mathrm{S}] \gg K_m$, the rate of the reaction is **zero order** with respect to substrate concentration (is independent of substrate concentration) but first order with respect to enzyme concentration.

If we are interested in **measuring the amount of enzyme** in a reaction mixture, we should, if possible, work at substrate concentrations so that the observed velocity approximates V_m. At these substrate concentrations, enzyme is directly rate limiting to the observed velocity. Conversely, if we are interested in **measuring substrate concentration** by measuring initial velocity, we must be at substrate concentrations less than K_m in order to have a rate directly proportional to substrate concentration.

16.2.1.3 Determination of Michaelis Constant (K_m) and V_m

To properly design an experiment in which velocity is zero order with respect to substrate and first order with respect to enzyme concentration, or conversely an experiment in which we would like to measure rates that are directly proportional to substrate concentration, we must know the K_m. The most popular method for determining K_m is the use of a **Lineweaver–Burk plot**. The reciprocal of the Michaelis–Menten equation is:

$$\frac{1}{v_0} = \frac{K_m}{V_m[S]} + \frac{1}{V_m} \quad [17]$$

This equation is that of a straight line $y = mx + b$ where m = slope and b = y-intercept. A plot of substrate concentration vs. initial velocity as shown in Fig. 16-2 can be transformed to a linear form via use of the reciprocal of Equation [17] and Fig. 16-3 (Lineweaver–Burk plot) results. The intercept of the plotted data on the y (vertical) axis is $1/V_m$ while the intercept on the x (horizontal) axis is $-1/K_m$. The slope of the line is K_m/V_m. Consequently, both K_m and V_m can be obtained using this method.

A disadvantage of the Lineweaver–Burk plot is that the data with the inherently largest error, collected at very low substrate concentrations and consequently low rates, tend to direct the drawing of a best fit line. An alternative method of plotting the data is the **Eadie Hofstee method**. The Michaelis–Menten equation can be rearranged to give:

$$v_0 = V_m - \frac{v_0 K_m}{[\mathrm{S}]} \quad [18]$$

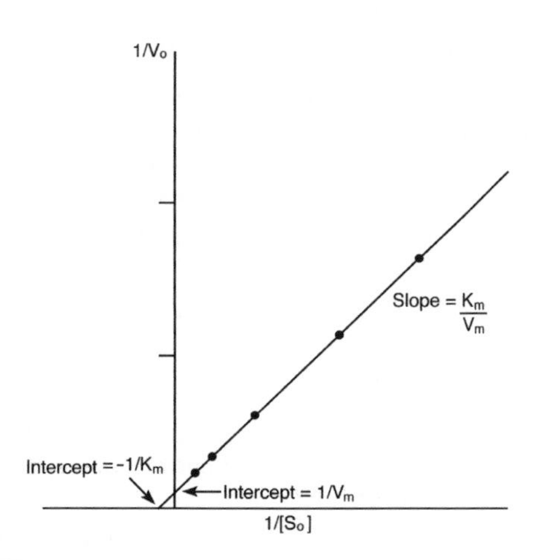

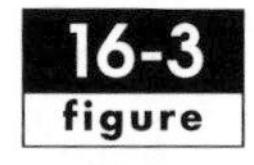
Plot of substrate-velocity data by the Lineweaver–Burk method.

Equation [18] is also the equation of a straight line and when v_0 vs. $v_0/[S]$ are plotted, the slope of the line is $-K_m$, the y-intercept is V_m, and the x-intercept is V_m/K_m. A more even spacing of the data is achieved by this method than by a Lineweaver–Burk plot.

16.2.2 Factors that Affect Enzyme Reaction Rate

The velocity of an enzyme-catalyzed reaction is affected by a number of factors, including enzyme and substrate concentrations, temperature, pH, ionic strength, and the presence of inhibitors and activators.

16.2.2.1 Effect of Enzyme Concentration

The velocity of an enzyme-catalyzed reaction will depend on the enzyme concentration in the reaction mixture. The expected relationship between enzyme activity and enzyme concentration is shown in Fig. 16-4. Doubling the enzyme concentration will double the rate of the reaction. If possible, determination of enzyme activity should be done at concentrations of substrate much greater than K_m. Under these conditions, a zero-order dependence of the rate with respect to substrate concentration and a first-order relationship between rate and enzyme concentration exist. It is critical that the substrate concentration is saturating during the entire period the reaction mixture is sampled and the amount measured of product formed or substrate disappearing is linear over the period during which the reaction is sampled. The activity of the enzyme is obtained as the slope of the linear part of the line of a plot of product or substrate concentration vs. time.

If a large number of samples is to be assayed, a single aliquot is often taken at a single time. This can be risky and will give good results only if the time at which the sample is taken falls on the linear portion of a plot of substrate concentration or product concentration vs. time of reaction (see Fig. 16-5). The plot becomes nonlinear if the substrate concentration falls below the concentration needed to saturate the enzyme, if the increase in concentration of product produces a significant amount of back reaction, or if the enzyme loses activity during the time of the

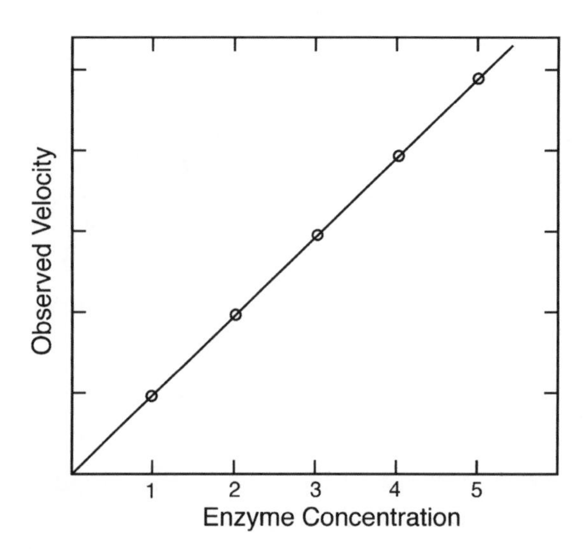

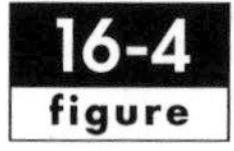
Expected effect of enzyme concentration on observed velocity of an enzyme-catalyzed reaction.

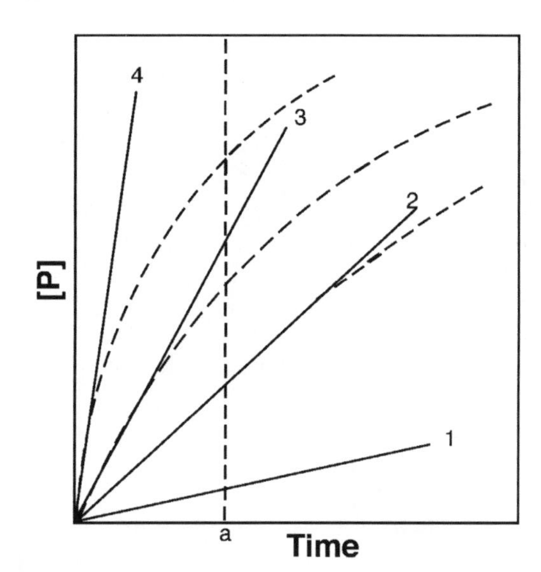

Effect of enzyme concentration on time course of an enzyme-catalyzed reaction. The *dashed lines* are experimentally determined data with enzyme concentration increasing from 1 to 4. The *solid lines* are tangents drawn from the initial slopes of the experimental data. If a single time point, *a*, is used for data collection, a large difference between actual data collected and that predicted from initial rates is seen.

assay. Normally, one designs an experiment in which enzyme concentration is estimated such that no more than 5–10% of the substrate has been converted to product within the time used for measuring the initial rate. In the example shown in Fig. 16-5, by sampling at the single point, a, an underestimation of the rate is made for curves 3 and 4. A better method of estimating rates is to measure initial rates of the reactions, in which the change in substrate or product concentration is determined at times as close as possible to time zero. This is shown in Fig. 16-5 by the solid lines drawn tangent to the slopes of the initial parts of the curves. The slope of the tangent line gives the initial rate.

Sometimes it is not possible to carry out enzyme assays at $[S] \gg K_m$. The substrate may be very expensive or relatively insoluble or K_m may be large (i.e., $K_m > 100\,mM$). Enzyme concentration can also be estimated at substrate concentrations much less than K_m. When substrate concentration is much less than K_m, the substrate term in the denominator of the Michaelis–Menten equation can be ignored and $v = (V_m[S])/K_m$ which is the equation for a first-order reaction with respect to substrate concentration. Under these conditions, a plot of product concentration vs. time gives a nonlinear plot (Fig. 16-6). A plot of $\log([S_o]/[S])$ vs. time gives a straight line relationship

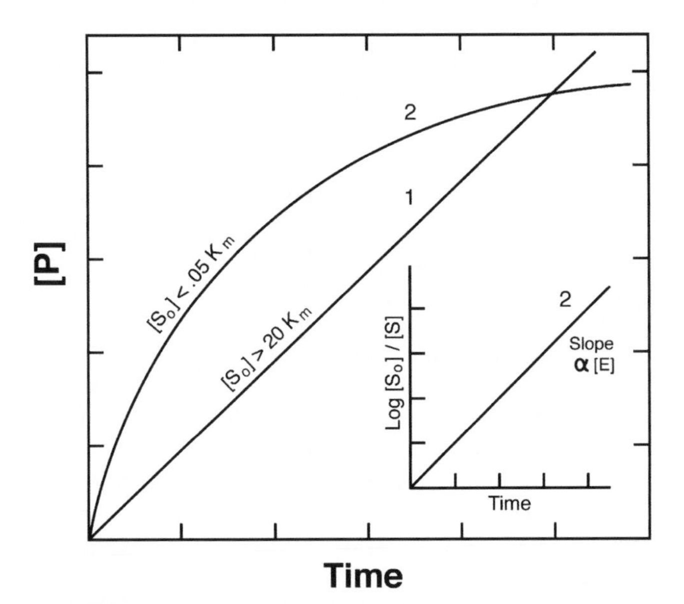

16-6 figure Product concentration $[P]$ formed as a function of time for an enzyme-catalyzed reaction. Line 1 is linear indicating a zero-order reaction with respect to substrate concentration $[S]$. The slope of line 1 is directly related to enzyme concentration. Line 2 is nonlinear. A replot of line 2 data, plotting $\log [S_0]/[S]$ vs. time is linear (*Insert*), indicating the reaction is first order with respect to substrate concentration. The slope of the replot is directly related to enzyme concentration.

(Fig. 16-6, inset). The slope of the line of the log plot is directly related to the enzyme concentration. When the slope of a series of these log plots is further plotted as a function of enzyme concentration, a straight line relationship should result. If possible, the reaction should be followed continuously or aliquots removed at frequent time intervals and the reaction be allowed to proceed to greater than 10% of the total reaction.

16.2.2.2 Effect of Substrate Concentration

The substrate concentration velocity relationship for an enzyme-catalyzed reaction in which enzyme concentration is constant is shown in Fig. 16-2. As noted before, the rate of the reaction is first order with respect to substrate concentration when $[S] \ll K_m$. At $[S] \gg K_m$, the reaction is zero order with respect to substrate concentration and first order with respect to [E]. At substrate concentrations between the first-order and zero-order regions, the enzyme-catalyzed reaction is mixed order with respect to substrate concentration. However, when initial rates are obtained, a linear relationship between v_0 and E_o should be seen.

16.2.2.3 Environmental Effects

16.2.2.3.1 Effect of Temperature on Enzyme Activity

Temperature can affect observed enzyme activity in several ways. Most obvious is that temperature can affect the stability of enzyme and also the rate of the enzyme-catalyzed reaction. Other factors in enzyme-catalyzed reactions that may be considered include the effect of temperature on the solubility of gases that are either products or substrates of the observed reaction and the effect of temperature on pH of the system. A good example of the latter is the common buffering species Tris (tris [hydroxymethyl] aminomethane), for which the pK_a changes 0.031 per 1°C change.

Temperature affects both the stability and the activity of the enzyme, as shown in Fig. 16-7. At relatively low temperatures, the enzyme is stable. However, at higher temperatures, denaturation dominates, and a markedly reduced enzyme activity represented by the negative slope portion of line 2 is observed. Line 1 of Fig. 16-7 shows the effect of temperature on the velocity of the enzyme-catalyzed reaction. The velocity is expected to increase as the temperature is increased. As shown by line 1, the velocity approximately doubles for every 10°C rise in temperature. The net effect of increasing temperature on the rate of conversion of substrate to product (line 1) and on the rate of the denaturation of enzyme (line 3) is line 2 of Fig. 16-7. The temperature optimum of the enzyme is at the maximum point of line 2. The temperature optimum is not a unique characteristic of the

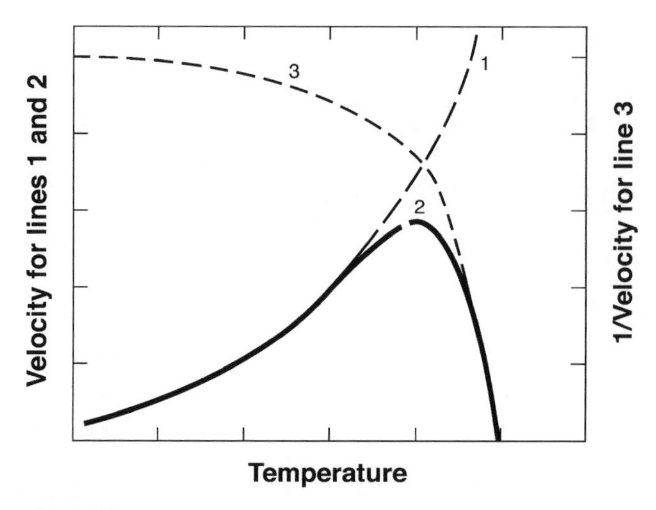

16-7 figure Effect of temperature on velocity of an enzyme-catalyzed reaction. Temperature effect on substrate to product conversion is shown by line 1. Line 3 shows effect of temperature on rate of enzyme denaturation (right-hand y-axis is for line 3). The net effect of temperature on the observed velocity is given by line 2 and the temperature optimum is at the maximum of line 2.

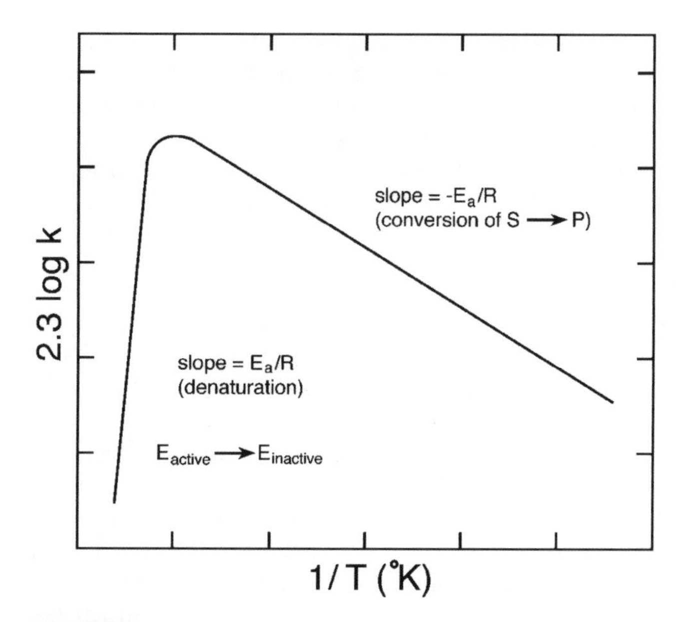

16-8 figure Effect of temperature on rate constant of an enzyme-catalyzed reaction. The data are plotted 2.3 log k vs. $1/T(°K)$ according to the Arrhenius equation, $k = Ae^{-E_a/RT}$.

enzyme. The optimum applies instead to the entire system because type of substrate, pH, salt concentration, substrate concentration, and time of reaction can affect the observed optimum. For this reason, investigators should fully describe a system in which the effects of temperature on observed enzyme activity are reported.

The data of line 2 of Fig. 16-7 can be plotted according to the **Arrhenius equation**:

$$k = Ae^{-E_a RT} \quad [19]$$

which can be written:

$$\log k = \log A - \frac{E_a}{2.3RT} \quad [20]$$

where:

k = specific rate constant at some temperature, T (K)
E_a = activation energy, the minimum amount of energy a reactant molecule must have to be converted to product
R = gas constant
A = frequency factor (preexponential factor)

The positive slope on the left side (high temperature) of the **Arrhenius plot** (Fig. 16-8) gives a measure of the **activation energy** (E_a) for the denaturation of the enzyme. Note that a small change in temperature has a very large effect on the rate of denaturation. The slope on the right side of Fig. 16-8 gives a measure of the E_a for the transformation of substrate to product catalyzed by the enzyme. If the experiment is carried out under conditions in which V_m is measured ($[S] \gg K_m$), then the activation energy observed will be for the catalytic step of the reaction.

16.2.2.3.2 Effect of pH on Enzyme Activity The observed rate of an enzyme-catalyzed reaction is greatly affected by the pH of the medium. Enzymes have pH optima and commonly have bell-shaped curves for activity vs. pH (Fig. 16-9). This pH effect is a manifestation of the effects of pH on enzyme stability and on rate of substrate to product conversion and may also be due to changes in ionization of substrate.

The rate of substrate to product conversion is affected by pH because pH may affect binding of substrate to enzyme and the ionization of catalytic groups such as carboxyl or amino groups that are part of the enzyme's active site. The stability of the tertiary or quarternary structure of enzymes is also pH dependent and affects the velocity of the enzyme reaction, especially at extreme acidic or alkaline pHs. The pH for maximum stability of an enzyme does not necessarily coincide with the pH for maximum activity of that same enzyme. For example, the proteolytic enzymes trypsin and chymotrypsin are stable at pH 3, while they have maximum activity at pH 7–8.

To establish the pH optimum for an enzyme reaction, the reaction mixture is buffered at different pHs and the activity of the enzyme is determined. To

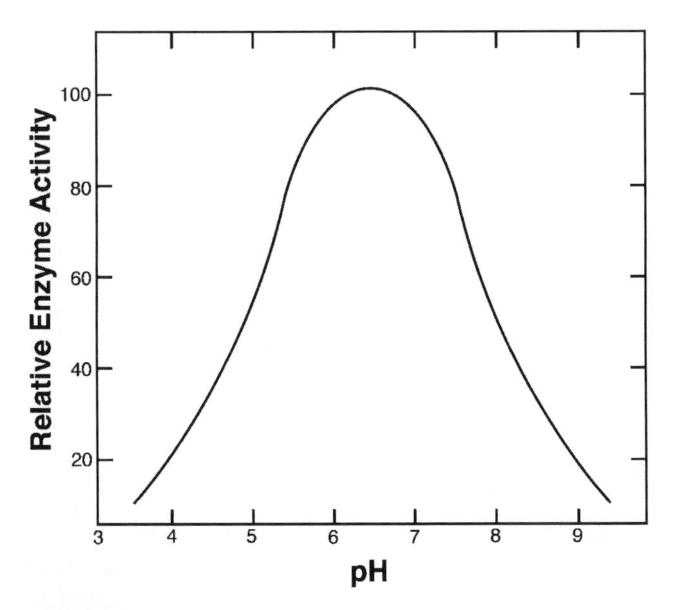

Typical velocity-pH curve for an enzyme-catalyzed reaction. The maximum on the curve is the optimum for the system and can vary with temperature, specific substrate, and enzyme source.

determine pH enzyme stability relationships, aliquots of the enzyme are buffered at different pH values and held for a specified period of time (e.g., 1 h). The pH of the aliquots is then adjusted to the pH optimum and each aliquot is assayed. The effect of pH on enzyme stability is thus obtained. These studies are helpful in establishing conditions for handling the enzyme and also may be useful in establishing methods for controlling enzyme activity in a food system. Note that pH stability and the pH optimum for the enzyme activity are not true constants. That is to say, these may vary with particular source of enzyme, the specific substrate used, the temperature of the experiment, or even the buffering species used in the experiment. In the use of enzymes for analysis, it is not necessary that the reaction be carried out at the pH optimum for activity, or even at a pH at which the enzyme is most stable, but it is critical to maintain a fixed pH during the reaction (i.e., use buffer) and to use the same pH in all studies to be compared.

16.2.2.4 Activators and Inhibitors

16.2.2.4.1 Activators Some enzymes contain, in addition to a protein portion, small molecules that are activators of the enzyme. Some enzymes show an absolute requirement for a particular inorganic ion for activity while others show increased activity when small molecules are included in the reaction medium. These small molecules can play a role in maintaining the conformation of the protein, or they may form an essential component of the active site, or they may form part of the substrate of the enzyme.

In some cases, the activator forms a nearly irreversible association with the enzyme. These nonprotein portions of the enzyme are called **prosthetic groups**. The amount of enzyme activator complex formed is equal to the amount of activator present in the mixture. In these cases, activator concentration can be estimated up to concentrations equal to total enzyme concentration by simply measuring enzyme activity.

In most cases, dissociation constants for an enzyme activator complex are within the range of enzyme concentration. Dissociable nonprotein parts of enzymes are categorized as **coenzymes**. When this type of activator is added to enzyme, a curvilinear relationship similar to a Michaelis–Menten plot results, making difficult the determination of an unknown amount of activator. A reciprocal plot analogous to a Lineweaver–Burk plot can be constructed using standards and unknown activator concentrations estimated from such a plot.

One food-related enzyme reaction involving an activator is measuring pyridoxal-phosphate, a form of vitamin B6. This reaction measures the reactivation of coenzyme free yeast aminotransferase by coupling the transamination reaction with malate dehydrogenase. This is possible when malate dehydrogenase, NADH, alpha-ketoglutarate, aspirate, and the aminotransferase are in excess, and the pyridoxal 5-phosphate added is rate limiting.

$$\text{alpha-ketoglutarate} + \text{aspartate} \underset{}{\overset{\text{aminotransferase}}{\rightleftarrows}} \text{glutamate} + \text{oxalacetate} \quad [21]$$

$$\text{oxalacetate} + \text{NADH} + \text{H}^+ \underset{}{\overset{\text{malate dehydrogenase}}{\rightleftarrows}} \text{malate} + \text{NAD}^+ \quad [22]$$

Another example of an essential activator is the pyridine coenzyme NAD^+. NAD^+ is essential for the oxidation of ethanol to acetaldehyde by alcohol dehydrogenase:

$$\text{ethanol} + \text{NAD}^+ \underset{}{\overset{\text{alcohol dehydrogenase}}{\rightleftarrows}} \text{acetaldehyde} + \text{NADH} + \text{H}^+ \quad [23]$$

In the reaction, NAD^+ is reduced to NADH and can be considered a second substrate. Another example of an activator of an enzyme is the chloride ion with α-amylase. In this case, α-amylase has some activity in the absence of chloride. With saturating levels of chloride, the α-amylase activity increases about fourfold. Other anions, including F^-, Br^-, and I^-, also activate α-amylase. These anions must not be in the reaction mixture if α-amylase stimulation is to be used as a method of determining chloride concentration.

16.2.2.4.2 Inhibitors An enzyme inhibitor is a compound that when present in an enzyme-catalyzed reaction medium decreases the enzyme activity. Enzyme inhibitors can be categorized as **irreversible** or **reversible** inhibitors. Enzyme inhibitors include inorganic ions, such as Pb^{2+} or Hg^{2+}, which can react with sulfhydryl groups on enzymes to inactivate the enzyme, compounds that resemble substrate, and naturally occurring proteins that specifically bind to enzymes (such as protease inhibitors found in legumes).

1. **Irreversible inhibitors**. When the dissociation constant of the inhibitor enzyme complex is very small, the decrease in enzyme activity observed will be directly proportional to the inhibitor added. The speed at which the irreversible combination of enzyme and inhibitor reacts may be slow, and the effect of time on the reduction of enzyme activity by the addition of inhibitor must be determined to ensure complete enzyme–inhibitor reaction. For example, the amylase inhibitor found in many legumes must be preincubated under specified conditions with amylase prior to measurement of residual activity to accurately estimate inhibitor content (1). Irreversible inhibitors decrease V_m as the amount of total active enzyme is reduced.
2. **Reversible inhibitors**. Most inhibitors exhibit a dissociation constant such that both enzyme and inhibitor are found free in the reaction mixture. Several types of reversible inhibitors are known: **competitive, noncompetitive**, and **uncompetitive**.

Competitive inhibitors usually resemble the substrate structurally and compete with substrate for binding to the active site of the enzyme, and only one molecule of substrate or inhibitor can be bound to the enzyme at one time. An inhibitor can be characterized as competitive by adding a fixed amount of inhibitor to reactions at various substrate concentrations and by plotting the resulting data by the Lineweaver–Burk method and noting the effect of inhibitor relative to that of control reactions in which no inhibitor is added. If the inhibitor is competitive, the slope and x-intercept of the plot with inhibitor are altered while the y-intercept ($1/V_m$) is unaltered. It can be shown that the ratio of the **uninhibited initial velocity** (ν_o) to the **inhibited initial velocity** (ν_i) gives:

$$\frac{\nu_0}{\nu_1} = \frac{[I]K_m}{K_i(K_m + [S])} \quad [24]$$

where:

K_i = dissociation constant of the enzyme–inhibitor complex

$[I]$ = concentration of competitive inhibitor

Thus, a plot of ν_o/ν_i vs. inhibitor concentration will give a straight line relationship. From this plot the concentration of a competitive inhibitor can be found (2).

A **noncompetitive inhibitor** binds to enzyme independent of substrate and is bound outside the active site of the enzyme. A noncompetitive inhibitor can be identified by its effect on the rate of enzyme-catalyzed reactions at various substrate concentrations and the data plotted by the Lineweaver–Burk method. A noncompetitive inhibitor will affect the slope and the y-intercept as compared to the uninhibited system while the x-intercept, $1/K_m$, is unaltered. Analogous to competitive inhibitors, a standard curve of ν_o/ν_i vs. inhibitor concentration may be prepared and used to determine the concentration of a noncompetitive inhibitor (2).

Uncompetitive inhibitors bind only to the enzyme–substrate complex. Uncompetitive inhibition is noted by adding a fixed amount of inhibitor to reactions at several substrate concentrations and plotting the data by the Lineweaver–Burk method. An uncompetitive inhibitor will affect both the x- and y-intercepts of the Lineweaver–Burk plot as compared to the uninhibited system, while maintaining an equal slope to the uninhibited system (i.e., a parallel line will result). A plot of ν_o/ν_i vs. inhibitor concentration can be prepared to use as a standard curve for the determination of the concentration of an uncompetitive inhibitor (2).

16.2.3 Methods of Measurement

16.2.3.1 Overview

For practical enzyme analysis, it is necessary to be familiar with the methods of measurement of the reaction. Any physical or chemical property of the system that relates to substrate or product concentration can be used to follow an enzyme reaction. A wide variety of methods are available to follow enzyme reactions, including **absorbance spectrometry, fluorimetry, manometric methods, titration, isotope measurement, chromatography, mass spectrometry**, and **viscosity**. A good example of the use of spectrophotometry as a method for following enzyme reactions is use of the spectra of the pyridine coenzyme NAD(H) and NADP(H), in which there is a marked change in absorbance at 340 nm upon oxidation–reduction (Fig. 16-10). Many methods depend on the increase or decrease in absorbance at 340 nm when these coenzymes are products or substrates in a coupled reaction.

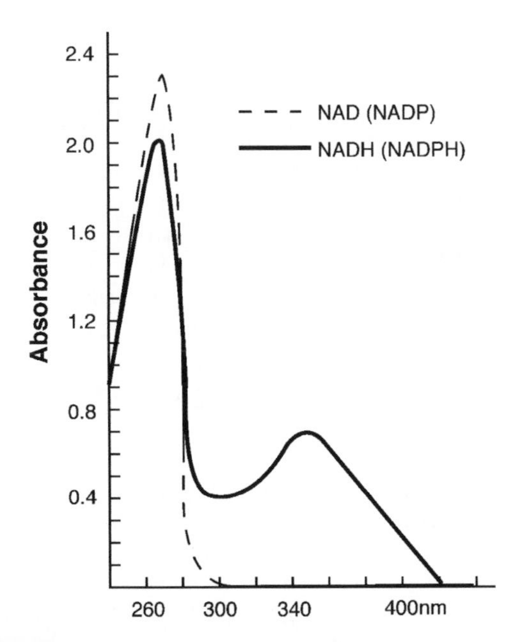

Absorption curves of NAD(P) and NAD(P)H; λ = wavelength. Many enzymatic analysis methods are based on the measurement of an increase or decrease in absorbance at 340 nm due to NAD(H) or NADP(H).

An example of using several methods to measure the activity of an enzyme is in the assay of α-amylase activity (3). α-Amylase cleaves starch at α-1,4 linkages in starch and is an endoenzyme. An endoenzyme cleaves a polymer substrate at internal linkages. This reaction can be followed by a number of methods, including reduction in viscosity, increase in reducing groups upon hydrolysis, reduction in color of the starch iodine complex, and polarimetry. However, it is difficult to differentiate the activity of α-amylase from β-amylase using a single assay. α-Amylase cleaves maltose from the nonreducing end of starch. While a marked decrease in viscosity of starch or reduction in iodine color would be expected to occur due to α-amylase activity, β-amylase can also cause changes in viscosity and iodine color if in high concentration. To establish whether α-amylase or β-amylase is being measured, the analyst must determine the change in number of reducing groups as a basis of comparison. Because α-amylase is an endoenzyme, hydrolysis of a few bonds near the center of the polymeric substrate will cause a marked decrease in viscosity, while hydrolysis of an equal number of bonds by the exoenzyme, β-amylase, will have little effect on viscosity.

In developing an enzyme assay, it is wise to first write out a complete, balanced equation for the particular enzyme-catalyzed reaction. Inspection of the products and substrates for chemical and physical properties that are readily measurable with available equipment will often result in an obvious choice of method for following the reaction in the laboratory.

If one has options in methodology, one should select the method that is able to monitor the reaction continuously, is most sensitive, and is specific for the enzyme-catalyzed reaction.

16.2.3.2 Coupled Reactions

Enzymes can be used in assays via coupled reactions. **Coupled reactions** involve using two or more enzyme reactions so that a substrate or product concentration can be readily followed. In using a coupled reaction, there is an **indicator reaction** and a **measuring reaction**. For example:

$$\underset{\text{measuring reaction}}{\mathrm{S1} \xrightarrow{\mathrm{E1}} \mathrm{P1}} \qquad [25]$$

$$\underset{\text{indicating reaction}}{\mathrm{P1} \xrightarrow{\mathrm{E2}} \mathrm{P2}} \qquad [26]$$

The role of the indicating enzyme (E2) is to produce P2, which is readily measurable and, hence, is an indication of the amount of P1 produced by E1. Alternatively the same sequence can be used in measuring S1, the substrate for E1. When a coupled reaction is used to measure the activity of an enzyme (e.g., E1 above), it is critical that the indicating enzyme E2 not be rate limiting in the reaction sequence: the measuring reaction must always be rate determining. Consequently, E2 activity should be much greater than E1 activity for an effective assay. Coupled enzyme reactions can have problems with respect to pH of the system if the pH optima of the coupled enzymes are quite different. It may be necessary to allow the first reaction (e.g., the measuring reaction catalyzed by E1 above, Equation [25]) to proceed for a time and then arrest the reaction by heating to denature E1. The pH is adjusted, the indicating enzyme (E2, Equation [26]) added, and the reaction completed. If an endpoint method is used with a coupled system, the requirements for pH compatibility are not as stringent as for a rate assay because an extended time period can be used to allow the reaction sequence to go to completion.

16.3 APPLICATIONS

As described previously, certain information is needed prior to using enzyme assays analytically. In general, knowledge of K_m, time course of the reaction, the enzyme's specificity for substrate, the pH optimum and pH stability of the enzyme, and effects of temperature on the reaction and stability of the enzyme are desirable. Many times this information is available from the literature. However, a few preliminary experiments may be necessary, especially in the case of experiments in which velocities are measured. A time course to establish linearity of product formation or substrate consumption in the reaction is a necessity. An experiment to show linearity of velocity of the enzyme reaction to enzyme concentration is recommended (see Fig. 16-5).

16.3.1 Substrate Assays

The following is not an extensive compendium of methods for the measurement of food components by enzymatic analysis. Instead, it is meant to be representative of the types of analyses possible. The reader can consult handbooks published by the manufacturers of enzyme kits (e.g., Megazyme; https://secure.megazyme.com/Dynamic.aspx?control=CSCatalogue&categoryName=AssayKits), the review article by Whitaker (2), a book by Henniger (4), and the series by Bergmeyer (5) for a more comprehensive guide to enzyme methods applicable to foods.

16.3.1.1 Sample Preparation

Because of the specificity of enzymes, sample preparation prior to enzyme analysis is often minimal and may involve only extraction and removal of solids by filtration or centrifugation. Regardless, due to the wide variety of foods that might be encountered by the analyst using enzyme assays, a check should be made of the extraction and enzyme reaction steps by standard addition of known amounts of analyte to the food and extract, and measuring recovery of that standard. If the standard additions are fully recovered, this is a positive indication that the extraction is complete, that sample does not contain interfering substances that require removal prior to the enzymatic analysis, and that the reagents are good. In some cases, interfering substances are present but can be readily removed by precipitation or adsorption. For example, polyvinylpolypyrrolidone (PVPP) powder can be used to decolorize juices or red wines. With the advent of small syringe minicolumns (e.g., C18, silica, and ion exchange cartridges), it is also relatively easy and fast to attain group separations to remove interfering substances from a sample extract.

16.3.1.2 Total Change/Endpoint Methods

While substrate concentrations can be determined in rate assays when the reaction is first order with respect to substrate concentration ($[S] \ll K_m$), substrate concentration can also be determined by the total change or endpoint method. In this method, the enzyme-catalyzed reaction is allowed to go to completion so that concentration of product, which is measured, is directly related to substrate. An example of such a system is the measurement of glucose using glucose oxidase and peroxidase, described below.

In some cases, an equilibrium is established in an endpoint method in which there is a significant amount of substrate remaining in equilibrium with product. In these cases, the equilibrium can be altered. For example, in cases in which a proton-yielding reaction is used, alkaline conditions (increase in pH) can be used. Trapping agents can also be used, in which product is effectively removed from the reaction, and by mass action the reaction goes to completion. Examples include the trapping of ketones and aldehydes by hydrazine. In this way, the product is continually removed and the reaction is pulled to completion. The equilibrium also can be displaced by increasing cofactor or coenzyme concentration.

Another means of driving a reaction to completion is a regenerating system (5). For example, in the measurement of glutamate, with the aid of glutamate dehydrogenase, the following can be done:

$$\text{glutamate} + NAD^+ + H_2O \xrightleftharpoons{\text{glutamate dehydrogenase}} \alpha\text{-ketoglutarate} + NADH + NH_4^+ \quad [27]$$

$$\text{pyruvate} + NADH + H^+ \xrightarrow{\text{lactate dehydrogenase}} NAD^+ + \text{lactate} \quad [28]$$

In this system, NADH is recycled to NAD^+ via lactate dehydrogenase until all the glutamate to be measured is consumed. The reaction is stopped by heating to denature the enzymes present, a second aliquot of glutamate dehydrogenase and NADH is added, and the α-ketoglutarate (equivalent to the original glutamate) measured via decrease in absorbance at 340 nm. An example in which the same equilibrium is displaced in the measurement of glutamate is as follows:

$$\text{glutamate} + NAD^+ + H_2O \xrightleftharpoons{\text{glutamate dehydrogenase}} \alpha\text{-ketoglutarate} + NADH + NH_4^+ \quad [29]$$

$$NADH + INT \xrightarrow{\text{diaphorase}} NAD^+ + \text{formazan} \quad [30]$$

Iodonitrotetrazolium chloride (INT) is a trapping reagent for the NADH product of the glutamate dehydrogenase catalyzed reaction. The formazan formed is measurable colorimetrically at 492 nm.

16.3.1.3 Specific Applications

16.3.1.3.1 Measurement of Sulfite Sulfite is a food additive that can be measured by several techniques, including titration, distillation followed by titration, gas chromatography, and colorimetric analysis. Sulfite also can be specifically oxidized to sulfate by the commercially available enzyme sulfite oxidase (SO):

$$SO_3^{2-} + O_2 + H_2O \xrightarrow{SO} SO_4^{2-} + H_2O_2 \quad [31]$$

The H_2O_2 product can be measured by several methods including use of the enzyme NADH-peroxidase:

$$H_2O_2 + NADH + H^+ \xrightarrow{NADH-peroxidase} 2H_2O + NAD^+ \quad [32]$$

The amount of sulfite in the system is equal to the NADH oxidized, which is determined by decrease in absorbance at 340 nm. Ascorbic acid can interfere with the assay but can be removed by using ascorbic acid oxidase (6).

16.3.1.3.2 Colorimetric Determination of Glucose The combination of the enzymes glucose oxidase and peroxidase can be used to specifically measure glucose in a food system (7) (see also Chap. 10, Sect. 10.3.4.3.3). Glucose is preferentially oxidized by glucose oxidase to produce gluconolactone and hydrogen peroxide. The hydrogen peroxide plus *o*-dianisidine in the presence of peroxidase produces a yellow color that absorbs at 420 nm Equations [33] and [34]. This assay is normally carried out as an endpoint assay and there is stoichiometry between the color formed and the amount of glucose in the extract, which is established with a standard curve. Because glucose oxidase is quite specific for glucose, it is a useful tool in determining the amount of glucose in the presence of other reducing sugars.

$$\beta\text{-D-glucose} + O_2 \xrightarrow{\text{glucose oxidase}} \delta\text{-gluconolactone} + H_2O_2 \quad [33]$$

$$H_2O_2 + \text{o-diansidine} \xrightarrow{\text{peroxidase}} H_2O + \underset{\text{(colored)}}{\text{oxidized dye}} \quad [34]$$

16.3.1.3.3 Starch/Dextrin Content Starch and dextrins can be determined by enzymatic hydrolysis using amyloglucosidase, an enzyme that cleaves α-1,4 and α-1,6 bonds of starch, glycogen, and dextrins, liberating glucose (see Chap. 10). The glucose formed can be subsequently determined enzymatically. Glucose can be determined by the previously described colorimetric method, in which glucose is oxidized by glucose oxidase and coupled to a colored dye via reaction of the glucose oxidase product, hydrogen peroxide, with peroxidase. An alternative method of measuring glucose is by coupling hexokinase (HK) and glucose-6-phosphate dehydrogenase (G6PDH) reactions:

$$\text{glucose} + \text{ATP} \xrightarrow{HK} \text{glucose-6-phosphate} + \text{ADP} \quad [35]$$

$$\text{glucose-6-phosphate} + NADP^+ \xrightarrow{G6PDH} \text{6-phosphogluconate} + NADPH + H^+ \quad [36]$$

The amount of NADPH formed is measured by absorbance at 340 nm and is a stoichiometric measure of the glucose originating in the dextrin or starch hydrolyzed by amyloglucosidase. The amount of starch determined by this method is calculated as follows:

$$c = \frac{V\,\text{MW}}{\varepsilon\, b\, v 1000} \times \Delta A_{340} \quad [37]$$

where:

c = starch in sample solution (g/L)
V = volume (ml) of reaction mixture
MW = molecular weight of starch (because this method measures glucose derived from starch, use 162.1, $MW_{glucose} - MW_{water}$)
ε = absorption coefficient of NADPH at 340 nm ($6.3\ \text{L mmol}^{-1}\ \text{cm}^{-1}$)
b = light pathlength of cuvette (1 cm)
v = volume of sample (ml)
ΔA_{340} = A_{340}, sample − A_{340}, reagent blank

Note that HK catalyzes the phosphorylation of fructose as well as glucose. The determination of glucose is specific because of the specificity of the second reaction, catalyzed by G6PDH, in which glucose-6-phosphate is the substrate.

This assay sequence can be used to detect the dextrins of corn syrup used to sweeten a fruit juice product. A second assay would be needed, however, without treatment with amyloglucosidase to account for the glucose in the product. The glucose determined in that assay would be subtracted from the result of the assay in which amyloglucosidase is used.

The same HK-G6PDH sequence used to measure glucose can also be used to measure other carbohydrates in foods. For example, lactose and sucrose can be determined via specific hydrolysis of these disaccharides by β-galactosidase and invertase, respectively, followed by the use of the earlier described HK-G6PDH sequence.

16.3.1.3.4 Determination of D*-Malic Acid in Apple Juice* Two stereoisomeric forms of malic acid exist. L-Malic acid occurs naturally, while the D form is normally not found in nature. Synthetically produced malic acid is a mixture of these two isomers. Consequently, synthetic malic acid can be detected by a determination of D-malic acid. One means of detecting the malic acid is through the use of the enzyme decarboxylating D-malate dehydrogenase (DMD) (8). DMD catalyzes the conversion of D-malic acid as follows:

$$\text{D-malic acid} + NAD^{+} \xrightarrow{DMD} \text{pyruvate} + CO_2 + NADH + H^{+} \quad [38]$$

The reaction can be followed by the measurement of NADH photometrically. Because CO_2 is a product of this reaction and escapes, the equilibrium of the reaction lies to the right and the process is irreversible. This assay is of value because the addition of synthetic D/L malic acid can be used to illegally increase the acid content of apple juice and apple juice products.

16.3.2 Enzyme Activity Assays

16.3.2.1 Peroxidase Activity

Peroxidase is found in most plant materials and is reasonably stable to heat. A heat treatment that will destroy all peroxidase activity in a plant material is usually considered to be more than adequate to destroy other enzymes and most microbes present. In vegetable processing, therefore, the adequacy of the blanching process can be monitored by following the disappearance of peroxidase activity (9). Peroxidase catalyzes the oxidation of guaiacol (colorless) in the presence of hydrogen peroxide to form tetraguaiacol (yellow brown) and water Equation [39]. Tetraguaiacol has an absorbance maximum around 450 nm. Increase in absorbance at 450 nm can be used to determine the activity of peroxidase in the reaction mixture.

$$H_2O_2 + \text{guaiacol} \xrightarrow{\text{peroxidase}} \underset{\text{(colored)}}{\text{tetraguaiacol}} + H_2O \quad [39]$$

16.3.2.2 Lipoxygenase

Recently it has been pointed out that lipoxygenase may be a more appropriate enzyme to measure the adequacy of blanching of vegetables than peroxidase (10). Lipoxygenase refers to a group of enzymes that catalyzes the oxidation by molecular oxygen of fatty acids containing a *cis, cis,* 1,4-pentadiene system producing conjugated hydroperoxide derivatives:

$$(\text{—CH} = \text{CH-CH}_2\text{-CH} = \text{CH–}) + O_2 \xrightarrow{\text{lipoxygenase}} \underset{\text{(conjugated)}}{(\text{—COOH–CH} = \text{CH–CH} = \text{CH–})} \quad [40]$$

A variety of methods can be used to measure lipoxygenase activity in plant extracts. The reaction can be followed by measuring loss of fatty acid substrate, oxygen uptake, occurrence of the conjugated diene at 234 nm, or the oxidation of a cosubstrate such as carotene (11). All these methods have been used, and each has its advantages. The oxygen electrode method is widely used and replaces the more cumbersome manometric method. The electrode method is rapid and sensitive and gives continuous recording. It is normally the method of choice for crude extracts, but secondary reactions involving oxidation must be corrected for or eliminated. Zhang et al. (12) have reported the adaptation of the O_2 electrode method to the assay of lipoxygenase in green bean homogenates without extraction. Due to the rapidity of the method (<3 min including the homogenization), on-line process control using lipoxygenase activity as a control parameter for optimization of blanching of green beans is a real possibility. The formation of conjugated diene fatty acids with a chromophore at 234 nm can also be followed continuously. However, optically clear mixtures are necessary. Bleaching of carotenoids has also been used as a measure of lipoxygenase activity. However, the stoichiometry of this method is uncertain, and all lipoxygenases do not have equal carotenoid bleaching activity. Williams et al. (10) have developed a semiquantitative spot test assay for lipoxygenase in which I^- is oxidized to I_2 in the presence of the linoleic acid hydroperoxide product and the I_2 detected as an iodine starch complex.

16.3.2.3 Phosphatase Assay

Alkaline phosphatase is a relatively heat stable enzyme found in raw milk. The thermal stability of alkaline phosphatase in milk is greater than the nonspore forming microbial pathogens present in milk. The phosphatase assay has been applied to dairy products to determine whether pasteurization has been done properly and to detect the addition of raw milk to pasteurized milk. A common phosphatase test is based on the phosphatase-catalyzed hydrolysis of disodium phenyl phosphate liberating phenol (13). The phenol product is measured colorimetrically after reaction with CQC (2,6-dichloroquinonechloroimide) to form a blue indophenol. The indophenol is extracted into *n*-butanol and measured at 650 nm. This is an example of a physical separation of product to allow the ready measurement of an enzyme reaction. More recently, a rapid fluorometric assay was developed and commercialized for measurement of alkaline phosphatase in which the rate of fluorophore production can be monitored directly without butanol extraction used to measure indophenol when phenylphosphate is used as substrate (14). The fluorometric assay was shown to give greater repeatability compared to the standard assay in which phenylphosphate is used as substrate and was capable of detecting 0.05% raw milk in a pasteurized milk sample. Similar chemistry has been applied to the measurement of acid phosphatase activity in meats as a means of ensuring adequate cooking via correlation of enzyme activity to endpoint temperature (15).

16.3.2.4 α-Amylase Activity

Amylase activity in malt is a critical quality parameter. The amylase activity in malt is often referred to as diastatic power and refers to the production of reducing substances by the action of α- and β-amylases on starch. The measurement of diastatic power involves digestion of soluble starch with a malt infusion (extract) and following increase in reducing substances by measuring reduction of Fehling's solution or ferricyanide. Specifically measuring α-amylase activity (often referred to as **dextrinizing activity**) in malt is more complicated and is based on using a limit dextrin as substrate. **Limit dextrin** is prepared by action of β-amylase (free of α-amylase activity) on soluble starch. The β-amylase clips maltose units off the nonreducing end of the starch molecule until an α-1,6- branch point is encountered. The resulting product is a β-limit dextrin that serves as the substrate for the endo cleaving α-amylase. A malt infusion is added to the previously prepared limit dextrin substrate and aliquots removed periodically to a solution of dilute iodine. The α-amylase activity is measured by changed color of the starch iodine complex in the presence of excess β-amylase used to prepare the limit dextrin. The color is compared to a colored disc on a comparator. This is continued until the color is matched to a color on a comparator. The time to reach that color is **dextrinizing time** and is a measure of α-amylase activity, a shorter time representing a more active preparation.

Because α-amylase is an endoenzyme, when it acts on a starch paste the viscosity of the paste is dramatically reduced, greatly influencing flour quality. Consequently, α-amylase activity is of great importance in whole wheat. Wheat normally has small amounts of α-amylase activity, but when wetted in the field, preharvest sprouting (pregermination) can occur in wheat, with a dramatic increase in α-amylase activity. Preharvest sprouting cannot be easily detected visually, so measurement of α-amylase activity can be used as a sensitive estimate of preharvest sprouting. The **falling number method** is a procedure in which ground wheat is heated with water to form a paste, and the time it takes for a plunger to fall through the paste is recorded (16). Accordingly, the time in seconds (the falling number) is inversely related to the α-amylase activity and the degree of preharvest sprouting. This method of measuring enzyme activity is a good example of using change in physical property of a substrate as a means of estimation of enzyme activity.

16.3.2.5 Rennet Activity

Rennet, an extract of bovine stomach, is used as a coagulating agent in cheese manufacture. Most rennet activity tests are based on noting the ability of a preparation to coagulate milk. For example, 12% nonfat dry milk is dispersed in a 10 m*M* calcium chloride solution and warmed to 35°C. An aliquot of the rennet preparation is added and the time of milk clotting observed visually. The activity of the preparation is calculated in relationship to a standard rennet. As opposed to coagulation ability, rennet preparations can also be evaluated for proteolytic activity by measuring the release of a dye from azocasein (casein to which a dye has been covalently attached). In this assay, the rennet preparation is incubated with 1% azocasein. After the reaction period, the reaction is stopped by addition of trichloroacetic acid. The trichloroacetic acid precipitates the protein that is not hydrolyzed. The small fragments of colored azocasein produced by the hydrolysis of the rennet are left in solution and absorbance read at 345 nm (17, 18). This assay is based on the increase in solubility of a substrate upon cleavage by an enzyme.

16.3.3 Biosensors/Immobilized Enzymes

The use of immobilized enzymes as analytical tools is currently receiving increased attention. An immobilized enzyme in concert with a sensing device is an example of a biosensor. A biosensor is a device comprised of a biological sensing element (e.g., enzyme, antibody, etc.) coupled to a suitable transducer (e.g., optical, electrochemical, etc.). Immobilized enzymes, because of their stability and ease of removal from the reaction, can be used repeatedly, thus eliminating a major cost in enzyme assays. The most widely used enzyme electrode is the glucose electrode in which glucose oxidase is combined with an oxygen electrode to determine glucose concentration (19–23). When the electrode is put into a glucose solution, the glucose diffuses into the membrane where it is converted to gluconolactone by glucose oxidase with the uptake of oxygen. The oxygen uptake is a measure of the glucose concentration. The Clark polarographic electrode can be used for measurement of the oxygen. More recently oxygen sensors based on fiber optics and fluorescence quenching have been made commercially available and have the advantage of not needing the maintenance of Clark electrodes. Glucose can also be measured by the action of glucose oxidase with the detection of hydrogen peroxide, in which the hydrogen peroxide is detected amperometrically at a polarized electrode (22). Similar systems have been commercialized in which lactate, ethanol, sucrose, lactose, and glutamate can be measured. In the case of some of these sensors multiple enzymes are immobilized. For example for sucrose analysis, invertase, mutarotase, and glucose oxidase are immobilized on the same membrane. A large number of other enzyme electrodes have been reported. For example, a glycerol sensor, in which glycerol dehydrogenase was immobilized, has been developed for the determination of glycerol in wine (23). NADH produced by the enzyme was monitored with a platinum electrode.

16.4 SUMMARY

Enzymes, due to their specificity and sensitivity, are valuable analytical devices for quantitating compounds that are enzyme substrates, activators, or inhibitors. In enzyme-catalyzed reactions, the enzyme and substrate are mixed under specific conditions (pH, temperature, ionic strength, substrate concentration, and enzyme concentrations). Changes in these conditions can affect the reaction rate of the enzyme and thereby the outcome of the assay. The enzymatic reaction is followed by measuring either the amount of product generated or the disappearance of the substrate. Applications for enzyme analyses will increase as a greater number of enzymes are purified and become commercially available. In some cases, gene amplification techniques will make enzymes available that are not naturally found in great enough amounts to be used analytically. The measurement of enzyme activity is useful in assessing food quality and as an indication of the adequacy of heat processes such as pasteurization and blanching. In the future, as in-line process control (to maximize efficiencies and drive quality developments) in the food industry becomes more important, immobilized enzyme sensors, along with microprocessors, will likely play a prominent role.

16.5 STUDY QUESTIONS

1. The Michaelis–Menten equation mathematically defines the hyperbolic nature of a plot relating reaction velocity to substrate concentration for an enzyme-mediated reaction. The reciprocal of this equation gives the Lineweaver–Burk formula and a straight-line relationship as shown below.

$$\frac{1}{v_0} = \frac{K_m}{V_m}\frac{1}{[S]} + \frac{1}{V_m}$$

   (a) Define what v_o, K_m, V_m, and [S] refer to in the Lineweaver–Burk formula.
   (b) Based on the components of the Lineweaver–Burk formula, label the y-axis, x-axis, slope, and y-intercept on the plot.
   (c) What factors that control or influence the rate of enzyme reactions affect K_m and V_m?

   (a) v_o

   K_m

   V_m

   [S]

   (b)

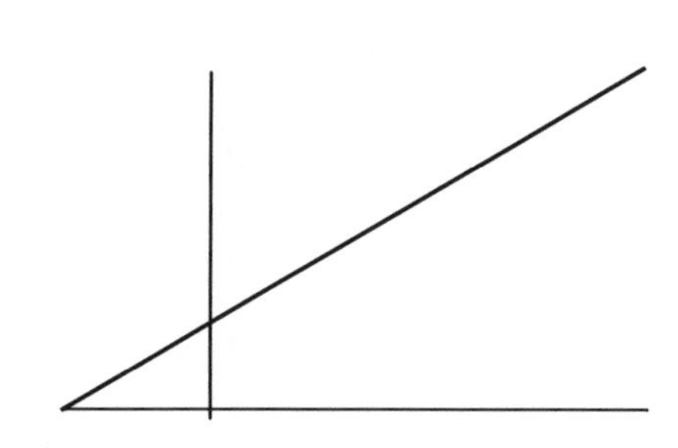

   (c) K_m

   V_m

2. Explain, on a chemical basis, why extremes of pH and temperature can reduce the rate of enzyme-catalyzed reactions.
3. Differentiate among competitive, noncompetitive, and uncompetitive enzyme inhibitors.
4. You believe that the food product you are working with contains a specific enzyme inhibitor. Explain how you would quantitate the amount of enzyme inhibitor (I) present in an extract of the food. The inhibitor (I) in question can be purchased commercially in a purified form from a chemical company. The inhibitor is known to inhibit the specific enzyme E, which reacts with the substrate S to generate product P, which can be quantitated spectrophotometrically.
5. What methods can be used to quantitate enzyme activity in enzyme-catalyzed reactions?
6. What is a coupled reaction, and what are the concerns in using coupled reactions to measure enzyme activity? Give a specific example of a coupled reaction used to measure enzyme activity.
7. Explain how D-malic acid can be quantitated by an enzymatic method to test for adulteration of apple juice.
8. Why is the enzyme peroxidase often quantitated in processing vegetables?
9. Explain the purpose of testing for phosphatase activity in the dairy industry, and explain why it can be used in that way.
10. The falling number value often is one of the quality control checks in processing cereal-based products. What is the falling number test, and what information does it provide? What other tests could be used to assay this quality factor?
11. Explain how glucose can be quantitated using a specific immobilized enzyme.

16.6 REFERENCES

1. Powers JR, Whitaker JR (1977) Effect of several experimental parameters on combination of red kidney bean (*Phaseolus vulgaris*) α-amylase inhibitor with porcine pancreatic α-amylase. J Food Biochem 1:239
2. Whitaker JR (1985) Analytical uses of enzymes. In: Gruenwedel D, Whitaker JR (eds) Food Analysis. Principles and Techniques, vol 3. Biological Techniques, Marcel Dekker, New York, pp 297–377
3. Bernfeld P (1955) Amylases, α and β. Methods Enzymol 1:149
4. Henniger G (2004) Nonindustrial enzyme usage: Enzymes in food analysis. In: Aehle W (ed) Enzymes in Industry, Production and Applications, 2nd edn, Wiley-VCH Verlag GmbH & Co. Weinheim, Germany, pp 322–334
5. Bergmeyer HU (1983) Methods of Enzymatic Analysis. Academic Press, New York
6. Beutler H (1984) A new enzymatic method for determination of sulphite in food. Food Chem 15:157
7. Raabo E, Terkildsen TC (1960) On the enzyme determination of blood glucose. Scand J Clin Lab Invest 12:402
8. Beutler H, Wurst B (1990) A new method for the enzymatic determination of D-malic acid in foodstuffs. Part I: Principles of the Enzymatic Reaction. Deutsche Lebensmittel-Rundschau 86:341
9. USDA (1975) Enzyme inactivation tests (frozen vegetables). Technical inspection procedures for the use of USDA inspectors. Agricultural Marketing Service, U.S. Department of Agriculture, Washington, DC
10. Williams DC, Lim MH, Chen AO, Pangborn, RM, Whitaker JR (1986) Blanching of vegetables for freezing – Which indicator enzyme to use. Food Technol 40(6): 130.
11. Surrey K (1964) Spectrophotometric method for determination of lipoxidase activity. Plant Physiology 39:65
12. Zhang Q, Cavalieri, RP, Powers JR, Wu J (1991) Measurement of lipoxygenase activity in homogenized green bean tissue. J Food Sci 56:719
13. Murthy GK, Kleyn DH, Richardson T, Rocco RM (1992) Phosphatase methods. In: Richardson GH (ed) Standard methods for the examination of dairy products, 16th edn. American Public Health Association, Washington, DC, p. 413
14. Rocco R (1990) Fluorometric determination of alkaline phosphatase in fluid dairy products: Collaborative study. J Assoc Off Anal Chem 73:842
15. Davis CE (1998) Fluorometric determination of acid phosphatase in cooked, boneless, nonbreaded broiler breast and thigh meat. J AOAC Int 81:887
16. AACC International (2010) Approved methods of analysis, 11th edn. (online), American Association of Cereal Chemists, St. Paul, MN
17. Christen GL, Marshall RT (1984) Selected properties of lipase and protease of *Pseudomonas fluorescens* 27 produced in 4 media. J Dairy Sci 67:1680
18. Kim SM, Zayas JF (1991) Comparative quality characteristics of chymosin extracts obtained by ultrasound treatment. J Food Sci 56:406
19. Reyes J, Cavalieri RP (2003) Biosensors. In: Heldman DR (ed) Encyclopedia of agricultural, food, and biological engineering, Marcel Dekker, New York, pp 119–123
20. Guilbault GG, Lubrano GJ (1972) Enzyme electrode for glucose. Anal Chim Acta 60:254
21. Borisov SM, Wolbeis OS (2009) Optical biosensors. Chem Rev 108:423
22. Shimizu Y, Morita K (1990) Microhole assay electrode as a glucose sensor. Anal Chem 62:1498
23. Matsumoto K (1990) Simultaneous determination of multicomponent in food by amperometric FIA with immobilized enzyme reactions in a parallel configuration. In: Schmid RD (ed) Flow injection analysis (FIA) based on enzymes or antibodies, GBF monographs, vol 14. VCH Publishers, New York, pp 193–204

Immunoassays

Y-H. Peggy Hsieh

Department of Nutrition, Food and Exercise Sciences, Florida State University, Tallahassee, FL 32306–1493, USA
yhsieh@fsu.edu

17.1 Introduction 303
17.1.1 Definitions 303
17.1.2 Binding Between Antigen and Antibody 304
17.1.3 Types of Antibodies 304
17.2 Theory 305
17.3 Enzyme Immunoassay Variations 306
17.3.1 Overview 306
17.3.2 Noncompetitive Immunoassay 307
17.3.2.1 Sandwich ELISA 308
17.3.3 Competitive Immunoassays 308
17.3.3.1 Bound Hapten Format 309
17.3.3.2 Bound Antibody Format 309
17.3.3.3 Standard Curve 309
17.3.4 Indirect Immunoassays 310
17.3.5 Western Blots 311
17.4 Lateral Flow Strip Assay 312
17.4.1 Overview 312
17.4.2 Procedure 312
17.4.3 Applications 313
17.5 Immunoaffinity Purification 313
17.6 Applications 313
17.7 Summary 314
17.8 Study Questions 314
17.9 Acknowledgement 315
17.10 References 315

S.S. Nielsen, *Food Analysis*, Food Science Texts Series, DOI 10.1007/978-1-4419-1478-1_17,

17.1 INTRODUCTION

Immunochemistry is a relatively new science that has developed rapidly in the last few decades. One of the most useful analytical developments associated with this new science is immunoassay. Originally immunoassays were developed in medical settings to facilitate the study of immunology, particularly the antibody–antigen interaction. Immunoassays now are finding widespread applications outside the clinical field because they are appropriate for a wide range of analytes ranging from proteins to small organic molecules. In the food analysis area, immunoassays are widely used for chemical residue analysis, identification of bacteria and viruses, and detection of proteins in food and agricultural products. Protein detection is important for determination of allergens and meat species content, seafood species identification, and detection of genetically modified plant tissues. While immunoassays of all formats are too numerous to cover completely in this chapter, there are several procedures that have become standard for food analysis because of their specificity, sensitivity, and simplicity.

17.1.1 Definitions

Immunoassays are analytical techniques based on the specific and high affinity binding of antibodies with particular target antigens. To fully understand immunoassays some of these terms need to be defined. The two essential elements of any immunoassay are antigens and antibodies. In an immunoassay, antigens and antibodies are used either as target molecules or capture molecules. In other words, a particular antigen can be used to capture its specific antibody, or a specific antibody can be used to trap the target antigen in a sample. An **antigen** is any molecule that induces the formation of antibodies and can bind to these antibodies. **Antibodies** are **immunoglobulin** (Ig) **proteins** produced by animals in response to an antigen. These antibody proteins are secreted by the activated B cells in immune system and bind the particular antigen responsible for their induction. Generally a molecule must be greater than 5000 dalton, abbreviated as Da (unit of molecular mass), to be perceived as an antigen by a mammalian immune system. Almost all proteins are large molecules and have the ability to induce antibody formation in the body of humans and animals. However, many of the molecules analyzed in food are not as large as proteins but are small molecules such as toxins, or antibiotics and chemical residues (e.g., pesticides). When animals are injected with small molecules, they do not develop antibodies against these molecules. To induce specific antibodies to recognize and bind the small target molecule, the solution is to covalently link the small molecule, or some appropriate derivative of the small molecule, to a larger carrier molecule. The small molecule that must be linked to a large carrier protein before it can be used as an immunogen to induce antibodies is called a **hapten**. The carrier protein-linked hapten is called a **conjugate antigen**. Haptens react specifically with the appropriate antibodies, but are not immunogenic. The most common molecules used as carriers are proteins that are fairly soluble for simplicity in chemical linking and foreign to the animal to properly stimulate an immune response. Typical carrier molecules include albumin proteins from a different species, such as bovine serum albumin and hemocyanins that are obtained from crustaceans. Of course, when a conjugate antigen is used for immunization of an animal, its immune system is stimulated to produce antibodies that bind not only the externally attached hapten, but also the exposed exterior of the covalently linked foreign protein.

There are five major classes of antibodies, IgA, IgE, IgG, IgM, IgD, according to their heavy chain structure. Animal blood contains trace amounts of IgA and IgD. IgM is a very large molecule and can be regarded as a precursor of IgG. IgE is only associated with allergic response in humans and animals. Among these five classes of antibodies, IgG has the highest concentration in blood and is the most important class used in food immunoassay. Since the antibody and antigen are central to any immunoassay, it is useful to better understand the basic structure of the antibody and how it binds the antigen. Figure 17-1 is an idealized diagram of an antibody IgG. The IgG is a Y-shaped molecule made up of four polypeptide chains that are linked by inter- and intradisulfide bonds. Two of the polypeptide chains are identical and roughly twice as large as the other two identical polypeptide chains. Because of their relative sizes, the former pair is known as **heavy chains** and the latter pair as **light chains**. Overall, an IgG antibody is a very large protein of approximately 150,000 Da.

Antigen is bound by two identical binding sites made up of the end portions (*N* terminals) of a heavy and light chain at the top of the Y. These two fragments capable of binding with antigen are called **Fab (fragment antigen binding)**. The third fragment with no antigen binding capability is called **Fc (fragment crystalizable)**, because it can be crystallized. Different antibodies produced by different B cells can have many variations in amino acid sequences near the binding sites for both the heavy and light chains. This leads to a tremendous diversity of binding sites for different antibodies. For example, a mouse has 10^7–10^8

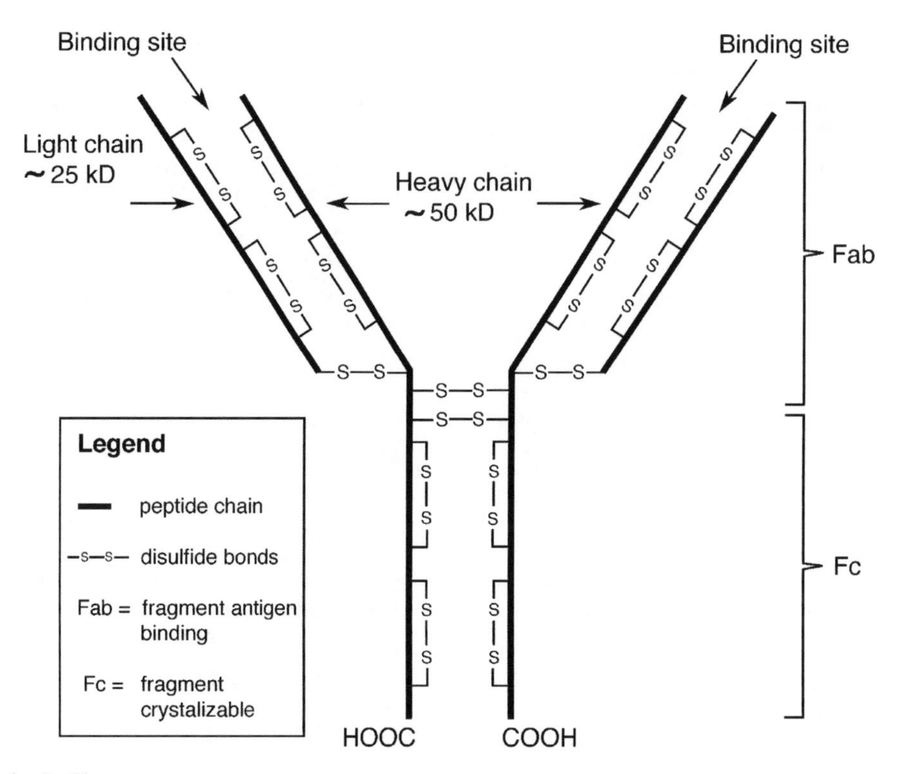

17-1 figure Antibody (IgG) structure.

different antibodies (and at least this number of different B cells), each with a unique binding site. The rest of the antibody (away from the binding site) is quite consistent, and small variations in this region on the heavy chains result in different antibody classes.

17.1.2 Binding Between Antigen and Antibody

Antibodies can develop remarkably strong binding affinities for their antigens. These affinities are among the strongest noncovalent interactions known between molecules. The binding strength (affinity) between antibody and antigen is one of the most important factors that determine the sensitivity of an immunoassay. The antibody binds to the outside of the antigen molecule in a specific region. This specific region bound by a single antibody binding site is known as an **epitope**. Two types of epitopes on an antigen can be formed. A **linear epitope** is formed by a continuous sequence of amino acid residues, and a **conformational epitope** is formed by noncontinuous amino acid sequences that are folded into close proximity from neighboring or overlapping peptide chain on the surface of the antigen (Fig. 17-2). If this 3-D conformation of the antigen is altered by some kind of the environmental conditions, such as heating or pH changes, the conformational epitope will be destroyed, which means that the antigen cannot bind to the antibody. Moreover, the binding of the antibody to the antigen does not involve covalent bonding, but the same interactions that are responsible for the tertiary structure of proteins. These interactions include electrostatic, hydrogen bonds, hydrophobic interaction, and Van der Waals. While the latter interactions, Van der Waals, are the weakest, they often can be the most important because every atom can contribute to the antibody–antigen binding as long as the atoms are very close to each other (generally about 0.3–0.4 nm). This requirement for very close proximity is why antibody to antigen binding is considered something like a lock and key interaction, where the surfaces of the antibody binding site and the antigen epitope are mirror complements of each other.

17.1.3 Types of Antibodies

A major variable in an immunoassay is the type of antibody used. When serum antibody is used from any animal, there are many different antibodies that bind different epitopes on the antigen. This collection of different antibodies is known as **polyclonal antibodies**. Scientists knew that individual B cells produced antibodies with only one binding site, but were unable to culture B cells outside of the animal. However, in 1975, Köhler and Milstein (1) successfully fused cancer, or myeloma cells, with B cells. The

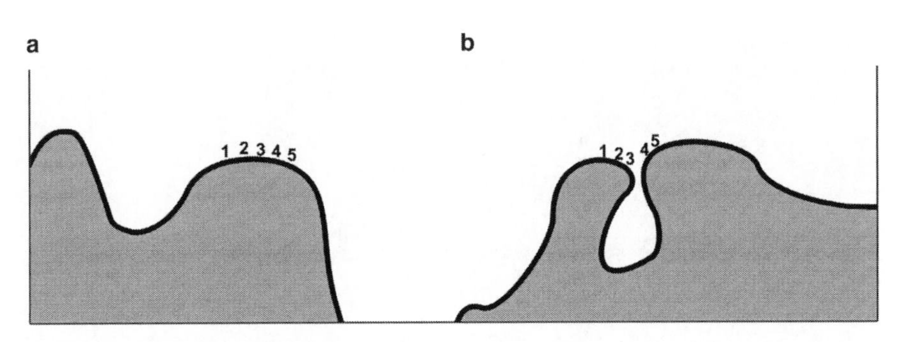

17-2 figure Linear epitope (**a**) and conformation-dependent epitope (**b**).

new fused cells, or **hybridomas**, retained the properties of both of the parent cells. That is, they could be cultured, like cancer cells, and produced antibodies like the B cells. Hybridomas thus can be cloned and cultured individually to produce different antibodies with different epitopes. Antibodies produced with this procedure became known as **monoclonal antibodies**. Monoclonal antibodies produced from a single hybridoma are identical in every way and bind antigen with only one type of binding site; that is, a single epitope is bound, therefore, they can be used as standard reagents in immunoassays. Moreover, the hybridomas were "immortalized" by the procedure and with proper care could produce as much identical antibody as required. It did not take the scientific community long to appreciate the tremendous advantages of these monoclonal antibodies, so Köhler and Milstein were awarded the Nobel Prize for their work on this in 1984. While monoclonal antibodies are initially much more expensive to produce, there is the possibility for limitless identical antibody production, often from nonanimal sources such as large-scale production of the hybridomas in cell growth chambers. These advantages outweigh the initial development costs for many immunoassay manufacturers. Detail procedures regarding the antibody production and characterization can be found in the book by Howard and Bethell (2)

17.2 THEORY

Based on the specific antibody–antigen affinity, various types of immunoassays have been developed to use either antibody as the capture molecule to search the target antigen, or use the antigen as the capture molecule to trap the antibody in a complex sample. The basis of every immunoassay is the detection and measurement of the primary antigen–antibody reaction. In its simplest form, antibody capture of antigen can involve a simple precipitation and be detected visually. Since all antibodies have at least two identical binding sites, they can crosslink epitopes from two identical antigens. If other antigen epitopes are further cross-linked by different antibodies, a large, insoluble network can result which is seen as a precipitate. The immunoprecipitation techniques, including immunodiffusion and agglutination, formed the basis of early development of immunoassay techniques and have been used widely for protein and cell identification using antisera. However, these methods only work for antigens with multiple epitopes.

To measure the quantity of soluble antibody-bound antigen molecules in a solution, all immunoassays require two things. The first is that there must be some method to separate or differentiate free antigen from bound antigen. Secondly, these antibody-bound antigens must be quantifiable at low concentrations for maximum sensitivity. Detection at very low concentrations has required very active labels. One of the first successful immunoassay procedures was developed by Yalow and Berson (3) in 1960. This procedure used radioactive iodine, I^{131}, a "hot" radioisotope with a half-life of only 8 days, as a label to reveal the primary antibody–antigen complex. This radioactive label allowed for the second requirement of immunoassays: quantification at low concentrations. Yalow and Berson used paper chromato-electrophoresis to separate their antibody-bound antigen from free antigen, fulfilling the first requirement of an immunoassay. With all the variations in separation and detection techniques in the early stage of immunoassay development, however, the radioactive iodine labeling remained and these assays became known as **radioimmunoassays** (RIA).

One of the techniques for the separation of unbound from bound molecules in immunoassays involves immobilizing protein on a hydrophobic solid surface. Proteins have large regions that contain hydrophobic amino acid groups that prefer not to be exposed to water. These nonpolar hydrophobic groups include hydrocarbons and aromatic groups that prefer to interact with similar groups, rather than a polar solvent such as water. In aqueous conditions, these regions will bind to other hydrophobic

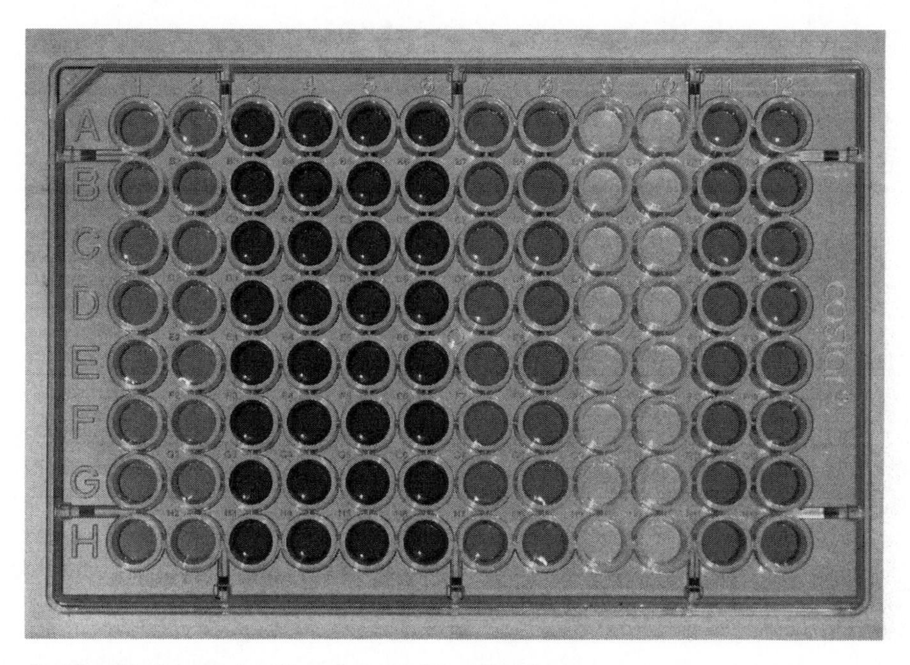

17-3 figure Picture of a 96-well plastic microtiter plate used for ELISA.

surfaces excluding water. Surfaces commonly introduced in immunoassays to take advantage of this type of binding include charcoal, nitrocellulose, and plastic. Plastic surface in many forms is used commonly for immunoassays. Among the most popular are microtiter plates made of plastics such as polystyrene or polyvinyl. These microtiter plates typically are formatted to contain 96 individual wells, each with a maximal capacity of about 300 µl of liquid (Fig. 17-3). To differentiate the wells, the vertical rows are labeled A to H and the columns numbered 1–12. It is important to realize that proteins bind to the bottom and sides of the wells in these plates randomly through hydrophobic interactions. The hydrophobic interactions between proteins and these surfaces increase at lower temperatures (due to less molecular motion) and with increased ionic strength of the solvent (increasing solvent polarity). Other forms of solid surface commonly used in immnoassays include plastic vials, magnetic beads, and nitrocellulose membranes or strips.

17.3 ENZYME IMMUNOASSAY VARIATIONS

17.3.1 Overview

Every immunoassay technique developed is based on the selection of an **amplification method** that will improve the sensitivity of assays. While RIAs worked well, they were confined to specially equipped laboratories because of the dangers associated with the use of radioactive material. Immunoassays did not develop for more general use, including field use, until enzyme labels were developed. Any immunoassay that uses an enzyme label to reveal the primary antibody–antigen binding is called an enzyme immunoassay. Pioneers in this development were Engvall and Perlmann (4) who in 1971 developed a type of enzyme immunoassay that they called an **enzyme-linked immunosorbent assay**, or ELISA. ELISA assays involve the binding of a soluble antigen or antibody to a solid support (immunosorbent), typically in the form of a 96-well plastic microtiter plate. The bound and unbound molecules can be separated by a washing step of the plate. Therefore, they also helped to popularize the use of hydrophobic plastics to immobilize proteins and to separate unbound molecules. Similar immunoassays using a solid support other than plastic microtiter plates also have been developed and are commonly used. Examples are the **dot blot** assay and **Western blot** assay which use nitrocellulose membranes. These developments have expanded the use of immunoassays to a wider range of applications.

The enzyme label used in an immunoassay converts a colorless substrate to a colored soluble product in the solution, thus generating a detectable signal for the assay. The amount of target protein antigen present in the sample extract is determined from the intensity of color developed in the immunoassay. The ideal enzyme for an ELISA or enzyme immunoassay is one that is stable, easily linked to antibodies or antigens, and rapidly catalyzes a noticeable change with a simple substrate. With the many enzymes available, two enzymes, **horseradish peroxidase** and **alkaline phosphatase**, by far are the most commonly used in immunoassays. Other enzymes used include β-galactosidase, glucose oxidase and glucose-6-phosphate dehydrogenase. The use of an enzyme to generate color signal also contributes to the sensitivity of the assay because a single enzyme molecule present at the end of the test converts many substrate molecules to detectable colored product, thus amplifying the signal generated by the assay. Minimal laboratory equipment is required to perform ELISA. The color generated from an assay can be visualized to determine the result in a qualitative assay or quantitated spectrophotometrically. The type of spectrophotometer used to quantitatively monitor color development caused by the enzyme action is called a **microplate reader**. Automated **microplate washer** also is available although most assays can be washed manually.

All ELISA protocols include the following five steps: (1) coating of antibody or antigen on a solid phase; (2) blocking the remaining uncoated surface on the solid phase with a blocking buffer containing a nonspecific protein such as bovine serum albumin (this is to minimize the nonspecific reactions and also protect the adsorbed antigen or antibody from surface denaturation); (3) incubating with different immunoassay reagents at a specified temperature and time; (4) washing the coated surface to separate free, unbound molecules from bound molecules; and (5) detecting the color developed from the assay visually or spectrophotometrically. Specific procedures vary with the different variations of ELISA. It is important to include both positive and negative controls in an assay along with any analyzed food sample because materials in the food extract can vary widely and these other components can have an effect on the competition for the antibody binding site. This is to ensure that the immunoassay works well (positive control shows positive signal) and that there is no contamination or nonspecific reactions in the assay system (negative control shows negative signal).

All immunoassay signals can be detected directly or indirectly. In the **direct enzyme immunoassay**, the enzyme label is directly linked to the primary detection molecule, therefore, more purified immunoreagents need to be used for the enzyme conjugation procedure. In contrast, the **indirect enzyme immunoassay** uses a commercially available intermediate reagent to link the capture molecule with an enzyme-conjugated secondary molecule. Although an additional step is involved, indirect assays require less immunoreagents, and in many cases could be more sensitive because more enzyme molecules can be linked to the detection antigen or antibody. Furthermore, both direct and indirect enzyme immunoassays can be configured in two types of formats: **competitive** and **noncompetitive**. Noncompetitive ELISA is commonly employed to analyze large molecules such as proteins in a food sample, while competitive ELISA is competitive in nature and mainly used for small molecule analysis. The amount of color development for the noncompetitive ELISA is directly related to the amount of antigen present in the sample. With any competitive ELISA format, there is an inverse relationship between the amount of color developed and the amount of antigen present in the sample (Fig. 17-4).

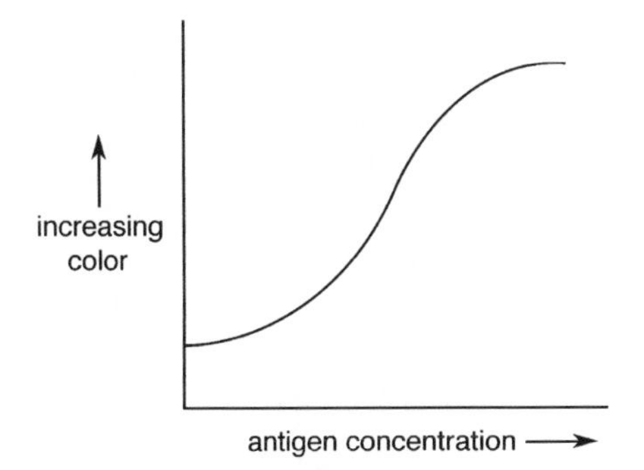

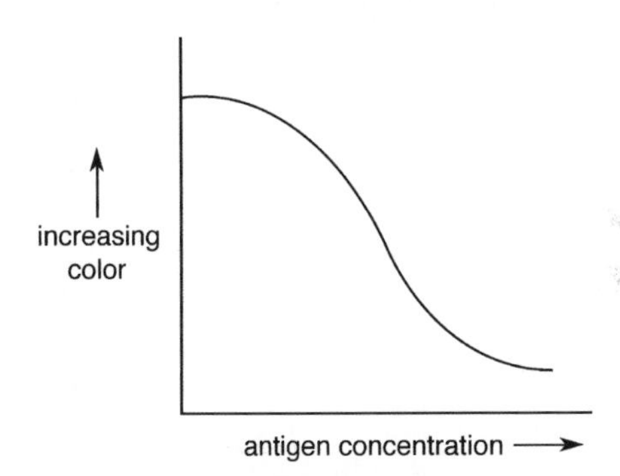

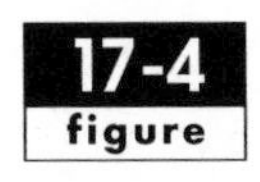
17-4 figure Relationship between color development and antigen concentration for different immunoassay formats.

The most commonly used ELISA variations in both competitive and noncompetitive ELISA formats are described in the following sections. In addition, another type of important enzyme immunoassay, Western blot, which is similar to ELISA but allows the determination of the molecular size of the antigentic protein, is also included in this section.

17.3.2 Noncompetitive Immunoassay

Non-competitive ELISA variations involve the revealing of the amount of primary antibody–antigen complex immobilized on the solid-phase by the amount of enzyme linked to the detection antigen or antibody molecules to produce a colored product in the assay solution. Therefore, at the end of the assay the color intensity is positively related to the amount of the target molecules. The absence of the target molecules produces no color, and the presence of high concentration of the target molecules produces strong color. This type of ELISA is used often to detect proteins in a food sample because protein molecules are large enough to link one or more antibodies or to an additional enzyme label on the surface of the protein.

17.3.2.1 Sandwich ELISA

One of the most popular formats for a noncompetitive enzyme immunoassay is the antibody **sandwich immunoassay**. A direct sandwich ELISA model is demonstrated in Fig. 17-5. The "meat" in the antibody sandwich is the target antigen. In food analysis this can involve identifying a protein adulterant, such as undeclared pork meat in a beef product; or a protein allergen such as peanut protein; or wheat protein in a product that would be a problem for people suffering from celiac disease.

Generally an antibody that binds to the antigen is first immobilized onto a solid phase. The most common immobilization of the antibody is simply binding it to a hydrophobic surface such as plastic. Excess antibody is removed by washing with a washing solution or simply water followed by a blocking step, and then the test is ready for analysis of a food extract. The immobilized antibody is called a **capture antibody**. The food extract being tested contains many compounds that might act as antigens. However, the antibody was prepared by immunization of an animal with a specific, purified protein antigen, and only this protein antigen in the food solution will bind to the capture antibody. Now the antigen and the capture antibody are immobilized and the remaining unbound molecules can be washed away. After the washing step, another antibody labeled with an enzyme is introduced. This antibody, called the **detection antibody**, also recognizes the antigen. Again excess detection antibody is washed away, then colorless enzyme substrate is added to develop a color if bound enzyme is present. Enzyme will only be present if the detection antibody has been immobilized by binding to antigen. The greater the color development, the greater the amount of antigen present. That is, there is a direct proportionality between the amount of color seen in the final step and the amount of antigen present in the extracted food sample. To increase the sensitivity of a sandwich immunoassay, one can use more antibodies for capture of the antigen or link more enzyme molecules to the assay system through the use of an intermediate reagent (Fig. 17-5). This immunoassay format can be made very sensitive and remarkably specific since two antibodies must detect the antigen.

When polyclonal antibodies are used in the sandwich immunoassay, the polyclonal antibody solution is divided into two parts. One part is bound to plastic to become the capture antibody. The second portion of the polyclonal antibody solution is conjugated to an enzyme and becomes the detection antibody. Monoclonal antibodies also can be used, but care must be exercised since a single type of monoclonal antibody cannot be used for both the capture and detection antibodies since only one unique epitope is recognized by any monoclonal antibody. In other words, the antigen must be able to bind two antibodies at the same time and therefore must use at least two distinct epitopes recognized by different monoclonal antibodies that recognize two distinct antigen epitopes.

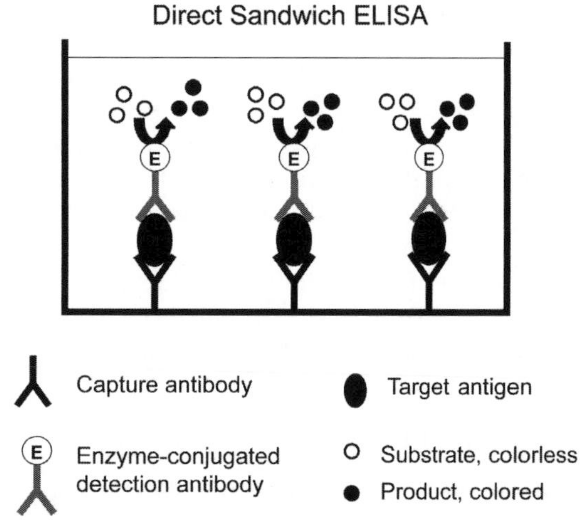

Direct Sandwich ELISA

Capture antibody
Target antigen
Enzyme-conjugated detection antibody
Substrate, colorless
Product, colored

17-5 figure Sandwich immunoassay.

17.3.3 Competitive Immunoassays

A problem in developing an immunoassay for detecting a small molecule is that a sandwich immunoassay format will not work since two different epitopes are required for both antibodies to bind. A small molecule represents only one epitope or even only part of one epitope. The **competitive immunoassay** format (Fig. 17-6) was, therefore, developed to solve this problem. The first requirement in a competitive ELISA involves immobilizing the small molecule, often as a hapten, or immobilizing the antibody. Subsequent procedures involve the competition between the free small antigen (from a sample) and the hapten (as an added reagent) for the binding of limited amount of the specific antibody. To bind the hapten to a surface such as nitrocellulose or plastic, it can again first be

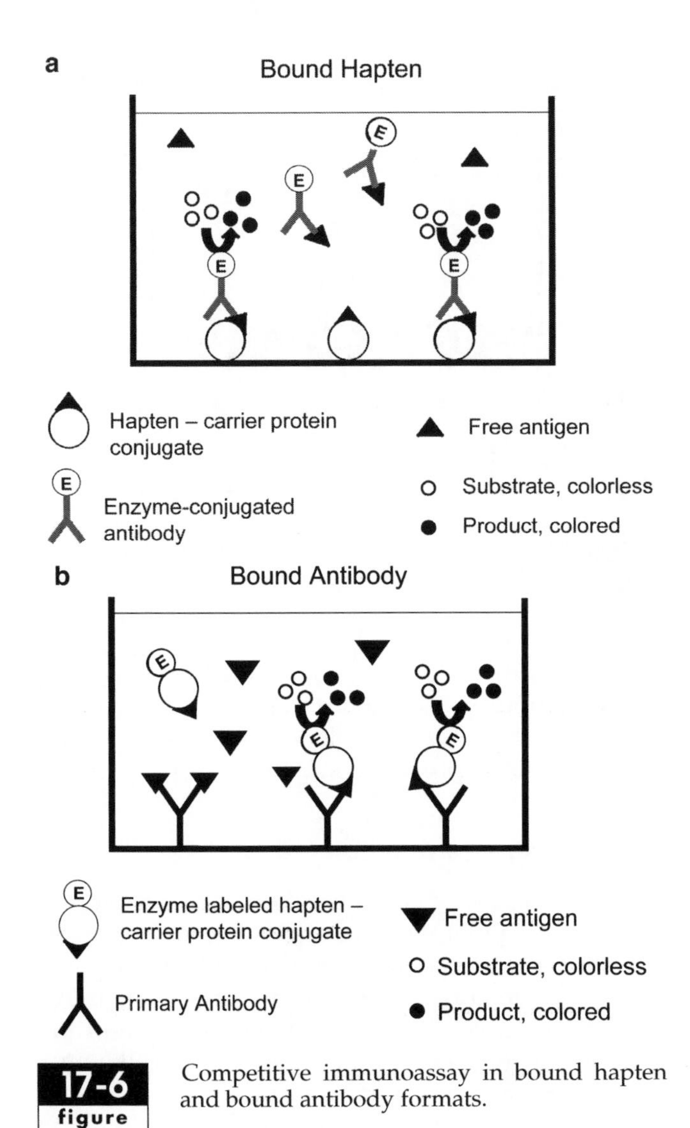

17-6 figure Competitive immunoassay in bound hapten and bound antibody formats.

linked to a protein that binds to these hydrophobic surfaces. However, the protein used for binding the hapten to the surface is different than the protein used for injection of the animal, since the animal also has developed antibodies against the carrier protein used for injection, and only the hapten-specific antibodies are desired for the competitive immunoassay.

To increase the sensitivity of a competitive immunoassay, the amount of limiting antibody should be reduced. Note that this is the reverse of what one would do to increase the sensitivity of a sandwich immunoassay. Theoretically the most sensitive competitive immunoassay would be between one antibody binding site and one hapten, with either of the two labeled with an enzyme. It is for this reason that the ability to detect the presence of the enzyme is so important for a competitive immunoassay. The more sensitive the system is to detect the enzyme, the more sensitive the competitive immunoassay. Two competitive ELISA procedures are described here.

17.3.3.1 Bound Hapten Format

In the bound hapten competitive immunoassay format (Fig. 17-6, top), the protein-bound hapten is first immobilized to a solid surface by the same hydrophobic interactions used to bind antibody. Excess material is washed away. Next a competition is created between the protein bound hapten and the free small molecule in a food extract, both competing for binding to the limited binding sites on the antibody labeled with a bound enzyme. It is important to realize that the free small molecule in the food extract is not completely identical to the immobilized hapten since the latter is covalently linked to a protein. However, if properly designed, the free molecule in the food extract is so chemically similar to the bound hapten that the competition for the limited number of antibody binding sites is nearly equal. The antibody bound to immobilized hapten remains after a subsequent washing step. The more small molecules in the food extract, the more antibody is bound to these free small molecules, and this unbound antibody (and its attached enzyme) will be washed away in the subsequent washing procedure. Finally, the amount of bound antibody is identified by adding the enzyme substrate and observing the amount of color developed. Therefore, there is an inverse relationship between the amount of small molecules or analyte in the food and the amount of color developed in the final step.

17.3.3.2 Bound Antibody Format

The other variation for a competitive immunoassay is to bind a limited amount of antibody to the plastic and create a competition between hapten bound to enzyme and free small molecules in the food extract (Fig. 17-6, bottom). It is generally believed that this second format is somewhat superior to the first format for sensitivity although it can require the use of more antibody reagent. Again after a washing step, the final procedure is a color development to determine the amount of bound hapten–enzyme. This competitive format also results in an inverse relationship between amount of color and free small molecules in the food extract.

17.3.3.3 Standard Curve

If one examines the data for a competitive immunoassay over a wide range of concentrations, the data always fit a sigmoidal curve (Fig. 17-7). Since all types of competitive immunoassays involve a reduction in absorbance with respect to a control (containing no small molecule or analyte), data often are presented as a ratio of sample absorbance to the absorbance of the control. This ratio is the left-side y-axis in Fig. 17-7.

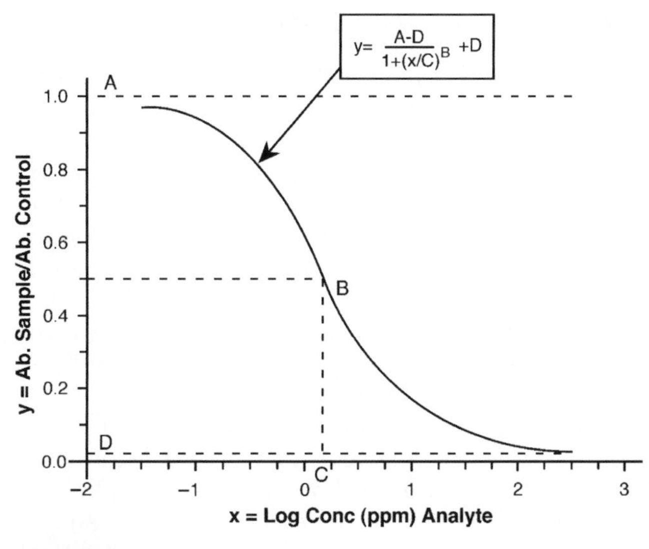

17-7 figure Typical competitive immunoassay standard curve.

As analyte concentration (usually presented on the x-axis using a logarithmic scale) becomes very low, the data curve approaches the maximum absorbance in an asymptotic manner. At the other extreme, very high amounts of analyte prevent antibody binding and the data curve approaches a very low absorbance value in an asymptotic manner. Ideally the low absorbance value is zero, but often it is found experimentally to be somewhat higher than zero. With polyclonal antibodies the main reason for this nonzero bottom limit is that some of the antibodies bind so strongly to hapten that they will not release in the presence of even very large amounts of free small molecules (analyte). The equation that describes the sigmoidal nature of the competitive immunoassay data is:

$$y = \left[(A - D)/(1 + (x/C)^B\right] + D \quad [1]$$

where:

- A = upper limit on y-axis (1.0 for Fig. 17-7)
- D = lower limit on y-axis (0.02 for Fig. 17-7)
- C = x coordinate representing y point half way between A and D, or inflection point of sigmoidal curve [$C = 2$ ($y = 0.51$ and $x = \log 2.0$) for Fig. 17-7]
- B = describes how rapidly the curve makes its transition from A to D ($B = 1.2$ for Fig. 17-7)

The final value, B, often has a magnitude near 1. Smaller values of B indicate a longer, shallower slope while larger values signify a shorter but steeper slope. Note that the value C is a measure of the sensitivity of the assay (2 ppm for Fig. 17-7). C is roughly the concentration of x required to reduce the control absorbance by half. The value of 50% absorbance reduction, known as **median inhibition concentration** or IC_{50} (concentration that reduce the effect by 50%), for different analytes is commonly quoted for immunoassay procedures.

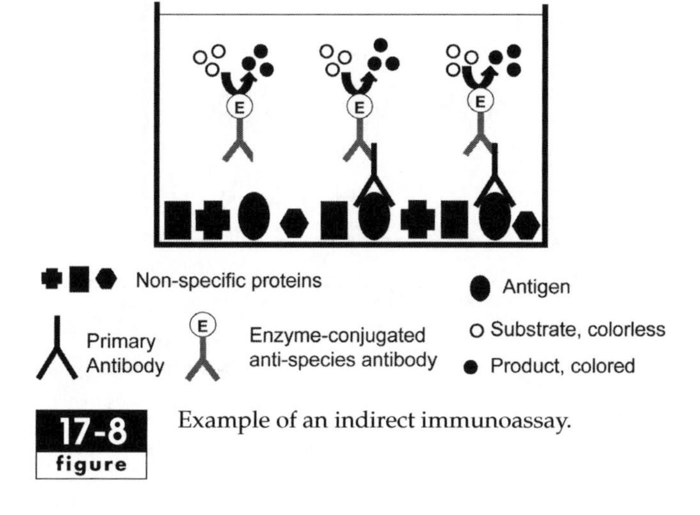

17-8 figure Example of an indirect immunoassay.

17.3.4 Indirect Immunoassays

The immunoassays described above use a **direct** format, meaning that the detecting molecules are linked to a label such as an enzyme, radioisotope, or florescent compound to directly measure the amount of antibody or antigen. The other popular variation of this format is the **indirect** immunoassay, meaning that they measure the amount of antibody or antigen indirectly, most often with an antispecies antibody. A simple form of indirect noncompetitive antibody-captured ELISA is demonstrated in Fig. 17-8. The method is often used in the early stage of immunoassay development to detect specific antibodies in antisera or screen hybridoma supernatants for searching desired antibodies. Theoretically, the antibody-captured ELISA can be made either as a direct ELISA or an indirect ELISA. However, because the target molecule is the specific antibody which appears in the biological fluid in very low quantity, it usually is impossible to obtain enough quantity of the specific antibody to prepare the enzyme-conjugated reagent for subsequent detection. Therefore, antibody-captured ELISA is almost always configured in an indirect assay format. The soluble antigen is adsorbed (coating) onto the surface of the wells and incubated. After blocking, diluted samples of antisera or hybridoma supernatants are then added to the wells and incubated to allow the immobilized antigen to bind specific antibodies in the sample. After washing away any unbound molecules, those bound antibodies can be detected by adding an enzyme-linked secondary antispecies antibody, which can easily bind to the constant region of the primary antibody. After another incubation and washing steps, a solution containing substrate is added to generate color in the solution. The color is positively related to the amount of target antibody present in the sample.

The secondary antispecies antibody used in the indirect ELISA does not have the specificity to bind the antigen, but only recognizes the primary antibody, thus to make a link of the enzyme label to the bound immunomolecules without interfering with the primary antigen–antibody binding. Since antibodies are proteins, they can act as antigens in another animal species. For example, rabbit antibodies injected into a goat can stimulate the goat's immune system to produce goat antibodies that bind to the rabbit antibodies. In this way, goat antirabbit antibodies can be produced to bind any antibody produced in a rabbit. There are many advantages to these antispecies antibodies. For example, when antispecies antibodies are used in the above antibody-captured ELISA format (Fig. 17-8), there is no need to label the primary antibody with an enzyme. After excess material is washed away, goat anti-rabbit antibody labeled with an enzyme can be added to detect the presence of any primary antibody which is produced in rabbit antiserum and binds to the antigen coated on the microtiter plate. Although this procedure adds an additional step, there are many other advantages. First, antispecies antibodies of all types are commercially available from many manufacturers. Also, these antispecies antibodies come with a variety of labels such as different enzymes, radioisotopes, or fluorescent compounds. These different label options become very useful for immunoassay development, or in the use of antibodies for detection in other systems such as examination of tissue under a microscope or proteins separated using electrophoretic techniques followed by an immunoassay. Since the antibody is a very large protein it has many sites for attachment of a labeled antispecies antibody. This multiplies the labels per antibody, increasing the ability to detect the antibody and resulting in increased sensitivity in an immunoassay since less primary antibody reagent can be used.

17.3.5 Western Blots

As one of the immunoassays that uses the specificity of the antigen-antibody interaction to indicate the presence of particular proteins in a sample, **Western blot** is a laboratory based method that combines two techniques: **polyacrylamide gel electrophoresis** (PAGE) and **immunoassay**. In the first part of a Western blot, proteins in a complex mixture are separated by PAGE according to their molecular mass. In the second part, the separated proteins are subjected to an immunoassay to detect the presence of antigenic proteins. Using this combination of techniques makes it possible to identify target proteins and confirm their identity by molecular mass. The detection reagent in a typical Western blot is an enzyme-labeled antibody conjugate, used in a direct or indirect immunoassay. If the original protein mixture was labeled with a radioactive material, then autoradiography is used to visualize the radioactive signal. The antibody used largely determines the specificity and sensitivity of the method. Specific proteins in picogram quantities can be detected in a highly sensitive Western blot.

To prepare protein samples for the initial separation by PAGE, they are typically boiled in a buffer solution containing a reducing agent (usually mercaptoethanol) and detergent (e.g., sodium dodecyl sulfate), to unfold the protein peptide chains. The treated sample is applied to a polyacrylamide gel and separated by electrophoresis based on **molecular mass** (see Chap. 15, Sect. 15.2.5). To prepare for the immunoassay portion of the method, the separated protein bands are transferred from the polyacrylamide gel to a **nitrocellulose membrane**. This membrane is then incubated with a solution of antibody–enzyme conjugate. After washing away the excess conjugate from the membrane, the enzyme–substrate is added. A colored band forms at the site on the membrane where the protein that reacted with the antibody was immobilized. The enzyme–substrate used in Western blot is different than the ones used in ELISA because the intent is to form an insoluble colored product that stays on the membrane. The color intensity and width of the protein band together indicate the concentration of the target protein in the sample extract. The molecular mass of the protein bound by the antibody can be estimated by its position relative to standard proteins of known molecular weight.

The Western blot procedure is technically complex, so it requires highly trained personnel working in a laboratory setting. However, the Western blot method is especially well suited to analysis of food samples that have been subjected to processing conditions. The Western blot method requires that the antibody used binds to a linear epitope on the protein, or is reactive to **denatured protein** (which would occur under the denaturing conditions in PAGE sample preparation). Because the Western blot method is highly sensitive and specific, and the antibodies used recognize denatured proteins, it is uniquely well suited to detect the presence of processed food proteins at low concentrations.

17.4 LATERAL FLOW STRIP ASSAY

17.4.1 Overview

The **lateral flow strip** (LFS) assay is a simple immunoassay format, used to determine if the target protein concentration is above or below a specified threshold (Fig. 17-9). The home pregnancy test is the best known LFS method. Results with LFS usually can be visualized in 10–20 min, and no washing steps are required to separate bound and unbound molecules. The characteristics of LFS methods – simplicity, low cost, ease-of-use, and reliability – make them ideal for use outside a laboratory setting, i.e., **field testing,** where supplies and equipment are limited.

Just as with other immunoassays, LFS methods are configured in a **competitive assay format** for detection of small molecules such as toxins or chemical residues, or in a **sandwich immunoassay format** for detection of large molecules. While the color in an enzyme immunoassay comes from an enzyme and its substrate creates a colored reaction product, the LFS methods instead use very small, spherical, colored particles (colloidal gold or colored latex) attached to antibodies to generate a positive colored signal. The capture antibody is immobilized in a zone on a porous membrane (usually nitrocellulose). By capillary action of the membrane, the test sample travels past the zone of immobilized antibody. The target protein in the test sample binds to the capture antibody. This type of sample movement and separation explains why LFS is called an **immunochromatographical assay**.

Because ELISA involves a signal amplification activity of the enzyme, it is generally regarded as inherently more sensitive than LFS methods. However, protein concentrations of less than a part per billion can be detected with a highly sensitive LFS.

17.4.2 Procedure

Figure 17-9 depicts a typical LFS sandwich immunoassay, and shows the various regions of the test strip. A primary antibody capable of binding the target protein is coated onto the surface of very small, colored particles (usually 20–40-nm diameter; colloidal gold or colored latex). These antibody-coated colored particles are dried in a porous pad. When these particles come in contact with liquid samples, they get reconstituted and are able to flow with the sample, moving across the **sample pad** of the strip by capillary action. Any large particulates in the sample liquid can be filtered out by a fiber filter placed at the front of the strip in the sample pad area. A second antibody, also capable of binding to the target protein, is immobilized at the **test line** (zone) on the surface of the fibers of the porous nitrocellulose membrane. A **control line** above

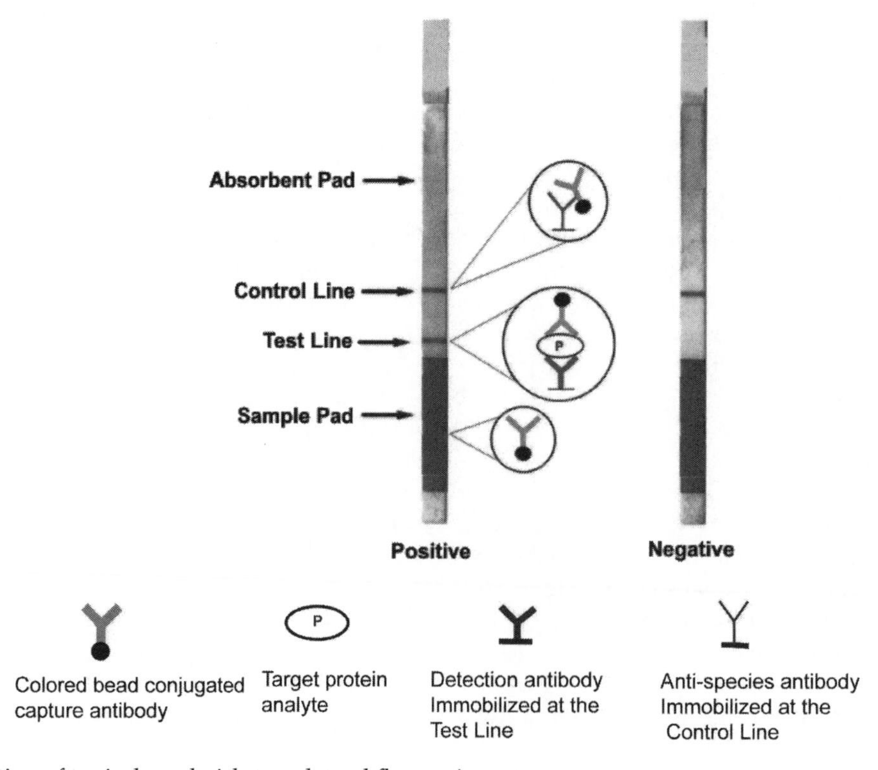

Illustration of typical sandwich-type lateral flow strip test.

the test line serves to indicate that the test ran appropriately. Sample gets drawn through the test strip by an **absorbent pad** placed at the back of the strip.

Both liquid and solid samples can be analyzed by the LFS assay. A solid sample must be dissolved or dispersed in a liquid solution to extract the target-protein (antigen) from the sample. To do the assay, the strip is put in contact with the sample solution at the end of the sample pad. The sample solution is drawn into the test strip by capillary action. The sample first passes through the filter, and the target protein is bound by the colored particle–antibody conjugate. Then the antigen–antibody-colored particle conjugate complex gets drawn into the membrane and is captured by the immobilized antibody in the test zone. A colored line becomes evident as more colored complexes are captured. The color intensity generally correlates with the amount of antigen present in the sample. A dedicated device can be used to measure the color intensity of the test line if a quantitative result is required. Samples containing no target protein show no color at the test line. The control zone contains an antispecies antibody capable of binding excess antibody-colored particle conjugates that pass through the test zone. Therefore, a colored line is formed for any complete test. If no colored line is formed at the control zone, the assay is invalid and needs to be repeated on a new strip.

17.4.3 Applications

Current generation of LFS methods are made into versatile formats and are useful in applications for which their primary attributes (i.e., speed, simplicity, low cost, etc.) are important aspects of the testing (e.g., **field testing**). LFS methods have been developed for qualitative and to some extent quantitative monitoring of food allergens, foodborne pathogens, food toxins, hormones, and certain food protein ingredients to ensure food safety and quality. They also have been developed to detect the presence of genetically modified organisms in processed foods for consumers' interest, and detect prohibited ruminant proteins in ruminant feedstuffs for the surveillance of mad cow disease. These easy-to-use rapid tests allow food processors and regulators to comply with regulations governing the labeling of food and feed products with minimal user training and facilities. The strengths, weaknesses, opportunities, and threats of LFS methods have been thoroughly discussed in a recent review article (5).

17.5 IMMUNOAFFINITY PURIFICATION

Besides the use of antibodies in immunoassays as described above, often antibodies are used in food analyses as complements to other analytical methods. This is due to the remarkable specificity of antibodies and their strong binding to antigen. The most common example of this is **immunoaffinity** purification, which is an antigen capture technique. Basically the antibody is immobilized on some support, most often using a covalent linking method so that there is no concern with "bleeding" of the antibody in later steps. The antibody can be bound to a solid phase such as agarose or silica gel. These antibody-bound solid phases can be used later for purification of antigen via a chromatography method or by the use of these phases on the surface of magnetic beads that are separated using a magnet. A simple purification sequence would involve exposing the antibody-bound solid to a food extract to first bind antigen, then washing the solid phase free of all unbound material, and finally releasing the pure antigen. Even though antibodies have such remarkably strong binding constants, they can be treated to release antigen by simple procedures such as changes in pH or solvent. Since the antibody is a protein, pH changes or solvent changes result in denaturation that changes the conformation of the binding site, releasing antigen. If these changes are carefully selected, denaturation can be reversed by reestablishing moderate conditions so that the valuable antibody-bound solid phase can be reused repeatedly. For sensitive antigens, like enzymes, these elution conditions also can be a concern.

These immunoaffinity purification procedures have been used for small molecules like toxins (e.g., aflatoxins) and even materials as large as cells. Different microorganisms contain unique cell surface antigens that can be selectively bound to aid in purification and differentiation.

17.6 APPLICATIONS

Immunoassays are a well-developed area in food analysis and there are several good textbooks available. For all sorts of laboratory techniques, Harlow and Lane (6) wrote one of the best books. The theory and practice of immunoassays is well handled in several books (7–9). There are even entire journals, such as *Food and Agricultural Immunology*, devoted to describing methods for preparing food immunoassays.

Because of the simplicity, sensitivity, and specificity of immunoassays, they are used widely as screening tests for pesticide (10, 11) and drug residues (12, 13) in food (see Chap. 18). Besides chemical analysis, immunoassay techniques are used in microbiology to rapidly detect food-borne pathogens (14, 15) and bacterial toxins (16). Immunoassays also are commonly used for meat and fish species identification (17). Since immunoassays can easily be developed to detect trace amounts of specific proteins, they are among a number of methods used to detect hidden food allergens (18, 19) and genetically modified organisms in foods (20). In fact, immunoassay can be developed to detect almost any organic substances in a food system. The flexibility is limited only by the availability and quality of the specific antibody used. Immunoassays are being automated for higher analytical throughput and for improved data quality (21). The research in the immunoassay area continues to develop specific antibodies, especially monoclonal antibodies for providing the ultimate specificity for the antibody–antigen interaction, different solid phase configurations to provide simpler means to carry out the assay, develop new detection systems to further improve the assay sensitivity, and alternative methodologies, such as immunosensors and immunoarray chips for multiple analytes detection and pattern recognition (15, 22).

17.7 SUMMARY

Almost any organic molecule in food can be determined using immunoassays as long as the specific antibodies are available. Both polyclonal antibodies and monoclonal antibodies or a combination of them can be used in an immunoassay. The remarkable selectivity and specificity of these assays are the result of the strong binding affinity between antibodies and their antigens. While the precise protocols of immunoassays can vary a great deal, all immunoassays use either a noncompetitive or a competitive format. The competitive format is the only one that can be used for quantification of small (about 1000 g/mol or less) molecules. ELISA has become the most popular immunoassay that uses an enzyme as the label to reveal the primary antibody–antigen binding through a color reaction catalyzed by the enzyme. In a noncompetitive ELISA with enzyme-derived color development, the more antibody bound molecules (analyte) in the food sample the more color develops, while in a competitive ELISA the reverse is true. The most common labels used for food immunoassays are enzymes and the most two common enzymes used are alkaline phosphatase and horseradish peroxidase.

Two other commonly used immunoassays are Western blot and LFS assay. Western blot is a laboratory-based immunoassay that combines PAGE and immunoassay to reveal the presence and the molecular mass of the antigenic protein on a membrane. Western blot is the most commonly used method for identifying and characterizing an antigen. A positive result also indicates the antigen conferring a linear epitope with the antibody. The LFS, on the other hand, is by far the simplest form of immunoassay. The assay is easy to use and is ideal for outside-the-laboratory applications where access to equipment and supplies is limited. While the general immunoassay procedure of the Western blot is similar to ELISA, the LFS assay is a one-step assay based on one-direction movement of the antigen in a sample solution towards immobilized capturing antibodies at different zones. The separation of bound and unbound molecules by washing steps is not required.

Food immunoassays can be prepared using very simple and rapid formats, making them ideal for kits used in the field. Such kits are commonly used for food testing at parts per million or lower levels of detection. While every effort is made to control the specificity of these field tests, they can suffer from false positives and false negatives. For this reason, immunoassay kits are used most often as rapid screening tests, while food samples that test positive for the target analyte by immunoassay are often confirmed using another, more laborious method.

Besides being the required constituent in immunoassays, antibodies also can be used to purify specific compounds in food for other analysis methods. These immunoaffinity purification methods allow for rapid purification of analytes from complex food matrices.

17.8 STUDY QUESTIONS

1. What is the relationship between an antigen and an antibody?
2. What is an epitope? What are the two types of epitopes?
3. What is the difference between monoclonal and polyclonal antibodies?
4. All immunoassays have two conditions that they must satisfy; what are they?
5. What is a hapten and what is a conjugated antigen?
6. What are the five general steps for an ELISA procedure?
7. What is the rationale for the blocking step in ELISA protocols?
8. What is the difference between direct and indirect immunoassays? What are the advantages and disadvantages associated with each type of assay format?
9. Two common immunoassays are the sandwich assay and the competitive assay. Which molecules are best detected by each? Why?
10. Explain why the concentration of antigen required to reduce the absorbance by 50% (usually very close to the value C for the equation of the sigmoidal curve) is such a useful value to determine for a competitive immunoassay.
11. Give four common applications of immunoassays in food analysis.
12. What is a Western blot? What is the major difference in reaction signals between Western blot and ELISA?

13. Compare and contrast ELISA and LFS, by identifying the similarities and differences in their characteristics, principles, and applications.
14. Describe, in general terms, how you would use immunoaffinity purification to isolate a protein for which you have developed antibodies.
15. All commercial potatoes contain the toxic glycoalkaloids α-solanine and α-chaconine. Both of these glycoalkaloids have the same large alkaloid portion, known as solanidine. Therefore polyclonal antibodies can be developed in rabbits against solanidine by chemically linking it to a foreign protein (foreign to the rabbit) and injecting the protein-bound hapten (solanidine linked using a succinic acid derivative) into rabbits. The antibodies that develop in the rabbit against the hapten bind to the alkaloid portion of both toxic glycoalkaloids. The rabbit antiserum containing primary polyclonal antibodies can be made highly specific to solanidin without cross reactivity with other similar molecules by removing all cross-reactive antibody components using an immunoaffinity column procedure.

    To develop an appropriate competitive ELISA, solanidine is again linked to a protein, but this time a different protein, and this conjugate is used to coat plastic microtiter plates. After excess conjugate is washed away the plates are ready for the competitive ELISA procedure.

    The glycoalkaloids in potatoes are extracted with methanol and this extract is further diluted with water for use in the ELISA procedure. A standard curve is prepared by diluting standard solutions of α-chaconine at low, medium, and high concentrations with similar aqueous methanol solutions. In addition a negative control is prepared using methanol and water at similar concentrations to the diluted potato extracts and standards, but without any glycoalkaloid present. Now the various extracts, standards, and negative controls are placed in individual wells with equivalent amounts of diluted rabbit serum containing the specific polyclonal antibodies. After incubation for 30 min at room temperature, all of the wells on the plate are again washed. Next a solution of commercially available goat antirabbit antibody conjugated to peroxidase is added to each well. After another 30 min of incubation, the wells are again thoroughly washed.

    Finally, phenylenediamine substrate solution is added to each well along with peroxide and again the plate is incubated for 30 min. After 30 min, the plate is rapidly read (in under 1 min) using an ELISA plate reader. The wells all contain differing amounts of yellow color.

    (a) Tomatidine is a glycoalkaloid found in tomatoes and contains the alkaloid portion tomatine. Would the polyclonal antibodies detect tomatidine?
    (b) Why is the protein that the hapten is attached to different for the ELISA procedure than for the injection?
    (c) Is the ELISA protocol direct or indirect?
    (d) Which wells would you expect to contain the most color, standards, potato extracts, or negative controls?
    (e) Would you be concerned if a potato extract gave almost no color at the end of the ELISA procedure?

17.9 ACKNOWLEDGEMENT

The author of this chapter wishes to acknowledge Dr. Peter Sporns, who wrote the immunoassay chapter for the 3rd edition of this book. Some of the text along with ideas for the content and organization of the current chapter came from Dr. Sporns' chapter, with his permission. Also ideas for the content and organization of the text on Western blot and lateral flow strip assays came from the chapter "Agricultural Biotechnology (GMO) Methods of Analysis," by Anne Bridges, Kimberly Magin and James Stave, in the 3rd edition of this book. Their contribution is recognized and appreciated.

17.10 REFERENCES

1. Köhler G, Milstein C (1975) Continuous cultures of fused cells secreting antibody of predefined specificity. Nature 256:495–497
2. Howard GC, Bethell DR (2001) Basic methods in antibody production and characterization. CRC, Boca Raton, FL
3. Yalow RS, Berson SA (1960) Immunoassay of endogenous plasma insulin in man. J Clin Invest 39:1157–1175
4. Engvall E, Perlmann P (1971) Enzyme-linked immunosorbent assay, ELISA III. Quantitation of specific antibodies by enzyme-labeled anti-immunoglobulin in antigen-coated tubes. J Immunol 109:129–135
5. Posthuma-Trumpie GA, Korf J, van Amerongen A (2009) Lateral flow (immuno) assay: its strengths, weaknesses, opportunities and threats. A literature survey. Anal Bioanal Chem 393:569–82
6. Harlow E, Lane D (1999) Using antibodies: a laboratory manual. Cold Spring Harbor Laboratory Press, Cold Spring Harbor, New York
7. Tijssen P (1985) In practice and theory of enzyme immunoassays – Volume 15 in the laboratory techniques in biochemistry and molecular biology series. Elsevier, Amsterdam
8. Deshpande SS (1996) Enzyme immunoassays: from concept to product development. Chapman and Hall, New York
9. Gabaldón JA, Maquieriera A, Puchades R (1999) Current trends in immunoassay-based kits for pesticide analysis. Crit Rev Food Sci Nutr 39:519–538
10. Wild D (2005) The immunoassay handbook. Elsevier, Amsterdam
11. Morozova VS, Levashova AI, Eremin SA (2005) Determination of pesticides by enzyme immunoassay. J Anal Chem 60:202–217
12. Mitchell JM, Griffiths MW, McEwen SA, McNab WB, Yee AJ (1998) Antimicrobial drug residues in milk and meat: causes, concerns, prevalence, regulations, tests and test performance. J Food Prot 61:742–756

13. Raig M, Toldrá F (2008) Veterinary drug residues in meat: concerns and rapid methods for detection. Meat Sci 78:60–67
14. Swaminathan B, Feng P (1994) Rapid detection of food-borne pathogenic bacteria. Annu Rev Microbiol 48: 401–426
15. Banada PP, Bhunia AK (2008) Antibodies and immunoassays for detection of bacterial pathogens. Ch. 21, In: Zourob M, Elwary S, Turner A (eds) Principles of bacterial detection: biosensors, recognition receptors and microsystems. Springer, New York.
16. Pimbley DW, Patel PD (1998) A review of analytical methods for the detection of bacterial toxins. J Appl Microbiol 84: 98S–109S
17. Hsieh Y-HP (2005) Meat species identification. In: Hui YH (ed) Handbook: food science, technology and engineering. CRC, Boca Raton, FL, pp 30–1–30–19
18. Owusu-Apenten RK (2002) Determination of trace protein allergens in foods. Ch. 11, In: Food protein analysis. Quantitative effects on processing. Marcel Dekker, New York, pp 297–339
19. Poms RE, Klein CL, Anklam E (2004) Methods for allergen analysis in food: a review. Food Addit Contam 21:1–31
20. Ahmed FE (2002) Detection of genetically modified organisms in foods. Trends Biotechnol 20:215–223
21. Bock JL (2000) The new era of automated immunoassay. Am J Clin Pathol 113:628–646
22. Corgier BP, Marquette CA, Blum LJ (2007) Direct electrochemical addressing of immunoglobulins: Immunochip on screen-printed microarray. Biosens Bioelectron 22:1522–1526

Analysis of Food Contaminants, Residues, and Chemical Constituents of Concern

*Baraem Ismail**

Department of Food Science and Nutrition, University of Minnesota, St. Paul, MN 55108-6099, USA
bismailm@umn.edu

and

Bradley L. Reuhs and S. Suzanne Nielsen

Department of Food Science, Purdue University, West Lafayette, IN 47907-2009, USA
breuhs@purdue.edu; nielsens@purdue.edu

S.S. Nielsen, *Food Analysis*, Food Science Texts Series, DOI 10.1007/978-1-4419-1478-1_18,

18.1 Introduction: Current and Emerging Food Hazards 319
18.2 Analytical Approach 320
 18.2.1 Choice of Analytical Method 320
  18.2.1.1 Qualitative or Semiquantitative Methods 320
  18.2.1.2 Quantitative Methods 321
 18.2.2 Sample Preparation 321
  18.2.2.1 Introduction 321
  18.2.2.2 Sample Homogenization 322
  18.2.2.3 Extraction and Cleanup 322
   18.2.2.3.1 Introduction 322
   18.2.2.3.2 Solid-Phase Microextraction 322
   18.2.2.3.3 Solid-Phase Extraction (QuEChERS) 323
   18.2.2.3.4 Microwave-Assisted Solvent Extraction 323
   18.2.2.3.5 Accelerated-Solvent Extraction 323
  18.2.2.4 Derivatization 323
18.3 Pesticide Residue Analysis 324
 18.3.1 Introduction 324
 18.3.2 Types of Analytical Methods 324
 18.3.3 Analytical Techniques Used for the Detection, Identification, and/or Quantification 325
  18.3.3.1 Biochemical Techniques 325
  18.3.3.2 Chromatographic Techniques 326
   18.3.3.2.1 Thin Layer Chromatography 326
   18.3.3.2.2 Gas Chromatography 326
   18.3.3.2.3 High-Performance Liquid Chromatography 326
  18.3.3.3 Mass Spectrometry Detection 327
   18.3.3.3.1 Gas Chromatography-Mass Spectrometry 327
   18.3.3.3.2 High-Performance Liquid Chromatography-Mass Spectrometry 327
18.4 Mycotoxin Analysis 327
 18.4.1 Introduction 327
 18.4.2 Sampling 330
 18.4.3 Detection and Determination 330
  18.4.3.1 Rapid Methods of Detection 331
   18.4.3.1.1 TLC 331
   18.4.3.1.2 Immunoassays 331
  18.4.3.2 Quantitative and Confirmative Chemical Methods 332
   18.4.3.2.1 HPLC 332
   18.4.3.2.2 GC 332
   18.4.3.2.3 Capillary Electrophoresis 332
  18.4.3.3 Other Methods of Analysis 332
18.5 Antibiotic Residue Analysis 332
 18.5.1 Introduction 332
 18.5.2 Detection and Determination 333
  18.5.2.1 Screening Methods 333
  18.5.2.2 Determinative and Confirmatory Methods 335
18.6 Analysis of GMOs 335
 18.6.1 Introduction 335
 18.6.2 Protein Methods 336
 18.6.3 DNA Methods 337
  18.6.3.1 DNA Extraction 337
  18.6.3.2 PCR Amplification 338
  18.6.3.3 DNA Analysis 338
 18.6.4 Method Comparison 338
18.7 Allergen Analysis 339
 18.7.1 Introduction 339
 18.7.2 Protein Methods 340
  18.7.2.1 General Considerations 340
  18.7.2.2 Protein-Based Analytical Techniques 340
 18.7.3 DNA Methods 341
18.8 Analysis of Other Chemical Contaminants and Undesirable Constituents 341
 18.8.1 Introduction 341
 18.8.2 Sulfites 341
 18.8.3 Melamine 342
 18.8.4 Packaging Material Residues 343
  18.8.4.1 Bisphenol A 343
  18.8.4.2 4-Methylbenzophenone 343
 18.8.5 Furans 343
 18.8.6 Acrylamide 343
 18.8.7 Benzene 343
 18.8.8 3-Monochloropropane 1,2-Diol (3-MCPD) 344
 18.8.9 Perchlorate 344
18.9 Summary 344
18.10 Study Questions 345
18.11 Acknowledgements 345
18.12 References 346

18.1 INTRODUCTION: CURRENT AND EMERGING FOOD HAZARDS

The food chain that starts with farmers and ends with consumers can be complex, involving multiple stages of production and distribution (planting, harvesting, breeding, transporting, storing, importing, processing, packaging, distributing to retail markets, and shelf storing) (Fig. 18-1). Various practices can be employed at each stage in the food chain, which may include pesticide treatment, agricultural bioengineering, veterinary drug administration, environmental and storage conditions, processing applications, economic gain practices, use of food additives, choice of packaging material, etc. Each of these practices can play a major role in food quality and safety, due to the possibility of contamination with or introduction (intentionally and nonintentionally) of hazardous substances or constituents. Legislation and regulation to ensure food quality and safety are in place and continue to develop to protect the stakeholders, namely farmers, consumers, and industry. [Refer to reference (1) for information on regulations of food contaminants and residues.]

The severity of an adverse effect associated with a food hazard is usually directly related to the dose. In many cases there is a threshold level (**tolerance level**) below which no adverse effects are observed. The US Environmental Protection Agency (EPA) establishes these tolerance levels, and the **Food and Drug Administration** (FDA) and **United States Department of Agriculture** (USDA) enforces them. However, food safety incidents, microbial and chemical in nature, continue to occur. Pesticide residues, mycotoxins, veterinary drug residues, some food additives, food adulterants, packaging hazardous chemicals, and environmental contaminants are of concern. The Rapid Alert System for Food and Feed (RASFF) reported that in the period between July 2003 and June 2007 a total of 12,641 alert notifications (2) categorized as follows: chemical (44%), mycotoxins (29%), microbial (17%), and other hazards (10%) (Fig. 18-2). Within the chemical category, the most frequently reported hazards include allergens (e.g., histamine and sulfite), heavy metals (e.g., mercury, lead, and cadmium), pesticides (e.g., omethoate, dimethoate, and isophenfos-methyl), and veterinary drugs (e.g., β-lactam, nitrofurans, sulfonamide, and chloramphenicol). Microbial contaminants include molds, viruses, and bacteria (discussion and methods of analysis for this category is beyond the scope of this chapter). Examples of current and emerging chemical hazards include fraud and food adulterants (e.g., melamine), packaging chemicals (e.g., bisphenol A and 4-methylbenzophenone), degradation metabolites (e.g., acrylamide and furan), and other chemical contaminants (e.g., 3-monochloropropane-1,2-diol, benzene, and perchlorate). Another category that can be of concern includes the genetically modified organisms (GMO) and their products. Introduction and usages of GMO in food products resulted in the development of legal requirements of safety and labeling.

Given the extent of concerns cited above, there is a strong need for adequate and reliable methods of detection and analysis to ensure food quality, safety, and fair trade. There are several well-established and reliable methods of analysis to detect food hazards. Development and validation of methods to detect and analyze emerging food hazards are currently underway. This chapter will cover some of the screening methods and quantitative methods that are commonly used for the detection and quantification of several food hazards, in addition to some recently developed methods for the detection of newly identified and emerging food hazards.

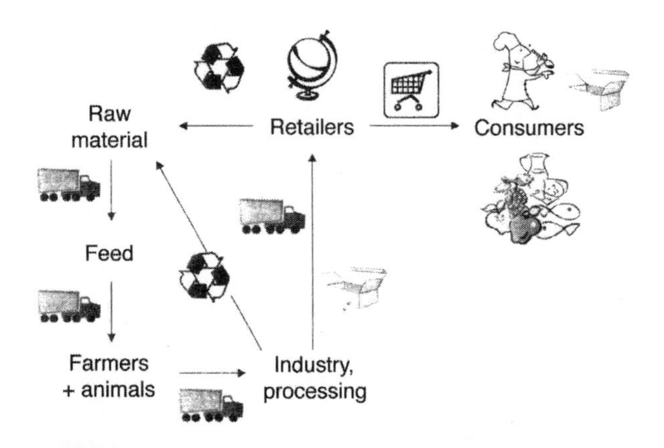

18-1 figure An illustration of the food chain. [From (2), with permission.]

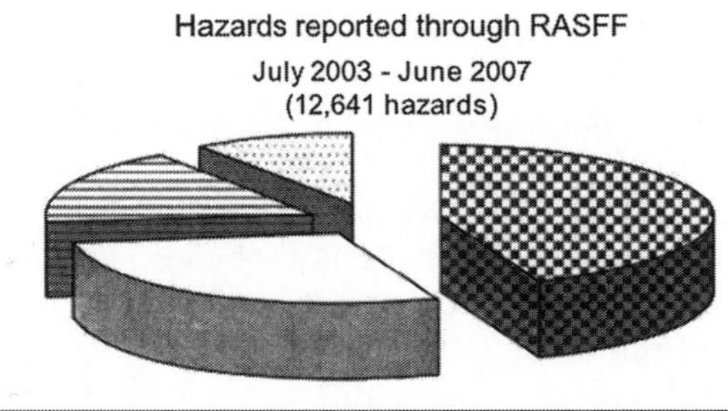

Chemical
Mycotoxin
Microbiological
Others, including fraud, biological hazards, labeling, chemical hazards, quality, hygiene, defective packaging and others

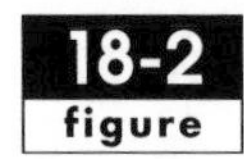

Categorized food chain hazards as reported through the Rapid Alert System for Food and Feed (RASEF) in the period of July 2003 to June 2007. [From (2), with permission.]

18.2 ANALYTICAL APPROACH

As with the analysis of any food constituent, there is an array of methodological approaches and techniques to choose from for the analysis of food hazards. Multiple factors influence the method of choice, including, but not limited to, the objectives, time, cost, effort, reliability of the method, complexity of the food matrix, suspected level of the contaminant, and availability of analytical instrumentation. The objectives can vary in complexity from mere detection of several suspected contaminants belonging to the same family to the determination of the exact level of a particular contaminant, or in more complex cases, detection of unknown adulterants. The general trend for regulatory institutions and industry is to implement inexpensive and rapid **screening methods**. However, depending on the objective of the analysis, **quantitative methods** that require sophisticated equipment might be needed. In this case, the industry may choose to send their samples to specialized laboratories. Once a method of analysis is chosen, appropriate consideration with regard to sampling and sample preparation needs to be made.

18.2.1 Choice of Analytical Method

The choice of analytical method for the detection and determination of food hazards must take into account not only the complexity of the food matrix, but also the characteristics of the analyte such as polarity, hydrophobicity, volatility, thermal stability, and chemical reactivity. The complexity of the food matrix and the characteristics of the analyte significantly influence the choice of extraction, separation, and detection techniques, as will be discussed in subsequent sections. Accuracy, precision, specificity, and sensitivity of the analytical method (see Chap. 4) are also important considerations. There are some official methods for the analysis of contaminants and residues, others are validated, and some are currently being developed and validated. Methods for the analysis of food hazards can be either qualitative, semiquantitative, or quantitative. A list of analytical methods for pesticide, mycotoxin, and veterinary drug residues is given in Table 18-1.

18.2.1.1 Qualitative or Semiquantitative Methods

Qualitative and **semiquantitative** methods, also known as **screening methods**, are usually used to assay a large number of samples for the presence of one or more contaminants belonging to the same family (e.g., antibiotic residues, see Sect. 18.5.2.1) in a relatively short time. These methods are generally robust, less sensitive to small changes in experimental and/or environmental conditions, and are not limited to a highly controlled lab environment. While qualitative methods detect the presence of certain contaminants, semiquantitative methods provide an estimate of the concentration of a detected contaminant or residue. The principal benefits of these methods are their low cost, relative speed, and simplicity. These methods

18-1 table **Summary of Analyses for Pesticide, Mycotoxin, and Antibiotic Residues in Foods**

Contaminant	*Quantitative*		*Semiquantitative or Qualitative (Screening Methods)*
Pesticides	Multiresidue (MRMs) GC (mostly) HPLC	Single-residue (SRMs) GC (mostly) HPLC Immunoassay	TLC Enzyme inhibition Immunoassay
Mycotoxins	HPLC (mostly) GC Capillary electrophoresis Immunoassays		TLC Immunoassay
Antibiotics	HPLC (mostly) GC Immunoassays		Microbial growth inhibition Receptor assays Enzyme substrate assays Immunoassays

HPLC, high-performance liquid chromatography; *GC*, gas chromatography; *TLC*, thin-layer chromatography.

include techniques such as **thin-layer chromatography** (*TLC*), **enzyme inhibition**, and **immunoassay** (see later sections of this chapter).

18.2.1.2 Quantitative Methods

For the quantitative analysis of chemical food contaminants and residues, **gas chromatography** (GC; Chap. 29) and **high-performance liquid chromatography** (HPLC; Chap. 28) are the two main analytical methodologies employed. As compared to HPLC, GC provides better separation efficiency and has been traditionally combined with more selective detectors for the analysis of food contaminants and residues. The combination of GC with mass spectrometry (MS), and the availability of relatively affordable benchtop GC-MS instruments, gave preference to GC analysis for multicomponent contaminant and residue analysis, in spite of having to derivatize polar analytes. However, thermally labile and/or large analytes that cannot be easily volatilized, such as mycotoxins, polar pesticides, and most of the veterinary drug residues, currently must be analyzed using HPLC. In recent years, major advances in HPLC-mass spectrometry (LC-MS) have facilitated direct, selective, and sensitive analysis of the polar analytes. For example, LC-MS is gradually replacing microbial and immunochemical methods used for the analysis of veterinary drugs (3). Additionally, LC-MS is being used for multiclass, multiresidue analysis of pesticides (4), due to the transition from the use of persistent and less polar compounds to the more readily degradable, more polar, thermolabile pesticides. The use of GC-MS or LC-MS provides simultaneous quantitation and structural identification of a wide range of compounds and the possibility of spectrometrically resolving coeluting peaks.

Immunoassay analytical techniques also are used for the detection and quantification of both single and multiple contaminants or residues. Detailed description of the various immunoassay techniques is provided in Chap. 17. Pesticides, antibiotics, and mycotoxins are typical examples of contaminants that can be rapidly and efficiently determined using immunoassays. Of the various immunoassay techniques, **enzyme-linked immunosorbent assays** (ELISA) and **immunosensor** techniques provide a highly sensitive detection of toxic analytes, whereas **immunoaffinity chromatography** (Chap. 17, Sect. 17.5) is used for concentration and clean up of the analyte of interest. In addition to being highly sensitive, immunoassays are simple, fast, and cost effective as compared to other methods used for the quantification of contaminants and residues. However, the antibodies used in the assay may have **cross-reactivity** (affinity) for related chemical structures.

18.2.2 Sample Preparation

18.2.2.1 Introduction

Food samples are usually too dilute (e.g., beverages) or too complex (e.g., meat) for direct analysis of trace contaminants and residues. Therefore, sample preparation, including homogenization, extraction, fractionation/clean-up, concentration and/or derivatization, normally precedes the analysis of food contaminants and residues (Fig. 18-3). Analysts are faced with continuous decrease of legislative limits for food contaminants and residues, resulting in the need for more precise and accurate measurements. Sample preparation techniques for the analysis of food contaminants and residues are continuously improved to guarantee high recovery and reproducibility. Fortunately, there also have been significant advances in analytical methods and equipment. The advanced technology of mass spectrometers (see Chap. 26), specifically, their enhanced "recognition features" when used in tandem, allows them to take over the "selectivity" of classical sample preparation methods. However, for quantification purposes, cleaner extracts are preferred. In both target and multiresidue methods, isotope dilution and addition of internal standard provide the most accurate results, compensating for sample matrix effects and ion suppression in MS. Also, faster and more efficient extraction methods are now available for the analysis of food contaminants and residues, as will be discussed in subsequent sections.

Analytical chemists often focus on perfecting the analytical technique (e.g., chromatographic analysis), overlooking the importance of sampling, sample storage, and sample preparation. Sampling and sample

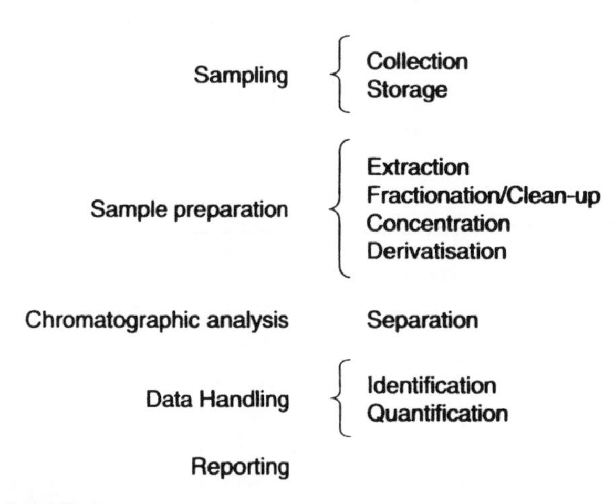

Flow diagram for the analysis of contaminants and residues. [From (6), with permission.]

preparation are labor intensive and time consuming, but are essential prerequisites for acquiring meaningful analytical data. There is a high probability of error and contamination during sampling and sample preparation, and these cannot be corrected at any point during the analysis. Therefore, an adequate plan for acquiring, storing, and preparing samples should be implemented and validated by statistical analysis (see Chap. 5 for more details). It is important to note that obtaining a sample that is representative of the level of trace substances in a heterogeneous mixture is not an easy task. It is crucial that the level of contaminants or residues determined in the acquired samples reflects the true concentration in the entire lot. Sampling for the analysis of a particular contaminant or residue will be briefly discussed in relevant sections of this chapter.

18.2.2.2 Sample Homogenization

Food contaminants and residues in many cases are unequally distributed in a food system. Removal of inedible parts (outer skins, stems, and cores), if applicable, and homogenization are necessary for reliable and accurate analysis. For instance, most pesticides are not translocated within plants and are expected to be located on the surface of fresh produce. Thus, inedible outer skin should be carefully removed, followed by homogenization. Homogenization can be achieved by chopping or grinding, followed by blending and mixing. Apart from homogenization, grinding reduces the structural features of the sample and enhances extraction efficiency. Contaminating the sample or exposing it to unnecessary heat, which can cause volatilization or degradation of contaminants or residues, should be avoided. Samples warm up with grinding, and some are difficult to grind at ambient temperatures because they are soft and can adhere in lumpy masses. Therefore, cryogrinding (freeze-grinding), which is the grinding of chilled samples, is preferred. The samples are normally chilled by dry ice, liquid carbon dioxide, or liquid nitrogen.

18.2.2.3 Extraction and Cleanup

18.2.2.3.1 Introduction Almost all food samples, with the exception of samples directly soluble in an organic solvent (e.g., vegetable oil), require a solvent extraction step (Chap. 27, Sect. 27.2; Chap. 29, Sect. 29.2.2.4) to isolate the target analytes from the matrix. Defatting of lipid-rich food matrices is often required, using hexane or isooctane, prior to the extraction of target contaminants. Extraction of contaminants and residues is traditionally done by solubilizing them in a suitable organic solvent, generally acetonitrile or acetone. **Anhydrous salt** (NaCl or Na_2SO_4) can be added to absorb water. In some cases, water is added so that the crude extract can be purified using a subsequent partitioning step with a second, water-immiscible solvent. When the extraction process is complete, the solvent is separated from insoluble solids by filtration.

Often, the crude sample extract is purified and concentrated before the separation/determination steps. The degree of cleanup required depends on the instrumental detector to be used and the chromatography mode in the automated separation stage of the analysis. The objective of clean up is to separate the target analytes from coextractives that can interfere with their detection. Often, the preliminary partitioning step is followed by a preparative chromatography step (see Chap. 27 for basic information on chromatography). The crude water–acetone or water–acetonitrile extract is partitioned with a relatively nonpolar organic solvent. The organic phase is reduced in volume by heating. The residues then can be further purified by column chromatography (either adsorption or size-exclusion chromatography). Separate fractions of column eluate can be analyzed. The choice of packing material is both analyte and matrix dependent. In addition to traditional column chromatography, immunoaffinity chromatography can be used for sample cleanup. For example, monoclonal antibody affinity chromatography is used as a one-step column cleanup prior to fluorometric determination of aflatoxin in milk (5).

In recent years, great efforts have been made to develop extraction techniques that are faster, more efficient, and require less solvent. Extraction and purification techniques have been developed based on the characteristics of the target analyte(s), which can be categorized into three classes: (1) **volatile organic compounds** (VOC) that can be analyzed via headspace techniques, after derivatization if needed, (2) **semivolatile organic compounds** (SVOC) that are GC amendable (thermally stable) (e.g., most of the pesticides), and (3) the **nonvolatile** (thermally labile) **organic compounds** (NVOC), which are mostly analyzed by HPLC after extraction. Some of the widely used extraction techniques will be covered briefly in subsequent sections. References (6–10) contain more details on the various extraction techniques.

18.2.2.3.2 Solid-Phase Microextraction VOCs are best analyzed by headspace techniques, which involve the injection of only the vapor phase, resulting in clean chromatograms with no contamination of nonvolatile compounds generally present in liquid extracts. Headspace sampling applications include the analysis of pesticides, furan, and residues from packaging materials. One method for headspace sampling is **solid-phase microextraction** (SPME) (see Chap. 29, Fig. 29-2), which employs partition [using polymers such as polydimethylsiloxane (PDMS) or polyacrylate] and adsorption (using divinylbenzenestyrene copolymers or mixtures of carboxen and PDMS) principles. The technique involves partitioning of the analyte between the sample matrix and the headspace, and partitioning between the headspace and the polymer-coated fiber. After equilibration, the fiber/needle assembly is transferred to the injector port of a GC where the analytes are thermally desorbed onto the column (see also Chap. 29, Section 29.2.2.5). SPME also can be used for SVOC and NVOC, where the polymer-coated fiber is directly immersed in the aqueous samples or placed inside a hollow cellulose membrane.

18.2.2.3.3 Solid-Phase Extraction (QuEChERS) Solid-phase extraction (SPE) is discussed in Chap. 29, Sect. 29.2.2.5. The recently developed **QuEChERS method** (11), which stands for quick, easy, cheap, effective, rugged, and safe, is a dispersive SPE (DSPE) technique that is by far the best method to extract multiple pesticide residues (Sect. 18.3.2). A single sample preparation method and one chromatography method linked to MS (preferably in tandem) to determine as many pesticides as possible is very much desired. The advantages of the QuEChERS method (outlined in Fig. 18-4) include speed, ease, minimal solvent use, and lower cost as compared to conventional SPE. The types of adsorbents and solvents, as well as the pH and polarity of the solvents, can be adjusted based on the sample matrix and types of analytes. QuEChERS kits are now commercially available (e.g., Sigma Aldrich/Supleco, Restek, and United Chemical Technologies).

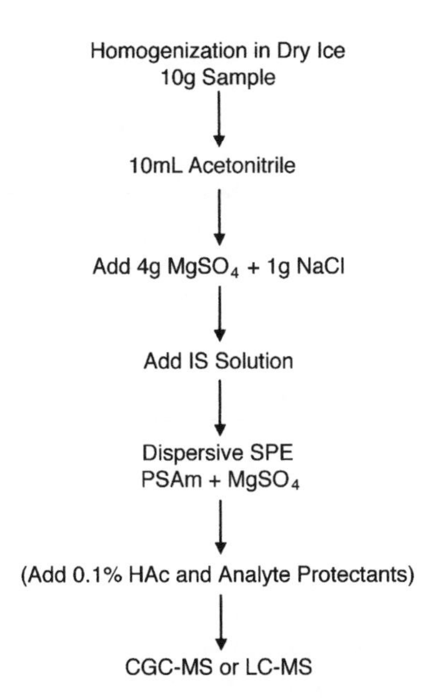

18-4 figure Flow diagram for the QuEChERS procedure. [From (6), with permission.]

18.2.2.3.4 Microwave-Assisted Solvent Extraction **Microwave-assisted solvent extraction** (MASE), which utilizes electromagnetic radiation to desorb organics from their solid matrix, can decrease significantly the extraction time and the quantity of solvent required to efficiently extract target analytes. The efficiency of MASE is attributed to the elevated temperatures that exceed the boiling point of the solvent(s) and the rapid transfer of the analyte(s) to the solvent phase. Commercial systems are available that incorporate the capacity for simultaneously extracting multiple samples within closed, lined (perfluoroalkoxy), pressurized vessels (up to 8–12), using microwave absorbing solvents. Disadvantages of MASE include lack of selectivity and loss of thermolabile analytes.

18.2.2.3.5 Accelerated-Solvent Extraction **Pressurized liquid extraction** (PLE) includes a number of extraction techniques, the most common being the **accelerated-solvent extraction** (*ASE*). ASE utilizes limited quantities of organic solvents at elevated temperature (up to 200°C) and pressure (1500–2000 psi) to statically extract solid samples for short periods of time (often <10 min) (12). Disadvantages of ASE include diluting effect and lack of selectivity, requiring further cleanup and concentration.

18.2.2.4 Derivatization

The chemical structure of analytes may need to be modified to become suitable for separation and detection by specific chromatographic techniques and detected via available and/or desirable detectors. The process of structural modification through multiple chemical reactions is known as **derivatization** (see Chap. 27, Sect. 27.3.4.2.1; Chap. 28, Sect. 28.2.4.7; Chap. 29, Sect. 29.2.3). Derivatization can be used to enhance the thermal stability or volatility of the analyte(s), to modify chromatographic behavior, to increase the selectivity of the detection, or to increase the sensitivity of the detection. Many types of derivatization reactions have been used in the analysis of contaminants and residues, such as selected pesticide residues as discussed in reference (13) and the antibiotic nitrofurans as discussed in reference (6).

18.3 PESTICIDE RESIDUE ANALYSIS

18.3.1 Introduction

A pesticide is any substance or a mixture of substances formulated to destroy or control any organism that competes with humans for food, destroys property, or spreads disease. "Pests" include weeds, microorganisms (e.g., fungi or bacteria), insects, and even mammals. Currently, more than 800 active substances, belonging to different chemical groups (4), are formulated in pesticide products (~500 registered pesticide products). The main classes of pesticides include: (1) **herbicides** [for weed control, e.g., triazine (e.g., atrazine)]; (2) **insecticides** [e.g., organochloride (OC) (e.g., dichlorodiphenyltrichloroethane, DDT); organophosphates (OP) (e.g., malathion, dimethoate, omethoate), and methylcarbamates (e.g., aldicarb)]; and (3) **fungicides** [e.g., phthalimide (e.g., Captan)]. Other types of pesticides may include acaricides, molluscicides, nematicides, pheromones, plant growth regulators, repellents, and rodenticides. A glossary of pesticide chemicals (14) is published by FDA.

Pesticides remain indispensable for feeding the world population and protecting it from diseases. If pesticides were not applied, an estimated one-third of the crop production would be lost. However, pesticides may have adverse effects on human health, including, but not limited to, cancer, acute neurologic toxicity, chronic neurodevelopment impairment, and dysfunction of the immune, reproductive, and endocrine systems. At present, due to the possible toxic effects of pesticides on human health and on the environment, there are strict regulations for their registration (including establishing of tolerance levels) and use all over the world. Pesticide residues may occur in food as a result of the direct application to a crop or a farm animal or as a postharvest treatment of food commodities. Pesticide residues also can occur in meat, milk, and eggs as a result of the farm animal consumption of feed from treated crops. Additionally, pesticide residues can occur in food from environmental contamination and spray drift. **Risk assessment studies**, required by the EPA, are done to determine the nature and extent of toxic effects, and to establish the level at which no adverse effects are observed (**no observed adverse effect level**, NOAEL). The **acceptable daily intake** (ADI) for humans may be calculated from the NOAEL using a safety factor, usually of 100 (i.e., ADI = NOAEL/100). In conducting risk assessment studies, the total aggregate pesticide exposure of various subpopulations and the risks associated with such exposures must be considered. Dietary exposure to a pesticide depends on both the actual residue in or on foods and the food consumption pattern. To determine the health risks associated with chronic intake, it is necessary to determine the quantity of the pesticides likely to be consumed over a prolonged period of time and compare this estimate with the ADI.

Based on risk assessment studies, **tolerance levels** (or **maximum residue level**, *MRL*) of pesticides (the active ingredients as well as toxic metabolites, and transformation products) in food are set and enforced by government agencies, as mentioned in Sect. 18.1, to protect stakeholders and regulate international trade. In fact, tolerance levels must be established prior to registration. The **registration** of a specific pesticide for its commercial production and application requires a complete set of analytical data to prove not only its efficacy but also its safety. In general, the tolerance levels in foods range between 0.01 and 10 mg/kg, depending on the commodity and the pesticide used. Low tolerance levels have necessitated the development of more accurate and sensitive analytical methods to meet the requirements in food.

18.3.2 Types of Analytical Methods

Several factors make the analysis of pesticide residues in food a complex task. These factors include the complexity of the food matrix, the significantly greater abundance of matrix components relative to that of the target pesticides, the possibility of having pesticide levels as low as pictogram or femtogram amounts, and the considerable differences in the physical and chemical properties of pesticides.

Analytical methods employed for the analysis of pesticides can be categorized into either **single-residue methods** (SRM) or **multiple-residue methods** (MRM). SRMs are designed to measure a single analyte and, often, its toxic metabolites and transformation products. The majority of SRMs have been developed for purposes of registration and setting tolerance levels or for investigating the metabolism and environmental fate of a specific pesticide. Sampling, extraction, purification, and determination in SRMs are optimized for the target pesticide. The majority of the SRMs currently in use, which have undergone EPA review or have been published in peer-reviewed scientific journals, are found in **Volume II of the *Pesticide Analytical Manual*** (*PAM* II) (15). For purposes of monitoring quality and safety, given the large number of pesticides and the considerable differences in their physical and chemical characteristics, MRMs are much preferred. Pesticides can be acidic, basic, or neutral, polar or nonpolar, and volatiles or nonvolatiles. Therefore, a MRM that can determine various pesticides in a single run is the most efficient analytical approach. Currently used MRMs are found in **Volume I of the**

Pesticide Analytical Manual (*PAM* I) (16). AOAC International (17) also has developed an MRM for pesticide residues, the AOAC Pesticide Screen (AOAC Method 970.52). The identification and quantification in MRMs currently used by the FDA and USDA are based on GC and HPLC analysis.

Prior to chromatographic analysis, sampling, extraction, and fractionation/cleanup in MRMs are optimized to ensure an efficient transfer of most if not all of the pesticide residues present in the sample matrix to the organic phase. To achieve efficient transfer of as much of the pesticide residues as possible from the sample matrix into the organic phase, partitioning using water-miscible solvent is performed. This is followed by partitioning with nonpolar solvent that is miscible with the polar solvent, yet immiscible with water. After the extraction step, a cleanup step is required to minimize matrix effects and maximize sensitivity and selectivity. Cleanup techniques involve using adsorption columns, in particular Florisil, alumina and silica gels, with solvent mixtures of low polarity for elution. Commercial kits are available for pesticide cleanup; some examples are provided in the top half of Table 18-2. Reference (18) includes detailed descriptions of different MRM extraction and cleanup protocols specific for pesticide analysis.

18.3.3 Analytical Techniques Used for the Detection, Identification, and/or Quantification

A wide range of analytical techniques can be used in pesticide analysis. This section highlights some of the current analytical techniques used for the detection and determination of pesticide residues, as well as recent developments.

18.3.3.1 Biochemical Techniques

Biochemical techniques, such as enzyme inhibition assays and immunoassays, are widely used for the detection of pesticides. Commercially available test kits include enzyme inhibition assays (e.g., see lower half of Table 18-2). The principle of these assays is based on the inhibition of a certain enzyme, essential for vital functions in insects, by pesticides present in the sample. If no pesticides are present, the enzyme will be active and act on the substrate to cause a change in color. If no change in color occurs, then the test is positive, and further confirmation can be performed following more sophisticated analysis, such as HPLC and GC, to identify and quantify the specific pesticides present.

18-2 table **Example Commercial Test Kits for Pesticide Residue Analysis**

Assay Type	*Example Matrices*	*Pesticide*	*Commercial Source*
Cleanup kits			
Solid phase extraction – chromatography	Fruits and vegetables, including those with fats, waxes, and pigments	Multiclass, multiresidue pesticides	QuEChERS® (Agilent Technologies, Inc., 5301 Stevens Creek Blvd, Santa Clara, CA)
Solid phase extraction – chromatography	A variety of food commodities including fruits and vegetables	Multiclass, multiresidue pesticides	DisQuE® Dispersive Sample Preparation Kit (Waters Corporation, 34 Maple Street, Milford, MA)
Enzyme inhibition test kits			
Insect esterase activity with color change	Water. Feeds, grains, fruits, tubers, leafy vegetables, *Brassica* vegetables, dairy products, and meats	Organophosphates (malathion, diazinon, parathion, etc.) and metabolites, *N*-methylcarbamates (aldicarb, carbofuran, etc.)	Cidelite® (Charm Sciences, Inc., 659 Andover Street, Lawrence, MA)
Insect cholinesterase activity with color change	General	General	Agri-Screen Ticket® (Neogen Corporation, 620 Lesher Place, Lansing, MI)

While enzyme inhibition assays are mostly used as screening methods, with limited sensitivity and selectivity, immunoassays can be tailored to particular purposes ranging from simple screening tests [field-portable (19)] to quantitative laboratory tests. Immunoassays are simple and sensitive, have high throughput, and are cost effective as compared with other conventional methods. Additionally, extensive cleanup of extracts is not necessary, unless cross-reactivity exists. Therefore, this technique can be very useful for monitoring programs when there are a large number of samples. Immunoassays can be either class- or compound-specific with different applications. ELISA accounts for almost 90% of the immunoassays used for pesticide residue analysis. The antibodies for ELISA testing of acylodien insecticides and triazine herbicides are class-specific and have broad reactivity, enabling the detection of closely related compounds. For this type of assay, the hapten moiety of the hapten–protein conjugate (see Chap. 17, Sect. 17.3.3) is common for a group of chemicals. Immunoassays for atrazine (20), chlorpyrifos (21), and flufenoxuron (22) are compound specific, with hapten–protein conjugate designed to preserve the unique portion of the target analyte. Therefore, these compound-specific assays can be applied in quantitative analyses. However, in some instances there is the possibility of under- or overestimation of the true pesticide concentrations due to matrix interference and cross-reactivity. References (18) and (19) include more information on the application of immunoassays in the analysis of pesticides.

18.3.3.2 Chromatographic Techniques

18.3.3.2.1 Thin Layer Chromatography **Thin-layer chromatography** (TLC) (Chap. 27, Sect. 27.3.4.2) can be used for screening purposes in the analysis of pesticides. Because of its low resolving capacity, low precision, and limited detection relative to GC and HPLC, it is not used as a quantitative method. However, TLC can be used as a semiquantitative method that precedes more accurate detection and quantification. An example application is the detection and estimation of pesticides that inhibit insect enzymes such as cholinesterases. Several OP and carbamate insecticides are capable of inhibiting this group of enzymes. Once a crude extract is separated by TLC, the plate is sprayed with a solution containing the enzyme(s), followed by a solution containing a specific substrate, which releases a colored product upon hydrolysis. The lack of color change indicates enzyme inhibition, due to the presence of pesticide residues, and the zone of inhibition is proportional to the quantity of pesticide present.

18.3.3.2.2 Gas Chromatography Recently, with the development of fused silica capillary columns (Chap. 29, Sect. 29.3.4.2), a large number of pesticides with similar physical and chemical properties can be separated and detected. In general, GC is the preferred method for the determination of volatile and thermally stable pesticides, such as the OC and OP classes. Choice of columns and detectors is made based on the nature of the pesticides. For example, 5% diphenyl, 95% dimethylpolysiloxane stationary phase columns are commonly used in MRMs.

Pesticides often contain heteroatoms, such as O, S, N, Cl, Br, and F, in a single molecule. Therefore, element-selective detectors are often used, such as a flame photometric detector (FPD), which is suited for the detection of P-containing compounds. The FPD (see Chap. 29, Sect. 29.3.5.4) is widely used for the detection of OP pesticides in various crops, without extensive cleanup required. For the determination of OC, the electron capture detector (ECD, see Chap. 29, Sect. 29.3.5.3) is used extensively due to its high sensitivity to organic halogen compounds. With a MRM approach for multiclass detection, and using these selective detectors, several GC injections are required, which is a limitation for conventional GC analysis. Additionally, identification in conventional GC analysis is highly dependent on retention time, which is not an absolute confirmation of identity, due to matrix interferences. Coupling of fused capillary columns to the highly specific and sensitive MS detection enhances not only the confirmation process, but also the quantitative determination (see Sect. 18.3.3.3).

18.3.3.2.3 High-Performance Liquid Chromatography Development of HPLC analysis for the separation and detection of pesticides became a necessity as the number of pesticides with poor volatility, relatively high polarity, and thermal instability increased. Classes such as *N*-methyl carbamate (NMC), urea herbicides, benzoylurea insecticides, and benzimidazole fungicides are typically analyzed by HPLC. These compounds are often analyzed by reversed-phase chromatography (Chap. 28, Sect. 28.3.2) with C18 or C8 columns and aqueous mobile phase, followed by UV absorption, UV diode array, fluorescence, or MS detection (Chaps. 22 and 26). Following exhaustive cleanup, phenylurea herbicides can be determined quantitatively, with enhanced selectivity and sensitivity by HPLC separation and UV detection at 254 nm (23). For the detection of benzimidazole fungicides, fluorescence is more sensitive than UV detection (4). When the sensitivity of UV and fluorescence detection is poor, postcolumn derivatization can be employed. For example, NMCs are determined following a postcolumn hydrolysis and derivatization coupled with fluorescence detection (4). Postcolumn derivatization, however, requires special equipment such a mixing chamber and a reactor, which might not always be available. Additionally, interference from other compounds having fluorescence properties is a major disadvantage.

Analysis of pesticides in complex systems following conventional HPLC analysis using fluorescence or UV is often inadequate. Even the use of diode array detection might not be specific enough to resolve spectral differences, which are often too small. Utilization of MS detection widened the scope of HPLC analysis of pesticides. LC-MS is becoming one of the most powerful techniques for the analysis of polar, ionic, and thermally labile pesticides (see Sect. 18.3.3.3).

18.3.3.3 Mass Spectrometry Detection

Details on MS instrumentation, modes of ionization, and mass analyzers are provided in Chap. 26. The subsequent sections will provide sample applications of GC-MS and LC-MS used in pesticide residue analysis. References (4), (18), (24), and (25) describe more application examples and details on method development.

18.3.3.3.1 Gas Chromatography-Mass Spectrometry The most common GC-MS technique for the analysis of pesticide residue involves single quadrupole instruments with electron impact (EI) ionization. The selected ion monitoring (SIM) option enhances selectivity, increases sensitivity, and minimizes interferences from coextracted compounds. For example, determination of 39 OP compounds by a single chromatographic run (using capillary GC) was achieved using GC-MS with SIM (26). Ion-trap detectors (ITD) are also used in the analysis of pesticide residues. Using ITD, the analyzing in full-scan mode provides higher sensitivity as compared to single quadruple analysis and allows for confirmation by library searches (NIST library spectrum search). Figure 18-5 illustrates how a targeted pesticide (propyzamide) was detected using GC-ITD MS. Additionally, the ITD enables tandem MS analysis (MS/MS) by means of collision-induced dissociation (CID). The use of tandem MS improves selectivity and significantly reduces background without loss of identification capability, thus enabling the analysis of pesticides at trace levels in the presence of many interfering compounds. Figure 18-6 provides an illustration of enhanced compound identity confirmation upon the use of MS/MS vs. only MS. For the determination of multiclass residues, triplequadrupole (QqQ) MS is becoming a powerful and fast analytical tool, requiring minimum sample preparation. The chromatogram separation becomes less important when the analysis is carried out by GC/QqQ-MS/MS, since the analyzer is able to monitor simultaneously a large number of coeluting compounds. Reliable quantitation and confirmation can be easily achieved, even at trace concentration levels. Time-of-flight (TOF) MS instruments also are gaining popularity for the simultaneous analysis of multiclass pesticides. For example, rapid analysis of 98 pesticides in fruit-based baby foods (at levels $\geq$0.01 mg/kg) was achieved using a GC/TOF-MS technique (27).

18.3.3.3.2 High-Performance Liquid Chromatography-Mass Spectrometry Similar to GC-MS, the use of LC-MS in the analysis of pesticide has recently undergone great development. Tandem MS methods, using atmospheric-pressure ionization (API), atmospheric-pressure chemical ionization (APCI), and electrospray ionization (ESI), have been developed for many pesticides such as OP, carbamate, and sulfonylurea pesticides (25). Specifically, the ESI coupled with MS/MS was found to have high sensitivity and selectivity for a wide range of pesticides in foods. Because of HPLC solvent interferences, single quadrupole instruments are not as widely used in LC-MS as compared to GC-MS for pesticide analysis. When using LC-MS for pesticide residue analysis, QqQ are the most widely used mass analyzers. However, recently, LC/TOF-MS has gained popularity due to its high speed, sensitivity, and selectivity (28).

18.4 MYCOTOXIN ANALYSIS

18.4.1 Introduction

Molds, which are filamentous fungi, can develop on food commodities and produce various types of chemical toxins, collectively known as **mycotoxins**. The main producers of mycotoxins are fungal species belonging to the genera *Aspergillus, Fusarium,* and *Penicillin*. Crops can be directly infected with fungal growth and subsequent mycotoxin contamination as a result of environmental factors such as temperature, humidity, weather fluctuations, mechanical damage of kernels, and pest attack. In addition, plant stress due to extreme soil dryness or lack of a balanced nutrient absorption can induce fungal growth. Fungal infection and mycotoxin contamination can occur at any stage of the food chain (Sect. 18.1). The United Nation's Food and Agriculture Organization estimated that up to 25% of the world's total crops per year are affected with "unacceptable" levels of mycotoxins (29).

The presence of mold on the surface of a food product does not confirm the presence of mycotoxins,

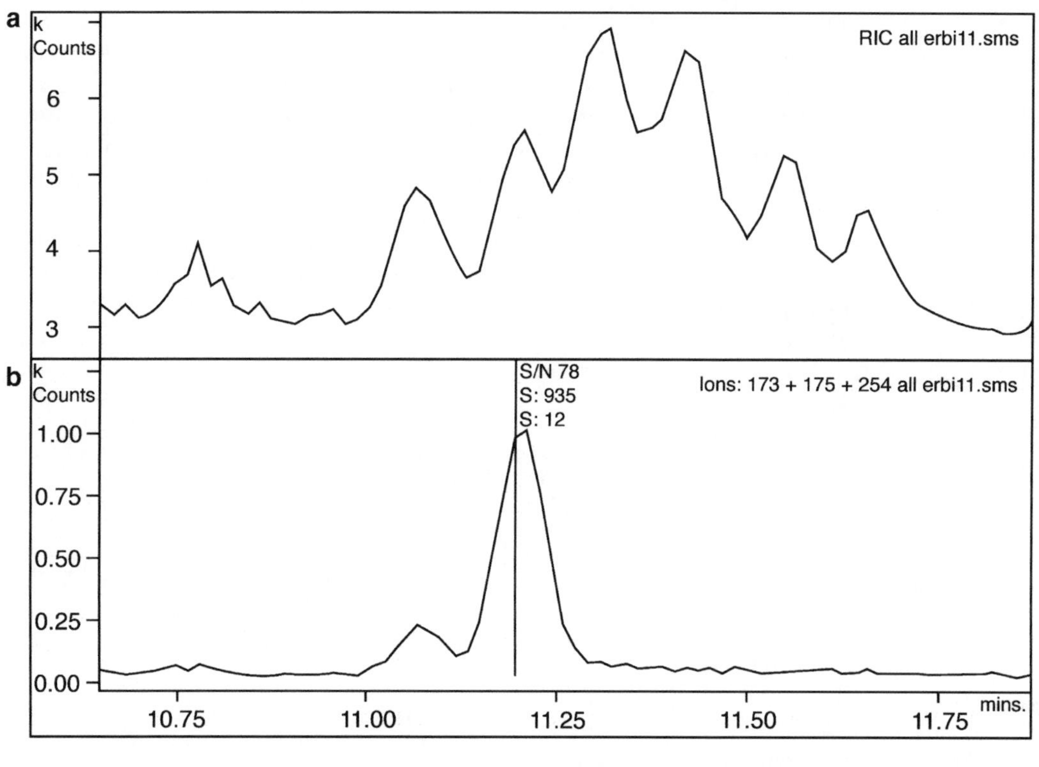

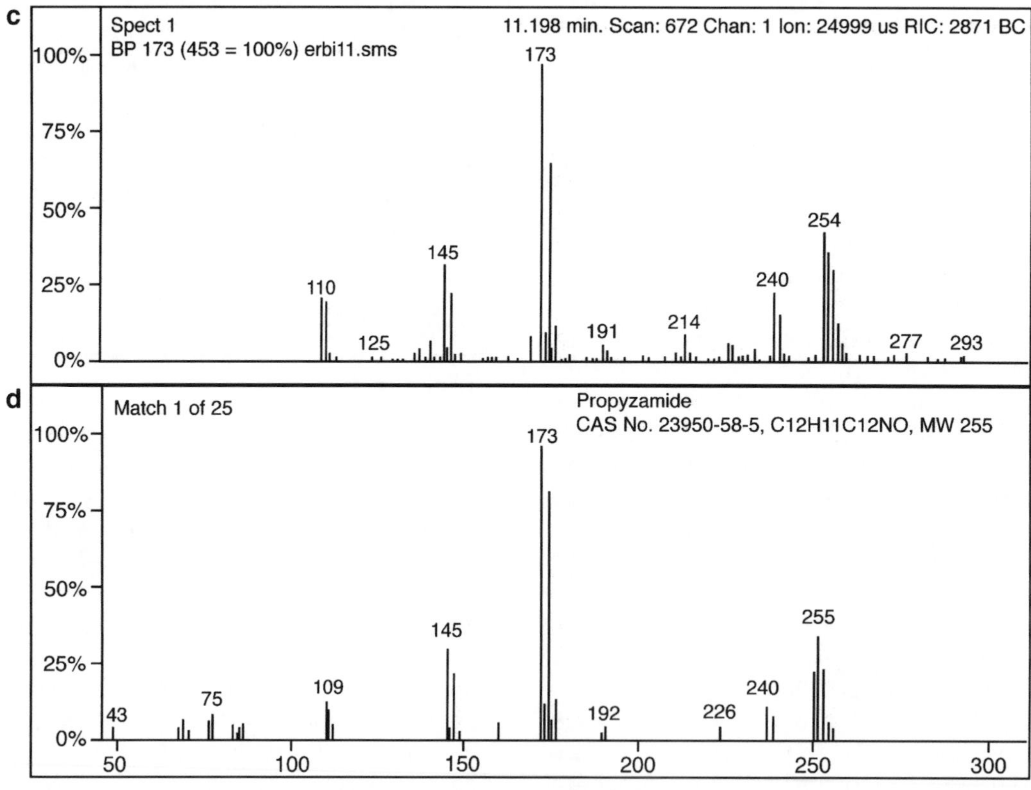

18-5 figure

Detection of propyzamide using GC-ITD MS. (**a**) Total ion chromatogram from retention time 10–12 min of an apricot sample fortified with 10 ng/g propyzamide. (**b**) Extracted ion chromatograms of the most characteristic propyzamide ions. (**c**) Mass spectrum of the eluate at 11.19 min. (**d**) NIST library spectrum of propyzamide. [Adapted from (4), with permission.]

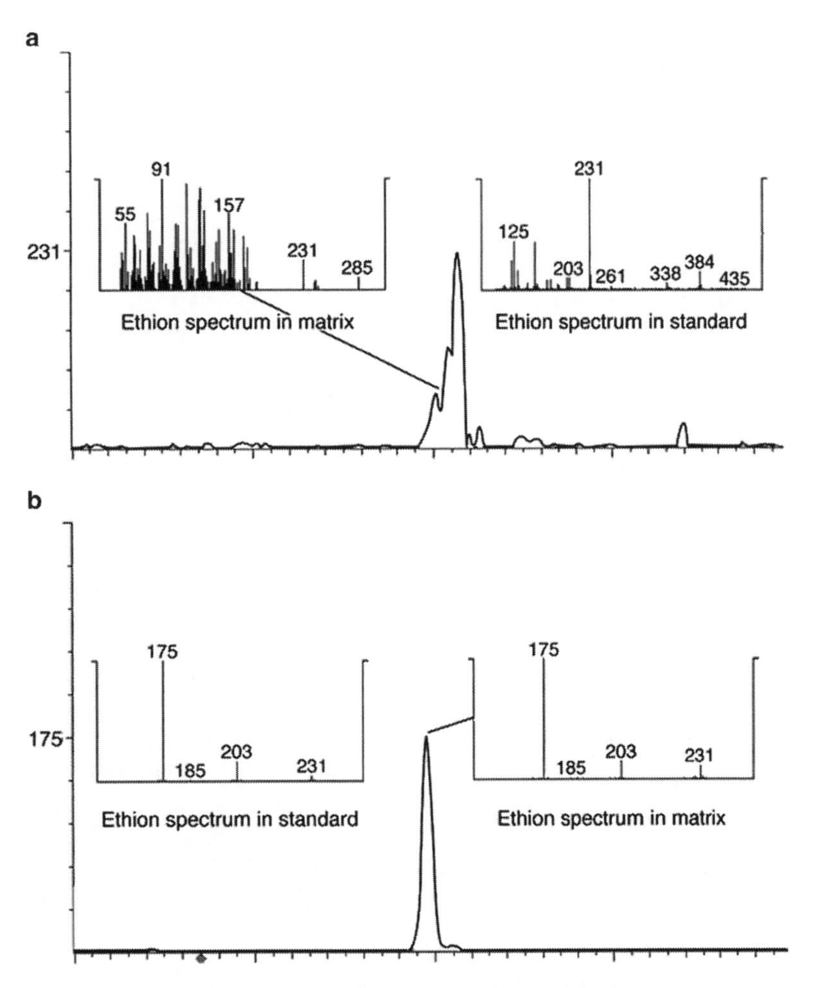

18-6 figure Pesticide residue analysis in vegetable extract using (**a**) EI full scan, showing noisy baseline and no spectral match vs. (**b**) EI MS/MS showing clean baseline, symmetrical peaks and excellent spectral match. (Used with permission from Varian, Inc., Palo Alto, CA.)

since their production varies greatly among different molds and in response to various environmental and nutritional conditions. Similarly, the absence of a visible mold infestation does not confirm the absence of mycotoxins. While molds can be destroyed by natural causes or processing, mycotoxins can survive. Mycotoxin contamination may occur in food as a result of the direct mold infection of plant-origin commodities, such as cereals, dried fruit, spices, grape, coffee, cocoa, fruit juices (especially apple based). Mycotoxins also can occur in milk, eggs, and to a minor extent in meat, as a result of the farm animal consumption of feed from contaminated crops. Additionally, humans and animals can be exposed to myctoxins through heavily contaminated dust, as in the case of harbors and warehouses.

More than 300 mycotoxins, belonging to various chemical classes, are known. However, the major classes of mycotoxin with a toxicological impact on human health include **aflatoxins** (B1, B2, M1, M2, G1, and G2), **ochratoxins** (e.g., ochratoxin A, OTA), **trichothecenes** [e.g., deoxynivalenol (DON), T2, and HT-2], **fumonisins** (FBs, e.g., FB1, FB2, FB3, etc.), *patulin* (a mycotoxin that occurs mainly in apples and apple products), and **zearalenone** (ZEA). The chemical classification and occurrence of mycotoxins (the major as well as minor ones) in raw commodities and processed foods is described in reference (29). The toxic effects of the aforementioned mycotoxins include, but are not limited to, genotoxicity, carcinogenicity, mutagenicity, and immunotoxicity. Genotoxic compounds have a probability of inducing an effect at any dose; therefore, no threshold dose should be considered and they should not be present in food. For example, there is no postulated safe dose for aflatoxin B1, which is genotoxic and is the most potent naturally occurring carcinogenic substance known. However, to provide risk managers with necessary basis for making decisions, threshold doses and "safe" **total daily intake** or **total weekly intake** (TDI or TWI) have been set for other mycotxoxins (Table 18-3). Based on the threshold doses, tolerance levels for mycotoxin were set in

18-3 table **Total Daily Intake (TDI)/Total Weekly Intake (TWI) for the Major Mycotoxins**

Mycotoxin[a]	*TDI (ng/kg bw per Day)*	*Organization*[b]
OTA	4	Health Canada (1989, 1996)
	5	Nordic Council (1991)
	5	EU (1998)
	14	JECFA (1996, 2001)
	120 (TWI)	EFSA (2006 [81])
FBs	2000	EU (2000)
	2000	JECFA (2000)
	400	Health Canada (2001)
DON	3000	Health Canada (1985)
	1000	Health Canada (2001)
	1000	Nordic Council (1998)
	1000	EU (1999)
	1000	JECFA (2000)
ZEA	100	Health Canada (1987)
	100	Nordic Council (1998)
	500	JECFA (2000)
	200	EU (2000)
Patulin	400	JECFA (1996)
	400	EU (2000)
	400	Health Canada (1996)

From (29), used with permission.
[a] OTA, ochratoxin; FBs, fumonisins; DON, deoxynivalenol; ZEA, zearalenone.
[b] EU, European Union; JECFA, Joint FAO/WHO Expert Committee on Food Additives.

the USA (e.g., 0–35 μg/kg for aflatoxins, 2.5–50 μg/kg for OTA, 300–2000 μg/kg for DON, and 5–100 μg/kg for patulin).

To limit human exposure to mycotoxins, a key critical control point is avoiding the processing of raw materials with unacceptable mycotoxins levels. Therefore, implementing periodic testing is a necessity, including the adoption of reliable sampling procedures and validated methods of analysis. Organizations such as the AOAC International, American Oil Chemists' Society (AOCS), AACC International, and the International Union of Pure and Applied Chemistry (IUPAC) have method validation programs for mycotoxin analysis.

18.4.2 Sampling

Compared to the analysis of other types of residues, the sampling step in the analysis of mycotoxins is by far the largest contributor to the total error. The possible variability is associated with the level and distribution of mycotoxins in the food commodity. An unevenly distributed 0.1% of the lot is usually highly contaminated, resulting in an overall level above the tolerance limit. Because of this heterogeneous distribution, an appropriate sampling plan is needed to assure that the concentration in a sample is the same as that in the whole lot. An incorrect sampling protocol can easily lead to false conclusions, often false negative, leading to undesirable health, economic, and trade impacts.

Recently, the Commission Regulation 401/2006 (30) provided specifications for the sampling of regulated mycotoxins in various food commodities including cereals, dried fruits, nuts, spices, milk and derived products, fruit juice, and solid apple products. Also, AOAC International, AOCS, and the FDA in collaboration with the USDA have developed detailed sampling plans for separate commodities. In general, an acceptable plan involves obtaining a large number of samples from multiple locations throughout the lot, creating a composite sample, grinding or slurring the composite sample (to reduce particle size and/or increase homogeneity), and subsampling for laboratory analysis. The number and size of collected samples and laboratory subsamples are dependent on the matrix and the size of the lot. For example, in sequential sampling of raw shelled peanuts (with a tolerance level of less than 15 μg/kg for aflatoxins), a bulk sample of approximately 70 kg is randomly accumulated (at a rate of one incremental portion per 225 kg of lot weight). This bulk sample is divided randomly into three 21.8-kg samples using a Dickens mechanical rotating divider, and then ground separately. A subsample (1100 g) is taken from one of the samples, formed into a slurry, and analyzed for aflatoxin, in duplicate. If the average of the two determinations is ≤8 μg/kg, the lot is passed and no further testing is performed. If the average is ≥45 μg/kg, the shipment is rejected. For averages >8 μg/kg and <45 μg/kg, a second 21.8-kg sample is analyzed in duplicate and the average of the four results is used to decide whether to accept (≤12 μg/kg) or reject (≥23 μg/kg) the lot. If the average falls between the second set of determining values, the third 21.8-kg sample is analyzed. If the average of the six determinations is <15 μg/kg, the lot is accepted.

18.4.3 Detection and Determination

Post sampling, sample preparation commonly includes extraction, cleanup, and concentration. Sample preparation steps (discussed in Sect. 18.2.2) that are specific for mycotoxin analysis have been developed. ASE often is used for mycotoxin extraction, while SPE often is used for cleanup, especially when multimycotoxin analysis is needed. For a specific mycotoxin, immunoaffinity LC (IAC) is used for cleanup.

Tremendous effort has been made to develop and optimize qualitative and quantitative methods for mycotoxin analysis. A list of official methods of mycotoxin analysis is included in reference (29). In addition, the ACS symposium series 1001 (31) provides details on newly developed methods of analysis for several mycotoxins. Mycotoxin analysis kits, both for cleanup and for detection, are commercially available (Table 18-4). The sections below briefly discuss some of the current and newly developed analytical techniques used for the detection and determination of mycotoxins in food.

18.4.3.1 Rapid Methods of Detection

18.4.3.1.1 TLC A large number of TLC methods for the analysis of mycotoxins have been accepted by AOAC International, including DON in barley and wheat, aflatoxin in peanuts and corn, aflatoxin M1 in milk and cheeses, OTA in barley and green coffee, and ZEA in corn. Conventional TLC techniques are commonly used for screening purposes, with detection limits reaching 2 ng/g. When results are positive, confirmatory and more sensitive quantitative analysis follows. The overall performance of TLC for the analysis of mycotoxin is improved when used in combination with IAC. Additionally, a microcomputer interfaced with a fluorodensitometer and semiconductor-based detection improves the data handling.

18.4.3.1.2 Immunoassays Three main types of immunoassays can be used for the analysis of mycotoxins: **radioimmunoassay** (RIA), ELISA (see Chap. 17), and **fluorescence polarization immunoassay** (FPIA). The use of RIA, where the mycotoxin (e.g., aflatoxin) is radiolabeled, has been gradually replaced by ELISA. Since all mycotoxins are relatively small molecules (MW<1000), competitive ELISA is used (see Chap. 17, Sect. 17.3.3). In FPIA, fluorecein-labeled mycotoxin competes with unlabelled mycotoxin analyte in a sample for binding with the antibodies. The greater the abundance of a specific mycotoxin analyte in the sample, the less fluorescein-labeled mycotoxin is bound to the antibodies, resulting in lower emission of plane-polarized light. FPIA has gained popularity since it involves no coating of the plate and less analysis time as compared to ELISA. Results of FPIA were comparable to that of ELISA for the analysis of DON, ZEA, and OTA (29).

Immuno-based testing methods, including membrane-based immunoassays, lateral flow strip (LFS) assays, and biosensors also have been developed for online control or field testing of mycotoxins. Membrane-based immunoassays are based on the principle of direct competitive ELISA, with the anti-mycotoxin antibody being coated on the membrane surface. For example, test kits based on membrane-baed immunoassays have been validated for OTA determination in wheat, rye, maize, and barely (32). LFS that is an immunochromatographic test (see Chap. 17, Sect. 17.4) is rapid and capable of detecting many mycotoxins simultaneously. For example, the AgraStrip® LFS kit (Romer Labs Inc., Union,

18-4 table Example Commercial Test Kits for Mycotoxin Residue Analysis

Assay Type	*Example Matrices*	*Mycotoxin*	*Commercial Source*
Antibodies in a competitive direct ELISA assay	Corn, cornmeal, corn gluten meal, corn/soy blend, wheat, rice, soy, barley, oats, whole cottonseed, cottonseed meal, raw peanuts, peanut butter, and mixed feeds	Aflatoxin, deoxynivalenol (DON), fumonisin, ochratoxin, T-2 toxin, and zearalenone	Agri-Screen® series (Neogen Corporation, 620 Lesher Place, Lansing, MI)
Lateral flow receptor	Milk, grain, feedstuffs, wine, and grape juice	Aflatoxin M_1 and M_2, DON, fumonisin, ochratoxin, T-2 toxin, and zearalenone	Charm ROSA® series (Charm Sciences, Inc., 659 Andover Street, Lawrence, MA)
Immunoaffinity columns with HPLC or fluorometry	Milk, milk products, wheat and grains, coffee, nuts, and other foods	Citrinin, aflatoxins M_1, B_1, B_2, G_1, and G_2, DON, fumonisin, ochratoxin, T-2 and HT-2 toxins, and zearalenone	AflaTest® WB and sister products (VICAM, 34 Maple St. Milford, MA)
Cleanup columns for HPLC, GC, or TLC analysis	Fruits and fruit juices, wheat, oats, corn, nuts, fruit, coffee, wine	Patulin, ergot alkaloids (ergometrine, ergotamine, ergosine, etc.), citrinin, moniliformin, aflatoxins, sterigmatocystin, zearalenone, ochratoxin, fumonisin, DON, and T2	Mycosep® and sister products (Romer Labs, Inc., 1301 Stylemaster Drive, Union, MO)

MO) can qualitatively determine the presence of aflatoxins B1, B2, G1, and G2, in 5 min. However, any positive test requires confirmation by a reference method such as HPLC (see Sect. 18.4.3.2.1). Biosensors, on the other hand, are compact analytical devices that use biological components such as nucleic acids, enzymes, antibodies, or cells, associated with a transduction system. The transduction system processes the signal produced by the interaction between the target molecule and the biological component. The use of biosensors for the detection of several mycotoxins, such as aflatoxins DON and OTA, is rapidly increasing (29).

18.4.3.2 Quantitative and Confirmative Chemical Methods

18.4.3.2.1 HPLC For quantitative determination, HPLC is the methodology of choice for most mycotoxins, specifically, aflatoxins, DON, OTA, ZEA, FBs, and patulin. Pre- or postcolumn derivatization (with trifluoroacetic acid or iodine) is required for the fluorescence detection of aflatoxins OTA, ZEA, and FBs. However, direct UV detection is used for DON and patulin. Reversed-phase chromatographic separation is normally employed for multimycotoxin analysis. A multifunctional column, which consists of a mixture of reversed-phase, size exclusion, and ion exchange stationary phases, also can be used (e.g., AOAC Method 49.2.19A for aflatoxin analysis). Methods using IAC-HPLC are reported for the determination of aflatoxins in peanut butter, pistachio paste, fig past, and paprika powder (33). Validated HPLC methods for the analysis of various mycotoxins are periodically reported in Methods Committee Reports published in *Journal of AOAC International*.

As is the case with pesticide residue analysis, coupling of HPLC with MS analysis, especially LC-MS/MS, provides greater sensitivity and selectivity and allows for simultaneous analysis of multiclass mycotoxins. Additionally, the use of LC-MS/MS allows for the detection of conjugated mycotoxins (i.e., the toxin bound to a polar compound such as glucose or another sugar), which are referred to as masked mycotoxins because they escape routine detection. A method has been recently developed for the determination of 39 mycotoxins, including conjugated DON, FBs, ZEA, and aflatoxins, in wheat and maize, using LC-ESI-triplequadrupole MS, without cleanup (34).

18.4.3.2.2 GC GC is not widely used for the detection of mycotoxins, except in the case of trichothecenes. Trichothecenes do not strongly absorb in the UV–Vis range and are nonfluorescent; therefore, GC methods were developed for their determination. Capillary column GC is commonly employed for the simultaneous detection of different trichothecenes, e.g., DON, T2, and HT-2, using trifluoroacetyl, heptafluorobutyry, or trimethylsilyl derivatization coupled with electron capture detection. GC is often linked to MS for peak confirmation. GC-MS also can be used also for the confirmation of patulin in apple juice (29). Validated and accepted methods for the determination of trichothecenes using GC are reported as AOAC Official Methods and by the American Society of Brewing Chemists.

18.4.3.2.3 Capillary Electrophoresis **Capillary electrophoresis** (CE, see Chap. 15, Sect. 15.2.5.3), generally used as a chromatographic technique, can be employed to separate mycotoxins from matrix components using electrical potential. Methods are available for the determination of patulin in apple juice (35), simultaneous determination of ochratoxins (A and B) and aflatoxins (36), and for the determination of ZEA using cyclodextrins (for enhancing the native fluorescence) (37).

18.4.3.3 Other Methods of Analysis

Other methods for the detection of mycotoxins have been reported such as **near** and **mid infrared** (NIR and MIR) spectroscopy, especially **Fourier transform infrared** (FTIR) spectroscopy (a type of MIR) (see Chap. 23). FTIR spectrometry was used for the detection of mycotoxins in infected corn by gathering information from MIR absorption spectra (38). Rapid and accurate identification of mycotoxigenic fungi and their mycotoxins (e.g., FB1 and DON) was achieved using MIR and NIR spectroscopy (39, 40). Calibration for these methods is based on reference HPLC or GC methods. Testing and validation of IR methods for the detection of mycotoxins are currently being pursued.

18.5 ANTIBIOTIC RESIDUE ANALYSIS

18.5.1 Introduction

Animals intended for human consumption may not only be given drugs (e.g., antibiotics, antifungals, tranquilizers, and anti-inflammatory drugs) at therapeutic levels to combat diseases, but also may be given drugs (mainly antibiotics) at subtherapeutic levels. Such low-level subtherapeutic drug treatment can have the following benefits: (1) reduce the incidence of infectious diseases caused by bacteria and protozoa, (2) increase the rate of weight gain, and (3) decrease the amount of feed needed to obtain weight gain. The **Center for**

Veterinary Medicine (*CVM*), a branch of the FDA, regulates the manufacture and distribution of drugs given to animals, including animals from which human foods are derived. The use of drugs in feeds is strictly regulated and monitored by the CVM, including an approval process and restrictions on the following: (1) species that may be treated, (2) dosage level and duration, (3) withdrawal period (time between the last drug treatment to the animal and the slaughter or use of milk or eggs by humans), and (4) tolerance level (total residues, including metabolites, conjugates, and residues bound to macromolecules, in uncooked edible tissues). Most drug residues of concern for human food are antibiotic residues, so that is the focus of this section. Some of the major families/types of antibiotics, referred to in later sections, include β-lactams, sulfa drugs, cephalosporins, tetracyclines, and chloramphenicol.

Residual levels of any antibiotics given to animals used for human consumption are of concern for a variety of reasons, including the fact that some consumers are allergic to certain antibiotics, and the possibility that some microbes may develop antibiotic resistance due to constant exposure. Recent data suggested that some antibiotics can be carcinogenic as well, such as nitrofuran compounds (41). Also, on a practical level for dairy products such as cheese made with starter cultures, antibiotic residues would reduce the intended microbial growth and therefore reduce acid production. For all these reasons and more, the FDA has strict regulations on antibiotic residues in human food. Because of these regulations, antibiotic-contaminated meat and milk (including milk products) are considered to be adulterated. The FSIS monitors antibiotic residues in meat products from cattle, swine, lamb, goat, and poultry. In 2004, 38 violative residues of penicillin, tilmicosin, neomycin, and gentamicin were reported (42). FDA's Center for Food Safety and Nutrition monitors antibiotic residues in milk and its products, including milk from pick-up tanks, pasteurized fluid milk, cheeses, etc. In 2003, out of four million samples tested, 3000 were reported to have antibiotic residues including β-lactam and sulfonamide (43).

Samples are tested for antibiotic residues using rapid screening methods. A positive result from a screening test suggests that one or more types of antibiotic are present, so further testing is required to identify and quantify the specific antibiotics present.

18.5.2 Detection and Determination

As with any residue analysis, sample preparation is often required to purify and concentrate the analytes of interest. Often, procedures such as defatting, protein hydrolysis (in case of meat or egg samples), protein precipitation (in case of dairy samples), and aqueous wash (in case of honey to remove excess sugar) precede the extraction of antibiotic residues. For the extraction of many antibiotics, liquid–liquid extraction and SPE are commonly used, followed by a partial purification step often using ion-exchange cleanup systems that take advantage of the acid/base character of the antibiotics. Sample preparation steps, including extraction and isolation (as discussed in Sect. 18.2.2) specific for drug residues, are reviewed in reference (44).

Tolerance levels are established for some antibiotics, while others have a zero tolerance (as is the case for nitrofurans and chloramphenicol). Chloramphenicol, for instance, is an antibiotic of considerable current concern in the USA, European Union, and other countries. It is used in some parts of the world in producing shrimp and has been found in imported seafood products (e.g., shrimp, crayfish, and crab). Because of the adverse health effects on humans, the FDA has banned the use of chloramphenicol in animals raised for food production and set a zero tolerance in human food [21 CFR 522.390 (3)]. Therefore, the analytical methods need to be as sensitive and selective as possible.

A wide variety of analytical methods have been developed and optimized for the analysis of antibiotics, categorized as screening or determinative and confirmatory. Reference (3) lists the methods currently used in the analysis of several antibiotics, and reference (45) lists the AOAC approved antibiotic test kits. One excellent resource for antibiotic test methods, specific for milk and dairy products, is reference (46). The sections below briefly discuss some of the current and newly developed analytical techniques used for the detection and determination of antibiotics in susceptible food commodities.

18.5.2.1 Screening Methods

Some of the major categories for rapid screening assays, some of which are quantitative, are: (1) microbial growth inhibition, (2) receptor assays, (3) enzyme substrate assays, and (4) immunoassays. Some screening methods are specific to individual antibiotics, some to a class of antibiotics, and some have no specificity. Screening assays for antibiotic residues in test samples initially relied mostly on inhibition of microbial growth, but now many use other principles for detection. Table 18-5 includes examples of screening assays, including the specific antibiotics detected.

For **microbial growth inhibition** assays, one commonly measures turbidity, zone of inhibition, or acid production. In a **turbidity assay**, an indicator organism growing in a clear liquid culture will cause an

18-5 table Example Commercial Test Kits for Antibiotic Residue Analysis

Assay Type	*Example Matrices*	*Antibiotics Detected*	*Commercial Source*
Bacterial inhibition	Raw cow milk. Goat and sheep milk. Other dairy products.	Penicillin	Not applicable
Competitive radio-receptor binding assay	Raw cow milk and liquid milk. Cream, condensed milk, and dairy powders. Meat and honey.	β-lactams (penicillin G, amoxicillin, ampicillin, ceftiofur, cephapirin, etc.), cephalosporins, sulfa drugs, tetracycline, macrolides, aminoglycosides	Charm II® (Charm Sciences, Inc., 659 Andover Street, Lawrence, MA)
Competitive binding assay with enzyme-based color development	Raw cow milk from individual cow and bulk tank testing.	β-lactams, tetracycline, chlortetracycline, oxytetracycline	SNAP® test kit (IDEXX Laboratories, Inc., One IDEXX Drive, Westbrook, ME)
Competitive ELISA	Shrimp	Chloramphenicol	Veratox® (Neogen Corporation, 620 Lesher Place, Lansing, MI)
Lateral flow receptor	Raw cow milk and liquid milk. Cream and condensed milk.	Sulfadimethoxine and sulfamethazine	Charm ROSA MRL Sulfamethoxine and Sulfamethazine

increased turbidity; growth is inhibited, so turbidity is reduced if antibiotics are present. In a **zone of inhibition assay**, the test material diffuses through an agar-based nutrient medium that has been uniformly inoculated with spores of a susceptible organism. Any antibiotics present in the test material will inhibit the germination and growth of the organism, creating clear zones. In the **acid production assays**, the acid produced when microbes grow causes a color change in the medium. No color change means that the test sample contained an inhibitory substance. While microbial growth inhibition assays are more time consuming than many newer screening tests, they are inexpensive, applicable to testing of large numbers of sample, and provide some sensitivity to multiple antibiotic categories (46). AOAC Official Methods include a nonspecific microbiological method for antibiotics and numerous microbiological methods for specific antibiotics (17).

An example of a **receptor assay** is the Charm II® test (Charm Science, Lawrence, MA), which has different versions, designed to detect different groups of antibiotics. The assay involves a competition between labeled antibiotics (labeled using ^{14}C or ^{3}H, depending on the specific type of Charm II® test system) and antibiotic residues in a milk sample for a limited number of specific binding sites on the surface of bacteria added to the test sample. The greater the concentration of antibiotic residue in the milk sample, the less radiolabeled tracer will become bound to the microorganism. The assay is started by adding to the milk sample the radiolabeled tracer antibiotic and the binding receptor from a susceptible microorganism. This sample is incubated, centrifuged, fat is removed, the microbial plug is resuspended in a scintillation fluid, and the radiolabeled tracer is measured (46). The method can be applied not only to milk and certain dairy products, but also to honey and meat.

Enzyme substrate assays involve measuring inhibition of an enzyme acting on a substrate, caused by the presence of an antibiotic. An example used for testing raw milk is the Penzyme® III commercial kit (Neogen, Lansing, MI), which is specific for β-lactam antibiotics, which inhibit D,D-carboxypeptidase on an equimolar basis. When this enzyme acts on a specific substrate, it causes release of D-alanine, which can be measured in additional steps of the assay resulting in a color change (46).

Some **immunoassays** for screening antibiotic residues are of ELISA type and some use lateral flow strips (see Chap. 17, Sects. 17.3 and 17.4, respectively). The Charm ROSA (Rapid One Step Assay) MRL assay, intended for milk and cream testing, is an example using a lateral flow strip. The receptor-gold used on the test strip is specially formulated to reduce false positives. Another example of an immunoassay marketed for testing milk is the SNAP® kit marketed for testing milk (IDEXX Laboratories, Inc., Westbrook, ME). The assay sets up a competition between residual antibiotics in a milk sample and enzyme-labeled antibiotics in the test kit. The enzyme acts on a substrate to cause a color change; any antibiotics in the milk will result in a decrease in color development.

An example of a competitive ELISA, used for the detection of a specific antibody, is the Veratox® assay (Neogen Corporation, Lansing, MI) for the antibiotic chloramphenicol.

18.5.2.2 Determinative and Confirmatory Methods

Quantitative determination of antibiotic residues in food products follows the same general steps as for other trace analytes. After sample preparation steps (including pretreatment, extraction, and purification), the partially purified extract is subjected to chromatographic separation, detection, and quantification. The most commonly used chromatography system is HPLC (mostly using reversed-phase separation mode) coupled with UV detection, using variable wavelength or diode array detection. HPLC coupled with fluorescence, chemiluminescence, or postcolumn reaction detectors also have been used in analysis of antibiotic residues. For confirmatory and identification purposes at trace levels, LC-MS and LC-MS/MS are increasingly used in the analysis of antibiotics.

The FDA provides regulatory LC-MS/MS methods to detect fluoroquinolones in honey (47, 48) and LC-MS methods for chlorampenicol and related compounds in honey. For shrimp, crab, and crawfish analysis of chlorampenicol and related compounds, the FDA has regulatory LC-MS/MS methods (49), and LC-MS methods not intended for regulatory purposes at this time (50). Additionally, LC-MS/MS has been used for confirmation of β-lactam residues in milk (51). The detection of 14 different types of sulfonamides, in milk (52) and in condensed milk and soft cheeses (53), at levels below 10 ng/ml, also was achieved using LC-MS/MS. LC-MS/MS has been compared with ultra-high-performance liquid chromatography/quadrupole time-of-flight MS (UHPLC/Q-Tof MS) for analysis of macrolide antibiotic residues in a variety of foods (54). LC-MS/MS gave a lower limit of detection and better precision, but the UPLC/Q-Tof MS provided better confirmation of positive findings.

18.6 ANALYSIS OF GMOS

18.6.1 Introduction

Agriculturally important plants may be genetically modified by the insertion of DNA from a different organism (**transgene**) into the plant's genome, conferring novel traits that it would not have otherwise, such as herbicide tolerance or insect resistance. The modified plant is termed a **genetically modified organism**, or **GMO**. Current GMO production relates primarily to four crops: **soybeans, corn, cotton**, and **canola** (a cultivar of rapeseed). Although protection from pests and herbicide tolerance are the most common GMO traits, other GMO crops include plants that have been modified to improve postharvest quality or enhance the nutritional makeup of the food. Examples include vegetables with an extended shelf life and "golden rice," which produces the precursor to vitamin A. What is the source of the transgenes and how are GMOs produced?

The enzyme 5-enolpyruvyl-shikimate-3-phosphate synthase (EPSPS) is a component of the metabolic pathway leading to the biosynthesis of the aromatic amino acids phenylalanine, tyrosine, and tryptophan. The EPSPS that is naturally produced by the Gram-negative soil bacterium *Agrobacterium* sp. strain CP4 (CP4 EPSPS) is unaffected by glyphosate herbicides, such as **Roundup**®. Introduction of the *cp4 epsps* gene into soybeans (**transformation**) confers glyphosate-containing-herbicide tolerance to the plants, because it functions in place of the EPSPSs normally produced by the crop plants, which are sensitive to glyphosate herbicides. Similarly, different strains of *Bacillus thuringiensis* (Bt), a Gram-positive soil bacterium, carry various *cry* genes that encode the Cry proteins, also known as **Bt toxins**. These toxins specifically target certain insects, and the introduction of *cry* genetic material into a plant will protect the crop from the insects. For example, the Bt Cry34Ab1 and Bt Cry35Ab1 toxins provide the corn plant with protection against western and northern corn rootworm larvae, but have no effect on other insects or animals.

There are multiple methods to achieve plant transformation. One involves the use of *Agrobacterium* itself. In nature, *Agrobacterium tumefaciens* is a plant pathogen that is able to cause a tumor-like growth on plants, termed crown-gall disease, by the horizontal gene transfer of DNA segments (T-DNA) into the plant via a tumor-inducing (Ti) plasmid. To construct a GMO, scientists simply replace the T-DNA with the DNA of the target gene, such as *cp4 epsps*, and accessory DNA needed to express the gene properly in the plant, such as the promoter sequence. Individual plant cells are then exposed to the bacteria, selected for transformation, and grown to full plants for the seeds. Other techniques involve a similar process, but use different vectors for the gene transfer. These include the "gene gun" system, in which tiny gold particles are coated with the DNA and shot into the plant cells. As above, transformed cells are selected and grown to adult, seed-producing plants.

Despite the prevalence of GMO crops (over 80% of US soybeans and 60% of US corn in 2006), there are many reservations about the practice. These range from concerns about limiting crop variation (i.e., monocropping) to fears of unintended impacts on other plants, insects, wildlife, and nearby communities. The possibility of GMO products acting as allergens is also a consideration. The debate about the use of GMO crops has resulted in a range of responses. Of course, all new products must be evaluated for toxicological, allergenic, nutritional, and other characteristics that could impact the inclusion of the resulting agricultural product in foods. There are specific guidelines put forth by various governments as to the nature of the analyses used, but all involve various testing regimes. However, in some places, such as the European Union (EU), measures to satisfy public concerns about GMOs include strict labeling guidelines. Therefore, even US producers must be able to accurately test ingredients and other products intended for sale in EU countries. Fortunately, many companies now produce high-quality test kits that are easy to use. Many of the kits are specific for a certain GMO protein or gene, such as the *cp4 epsps* gene that encodes the Roundup Ready phenotype, or the CP4 EPSPS protein itself. These kits generally fall into two categories, based on the methodology: **immunoassays** that are specific for the proteins and **polymerase chain reaction** (PCR) kits that specifically amplify the DNA of the GMO gene so that it can be detected. In addition, PCR kits are available that are directed at DNA sequences that are shared by many GMOs, such as the promoter sequence that is commonly used. Further reading on the GMO and GMO detection topics is listed in references (55–57).

18.6.2 Protein Methods

The protein methods involve immunoassays of various types (see Chap. 17 for details of methods). Immunoassays are highly specific because they are based on the use of **antibodies** for detection. Although the traditional **ELISA** is used for GMO detection, an easier and quicker type of assay involves the use of lateral flow strips. These are **single-step lateral flow immunochromatographic assays** that can be performed quickly, although, unlike ELISA analyses, they are not quantitative. Formerly, western blots, which are immunoassays that follow an electrophoresis step, were also used in GMO protein detection, but with the commercialization of ELISA and lateral flow assays the application of Western blot technology to this problem is no longer practical. Only the ELISA and lateral flow assays will be discussed below.

The **sandwich ELISA** technique is used commonly to test for GMOs. Table 18-6 includes an

18-6 table **Example Commercial Test Kits for GMO Analysis**

Assay Type	*Example Matrices*	*GMO, (Protein), and Trait*	*Commercial Source*
Single-step lateral flow immunochromatographic assay	Corn and soybeans products	Roundup Ready® (CP4 EPSPS protein). Confers tolerance to Roundup® herbicides	Reveal® (Neogen Corporation, 620 Lesher Place, Lansing, MI)
Single-step lateral flow immunochromatographic assay	Corn products and other matrices	StarLink® (Bt-Cry9C and PAT proteins). European corn borer resistance and phosphinothricin (PPT) herbicide tolerance	AgraStrip® GMO ST (Romer Labs, Inc., 1301 Stylemaster Drive, Union, MO)
Single-step lateral flow immunochromatographic assay	Cotton and seeds	BollGard® II (Bt-Cry 2A and Bt-Cry 1Ab/1Ac proteins) control of most leaf- and boll-feeding worm species	FlashKits® Cotton (Agdia Biofords, 5 Rue Henri Desbrueres, Evry, France
Enzyme linked immunosorbent assay (ELISA)	Leaf tissue and single seed	YieldGard Plus, YieldGard Plus/RR2, YieldGard VT Triple, YieldGard Rootworm, and Yieldgard VT Rootworm/RR2 (Cry1Ab and Cry3Bb1 proteins). Control of corn borer and corn rootworm larvae	QualiPlate® Kit for Cry1Ab & Cry3Bb1 (Envirologix, Inc., 500 Riverside Industrial Parkway, Portland, ME)
PCR kit	Most food samples and raw material containing soybean	Roundup Ready® Soya. (*cp4 epsps* gene). Tolerance to Roundup® herbicides	LightCycler® GMO Soya Quantification (Roche Diagnostics, 9115 Hague Road, Indianapolis, IN
PCR kit	Most food samples and raw material containing corn	NaturGard® KnockOut® (*cry 1ab* and *bar* genes). European corn borer resistance and PPT herbicide tolerance	LightCycler® GMO Maize Quantification (Roche Diagnostics)

example ELISA test kit for Cry1Ab and Cry3Bb1, which confer resistance to corn borer and corn rootworm larvae, respectively. The ELISA plate kits include the following components necessary for the effective quantification of the GMO proteins present in a sample: (1) 96-well antibody-coated microwell plates, (2) enzyme conjugate (antibodies with enzyme attached), (3) reagents (including enzyme substrate), and (4) standards (including positive control). The positive controls provided with the kit are used both for determination of when the experiments should be stopped and for the quantitative comparison with the samples. Kits that are intended for the detection of more than one protein include multiple conjugates (in the same solution) and substrates (in separate solutions) so that each protein can be separately detected in different wells in one assay. The plates are coated with antibodies to both proteins so that the initial step is universal. These assay kits have detection thresholds as low as 1–10 parts per billion, although, with technical support help from the kit producer, they can be diluted to achieve higher detection thresholds. With the use of multiple plates, hundreds of samples can be tested in less than 2 h. Typically, 44 paired samples and four paired standards are analyzed on each plate.

The lateral flow strip is a self-contained immunochromatographic assay that can easily be used on site. Lateral flow strip tests are calibrated to detect the GMO protein at certain concentration level or above. The GMO lateral flow test kit assays can be performed with no special equipment and no significant level of expertise, making it the assay of choice for qualitative field tests. In addition to the test kit, a sample extraction jar and pipettes are all the equipment that is necessary. Therefore, the lateral flow strips can be used to test for GMO material in the growing field, at storage areas, transit points, or at any point of transport or processing. Positive test results can be read in as little as 5 min. Examples of positive and negative lateral flow strip results are shown in Fig. 18-7. Note that the "**control line**" appears on both strips, whereas the "**test line**" is present only in the positive strip.

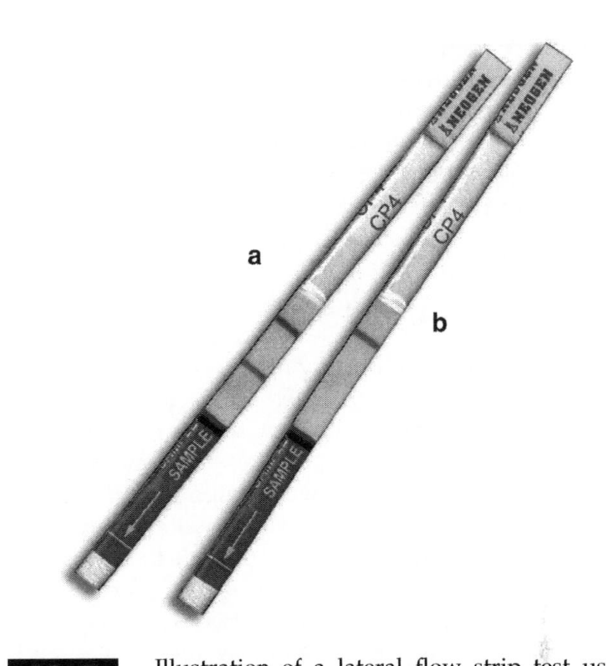

18-7 figure Illustration of a lateral flow strip test used for GMO Testing in Roundup Ready modified foods, showing a positive result (**a**) and a negative result (**b**). (Used with permission from Neogen Corporation, Lansing, MI.)

18.6.3 DNA Methods

As an alternative to protein identification, the detection of the transgene DNA itself is an effective method of testing for GMO material in a sample. The analysis involves three distinct steps: (1) the **extraction** of the DNA from the sample, (2) the **amplification** of the DNA by PCR, and (3) the **identification and quantification** of the amplified DNA (Note: for **real-time quantitative PCR analyses**, amplification and detection/quantification occur at the same time). Although all three steps are important, the PCR amplification is critical to the specificity and success of the analysis. Numerous PCR kits are available from various sources, and examples are listed in Table 18-6. In addition to the gene-specific PCR kits listed, other screening kits are available that amplify the commonly used promoters and transcription terminators for the general assay for GMO material without identifying the specific GMO.

18.6.3.1 DNA Extraction

To facilitate the analysis, extraction of the DNA from the food matrix should be performed prior to significant processing of the material. Therefore, the testing of raw ingredients is preferable and more common than testing highly processed foods. It is important to note that the application of extreme levels of sheer and heat can damage the DNA, rendering the subsequent PCR and detection ineffective. Another important consideration is the specific food matrix to be analyzed. The types of extraction protocols differ somewhat, but all include some method of disrupting the matrix to release the DNA, and in a food matrix this will usually be accomplished by grinding the sample to a fine powder. This is followed by dispersal of the ground material into an extraction solution and removal of unwanted components. For example, lipids may be removed by a detergent and protein by the addition of a protease. A final step may involve the precipitation of the DNA with cold alcohol, such as ethanol or isopropanol. The exact protocol used will depend on the specific matrix.

18.6.3.2 PCR Amplification

PCR is a method of amplifying the copy number of a specific sequence of DNA. PCR is a cyclic method that exponentially increases the copy number, by means of enzymatic replication, and it can be repeated numerous times to yield millions of copies of the target sequence. It uses thermal cycling to alternately replicate the target sequence and then melt the DNA into single strands in order to repeat the process. The method is based on the use of two synthetic DNA fragments that are complementary to opposite ends of the target sequence. These fragments are termed the **primers**, which are commonly 18–35 bases in length, and they can be produced only when the target sequence is known. If a general identification of any GMO material is desired, then the primers used would be complementary to the promoter sequence, which is common to all transgenic crop plants that are normally grown for industrial food production. If a specific GMO product is to be identified, then the primers would consist of a sequence that includes the trangene DNA and plant DNA. This is necessary to avoid the detection of bacterial DNA that originates from bacteria that may have been on or in the plants (both the Bt toxin and Roundup® tolerance transgenes are from soil bacteria).

In addition to the specific primers, the PCR mixture also includes a heat-stable **DNA polymerase**, such as Taq polymerase (from *Thermus aquaticus*), the **nucleotide bases** that comprise DNA, in the form of **deoxynucleoside triphosphates** (dNTPs), and a buffer solution to maintain the optimal conditions for the reactions. All of these components are present in the commercial kits (see Table 18-6). The reaction vials, which typically contain 20–200 μl of the mixture solution, are then placed in a **thermal cycler**. Once the PCR system is started, the mixture is first put through a high-temperature cycle to melt the DNA (separate it into single strands), and then a lower temperature cycle to allow the primers to anneal to the single stranded target DNA, and finally an intermediate temperature cycle to allow the DNA polymerase to synthesize a new DNA strand complementary to the target strand by adding dNTPs, starting at the primers. The process is then repeated, usually for 30–50 cycles, which is sufficient to produce millions of copies of the DNA.

18.6.3.3 DNA Analysis

When sufficient DNA has been generated by PCR, the sample can be analyzed by agarose gel electrophoresis. The sample and standards migrate through the gel, and after the run, the gel is stained and the presence and abundance of DNA can be identified, by comparison to the position and degree of staining of the standards.

Some of the newer kits contain reagents to specifically label the double-stranded DNA with tags, such as Fluorescein. These kits are meant to be used with special PCR equipment that deposits the finished, labeled mixture into a capillary where the intensity of the fluorescence, which is proportional to the abundance of DNA, is determined by high-sensitivity fluorescence spectroscopy, rendering the electrophoresis step unnecessary. In this case the specificity is contingent on the sequence of the primers, which directly determine the nature of the double-stranded DNA present. These kits also include standards and other items necessary for accurate quantification of the target DNA abundance in the sample. If real-time PCR is used, the fluorescence is read in the reaction tube after each cycle of amplification, yielding a curve containing multiple data points rather than a single data point for each sample.

18.6.4 Method Comparison

The example applications are listed in Table 18-6. These include ELISA assays, single-step lateral flow immunochromatographic assays, and the PCR kits for Bt toxin and Roundup Ready® soybeans. The latter kit, the LightCycler® GMO Soya Quantification kit (Roche Diagnostics), is an example of the complete assay kits that are intended for use with special equipment, in this case the Roche LightCycler System®. When used with this system and the software provided, many samples can be analyzed in less than 1 h, including the production of a final report, with very little labor after extraction and system set up. The obvious disadvantage to the latter system is the need for a well-equipped laboratory, the special equipment, and a significant level of expertise. In contrast, the lateral flow immunochromatographic strips can be used by anybody with minimal training, they are a complete analysis without further handling, and they can be used on site, without any large or expensive equipment. The major disadvantage of the lateral flow strips is that they are not quantitative. The ELISA tests are immunoassays that are quantitative, but they require some expertise and a limited access to laboratory space and equipment. Consequently, each analysis should be tailored to the information needed and the requirements of the methodology.

18.7 ALLERGEN ANALYSIS

18.7.1 Introduction

Food allergens are the food proteins that trigger an allergic response. Symptoms of an allergic response include hives, face and tongue swelling, difficulty in breathing and can include the severe, life-threatening allergic reaction called anaphylactic shock. It is important to note that food allergy, which triggers an immune system reaction, is distinct from other adverse responses to food, such as food intolerance (e.g., lactose intolerance), pharmacologic reactions (due mainly to food additives such as sulfites and benzoate), and toxin-mediated reactions (due to residues such as pesticides and mycotoxins).

A significant percentage of the population has food allergens, and the prevalence is rising. More than 160 foods can cause allergic reactions in people with food allergies, but over 90% of the food allergic reactions in the USA are caused by the eight most common allergenic foods, referred to as "**the big eight**": milk, eggs, fish, crustacean shellfish, tree nuts, peanuts, wheat, and soybeans (58). In the USA, the Food Allergen Labeling and Consumer Protection Act of 2004 targets these eight food allergens (59). It applies to all foods regulated by the FDA (both domestic and imported), requiring that labels list all ingredients by their common names, and identify the source of all ingredients derived from the eight most common food allergens. Food allergies are an issue around the world; numerous other countries have or are considering labeling regulations. Since there is no cure for food allergies, strict avoidance of food allergens is the only effective action. All of these facts point to the importance of both screening and quantitative methods for analysis of allergenic foods.

Methods available so far for the detection of food allergens are mainly based on protein or DNA detection, as discussed in subsequent sections. A commercially-available rapid screening method that is not based on protein or DNA detection is the swab and adenosine tri-phosphate (ATP)-sensitive detection method. This method is based on the detection of ATP present on the surface of multiple allergenic foods, e.g., peanut butter, whole egg, soybeans, and milk. It is used mainly to prevent food allergen cross contact during cleaning of processing equipment. Test kits for food allergen analysis (protein- or DNA-based, and others) are commercially available (Table 18-7). A review of food allergen analytical methods is given in (60).

18-7 table **Example Commercial Test Kits for Allergen Analysis**

Assay Type	*Example Matrices*	*Allergen Proteins*	*Commercial Source*
Antibodies in a sandwich ELISA assay	Cookies, crackers, chocolate bars, cereals, pasta, salad dressing, cake mix, ice cream, various food ingredients, milk (for soy), granola bars, juices, and environmental surfaces and processing equipment	Almond, egg, grain prolamins (wheat gliadin, rye secalin, and barley hordein), hazelnut, milk, peanut, and soy	Alert® and Veratox® series (Neogen Corporation, 620 Lesher Place, Lansing, MI)
Antibodies in a sandwich ELISA assay	Raw materials and processed foods	Peanut, hazelnut, soy, and prolamins	AgraQuant® Allergen Kits (Romer Labs, Inc., 1301 Stylemaster Drive, Union, MO)
Swab and ATP sensitive detection	Environmental surfaces and processing equipment [Note: This test is not specific for the allergens, but quickly detects the presence of food residue (ATP) where allergen-containing food ingredients have been processed]	Food residue in general. The following allergenic foods contain ATP: peanut butter, soybeans, nonfat milk powder, whole egg powder, whole wheat flour, shrimp, sesame, almond	AllerGiene® (Charm Sciences, Inc., 659 Andover Street, Lawrence, MA)
DNA selection for PCR and gel electrophoresis	Most food products	Peanut, soy, and cow's milk	BIOKITS® DNA Selection Module (Tepnel Research Products, 550 West Avenue, Stamford, CT)

18.7.2 Protein Methods

18.7.2.1 General Considerations

Similar to the analysis of many other food constituents of concern present in small amounts, sampling adequacy and the detection limits are of concern with analysis for allergens. Another concern is the adequate extraction of the different allergens. Unlike the trace analytes of concern mentioned thus far, since the analytes are proteins, the extraction solution is normally a buffer at various pHs and salt concentration. Extraction buffers differ in their ability to extract food allergens, with some solutions extracting the same allergens in different concentrations, while others cannot extract all the allergens. For example, phosphate buffer fails to extract the major peanut allergen (Ara h 3); however, the efficiency of extracting this allergen is greatly enhanced upon the addition of salt (Fig. 18-8). Additionally, it is crucial that the extraction solution is compatible with the assay used (e.g., immunoassay) and does not alter the chemical structure of the analyte. The choice of extraction procedure should also take into consideration the food processing conditions employed. Upon processing, the solubility of the proteins can be reduced due to denaturation and aggregation, resulting in reduced protein recovery (61). Therefore, to obtain reliable and accurate results, it is crucial to select the right extraction solution for the target analyte(s).

18.7.2.2 Protein-Based Analytical Techniques

Protein-based methods used for the analysis of food allergens usually involve antibody-based assays (immunoassays) due to their specificity and sensitivity. Immunoassays target the offending allergen by use of monoclonal or polyclonal antibodies or a combination of both. Most of the commercially available test kits, which differ in specificity and the number of proteins they target, use polyclonal antibodies. Immunoassay-based methods used for the analysis of food allergens include ELISA (see Chap. 17, Sect. 17.3), Western blot (see Chap. 17, Sect. 17.3.5), biosensor immunoassays (antibodies immobilized on a biosensor chip), and dot immunoblotting (similar to Western blot; however, protein extract is spotted directly onto the nitrocellulose or PVDF membrane prior to enzyme-labeled protein specific antibody) (62).

Western blot and dot immunoblotting are mostly used for qualitative and screening purposes. The most commonly used immunoassays for the quantitative analysis of food allergens are ELISA methods, using a competitive (see Chap. 17, Sect. 17.3.3), or, more commonly, sandwich format (see Chap. 17, Sect. 17.3.2.1). The competitive ELISA is used for the small protein allergens, with a molecular weight less than 5 KDa. Numerous sandwich, as well as competitive ELISA methods have been developed for several food allergens (60). Reference (60) includes a list of commercially available sandwich and competitive ELISA test kits.

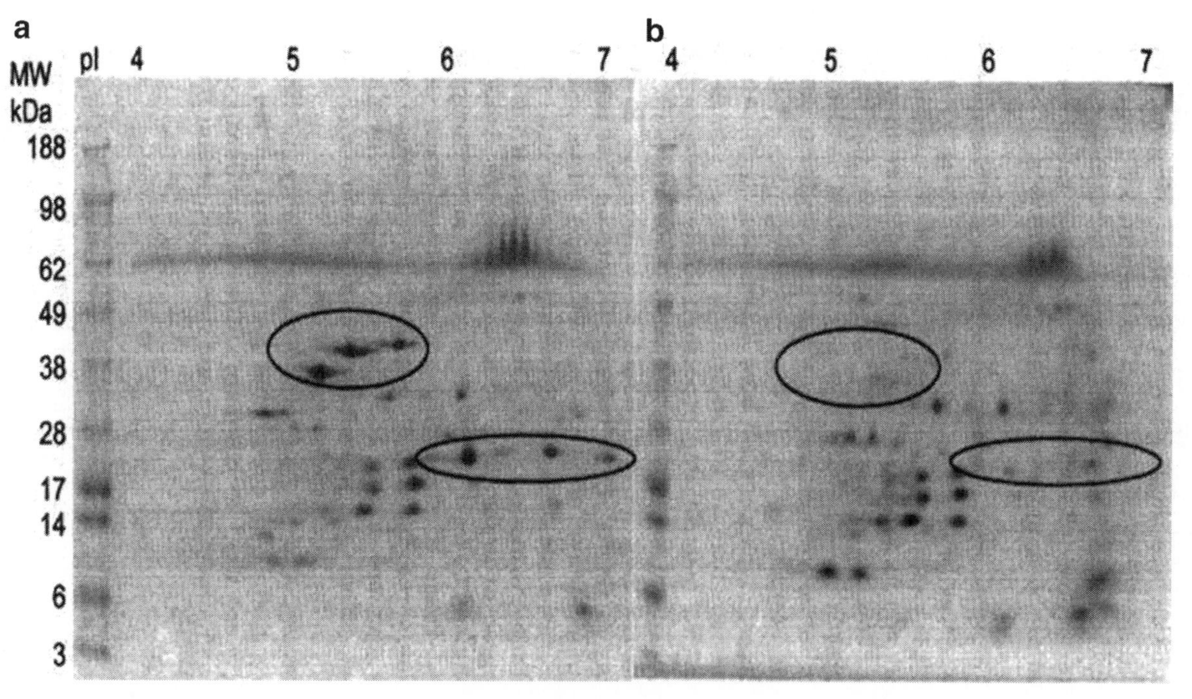

18-8 figure Evaluation, by two-dimensional electrophoresis, of the ability of (**a**) high salt buffer and (**b**) phosphate buffer to extract the peanut allergen Ara h 3 from raw peanuts. [From (61), with permission.]

Lateral flow test strips (dipstick assays), which are based on ELISA principle, are also commercially available for food allergen analysis. Dipstick tests are used for screening purposes, because they are very rapid, inexpensive, and do not require instrumentation.

18.7.3 DNA Methods

There are several advantages and disadvantages for the use of DNA-based methods in the analysis of food allergens. DNA-based methods do not target the allergen in the sample; therefore, the detection of the allergen-encoding DNA does not always correlate with the presence of the allergen, especially when the food has been fortified with purified protein. Upon processing, for example, production of soy protein isolate (which are purified proteins used as ingredients for fortification and functionality enhancement), protein and DNA could be separated resulting in false conclusions regarding the presence of the allergen in the sample. Regardless of these disadvantages, DNA-based methods are very specific and sensitive techniques with the advantage that targeted DNA is less affected by several processing and extraction conditions as compared to proteins.

DNA-based methods involve the extraction of the DNA (see Sect. 18.6.3.1) followed by amplification by PCR using a thermostable polymerase (see Sect. 18.6.3.2). The amplified sample is then visualized by fluorescence staining or by Southern blotting following agarose gel electrophoresis. This procedure normally provides qualitative data, or semiquantitative data if internal standards were used. Quantification can be achieved if real-time PCR [see Sect. 18.6.3.1 and reference (60)] or PCR-ELISA was used. PCR-ELISA method involves linking the amplified DNA fragment of an allergenic food to a specific protein-labeled DNA probe, which then is coupled with a specific enzyme-labeled antibody. Quantification of the DNA is based on the enzyme-substrate color producing reaction. Reference (60) includes a list of commercially available real-time PCR and PCR-ELISA test kits for the analysis of allergens; readers are referred to reference (60).

18.8 ANALYSIS OF OTHER CHEMICAL CONTAMINANTS AND UNDESIRABLE CONSTITUENTS

18.8.1 Introduction

Pesticide residues, mycotoxins, antibiotic residues, and allergens in foods have been of concern for many years. However, in any given time period, there are numerous additional chemical hazards that must be addressed. Many of these substances fall into the following categories: (1) banned (e.g., coumarin), or allowed in some countries but not others, (2) legally limited (e.g., sulfites, benzene), (3) intentional contaminants (e.g., melamine), (4) approved for use, but of concern (e.g., monosodium glutamate), or (5) natural constituents of concern (e.g., acrylamide, furan). This section includes the analysis of some such current and emerging hazards, covered in much less detail than most earlier sections of the chapter. In some cases the methods are well established, while in other cases both screening and quantitative methods are just now being developed. For some chemical substances, currently with only expensive and time-consuming methods of analysis, efforts will no doubt continue to develop more rapid and inexpensive methods. For most substances covered, a major question for the method of analysis is the limit of detection (LOD).

18.8.2 Sulfites

Although sulfites are classified as allergens, they are covered in this separate section because the nature of the chemical compound, symptoms, and methods of analysis are quite different than for other allergens. Sulfites and sulfiting agents are a group of chemical compounds that include sulfur dioxide (SO_2), sulfurous acid (HSO_3), and the following inorganic sulfite salts that can liberate SO_2: sodium (Na) and potassium (K) sulfite, Na and K bisulfite, and Na and K metabisulfite (63). In some foods they occur naturally, but in other foods they are added for a variety of reasons, including preventing microbial growth and browning. Sulfites naturally occur to some extent in all wines, but are commonly added to stop fermentation at the appropriate time, and to prevent spoilage and oxidation. Dried fruits and vegetable products are sometimes treated with sulfites to reduce browning. Shrimp, lobster, and related crustaceans can be treated with sulfite to prevent "blackspot." Some consumers are highly intolerant to sulfite residues in food, most commonly resulting in asthma attacks (64). Therefore, the FDA forbids sulfites from being added to foods intended for raw consumption (e.g., salad bar foods) and requires the phrase "contains sulfites" on the label of foods that contain greater than 10 ppm sulfites, whether naturally occurring or added during manufacturing [21 CFR 101.100 (a)(4)] (65). Many countries besides the USA have set strict limits on the residual levels of sulfites in various foods.

Reactions of sulfites with other food components make analysis challenging, often resulting in decreased levels during storage. Most methods of analysis detect free forms of sulfite plus some bound forms. However, none of the available methods, alone, measures all forms of sulfite in foods, which includes free inorganic sulfite plus the many sulfite bound forms. It is not known which forms of sulfite cause the adverse responses in sulfite-sensitive consumers, so the focus has been on measuring as much as possible of the residual sulfite, both free and bound forms (63).

One of the long-time, quantitative methods for sulfite analysis of foods is the **Monier-Williams procedure** (AOAC Method 990.28) (17), which measures "total" SO_2 (actually, free sulfite plus reproducible portion of bound sulfites, such as carbonyl addition products). The FDA refers to this method in regulations for labeling sulfite-containing foods. In this method, the test sample is heated with HCl, converting sulfite to SO_2. Nitrogen gas bubbled through the sample sweeps SO_2 through a condenser and a hydrogen peroxide solution, oxidizing SO_2 to H_2SO_4. The sulfite content of the sample is directly related to the amount of H_2SO_4 generated, measured by either a gravimetric or turbidimetric procedure.

Other analytical methods for sulfites include the following:

1. The "**Ripper**" **method**, long used by the wine industry as a rapid screening method, compared to the more time-consuming Monier-Williams method; sulfite is titrated with an iodide-iodate solution, using a starch endpoint indicator; measures "free" SO_2 (APHA Standard Method 4500-$SO_3{}^{-2}$ B) (66).
2. Enzymatic method; sulfite is oxidized to sulfate, generating hydrogen peroxide that is further reacted with NADH-peroxidase to produce NAD, which is measured by absorption at 340 nm (see Chap. 16, Sect. 16.3.1.3.1).
3. Ion Chromatography, using amperometric detector (AOAC Method 990.31) (17).
4. HPLC, with ultraviolet (67) or fluorometric (68) detection.

18.8.3 Melamine

Melamine, a trimer of cyanamide with 66% nitrogen, has been added illegally as an economic adulterant to foods and ingredients to increase the apparent protein content. The standard Kjeldahl and Dumas methods, which estimate protein by measuring nitrogen content, are misled by such a nitrogen-rich compound (see also Chap. 9). Melamine has several industrial uses and is also a metabolite of the pesticide cyromazine. In 2008, melamine was linked to the death of pets that consumed pet food containing contaminated wheat gluten. This was followed by the illness and death of infants in China who consumed infant formula with contaminated milk. The deaths were attributed to kidney failure, caused by melamine and cyanuric acid (a melamine-like compound) absorption in the bloodstream, then their interaction and crystallization to block and damage renal cells. Limits for acceptable melamine consumption have been set by various countries and organizations; the US FDA set a threshold of 1 ppm for melamine in infant formula and 2.5 ppm in other products (69–71).

The FDA has two major methods for quantifying melamine and cyanuric acid, with both methods using liquid chromatography with triple quadrupole tandem mass spectrometry (LC-MS/MS) (see also Chap. 26, Sect. 26.6). Both methods use a special type of normal-phase chromatography, referred to as hydrophilic interaction liquid chromatography (HILIC), primarily used to separate very polar compounds. Both methods involve extraction and multiple cleanup steps, separation on a zwitterionic HILIC LC column, and electrospray ionization in both the negative ion and positive ion modes. Method LIB 4421 (72), developed for testing infant formula (and milk-containing samples), has a limit of quantitation of 0.25 μg/g for both melamine and cyanuric acid. Method LIB 4421 (73) is an interim method for melamine and cyanuric acid in foods. It has a limit of quantitation for melamine of 25 μg/kg for tissue and liquid formula and 200 μg/kg for dry infant formula powder. Those limits for cyanuric acid are 50 and 200 μg/kg, respectively. The FDA also developed a GC-MS screening method, suitable for a variety of food matrixes, for the presence of melamine and related compounds, at a minimum reporting level of 10 μg/g (74).

Because the methods described above, especially the LC-MS/MS methods, are very time-consuming and expensive, numerous faster and less expensive methods have been developed to test for melamine, including those briefly described below. Comparisons of the various methods have been reported (75).

1. Ultrasound-assisted extractive electrospray ionization mass spectrometry (EESI-MS): Melamine-containing liquids are nebulized into a fine spray using ultrasounds; the spray is ionized by EESI and analyzed using mass spectrometry (76).
2. Low-temperature plasma probe (to ionize the sample) with mass spectrometry (LTP-MS/MS) (77).
3. Near-infrared (NIR) and mid-infrared (MIR): The MIR system was Fourier transform using either diffuse reflectance spectroscopy (FTIR-DRIFTS) or multibounce attenuated total reflectance (FTIR-ATR) to collect spectra (78).
4. Immunoassays, using commercially available plate ELISA or lateral flow strips.

18.8.4 Packaging Material Residues

The chemicals used to make food packaging materials, or their break down products, can sometimes migrate into the food, causing food safety concerns. The two chemical substances described below are currently of concern.

18.8.4.1 Bisphenol A

Bisphenol A (BPA), an organic compound with two phenol groups, is regulated by the FDA for use in food contact applications. BPA is primarily used to make polycarbonate plastic and can leach from the plastic under certain conditions. There is concern about negative health effects since low doses of BPA can mimic the body's own hormones (79). A first study of BPA effects on humans (80) showed that high levels of BPA in the urine were significantly associated with heart disease, diabetes, and abnormally high levels of certain liver enzymes. Canada has banned BPA in many products for babies and young children, the European Union has been reducing the limit for all Bisphenol A diglyceridyl ether (BADGE) derivatives, and the FDA is reevaluating BPA safety levels after presenting a draft assessment on BPA in 2008 (79).

Numerous companies/analytical laboratories report the analysis of BPA in foods by GC-MS, GC-MS/MS, LC-MS, and LC-MS/MS (reported LOD of <1 mg/kg for LC-MS/MS). Other analytical methods include use of commercially available immunoaffinity columns and ELISA.

18.8.4.2 4-Methylbenzophenone

Benzophenone and 4-methylbenzophenone (a metabolite of benzophenone) are chemical components of ink used for food packaging. Benzophenone is known to cause liver and kidney hypertrophy in animals, but data are not available for 4-methylbenzophenone. Because of their volatility, and if there is no functional barrier, these chemicals can migrate from the printed surface of cardboard boxes into food contained in the box. With the recent discovery of 4-methylbenzophenone in cereal, the European Food Safety Authority (81) reevaluated the toxicological data on benzophenone. EFSA concluded that there is no health risk from the short-term consumption of breakfast cereals contaminated with 4-methylbenzophenone at levels reported, but will be gathering further data since a health risk for children could not be excluded (81). Because of these concerns, analytical testing laboratories offer testing services for 4-methylbenzophenone using LC-MS/MS methods. Reported LOD for known food matrices and known packaging matrices are 10 ppb and 1 ppm, respectively (82).

18.8.5 Furans

Furan, a known carcinogen, is a colorless, volatile liquid used in some chemical manufacturing industries. It has been found recently in a wide variety of foods, although it is likely that it has been present for many years and only just recently detected due to the availability of more sensitive analytical techniques. Furan seemingly forms during traditional heat treatment techniques (e.g., retorting foods in cans and jars) (83). In 2005 the FDA set out a plan to measure the furan levels in various foods to estimate dietary exposure and to identify mechanisms responsible for furan formation in foods (84). The FDA developed a headspace (HS) GC-MS method for furan analysis, using standard addition for quantification (85). This method has been the basis for numerous comparison studies using slight modifications for specific food types (86) and faster throughput (e.g., using headspace SPME), achieving LOD of 0.02–0.12 ng/g (87).

18.8.6 Acrylamide

Like furan, acrylamide was only recently identified in foods, detectable with improved analytical methods, and formed with certain types of heat treatment. The uses of acrylamide include wastewater treatment, gel electrophoresis (see Chap. 15, Sect. 15.2.5), and papermaking. Acrylamide is a known neurotoxin and carcinogen, and it poses numerous other health risks. Acrylamide was discovered in food by Swedish scientists in 2002 (88), reporting its presence in a variety of fried and oven-baked foods. Formation of acrylamide occurs in carbohydrate-rich foods cooked at high temperatures, seemingly when the amino acid asparagine and the carbonyl group of reducing sugars react. The FDA, Food and Agriculture Organization/World Health Organization (FAO/WHO), and many other agencies around the world studied the acrylamide issue and developed recommendations. The FAO/WHO set a NOAEL for acrylamide neuropathy at 0.5 mg/kg body weight/day. The FDA's action plan for acrylamide included updating and validating its LC-MS/MS method (89, 90). This LC-MS/MS method, along with a GC-MS method, is acknowledged as the major useful and authoritative method for acrylamide determination, and they have been compared (91).

18.8.7 Benzene

Benzene, a known carcinogen, is used to manufacture a wide range of industrial products and is released into the air from emissions of automobiles and from burning oil and coal. The EPA has set a maximum contaminant level (MCL) for benzene in drinking water of 5 ppb, the same level adopted by the FDA for an allowable level in bottled water. In 1990 the soft drink industry informed the FDA that benzene could form at low levels in some beverages that contained both ascorbic acid and benzoate salts. Further research showed that benzene formation in the presence of these ingredients was stimulated by elevated temperature and light. A reformulation of beverages was undertaken by many manufacturers to reduce or eliminate benzene formation. The FDA has continued to monitor the benzene level in beverages containing benzoate salts and ascorbic acid (92). The FDA method for determination of benzene in soft drinks and other beverages is headspace sampling followed by GC-MS (93). One version of the method uses cryogenic focusing of the sampled headspace and results in a LOD of 0.2 ppb; a second version incorporates GC separations on a special capillary column, giving a LOD of 0.02 ppb.

18.8.8 3-Monochloropropane 1,2-Diol (3-MCPD)

The most commonly found member of the chemical contaminant group known as chloropropanols is 3-MCPD, a known carcinogen and a suspected genotoxin. This chemical can be created when proteins (soy) are hydrolyzed by heat and food-grade acids to create **acid-hydrolyzed vegetable protein** (acid-HVP), specifically when hydrochloric acid reacts with residual fat in the protein source. The 3-MCPD is a well-known contaminant of acid-HVP, which is a common savory ingredient in soups, savory snacks, and gravy mixes (94). Because of the way acid-HVP and soy sauce are made and used, 3-MCPD is found in many types of Asian-style sauces. The levels of 3-MCPD formed can be reduced, but not completely eliminated, by modifying the manufacturing process for acid-HVP. Based on FDA regulations in the USA, acid-HVP that contains 3-MCPD at levels greater than 1 ppm is not generally recognized as safe (GRAS), so it is an unapproved food additive (94). Numerous other countries are concerned about the levels of 3-MCPD in acid-HVP and Asian-style sauces and have set limits ranging from 0.01 to 1 ppm (95, 96). The current method of choice is GC-MS, after a lengthy sample preparation that involves multiple extractions and derivatizations (97).

18.8.9 Perchlorate

Perchlorate is an industrial chemical contaminant used as a component of rocket fuel, but it is also found to occur naturally and may be generated under certain climatic conditions. At high doses, perchlorate can interfere with iodide uptake into the thyroid gland and lead to hypothyroidism. Perchlorate has been found in a variety of foods, including bottle water, milk, and lettuce (98,99). The FDA developed a rapid, accurate, sensitive, and specific assay for perchlorate, using ion chromatography-tanden mass spectrometry (IC-MS/MS) (100). The method involves extraction, cleanup by solid-phase extraction, and filtration prior to the IC-MS/MS determination. The method has been updated to allow determination of perchlorate in a wider variety of foods, with an LOD of 0.5–3 ppb, depending on the food product.

18.9 SUMMARY

Consumer concerns and governement regulations focused on the safety of foods dictate the need for analysis of various food contaminants, residues, and chemical constituents of concern. These compounds include pesticide residues, mycotoxins, antibiotic residues, GMOs, allergens, food adulterants, packaging material hazardous chemicals, environmental contaminants, and certain other chemicals. Both rapid screening methods and more time-consuming quantitative methods are required to meet the needs of industry and government, in an effort to ensure a safe and reliable food supply. A positive result from a screening method usually leads to further testing to confirm and quantify the presence of the compound of concern. Sampling and sample preparation can be a significant challenge due to the low levels of the chemicals and the complex food matrices. Sample preparation often includes homogenization, extraction, and cleanup, and sometimes requires derivatization. Screening methods increasingly utilize immune-based techniques, such as ELISA, LFS, immunosensors, and immunoaffinity chromatography columns. Some immunoassays can be considered quantitative, rather than just screening methods. Other screening methods commonly used include enzyme inhibition assays, thin layer chromatography, and inhibition of microbial growth. While GC is the most common chromatographic technique for quantitative analysis of pesticides, for many other compounds of concern covered in this chapter, the predominant chromatographic method is HPLC. Both GC and HPLC analysis are now commonly coupled with mass spectrometry detection, often using MS tandem systems. The testing for GMOs and allergens typically involves either protein-based methods (e.g., immunoassays) or DNA methods using PCR. Work continues to improve and develop various methods of analysis for chemical residues and compounds of concern, focusing largely on speed, cost, and reliability of screening methods, and detection limits of quantitative methods.

18.10 STUDY QUESTIONS

1. Explain the importance of each of the following steps in sample preparation for the analysis of contaminants and residues of concern:
   (a) Grinding/homogenization
   (b) Extraction
   (c) Cleanup/purification
   (d) Derivatization
2. In the analysis of contaminants and residues of concern, compare and contrast:
   (a) GC vs. LC analysis
   (b) MS vs. MS/MS analysis
   (c) LC with fluorescence or UV detection vs. LC-MS
   (d) TLC vs. automated chromatography (LC or GC)
   (e) Microplate ELISA vs. LFS
3. The "tolerance level" for residues of the pesticide chloropyrifos on corn grain is 0.05 ppm.
   (a) What is meant by "tolerance level"?
   (b) What federal agency sets that tolerance level?
   (c) What federal agency enforces that tolerance level?
   (d) What is "ppm" equivalent to in terms of (a) weight per volume, and (b) weight per weight units commonly used to express concentration?
   wt/vol:
   wt/wt:
   (e) In Volumes I and II of the *Pesticide Analytical Manual* you find described "multiresidue" methods and "single-residue" methods. You also have found numerous screening methods. Which one of these three types of methods (i.e., multiresidue, single-residue, screening) would you use to ensure your compliance with the tolerance level for this pesticide? Briefly explain the nature of this type of method, and why you chose this method over the other two types of methods.
4. Mycotoxins are of potential concern in corn, especially in certain growing and storage conditions.
   (a) Sampling is a major contributor to error in the analysis for mycotoxins. Why is sampling for mycotoxin analysis such a challenge?
   (b) Identify the most commonly used quantitative chromatographic method for mycotoxins. Justify the preference of this method.
5. Regarding antibiotic residues, briefly explain the following:
   (a) How might these get into foods?
   (b) What types of foods most likely contain them?
   (c) Why are these antibiotic residues a problem?
   (d) Why have techniques such as immunoassays largely replaced microbial growth inhibition assays for screening antibiotic residues?
   (e) What method is most commonly used for accurate quantitative determination and confirmation?
6. You want to identify GMOs in a soybean field where you do not have access to a sophisticated analytical laboratory.
   (a) What food constituent would you be analyzing, and which analytical technique would you use? Explain the principle of this technique.
   (b) Is it possible to use Western blot? Answer by referring to the principle of the Western blot technique.
7. You are analyzing unwanted chemical components in the foods/raw materials listed below. Identify one possible unwanted chemical component that would likely be of concern with this specific food, and state an appropriate method for quantitative analysis. Also give one appropriate screening method for each chemical component identified. (Note: For each food, give a different likely unwanted chemical component and a different quantitative method of analysis.)

	Likely unwanted chemical component	*Quantitative analysis method*	*Screening method*
(a) Oats			
(b) Peanuts			
(c) Milk			
(d) Wine			

8. Regarding GMOs:
   (a) Describe the source and function of the common trans genes.
   (b) Describe a method for transformation of the plants.
9. Describe the three basic methods for GMO detection and quantitation and describe the strengths and weaknesses of each.
10. For the analysis of food allergens, protein-based or DNA-based methods are commonly used.
    (a) Give an example of when you would choose a protein-based method over a DNA-based method. Justify your choice and provide the principle of the method of choice.
    (b) Give an example of when you would choose a DNA-based method over a protein-based method. Justify your choice and provide the principle of the method of choice.
11. Regarding the compounds described in Sect. 18.8:
    (a) Identify two compounds recently identified in foods, but likely present for many years. Why are they just now being raised as a concern?
    (b) Identify two compounds more likely analyzed by GC-MS than by LC-MS (or LC-MS/MS). Explain the preference of GC-MS over LC-MS in this case.
    (c) Regarding sulfites, explain how this "allergen" differs from the food allergens described in Sect. 18.7 in the response of sensitive humans, how they differ in the nature of quantitative methods, and why quantitative determination is relatively difficult for sulfites than for food allergens.

18.11 ACKNOWLEDGEMENTS

The authors of this chapter wish to acknowledge William D. Marshall, who wrote for earlier editions of the book the chapter "Analysis of Pesticide, Mycotoxin, and Drug Residues in Foods." Some of the ideas for the content and organization of the current chapter came from Dr. Marshall's chapter. Also, some ideas for the content and organization of the text on GMO analysis came from the chapter "Agricultural Biotechnology (GMO) Methods of Analysis," by Anne Bridges, Kimberly Magin, and James Stave, in the 3rd edition of this book. Their contribution is recognized and appreciated.

18.12 REFERENCES

1. Arvanitoyannis IS (2008) International regulations on food contaminants and residues, ch. 2. In: Pico Y (ed) Food contaminants and residue analysis. Comprehensive analytical chemistry, D. Barcelo (ed) vol 51. Elsevier, Oxford, UK
2. Nielen MWF, Marvin HJP (2008) Challenges in chemical food contaminants and residue analysis, ch. 1. In: Pico Y (ed) Food contaminants and residue analysis. Comprehensive analytical chemistry, D. Barcelo (ed) vol 51. Elsevier, Oxford, UK
3. Turnipseed SB, Andersen WC (2008) Veterinary drug residues, ch. 10. In: Pico Y (ed) Food contaminants and residue analysis. Comprehensive analytical chemistry, vol 51. Elsevier, Oxford, UK
4. Sannino A (2008) Pesticide residues, ch. 9. In: Pico Y (ed) Food contaminants and residue analysis. Comprehensive analytical chemistry, D. Barcelo (ed) vol 51. Elsevier, Oxford, UK
5. Hansen TJ (1990) Affinity column cleanup and direct fluorescence measurement of aflatoxin M_1 in raw milk. J Food Prot 53:75–77
6. Sandra P, David F, Vanhoenacker G (2008) Advanced sample preparation techniques for the analysis of food contaminants and residues, ch. 5. In: Pico Y (ed) Food contaminants and residue analysis. Comprehensive analytical chemistry, D. Barcelo (ed) vol 51. Elsevier, Oxford, UK
7. Handley A (ed) (1999) Extraction methods in organic analysis. Scheffield Academic, Scheffield, England
8. Mitra S (ed) (2003) Sample preparation techniques in analytical chemistry. Wiley, Hoboken, NJ
9. Ramos L, Smith RM (eds) (2007) Advances in sample preparation, part I. J Chromatogr A 1152:1
10. Ramos L, Smith RM (eds) (2007) Advances in sample preparation, part II. J Chromatogr A 1153:1
11. Anastassiades M, Lehotay SJ, Stajnbaher D, Schenck FJ (2003) Fast and easy multiresidue method employing acetonitrile extraction/partitioning and "dispersive solid-phase extraction" for the determination of pesticide residues in produce. J AOAC Int 86:412–431
12. Richer BE, Jones BA, Ezzel JL, Porter NL, Avdalovic N, Pohl C (1996) Accelerated solvent extraction: a technique for sample preparation. Anal Chem 68:1033–1039
13. Cairns T, Sherma J (1992) Emerging strategies for pesticide analysis. In: Modern methods for pesticide analysis, 9th edn. CRC, Boca Raton, FL
14. FDA (2005) FDA glossary of pesticide chemicals. http://vm.cfsan.fda.gov/~acrobat/pestglos.pdf
15. FDA (2002) Pesticide analytical manual volume II (updated January, 2002). http://www.fda.gov/Food/ScienceResearch/LaboratoryMethods/PesticideAnalysisManualPAM/ucm113710.htm
16. FDA (1999) Pesticide analytical manual volume I (PAM), 3rd edn (1994, updated October, 1999). http://www.fda.gov/Food/ScienceResearch/LaboratoryMethods/PesticideAnalysisManualPAM/ucm111455.htm
17. AOAC International (2007) Official methods of analysis, 18th edn., 2005; current through revision 2, 2007 (on-line). AOAC International, Gaithersburg, MD
18. Tadeo JL (2008) Analysis of pesticides in food and environmental samples. CRC/Taylor and Francis Group, Boca Ranton, FL
19. Lee NA, Kennedy IR (2001) Environmental monitoring of pesticides by immunoanalytical techniques: Validation, current status, and future perspectives. J AOAC Int 84:1393–1406
20. Gierish T (1993) A new monoclonal antibody for the sensitive detection of atrazine with immunoassay in microtiter plate and dipstick format. J Agric Food Chem 41:1006–1011
21. Manclus JJ, Primo J, Montoya A (1996) Development of enzyme-linked immunosorbent assays for the insecticide chlorpyrifos. I. Monoclonal antibody production and immunoassay design. J Agric Food Chem 44:4052–4062
22. Wang S, Allan RD, Skerritt JH, Kennedy IR (1999) Development of a compound-specific ELISA for flufenoxuron and an improved class-specific assay for benzoylphenylurea insect growth regulators. J Agric Food Chem 47:3416–3424
23. Sannino A (1998) Determination of phenylurea herbicide residues in vegetables by liquid chromatography after gel permeation chromatography and florisil cartridge cleanup. J AOAC Int 81:1048–1053
24. Soderberg D (2005) Committee on residues and related topics; pesticides and other chemical contaminants. General referee reports. J AOAC Int 88:331–345
25. Pico Y, Blasco C, Font G (2004) Environemental and food applications of LC tandem mass spectrometry in pesticide-residues analysis: an overview. Mass Spectrom Rev 23:45–85
26. Sannino A, Ambriani PM, Bandini M, Bolzoni L (1995) Multiresidue method for determination of organophosphorus insecticide residues in fatty processed foods by gel permeation chromatography. J AOAC Int 78:1502–1512
27. Patel K, Fussell RJ, Goodall DM, Keelye BJ (2004) Evaluation of large volume-difficult matrix introduction-gas chromatography-time of flight-mass spectrometry (LV-DMI-GC-TOF-MS) for the determination of pesticides in fruit-based baby foods. Food Addit Contam 21:658–669
28. Gilbert-López B, García-Reyes JF, Ortega-Barrales P, Molina-Díaz A, Fernández-Alba AR (2007) Analyses of pesticide residues in fruit-based baby food by liquid chromatography/electrospray ionization time-of-flight mass spectrometry. Rapid Commun Mass Spectrom 21(13):2059–2071

29. Brera C, De Santis B, Debegnach F, Miraglia M (2008) Mycotoxins, ch. 12. In: Pico Y (ed) Food contaminants and residue analysis. Comprehensive analytical chemistry, D. Barcelo (ed) vol 51. Elsevier, Oxford, UK
30. Commission Regulation (EC) N 401/2006 of 23 February 2006. Laying down the methods of sampling and analysis for the official control of the levels of mycotoxins in food stuff
31. Siantar DP, Trucksess MW, Scott PM, Herman EM (eds) (2008) Food contaminants: mycotoxins and food allergens. ACS symposium series 1001. American Chemical Soceity, Washington, DC
32. De Saeger S, Sibanda L, Desmet A, Van Peteghem C (2002) A collaborative study to validate novel field immunoassay kits for rapid mycotoxin detection. Int J Food Microbiol 75:135–142
33. Garner RG, Whattam MM, Taylor PJ, Stow MW (1993) Analysis of United Kingdom purchased spices for aflatoxins using an immunoaffinity column clean-up procedure followed by high performance liquid chromatography. J Chromatogr A 648:485–490
34. Development and validation of a liquid chromatography/tandem mass spectrometric method for the determination of 39 mycotoxins in wheat and maize. Rapid Commun Mass Spectrom 20:2649–2659
35. Tsao R, Zhou T (2000) Micellar electrokinetic capillary electrophoresis for rapid analysis of patulin in apple cider. J Agric Food Chem 48:5231–5235
36. Peña R, Alcaraz MC, Arce L, Ríos A, Valcárcel M (2002) Screening of aflatoxins in feed samples using a flow system coupled to capillary electrophoresis. J Chromatogr A 967:303–314
37. Maragos CM, Appell M (2007) Capillary electrophoresis of the mycotoxin zearalenone using cyclodextrin-enhanced fluorescence. J Chromatogr A 1143:252–257
38. Greene RV, Gordon SH, Jackson MA, Bennett GA (1992) Detection of fungal contamination in corn: potential of FTIR-PAS and DRS. J Agric Food Chem 40: 1144–1149
39. Berardo N, Pisacane V, Battilani P, Scandolara A, Pietri A, Marocco A (2005) Rapid detection of kernel rots and mycotoxin in maize by near-infrared reflectance spectroscopy. J Agric Food Chem 53:8128–8134
40. Pettersson H, Åberg L (2003) Near infrared spectroscopy for determination of mycotoxin in cereals. Food Control 14:229–232
41. De la Calle MB, Anklam E (2005) Semicarbazide: occurrence in food products and state-of-the-art in analytical methods used for its determination. Anal Bioanal Chem 382:968–977
42. United States Department of Agriculture (2004) Red book. USDA Food Safety Inspection Service. http://www.fsis.usda.gov/Science/2004_Red_Book/index.asp
43. FDA (2004) National milk drug residue data base. Food and Drug Adminstration, Center for Food Safety and Nutrition. http://www.cfsa.fda.gov/~ear/milkrp03.html
44. Fedeniuk RW, Shand P (1998) Theory and methodology of antibiotic extraction from biomatrices. J Chromatogr A 812:3–15
45. Antibiotic and Veterinary Drug Residue Test Kits (2007) http://www.aoac.org/testkits/kits-antibiotics.HTM
46. Bulthaus M (2004) Detection of antibiotic/drug residues in milk and dairy products, ch. 12. In: Wehr HM, Frank JF (eds) Standard methods for the examination of dairy products, 17th edn. American Public Health Association, Washington, DC
47. FDA (2009) Fluoroquinolones (last updated 5/14/2009). http://www.fda.gov/Food/ScienceResearch/LaboratoryMethods/DrugChemicalResiduesMethodology/ucm071463.htm
48. FDA (2006) Preparation and LC/MS/MS analysis of honey for fluoroquinolone residues. 29 Sept 2006 (last updated 6/18/2009). http://www.fda.gov/Food/ScienceResearch/LaboratoryMethods/DrugChemicalResiduesMethodology/ucm071495.htm
49. Anonymous (2009) Code of federal regulations. Chloramphicol infection. 21 CFR 522.390 (3). US Government Printing Office, Washington, DC
50. FDA (2009) Analytical methods for residues of chloramphenicol and related compounds in foods. http://www.fda.gov/Food/ScienceResearch/LaboratoryMethods/DrugChemicalResiduesMethodology/ucm113126.htm http://www.fda.gov/Food/ScienceResearch/LaboratoryMethods/DrugChemicalResiduesMethodology/ucm071463.htm
51. Holstege DM, Puschner B, Whitehead G, Galey FD (2002) Screening and mass spectral confirmation of B-lactam antibiotic residues in milk using LC-MS/MS. J Agric Food Chem 50(2):406–411
52. Cavalier C, Curini R, Di Corcia A, Nazzari M, Samperi R (2003) A simple and sensitive liquid chromatography-mass spectrometry confirmatory method for analyzing sulfonamide antibacterials in milk and egg. J Agric Food Chem 51:558–566
53. Clark SB, Turnipseed SB, Madson MR (2005) Confirmation of sulfamethazine, sulfathiazole, and sulfadimethoxine residues in condensed milk and soft-cheese products by liquid chromatography/tandem mass spectrometry. J AOAC Int 88:736–743
54. Wang J, Leung D (2007) Analyses of macrolide antibiotic residues I eggs, raw milk, and honey using both ultra-performance liquid chromatography/quadrupole time-of-flight mass spectrometry and high-performance liquid chromatography/tandem mass spectrometry. Rapid Commun Mass Spectrom 21(19):3213–3222
55. Ahmed FE (2004) Testing of genetically modified organisms in foods. CRC, Boca Raton, FL
56. Heller KJ (2003) Genetically engineered food: methods and detection. Wiley-VCH, Weinheim, Germany
57. Jackson JF, Linskens HF (2009) Testing for genetic manipulation in plants (molecular methods of plant analysis). Springer, New York
58. FDA (2007) Food allergies: what you need to know. February 2007 (last updated 6/30/2009). http://www.fda.gov/Food/ResourcesForYou/Consumers/ucm079311.htm
59. FDA (2004) Food allergen labeling and consumer protection act of 2004 (Public Law 108–282, Title II). 2 Aug 2004. http://www.fda.gov/Food/LabelingNutrition/FoodAllergensLabeling/

GuidanceComplianceRegulatoryInformation/ucm106187.htm
60. Poms RE, Klein CL, Anklam E (2004) Methods for allergen analysis in food: a review. Food Addit Contam 21(1):1–31
61. Westphal CD (2008) Improvement of immunoassays for the detection of food allergens, ch. 29. In: Siantar DP, Trucksess MW, Scott PM, Herman EM (eds) Food contaminants: mycotoxins and food allergens. ACS symposium series 1001. American Chemical Society, Washington, DC
62. Yman IM, Eriksson A, Johansson MA, Hellenas K-E (2006) Food allergen detection with biosensor immunoassays. J AOAC Int 89(3):856–861
63. Taylor SL, Bush RK, Nordlee JA (2003) Sulfites, ch. 24. In: Metcalfe DD, Simon RA (eds) Food allergy: adverse reactions to foods and food additives, 3rd edn. Blackwell, Malden, MA, pp 324–341
64. Montalbano MM, Bush RK (2003) Asthna and food additives, ch. 22. In: Metcalfe DD, Simon RA (eds) Food allergy: adverse reactions to foods and food additives, 3rd edn. Blackwell, Malden, MA, pp 299–309
65. Anonymous (2009) Code of federal regulations. Food; exemptions from labeling. 21 CFR 101.100 (a)(4). US Government Printing Office, Washington, DC
66. Eaton AD, Clesceri LS, Rice EW, Greenberg AE (eds) (2005) Standard methods for the examination of water and wastewater, 21st edn. Method 4500-SO_3^{-2} B. American Public Health Association, Washington, DC
67. McFeeters RF, Barish AO (2003) Sulfite analysis of fruits and vegetables by high-performance liquid chromatography (HPLC) with ultraviolet spectrophotometric detection. J Agric Food Chem 51:1513–1517
68. Chung SWC, Chan BTP, Chan ACM (2008) Determination of free and reversibly-bound sulfite in selected foods by high-performance liquid chromatography with fluorometric detection. J AOAC Int 91(1):98–102
69. FDA (2008) Dear colleague. Letter to the United States Food Manufacturing Industry, regarding melamine. 10 Oct 2008 (last updated 6/18/2009). http://www.fda.gov/Food/FoodSafety/FoodContaminantsAdulteration/ChemicalContaminants/Melamine/ucm164514.htm
70. FDA (2008) Update: interim safety and risk assessment of melamine and its analogues in food for humans. 28 Nov 2008 (last updated 6/18/2009). http://www.fda.gov/Food/FoodSafety/FoodContaminantsAdulteration/ChemicalContaminants/Melamine/ucm164520.htm
71. FDA (2007) Interim melamine and analogues safety/rick assessment. 25 May 2007. http://www.fda.gov/Food/FoodSafety/FoodContaminantsAdulteration/ChemicalContaminants/Melamine/ucm164658.htm
72. Turnipsee S, Casey C, Nochetto C, Heller DN (2008) Determination of melamine and cyanuric acid residues in infant formula using LC-MS/MS. Laboratory information bulletin no. 4421, vol 24, October 2008. Food and Drug Administration, Washington, DC. http://www.fda.gov/Food/ScienceResearch/LaboratoryMethods/DrugChemicalResiduesMethodology/ucm071637.htm
73. Smoker M, Krynitsky AJ (2008) Interim method for determination of melamine and cyanuric acid residues in foods using LC-MS/MS: version 1.0. Laboratory information bulletin no. 4422, October 2008. Food and Drug Administration, Washington, DC. http://www.fda.gov/Food/ScienceResearch/LaboratoryMethods/DrugChemicalResiduesMethodology/ucm071673.htm
74. Litzau JJ, Mercer GE, Mulligan KJ (2008) GC-MS screen for the presence of melamine, ammeline, ammelide, and cyanuric acid. Laboratory information bulletin no. 4423, vol 24, October 2008. Food and Drug Administration, Washington, DC. http://www.fda.gov/Food/ScienceResearch/LaboratoryMethods/DrugChemicalResiduesMethodology/ucm071759.htm
75. Kim B, Perkins LB, Bushway RJ, Besbit S, Fan T, Sheridan R, Greene V (2008) Determination of melamine in pet food by enzyme immunoassay, high-performance liquid chromatography with diode array detection, and ultra-performance liquid chromatography with tandem mass spectrometry. J AOAC Int 91(2):408–413
76. Zhu L, Gamex G, Chen H, Chingin K, Zenobi R (2008) Rapid detection of melamine in untreated milk and wheat gluten by ultrasound-assisted extractive electrospray ionization mass spectrometry (EESI-MS). Chem Commun 559:doi: 10.1039/b818541g
77. Huang G, Ouyang Z, Cooks RG (2008) High-throughput trace melamine analysis in complex mixtures. Chem Commun 556:doi: 10.1039/b818059h
78. Mauer LJ, Chernyshova AA, Hiatt A, Deering A, Davis R (2009) Melamine detection in infant formula powder using near- and mid-infrared spectroscopy. J Agric Food Chem 57(10):3974–3980
79. FDA (2008) Draft assessment of bisphenol A for use in food contact applications. Food and Drug Administration, Washington, DC. http://www.fda.gov/ohrms/dockets/ac/08/briefing/2008–0038b1_01_02_FDA%20BPA%20Draft%20Assessment.pdf
80. Lang IA, Galloway TS, Scarlett A, Henley WE, Depledge M, Wallace RB, Melzer D (2008) Association of urinary bisphenol A concerntration with medical disorders and laboratory abnormalities in adults. JAMA 300(300):1303. PMID 18799442. http://jama.ama-assn.org/cgi/content/full/300.11.1303
81. European Food Safety Authority (EFSA) (2009) Toxicological evaluation of benzophenone. Scientific opinion of the panel on food contact materials, enzymes, flavourings and processing aids (CEF). Question no. EFSA-Q-2009–411. Adopted 14 May 2009. EFSA J 1104:2–3. http://www.efsa.europa.eu/cs/BlobServer/Scientific_Opinion/cef_ej1104_benzophenone_op_sum.pdf?ssbinary=true
82. Silliker (2009) 4-Methylbenzophone testing services. Silliker, Inc., Homewood, IL. http://www.silliker.com/usa/html/labservices/4methylbenzophenone.php
83. Vranová J, Ciesarová Z (2009) Furan in food – a review. Czech J Food Sci 27 (1):1–10
84. FDA (2005) FDA action plan for furan in food. 1 Sept 2005. Updated as of 5/25/2009. http://www.fda.gov/ohrms/dockets/ac/08/briefing/2008–0038b1_01_02_FDA%20BPA%20Draft%20Assessment.pdf

85. FDA (2004) Determination of furan in foods. 7 May 2004 (updated 2 June 2005 and 27 Oct 2006). http://www.fda.gov/Food/FoodSafety/FoodContaminantsAdulteration/ChemicalContaminants/Furan/ucm078400.htm
86. Nyman PJ, Morehourse KM, Perfetti GA, Diachenko GW, Holcomb JR (2008) Single-laboratory validation of a method for the determination of furan in foods using headspace gas chromatography/mass spectrometry, part 2, low-moisture snack foods. J AOAC Int 91(2):414–421
87. Altaki MS, Santos FJ, Galceran MT (2007) Analysis of furan I foods by headspace solid-phase microextraction – gas chromatography – ion trip mass spectrometry. J Chromatogr A 1146(1):103–109
88. Tareke E, Rydberg P, Karlsson P, Eriksson S, Törnqvist M (2002) Analysis of acrylamide, a carcinogen formed in heated foodstuffs. J Agric Food Chem 50(17):4998–5006
89. FDA (2004) FDA action plan for acrylamide in food (March 2004). http://www.fda.gov/Food/FoodSafety/FoodContaminantsAdulteration/ChemicalContaminants/Acrylamide/ucm053519.htm
90. FDA (2002) Detection and quantitation of acrylamide in foods. 20 June 2002 (updated 23 July 2002 and 24 Feb 2003). http://www.fda.gov/Food/FoodSafety/FoodContaminantsAdulteration/ChemicalContaminants/Acrylamide/ucm053537.htm
91. Zhang Y, Zhang G, Zhang Y (2005) Occurance and analytical methods of acryalmide I heat-treated foods. Review and recent developments. J Chromatogr A 1075(1–2):1–21
92. FDA (2006) Questions and answers on the occurrence of benzene in soft drinks and other beverages. 19 May 2006 (last updated 5/15/2009). http://www.fda.gov/Food/FoodSafety/FoodContaminantsAdulteration/ChemicalContaminants/Benzene/ucm055131.htm
93. FDA (2009) Determination of benzene in soft drinks and other beverages (22 June 2009). http://www.fda.gov/Food/FoodSafety/FoodContaminantsAdulteration/ChemicalContaminants/Benzene/ucm055131.htm
94. Australia New Zealand Food Authority (2001) Maximum limit for chloropropanols in soy and oyster sauces. Draft assessment. Proposal P243 (10 Oct 2001, updated 0/502). http://www.foodstandards.gov.au/_srcfiles/P243%20DraftFAR.pdf
95. FDA (2008) Compliance Policy Guide Sec. 500.500: guidance levels for 3-MCPD (3-cholro-1,2-propanediol) in acid-hydrolyzed protein and Asian-style sauces. 14 March 2008 (last updated 7/30/2009). http://www.fda.gov/ICECI/ComplianceManuals/CompliancePolicyGuidanceManual/ucm074419.htm
96. International Life Sciences Institute (ILSI Europe) (2009) Workshop on 3-MCPD esters in food products. 5–6 Feb 2009. Summary report. Brussels, Belgium. http://www.bezpecnostpotravin.cz/UserFiles/File/Kvasnickova/Report3MCPDReport.pdf
97. Deutsche Gesellschaft fur Fettwissenschaft (DGF) (2009) DGF standard methods. Section C – Fats C-III 19 (09). Ester-bound 3-chloropropane-1,2-diol (3-MCPD esters) and 3-MCPD forming substances. http://www.dgfett.de/methods/c_iii_18(09)_e_3mcpd_ester.pdf
98. FDA (2007) FDA's role in measuring and assessing perchlorate levels in food and beverages. 25 April 2007 (last updated 7/6/2009). http://www.fda.gov/NewsEvents/Testimony/ucm154025.htm
99. FDA (2009) Perchlorate questions and answers (updated 5/25/2009). http://www.fda.gov/Food/FoodSafety/FoodContaminantsAdulteration/ChemicalContaminants/Perchlorate/ucm077572.htm
100. FDA (2005) Rapid determination of perchlorate anion in foods by ion chromatography-tandem mass spectrometry. Revision 2: 12 April 2005 (last updated 5/14/2009). http://www.fda.gov/Food/FoodSafety/FoodContaminantsAdulteration/ChemicalContaminants/Perchlorate/ucm077793.htm

Analysis for Extraneous Matter

Hulya Dogan, Bhadriraju Subramanyam, and John R. Pedersen*

Department of Grain Science and Industry, Kansas State University, Manhattan, KS 66506, USA
dogan@k-state.edu
sbhadrir@k-state.edu

19.1 Introduction 353
  19.1.1 Federal Food, Drug, and Cosmetic Act 353
  19.1.2 Good Manufacturing Practices 353
  19.1.3 Defect Action Levels 353
  19.1.4 Purposes of Analyses 353
19.2 General Considerations 353
  19.2.1 Definition of Terms 353
    19.2.1.1 Extraneous Materials 354
    19.2.1.2 Filth 354
    19.2.1.3 Heavy Filth 354
    19.2.1.4 Light Filth 354
    19.2.1.5 Sieved Filth 354
  19.2.2 Diagnostic Characteristics of Filth 354
19.3 Official and Approved Methods 354
19.4 Basic Analysis 355
  19.4.1 Sieving Method 356
  19.4.2 Sedimentation Method 356
  19.4.3 Flotation Methods 356
    19.4.3.1 Cracking Flotation Method 357
    19.4.3.2 Light Filth Floatation Method 357

S.S. Nielsen, *Food Analysis*, Food Science Texts Series, DOI 10.1007/978-1-4419-1478-1_19,

19.4.4 Objectivity/Subjectivity of Methods 358
19.5 Other Techniques 358
19.5.1 Overview 358
19.5.2 X-Ray Radiography 359
19.5.3 X-Ray Microtomography 359
19.5.4 Electrical Conductance Method 360
19.5.5 Impact-Acoustic Emission 360
19.5.6 Microscopy Techniques 360
19.5.7 Near-Infrared Spectroscopy 361
19.5.8 Enzyme-Linked Immunosorbent Assays 361
19.6 Comparison of Methods 361
19.7 Isolation Principles Applied to Food Processing 363
19.8 Summary 364
19.9 Study Questions 364
19.10 Acknowledgement 364
19.11 References 364

19.1 INTRODUCTION

Analysis for extraneous matter is an important element both in the selection of raw materials for food manufacturing and for monitoring the quality of processed foods. The presence of extraneous material in a food product is unappealing and can pose a serious health hazard to the consumer. It also represents lack of good manufacturing practices and sanitary conditions in production, storage, or distribution. The presence of extraneous materials in the product ingredients may render the final product adulterated and not suitable for human food.

19.1.1 Federal Food, Drug, and Cosmetic Act

The **Federal Food, Drug, and Cosmetic Act** (FD&C Act) of 1938 with Amendments administered and enforced by the US Food and Drug Administration (FDA) (1) defines a food as **adulterated** "if it consists in whole or in part of any filthy, putrid, or decomposed substance, or if it is otherwise unfit for food [Section 402 [21 USC 342] (a)(3)]; or if it has been prepared, packed, or held under unsanitary conditions whereby it may have become contaminated with filth, or whereby it may have been rendered injurious to health" [Section 402 [21 USC 342] (a)(4)]. The filthy, putrid, or decomposed substances referred to in the law include the extraneous matter addressed in this chapter. In addition, extraneous matter includes adulterants that may be encountered in processing systems, such as lubricants, metal particles, or other contaminants (animate or inanimate) that may be introduced into a food intentionally or because of a poorly operated food processing system. These aspects are not covered in this chapter.

19.1.2 Good Manufacturing Practices

The **Current Good Manufacturing Practice in Manufacturing, Packing, or Holding Human Food** (cGMPs) was published in 1969 by the Food and Drug Administration (FDA) (21 CFR Part 110) to provide guidance for compliance with the FD&C Act (2) (see also Chap. 2). That regulation provides guidelines for operating a food processing facility in compliance with Section 402 (a)(4), and these guidelines have not been revised since 1986. Currently, the cGMPs are being amended to make the compliance guidelines more risk-based. Paramount to complying with the FD&C Act and cGMPs is the thorough inspection of raw materials and routine monitoring of food processing operations to ensure protection of the consuming public from harmful or filthy food products.

19.1.3 Defect Action Levels

Most of our foods are made from or consist in part of ingredients that are obtained from plants or animals and are mechanically stored, handled, and transported in large quantities. It would be virtually impossible to keep those materials completely free of various forms of contaminants. In recognition of that, the FDA (3) has established **defect action levels** (DALs) that reflect current maximum levels for natural or unavoidable defects in food for human use that present no health hazard. They reflect the maximum levels that are considered unavoidable under good manufacturing practices and apply mainly to contaminants that are unavoidably carried over from raw agricultural commodities into the food processing system. The manner in which foods are manufactured may lead to their contamination with extraneous materials if strict controls in processing are not maintained. This latter type of contamination leads to food safety issues, and DALs are not used to determine compliance. Other actionable levels of contaminants may be found in the FDA *Compliance Policy Guide* (*CPG*) *Manual* (4).

The most current information of FDA laws and regulations relevant to extraneous matter, including cGMPs, DALs, and CPGs, can be found on the Internet:

Federal Food, Drug, and Cosmetic Act (FD&C Act) – http://www.fda.gov/opacom/laws/fdcact/fdctoc.html

Current Good Manufacturing Practices (cGMPs) – http://www.gmp1st.com/fdreg.htm

Food Defect Action Levels (DALs) – http://www.cfsan.fda.gov/~dms/dalbook.html

Compliance Policy Guidance (CPG) – http://www.fda.gov/ora/compliance_ref/cpg/

19.1.4 Purposes of Analyses

The major purposes for conducting analyses for extraneous matter in foods are to ensure the protection of the consuming public from harmful or filthy food products, to meet regulatory requirements of the FD&C Act Sections 402 (a)(3) and 402 (a)(4), and to comply with DALs.

19.2 GENERAL CONSIDERATIONS

19.2.1 Definition of Terms

Terms used by AOAC International (AOAC Method 970.66) to classify or characterize various types of extraneous materials are defined as follows.

19.2.1.1 Extraneous Materials

Any foreign matter in product associated with objectionable conditions or practices in production, storage, or distribution; included are various classes of filth, decomposed material (decayed tissues due to parasitic or nonparasitic causes), and miscellaneous matter such as sand and soil, glass, rust, or other foreign substances. Bacterial counts are not included.

19.2.1.2 Filth

Any objectionable matter contributed by animal contamination such as rodent, insect, or bird matter, or any other objectionable matter contributed by unsanitary conditions.

19.2.1.3 Heavy Filth

Heavier material separated from products by sedimentation based on different densities of filth, food particles, and immersion liquids. Examples of such filth are sand, soil, insect and rodent excreta pellets and pellet fragments, and some animal excreta pellets.

19.2.1.4 Light Filth

Lighter filth particles that are oleophilic and are separated from product by floating them in an oil–aqueous liquid mixture. Examples are insect fragments, whole insects, rodent hairs and fragments, and feather barbules.

19.2.1.5 Sieved Filth

Filth particles of specific size ranges separated quantitatively from product by use of selected sieve mesh sizes.

19.2.2 Diagnostic Characteristics of Filth

There are certain qualities characteristic to extraneous materials that serve as proof of presence of foreign or objectionable matter in food. Examples include specific diagnostic characteristics of **molds** (i.e., parallel hyphal walls, septation, granular appearance of cell contents, branching of hyphae, blunt ends of hyphal filaments, nonrefracted appearance of hyphae); diagnostic characteristics of **insect fragments** (i.e., recognizable shape, form, or surface sculpture, an articulation or joint, setae or setal pits, sutures), **rodent hairs** (i.e., pigment patterns and structural features), **feather barbules** (i.e., structural features); diagnostic characteristics of **insect-damaged grains** (IDK) and packaging materials; and chemical identification of **animal urine** and **excrement**. These diagnostic characteristics are outlined by AOAC International (formerly Association of Official Analytical Chemists) for positive identification of extraneous matter or filth (5).

The AACC International (formerly American Association of Cereal Chemists) publishes a methods book that includes a section on extraneous matter, containing descriptive material helpful in identifying insect and rodent contaminants (6). Several microscopic and radiographic illustrations are provided by the AACC International as authentic reference materials to help analysts to identify filth. AACC Method 28–95, "Insect, Rodent Hair, and Radiographic Illustrations," provides a series of colored pictures representative of insect fragments commonly found in cereal products and pictorial examples of rodent hair structure.

Kurtz and Harris (7) provide a virtual parts catalog of insect fragments with a series of micrographs. Gentry et al. (8), an updated version of the Kurtz and Harris publication, includes colored micrographs of common insect fragments. Also included in AACC 28–95 are radiographic examples of grain kernels that contain internal insect infestation. AACC Method 28-21A, "X-ray Examination for Internal Insect Infestation," provides an outline of the apparatus and procedure for X-ray examination of internal insect infestation in grain (9).

19.3 OFFICIAL AND APPROVED METHODS

There are various laboratory methods for separating (isolating) extraneous materials from foods and for identifying and enumerating them. The FDA and the AOAC International have published reference articles, books, and methods on analysis of extraneous materials. The most authoritative source, and that generally considered official by the FDA, is the *Official Methods of Analysis of AOAC International*, Chap. 16, "Extraneous Materials: Isolation" (5). This chapter includes methods for extraneous matter isolation in various food categories (Table 19-1). The AOAC International "Extraneous Materials: Isolation" chapter contains a subchapter dealing with **molds**. This includes identification of molds and methods for isolation of molds in fruits and fruit products and vegetables and vegetable products.

The AACC International (6) has established methods for isolating and identifying extraneous matter in cereal grains and their products (AACC Method 28-00, listed in Table 19-2). In most instances, the AACC methods are based on FDA or AOAC methods, but the format is slightly different. The AACC presents each procedure in an outline form that includes the scope, apparatus, and reagents required and the procedure in itemized steps while the AOAC uses a narrative paragraph form (Table 19-2).

19-1 table Official Methods of AOAC International for Analysis of Extraneous Materials

Section	Title
16	Extraneous materials: isolation
16.1	General
16.2	Beverages and beverage materials
16.3	Dairy products
16.4	Nuts and nut products
16.5	Grains and their products
16.6	Baked goods
16.7	Breakfast cereals
16.8	Eggs and egg products
16.9	Poultry, meat, and fish and other marine products
16.10	Fruits and fruit products
16.11	Snack food products
16.12	Sugars and sugar products
16.13	Vegetables and vegetable products
16.14	Spices and other condiments
16.15	Miscellaneous
16.16	Animal excretions
16.17	Mold
16.18	Fruits and fruit products
16.19	Vegetables and vegetable products

A valuable resource on analysis for extraneous matter is *Principles of Food Analysis for Filth, Decomposition and Foreign Matter*, FDA Technical Bulletin No. 1 (10). The FDA *Training Manual for Analytical Entomology in the Food Industry* (11) is prepared to facilitate the orientation of food analysts to the basic techniques they will need for filth analysis. A recent, more advanced resource is *Fundamentals of Microanalytical Entomology: A Practical Guide to Detecting and Identifying Filth in Foods* (12). Most chapter authors of this resource are, or have been, FDA personnel "involved in the forensic aspect of piecing together the etiological puzzles of how insect filth gets into processed food products" (12). The authors share their experience gained in gathering and developing evidence used to document violations of the law that the FDA is mandated to enforce.

19.4 BASIC ANALYSIS

Various methods for isolation of extraneous matter are suggested in Sect. 19.3, which define different types of filth: separation on the basis of differences in **density**, **affinity for oleophilic solvents**, **particle size**; **diagnostic characteristics** for identification of filth; and **chemical identification** of contaminants. Since all methods of analysis for extraneous matter for all categories of food cannot be discussed in this chapter, only the underlying principles of the methods are summarized below. Readers may need to refer to the specific AOAC methods cited for detailed instructions of the procedures.

19-2 table Approved Methods of the AACC International for Analysis of Extraneous Materials

Number	Title
28	Extraneous matter
28-01.01	Apparatus or materials for extraneous matter methods
28-02.01	Reagents for extraneous matter methods
28-03.02	Special techniques for extraneous matter methods
28-06.01	Cinder and sand particles in farina – counting method
28-07.01	Cinder and sand particles in farina – gravimetric method
28-10.02	Macroscopic examination of external contamination in whole grains
28-19.01	External filth and internal insect infestation in whole corn
28-20.02	Microscopic examination of external contamination in whole grains
28-21.02	X-ray examination for internal insect infestation
28-22.02	Cracking-flotation test for internal insects in whole grains
28-30.02	Macroscopic examination of materials hard to hydrate
28-31.02	Pancreatin sieving method, for insect and rodent filth in materials hard to hydrate
28-32.02	Sieving method, for materials hard to hydrate
28-33.02	Pancreatin nonsieving method for insect and rodent filth in materials easy to hydrate
28-40.01	Acid hydrolysis method for insect fragments and rodent hairs – wheat–soy blend
28-41.03	Acid hydrolysis method for extracting insect fragments and rodent hairs – light filth in white flour
28-43.01	Glass plate method, for insect excreta
28-44.01	Iodine method, for insect eggs in flour
28-50.01	Decantation method, for rodent excreta
28-51.02	Flotation method, for insect and rodent filth
28-60.02	Tween-versene method, for insect fragments and rodent hairs in rye flour
28-70.01	Defatting-digestion method, for insect fragments and rodent hairs
28-75.02	Sieving method, for light filth in starch
28-80.01	Flotation method, for insect and rodent filth in popped popcorn
28-85.01	Ultraviolet light examination, for rodent urine
28-86.01	Xanthydrol test, for urea
28-87.01	Urease-bromthymol blue test paper, for urea
28-93.01	Direction of insect penetration into food packaging
28-95.01	Insect, rodent hair, and radiographic illustrations

The AOAC and the AACC methods for analysis of extraneous materials involve the use of one or more of the following basic methods: filtration, sieving, wet sieving, gravimetry, sedimentation/flotation, cracking flotation, heat, acid or enzyme digestion, macroscopic and microscopic methods, and mold counts.

19-1 figure

Fisher USA Standard test sieve. Range of mesh sizes is available for various particle size separations.

19.4.1 Sieving Method

Separation is based on difference in particle size of product and contaminant using **standard test sieves** (Fig. 19-1). For instance, the insects are (larger) separated from spices (smaller) using a 20 mesh sieve, and wheat grains (larger) are separated from insects (smaller) using a 10 or 12 mesh sieve. Then the contaminant is identified using a **widefield stereomicroscope**.

19.4.2 Sedimentation Method

Separation is based on different densities of product, contaminant, and immersion fluid. Specific gravity of immersion solution (carbon tetrachloride/chloroform) allows heavier shell, sand, glass, metal, or excreta contaminants to settle; less dense product floats. Apparent lower **specific gravity** of internally infested wheat kernels, for instance, allows them to float while sound wheat kernels settle in 1.19 specific gravity solution. Contaminants are then identified using a microscope.

Analysis of high-fat-containing samples such as nuts requires defatting using petroleum ether prior to filth analysis (AOAC Method 968.33A). The chloroform and chloroform:carbon tetrachloride solvents allow pieces of shell, sand, and soil to settle at the bottom of the beaker on the basis of specific gravity and cause the defatted nut meats to float and be decanted. Essentially the same procedure is suggested to isolate pieces of rodent excreta from corn grits, rye and wheat meal, whole wheat flour, farina, and semolina in AOAC Method 941.16A. It should be noted that the use of the more toxic solvents such as carbon tetrachloride, chloroform, and petroleum ether is avoided in most contemporary analytical methods.

19.4.3 Flotation Methods

Flotation methods are designed to isolate microscopic filth by floating the filth upwards, typically in an **oil/water-phased system**. Insect fragments, mites, and hairs are lipophilic and likely to be in the oil phase, thus they float to the surface with the oils. Plant tissues and most related tissues are hydrophilic, and they tend to stay in the water phase. Therefore, separation is based on the principle of affinity for oleophilic solvents. Gravity further helps this process, and larger particles sink. To accomplish the separation of filth from food, a number of solution systems are used to ensure that the majority of the product sinks, while the oils with trapped filth float. The oil phase is trapped off with a **Wildman trap flask** (Fig. 19-2), filtered, collected on a filter paper, and examined microscopically to determine the amount and kinds of filth present (13).

Flotation is a common method used to determine insect fragments, rodent hairs, and other forms of light filth in wheat flour (AOAC Method 972.32). The acid digestion is used to break down the starch in the flour and allows the other flour constituents to more cleanly separate from the dilute acid solution. Although the AOAC method calls for digestion by autoclaving, AACC Method 28–41B provides for an alternative hotplate digestion, which might be more convenient for some laboratories. The oleophilic property of insect fragments, rodent hairs, and feather barbules allows them to be coated by the mineral oil and trapped in the oil layer for separation and collection on ruled filter paper. The heavier sediments of the digestion are washed and drained from the funnel. Fragments and rodent hairs are reported on the basis of 50 g of flour.

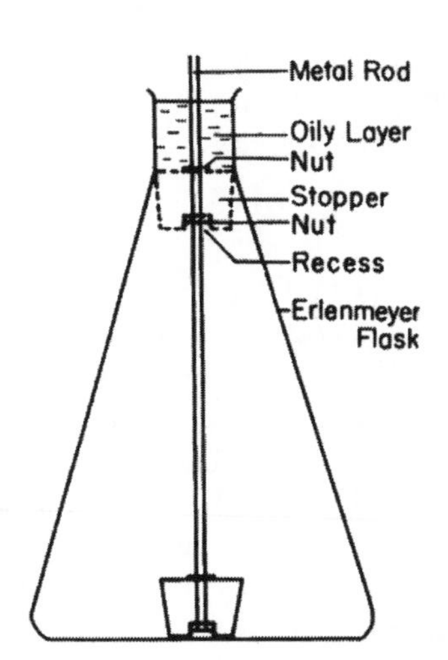

19-2 figure

Wildman trap flask. Stopper on shaft is lifted up to neck of flask to trap off floating layer. [Adapted from (5), AOAC Method 945.75, Extraneous Materials in Products.]

19.4.3.1 Cracking Flotation Method

Internal infesting insects (such as in grains) can be determined using an oleophilic method. First any external insects are removed by sieving. Grain sample is coarsely cracked to free insects from kernels. Cracked grain sample is digested in 3–5% HCl solution and sieved with water to remove hydrolyzed starch and acid. The sample is transferred to a Wildman trap flask and boiled in 40% alcohol solution to deaerate. Tween 80 (polyoxyethylene sorbitan monooleate) and Na_4EDTA (tetrasodium salt of ethylenediaminetetraacetic acid) solutions are added to cause light bran particles to remain in solution during oil extraction of insect material. Light mineral oil is added to the solution to form a floating layer in which insect material is attracted due to its oleophilic nature. The oil layer is filtered through a ruled paper to collect the contaminating insects. The filter paper is examined microscopically.

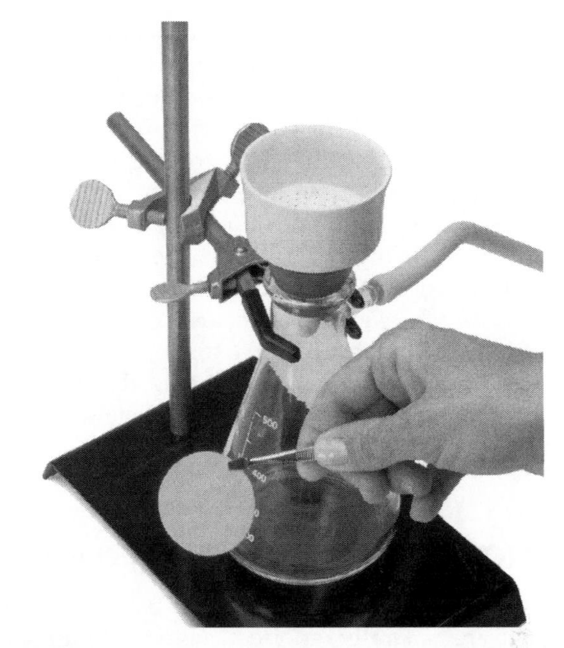

19-3 figure Filter flask and funnel. Funnel has a collar (partially raised) that holds ruled filter paper in place on the funnel base for trapping filth for examination. Suction is applied with a water aspirator. (http://www.whatman.com)

19.4.3.2 Light Filth Floatation Method

Oleophilic filth is defined as light filth. Examples of light filth include insects, insect fragments, hairs, and feather barbules, which can be detected in a food product by separating them from the food in the oil phase of an oil/aqueous mixture. The analysis of light filth is accomplished through a series of steps, starting with a pretreatment that removes fats, oils, soluble solids, and fine particulate matter to enhance the wettability of the food. The second step requires mixing the food with a water and oil mixture. The food will remain in the aqueous phase and the light filth will rise to the top with the oil phase. In the third step, the extract with filth elements is poured onto a ruled filter paper using a filter flask and funnel (Fig. 19-3) and examined line by line under a stereomicroscope (Fig. 19-4). After identification and enumeration, the results are reported to provide the following information: (a) whole or equivalent insects (adults, pupae, maggots, larvae, cast skins), (b) insect fragments, identified, (c) insect fragments, unidentified, (d) aphids, scale insects, mites, spiders, etc. and their fragments, (e) rodent hairs (state the length of the hairs).

Insect fragments, rodent hairs, and other light filth can be isolated from flour samples by an acid hydrolysis method. Sediment products of the digestion are allowed to settle out in a separatory funnel and are drained away. The remaining oil layer is filtered through a ruled filter paper and the contaminants identified microscopically. However, certain grain products such as whole wheat flour contain amounts of bran particles that may result in excessive amounts of material being trapped in the oil layer, making it difficult to identify particles of filth.

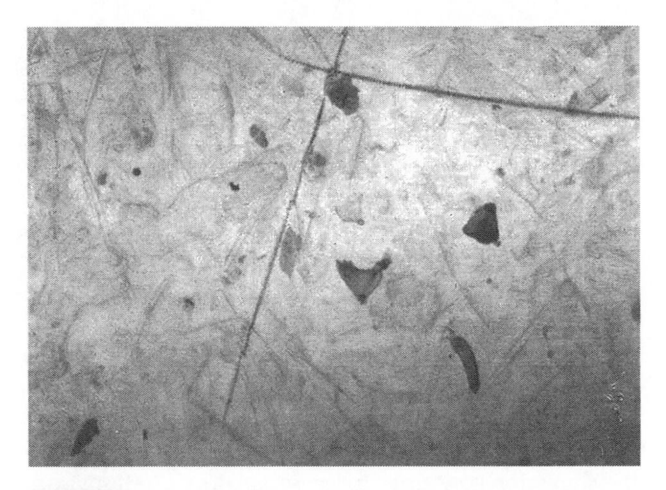

19-4 figure Microscopic view of insect fragments and rodent hairs on a filter paper.

The "Tween-Versene Method for Insect Fragments and Rodent Hairs in Rye Flour" (AACC Method 28–60A) utilizes two chemical agents that tend to suppress bran accumulation in the heptane recovery layer. Tween 80 (polyoxyethylene sorbitan monooleate) is a nonionic agent that appears to have certain surface active properties that make it a useful adjunct to Na_4EDTA (tetrasodium salt of ethylenediaminetetraacetic acid). In the presence of Tween 80, Na_4EDTA appears to be a depressor for food materials (such as bran and other light plant matter), which otherwise tend to float. It has been suggested that the chelating properties of Na_4EDTA may result in its adsorption onto the surfaces of food particles along with the surfactant Tween 80, thereby preventing an attraction of food particles to oils used to isolate light filth. By preventing plant material from being collected in the heptane layer that is trapped off, contaminants such as oleophilic insect parts (exoskeleton) that are contained in the separating oil are much easier to distinguish and identify. AACC Method 28–95 provides a description of insect fragments and rodent hair characteristics with illustrations (6).

19.4.4 Objectivity/Subjectivity of Methods

Insect parts, rodent hairs, and feather barbules in food products are generally reported as the total number of filth elements counted of each kind encountered per sample unit. They are identified on the basis of objective criteria. However, identifying insect fragments is not a simple task. Training and supervised practice are required to achieve competence and consistency. Some fragments are easily identified on the basis of structural shape and form. Mandibles, for example, are quite distinctive in their shape and configuration; certain species of insects can be determined on the basis of this one structure. In other instances, fragments may be mere chips of insect cuticle that have neither distinctive shape nor form but can be identified as being of insect origin if they have one or more of the characteristics given in Sect. 19.2.2. Experienced analysts should rarely misinterpret fragments.

Isolation of extraneous material from a food product so that it can be identified and enumerated can be a very simple procedure or one that requires a series of several rather involved steps. In the process of isolating fragments from flour by the acid hydrolysis method, for instance, the sample is transferred from the digestion container to the separatory container and then to the filter paper for identification and enumeration. At each of those transfers there is an opportunity for loss of fragments. Although the analyst may have made every effort to maintain the isolation "quantitative," there are opportunities for error. Both fragment loss and analyst variation are minimized by common use of standard methods and procedures and by proper training and supervised practice.

Another concern involves the significance of **insect fragment counts** (as well as particles of sand, pieces of rodent excreta, rodent hairs, etc.) in relation to **fragment or particle size**. Fragment counts are reported on a numerical basis; they do not reflect the total contaminant biomass that is present. A small fragment is counted the same as a large fragment. The size of the fragment may be a reflection of the process to which a common raw material (e.g., wheat) has been subjected; a more vigorous process produces more and smaller fragments than a less vigorous process. The state of insects may also be a factor. Dead (dried) forms produce greater numbers of fragments than live forms. These factors have been of concern to food processors for some time and have prompted the search for more objective means of determining insect contamination.

19.5 OTHER TECHNIQUES

19.5.1 Overview

Methods described in Sect. 19.3 are directed primarily at routine quality control efforts to determine if the level of natural or unavoidable defects is below the defect action level. To a certain extent, those routine methods can be used to identify the source of contaminants in processed foods. For example, the identification of certain insect fragments can indicate infestation in the raw commodity rather than in the processing system. However, other more sophisticated techniques offer opportunities to pinpoint the nature and source of other contaminants that may exist unavoidably or due to mistakes, accidents, material or equipment failures, or intentional adulteration.

The detection of insects in stored grain and the quantitation of insect parts present in grain products represent a serious and continuing problem for the grain industry. Approved methods of detection primarily involve visual and microscopic inspection and X-ray analysis, which require trained personnel and are time-consuming, difficult to standardize, and expensive. The assays for insect contamination are preferred to be highly specific, sensitive, rapid, and inexpensive. Moreover, it ideally should be employable by persons having minimal training, particularly in nonlaboratory settings such as at grain elevator and processing sites.

There are several attempts to develop rapid and efficient methods including the use of nuclear magnetic resonance, sound amplification, and infrared spectrometry as alternatives to presently used chemical techniques mentioned in preceding sections. Most of these techniques are expensive and challenging due to difficulty in quantification and identification of specific infestations. Immunological assays, which have found widespread use in clinical diagnostic settings and also in home use, have been explored to detect insect contamination. These methods are described below as they relate to detecting an infestation.

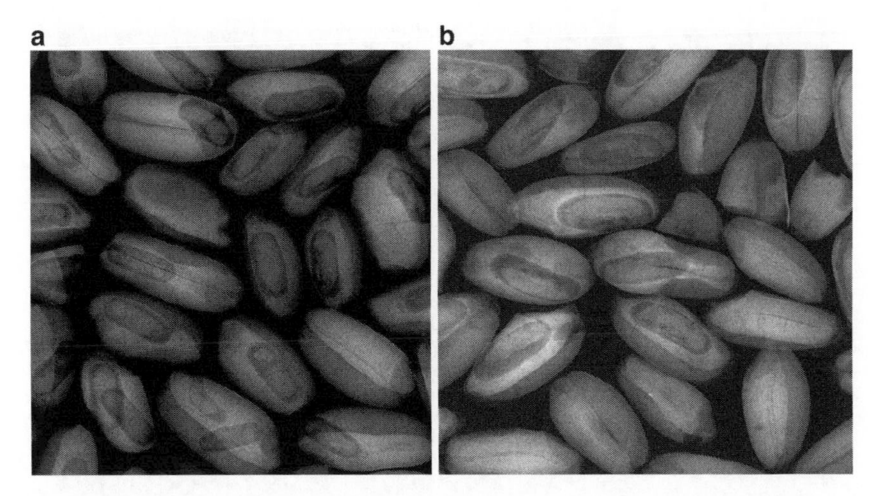

19-5 figure X-ray radiograph of infested wheat: (**a**) Lesser grain borer pupae, (**b**) Rice weevil pupae. (Courtesy of Moses Khamis.)

19.5.2 X-Ray Radiography

X-ray radiography is widely used as a test reference method (14). Grain processors use it as a means of inspecting wheat for internal insect infestation, which is the main source of insect fragments in processed cereal products (Fig. 19-5).The existing X-ray techniques enable the classification of at least four stages of insect development by measuring the area occupied by the insect, and an accurate classification is also possible based on visible insect morphology (15). The use of real-time digital imaging instead of X-ray radiographs to discriminate the infested kernels significantly shortened the X-ray procedures. Conventional film observations give, however, better accuracy (3% error rate) than the digital images (11.7%) for infestation by third-larval instars, while the error is less than 1% for both methods with a more advanced stage of larval development (16).

19.5.3 X-Ray Microtomography

X-ray microtomography (XMT) is an emerging 3-D imaging technique that operates on the same basic principles as medical computed tomography (CT) scanners, but has much higher resolution. It is very effective in characterizing various internal structural features, which are not possible with conventional 2-D imaging methods. Conventional imaging techniques such as light microscopy, scanning electron microscopy (SEM), and digital video imaging have some limitations: They are destructive in nature, as sample preparation involves cutting to expose the cross section to be viewed. High-resolution XMT has a wide range of applications in science and engineering for which accurate 3-D imaging of internal structure of objects is crucial.

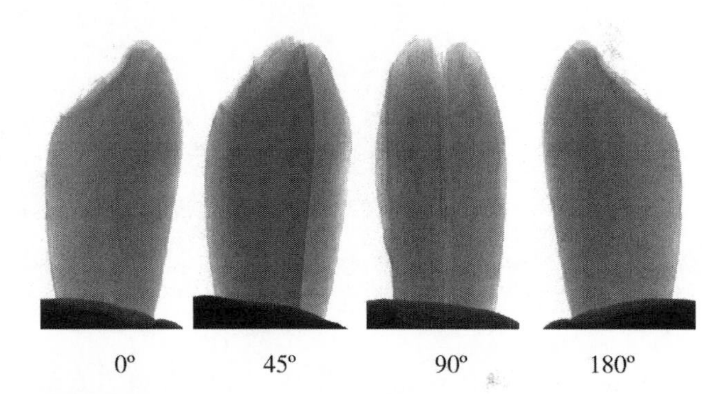

19-6 figure Shadow images of a wheat kernel at various rotation angles.

Microtomographical techniques involve targeting the specimen with a polychromatic X-ray beam with high spatial coherence. The X-rays not absorbed by the specimen fall on specifically designed X-ray scintillators that produce visible light, which is then recorded by a charge-coupled device (CCD) camera. Scans are done by rotating the specimen between a fixed X-ray source and detector, around the axis perpendicular to the X-ray beam, while collecting radiographs of the specimen at small angular increments in the range 0–360°. The radiographs are reconstructed into a series of 2-D slices. The series of 2-D slices are then reconstructed into a 3-D image. The resulting XMT data can be visualized by 3-D rendering or 2-D slices derived from virtual model, using software that allows reconstruction of cross sections at various depth increments and along any desired orientation of the plane of cut (17). XMT is able to capture several features of the internal structures of grain kernels, which are not possible with the conventional imaging methods. Figure 19-6 is demonstration of shadow images

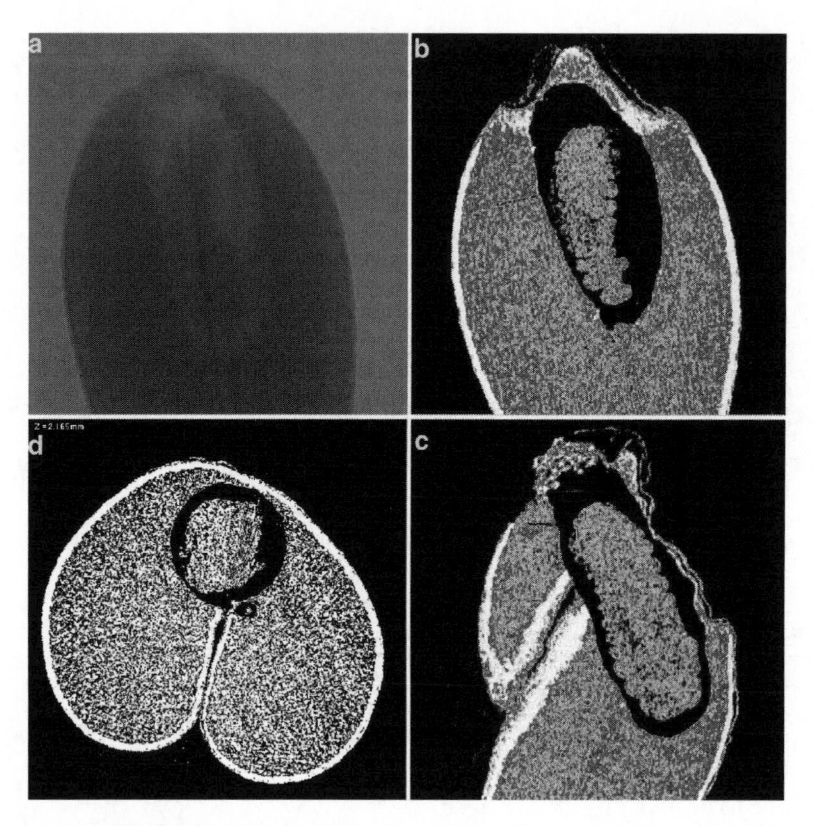

19-7 figure X-ray microtomography (XMT) images: (**a**) shadow image, (**b**) sagittal view, (**c**) axial view, and (**d**) coronal view of lesser grain borer in a wheat kernel.

taken at several step angles during scanning. A typical scan creates 200–400 of those images that are then used to create axial, sagittal, and coronal views shown in Fig. 19-7.

19.5.4 Electrical Conductance Method

The **electrical conductance** method is based on monitoring the conductance signals for each single kernel during milling in a Single Kernel Characteristics System (SKCS), which is commonly used for wheat hardness determination (15). This method is highly accurate for detecting older developmental stages of insects: the percentage of properly classified cases for small, medium, and large larvae and pupae is 24.5, 62.2, 87.5, and 88.6, respectively. The accuracy of this method depends also on insect species (rice weevil and lesser grain borer) and wheat type (soft or hard red winter wheat).

19.5.5 Impact-Acoustic Emission

Impact-acoustic emissions are used as a nondestructive, real-time method for detection of damaged grains and shelled nuts (18). Kernels are impacted onto a steel plate and the resulting acoustic signal is analyzed to detect damage using different methods: modeling of the signal in the time domain, computing time-domain signal variances and maximums in short-time windows, analysis of the frequency spectrum magnitudes, and analysis of a derivative spectrum. Features were used as inputs to a stepwise discriminant analysis routine, which selected a small subset of features for accurate classification using a neural network. Pearson et al. (18) reported that impact-acoustic emissions is a feasible and promising method for detection of IDK, sprout damage, and scab damage. More study is needed to improve accuracy on kernels infested with insects that have not yet emerged from the kernels. The computational cost of classifying a kernel using this technique is very low, allowing inspection of large numbers of wheat kernels very rapidly, ~40 kernels/s. Grain inspectors usually use a 100 g (3000 kernel) sample to inspect for IDK. This takes an inspector approximately 20 min to analyze manually, but can be accomplished in about 75 s with an acoustic system.

19.5.6 Microscopy Techniques

Microscopy techniques including light microscopy, fluorescence microscopy, and scanning electron microscopy (SEM) are used to study the structure/function relationships of food, but also can be applied to questions of extraneous matter. For example, SEM with energy dispersive spectroscopy (EDS) can be used to determine the nature of metals in products that may be due to equipment failure or intentional adulteration due to tampering (19). Light microscopy in a polarized mode can be used to distinguish between plastics, glass, and other fiber or crystalline contaminants (20).

19.5.7 Near-Infrared Spectroscopy

Near-infrared spectroscopy (NIRS) is a relatively fast, accurate, and economical technique available to the grain industry for compositional analysis such as water, oil, fiber, starch, and protein in grains and seeds. It also has relatively recent applications in analysis of extraneous matter. NIRS has been used to identify several coleopteran species (21), and to detect parasitized weevils in wheat kernels (22) and external and internal insect infestation in wheat (23–25). Berardo et al. (26) reported that NIRS predicts the percentage of *Fusarium verticillioides* infection in maize kernels and the content of ergosterol and fumosin B1 in meals. In the same way, promising results were obtained when NIRS methodology was applied to detect scab-damaged kernels (27) and estimate deoxynivalenol, ergosterol, and fumonisin in single kernels of wheat (21) and corn (28).

Near-infrared spectroscopy used with a single kernel characterization system is able to detect later stages of internal insect infestation in wheat with a 95% confidence (25). In contrast to other procedures, this system is capable of being automated and incorporated into the current grain inspection process. NIRS also has been compared with the current standard insect fragment flotation method for its ability to detect insect fragments in flour (29). Fragment counts with both techniques were correlated; however, the flotation method was more sensitive below the FDA DAL of 75 insect fragments/50 g of wheat flour. NIR spectroscopy was able to predict accurately whether flour samples contained less than or more than 130 fragments/50 g.

19.5.8 Enzyme-Linked Immunosorbent Assays

To develop an optimal immunological assay for an insect contamination of foodstuffs, antibodies are required that are directed against an insect-specific antigen, preferably protein, likely to be present in any life stage of the contaminating insect or in insect remains. Antigens and antibodies are two key parts of any immunoassay (see also Chap. 17).

For an immunoassay with broad specificity it is required to use an insect-specific protein such as myosin. Myosin is ubiquitous in insects; it is present in large quantities in adult insect tissue and is also present in appreciable quantities in other life stages (30). An **enzyme-linked immunosorbent assay** (ELISA) method has been developed to measure quantitatively the amount of insect material in a sample (30). It is also possible to develop an immunoassay specific for a particular species of insect contamination using antibodies having a unique species specificity. Kitto et al. (31) developed such techniques (patented in 1992) for detecting the amount of insect contamination in foodstuffs. The method comprises the following steps:

1. Preparing an aqueous solution or suspension of a homogenized grain sample
2. Substantially affixing at least a portion of solution or suspension to a solid surface
3. Applying to solid surface a specifically binding insect antigen (or antibody) and enzyme to form an antibody-enzyme conjugate, resulting in formation of a colored product when the enzyme reacts with a substrate
4. Washing unbound conjugate from the solid surface
5. Incubating the solid surface with an enzyme substrate under conditions allowing colored product to be formed when enzyme is present
6. Correlating amounts of color formed with an amount of insect contamination

Recent research (32) showed that the myosin in fourth instars of the lesser grain borer developing within kernels of wheat degraded within the first 2 weeks when larvae were killed with phosphine, a fumigant commonly used to manage insect infestations in stored grain. Myosin degradation resulted in underestimating insect fragment estimates by about 58%.

19.6 COMPARISON OF METHODS

A number of methods that have been developed to detect insects in commodity samples (Table 19-3) are described here in general terms:

1. Density separation based on infested kernels being lighter weight and floating in a liquid
2. Staining kernels to detect weevil egg plugs
3. Detection of carbon dioxide or uric acid produced by the internally feeding insects
4. Detection of insects hidden inside kernels using near-infrared spectroscopy (NIRS)
5. Detection by use of nuclear magnetic resonance (NMR)
6. Detection by X-ray images and digital image analysis techniques

Insect Detection Methods Applicable for Commodity Samples (Adapted from (35))

Test Method	*Applicability*	*Comments*
Visual inspection	Whole grains, milled products	Qualitative, only high-level infestation detected
Sampling and sieving	Whole grains, milled products	Commonly practiced, hidden infestation not detected
Heat extraction	Whole grains	Adults and larvae detected
Acoustics	Whole grains	*Feeding sounds*: Active stages detected *Impact-acoustic emissions*: Nondestructive, real time; detect insect, sprout, and scab damage
Breeding out	Whole grains	Time consuming
Imaging techniques		
X-ray method	Whole grains	Non-destructive, highly accurate, able to detect both live and dead insects inside grain kernels; cannot detect insect eggs, prohibitive capital cost
Near infrared spectroscopy	Whole grains, milled products	Rapid, sensitive, can be automated, no sample preparation; cannot detect low levels of infestation, sensitive to moisture content, calibration of equipment is complex and frequent
Nuclear magnetic resonance	Whole grains	Less sensitive
Serological techniques	Whole grains, milled products	Highly sensitive, species specific; shows infestation from unknown past to till date
Uric acid determination	Whole grains, milled products	Shows infestation from unknown past to till date
CO_2 analysis	Whole grains	Simple, time consuming; indicates current level of infestation; not suitable for grains having >15% moisture
Specific gravity methods	Whole grains	Simple and quick, not suitable for oats and maize
Cracking and floatation method	Whole grains	Variable results noted
Fragment count	Whole grains, milled products	Highly variable results noted; shows infestation from unknown past to till date
Staining techniques		
Egg plugs	Whole grains	Specific for *Sitophilus* spp.
Ninhydrin method	Whole grains	Eggs and early larvae not indicated

7. Acoustical sensors to hear sounds from insects feeding inside kernels
8. Enzyme-linked immunosorbent assays (ELISA) to detect myosin in insect muscle

Some of the recent methods have been developed by adapting the single-kernel characterization system (SKCS), computed tomography (CT), acoustic-impact emissions, and use of a electrical conductive roller mill (15,33,34).

The choice of method depends on several factors: (a) type of infestation (inside or outside food grains, in the surrounding premises or inside bulk grain), (b) required level of inspection (macroscopic vs. microscopic, qualitative vs. quantitative), (c) availability of equipment and facilities, and (d) required sensitivity (35). Most of the methods aim to detect the presence of live insects directly or indirectly. External insects are detected by visual inspection, sampling, sieving, and heat-extraction methods, while internal (hidden) insects are detected by radiography, staining techniques to identify egg plugs, and near-infrared and fragment count methods. Determination of uric acid or CO_2 level serves as an indirect way of detecting and estimating internally feeding insects, and these methods may be suitable if infestations are restricted to one insect species. Depending on some storage conditions, grain may contain molds and insects, and in such cases CO_2 produced by molds may interfere in accurately detecting or estimating insects. Fragment count and ELISA methods can be used for the detection of both living and dead insects. In general, problems encountered with these detection methods are that the most accurate methods, such as X-ray and computed tomography (CT), are laborious and expensive, while rapid, automated methods may not be suitable for detecting eggs and young larvae (32).

19.7 ISOLATION PRINCIPLES APPLIED TO FOOD PROCESSING

Examination of stored-product insects often requires extracting them from the commodity. An intensive summary of literature survey on insect extraction and detection methods can be found in reference (36). Isolation principles, such as particle size and density, discussed in the preceding sections are designed to identify extraneous materials in finished food products, monitoring quality, and compliance with DALs. In addition, some of these principles of isolation are used in a **proactive** way during processing to prevent extraneous matter from being incorporated into finished food products.

Wheat that contains hidden internal insect infestation is the primary source of insect fragments in processed cereal products. The current DAL for internal insect infestation in wheat is **32 IDK** per 100 g of wheat (3). IDK are those **visually** determined to have insect tunneling or emergence holes. Most processors rely on much lower levels of IDK ($\leq$6 IDK/100 g) to produce flour that meets customer tolerances and the FDA's DAL for insect fragments in flour. In addition, to prevent adulteration of flour with filth, entoleters and infestation destroyers in the milling process break up insect-damaged kernels, and these broken kernels along with the insect fragments are aspirated out of the milling stream. As previously indicated, **X-ray radiography** is used by some as a means for selecting grains for processing or for research purposes to age-grade internally developing stages of stored-grain insects. More recently, NIR spectroscopy has provided a new tool for assessing internal insect infestation in wheat. By selectively milling only wheat that has minimal or no evidence of internal insect infestation, grain processors can effectively limit insect fragments in their products. In like manner, bakers and other users of processed grain products can selectively monitor for insect fragments in their raw materials using one of the approved methods for extraction and enumeration of fragments or by sending samples to a private laboratory for fragment analysis.

Most food processing systems that deal with agricultural products generally apply some type of cleaning operations as an initial step. In flour milling, for example, wheat is passed through a system called the "cleaning house," which consists of a series of machines that apply the principles of particle size and density separation. Sieves remove contaminants larger than wheat kernels as well as finer contaminants such as sand. In addition, air (aspiration) is used to remove plant material that is lighter than the grain. Current equipment to remove stones and other dense materials the same size as grain kernels use air passed upward through an inclined, tilted table. This causes the grain to "float" off the side of the table and the heavier material to continue and "tail" over the end of the table. In earlier systems, grains were passed through washers in which water separated the grain from heavier material (such as stones) much like fluming of potatoes or fruit does. Impact with rotating disks and steel pegs (entoleters and infestation destroyers) or grinding operations are used prior to milling to break open kernels of wheat containing internal insect infestation. As a means of reducing insect fragments in the finished product, this process is followed by aspiration to lift out any light insect contaminants released in the operation.

As a final step in wheat milling, flour is generally passed through sieves fine enough to remove insect eggs and any other contaminants that might be present. This is to assure that when flour leaves the mill it is free of any viable form of insect contamination (37). Where flour is used in large quantities, such as commercial bakeries, prior to use, flour is again sieved to ensure that no contamination has occurred in transport and storage of the flour.

Metal contamination has been a major concern of all food processors. Although metal detection methods are not specifically among the isolation techniques represented in AOAC International or AACC International methods manuals, they serve the purpose of isolating contaminants from food products. Magnets of various types have been used on raw materials and processing systems to prevent the passage of metal into handling and processing equipment where both equipment damage and product contamination are concerns. **Metal detectors** are employed in many food processing operations and on finished product packaging lines to detect ferrous and nonferrous metal fragments and to prevent contaminated products from entering consumer food channels.

Recent X-ray technology suggests that X rays may have an advantage over other methods for detecting metal and that they can also be used to detect glass, wood, plastic, and bone chips in foods. Detection of these extraneous materials also can be automated with rejection systems in packaging lines (38).

19.8 SUMMARY

Extraneous matter in raw ingredients and in processed foods might be unavoidable in the array of foods that are stored, handled, processed, and transported. DALs are established for amounts considered unavoidable and of no health hazard. A variety of methods are available to isolate extraneous matter from foods. Those methods largely prescribed by AOAC International employ a series of physical and chemical means to separate the extraneous material for identification and enumeration. Major concerns in the analysis of food products for extraneous matter are the objectivity of methods and the availability of adequately trained analysts. Some "principles" of isolation are applied in a proactive way in food processing operations.

Currently available methods (both macroscopic and microscopic) show varying degrees of efficiency in analysis of extraneous matter and filth in foods. Some techniques are time-consuming, require trained personnel, and are difficult to implement in real time. Some techniques have not been found feasible to be implemented in food inspection systems because of their cost, unreliability, and the varying degrees of success obtained in detecting infestations. Macroscopic and microscopic procedures for characterizing defects in foods tend to supplement each other and together provide a comprehensive evaluation of defects in the product. It is important that the analyst realize the close association of complementary methods for use as a joint approach in solving analytical problems.

19.9 STUDY QUESTIONS

1. Indicate why the FDA has established DALs.
2. Explain why practicing cGMPs has no impact on DALs.
3. List three major reasons for conducting analysis for extraneous matter in foods.
4. What two resources provide methods for separating extraneous matter from cereal grains and their products?
5. There are several basic principles involved in separating (isolating) extraneous matter from foods. List five of these principles and give an example of each principle.
6. Briefly describe the major constraint(s) to currently accepted methods for analyses of extraneous matter in foods.
7. Explain how some of the more recent analytical techniques can assist in identifying sources of extraneous matter in foods.
8. What are some likely sources of error with the various analytical methods?

19.10 ACKNOWLEDGEMENT

This contribution is paper number 10-062-B of the Kansas Agricultural Experiment Station, Kansas State University, Manhattan, KS 66506.

19.11 REFERENCES

1. FDLI (1993) Federal food drug and cosmetic act, as amended. In: Compilation of food and drug laws. The Food and Drug Law Institute, Washington, DC
2. FDA (2009) Current good manufacturing practice in manufacturing, packing, or holding human food. Part 110, Title 21: food and drugs. In: Code of federal regulations. Office of the Federal Register National Archives and Records Administration, Washington, DC
3. FDA (1995) The food defect action levels – current levels for natural or unavoidable defects for human use that present no health hazard (revised 1998). Department of Health and Human Services, Food and Drug Administration, Washington, DC
4. FDA (2000) Compliance policy guide manual. Food and Drug Administration, Office of Regulatory Affairs, Washington, DC
5. AOAC International (2007) Extraneous materials: isolation. In: Horwitz W, Latimer G (eds) Official methods of analysis, 18th edn., 2005; Current through revision 2, 2007 (on-line). AOAC International, Gaithersburg, MD
6. AACC International (2010) AACC method 28 extraneous matter. In: Approved methods of the American Association of Cereal Chemists, 11th edn. American Association of Cereal Chemists, St. Paul, MN
7. Kurtz OL, Harris KL (1962) Micro-analytical entomology for food sanitation control. Association of Official Analytical Chemists, Washington, DC
8. Gentry JW, Harris KL (eds) (1991) Microanalytical entomology for food sanitation control, vols 1 and 2. Association of Official Analytical Chemists, Melbourne, FL
9. AACC (2000) X-ray examination for internal insect infestation, AACC method 28–21A. In: Approved methods of the American Association of cereal chemists, 10th edn. American Association of Cereal Chemists, St. Paul, MN
10. FDA (1981) Principles of food analysis for filth, decomposition, and foreign matter. FDA technical bulletin no. 1, Gorham JR (ed). Association of Official Analytical Chemists, Arlington, VA
11. FDA (1978) Training manual for analytical entomology in the food industry. FDA technical bulletin no. 2, Gorham JR (ed). Association of Official Analytical Chemists, Arlington, VA
12. Olsen AR (ed) (1995) Fundamentals of microanalytical entomology – a practical guide to detecting and identifying filth in foods. CRC Press, Boca Raton, FL
13. FDA (1998) Introduction and apparatus for macroanalytical methods. In: FDA technical bulletin number 5, macroanalytical procedures manual (MPM), FDA, Washington, DC
14. Pedersen JR (1992) Insects: identification, damage, and detection, ch. 12. In: Sauer DB (ed) Storage of cereal

grains and their products. American Association of Cereal Chemists, St. Paul, MN, pp 635–689
15. Pearson TC, Brabec DL, Schwartz CR (2003) Automated detection of internal insect infestations in whole wheat kernels using a Perten SKCS 4100. Appl Eng Agric 19:727–733
16. Haff RP, Slaughter DC (2004) Real-time X-ray inspection of wheat for infestation by the granary weevil, *Sitophilus granarius* (L.). Trans ASAE 47:531–537
17. Dogan H (2007) Non-destructive Imaging of agricultural products using X-ray microtomography. Proc Microsc Microanal Conf 13(2):512–513
18. Pearson TC, Cetin AE, Tewfik AH, Haff RP (2007) Feasibility of impact-acoustic emissions for detection of damaged wheat kernels. Digital Signal Process 17:617–633
19. Goldstein JI, Newbury DE, Echlin P, Joy DC, Romig AD Jr, Lyman CE, Fiori C, Lifshin E (1992) Scanning electron microscopy and X-ray microanalysis. A text for biologists, materials scientists, and geologists, 2nd edn. Plenum, New York
20. McCrone WC, Delly JG (1973) The particle atlas, 2nd edn. Ann Arbor Science, Ann Arbor, MI
21. Dowell FE, Ram MS, Seitz LM (1999) Predicting scab, vomitoxin, and ergosterol in single wheat kernels using near-infrared spectroscopy. Cereal Chem 76(4):573–576
22. Baker JE, Dowell FE, Throne JE (1999) Detection of parasitized rice weevils in wheat kernels with near-infrared spectroscopy. Biol Control 16:88–90
23. Ridgway C, Chambers J (1996) Detection of external and internal insect infestation in wheat by near-infrared reflectance spectroscopy. J Sci Food Agric 71:251–264
24. Ghaedian AR, Wehling RL (1997) Discrimination of sound and granary-weevil-larva-infested wheat kernels by near-infrared diffuse reflectance spectroscopy. J AOAC Int 80:997–1005
25. Dowell FD, Throne JE, Baker JE (1998) Automated nondestructive detection of internal insect infestation of wheat kernels by using near-infrared reflectance spectroscopy. J Econ Entomol 91:899–904
26. Berardo N, Pisacane V, Battilani P, Scandolara A, Pietro A, Marocco A (2005) Rapid detection of kernel rots and mycotoxins in maize by near-infrared reflectance spectroscopy. J Agric Food Chem 53:8128–8134
27. Delwiche SR, Hareland GA (2004) Detection of scab-damaged hard red spring wheat kernels by near-infrared reflectance. Cereal Chem 81(5):643–649
28. Dowell FE, Pearson TC, Maghirang EB, Xie F, Wicklow DT (2002) Reflectance and transmittance spectroscopy applied to detecting fumonisin in single corn kernels infected with *Fusarium verticillioides*. Cereal Chem 79(2):222–226
29. Perez-Mendoza P, Throne JE, Dowell FE, Baker JE (2003) Detection of insect fragments in wheat flour by near-infrared spectroscopy. J Stored Prod Res 39:305–312
30. Quinn FA, Burkholder WE, Kitto GB (1992) Immunological technique for measuring insect contamination of grain. J Econ Entomol 85:1463–1470
31. Kitto GB, Quinn FA, Burkholder W (1992) Techniques for detecting insect contamination of foodstuffs. US Patent 5,118,610
32. Atui MB, Flin PW, Lazzari SMN, Lazzari FA (2007) Detection of *Rhyzopertha dominica* larvae in stored wheat using ELISA: the impact of myosin degradation following fumigation. J Stored Prod Res 43:156–159
33. Toews MD, Pearson TC, Campbell JF (2006) Imaging and automated detection of *Sitophilus oryzae* (Coleoptera: Curculionidae) pupae in hard red winter wheat. J Econ Entomol 99(2):583–592
34. Pearson TC, Brabec DL (2007) Detection of wheat kernels with hidden insect infestations with an electrically conductive roller mill. Appl Eng Agric 23(5):639–645
35. Rajendran S (2005) Detection of insect infestation in stored foods. In: Taylor SL (ed) Advances in food and nutrition research, vol 49. Elsevier Academic, UK, pp 163–232
36. Hagstrum DW, Subramanyam Bh (2006) Fundamentals of stored product entomology. AACC International, St. Paul, MN
37. Mills R, Pedersen J (1990) A flour mill sanitation manual. Eagan, St. Paul, MN
38. FMC FoodTech (2001) X-ray technology. Solutions 2:20, 21

Determination of Oxygen Demand

Yong D. Hang

Department of Food Science and Technology, Cornell University, Geneva, NY 14456, USA
ydh1@cornell.edu

20.1 Introduction 369
20.2 Methods 369
20.2.1 Biochemical Oxygen Demand 369
20.2.1.1 Principle 369
20.2.1.2 Procedure 369
20.2.1.3 Applications and Limitations 369
20.2.2 Chemical Oxygen Demand 370
20.2.2.1 Principle 370
20.2.2.2 Procedure 370
20.2.2.3 Applications and Limitations 370
20.3 Comparison of BOD and COD 370
20.4 Sampling and Handling Requirements 371
20.5 Summary 371
20.6 Study Questions 371
20.7 Practice Problems 372
20.8 References 372

S.S. Nielsen, *Food Analysis*, Food Science Texts Series, DOI 10.1007/978-1-4419-1478-1_20,

20.1 INTRODUCTION

Oxygen demand is a commonly used parameter to evaluate the potential effect of organic pollutants on either a wastewater treatment process or a receiving water body. Because microorganisms utilize these organic materials, the concentration of dissolved oxygen is greatly depleted from the water. The oxygen depletion in the environment can have a detrimental effect on fish and plant life.

The two main methods used to measure the oxygen demand of water and wastewater are **biochemical oxygen demand** (BOD) and **chemical oxygen demand** (COD). This chapter briefly describes the principles, procedures, applications, and limitations of each method. Methods described in this chapter are adapted from *Standard Methods for the Examination of Water and Wastewater*, published by the American Public Health Association (APHA) (1). The book includes step-by-step procedures with equipment for BOD, COD, and other tests for water and wastewater.

20.2 METHODS

20.2.1 Biochemical Oxygen Demand

20.2.1.1 Principle

The **biochemical oxygen demand** (BOD) determination is a measure of the amount of oxygen required by microorganisms to oxidize the biodegradable organic constituents present in water and wastewater. The method is based on the direct relationship between the concentration of organic matter and the amount of oxygen used to oxidize the pollutants to water, carbon dioxide, and inorganic nitrogenous compounds. The **oxygen demand** of water and wastewater is proportional to the amount of **organic matter** present. The BOD method measures the biodegradable carbon (carbonaceous demand) and, under certain circumstances, the biodegradable nitrogen (nitrogenous demand).

20.2.1.2 Procedure

Place a known amount of a water or wastewater sample that has been seeded with an effluent from a biological waste treatment plant in an airtight BOD bottle and measure the initial dissolved oxygen immediately. Incubate the sample at 20 °C and, after 5 days, measure the dissolved oxygen content again (APHA Method 4500–0). The dissolved oxygen content can be determined by the membrane electrode method (APHA Method 4500-O G) or the azide modification (APHA Method 4500-0 C), permanganate modification (APHA Method 4500-0 D), alum flocculation modification (4500-0 E) or copper sulfate–sulfamic acid flocculation modification (APHA Method 4500-0 F) of the iodometric method (APHA Method 4500-0 B) to minimize interference by nitrite or ferrous or ferric iron. The iodometric method is a titrimetric procedure that is based on the oxidizing property of dissolved oxygen while the membrane electrode method is based on the diffusion rate of dissolved oxygen across a membrane. A dissolved oxygen meter with an oxygen-sensitive membrane electrode made by Fisher, Orion, YSI, or other companies is used to measure the diffusion current, which is linearly proportional to the concentration of dissolved oxygen under steady-state conditions. It is important to change frequently and calibrate the membrane electrode to eliminate the effect of interfering gases such as hydrogen sulfide. The azide modified iodometric procedure, for example, is used to remove interference of nitrite, which is the most commonly interfering material in water and wastewater. The alum flocculation modification method is commonly used to minimize the interference caused by the presence of suspended solids. The BOD value, which is expressed as mg/L, can be calculated from the difference in the initial dissolved oxygen and the content of dissolved oxygen after the incubation period according to the following equation (APHA Method 5210 B):

$$\text{BOD (mg/L)} = 100/P \times (\text{DOB} - \text{DOD}) \qquad [1]$$

where:

DOB = initial oxygen in diluted sample (mg/L)
DOD = oxygen in diluted sample after 5-day incubation, (mg/L)
P = ml sample × 100/capacity of bottle

20.2.1.3 Applications and Limitations

The BOD test is used most widely to measure the organic loading of waste treatment processes, to determine the efficiency of treatment systems, and to assess the effect of wastewater on the quality of receiving waters. The 5-day BOD test has some drawbacks because of the following:

1. The procedure requires an incubation time of at least 5 days.
2. The BOD method does not measure all the organic materials that are biodegradable.
3. The test is not accurate without a proper seeding material.
4. Toxic substances such as chlorine present in water and wastewater may inhibit microbial growth.

20.2.2 Chemical Oxygen Demand

20.2.2.1 Principle

The **chemical oxygen demand** (COD) determination is a rapid way to estimate the quantity of oxygen used to oxidize the organic matter present in water and wastewater. Most organic compounds are destroyed by refluxing in a strong acid solution with a known quantity of a strong oxidizing agent such as **potassium dichromate**. The excess amount of potassium dichromate left after digestion of the organic matter is measured. The amount of organic matter that is chemically oxidizable is directly proportional to the potassium dichromate consumed.

20.2.2.2 Procedure

A known quantity of sample of water or wastewater is refluxed at elevated temperatures for up to 2 h with a known quantity of potassium dichromate and sulfuric acid using an open reflux method (APHA Method 5220 B), a closed reflux titrimetric method (APHA Method 5220 C), or a closed reflux colorimetric method (APHA Method 5220 D). The amount of potassium dichromate left after digestion of the organic matter is titrated with a standard ferrous ammonium sulfate (FAS) solution using orthophenanthroline ferrous complex as an indicator. The amount of oxidizable organic matter, determined as oxygen equivalent, is proportional to the potassium dichromate used in the oxidative reaction. The COD value can be calculated from the following equation (APHA Method 5220 B):

$$\text{COD (mg/L)} = (A - B) \times M \times 8000/D \qquad [2]$$

where:

A = ml FAS used for blank

B = ml FAS used for sample

M = molarity of FAS

D = ml sample used

8000 = milliequivalent weight of oxygen × 1000 ml/L

20.2.2.3 Applications and Limitations

Potassium dichromate is widely used for the COD method because of its advantages over other oxidizing compounds in oxidizability, applicability to a wide variety of waste samples, and ease of manipulation. The dichromate reflux method can be used to measure the samples with COD values of greater than 50 mg/L.

The COD test measures carbon and hydrogen in organic constituents but not nitrogenous compounds. Furthermore, the method does not differentiate between biologically stable and unstable compounds present in water and wastewater. The COD test is a very important procedure for routinely monitoring industrial wastewater discharges and for the control of waste treatment processes. The test is faster and more reproducible than the BOD method. The obvious disadvantages of the COD method are as follows:

1. Aromatic hydrocarbons, pyridine, and straight-chain aliphatic compounds are not readily oxidized.
2. The method is very susceptible to interference by **chloride**, and thus the COD of certain food processing waste effluents such as pickle and sauerkraut brines cannot be readily determined without modification. This difficulty may be overcome by adding **mercuric sulfate** to the sample prior to refluxing. Chloride concentrations greater than 500–1000 mg/L may not be corrected by the addition of mercuric sulfate. A chloride correction factor can be developed for a particular waste by the use of proper blanks.

20.3 COMPARISON OF BOD AND COD

The BOD and COD analyses of water and wastewater can result in different values because the two methods measure different materials. As shown in Table 20-1, the COD value of a waste sample is usually higher than its BOD because of the following:

1. Many organic compounds that can be chemically oxidized cannot be biochemically oxidized. For example, cellulose cannot be determined by the BOD method but can be measured by the COD test.
2. Certain inorganic compounds such as ferrous iron, nitrites, sulfides, and thiosulfates are readily oxidized by potassium dichromate. This inorganic COD introduces an error when computing the organic matter of water and wastewater.

Oxygen Demand of Tomato Processing Wastes

Item	*1973*	*1974*	*1975*
BOD (mg/L)	2400	1300	1200
COD (mg/L)	5500	3000	2800
TOC (mg/L)	2000	1100	1000

From (2).

20-2 table **COD and BOD Values of Selected Fruit and Vegetable Processing Wastes**

Product	*COD (mg/L)*	*BOD (mg/L)*	*Mean Ratio (BOD/COD)*
Apples	395–37,000	240–19,000	0.55
Beets	445–13,240	530–6,400	0.57
Carrots	1750–2910	817–1,927	0.52
Cherries	1200–3795	600–1,900	0.53
Corn	3400–10,100	1587–5341	0.50
Green beans	78–2200	43–1,400	0.55
Peas	723–2284	337–1350	0.61
Sauerkraut	470–65,000	300–41,000	0.66
Tomatoes	652–2305	454–1575	0.72
Wax beans	193–597	55–323	0.58
Wine	495–12,200	363–7645	0.60

From (2).

3. The BOD test can give low values because of a poor seeding material. The COD test does not require an inoculum.
4. Some aromatics and nitrogenous (ammonium) compounds are not oxidized by the COD method. Other organic constituents such as cellulose or lignin, which are readily oxidized by potassium dichromate, are not biologically degraded by the BOD method.
5. Toxic materials present in water and wastewater that do not interfere with the COD test can affect the BOD results.

The COD has value for specific wastes since it is possible to obtain a direct correlation between COD and BOD values. Table 20-2 shows the COD and BOD values of waste effluents from fruit and vegetable processing factories. The BOD/COD ratios of these processing waste effluents varied considerably and ranged from 0.50 to 0.72 (2). The **BOD/COD ratio** can be a useful tool for rapid determination of the biodegradability of organic matter present in the wastes. A low BOD/COD ratio indicates the presence of a large amount of nonbiodegradable organic matter. Samples of wastewater with high BOD/COD ratios have a small amount of organic matter that is nonbiodegradable.

20.4 SAMPLING AND HANDLING REQUIREMENTS

Samples of water and wastewater collected for oxygen demand determinations must be analyzed as soon as possible or stored under properly controlled conditions until analyses can be made.

Samples for the BOD test can be kept at low temperatures (4°C or below) for up to 48 h. Chemical preservatives should not be added to water and wastewater because they can interfere with BOD analysis.

Untreated wastewater samples for the COD test must be collected in glass containers and analyzed promptly. The COD samples can be stored at 4 °C or below for up to 28 days if these are acidified with a concentrated mineral acid (sulfuric acid) to a pH value of 2.0 or below.

20.5 SUMMARY

Oxygen demand is most widely used to determine the effect of organic pollutants present in water and wastewater on receiving streams and rivers. The two important methods used to measure oxygen demand are BOD and COD.

The BOD test measures the amount of oxygen required by microorganisms to oxidize the biodegradable organic matter present in water and wastewater. The COD method determines the quantity of oxygen consumed during the oxidation of organic matter in water and wastewater by potassium dichromate.

Of the two methods used to measure oxygen demand, the BOD test has the widest application in measuring waste loading to treatment systems, in determining the efficiency of treatment processes, and in evaluating the quality of receiving streams and rivers because it most closely approximates the natural conditions of the environment. The COD test can be used to monitor routinely the biodegradability of organic matter in water and wastewater if a relationship between COD and BOD has been established.

20.6 STUDY QUESTIONS

1. In your new job as supervisor of a lab that has previously been using the BOD method to determine oxygen demand of wastewater, you have decided to change to the COD method.
   (a) Differentiate the basic principle and procedure of the BOD and COD methods for your lab technicians.
   (b) In what case would they be instructed to use mercuric sulfate in the COD assay?
   (c) You realize there are advantages and disadvantages of the two potential methods, BOD and COD. Give two advantages and two disadvantages for the BOD method as compared to the COD method.
2. In each case described below, indicate if you would expect the COD value to be higher or lower than results from a BOD test. Explain your answer.
   (a) Poor seed material in BOD test
   (b) Sample contains toxic materials

(c) Sample high in aromatics and nitrogenous compounds
(d) Sample high in nitrites and ferrous iron
(e) Sample high in cellulose and lignin

20.7 PRACTICE PROBLEMS

1. Determine the BOD value of a sample given the following data (see Equation [1]):

   DOB = 9.0 mg/L
   DOD = 6.6 mg/L
   P = 15 ml
   Capacity of bottle = 300 ml

2. Determine the COD value of a sample given the following data (see Equation [2]):

   ml FAS for blank = 37.8 ml
   ml FAS for sample = 34.4 ml
   Molarity of FAS = 0.025 M
   Sample = 5 ml

Answers

1. BOD = 48 mg/L

   Calculation:

$$\begin{aligned}\text{BOD(mg/L)} &= 100/P \times (9.0\,\text{mg/L} - 6.6\,\text{mg/L})\\ &= 100/P \times 2.4\\ &= 240/P\\ &= 240/(15\,\text{mL} \times 100/300\,\text{ml})\\ &= 240/5\\ &= 48\end{aligned}$$

2. COD = 136 mg/L

   Calculation:

$$\begin{aligned}\text{COD (mg/L)} &= (37.8\,\text{ml} - 34.4\,\text{ml}) \times 0.025 \times 8000/D\\ &= 3.4 \times 0.025 \times 8000/D\\ &= 680/D\\ &= 680/5\\ &= 136\end{aligned}$$

20.8 REFERENCES

1. Eaton AD, Clesceri LS, Rice EW, Greenberg AE (eds) (2005) Standard methods for the examination of water and wastewater, 21st edn. American Public Health Association (APHA), Washington, DC
2. Splittstoesser DF, Downing DL (1969) Analysis of effluents from fruit and vegetable processing factories. NY State Agr Exp Sta Res Circ 17. Geneva, New York

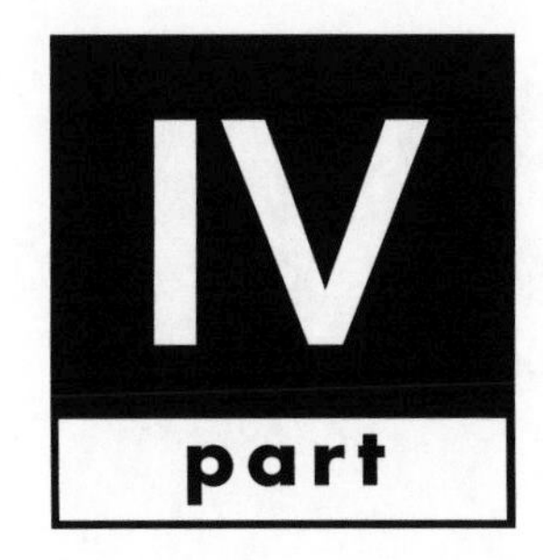

Spectroscopy

Basic Principles of Spectroscopy

Michael H. Penner

Department of Food Science and Technology, Oregon State University, Corvallis, OR 97331-6602, USA

mike.penner@orst.edu

21.1 Introduction 377
21.2 Light 377
 21.2.1 Properties 377
 21.2.2 Terminology 378
 21.2.3 Interference 378
21.3 Energy States of Matter 379
 21.3.1 Quantum Nature of Matter 379
 21.3.2 Electronic, Vibrational, and Rotational Energy Levels 380
 21.3.3 Nuclear Energy Levels in Applied Magnetic Fields 382
21.4 Energy Level Transitions in Spectroscopy 382
 21.4.1 Absorption of Radiation 382
 21.4.2 Emission of Radiation 383
21.5 Summary 384
21.6 Study Questions 384
21.7 Resource Materials 385

S.S. Nielsen, *Food Analysis*, Food Science Texts Series, DOI 10.1007/978-1-4419-1478-1_21,

21.1 INTRODUCTION

Spectroscopy deals with the production, measurement, and interpretation of spectra arising from the **interaction of electromagnetic radiation with matter**. There are many different spectroscopic methods available for solving a wide range of analytical problems. The methods differ with respect to the species to be analyzed (such as molecular or atomic spectroscopy), the type of radiation–matter interaction to be monitored (such as absorption, emission, or diffraction), and the region of the electromagnetic spectrum used in the analysis. Spectroscopic methods are very informative and widely used for both quantitative and qualitative analyses. Spectroscopic methods based on the absorption or emission of radiation in the **ultraviolet** (UV), **visible** (Vis), **infrared** (IR), and radio (**nuclear magnetic resonance, NMR**) frequency ranges are most commonly encountered in traditional food analysis laboratories. Each of these methods is distinct in that it monitors different types of molecular or atomic transitions. The basis of these transitions is explained in the following sections.

21.2 LIGHT

21.2.1 Properties

Light may be thought of as particles of energy that move through space with wavelike properties. This image of light suggests that the energy associated with a ray of light is not distributed continuously through space along the wave's associated electric and magnetic fields but rather that it is concentrated in discrete packets. Light is therefore said to have a dual nature: **particulate** and **wavelike**. Phenomena associated with light propagation, such as interference, diffraction, and refraction, are most easily explained using the wave theory of electromagnetic radiation. However, the interaction of light with matter, which is the basis of absorption and emission spectroscopy, may be best understood in terms of the particulate nature of light. Light is not unique in possessing both wavelike and particulate properties. For example, fundamental particles of matter, such as electrons, protons, and neutrons, are known to exhibit wavelike behavior.

The wave properties of electromagnetic radiation are described in terms of the wave's frequency, wavelength, and amplitude. A graphical representation of a plane-polarized electromagnetic wave is given in Fig. 21-1. The wave is plane polarized in that the oscillating electric and magnetic fields making up the wave are each limited to a single plane. The **frequency** (ν, *the lower case Greek letter nu*) of a wave is defined as the number of oscillations the wave will make at a given point per second. This is the reciprocal of the **period** (p) of a wave, which is the time in seconds required for successive maxima of the wave to pass a fixed point. The **wavelength** (λ) represents the distance between successive maxima on any given wave. The units used in reporting wavelengths will depend on the region of electromagnetic radiation used in the analysis. Spectroscopic data sometimes are reported with respect to **wavenumbers** ($\bar{\nu}$), which are reciprocal wavelengths in units of cm^{-1}. Wavenumbers are encountered most often in IR spectroscopy. The **velocity of propagation** (v_i) of an electromagnetic wave, in units of distance per second, in any given medium "i" can be calculated by taking the product of the frequency of the wave, in cycles per second, and its wavelength in that particular medium.

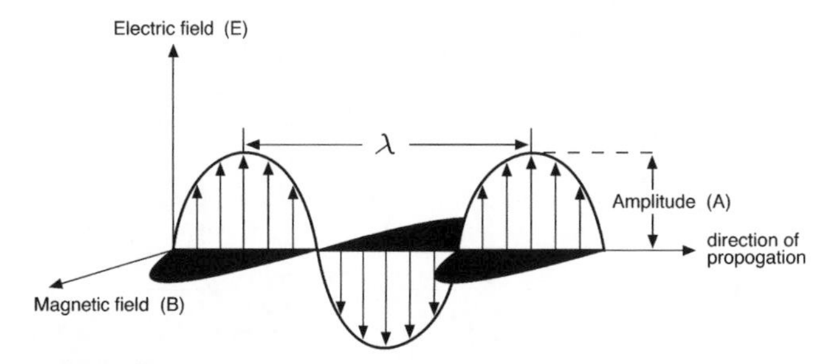

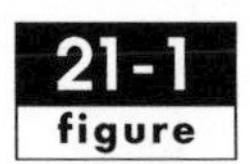

Representation of plane-polarized electromagnetic radiation propagating along the x-axis. The electric and magnetic fields are in phase, perpendicular to each other, and to the direction of propagation.

$$v_i = \nu\lambda_i \quad [1]$$

where:

v_i = velocity of propagation in medium i
ν = frequency (of associated wave)
λ_i = wavelength in medium i

The frequency of an electromagnetic wave is determined by the source of the radiation, and it remains constant as the wave traverses different media. However, the velocity of propagation of a wave will vary slightly depending on the medium through which the light is propagated. The wavelength of the radiation will change in proportion to changes in wave velocity as defined by Equation [1]. The **amplitude of the wave** (A) represents the magnitude of the electric vector at the wave maxima. The **radiant power** (P) and **radiant intensity** (I) of a beam of radiation are proportional to the square of the amplitude of the associated waves making up that radiation. Figure 21-1 indicates that electromagnetic waves are composed of **oscillating magnetic** and **electric fields**, the two of which are mutually perpendicular, in phase with each other, and perpendicular to the direction of wave propagation.

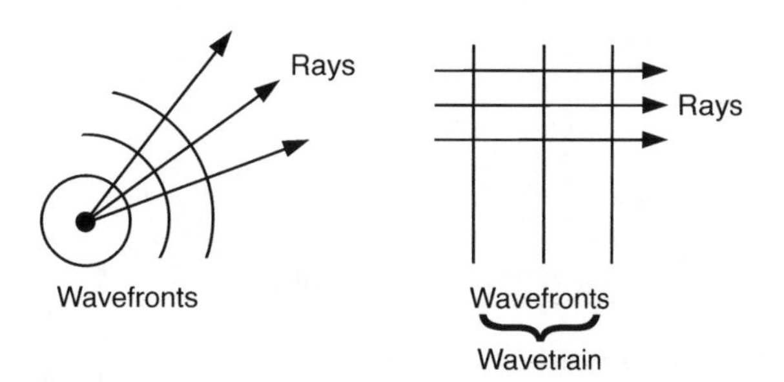

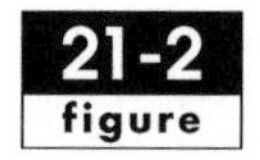

Wavefronts, wave-trains, and rays. [From Hugh D. Young, *University Physics* (8th edn), p 947, © 1992 by Addison-Wesley, Reading, MA. Courtesy of the publisher.]

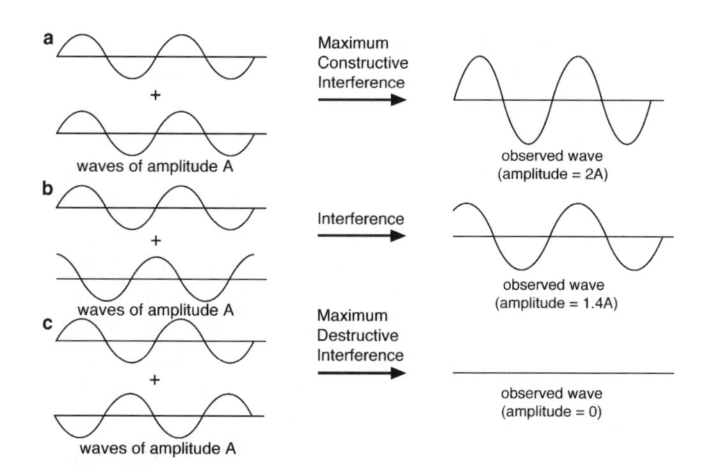

21-3 figure
Interference of identical waves that are (**a**) in phase, (**b**) 90° out of phase, and (**c**) 180° out of phase.

As drawn, the waves represent changes in the respective field strengths with time at a fixed location or changes in the respective field strengths over distance at a fixed time. The electrical and magnetic components of the waves are represented as a series of vectors whose lengths are proportional to the magnitude of the respective field. It is the oscillating electric field that is of most significance to spectroscopic phenomena such as absorption, transmission, and refraction. However, a purely electric field, without its associated magnetic field, is impossible.

21.2.2 Terminology

The propagation of electromagnetic waves is often described in terms of wavefronts or trains of waves (Fig. 21-2). A **wavefront** represents the locus of a set of points all of which are in phase. For a point source of light, a concentric ring that passes through the maxima of adjacent light rays will represent a wavefront. The entire ring need not be drawn in all cases, such that wavefronts may represent planes of light in cases where the observation is sufficiently removed from the point source that the curved surface appears planar. Wavefronts are most typically drawn by connecting maxima, minima, or both for adjacent rays. If maxima are used for depicting wavefronts, then each of the wavefronts will be separated by one wavelength. A **train of waves**, or **wave-train**, refers to a series of wavefronts all of which are in phase, that is, each individual wave will have a maximum amplitude at the same location in space. A wave-train also may be represented by a series of light rays. Rays of light are used generally with reference to the corpuscular nature of light, representing the path of photons. A wave-train would indicate that a series of photons, all in phase, followed the same path.

21.2.3 Interference

Interference is the term used to describe the observation that when two or more wave-trains cross one another, they result in an instantaneous wave, at the point of intersection, whose amplitude is the algebraic sum of the amplitudes of the individual waves at the point of intersection. The law describing this wave behavior is known as the **principle of superposition**. Superposition of sinusoidal waves is illustrated in Fig. 21-3. Note that in all cases, the effective amplitude of the perceived wave at the point in question is the combined effect of each of the waves that crosses that point at any given instant. In spectroscopy, the amplitude of most general interest is that corresponding to the magnitude of the resulting electric field intensity. **Maximum constructive interference** of two waves occurs when the waves are completely in phase (i.e., the maxima of one wave align with the maxima of the other wave), while **maximum destructive interference** occurs when waves are 180° out of phase (the maxima of one wave align with the minima of the other wave). This concept of interference is fundamental to the interpretation of diffraction data, which represents a specialized segment of qualitative spectroscopy. Interference phenomena also are widely used in the design of spectroscopic instruments that require the dispersion or selection of radiation, such as those instruments employing grating monochromators or interference filters, as described in Chap. 23.

Interference phenomena are best rationalized by considering the wavelike nature of light. However, phenomena such as the absorption and emission of radiation are more easily understood by considering the particulate nature of light. The particles of energy that move through space with wavelike properties are called **photons**. The energy of a photon can be defined in terms of the frequency of the wave with which it is associated Equation [2].

$$E = h\nu \qquad [2]$$

where:

E = energy of a photon
h = Planck's constant
ν = frequency (of associated wave)

This relationship indicates that the photons making up **monochromatic light**, which is electromagnetic radiation composed of waves having a single frequency and wavelength, are all of equivalent energy. Furthermore, just as the frequency of a wave is a constant determined by the radiation source, the energy of associated photons also will be unchanging. The brightness of a beam of monochromatic light, when expressed in terms of the particulate nature of light, will be the product of the photon flux and the energy per photon. The **photon flux** refers to the number of photons flowing across a unit area perpendicular to the beam per unit time. It follows that to change the brightness of a beam of monochromatic light will require a change in the photon flux. In spectroscopy, the term *brightness* is generally not used, but rather one refers to the **radiant power** (P) or the **radiant intensity** (I) of a beam of light. Radiant power and radiant intensity often are used synonymously when referring to the amount of radiant energy striking a given area per unit time. In terms of International Scientific (SI) Units (time, seconds; area, meters; energy, joules), radiant power equals the number of joules of radiant energy impinging on a 1 m^2 area of detector per second. The basic interrelationships of light-related properties and a general scheme of the electromagnetic spectrum are presented in Table 21-1 and Fig. 21-4, respectively.

21.3 ENERGY STATES OF MATTER

21.3.1 Quantum Nature of Matter

The energy content of matter is quantized. Consequently, the potential or internal energy content of an atom or molecule does not vary in a continuous manner but rather in a series of discrete steps. Atoms and molecules, under normal conditions, exist predominantly in the **ground state**, which is the state of lowest energy. Ground state atoms and molecules can gain energy, in which case they will be elevated to one of their higher energy states, referred to as **excited states**. The quantum nature of atoms and molecules puts limitations on the energy levels that are available to these species. Consequently, there will be specific "allowed" **internal energy levels** for each atomic or molecular species. Internal energy levels not corresponding to an allowed value for that particular species are unattainable. The set of available energy levels for any given atom or molecule will be distinct for that species. Similarly, the potential energy spacings between allowed internal energy levels will be characteristic of a species. Therefore, the set of potential energy spacings for a species may be used qualitatively as a distinct fingerprint. Qualitative absorption and emission spectroscopy make use of this phenomenon in that these techniques attempt to determine an unknown compound's relative energy spacings by measuring transitions between allowed energy levels.

Properties of Light

Symbols/Terms	*Relationship*	*Frequently Used Units*
λ = wavelength	$\lambda_i \nu = v_i$ ($v_i = c$ *in a vacuum*)	nm (nanometers, 10^{-9}m) Å (Ångstrom units, 10^{-10} m) μm (microns, 10^{-6} m) mμ (millimicrons, 10^{-9} m)
ν = frequency		Hz (hertz, 1 Hz = 1 oscillation per second)
c = speed of light		2.9979×10^8 m s^{-1} in vacuum
$\bar{\nu}$ = wavenumber	$= 1/\lambda$	cm^{-1} kK (kilokayser, 1 kK = 1,000 cm^{-1})
p = period	$p = 1/\nu$	s
E = energy	$E = h\nu$ $= hc/\lambda$ $= hc\bar{\nu}$	J (1 joule = 1 kg m^2 s^{-2}) cal (calorie, 1 cal = 4.184 J) erg (1 erg = 10^{-7} J) eV (1 electron volt = 1.6022×10^{-19} J)
h = Planck's constant		6.6262×10^{-34} J s
P = radiant power	Amount of energy striking a given unit area per unit time	(Joules)$(m^2)^{-1}(s)^{-1}$

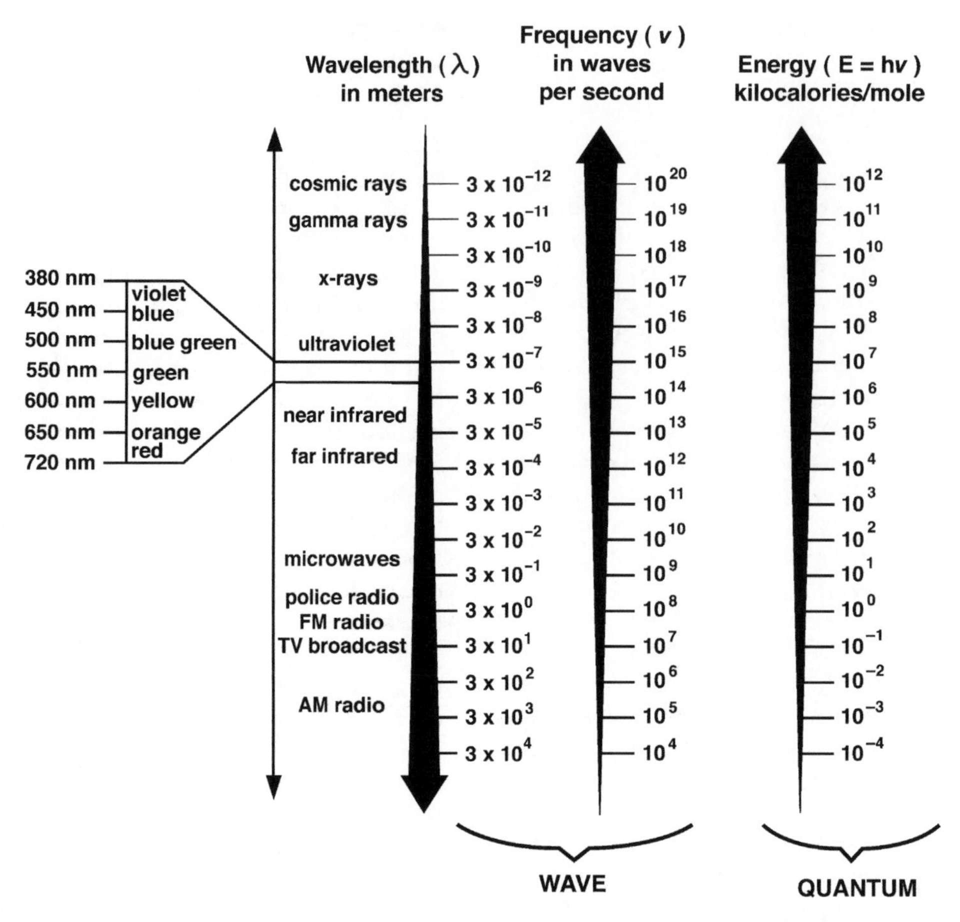

The electromagnetic spectrum. [From (10), p 3. Courtesy of Milton Roy Company, Rochester, NY, a subsidiary of Sundstrand Corporation.]

21.3.2 Electronic, Vibrational, and Rotational Energy Levels

The relative **potential energy** of an atom or molecule corresponds to the energy difference between the energy state in which the species exists and that of the ground state. Figure 21-5 is a partial molecular energy level diagram depicting potential energy levels for an organic molecule. The lowest energy state in the figure, bottom line in bold, represents the ground state. There are three **electronic energy states** depicted, each with its corresponding vibrational and rotational energy levels. Each of the electronic states corresponds to a given **electron orbital**. Electrons in different orbitals are of different potential energy. When an electron changes orbitals, such as when absorbing or emitting a photon of appropriate energy, it is termed an **electronic transition** since it is the electron that is changing energy levels. However, any change in the potential energy of an electron will, by necessity, result in a corresponding change in the potential energy of the atom or molecule that the electron is associated with.

Atoms are like **molecules** in that only specific energy levels are allowed for atomic electrons. Consequently, an energy level diagram of an atom would consist of a series of electronic energy levels. In contrast to molecules, the electronic energy levels of atoms have no corresponding vibrational and rotational levels and, hence, may appear less complicated. Atomic energy levels correspond to allowed electron shells (orbits) and corresponding subshells (i.e., 1s, 2s, 2p, etc.). The magnitude of the energy difference between the ground state and first excited states for valence electrons of atoms and bonding electrons of molecules is generally of the same range as the energy content of photons associated with UV and Vis radiation.

The wider lines within each electronic state of Fig. 21-5 depict the species' **vibrational energy levels**. The atoms that comprise a molecule are in constant motion, vibrating in many ways. However, in all cases the energy associated with this vibrational motion corresponds to defined quantized energy levels. The energy differences between neighboring vibrational energy levels are much smaller than those between

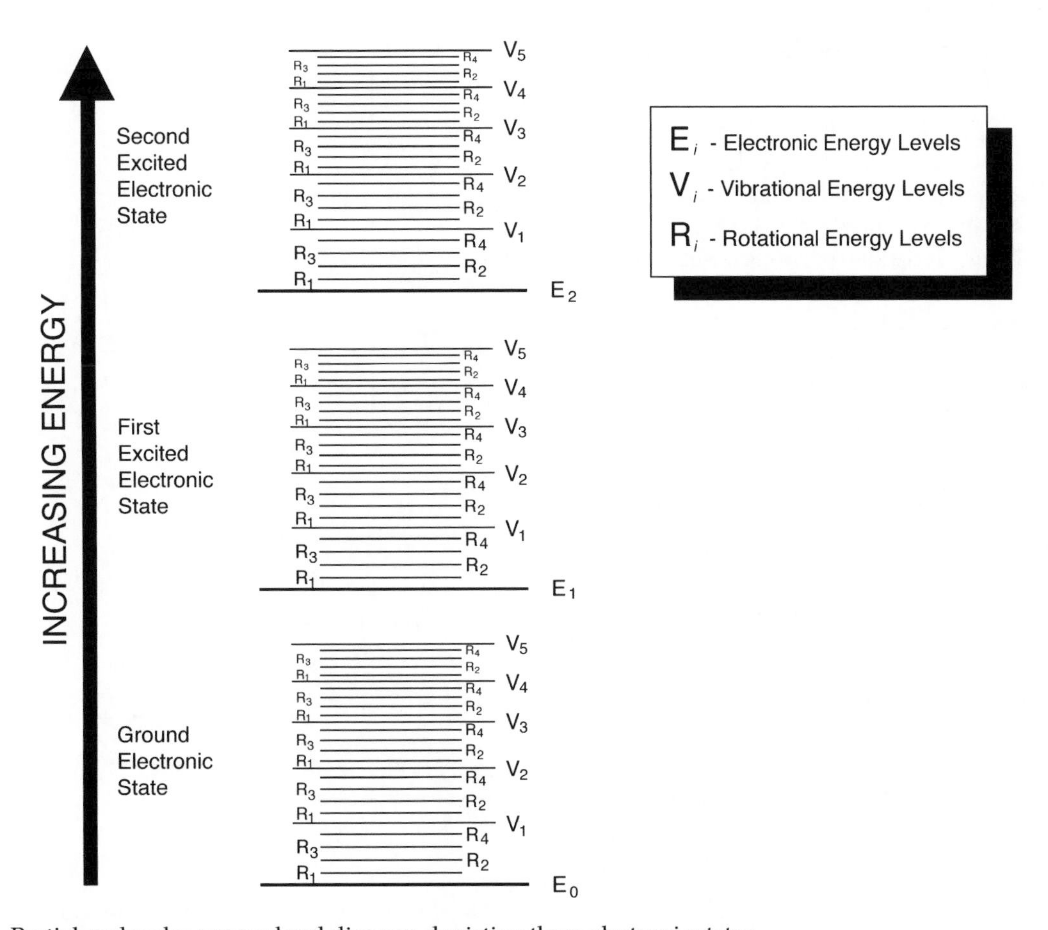

21-5 figure Partial molecular energy level diagram depicting three electronic states.

adjacent electronic energy levels. Therefore, it is common to consider that several vibrational energy levels are superimposed on each of the molecular electronic energy levels. Energy differences between allowed vibrational energy levels are of the same magnitude as the energy of photons associated with radiation in the IR region. Vibrational energy levels would not be superimposed on an atomic potential energy level diagram since this vibrational motion does not exist in a single atom. In this respect, the potential energy diagram for an atom is less complex than that for a molecule, the atomic energy level diagram having fewer energy levels.

The potential energy of a molecule also is quantized in terms of the energy associated with the rotation of the molecule about its center of gravity. These **rotational energy levels** are yet more closely spaced than the corresponding vibrational levels, as depicted by the narrow lines within each electronic state shown in Fig. 21-5. Hence, it is customary to consider several rotational energy levels superimposed on each of the permitted vibrational energy levels. The energy spacings between rotational energy levels are of the same magnitude as the energy associated with photons of microwave radiation. Microwave spectroscopy is not commonly used in food analysis laboratories; however, the presence of these different energy levels will impact the spectrum observed in other forms of spectroscopy, as will be discussed later. Similar to the situation of vibrational energy levels, rotational energy levels are not of consequence to atomic spectroscopy.

In summation, the internal energy of an **atom** is described in terms of its **electronic energy levels**, while the internal energy of a **molecule** is dependent on its **electronic**, **vibrational**, and **rotational energies**. The algebraic form of these statements follows.

$$E_{\text{atom}} = E_{\text{electronic}} \quad [3]$$

$$E_{\text{molecule}} = E_{\text{electronic}} + E_{\text{vibrational}} + E_{\text{rotational}} \quad [4]$$

The spectroscopist makes use of the fact that each of these associated energies is quantized and that different species will have somewhat different energy spacings.

21.3.3 Nuclear Energy Levels in Applied Magnetic Fields

NMR spectroscopy makes use of yet another type of quantized energy level. The energy levels of importance to NMR spectroscopy differ with respect to those described above in that they are relevant only in the presence of an applied external **magnetic field**. The basis for the observed energy levels may be rationalized by considering that the nuclei of some atoms behave as tiny bar magnets. Hence, when the atoms are placed in a magnetic field, their nuclear magnetic moment will have a preferred orientation, just as a bar magnet would behave. The NMR-sensitive nuclei of general relevance to the food analyst have two permissible orientations. The energy difference between these allowed orientations depends on the effective magnetic field strength that the nuclei experience. The effective magnetic field strength will itself depend on the strength of the applied magnetic field and the chemical environment surrounding the nuclei in question. The applied magnetic field strength will be set by the spectroscopist, and it is essentially equivalent for each of the nuclei in the applied field. Hence, differences in energy spacings of NMR-sensitive nuclei will depend solely on the identity of the nucleus and its environment. In general, the energy spacings between permissible nuclear orientations, under usable external magnetic field strengths, are of the same magnitude as the energy associated with radiation in the radiofrequency range.

21.4 ENERGY LEVEL TRANSITIONS IN SPECTROSCOPY

21.4.1 Absorption of Radiation

The **absorption of radiation** by an atom or molecule is that process in which energy from a photon of electromagnetic radiation is transferred to the absorbing species. When an atom or molecule absorbs a photon of light, its internal energy increases by an amount equivalent to the amount of energy in that particular photon. Therefore, in the process of absorption, the species goes from a lower energy state to a more **excited state**. In most cases, the species is in the **ground state** prior to absorption. Since the absorption process may be considered quantitative (i.e., all of the photon's energy is transferred to the absorbing species), the photon being absorbed must have an energy content that exactly matches the energy difference between the energy levels across which the transition occurs. This must be the case due to the quantized energy levels of matter, as discussed previously. Consequently, if one plots photon energy vs. the relative absorbance of radiation uniquely composed of photons of that energy, one observes a characteristic **absorption spectrum**, the shape of which is determined by the relative absorptivity of photons of different energy. The **absorptivity** of a compound is a wavelength-dependent proportionality constant that relates the absorbing species concentration to its experimentally measured absorbance under defined conditions. A representative absorption spectrum covering a portion of the UV radiation range is presented in Fig. 21-6. The independent variable of an absorption spectrum is most commonly expressed in terms of the wave properties (wavelength, frequency, or wavenumbers) of the radiation, as in Fig. 21-6, rather than the energy of the associated photons.

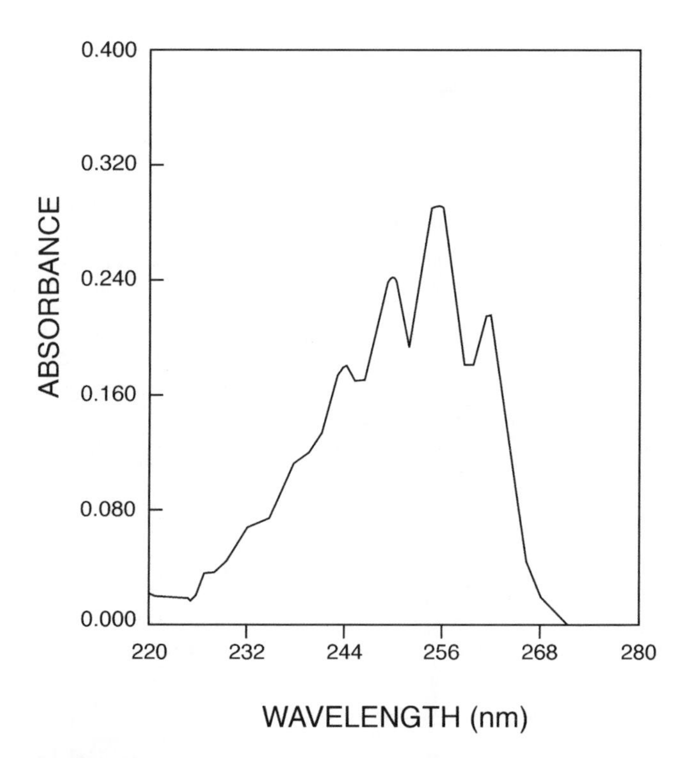

21-6 figure Absorption spectrum of a 0.005 *M* benzene in water solution.

Various molecular transitions resulting from the absorption of photons of different energy are shown schematically in Fig. 21-7. The transitions depicted represent those that may be induced by absorption of UV, Vis, IR, and microwave radiation. The figure also includes transitions in which the molecule is excited from the ground state to an exited electronic state with a simultaneous change in its vibrational or rotational energy levels. Although not shown in the figure, the absorption of a photon of appropriate energy also may cause simultaneous changes in electronic, vibrational, and rotational energy levels. The ability of molecules to have simultaneous transitions between the different energy levels tends to broaden the peaks in the UV–Vis absorption spectrum of molecules relative to those peaks observed in the absorption spectrum of atoms. This would be expected when one considers that vibrational and rotational energy levels are absent in an atomic energy level diagram. The depicted transitions between vibrational energy levels, without associated electronic transitions, are induced by radiation in the IR region. Independent transitions between allowed rotational energy levels also are depicted, these resulting from the absorption of photons of microwave radiation. A summary of transitions relevant to atomic and molecular absorption spectroscopy, including corresponding wavelength regions, is presented in Table 21-2.

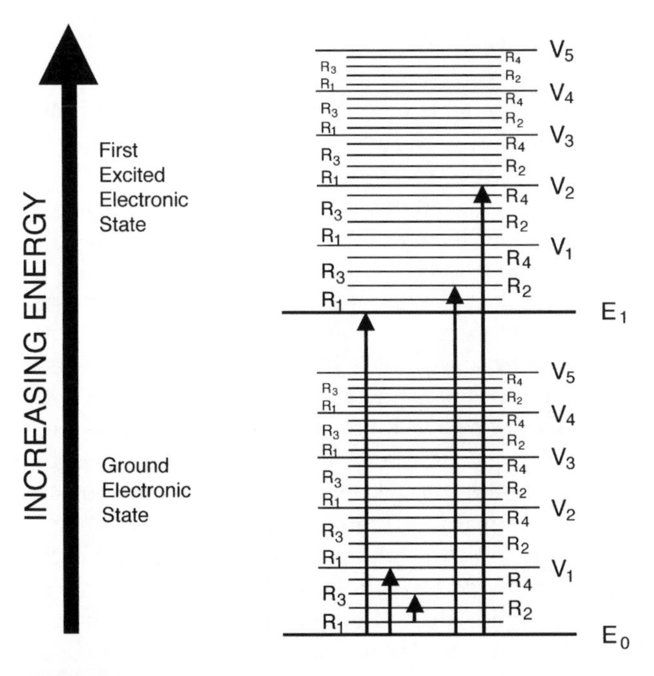

21-7 figure Partial molecular energy level diagram including electronic, vibrational, and rotational transitions.

21.4.2 Emission of Radiation

Emission is essentially the reverse of the absorption process, occurring when energy from an atom or molecule is released in the form of a photon of radiation. A molecule raised to an excited state will typically remain in the excited state for a very short period of time before relaxing back to the ground state. There are several **relaxation processes** through which an excited molecule may dissipate energy. The most common relaxation process is for the excited molecule to dissipate its energy through a series of small steps brought on by collisions with other molecules. The energy is thus converted to kinetic energy, the net result being the dissipation of the energy as heat. Under normal conditions, the dissipated heat is not enough to measurably affect the system. In some cases, molecules excited by the absorption of UV or Vis light will lose a portion of their excess energy through the emission of a photon. This emission process is referred to as either **fluorescence** or **phosphorescence**, depending on the nature of the excited state. In molecular

21-2 table **Wavelength Regions, Spectroscopic Methods, and Associated Transitions**

Wavelength Region	*Wavelength Limits*	*Type of Spectroscopy*	*Usual Wavelength Range*	*Types of Transitions in Chemical Systems with Similar Energies*
Gamma rays	0.01–1 Å	Emission	<0.1 Å	Nuclear proton/neutron arrangements
X-rays	0.1–10 nm	Absorption, emission, fluorescence, and diffraction	0.1–100 Å	Inner-shell electrons
Ultraviolet	10–380 nm	Absorption, emission, and fluorescence	180–380 nm	Outer-shell electrons in atoms, bonding electrons in molecules
Visible	380–750 nm	Absorption, emission, and fluorescence	380–750 nm	Same as ultraviolet
Infrared	0.075–1000 μm	Absorption	0.78–300 μm	Vibrational position of atoms in molecular bonds
Microwave	0.1–100 cm	Absorption	0.75–3.75 mm	Rotational position in molecules
		Electron spin resonance	3 cm	Orientation of unpaired electrons in an applied magnetic field
Radiowave	1–1000 m	Nuclear magnetic resonance	0.6–10 m	Orientation of nuclei in an applied magnetic field

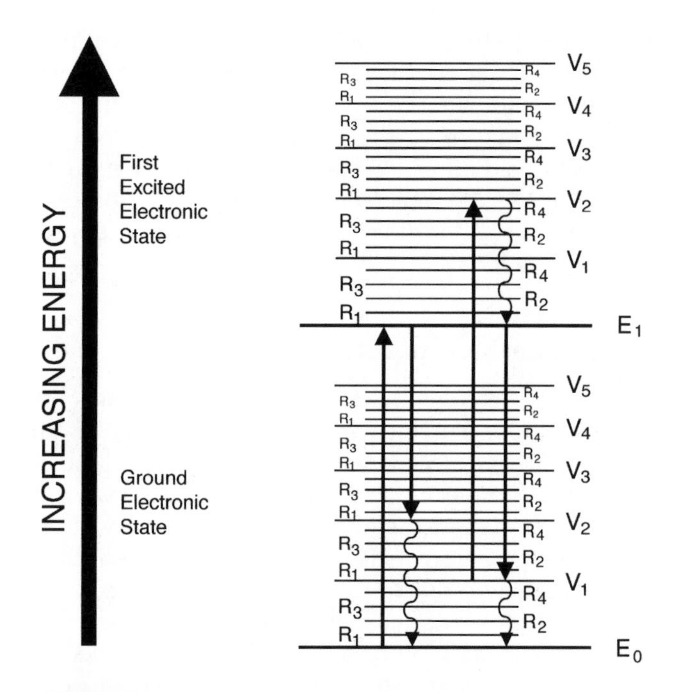

21-8 figure Partial molecular energy level diagram including absorption, vibrational relaxation, and fluorescence relaxation.

fluorescence spectroscopy, the photons emitted from the excited species generally will be of lower energy and longer wavelength than the corresponding photons that were absorbed in the excitation process. The reason is that, in most cases, only a fraction of the energy difference between the excited and ground states is lost in the emission process. The other fraction of the excess energy is dissipated as heat during vibrational relaxation. This process is depicted in Fig. 21-8, which illustrates that the excited species undergoes vibrational relaxation down to the lowest vibrational energy level within the excited electronic state, and then undergoes a transition to the ground electronic state through the emission of a photon. The photon emitted will have an energy that equals the energy difference between the lowest vibrational level of the excited electronic state and the ground electronic state level it descends to. The fluorescing molecule may descend to any of the vibrational levels within the ground electronic state. If the fluorescence transition is to an excited vibrational level within the ground electronic state, then it will quickly return to the ground state (lowest energy level) via vibrational relaxation. In yet other cases, an excited species may be of sufficient energy to initiate some type of photochemistry that ultimately leads to a decrease in the system's potential energy. In all cases, the relaxation process is driven by the tendency for a species to exist at its lowest permissible internal energy level. The relaxation process that dominates a system will be the one that minimizes the lifetime of the excited state. Under normal conditions, the relaxation process is so rapid that the population of molecules in the ground state is essentially unchanged.

21.5 SUMMARY

Spectroscopy deals with the interaction of electromagnetic radiation with matter. Spectrochemical analysis, a branch of spectroscopy, encompasses a wide range of techniques used in analytical laboratories for the qualitative and quantitative analysis of the chemical composition of foods. Common spectrochemical analysis methods include UV, Vis, and IR absorption spectroscopy; molecular fluorescence spectroscopy; and NMR spectroscopy. In each of these methods, the analyst attempts to measure the amount of radiation either absorbed or emitted by the analyte. All of these methods make use of the facts that the energy content of matter is quantized and that photons of radiation may be absorbed or emitted by matter if the energy associated with the photon equals the energy difference for allowed transitions of that given species. The above methods differ from each other with respect to the radiation wavelengths used in the analysis or the molecular vs. atomic nature of the analyte.

21.6 STUDY QUESTIONS

1. Which phenomena associated with light are most readily explained by considering the wave nature of light? Explain these phenomena based on your understanding of interference.
2. Which phenomena associated with light are most readily explained by considering the particulate nature of light? Explain these phenomena based on your understanding of the quantum nature of electromagnetic radiation.
3. What does it mean to say that the energy content of matter is quantized?
4. Molecular absorption of radiation in the UV–Vis range results in transitions between what types of energy levels?
5. Molecular absorption of radiation in the IR range results in transitions between what types of energy levels?
6. Why is an applied magnetic field necessary for NMR spectroscopy?
7. How do the allowed energy levels of molecules differ from those of atoms? Answer with respect to the energy level diagram depicted in Fig. 21-5.
8. In fluorescence spectroscopy, why is the wavelength of the emitted radiation longer than the wavelength of the radiation used for excitation of the analyte?

21.7 RESOURCE MATERIALS

1. Ball DW (2001) The basics of spectroscopy. Society of Photo-optical Instrumentation Engineers, Bellingham, WA
2. Currell G (2000) Analytical instrumentation – performance characteristics and quality. Wiley, New York, pp 67–91
3. Duckett S (2000) Foundations of spectroscopy. Oxford University Press, New York
4. Hargis LG (1988) Analytical chemistry – principles and techniques. Prentice Hall, Englewood Cliffs, NJ
5. Harris DC (2006) Quantitative chemical analysis, 7th edn. WH Freeman, New York
6. Harris DC, Bertolucci MD (1989) Symmetry and spectroscopy. Dover, Mineola, NY
7. Harwood LM, Claridge TDW (1997) Introduction to organic spectroscopy. Oxford University Press, New York
8. Ingle JD Jr, Crouch SR (1988) Spectrochemical analysis. Prentice Hall, Englewood Cliffs, NJ
9. Meyers RA (ed) (2000) Encyclopedia of analytical chemistry: applications, theory, and instrumentation. 5:2857–4332
10. Milton Roy educational manual for the SPECTRONIC® 20 & 20D spectrophotometers (1989) Milton Roy Co., Rochester, NY
11. Ramette RW (1981) Chemical equilibrium and analysis. Addison-Wesley, Reading, MA
12. Robinson JW, Frame EMS, Frame GM II (2005) Undergraduate instrumental analysis, 6th edn. Marcel Dekker, New York
13. Young HD, Freedman RA (2000) Sears and Zemansky's university physics, 10th edn. Addison-Wesley Longman, Reading, MA
14. Skoog DA, Holler FJ, Crouch SR (2007) Principles of instrumental analysis, 6th edn. Brooks/Cole, Pacific Grove, CA
15. Tinoco I Jr, Sauer K, Wang JC, Puglisi E (2001) Physical chemistry — principles and applications in biological sciences, 4th edn. Prentice Hall, Englewood Cliffs, NJ

Ultraviolet, Visible, and Fluorescence Spectroscopy

Michael H. Penner

Department of Food Science and Technology, Oregon State University, Corvallis, OR 97331-6602, USA
mike.penner@orst.edu

22.1 Introduction 389
22.2 Ultraviolet and Visible Absorption Spectroscopy 389
22.2.1 Basis of Quantitative Absorption Spectroscopy 389
22.2.2 Deviations from Beer's Law 391
22.2.3 Procedural Considerations 391
22.2.4 Calibration Curves 393
22.2.5 Effect of Indiscriminant Instrumental Error on the Precision of Absorption Measurements 394
22.2.6 Instrumentation 394
22.2.6.1 Light Source 394
22.2.6.2 Monochromator 395
22.2.6.3 Detector 396
22.2.6.4 Readout Device 397
22.2.7 Instrument Design 398
22.2.8 Characteristics of UV–Vis Absorbing Species 399
22.3 Fluorescence Spectroscopy 400
22.4 Summary 402
22.5 Study Questions 402
22.6 Practice Problems 403
22.7 Resource Materials 404

S.S. Nielsen, *Food Analysis*, Food Science Texts Series, DOI 10.1007/978-1-4419-1478-1_22,

22.1 INTRODUCTION

Spectroscopy in the ultraviolet–visible (UV–Vis) range is one of the most commonly encountered laboratory techniques in food analysis. Diverse examples, such as the quantification of macrocomponents (total carbohydrate by the phenol-sulfuric acid method), quantification of microcomponents, (thiamin by the thiochrome fluorometric procedure), estimates of rancidity (lipid oxidation status by the thiobarbituric acid test), and surveillance testing (enzyme-linked immunoassays), are presented in this text. In each of these cases, the analytical signal for which the assay is based is either the emission or absorption of radiation in the UV–Vis range. This signal may be inherent in the analyte, such as the absorbance of radiation in the visible range by pigments, or a result of a chemical reaction involving the analyte, such as the colorimetric copper-based Lowry method for the analysis of soluble protein.

Electromagnetic radiation in the UV–Vis portion of the spectrum ranges in wavelength from approximately 200 to 700 nm. The **UV range** runs from 200 to 350 nm and the **Vis range** from 350 to 700 nm (Table 22-1). The UV range is colorless to the human eye, while different wavelengths in the visible range each have a characteristic color, ranging from violet at the short wavelength end of the spectrum to red at the long wavelength end of the spectrum. Spectroscopy utilizing radiation in the UV–Vis range may be divided into two general categories, **absorbance** and **fluorescence** spectroscopy, based on the type of radiation–matter interaction that is being monitored. Each of these two types of spectroscopy may be subdivided further into **qualitative** and **quantitative** techniques.

22-1 table **Spectrum of Visible Radiation**

Wavelength (nm)	*Color*	*Complementary Hue*[a]
<380	Ultraviolet	
380–420	Violet	Yellow-green
420–440	Violet-blue	Yellow
440–470	Blue	Orange
470–500	Blue-green	Red
500–520	Green	Purple
520–550	Yellow-green	Violet
550–580	Yellow	Violet-blue
580–620	Orange	Blue
620–680	Red	Blue-green
680–780	Purple	Green
>780	Near-infrared	

[a]Complementary hue refers to the color observed for a solution that shows maximum absorbance at the designated wavelength assuming a continuous spectrum "white" light source.

In general, quantitative absorption spectroscopy is the most common of the subdivisions within UV–Vis spectroscopy.

22.2 ULTRAVIOLET AND VISIBLE ABSORPTION SPECTROSCOPY

22.2.1 Basis of Quantitative Absorption Spectroscopy

The objective of quantitative absorption spectroscopy is to determine the concentration of analyte in a given sample solution. The determination is based on the measurement of the amount of light absorbed from a reference beam as it passes through the sample solution. In some cases, the analyte may naturally absorb radiation in the UV–Vis range, such that the chemical nature of the analyte is not modified during the analysis. In other cases, analytes that do not absorb radiation in the UV–Vis range are chemically modified during the analysis, converting them to a species that absorbs radiation of the appropriate wavelength. In either case, the presence of analyte in the solution will affect the amount of radiation transmitted through the solution and, hence, the relative transmittance or absorbance of the solution may be used as an index of analyte concentration.

In actual practice, the solution to be analyzed is contained in an absorption cell and placed in the path of radiation of a selected wavelength(s). The amount of radiation passing through the sample is then measured relative to a reference sample. The relative amount of light passing through the sample is then used to estimate the analyte concentration. The process of absorption may be depicted as in Fig. 22-1. The radiation incident on the absorption cell, P_0, will have

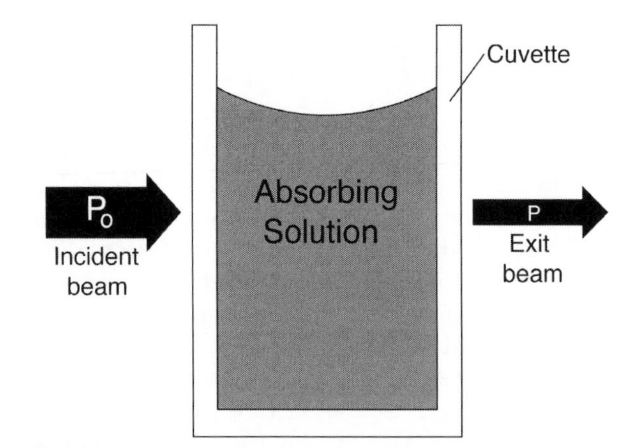

22-1 figure Attenuation of a beam of radiation as it passes through a cuvette containing an absorbing solution.

significantly greater radiant power than the radiation exiting the opposite side of the cell, P. The decrease in radiant power as the beam passes through the solution is due to the capture (absorption) of photons by the absorbing species. The relationship between the power of the incident and exiting beams typically is expressed in terms of either the transmittance or the absorbance of the solution. The **transmittance** (T) of a solution is defined as the ratio of P to P_0 as given in Equation [1]. Transmittance also may be expressed as a percentage as given in Equation [2].

$$T = P/P_0 \quad [1]$$

$$\%T = (P/P_0) \times 100 \quad [2]$$

where:

T = transmittance
P_0 = radiant power of beam incident on absorption cell
P = radiant power of beam exiting the absorption cell
$\%T$ = percent transmittance

The terms T and $\%T$ are intuitively appealing, as they express the fraction of the incident light absorbed by the solution. However, T and $\%T$ are not directly proportional to the concentration of the absorbing analyte in the sample solution. The nonlinear relationship between transmittance and concentration is an inconvenience since analysts are generally interested in analyte concentrations. A second term used to describe the relationship between P and P_0 is **absorbance** (A). Absorbance is defined with respect to T as shown in Equation [3].

$$A = \log(P_0/P) = -\log T = 2 - \log \%T \quad [3]$$

where:

A = absorbance
T and $\%T$ = as in Equations [1] and [2], respectively

Absorbance is a convenient expression in that, under appropriate conditions, it is directly proportional to the concentration of the absorbing species in the solution. Note that, based on these definitions for A and T, the absorbance of a solution *is not* simply unity minus the transmittance. In quantitative spectroscopy, the fraction of the incident beam that is not transmitted does not equal the solution's absorbance (A).

The relationship between the absorbance of a solution and the concentration of the absorbing species is known as **Beer's law** (Equation [4]).

$$A = abc \quad [4]$$

where:

A = absorbance
c = concentration of absorbing species
b = path length through solution (cm)
a = absorptivity

There are no units associated with absorbance, A, since it is the log of a ratio of beam powers. The concentration term, c, may be expressed in any appropriate units (M, mM, mg/ml, %). The path length, b, is in units of cm. The **absorptivity**, a, of a given species is a proportionality constant dependent on the molecular properties of the species. The absorptivity is wavelength dependent and may vary depending on the chemical environment (pH, ionic strength, solvent, etc.) the absorbing species is experiencing. The units of the absorptivity term are $(\text{cm})^{-1}$ $(\text{concentration})^{-1}$. In the special case where the concentration of the analyte is reported in units of molarity, the absorptivity term has units of $(\text{cm})^{-1}$ $(\text{M})^{-1}$. Under these conditions, it is designated by the symbol ε, which is referred to as the **molar absorptivity**. Beer's law expressed in terms of the molar absorptivity is given in Equation [5]. In this case, c refers specifically to the molar concentration of the analyte.

$$A = \varepsilon bc \quad [5]$$

where:

A and b = as in Equation [4]
ε = molar absorptivity
c = concentration in units of molarity

Quantitative spectroscopy is dependent on the analyst being able to accurately measure the fraction of an incident light beam that is absorbed by the analyte in a given solution. This apparently simple task is somewhat complicated in actual practice due to processes other than analyte absorption that also result in significant decreases in the power of the incident beam. A pictorial summary of reflection and scattering processes that will decrease the power of an incident beam is given in Fig. 22-2. It is clear that these processes must be accounted for if a truly quantitative estimate of analyte absorption is necessary. In practice, a reference cell is used to correct for these processes. A **reference cell** is one that, in theory, exactly matches the sample absorption cell with the exception that it contains no analyte. A reference cell often is prepared by adding distilled water to an absorption cell. This reference cell is then placed in the path of the light beam, and the power of the radiation exiting the reference cell is measured and taken as P_0 for the sample cell. This procedure assumes that all processes except the selective absorption of radiation by the analyte are equivalent for the sample and reference cells.

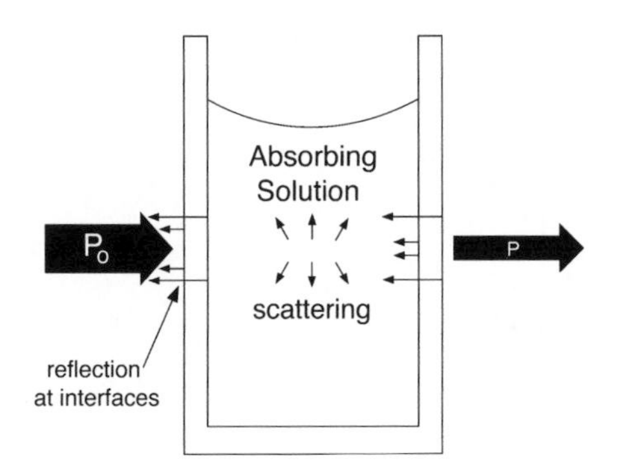

Factors contributing to the attenuation of a beam of radiation as it passes through a cuvette containing an absorbing solution.

The absorbance actually measured in the laboratory approximates Equation [6].

$$A = \log(P_{\text{solvent}}/P_{\text{analyte solution}}) \cong \log(P_0/P) \quad [6]$$

where:

P_{solvent} = radiant power of beam exiting cell containing solvent (blank)
$P_{\text{analyte solution}}$ = radiant power of beam exiting cell containing analyte solution
P_0 and P = as in Equation [1]
A = as in Equation [3]

22.2.2 Deviations from Beer's Law

It should never be assumed that Beer's law is strictly obeyed. Indeed, there are several reasons for which the predicted linear relationship between absorbance and concentration may not be observed. In general, Beer's law is applicable only to dilute solutions, up to approximately 10 m*M* for most analytes. The actual concentration at which the law becomes limiting will depend on the chemistry of the analyte. As analyte concentrations increase, the intermolecular distances in a given sample solution will decrease, eventually reaching a point at which neighboring molecules mutually affect the charge distribution of the other. This perturbation may significantly affect the ability of the analyte to capture photons of a given wavelength; that is, it may alter the analyte's absorptivity (a). This causes the linear relationship between concentration and absorption to break down since the absorptivity term is the constant of proportionality in Beer's law (assuming a constant path length, b). Other chemical processes also may result in deviations from Beer's law, such as the reversible association–dissociation of analyte molecules or the ionization of a weak acid in an unbuffered solvent. In each of these cases, the predominant form of the analyte may change as the concentration is varied. If the different forms of the analyte, for example, ionized vs. neutral, have different absorptivities (a), then a linear relationship between concentration and absorbance will not be observed.

A further source of deviation from Beer's law may arise from limitations in the instrumentation used for absorbance measurements. Beer's law strictly applies to situations in which the radiation passing through the sample is monochromatic, since under these conditions a single absorptivity value describes the interaction of the analyte with all the radiation passing through the sample. If the radiation passing through a sample is polychromatic and there is variability in the absorptivity constants for the different constituent wavelengths, then Beer's law will not be obeyed. An extreme example of this behavior occurs when radiation of the ideal wavelength and stray radiation of a wavelength that is not absorbed at all by the analyte simultaneously pass through the sample to the detector. In this case, the observed transmittance will be defined as in Equation [7]. Note that a limiting absorbance value will be reached as $P_0 \gg P$, which will occur at relatively high concentrations of the analyte.

$$A = \log(P_0 + P_s)/(P + P_s) \quad [7]$$

where:

P_s = radiant power of stray light
A = as in Equation [3]
P and P_0 = as in Equation [1]

22.2.3 Procedural Considerations

In general, the aim of quantitative measurements is to determine the concentration of an analyte with optimum precision and accuracy, in a minimal amount of time and at minimal cost. To accomplish this, it is essential that the analyst consider potential errors associated with each step in the methodology of a particular assay. Potential sources of error for spectroscopic assays include inappropriate sample preparation techniques, inappropriate controls, instrumental noise, and errors associated with inappropriate conditions for absorbance measurements (such as extreme absorbance/transmittance readings).

Sample preparation schemes for absorbance measurements vary considerably. In the simplest case, the analyte-containing solution may be measured directly following homogenization and clarification. Except for special cases, homogenization is required prior to any analysis to ensure a representative sample. Clarification of samples is essential prior to taking absorbance readings in order to avoid the apparent absorption due to scattering of light by turbid solutions. The **reference solution** for samples in this simplest case will be the sample solvent, the solvent being water or an aqueous buffer in many cases. In more complex situations, the analyte to be quantified may need to be chemically modified prior to making absorbance measurements. In these cases, the analyte that does not absorb radiation in an appropriate spectral range is specifically modified, resulting in a species with absorption characteristics compatible with a given spectrophotometric measurement. Specific reactions such as these are used in many colorimetric assays that are based on the absorption of radiation in the Vis range. The reference solution for these assays is prepared by treating the sample solvent in a manner identical with that of the sample. The reference solution therefore will help to correct for any absorbance due to the modifying reagents themselves and not the modified analyte.

A **sample-holding cell** or **cuvette** should be chosen after the general spectral region to be used in a spectrophotometric measurement has been determined. Sample holding cells vary in composition and dimensions. The sample holding cell should be composed of a material that does not absorb radiation in the spectral region being used. Cells meeting this requirement for measurements in the **UV range** may be composed of **quartz** or **fused silica**. For the **Vis range**, cells made of **silicate glass** are appropriate and inexpensive **plastic** cells also are available for some applications. The dimensions of the cell will be important with respect to the amount of solution required for a measurement and with regard to the path length term used in Beer's law. A typical absorption cell is 1 cm^2 and approximately 4.5-cm long. The path length for this traditional cell is 1 cm, and the minimum volume of solution needed for standard absorption measurements is approximately 1.5 ml. Absorption cells with path lengths ranging from 1 to 100 mm are commercially available. Narrow cells, approximately 4 mm in width, with optical path lengths of 1 cm, are also available. These narrow cells are convenient for absorbance measurements when limiting amounts of solution, less than 1 ml, are available.

In many cases, the analyst will need to **choose an appropriate wavelength** at which to make absorbance measurements. If possible, it is best to choose the wavelength at which the analyte demonstrates maximum absorbance and where the absorbance does not change rapidly with changes in wavelength (Fig. 22-3). This position usually corresponds to the apex of the highest absorption peak. Taking measurements at this apex has two advantages: (1) maximum sensitivity, defined as the absorbance change per unit change in analyte concentration, and (2) greater adherence to Beer's law since the spectral region making up the radiation beam is composed of wavelengths with relatively small differences in their molar absorptivities for the analyte being measured (Fig. 22-3). The latter

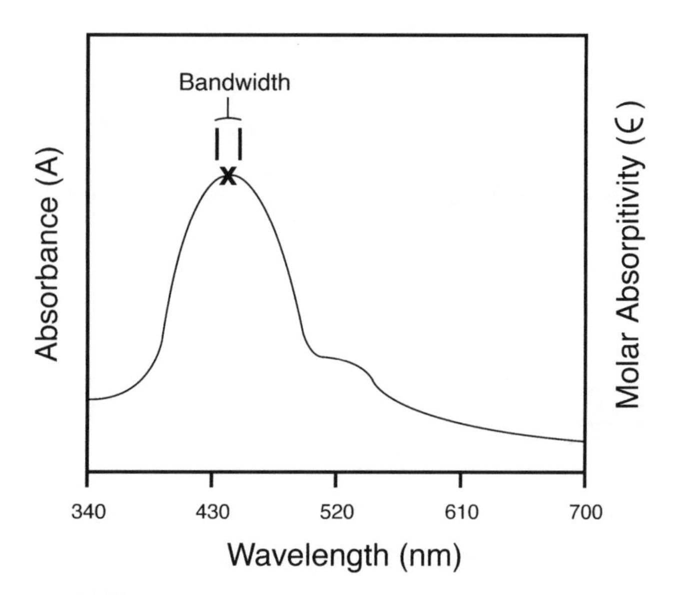

22-3 figure Hypothetical absorption spectrum between 340 and 700 nm. The effective bandwidth of the radiation used in obtaining the spectrum is assumed to be approximately 20 nm. Note that at the point indicated there is essentially no change in molar absorptivity over this wavelength range.

point is important in that the radiation beam used in the analysis will be composed of a small continuous band of wavelengths centered about the wavelength indicated on the instrument's wavelength selector.

The actual **absorbance measurement** is made by first calibrating the instrument for 0% and then 100% transmittance. The 0% transmittance adjustment is made while the photodetector is screened from the incident radiation by means of an occluding shutter, mimicking infinite absorption. This adjustment sets the base level current or "dark current" to the appropriate level, such that the readout indicates zero. The 100% transmittance adjustment then is made with the occluding shutter open and an appropriate reference cell/solution in the light path. The reference cell itself should be equivalent to the cell that contains the sample (i.e., a "matched" set of cells is used). In many cases, the same cell is used for both the sample and reference solutions. The reference cell generally is filled with solvent, that often being distilled/deionized water for aqueous systems. The 100% T adjustment effectively sets $T = 1$ for the reference cell, which is equivalent to defining P_0 in Equation [1] as equivalent to the radiant power of the beam exiting the reference cell. The 0% T and 100% T settings should be confirmed as necessary throughout the assay. The sample cell that contains analyte then is measured without changing the adjustments. The adjustments made with the reference cell will effectively set the instrument to give a sample readout in terms of Equation [6]. The readout for the sample solution will be between 0 and 100% T. Most modern spectrophotometers allow the analyst to make readout measurements either in absorbance units or as percent transmittance. It is generally most convenient to make readings in absorbance units since, under optimum conditions, absorbance is directly proportional to concentration. When making measurements with an instrument that employs an analog swinging needle type of readout, it may be preferable to use the linear percent transmittance scale and then calculate the corresponding absorbance using Equation [3]. This is particularly true for measurements in which the percent transmittance is less than 20.

22.2.4 Calibration Curves

In most instances, it is advisable to use calibration curves for quantitative measurements. In food analysis, there are a large number of empirical assays for which calibration curves are essential. The calibration curve is used to establish the relationship between analyte concentration and absorbance. This relationship is established experimentally through the analysis of a series of samples of known analyte concentration. The standard solutions are best prepared with the same reagents and at the same time as the unknown. The concentration range covered by the standard solutions must include that expected for the unknown. Typical calibration curves are depicted in Fig. 22-4. **Linear calibration curves** are expected for those systems that obey Beer's law. **Nonlinear calibration curves** are used for some assays, but linear relationships generally are preferred due to the ease of processing the data. Nonlinear calibration curves may be either due to concentration-dependent changes in the chemistry of the system or due to limitations inherent in the instruments used for the assay. The nonlinear calibration curve in Fig. 22-4b reflects the fact that the **calibration sensitivity**, defined as change in absorbance per unit change in analyte concentration, is not constant. For the case depicted in Fig. 22-4b, the assay's concentration-dependent decrease in sensitivity obviously begins to limit its usefulness at analyte concentrations above 10 m*M*.

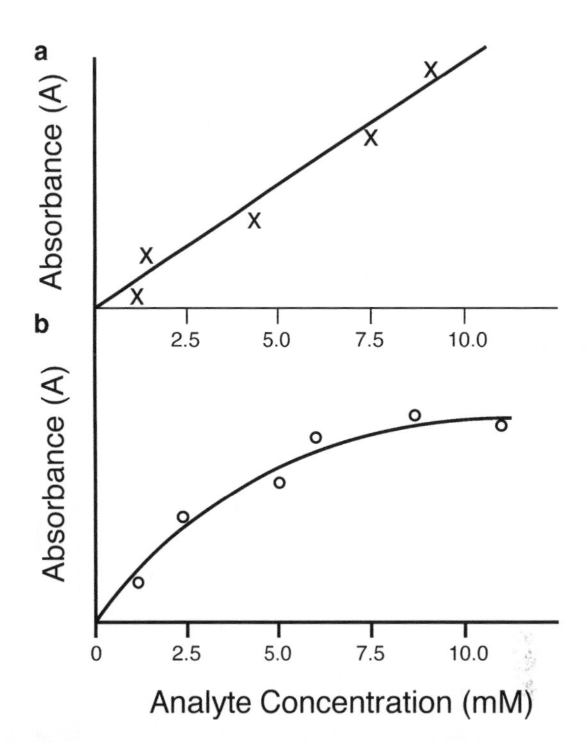

22-4 figure Linear (**a**) and nonlinear (**b**) calibration curves typically encountered in quantitative absorption spectroscopy.

In many cases, truly representative calibration standards cannot be prepared due to the complexity of the unknown sample. This scenario must be assumed when insufficient information is available on the extent of interfering compounds in the unknown. **Interfering compounds** include those that absorb radiation in the same spectral region as the analyte, those that influence the absorbance of the analyte, and those compounds that react with modifying reagents that are supposedly specific for the analyte. This means that calibration curves are potentially in error if the unknown and the standards differ with respect to pH, ionic strength, viscosity, types of impurities, and the like. In these cases, it is advisable to calibrate the assay system by using a **standard addition protocol**. One such protocol goes as follows: To a series of flasks add a constant volume of the unknown (V_u) for which you are trying to determine the analyte concentration (C_u). Next, to each individual flask add a known volume (V_s) of a standard analyte solution of concentration C_s, such that each flask receives a unique volume of standard. The resulting series of flasks will contain identical volumes of the unknown and different volumes of the standard solution. Next, dilute all flasks to the same total volume, V_t. Each of the flasks then is assayed, with each flask treated identically. If Beer's law is obeyed, then the measured absorbance of each flask will be proportional to the total analyte concentration as defined in Equation [8].

$$A = k[(V_s C_s + V_u C_u)/(V_t)] \quad [8]$$

where:

V_s = volume of standard
V_u = volume of unknown
V_t = total volume
C_s = concentration of standard
C_u = concentration of unknown
k = proportionality constant (path length × absorptivity)

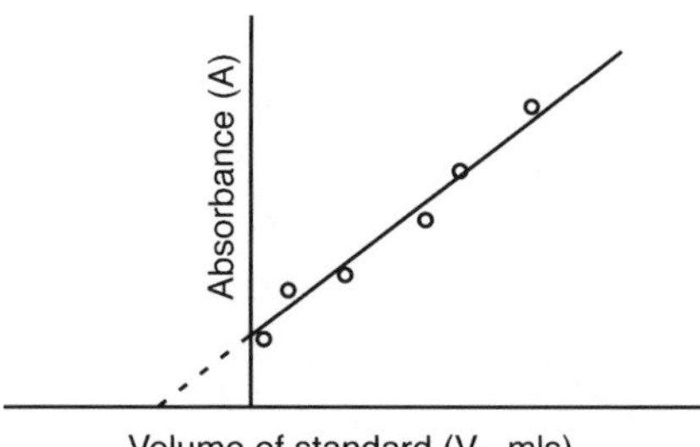

Calibration curve for the determination of the analyte concentration in an unknown using a standard addition protocol. *A*, absorbance; V_s, volume of standard analyte solution; as discussed in text.

The results from the assays are then plotted with the volume of standard added to each flask (V_s) as the independent variable and the resulting absorbance (A) as the dependent variable (Fig. 22-5). Assuming Beer's law, the line describing the relationship will be as in Equation [9], in which all terms other than V_s and A are constants. Taking the ratio of the slope of the plotted line Equation [10] to the line's intercept Equation [11] and rearranging gives Equation [12], from which the concentration of the unknown, C_u, can be calculated since C_s and V_u are experimentally defined constants.

$$A = kC_sV_s/V_T + V_uC_uk/V_t \quad [9]$$

$$\text{Slope} = kC_s/V_t \quad [10]$$

$$\text{Intercept} = V_uC_uk/V_t \quad [11]$$

$$C_u = (\text{measured intercept}/\text{measured slope})(C_s/V_u) \quad [12]$$

where:

V_s, V_u, V_t, C_s, C_u, and k = as in Equation [8]

22.2.5 Effect of Indiscriminant Instrumental Error on the Precision of Absorption Measurements

All spectrophotometric assays will have some level of **indiscriminant error** associated with the absorbance/transmittance measurement itself. Indiscriminant error of this type often is referred to as **instrument noise**. It is important that the assay be designed such that this source of error is minimized, the objective being to keep this source of error low relative to the variability associated with other aspects of the assay, such as sample preparation, subsampling, reagent handling, and so on. Indiscriminant instrumental error is observed with repeated measurements of a single homogeneous sample. The relative concentration uncertainty resulting from this error is not constant over the entire percent transmittance range (0–100%). Measurements at intermediate transmittance values tend to have lower relative errors, thus greater relative precision, than measurements made at either very high or very low transmittance. **Relative concentration uncertainty** or relative error may be defined as S_c/C, where S_c is the sample standard deviation and C is the measured concentration. Relative concentration uncertainties of from 0.5 to 1.5% are to be expected for absorbance/transmittance measurements taken in the optimal range. The optimal range for absorbance measurements on simple, less expensive spectrophotometers is from approximately 0.2 to 0.8 absorbance units, or 15–65% transmittance. On more sophisticated instruments, the range for optimum absorbance readings may be extended up to 1.5 or greater. To be safe, it is prudent to always make absorbance readings under conditions at which the absorbance of the analyte solution is less than 1.0. If there is an anticipated need to make measurements at absorbance readings greater than 1.0, then the relative precision of the spectrophotometer should be established experimentally by repetitive measurements of appropriate samples. Absorbance readings outside the optimal range of the instrument may be used, but the analyst must be prepared to account for the higher relative error associated with these extreme readings. When absorbance readings approach the limits of the instrumentation, then relatively large differences in analyte concentrations may not be detected.

22.2.6 Instrumentation

There are many variations of spectrophotometers available for UV–Vis spectrophotometry. Some instruments are designed for operation in only the visible range, while others encompass both the UV and Vis range. Instruments may differ with respect to design, quality of components, and versatility. A basic spectrophotometer is composed of five essential components: the **light source**, the **monochromator**, the **sample/reference holder**, the **radiation detector**, and a **readout device**. A power supply is required for instrument operation. A schematic depicting component interrelationships is shown in Fig. 22-6.

22.2.6.1 Light Source

Light sources used in spectrophotometers must continuously emit a strong band of radiation encompassing the entire wavelength range for which the instrument is designed. The power of the emitted radiation must be sufficient for adequate detector response, and it should not vary sharply with changes in wavelength or drift significantly over the experimental time scale.

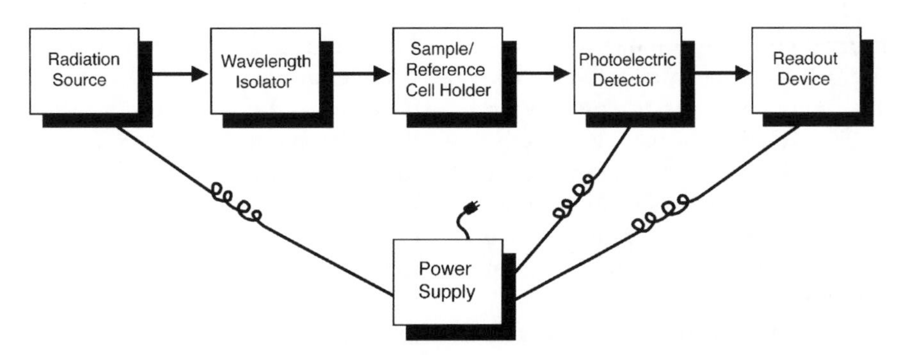

22-6 figure Arrangement of components in a simple single-beam, UV–Vis absorption spectrophotometer.

The most common radiation source for Vis spectrophotometers is the **tungsten filament lamp**. These lamps emit adequate radiation covering the wavelength region from 350 to 2500 nm. Consequently, tungsten filament lamps also are employed in near-infrared spectroscopy. The most common radiation sources for measurements in the UV range are **deuterium electrical-discharge lamps**. These sources provide a continuous radiation spectrum from approximately 160 nm to 375 nm. These lamps employ quartz windows and should be used in conjunction with quartz sample holders, since glass significantly absorbs radiation below 350 nm.

22.2.6.2 Monochromator

The component that functions to isolate the specific, narrow, continuous group of wavelengths to be used in the spectroscopic assay is the **monochromator**. The monochromator is so named because light of a single wavelength is termed **monochromatic**. Theoretically, **polychromatic radiation** from the source enters the monochromator and is dispersed according to wavelength, and **monochromatic radiation** of a selected wavelength exits the monochromator. In practice, light exiting the monochromator is not of a single wavelength, but rather it consists of a narrow continuous band of wavelengths. A representative monochromator is depicted in Fig. 22-7. As illustrated, a typical monochromator is composed of **entrance** and **exit slits**, **concave mirror(s)**, and a **dispersing element** (the grating in this particular example). Polychromatic light enters the monochromator through the entrance slit and is then culminated by a concave mirror. The culminated polychromatic radiation is then dispersed, dispersion being the physical separation in space of radiation of different wavelengths. The radiation of different wavelengths is then reflected from a concave mirror that focuses the different wavelengths of light sequentially along the focal plane. The radiation that aligns with the exit slit in the focal plane thus

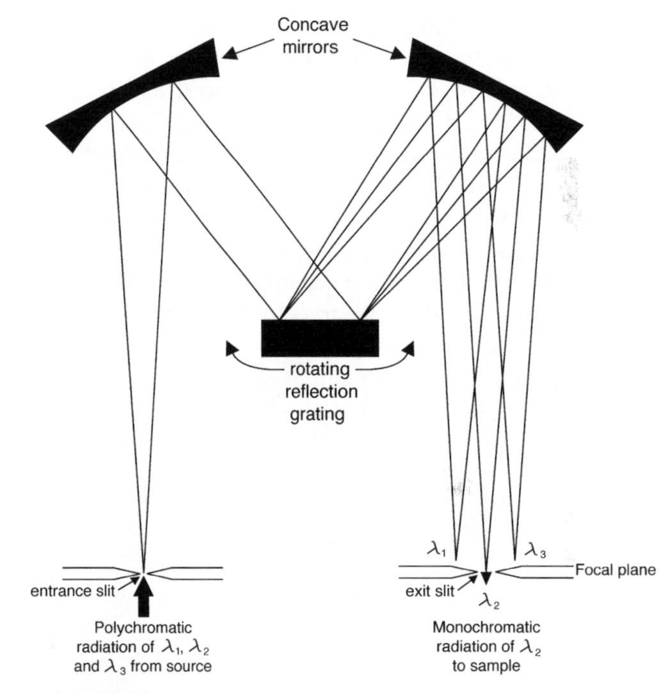

22-7 figure Schematic of a monochromator employing a reflection grating as the dispersing element. The concave mirrors serve to culminate the radiation into a beam of parallel rays.

is emitted from the monochromator. The radiation emanating from the monochromator will consist of a narrow range of wavelengths presumably centered around the wavelength specified on the wavelength selection control of the instrument.

The size of the wavelength range passing out of the exit slit of the monochromator is termed the **bandwidth** of the emitted radiation. Many spectrophotometers allow the analyst to adjust the size of the monochromator exit slit (and entrance slit) and, consequently, the bandwidth of the emitted radiation. Decreasing the exit slit width will decrease the associated bandwidth and the radiant power of the emitted beam. Conversely, further opening of the exit slit will result in a beam of greater radiant power, but one that has a larger bandwidth. In some cases where resolution is critical, such as some qualitative work, the narrower slit width may be advised. However, in most quantitative work a relatively open slit may be used since adsorption peaks in the UV–Vis range generally are broad relative to spectral bandwidths. Also, the signal-to-noise ratio associated with transmittance measurements is improved due to the higher radiant power of the measured beam.

The effective bandwidth of a monochromator is determined not only by the slit width but also by the quality of its dispersing element. The dispersing element functions to spread out the radiation according to wavelength. **Reflection gratings**, as depicted in Fig. 22-8, are the most commonly used dispersing elements in modern spectrophotometers. Gratings sometimes are referred to as **diffraction gratings** because the separation of component wavelengths is dependent on the different wavelengths being diffracted at different angles relative to the grating normal. A reflection grating incorporates a reflective surface in which a series of closely spaced grooves has been etched, typically between 1200 and 1400 grooves per millimeter. The grooves themselves serve to break up the reflective surface such that each point of reflection behaves as an independent point source of radiation.

Referring to Fig. 22-8, lines 1 and 2 represent rays of parallel monochromatic radiation that are in phase and that strike the grating surface at an angle i to the normal. Maximum constructive interference of this radiation is depicted as occurring at an angle r to the normal. At all other angles, the two rays will partially or completely cancel each other. Radiation of a different wavelength would show maximum constructive interference at a different angle to the normal. The wavelength dependence of the diffraction angle can be rationalized by considering the relative distance the photons of rays 1 and 2 travel and assuming that maximum constructive interference occurs when the waves associated with the photons are completely in phase. Referring to Fig. 22-8, prior to reflection, photon 2 travels a distance CD greater than photon 1. After reflection, photon 1 travels a distance AB greater than photon 2. Hence, the waves associated with photons 1 and 2 will remain in phase after reflection only if the net difference in the distance traveled is an integral multiple of their wavelength. Note that for a different angle r the distance AB would change and, consequently, the net distance CD–AB would be an integral multiple of a different wavelength. The net result is that the component wavelengths are each diffracted at their own unique angles r.

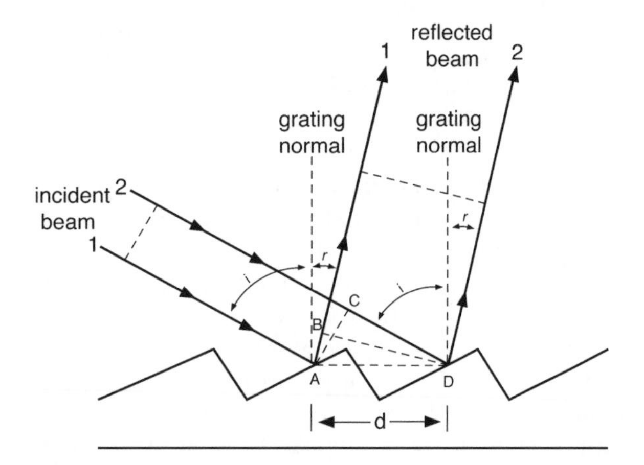

22-8 figure Schematic illustrating the property of diffraction from a reflection grating. Each reflected point source of radiation is separated by a distance d.

22.2.6.3 Detector

In a spectroscopic measurement, the light transmitted through the reference or sample cell is quantified by means of a **detector**. The detector is designed to produce an electric signal when it is struck by photons. An ideal detector would give a signal directly proportional to the radiant power of the beam striking it; would have a high signal-to-noise ratio; and would have a relatively constant response to light of different wavelengths, such that it was applicable to a wide range of the radiation spectrum. There are several types and designs of radiation detectors currently in use. The most commonly encountered detectors are the **phototube**, the **photomultiplier tube**, and the **photodiode detectors**. All of these detectors function by converting the energy associated with incoming photons into electrical current. The **phototube** consists of a semicylindrical cathode covered with a photoemissive surface and a wire anode, the electrodes being housed under vacuum in a transparent tube (Fig. 22-9a). When photons strike the photoemissive surface of the cathode, there is an emission of electrons, and the freed electrons then are collected at the anode. The net result of this process is that a measurable current is created. The number of electrons emitted from the cathode and the subsequent current through the system are directly proportional to the number of photons, or radiant power of the beam, impinging on the photoemissive surface. The **photomultiplier tube** is of similar design. However, in the photomultiplier tube there is an amplification of the number of electrons collected at the anode per photon striking the photoemissive surface of the cathode

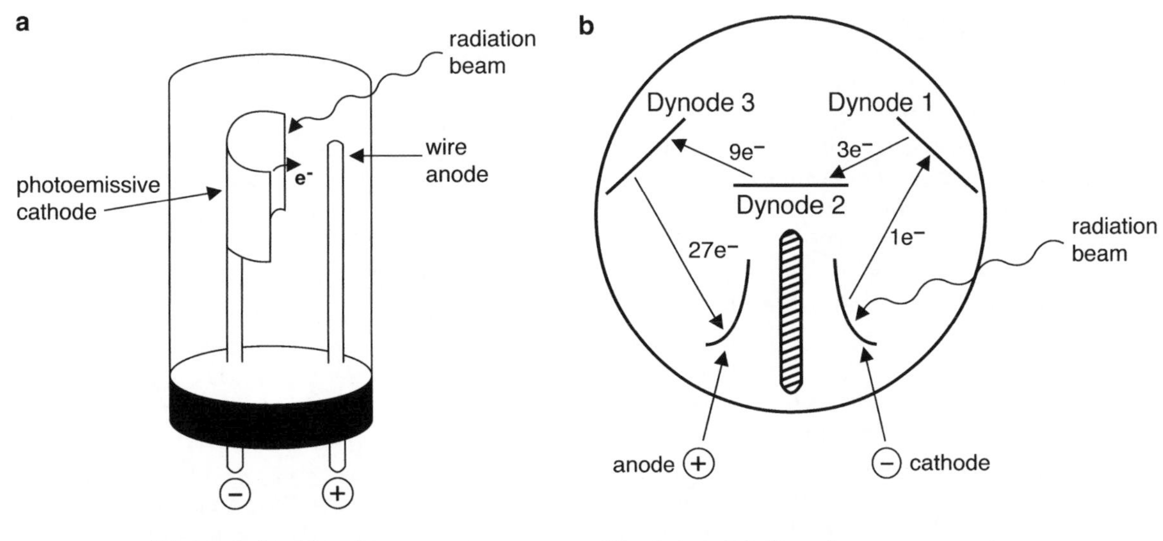

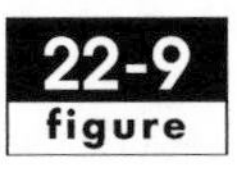

Schematic diagram of a typical phototube design (**a**) and the cathode–dynode–anode arrangement of a representative photomultiplier tube (**b**).

(Fig. 22-9b). The electrons originally emitted from the cathode surface are attracted to a dynode with a relative positive charge. At the dynode, the electrons strike the surface, causing the emission of several more electrons per original electron, resulting in an amplification of the signal. Signal amplification continues in this manner, as photomultiplier tubes generally contain a series of such dynodes, with electron amplification occurring at each dynode. The cascade continues until the electrons emitted from the final dynode are collected at the anode of the photomultiplier tube. The final gain may be as many as 10^6–10^9 electrons collected per photon.

Photodiode detectors now are used commonly in UV–Vis spectrophotometers. These are solid-state devices in which the light-induced electrical signal is a result of photons exciting electrons in the semiconductor materials from which they are fabricated, most commonly silicon. Spectrophotometers using photodiode detectors may contain a single diode detector or a linear array of diodes (diode array spectrophotometers). If a single photodiode detector is used, then the arrangement of components is generally as depicted in Fig. 22-6. If an array of photodiode detectors is used, then the light originating from the source typically passes into the sample prior to it being dispersed. The light transmitted through the sample is subsequently dispersed onto the diode array, with each diode measuring a narrow band of the resulting spectrum. This design allows one to simultaneously measure multiple wavelengths, allowing nearly instantaneous collection of an entire absorption spectrum. Diode-based detectors are generally reported to be more sensitive than phototubes but less sensitive than photomultiplier tubes.

22.2.6.4 Readout Device

The signal from the detector is generally amplified and then displayed in a usable form to the analyst. The final form in which the signal is displayed will depend on the complexity of the system. In the simplest case, the analog signal from the detector is displayed on an **analog meter** through the position of a needle on a meter face calibrated in percent transmission or absorbance. Analog readouts are adequate for most routine analytical purposes; however, analog meters are somewhat more difficult to read and, hence, the resulting data are expected to have somewhat lower precision than that obtained on a digital readout (assuming the digital readout is given to enough places). **Digital readouts** express the signal as numbers on the face of a meter. In these cases, there is an obvious requirement for signal processing between the analog output of the detector and the final digital display. In virtually all cases, the signal processor is capable of presenting the final readout in terms of either absorbance or transmittance. Many of the newer instruments include microprocessors capable of more extensive data manipulations on the digitized signal. For example, the readouts of some spectrophotometers may be in concentration units, provided the instrument has been correctly calibrated with appropriate reference standards.

22.2.7 Instrument Design

The optical systems of spectrophotometers fall into one of the two general categories: They are either single-beam or double-beam instruments. In a **single-beam instrument**, the radiant beam follows only one path, that going from the source through the sample to the detector (Fig. 22-6). When using a single-beam instrument, the analyst generally measures the transmittance of a sample after first establishing 100% T, or P_0, with a reference sample or blank. The blank and the sample are read sequentially since there is but a single light path going through a single cell-holding compartment. In a **double-beam instrument**, the beam is split such that one-half of the beam goes through one cell-holding compartment and the other half of the beam passes through the second. The schematic of Fig. 22-10 illustrates a double-beam optical system in which the beam is split in time between the sample and reference cell. In this design, the beam is alternately passed through the sample and reference cells by means of a rotating sector mirror with alternating reflective and transparent sectors. The double-beam design allows the analyst to simultaneously measure and compare the relative absorbance of a sample and a reference cell. The advantage of this design is that it will compensate for deviations or drifts in the radiant output of the source since the sample and reference cells are compared many times per second. The disadvantage of the double-beam design is that the radiant power of the incident beam is diminished because the beam is split. The lower energy throughput of the double-beam design is generally associated with inferior signal-to-noise ratios. Computerized single-beam spectrophotometers now are available that claim to have the benefits of both the single- and double-beam designs. Their manufacturers report that previously troublesome source and detector drift and noise problems have been stabilized such that simultaneous reading of the reference and sample cell is not necessary. With these instruments, the reference and sample cells are read sequentially, and the data are stored and then processed by the associated computer.

The Spectronic® 20 is a classic example of a simple single-beam visible spectrophotometer (Fig. 22-11). The white light emitted from the source passes into the monochromator via its entrance slit; the light is then dispersed into a spectrum by a diffraction grating; and a portion of the resulting spectrum then leaves the monochromator via the exit slit. The radiation emitted from the monochromator passes through a sample compartment and strikes the silicon photodiode detector, resulting in an electrical signal proportional to the intensity of impinging light. The lenses depicted in Fig. 22-11 function in series to focus the light image on the focal plane that contains the exit slit. To change

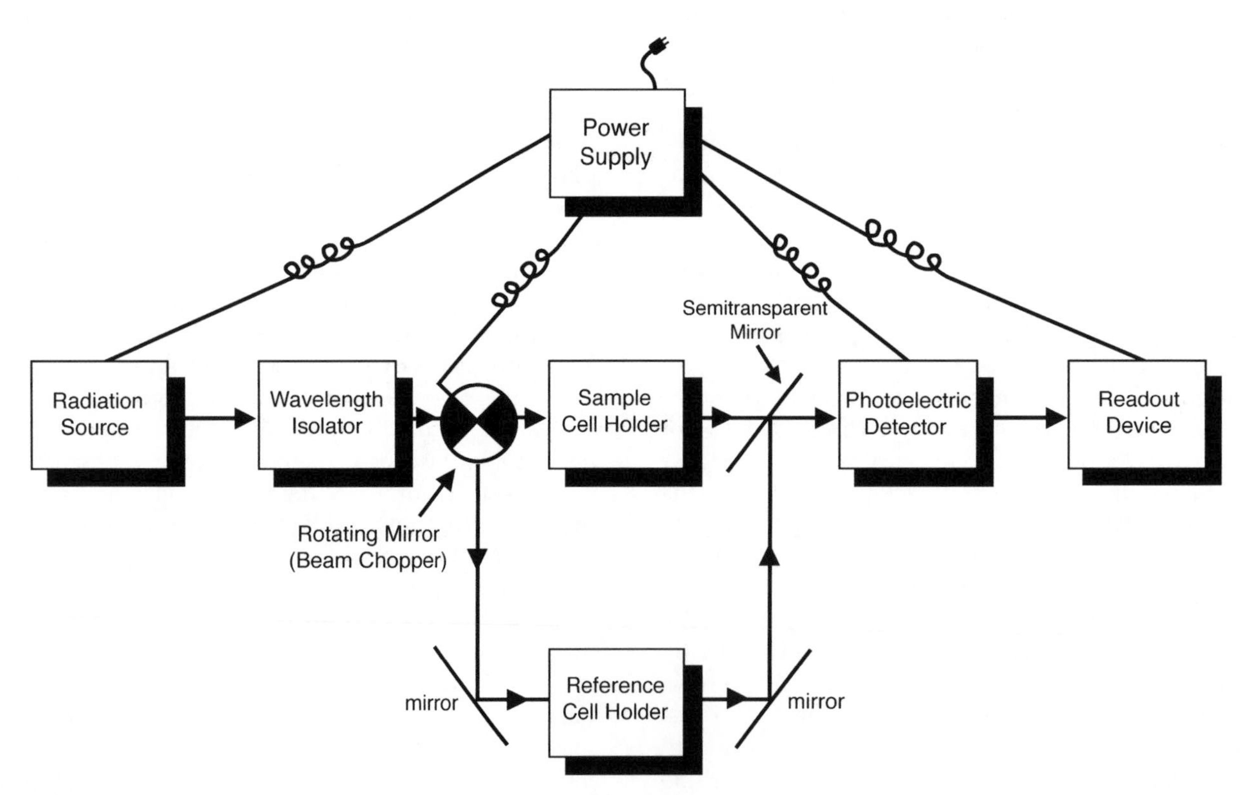

22-10 figure Arrangement of components in a representative double-beam UV–Vis absorption spectrophotometer. The incident beam is alternatively passed through the sample and reference cells by means of a rotating beam chopper.

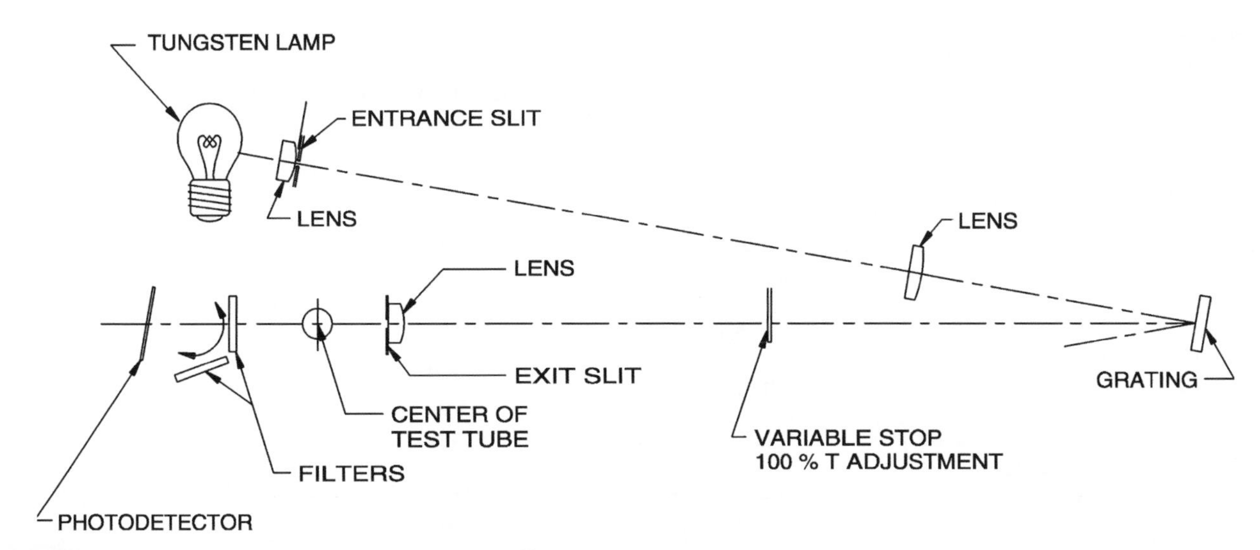

22-11 figure Optical system for the SPECTRONIC® 20 Spectrophotometer. (Courtesy of Thermo Spectronic, Rochester, NY, Thermo Electron Spectroscopy.)

the portion of the spectrum exiting the monochromator, one rotates the reflecting grating by means of the wavelength cam. A shutter automatically blocks light from exiting the monochromator when no sample/reference cell is in the instrument; the zero percent *T* adjustment is made under these conditions. The light control occluder is used to adjust the radiant power of the beam exiting the monochromator. The occluder consists of an opaque strip with a V-shaped opening that can be physically moved in or out of the beam path. The occluder is used to make the 100% *T* adjustment when an appropriate reference cell is in the instrument.

22.2.8 Characteristics of UV–Vis Absorbing Species

The absorbance of UV–Vis radiation is associated with electronic excitations within atoms and molecules. Commonly encountered analytical methods based on UV–Vis spectroscopy do not use UV radiation below 200 nm. Hence, the excitations of interest in traditional UV–Vis spectroscopy are the result of unsaturation and/or the presence of nonbonded electrons in the absorbing molecules. The UV-absorption characteristics of several functional groups common to food constituents are tabulated in Table 22-2. The presented wavelengths of maximum absorbance and molar absorptivities are only approximate since the environment to which the functional group is exposed, including neighboring constituents and solvents, will have an influence on the electronic properties of the functional group.

The type of information contained in Table 22-2 will likely be useful in determining the feasibility of UV–Vis spectroscopy for specific applications. For example, it is helpful to know the absorption characteristics of carboxyl groups if one is considering the feasibility of using UV–Vis absorption spectroscopy as a detection method to monitor nonderivatized organic acids eluting from liquid chromatography columns. With respect to this particular organic acids question, the table indicates that organic acids are likely to absorb radiation in the range accessible to most UV–Vis detectors (>200 nm). However, the table also indicates that sensitivity of such a detection method is likely to be limited due to the low molar absorptivity of carboxyl groups at such wavelengths. This explains why high-performance liquid chromotography methods for organic acid quantification sometimes make use of UV–Vis detectors tuned to $\sim$210 nm (e.g., Resource Material 3), and why there are research efforts aimed at developing derivitization methods to enhance the sensitivity of UV–Vis-based methods for the quantification of organic acids (Resource Material 10).

The data of Table 22-2 also illustrate the effect of conjugation on electronic transitions. Increased conjugation leads to absorption maxima at longer wavelengths due to the associated decrease in the electronic energy spacing within a conjugated system (i.e., lower energy difference between the ground and excited state). The aromatic compounds included in the table were chosen due to their relevance to protein quantification: benzene/phenylalanine, phenol/tyrosine, and indole/tryptophan (Chap. 9, Sect. 9.2.8). The table

22-2 table **Representative Absorption Maxima Above 200 nm for Select Functional Groups**

Chromophore	*Example*	λ_{max}[a]	ε_{max}[b]	*Resource Material*
Nonconjugated systems				
R–CHO	Acetaldehyde	290	17	4
R_2–CO	Acetone	279	15	4
R–COOH	Acetic acid	208	32	4
R–$CONH_2$	Acetamide	220	63	20
R–SH	Mercaptoethane	210	1200	16
Conjugated systems				
R_2C=CR_2	Ethylene	<200	–	20
R–CH=CH–CH=CH–R	1,3 Butadiene	217	21,000	20
R–CH=CH–CH=CH–CH=CH–R	1,3,5 Hexatriene	258	35,000	20
–11 conjugated double bonds–	β-Carotene	465	125,000	12
R_2C=CH–CH=O	Acrolein (2-propenal)	210	11,500	20
		315	14	20
HOOC–COOH	Oxalic acid	250	63	20
Aromatic compounds[c]				
C_6H_6	Benzene	256	200	20
C_6H_5OH	Phenol	270	1450	20
C_8H_7N	Indole	278	2500	NIST database[d]

[a] λ_{max}, Wavelength (in nm), of a maximum absorbance greater than 200 nm.
[b] ε_{max}, Molar absorptivity, units of $(cm)^{-1}$ $(M)^{-1}$.
[c] Spectra of the aromatic compounds generally contain an absorption band(s) of higher intensity at a lower wavelength [e.g., phenol has an absorption maxima of ~210 nm with a molar absorptivity of ~6200 $(cm)^{-1}$ $(M)^{-1}$; Values from (20)]. Only the absorption maxima corresponding to the longer wavelengths are included in the table.
[d] NIST Standard Reference Database 69: *NIST Chemistry WebBook.* (The presented values were estimated from the UV–Vis spectrum for indole presented online: http://webbook.nist.gov/cgi/cbook.cgi?Name = indole&Units = SI&cUV = on. The Web site contains UV–Vis data for many compounds that are of potential interest to food scientists.)

indicates that typical proteins will have an absorption maximum at approximately 278 nm (high molar absorptivity of the indole side chain of tryptophan), as well as another peak at around 220 nm. This latter peak corresponds to the amide/peptide bonds along the backbone of the protein, the rationale being deduced from the data for the simple amide included in the table (i.e., acetamide).

22.3 FLUORESCENCE SPECTROSCOPY

The technique of fluorescence spectroscopy is generally 1–3 orders of magnitude more sensitive than corresponding absorption spectroscopy methods. In **fluorescence spectroscopy**, the signal being measured is the electromagnetic radiation that is emitted from the analyte as it relaxes from an excited electronic energy level to its corresponding ground state. The analyte is originally activated to the higher energy level by the absorption of radiation in the UV or Vis range. The processes of activation and deactivation occur simultaneously during a fluorescence measurement. For each unique molecular system, there will be an optimum radiation wavelength for sample excitation and another, of longer wavelength, for monitoring fluorescence emission. The respective wavelengths for excitation and emission will depend on the chemistry of the system under study.

The instrumentation used in fluorescence spectroscopy is composed of essentially the same components as the corresponding instrumentation used in UV–Vis absorption spectroscopy. However, there are definite differences in the arrangement of the optical systems used for the two types of spectroscopy (compare Figs. 22-6 and 22-12). In fluorometers and spectrofluorometers, there is a need for two wavelength selectors, one for the **excitation beam** and one for the **emission beam**. In some simple fluorometers, both wavelength selectors are filters such that the excitation and emission wavelengths are fixed. In more sophisticated spectrofluorometers, the excitation and emission wavelengths are selected by means of grating monochromators. The photon detector of fluorescence instrumentation is generally arranged such that the emitted radiation that strikes the detector is traveling at an angle of 90° relative to the axis of the excitation beam. This detector placement minimizes signal

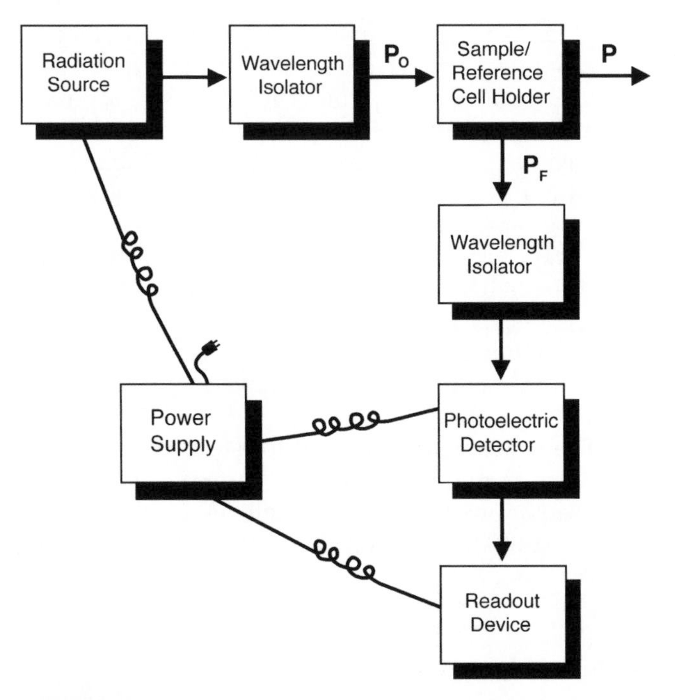

Schematic diagram depicting the arrangement of the source, excitation and emission wavelength selectors, sample cell, photoelectric detector, and readout device for a representative fluorometer or spectrofluorometer.

interference due to transmitted source radiation and radiation scattered from the sample.

The **radiant power** of the fluorescence beam (P_F) emitted from a fluorescent sample is proportional to the change in the radiant power of the source beam as it passes through the sample cell Equation [13]. Expressing this in another way, the radiant power of the fluorescence beam will be proportional to the number of photons absorbed by the sample.

$$P_F = \varphi(P_0 - P) \quad [13]$$

where:

P_F = radiant power of beam emitted from fluorescent cell
φ = constant of proportionality
P_0 and P = as in Equation [1]

The constant of proportionality used in Equation [13] is termed the **quantum efficiency** (φ), which is specific for any given system. The quantum efficiency equals the ratio of the total number of photons emitted to the total number of photons absorbed. Combining Equation [3] and Equation [5] allows one to define P in terms of the analyte concentration and P_0, as given in Equation [14].

$$P = P_0 10^{-\varepsilon bc} \quad [14]$$

where:

P_0 and P = as in Equation [1]
$\varepsilon, b,$ and c = as in Equation [5]

Substitution of Equation [14] into Equation [13] gives an expression that relates the radiant power of the fluorescent beam to the analyte concentration and P_0, as shown in Equation [15]. At low analyte concentrations, $\varepsilon bc < 0.01$, Equation [15] may be reduced to the expression of Equation [16]. Further grouping of terms leads to the expression of Equation [17], where k incorporates all terms other than P_0 and c.

$$P_F = \varphi P_0(1-10^{-\varepsilon bc}) \quad [15]$$

$$P_F = \varphi P_0 \, 2.303 \, \varepsilon bc \quad [16]$$

$$P_F = kP_0 c \quad [17]$$

where:

k = constant of proportionality
P_F = as in Equation [13]
c = as in Equation [5]

Equation [17] is particularly useful because it emphasizes two important points that are valid for the conditions assumed when deriving the equation, particularly the assumption that analyte concentrations are kept relatively low. First, the fluorescent signal will be directly proportional to the analyte concentration, assuming other parameters are kept constant. This is very useful because a linear relationship between signal and analyte concentration simplifies data processing and assay troubleshooting. Second, the sensitivity of a fluorescent assay is proportional to P_0, the power of the incident beam, the implication being that the sensitivity of a fluorescent assay may be modified by adjusting the source output.

Equations [16] and [17] will eventually break down if analyte concentrations are increased to relatively high values. Therefore, the **linear concentration range** for each assay should be determined experimentally. A representative calibration curve for a fluorescence assay is presented in Fig. 22-13. The nonlinear portion of the curve at relatively high analyte concentrations results from decreases in the fluorescence yield per unit concentration. The fluorescence yield for any given sample is also dependent on its environment. Temperature, solvent, impurities, and pH may influence this parameter. Consequently, it is imperative that these environmental parameters be accounted for in the experimental design of fluorescence assays. This may be particularly important in the preparation of appropriate reference standards for quantitative work.

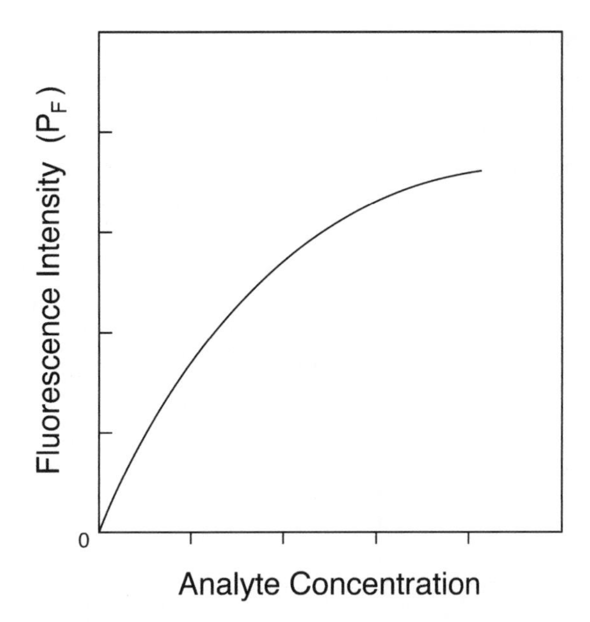

Relationship between the solution concentration of a fluorescent analyte and that solution's fluorescence intensity. Note that there is a linear relationship at relatively low analyte concentrations that eventually goes nonlinear as the analyte concentration increases.

22.4 SUMMARY

UV and Vis absorption and fluorescence spectroscopy are used widely in food analysis. These techniques may be used for either qualitative or quantitative measurements. Qualitative measurements are based on the premise that each analyte has a unique set of energy spacings that will dictate its absorption/emission spectrum. Hence, qualitative assays generally are based on the analysis of the absorption or emission spectrum of the analyte. In contrast, quantitative assays most often are based on measuring the absorbance or fluorescence of the analyte solution at one wavelength. Quantitative absorption assays are based on the premise that the absorbance of the test solution will be a function of the solution's analyte concentration.

Under optimum conditions, there is a direct linear relationship between a solution's absorbance and its analyte concentration. The equation describing this linear relationship is known as Beer's law. The applicability of Beer's law to any given assay always should be verified experimentally by means of a calibration curve. The calibration curve should be established at the same time and under the same conditions that are used to measure the test solution. The analyte concentration of the test solution then should be estimated from the established calibration curve.

Molecular fluorescence methods are based on the measurement of radiation emitted from excited analyte molecules as they relax to lower energy levels. The analytes are raised to the excited state as a result of photon absorption. The processes of photon absorption and fluorescence emission occur simultaneously during the assay. Quantitative fluorescence assays are generally 1–3 orders of magnitude more sensitive than corresponding absorption assays. Like absorption assays, under optimal conditions there will be a direct linear relationship between the fluorescence intensity and the concentration of the analyte in the unknown solution. Most molecules do not fluoresce and, hence, cannot be assayed by fluorescence methods.

The instrumentation used for absorption and fluorescence methods has similar components, including a radiation source, wavelength selector(s), sample holding cell(s), radiation detector(s), and a readout device.

22.5 STUDY QUESTIONS

1. Why is it common to use absorbance values rather than transmittance values when doing quantitative UV–Vis spectroscopy?
2. For a particular assay, your plot of absorbance vs. concentration is not linear. Explain the possible reasons for this.
3. What criteria should be used to choose an appropriate wavelength at which to make absorbance measurements, and why is that choice so important?
4. In a particular assay, the absorbance reading on the spectrophotometer for one sample is 2.033 and for another sample is 0.032. Would you trust these values? Why or why not?
5. Explain the difference between electromagnetic radiation in the UV and Vis ranges. How does quantitative spectroscopy using the UV range differ from that using the Vis range?
6. What is actually happening inside the spectrophotometer when the analyst "sets" the wavelength for a particular assay?
7. Considering a typical spectrophotometer, what is the effect of decreasing the exit slit width of the monochromator on the light incident to the sample?
8. Describe the similarities and differences between a phototube and a photomultiplier tube. What is the advantage of one over the other?
9. Your lab has been using an old single-beam spectrophotometer that must now be replaced by a new spectrophotometer. You obtain sales literature that describes single-beam and double-beam instruments. What are the basic differences between a single-beam and a double-beam spectrophotometer, and what are the advantages and disadvantages of each?
10. Explain the similarities and differences between UV–Vis spectroscopy and fluorescence spectroscopy with regard to instrumentation and principles involved. What is the advantage of using fluorescence spectroscopy?

22.6 PRACTICE PROBLEMS

1. A particular food coloring has a molar absorptivity of $3.8 \times 10^3\ cm^{-1}\ M^{-1}$ at 510 nm.
   (a) What will be the absorbance of a $2 \times 10^{-4}\ M$ solution in a 1-cm cuvette at 510 nm?
   (b) What will be the percent transmittance of the solution in (a)?
2.
   (a) You measure the percent transmittance of a solution containing chromophore X at 400 nm in a 1-cm path length cuvette and find it to be 50%. What is the absorbance of this solution?
   (b) What is the molar absorptivity of chromophore X if the concentration of X in the solution measured in question 2(a) is 0.5 m*M*?
   (c) What is the concentration range of chromophore X that can be assayed if, when using a sample cell of path length 1, you are required to keep the absorbance between 0.2 and 0.8?
3. What is the concentration of compound Y in an unknown solution if the solution has an absorbance of 0.846 in a glass cuvette with a path length of 0.2 cm? The absorptivity of compound Y is $54.2\ cm^{-1}\ (mg/ml)^{-1}$ under the conditions used for the absorption measurement.
4.
   (a) What is the molar absorptivity of compound Z at 295 nm and 348 nm, given the absorption spectrum shown in Fig. 22-14 (which was obtained using a UV–Vis spectrophotometer and a 1 m*M* solution of compound Z in a sample cell with a path length of 1 cm)?
   (b) Now you have decided to make quantitative measurements of the level of compound Z in different solutions. Based on the above spectrum, which wavelength will you use for your measurements? Give two reasons why this is the optimum wavelength.

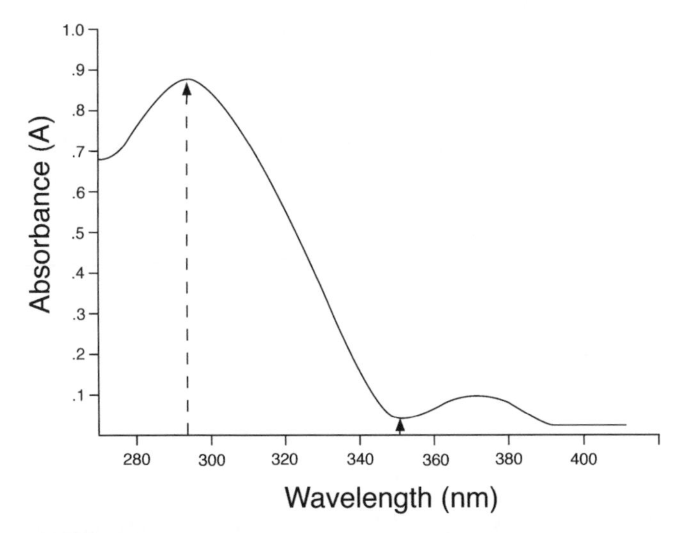

22-14 figure Absorption spectrum of compound Z, to be used in conjunction with problems 4(a) and 4(b).

Answers

1. (a) = 0.76, (b) = 17.4

This problem requires a knowledge of the relationship between absorbance and transmittance and the ability to work with Beer's law.

Given: molar absorptivity $= 3.8 \times 10^3\ cm^{-1}\ M^{-1}$

(a) Use Beer's Law $\rightarrow A = \varepsilon bc$ (see Equation [5] of text)

where:

$$\varepsilon = 3.8 \times 10^3\ cm^{-1}\ M^{-1}$$
$$b = 1\ cm$$
$$c = 2 \times 10^{-4}\ M$$

Plugging into Beer's Law gives the Answer $\rightarrow$ Absorbance = 0.76

(b) Use definition of absorbance $\rightarrow A = -\log T$ (see Equation [3] of text)

where:

$T = P/P_0$

Rearranging Equation [3]:

$$-A = \log T$$
$$10^{-A} = T$$
$$A = 0.76 \text{ [from part (a) of question]}$$
$$10^{-.76} = 0.1737 = T$$
$$\%T = 100 \times T \text{ (combining Equations [1] and [2] of text)}$$

Answer $\rightarrow \%T = 17.4$

2. (a) = 301, (b) = $602\ cm^{-1}\ M^{-1}$, (c) = $0.33 \times 10^{-3}\ M$ to $1.33 \times 10^{-3}\ M$

This problem again requires knowledge of the relationship between absorbance and transmittance and the manipulation of Beer's law. Care must be taken in working with the appropriate concentration units.

(a) $T = 0.5$

Use $A = -\log T = -\log 0.5 = 0.301$

Answer $\rightarrow$ 0.301

(b) Given that the solution in part (a) is 0.5 m*M* (equivalent to $5 \times 10^{-4}\ M$)

Rearranging Beer's Law:

$$\varepsilon = A/(bc)$$
$$\varepsilon = 0.301/[(1\ cm) \times (5 \times 10^{-4}\ M)]$$

Answer $\rightarrow \varepsilon = 602\ cm^{-1}\ M^{-1}$

(c) To answer the problem, find the concentration that will give an absorbance of 0.200 (lower limit) and the concentration that will give an absorbance of 0.800 (upper limit). In both cases, use Beer's law to determine the appropriate concentrations.

where:

$$c = A/\varepsilon b$$

$$\text{Lowest concentration} = 0.2/[(602\ cm^{-1}\ M^{-1})(1\ cm)] = 3.3 \times 10^{-4}\ M \text{ (i.e., 0.33 m}M\text{)}$$

$$\text{Highest concentration} = 0.8/[(602\ cm^{-1}\ M^{-1})(1\ cm)] = 1.3 \times 10^{-3}\ M \text{ (i.e., 1.33 m}M\text{)}$$

3. 0.078 mg/ml

   This problem illustrates (1) that concentration need not be expressed in units of molarity and (2) that the path length of the cuvette must be considered when applying Beer's law. In the present problem, the analyte concentration is given in mg/ml, thus the absorptivity must be in analogous units.

   Apply: $c = A/\varepsilon b$
   where:

$$A = 0.846$$
$$\varepsilon = 54.2\,\text{cm}^{-1}\,(\text{mg/ml})^{-1}$$
$$b = 0.2\,\text{cm}$$

   Answer → 0.078 mg/ml

4. (a) = 860 at 295 nm, 60 at 348 nm; (b) = 295 nm; optimum sensitivity and more likely to adhere to Beer's law.

   This problem presents the common situation in which one wants to use absorbance spectroscopy for quantitative measurements but is unsure what wavelength to choose for the measurements. Furthermore, the absorptivity of the analyte at the different wavelengths of interest is unknown. A relatively simple way to obtain the necessary information is to do determine the absorption spectrum of the analyte at a known concentration.

   (a) The arrows on the provided spectrum indicate the points on the spectrum corresponding to 295 nm and 348 nm. The problem notes that the absorption spectrum was obtained using a 1 m*M* solution (i.e., $1 \times 10^{-3}\,M$ solution) of the analyte and that the path length of the cuvette was 1 cm. The answer to the problem is thus determined by taking the absorbance of the analyte at the two wavelengths in question and then plugging the appropriate data into Beer's law. It is somewhat difficult to get an exact absorbance reading from the presented spectrum, but we can estimate that the absorbance of the 1 m*M* solution is ~0.86 at 295 nm and ~0.06 at 348 nm.

   Using $\varepsilon = A/bc$
   Answer →

$$\text{At 295 nm } \varepsilon = 0.86/[(1\,\text{cm})\,(0.001\,M)]$$
$$= 860\,\text{cm}^{-1}\,M^{-1}$$
$$\text{At 348 nm } \varepsilon = 0.06/[(1\,\text{cm})\,(0.001\,M)]$$
$$= 60\,\text{cm}^{-1}\,M^{-1}$$

   (b) In general, analysts strive to obtain maximum sensitivity for their assays, where sensitivity refers to the change in assay signal per unit change in analyte concentration (the assay signal in this case is absorbance). The absorbance values for the analyte at the different wavelengths, taken from the absorption spectrum, and/or the relative absorptivity values for the analyte at the different wavelengths provide a good approximation of the relative sensitivity of the assay at different wavelengths. (It is an approximation because we have not determined the variability/precision of the measurements at the different wavelengths.) It can be seen from the given spectrum that absorbance "peaks" were at ~298 nm and ~370 nm. The sensitivity of the assay, relative to neighboring wavelengths, is expected to be maximum at these absorbance peaks. The peak at 295 nm is significantly higher than that at 370 nm, so the sensitivity of the assay is expected to be significantly higher at 295 nm. Thus, this would be the optimum wavelength to use for the assay. A second reason to choose 295 nm is because it appears to be in the middle of the "peak" and, thus, small changes in wavelength due to instrumental/operator limitations are not expected to appreciably change the absorptivity values. Therefore, the assay is more likely to adhere to Beer's law.

   There are situations in which an analyst may choose to not use the wavelength corresponding to an overall maximum absorbance. For example, if there are known to be interfering compounds that absorb at 295 nm, then an analyst may choose to do take absorbance measurements at 370 nm.

22.7 RESOURCE MATERIALS

1. Brown CW (2005) Ultraviolet, visible, near-infrared spectrophotometers. In: Cazes J (ed) Ewing's analytical instrumentation handbook, 3rd edn. Marcel Dekker, New York
2. Currell G (2000) Analytical instrumentation – performance characteristics and quality. Wiley, New York, pp 67–91
3. DeBolt S, Cook DR, Ford CM (2006) l-Tartaric acid synthesis from vitamin C in higher plants. Proc Natl Acad Sci 103:5608–5613
4. Feinstein K (1995) Guide to spectroscopic identification of organic compounds. CRC, Boca Raton, FL
5. Hargis LG (1988) Analytical chemistry – principles and techniques. Prentice-Hall, Englewood Cliffs, NJ
6. Harris DC (2006) Quantitative chemical analysis, 7th edn. W.H. Freeman, New York
7. Harris DC, Bertolucci MD (1989) Symmetry and spectroscopy. Dover, Mineola, NY
8. Ingle JD Jr, Crouch SR (1988) Spectrochemical analysis. Prentice-Hall, Englewood Cliffs, NJ
9. Milton Roy educational manual for the SPECTRONIC® 20 & 20D spectrophotometers (1989) Milton Roy Co., Rochester, NY
10. Miwa H (2000) High-performance liquid chromatography determination of mono-, poly- and hydroxycarboxylic acids in foods and beverages as their 2-nitrophenylhydrazides. J Chromatogr A 881:365–385
11. Owen T (2000) Fundamentals of UV–visible spectroscopy, Agilent Technologies. http://www.chem.agilent.com/scripts/LiteraturePDF.asp?iWHID=20660
12. Pavia DL, Lampman GM, Kriz GS Jr (1979) Introduction to spectroscopy: a guide for students of organic chemistry. W.B. Saunders, New York
13. Perkampus H-H (1994) UV–Vis spectroscopy and its applications. Springer, Berlin, Germany
14. Ramette RW (1981) Chemical equilibrium and analysis. Addison-Wesley, Reading, MA

15. Robinson JW, Frame EMS, Frame GM II (2005) Undergraduate instrumental analysis, 6th edn. Marcel Dekker, New York
16. Shriner RL, Fuson RC, Curtin DY, Morrill TC (1980) The systematic identification of organic compounds – a laboratory manual, 6th edn. Wiley, New York
17. Skoog DA, Holler FJ, Crouch SR (2007) Principles of instrumental analysis, 6th edn. Brooks/Cole, Pacific Grove, CA
18. Tinoco I Jr, Sauer K, Wang JC, Puglisi D (2001) Physical chemistry: principles and applications in biological sciences, 4th edn. Prentice-Hall, Englewood Cliffs, NJ
19. Thomas MJK, Ando DJ (1996) Ultraviolet and visible spectroscopy, 2nd edn. Wiley, New York
20. Yadav LDS (2005) Organic spectroscopy. Kluwer Academic, Boston, MA

Infrared Spectroscopy

Randy L. Wehling
Department of Food Science and Technology, University of Nebraska, Lincoln, NE 68583-0919, USA
rwehling1@unl.edu

23.1 Introduction 409
23.2 Principles of IR Spectroscopy 409
23.2.1 The IR Region of the Electromagnetic Spectrum 409
23.2.2 Molecular Vibrations 409
23.2.3 Factors Affecting the Frequency of Vibration 409
23.3 Mid-IR Spectroscopy 410
23.3.1 Instrumentation 410
23.3.1.1 Dispersive Instruments 410
23.3.1.2 Fourier Transform Instruments 410
23.3.1.3 Sample Handling Techniques 411
23.3.2 Applications of Mid-IR Spectroscopy 411
23.3.2.1 Absorption Bands of Organic Functional Groups 411
23.3.2.2 Presentation of Mid-IR Spectra 411
23.3.2.3 Qualitative Applications 412
23.3.2.4 Quantitative Applications 412
23.3.3 Raman Spectroscopy 412
23.4 Near-Infrared Spectroscopy 413
23.4.1 Principles 413
23.4.1.1 Principles of Diffuse Reflection Measurements 413

S.S. Nielsen, *Food Analysis*, Food Science Texts Series, DOI 10.1007/978-1-4419-1478-1_23,

23.4.1.2 Absorption Bands in the NIR Region 414
23.4.2 Instrumentation 415
23.4.3 Quantitative Methods Using NIR Spectroscopy 416
23.4.3.1 Calibration Methods Using Multiple Linear Regression 417
23.4.3.2 Calibration Development Using Full Spectrum Methods 417
23.4.4 Qualitative Analysis by NIR Spectroscopy 417
23.4.5 Applications of NIR Spectroscopy to Food Analysis 417
23.5 Summary 418
23.6 Study Questions 418
23.7 References 419

23.1 INTRODUCTION

Infrared (IR) spectroscopy refers to measurement of the absorption of different frequencies of IR radiation by foods or other solids, liquids, or gases. IR spectroscopy began in 1800 with an experiment by Herschel. When he used a prism to create a spectrum from white light and placed a thermometer at a point just beyond the red region of the spectrum, he noted an increase in temperature. This was the first observation of the effects of IR radiation. By the 1940s, IR spectroscopy had become an important tool used by chemists to identify functional groups in organic compounds. In the 1970s, commercial near-IR reflectance instruments were introduced that provided rapid quantitative determinations of moisture, protein, and fat in cereal grains and other foods. Today, IR spectroscopy is used widely in the food industry for both qualitative and quantitative analysis of ingredients and finished foods.

In this chapter, the techniques of mid- and near-IR spectroscopy are described, including the principles by which molecules absorb IR radiation, the components and configuration of commercial IR spectrometers, sampling methods for IR spectroscopy, and qualitative and quantitative applications of these techniques to food analysis.

23.2 PRINCIPLES OF IR SPECTROSCOPY

23.2.1 The IR Region of the Electromagnetic Spectrum

Infrared radiation is electromagnetic energy with **wavelengths** (λ) longer than visible light but shorter than microwaves. Generally, wavelengths from 0.8 to 100 micrometers (μm) can be used for IR spectroscopy and are divided into the near-IR (0.8–2.5 μm), the mid-IR (2.5–15 μm), and the far-IR (15–100 μm) regions. One μm is equal to 1×10^{-6} m. The near- and mid-IR regions of the spectrum are most useful for quantitative and qualitative analysis of foods.

IR radiation also can be measured in terms of its **frequency**, which is useful because frequency is directly related to the energy of the radiation by the following relationship:

$$E = h\nu \qquad [1]$$

where:

E = energy of the system
h = Planck's constant
ν = frequency, in hertz

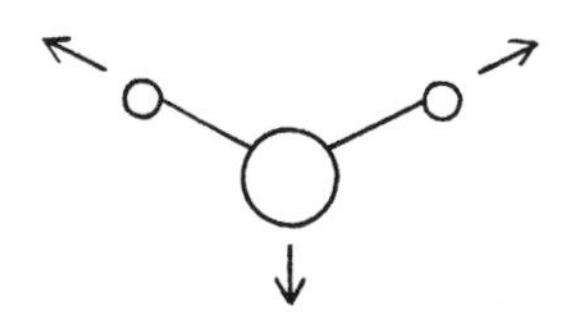

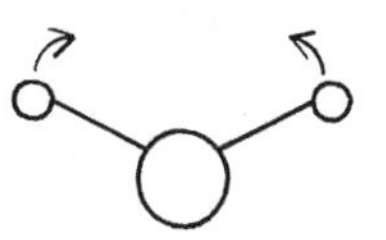

Vibrational modes of the water molecule. Frequencies of the fundamental vibration for the symmetrical stretch, asymmetric stretch, and scissoring motion are 3652 cm^{-1}, 3756 cm^{-1}, and 1596 cm^{-1}, respectively.

Frequencies are commonly expressed as **wavenumbers** ($\bar{\nu}$, in reciprocal centimeters, cm^{-1}). Wavenumbers are calculated as follows:

$$\bar{\nu} = 1/(\lambda \text{ in cm}) = 10^4/(\lambda \text{ in } \mu\text{m}) \qquad [2]$$

23.2.2 Molecular Vibrations

A molecule can absorb IR radiation if it vibrates in such a way that its charge distribution, and therefore its electric dipole moment, changes during the vibration. Although there are many possible vibrations in a polyatomic molecule, the most important vibrations that produce a change in dipole moment are stretching and bending (scissoring, rocking, twisting, wagging) motions. Examples of these vibrations for the water molecule are shown in Fig. 23-1. Note that the stretching motions vibrate at higher frequencies than the scissoring motion. Also, asymmetric stretches are more likely to result in a change in dipole moment, with corresponding absorption of IR radiation, than are symmetric stretches.

23.2.3 Factors Affecting the Frequency of Vibration

A molecular vibration can be thought of as a harmonic oscillator, with the energy level for any molecular vibration given by the following equation:

$$E = (v + 1/2)\,(\mathrm{h}/2\,\pi)\sqrt{k/\frac{m_1 m_2}{m_1 + m_2}} \qquad [3]$$

where:

v = vibrational quantum number (positive integer values, including zero, only)

h = Planck's constant

k = force constant of the bond

m_1 and m_2 = masses of the individual atoms involved in the vibration

Note that the vibrational energy, and therefore the frequency of vibration, is directly proportional to the strength of the bond and inversely proportional to the mass of the molecular system. Thus, different chemical functional groups will vibrate at different frequencies. A vibrating molecular functional group can absorb radiant energy to move from the lowest ($v = 0$) vibrational state to the first excited ($v = 1$) state, and the frequency of radiation that will make this occur is identical to the initial frequency of vibration of the bond. This frequency is referred to as the **fundamental absorption**. Molecules also can absorb radiation to move to a higher ($v = 2$ or 3) excited state, such that the frequency of the radiation absorbed is two or three times that of the fundamental frequency. These absorptions are referred to as **overtones**, and the intensities of these absorptions are much lower than the fundamental since these transitions are less favored. Combination bands can also occur if two or more different vibrations interact to give bands that are sums of their fundamental frequencies. The overall result is that each functional group within the molecule absorbs IR radiation in distinct wavelength bands rather than as a continuum.

23.3 MID-IR SPECTROSCOPY

Mid-IR spectroscopy measures a sample's ability to absorb light in the 2.5–15 μm (4000–650 cm^{-1}) region. Fundamental absorptions are primarily observed in this spectral region.

23.3.1 Instrumentation

Two types of spectrometers are used for mid-IR spectroscopy: dispersive instruments and Fourier transform (FT) instruments. Almost all newer instruments are of the FT type.

23.3.1.1 Dispersive Instruments

Dispersive instruments use a **monochromator** to disperse the individual frequencies of radiation and sequentially pass them through the sample so that the absorption of each frequency can be measured. IR spectrometers have components similar to ultraviolet-visible (UV-Vis) spectrometers, including a radiation source, a monochromator, a sample holder, and a detector connected to an amplifier system for recording the spectra. Most dispersive IR spectrometers are double-beam instruments.

A common mid-IR source is a coil of **Nichrome wire** wrapped around a ceramic core that glows when an electrical current is passed through it. A **Globar**, which is a silicon carbide rod across which a voltage is applied, also can be used as a more intense source. Older spectrometers used sodium chloride prisms to disperse the radiation into monochromatic components, but more modern instruments use a diffraction grating to achieve this effect. Detectors include the **thermocouple detector**, whose output voltage varies with temperature changes caused by varying levels of radiation striking the detector, and the **Golay detector**, in which radiation striking a sealed tube of xenon gas warms the gas and causes pressure changes within the tube. However, most current instruments use either **pyroelectric** detectors, such as deuterated triglycine sulfate (DTGS) crystals, or solid-state **semiconductor detectors**. Variation in the amount of radiation striking a DTGS detector causes the temperature of the detector to change, which results in a change in the dielectric constant of the DTGS element. The resulting change in capacitance is measured. Semiconductor detectors, such as those made from a mercury:cadmium:telluride (MCT) alloy, have conductivities that vary according to the amount of radiation striking the detector surface. These detectors respond faster and to smaller changes in radiation intensity than other detectors; however, they typically require cryogenic cooling. DTGS and MCT detectors are the most commonly used detectors in Fourier Transform instruments discussed in Sect. 23.3.1.2.

23.3.1.2 Fourier Transform Instruments

In **Fourier transform** (FT) instruments, the radiation is not dispersed, but rather all wavelengths arrive at the detector simultaneously and a mathematical treatment, called an **FT**, is used to convert the results into a typical IR spectrum. Instead of a monochromator, the instrument uses an **interferometer**. In a Michelson interferometer, which is the most commonly used design, an IR beam is split and then recombined by reflecting back the split beams with mirrors (Fig. 23-2). If the pathlength of one beam is varied by moving its mirror, the two beams will interfere either constructively or destructively as they are combined, depending on their phase difference. Therefore, the intensity of the radiation reaching the detector varies as a function of the optical path difference, and the

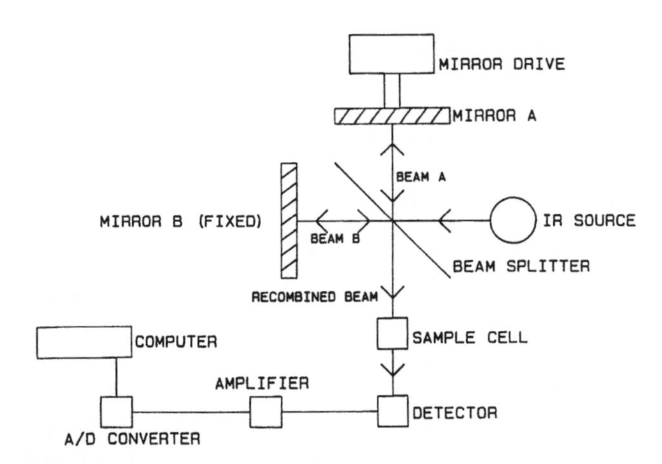

23-2 figure Block diagram of an interferometer and associated electronics typically used in an FTIR instrument.

pattern of energy intensity obtained as a function of optical path difference is referred to as an **interferogram**. When a sample is placed in the recombined beam ahead of the detector, the molecules in the sample absorb at their characteristic frequencies, and thus the radiation reaching the detector is modified by the presence of the sample. This interferogram showing intensity vs. pathlength is then converted by Fourier transformation into an IR spectrum giving absorbance vs. frequency. A computer allows the mathematical transformation to be completed rapidly. Because all wavelengths are measured at once, FT instruments can acquire spectra more rapidly, with a greatly improved signal-to-noise ratio, as compared to dispersive instruments.

23.3.1.3 Sample Handling Techniques

Liquids may be measured by **transmission IR spectroscopy**. Because absorptivity coefficients in the mid-IR are high, cells with pathlengths of only 0.01–1.0 mm are commonly used. Quartz and glass absorb in the mid-IR region, so cell windows are made of non-absorbing materials such as halide or sulfide salts. Halide salts are soluble in water, and care must be taken when selecting cells for use with aqueous samples. Cells are also available with windows made from more durable and less soluble materials, such as zinc selenide, but are more expensive than those with halide salt windows. Transmission spectra of solids can be obtained by finely grinding a small amount of the sample with potassium bromide, pressing the mixture into a pellet under high pressure, and inserting the pellet into the IR beam. An alternative technique is to disperse a finely divided solid in Nujol mineral oil to form a mull.

Attenuated total reflectance (ATR) **cells** are available for obtaining spectra from solid samples that are too thick for transmission measurements, pastes such as peanut butter, and liquids including viscous liquids. ATR measures the total amount of energy reflected from the surface of a sample in contact with an IR transmitting crystal. IR radiation passes through the crystal to the sample, where the radiation penetrates a short distance into the sample before it is reflected back into the transmitting crystal. Therefore, the intensity of the reflected radiation is decreased at wavelengths where the sample absorbs radiation, allowing a spectrum to be obtained that is similar to a transmission spectrum. Similarly, **internal reflectance cells** also are available for use with liquid samples, where the IR radiation penetrates a few micrometers into the liquid before being reflected back into an IR transmitting crystal in contact with the liquid. These types of cells are especially useful for samples such as aqueous liquids that absorb strongly in the mid-IR region.

Transmission spectra can be obtained from gas samples using a sealed 2–10-cm glass cell with IR transparent windows. For trace analysis, multiple-pass cells are available that reflect the IR beam back and forth through the cell many times to obtain pathlengths as long as several meters. FTIR instruments also can be interfaced to a gas chromatograph, to obtain spectra of compounds eluting from the chromatography column.

Commercial instruments are also available in which a **microscope** is interfaced to a FTIR spectrometer. The IR beam can be focused through the microscope onto a thin specimen mounted on a microscope slide. The IR spectrum then can be obtained from a very small area of the sample that measures only a few micrometers on each side. By moving the microscope stage, a profile of spectra across the sample can be obtained and used to evaluate the homogeneity of the sample. Mid-infrared imaging instruments are also available that use an array detector, similar to a digital camera, to obtain an image of the sample at different frequencies.

23.3.2 Applications of Mid-IR Spectroscopy

23.3.2.1 Absorption Bands of Organic Functional Groups

The wavelength bands where a number of important organic functional groups absorb radiation in the mid-IR region are shown in Table 23-1.

23.3.2.2 Presentation of Mid-IR Spectra

Spectra are normally presented with either wavenumbers or wavelengths plotted on the *x*-axis and either percent transmittance or absorbance plotted on the

23-1 table Mid-IR Absorption Frequencies of Various Organic Functional Groups

Group	*Absorbing Feature*	*Frequency* (cm^{-1})
Alkanes	–CH stretch and bend	3000–2800
	$-CH_2$ and $-CH_3$ bend	1470–1420 and 1380–1340
Alkenes	Olefinic –CH stretch	3100–3000
Alkynes	Acetylenic –CH stretch	3300
Aromatics	Aromatic –CH stretch	3100–3000
	–C=C– stretch	1600
Alcohols	–OH stretch	3600–3200
	–OH bend	1500–1300
	C–O stretch	1220–1000
Ethers	C–O asymmetric stretch	1220–1000
Amines	Primary and secondary –NH stretch	3500–3300
Aldehydes and ketones	–C=O stretch	1735–1700
	–CH (doublet)	2850–2700
Carboxylic acids	–C=O stretch	1740–1720
Amides	–C=O stretch	1670–1640
	–NH stretch	3500–3100
	–NH bend	1640–1550

y-axis. The mid-IR spectrum of polystyrene is shown in Fig. 23-3 and is typical of the common method of presentation of IR spectra.

23.3.2.3 Qualitative Applications

The center frequencies and relative intensities of the absorption bands can be used to identify specific functional groups present in an unknown substance. A substance also can be identified by comparing its mid-IR spectrum to a set of **standard spectra** and determining the closest match. Spectral libraries are available from several sources, but probably the largest collection of standards is the Sadtler Standard Spectra (Sadtler Division of Bio-Rad, Inc., Philadelphia, PA). Standard spectra are now commonly stored in digital format to allow searching by computer algorithm to determine the best match with an unknown compound. Common food applications include the identification of flavor and aroma compounds, particularly when FTIR measurements are coupled with gas chromatography. IR spectra also are useful for obtaining positive identification of packaging films.

23.3.2.4 Quantitative Applications

IR spectroscopic measurements obey **Beer's law**, although deviations may be greater than in UV-Vis spectroscopy due to the low intensities of IR sources, the low sensitivities of IR detectors, and the relative narrowness of mid-IR absorption bands. However, quantitative measurements can be successfully made. Perhaps the most extensive use of this technique is in the **infrared milk analyzers**, which have the ability to analyze hundreds of samples per hour. The fat, protein, and lactose contents of milk can be determined simultaneously with one of these instruments. The ester carbonyl groups of lipid absorb at 5.73 μm $(1742\,cm^{-1})$, the amide groups of protein at 6.47 μm $(1348\,cm^{-1})$, and the hydroxyl groups of lactose at 9.61 μm $(1046\,cm^{-1})$. These automated instruments homogenize the milk fat globules to minimize light scattering by the sample, and then pump the milk into a flow-through cell through which the IR beam is passed. In some instruments, the monochromator uses simple optical interference filters that pass only a single wavelength of radiation through the sample, and the filter is selected depending on which constituent the operator wishes to measure. The instrument is calibrated using samples of known concentration to establish the slope and intercept of a Beer's law plot. Newer analyzers use an FTIR instrument to measure the absorbance at many wavelengths simultaneously, and then apply a **multiple linear regression** (MLR) or **partial least squares** (PLS) **regression** equation to predict the concentration of each constituent from the absorbance values at selected wavelengths. Regression techniques are described in more detail in Sect. 23.4. Official methods have been adopted for the IR milk analyzers, and specific procedures for operation of these instruments are given (1,2).

Commercial instruments also are available for measuring the fat content of emulsified meat samples by IR spectroscopy. A number of additional applications have been reported in recent years, particularly with respect to the characterization of fats and oils, including measurement of the degree of unsaturation, determination of *cis* and *trans* contents, and identification of the source of olive oils (3,4).

23.3.3 Raman Spectroscopy

Raman spectroscopy is a vibrational spectroscopic technique that is complementary to IR measurements (5). When a photon of light collides with a molecule, the collision can result in the photon being scattered. The collision can either be elastic, where kinetic energy is conserved, or inelastic, where energy is lost. When the photon interacts with the molecule, the energy of the molecule is raised to an unstable virtual state. When most molecules return to their ground state, they return to the lowest vibrational state, and the scattered photon has the same energy as the incident light. This is known as **Rayleigh scattering**. However, a few molecules may return to a higher vibrational state. For those molecules, the scattered light will have less energy (lower frequency) than the incident light.

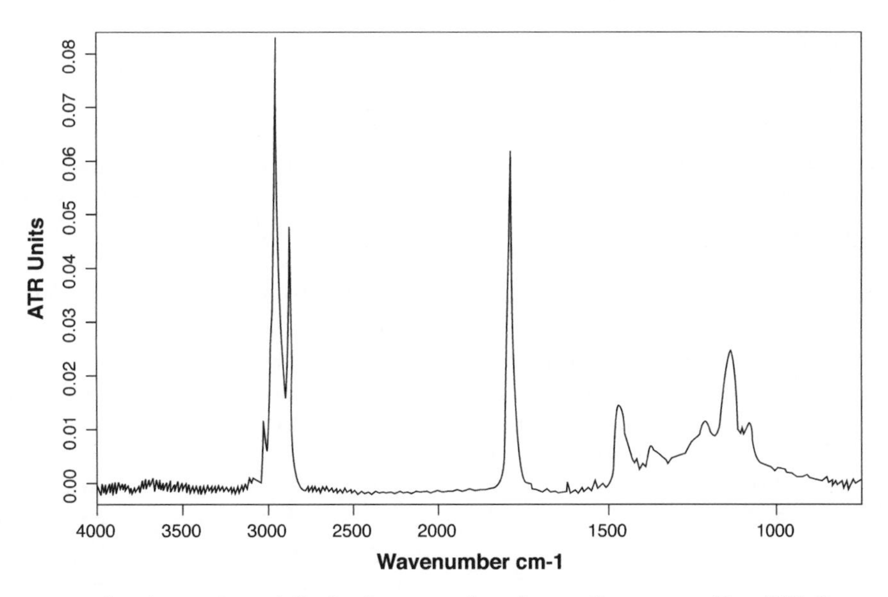

23-3 figure

Mid-IR spectra of native and partially-hydrogenated soybean oils measured by ATR. Frequency in wavenumbers is plotted on the *x*-axis, with intensity on the *y*-axis. The bands just above 3000 cm^{-1} indicate the presence of unsaturated hydrocarbons in the molecules, while the strong –CH bands just below 3000 cm^{-1} indicate that saturated hydrocarbons are present in large amounts. The sharp band between 1700 and 1750 cm^{-1} arises from the carbonyls in the ester linkages between the fatty acids and glycerol backbone. Other –CH bands can be observed below 1500 cm^{-1}, including a –CH band exclusively associated with trans double bonds at 960 cm^{-1}.

The difference in frequency between the incident and scattered light is equal to the frequency of vibration of the molecule. This is known as **Raman scattering**. For Raman scattering to occur, a molecule must undergo a change in polarizability, but does not need to undergo a change in dipole moment. Thus, symmetrical vibrations that cannot be observed by IR spectroscopy can be observed by Raman. Raman is complementary to IR spectroscopy in that some vibrations are only Raman active, some are only IR active, and some are both.

In Raman spectroscopy, the intensity of the scattered light depends on the intensity of the source; therefore, a laser is most commonly used as the source of radiation. To minimize interference from fluorescence, near-infrared lasers are frequently used. The intensity of the scattered light is measured at different frequencies from the incident light, and a plot of $\Delta\nu$ in cm^{-1} vs. intensity is made to obtain the Raman spectrum.

Quantitative analysis can be performed with Raman spectroscopy, as the intensity of light scattered at a specific frequency is directly proportional to the concentration of scattering molecules present. Raman spectroscopy is applicable to both solid and liquid samples, often with minimal sample preparation for solids. Water is a very weak Raman scatterer, allowing low concentrations of organic molecules in aqueous solution to be measured. Conversely, water is a very strong IR absorber, which increases the complexity of applying IR techniques to aqueous samples. Finally, the Raman spectrum of a compound is often less complex than its mid or near-IR spectrum. These factors make Raman spectroscopy a useful alternative to IR spectroscopy. Applications of Raman spectroscopy to food analysis have been reviewed in the literature (6).

23.4 NEAR-INFRARED SPECTROSCOPY

Measurements in the **near-IR** (NIR) spectral region (0.7–2.5 μm, equal to 700–2500 nm) are more widely used for quantitative analysis of foods than are mid-IR measurements. Several commercial instruments are available for compositional analysis of foods using NIR spectroscopy. A major advantage of NIR spectroscopy is its ability to measure directly the composition of solid food products by use of diffuse reflection techniques.

23.4.1 Principles

23.4.1.1 Principles of Diffuse Reflection Measurements

When radiation strikes a solid or granular material, part of the radiation is reflected from the sample surface. This mirror-like reflection is called **specular reflection**, and gives little useful information about the sample. Most of the specularly reflected radiation is directed back toward the energy source. Another portion of the radiation will penetrate through the surface of the sample and be reflected off several sample particles before it exits the sample. This is referred to as **diffuse reflection**, and this diffusely reflected radiation emerges from the surface at random angles through 180°. Each time the radiation interacts with a sample particle, the chemical constituents in the sample can absorb a portion of the radiation. Therefore, the diffusely reflected radiation contains information about the chemical composition of the sample, as indicated by the amount of energy absorbed at specific wavelengths.

The amount of radiation penetrating and leaving the sample surface is affected by the size and shape of the sample particles. Compensation for this effect may be achieved by grinding solid or granular materials with a sample preparation mill to a fine, uniform particle size, or by applying mathematical corrections when the instrument is calibrated (7).

23.4.1.2 Absorption Bands in the NIR Region

The absorption bands observed in the NIR region are primarily overtones and combinations. Therefore, the absorptions tend to be weak in intensity. However, this is actually an advantage, since absorption bands that have sufficient intensity to be observed in the NIR region arise primarily from functional groups that have a hydrogen atom attached to a carbon, nitrogen, or oxygen, which are common groups in the major constituents of food such as water, proteins, lipids, and carbohydrates. Table 23-2 lists the absorption bands associated with a number of important food constituents.

The absorption bands in the NIR region tend to be broad and frequently overlap, yielding spectra that are quite complex. However, these broad bands are especially useful for quantitative analysis. Typical NIR spectra of wheat, dried egg white, and cheese are shown in Fig. 23-4. Note that strong absorption bands associated with the −OH groups of water are centered at ca. 1450 and 1940 nm. These bands are the dominant features in the spectrum of cheese, which contains 30–40% moisture, and they are still prominent even in the lower moisture wheat and egg white samples. Bands arising from the −NH groups in protein

23-2 table Near-IR Absorption Bands of Various Food Constituents

Constituent	*Absorber*	*Wavelength (nm)*
Water	−OH stretch/deformation combination	1920–1950
	−OH stretch	1400–1450
Protein – peptides	−NH deformation	2080–2220 and 1560–1670
Lipid	Methylene −CH stretch	2300–2350
	$-CH_2$ and $-CH_3$ stretch	1680–1760
Carbohydrate	C−O, O−H stretching combination	2060–2150

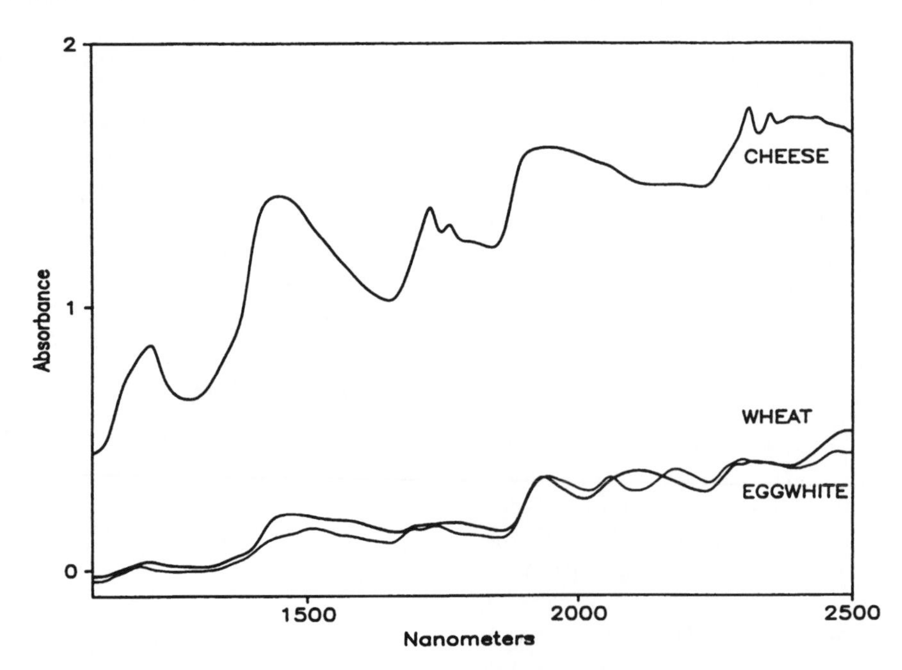

23-4 figure NIR spectra of cheese, wheat, and dried egg white plotted as $\log(1/R)$ vs. wavelength in nm.

can be observed at 2060 nm and 2180 nm in the egg white spectrum but are partially obscured by a starch absorption band, centered at 2100 nm, in the wheat sample. Relatively sharp absorption bands arising from −CH groups in lipid can be observed at 2310 and 2350 nm, and another band from these groups is seen around 1730 nm. The 1730-nm band overlaps a weak protein absorption. The lipid bands are distinctly observable in the cheese spectrum.

23.4.2 Instrumentation

A commercial NIR spectrometer is shown in Fig. 23-5. The radiation source in most NIR instruments is a tungsten-halogen lamp with a quartz envelope, similar to a projector lamp. These lamps emit significant amounts of radiation in both the visible and NIR spectral regions. Semiconductor detectors are most commonly used in NIR instruments, with silicon detectors used in the 700–1100-nm range, and lead sulfide used in the 1100–2500-nm region. In situations for which a rapid response to changing light intensity is needed, such as in online monitoring, indium–gallium–arsenide (InGaAs) detectors can be used. Many InGaAs detectors are limited to a maximum wavelength of 1700 nm, although commercial InGaAs detectors with a range extended to longer wavelengths are now available. Most commercial NIR instruments use monochromators, rather than interferometers, although some commercial instruments are now using FT technology. Monochromator-based instruments may be of the scanning type, in which a grating is used to disperse the radiation by wavelength, and the grating is rotated to impinge a single wavelength (or more appropriately, a narrow band of wavelengths) onto a sample at any given time. Using this arrangement, it takes several seconds to collect a spectrum from a sample over the entire NIR region. Some rapid scanning instruments impinge light over the entire NIR region onto the sample, with the reflected or transmitted light then directed onto a fixed grating that disperses the light by wavelength, and also focuses it onto a multichannel array detector that measures all wavelengths at once. These instruments can obtain a spectrum from a sample in less than 1 s.

Instruments dedicated to specific applications can use optical interference filters to select 6–20 discrete wavelengths that can be impinged on the sample. The filters are selected to obtain wavelengths that are known to be absorbed by the sample constituents. The instrument inserts filters one at a time into the light beam to direct individual wavelengths of radiation onto the sample.

Either **reflection** or **transmission** measurements may be made in NIR spectroscopy, depending on the type of sample. In the reflection mode, used primarily for solid or granular samples, it is desirable to measure only the diffusely reflected radiation that contains information about the sample. In many instruments, this is accomplished by positioning the detectors at a 45° angle with respect to the incoming IR beam, so that the specularly reflected radiation is not measured (Fig. 23-6a). Other instruments use an integrating sphere, which is a gold-coated metallic sphere with the detectors mounted inside (Fig. 23-6b). The sphere collects the diffusely reflected radiation coming at various angles from the sample and focuses it onto the detectors. The specular component escapes from the

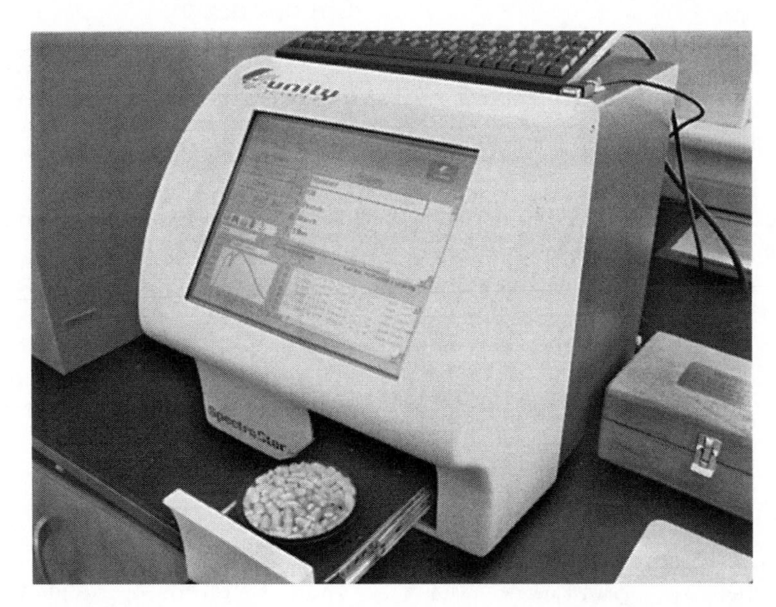

A modern commercial NIR instrument. (Photograph courtesy of the University of Nebraska-Lincoln.)

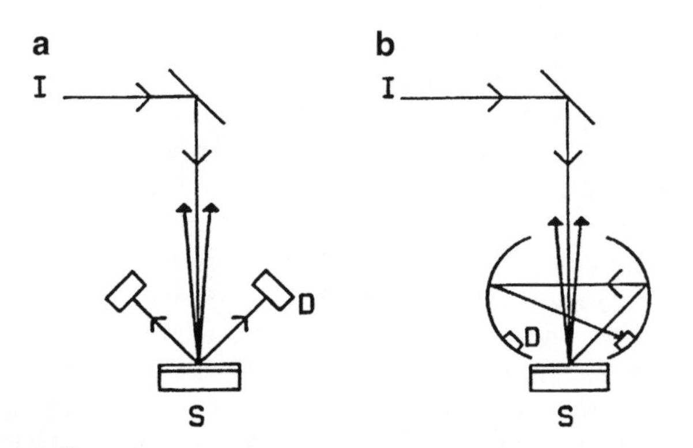

23-6 figure Typical instrument geometries for measuring diffuse reflectance from solid food samples. Radiation from the monochromator (I) is directed by a mirror onto the sample (S). Diffusely reflected radiation is measured directly by detectors (D) placed at a 45° angle to the incident beam (**a**) or is collected by an integrating sphere and focused onto the detectors (**b**). In both cases, the specularly reflected radiation is not measured.

sphere through the same port by which the incident beam enters and strikes the sample.

Samples are often prepared by packing the food tightly into a cell against a quartz window, thereby providing a smooth, uniform surface from which reflection can occur. Quartz does not absorb in the NIR region. At each wavelength, the intensity of light reflecting from the sample is compared to the intensity reflected from a nonabsorbing reference, such as a ceramic or fluorocarbon material. Reflectance (R) is calculated by the following formula:

$$R = I/I_0 \quad [4]$$

where:

I = intensity of radiation reflected from the sample at a given wavelength

I_0 = intensity of radiation reflected from the reference at the same wavelength

Reflectance data are expressed most commonly as $\log(1/R)$, an expression analogous to absorbance in transmission spectroscopy. Reflectance measurements also are expressed sometimes as differences, or derivatives, of the reflectance values obtained from adjacent wavelengths:

$$(\log R_{\lambda_2} - \log R_{\lambda_1}) \quad [5]$$

or

$$(2\log R_{\lambda_2} - \log R_{\lambda_1} - \log R_{\lambda_3}) \quad [6]$$

These derivative values are measures of the changes in slope of the spectrum, where λ_1, λ_2, and λ_3 are adjacent wavelengths typically separated by 5–20 nm, with the higher numbers representing longer wavelengths.

Transmission measurements also can be made in the NIR region, and this is usually the method of choice for liquid samples. A liquid is placed in a quartz cuvette and the absorbance measured at the wavelengths of interest. Transmission measurements also can be taken from solid samples, but generally only in the 700–1100-nm range. In this wavelength region, the absorption bands are higher overtones that are very weak, allowing the radiation to penetrate through several millimeters of a solid sample. The use of transmission measurements can minimize the degree of sample preparation needed. Since the IR beam passes through the entire sample, the need for a smooth, homogeneous sample surface is reduced.

NIR energy can be transmitted through a fiber optic cable some distance from the monochromator or interferometer, allowing reflection or transmission measurements to be made remotely from the instrument. This is very useful if we wish to take measurements in a processing plant environment. Commercial probes are available that can be inserted directly into bulk granular materials, or inserted into a pipe carrying a liquid.

As with mid-IR, NIR imaging instruments are also now commercially available. These instruments use an array detector so that a digital image of a food sample can be obtained at various wavelengths, or a spectrum can be obtained from a single pixel in a digital image. This technique is often referred to as hyperspectral imaging, and holds much potential for evaluating sample heterogeneity, or identifying small features or contaminants on an intact food sample.

23.4.3 Quantitative Methods Using NIR Spectroscopy

NIR instruments can be calibrated to measure various constituents in food and agricultural commodities. Because of the overlapping nature of the NIR absorption bands, it is usually necessary to take measurements at two or more wavelengths to quantitate a food component reliably. Multivariate statistical techniques are used to relate the spectral data collected at multiple wavelengths to the concentration of the component of interest in the food. The simplest statistical technique used is MLR, which applies an equation of the following form to predict the amount of a constituent present in the food from the spectral measurements:

$$\%\,\text{constituent} = z + a\log(1/R_{\lambda 1}) + b\log(1/R_{\lambda 2}) + c\log(1/R_{\lambda 3}) + \ldots \quad [7]$$

where each term represents the spectral measurement at a different wavelength multiplied by a corresponding coefficient. Each coefficient and the intercept (z)

are determined by the multivariate regression analysis. Absorbance or derivatized reflectance data also can be used in lieu of the $\log(1/R)$ format. Use of derivatized reflectance data has been found to provide improved results in some instances, particularly with samples that may not have uniform particle sizes. Other mathematical techniques also are available that can be applied to the reflectance data to correct for the effects of nonuniform particle size (7).

23.4.3.1 Calibration Methods Using Multiple Linear Regression

The first step in calibrating an NIR instrument is to select a set of calibration, or training, samples. The samples should be representative of the products that will be analyzed, contain the constituent of interest at levels covering the range that is expected to be encountered, and have a relatively uniform distribution of concentrations across that range. The calibration samples are analyzed by the classical analytical method normally used for that constituent, and spectral data also are obtained on each sample with the NIR instrument at all available wavelengths. All data are stored into computer memory. MLR is then used to select the optimum wavelengths for measurement and the associated coefficients for each wavelength. Wavelengths are selected based on statistical significance by using a step forward or reverse stepwise regression procedure or by using a computer algorithm that tests regressions using all possible combinations of two, three, or four wavelengths to determine the combination that provides the best results. Most calibrations will use between two and six wavelengths, and one should always check to make certain that the wavelengths chosen on the basis of statistical significance also make sense from a spectroscopic standpoint. Calibration results are evaluated by comparing the multiple correlation coefficients, *F*s of regression, and standard errors for the various equations developed. It is desirable to maximize the correlation coefficient (generally R should be >0.9) and minimize the standard error. A calibration always should be tested by using the instrument to predict the composition of a set of test samples that are completely independent of the calibration set and comparing the results obtained to the classical method.

23.4.3.2 Calibration Development Using Full Spectrum Methods

Calibration techniques such as **PLS regression** and **principal components regression** (PCR) have been developed that use information from all wavelengths in the entire NIR spectrum, rather than a few selected wavelengths, to predict sample composition. PLS and PCR use data reduction techniques to extract from a large number of variables (i.e., reflectance or absorbance measurements at many wavelengths) a much smaller number of new variables that account for most of the variability in the samples. These new variables then can be used to develop a regression equation to predict the amount of a constituent in samples of a food. In PLS and PCR methods, it is not necessary to eliminate spectral information, as it is in MLR where only a limited number of wavelengths are used. PLS and PCR methods are reported to yield improved results for some samples (8).

Artificial neural networks have also recently been used to predict composition from NIR spectra. Neural networks may have some advantages over the linear regression techniques for dealing with highly complex samples, or samples from diverse geographic regions.

23.4.4 Qualitative Analysis by NIR Spectroscopy

NIR spectroscopy also can be used to classify a sample into one of two or more groups, rather than to provide quantitative measurements. **Discriminant analysis** techniques can be used to compare the NIR spectrum of an unknown sample to the spectra of samples from different groups. The unknown sample then is classified into the group to which its spectrum is most similar. While this technique has been more widely used in the chemical and pharmaceutical industries for raw material identification, it has also been used for food applications, including the classification of wheat as hard red spring or hard red winter (9), the identification of orange juice samples from different sources (10), authentication of the source of olive oils (11), and discrimination of beef by breed and muscle type (12).

23.4.5 Applications of NIR Spectroscopy to Food Analysis

Theory and applications of NIR spectroscopy to food analysis have been discussed in several publications (13–16). The technique is widely used throughout the grain, cereal products, and oilseed processing industries. NIR techniques using measurements from ground or whole grain samples have been adopted as approved methods of analysis by AACC International (17) for measuring protein in barley, oats, rye, triticale, wheat, and wheat flour, as well as moisture, protein, and oil in soybeans. These approved methods describe the instruments available for making these measurements, including a list of current manufacturers with contact information in Method 39-30, as well as the proper techniques for preparing samples and calibrating the instruments. NIR instruments now are used by the official grain inspection agencies in both the US and Canada for measuring protein, moisture, and oil in cereals and oilseeds.

Components, such as protein and dietary fiber, can be determined successfully in a number of cereal-based foods using NIR spectroscopy (18–20). Modern instruments and calibration techniques allow a wide variety of products, such as cookies, granola bars, and ready-to-eat breakfast foods, to be analyzed using the same calibration.

NIR spectroscopy also can be used for numerous other commodities and food products. The technique has been used successfully to evaluate composition and quality of red meats and processed meat products (21–23), poultry (24), and fish (25). NIR spectroscopy is useful also for analyzing a number of dairy products and nondairy spreads, including measuring moisture in butter and margarine (26); moisture, fat, and protein in cheese (27, 28); and lactose, protein, and moisture in milk and whey powders (29). NIR techniques also have shown promise for measuring total sugars and soluble solids in fruits, vegetables, and juices (30–32), are being used commercially for monitoring the sugar content in corn sweeteners (33), and can be used to quantitate sucrose and lactose in chocolate (34).

NIR spectroscopy also is showing potential for measuring specific chemical constituents in a food that affect its end-use quality, for monitoring changes that occur during processing or storage, and for directly predicting processing characteristics of a commodity that are related to its chemical composition. Examples include determining the amylose content in rice starch, an important determinant of rice quality (35, 36), monitoring peroxide value in vegetable oils (37), monitoring degradation of frying oils (38), and predicting corn processing quality (39, 40).

These are only a few examples of current applications. If a substance absorbs in the NIR region, and is present at a level of a few tenths of a percent or greater, it has potential for being measured by this technique. The primary advantage of NIR spectroscopy is that once the instrument has been calibrated, several constituents in a sample can be measured rapidly (from 30 s to 2 min) and simultaneously. To measure multiple constituents, a calibration equation for each constituent is stored into the memory of the instrument. Measurements are taken at all wavelengths needed by the calibrations, and each equation then is solved to predict the constituents of interest. No sample weighing is required, and no hazardous reagents are used or chemical waste generated. It is also adaptable for **online measurement systems** (41). Disadvantages include the high initial cost of the instrumentation, which may require a large sample load to justify the expenditure, and the fact that specific calibrations may need to be developed for each product measured. Also, the results produced by the instrument can be no better than the data used to calibrate it, which makes careful analysis of the calibration samples of highest importance.

23.5 SUMMARY

IR spectroscopy measures the absorption of radiation in the near- ($\lambda = 0.8$–$2.5\,\mu m$) or mid- ($\lambda = 2.5$–$15\,\mu m$) IR regions by molecules in food or other substances. IR radiation is absorbed as molecules change their vibrational energy levels. Mid-IR spectroscopy is especially useful for qualitative analysis, such as identifying specific functional groups present in a substance. Different functional groups absorb different frequencies of radiation, allowing the groups to be identified from the spectrum of a sample. Quantitative analysis also can be achieved by mid-IR spectroscopy, with milk analysis being a major application. NIR spectroscopy is used most extensively for quantitative applications, using either transmission or diffuse reflection measurements that can be taken directly from solid foods. By using multivariate statistical techniques, NIR instruments can be calibrated to measure the amounts of various constituents in a food sample based on the amount of IR radiation absorbed at specific wavelengths. NIR spectroscopy requires much less time to perform quantitative analysis than do many conventional wet chemical or chromatographic techniques.

23.6 STUDY QUESTIONS

1. Describe the factors that affect the frequency of vibration of a molecular functional group and thus the frequencies of radiation that it absorbs. Also, explain how the fundamental absorption and overtone absorptions of a molecule are related.
2. Describe the essential components of an FT mid-IR spectrometer and their function, and compare the operation of the FT instrument to a dispersive instrument. What advantages do FT instruments have over dispersive IR spectrophotometers?
3. Describe the similarities and differences between mid-infrared spectroscopy and Raman spectroscopy.
4. Of the three antioxidants butylated hydroxytoluene (BHT), butylated hydroxyanisole (BHA), and propyl gallate, which would you expect to have a strong IR absorption band in the 1700–1750 cm^{-1} spectral region? Look up these compounds in a reference book if you are uncertain of their structure.
5. Describe the two ways in which radiation is reflected from a solid or granular material. Which type of reflected radiation is useful for making quantitative measurements on solid samples by NIR spectroscopy? How are NIR

reflectance instruments designed to select for the desired component of reflected radiation?
6. Describe the steps involved in calibrating an NIR reflectance instrument to measure the protein content of wheat flour. Why is it usually necessary to make measurements at more than one wavelength?

23.7 REFERENCES

1. AOAC International (2007) Official methods of analysis, 18th edn., 2005; Current through revision 2, 2007 (On-line). AOAC International, Gaithersburg, MD
2. Lahner BS (1996) Evaluation of Aegys MI 600 Fourier transform infrared milk analyzer for analysis of fat, protein, lactose, and solids nonfat: A compilation of eight independent studies. J AOAC Int 79:1388
3. van de Voort FR, Ismail AA, Sedman J (1995) A rapid, automated method for the determination of *cis* and *trans* content of fats and oils by Fourier transform infrared spectroscopy. J Am Oil Chem' Soc 72:873
4. Gurdeniz G, Toklati F, Ozen B (2007) Differentiation of mixtures of monovarietal olive oils by mid-infrared spectroscopy and chemometrics. Eur J Lipid Sci Technol 109:1194–1202
5. An introduction to Raman for the infrared spectroscopist. Inphotonics Technical Note No. 11, Norwood, MA (www.http://inphotonics.com/technote11.pdf)
6. Thygesen LG, Lokke MM, Micklander E, Engelsen SB (2003) Vibrational microspectroscopy of food. Raman vs. FT-IR. Trends Food Sci Technol 14:50–57
7. Martens H, Naes T (2001) Multivariate calibration by data compression. Ch. 4, In: Williams PC, Norris KH (eds) Near infrared technology in the agricultural and food industries, 2nd edn. American Association of Cereal Chemists, St. Paul, MN, p 75
8. Martens H, Naes T (2001) Multivariate calibration by data compression. Ch. 4, In: Williams PC, Norris KH (eds) Near infrared technology in the agricultural and food industries, 2nd edn. American Association of Cereal Chemists, St. Paul, MN, pp 59–100
9. Delwiche SR, Chen YR, Hruschka WR (1995) Differentiation of hard red wheat by near-infrared analysis of bulk samples. Cereal Chem 72:243
10. Evans DG, Scotter CN, Day LZ, Hall MN (1993) Determination of the authenticity of orange juice by discriminant analysis of near infrared spectra. A study of pretreatment and transformation of spectral data. J Near Infrared Spectrosc 1:33
11. Bertran E, Blanco M, Coello J, Iturriaga H, Maspoch S, Montoliu I (2000) Near infrared spectrometry and pattern recognition as screening methods for the authentication of virgin olive oils of very close geographical origins. J Near Infrared Spectrosc 8:45
12. Alomar D, Gallo C, Castaneda M, Fuchslocher R (2003) Chemical and discriminant analysis of bovine meat by near infrared reflectance spectroscopy (NIRS). Meat Sci 63:441
13. Osborne BG, Fearn T, Hindle PH (1993) Practical NIR spectroscopy with application in food and beverage analysis. Longman, Essex, UK
14. Williams PC, Norris KH (eds) (2001) Near-infrared technology in the agricultural and food industries, 2nd edn. American Association of Cereal Chemists, St. Paul, MN
15. Ozaki Y, McClure WF, Christy AA (2006) Near-infrared spectroscopy in food science and technology. Wiley, Hoboken, NJ
16. Woodcock T, Downey G, O'Donnel CP (2008) Review: better quality food and beverages: the role of near infrared spectroscopy. J Near Infrared Spectrosc 16:1
17. AACC International (2010) Approved methods of analysis, 11th edn (On-line). The American Association of Cereal Chemists, St. Paul, MN
18. Kays SE, Windham WR, Barton FE (1998) Prediction of total dietary fiber by near-infrared reflectance spectroscopy in high-fat and high-sugar-containing cereal products. J Agric Food Chem 46:854
19. Kays SE, Barton FE (2002) Near-infrared analysis of soluble and insoluble dietary fiber fractions of cereal food products. J Agric Food Chem 50:3024
20. Kays WE, Barton FE, Windham WR (2000) Predicting protein content by near infrared reflectance spectroscopy in diverse cereal food products. J Near Infrared Spectrosc 8:35
21. Oh EK, Grossklaus D (1995) Measurement of the components in meat patties by near infrared reflectance spectroscopy. Meat Sci 41:157
22. Geesink GH, Schreutelkamp FH, Frankhuizen R, Vedder HW, Faber NM, Kranen RW, Gerritzen MA (2003) Prediction of pork quality attributes from near infrared reflectance spectra. Meat Sci 65:661
23. Naganathan GK, Grimes LM, Subbiah J, Calkins CR, Samal A, Meyer G (2008) Visible/near-infrared hyperspectral imaging for beef tenderness prediction. Comput Electron Agric 64:225
24. Windham WR, Lawrence KC, Feldner PW (2003) Prediction of fat content in poultry meat by near-infrared transmission analysis. J Appl Poult Res 12:69
25. Solberg C, Fredriksen G (2001) Analysis of fat and dry matter in capelin by near infrared transmission spectroscopy. J Near Infrared Spectrosc 9:221
26. Isakkson T, Nærbo G, Rukke EO (2001) In-line determination of moisture in margarine, using near infrared diffuse transmittance. J Near Infrared Spectrosc 9:11
27. Pierce MM, Wehling RL (1994) Comparison of sample handling and data treatment methods for determining moisture and fat in Cheddar cheese by near-infrared spectroscopy. J Agri Food Chem 42:2830
28. Rodriquez-Otero JL, Hermida M, Cepeda A (1995) Determination of fat, protein, and total solids in cheese by near infrared reflectance spectroscopy. J AOAC Int 78:802
29. Wu D, He Y, Feng S (2008) Short-wave near-infrared spectroscopy analysis of major compounds in milk powder and wavelength assignment. Anal Chim Acta 610:232

30. Tarkosova J, Copikova J (2000) Determination of carbohydrate content in bananas during ripening and storage by near infrared spectroscopy. J Near Infrared Spectrosc 8:21
31. Segtman VH, Isakkson T (2000) Evaluating near infrared techniques for quantitative analysis of carbohydrates in fruit juice model systems. J Near Infrared Spectrosc 8:109
32. Camps C Christen D (2009) Non-destructive assessment of apricot fruit quality by portable visible-near infrared spectroscopy. LWT – Food Sci Technol 42:1125
33. Psotka J, Shadow W (1994) NIR analysis in the wet corn refining industry – A technology review of methods in use. Int Sugar J 96:358
34. Tarkosova J, Copikova J (2000) Fourier transform near infrared spectroscopy applied to analysis of chocolate. J Near Infrared Spectrosc 8:251
35. Villareal CP, De la Cruz NM, Juliano BO (1994) Rice amylose analysis by near-infrared transmittance spectroscopy. Cereal Chem 71:292
36. Delwiche SR, Bean MM, Miller RE, Webb BD, Williams PC (1995) Apparent amylose content of milled rice by near-infrared reflectance spectrophotometry. Cereal Chem 72:182
37. Yildiz G, Wehling RL, Cuppett SL (2001) Method for determining oxidation of vegetable oils by near-infrared spectroscopy. J Am Oil Chem Soc 78:495
38. Ng CL, Wehling RL, Cuppett SL (2007) Method for determining frying oil degradation by near-infrared spectroscopy. J Agric Food Chem 55:593
39. Wehling RL, Jackson DS, Hooper DG, Ghaedian AR (1993) Prediction of wet-milling starch yield from corn by near-infrared spectroscopy. Cereal Chem 70:720
40. Paulsen MR, Singh M (2004) Calibration of a near-infrared transmission grain analyzer for extractable starch in maize. Biosyst Eng 89:79
41. Psotka J (2001) Challenges of making accurate online near-infrared measurements. Cereal Foods World 46:568

Atomic Absorption Spectroscopy, Atomic Emission Spectroscopy, and Inductively Coupled Plasma-Mass Spectrometry

Dennis D. Miller and Michael A. Rutzke*

Department of Food Science, Cornell University, Ithaca, NY 14853-7201, USA
ddm2@cornell.edu
mar9@cornell.edu

24.1 Introduction 423
24.2 General Principles 423
24.2.1 Energy Transitions in Atoms 423
24.2.2 Atomization 424
24.3 Atomic Absorption Spectroscopy 425
24.3.1 Principles of Flame Atomic Absorption Spectroscopy 425
24.3.2 Principles of Electrothermal Atomic Absorption Spectroscopy (Graphite Furnace AAS) 426
24.3.3 Instrumentation for Atomic Absorption Spectroscopy 426
24.3.3.1 Radiation Source 426
24.3.3.2 Atomizers 427

S.S. Nielsen, *Food Analysis*, Food Science Texts Series, DOI 10.1007/978-1-4419-1478-1_24,

24.3.3.3 Monochromator 428
24.3.3.4 Detector/Readout 428
24.3.4 General Procedure for Atomic Absorption Analysis 428
24.3.4.1 Operation of a Flame Atomic Absorption Instrument 428
24.3.4.2 Calibration 429
24.3.4.2.1 Selection of Standards 429
24.3.4.2.2 Sensitivity Check 429
24.3.5 Interferences in Atomic Absorption Spectroscopy 429
24.3.5.1 Spectral Interference 429
24.3.5.1.1 Absorption of Source Radiation 429
24.3.5.1.2 Background Absorption of Source Radiation 429
24.3.5.2 Nonspectral Interferences 430
24.3.5.2.1 Transport Interferences 430
24.3.5.2.2 Solute Volatilization Interferences 430
24.3.5.2.3 Ionization Interference 430
24.4 Atomic Emission Spectroscopy 430
24.4.1 Principles of Flame Emission Spectroscopy 431
24.4.2 Principles of Inductively Coupled Plasma-Atomic Emission Spectroscopy 431
24.4.3 Instrumentation for Flame Emission Spectroscopy 431
24.4.4 Instrumentation for ICP-AES 431
24.4.4.1 Argon Plasma Torch 432
24.4.4.1.1 Characteristics of an Argon Plasma Torch 432
24.4.4.1.2 Sample Introduction into the Plasma and Sample Excitation 433
24.4.4.1.3 Radial and Axial Viewing 433
24.4.4.2 Detectors and Optical Systems 434
24.4.5 General Procedure for Analysis by ICP-AES 435
24.4.6 Interferences in ICP-AES 435
24.5 Comparison of AAS and ICP-AES 436
24.6 Applications of Atomic Absorption and Emission Spectroscopy 436
24.6.1 Uses 436
24.6.2 Practical Considerations 437
24.6.2.1 Reagents 437
24.6.2.2 Standards 437
24.6.2.3 Labware 437
24.7 Inductively Coupled Plasma-Mass Spectrometry 437
24.8 Summary 439
24.9 Study Questions 440
24.10 Practice Problems 440
24.11 References 442

24.1 INTRODUCTION

Atomic spectroscopy has played a major role in the development of our current database for mineral nutrients and toxicants in foods. When atomic absorption spectrometers became widely available in the 1960s, the development of **atomic absorption spectroscopy** (AAS) methods for accurately measuring trace amounts of mineral elements in biological samples paved the way for unprecedented advances in fields as diverse as food analysis, nutrition, biochemistry, and toxicology (1). The application of plasmas as excitation sources for **atomic emission spectroscopy** (AES) led to the commercial availability of instruments for **inductively coupled plasma - atomic emission spectroscopy** (ICP-AES) beginning in the late 1970s. This instrument has further enhanced our ability to measure the mineral composition of foods and other materials rapidly, accurately, and precisely. More recently, plasmas have been joined with **mass spectrometers** (MS) to form inductively coupled plasma-mass spectrometer ICP-MS instruments that are capable of measuring mineral elements with extremely low detection limits. These three instrumental methods have largely replaced traditional wet chemistry methods for mineral analysis of foods, although traditional methods for calcium, chloride, iron, and phosphorus remain in use today (see Chap. 12).

In theory, virtually all of the elements in the periodic chart may be determined by AAS or AES. In practice, atomic spectroscopy is used primarily for determining mineral elements. Table 24-1 lists mineral elements of interest in foods. The USDA database for Ca, Fe, Na, and K in foods is reasonably good. The database for the trace elements and toxic heavy metals is incomplete and should be expanded.

This chapter deals with the basic principles that underlie analytical atomic spectroscopy and provides an overview of the instrumentation available for measuring atomic absorption and emission. A brief section on inductively coupled plasma mass spectrometry is also included. In addition, some practical problems associated with the use of these technologies are addressed. Readers interested in a more thorough treatment of these topics are referred to two excellent monographs available from PerkinElmer, Inc. [Beaty and Kerber (3); and Boss and Fredeen (4)] and to a book by Thomas (5). A more general source is *Principles of Instrumental Analysis* (5th Ed.) by Skoog et al. (6).

24-1 table **Elements in Foods Classified According to Nutritional Essentiality, Potential Toxic Risk, and Inclusion in USDA Nutrient Database for Standard Reference**

Essential Nutrient	*Toxicity Concern*	*USDA Nutrient Database*
Calcium	Lead	Calcium
Phosphorous	Mercury	Iron
Sodium	Cadium	Magnesium
Potassium	Nickel	Phosphorous
Chlorine	Arsenic	Potassium
Magnesium	Thalliuim	Sodium
Iron		Zinc
Iodine		Copper
Zinc		Manganese
Copper		Selenium
Selenium		
Chromium		
Manganese		
Arsenic		
Boron		
Molybdenum		
Nickel		
Silicon		

Compiled based on References (1) and (2).

24.2 GENERAL PRINCIPLES

AAS quantifies the absorption of electromagnetic radiation by well-separated neutral atoms in the gaseous state, while AES measures emission of radiation from atoms excited by heat or other means. Atomic spectroscopy is particularly well suited for analytical measurements because atomic spectra consist of discrete lines, and every element has a unique spectrum. Therefore, individual elements can be identified and quantified accurately and precisely even in the presence of atoms of other elements.

24.2.1 Energy Transitions in Atoms

Atomic absorption spectra are produced when **ground state atoms** (or ions) **absorb energy** from a radiation source. Atomic emission spectra are produced when **excited neutral atoms emit energy** on returning to the ground state or a lower energy state. Absorption of a photon of radiation causes an outer shell electron to jump to a higher energy level, moving the atom into an **excited state**. The excited atom may fall back to a lower energy state, releasing a photon in the process. Atoms absorb or emit radiation of discrete wavelengths because the allowed energy levels of electrons in atoms are fixed (not random). The energy change associated with a transition between two energy levels is directly related to the frequency of the absorbed radiation:

$$E_e - E_g = h\nu \quad [1]$$

where:

E_e = energy in excited state

E_g = energy in ground state

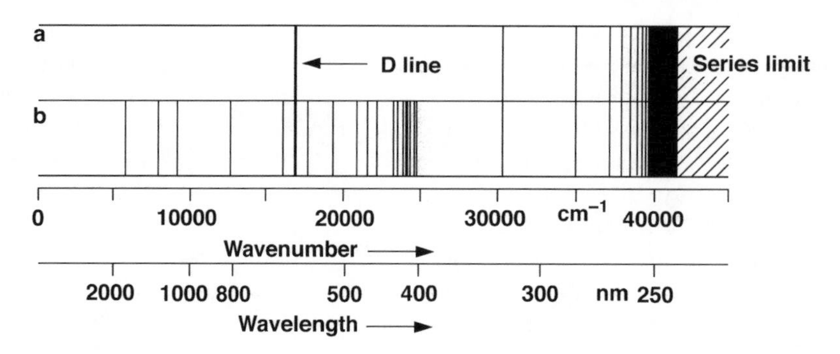

24-1 figure Spectra for sodium. The upper spectrum (**a**) is the absorption spectrum and the lower (**b**) is the emission spectrum. [From Welz, G. 1985. Atomic Absorption Spectrometry. VCH, Weinheim, Germany, reprinted with permission of VCH Publishers (1985).]

h = Planck's constant

ν = frequency of the radiation

Rearranging, we have:

$$\nu = (E_e - E_g)/h \quad [2]$$

or, since $\nu = c/\lambda$

$$\lambda = hc/(E_e - E_g) \quad [3]$$

where:

c = speed of light

λ = wavelength of the absorbed or emitted light

The above relationships clearly show that for a given electronic transition, radiation of a discrete wavelength is either absorbed or emitted. Each element has a unique set of allowed transitions and therefore a unique spectrum. The absorption and emission spectra for sodium are shown in Fig. 24-1. For **absorption**, transitions involve primarily the excitation of electrons in the ground state, so the number of transitions is relatively small. **Emission**, on the other hand, occurs when electrons in various excited states fall to lower energy levels including, but not limited to, the ground state. Therefore, the emission spectrum has more lines than the absorption spectrum. When a transition is from or to the ground state, it is termed a **resonance transition**, and the resulting spectral line is called a **resonance line**. See Chap. 21 for a more detailed discussion of atomic and molecular energy transitions.

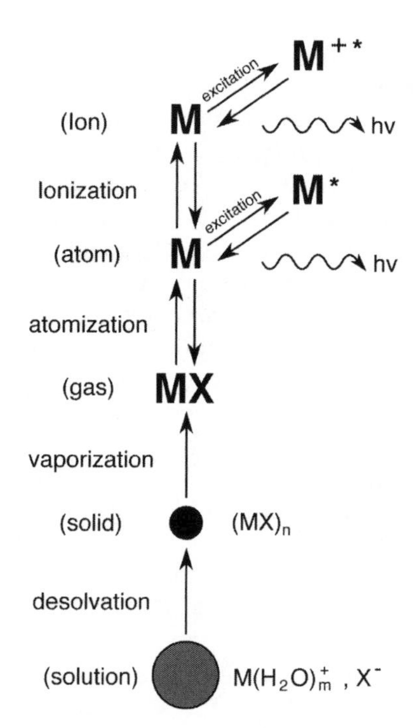

24-2 figure A schematic representation of the atomization of an element in a flame or plasma. The *large circle* at the bottom represents a tiny droplet of a solution containing the element (M) as part of a compound. [From (4), used with permission. Courtesy of the PerkinElmer Corporation, Shelton, CT.]

24.2.2 Atomization

Atomic spectroscopy requires that atoms of the element of interest be in the **atomic state** (not combined with other elements in a compound) and that they be well separated in space. In foods, virtually all elements are present as compounds or complexes and, therefore, must be converted to neutral atoms (atomized) before atomic absorption or emission measurements can be made. **Atomization** involves separating particles into individual molecules (vaporization) and breaking molecules into atoms. It is usually accomplished by exposing the analyte (the substance being measured) to high temperatures in a flame or plasma although other methods may be used. A solution containing the analyte is introduced into the flame or plasma as a fine mist. The solvent quickly evaporates, leaving solid particles of the analyte that vaporize and decompose to atoms that may absorb radiation (atomic absorption) or become excited and subsequently emit radiation (atomic emission). This process is shown schematically in Fig. 24-2.

Methods for Atomization of Analytes

Source of Energy for Atomization	Approximate Atomization Temperature (°C)	Analytical Method
Flame	1700–3150	AAS, AES
Electrothermal	1200–3000	AAS (graphite furnace)
Inductively Coupled Argon Plasma	6000–7000	ICP-AES, ICP-MS

Three methods for atomizing samples are summarized in Table 24-2.

24.3 ATOMIC ABSORPTION SPECTROSCOPY

AAS is an analytical method based on the absorption of ultraviolet or visible radiation by free atoms in the gaseous state. It is a relatively simple method and was the most widely used form of atomic spectroscopy in food analysis for many years. It is being gradually replaced by the more versatile inductively coupled plasma (ICP) spectroscopy and inductively coupled plasma-mass spectrometry. Two types of atomization are commonly used in AAS: **flame atomization** and **electrothermal (graphite furnace) atomization**.

24.3.1 Principles of Flame Atomic Absorption Spectroscopy

A schematic diagram of a flame atomic absorption spectrometer is shown in Fig. 24-3. In flame AAS, a nebulizer–burner system is used to convert a solution of the sample into an atomic vapor. It is important to note that the sample must be in solution (usually an aqueous solution) before it can be analyzed by flame AAS. The sample solution is **nebulized** (dispersed into tiny droplets), mixed with fuel and an oxidant, and burned in a flame produced by oxidation of the fuel by the oxidant. Atoms and ions are formed within the flame as analyte compounds are decomposed by the high temperatures (Fig. 24-2). The flame itself serves as the sample compartment. The temperature of the flame is important because it will affect the efficiency of converting compounds to atoms and ions and because it influences the distribution between atoms and ions in the flame. Atoms and ions of the same element produce different spectra so they absorb radiation of different wavelengths. Therefore, it is desirable to choose a flame temperature that will maximize atomization and minimize ionization because the radiation coming from the lamp has emission lines specific to the corresponding atoms, not ions. This means that absorption efficiency will be decreased when atoms become ionized. Both atomization efficiency and ionization increase with increasing flame temperature, so choice of the optimal flame is not a simple matter. Flame characteristics may be manipulated by choice of oxidant and fuel and by adjustment of the oxidant/fuel ratio. The most common **oxidant–fuel combinations** are air-acetylene and nitrous oxide-acetylene. Also, adding cesium, an element with low ionization energy, to the sample will suppress ionization of other elements in the sample. The instrument instruction manual or the literature should be consulted for recommended flame characteristics.

Once the sample is atomized in the flame, its quantity is measured by determining the attenuation of a beam of radiation passing through the flame. For the measurement to be specific for a given element, the radiation source is chosen so that the emitted radiation contains an emission line that corresponds to one of the most intense lines in the atomic spectrum of the element being measured. This is accomplished by fabricating lamps in which the element to be determined serves as the cathode. Thus, the radiation emitted from the lamp is the emission spectrum of the element. The emission line of interest is isolated by passing the beam through a monochromator so that only radiation of a very narrow band width reaches the detector. Usually, one of the strongest spectral lines is chosen; for example, for sodium the monochromator is set to pass radiation with a wavelength of 589.0 nm. The principle of this process is illustrated in Fig. 24-4. Note that the intensity of the radiation leaving the flame is less than the intensity of radiation coming from the source. This is because sample atoms in the flame absorb some of the radiation. Note also that the line width of the radiation from the source is narrower than the corresponding line width in the absorption spectrum. This is because the higher temperature of the flame causes a broadening of the line width.

The amount of radiation absorbed by the sample is given by **Beer's law**:

$$A = \log(I_o/I) = abc \quad [4]$$

where:

A = absorbance
I_o = intensity of radiation incident on the flame
I = intensity of radiation exiting the flame
a = molar absorptivity
b = path length through the flame
c = concentration of atoms in the flame

Clearly, absorbance is directly related to the concentration of atoms in the flame.

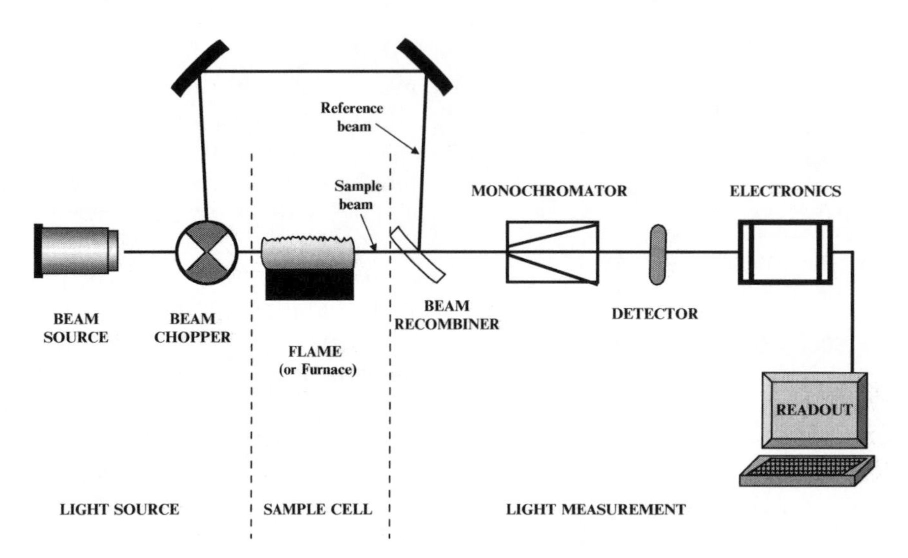

24-3 figure Schematic representation of a double-beam atomic absorption spectrophotometer. [Adapted from (4).]

24.3.2 Principles of Electrothermal Atomic Absorption Spectroscopy (Graphite Furnace AAS)

Electrothermal AAS is identical to flame AAS except for the atomization process. **Electrothermal atomization** involves heating the sample to a temperature (2000–3000°C) that produces volatilization and atomization. This is accomplished in a tube or cup positioned in the light path of the instrument so that absorbance is determined in the space directly above the surface where the sample is heated. The advantages of electrothermal atomization are that it can accommodate smaller samples than are required for flame atomic absorption and that limits of detection are lower. Disadvantages are the added expense of the electrothermal furnace, lower sample throughput, more difficult operation, and lower precision. In addition, matrix interferences are more of a problem with electrothermal atomization.

24.3.3 Instrumentation for Atomic Absorption Spectroscopy

Atomic absorption spectrometers consist of the following components (Fig. 24-3):

1. **Radiation source**, a hollow cathode lamp (HCL) or an electrode-less discharge lamp (EDL)
2. **Atomizer**, usually a nebulizer–burner system or an electrothermal furnace
3. **Monochromator**, usually an ultraviolet-visible (UV-Vis) grating monochromator
4. **Detector**, a photomultiplier tube (PMT) or a solid-state detector (SSD)
5. **Computer**

In **double-beam instruments**, which are by far the most common, the beam from the light source is split by a rotating mirrored chopper into a reference beam and a sample beam. The reference beam is diverted around the sample compartment (flame or furnace) and recombined before passing into the monochromator. The electronics are designed to produce a ratio of the reference and sample beams. This way, fluctuations in the radiation source and the detector are canceled out, yielding a more stable signal.

24.3.3.1 Radiation Source

The radiation source in atomic absorption spectrometers may be either a HCL or an EDL. **Hollow cathode lamps** consist of a hollow tube filled with argon or neon, an anode made of tungsten, and a cathode made of the metallic form of the element being measured (Fig. 24-5). When voltage is applied across the electrodes, the lamp emits radiation characteristic of the metal in the cathode; if the cathode is made of iron, an iron spectrum is emitted. When this radiation passes through a flame containing the sample, iron atoms in the flame will absorb some of it because it contains radiation of exactly the right energy for exciting iron atoms. This makes sense when we remember that for a given electronic transition, either up or down in energy, the energy of an emitted photon is exactly the same as the energy of an absorbed photon. Of course, this means that it is necessary to use a different lamp for each element analyzed (there are a limited number of multielement lamps available that contain cathodes made of more than one element). HCLs for about 60 metallic elements may be purchased from commercial sources, which means that atomic absorption may be used for the analysis of up to 60 elements. Some

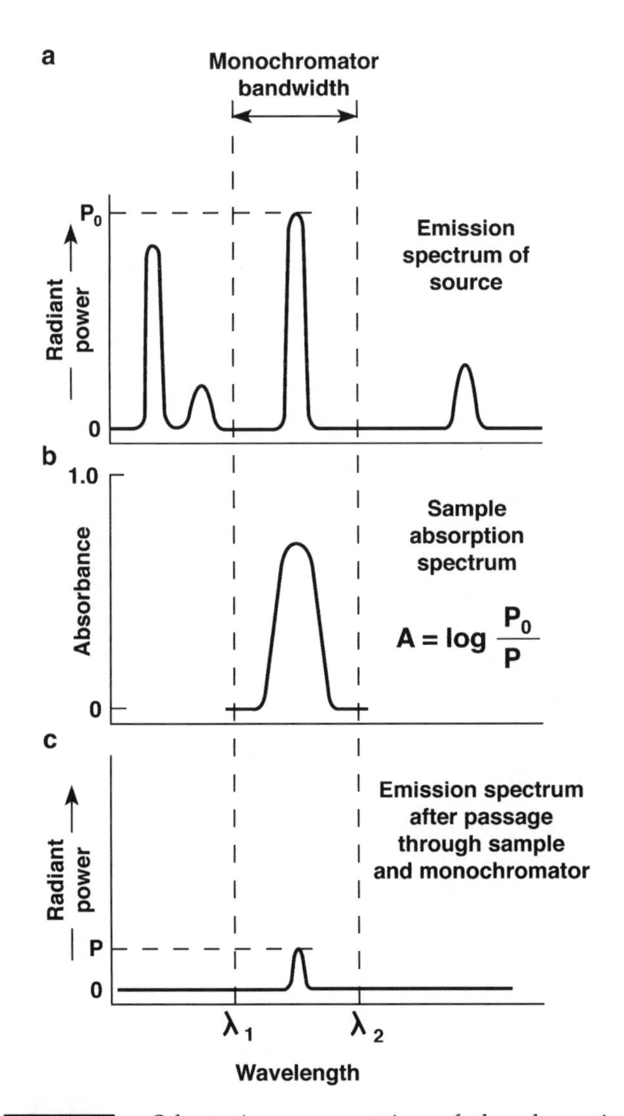

24-4 figure Schematic representation of the absorption of radiation by a sample during an atomic absorption measurement. The spectrum of the radiation source is shown in (**a**). As the radiation passes through the sample (**b**), it is partially absorbed by the element of interest. Absorbance is proportional to the concentration of the element in the flame. The radiant power of the radiation leaving the sample is reduced because of absorption by the sample (**c**). [From (6), used with permission. Illustration from Principles of Instrumental Analysis, 3rd ed., by Douglas A. Skogg. © 1985 Reprinted with permission of Brooks/Cole, a division of Thompson Learning.]

manufacturers are now recommending **electrode-less discharge lamps** for more volatile elements such as arsenic, mercury, and cadmium (7). Like HCLs, EDLs consist of a hollow glass vessel containing an inert filler gas plus the element of interest. However, the discharge is produced by a radio frequency generator coil rather than an electric current (8).

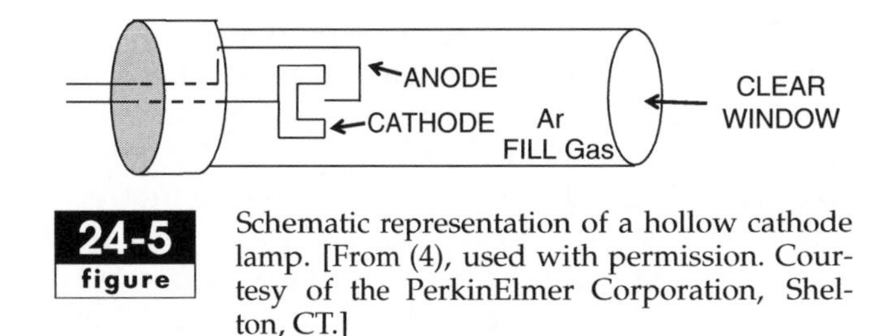

24-5 figure Schematic representation of a hollow cathode lamp. [From (4), used with permission. Courtesy of the PerkinElmer Corporation, Shelton, CT.]

Radiation reaching the monochromator comes from two sources, the attenuated beam from the HCL or EDL and excited atoms in the flame. Instruments are designed to discriminate between these two sources either by modulating the lamp so that the output fluctuates at a constant frequency or by positioning a chopper perpendicular to the light path between the source and the flame (Fig. 24-3). A **chopper** is a disk with segments removed. The disk is rotated at a constant speed so that the light beam reaching the flame is either on or off at regular intervals. The flame also produces radiation but flame radiation is continuous. Therefore, the radiation reaching the detector consists of the sum of an alternating and a continuous signal. Instrument electronics subtract the continuous signal and send only the alternating signal to the readout. This effectively eliminates the contribution of emissions from elements in the flame to the final signal.

24.3.3.2 Atomizers

Several types of atomizers are used in AAS: These include **flame**, **electrothermal**, **cold vapor technique for mercury**, and **hydride generation**.

The **flame atomizer** consists of a **nebulizer** and a **burner**. The nebulizer is designed to convert the sample solution into a fine mist or aerosol. This is accomplished by aspirating the sample through a capillary into a chamber through which oxidant and fuel are flowing. The chamber contains baffles which remove larger droplets, leaving a very fine mist. Only about 1% of the total sample is carried into the flame by the oxidant–fuel mixture. The larger droplets fall to the bottom of the mixing chamber and are collected as waste. The burner head contains a long, narrow slot that produces a flame that may be 5–10 cm in length. This gives a long path length that increases the sensitivity of the measurement.

Flame characteristics may be manipulated by adjusting oxidant/fuel ratios and by choice of oxidant and fuel. **Air-acetylene** and **nitrous oxide-acetylene** are the most commonly used oxidant–fuel mixtures although other oxidants and fuels may be used for some elements. There are three types of flames:

(1) **Stoichiometric**. This flame is produced from stoichiometric amounts of oxidant and fuel so the fuel is completely burned and the oxidant is completely consumed. It is characterized by yellow fringes. (2) **Oxidizing**. This flame is produced from a fuel–lean mixture. It is the hottest flame and has a clear blue appearance. (3) **Reducing**. This flame is produced from a fuel-rich mixture. It is a relatively cool flame and has a yellow color. Analysts should follow guidelines for the proper type of flame for each element.

Flame atomizers have the advantage of being stable and easy to use. However, sensitivity is relatively low because much of the sample never reaches the flame and the residence time of the sample in the flame is short.

Electrothermal atomizers are typically cylindrical graphite tubes connected to an electrical power supply. They are commonly referred to as **graphite furnaces**. The sample is introduced into the tube through a small hole using a microliter syringe (sample volumes normally range from 0.5 to 10 μL). During operation, the system is flushed with an inert gas to prevent the tube from burning and to exclude air from the sample compartment. The tube is heated electrically. Through a stepwise increase in temperature, first the sample solvent is evaporated, then the sample is ashed, and finally the temperature is rapidly increased to 2,000–3,000°C to quickly vaporize and atomize the sample.

The **cold vapor technique** works only for mercury, because mercury is the only mineral element that can exist as free atoms in the gaseous state at room temperature. In this technique, mercury compounds in a sample are reduced to elemental mercury by the action of stannous chloride, a strong reducing agent. The elemental mercury is then carried in a stream of air or argon into an absorption cell and atomic absorption is measured the same way as it is in flame ionization and electrothermal instruments. This method has the advantage of very high sensitivity because all of the mercury in the sample can be transferred to the absorption cell and measured. See reference (3) for a more detailed description of this technique.

In the **hydride generation technique**, volatile hydrides of elements are formed by reacting samples with sodium borohydride. The hydrides then are carried into an absorption cell and heated to decompose them into free atoms. Then atomic absorption measurements are carried out in the same manner as with other atomization techniques. As with the cold mercury vapor technique, sensitivity is high because there is very little sample loss. However, this technique is limited to a relatively few elements that are capable of forming volatile hydrides. These include As, Pb, Sn, Bi, Sb, Te, Ge, and Se. See reference (3) for a more detailed explanation of this technique.

24.3.3.3 Monochromator

The **monochromator** is positioned in the optical path between the flame or furnace and the detector (Fig. 25-3). Its purpose is to isolate the resonance line of interest from the rest of the radiation coming from the flame or furnace and the lamp so that only radiation of the desired wavelength reaches the detector. Typically, monochromators of the grating type are used (see Chap. 22).

24.3.3.4 Detector/Readout

Two types of detectors are used in AA spectrometers, **photomultiplier tubes** and **solid-state detectors**. Detectors convert the radiant energy reaching it into an electrical signal. This signal is processed to produce either an analog or a digital readout. Modern instruments are interfaced with computers for data collection, manipulation, and storage (see Chap. 22).

24.3.4 General Procedure for Atomic Absorption Analysis

While the basic design of all atomic absorption spectrometers is similar, operation procedures do vary from one instrument to another. Therefore, it is always good practice to carefully review operating procedures provided by the manufacturer before using the instrument. Most manuals have detailed procedures for the operation of the instrument as well as tables listing standard conditions (wavelength and slit width requirements, interferences and steps for avoiding them, flame characteristics, linear range, and suggestions for preparing standards) for each element. Be certain to pay close attention to safety precautions recommended by the manufacturer. *ACETYLENE IS AN EXPLOSIVE GAS*, and great care must be taken to avoid dangerous and damaging explosions.

24.3.4.1 Operation of a Flame Atomic Absorption Instrument

The following is a generalized procedure that will be similar but not identical to procedures found in instrument operating manuals:

1. Turn the lamp current control knob to the off position.
2. Install the required lamp in the lamp compartment.
3. Turn on main power and power to lamp. Set lamp current to the current shown on the lamp label.
4. Select required slit width and wavelength and align light beam with the optical system.

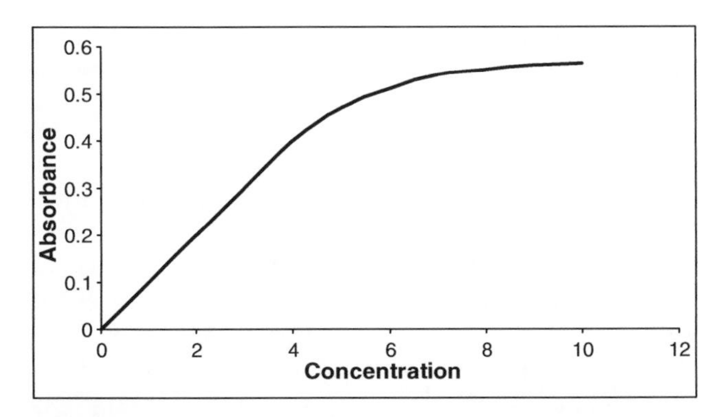

24-6 figure A plot of absorbance vs. concentration showing nonlinearity above a certain concentration. [From (3), used with permission. Courtesy of the PerkinElmer Corporation, Shelton, CT.]

5. Ignite flame and adjust oxidant and fuel flow rates.
6. Aspirate distilled water. Aspirate blank and zero instrument.
7. Aspirate standards and sample.
8. Aspirate distilled water.
9. Shut down instrument.

24.3.4.2 Calibration

According to Beer's law, absorbance is directly related to concentration. However, a plot of absorbance vs. concentration will deviate from linearity when concentration exceeds a certain level (Fig. 24-6). Therefore, it is always necessary to calibrate the instrument using appropriate standards that closely resemble the sample. This may be done by running a series of standards and plotting absorbance vs. concentration or, in the case of most modern instruments, programming the instrument to read in units of concentration.

24.3.4.2.1 Selection of Standards The first step in calibration is to select the number and concentrations of standards to use. It is best to use standards that are at least 99.999% pure when preparing multielement standards. When operating in the linear range, only one standard is needed. The linear range may be determined by running a series of standards of increasing concentration and plotting absorbance vs. concentration. Operating manuals should contain values for linear ranges. The concentration of the standard should be higher than that of the most concentrated sample. If the range of concentration exceeds the linear range, multiple standards must be used or the sample diluted. Again, the concentration of the most concentrated standard should exceed the concentration of the most concentrated sample.

24.3.4.2.2 Sensitivity Check Because many factors can influence the operating efficiency of an instrument, it is a good idea to check instrument output using a standard of known concentration. Operating manuals should have values for characteristic concentrations for each element. For example, manuals for PerkinElmer atomic absorption spectrophotometers state that a 5.0 mg/L aqueous solution of iron "will give a reading of *approximately* 0.2 absorbance units." If the measured absorbance reading deviates significantly from this value, appropriate adjustments (e.g., flame characteristics, lamp alignment, etc.) should be made.

24.3.5 Interferences in Atomic Absorption Spectroscopy

With any analytical technique, it is important to be on the lookout for possible interferences. Atomic spectroscopy techniques are powerful partly because measurements of individual elements can usually be made without laborious separations. There are two main reasons for this. First, as mentioned previously, a single narrow emission line is used for the measurement. Second, these are relative techniques; that is, quantitative results for an unknown sample are possible only through comparison with a standard of known concentration. If there are matrix-effect problems, they can often be overcome by using the same matrix for the standard or by employing the method of additions approach.

The following is a brief discussion of common interference problems in AAS. See your instrument manual for a thorough discussion of interference problems in AAS and reference (6) for a list of interferences for each element. Two types of interferences are encountered in AAS: **spectral interference** and **nonspectral interference**.

24.3.5.1 Spectral Interference

24.3.5.1.1 Absorption of Source Radiation An element in the sample other than the element of interest may absorb at the wavelength of the spectral band being used. Such interference is rare because emission lines from HCLs are so narrow that only the element of interest is capable of absorbing the radiation. One example where this problem does occur is with the interference of iron in zinc determinations. Zinc has an emission line at 213.856 nm, which overlaps the iron line at 213.859 nm. The problem may be solved by choosing an alternative emission line for measuring zinc or by narrowing the monochromator slit width.

24.3.5.1.2 Background Absorption of Source Radiation Particulates present as a result of incomplete atomization may scatter source radiation, thereby attenuating the radiation reaching the detector. This problem may be overcome by going to a higher flame temperature to ensure complete atomization of the sample. Some instruments are equipped with automatic background correction devices.

24.3.5.2 Nonspectral Interferences

24.3.5.2.1 Transport Interferences These result when something in the sample solution affects the rate of aspiration, nebulization, or transport into the flame. Transport interferences are rarely a problem with graphite furnace instruments but may cause substantial errors in flame AAS. Such factors as viscosity, surface tension, vapor pressure, and density of the sample solution can influence the rate of transport of sample into the flame. Transport interferences often can be overcome by matching as closely as possible the physical properties of the sample and the standards. For example, use the same solvent for the sample and the standard, or add the interferant in the sample (e.g., sugar) to the standard. The method of additions also may be used to overcome transport interferences.

24.3.5.2.2 Solute Volatilization Interferences Matrix interferences result from the reduction or enhancement of the emitted signal due to differences between the composition of the standard and the composition of the sample. Flame atomic absorption is prone to solute vaporization interferences, which are changes in the lateral migration of an analyte due to the matrix. For example it has been observed in flame absorption and emission that alkaline elements are depressed by elevated levels of aluminum and phosphorus (9). Chemical interferences occur in flames when calcium is determined in a matrix that varies in phosphate concentration. This is because phosphate reduces the number of free calcium atoms in the flame by converting calcium phosphate to calcium pyrophosphate, which is not decomposed by the flame.

24.3.5.2.3 Ionization Interference Ionization of analyte atoms in the flame may cause a significant interference. **Easily ionized elements** (EIE) like K, Na, Li, Cs, produce large changes in emission intensities in the flame. (Remember that absorption and emission lines of atoms and ions of the same element are different and that atomic absorption spectrometers are tuned to measure *atomic* absorption, not *ionic* absorption. Therefore, any factor that reduces the concentration of atoms in the flame will lower the absorbance reading.) The ionization of atoms results in an equilibrium situation:

$$M \leftrightarrows M^{+} + e^{-} \qquad [5]$$

Ionization increases with increasing flame temperature and normally is not a problem in air-acetylene flames because the temperature is not high enough. It can be a problem in nitrous oxide-acetylene flames with elements that have ionization potentials of 7.5 eV or less. Ionization is suppressed by the presence of EIE, such as potassium, through mass action. When potassium ionizes, it increases the concentration of electrons in the flame and shifts the above equilibrium to the left. Reagents added to reduce ionization are called **ionization suppressors**.

24.4 ATOMIC EMISSION SPECTROSCOPY

In contrast to AAS, the source of radiation in AES is the excited atoms or ions in the sample rather than an external source. Figure 24-7 shows a simplified diagram of an atomic emission spectrometer. (*Note*: In recent years, some instrument manufacturers have started calling AES instruments "optical emission spectrometers" (OES) because they measure light emitted when excited atoms return to the ground state. In this chapter, we will use AES rather than OES but the two terms are virtually interchangeable.)

As with AAS, the sample must be atomized to produce usable spectra for quantitative analysis. The difference is that in emission spectroscopy, sufficient heat is applied to the sample to excite atoms to higher energy levels. Aside from the external radiation source required for AAS, instrumentation for AES is similar. In fact, many instruments may be operated in either the absorption or emission mode.

Emissions are produced when electrons from excited neutral atoms move back to lower energy

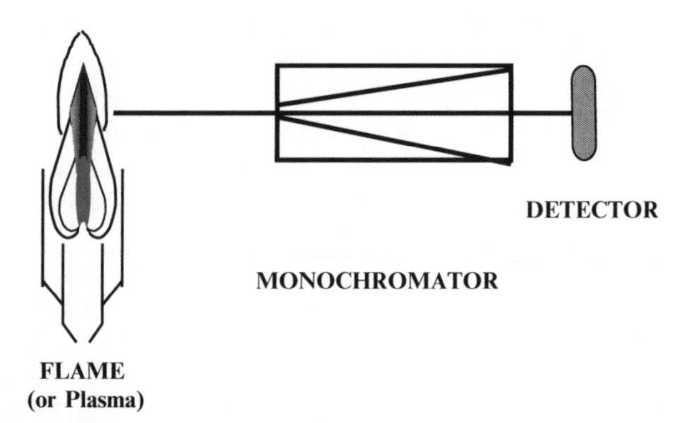

24-7 figure A simplified diagram of an atomic emission spectrometer. [Adapted from (4).]

states. Emissions have wavelengths characteristic of individual elements because, as discussed previously, the allowed energy levels for electrons are unique for each element. Energy for excitation may be produced by several methods, including **heat** (usually from a flame), **light** (from a laser), **electricity** (arcs or sparks), or **radio waves** (inductively coupled plasma). Emissions are passed through monochromators or filters prior to detection by PMTs, charge-coupled devices, or charge injection devices.

The two most common forms of AES used in food analysis are **flame emission spectroscopy** and **inductively coupled plasma-atomic emission spectroscopy** (ICP-AES).

24.4.1 Principles of Flame Emission Spectroscopy

Flame emission spectroscopy employs a nebulizer–burner system to atomize and excite the sample. The instrument may be either a spectrophotometer (which uses a monochromator to isolate the desired emission line) or a photometer (which uses a filter to isolate emission lines). Flame emission is most useful for elements with relatively low excitation energies. These include Na, K, and Ca.

24.4.2 Principles of Inductively Coupled Plasma-Atomic Emission Spectroscopy

Inductively coupled plasma emission spectroscopy has become widely available in the last three decades or so. In ICP-AES, a plasma is used as the atomization and excitation source. A **plasma** is defined as gaseous mixture containing significant concentrations of cations and electrons. Temperatures in plasmas are very high (in the neighborhood of 5000–10000K) resulting in very effective atomization. Fortunately, excessive ionization of sample atoms is not a problem, probably because of the high concentration of electrons contributed by the ionization of the argon. See references (4, 7, 10, 11) for detailed descriptions of ICP-ES.

24.4.3 Instrumentation for Flame Emission Spectroscopy

Flame emission spectrometers consist of the following components:

1. **Atomization-excitation source**, usually a nebulizer that feeds into a laminar flow burner
2. **Monochromator** or **filter**. (Instruments with monochromators are more versatile because any wavelength in the UV-Vis spectrum can be selected, whereas a given filter is designed to pass only one narrow wavelength band.)
3. **Detector**
4. **Readout device**

A comparison of the components of atomic absorption and flame emission spectrometers quickly reveals their similarities. In emission spectrometers, the flame is the radiation source, so the HCL and the chopper are not required. Many modern atomic absorption instruments also can be operated as flame emission spectrometers. Specialized instruments specifically designed for the analysis of sodium, potassium, lithium, and calcium in biological samples are made by some manufacturers. These instruments are called **flame photometers**. They employ interference filters to isolate the spectral region of interest. Low flame temperatures are used so that only easily excited elements such as the alkali and alkaline earth metals produce emissions. This results in a simpler spectrum and reduces interference from other elements that may be present.

24.4.4 Instrumentation for ICP-AES

There are three basic types of ICP-AES instruments available today: the **simultaneous PMT spectrometer**, the **sequential PMT spectrometer**, and the **echelle spectrometer**. All three instruments are capable of determining multiple elements in the same sample. The PMT spectrometers use PMTs as detectors. Echelle spectrometers use either **charge injection devices** (CIDs) or **charge-coupled devices** (CCDs) to detect the emissions. The simultaneous PMT spectrometer analyzes for a limited number of elements simultaneously, while the sequential PMT spectrometer measures multiple elements sequentially in rapid succession. Echelle spectrometers also measure multiple elements simultaneously but they can detect a much larger number of elements than instruments with PMT detectors.

Inductively coupled plasma-atomic emission spectrometers consist of the following components (see Fig. 24-8):

1. **Argon plasma torch**
2. **Monochromator**, **polychromator**, or **echelle optical system**
3. **Detector(s):** one photomultiplier tube for sequential spectrometers, multiple photomultiplier tubes for simultaneous PMT spectrometers, or CID or CCD detectors for echelle spectrometers
4. **Computer** for data collection and treatment

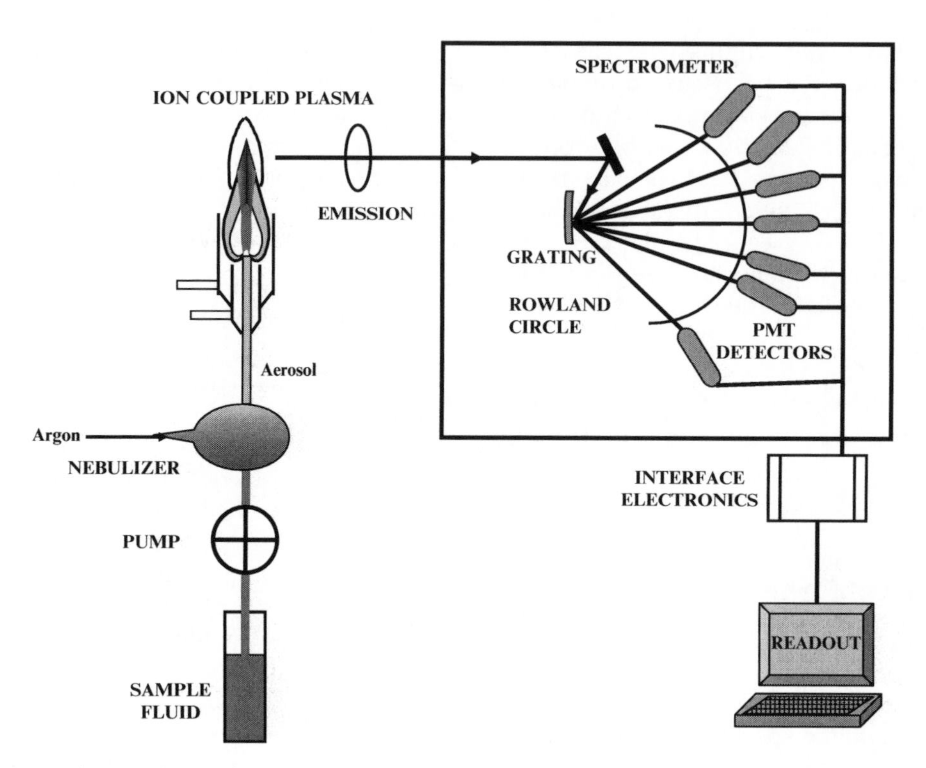

Schematic of an inductively coupled plasma-atomic emission simultaneous spectrometer. *PMT*, photomultiplier tube.

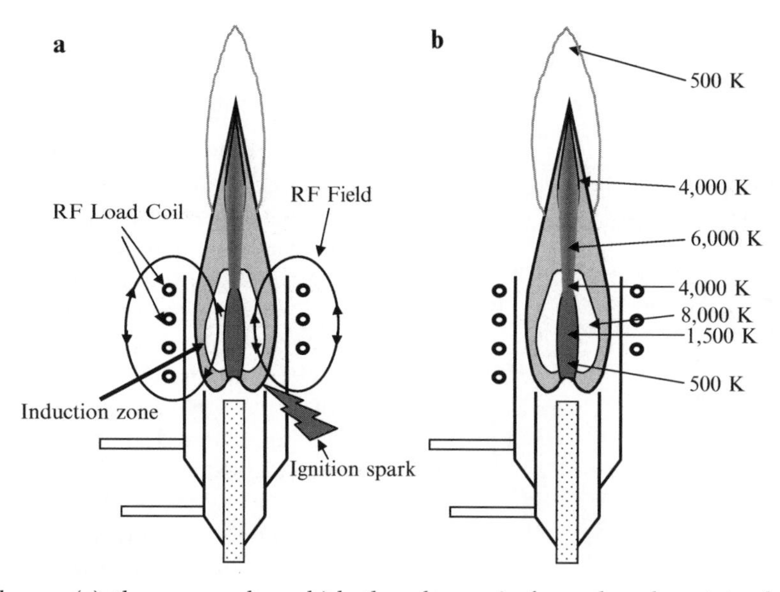

The ICP plasma (**a**) the process by which the plasma is formed and sustained and (**b**) the temperature distribution of the plasma.

24.4.4.1 Argon Plasma Torch

24.4.4.1.1 Characteristics of an Argon Plasma Torch The heart of the ICP-AES instrument is a plasma torch (Fig. 24-9). It operates at extremely high temperatures, causing complete atomization of the sample. The torch consists of two concentric quartz tubes centered in a copper coil. During operation of the torch, a steam of argon gas flows through the tubes and radio frequency (RF) power is applied to the copper coil, creating an oscillating magnetic field inside the tube. The plasma is started by applying the RF power and ionizing the argon gas with an electric spark to form argon ions and electrons. The oscillating magnetic field couples with the electrons and argon ions forcing them to flow in an annular (ring-shaped) path. The argon is heated by a process known as inductive coupling, to temperatures ranging as high as 10,000K. Heating does not involve burning fuel to directly heat and atomize the sample, as is the case with flame atomic absorption and emission spectroscopy (argon is a noble gas and will not combust). Rather, heating is accomplished by transferring RF energy to free electrons in a manner similar to the transfer of microwave energy to water in a microwave oven. These high-energy electrons in turn collide with argon atoms and ions causing a rapid increase in temperature. The process continues until about 1% of the argon atoms are ionized. At this point the plasma is stable and self-sustaining for as long as the RF field is applied at constant power.

The extremely high temperatures and the inert atmosphere of argon plasmas are ideal for atomizing and exciting analytes. The low oxygen content reduces the formation of oxides, which is sometimes a problem with flame methods. The nearly complete atomization of the sample minimizes chemical interferences. The relatively uniform temperatures in the plasma (compared to non-uniform temperatures in flames) and the relatively long residence time in the plasma give good linear responses over a wide concentration range (up to several orders of magnitude).

24.4.4.1.2 Sample Introduction into the Plasma and Sample Excitation In almost all cases, samples are introduced as aerosols carried by a stream of argon in the injector tube inside the annulus of the plasma at the base of the RF load coil. The analyte is atomized and excited as shown in Fig. 24-2.

24.4.4.1.3 Radial and Axial Viewing Emissions from ICP torches can be viewed either radially or axially. In the radial view, the optics are aligned perpendicular to the torch (Fig. 24-10a). In the axial view, the light is viewed by looking down the center of the torch (Fig. 24-10b). The advantage of axial viewing is that detection limits are lower.

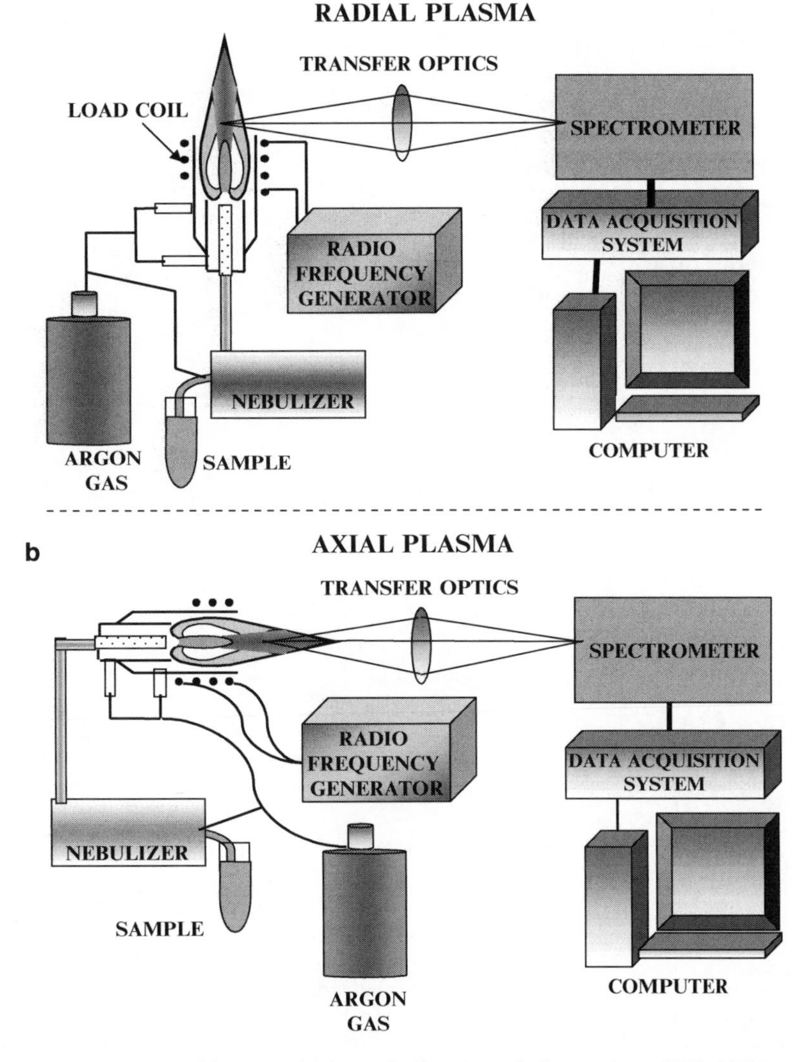

24-10 figure Major components and typical layout of (**a**) a radially viewed plasma in an ICP-AES instrument and (**b**) an axially viewed plasma in an ICP-AES instrument.

24.4.4.2 Detectors and Optical Systems

Older ICP-AES instruments use PMT detectors (Fig. 24-8). While PMT detectors are extremely reliable and durable, a separate detector is required for each wavelength and this limits the number of elements that can be determined simultaneously because PMTs are relatively large and bulky. Modern ICP-AES instruments are equipped with solid-state detectors that are capable of measuring continuous emission spectra. There are two types of solid-state detectors, the **charge injection device** (CID) and the **charge-coupled device** (CCD). Both detectors have multiple rows of pixels that are sensitive to light. Both types generate and store a charge when struck by radiation, and the magnitude of the charge is proportional to the intensity of the incident radiation. The primary difference is how the signal is read off of the chip. In the CCD detectors, the pixels from each row are read sequentially. For example, after the first pixel of a row is read, the charge collected by pixel two is then transferred to pixel one to be read. Then the charge from pixel three is transferred to pixel two and so on. This transfer process continues until all the pixels in a row are measured. Each row of pixels is read independently. The size of each pixel is about 27 by 27 μm. In CID detectors, the data from each individual pixel are collected and read independently. This has the advantage of collecting the signals at their optimal signal-to-noise ratio.

Emissions coming from the plasma torch are passed through an **echelle optical system** which produces a two-dimensional spectrum and focuses it on the detector (Fig. 24-11). These echelle spectrometers have several advantages over a PMT-based spectrometer. Most important is the ability to measure the continuous emission spectra from about 166 nm to about 840 nm. This gives the analyst much more flexibility to select the most ideal wavelengths to use for the analysis. In addition, background subtraction can be done more accurately without increasing the analysis time. The older PMT-based spectrometers were constructed to meet the specific needs of a customer. For example, if a customer wanted to be able to analyze food samples for Ca, Na, Zn, Fe, and Cu, the instrument would be configured to focus emission lines from each of these elements on separate PMTs arranged in a circle (Fig. 24-8). Once the instrument was assembled and installed it was very difficult to reconfigure the instrument for additional applications. Also, the PMT-based spectrometers were quite large, measuring about 170-cm long, 90-cm deep, and 150-cm high. These dimensions were necessary to house the Rowland circle optics, which used a concave grating to disperse light in a one-dimensional arc over a distance ranging from 50 to 100 cm (Fig. 24-8). These PMT-based ICP-AESs were limited to measuring fewer than 70 elements simultaneously. This was primarily due to the size of the PMT detectors. The new echelle instruments are very compact and much smaller. The dimensions are 85-cm long, 75-cm deep, and 60-cm tall. These instruments are about one-sixth the size of a PMT instrument. These instruments use a compact echelle optical design which uses two dispersing elements in series. The first dispersing

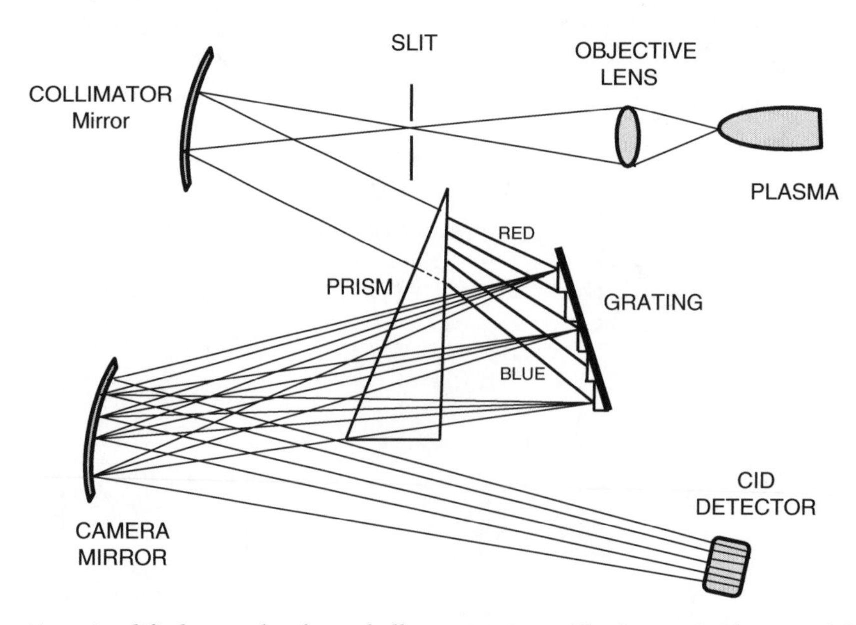

This figure is a simplified example of an echelle spectrometer. The important feature of this spectrometer is that combines two light dispersing elements in series. The first is a prism which disperses the light in the *X*-direction without overlapping orders. Second is a grating which disperses the light in the *Y*-direction, producing a two-dimensional image on the detector.

element is a prism, which disperses the light without any wavelength overlap. The light then strikes a low-density ruled grating (about 53 groves per mm) which separates the light further at high orders, producing a two-dimensional display on a small solid-state detector. These new echelle spectrometer-based ICPs now have the same detection limits as the former PMT-based systems, with the flexibility of being able to measure hundreds of lines simultaneously.

24.4.5 General Procedure for Analysis by ICP-AES

As is the case with atomic absorption spectrometers, operating procedures for atomic emission spectrometers vary somewhat from instrument to instrument. ICP atomic emission spectrometers are controlled by computers. The software contains **methods** that specify instrument operating conditions. The computer may be programmed by the operator, or, in some cases, default conditions may be used. Once the method is established, operation is highly automated.

24.4.6 Interferences in ICP-AES

Generally, interferences in ICP-AES analyses are less of a problem than with AAS, but they do exist and must be taken into account. **Spectral interferences** are the most common. Samples containing high concentrations of certain ions may cause an increase (shift) in background emissions at some wavelengths. This will cause a positive error in the measurement, referred to as **background shift interference** (Fig. 24-12). Correction is relatively simple. A pure solution of the interfering element at the concentration of this element in the sample is prepared and an emission measurement is made at a wavelength above and below the emission line of the analyte. An average of these two emissions is then subtracted from the emission of the analyte. (*Note*: In the example in Fig. 24-12, the intensity of the emission from the aluminum varies with the wavelength and this is why it is necessary to make measurements above and below the emission line for lead. If the intensity does not vary with wavelength, then only one measurement near the wavelength of the analyte's emission line is required for correction.) Alternatively, another emission line in a region where there is no background shift could be chosen. Another type of spectral interference, called **spectral overlap interference**, occurs when the resolution of the instrument is insufficient to prevent overlap of the emission line of one element with that of another. (*Note*: Even though we refer to emissions as "lines", the actual emissions have some width, and the wider the line, the lower the resolution. This is not a problem when the spectral lines of all the elements in a sample are sufficiently separated so there is no overlap, but if there is overlap, there is interference.) For example, when determining Cd in a sample containing high concentrations of Fe, some of the emission from one of the Fe lines overlaps the preferred Cd emission line (see Fig. 24-13). This will cause an apparent increase in the measured concentration of Cd. These interferences

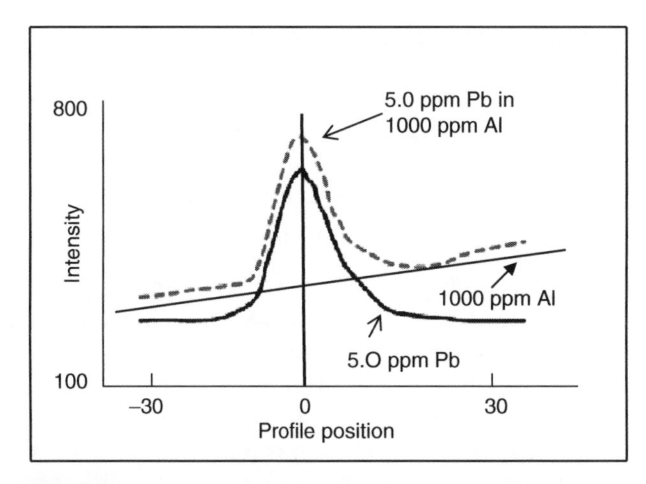

24-12 figure Background shift interference of aluminum on lead. As shown by the sloping solid line in the figure, aluminum increases the background in the wavelength region around the preferred emission line for lead (220.35 nm). Note that a profile position of 0 is the center of the preferred emission line for the analyte (220.35 nm in this case). The units on the *x*-axis are arbitrary units. The units on the *y*-axis are also arbitrary but reflect the intensity of the emissions striking the detector.

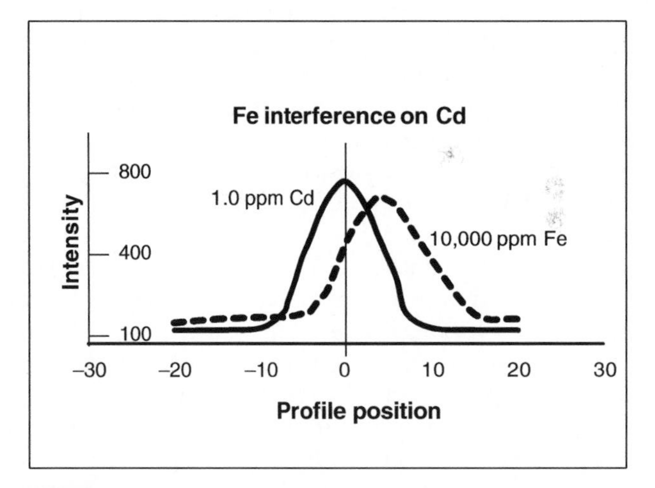

24-13 figure Spectral overlap interference of iron on cadmium. The overlapping spectra indicate that the resolution of the instrument is insufficient to separate the emissions from iron and cadmium at the wavelength chosen for cadmium measurement (226.502 nm). Therefore, if a sample being analyzed for cadmium also contains iron, there will be interference.

can be corrected by running a solution of pure Fe and measuring the apparent Cd concentration in the pure Fe sample which, presumably, has no Cd in it. Next, the sample is analyzed for Cd and Fe and the contribution of Fe to the Cd signal is subtracted, thereby giving an accurate estimate of the true Cd concentration.

24.5 COMPARISON OF AAS AND ICP-AES

AAS and ICP-AES have many advantages in common. Both are capable of measuring trace metal concentrations in complex matrices with excellent precision and accuracy. Sample preparation is relatively simple. For most applications, sample preparation for both techniques involves destruction of organic matter by ashing, followed by dissolution of the ash in an aqueous solvent, usually a dilute acid. In comparison with traditional wet chemistry methods, measurements with AAS and ICP-AES are extremely rapid.

ICP-AES instruments are capable of determining concentrations of multiple elements in a single sample with a single aspiration. This offers a significant speed advantage over AAS when the objective is to quantify several elements in a given sample. ICP-AES may also offer an advantage over AAS when analyzing for elements in refractory compounds. **Refractory compounds** are compounds that are unusually stable at high temperatures and may not be fully atomized in the flame of an AAS. Most refractory compounds are readily atomized in the much higher temperatures of a plasma torch.

Another advantage of ICP-AES over AAS is that ICP-AES has a much larger analytical working range. Analytical working range refers to the range of concentrations over which an instrument can be operated without recalibration. The analytical working range for ICP-AES is 4–6 orders of magnitude. For example, it is possible to measure samples varying in concentration from 1 μg/L to 1 g/L without having to recalibrate the instrument. The working range for an AAS instrument is about 3 orders of magnitude, or, compared to the example above, from 1 μg/L to 1 mg/L.

One way of comparing analytical methods is to compare limits of detection. **Limits of detection** may be defined qualitatively as the lowest concentration of the element that can be distinguished from the blank at a given level of confidence, usually 98%.

Table 24-3 lists limits of detection for the various methods for elements that may be of interest to food analysts. It should be noted that these are approximate values, and limits of detection will vary depending on the sample matrix, the stability of the instrument, and other factors. A two or threefold difference in limit of detection is probably not meaningful, but an order of magnitude difference probably is.

Approximate Detection Limits (μg/L) for Selected Elements When Analyzed with Various Instruments[a]

Element	*Flame AA*[b]	*GFAA*[c]	*ICP-AES*[d]	*ICP-MS*[e]
Al	45	0.1	1	0.005
Ca	1.5	0.01	0.05	0.0002
Cd	0.8	0.002	0.1	0.00009
Cu	1.5	0.014	0.4	0.0002
Fe	5	0.06	0.1	0.0003
Hg	300	0.6	1	0.01
K	3	0.005	1	0.0002
Mg	0.15	0.004	0.04	0.0003
Mn	1.5	0.005	0.1	0.00007
Na	0.3	0.005	0.5	0.0003
Pb	15	0.05	1	0.00004
Se	100	0.05	4	0.0007
Zn	1.5	0.02	0.2	0.0003

Adapted from (7), by PerkinElmer, Shelton, CT.
[a]Detection limit is defined as lowest concentration of the element in a solution that can be detected with 98% confidence.
[b]Flame atomic absorption.
[c]Graphite furnace atomic absorption.
[d]Inductively coupled plasma-atomic emission spectroscopy, axial view, and simultaneous multielement conditions.
[e]Inductively coupled plasma-mass spectrometry.

24.6 APPLICATIONS OF ATOMIC ABSORPTION AND EMISSION SPECTROSCOPY

24.6.1 Uses

Atomic absorption and emission spectroscopy are widely used for the quantitative measurement of mineral elements in foods. In principle, any food may be analyzed with any of the atomic spectroscopy methods discussed. In most cases, it is necessary to **ash** the food to destroy organic matter and to dissolve the ash in a suitable solvent (usually water or dilute acid) prior to analysis (see Chap. 7 for details on ashing methodology). Proper ashing is critical to accuracy. Some elements may be volatile at temperatures used in **dry ashing** procedures. Volatilization is less of a problem in **wet ashing**, except for the determination of boron, which is recovered better using a dry ashing method. However, ashing reagents may be contaminated with the analyte. It is therefore wise to carry blanks throughout the ashing procedure.

Some liquid products may be analyzed without ashing, provided appropriate precautions are taken to avoid interferences. For example, vegetable oils may be analyzed by dissolving the oil in an organic solvent such as acetone or ethanol and aspirating the solution directly into a flame atomic absorption spectrometer. Milk samples may be treated with trichloroacetic acid to precipitate the protein; the resulting supernatant is analyzed directly. A disadvantage of this approach is that the sample is diluted in the process and the analyte can become entrapped or complexed to the precipitated proteins. This may be a problem when analytes are present in low concentrations. An alternative approach is to use a graphite furnace for atomization. For example, an aliquot of an oil may be introduced directly into a graphite furnace for atomization. The choice of method will depend on several factors, including instrument availability, cost, precision/sensitivity, and operator skill.

24.6.2 Practical Considerations

24.6.2.1 Reagents

Since concentrations of many mineral elements in foods are at the trace level, it is essential to use highly pure chemical reagents and water for preparation of samples and standard solutions. Only reagent grade chemicals should be used. Water may be purified by distillation, deionization, or a combination of the two. **Reagent blanks** should always be carried through the analysis. The purest reagents are prepared by sub-boiling distillation.

24.6.2.2 Standards

Quantitative atomic spectroscopy depends on comparison of the sample measurement with appropriate standards. Ideally, standards should contain the analyte metal in known concentrations in a solution that closely approximates the sample solution in composition and physical properties. If multielement standards are used the individual elements that compose the multielement standard must be checked for other elements that may be present as contaminants, and the final concentration needs to include all contaminants. A series of standards of varying concentrations should be run to generate a calibration curve. Because many factors can affect the measurement, such as flame temperature, aspiration rate, and the like, it is essential to run standards frequently, preferably right before and/or right after running the sample. Standard solutions may be purchased from commercial sources, or they may be prepared by the analyst. Obviously, standards must be prepared with extreme care since the accuracy of the analyte determination depends on the accuracy of the standard. Perhaps the best way to check the accuracy of a given assay procedure is to analyze a reference material of known composition and similar matrix. Standard reference materials may be purchased from the United States National Institute of Standards and Technology (formerly the National Bureau of Standards) (11).

24.6.2.3 Labware

Vessels used for sample preparation and storage must be clean and free of the elements of interest. Plastic containers are preferable because glass has a greater tendency to adsorb and later leach metal ions. All labware should be thoroughly washed with a detergent, carefully rinsed with distilled or deionized water, soaked in an acid solution (1 *N* HCl is sufficient for most applications), and rinsed again with distilled or deionized water.

24.7 INDUCTIVELY COUPLED PLASMA-MASS SPECTROMETRY

As described above, atomic absorption and emission spectrometers are designed to quantify mineral elements in a sample by measuring either the absorption or emission of radiation by the element of interest at a wavelength that is unique to that element. Another approach is to directly measure the actual number of atoms (as ions) of the element in the sample. This is possible with instruments that combine an inductively coupled plasma torch with a mass spectrometer to make an **inductively coupled plasma-mass spectrometer** (ICP-MS) (Fig. 24-14). See references (5) and (12) for an in-depth discussion of the principles and capabilities of ICP-MS.

The principles of mass spectrometry are described in detail in Chap. 26. Briefly, mass spectrometers are instruments that separate ions according to their mass to charge ratio (m/z) and then accurately count these ions as they strike a detector. This requires that the element or compound of interest be in the gaseous state and that it be ionized. Inductively coupled plasmas are very efficient at atomizing and ionizing virtually all elements in the periodic chart. The technical difficulty is in transferring the ions generated by a plasma operating at atmospheric pressure into the mass spectrometer which operates in a high vacuum. This problem is resolved by using an interface made up of two water-cooled cones placed one behind the other with a partial vacuum between them. This process allows the ionized minerals from the plasma to enter the high vacuum of the MS while removing most of the argon gas coming from the torch by pumping it away with a vacuum pump.

Inductively coupled plasma mass spectrometers are based on several advanced technologies and, as such, are highly sophisticated instruments. However, several companies have begun marketing instruments that can be successfully operated by analysts from a variety of fields, making ICP-MS a practical tool for mineral analyses in foods.

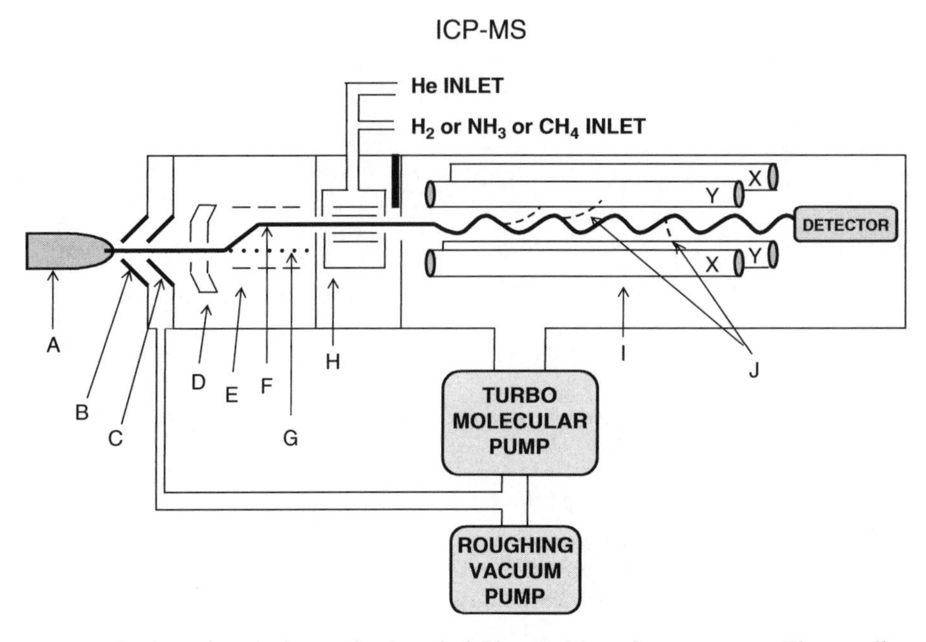

A simplified diagram of an Inductively Coupled Plasma-Mass Spectrometer. The axially oriented plasma (A) generates ions that are swept into the mass spectrometer. The ion beam first passes through an interface made up of two cones placed one behind the other (B and C). The first is called a sampling cone (B); it is water cooled to prevent the heat from the plasma from damaging it. Next the ions, light, and gases pass through to a skimmer cone (C). There is a partial vacuum between the cones and this vacuum removes most of the argon gas and neutrals (unionized atoms and molecules) by pumping them away with a vacuum pump. The ions, electrons, light, and remaining neutrals generated in the plasma then pass through extraction lenses (D) which are negatively charged to repel electrons and attract and accelerate the positive ions. These positively charged ions travel through to the off-axis ion omega lenses (E). These lenses separate the ions from photons and neutrals by bending the ion beam (F). The paths of the photons and neutrals (G) are unaffected and they collide with the wall of the wall of the instrument. This process is necessary to prevent the photons and neutrals from striking the detector and creating a false signal. The ion beam then passes into the collision reaction cell (H). At this location gasses such as H_2, He, NH_3 or CH_4 may be introduced to remove polyatomic interferences. The ion beam then passes into the quadrupole (I). The quadrupole consists of four parallel rods. One paired set of rods (X) has a direct current voltage applied to it and the opposite pair (Y) has a radio frequency voltage applied to it. When specific rf and dc voltages are applied to the rods, a single ion with a specific m/z will be electrostatically steered between the roods and move toward the detector. All other ions with a different m/z will be rejected (J). The rf voltage is then changed to allow an ion with a different m/z to pass through to the detector. This process is repeated until all of the elements of interest are measured.

In ICP-MS mineral analyses, samples are prepared by ashing the food or other matrix and dissolving the ash in a dilute acid in the same way they are prepared for AAS and AES analyses. The solution is then aspirated into the plasma torch in the same way it is introduced into an ICP-AES instrument, but instead of having a monochromator or polychromator and a device for separating and detecting **light** of specific wavelengths, the ICP-MS uses a mass spectrometer to separate and detect **ions** of the element or compound.

The identification and quantitation of a mineral element in a food by ICP-MS depends on the existence of a unique **mass-to-charge ratio** (m/z) for the element. If two elements have the same m/z, then there will be interferences. Fortunately, many elements have more than one isotope and this can be used to good advantage when two m/z ratios overlap. For example, iron has four naturally occurring stable isotopes with masses of 54, 56, 57, and 58 **atomic mass units** (amu). The natural abundances of these isotopes are: 5.8%, 91.75%, 2.1%, and 0.28% for ^{54}Fe, ^{56}Fe, ^{57}Fe, and ^{58}Fe, respectively. Nickel has five stable isotopes with the following abundances: ^{58}Ni, 68.04%; ^{60}Ni, 26.22%; ^{61}Ni, 1.14%; ^{62}Ni, 3.63%; ^{64}Ni, 0.93%. This means that the signal for ^{58}Fe and ^{58}Ni will overlap, resulting in an isobaric interference because they both have an m/z of 58. Fortunately, with the exception of indium, all of the elements have at least one isotope with a unique mass, and this isotope can be used to determine the concentration of the element in the sample. Thus, to determine the concentration of Fe or Ni in the example above, the analyst would select ^{56}Fe for the determination of Fe and select ^{60}Ni for the determination of Ni in the sample. (It is best to select the isotope with the highest natural abundance because the measurement precision is higher for more abundant isotopes.)

Another source of interference occurs when ions of a particular element exist with **multiple charges**. Most ions in a plasma have a charge of +1 but doubly charged ions may form. Fortunately, ions with charges greater than one have negligible impact with the possible exception of ^{138}Ba which may lose 1 or 2 electrons to produce singly and doubly charged ions with m/z ratios of 138 and 69, respectively. ^{69}Ga has an m/z of 69 so the presence of barium in a sample could increase the apparent concentration of gallium.

Another possible source of interference comes from the formation of molecular species in the plasma, sometimes referred to as **polyatomic interferences**. For example, argon can react with oxygen to produce ArO^+ which has an m/z of 56, the same m/z as $^{56}Fe^+$. Also the high temperature of the plasma can cause inert gases to form dimers, for example, two Ar atoms may combine to form $ArAr^+$ which has the same m/z as selenium at mass 78. Also, Ar may react with Cl from the sample to form $ArCl^+$ which has an m/z of 75, the same as arsenic. Fortunately modern ICP-MS instruments are equipped with devices called *collision reaction cells* (CRC) that can eliminate or compensate for these types of interferences.

The major *advantage* of ICP-MS is its extremely low detection limits, allowing the measurement of concentrations in the parts per trillion range. Another advantage is that ICP-MS is capable of measuring Se, Ni, Co, Cu, As, Pb, Hg, and Tl with excellent sensitivity. It is difficult to measure these elements with either AAS or ICP-AES. A *disadvantage* is the high cost of purchasing and maintaining ICP-MS instruments.

24.8 SUMMARY

Atomic spectroscopy is used to determine the concentration of elements in a sample. There are several types of atomic spectroscopy. AAS, AES, and ICP-MS are the most common.

For most applications these three techniques require the sample to be in a dilute acid solution. Therefore sample preparation usually involves destroying the organic matter by dry ashing (heating the sample to about 500°C to burn off the organic matter) or wet ashing (heating the sample in nitric acid and perchloric acid) followed by dissolution of the ash in dilute acid. Once the sample is dissolved, it is introduced into the instrument by aspirating the solution into a flame or plasma where it encounters very high temperatures (2000–3000K for flames and 6000–10000K for plasmas). Nearly all compounds or salts introduced into the flame or plasma are decomposed to atoms. These atoms are well separated, are in the gaseous state, and are mostly neutral in a flame but a high fraction of them lose an electron and become charged in a plasma. The final step is to measure quantitatively the concentration of the elements present in flame or plasma. This quantitation is achieved by either optical spectroscopy or mass spectrometry.

Optical spectroscopy depends on the absorption (AAS) or emission (AES) of electromagnetic radiation (light) by the atoms in the gas state. Electrons in atoms occupy orbitals of discrete energy levels that are unique for each element. Electrons in the ground state can be pushed into a higher energy level by light or by heat.

In AAS light is used to push electrons into a higher energy level; only that light with a wavelength corresponding to the energy difference between the ground state and some excited state will be absorbed. These differences in energy are unique for each element. For example, when light with a wavelength corresponding to the energy level differences in iron is passed through a collection of atoms that includes iron, only the iron atoms will absorb the light. Furthermore, the amount of light absorbed is directly related to the concentration of the iron atoms in the sample. By measuring the absorbance of light of a particular wavelength by an atomized sample, we can determine the concentration of an element even when many other elements are present.

In AES, the optical approach involves exciting the electrons in an element by heating and then measuring the intensity of the light that is emitted when the electrons fall back to the ground state or a lower energy state. Again, the energy differences between excited and lower energy states are unique for each element. Therefore, each element will emit light of a unique wavelength. AES instruments are designed to separate the light that is emitted from excited atoms and to quantitatively measure the intensity of the emitted light.

In contrast, ICP-MS instruments are designed to measure (count) ions directly. This requires that the atoms in a plasma be separated according to their masses. To accomplish this, atoms in the plasma are ionized and then introduced into a mass spectrometer which accelerates the ions, separates them according to their mass-to-charge ratio (m/z), and focuses the ion beam on a detector which counts the actual number of ions that strike it. The resolution of modern mass spectrometers is sufficient to separate ions with m/z differences of about 1 amu, giving the ICP-MS the ability to measure the concentration of elements in complex samples with high sensitivity and specificity.

Atomic spectroscopy is a powerful tool for the quantitative measurement of elements in foods. The development of these technologies over the past five or six decades has had a major impact on several fields, including food science and technology, food safety and toxicology, nutrition, biochemistry, and biology.

Today, accurate and precise measurements of a large number of mineral nutrients and nonnutrients in foods can be made rapidly and with minimal sample preparation using commercially available instrumentation.

24.9 STUDY QUESTIONS

1. Explain the significance of energy transitions in atoms and of atomization for the techniques of AAS and AES.
2. Describe the similarities and differences between AAS and AES for mineral analysis.
3. A new employee in your laboratory is somewhat familiar with the application, principles involved, instrumental components, and quantitation procedure for UV-Vis spectroscopy. The employee must now learn to do analyses using AAS. Explain to the new employee the (a) applications, (b) principles, (c) instrumental components and their arrangement (use diagrams and explain differences), and (d) quantification procedure for AAS by comparing and contrasting these same items for UV-Vis spectroscopy. (Assume you are talking about double-beam systems.)
4. What would be the advantages of having an atomic absorption unit that had a graphite furnace (vs. a flame)?
5. The analytical laboratory in your company plans to purchase an inductively coupled plasma-atomic emission spectrometer.
   (a) Explain the instrumentation and principle of its operation to analyze foods for specific minerals.
   (b) Explain how AAS differs in instrumentation and principle of operation from what you described previously for ICP-AES.
   (c) What are the advantages of ICP-AES over AAS?
   (d) The analytical lab in your company handles a large number of samples and analyzes them for multiple elements. Would you request purchase of a simultaneous PMT spectrometer, a sequential PMT spectrometer, or a CID spectrometer? Explain your answer.
   (e) For most types of food samples other than clear liquids, what type of sample preparation and treatment is generally required before using ICP-AES, AES, or AAS for analysis?
6. You are training a new technician in your laboratory on mineral analysis by AAS and ICP-AES. Briefly describe the purpose of each of the following items associated with those instruments/analysis.
   (a) HCL in AAS
   (b) Plasma in ICP-AES
   (c) Monochromator in ICP-AES
   (d) Nebulizer in AAS and ICP-AES
7. In the quantitation of Na by atomic absorption, KCl or LiCl was not added to the sample. Would you likely over- or underestimate the true Na content? Explain why either KCl or LiCl is necessary to obtain accurate results.
8. Give five potential sources of error in sample preparation prior to atomic absorption analysis.
9. You are performing iron analysis on a milk sample using AAS. Your results for the blank are high. What could be causing this problem and what is a possible remedy?
10. The detection limit for calcium is lower for ICP emission spectroscopy than it is for flame AAS. How is the detection limit determined, and what does it mean?
11. As the manager of the quality assurance laboratory for your company, you ask one of your technicians to find the AOAC methods for sodium determination in a specific food product. Your technician finds the following methods listed: Volhard titration, ion selective electrode, and ICP-AES. Your technician asks you about the differences between these methods. To answer the question, differentiate the principles involved, and explain why your lab might choose to use one method over the other. (See also Chap. 12.)
12. When analyzing a sample for mineral elements using ICP-MS, the instrument is programmed to count the number of ions with a specific m/z striking the detector. You decide to determine the concentrations of potassium and calcium in a sample of wheat flour. What m/z ratio would you use for potassium? For calcium? Why? (Hint, study the masses of all the naturally occurring and stable isotopes for the 2 elements and for argon (see table) and select isotopes with no interferences.) Why is it important to know the masses of argon isotopes as well as potassium and calcium?

Isotope	*Natural Abundance (%)*
^{36}Ar	0.34
^{38}Ar	0.063
^{40}Ar	99.6
^{39}K	93.2
^{40}K	0.012
^{41}K	6.73
^{40}Ca	96.95
^{42}Ca	0.65
^{43}Ca	0.14
^{44}Ca	2.086
^{46}Ca	0.004

24.10 PRACTICE PROBLEMS

1. Your company manufactures and markets an enriched all-purpose flour product. You purchase a premix containing elemental iron powder, riboflavin, niacin, thiamin, and folate which you mix with your flour during milling. You specified to the supplier that the premix be formulated so that when added to flour at a specified rate, the concentration of added iron is 20 mg/lb flour. However, you have reason to believe that the iron concentration in the premix is too low so you decide to analyze your enriched flour using your new atomic absorption spectrometer. Current FDA guidelines for

enriched flour specify that the iron concentration should be 20 mg/lb: http://www.accessdata.fda.gov/scripts/cdrh/cfdocs/cfcfr/CFRSearch.cfm?fr=137.165 (accessed 6/18/2009). You follow the following protocol to determine the iron concentration.

(a) Weigh out 10.00 g of flour, in triplicate (each replicate should be analyzed separately).
(b) Transfer the flour to an 800-ml Kjeldahl flask.
(c) Add 20 ml of deionized water, 5 ml of concentrated H_2SO_4, and 25 ml of concentrated HNO_3
(d) Heat on a Kjeldahl burner in a hood until white SO_3 fumes form.
(e) Cool, add 25 ml of deionized water, and filter quantitatively into a 100-ml volumetric flask. Dilute to volume.
(f) Prepare iron standards with concentrations of 2, 4, 6, 8, and 10 mg/l.
(g) Install an iron hollow cathode lamp in your atomic absorption spectrometer and turn on the instrument and adjust it according to instructions in the operating manual.
(h) Run your standards and each of your ashed samples and record the absorbances.

Calculate the iron concentration in each of your replicates, express as mg Fe/lb flour.

Table. Absorbance data for iron standards and flour samples

Sample	*Fe Conc. (mg/L)*	*Absorbance*	*Corrected Absorbance*
Reagent Blank	–	0.01	–
Standard 1	2.0	0.22	0.21
Standard 2	4.0	0.40	0.39
Standard 3	6.0	0.63	0.62
Standard 4	8.0	0.79	0.78
Standard 5	10.0	1.03	1.02
Flour Sample 1	?	0.28	0.27
Flour Sample 2	?	0.29	0.28
Flour Sample 3	?	0.26	0.25

2. Describe a procedure for determining calcium, potassium, and sodium in infant formula using an ICP-AES. Note: Concentrations of Ca, K, and Na in infant formula are around 700, 730, and 300 mg, respectively.

Answers

1.

(a) Enter the data for the standards into Excel. Using the scatter plot function, plot the standard curve and generate a trend line using linear regression. Include the equation for the line and the R^2 value. Your results should look like the standard curve shown.
(b) Using the equation, calculate the iron concentration in the solution in the volumetric flask for each of your samples. Your answers should be 2.68, 2.79, 2.48 mg/L for samples 1, 2, and 3 respectively. The mean is 2.65 mg/L; the standard deviation is 0.16.

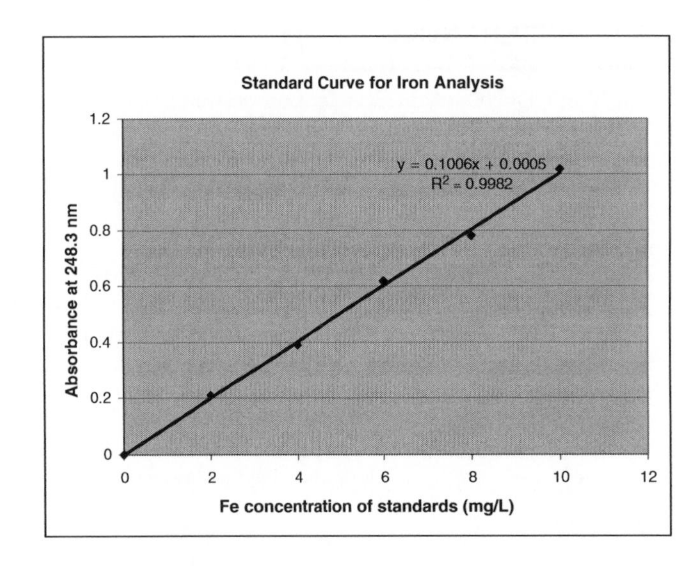

(c) Now determine the iron concentration in the flour. Recall that you transferred the solution from the Kjeldahl flask quantitatively into the 100-ml volumetric flask. Therefore all of the iron in the flour sample should be in the volumetric flask. The mean concentration is 2.65 mg/L. The volume is 0.1.L Therefore, the amount of iron in the 10 g of flour is 0.265 mg. To convert this to mg/lb, multiply by 454/10: 0.265 mg/10 g × 454 g/lb = 12 mg Fe/lb flour.
(d) Your suspicions are confirmed; your supplier shorted you on iron in the premix. You need to correct this as soon as possible because your flour does not conform to the FDA's standard of identity for enriched flour and you may be subject to legal action by the FDA.

2. Consult AOAC Method 984.27 [AOAC International. 2007. *Official Methods of Analysis*, 18th ed., 2005; Current through Revision 2, 2007 (On-line). AOAC International, Gaithersburg, MD.] The following approach may be used:

(a) Shake can vigorously.
(b) Transfer 15.0 ml of formula to a 100-ml Kjeldahl flask. (Carry two reagent blanks through with sample.)
(c) Add 30 ml of $HNO_3 - HClO_4$ (2:1).
(d) Leave samples overnight.
(e) Heat until ashing is complete (follow AOAC procedure carefully –mixture is potentially explosive.)
(f) Transfer quantitatively to a 50-ml vol flask. Dilute to volume.
(g) Calibrate instrument. Choose wavelengths of 317.9 nm, 766.5 nm, and 589.0 nm for Ca, K, and Na, respectively. Prepare calibration standards containing 200, 200, and 100 ng/ml for Ca, K, and Na, respectively.
(h) The ICP-AES computer will calculate concentrations in the samples as analyzed.

To convert to concentrations in the formula, use the following equation:
Concentration in formula = Concentration measured by ICP × (50 ml/15 ml)

24.11 REFERENCES

1. Miller DD (2008) Minerals. In: Damodaran S, Parkin KL, Fennema OR (eds) Food Chemistry, 4th edn. CRC Press Taylor and Francis Group, Boca Raton, FL
2. U.S. Department of Agriculture, Agricultural Research Service (2009) USDA Nutrient Database for Standard Reference. Nutrient Data Laboratory Home Page, http://www.nal.usda.gov/fnic/foodcomp/search/. Accessed April 2009
3. Beaty RD, Kerber JD (1993) Concepts, instrumentation and techniques in atomic absorption spectrophotometry. PerkinElmer Corporation, Norwalk, CT
4. Boss CB, Fredeen KJ (1989) Concepts, instrumentation and techniques in inductively coupled plasma atomic emission spectrometry. PerkinElmer Corporation, Norwalk, CT
5. Thomas R (2008) Practical guide to ICP-MS: a tutorial for beginners, 2nd edn. CRC, Taylor and Francis Group, Boca Raton, FL
6. Skoog DA, Holler FJ, Nieman TA (1998) Principles of instrumental analysis, 5th edn. Saunders College Publishing, Philadelphia, PA
7. Perkin E (2009) Atomic spectroscopy: guide to selecting the appropriate technique and system. http://las.perkinelmer.com/content/relatedmaterials/brochures/bro_atomicspectroscopytechniqueguide.pdf. Accessed May 2009
8. Ganeev A, Gavare Z, Khutorshikov VI, Hhutorshikov SV, Revalde G, Skudra A, Smirnova GM, Stankov NR (2003) High-frequency electrodeless discharge lamps for atomic absorption spectrometry. Spectrochimica Acta Part B 58:879–889
9. West AC, Fassel VA, Kniseley RN (1973) Lateral diffusion interferences in flame atomic absorption and emission spectrometry. Anal Chem 45:1586–1594
10. Boss CB, Fredeen KJ (1989) ICP-AES methodology. In: Concepts, Instrumentation, and Techniques in Inductively Coupled Plasma Atomic Emission Spectrometry. PerkinElmer Corporation, Norwalk, CT
11. National Institute of Standards and Technology (2009) http://ts.nist.gov/measurementservices/referencematerials/index.cfm. Accessed May 2009
12. Anon (2009) Inductively coupled plasma-mass spectrometry: a primer. Agilent Technologies, Wilmington, DE. http://www.agilenticpms.com/primer. Accessed June, 2009

Nuclear Magnetic Resonance

*Bradley L. Reuhs**
Department of Food Science, Purdue University, West Lafayette, IN 47907-2009, USA
breuhs@purdue.edu

and

Senay Simsek
Department of Plant Sciences, North Dakota State University, Fargo, ND 58108, USA
senay.simsek@ndsu.edu

25.1 Introduction 445
25.2 Principles of NMR Spectroscopy 445
 25.2.1 Magnetic Field 445
 25.2.2 Radio Frequency Pulse and Relaxation 446
 25.2.3 Chemical Shift and Shielding 447
 25.2.4 1D-NMR Experiment 448
 25.2.5 Coupling and 2D-NMR 449
25.3 NMR Spectrometer 449
25.4 Applications 450
 25.4.1 NMR-Related Techniques and General Applications 450
  25.4.1.1 Liquids 450
  25.4.1.2 Solids 452

S.S. Nielsen, *Food Analysis*, Food Science Texts Series, DOI 10.1007/978-1-4419-1478-1_25,

25.4.1.3 Magnetic Resonance Imaging 452
25.4.1.4 Relaxometry 453
25.4.2 Specific Food Application Examples 453
25.4.2.1 Oil/Fat 453
25.4.2.1.1 Fatty Acid Profile 453
25.4.2.1.2 Verification of Vegetable Oil Identity 453
25.4.2.1.3 Monitoring of Oxidation 454
25.4.2.1.4 Solid Fat Content 454
25.4.2.2 Water 454
25.4.2.3 Ingredient Assays 454
25.5 Summary 455
25.6 Study Questions 455
25.7 References 455
25.8 Resource Materials 456

25.1 INTRODUCTION

Nuclear magnetic resonance (NMR) **spectroscopy** is a powerful analytical technique with a wide variety of applications. It may be used for complex structural studies, for protocol or process development, or as a simple quality assay for which structural information is important. It is nondestructive, and high-quality data may be obtained from milligram, even microgram, quantities of sample. Whereas other spectroscopy techniques may be used to determine the nature of the functional groups present in a sample, only NMR spectroscopy can provide the data necessary to determine the complete structure of a molecule. The applicability of NMR to food analysis has increased over the last three decades. In addition to improved instrumentation and much lower costs, very complex and specialized NMR techniques can now be routinely performed by a student or technician. These experiments can be set up with the click of a button/icon, as all the basic parameters are embedded into default experiment files listed in the data/work station software, and the results are obtained in a short time.

NMR instruments may be configured to analyze samples in solutions or in the solid state. In fact, these two types of analyses can be used in tandem to follow the fate of a given molecule within a specific system. For example, as a fruit ripens, much of the pectin will be released from the solid matrix around the plant cells into solution in the ripe fruit liquid. The development of this process can be followed by liquids vs. solids analyses during the ripening time. As ripening progresses, the NMR signals associated with pectin will decrease in the solids NMR analyses and increase in the liquids NMR spectra.

Other food applications of NMR spectroscopy include structural analysis of food components, such as fiber, to correlate the structure to the rheological properties. Routine analyses can be used to determine the quality of a product or to test the purity of ingredients. Related techniques, such as **NMR relaxometry**, can be used to monitor processing operations; for example, relaxometry can be used to follow the solubilization of powdered ingredients in water to optimize processing parameters. **Magnetic resonance imaging** (MRI) is a nondestructive technique that can be used to image product quality and changes during processing and storage. For example, MRI is used to image the freezing process, with the goal of increasing shelf life. When combined with rheological analyses, sauce and paste flow in a processing system can be measured.

This chapter will cover the basic principles and applications of NMR spectroscopy, as well as a brief description of relaxometry and MRI. Specific applications to food analysis will be highlighted.

25.2 PRINCIPLES OF NMR SPECTROSCOPY

25.2.1 Magnetic Field

NMR differs from most other forms of spectroscopy in that it is the **atomic nuclei** that are the subject of study, and the energy absorbed and emitted by the nuclei is in the radio frequency range. Many nuclei possess an angular momentum, which means that they have a characteristic spin quantum number (I), and may be analyzed using NMR. The most common nuclei analyzed by NMR are the proton (H) and the ^{13}C isotope of carbon, as well as ^{19}F and ^{31}P, all of which have a spin $I = 1/2$. Nuclei with other spin quantum numbers will not be considered in this chapter, and the theoretical discussion will focus on the proton. These nuclei are also charged, and a spinning charge generates a magnetic field. Simply put, the nuclei behave like tiny magnets that interact with an applied, external magnetic field.

Once the nuclei are placed within a strong **external magnetic field** (B_0), the spin of the nuclei will align with that field (Fig. 25-1). Because of quantum mechanical constraints (nuclei of spin I have $2I + 1$ possible orientations in the external magnetic field), there are only two orientations that the spin 1/2 nuclei can adopt: either aligned with the applied magnetic field (parallel or spin $+1/2$) or aligned against the field (anti-parallel or spin $-1/2$). The parallel orientation has a slightly lower energy associated with it and, therefore, has a slightly higher population. It is this excess of nuclei in the spin $+1/2$ state that produces the net magnetization that is manipulated during an NMR experiment. The spin of the nuclei is not around the center axis, but comparable to a gyration (Fig. 25-1). The motion of a spinning charged particle in an external magnetic field is similar to that of a spinning gyroscope in a gravitational field. This type of motion is known as **precession**, and there is a specific precessional orbit and frequency, the **Larmor frequency**, which is related to the magnetic properties of the nuclei. The magnitude and direction of the local magnetic field describe the magnetic moment or **magnetic dipole** of the system. However, due to the precession and the lower energy state excess of nuclei, there is a net vector parallel to the applied field (Fig. 25-1).

All nuclei of the same element, H for example, will have a nearly identical Larmor frequency in a magnetic field. The specific frequency is dependent on the strength of the external magnetic field, and the Larmor frequency of H in this field defines the NMR instrument. For example, a proton has a Larmor frequency of 500 MHz in an 11.7 T magnet, so the instrument is termed a "500 MHz NMR spectrometer." The strength of the magnet not only governs the Larmor

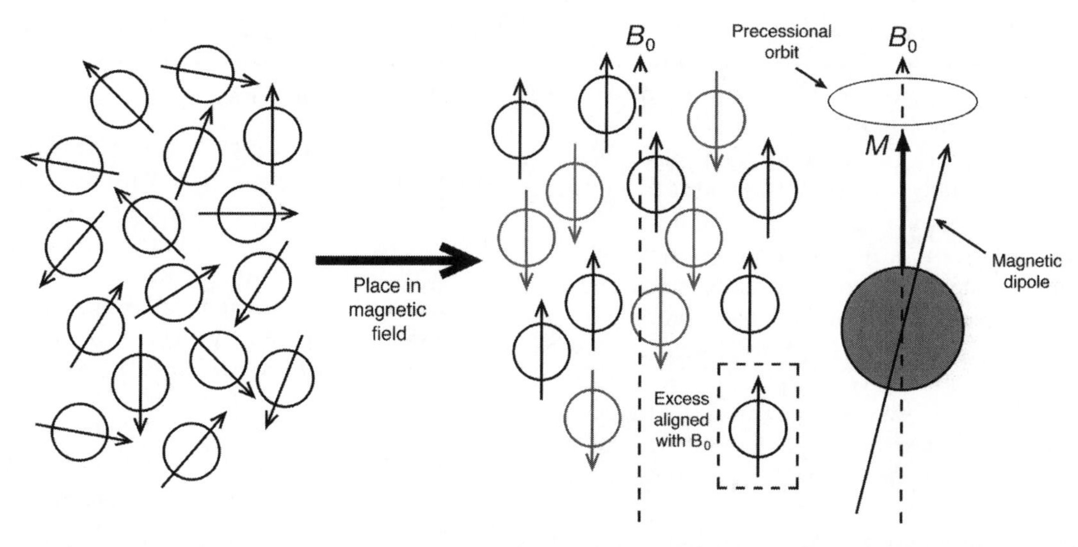

25-1 figure

Nuclear spin and magnetic vectors are randomly ordered outside of the NMR magnet. However, once placed in an applied magnetic field, the NMR magnet, the nuclei align either with the applied field, B_0, (parallel) or against it (antiparallel). There is a slight excess in the population aligned parallel to B_0. Although the magnetic dipole tracks a precessional orbit, the net magnetization (M) is aligned with B_0.

frequency of the nuclei, but also determines the degree of excess nuclei in the parallel orientation. The excess of nuclei in the parallel orientation increases with an increase in the external magnetic field strength, and this in turn impacts the signal intensity of the NMR experiment (i.e., more protons will be detected in a higher field strength instrument). This is one reason researchers seek ever more powerful magnets for NMR. Field strength impacts sensitivity and the signal/noise (signal-to-noise) ratio, and, therefore, the information obtained from the NMR experiment.

The development of more powerful NMR instruments was contingent on the advances in the production of cryo- or superconducting magnets, in which the magnet coil is held at the temperature of liquid helium (around 3K). Currently, there are instruments as powerful as 900 MHz available in many university and industry facilities. In addition to an increase in the sensitivity and the signal/noise ratio, superconducting magnets also have the advantage that once charged, they maintain the charge for years without the input of additional energy, due to the low temperature. The major disadvantage is the need for periodic addition of liquid N_2 and liquid He, which in the case of the latter can be quite expensive, particularly with the very large magnets associated with the high field strength instruments.

25.2.2 Radio Frequency Pulse and Relaxation

Early NMR instruments relied on electromagnets and a simple **radio frequency** (RF) transmitter, and the analyses were performed by a sweep through a range of energy levels. The collected spectra contain the frequency information; hence, it is termed frequency-domain NMR. Although this enabled the development of NMR spectroscopy, it would not facilitate the modern NMR experiment. One of the major developments in NMR technology was the **RF pulse**, in which a large range of frequencies is excited by a short pulse of RF energy around a centered carrier frequency, which is at the Larmor, or resonance, frequency of the nuclei under study. This pulse simultaneously excites all of the protons in the sample. The NMR data for all the protons is collected during a short time after the pulse is applied; therefore, it is termed time-domain NMR. The excitation of a range of radio frequencies by a pulse is similar to the excitation of a range of audio frequencies when a clapper strikes a bell, and the size and construction of the bell determines the frequency range that is emitted. In NMR, the carrier frequency, transmitter power, and duration of the RF pulse determine the frequency range of the pulse.

Once the sample is placed in the magnet, the protons align parallel or antiparallel to the applied, external magnetic field, $\boldsymbol{B_0}$, with an excess in parallel orientation. The net magnetization of the nuclei in the parallel orientation is aligned with the z-axis in an xyz graphical representation of the system (Fig. 25-2a), although the precession of each nucleus is random. However, after a pulse of RF energy is applied to the system, the nuclei precess coherently and individual nuclei absorb energy and shift to a higher energy state. The pulse, which is applied by a transmitter coil perpendicular to the z-axis ($\boldsymbol{B_1}$), tilts the net magnetization vector away from the z-axis and toward the ***xy* plane** (Fig. 25-2b).

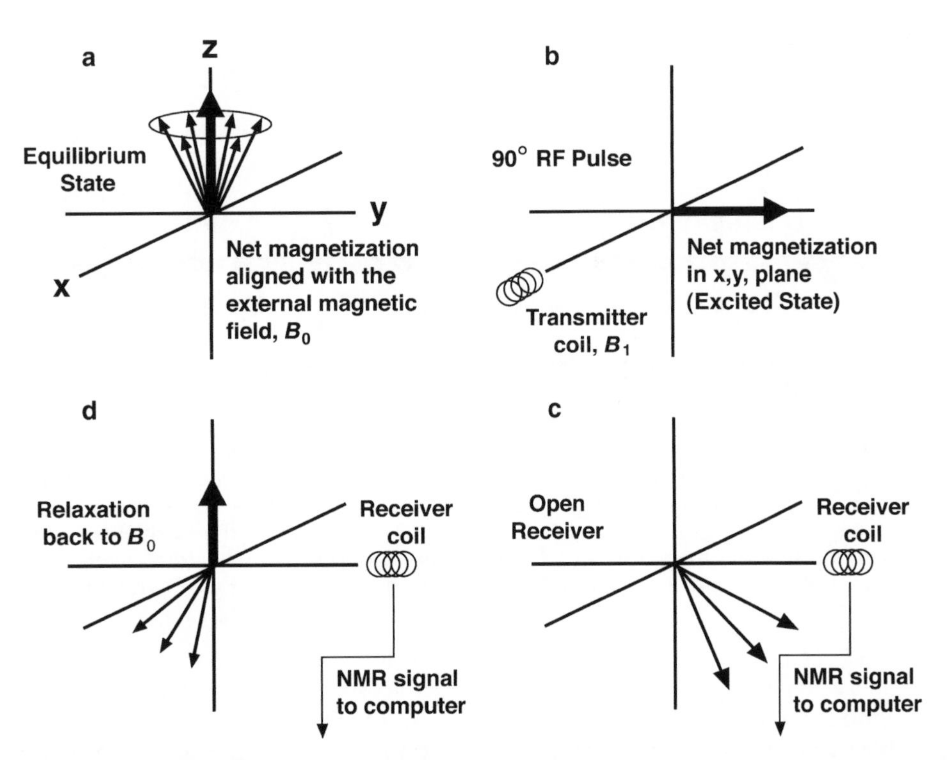

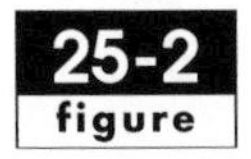

(**a**) Prior to the RF energy pulse, the net magnetization (M) composed of all the component vectors is in the equilibrium state, aligned with B_0. (**b**) The 90° RF pulse, which covers the resonance frequencies of all relevant nuclei in the sample and originates perpendicular to the z-axis (B_1), causes the nuclei to move to a higher energy state, and the net magnetization tilts into the xy plane. (**c**) Once in the xy plane, the net magnetic vector separates into the component vectors for each unique population of nuclei. As these oscillate in the xy plane, they emit RF signals that are detected by the NMR instrument after passing through the receiver coil, which is located perpendicular to both B_0 and the transmitter coil. (**d**) As the component vectors continue to oscillate in the xy plane (and emit RF signals), the nuclei begin to relax back to the equilibrium state. The NMR instrument may be set up to repeat this process, with additional pulses, numerous times; the collected data are then added together to improve the signal/noise ratio and resolution.

Although the parameters that define a pulse include the transmitter power and the pulse duration, a specific pulse used in an NMR experiment is usually described by the degree to which the net magnetization is tilted. The most common pulse is the 90° pulse, which tilts the net magnetization exactly into the xy plane where the receiver coil is located, thereby maximizing the resulting signals. Many NMR experiments use a series of pulses, termed a **pulse sequence**, to manipulate the magnetization. Complex pulse sequences are essential for the two-dimensional (2D) NMR experiments (Sect. 25.2.5) that are required for complete structural analyses of complex molecules.

Once the net magnetization has been tilted into the xy plane by a 90° pulse, the magnetization begins to decay back to the z-axis. This process is termed **NMR relaxation**, and it involves both **spin–lattice** ($T1$) and **spin–spin** ($T2$) relaxation. ***T*1 relaxation** is associated with the interaction of the magnetic fields of the excited-state nuclei with the magnetic fields of other nuclei within the "lattice" of the total sample. ***T*2 relaxation** involves the interactions of neighboring nuclei that lead to a diminishment in the energy state of the excited-state nuclei and the loss of phase coherence. The mechanisms behind relaxation are complex, but the process can be utilized for some specific NMR experiments that, among other things, take advantage of the fact that samples in different forms, liquids vs. solids for example, relax at different rates.

25.2.3 Chemical Shift and Shielding

The total H population in the sample determines the net magnetization of the system in the external magnetic field, $\mathbf{B_0}$. The exact frequency of a unique population of protons (i.e., all protons in a specific chemical location in the molecule), however, is also dependent on the immediate environment of the nucleus, principally the density of the electron cloud surrounding the nuclei, which determines the electronic environment of the nuclei. This is referred to as the "**shielding**" effect, because the electrons create a secondary, induced magnetic field that opposes the applied field, shielding the nuclei from the applied field. The resulting frequency differences are so small, relative to the Larmor frequency, that they are

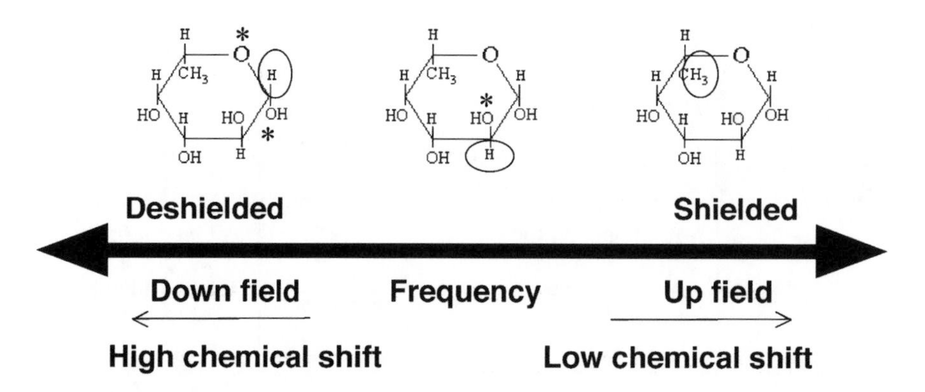

25-3 figure The shielding effect is responsible for the small, but detectable, differences in the resonance frequencies of nuclei such as protons. Protons that are not close to an electronegative group in the molecule, such as the protons in the methyl group in fucose (*top right*), a common 6-deoxy sugar, will be shielded by the electrons surrounding it, and it will have a low chemical shift, upfield in the NMR spectrum. Protons near one oxygen atom (Indicated by an Asterisk), such as the ring protons in sugars (*top middle*), will be intermediately shielded and have a chemical shift towards the middle of the spectrum. And a proton that is near two oxygen atoms, such as the anomeric proton in sugars (*top left*), will be relatively deshielded and have a high chemical shift downfield in the spectrum.

commonly reported in parts per million (ppm). However, they are large enough to be clearly detected and resolved during an NMR experiment. The frequency differences that result from the differences in the electronic environments yield the **chemical shift** of the nuclei. Following the processing of the NMR data, these differences result in a series of resonance signals, each representing a unique proton population, along the x-axis in a one-dimensional (1D)-NMR spectrum. Those protons that have a relatively dense electron cloud are considered shielded, since the electron cloud works in opposition to the external magnetic field, and the resonances will be found on the right, or upfield, side of the spectrum, at a lower chemical shift. As deshielding increases, the resonances are shifted further to the left, or downfield, at a progressively higher chemical shift (Fig. 25-3).

One of the most important determinants of the chemical shift for a specific population of protons (i.e., all the protons in the sample that are in an identical molecular location) is the proximity to an electronegative group or atom, such as O (Fig. 25-3). For example, protons that are not located near any electronegative groups or atoms, such as those in the methyl group of a 6-deoxy sugar, are heavily shielded and the resonances will be found on the far right side of a proton NMR spectrum (Fig. 25-3). In contrast, the proton on the C1 of a typical sugar (the anomeric proton) is near two O atoms, the O in the –OH group (or the O in the linkage to the next sugar in a polymer, such as starch) and the O that forms part of the hemiacetal ring structure of the sugar. Consequently, the resonances associated with the highly deshielded anomeric protons are found on the left side of the proton spectrum. Protons near one O, such as the ring protons in a typical sugar, will be partially shielded, and the resonances associated with these protons will be in the middle region of the spectrum.

25.2.4 1D-NMR Experiment

For solution NMR spectroscopy, the sample is dissolved in solvents produced for NMR analysis, such as D_20 (where D = deuterium; this is to avoid overloading the NMR signal with solvent protons), and pipetted into a NMR tube, which is then capped and placed into the magnet. The net magnetization of the nuclei in the parallel orientation in the external magnetic field, B_0, is aligned with the z-axis; this is the equilibrium state. The 90° pulse at the Larmor frequency of H from the transmitter in the x-axis tilts the net magnetic vector into the xy plane. There the component vectors (those representing each unique population of protons) will oscillate at their specific NMR resonance frequencies, separating the net vector into numerous component magnetic vectors, which then induce radio signals (the NMR signal) in the receiver coil located in the y-axis (Fig. 25-2c). At the same time, the magnitude of the vectors will decay in the xy plane (relaxation) and return to the z-axis (Fig. 25-2d). This process is termed one "**scan**."

The result of these actions is a combined signal for all protons in the sample, which rapidly decreases over time. This is termed a **free induction decay** (FID) (Fig. 25-4a), and it contains all the frequency and intensity information (as well as phase, which will not be considered) for each of the unique populations of nuclei in the sample. **Fourier transformation**, a mathematical operation that converts one function

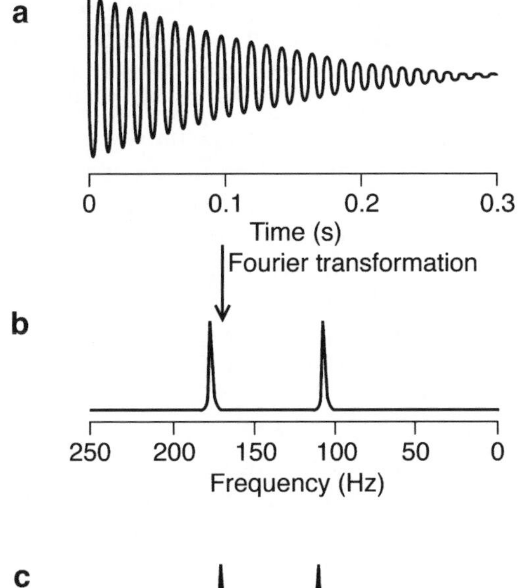

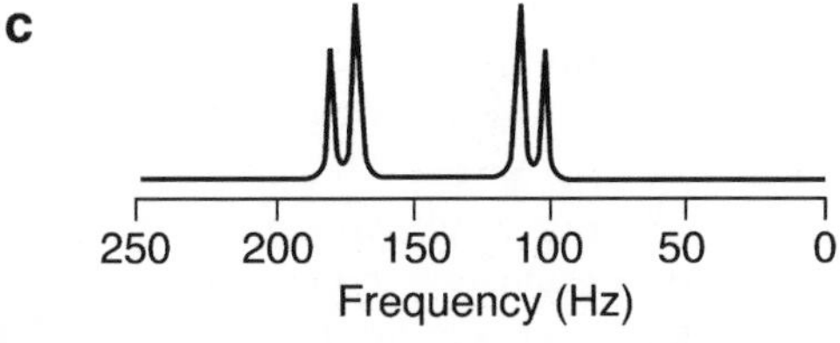

25-4 figure After the 90° RF pulse moves the magnetization to a higher energy state in the *xy* plane, the receiver is turned on and it collects the RF signals emitted by oscillating nuclei; the emitted RF signals rapidly decay over a short time as the magnetization relaxes back to the equilibrium state (see Fig. 25-2). (**a**) The result is NMR data that contains all of the frequency and intensity information for the nuclei under analysis, and which diminishes as the signals decay; this data is termed a "free induction decay" (FID), and it represents the time domain information obtained from the NMR experiment. (**b**) Once the FID has been processed by Fourier Transformation, an NMR spectrum is obtained; this represents the frequency domain NMR information. The resonance "peaks" found on the *x*-axis are due to unique populations of protons, two in this case. (**c**) If the two protons are coupled through the molecular bonds, they will each have the effect of splitting the resonance peaks into two distinct peaks. The degree of splitting, reported in Hz, is indicative of the strength of the coupling effect.

of a variable into another, is then applied to convert the time-domain FID to the frequency-domain **NMR spectrum** (an *xy* plot) (Fig. 25-4b). The frequency information for each unique proton population is presented on the *x*-axis of the NMR spectrum. The signal intensity is related to the *y*-axis; however, this is not labeled because it has no units and, moreover, the linewidth differences of each resonance make the *y*-value imprecise. The only method to compare the signal intensity and, consequently, the relative abundance of specific protons in a 1D-NMR spectrum is to compare the integration values of the resonance peaks.

In practice, the NMR spectrometer is set up so that the pulse is applied numerous times, usually in increments of 16 scans. For samples that are present in a high concentration, 16 or 32 scans per experiment is common, and 256 or 512 scans are commonly used for more dilute samples. After each scan, the new data are added to the data already collected. The result of adding the data from numerous scans is a significant improvement in the signal/noise ratio and resolution.

25.2.5 Coupling and 2D-NMR

Another essential concept to consider is "**coupling**." Coupling is a result of the influence of electrons in covalent bonds on the local magnetic field of nearby nuclei. Through the intervening bonds, two nearby nuclei will affect the chemical shift of one another, resulting in the splitting of the resonances from each unique population of nuclei into two distinct resonances (Fig. 25-4c). The coupling strength is affected by both the proximity of the nuclei to one another and the geometry of the intervening bonds. For example, protons that have a *trans* relationship ("across" from each other) have a much stronger coupling than those that have a *cis* relationship ("on the same side" as one another). Thus, the use of coupling data yields information about the geometry of a specific molecule.

A more important impact of coupling is that complex 2D-NMR experiments have been designed to take advantage of the coupling phenomenon, to produce the data necessary for the complete structure determination of a molecule. 2D-NMR experiments are essentially a series of 1D experiments in which the pulse sequence includes several pulses and a variable parameter, such as a delay between two of the pulses. The computer collects all the spectra and plots them out as a 2D plot, in which "**cross peaks**" show the coupling correlations of nearby nuclei.

25.3 NMR SPECTROMETER

The NMR spectrometer consists of a powerful cryomagnet into which the sample is placed, a set of electronics for transmitting and collecting radio signals, and a data/work station (Fig. 25-5). Modern instruments use superconducting magnets that are cooled to a very low temperature by a jacket of liquid helium, which has a boiling point of 4.2K. This jacket is, in turn, surrounded by an outer jacket of liquid nitrogen, which is cheaper and easier to work with than liquid helium. The core of the superconducting magnet consists of coil windings of thin wires made

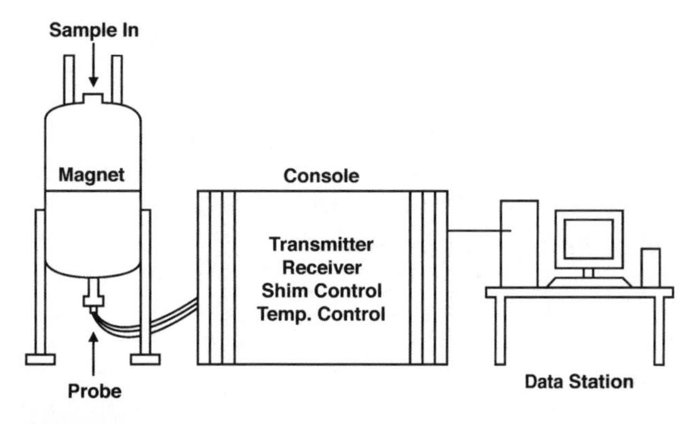

25-5 figure A diagram of an NMR spectrometer. The instrument consists of a superconducting cryomagnet (the NMR magnet), an electronics console, and a data/work station that also controls all the functions of the instrument.

of superconducting alloys, such as niobium–titanium or niobium–tin. The coil (i.e., the magnet) and the coolants are contained in an insulated dewar, termed a cryostat, which includes a vacuum chamber around the liquid jackets.

Once the magnet is cooled to the operating temperature, by the addition of the coolants, and energized by an external power supply, the magnet will maintain its charge and magnetic field for years. One of the most important aspects of maintenance for NMR instruments is the routine filling, or topping off, of the coolants. A typical research instrument is filled with liquid nitrogen on a weekly basis, and liquid helium is added monthly. Failure to maintain the coolant levels will result in the "**quenching**" of the magnet, in which the coolants boil off violently and the magnet loses its charge. Should this happen, the magnet will need to be refilled and recharged, at the very least, and may also require expensive repairs.

Down the center of the magnet, but external to the dewar (i.e., at room temperature) is a tubular space, the **magnet bore**, that serves several functions. A multifunction device, termed a **probe**, is placed inside the magnet bore from the bottom. Inside the probe, just inside of the main magnet coil, are small, secondary magnetic coils that receive power from the NMR instrument hardware. These small coils are manipulated by the operator to make fine adjustments to the magnetic field, to optimize the magnetic field; hence, they are called shims (in construction, a shim is a small piece of wood or metal that is placed between two layers of building material to obtain a better fit, such as when leveling a door frame). The probe also contains the coils for transmitting and receiving the RF energy. Thus, the probe is the central piece of hardware for the NMR experiment. Finally, at the top of the bore, and at the top of the whole system, is the sample insertion point. The sample tube, which is in a holder, is lowered down through the bore by a diminishing stream of forced air until it gently comes to rest at the top of the probe; there the sample is correctly aligned with the magnetic field and the probe hardware. One of the most common mistakes for an inexperienced operator is to drop the sample holder and sample into the magnet bore without the airstream flowing, which results in a rapid descent of the sample holder and sample, breaking the NMR tube and damaging the probe (as well as angering the instrument shop manager).

Connected to the magnet probe by several cables is an electronics console that includes the transmitter, the receiver, and other systems that control the NMR instrument, such as the sample temperature control unit. The transmitter includes systems to produce the pulse at the correct frequency for each nucleus that may be observed. For example, a 500-MHz NMR spectrometer requires a transmitter at 500 MHz for ^{1}H-NMR analyses and a second transmitter system at 125 MHz for ^{13}C-NMR. The console also houses the receiver electronics, which process the NMR signal from the probe receiver coils, the electronics that control the shim electromagnets, and the probe temperature control system. All of this is managed by a computer data/work station and some NMR-specific peripherals.

25.4 APPLICATIONS

Table 25-1 lists recent applications of the NMR and related techniques to food analysis. The following discussion describes general applications and some specific applications to food research. See the references in Table 25-1 for detailed information on the various techniques.

25.4.1 NMR-Related Techniques and General Applications

25.4.1.1 Liquids

Liquids NMR is used for relatively pure samples that readily go into solution in any of the many solvents produced for NMR analyses. Common samples include carbohydrates, proteins, lipids, phenolics, and many other classes of organic compounds. The experiment described in Sect. 25.2.4 is an example of a typical liquid 1D-NMR spectroscopy analysis. The result is a 1D-NMR spectrum with the resonances plotted along the *x*-axis. This is the simplest application of NMR spectroscopy, but it can be very informative. For example, plant- and yeast-derived β-glucans are currently an important topic of research, as they have many health-related benefits, and they are often discarded

Recent Food Applications of Magnetic Resonance Spectroscopy and Related Techniques

Food	*Analysis*	*NMR Method*	*Refs.*
Potato	Effect of microwave heating	MRI	(16)
Pear	Spatial and longitudinal evaluation of core breakdown	MRI	(13)
Bread products	Structural analysis and water status	Single point imaging	(18)
Emulsions	Charge effects of protein interactions at the interface	^{1}H-NMR	(2)
Cheese	Changes in water and fat from temperature and age	2D Laplace inversion, 2D-NMR	(8)
Sunflower seeds	Moisture:oil:solid matter ratios and fatty acid content	Spin–spin, and spin–lattice relaxometry	(1)
Herbs and spices	Antioxidant activity	Electron spin resonance	(4)
Dough	Assessment of dough porosity during proofing	Spin-echo and gradient-echo	(9)
Monoglycerides	Phase determination and behavior	Solid-state NMR	(11)
Beer	Characterization of beer components	NMR spectroscopy	(5)
Meat	Assessment of meat quality	NMR and MRI	(19)
Starch	Amylopectin retrogradation	^{1}H NMR relaxation and CP-MAS spectrometry	(6)
Meat – salt	Interactions of sodium ions with meat	^{23}Na and ^{35}Cl NMR	(7)
Multicomponent foods	Water distribution, food spoilage	NMR relaxometry	(8)
Apple	Texture – mealiness	MRI	(10)
Dairy products	Structural changes in casein micelles, water retention, and macroscopic changes	NMR diffusometry and MRI	(15)
Olive oil	Detection of adulteration by other oils	^{13}C-NMR	(12)
Vegetable oil	Quality control, oil stability, polar compound determination, and color determination	^{13}C-NMR	(20)
Wine	Chemical composition, phenolic compound composition	^{1}H-NMR	(14)
Starch	Chemicophysical properties	CP-MAS-NMR	(3)
Soybean cellulose and pectin	Enzymatic degradation	CP-MAS-NMR	(3)
Wheat bran	Cell wall polysaccharides	CP-MAS-NMR	(3)
Potato	Polyphenol compound analysis	CP-MAS-NMR	(3)
Wheat straw	Degradation, composition and structure of lignocellulosic fibers	CP-MAS-NMR	(3)
Mung bean	Xyloglucan content	^{13}C-CP-MAS-NMR	(3)
Wheat glutenin subunits	Hydration, plasticization, and disulphide bonds	CP-MAS-NMR	(3)
Durum wheat flour	Geographical source	^{1}H-HR-MAS-NMR	(3)
Wine, fruit juice, and olive oil	Detection of adulteration	Specialized NMR techniques	(17)

NMR, nuclear magnetic resonance spectroscopy; *MRI*, magnetic resonance imaging; *CP-MAS*, cross polarization-magic angle spinning NMR.

as waste from many food processing systems. NMR spectroscopy is the best analytical tool available to determine the purity and identity of the β-glucans as various food processors, particularly the cereal, baking, and brewing industries, work to extract these valuable byproducts from the waste stream in a cost-effective manner. Figure 25-6 shows a 1D-NMR spectrum of 1,3–1,4 mixed linkage β-glucans from cereal (oat) processing waste. From this spectrum, both the purity and relative ratio of 1,3 to 1,4 linkages could be determined.

If additional structural information is needed, there are many powerful 2D- and 3D-NMR analyses available for the assignment of the chemical shifts of each ^{1}H and ^{13}C atom in an organic molecule. Once assigned, other experiments enable an assessment of the relative proximity of these nuclei through molecular bonds and through space. 2D-NMR, therefore, can be applied to any sample for which structural information is required, such as a health-related fiber or a new sweetener. This information may be critical if a company or researcher wishes to file for a patent.

NMR spectroscopy also is a valuable assay tool in batch ingredient analysis for quality assurance. In such assays, the structural assignments of the spectra would not be as important as the consistency of

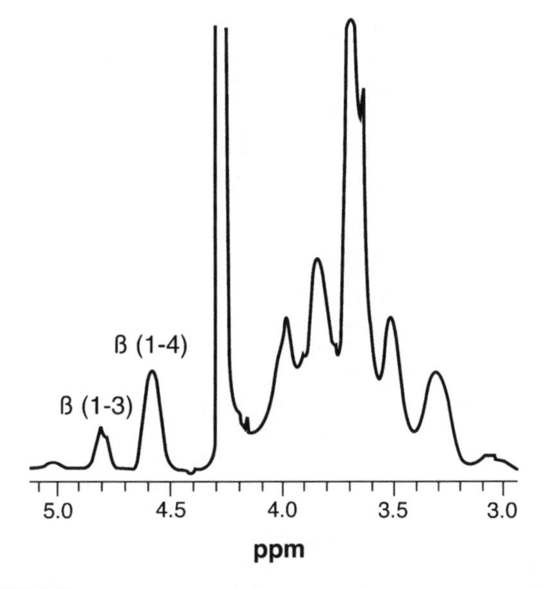

A 1D-H-NMR spectrum of 1,3–1,4 mixed linkage β-glucans from oat processing waste. Both the purity of the sample and the ratio of 1,3 to 1,4 linkages could be determined from the spectrum.

the spectra compared to a spectrum of a high-quality, control product. This application can be used with many types of ingredients, because NMR solvents are available for compounds with a range of solubility properties.

25.4.1.2 Solids

The principles that underlie solid-state NMR are similar to those discussed in Sect. 25.2; however, due to the fact that the sample is not freely tumbling about in solution there is a "directional" aspect (anisotropic or orientation-dependent interactions) to the solid-state analysis. The anisotropic nature of solids results in very broad signals and yields spectra that lack the structural information obtained from samples in solution. One method to overcome this problem is **magic angle spinning** (MAS) experiments, in which the line broadening due to the anisotropy is countered when the sample holder is spun at a specific "magic" angle relative to the external field, B_0, yielding much narrower lines. MAS is often combined with cross polarization enhancement (CP-MAS), in which the magnetization from more easily detected nuclei is transferred to those that are less easily detected (such as from H to ^{13}C).

Solid-state NMR analyses can be applied to many types of samples, such as powders and fresh vegetable tissue. Solid-state ^{13}C-CPMAS-NMR techniques can be used to monitor the chemical composition and the physicochemical properties in the solid portion of an intact food sample. This has been applied to composition studies of different mushroom species, and solid-state ^{13}C-CPMAS-NMR spectroscopy showed significant differences in the ratio of carbohydrate to protein resonances between different species. Also, high-resolution ^{1}H-MAS-NMR techniques enabled food researchers to discriminate between durum wheat flours from Southern Italy, which differ in composition depending on the region of origin. A similar application was used to correlate composition with origin in a study of Parmesan cheese.

25.4.1.3 Magnetic Resonance Imaging

Magnetic resonance imaging (MRI) is unique in that the sample can be placed into the magnet in the native form, and 2D or 3D images of the sample can be generated. MRI involves variations in field strength and the center frequency of the pulses over time and space, along with the application of field gradients in different geometric positions relative to the magnet bore (B_0). The end result is a spatial "encoding" of the sample protons with different phase and frequency values. After multidimensional Fourier transformations of multiple FIDs from different spatial "slices" of the sample, an image of the sample is produced that contains information about the state of the tissue or other material under study.

The sample can be a medical patient, a small test animal, a diseased plant stem, a ripened fruit, or even a complex food product in various steps of processing or its final form. For example, a packaged product could be analyzed over time in the package to track water movement or loss. The MRI analyses would not affect the product. There are many potential applications for MRI in the food industry; for example, it can be used to image the freezing process in frozen food production, with the goal of increasing shelf life. MRI also may be applied to analyses of the composition and characteristics of pastes and sauces, to locate voids in products, or to examine the fat distribution in meats or the water/fat distribution in emulsions. It can also provide detailed information about the thickness of a filling or coating, structural changes, including water loss, in a product via heat transfer (cooking), or changes associated with hydration of a food product during processing. When combined with rheological analyses, sauce and paste flow in a processing system may be monitored.

MRI images of Clementine fruit are shown in Fig. 25-7. One image shows freeze damage to the interior pericarp region of the fruit. The other image shows the presence of an unwanted seed. Such problems often are undetected by a simple visual inspection of the fruit.

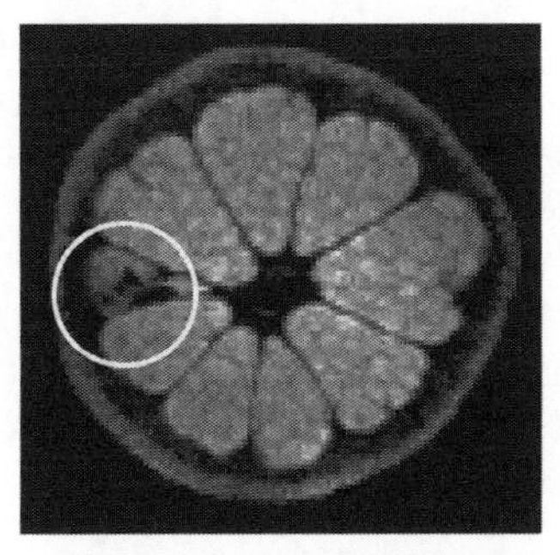

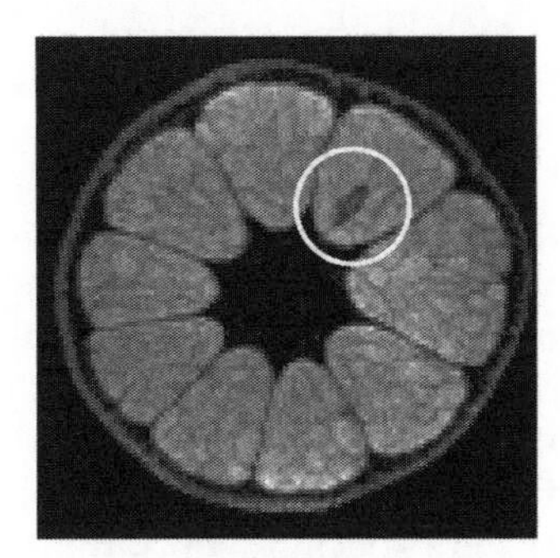

25-7 figure MRI images (18 mm slice thickness) of clementine citrus fruit with defects. Freeze damage is shown in the image on the top and an unwanted seed is shown in the image on the bottom. (Images courtesy of Michael McCarthy, Aspect AI Ltd., Netanya, Israel.)

As the high costs of purchasing and maintaining a large-bore MRI instrument decrease over time, as they did for NMR spectroscopy, and as smaller bore instruments become more common, this important tool should become available to even small food companies and food science departments. These instruments may become a common sight in even modest research and development laboratories over the next decade.

25.4.1.4 Relaxometry

In the plastics industry, small molecules are mixed with the large polymers to make the system more fluid. These small molecules are termed plasticizers, and in food processing the natural equivalent is water. The amount of water available to act as a plasticizer is a very important factor for food quality. An increase or a reduction in the amount of plasticizer can affect the glass transition process that, in turn, affects the quality of the final product. Water exists in several states in food, and the interaction of water molecules with food components can be investigated by the measurement of NMR relaxation. This includes both the spin–lattice ($T1$) and spin–spin ($T2$) relaxation times (Sect. 25.2.2) of the water protons. The relaxation times are related to the magnetic interactions of water protons with the surrounding environment, and the effective relaxation time is related to the extent of the association between the water molecules and immobilized or slowly moving macromolecules. In general, as the macromolecular content increases, the relaxation times of the water protons also increase.

25.4.2 Specific Food Application Examples

High-resolution NMR spectroscopy has been used for the analysis of complex systems such as food samples, biofluids, and biological tissues because it provides information on a wide range of compounds found in the food matrix in a single experiment. NMR spectrometry is nondestructive and offers advantages in the simplicity of sample preparation and rapidity of analysis. The short time frame needed to obtain NMR spectra (minutes), coupled with automation, enables the analysis of many samples with minimal operator input. There are two basic types of analysis in the application of NMR to the food industry: (1) identification of distinct resonances and, therefore, specific compounds, and (2) use of chemometric profile analysis, in which the whole spectral profile is used without assigning particular resonances.

25.4.2.1 Oil/Fat

25.4.2.1.1 Fatty Acid Profile Physical and chemical properties of fats, oils, and their derivatives are mainly influenced by their fatty acid profile. Even though gas chromatography (GC) is usually used for determining the fatty acid profile (Chap. 8, Sect. 8.3.6; Chap. 14, Sect. 14.6.2), the common unsaturated fatty acids, such as oleic, linoleic, and linolenic acids in an oil or fat sample, can be quantified using ^{1}H-NMR, by integration of select signals in the spectra. Although GC provides accurate information about complete fatty acid profile, it lacks information about the fatty acid distribution on the glycerol anchors, which is important to determine the functionality of the ingredient in processing, such as the crystallization point or how it plasticizes the dough in a baked product. For example, the correct type of fat is essential for quality pie crusts or croissants. The fatty acid distribution on the glycerol anchors can be obtained from ^{13}C-NMR analysis. There are two groups of resonances in the carbonyl region of the spectrum; the first is due to fatty acids in positions 1,3, and the second is from fatty acids in position 2 of the glycerol moiety.

25.4.2.1.2 Verification of Vegetable Oil Identity Even though different oils or fats may be purposely mixed for specific reasons, the adulteration of high-value oils with oils of lesser value is an issue of economic and commercial importance. This is primarily a problem with olive oil, because it is expensive and has superior nutritional value. Accordingly, many studies from major olive oil-producing Mediterranean countries, such as Greece, Italy, and Spain, deal with identifying lower-value oils, such as hazelnut oil, used for adulterating olive oil. The adulteration problem is complicated by the fact that the lower-value oils usually have fatty acid profiles similar to olive oil. Among the methods used for analyzing potentially adulterated olive oil are ^{13}C-NMR and ^{1}H-NMR spectrometry. For example, NMR is utilized in conjunction with multivariate statistical analyses of specific resonances in NMR spectra of olive oil diluted with hazelnut or sunflower oil. These methods also can be used to identify the variety and geographical origin of the oil.

25.4.2.1.3 Monitoring of Oxidation The oxidation of vegetable oils is a significant quality problem and can lead to further deterioration of the oil. Highly unsaturated fatty acids, with *bis*-allylic methylene groups, are particularly susceptible to oxidation. Primary and secondary oxidation products, such as hydroperoxides and aldehydes, are easily detected by ^{1}H-NMR analyses. ^{1}H-NMR is especially useful for such analyses because the samples do not require any additional treatments, such as derivatization, that could cause degradation.

25.4.2.1.4 Solid Fat Content While most analyses discussed in this chapter depend on high-resolution NMR instruments, a benchtop, low-resolution pulsed NMR instrument can be used to determine the solid fat content (SFC) of a sample (see also Chap. 14, Sect. 14.3.11). For example, the amount of solid triacylglycerols in the oil or fat at different temperatures can be determined. This method is based on the difference in relaxation times between solids and liquids, and after a delay only the NMR signal of the liquid fat is measured. The solid content is then estimated. Crystallization mechanisms of fat blends also can be studied using SFC measurements.

25.4.2.2 Water

Glass transition is an important property of foods, and the glass transition temperature (Tg), which is dependent on water content, impacts both the processing and the storage of food products. Tg can be determined with an **NMR state diagram**, which is a curve relating NMR relaxation time to glass transition temperature at different moisture contents. This information is important because processing and storage temperatures above Tg at any point during production and distribution of a product are associated with more rapid deterioration. Spin–spin relaxation time (T2) is commonly used as an indication of proton mobility, which is different above and below the Tg of a given product. Although the differential scanning calorimeter (DSC) (Chap. 31, Sect. 31.3.2) is most commonly used for simple Tg analyses, the ability to generate NMR state diagrams increases the value of NMR for many applications.

25.4.2.3 Ingredient Assays

Adulteration in fruit juice is not easy to detect by taste or color. For example, orange juice can be blended with relatively inexpensive grapefruit juice, but the presence of the grapefruit juice in a commercially available orange juice product poses serious health risks for consumers with certain medical conditions. Grapefruit juice has a number of coumarin-like flavonoids and other powerful CYP450 inhibitors that negatively impact the metabolism of many prescribed drugs. Therefore, the detection and prevention of this kind of adulteration is especially important. NMR-based chemometric approaches using Independent Component Analysis, a variant of Principle Component Analysis, are now applied to this problem. Selected regions of the ^{1}H NMR spectra, which are known to contain distinguishing flavonoid glycoside signals, are accurately analyzed in a relatively short time. Another common issue with juice preparation is the differentiation between freshly squeezed juices and those produced from pulp washes, which can be added to fresh-squeezed orange juice to reduce production costs. ^{1}H NMR, in combination with Principal Component Analyses, can easily and accurately distinguish the fresh-squeezed and pulp-wash orange juice.

NMR is also used in monitoring batch-to-batch quality and production site differences in beer. Large multinational breweries prepare their beers at many different geographic locations and require methods for quality control at a detailed molecular level. NMR can be used in conjunction with principal component analysis to distinguish beer from different production sites based on lactic acid, pyruvic acid, dextran, adenosine, inosine, uridine, tyrosine, and 2-phenylethanol content. Quantifying these compounds allows the producers to identify production sites where there is greater variability in these compounds (and therefore poorer quality control).

NMR methods are used by other producers to improve quality control in soft drink production, juice production, and vegetable oil manufacturing. Similar methods also are used to monitor the quality of functional foods and neutraceuticals (food extracts with positive medicinal effects) that are harvested from different geographic locations.

25.5 SUMMARY

Nuclear magnetic resonance technology provides powerful research instrumentation for a variety of applications, from structural elucidation of complex molecules, to 3D-imaging of fresh tissue, to simple ingredient assays for quality assurance. NMR differs from most other forms of spectroscopy because the nucleus is the subject of study and the excitation step uses radio-frequency electromagnetic energy. The proton (H) and the ^{13}C isotope are the most commonly studied nuclei, and each has a characteristic charge and spin, which results in a small, local magnetic field. NMR analyses require an external magnetic field, which causes the local magnetic fields of the nuclei to align in a parallel or antiparallel orientation. There is a slight excess in the parallel orientation (in the *z*-axis aligned with B_0), and it is the net magnetic vector of this population that is detected during an NMR experiment. A pulse of RF energy moves this net magnetism into the *xy* plane, where a reemitted radio signal (the NMR signal) is detected. This signal, which decays quickly, contains the intensity and frequency information for all the nuclei in the sample, and the resulting FID is converted by Fourier transformation into the NMR spectrum, which shows the various resonances spread along the *x*-axis based on differences in frequency.

The NMR instrument consists of a cryomagnet with the transmitter and receiver antennae in the central bore, an electronics console with the transmitter and receiver hardware, and a data/work station that controls all the functions of the instrument. In addition to NMR spectrometers, with both solids and liquids applications, there are other related instruments, such as MRI, that are based on the same principles, but yield different information.

Among the common applications of NMR to food science are structural studies that examine the correlation between chemical structure and health benefits or functionality of food ingredients, studies of the effects of processing on food properties and quality, composition studies of food ingredients or even fresh vegetable tissue, imaging of food products, and determination of SFC or ingredient purity.

25.6 STUDY QUESTIONS

1. Explain the basic principles associated with NMR spectroscopy, including the function of the magnet and the concept of nuclear spin.
2. Describe the interaction of the net magnetization with the RF pulse (90°) and the subsequent NMR signals.
3. Explain the concept of shielding and chemical shift.
4. Describe the FID and the NMR spectrum, including the concepts of time domain, frequency domain, and data transformation.
5. List the components of the NMR spectrometer and their functions.
6. What kinds of samples are analyzed by (a) liquids NMR and (b) solid-state NMR?
7. What kind of final data does one obtain with an MRI? List two applications of MRI.
8. What is the primary use of relaxometry in food analysis?
9. List the general types of food applications of NMR and give an example of each.

25.7 REFERENCES

1. Albert G, Pusiol DJ, Zuriaga MJ (2003) Spin–spin and spin–lattice relaxometry in sunflower seeds. In: Belton PS, Gill AM, Webb GA, Rutledge D (eds) Magnetic resonance in food science: latest developments. The Royal Society of Chemistry, Cambridge, UK, pp 93–100
2. Areas JAG, Watts A (2003) Charge effects of protein interactions at lipid–water interface probed by ^{2}H-NMR. In: Belton PS, Gill AM, Webb GA, Rutledge D (eds) Magnetic resonance in food science: latest developments. The Royal Society of Chemistry, Cambridge, UK, pp 70–76
3. Bertocchi F, Paci M (2008) Applications of high-resolution solid-state NMR spectroscopy in food science. J Agric Food Chem 56(20):9317–9327
4. Dillas SM, Canadanovic-Brunet JM, Cetkovic GS, Tumbas VT (2003) Antioxidative activity of some herbs and spices – a review of ESR studies. In: Belton PS, Gill AM, Webb GA, Rutledge D (eds) Magnetic resonance in food science: latest developments. The Royal Society of Chemistry, Cambridge, UK, pp 110–120
5. Duarte IF, Spraul M, Godejohann M, Braumann U, Gil AM (2003) Application of NMR and hyphenated NMR spectroscopy for the study of beer components. In: Belton PS, Gill AM, Webb GA, Rutledge D (eds) Magnetic resonance in food science: latest developments. The Royal Society of Chemistry, Cambridge, UK, pp 151–160
6. Farhat IA, Ottenhof MA, Marie V, de Bezenac E (2003) ^{1}H-NMR relaxation study of amylopectin retrogradation. In: Belton PS, Gill AM, Webb GA, Rutledge D (eds) Magnetic resonance in food science: latest developments. The Royal Society of Chemistry, Cambridge, UK, pp 172–179
7. Foucat L, Donnat JP, Renou JP (2003) ^{23}Na and ^{35}Cl NMR studies of the interactions of sodium and chloride ions with meat products. In: Belton PS, Gill AM, Webb GA, Rutledge D (eds) Magnetic resonance in food science: latest developments. The Royal Society of Chemistry, Cambridge, UK, pp 180–185
8. Godefroy S, Creamer LK, Watkinson PJ, Callaghan PT (2003) The use of 2D Laplace inversion in food material. In: Belton PS, Gill AM, Webb GA, Rutledge D (eds) Magnetic resonance in food science: latest developments. The Royal Society of Chemistry, Cambridge, UK, pp 85–92
9. Grenier A, Lucas T, Davenel A, Collewet G, Le Bail A (2003) Comparison of two sequences: spin-echo and

gradient echo for the assessment of dough porosity during proving. In: Belton PS, Gill AM, Webb GA, Rutledge D (eds) Magnetic resonance in food science: latest development. The Royal Society of Chemistry, Cambridge, UK, pp 136–143
10. Hills BP, Meriodeau L, Wright KM (2003) NMR studies of complex foods in off-line and on-line situations. In: Belton PS, Gill AM, Webb GA, Rutledge D (eds) Magnetic resonance in food science: latest developments. The Royal Society of Chemistry, Cambridge, UK, pp 186–198
11. Hughes E, Frossard P, Sagalowicz L, Appolonia Nouzille C, Raemy A, Watzke H (2003) Solid state NMR of lyotropic food systems. In: Belton PS, Gill AM, Webb GA, Rutledge D (eds) Magnetic resonance in food Science: latest developments. The Royal Society of Chemistry, Cambridge, UK, pp 144–150
12. Kyricou I, Zervou M, Petrakis P, Mavromoustakos T (2003) An effort to develop an analytical method to detect adulteration of olive oil by hazelnut oil. In: Belton PS, Gill AM, Webb GA, Rutledge D (eds) Magnetic resonance in food science: latest developments. The Royal Society of Chemistry, Cambridge, UK, pp 223–230
13. Lammertyn J, Dresselaers T, Van Hecke P, Jancsok P, Wevers M, Nicolai BM (2003) Application and comparison of MRI and X-Ray CT to follow spatial and longitudinal evolution of core breakdown in pears. In: Belton PS, Gill AM, Webb GA, Rutledge D (eds) Magnetic resonance in food science: latest developments. The Royal Society of Chemistry, Cambridge, UK, pp 46–53
14. Maraschin RP, Ianssen C, Arsego JL, Capel LS, Dias PF, Cimadon AMA, Zanus C, Caro MSB, Maraschin M (2003) Solid phase extraction and ^{1}H-NMR analysis of Brazilian Cabernet Sauvignon wines – a chemical composition correlation study. In: Belton PS, Gill AM, Webb GA, Rutledge D (eds) Magnetic resonance in food science: latest developments. The Royal Society of Chemistry, Cambridge, UK, pp 255–260
15. Mariette F (2003) NMR relaxometry and MRI for food quality control: application to dairy products and processes. In: Belton PS, Gill AM, Webb GA, Rutledge D (eds) Magnetic resonance in food science: latest developments. The Royal Society of Chemistry, Cambridge, UK, pp 209–222
16. Nott KP, Shaarani SM, Hall LD (2003) The effect of microwave heating on potato texture studied with magnetic resonance imaging. In: Belton PS, Gill AM, Webb GA, Rutledge D (eds) Magnetic resonance in food science: latest developments. The Royal Society of Chemistry, Cambridge, UK, pp 38–45
17. Ogrinc N, Kosir IJ, Spangenberg JE, Kidric J (2003) The application of NMR and MS methods for detection of adulteration of wine, fruit juices, and olive oil: a review. Anal Bioanal Chem 376(4):424–430
18. Ramos-Cabrer P, van Duynhoven JPM, Blezer ELA, Nicolay K (2003) Single point imaging of bread-based food products. In: Belton PS, Gill AM, Webb GA, Rutledge D (eds) Magnetic resonance in food science: latest developments. The Royal Society of Chemistry, Cambridge, UK
19. Renou JP, Bielicki G, Bonny JM, Donnat JP, Foucat L (2003) Assessment of meat quality by NMR. In: Belton PS, Gill AM, Webb GA, Rutledge D (eds) Magnetic resonance in food science: latest developments. The Royal Society of Chemistry, Cambridge, UK, pp 161–171
20. Zamora R, Gomez G, Hidalgo FJ (2003) Quality control of vegetable oils by ^{13}C NMR spectroscopy. In: Belton PS, Gill AM, Webb GA, Rutledge D (eds) Magnetic resonance in food science: latest developments. The Royal Society of Chemistry, Cambridge, UK, pp 231–238

25.8 RESOURCE MATERIALS

1. Berger S, Braun S (2004) 200 and more NMR experiments: a practical course. Wiley–VCH, Weinheim, Germany
2. Farhat IA, Belton PS, Webb GA (2007) Magnetic resonance in food science: from molecules to man. The Royal Society of Chemistry, Cambridge, England
3. Günther H (1995) NMR spectroscopy, 2nd edn. Wiley, New York
4. Hills B (1998) Magnetic resonance imaging in food science. Wiley, New York
5. Jacobsen NE (2007) NMR spectroscopy explained: simplified theory, applications and examples for organic chemistry and structural biology. Wiley, New York
6. McMurry JE (2007) Organic chemistry, 7th edn. Brooks-Cole, Salt Lake City, UT
7. Skoog DA, Holler FJ, Crouch SR (2007) Principles of instrumental analysis, 6th edn. Brooks-Cole, Salt Lake City, UT
8. Webb GA, Belton PS, Gil AM, Delgadillo I (eds) (2001) Magnetic resonance in food science: a view to the future. The Royal Society of Chemistry, Cambridge, England

Mass Spectrometry

J. Scott Smith
Food Science Institute, Kansas State University, Manhattan, KS 66506-1600, USA
jsschem@ksu.edu

and

Rohan A. Thakur
Taylor Technology, Princeton, NJ 08540, USA
rt@taytech.com

26.1 Introduction 459
26.2 Instrumentation: The Mass Spectrometer 459
  26.2.1 Overview 459
  26.2.2 Sample Introduction 459
  26.2.3 Ionization 460
  26.2.4 Mass Analyzers 460
26.3 Interpretation of Mass Spectra 462
26.4 Gas Chromatography–Mass Spectrometry 464
26.5 Liquid Chromatography–Mass Spectrometry 465
  26.5.1 Electrospray Interface 466

S.S. Nielsen, *Food Analysis*, Food Science Texts Series, DOI 10.1007/978-1-4419-1478-1_26,

26.5.2 Atmospheric Pressure Chemical Ionization 466
26.5.3 Tandem Mass Spectrometry 467
26.6 Applications 467
26.7 Summary 470
26.8 Study Questions 470
26.9 Resource Materials 470
26.10 References 470

26.1 INTRODUCTION

Mass spectrometry (MS) is unique among the various spectroscopy techniques in both theory and instrumentation. As you may recall, spectroscopy involves the interaction of electromagnetic radiation or some form of energy with molecules. The molecules absorb the radiation and produce a spectrum either during the absorption process or as the excited molecules return to the ground state. MS works by placing a charge on a molecule, thereby converting it to an ion in a process called **ionization**. The generated ions are then resolved according to their **mass-to-charge ratio** (m/z) by subjecting them to electrostatic fields (**mass analyzer**) and finally detected. An additional stage of ion fragmentation may be included before detection to elicit structural information in a technique known as **tandem MS**. The result of ion generation, separation, fragmentation, and detection is manifested as a mass spectrum that can be interpreted to yield molecular weight or structural information. The uniqueness of this process allows the method to be used for both detection and identification of an unknown compound.

Because of recent advances in instrument design, electronics, and computers, MS has become routine in many analytical labs. Probably the most common application is the interfacing of the mass spectrometer (MS) with **gas chromatography** (GC) (see Chap. 29), in which the MS is used to confirm the identity of compounds as they elute off the GC column. The use of **high-performance liquid chromatography** (HPLC) (see Chap. 28) with the MS also has recently become more routine due to advances made in the interconnecting interfaces.

26.2 INSTRUMENTATION: THE MASS SPECTROMETER

26.2.1 Overview

The MS performs three basic functions. There must be a way to **ionize** the molecules, which occurs in the ion source by a variety of techniques such as electron impact, matrix-assisted-laser-desorption, or atmospheric pressure ionization. The charged molecular ion and its fragments must be **separated** according to their m/z, and this occurs in the mass analyzer section (e.g., quadrupoles, ion traps, time-of-flight (TOF), Fourier transform). Finally the separated, charged fragments must be **monitored** by a detector. The block diagram in Fig. 26-1 represents the various components of an MS.

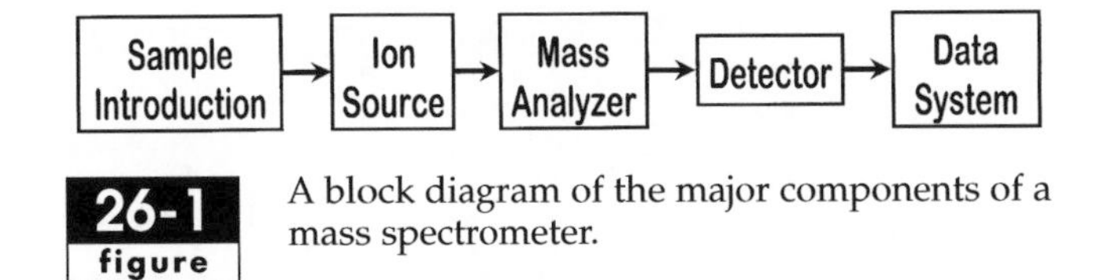

26-1 figure A block diagram of the major components of a mass spectrometer.

Sample introduction can be **static**, **direct insertion probes**, or **dynamic**, which involves interfacing chromatographic equipment such as gas or liquid chromatography (LC). Interface methods include **heated capillary transfer lines** (GC–MS), and LC–MS techniques such as **electrospray** and **atmospheric pressure ionization interfaces**. Common MS interfaces will be discussed in more detail in Sects. 26.4 and 26.5.

Figure 26-2 depicts the interior of a typical **quadrupole MS**. The region between ion generation and detection is maintained by different vacuum pumps. Each successive region from the source is kept at lower vacuum than the preceding region, with the mass analyzer/detector being in the region of the strongest vacuum ($\approx 10^{-6}$ Torr). A vacuum is necessary to avoid ion–molecule reactions between the charged ions and other gaseous molecules before they reach the detector, thereby increasing both sensitivity and resolution.

26.2.2 Sample Introduction

The initial step in operating the MS is to get the sample into the ion source chamber. Pure compounds that are gases or volatile liquids are injected directly into the source region. This requires no special equipment or apparatus and is much the same as injecting a sample into a GC. Thus, this **static method** of introducing the sample to the source is called **direct injection**. With solids that are at least somewhat volatile, the **direct insertion probe method** is used, in which the sample is placed in a small cup at the end of a stainless steel rod or probe. The probe is inserted into the ion source through one of the sample inlets, and the source is heated until the solid vaporizes. The mass spectrum is then obtained on the vaporized solid material as with the direct injection method.

Both direct injection and direct insertion probe methods work well with pure samples, but their use is very limited when analyzing complex mixtures of several compounds. For mixtures, sample introduction is a **dynamic method** in which the sample must be separated into the individual compounds and then analyzed by the MS. This is done typically by GC or HPLC connected to an MS by an **interface** (see Sects. 26.4 and 26.5). The interface removes excess GC carrier gas or HPLC solvent that would otherwise overwhelm the vacuum pumps of the MS.

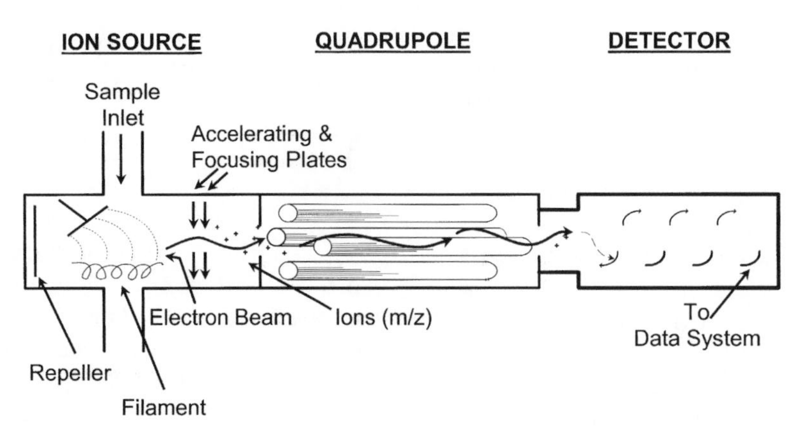

26-2 figure Schematic of a typical mass spectrometer. The sample inlets (interfaces) at the top and bottom can be used for direct injection or interfacing to a GC.

26.2.3 Ionization

In GC–MS analysis, once in the ion source, the compound is exposed to a beam of electrons emitted from a filament composed of rhenium or tungsten metal. When a direct current is applied to the filament (usually 70 electron volts, eV), it heats and emits electrons that move across the ion chamber toward the positive electrode. As the electrons pass through the source region, they come in close proximity to the sample molecule and extract an electron, forming an ionized molecule. Once ionized, the molecules contain such high internal energies that they can further fragment into smaller molecular fragments. This entire process is called **electron impact** (EI) **ionization**, although the emitted electrons rarely hit a molecule.

The eventual outcome of the ionization process is both negatively and positively charged molecules of various sizes unique to each compound. When the repeller plate at the back of the source is positively charged, it repels the positive fragments toward the quadrupole mass analyzer. Thus, we look only at the **positive fragments**, although negative fragments are sometimes analyzed. As the positively charged fragments leave the ion source, they pass through holes in the accelerating and focusing plates. These plates serve to increase the energy of the charged molecules and to focus the beam of ions, so that a maximum amount reaches the mass analyzer.

26.2.4 Mass Analyzers

The heart of an MS is the **mass analyzer**. It performs the fundamental task of separating the charged fragments based on their m/z, and dictates the mass range, accuracy, resolution, and sensitivity. There are seven basic types of mass analyzers: **quadrupoles** (Q), **ion traps** (IT), **time-of-flight** (TOF), **magnetic sectors**, **isotope ratio MS**, **Fourier-transform-based ion cyclotrons** (FT-ICR) and **Orbitraps** (OT), and **accelerator mass spectrometers** (AMS). In addition to these detectors, recently, **inductively coupled plasma** (ICP) torches have been coupled to mass spectrometers (ICP–MS, see Chap. 24). Combinations of these basic mass analyzer types greatly enhance the information derived from routine MS such as the **triple stage quadruple** (TSQ), IT–FT, IT–OT, Q–TOF, and **quadrupole–ion trap** (Q–Trap).

Over the years, four types of MS have become ubiquitous: quadrupole–MS (single quadrupole and TSQ) have found routine use in quantitative analysis; IT–MS have made qualitative analysis easier through the advantages of **multiple stages of MS** (MS^n); while TOF/Q–TOF and FT-based MS have enabled **high resolution accurate mass** applications. The other less common types of mass analyzers such as the **magnetic sector** continue to find use for highly specialized applications, such as dioxin analysis, which requires ultrahigh resolution to separate away isobaric interferences. **Isotope ratio** and **AMS** are popular in geochemistry and nutrition sciences because of their extreme specificity, which allows them to detect far below the detection limits of the more commonly used mass analyzer types.

As the name implies, the magnetic sector analyzer uses a magnetic field to separate the ions based on their m/z. As shown in Fig. 26-3, ions produced in the ion source region of the instrument are accelerated down a curved tube that runs through a magnet. Through a combination of magnetic field strength and ion velocity, the ions take on a curved path to the detector. Ions having a curved path that keeps them in the center of the analyzer tube will reach the detector opening. Ions that hit the sides of the tube are pumped away and do not reach the detector.

To detect multiple ions and produce the typical mass spectra, the magnet field strength is changed so that all possible m/z ratios are seen. This is achieved

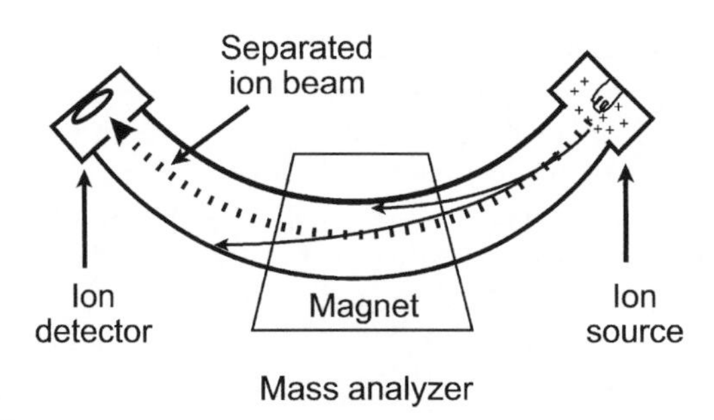

26-3 figure Diagram of a magnetic sector mass analyzer.

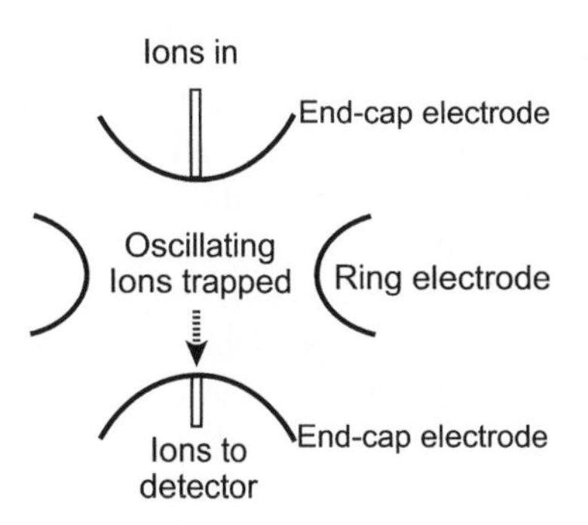

26-4 figure Diagram of an ion trap mass analyzer.

by ramping up the current to the electromagnet over the analysis time, thereby varying the field strength. Selecting the current range determines the mass range to be analyzed. If the field strength is changed very quickly, then all of the ions are detected almost instantaneously.

Quadrupole mass analyzers are based on the ion-focusing work of the Greek electrical engineer Christophilos. The word "quadrupole" is derived from the Latin word for "fourfold" *(quadruplus)*, and "pole," to describe the array of four rods that are used. The four rods are used to generate two equal but out-of-phase electric potentials: one is alternating current (AC) frequency of applied voltage that falls in radiofrequency (RF) range and the other one is direct current (DC). The potential difference can be varied to create an oscillating electrical field between two of the opposite rods, resulting in their having equal but opposite charges.

When, for example, a positive-charged ion enters the quadrupole field, it will be instantly attracted toward a rod maintained at negative potential, and if the potential of that rod changes before the ion impacts, it will be deflected (i.e., change direction). Thus, every stable ion (i.e., ion with stable flight path) entering the quadrupolar region traces a sine wave-type pattern on its way to the detector. By adjusting the potentials on the rods, selected ions, a mass range, or only a single ion can be made stable and detected. The unstable ions impact one of the four rods, releasing them from the influence of the oscillating field, and are pumped away by the vacuum pumps. A quadrupole mass analyzer is commonly referred to as a **mass filter**, because, in principle, the device filters ions that achieve stability from those that do not.

Ion traps are essentially multidimensional quadrupole mass analyzers that store ions (trap) and then eject these trapped ions according to their m/z ratios. Once the ions are trapped, multiple stages of MS (MS^n) can be achieved, mass resolution can be increased, and sensitivity can be improved. The major difference between an ion trap and a quadrupole mass analyzer is that in an ion trap the unstable ions are ejected and detected while the stable ions are trapped (MS in time), whereas in a quadrupole, the ions with a stable flight path reach the detector, and the unstable ions hit the rods and are pumped away (MS in space).

Figure 26-4 shows the cross-sectional view of a 3-D ion trap mass analyzer. It consists of a ring electrode sandwiched between a perforated-entrance, end-cap electrode and a perforated-exit, end-cap electrode. An AC (RF) voltage and variable amplitude is applied to the ring electrode, producing a 3-D quadrupole field within the mass analyzer cavity.

Ions formed in the source are electronically injected into the ion trap, where they come under the influence of a time-varying RF field. The ions are trapped within the mass analyzer cavity, and the applied RF voltage drives ion motion in a figure eight toward the end-caps. Thus, for an ion to be trapped it must have a stable trajectory in both the axial and radial directions. To detect the ions, the frequency applied to the ring electrode is changed and the ion trajectories are made unstable. This results in the ions being ejected through the perforated end-caps, thus enabling them to reach the detector.

Helium is continuously infused into the ion trap cavity, and primarily serves as a dampening gas. Being lighter than any ions entering the trap (low-mass cut-off is normally m/z 50), it nonelastically collides with the ions entering the trap. This results in the helium molecules (not ions) absorbing kinetic energy from the ions entering the trap, effectively dampening and focusing the ions to the center of the trap, where they can be trapped more efficiently. This process also improves sensitivity and resolution (i.e., the ability to resolve a peak from its adjacent neighbors) in the trap. Recent developments in ion trap technology have resulted in **2-D ion traps**, which substantially increase ion-trapping volume by spreading the ion cloud in a quadrupole-like assembly. This allows for a higher number of ions to be stored simultaneously (increased space-charge capacity), thereby significantly increasing analytical sensitivity.

Time-of-flight mass analyzers separate ions according to the time required to reach the detector, while traveling over a known distance. Ions are pulsed from the source with the same kinetic energy, which causes ions of different m/z ratios to acquire different velocities (lighter ions travel faster while heavier ions travel slower). The difference in velocities translates to difference in time reaching the detector, upon which the mass spectrum is generated. Theoretically, TOF instruments have no upper mass range, which makes them useful for the analysis of biopolymers and large molecules, and have fast duty cycles since they technically transmit all m/z ions (full scan mode). The use of reflectrons (ion mirrors) can quickly increase mass resolution of TOF instruments by increasing ion drift path length by bouncing ions in a V or W pattern without drastically increasing the instrument footprint.

Fourier-transform-based mass analyzers deconvolute image currents produced by ion motion (harmonic oscillations or cyclotron motion) into mass spectra. A Fourier-transform ion cyclotron resonance mass analyzer traps ions in a magnetic field (Penning traps), while a Fourier-transform Orbitrap mass analyzer traps ions in an electric field (Kingdon trap). Both analyzer types are unique from the previously listed mass analyzers, because the ions themselves are never resolved in space or time, nor are they detected by impinging upon a detector. Rather, the frequency (cyclotron motion) is measured as a function of the applied electric (Orbitrap) or magnetic field (ICR). This results in sub part-per-million (ppm) mass accuracy measurements allowing determination of elemental composition. Extremely high resolution (>1 M at m/z 400) can be achieved with this type of mass analyzer.

26.3 INTERPRETATION OF MASS SPECTRA

As previously indicated, a **mass spectrum** is a plot (or table) of the intensity of various mass fragments (m/z) produced when a molecule is subjected to one of the many types of ionization techniques. The electron beam generated by a heated filament is used to ionize the molecules in classic GC–MS. It is usually kept at a constant potential of 70 eV because this produces sufficient ions without too much fragmentation, which would result in a loss of the higher molecular weight ions. Another advantage of using 70 eV for ionization is that the resulting mass spectra are usually very similar regardless of the make and model of the instrument. This allows for computer-assisted mass spectral matching to database libraries that help in unknown compound identification. In fact, most MSs now come with a MS spectral database and the required matching software.

Typical mass spectra include only positive fragments that usually have a charge of +1. Thus, the **mass-to-charge ratio** is the molecular mass of the fragment divided by +1, which equals the mass of the fragment. As yet, the mass-to-charge ratio unit has no name and is currently abbreviated by the symbol m/z (older books use m/e). In some publications, the unit of mass-to-charge ratio is allowed to be called the **Thomson** after the late J.J. Thomson, who constructed one of the first instruments for the determination of the m/z of ions.

A mass spectrum for butane is illustrated in Fig. 26-5. The relative abundance is plotted on the y-axis and the m/z is plotted on the x-axis. Each line on the bar graph represents an m/z fragment with the

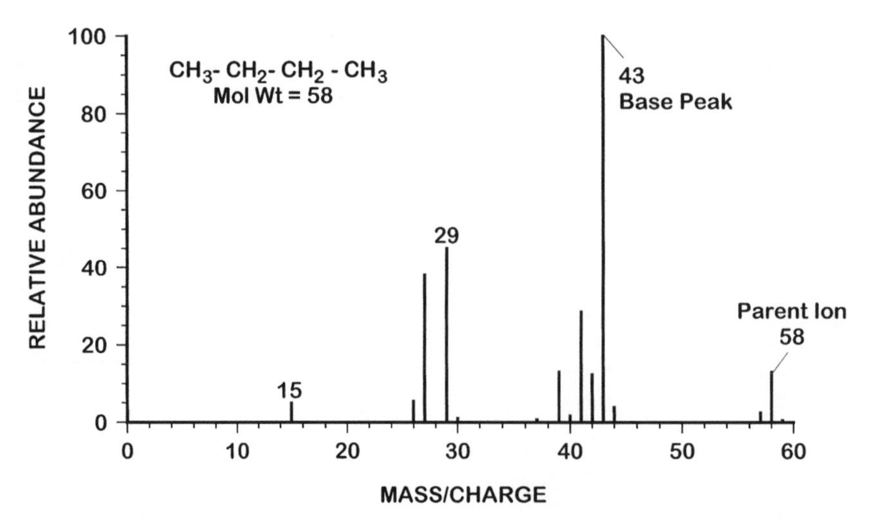

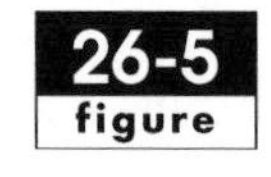

Mass spectrum of butane obtained by electron impact ionization.

abundance unique to a specific compound. The spectrum always contains what is called the **base peak** or **base ion**. This is the fragment (m/z) that has the highest abundance or intensity. When the signal detector is processed by the computer, the m/z with the highest intensity is taken to be 100%, and the abundance of all the other m/z ions is adjusted relative to the base peak. The base peak always will be presented as 100% relative abundance. Butane has the base peak at an m/z of 43.

Another important fragment is the **molecular ion** or **parent ion**, designated by the symbol $M^{+\bullet}$. This peak has the highest mass number and represents the positively charged intact molecule with a m/z equal to the molecular mass. The harsher ionizing techniques such as the EI shown here (Fig. 25-5) produces an ion (radical cation) at m/z 58 by stripping an electron. Because the mass of a single electron can be considered insignificant, the molecular ion produced by EI-type ionization is indicative of the molecular weight of that compound. All other molecular fragments originate from this charged species, so it is easy to see why it is called the molecular or parent ion. It is not always present, because, sometimes, the parent ion decomposes before it has a chance to traverse the mass analyzer. However, a mass spectrum is still obtained, and this becomes a problem only when determining the molecular mass of an unknown. The remainder of the mass spectrum is a consequence of the stepwise cleavage of large fragments to yield smaller ones termed **product ions** or **daughter ions**. The process is relatively straightforward for alkanes, such as butane, making possible identification of many of the fragments.

As indicated previously, the initial step in EI ionization is the abstraction of an electron from the molecule as electrons from the beam pass in close proximity. The equation below illustrates the first reaction that produces the positively charged product ion.

$M + e$	$\Rightarrow M^{+\bullet}$	$+2e$	
(from electron beam)	(molecular ion)	(one electron from the electron beam and one from the molecular ion, M)	[1]

The M symbolizes the unionized molecule as it reacts with the electron beam and forms a radical cation. The cation will have an m/z equal to the molecular weight. The parent ion then sequentially fragments in a unimolecular fashion. (Note that the product ion is often written as M^+, for which the free electron, symbolized by the dot, is assumed. Regardless, the molecule has lost one electron and still retains all the protons; thus, the net charge must be positive.) The reactions of butane as it forms several of the predominant product ion (daughter) fragments are shown below.

$$CH_3-CH_2-CH_2-CH_3+e \Rightarrow CH_2-CH_2-CH_2-CH_3{}^{+\bullet} \quad (m/z=58)+2e \qquad [2]$$

$$CH_3-CH_2-CH_2-CH_3{}^{+\bullet} \Rightarrow CH_3-CH_2-CH_2-CH_2{}^{+} \quad (m/z=57)+{}^{\bullet}H \qquad [3]$$

$$CH_3-CH_2-CH_2-CH_3{}^{+\bullet} \Rightarrow CH_3-CH_2-CH_2{}^{+} \quad (m/z=43)+{\bullet}CH_3 \qquad [4]$$

$$CH_3-CH_2-CH_2-CH_3{}^{+\bullet} \Rightarrow CH_3-CH_2{}^{+\bullet} \quad (m/z=29)+{\bullet}CH_3-CH_3 \qquad [5]$$

$$CH_3-CH_2 \Rightarrow CH_3{}^{+\bullet} \quad (m/z=15)+{\bullet}CH_2 \qquad [6]$$

Many of the fragments for butane result from direct cleavage of the methylene groups. With alkanes, you will always see fragments in the mass spectrum that are produced by the sequential loss of CH_2 or CH_3 groups.

Close examination of the butane mass spectrum in Fig. 26-5 reveals a peak that is 1 m/z unit larger than the molecular ion at $m/z = 58$. This peak is designated by the symbol $M+1$ and is due to the naturally occurring isotopes. The most abundant isotope of carbon has a mass of 12; however, a small amount of ^{13}C is also present (1.11%). Any ions that contained a ^{13}C or a deuterium isotope would be 1 m/z unit larger, although the relative abundance would be low.

Another example of MS fragmentation patterns is shown for methanol in Fig. 26-6. Again, the fragmentation pattern is straightforward. The molecular ion $(CH_3-OH^{+\bullet})$ is at a m/z of 32, which is the molecular weight. Other fragments include the base peak at a m/z of 31 due to CH_2-OH^+, the CHO^+ fragment at a m/z of 29, and the $CH_3{}^+$ fragment at a m/z of 15.

So far, only EI types of ionization have been discussed. Another common fragmentation method is **chemical ionization** (CI). In this technique, a gas is ionized, such as methane (CH_4), which then directly ionizes the molecule. This method is classified as a **soft ionization** because only a few fragments are produced. The most important use of CI is the determination of the molecular ion since there is usually a fragment that is 1 m/z unit larger than that obtained with EI. Thus, a mass spectrum of butane taken by the CI method would have a quasimolecular (parent) ion at $m/z = 59(M+H)$. Many LC–MS interfaces use CI methods and so it is common to see $(M+H)^+$ ions, where they are often called precursor ions (i.e., they produce the ion fragments or daughters). As can be seen, the reactions of the cleavage process can be quite involved. Many of the reactions are covered in detail in the books by McLafferty and Turecek and by Davis and Frearson listed in Sect. 26.9.

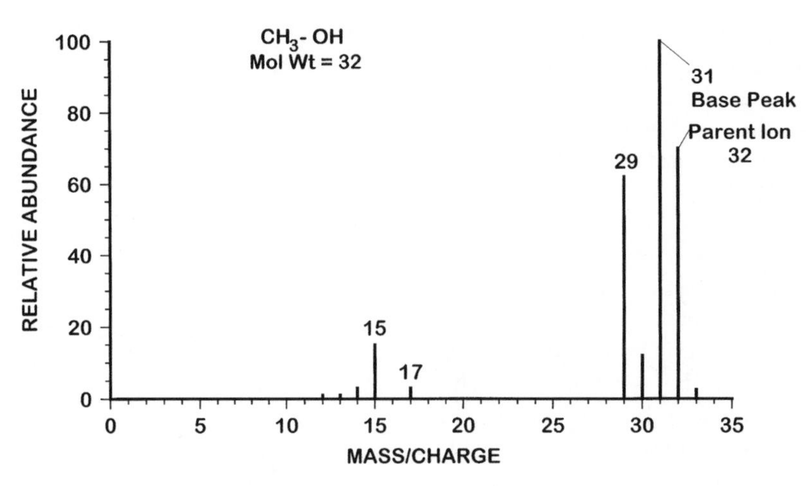

26-6 figure Mass spectrum of methanol obtained by electron impact ionization.

26.4 GAS CHROMATOGRAPHY–MASS SPECTROMETRY

Although samples can be introduced directly into the MS ion source, many applications require chromatographic separation before analysis. The rapid development of **gas chromatography–mass spectrometry** (GC–MS) have allowed for the coupling of the two methods for routine separation problems (see Chap. 29). An MS coupled to GC allows the peaks to be identified or confirmed, and, if an unknown is present, it can be identified using a computer-assisted search of a library containing known MS spectra. Another critical function of GC–MS is to ascertain the purity of each peak as it elutes from the column. Does the material eluting in a peak contain one compound or is it a mixture of several compounds that just happen to coelute with the same retention time?

In most cases, a **capillary GC column** is connected directly to the MS source via a **heated capillary transfer line**. The transfer line is kept hot enough so as to avoid condensation of the volatile component eluting from the GC column on its way into the low-pressure MS source. The sample flows through the GC column into the interface and then on to be processed by the MS. A computer is used to store and process the data from the MS.

An example of the power of GC–MS is shown below in the separation of the methyl esters of several long-chain fatty acids (Fig. 26-7). Long-chain fatty acids must have the carboxylic acid group converted or blocked with a methyl group to make them volatile. Methyl esters of palmitic (16:0), oleic (18:1), linoleic (18:2), linolenic (18:3), stearic (18:0), and arachidic (20:0) acids were injected onto a column that was supposed to be able to separate all the naturally occurring fatty acids. However, the GC tracing showed only four peaks, when it was known that six different methyl esters were in the sample. The logical explanation is that one or two of the peaks contain a mixture of methyl esters resulting from poor resolution on the GC column.

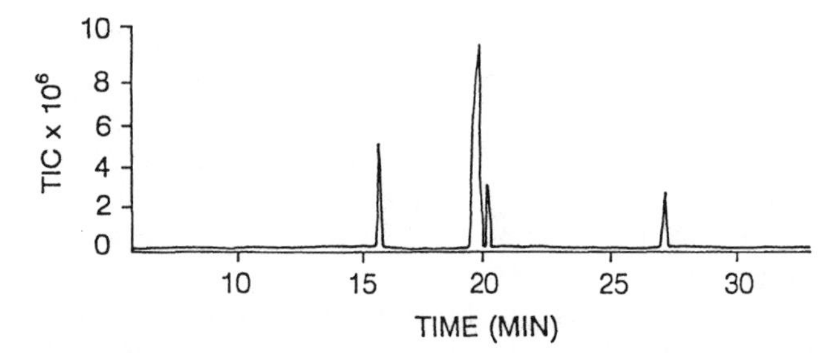

26-7 figure Total ion current GC chromatogram of the separation of the methyl esters of six fatty acids. Detection is by electron impact ionization using a direct capillary interface.

The purity of the peaks is determined by running the GC–MS and taking mass spectra at very short increments of time (1 s or less). If a peak is pure, then the mass spectra taken throughout the peak should be the same. In addition, the mass spectrum can be compared with the library of spectra stored in the computer.

The **total ion current** (TIC) **chromatogram** of the separation of the fatty acid methyl esters is shown in Fig. 26-7. There are four peaks eluting off the column between 15.5 and 28 min. The first peak at 15.5 min has the same mass spectrum throughout, indicating that only one compound is eluting. A computer search of the MS library gives an identification of the peak to the methyl ester of palmitic acid. The mass spectra shown in Fig. 26-8 compare the material eluting from the column to the library mass spectrum.

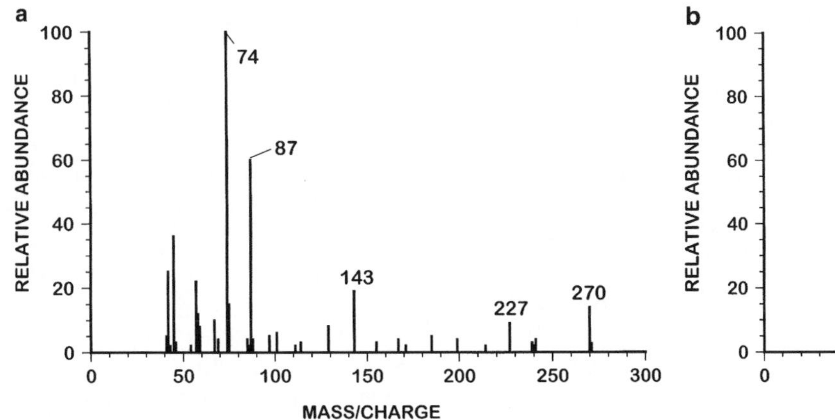

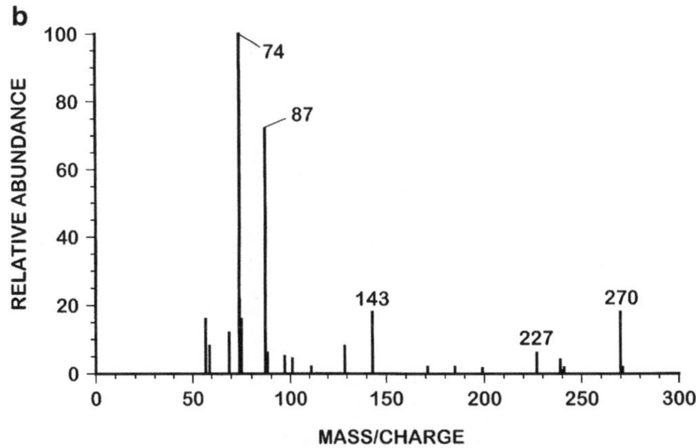

26-8 figure Mass spectra of (**a**) the peak at 15.5 min in the TIC chromatogram shown in Fig. 26-7 and (**b**) the methyl ester of palmitic acid from a computerized MS library.

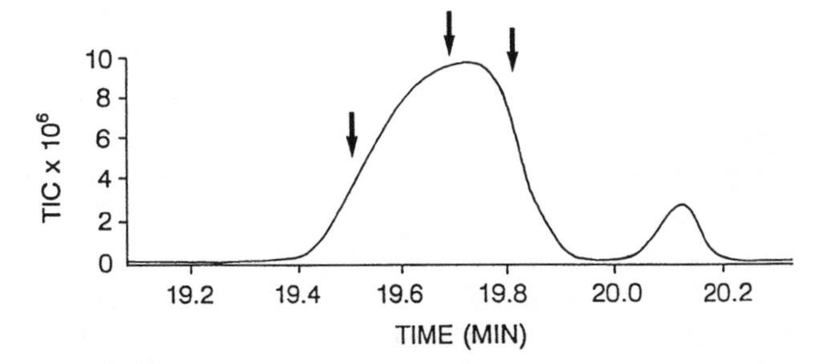

26-9 figure Enlargement of the region 19.2–20.2 min from the TIC chromatogram shown in Fig. 26-7. Arrows indicate where mass spectra were obtained.

Most of the fragments match, although the GC–MS scan does have many small fragments not present on the library mass spectrum. This is common background noise and usually does not present a problem. The data from the rest of the chromatogram indicate that the peaks at both 20 and 27 min contain only one component. The computer match identifies the peak at 20 min as stearic acid, methyl ester, and the peak at 27 min as arachidic acid, methyl ester. However, the peak located at 19.5 min is shown to have several different mass spectra, indicating impurity or coeluting compounds.

In Fig. 26-9, the region around 19 min has been enlarged. The arrows indicate where different mass spectra were obtained. The computer identified the material in the peak at 19.5 min as linoleic acid, methyl ester; the material at 19.7 min as oleic acid, methyl ester; and the material at 19.8 min as linolenic acid, methyl ester. Thus, as we originally suspected, several of the methyl esters were coeluting off the GC column. This example illustrates the tremendous power of GC–MS used in both a quantitative and a qualitative manner.

26.5 LIQUID CHROMATOGRAPHY–MASS SPECTROMETRY

Years ago the only way to obtain a decent mass spectrum of material from HPLC separations was to collect fractions, evaporate off the solvent, and introduce the sample into a conventional MS by direct injection or direct probe. Although this method was sometimes adequate, the direct on-line coupling of the two instruments was a tremendous advantage in terms of time and ease of operation.

For a **high-performance liquid chromatography–mass spectrometry** (LC–MS) interface, the same overall requirements must be met as for GC–MS. There must be a way to remove the excess solvent, while converting a fraction of the liquid effluent into the gas phase, making it amenable for MS analysis. Furthermore most compounds analyzed by HPLC are either nonvolatile or thermally labile, making the task of liquid-to-gas phase transition even more challenging, especially while maintaining compound integrity.

How does LC–MS work? A modern LC–MS ionization interface converts liquid (LC eluent) into gas phase ions (sampled by the MS) by a process of desolvation in the presence of a highly charged electrical field at atmospheric pressure. The energy applied to evaporate the solvent (thermal and electrical) is almost completely used in the desolvation process, and does not contribute to degradation (usually thermal) of any labile species present in the LC eluant. Of the many different types of LC–MS ionization interfaces developed over the years, i.e., Moving Belt, Thermospray, and Particle Beam, it was the development of the atmosphere pressure-based ionization interfaces, **electrospray** (ESI), and **atmospheric pressure chemical ionization** (APCI) that made LC–MS a routine technique. Recently, a technique known as **atmospheric**

pressure photoionization (APPI) has been developed as a complementary technique to APCI.

26.5.1 Electrospray Interface

Electrospray, the most popular LC–MS technique in use today, was developed by John Fenn and coworkers in 1984. For his effort, John Fenn was awarded the 2002 Nobel Prize in chemistry, which he shared with Koichi Tanaka and Kurt Wuthrich. Electrospray is a highly sensitive technique with limits of detection in the low femtogram range and a linear dynamic range between 3 and 4 orders of magnitude. Normally, polar compounds are amenable to ESI analysis with the type of ion produced depending on the initial charge. That is, positively charged compounds yield positive ions, while negatively charged compounds such as those containing free carboxylic acid functional groups will produce negative ions.

The ESI source as depicted in Fig. 26-10 consists of a nozzle that contains a fused-silica capillary sample tube (serves to transfer the LC effluent) coaxially positioned within a metal capillary tube to which a variable electrical potential can be applied against a counter-electrode, which is usually the entrance to the MS. Compressed nitrogen gas at high velocity is coaxially introduced to aid in the nebulization of the LC effluent as it exits the tip of the metal capillary tube. The relative velocity difference between the streams of nitrogen gas (fast moving) and LC effluent (slow moving) at the ESI tip, results in producing a fine spray of highly charged droplets. At nanoflow rates (<1 μL/min), the force of the electrical field is strong enough to break-up the LC effluent into fine droplets without the use of nebulizing gas, in a process known as nanospray. For conventional HPLC flow-rates (1–1000 μL/min), the sheer volume of liquid requires an initial droplet size reduction through the use of nebulizing nitrogen gas, creating the required microdroplets, which can now be influenced by the prevailing electrical field.

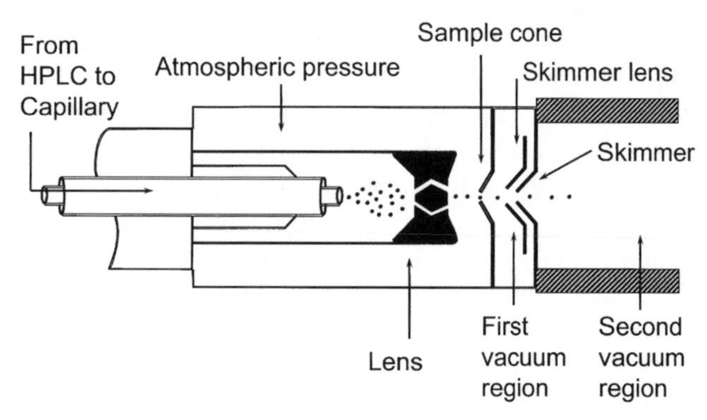

26-10 figure Schematic of an electrospray LC–MS interface.

At this point, the repulsive forces due to the accumulation of "like" charges inside the rapidly reducing microdroplet volume, creates an imbalance with the forces of surface tension that are trying to conserve the spherical structure of the microdroplet. The positive charge is drawn out, but cannot escape the surface of the liquid, and forms what is known as a **Taylor cone**. Further reduction of the diameter of the droplets causes the Taylor cone to stretch to a critical point, at which the charge escapes the liquid surface, and is emitted as a gas phase ion in a process known as a **coulombic explosion**. Two theories describe the ion formation process: one proposed by Dole in the late 1960s is known as the "single ion in droplet" theory, which describes a process of droplet fissure until the microdroplet contains one charge that it emits into the gas phase, while the theory proposed by Iribarne and Thompson in the late 1970s describes the spontaneous emission of multiple gas phase ions from the droplet surface upon the droplet reaching a certain critical diameter. Both theories currently prevail with a common emphasis on droplet size reduction as a prerequisite for the emission of gas phase ions.

One of the many advantages of the ESI process is its ability to generate multiple-charged ions and tolerate conventional HPLC flow rates. Proteins and other large polymers (e.g., between 2000 and 70,000 Da) can be easily analyzed on LC–MS systems having a mass limit of m/z 2000, due to multiple charging phenomenon. For example, Interleukin-8 (rat) has a mass of 7845.3 Da, but develops up to +8 charges, with the most abundant ion appearing at the +4 charge state (i.e., 7849.2 ÷ 4 charges = 1962.3 m/z), and thus can be analyzed on an LC–MS system having a mass limit of m/z 2000. Powerful software can process in excess of +50 charge states, to yield the molecular ion information for larger proteins. A limitation of the ESI process is the phenomenon of ion suppression or matrix effects, which usually causes a reduction in response for the analyte signal intensity in presence of matrix components. Matrix factor corrections are used to account for ion suppression effects, including the use of stable labeled internal standards for quantitative analysis. The ESI interface is more susceptible to matrix effects (usually ion suppression) than compared to the APCI interface discussed in Sect. 26.5.2.

26.5.2 Atmospheric Pressure Chemical Ionization

The **APCI interface** is normally used for compounds of low polarity and some volatility. It is harsher than ESI and is a gas-phase ionization technique. Therefore

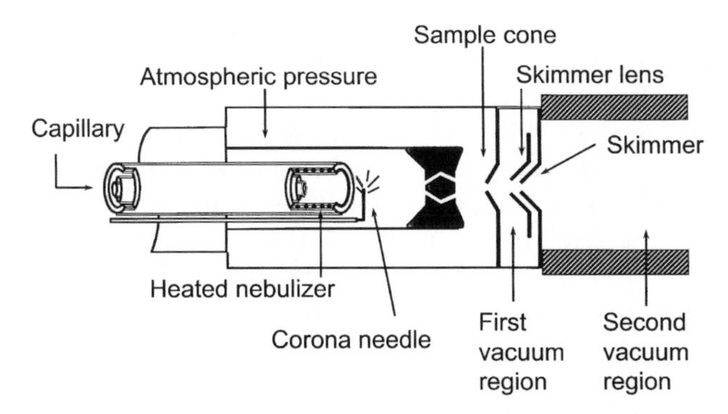

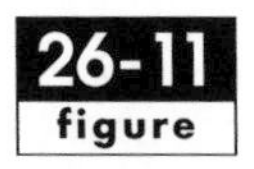
26-11 figure Schematic of an atmospheric pressure chemical ionization LC–MS interface.

gas-phase chemistries of the analyte and solvent vapor play an important part in the APCI process.

Figure 26-11 shows the schematic diagram of an APCI interface. The LC effluent-carrying fused-silica capillary tube protrudes about halfway inside a silicon-carbide (ceramic) vaporizer tube. The vaporizer tube is maintained at approximately 400–500 °C, and serves to vaporize the LC effluent. High voltage is applied to a corona needle positioned near the exit of the vaporizer tube. The high voltage creates a **corona discharge** that forms reagent ions from the mobile phase and nitrogen nebulizing gas. These ions react with the sample molecules (M) and convert them to ions. A common cascade of reactions occurring in the presence of water, nitrogen gas, and the high-voltage corona discharge is as follows:

$$e^- + N_2 \rightarrow N_2^{+\bullet} + 2e^- \quad [7]$$

$$N_2^{+\bullet} + H_2O \rightarrow N_2 + H_2O^{+\bullet} \quad [8]$$

$$H_2O^{+\bullet} + H_2O \rightarrow H_3O^+ + OH^\bullet \quad [9]$$

$$M + H_3O^+ \rightarrow (M+H)^+ + H_2O \quad [10]$$

The APCI interface is a robust interface and can handle high flow rates of up to 2 ml/min. It is unaffected by minor changes in buffer strength or composition, and is typically used to analyze molecules less than 2,000 Da. It does not facilitate multiple charges, and hence cannot be used to analyze large biomolecules/polymers. In terms of matrix effects, APCI usually shows "ion enhancement" rather than "ion suppression." This is due to the matrix components enriching the plasma generation process, thereby enhancing the efficiency of the ionization process. As a result, there is an increase in response for the analyte signal in the presence of matrix components, requiring matrix factor correction through the appropriate use of stable labeled internal standards for quantitative analysis.

Improvements in LC–MS interfaces are continuously being made, although no universal interface for the various types of compounds is available yet. The APPI interface, which uses a krypton or xenon light source to generate a beam of photons instead of a corona discharge-generated plasma as in APCI, is an example of improvements made in LC–MS interfaces. Compounds having ionization potentials lower than the wavelength of the light source will be ionized. Since most HPLC solvents do not ionize at the wavelengths generated by the commonly used photon sources, improvements in the signal-to-noise ratio and hence detection limits are possible using APPI.

26.5.3 Tandem Mass Spectrometry

The use of **tandem MS** (MS/MS, MS^n) is used in GC–MS and is especially helpful in LC–MS, for which analysis is often limited by the lack of a suitable detector. Use of the MS/MS mode of detection is often the only way to detect and confirm compounds eluting off the column. For example, fatty acids can be measured directly by HPLC, but unless the concentrations are high, an ultraviolet or refractive index detector will not pick them up. LC–MS/MS will allow for the assay of trace amounts present in effluent. LC–MS/MS (TSQ) has become the most commonly used quantitative analysis of nonvolatile pesticides, amino acids, antibiotics, steroids, proteins, lipids, and sugars, and is often the preferred method for analysis for regulatory agencies.

26.6 APPLICATIONS

The use of MS in the field of food science is well established and growing rapidly as food exports from Asia increase yearly to the US and Europe. While GC–MS has been used for years, lower-priced LC–MS/MS instrumentation has become indispensible for the analysis of compounds such as chloramphenicol, nitrofurans, sulfonamides, tetracyclines, melamine, acrylamide, and malachite green in foods such as honey, fish, shrimp, and milk (see Chap. 18). Agencies such as the Food and Drug Administration (FDA), Center for Food Safety and Applied Nutrition (CFSAN), European Food Safety Authority (EFSA), Health Canada, and Japan Food Safety Commission use MS-based techniques to drive regulatory standards for banned substances, safe-guarding the food supply for human consumption.

To give an appreciation of the usefulness of LC–MS, several examples are provided below. It is important to remember that there are a wide variety of methods now available to analyze just about any type of sample in a variety of matrices.

Due to the prevalence of consumption of caffeine-containing drinks throughout the world, the analysis of this small bioactive compound has been of interest for many years. Over 20 years ago, HPLC methods were published showing that caffeine and other alkaloids, theobromine and theophylline, could be analyzed by HPLC using an ultraviolet detector. While HPLC–UV analysis is quite acceptable, the use of LC–MS can verify and enhance identification in a variety of complex food systems.

Figure 26-12 illustrates a reversed-phase HPLC column separation and MS spectrum obtained using the ESI interface coupled with MS/MS. An aqueous coffee extract was filtered and separated by HPLC using an acetic acid–acetonitrile mobile phase. For comparative purposes, a separate HPLC separation was achieved with the same HPLC column except detection was with a UV detector. The HPLC chromatogram in Fig. 26-12a shows the TIC, an indicator of total ions and thus compounds eluting, and matches the HPLC-UV chromatogram (not shown). Figure 26-12b is the selected ion trace of ions $m/z = 180.7–181.7$, which would correspond to the protonated molecular ions $(M + H)^+$ of theobromine and theophylline at 181.2 (both compounds are isomers and have identical masses of 180.2). The chromatogram in Fig. 26-12c is the selected ion trace of $m/z = 194.2–196.2$, which corresponds to the protonated molecular ion of caffeine (195.2). The MS/MS of caffeine and theobromine are presented in Fig. 26-12d, e and show the protonated molecular ions for both caffeine and theobromine, and several ion fragment m/z.

The recent **melamine** food contamination issue of 2007–2008 is another excellent example of the use of LC–MS and MS/MS in both as a detective role and later as an official analytical method. Melamine is a six-membered cyclic nitrogenous ring compound with three amines attached to the carbons in the ring and

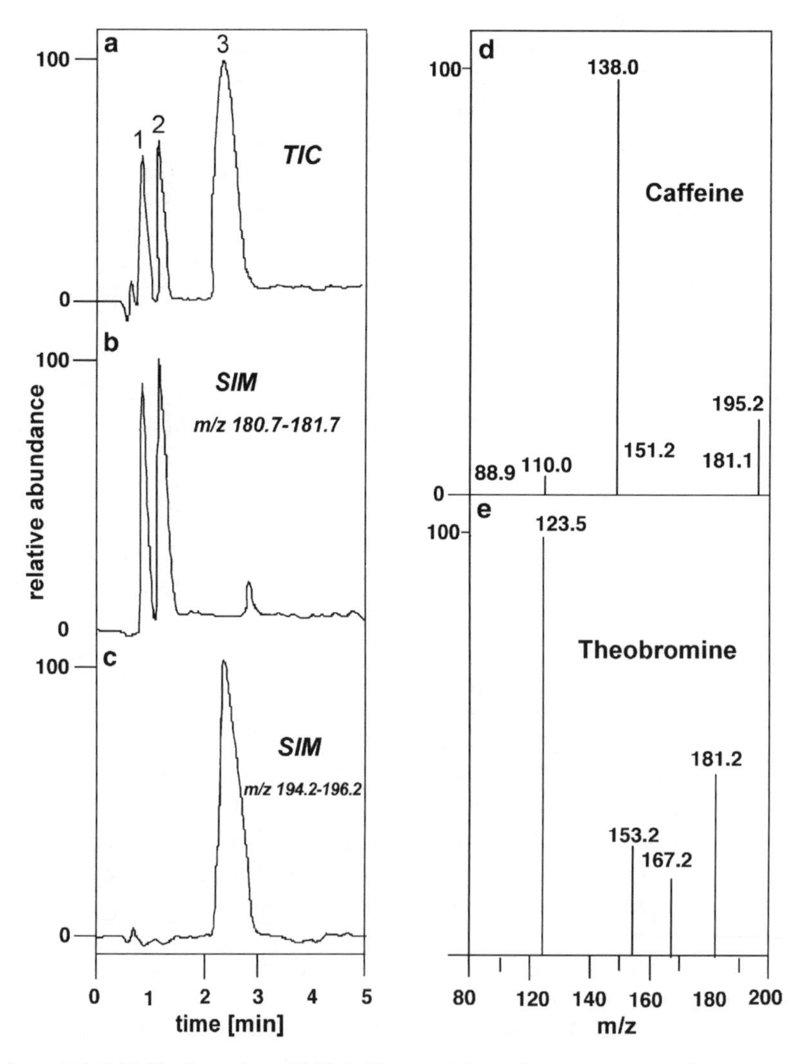

Reversed-phase LC–MS (Parts **a–c**) and MS/MS separation of an aqueous coffee extract: (**a**) TIC, total ion current of extract; 1-theobromine, 2-theophylline, 3-caffeine; (**b**) selected ion trace, $m/z = 180.7–181.7$; (**c**) selected ion trace, $m/z = 194.2–196.2$; (**d**) MS–MS ionization of caffeine; (**e**) MS/MS of theobromine. [From Reference (1), used with permission of Elsevier, New York.]

contains 67% nitrogen (Fig. 26-13). Since most major test for protein measure nitrogen, the compound was used as an economic adulterant in wheat gluten and dried milk powder to artificially increase the apparent protein content. Due to the polar nature of the amine groups, melamine is not volatile and thus cannot be analyzed with GC unless a derivative is synthesized, thus LC–MS is a preferred method. Methods have been presented that entail the separation of melamine by HPLC with detection by ESI. ESI produces a strong protonated molecular ion at $m/z = 127.1$ with MS/MS fragments of 127.1 and 85.1. One of the FDA suggested analytical methods entail monitoring of all these ions, which produces very good specificity and sensitivity.

The area of bioactive foods components has grown dramatically over the last 10 years in part due to the availability of LC–MS. Many bioactive compounds in fruits, vegetables, and spices are polar and are not volatile. Thus identification and evaluations were very difficult without HPLC methods coupled with MS.

A good example of the use of LC–MS is in the measurement of the polyphenolic flavonoids called the catechins. These antioxidant compounds present in catechu, green tea, cocoas, and chocolates appear to have several beneficial biological effects including enhanced heart and blood vessel health.

Figure 26-14 shows a separation of the major green tea catechins by HPLC with ESI–MS detection. At the bottom of the figure is depicted the chromatogram showing the separation of the catechins by HPLC. The ESI–MS detector was used to monitor all ions from m/z 120–2,200 (TIC mode). The top left panel show the MS obtained for the epicatechin peak eluting at about 12.1 min. As typical with ESI, there are few fragments, though the M + H molecular ion at 291.3 m/z is predominant. Further ionization (MS/MS) of the epicatechin yields two major protonated fragments, m/z 273.3 (loss of OH from the C ring) and 139.3 (oxidation and cleavage of the A ring). With these data it is possible to elucidate isomers and also possible degradation pathways.

mol mass =126.12

26-13 figure Chemical structure of melamine.

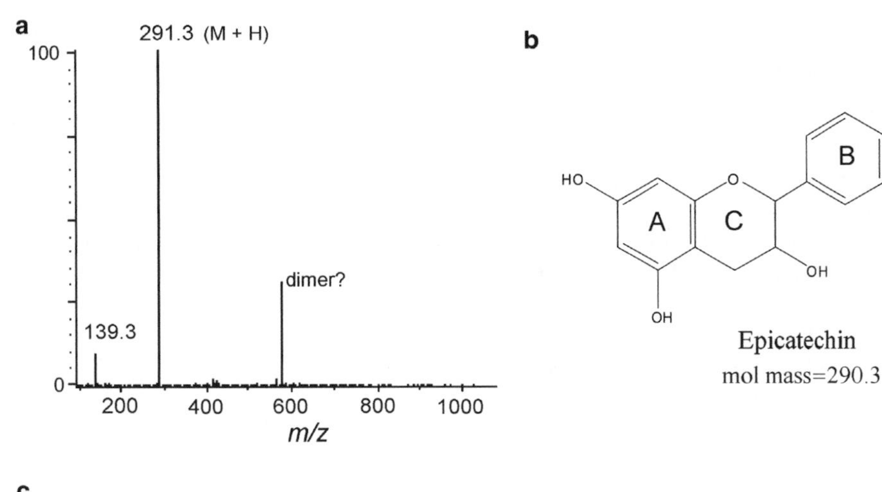

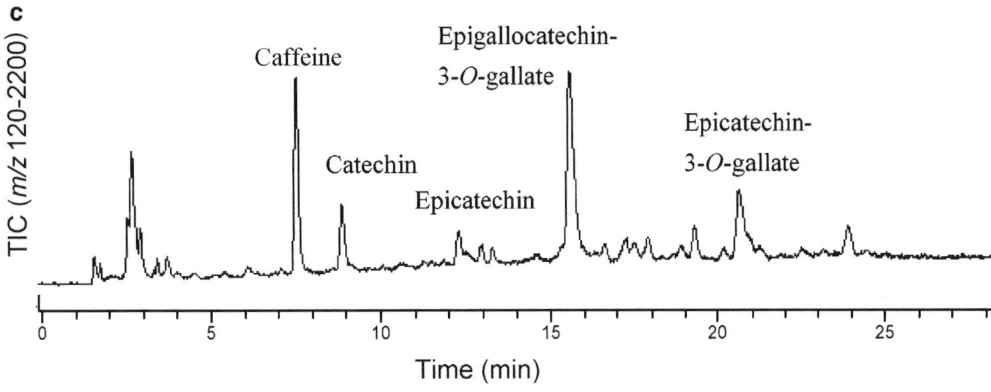

LC–ESI–MS of green tea leaf extracts. (**a**) ESI mass spectra of epicatechin. (**b**) chemical structure of epicatechin. (**c**) TIC scanned from m/z 120 to 2200. [From Reference (2), used with permission of American Chemical Society, Washington, DC.]

There are many different applications of MS in food science. In considering the application of GC–MS or LC–MS in food systems, note that if a compound can be separated by a GC or LC method, then chances are good that a MS can be used. For years, the MS was used only in a qualitative manner, to check the purity of eluting peaks or for compound identification. With smaller units, the use of MS as a universal detector has gained wide acceptance. The advantage of utilizing the MS as a detector is that only certain ions need to be monitored, which makes it a selective detector. This technique currently is used extensively for screening of pesticide residues and highly sensitive quantitative analysis of banned substances.

26.7 SUMMARY

MS is fairly simple when examined closely. The basic requirements are to (1) somehow get the sample into an ionizing chamber where ions are produced, (2) separate the ions formed by magnets, quadrupoles, drift-tubes, electric fields, (3) detect the m/z, and (4) output the data to some type of computer.

Since the qualitative and quantitative aspects of MSs are so powerful, they are routinely coupled to GCs and HPLCs. The interface for GCs is versatile and easy to use; however, the extensive sample preparation required for GC–MS analysis makes its utility cumbersome. Far simpler sample preparation procedures, wider ionization ranges for different classes of compounds, faster analysis times, routine high sensitivity, access to accurate mass capability, and the advent of ultrahigh-pressure chromatography (UPLC) has greatly increased the adoption of LC–MS as an analytical technique for food analysis.

LC–MS is now becoming routine, although it is still difficult to compare results when using different types of interfaces and it remains susceptible to ion suppression effects. Matrix effects or ion suppression is the difference in response observed for the target compound as an influence of matrix components on the ionization process. Matrix effects are commonly observed for ESI, while APCI is less susceptible due to its inherently different ionization process. The use of LC–MS has grown rapidly, since the MS is a universal detector for both qualitative and quantitative information.

26.8 STUDY QUESTIONS

1. What are the basic components of a MS?
2. What are the unique aspects of data that a MS provides? How is this useful in the analysis of foods?
3. What is EI ionization? What is CI ionization?
4. What is the base peak on a mass spectrum? What is the molecular ion peak?
5. What are the major ions (fragments) expected in the EI mass spectrum of ethanol ($CH_3 - CH_2 - OH$)?
6. What are the major differences in how ionization occurs in the electrospray versus the APCI interface? What is ion suppression?
7. What are the major differences between the quadrupole, ion trap, time of flight, and Fourier transform mass analyzer? What are the advantages of using each analyzer? What is especially unique about a Fourier transform-based mass analyzer?

26.9 RESOURCE MATERIALS

1. Barker J (1999) Mass spectrometry, 2nd edn. Wiley, New York. One of the best introductory texts on mass spectrometry in its second edition. The author starts at a very basic level and slowly works through all aspects of MS, including ionization, fragmentation patterns, GC–MS, and LC–MS.
2. Ho C-T, Lin J-K, Shahidi F (eds) (2009) Tea and tea products. CRC, Boca Raton, FL. A good review of current literature on the chemistry and health-promoting properties of tea. Includes several chapters that discuss analytical methods for analyzing bioactive compounds and flavonoids in teas.
3. Lee TA (1998) A beginner's guide to mass spectral interpretation. Wiley, New York. A good basic introduction to Mass Spectrometry with many practical examples.
4. McLafferty FW, Turecek F (1993) Interpretation of mass spectra, 4th edn. University Science Books, Sausalito, CA. The fourth edition of an essential classic book on how molecules fragment in the ion source. Contains many examples of different types of molecules.
5. Niessen WMA, van der Greef J (1992) Liquid chromatography–mass spectrometry. Marcel Dekker, New York. A thorough, though somewhat dated, review of all the LC–MS methods and interfaces.
6. Silverstein RM, Webster FX (1997) Spectrometric identification of organic compounds, 6th edn. Wiley, New York. An introductory text for students in organic chemistry. Chapter 2 presents mass spectrometry in an easily readable manner. Contains many examples of the mass spectra of organic compounds.

26.10 REFERENCES

1. Huck CW, Guggenbichler W, Bonn GK (2005) Analysis of caffeine, theobromine and theophylline in coffee by near infrared spectroscopy (NIRS) compared to high-performance liquid chromatography (HPLC) coupled to mass spectrometry. Analytica Chimica Acta 538 (1–2):195–203
2. Shen D, Wu Q, Wang M, Yang Y, Lavoie EJ, Simon JE (2006) Determination of the predominant catechins in *Acacia catechu* by liquid chromatography/electrospray ionization-mass spectrometry. J Agric Food Chem 54(9):3219–3224

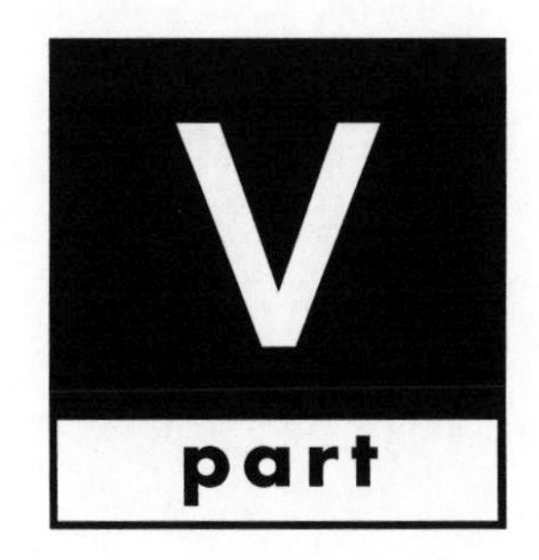

Chromatography

Basic Principles of Chromatography

*Baraem Ismail**
Department of Food Science and Nutrition, University of Minnesota, St. Paul, MN 55108-6099, USA
bismailm@umn.edu

and

S. Suzanne Nielsen
Department of Food Science, Purdue University, West Lafayette, IN 47907-2009, USA
nielsens@purdue.edu

27.1 Introduction 475
27.2 Extraction 475
27.2.1 Batch Extraction 475
27.2.2 Continuous Extraction 475
27.2.3 Countercurrent Extraction 475
27.3 Chromatography 475
27.3.1 Historical Perspective 475
27.3.2 General Terminology 476

S.S. Nielsen, *Food Analysis*, Food Science Texts Series, DOI 10.1007/978-1-4419-1478-1_27,

27.3.3 Gas Chromatography 476
27.3.4 Liquid Chromatography 477
27.3.4.1 Paper Chromatography 477
27.3.4.2 Thin-Layer Chromatography 478
27.3.4.2.1 General Procedures 478
27.3.4.2.2 Factors Affecting Thin-Layer Separations 478
27.3.4.3 Column Liquid Chromatography 479
27.3.5 Supercritical Fluid Chromatography 480
27.4 Physicochemical Principles of Chromatographic Separation 481
27.4.1 Adsorption (Liquid–Solid) Chromatography 481
27.4.2 Partition (Liquid–Liquid) Chromatography 482
27.4.2.1 Introduction 482
27.4.2.2 Coated Supports 483
27.4.2.3 Bonded Supports 483
27.4.3 Ion-Exchange Chromatography 483
27.4.4 Size-Exclusion Chromatography 485
27.4.5 Affinity Chromatography 488
27.5 Analysis of Chromatographic Peaks 489
27.5.1 Separation and Resolution 490
27.5.1.1 Developing a Separation 490
27.5.1.2 Chromatographic Resolution 491
27.5.1.2.1 Introduction 491
27.5.1.2.2 Column Efficiency 492
27.5.1.2.3 Column Selectivity 494
27.5.1.2.4 Column Capacity Factor 494
27.5.2 Qualitative Analysis 495
27.5.3 Quantitative Analysis 495
27.6 Summary 496
27.7 Study Questions 497
27.8 Acknowledgments 498
27.9 References 498

27.1 INTRODUCTION

Chromatography has a great impact on all areas of analysis and, therefore, on the progress of science in general. Chromatography differs from other methods of separation in that a wide variety of materials, equipment, and techniques can be used. [Readers are referred to references (1–19) for general and specific information on chromatography.]. This chapter will focus on the principles of chromatography, mainly **liquid chromatography** (LC). Detailed principles and applications of **gas chromatography** (GC) will be discussed in Chap. 29. In view of its widespread use and applications, high-performance liquid chromatography (HPLC) will be discussed in a separate chapter (Chap. 28). The general principles of extraction are first described as a basis for understanding chromatography.

27.2 EXTRACTION

In its simplest form, extraction refers to the transfer of a solute from one liquid phase to another. Extraction in myriad forms is integral to food analysis – whether used for preliminary sample cleanup, concentration of the component of interest, or as the actual means of analysis. Extractions may be categorized as **batch**, **continuous**, or **countercurrent** processes. (Various extraction procedures are discussed in detail in other chapters: traditional solvent extraction in Chaps. 8, 18, and 29; accelerated solvent extraction in Chap. 18; solid-phase extraction in Chaps. 18 and 29; and solid-phase microextraction and microwave-assisted solvent extraction in Chap. 18).

27.2.1 Batch Extraction

In **batch extraction** the solute is extracted from one solvent by shaking it with a second, immiscible solvent. The solute **partitions**, or distributes, itself between the two phases and, when equilibrium has been reached, the **partition coefficient**, K, is a constant.

$$K = \frac{\text{Concentration of solute in phase 1}}{\text{Concentration of solute in phase 2}} \quad [1]$$

After shaking, the phases are allowed to separate, and the layer containing the desired constituent is removed, for example, in a separatory funnel. In batch extraction, it is often difficult to obtain a clean separation of phases, owing to emulsion formation. Moreover, partition implies that a single extraction is usually incomplete.

27.2.2 Continuous Extraction

Continuous liquid–liquid extraction requires special apparatus, but is more efficient than batch separation. One example is the use of a Soxhlet extractor for extracting materials from solids. Solvent is recycled so that the solid is repeatedly extracted with fresh solvent. Other pieces of equipment have been designed for the continuous extraction of substances from liquids, and different extractors are used for solvents that are heavier or lighter than water.

27.2.3 Countercurrent Extraction

Countercurrent distribution refers to a serial extraction process. It separates two or more solutes with different partition coefficients from each other by a series of partitions between two immiscible liquid phases. Liquid–liquid partition chromatography (Sect. 27.4.2), also known as countercurrent chromatography, is a direct extension of countercurrent extraction. Years ago the countercurrent extraction was done with a "Craig apparatus" consisting of a series of glass tubes designed such that the lighter liquid phase (**mobile phase**) was transferred from one tube to the next, while the heavy phase (**stationary phase**) remained in the first tube (4). The liquid–liquid extractions took place simultaneously in all tubes of the apparatus, which was usually driven electromechanically. Each tube in which a complete equilibration took place corresponded to one theoretical plate of the chromatographic column (refer to Sect. 27.5.2.2.1). The greater the difference in the **partition coefficients** of various substances, the better was the separation. A much larger number of tubes was required to separate mixtures of substances with close partition coefficients, which made this type of countercurrent extraction very tedious. Modern **liquid–liquid partition chromatography** (Sect. 27.4.2) that developed from this concept is much more efficient and convenient.

27.3 CHROMATOGRAPHY

27.3.1 Historical Perspective

Modern chromatography originated in the late nineteenth and early twentieth centuries from independent work by David T. Day, a distinguished American geologist and mining engineer, and Mikhail Tsvet, a Russian botanist. Day developed procedures for fractionating crude petroleum by passing it through Fuller's earth, and Tsvet used a column packed with chalk to separate leaf pigments into colored bands.

Because Tsvet recognized and correctly interpreted the chromatographic processes and named the phenomenon **chromatography**, he is generally credited with its discovery.

After languishing in oblivion for years, chromatography began to evolve in the 1940s due to the development of column partition chromatography by Martin and Synge and the invention of paper chromatography. The first publication on GC appeared in 1952. By the late 1960s, GC, because of its importance to the petroleum industry, had developed into a sophisticated instrumental technique, which was the first instrumental chromatography to be available commercially. Since early applications in the mid-1960s, HPLC, profiting from the theoretical and instrumental advances of GC, has extended the area of liquid chromatography into an equally sophisticated and useful method. SFC, first demonstrated in 1962, is finally gaining popularity. Modern chromatographic techniques, including automated systems, are widely utilized in the characterization and quality control of food raw materials and food products.

27.3.2 General Terminology

Chromatography is a general term applied to a wide variety of separation techniques based on the partitioning or distribution of a sample (**solute**) between a moving or mobile phase and a fixed or stationary phase. Chromatography may be viewed as a series of equilibrations between the mobile and stationary phase. The relative interaction of a solute with these two phases is described by the **partition** (K) or **distribution** (D) **coefficient** (ratio of concentration of solute in stationary phase to concentration of solute in mobile phase). The mobile phase may be either a gas (GC) or liquid (LC) or a supercritical fluid (SFC). The stationary phase may be a liquid or, more usually, a solid. The field of chromatography can be subdivided according to the various techniques applied (Fig 27-1), or according to the physicochemical principles involved in the separation. Table 27-1 summarizes some of the chromatographic procedures or methods that have been developed on the basis of different mobile–stationary phase combinations. Inasmuch as the nature of interactions between solute molecules and the mobile or stationary phases differ, these methods have the ability to separate different kinds of molecules. (The reader is urged to review Table 27-1 again after having read this chapter.)

27.3.3 Gas Chromatography

Gas chromatography is a column chromatography technique, in which the mobile phase is gas and the stationary phase is either an immobilized liquid or a solid packed in a closed tube. GC is used to separate thermally stable volatile components of a mixture. Gas chromatography, specifically gas–liquid chromatography, involves vaporizing a sample and injecting it onto

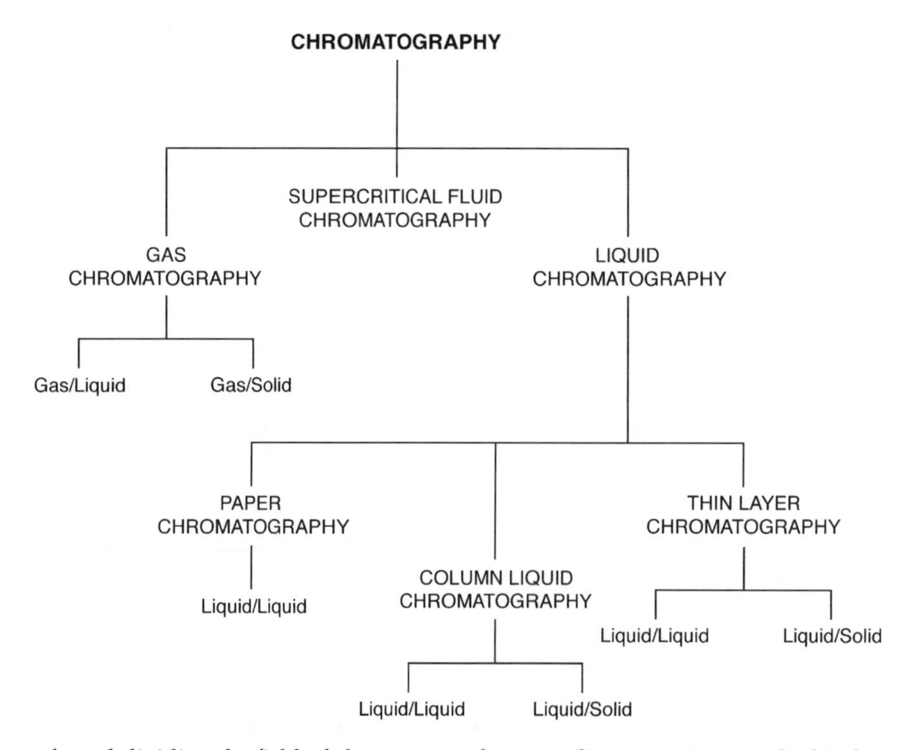

A scheme for subdividing the field of chromatography, according to various applied techniques.

Characteristics of Different Chromatographic Methods

Method	*Mobile/Stationary Phase*	*Retention Varies with*
Gas–liquid chromatography	Gas/liquid	Molecular size/polarity
Gas–solid chromatography	Gas/solid	Molecular size/polarity
Supercritical fluid chromatography	Supercritical fluid/solid	Molecular size/polarity
Reversed-phase chromatography	Polar liquid/nonpolar liquid or solid	Molecular size/polarity
Normal-phase chromatography	Less polar liquid/more polar liquid or solid	Molecular size/polarity
Ion-exchange chromatography	Polar liquid/ionic solid	Molecular charge
Size-exclusion chromatography	Liquid/solid	Molecular size
Hydrophobic-interaction chromatography	Polar liquid/nonpolar liquid or solid	Molecular size/polarity
Affinity chromatography	Water/binding sites	Specific structure

the head of the column. Under a controlled temperature gradient, the sample is transported through the column by the flow of an inert, gaseous mobile phase. Volatiles are then separated based on several properties, including boiling point, molecular size, and polarity. Physiochemical principles of separation are covered in Sect. 27.4. However, details of the chromatographic theory of separation as it applies specifically to GC, as well as detection and instrumentation of GC, are detailed in Chap. 29.

27.3.4 Liquid Chromatography

There are several **liquid chromatography** techniques applied in food analysis, namely paper chromatography, thin layer chromatography (TLC) (both of these techniques may be referred to as **planar chromatography**), and column liquid chromatography, all of which involve a liquid mobile phase and either a solid or a liquid stationary phase. However, the physical form of the stationary phase is quite different in each case. Separation of the solutes is based on their physicochemical interactions with the two phases, which is discussed in Sect. 27.4.

27.3.4.1 Paper Chromatography

Paper chromatography was introduced in 1944. In paper chromatography the stationary phase and the mobile phase are both liquid (**partition chromatography**, see Sect. 27.4.2). Paper generally serves as a support for the liquid stationary phase. The dissolved sample is applied as a small spot or streak one half inch or more from the edge of a strip or square of filter paper (usually cellulose), which is then allowed to dry. The dry strip is suspended in a closed container in which the atmosphere is saturated with the **developing solvent** (mobile phase), and the paper chromatogram is **developed**. The end closer to the sample is placed in contact with the solvent, which then travels up or down the paper by capillary action (depending on whether **ascending** or **descending** development is used), separating the sample components in the process. When the solvent front has traveled the length of the paper, the strip is removed from the developing chamber and the separated zones are detected by an appropriate method.

The stationary phase in paper partition chromatography is usually water. However, the support may be impregnated with a nonpolar organic solvent and developed with water or other polar solvents or water (**reversed-phase** paper chromatography). In the case of complex sample mixtures, a two-dimensional technique may be used. The sample is spotted in one corner of a square sheet of paper, and one solvent is used to develop the paper in one direction. The chromatogram is then dried, turned 90°, and developed again, using a second solvent of different polarity. Another means of improving resolution is the use of **ion-exchange** (Sect. 27.4.3) papers, that is, paper that has been impregnated with ion-exchange resin or paper, with derivatized cellulose hydroxyl groups (with acidic or basic moieties).

In paper and thin-layer chromatography, components of a mixture are characterized by their relative mobility (R_f) value, where:

$$R_f = \frac{\text{Distance moved by component}}{\text{Distance moved by solvent}} \quad [2]$$

Unfortunately, R_f values are not always constant for a given solute/sorbent/solvent, but depend on many factors, such as the quality of the stationary phase, layer thickness, humidity, development distance, and temperature.

27.3.4.2 Thin-Layer Chromatography

Thin-layer chromatography (TLC), first described in 1938, has largely replaced paper chromatography because it is faster, more sensitive, and more reproducible. The resolution in TLC is greater than in paper chromatography because the particles on the plate are smaller and more regular than paper fibers. Experimental conditions can be easily varied to achieve separation and can be scaled up for use in column chromatography, although thin-layer and column procedures are not necessarily interchangeable, due to differences such as the use of binders with TLC plates, vapor-phase equilibria in a TLC tank, etc. There are several distinct advantages to TLC: high sample throughput, low cost, the possibility to analyze several samples and standards simultaneously, minimal sample preparation, and that a plate may be stored for later identification and quantification.

TLC is applied in many fields, including environmental, clinical, forensic, pharmaceutical, food, flavors, and cosmetics. Within the food industry, TLC may be used for quality control. For example, corn and peanuts are tested for aflatoxins/mycotoxins prior to their processing into corn meal and peanut butter, respectively. Applications of TLC to the analysis of a variety of compounds, including lipids, carbohydrates, vitamins, amino acids, and natural pigments, are discussed in reference (5).

27.3.4.2.1 General Procedures TLC utilizes a thin (ca. 250 μm thick) layer of **sorbent** or **stationary phase** bound to an **inert support** in a planar configuration. The support is often a glass plate (traditionally, 20 cm × 20 cm), but plastic sheets and aluminum foil also are used. Precoated plates, of different layer thicknesses, are commercially available in a wide variety of sorbents, including chemically modified silicas. Four frequently used TLC sorbents are silica gel, alumina, diatomaceous earth, and cellulose. Modified silicas for TLC may contain polar or nonpolar groups, so both normal and reversed-phase (see Sect. 27.4.2.1) thin-layer separations may be carried out. **High-performance thin-layer chromatography** (HPTLC) simply refers to TLC performed using plates coated with smaller, more uniform particles. This permits better separations in shorter times.

If **adsorption** TLC is to be performed, the sorbent is first **activated** by drying for a specified time and temperature. Sample (in carrier solvent) is applied as a spot or streak 1–2 cm from one end of the plate. After evaporation of the carrier solvent, the TLC plate is placed in a closed **developing chamber** with the end of the plate nearest the spot in the solvent at the bottom of the chamber. Traditionally, solvent migrates up the plate (**ascending development**) by capillary action and sample components are separated. After the TLC plate has been removed from the chamber and solvent allowed to evaporate, the separated bands are made visible or detected by other means. Specific **chemical reactions (derivatization)**, which may be carried out either before or after chromatography, often are used for this purpose. Two examples are reaction with sulfuric acid to produce a dark charred area (a **destructive chemical method**) and the use of iodine vapor to form a colored complex (a **nondestructive method** inasmuch as the colored complex is usually not permanent). Common **physical detection methods** include the measurement of absorbed or emitted electromagnetic radiation (e.g., fluorescence) by means of autoradiography and the measurement of β-radiation from radioactively labeled compounds. **Biological methods** or biochemical inhibition tests can be used to detect toxicologically active substances. An example is measuring the inhibition of cholinesterase activity by organophosphate pesticides.

Quantitative evaluation of thin-layer chromatograms may be performed (1) in situ (directly on the layer) by using a **densitometer** or (2) after scraping a zone off the plate, eluting compound from the sorbent, and analyzing the resultant solution (e.g., by liquid scintillation counting).

27.3.4.2.2 Factors Affecting Thin-Layer Separations In both planar and column liquid chromatography, the nature of the compounds to be separated determines what type of stationary phase is used. Separation can occur by adsorption, partition, ion-exchange, size-exclusion, or multiple mechanisms (Sect. 27.4). Table 27-2 lists the separation mechanisms involved in some typical applications on common TLC sorbents.

Solvents for TLC separations are selected for specific chemical characteristics and **solvent strength** (a measure of interaction between solvent and sorbent; see Sect. 27.4.1). In simple adsorption TLC, the higher the solvent strength, the greater the R_f value of the solute. An R_f value of 0.3–0.7 is typical. Mobile phases have been developed for the separation of various compound classes on the different sorbents [see Table 7.1 in reference (15)].

In addition to the sorbent and solvent, several other factors must be considered when performing planar chromatography. These include the **type of developing chamber** used, **vapor phase conditions** (saturated vs. unsaturated), **development mode**

27-2 table Thin-Layer Chromatography Sorbents and Mode of Separation

Sorbent	*Chromatographic Mechanism*	*Typical Application*
Silica gel	Adsorption	Steroids, amino acids, alcohols, hydrocarbons, lipids, aflatoxins, bile acids, vitamins, alkaloids
Silica gel RP	Reversed phase	Fatty acids, vitamins, steroids, hormones, carotenoids
Cellulose, kieselguhr	Partition	Carbohydrates, sugars, alcohols, amino acids, carboxylic acids, fatty acids
Aluminum oxide	Adsorption	Amines, alcohols, steroids, lipids, aflatoxins, bile acids, vitamins, alkaloids
PEI cellulose[a]	Ion exchange	Nucleic acids, nucleotides, nucleosides, purines, pyrimidines
Magnesium silicate	Adsorption	Steroids, pesticides, lipids, alkaloids

Reprinted from (15) by permission of Wiley, New York.
[a]PEI cellulose refers to cellulose derivatized with polyethyleneimine (PEI).

(ascending, descending, horizontal, radial, etc.), and **development distance**. For additional reading refer to references (5), (7), and (16).

27.3.4.3 Column Liquid Chromatography

Column chromatography is the most useful method of separating compounds in a mixture. Fractionation of solutes occurs as a result of differential migration through a closed tube of stationary phase, and analytes can be monitored while the separation is in progress. In **column liquid chromatography**, the mobile phase is liquid and the stationary phase can be either solid or liquid supported by an inert solid. A system for **low-pressure** (i.e., performed at or near atmospheric pressure) column liquid chromatography is illustrated in Fig. 27-2.

Having selected a stationary and mobile phase suitable for the separation problem at hand, the analyst must first prepare the **stationary phase** (**resin**, **gel**, or **packing material**) for use according to the supplier's instructions. (For example, the stationary phase often must be **hydrated** or **preswelled** in the mobile phase). The prepared stationary phase then is **packed** into a column (usually glass), the length and diameter of which are determined by the amount of sample to be loaded, the separation mode to be used, and the degree of resolution required. Longer and narrower columns usually enhance resolution and separation. Adsorption columns may be either dry or wet packed; other types of columns are wet packed. The most common technique for wet packing involves making a slurry of the adsorbent with the solvent and pouring this into the column. As the sorbent settles, excess solvent is drained off and additional slurry is added. This process is repeated until the desired bed height is obtained. (There is a certain art to pouring uniform columns and no attempt is made to give details here.) If the packing solvent is different from the initial eluting solvent, the column must be thoroughly washed (**equilibrated**) with the starting mobile phase.

The sample to be fractionated, dissolved in a minimum volume of mobile phase, is applied in a layer at the top (or head) of the column. Classical or low-pressure chromatography utilizes only **gravity** flow or a **peristaltic pump** to maintain a flow of **mobile phase** (**eluent** or **eluting solvent**) through the column. In the case of a gravity-fed system, eluent is simply siphoned from a reservoir into the column. The flow rate is governed by the hydrostatic pressure, measured as the distance between the level of liquid in the reservoir and the level of the column outlet. If eluent is fed to the column by a peristaltic pump (see Fig. 27-2), then the flow rate is determined by the pump speed and, thus, regulation of hydrostatic pressure is not necessary.

The process of passing the mobile phase through the column is called **elution**, and the portion that emerges from the outlet end of the column is sometimes called the **eluate** (or effluent). Elution may be **isocratic** (constant mobile-phase composition) or a **gradient** (changing the mobile phase, e.g., increasing solvent strength or pH) during elution in order to enhance resolution and decrease analysis time (see also Sect. 27.5.1). As elution proceeds, components of the sample are selectively retarded by the stationary phase based on the strength of interaction with the stationary phase, and thus they are eluted at different times.

The column eluate may be directed through a detector and then into tubes, changed at intervals by

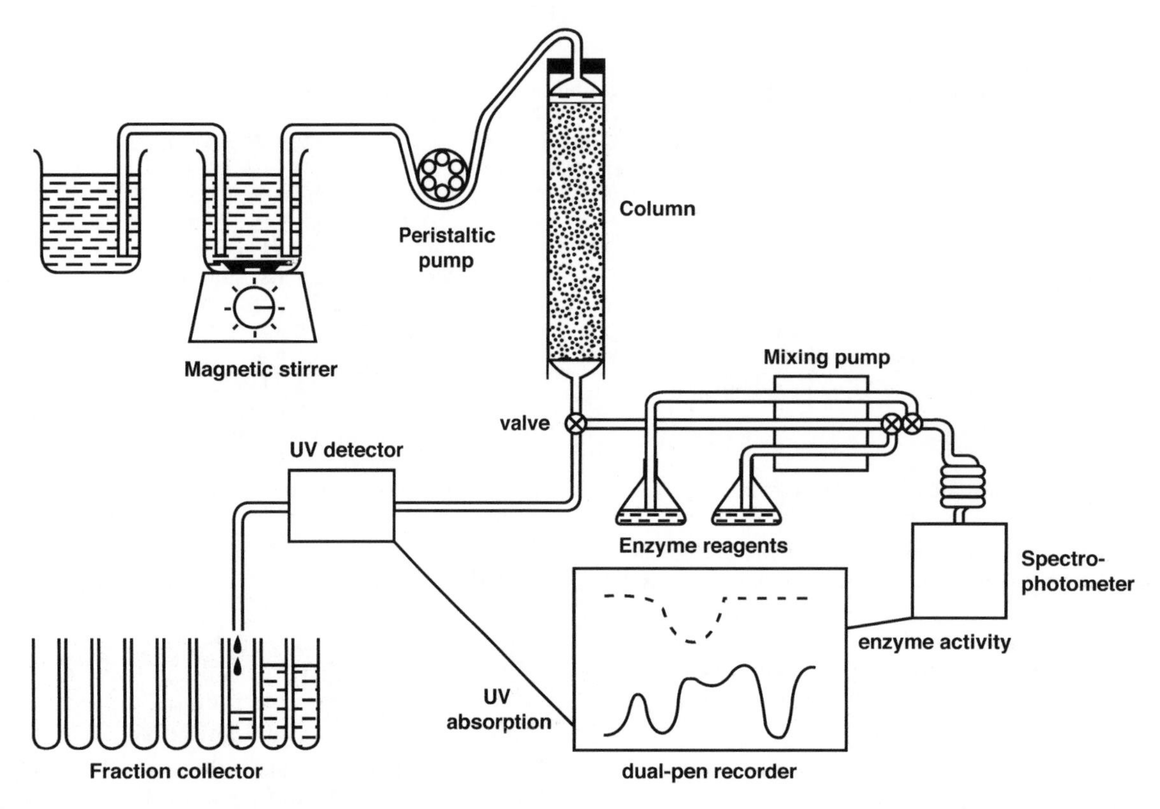

A system for low-pressure column liquid chromatography. In this diagram, the column effluent is being split between two detectors in order to monitor both enzyme activity (*at Right*) and UV absorption (*at Left*). The two tracings can be recorded simultaneously by using a dual-pen recorder. [Adapted from (12), with permission.]

a fraction collector. The detector response, in the form of an electrical signal, may be recorded (the **chromatogram**), using either a chart recorder or a computerized software, and used for qualitative or quantitative analysis, as discussed in more detail later. The fraction collector may be set to collect eluate at specified time intervals or after a certain volume or number of drops has been collected. Components of the sample that have been chromatographically separated and collected then can be further analyzed as needed.

27.3.5 Supercritical Fluid Chromatography

SFC refers to chromatography performed above the **critical pressure** (P_c) and **critical temperature** (T_c) of the mobile phase. A supercritical fluid (or compressed gas) is neither a liquid nor a typical gas. The combination of P_c and T_c is known as the **critical point**. A supercritical fluid can be formed from a conventional gas by increasing the pressure, or from a conventional liquid by raising the temperature. Carbon dioxide frequently is used as a mobile phase for SFC; however, it is not a good solvent for polar and high-molecular-weight compounds. A small amount of a polar, organic solvent such as methanol can be added to a nonpolar supercritical fluid to enhance solute solubility, improve peak shape, and alter selectivity. Other supercritical fluids that have been used in food applications include nitrous oxide, trifluoromethane, sulfur hexafluoride, pentane, and ammonia.

Supercritical fluids confer chromatographic properties intermediate to LC and GC. The **high diffusivity** and **low viscosity** of supercritical fluids mean decreased analysis times and improved resolution compared to LC. SFC offers a wide range of **selectivity** (Sect. 27.5.2) adjustment, by changes in **pressure** and **temperature** as well as changes in **mobile phase composition** and the **stationary phase**. In addition, SFC makes possible the separation of **nonvolatile**, **thermally labile compounds** that are not amenable to GC.

SFC can be performed using either **packed columns** or **capillaries**. Packed column materials are similar to those used for HPLC. Small particle, porous, high surface area, hydrated silica may serve as the stationary phase itself, or simply as a support for a bonded stationary phase (Chap. 28). Polymer-based packing has been used, but is less satisfactory owing to long solute retention times. Capillaries are generally coated with a polysiloxane (–Si–O–Si) film, which is then cross-linked to form a polymeric stationary phase that cannot be washed off by the mobile phase. Polysiloxanes containing different functional groups, such as methyl, phenyl, or cyano, may be used to vary the polarity of this stationary phase. Instrumentation for packed column SFC is similar to that used for HPLC with one major difference: A back pressure regulator is used to control the **outlet pressure** of the system. Without this device, the fluid would expand to a low-pressure, low-density gas. Besides the advantages of decreased analysis time and improved resolution, SFC offers the possibility to use a wide variety of detectors, including those designed for GC.

SFC has been used primarily for nonpolar compounds. Fats, oils, and other lipids are compounds to which SFC is increasingly applied. For example, the noncaloric fat substitute, Olestra®, was characterized by SFC-MS (mass spectroscopy). Other researchers have used SFC to detect pesticide residues, study thermally labile compounds from members of the *Allium* genus, fractionate citrus essential oils, and characterize compounds extracted from microwave packaging (3). Borch-Jensen and Mollerup (1) highlighted the use of packed column and capillary SFC for the analysis of food and natural products, especially fatty acids and their derivatives, glycerides, waxes, sterols, fat-soluble vitamins, carotenoids, and phospholipids.

27.4 PHYSICOCHEMICAL PRINCIPLES OF CHROMATOGRAPHIC SEPARATION

Several physicochemical principles (illustrated in Fig. 27-3) are involved in chromatography mechanisms employed to separate or fractionate various compounds of interest, regardless of the specific techniques applied (discussed in Sect. 27.3). The mechanisms described below apply mainly to liquid chromatography; GC mechanisms will be detailed in Chap. 29. Although it is more convenient to describe each of these phenomena separately, it must be emphasized that more than one mechanism may be involved in a given fractionation. For example, many cases of partition chromatography also involve adsorption.

27.4.1 Adsorption (Liquid–Solid) Chromatography

Adsorption chromatography is the oldest form of chromatography, originated with Tsvet in 1903 in the experiments that spawned modern chromatography. In this chromatographic mode, the stationary phase is a finely divided solid to maximize the surface area. The stationary phase (**adsorbent**) is chosen to permit differential interaction with the components of the sample to be resolved. The intermolecular forces thought to be primarily responsible for chromatographic adsorption include the following:

- Van der Waals forces
- Electrostatic forces
- Hydrogen bonds
- Hydrophobic interactions

Sites available for interaction with any given substance are heterogeneous. Binding sites with greater affinities, the most active sites, tend to be populated first, so that additional solutes are less firmly bound. The net result is that adsorption is a concentration-dependent process, and the **adsorption coefficient** is *not* a constant (in contrast to the **partition coefficient**). Sample loads exceeding the adsorptive capacity of the stationary phase will result in relatively poor separation.

Classic adsorption chromatography utilizes **silica** (slightly acidic), **alumina** (slightly basic), charcoal (nonpolar), or a few other materials as the stationary phase. Both silica and alumina possess surface hydroxyl groups, and Lewis acid-type interactions determine their adsorption characteristics. The elution order of compounds from these adsorptive stationary phases can often be predicted on the basis of their relative **polarities** (Table 27-3). Compounds with the most polar functional groups are retained most strongly on polar adsorbents and, therefore, are eluted last. Nonpolar solutes are eluted first.

One model proposed to explain the mechanism of liquid–solid chromatography is that **solute** and **solvent** molecules are competing for active sites on the adsorbent. Thus, as relative adsorption of the mobile phase increases, adsorption of the solute must decrease. Solvents can be rated in order of their strength of adsorption on a particular adsorbent, such as silica. Such a **solvent strength** (or polarity) **scale** is called a **eluotropic series**. A eluotropic series for alumina is listed in Table 27-4. Silica has a similar rank ordering. Once an adsorbent has been chosen, solvents can be selected from the eluotropic series for that adsorbent. Mobile phase polarity can be increased (often by admixture of more polar solvents) until elution of the compound(s) of interest has been achieved.

Adsorption chromatography separates aromatic or aliphatic nonpolar compounds, based primarily on the type and number of functional groups present. The labile, fat-soluble chlorophyll and carotenoid pigments from plants have been studied extensively by adsorption column chromatography. Adsorption chromatography also has been used for the analysis of fat-soluble vitamins. Frequently, it is used as a batch procedure for removal of impurities from samples prior to other analyses. For example, disposable solid-phase extraction cartridges (see Chap. 29) containing silica have been used for food analyses, such as lipids in soybean oil, carotenoids in citrus fruit, and vitamin E in grain.

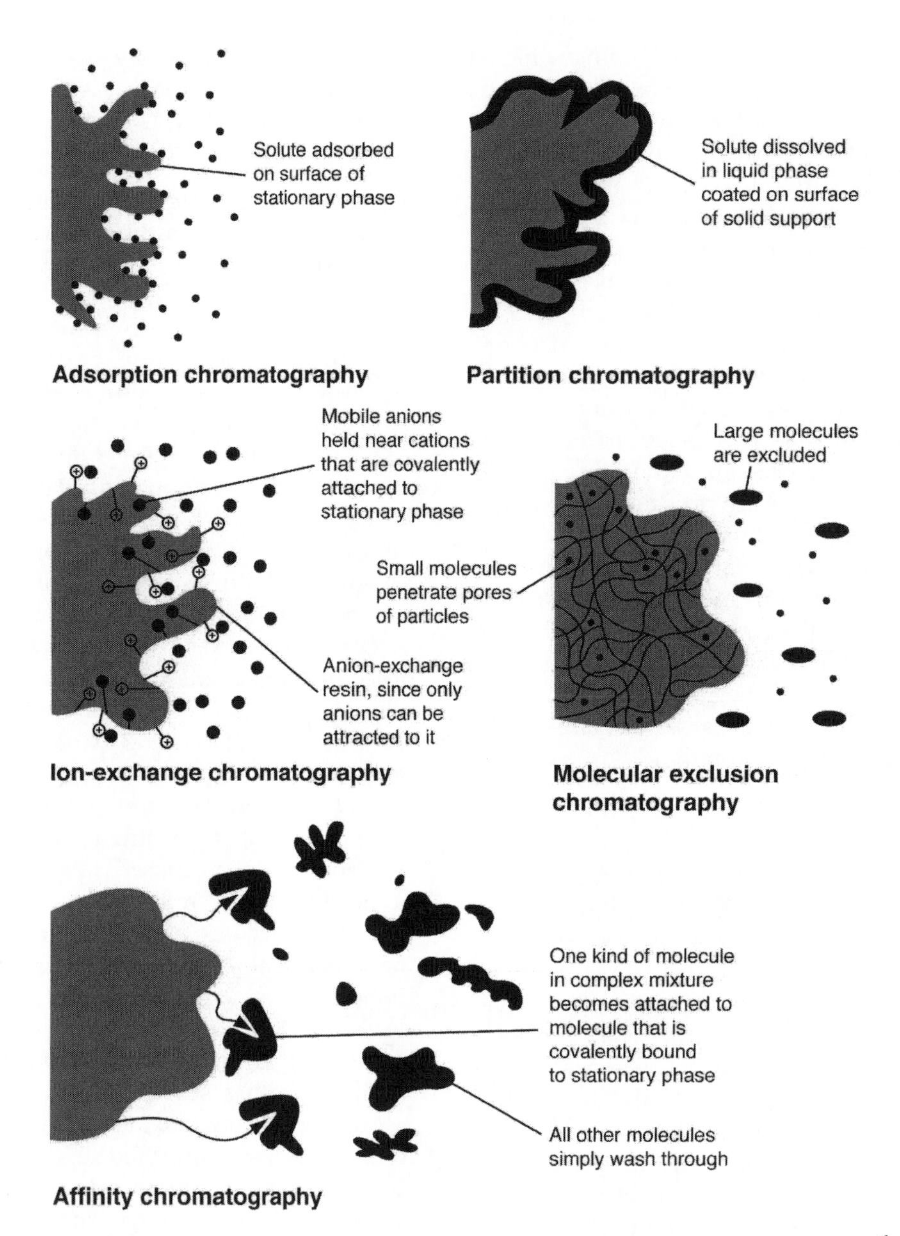

27-3 figure Physicochemical principles of chromatography. [From *Quantitative Chemical Analysis, 5th ed.*, by D.C. Harris. © 1999 by W.H. Freeman & Co. Used with permission.]

27.4.2 Partition (Liquid–Liquid) Chromatography

27.4.2.1 Introduction

In 1941, Martin and Synge undertook an investigation of the amino acid composition of wool, using a countercurrent extractor (Sect. 27.2.3) of 40 tubes with chloroform and water flowing in opposite directions. The efficiency of the extraction process was improved enormously when a column of finely divided inert support material was used to hold one liquid phase (stationary phase) immobile, while the second liquid, an immiscible solvent (mobile phase), flowed over it, thus providing intimate contact between the two phases. Solutes partitioned between the two liquid phases according to their partition coefficients, hence the name **partition chromatography**.

A partition system is manipulated by changing the nature of the two liquid phases, usually by combination of solvents or pH adjustment of buffers. Often, the more polar of the two liquids is held stationary on the inert support and the less polar solvent is used to elute the sample components (**normal-phase chromatography**). Reversal of this arrangement, using a **nonpolar stationary phase** and a **polar mobile phase**, has come to be known as **reversed-phase chromatography** (see Sect. 27.4.2.3).

Compounds Class Polarity Scale[a]

Fluorocarbons
Saturated hydrocarbons
Olefins
Aromatics
Halogenated compounds
Ethers
Nitro compounds
Esters ≈ ketones ≈ aldehydes
Alcohols ≈ amines
Amides
Carboxylic acids

From (9), used with permission.
[a]Listed in order of increasing polarity.

Eluotropic Series for Alumina

Solvent
1-Pentane
Isooctane
Cyclohexane
Carbon tetrachloride
Xylene
Toluene
Benzene
Ethyl ether
Chloroform
Methylene chloride
Tetrahydrofuran
Acetone
Ethyl acetate
Aniline
Acetonitrile
2-Propanol
Ethanol
Methanol
Acetic acid

From (9), used with permission.

Polar **hydrophilic** substances, such as amino acids, carbohydrates, and water-soluble plant pigments, are separable by **normal-phase** partition chromatography. **Lipophilic** compounds, such as lipids and fat-soluble pigments, and **polyphenols** may be resolved with **reversed-phase** systems. Liquid–liquid partition chromatography has been invaluable to carbohydrate chemistry. Column liquid chromatography on finely divided cellulose has been used extensively in preparative chromatography of sugars and their derivatives. Paper chromatography (Sect. 27.3.4.1) is a simple method for distinguishing between various forms of sugars (following normal-phase partition chromatography) or phenolic compounds (following reverse-phase partition chromatography) present in foods.

27.4.2.2 Coated Supports

In its simplest form, the stationary phase for partition chromatography consists of a liquid coating on a solid matrix. The solid support should be as inert as possible and have a large surface area in order to maximize the amount of liquid held. Some examples of solid supports that have been used are silica, starch, cellulose powder, and glass beads. All are capable of holding a thin film of water, which serves as the stationary phase. It is important to note that materials prepared for adsorption chromatography must be **activated** by drying them to remove surface water. Conversely, some of these materials, such as silica gel, may be used for partition chromatography if they are deactivated by impregnation with water or the desired stationary phase. One disadvantage of liquid–liquid chromatographic systems is that the liquid stationary phase is often stripped off. This problem can be overcome by chemically bonding the stationary phase to the support material, as described in the next section.

27.4.2.3 Bonded Supports

The liquid stationary phase may be covalently attached to a support by a chemical reaction. These **bonded phases** have become very popular for HPLC use, and a wide variety of both polar and nonpolar stationary phases is now available. Especially widely used is reversed-phase HPLC (see Chap. 28), with a nonpolar bonded stationary phase (e.g., silica covered with C_8 or C_{18} groups) and a polar solvent (e.g., water–acetonitrile). It is important to note that mechanisms other than partition may be involved in the separation using bonded supports. Bonded-phase HPLC columns have greatly facilitated the analysis of vitamins in foods and feeds, as discussed in Chap. 3 of reference (10). Additionally, bonded-phase HPLC is widely used for the separation and identification of polyphenols such as phenolic acids (e.g., *p*-coumaric, caffeic, ferulic, and sinapic acids) and flavonoids (e.g., flavonols, flavones, isoflavones, anthocyanidins, catechins, and proanthocyanidins).

27.4.3 Ion-Exchange Chromatography

Ion exchange is a separation/purification process occurring naturally, for example, in soils and is utilized in water softeners and deionizers. Three types of separation may be achieved: (1) ionic from nonionic, (2) cationic from anionic, and (3) mixtures of similarly charged species. In the first two cases, one substance binds to the ion-exchange medium, whereas the other substance does not. Batch extraction methods can be used for these two separations; however, chromatography is needed for the third category.

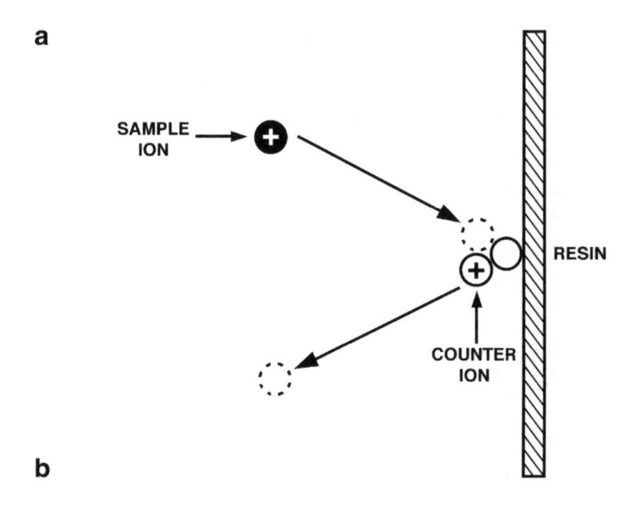

$$M^+ + (Na^+\ {}^-O_3S-\text{Resin}) \rightleftharpoons (M^+\ {}^-O_3S-\text{Resin}) + Na^+$$
M^+ = Cation

$$X^- + (Cl^-\ {}^+R_4N-\text{Resin}) \rightleftharpoons (X^-\ {}^+R_4N-\text{Resin}) + Cl^-$$
X^- = Anion

The basis of ion-exchange chromatography. (**a**) Schematic of the ion-exchange process; (**b**) ionic equilbria for cation- and anion-exchange processes. [From (9), used with permission.]

Ion-exchange chromatography may be viewed as a type of adsorption chromatography in which interactions between solute and stationary phase are primarily **electrostatic** in nature. The stationary phase (ion exchanger) contains fixed functional groups that are either negatively or positively charged (Fig. 27-4a). Exchangeable counterions preserve charge neutrality. A sample ion (or charged sites on large molecules) can exchange with the counterion to become the partner of the fixed charge. Ionic equilibrium is established as depicted in Fig. 27-4b. The functional group of the stationary phase determines whether cations or anions are exchanged. **Cation exchangers** contain covalently bound negatively charged functional groups, whereas **anion exchangers** contain bound positively charged groups. The chemical nature of these acidic or basic residues determines how stationary-phase ionization is affected by the mobile-phase pH.

The strongly acidic sulfonic acid moieties (RSO_3^-) of "strong"-cation exchangers are completely ionized at all pH values above 2. Strongly basic quaternary amine groups ($RNR'_3{}^+$) on "strong"-anion exchangers are ionized at all pH values below 10. Since maximum negative or positive charge is maintained over a broad pH range, the exchange or binding capacity of these stationary phases is essentially constant, regardless of mobile-phase pH. "Weak"-cation exchangers contain weakly acidic carboxylic acid functional groups, (RCO_2^-); consequently, their exchange capacity varies considerably between ca. pH 4 and 10. Weakly basic anion exchangers possess primary, secondary, or tertiary amine residues ($R\text{-}NHR'_2{}^+$), which are deprotonated in moderately basic solution, thereby losing their positive charge and the ability to bind anions. Thus, one way of eluting solutes bound to an ion-exchange medium is to change the mobile-phase pH. A second way to elute bound solutes is to increase the ionic strength (e.g., use NaCl) of the mobile phase, to weaken the electrostatic interactions.

Chromatographic separations by ion exchange are based upon differences in affinity of the exchangers for the ions (or charged species) to be separated. The factors that govern **selectivity** of an exchanger for a particular ion include the ionic valence, radius, and concentration; the nature of the exchanger (including its displaceable counterion); and the composition and pH of the mobile phase. To be useful as an ion exchanger, a material must be both ionic in nature and highly permeable. Synthetic ion exchangers are thus cross-linked polyelectrolytes, and they may be inorganic (e.g., aluminosilicates) or, more commonly, organic compounds. **Polystyrene**, made by cross-linking styrene with divinyl benzene (DVB), may be modified to produce either anion- or cation-exchange resins (Fig. 27-5). Polymeric resins such as these are commercially available in a wide range of particle sizes and with different degrees of cross-linking (expressed as weight percent of DVB in the mixture). The extent of cross-linking controls the rigidity and porosity of the resin, which, in turn, determines its optimal use. Lightly cross-linked resins permit rapid equilibration of solute, but particles swell in water, thereby decreasing charge density and selectivity (relative affinity) of the resin for different ions. More highly cross-linked resins exhibit less swelling, higher exchange capacity, and selectivity, but longer equilibration times. The small pore size, high charge density, and inherent hydrophobicity of the older ion-exchange resins have limited their use to small molecules [molecular weight (MW) <500].

Ion exchangers based on **polysaccharides**, such as cellulose, dextran, or agarose, have proven very useful for the separation and purification of large molecules, such as proteins and nucleic acids. These materials, called **gels**, are much softer than polystyrene resins, and thus may be derivatized with strong or with weak acidic or basic groups via OH moieties on the polysaccharide backbone (Fig. 27-6). They have much larger pore sizes and lower charge densities than the older synthetic resins.

Food-related applications of ion-exchange chromatography include the separation of amino acids, sugars, alkaloids, and proteins. Fractionation of amino acids in protein hydrolyzates was initially carried out

Styrene

Divinylbenzene

Crosslinked styrene-divinylbenzene copolymer

R = H, Plain polystyrene

R = $CH_2N^+(CH_3)_3Cl^-$, Anion-exchanger

R = $SO_3^-H^+$, Cation-exchanger

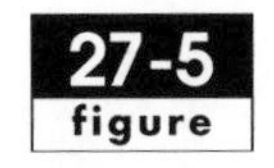

Chemical structure of polystyrene-based ion-exchange resins.

by ion-exchange chromatography; automation of this process led to the development of commercially produced amino acid analyzers (see Chap. 15). Many drugs, fatty acids, and the acids of fruit, being ionizable compounds, may be chromatographed in the ion-exchange mode. For additional details on the principles and applications of ion chromatography please refer to reference (18).

27.4.4 Size-Exclusion Chromatography

Size-exclusion chromatography (SEC), also known as **molecular exclusion, gel permeation** (GPC), and **gel-filtration chromatography** (GFC), is probably the easiest mode of chromatography to perform and to understand. It is widely used in the biological sciences for the resolution of macromolecules, such as proteins and carbohydrates, and also is used for the fractionation and characterization of synthetic polymers. Unfortunately, nomenclature associated with this separation mode developed independently in the literature of the life sciences and in the field of polymer chemistry, resulting in inconsistencies.

In the **ideal SEC** system, molecules are separated solely on the basis of their size; no interaction occurs between solutes and the stationary phase. In the event that solute/support interactions do occur, the separation mode is termed **nonideal SEC**. The stationary phase in SEC consists of a column packing material that contains pores comparable in size to the molecules to be fractionated. Solutes too large to enter the pores travel with the mobile phase in

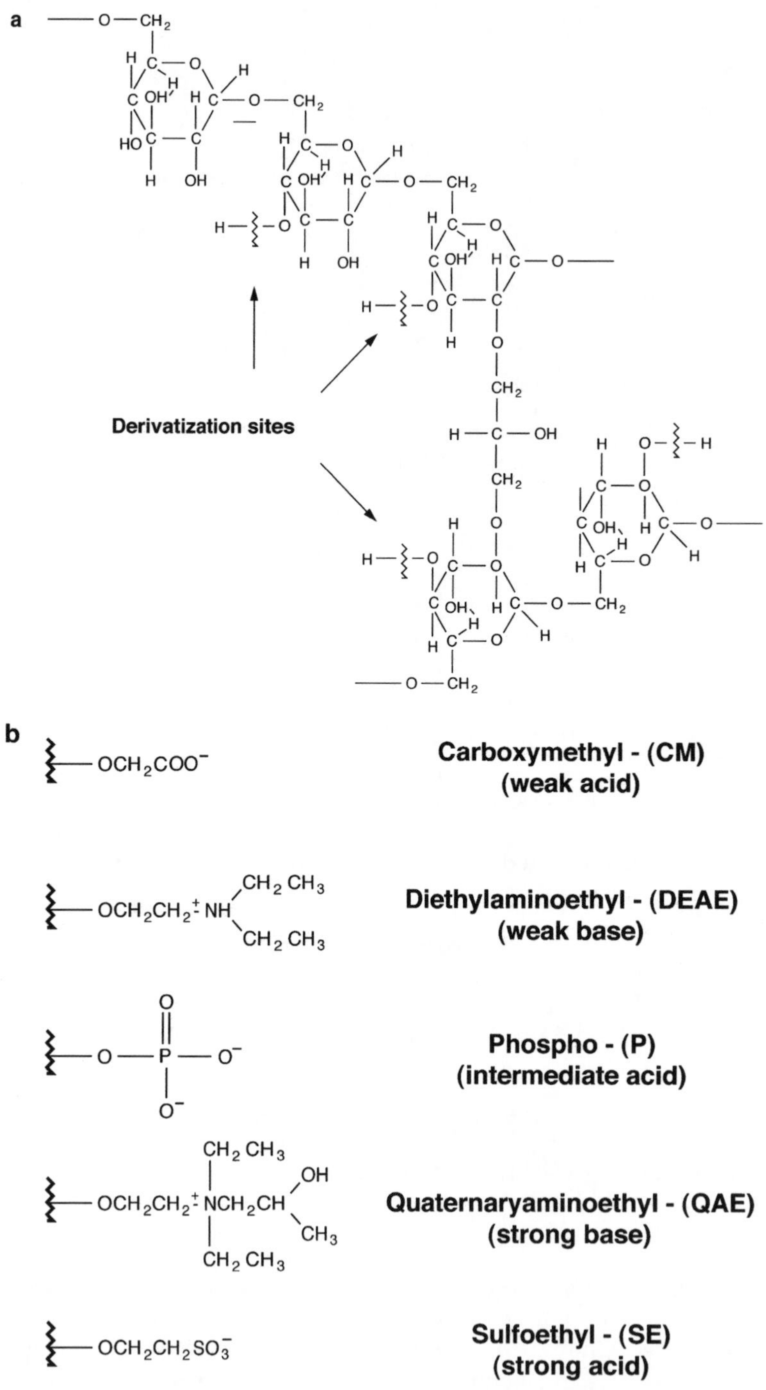

a

Derivatization sites

b

- OCH_2COO^- — **Carboxymethyl - (CM) (weak acid)**
- $OCH_2CH_2\overset{+}{N}H(CH_2CH_3)_2$ — **Diethylaminoethyl - (DEAE) (weak base)**
- $O{-}P(=O)(O^-){-}O^-$ — **Phospho - (P) (intermediate acid)**
- $OCH_2CH_2\overset{+}{N}(CH_2CH_3)_2CH_2CH(OH)CH_3$ — **Quaternaryaminoethyl - (QAE) (strong base)**
- $OCH_2CH_2SO_3^-$ — **Sulfoethyl - (SE) (strong acid)**

27-6 figure Chemical structure of one polysaccharide-based ion-exchange resin. (**a**): Matrix of cross-linked dextran ("Sephadex," Pharmacia Biotech, Inc., Piscataway NJ); (**b**): functional groups that may be used to impart ion-exchange properties to the matrix.

the **interstitial space** (between particles) **outside the pores**. Thus, the largest molecules are eluted first from an SEC column. The volume of the mobile phase in the column, termed the **column void volume**, V_o, can be measured by chromatographing a very large (totally excluded) species, such as Blue Dextran, a dye of MW $= 2 \times 10^6$.

As solute dimensions decrease, approaching those of the packing pores, molecules begin to diffuse into the packing particles and, consequently, are slowed down. Solutes of low molecular weight (e.g., glycyltyrosine) that have free access to all the available pore volume are eluted in the volume referred to as V_t. This value, V_t, which is equal to the column void volume,

V_o, plus the volume of liquid inside the sorbent pores, V_i, is referred to as the **total permeation volume** of the packed column ($V_t = V_o + V_i$). These relationships are illustrated in Fig. 27-7. Solutes are ideally eluted between the void volume and the total liquid volume of the column. Because this volume is limited, only a relatively small number of solutes (ca. 10) can be completely resolved by SEC under ordinary conditions.

The behavior of a molecule in a size-exclusion column may be characterized in several different ways. Each solute exhibits an **elution volume**, V_e, as illustrated in Fig. 27-7. However, V_e depends on column dimensions and the way in which the column was packed. The **available partition coefficient** is used to define solute behavior independent of these variables:

$$K_{av} = (V_e - V_o)/(V_t - V_o) \quad [3]$$

where:

K_{av} = available partition coefficient
V_e = elution volume of solute
V_o = column void volume
V_t = total permeation volume of column

The value of K_{av} calculated from experimental data for a solute chromatographed on a given SEC column defines the proportion of pores that can be occupied by that molecule. For a large, totally excluded species, such as Blue Dextran or DNA, $V_e = V_o$ and $K_{av} = 0$. For a small molecule with complete access to the internal pore volume, such as glycyltyrosine, $V_e = V_t$ and $K_{av} = 1$.

For each size-exclusion packing material, a plot of K_{av} vs. the logarithm of the molecular weight for a series of solutes, similar in molecular shape and density, will give an S-shaped curve (Fig. 27-8). In the case of proteins, K_{av} is actually better related to the **Stokes radius**, the average radius of the protein in solution. The central, linear portion of this curve describes the **fractionation range** of the matrix, wherein maximum separation among solutes of similar molecular weight is achieved. This correlation between solute elution behavior and molecular weight (or size) forms the basis for a widely used method for characterizing large molecules such as proteins and polysaccharides. A size-exclusion column is calibrated with a series of solutes of known molecular weight (or Stokes radius) to obtain a curve similar to that shown in Fig. 27-8. The value of K_{av} for the unknown is then determined, and an estimate of molecular weight (or size) of the unknown is made by interpolation of the calibration curve.

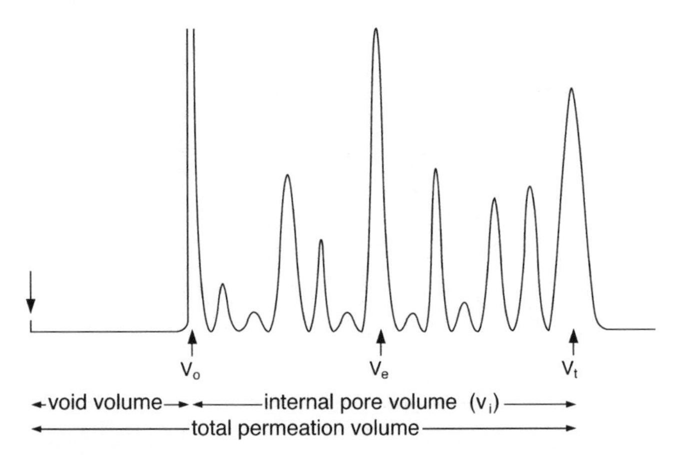

27-7 figure

Schematic elution profile illustrating some of the terms used in size-exclusion chromatography. [Adapted from (8), p. A271, with kind permission from Elsevier Science – NL, Sara Burgerhartstraat 25, 1055 KV Amsterdam, The Netherlands.]

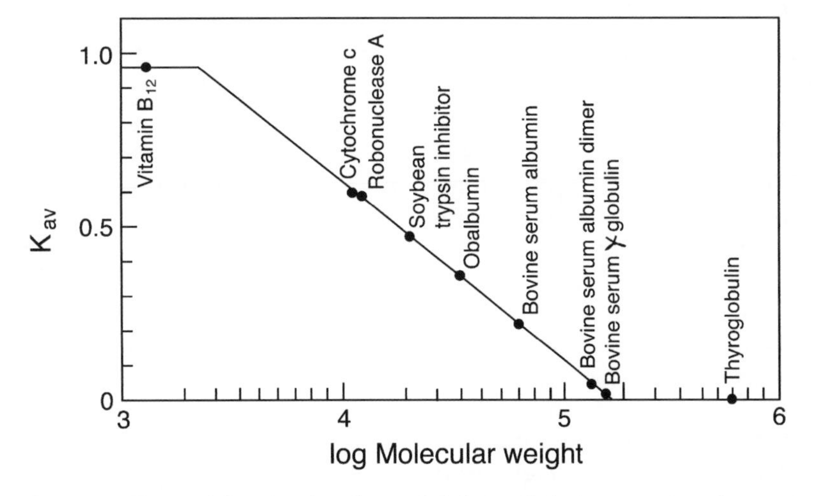

Relationship between K_{av} and log (molecular weight) for globular proteins chromatographed on a column of Sephadex G-150 Superfine. (Reproduced by permission of Pharmacia Biotech, Inc., Piscataway, NJ.)

Column packing materials for SEC can be divided into two groups: semirigid, **hydrophobic media**, and soft, **hydrophilic gels**. The former are usually derived from **polystyrene** and are used with organic mobile phases (GPC or nonaqueous SEC) for the separation of polymers, such as rubbers and plastics. Soft gels, **polysaccharide**-based packings, are typified by Sephadex, a cross-linked dextran (see Fig. 27-6a). These materials are available in a wide range of pore sizes and are useful for the separation of water-soluble substances in the molecular weight range $1–2.5 \times 10^7$. In selecting an SEC column packing, both the purpose of the experiment and size of the molecules to be separated must be considered. If the purpose of the experiment is group separation, where molecules of widely different molecular sizes need to be separated, a matrix is chosen such that the larger molecules, e.g., proteins, are eluted in the void volume of the column, whereas small molecules are retained in the total volume. A common example of group separation is buffer exchange and desalting. When SEC is used for separation of macromolecules of different sizes, the molecular sizes of all the components must fall within the fractionation range of the gel.

As discussed previously, SEC can be used, directly, to fractionate mixtures or, indirectly, to obtain information about a dissolved species. In addition to molecular weight estimations, SEC is used to determine the molecular weight distribution of natural and synthetic polymers, such as dextrans and gelatin preparations. Fractionation of biopolymer mixtures is probably the most widespread use of SEC, since the mild elution conditions employed rarely cause denaturation or loss of biological activity. It is also a fast, efficient alternative to dialysis for desalting solutions of macromolecules, such as proteins.

27.4.5 Affinity Chromatography

Affinity chromatography is unique in that separation is based on the specific, reversible interaction between a solute molecule and a ligand immobilized on the chromatographic stationary phase. While discussed here as a separate type of chromatography, affinity chromatography could be viewed as the ultimate extension of adsorption chromatography. Although the basic concepts of so-called biospecific adsorption were known as early as 1910, they were not perceived as potentially useful laboratory tools until ca. 1968.

Affinity chromatography usually involves immobilized biological materials as the stationary phase. These **ligands** can be antibodies, enzyme inhibitors, lectins, or other molecules that selectively and reversibly bind to complementary analyte molecules in the sample. Separation exploits the lock and key binding of biological systems. Although both ligands and the species to be isolated are usually biological macromolecules, the term affinity chromatography also encompasses other systems, such as separation of small molecules containing *cis*-diol groups via phenylboronic acid moieties on the stationary phase.

The principles of affinity chromatography are illustrated in Fig. 27-9. A ligand, chosen based on its specificity and strength of interaction with the molecule to be isolated (analyte), is immobilized on a suitable support material. As the sample is passed through this column, molecules that are complementary to the bound ligand are adsorbed while other sample components are eluted. Bound analyte is subsequently eluted via a change in the mobile-phase composition as will be discussed below. After reequilibration with the initial mobile phase, the stationary phase is ready to be used again. The ideal support for

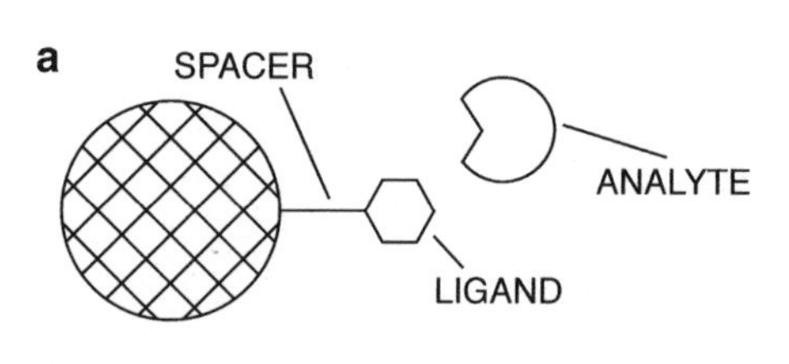

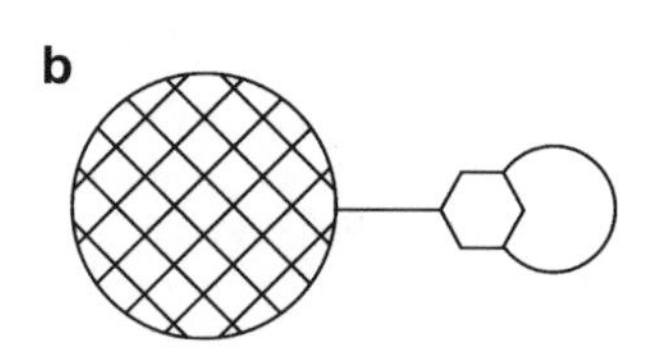

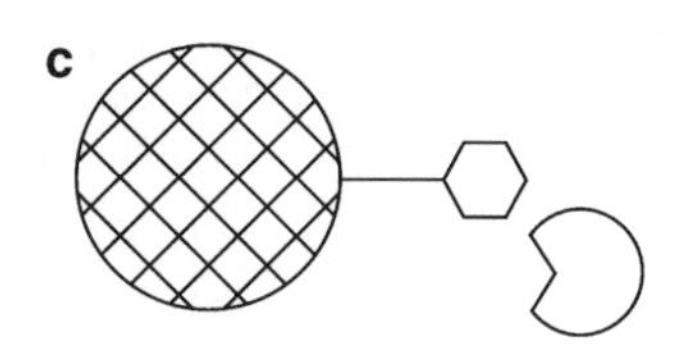

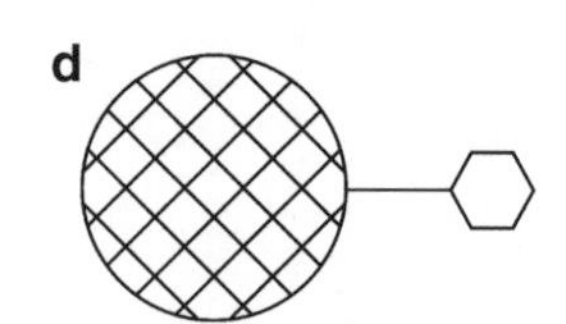

Principles of bioselective affinity chromatography. (**a**) The support presents the immobilized ligand to the analyte to be isolated. (**b**) The analyte makes contact with the ligand and attaches itself. (**c**) The analyte is recovered by the introduction of an eluent, which dissociates the complex holding the analyte to the ligand. (**d**) The support is regenerated, ready for the next isolation. [Reprinted from (8), p. A311, with kind permission from Elsevier Science NL, Sara Burgerhartstraat 25, 1055 KV Amsterdam, The Netherlands.]

affinity chromatography should be a porous, stable, high-surface-area material that does not adsorb anything itself. Thus, polymers such as agarose, cellulose, dextran, and polyacrylamide are used, as well as controlled-pore glass.

Affinity ligands are usually attached to the support or matrix by covalent bond formation, and optimum reaction conditions often must be found empirically. Immobilization generally consists of two steps: **activation** and **coupling**. During the activation step, a reagent reacts with functional groups on the support, such as hydroxyl moieties, to produce an activated matrix. After removal of excess reagent, the ligand is coupled to the activated matrix. (Preactivated supports are commercially available, and their availability has greatly increased the use of affinity chromatography.) The coupling reaction most often involves free amino groups on the ligand, although other functional groups can be used. When small molecules such as phenylboronic acid are immobilized, a **spacer arm** (containing at least four to six methylene groups) is used to hold the ligand away from the support surface, enabling it to reach into the binding site of the analyte.

Ligands for affinity chromatography may be either **specific** or **general** (i.e., group specific). Specific ligands, such as antibodies, bind only one particular solute. General ligands, such as nucleotide analogs and lectins, bind to certain classes of solutes. For example, the lectin concanavalin A binds to all molecules that contain terminal glucosyl and mannosyl residues. Bound solutes then can be separated as a group or individually, depending upon the elution technique used. Some of the more common general ligands are listed in Table 27-5. Although less selective, general ligands provide greater convenience.

Elution methods for affinity chromatography may be divided into **nonspecific** and **(bio)specific** methods. Nonspecific elution involves disrupting ligand analyte binding by changing the mobile-phase pH, ionic strength, dielectric constant, or temperature. If additional selectivity in elution is desired, for example, in the case of immobilized general ligands, a biospecific elution technique is used. Free ligand, either identical to or different from the matrix-bound ligand, is added to the mobile phase. This free ligand competes for binding sites on the analyte. For example, glycoproteins bound to a concanavalin A (lectin) column can be eluted by using buffer containing an excess of lectin. In general, the eluent ligand should display greater affinity for the analyte of interest than the immobilized ligand.

In addition to protein purification, affinity chromatography may be used to separate supramolecular structures such as cells, organelles, and viruses; concentrate dilute protein solutions; investigate binding mechanisms; and determine equilibrium constants.

27-5 table **General Affinity Ligands and Their Specificities**

Ligand	*Specificity*
Cibacron Blue F3G-A dye, derivatives of AMP, NADH, and NADPH	Certain dehydrogenases via binding at the nucelotide binding site
Concanavalin A, lentil lectin, wheat-germ lectin	Polysaccharides, glycoproteins, glycolipids, and membrane proteins containing sugar residues of certain configurations
Soybean trypsin inhibitor, methyl esters of various amino acids, D-amino acids	Various proteases
Phenylboronic acid	Glycosylated hemoglobins, sugars, nucleic acids, and other *cis*-diol-containing substances
Protein A	Many immunoglobulin classes and subclasses via binding to the F_c region
DNA, RNA, nucleosides, nucleotides	Nucleases, polymerases, nucleic acids

Affinity chromatography has been useful especially in the separation and purification of enzymes and glycoproteins. In the case of the latter, carbohydrate-derivatized adsorbents are used to isolate specific lectins, such as concanavalin A, and lentil or wheat-germ lectin. The lectin then may be coupled to agarose, such as concanavalin A- or lentil lectin-agarose, to provide a stationary phase for the purification of specific glycoproteins, glycolipids, or polysaccharides. For additional details on affinity chromatography please refer to reference (19).

27.5 ANALYSIS OF CHROMATOGRAPHIC PEAKS

Once the chromatographic technique (Sect. 27.3) and chromatographic mechanism (Sect. 27.4) have been chosen, the analyst has to ensure adequate separation of constituents of interests from a mixture, in a reasonable amount of time. After separation is achieved and chromatographic peaks are obtained, qualitative as well as quantitative analysis can be carried out. Basic principles of separation and resolution will be discussed in the subsequent sections. Understanding these principles allows the analyst to optimize separation and perform qualitative and quantitative analysis.

27.5.1 Separation and Resolution

This section will discuss separation and resolution as it pertains mainly to LC; separation and resolution optimization as it pertains specifically to GC will be discussed in Chap. 29.

27.5.1.1 Developing a Separation

There may be numerous ways to accomplish a chromatographic separation for a particular compound. In many cases, the analyst will follow a standard laboratory procedure or published methods. In the case of a sample that has not been previously analyzed, the analyst begins by evaluating what is known about the sample and defines the goals of the separation. How many components need to be resolved? What degree of resolution is needed? Is qualitative or quantitative information needed? Molecular weight (or molecular weight range), polarity, and ionic character of the sample will guide the choice of chromatographic separation mechanism (**separation mode**). Figure 27-10 shows that more than one correct choice may be possible. For example, small ionic compounds may be separated by ion-exchange, ion-pair reversed-phase (see Sect. 28.3.2.1), or reversed-phase LC. In this case, the analyst's choice may be based on convenience, experience, and personal preference.

Having chosen a separation mode for the sample at hand, one must select an appropriate stationary phase, elution conditions, and a detection method. Trial experimental conditions may be based on the results of a literature search, the analyst's previous experience with similar samples, or general recommendations from chromatography experts.

To achieve separation of sample components by all modes except SEC, one may utilize either isocratic or gradient elution. **Isocratic elution** is the most simple and widely used technique, in which solvent composition and flow rate are held constant. **Gradient elution** involves reproducibly varying mobile phase composition or flow rate (flow programming) during the LC analysis. Gradient elution is used when sample components possess a wide range of polarities, so that an isocratic mobile phase does not elute all components within a reasonable time. The change may be continuous or stepwise. Gradients of increasing ionic strength are extremely valuable in ion-exchange chromatography (see Sect. 27.4.3). Gradient elution is commonly used for desorbing large molecules, such as

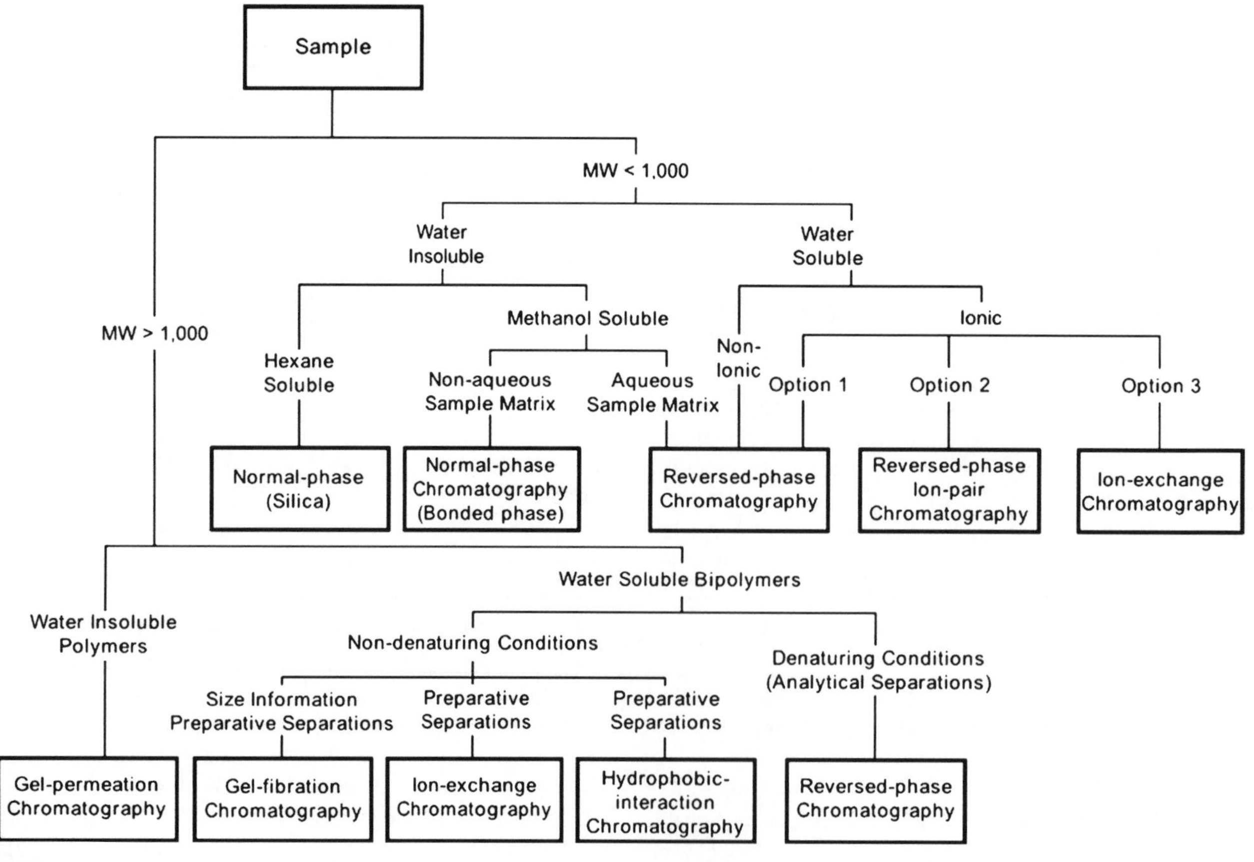

A schematic for choosing a chromatographic separation mode based on sample molecular weight and solubility. [From (11), used with permission.]

proteins, which can undergo multiple-site interaction with a stationary phase. Increasing the "strength" of the mobile phase (Sect. 27.4.1), either gradually or in a stepwise fashion, shortens the analysis time.

Method development may begin with an isocratic mobile phase, possibly of intermediate solvent strength; however, using gradient elution for the initial separation may ensure that some level of separation is achieved within a reasonable time period and nothing is likely to remain on the column. Data from this initial run allow one to determine if isocratic or gradient elution is needed, and to estimate optimal isocratic mobile phase composition or gradient range. The use of a gradient run does not presuppose that the final method will use gradient elution.

Once an initial separation has been achieved, the analyst can proceed to optimize resolution. This generally involves manipulation of mobile phase variables, including the nature and percentage of organic components, pH, ionic strength, nature and concentration of additives (such as ion-pairing agents), flow rate, and temperature. In the case of gradient elution, gradient steepness (slope) is another variable to be optimized. However, the analyst must be aware of the principles of chromatographic resolution as will be discussed in the following section.

27.5.1.2 Chromatographic Resolution

27.5.1.2.1 Introduction The main goal of chromatography is to segregate components of a sample into separate bands or peaks as they migrate through the column. A **chromatographic peak** is defined by several parameters including **retention time** (Fig. 27-11), **peak width**, and **peak height** (Fig. 27-12). The volume of the mobile phase required to elute a compound from an LC column is called the **retention volume**, V_R. The associated time is the **retention time**, t_R. Shifts in retention time and changes in peak width greatly influence **chromatographic resolution**.

Differences in column dimensions, loading, temperature, mobile phase flow rate, system dead-volume, and detector geometry may lead to discrepancies in retention time. By subtracting the time required for the mobile phase or a nonretained solute (t_m or t_o) to travel through the column to the detector, one obtains an **adjusted retention time**, t'_R (or volume) as depicted in Fig. 27-11. The adjusted retention time (or volume) corrects for differences in system dead-volume; it may be thought of as the time the sample spends adsorbed on the stationary phase.

The **resolution** of two peaks from each other is related to the **separation factor**, α. Values for α (Fig. 27-11) depend on temperature, the stationary phase, and mobile phase used. Resolution is defined as follows:

$$R_S = \frac{2\Delta t}{w_2 + w_1} \quad [4]$$

where:

R_S = resolution
Δt = difference between retention times of peaks 1 and 2
w_2 = width of peak 2 at baseline
w_1 = width of peak 1 at baseline

Figure 27-12 illustrates the measurement of peak width [part (a)] and the values necessary for calculating resolution [part (b)]. (Retention and peak or band width must be expressed in the same units, i.e., time or volume.)

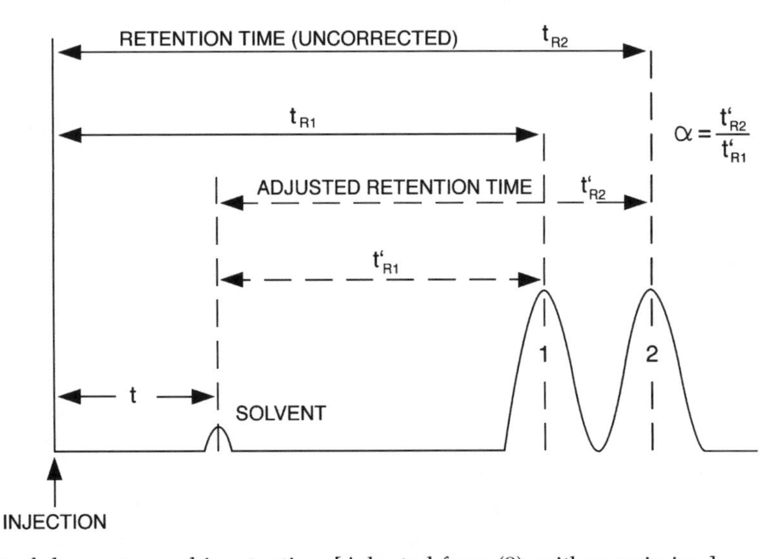

Measurement of chromatographic retention. [Adapted from (9), with permission.]

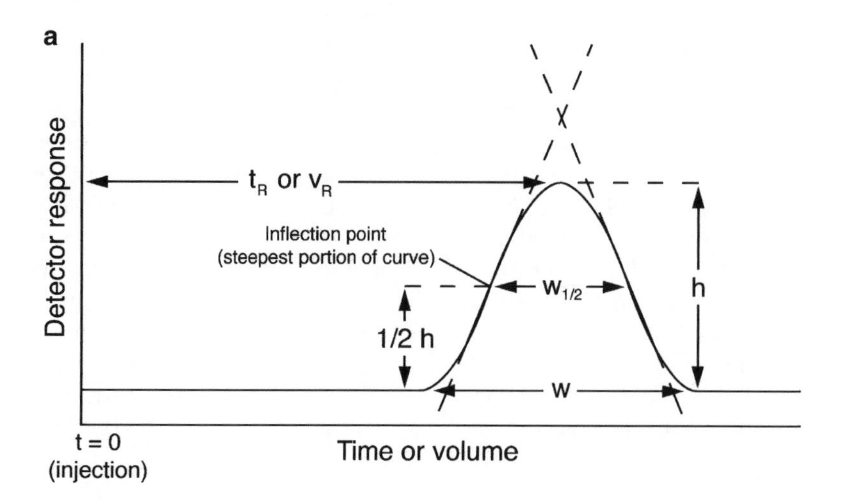

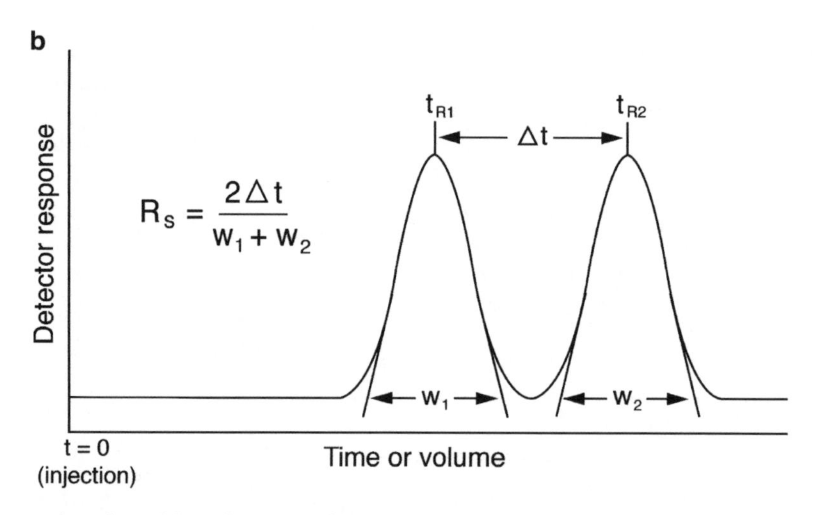

Measurement of peak width and its contribution to resolution. (**a**) Idealized Gaussian chromatogram, illustrating the measurement of w and $w_{1/2}$; (**b**) the resolution of two bands is a function of both their relative retentions and peak widths. [Adapted from (9), with permission.]

Chromatographic resolution is a function of column **efficiency**, **selectivity**, and the **capacity factor**. Mathematically, this relationship is expressed as follows:

$$R_s = \underbrace{1/4\sqrt{N}}_{a}\underbrace{\left(\frac{\alpha - 1}{\alpha}\right)}_{b}\underbrace{\left(\frac{k'}{k'+1}\right)}_{c} \quad [5]$$

where:

a = column **efficiency** term
b = column **selectivity** term
c = **capacity** term

These terms, and factors that contribute to them, will be discussed in the following sections.

27.5.1.2.2 Column Efficiency If faced with the problem of improving resolution, a chromatographer should first examine the **efficiency** of the column. An efficient column keeps the bands from spreading and gives narrow peaks. Column efficiency can be calculated by

$$N\left(\frac{t_R}{\sigma}\right)^2 = 16\left(\frac{t_R}{w}\right)^2 = 5.5\left(\frac{t_R}{w_{1/2}}\right)^2 \quad [6]$$

where:

N = number of theoretical plates
t_R = retention time
σ = standard deviation for a Gaussian peak
w = peak width at baseline ($w = 4\sigma$)
$w_{1/2}$ = peak width at half height

The measurement of t_R, w, and $w_{1/2}$ is illustrated in Fig. 27-12. (Retention volume may be used instead of t_R; in this case, band width is also measured in units of volume.) Although some peaks are not actually Gaussian in shape, normal practice is to treat them as if they were. In the case of peaks that are incompletely resolved or slightly asymmetric, peak width at half height is more accurate than peak width at baseline.

The value N calculated from the above equation is called the number of **theoretical plates**. The theoretical plate concept, borrowed from distillation theory, can best be understood by viewing chromatography as a series of equilibrations between mobile and stationary phases, analogous to countercurrent distribution. Thus, a column would consist of N segments (theoretical plates) with one equilibration occurring in each. As a first approximation, N is independent of retention time and is therefore a useful measure of column performance. One method of monitoring column performance over time is to chromatograph a standard compound periodically, under constant conditions, and to compare the values of N obtained. It is important to note that columns often behave as if they have a different number of plates for different solutes in a mixture. Different solutes have different partition coefficient and thus have distinctive series of equilibrations between mobile and stationary phases. Band broadening due to column deterioration will result in a decrease of N for a particular solute. Band broadening is a result of an extended time for a solute to reach equilibrium between mobile and stationary phases.

The number of theoretical plates is generally proportional to column length. Since columns are available in various lengths, it is useful to have a measure of column efficiency that is independent of column length. This may be expressed as follows:

$$\mathrm{HETP} = \frac{L}{N} \quad [7]$$

where:

HETP = height equivalent to a theoretical plate
L = column length
N = number of theoretical plates

The so-called **HETP** is sometimes more simply described as **plate height** (H). If a column consisted of discrete segments, HETP would be the height of each imaginary segment. Small plate height values (a large number of plates) indicate good efficiency of separation. Conversely, reduced number of plates results in poor separation due to the extended equilibrium time in a deteriorating column.

In reality, columns are not divided into discrete segments and equilibration is not infinitely fast. The plate theory is used to simplify the equilibration concept. The movement of solutes through a chromatography column takes into account the finite rate at which a solute can equilibrate itself between stationary and mobile phases. Thus, band shape depends on the rate of elution and is affected by solute diffusion. Any mechanism that causes a band of solute to broaden will increase HETP and decrease column efficiency. The various factors that contribute to plate height are expressed by the **Van Deemter equation**:

$$\mathrm{HETP} = \mathrm{A} + \frac{\mathrm{B}}{u} + \mathrm{C}u \quad [8]$$

where:

HETP = height equivalent to a theoretical plate
A, B, C = constants
u = mobile phase rate

The constants A, B, and C are characteristic for a given column, mobile phase, and temperature. The A term represents the **eddy diffusion** or multiple flowpaths. Eddy diffusion refers to the different microscopic flowstreams that the mobile phase can take between particles in the column (analogous to eddy streams around rocks in a brook). Sample molecules can thus take different paths as well, depending on which flowstreams they follow. As a result, solute molecules spread from an initially narrow band to a broader area within the column. Eddy diffusion may be minimized by good column packing techniques and the use of small diameter particles of narrow particle size distribution.

The B term of the Van Deemter equation, sometimes called the longitudinal diffusion term, exists because all solutes diffuse from an area of high concentration (the center of a chromatographic band) to one of low concentration (the leading or trailing edge of a chromatographic band). In LC, the contribution of this term to HETP is small except at low flow rate of the mobile phase. With slow flow rates there will be more time for a solute to spend on the column, thus its diffusion will be greater.

The C (mass transfer) term arises from the finite time required for solute to equilibrate between the mobile and stationary phases. Mass transfer is practically the partitioning of the solute into the stationary phase, which does not occur instantaneously and depends on the solute's partition and diffusion coefficients. If the stationary phase consists of porous particles (see Chap. 28, Sect. 28.2.3.2, Fig. 28-3), a sample molecule entering a pore ceases to be transported by the solvent flow and moves by diffusion only. Subsequently, this solute molecule may diffuse back to the mobile phase flow or it may interact with the stationary phase. In either case, solute molecules inside the pores are slowed down relative to those outside the pores and band broadening occurs. Contributions to HETP from the C term can be minimized by using porous particles of small diameter or pellicular packing materials (Chap. 28, Sect. 28.2.3.2.2).

As expressed by the Van Deemter equation, **mobile phase flow rate**, u, contributes to plate height in opposing ways – increasing the flow rate increases the equilibration point (Cu), but decreases longitudinal diffusion of the solute particles (B/u). A Van Deemter plot (Fig. 27-13) may be used to determine the mobile phase flow rate at which plate height is minimized and column efficiency is maximized. Flow rates above the optimum may be used to decrease analysis time if adequate resolution is still obtained. However,

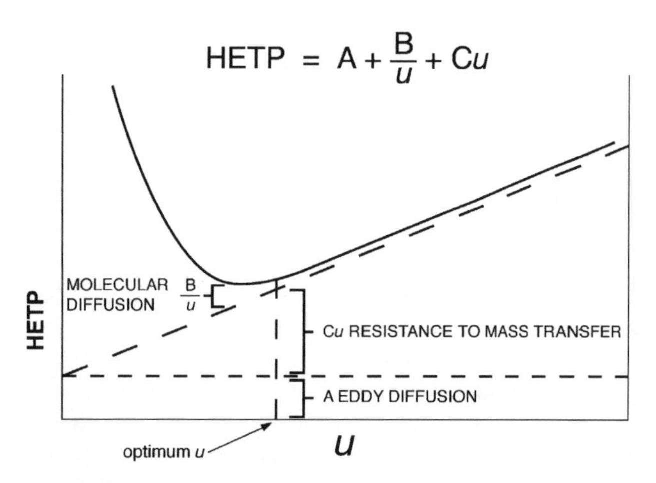

27-13 figure Van Deemter plot of column efficiency (HETP) vs. mobile phase rate (u). Optimum u is noted. (Courtesy of Hewlett-Packard Co., Analytical Customer Training, Atlanta, GA.)

at very high flow rates, there will be less time to approach equilibrium, which will lead to broadening of the band.

In addition to flow rate, temperature can affect the longitudinal diffusion and the mass transfer. Increasing the temperature causes enhanced movement of the solute between the mobile phase and the stationary phase, and within the column, thus leading to faster elution and narrower peaks.

27.5.1.2.3 Column Selectivity Chromatographic resolution depends on **column selectivity** as well as efficiency. Column selectivity refers to the distance, or relative separation, between two peaks and is given by

$$\alpha = \frac{t_{R2} - t_o}{t_{R1} - t_o} = \frac{t'_{R2}}{t'_{R1}} = \frac{K_2}{K_1} \qquad [9]$$

where:

α = separation factor

t_{R1} and t_{R2} = retention times of components 1 and 2, respectively

t_o (or t_m) = retention time of unretained components (solvent front)

t'_{R1} and t'_{R2} = adjusted retention times of components 1 and 2, respectively

K_1 and K_2 = distribution coefficients of components 1 and 2, respectively

Retention times (or volumes) are measured as shown in Fig. 27-11. The time, t_o, can be measured by chromatographing a solute that is not retained under the separation conditions (i.e., travels with the solvent front). When this parameter is expressed in units of volume, V_o or V_m, it is known as the **dead-volume** of the system. Selectivity is a function of the stationary and/or mobile phase. For example, selectivity in ion-exchange chromatography is influenced by the nature and number of ionic groups on the matrix but also can be manipulated via pH and ionic strength of the mobile phase. Good selectivity is probably more important to a given separation than high efficiency (Fig. 27-14), since resolution is directly related to selectivity but is quadratically related to efficiency; thus a fourfold increase in N is needed to double R_S (Equation [5]).

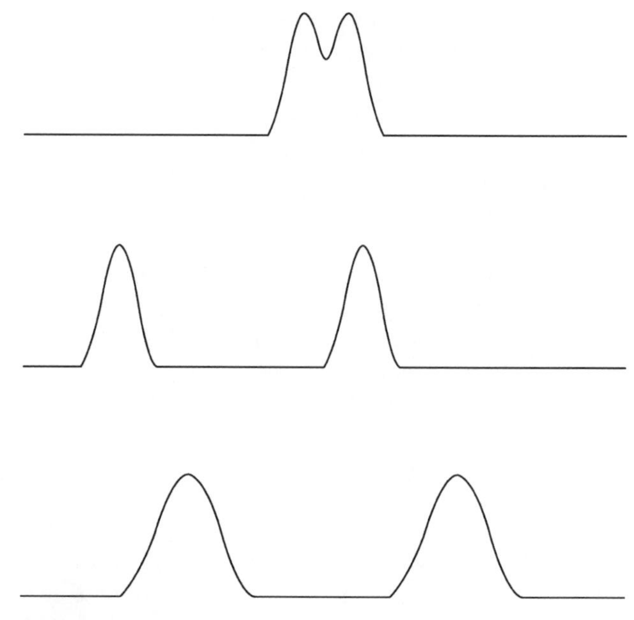

27-14 figure Chromatographic resolution: efficiency vs. selectivity. (**a**) Poor resolution; (**b**) good resolution due to high column efficiency; (**c**) good resolution due to column selectivity. [From (9), used with permission.]

27.5.1.2.4 Column Capacity Factor The **capacity** or **retention factor**, k', is a measure of the amount of time a chromatographed species (solute) spends in/on the stationary phase relative to the mobile phase. The relationship between capacity factors and chromatographic retention (which may be expressed in units of either volume or time) is shown below:

$$k' = \frac{KV_s}{V_m} = \frac{V_R - V_m}{V_m} = \frac{t_R - t_o}{t_o} \qquad [10]$$

where:

k' = capacity factor

K = distribution coefficient of the solute

V_s = volume of stationary phase in column

V_m = volume of mobile phase

V_R = retention volume of solute

t_R = retention time of solute

t_o = retention time of unretained components (solvent front)

Small values of k' indicate little retention, and components will be eluted close to the solvent front, resulting in poor separations. Overuse or misuse of the column may lead to the loss of some functional groups, thus resulting in small k' values. Large values of k' result in improved separation but also can lead to broad peaks and long analysis times. On a practical basis, k' values within the range of 1–15 are generally used. (In the equation for R_s, k' is actually the average of $k_1{}'$ and $k_2{}'$ for the two components separated.)

27.5.2 Qualitative Analysis

Once separation and resolution have been optimized, identification of the detected compounds can be achieved. (Various detection methods are outlined in Chaps. 28 and 29.) Comparing V_R or t_R to that of standards chromatographed under identical conditions often enables one to identify an unknown compound. When it is necessary to compare chromatograms obtained from two different systems or columns, it is better to compare **adjusted retention time**, t'_R (see Sect. 27.5.2.2). Different compounds may have identical retention times. In other words, even if the retention time of an unknown and a standard are equivalent, the two compounds might not be identical. Therefore, other techniques are needed to confirm peak identity. For example:

1. Spike the unknown sample with a known compound and compare chromatograms of the original and spiked samples to see which peak has increased. Only the height of the peak of interest should increase, with no change in retention time, peak width, or shape.
2. A diode array detector can provide absorption spectra of designated peaks (see Sects. 23.2.6 and 28.2.4.1). Although identical spectra do not prove identity, a spectral difference confirms that sample and standard peaks are different compounds.
3. In the absence of spectral scanning capability, other detectors, such as absorption or fluorescence, may be used in a ratioing procedure. Chromatograms of sample and standard are monitored at each of two different wavelengths. The ratio of peak areas at these wavelengths should be the same if sample and standard are identical.
4. Peaks of interest can be collected and subjected to additional chromatographic separation using a different separation mode.
5. Collect the peak(s) of interest and establish their identity by another analytical method (e.g., mass spectrometry, which can give a mass spectrum that is characteristic of a particular compound; see Chap. 26).

27.5.3 Quantitative Analysis

Assuming that good chromatographic resolution and identification of sample components have been achieved, quantification involves measuring peak height, area, or mass and comparing these data with those for standards of known concentration. When strip chart recorders were commonly used, measurement of the peak height, area (usually using: peak area = width at half height × height) (see Fig. 27-12a), or mass was done manually. This was followed by stand-alone integrators that produced a chromatogram on a strip of paper and digitally integrated the peaks by area for a tabular report. Nearly all chromatography systems (especially GC and HPLC systems) now use data analysis software, which recognizes the start, maximum, and end of each chromatographic peak, even when not fully resolved from other peaks. These values then are used to determine retention times and peak areas. At the end of each run, a report is generated that lists these data and postrun calculations, such as relative peak areas, areas as percentages of the total area, and relative retention times. If the system has been standardized, data from external or internal standards can be used to calculate analyte concentrations. Data analysis software to quantify peaks is not as common in low-pressure, preparative chromatography, in which postchromatography analysis of collected fractions is used to identify samples eluted. Examples of postchromatography analysis include the BCA (bicinchoninic acid) protein assay (Chap. 9, Sect. 9.2.7) and the phenol-sulfuric acid assay for carbohydrate (Chap. 10, Sect. 10.3.2). After obtaining the absorbance reading on a spectrophotometer for such assays, the results are plotted as fraction number on the x-axis and absorbance on the y-axis, to determine which fractions contain protein and/or carbohydrate.

Having quantified sample peaks, one must compare these data with appropriate standards of known concentration to determine sample concentrations. Comparisons may be by means of **external** or **internal** standards. Comparison of peak height, area, or mass of unknown samples with standards injected separately (i.e., **external standards**) is common practice. Standard solutions covering the desired concentration range (preferably diluted from one stock solution) are chromatographed, and the appropriate data (peak height, area, or mass) plotted vs. concentration to obtain a standard curve. An identical volume of sample is then chromatographed, and height, area, or mass of the sample peak is used to determine sample concentration via the standard curve (Fig. 27-15a). This absolute calibration method requires precise analytical technique and requires that detector sensitivity be constant from day to day if the calibration curve is to remain valid.

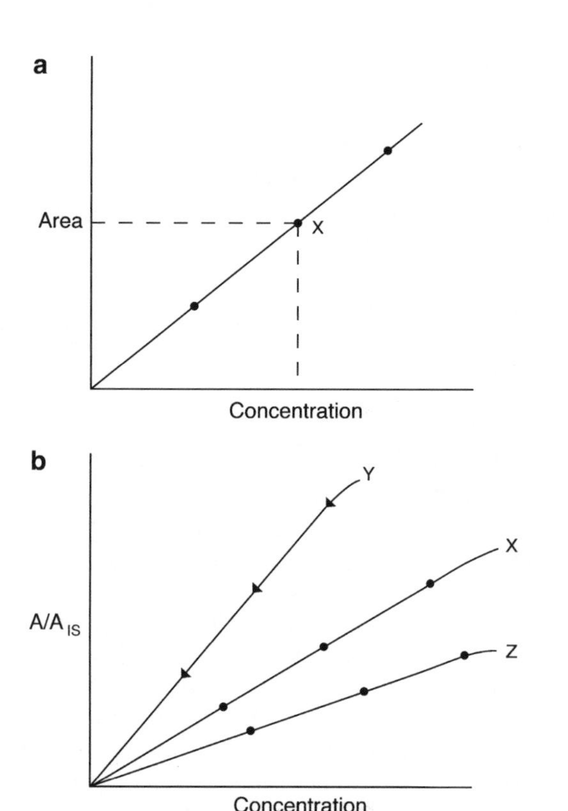

27-15 figure Calibration curves for quantification of a sample component, *x*. (**a**): External standard technique; (**b**): internal standard technique. [Adapted from (9), with permission.]

Use of the **internal standard** (relative or indirect) method can minimize errors due to sample preparation, apparatus, and operator technique. In this technique, a compound is utilized that is structurally related to, but is eluted independently of, compounds of interest in the sample to be analyzed. Basically, the amount of each component in the sample is determined by comparing the height, area, or mass of that component peak to the height, area, or mass of the internal standard peak. However, variation in detector response between compounds of different chemical structure must be taken into account. One way to do this is by first preparing a set of standard solutions containing varying concentrations of the compound(s) of interest. Each of these solutions is made to contain a known and constant amount of the internal standard. These standard solutions are chromatographed, and peak height, area, or mass is measured. Ratios of peak height, area, or mass (compound of interest/internal standard) are calculated and plotted against concentration to obtain calibration curves such as those shown in Fig. 27-15b. A separate response curve must be plotted for each sample component to be quantified. Next, a known amount of internal standard is added to the unknown sample, and the sample is chromatographed. Peak height, area, or mass ratios (compound of interest/internal standard) are calculated and used to read the concentration of each relevant component from the appropriate calibration curve. The advantages of using internal standards are that injection volumes need not be accurately measured and the detector response need not remain constant since any change will not alter ratios. The main disadvantage of the internal standard technique is the difficulty of finding a standard that does not interfere chromatographically with components of interest in the sample.

Standards should be included during each analytical session, since detector response may vary from day to day. Analyte recovery should be checked periodically. This involves addition of a known quantity of standard to a sample (usually before extraction) and determination of how much is recovered during subsequent analysis. During routine analyses, it is highly desirable to include a control or check sample, a material of known composition. This material is analyzed parallel to unknown samples. When the concentration of analyte measured in the control falls outside an acceptable range, data from other samples analyzed during the same period should be considered suspect. Carefully analyzed food samples and other substances are available from the National Institute of Standards and Technology (formerly the National Bureau of Standards) for use in this manner.

27.6 SUMMARY

Chromatography is a separation method based on the partitioning of a solute between a mobile phase and a stationary phase. The mobile phase may be liquid, gas, or a supercritical fluid. The stationary phase may be an immobilized liquid or a solid, in either a planar or column form. Based on the physicochemical characteristics of the analyte, and the availability of instrumentation, a chromatographic system is chosen to separate, indentify, and quantify the analyte. Chromatographic modes include adsorption, partition, ion exchange, size exclusion, and affinity chromatography. Factors to be considered when developing a separation include mobile phase variables (strength, pH, temperature, and flow rate), and column efficiency, selectivity, and capacity. Following detection, a chromatogram provides both qualitative and quantitative information via retention time and peak height area data.

For an introduction to the techniques of HPLC and GC, the reader is referred again to Chaps. 28 and 29 in this text or to the excellent 5th edition of *Quantitative Chemical Analysis* by D.C. Harris (6). The book by R.M. Smith (13) also contains information on basic concepts of chromatography and chapters devoted to TLC, LC, and HPLC, as well as an extensive discussion of GC. References (5), (7), (15), and (16) contain a wealth of information on TLC. SFC is discussed in detail by Caude and Thiébaut (2). *Chromatograph* (8), the standard work edited by E. Heftmann (2004 and earlier editions), is an excellent source of information on both fundamentals (Part A) and applications (Part B) of chromatography. Part B includes chapters on the chromatographic analysis of amino acids, proteins, lipids, carbohydrates, and phenolic compounds. In addition, *Fundamental* and *Applications Reviews* published (in alternating years) by the journal *Analytical Chemistry* relate new developments in all branches of chromatography, as well as their application to specific areas, such as food. Recent books and general review papers are referenced, along with research articles published during the specified review period.

27.7 STUDY QUESTIONS

1. Explain the principle of countercurrent extraction and how it developed into partition chromatography.
2. For each set of two (or three) terms used in chromatography, give a brief explanation as indicated to distinguish between the terms.

   a. Adsorption vs. partition chromatography

	Adsorption	Partition
Nature of stationary phase		
Nature of mobile phase		
How solute interacts with the phases		

   b. Normal-phase vs. reversed-phase chromatography

	Normal-phase	Reversed-phase
Nature of stationary phase		
Nature of mobile phase		
What elutes last		

   c. Cation vs. anion exchangers

	Cation exchanger	Anion exchanger
Charge on column		
Nature of compounds bound		

   d. Internal standards vs. external standards

Nature of Stds.	How stds. are handled in relation to samples	What is plotted on std. curve
Internal standard		
External standard		

   e. TLC vs. column liquid chromatography

	Thin layer	Column liquid
Nature and location of stationary phase		
Nature and location of mobile phase		
How sample is applied		
Identification of solutes separated		

   f. HETP vs. N vs. L (from the equation HETP $= L/N$)

3. State the advantages of TLC as compared to paper chromatography.
4. State the advantages of column liquid chromatography as compared to planar chromatography.
5. Explain how SFC differs from LC and GC, including the advantages of SFC.
6. What is the advantage of bonded supports over coated supports for partition chromatography?
7. You are performing LC using a stationary phase that contains a polar nonionic functional group. What type of chromatography is this, and what could you do to increase the retention time of an analyte?
8. You applied a mixture of proteins, in a buffer at pH 8.0, to an anion-exchange column. On the basis of some assays you performed, you know that the protein of interest adsorbed to the column.
   (a) Does the anion-exchange stationary phase have a positive or negative charge?
   (b) What is the overall charge of the protein of interest that adsorbed to the stationary phase?
   (c) Is the isoelectric point of the protein of interest (adsorbed to the column) higher or lower than pH 8.0?
   (d) What are the two most common methods you could use to elute the protein of interest from the anion-exchange column? Explain how each method works. (See also Chap. 15.)
9. Would you use a polystyrene- or a polysaccharide-based stationary phase for work with proteins? Explain your answer.
10. Explain how you would use SEC to estimate the molecular weight of a protein molecule. Include an explanation of what information must be collected and how it is used.
11. Explain the principle of affinity chromatography, why a spacer arm is used, and how the solute can be eluted.
12. What is gradient elution from a column, and why is it often advantageous over isocratic elution?
13. A sample containing compounds A, B, and C is analyzed via LC using a column packed with a silica-based C_{18} bonded phase. A 1:5 solution of ethanol and H_2O was used as the mobile phase. The following chromatogram was obtained.

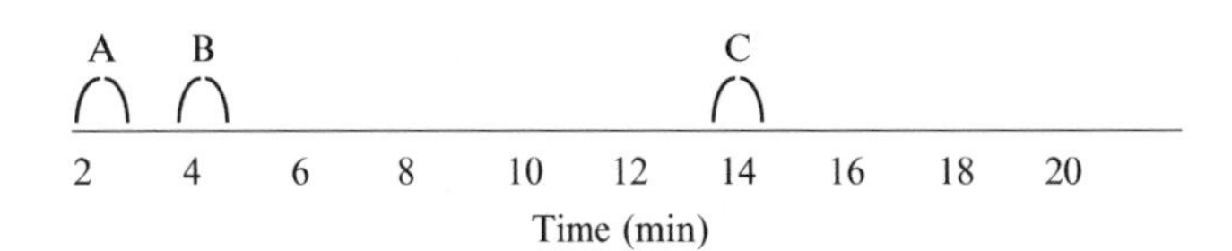

Assuming that the separation of compounds is based on their polarity,

(a) Is this normal- or reversed-phase chromatography? Explain your answer.
(b) Which compound is the most polar?
(c) How would you change the mobile phase so that compound C would elute sooner, without changing the relative positions of compounds A and B? Explain why this would work.
(d) What could possibly happen if you maintained an isocratic elution mode at low solvent strength?

14. Using the Van Deemter equation, HETP, and N, as appropriate, explain why the following changes may increase the efficiency of separation in column chromatography:
    (a) Changing the flow rate of the mobile phase
    (b) Increasing the length of the column
    (c) Reducing the inner diameter of the column
15. State the factors and conditions that lead to poor resolution of two peaks.
16. How can chromatographic data be used to quantify sample components?
17. Why would you choose to use an internal standard rather than an external standard? Describe how you would select an internal standard for use.
18. To describe how using internal standards works, answer the following questions.
    (a) What specifically will you do with the standards?
    (b) What do you actually measure and plot?
    (c) How do you use the plot?

27.8 ACKNOWLEDGMENTS

The authors of this chapter wish to acknowledge Dr. Bradley Reuhs, who was of great help in discussions about reorganizing the chromatography chapters and in the editing of this chapter.

27.9 REFERENCES

1. Borch-Jensen C, Mollerup J (1996) Applications of supercritical fluid chromatography to food and natural products. Semin Food Anal 1:101–116
2. Caude M, Thiébaut D (1999) Practical supercritical fluid chromatography and extraction. Harwood Academic, Amsterdam
3. Chester TL, Pinkston JD, Raynie DE (1996) Supercritical fluid chromatography and extraction (fundamental review). Anal Chem 68:487R–514R
4. Craig LC (1943) Identification of small amounts of organic compounds by distribution studies. Application to Atabrine. J Biol Chem 150:33–45
5. Fried B, Sherma J (1999) Thin-layer chromatography, 4th edn. Marcel Dekker, New York
6. Harris DC (1999) Quantitative chemical analysis, 5th edn. W.H. Freeman, New York
7. Hahn-Deinstrop E (2007) Applied thin-layer chromatography: best practice and avoidance of mistakes, 2nd edn. Wiley-VCH, Weinheim, Germany
8. Heftmann E (ed) (2004) Chromatography, 6th edn. Fundamentals and applications of chromatography and related differential migration methods. Part A: fundamentals and techniques. Part B: applications. J Chromatogr Library Ser vols 69A and 69B. Elsevier, Amsterdam
9. Johnson EL, Stevenson R (1978) Basic liquid chromatography. Varian Associates, Palo Alto, CA
10. Lawrence JF (ed) (1984) Food constituents and food residues: their chromatographic determination. Marcel Dekker, New York
11. Lough WJ, Wainer IW (eds) (1995) High performance liquid chromatography: fundamental principles and practice. Blackie Academic & Professional, Glasgow, Scotland
12. Scopes RK (1994) Protein purification: principles and practice, 3rd edn. Springer-Verlag, New York
13. Smith RM (1988) Gas and liquid chromatography in analytical chemistry, Wiley, Chichester, England
14. Snyder LR, Kirkland JJ (eds) (1979) Introduction to modern liquid chromatography, 2nd edn. Wiley, New York
15. Touchstone JC (1992) Practice of thin layer chromatography. Wiley, New York
16. Wall PE (2005) Thin-layer chromatography: a modern practical approach. The Royal Society of Chemistry, Cambridge, UK
17. Walters RR (1985) Report on affinity chromatography. Anal Chem 57:1099A–1113A
18. Weiss J (2004) Handbook of ion chromatography, 3rd edn. Wiley-VCH Verlag GmbH & Co. KGaA, Weinheim
19. Zachariou M (2008) Affinity chromatography: methods and protocols. Humana, Totowa, NJ

High-Performance Liquid Chromatography

Bradley L. Reuhs* and Mary Ann Rounds

Department of Food Science, Purdue University, West Lafayette, IN 47907-2009, USA

breuhs@purdue.edu

28.1 Introduction 501
28.2 Components of an HPLC System 501
28.2.1 Pump 501
28.2.2 Injector 502
28.2.3 Column 503
28.2.3.1 Column Hardware 503
28.2.3.1.1 Precolumns 503
28.2.3.1.2 Analytical Columns 503
28.2.3.2 HPLC Column Packing Materials 503
28.2.3.2.1 General Requirements 503
28.2.3.2.2 Silica-Based Column Packings 504
28.2.3.2.3 Porous Polymeric Column Packings 504

S.S. Nielsen, *Food Analysis*, Food Science Texts Series, DOI 10.1007/978-1-4419-1478-1_28,

28.2.4 Detector 505
28.2.4.1 UV-Vis Absorption Detectors 505
28.2.4.2 Fluorescence Detectors 505
28.2.4.3 Refractive Index Detectors 505
28.2.4.4 Electrochemical Detectors 505
28.2.4.5 Other HPLC Detectors 506
28.2.4.6 Coupled Analytical Techniques 506
28.2.4.7 Chemical Reactions 506
28.2.5 Data Station Systems 507
28.3 Applications in HPLC 507
28.3.1 Normal Phase 507
28.3.1.1 Stationary and Mobile Phases 507
28.3.1.2 Applications of Normal-Phase HPLC 508
28.3.2 Reversed Phase 508
28.3.2.1 Stationary and Mobile Phases 508
28.3.2.2 Applications of Reversed-Phase HPLC 508
28.3.3 Ion Exchange 509
28.3.3.1 Stationary and Mobile Phases 509
28.3.3.2 Applications of Ion-Exchange HPLC 509
28.3.3.2.1 Ion Chromatography 509
28.3.3.2.2 Ion Exchange Chromatography of Carbohydrates and Proteins 510
28.3.4 Size Exclusion 510
28.3.4.1 Column Packings and Mobile Phases 510
28.3.4.2 Applications of High Performance SEC 511
28.3.5 Affinity 511
28.4 Summary 511
28.5 Study Questions 512
28.6 Acknowledgments 512
28.7 References 512

28.1 INTRODUCTION

High-performance liquid chromatography (HPLC) developed during the 1960s as a direct offshoot of classic column liquid chromatography through improvements in the technology of columns and instrumental components (pumps, injection valves, and detectors). Originally, HPLC was the acronym for *high-pressure liquid chromatography*, reflecting the high operating pressures generated by early columns. By the late 1970s, however, *high-performance liquid chromatography* had become the preferred term, emphasizing the effective separations achieved. In fact, newer columns and packing materials offer high performance at moderate pressure (although still high pressure relative to gravity-flow liquid chromatography). HPLC can be applied to the analysis of any compound with solubility in a liquid that can be used as the mobile phase. Although most frequently employed as an **analytical** technique, HPLC also may be used in the **preparative** mode. There are many *advantages* of HPLC over traditional low pressure column liquid chromatography:

1. Speed (many analyses can be accomplished in 30 min or less)
2. A wide variety of stationary phases
3. Improved resolution
4. Greater sensitivity (various detectors can be employed)
5. Easy sample recovery (less eluent volume to remove)

Application of HPLC to the analysis of food began in the late 1960s, and its use increased with the development of column packing materials that would separate sugars. Using HPLC to analyze sugars was justified economically as a result of sugar price increases in the mid 1970s, which motivated soft drink manufacturers to substitute high-fructose corn syrup for sugar. Monitoring sweetener content by HPLC assured a good quality product. Other early food applications included the analysis of pesticide residues in fruits and vegetables, organic acids, lipids, amino acids, toxins (such as aflatoxins in peanuts), and vitamins (1). HPLC continues to be applied to these, and many more, food-related analyses today (2–5).

28.2 COMPONENTS OF AN HPLC SYSTEM

A schematic diagram of a basic HPLC system is shown in Fig. 28-1. The main components of this system – **pump, injector, column, detector**, and **data system** – are discussed briefly in the sections below. Also important are the mobile phase (**eluent**) reservoirs, and a fraction collector, which is used if further analysis of separated components is needed. Connecting tubing, tube fittings, and the materials out of which components are constructed also influence system performance and lifetime. References (1) and (6–10,15) include detailed discussions of HPLC equipment, with the book by Bidlingmeyer (1) especially appropriate for beginners. The unique organization of reference (8) is intended for those who may need to learn chromatography quickly in an industrial environment. Two useful books on HPLC troubleshooting are those written by Gertz (11) and Dolan and Snyder (12). In addition, much information on HPLC equipment, hardware, and troubleshooting hints may be found in publications such as *LC·MS, American Laboratory, Chemical & Engineering News*, and similar periodicals. Manufacturers are also a source of practical information on HPLC instrumentation and columns/stationary phase material.

28.2.1 Pump

The **HPLC pump** delivers the mobile phase through the system, typically at a flow rate of 0.4–1 ml/min, in a controlled, accurate, and precise manner. The majority of pumps currently used in HPLC (>90%) are reciprocating, piston-type pumps. The dual piston pump systems with ball check valves are the most efficient pumps available. One disadvantage of reciprocating pumps is that they produce a pulsating flow, requiring the addition of pulse dampers to suppress fluctuations. A mechanical **pulse damper** or **dampener** consists of a device (such as a deformable metal component or tubing filled with compressible liquid) that can change its volume in response to changes in pressure.

Gradient elution systems for HPLC are used to vary the mobile phase concentration during the run, by mixing mobile phase from two or more reservoirs. This is accomplished with low-pressure mixing, in which mobile phase components are mixed before entering the high-pressure pump, or high-pressure mixing, in which two or more independent, programmable pumps are used. For low-pressure gradient systems, a computer-controlled proportioning valve, followed by a mixing chamber at the inlet to the pumps is used, which results in extremely accurate and reproducible gradients. Gradient HPLC is extremely important for the effective elution of all components of a sample and for optimal resolution. It is routinely applied to all modes of HPLC except size-exclusion chromatography.

Many commercially available HPLC pumping systems and connecting lines are made of grade ANSI 316 stainless steel, which can withstand the pressures generated. Also it is resistant to corrosion by oxidizing agents, acids, bases, and organic solvents, although mineral acids and halide ions do attack stainless

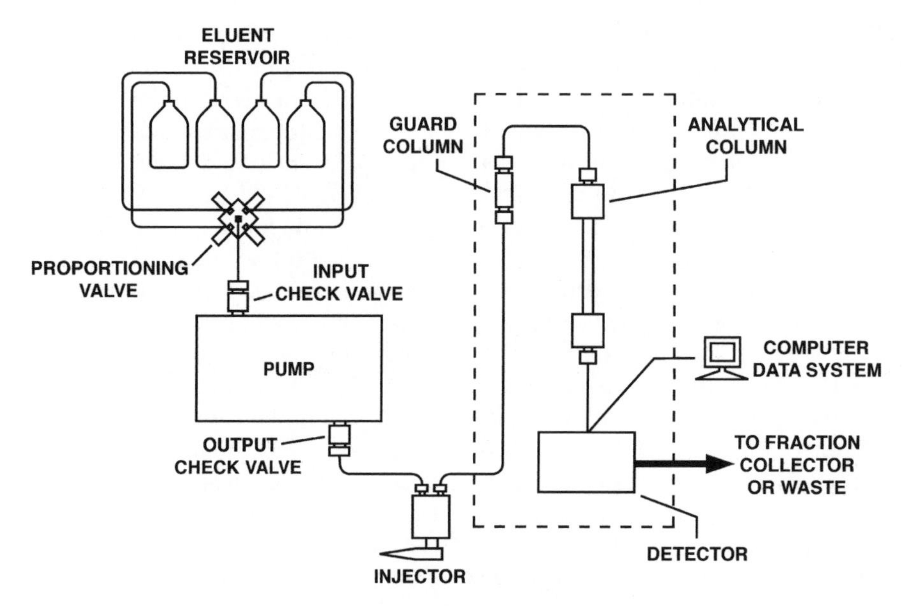

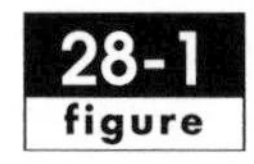
Schematic representation of a system for high-performance liquid chromatography (not drawn to scale). Column(s) and detector may be thermostatted, as indicated by the *dashed line*, for operation at elevated temperature.

steel. In other systems, all components that come into contact with the eluent are made of sturdy, inert polymers, and even employ sapphire pistons, which are resistant to extreme pH and high salt concentration. The latter systems can be used for all applications except normal phase, which uses organic solvents as the mobile phase. The polymer-based systems have facilitated a wider application of ion exchange HPLC.

All HPLC pumps contain moving parts such as check valves and pistons, and are quite sensitive to dust and particulate matter in the liquid being pumped. Therefore, it is advisable to filter the mobile phase using 0.45 or 0.22-μm filters prior to use. Degassing HPLC eluents, by the application of a vacuum or by sparging with helium, also is recommended to prevent the problems caused by air bubbles in a pump or detector.

28.2.2 Injector

The role of the injector is to place the sample into the flowing mobile phase for introduction onto the column. Virtually all HPLC systems use **valve injectors**, which separate sample introduction from the high-pressure eluent system. With the injection valve in the LOAD position (Fig. 28-2a), the sample is loaded into an **external, fixed-volume loop** using a syringe. Eluent, meanwhile, flows directly from the pump to the column at high pressure. When the valve is rotated to the INJECT position (Fig. 28-2b), the loop becomes part of the eluent flow stream and sample is carried onto the column. Such injectors are generally trouble free and afford good precision.

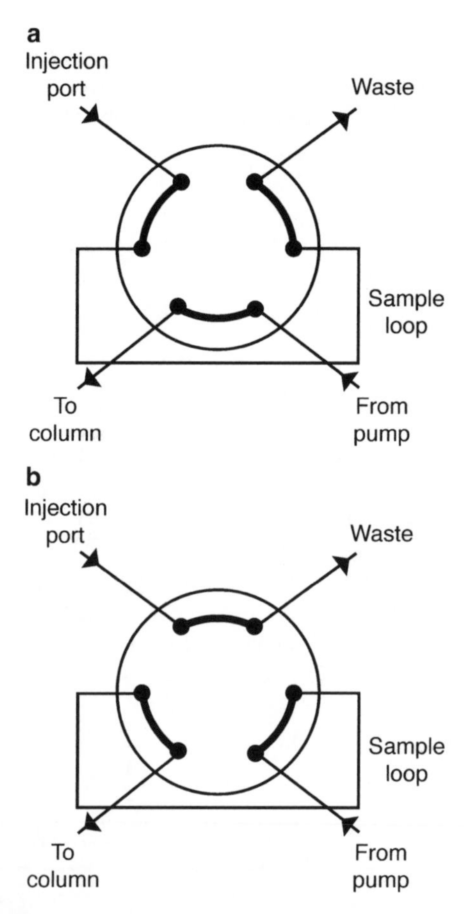

Valve-type injector. The valve allows the sample loop to be (**a**) isolated from the pump eluent stream (LOAD position) or (**b**) positioned in it (INJECT position). [from (9), used with permission.]

Changing the loop allows different volumes to be injected. Although injection volumes of 10–100 μl are typical, both larger (e.g., 1–10 ml) and smaller (e.g., ≤2 μl) sample volumes can be loaded by utilizing special hardware. An important advantage of the loop valve design is that it is readily adapted to automatic operation. Thus, automated sample injectors, or **autosamplers**, may be used to store and inject large numbers of samples. Samples are placed in uniform-size vials, sealed with a septum, and held in a (possibly refrigerated) tray. A computer actuated needle penetrates the septum to withdraw solution from the vial, and a mechanically or pneumatically operated valve introduces it onto the column. Autosamplers can reduce the tedium and labor costs associated with routine HPLC analyses and improve assay precision. However, because samples may remain unattended for 12–24 h prior to automatic injection, sample stability is a limiting factor for using this accessory.

28.2.3 Column

28.2.3.1 Column Hardware

An **HPLC column** is usually constructed of stainless steel tubing with terminators that allow it to be connected between the injector and detector of the system (Fig. 28-1). Columns also are made from glass, fused silica, titanium, and polyether ether ketone (PEEK) resin; the PEEK columns are essential for the high pH, high salt concentrations necessary for the powerful ion exchange HPLC systems. Many types and sizes of columns are commercially available, ranging from 5 cm × 50 cm (or larger) preparative columns down to wall-coated capillary columns.

28.2.3.1.1 Precolumns Auxiliary columns that precede the analytical HPLC column are termed **precolumns**. Short (≤5 cm) expendable columns, called **guard columns**, often are used to protect the analytical column from strongly adsorbed sample components. A guard column (or cartridge) is installed between the injector and analytical column via short lengths of capillary tubing (or a cartridge holder). They may be filled with either **pellicular** media (see Sect. 28.2.3.2.2) of the same bonded phase as the analytical column, or with **microparticulate** (≤10 μm) packing material identical to that of the analytical column. Microparticulate guard columns are usually purchased as prepacked, disposable inserts for use in a special holder, and cost much less than replacing an analytical column. A guard column (or cartridge) should be repacked or replaced before its binding capacity is exceeded and contaminants pollute the analytical column.

28.2.3.1.2 Analytical Columns The most commonly used **analytical** HPLC columns are 10, 15, or 25-cm long with an internal diameter of 4.6 or 5 mm (9). Short (3 cm) columns, packed with ≤3 μm particles, are gaining popularity for fast separations; for example, in method development or process monitoring. In recent years, the use of columns with smaller internal diameters (<0.5–2.0 mm), including wall-coated capillary columns, has increased. The advantages of using smaller diameter columns include a decreased consumption of mobile phase, an increased peak concentration, increased resolution, and the ability to couple HPLC with mass spectrometry (MS) (13).

Various names have been used for the reduced-volume columns. Dorsey et al. (14) refer to columns with internal diameters of 0.5–2.0 mm as **microbore**, while packed or open tubular columns having internal diameters of <0.5 mm are termed **microcolumns** or **capillary columns** (a capillary column is a narrow-bore open tubular column, in which the inner surface is coated with a thin layer of stationary phase). In the case of the packed columns, the microbore or microcolumns contain very small particle size packing material. Because of the extremely high operating pressures of these systems, they are often referred to as **ultra-HPLC** (UHPLC). To achieve good performance from microcolumns, it is essential to have an HPLC system with very low dead-volume, so that peak broadening outside the column does not destroy resolution achieved within the column. Pumps, injectors, and other hardware designed specifically for use with these columns are available from commercial suppliers.

28.2.3.2 HPLC Column Packing Materials

The development of a wide variety of column packing materials has contributed substantially to the success and widespread use of HPLC.

28.2.3.2.1 General Requirements A **packing material** serves, first of all, to form the chromatographic bed; however, in most modes of chromatography the column packing material serves as both **support** and the **stationary phase**. Requirements for HPLC column packing materials are good chemical stability, sufficient mechanical strength to withstand pressure generated during use, and the availability of a well-defined particle size, with a narrow particle size distribution (10). Two materials that meet the above criteria are porous silica and synthetic organic resins (see Sects. 28.2.3.2.2 and 28.2.3.2.3, respectively).

28.2.3.2.2 Silica-Based Column Packings **Porous silica** meets the above criteria quite well and can be prepared in a wide range of particle and pore sizes, with a narrow particle size distribution. Both **particle size** and **pore diameter** are important: Small particles reduce the distance a solute must travel between stationary and mobile phases, which facilitates equilibration and results in good column efficiencies (Chap. 27, Sect. 27.5.2.2.2). However, small particles also yield greater flow resistance and higher pressure at equivalent flow rates. Spherical particles of 3, 5, or 10-μm diameter are utilized in analytical columns. One-half or more of the volume of porous silica consists of the **pores** (10). Use of the smallest possible pore diameter will maximize **surface area** and **sample capacity**, which is the amount of sample that can be separated on a given column. Packing materials with a pore diameter of 50–100 Å and surface area of 200–400 m^2/g are used for low-molecular-weight (<500 Da) solutes. For increasingly larger molecules, such as proteins and polysaccharides, it is necessary to use wider pore materials (pore diameter ≥300 Å), so that internal surface is accessible to the solute (10).

Bonded phases (Fig. 28-3a) are made by covalently bonding hydrocarbon moieties to –OH groups (silanols) on the surface of silica particles (10, 16). Often, the silica is reacted with an organochlorosilane:

$$\equiv Si{-}OH + Cl{-}Si(R_1)(R_2){-}R_3 \longrightarrow \equiv Si{-}O{-}Si(R_1)(R_2){-}R_3 + HCl \quad [1]$$

Substituents R_1 and R_2 may be halides or methyl groups. The nature of R_3 determines whether the resulting bonded phase will exhibit normal-phase, reversed-phase, or ion-exchange chromatographic behavior. The main disadvantage of silica and silica-based bonded-phase column packings is that the silica skeleton slowly dissolves in aqueous solutions, and the rate of this process becomes prohibitive at pH<2 and >8.

A **pellicular packing material** (Fig. 28-3b) is made by depositing a thin layer or coating onto the surface of an inert, usually nonporous, microparticulate **core**. Core material may be either inorganic, such as silica, or organic, such as poly(styrene-divinylbenzene) or latex. Functional groups, such as ion-exchange sites, are then present at the surface only. The rigid core ensures good physical strength, whereas the thin stationary phase provides for rapid mass transfer and favorable column efficiency.

28.2.3.2.3 Porous Polymeric Column Packings Synthetic organic resins offer the advantages of good chemical stability and the possibility to vary interactive properties through direct chemical modification. Two major categories of porous polymeric packing materials exist.

Microporous or gel-type resins (Fig. 28-3c) are comprised of crosslinked copolymers in which the apparent porosity, evident only when the gel is in its swollen state, is determined by the degree of crosslinking. These gel-type packings undergo swelling and contraction with changes in the chromatographic mobile phase. Microporous polymers of less than ca. 8% crosslinking are not sufficiently rigid for HPLC use.

Macroporous resins are highly crosslinked (e.g., ≥50%) and consist of a network of microspheric gel beads joined together to form a larger bead (Fig. 28-3d). Large, permanent pores, ranging from 100 to 4000 Å or more in diameter, and large surface areas (≥100 m^2/g) are the result of interstitial spaces between the microbeads (16). Rigid microparticulate poly(styrene-divinylbenzene) packing materials of the macroporous type are popular for HPLC use. They are stable from pH 1 to 14 and are available in a variety of particle and pore sizes. These resins can be used in unmodified form for reversed-phase chromatography or functionalized for use in other HPLC modes.

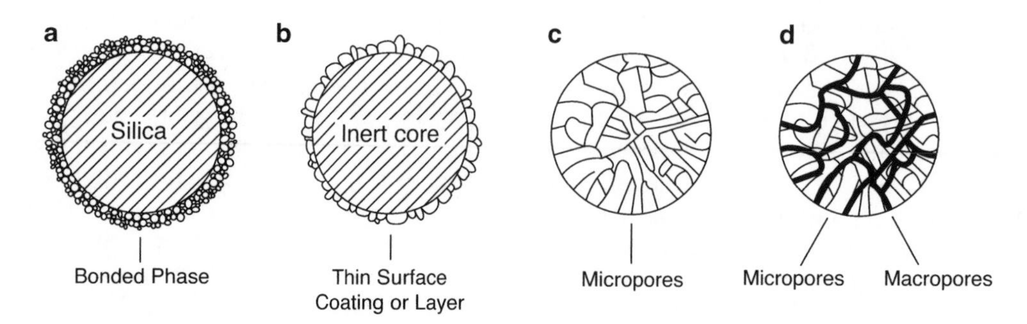

28-3 figure Some types of packing materials utilized in HPLC. (**a**) Bonded-phase silica; (**b**) pellicular packing; (**c**) microporous polymeric resin; (**d**) macroporous polymeric resin. [Adapted from (16), p. 621, by courtesy of Marcel Dekker, Inc.]

28.2.4 Detector

A **detector** translates sample concentration changes in the HPLC column effluent into electrical signals. Spectrochemical, electrochemical, or other properties of solutes may be measured by a variety of instruments, each of which has advantages and disadvantages. The choice of which to use depends on solute type and concentration, and on detector sensitivity, linear range, and compatibility with the solvent and elution mode to be used. Cost also may influence detector selection. One common feature of most HPLC detectors is the presence of a flow cell, through which the eluent flows as it is analyzed by the detector system. These flow cells are often delicate and easily polluted or damaged, so care must be taken when handling them.

The most widely used HPLC detectors are based on ultraviolet-visible (UV-Vis) and fluorescence spectrophotometry, refractive index determination, and electrochemical analysis (see Chap. 22 for detailed discussion of UV-Vis and fluorescence spectrophotometry). Many other methods, such as light scattering or mass spectrometry, also can be applied to the detection of analytes in HPLC eluents. More than one type of HPLC detector may be used in series, to provide increased **specificity** and **sensitivity** for multiple types of analytes. In one food-related application, a multidetector HPLC system equipped with a diode array absorption detector coupled to fluorescence and electrochemical detectors was used to monitor a wide variety of Maillard reaction products (2).

28.2.4.1 UV-Vis Absorption Detectors

Many HPLC analyses are carried out using a **UV-Vis absorption** detector, which can measure the absorption of radiation by chromophore-containing compounds. The three main types of UV-Vis absorption detectors are **fixed wavelength, variable-wavelength**, and **diode array spectrophotometers** (9). As its name implies, the simplest design operates at a single, **fixed wavelength**. A filter is used to isolate a single emission line (e.g., at 254 nm) from a source such as a mercury lamp. This type of detector is easy to operate and inexpensive, but of limited utility.

The most popular general purpose HPLC detector today is the **variable-wavelength** detector in which deuterium and tungsten lamps serve as sources of ultraviolet and visible radiation, respectively. Wavelength selection is provided by a **monochromator**, a device that acts somewhat like a prism to deflect light. An exit slit in the monochromator allows light from a limited range of wavelengths to pass through, and rotating the monochromator allows one to change the operating wavelength.

Diode array spectrophotometric detectors can provide much more information about sample composition than is possible with monochromatic detection. In this instrument, all the light from a deuterium lamp is spread out into a spectrum that falls across an array of photodiodes mounted on a silicon chip. These are read almost simultaneously by a microprocessor to provide the full absorption spectrum from 200 to 700 nm every 0.1 s, which may enable the components of a mixture to be identified. Although considerably more expensive than variable-wavelength detectors, they are useful in method development and in routine analyses in which additional evidence of peak identity, without further analysis, is needed.

28.2.4.2 Fluorescence Detectors

Some organic compounds can re-emit a portion of absorbed UV-Vis radiation at a longer wavelength (lower energy). This is known as **fluorescence**, and measurement of the **emitted light** provides another useful detection method. Fluorescence detection is both selective and very sensitive, providing up to 1000-fold lower detection limits than for the same compound in absorbance spectrophotometry (10). Although relatively few compounds are inherently fluorescent, analytes often are converted into fluorescent derivatives (see Sect. 28.2.4.7). Ideal for trace analysis, fluorescence detection has been used for the determination of various vitamins in foods and supplements, monitoring aflatoxins in stored cereal products, and the detection of aromatic hydrocarbons in wastewater.

28.2.4.3 Refractive Index Detectors

Refractive index (RI) detectors measure change in the **RI** of the mobile phase due to dissolved analytes, which provides a nearly universal method of detection. However, because a bulk property of the eluent is being measured, RI detectors are less sensitive than other types. Another disadvantage is that they cannot be used with gradient elution, as any change in eluent composition will alter its RI, thereby changing the baseline signal. RI detectors are widely used for analytes that do not contain UV-absorbing chromophores, such as carbohydrates and lipids, when the analytes are present at relatively high concentration.

28.2.4.4 Electrochemical Detectors

Electroanalytical methods used for HPLC detection are based either on electrochemical oxidation–reduction of the analyte or on changes in conductivity of the eluent. **Amperometric detectors** measure the change in current as the analyte is oxidized or reduced by the application of voltage across electrodes of the flow cell. This method is highly selective (nonreactive compounds give no response) and very sensitive. A major application of electrochemical detection has been for the routine determination of catecholamines, which are phenolic compounds of clinical importance that are present in blood and tissues at very low levels. The development of a triple-pulsed amperometric detector, which overcame the problem of electrode poisoning (accumulation of oxidized product on the electrode surface), has allowed electrochemical detection to be applied to the analysis of carbohydrates (see Sect. 28.3.3.2.2). Pulsed electrochemical detection also has excellent sensitivity for the quantification of flavor-active alcohols, particularly terpenols (2).

Analytes that are ionized and carry a charge can be detected by measuring the change in eluent **conductivity** between two electrodes. **Conductivity detection** has been used mainly to detect inorganic anions and cations and organic acids upon elution from weak ion-exchange columns. Its principal application has been the basis of **ion chromatography** (Sect. 28.3.3.2.1). An excellent overview of electrochemical detection is provided by Swedesh (8).

28.2.4.5 Other HPLC Detectors

Unfortunately, there is no *truly universal* HPLC detector with *high sensitivity*. Thus, there have been many attempts to find new principles that could lead to improved instrumentation. One interesting concept is the **evaporative light scattering detector**. The mobile phase is sprayed into a heated air stream, evaporating volatile solvents and leaving nonvolatile analytes as aerosols. These droplets or particles can be detected because they will scatter a beam of light (8). HPLC with light scattering detection has been applied to the analysis of wheat flour lipids. Also, light scattering detectors are quite useful for the characterization of polymers by size-exclusion chromatography. Improvements in laser applications brought about the development of **low-angle laser light scattering** (LALLS) and **multi-angle laser light scattering** (MALLS) detectors. With these detectors, there is no need to evaporate the mobile phase, as the laser beam is directed at the flow cell, and scattered laser light is then monitored by photo detectors set at specific angles to the cell. In MALLS there may be as many as 18 different photo detectors at discrete angles, each continuously collecting and analyzing the scattered light; from this data, the computer can determine the molecular weight of the eluting sample.

Radioactive detectors are widely used for pharmacokinetic and metabolism studies with radiolabeled drugs. Decay of a radioactive nucleus leads to excitation of a scintillator, which subsequently loses its excess energy by photon emission. Photons are counted by a photomultiplier tube and the number of counts per second is proportional to radiolabeled analyte (9).

A **chemiluminescent nitrogen detector** (CLND) allows nitrogen-containing compounds, such as amino acids, to be detected without using chemical derivatization (Sect. 28.2.4.7). This nitrogen-specific detection system has been used to quantify caffeine in coffee and soft drink beverages, and to analyze capsaicin in hot peppers (2).

28.2.4.6 Coupled Analytical Techniques

To obtain more information about the analyte(s), eluent from an HPLC system can be passed on to a second analytical instrument, such as infrared (IR), nuclear magnetic resonance (NMR), or MS [see Chaps. 23, 25, and 26, respectively, or reference (6)]. The coupling of spectrometers with liquid chromatography (LC) was initially slow to gain application, due to many practical problems. For example, in the case of **HPLC with mass spectrometric detection** (LC–MS), the liquid mobile phase affected the vacuum in the MS. This problem was addressed by the development of commercial interfaces that allow solvent to be evaporated so that only analyte is carried to the spectrometer. Two commonly used interface techniques are discussed in detail by Harris (6). The use of microbore or capillary HPLC columns with a low flow volume also facilitates direct coupling of the two instruments (13). LC–MS systems continue to improve, and the applications are expanding to nearly every class of relatively low molecular weight compounds, including bioactives and contaminants.

28.2.4.7 Chemical Reactions

Detection sensitivity or specificity may sometimes be enhanced by converting the analyte to a **chemical derivative** with different or additional characteristics. An appropriate reagent can be added to the sample prior to injection (i.e., **precolumn derivatization**) or combined with column effluent before it enters the detector (i.e., **postcolumn derivatization**). Automated amino acid analyzers utilize postcolumn derivatization, usually with ninhydrin, for reliable and reproducible analysis of amino acids. Precolumn derivatization of amino acids with *o*-phthalaldehyde or similar reagents permits highly sensitive HPLC determination of amino acids using fluorescence detection (Chap. 15, Sect. 15.3.1.2). In addition, fractions may be collected after passing through the detector and aliquots of each fraction analyzed by various means, including chemical/colorimetric assays, such as the Lowry protein assay (Chap. 9, Sect. 9.2.5) or a total carbohydrate assay (Chap. 10, Sect. 10.3.2). The results can then be plotted and overlaid with the detector plot, yielding very important information about the compounds eluting in various peaks.

28.2.5 Data Station Systems

A detector provides an electronic signal related to the composition of the HPLC column effluent. It is the job of the last element in the chain of HPLC instrumentation to display the chromatogram, and integrate the peak areas. **Data stations** and **software packages** are nearly ubiquitous with modern HPLC, and all come with very powerful tools for sample identification and quantitation. As an HPLC analysis progresses, the data from the HPLC detector(s) are digitized and saved to the hard drive of a dedicated computer. The operator can then manipulate the data, by assigning and integrating the peaks, for example, and then print out plots and tables for further assessment. Importantly, the software programs can be set up prior to the analysis to perform nearly all these functions, without further input from the operator. For example, retention times can be calculated relative to an internal standard in pesticide residue analysis, and the results compared to a stored database of standards that the software automatically accesses when the analysis is complete. The software will then assign and integrate the peaks, and construct a complete report that will be displayed when the file is opened.

The data stations are now more than simply for interpreting results: the software packages also include all the parameters needed to run the HPLC, including start and stop, injection of the sample, and developing the gradient via control of the proportioning pump systems. When combined with an autosampler, the data station can carry out the entire operation, on hundreds of samples, in the absence of an operator, and, with networking, deliver the analysis file/report to an office computer.

28.3 APPLICATIONS IN HPLC

The basic physicochemical principles underlying all liquid chromatographic separations – adsorption, partition, ion exchange, size exclusion, and affinity – are discussed in Chap. 27, and details will not be repeated here. The number of separation modes utilized in HPLC, however, is greater than that available in classic chromatography. Examples of HPLC applications in food analysis are given in Table 28-1. This is attributable to the success of bonded phases, initially developed to facilitate liquid–liquid partition chromatography (Chap. 27, Sect. 27.4.2). In fact, reversed-phase chromatography is the most widely used separation mode in modern HPLC.

28.3.1 Normal Phase

28.3.1.1 Stationary and Mobile Phases

In **normal-phase** HPLC, the **stationary phase** is a **polar adsorbant**, such as bare silica or silica to which polar nonionic functional groups – hydroxyl, nitro, cyano (nitrile), or amino – have been chemically attached. These bonded phases are moderately polar and the surface is more uniform, resulting in better peak shapes. The **mobile phase** for this mode consists of a **nonpolar solvent**, such as hexane, to which is added a more polar modifier, such as methylene chloride, to control solvent strength and selectivity. Solvent

28-1 table **Example Applications of HPLC in the Analysis of Various Food Constituents**

Analyte	*Separation Mode*	*Method of Detection*	*Chapter*	*Section*
Mono- and oligosaccharides	Ion exchange; normal- or reversed phase	Electrochemical; refractive index; postcolumn analysis	10	10.3.4.1
Vitamin E	Normal or reversed phase	Fluorescence; electrochemical; UV	11	11.2.5.2
Amino acids	Ion exchange; reversed-phase	Post or precolumn derivatization	15	15.3.1
Pesticides	Normal or reversed phase	UV; fluorescence; mass spectrometry	18	18.3.3.2.3; 18.3.3.3.2
Mycotoxins	Reversed phase; immunoaffinity	UV; fluorescence	18	18.4.3.2.1
Antibiotics	Reversed phase	UV	18	18.5.2.2
Melamine	Reversed phase, ion exchange	UV; mass spectrometry	18	18.8.3
Sulfites	Ion exchange	UV; electrochemical	18	18.8.2

strength refers to the way a solvent affects the migration rate of the sample. **Weak solvents** increase retention (large k' values) and **strong solvents** decrease retention (small k' values).

28.3.1.2 Applications of Normal-Phase HPLC

In the past, normal-phase HPLC was used for the analysis of fat-soluble vitamins, although reverse phase is currently applied more frequently for these analyses (see Table 28-1). Normal phase is currently used for the analyses of biologically active polyphenols from natural plant sources, such as grape and cocoa. It is also used for the analysis of relatively polar vitamins, such as vitamins A, D, E, and K (see Chap. 11), and also natural carotenoid pigments, which impart both color and health benefits to foods. Highly hydrophilic species, such as carbohydrates (see Chap. 10, Sect. 10.3.4.1), also may be resolved by normal-phase chromatography, using amino bonded-phase HPLC columns (10).

28.3.2 Reversed Phase

28.3.2.1 Stationary and Mobile Phases

More than 70% of all HPLC separations are carried out in the reversed-phase mode, which utilizes a **nonpolar stationary phase** and a **polar mobile phase**. **Octadecylsilyl** (ODS) **bonded phases**, with an octadecyl (C_{18}) chain [$-(CH_2)_{17}CH_3$], are the most popular reversed-phase packing materials, although shorter chain hydrocarbons [e.g., octyl (C_8) or butyl (C_4)] or phenyl groups are also used. Many silica-based, reversed-phase columns are commercially available. Differences in their chromatographic behavior result from variation in the type of organic group bonded to the silica matrix or the chain length of organic moiety.

Reversed-phase HPLC utilizes **polar mobile phases**, usually water mixed with methanol, acetonitrile, or tetrahydrofuran. Solutes are retained due to **hydrophobic interactions** with the **nonpolar stationary phase** and are eluted in order of increasing hydrophobicity (decreasing polarity). Increasing the polar (aqueous) component of the mobile phase increases solute retention (larger k' values) (see Chap. 27, Sect. 27.5.2.2.4), whereas increasing the organic solvent content of the mobile phase decreases retention (smaller k' values). Various additives can serve additional functions. For example, although ionic compounds often can be resolved without them, **ion-pair reagents** may be used to facilitate chromatography of ionic species on reversed-phase columns. These reagents are ionic surfactants, such as octanesulfonic acid, which can neutralize charged solutes and make them more lipophilic. This type of chromatography is referred to as **ion-pair reversed-phase**.

28.3.2.2 Applications of Reversed-Phase HPLC

Reversed-phase has been the HPLC mode most used for analysis of plant proteins. Cereal proteins, among the most difficult of these proteins to isolate and characterize, are now routinely analyzed by this method (10). Both water- and fat-soluble vitamins (Chap. 11) can be analyzed by reversed-phase HPLC (2–5), and the availability of fluorescence detectors has enabled researchers to quantitate very small amounts of the different forms of vitamin B_6 (vitamers) in foods and

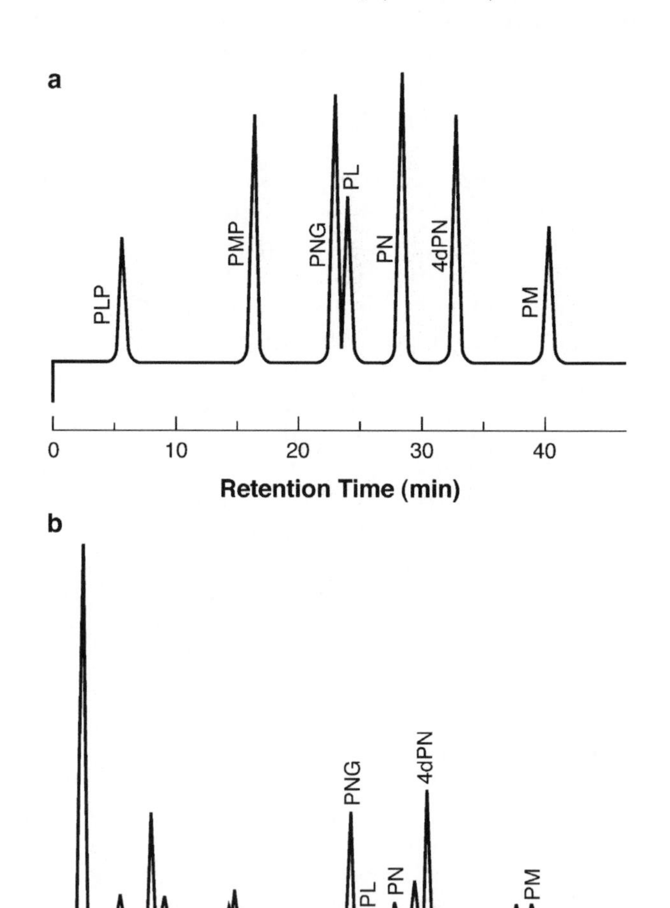

Analysis of vitamin B_6 compounds by reversed-phase HPLC with fluorescence detection. Some of the standard compounds (**a**) are present in a sample of rice bran extract (**b**). Sample preparation and analytical procedures are described in reference (17). *PL*, pyridoxal; *PLP*, pyridoxal phosphate; *PM*, pyridoxamine; *PMP*, pyridoxamine phosphate; *PN*, pyridoxine; *PNG*, pyridoxine β-D-glucoside. [Reprinted in part with permission from (17). Copyright 1991 American Chemical Society.]

biological samples. Figure 28-4 shows the separation of several of these vitamers in a rice bran extract achieved by reversed-phase ion-pair HPLC (17).

Reversed-phase ion-pair HPLC can be used to resolve carbohydrates on C_{18} bonded-phase columns (10), and the constituents of soft drinks (caffeine, aspartame, etc.) can be rapidly separated. Reversed-phase HPLC using a variety of detection methods, including RI, UV, and light scattering, has been applied to the analysis of lipids (2–5,10). Antioxidants, such as butylated hydroxyanisole (BHA) and butylated hydroxytoluene (BHT), can be extracted from dry foods and analyzed with simultaneous UV and fluorescence detection (3). Phenolic flavor compounds (such as vanillin) and pigments (such as chlorophylls, carotenoids, and anthocyanins) are also easily analyzed (2–5,10). A typical chromatogram of carotenoids present in a carrot extract is shown in Fig. 28-5. Reversed-phase ion-pair chromatography also is used for the separation of synthetic food colors (e.g., FD&C Red No. 40 and FD&C Blue No. 1) (5).

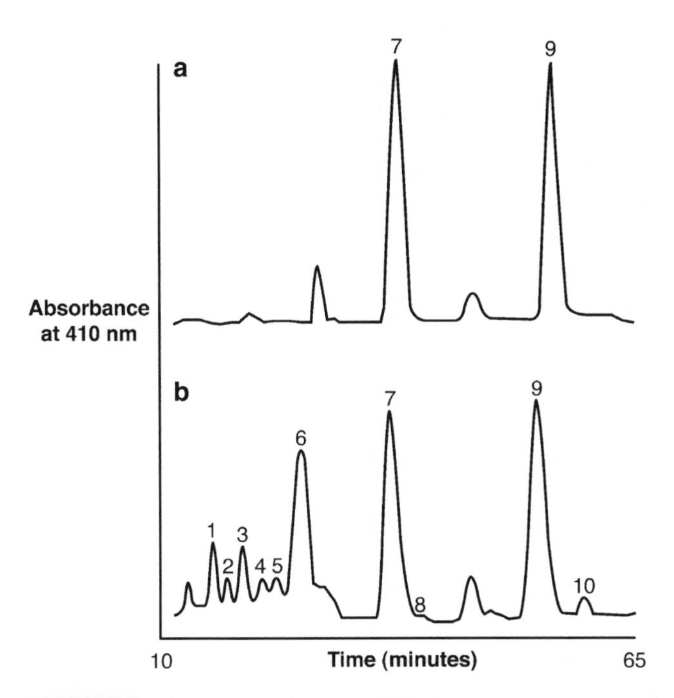

28-5 figure Reversed-phase HPLC separation of α-carotene (AC) and β-carotene (BC) isomers in (**a**) fresh and (**b**) canned carrots using a 5 μm C_{30} stationary phase. Peak 1, 13-*cis* AC; 2, unidentified *cis* AC; 3,13′-*cis* AC; 4, 15-*cis* BC; 5, Unidentified *cis* AC; 6, 13-*cis* BC; 7, all-*trans* AC; 8, 9-*cis* AC; 9, all-*trans* BC; 10, 9-*cis* BC. [Reprinted with permission from (18). Copyright 1997 American Chemical Society.]

28.3.3 Ion Exchange

28.3.3.1 Stationary and Mobile Phases

Packing materials for **ion-exchange HPLC** are usually **functionalized organic resins**, such as sulfonated or aminated poly(styrene-divinylbenzene) (Chap. 27, Sect. 27.4.3). **Macroporous resins** are most effective for HPLC columns due to their rigidity and permanent pore structure. **Pellicular packings** also are utilized, particularly in the CarboPac™ (Dionex) series, in which the nonporous, latex resin beads are coated with functionalized microbeads. The **mobile phase** in ion-exchange HPLC is usually an **aqueous buffer**, and solute retention is controlled by changing mobile phase ionic strength and/or pH. **Gradient elution** (gradually increasing ionic strength) is frequently employed.

28.3.3.2 Applications of Ion-Exchange HPLC

Ion-exchange HPLC has many applications, ranging from the detection of simple inorganic ions, to analysis of carbohydrates and amino acids, to the preparative purification of proteins oligosaccharides.

28.3.3.2.1 Ion Chromatography **Ion chromatography** is simply high-performance ion-exchange chromatography using a relatively **low-capacity stationary phase** (either anion- or cation-exchange) and, usually, a **conductivity detector**. All ions conduct an electric current; thus, measurement of electrical conductivity is an obvious way to detect ionic species. Because the mobile phase also contains ions, however, background conductivity can be relatively high. One step toward solving this problem is to use much lower capacity ion-exchange packing materials, so that more dilute eluents may be employed. In **nonsuppressed** or **single-column ion chromatography**, the detector cell is placed directly after the column outlet and eluents are carefully chosen to maximize changes in conductivity as sample components elute from the column. **Suppressed ion chromatography** utilizes an eluent that can be selectively removed by the use of ion-exchange membranes (10). Suppressed ion chromatography permits the use of more concentrated mobile phases and gradient elution. Ion chromatography can be used to determine inorganic anions and cations, transition metals, organic acids, amines, phenols, surfactants, and sugars. Some specific examples of ion chromatography applied to food matrices include the determination of organic and inorganic ions in milk; organic acids in coffee extract and wine; chlorine in

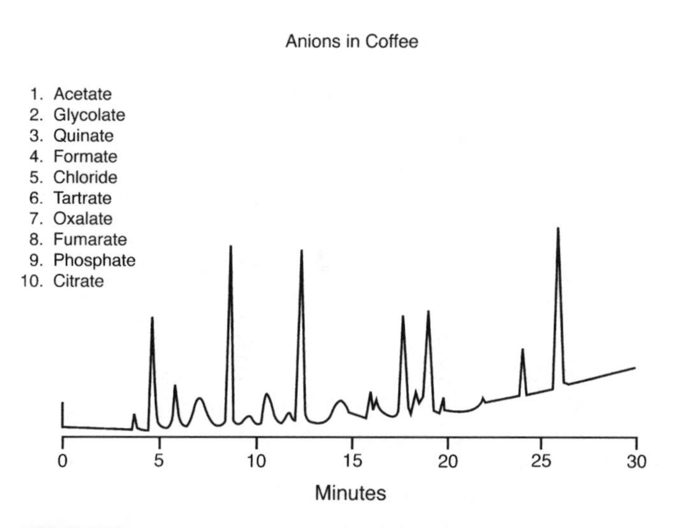

28-6 figure Ion-chromatographic analysis of organic acids and inorganic anions in coffee. Ten anions (listed) were resolved on an IonPac AS5A column (Dionex) using a sodium hydroxide gradient and suppressed conductivity detection. (Courtesy of Dionex Corp., Sunnyvale, CA.)

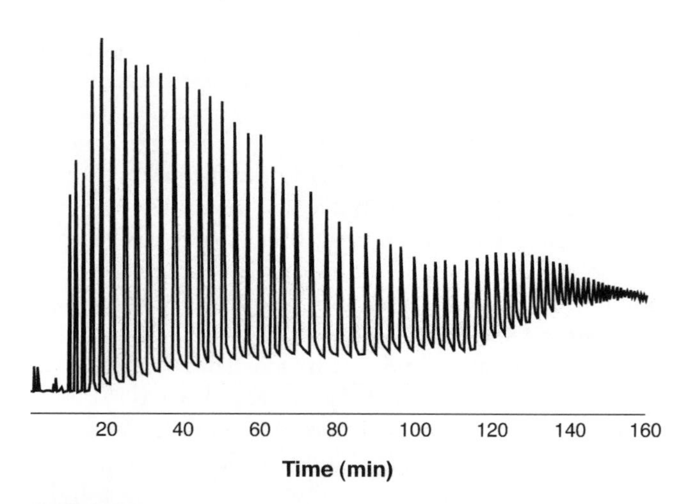

28-7 figure Anion exchange analysis of iso-amylase-treated waxy corn starch. The enzyme debranches the amylopectin, and the chromatogram represents the branch chain-length distribution, from four sugars in length and up. The analysis was performed with anion-exchange HPLC (Dionex^{TM}), with a pulsed amperometric detector.

infant formula; and trace metals, phosphates, and sulfites in foods. Figure 28-6 illustrates the simultaneous determination of organic acids and inorganic anions in coffee by ion chromatography.

28.3.3.2.2 Ion Exchange Chromatography of Carbohydrates and Proteins Both cation- and anion-exchange stationary phases have been applied to HPLC of carbohydrates. The advantage of separating carbohydrates by **anion exchange** is that retention and selectivity may be altered by changes in eluent composition. Carbohydrate analysis has benefited greatly by the development of a system that involves **anion-exchange HPLC** at high pH ($\geq$12), and detection by a **pulsed amperometric detector** (PAD). Pellicular column packings (see Sect. 28.2.3.2.2), consisting of nonporous latex beads coated with a thin film of strong anion exchanger, provide the necessary fast exchange, high efficiency, and resistance to strong alkali. These systems may be used in a variety of applications, from routine quality control to basic research. One common application is the determination of oligosaccharide distributions in corn syrups and other starch hydrolysates (Fig. 28-7).

Amino acids have been resolved on polymeric ion exchangers for more than 40 years (see Chap. 15, Sect. 15.3.1.2). Ion exchange is one of the most effective modes for HPLC of proteins and, recently, has been recognized as valuable for the fractionation of peptides.

28.3.4 Size Exclusion

Size-exclusion chromatography (SEC) fractionates solutes solely on the basis of size, with larger molecules eluting first. Due to the limited separation volume available in this chromatographic mode, as explained in Chap. 27 (Sect. 27.4.4), the peak capacity of a size-exclusion column is relatively small. Thus, the "high-performance" aspect of HPLC is not really applicable in the case of size exclusion. The main advantage gained from use of small particle packing materials is speed. Relatively small amounts of sample can be analyzed or separated and collected in $\leq$60 min, compared to $\leq$24 h separations using classic low pressure systems (10). A second advantage is that the sample concentration is higher and the relative volume is lower, so there is much less eluent to remove.

28.3.4.1 Column Packings and Mobile Phases

Size-exclusion packing materials or columns are selected so that matrix pore size matches the molecular weight range of the species to be resolved. Prepacked columns of microparticulate media are available in a wide variety of pore sizes. **Hydrophilic packings**, for use with water-soluble samples and aqueous mobile phases, may be surface-modified silica or methacrylate resins. **Poly(styrene-divinylbenzene) resins** are useful for nonaqueous size-exclusion chromatography of synthetic polymers.

The mobile phase in this mode is chosen for sample solubility, column compatibility, and minimal solute–stationary phase interaction. Otherwise, it has little effect on the separation. Aqueous buffers are used for biopolymers, such as proteins and polysaccharides, both to preserve biological activity and to prevent adsorptive interactions. Tetrahydrofuran or dimethylformamide is generally used for size-exclusion chromatography of other polymer samples, to ensure sample solubility.

28.3.4.2 Applications of High Performance SEC

Hydrophilic polymeric size-exclusion packings are used for the rapid determination of **average molecular weight** and **molecular weight range** of polysaccharides, including amylose, amylopectin, and other soluble gums such as xanthan, pullulan, guar, and water-soluble cellulose derivatives. **Molecular weight distribution** can be determined directly from high performance size-exclusion chromatography, if LALLS or MALLS is used for detection (8, 10). The application of aqueous size-exclusion chromatography to two commercially important polysaccharides, xanthan and carboxymethyl cellulose, is discussed in detail in reference (8).

SEC analysis has been useful to better understand numerous food components and systems. SEC analysis of tomato cell-wall pectin from hot-and cold-break tomato preparations (Fig. 28-8) showed that the cell-wall pectin was not differentially degraded by the different processing procedures. Size-exclusion HPLC

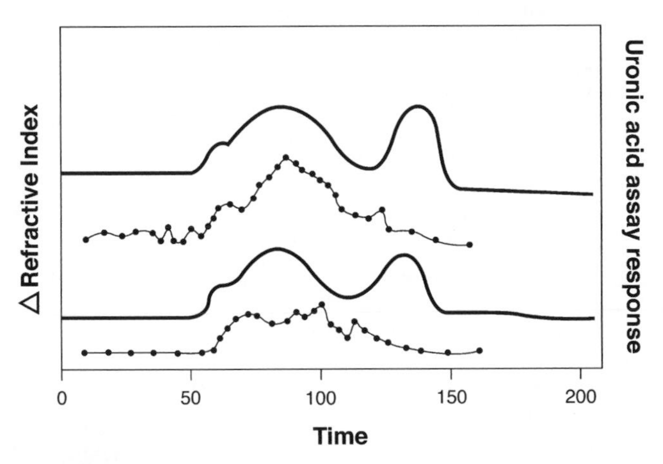

28-8 figure Analysis of tomato cell-wall pectin from hot- and cold-break tomato preparations by size-exclusion chromatography. The *solid lines* are from a refractive index detector response. The lines with markers result from a postchromatography analysis. Aliquots of collected fractions were analyzed by a colorimetric chemical assay that is specific for pectic sugars.

has been shown to be a rapid, one-step method for assessing soybean cultivars on the basis of protein content (proteins in the extracts of nondefatted flours from five soybean cultivars were separated into six common peaks, and cultivars could be identified by the percent total area of the fifth peak). A size-exclusion liquid chromatographic method also has been applied to the determination of polymerized triacylglycerols in oils and fats (2).

28.3.5 Affinity

Affinity chromatography is based on the principle that the molecules to be purified can form a selective but reversible interaction with another molecular species that has been immobilized on a chromatographic support. Although almost any material can be immobilized on a suitably activated support, the major ligands are proteins, including lectins (Chap. 15, Sect. 15.2.3.2.2), nucleic acids, and dyes (Chap. 27, Sect. 27.4.5). Affinity chromatography is used to purify many glycoproteins. Affinity chromatography using immobilized folate-binding protein is an effective tool in purifying sample extracts for HPLC analysis of folates in foods [e.g., reference (19)].

28.4 SUMMARY

HPLC is a chromatographic technique of great versatility and analytical power. A basic HPLC system consists of a pump, injector, column, detector, and data system. The pump delivers mobile phase through the system. An injector allows sample to be placed into the flowing mobile phase for introduction onto the column. The HPLC column consists of stainless steel or polymer hardware filled with a separation packing material. Various auxiliary columns, particularly guard columns, may be used prior to the analytical column. Detectors used in HPLC include UV-Vis absorption, fluorescence, RI, electrochemical, and light scattering, as well as coupled analytical systems, such as a mass spectrometer. Detection sensitivity or specificity sometimes can be enhanced by chemical derivatization of the analyte. Computer-controlled data station systems offer data collection and processing capabilities, and can run the instrument when an automated system is needed. A broad variety of column packing materials have contributed greatly to the widespread use of HPLC. These column packing materials may be categorized as silica-based (porous silica, bonded phases, pellicular packings) or polymeric (microporous, macroporous, or pellicular/nonporous). The success of silica-based bonded phases has expanded the applications of normal-phase and reversed-phase modes of separation in HPLC. Separations also are achieved with ion-exchange, size-exclusion, and affinity chromatography. HPLC is widely used for the analysis of small molecules and ions, such as sugars, vitamins, and amino acids, and is applied to the separation and purification of macromolecules, such as proteins and polysaccharides.

28.5 STUDY QUESTIONS

1. Why might you choose to use HPLC rather than traditional low-pressure column chromatography?
2. What is a guard column and why is it used?
3. Give three general requirements for HPLC column packing materials. Describe and distinguish among porous silica, bonded phases, pellicular, and polymeric column packings, including the advantages and disadvantages of each type.
4. What is the primary function of an HPLC detector (regardless of type)? What factors would you consider in choosing an HPLC detector? Describe three different types of detectors and explain the principles of operation for each.
5. You are performing HPLC using a stationary phase that contains a polar nonionic functional group. What type of chromatography is this, and what could you do to increase the retention time of an analyte?
6. Why are external standards commonly used for HPLC (unlike in GC, for which internal standards are more commonly used)?
7. Ion chromatography has recently become a widely promoted chromatographic technique in food analysis. Describe ion chromatography and give at least two examples of its use.
8. Describe one application each for ion-exchange and size-exclusion HPLC.

28.6 ACKNOWLEDGMENTS

Mary Ann Rounds (deceased) has been included as an author in this chapter, in recognition of the key role she played in writing the chapter on HPLC for the 1st–3rd editions of this textbook. Much of her writing was included in this chapter for the 4th edition. Dr. Reuhs, the primary author of this chapter, also wishes to acknowledge the involvement of Dr. Jesse F. Gregory, III, with this chapter for previous editions of the book. Dr. Baraem Ismail is acknowledged for her thoughtful suggestions on reorganizing the chromatography chapters, and her preparation of Table 28-1.

28.7 REFERENCES

1. Bidlingmeyer BA (1993) Practical HPLC methodology and applications. Wiley, New York
2. Chang SK, Holm E, Schwarz,J, Rayas-Duarte P (1995) Food (applications review). Anal Chem 67:127R–153R
3. Matissek R, Wittkowski R (eds) (1993) High performance liquid chromatography in food control and research. Technomic Publishing, Lancaster, PA
4. Nollet LML (ed) (2000) Food analysis by HPLC, 2nd edn. Marcel Dekker, New York
5. Macrae R (ed) (1988) HPLC in food analysis, 2nd edn. Academic, New York, NY
6. Harris DC (2006) Quantitative chemical analysis, 7th edn. W.H. Freeman and Co., New York
7. Hanai T (2004) HPLC: a practical guide. The Royal Society of Chemistry, Cambridge
8. Swadesh J (ed) (2000) HPLC: practical and industrial applications, 2nd edn. CRC, Boca Raton, FL
9. Lough WJ, Wainer IW (eds) (2008) High performance liquid chromatography: fundamental principles and practice. Springer, New York
10. Heftmann E (ed) (1992) Chromatography, 5th edn. Fundamentals and applications of chromatography and related differential migration methods. Part A: fundamentals and techniques. Part B: applications. J Chromatogr Libr Ser Vols. 51A and 51B. Elsevier, Amsterdam
11. Gertz C (1990) HPLC tips and tricks. LDC Analytical, Riviera Beach, FL
12. Dolan JW, Snyder LR (1989) Troubleshooting HPLC systems: a systematic approach to troubleshooting LC equipment and separations. Humana, Clifton, NJ
13. Ishii D (ed) (1988) Introduction to microscale high-performance liquid chromatography. VCH Publishers, New York
14. Dorsey JG, Cooper WT, Siles BA, Foley JP, Barth HG (1996) Liquid chromatography: theory and methodology (fundamental review). Anal Chem 68:515R–568R
15. LaCourse WR (2000) Column liquid chromatography: equipment and instrumentation (fundamental review). Anal Chem 72:37R–51R
16. Unger KK (1990) Packings and stationary phases in chromatographic techniques. Marcel Dekker, New York
17. Gregory JF, Sartain DB (1991) Improved chromatographic determination of free and glycosylated forms of vitamin B_6 in foods. J Agric Food Chem 39:899–905
18. Lessin WJ, Catignani GL, Schwartz SJ (1997) Quantification of *cis-trans*isomers of provitamin A carotenoids in fresh and processed fruits and vegetables. J Agric Food Chem 45:3728–3732
19. Pfeiffer C, Rogers LM, Gregory JF (1997) Determination of folate in cereal-grain food products using tri-enzyme extraction and combined affinity and reverse-phase liquid chromatography. J Agric Food Chem 45:407–413
20. Synder LR, Glajch JL, Kirkland JJ (1997) Practical HPLC method development, 2nd edn. Wiley, New York

Gas Chromatography

Michael C. Qian*
Department of Food Science and Technology, Oregon State University, Corvallis, OR 97331-6602, USA
michael.qian@oregonstate.edu

and

Devin G. Peterson and Gary A. Reineccius
Department of Food Science and Nutrition, University of Minnesota, St. Paul, MN 55108-6099, USA
dgp@umn.edu
greinecc@umn.edu

29.1 Introduction 515
29.2 Sample Preparation for Gas Chromatography 515
29.2.1 Introduction 515
29.2.2 Isolation of Solutes from Foods 515
29.2.2.1 Introduction 515
29.2.2.2 Headspace Methods 516
29.2.2.3 Distillation Methods 516

S.S. Nielsen, *Food Analysis*, Food Science Texts Series, DOI 10.1007/978-1-4419-1478-1_29,

29.2.2.4 Solvent Extraction 517
29.2.2.5 Solid-Phase Extraction 517
29.2.2.6 Direct Injection 519
29.2.3 Sample Derivatization 519
29.3 Gas Chromatographic Hardware and Columns 520
29.3.1 Gas Supply System 520
29.3.2 Injection Port 520
29.3.2.1 Hardware 520
29.3.2.2 Sample Injection Techniques 521
29.3.2.2.1 Split Injection 521
29.3.2.2.2 Splitless Injection 521
29.3.2.2.3 Temperature Programmed Injection 521
29.3.2.2.4 On-Column Injections 522
29.3.2.2.5 Thermal Desorption Injection 522
29.3.3 Oven 522
29.3.4 Column and Stationary Phases 522
29.3.4.1 Packed Columns 522
29.3.4.2 Capillary Columns 523
29.3.4.3 Gas–Solid (PLOT) Chromatography 524
29.3.5 Detectors 524
29.3.5.1 Thermal Conductivity Detector 524
29.3.5.1.1 Operating Principles 524
29.3.5.1.2 Applications 525
29.3.5.2 Flame Ionization Detector 525
29.3.5.2.1 Operating Principles 525
29.3.5.2.2 Applications 526
29.3.5.3 Electron Capture Detector 526
29.3.5.3.1 Operating Principles 526
29.3.5.3.2 Applications 526
29.3.5.4 Flame Photometric Detector and Pulsed Flame Photometric Detector 526
29.3.5.4.1 Operating Principles 526
29.3.5.4.2 Applications 527
29.3.5.5 Photoionization Detector 527
29.3.5.5.1 Operating Principles 527
29.3.5.5.2 Applications 528
29.3.5.6 Electrolytic Conductivity Detector 528
29.3.5.6.1 Operating Principles 528
29.3.5.6.2 Applications 528
29.3.5.7 Thermionic Detector 528
29.3.5.7.1 Operating Principles 528
29.3.5.7.2 Applications 528
29.3.5.8 Hyphenated Gas Chromatographic Techniques 528
29.3.5.9 Multidimensional Gas Chromatography 528
29.3.5.9.1 Conventional Two-Dimensional GC 529
29.3.5.9.2 Comprehensive Two-Dimensional GC 529
29.4 Chromatographic Theory 530
29.4.1 Introduction 530
29.4.2 Separation Efficiency 530
29.4.2.1 Carrier Gas Flow Rates and Column Parameters 530
29.4.2.2 Carrier Gas Type 531
29.4.2.3 Summary of Separation Efficiency 532
29.5 Applications of GC 532
29.5.1 Residual Volatiles in Packaging Materials 533
29.5.2 Separation of Stereoisomers 533
29.5.3 Headspace Analysis of Ethylene Oxide in Spices 533
29.5.4 Aroma Analysis of Heated Butter 534
29.6 Summary 535
29.7 Study Questions 535
29.8 References 536

29.1 INTRODUCTION

The first publication on gas chromatography (GC) was in 1952 (1), while the first commercial instruments were manufactured in 1956. James and Martin (1) separated fatty acids by GC, collected the column effluent, and titrated the individual fatty acids for quantitation. GC has advanced greatly since that early work and is now considered to be a mature field that is approaching theoretical limitations.

The types of analysis that can be done by GC are very broad. GC has been used for the determination of fatty acids, triglycerides, cholesterol and other sterols, gases, solvent analysis, water, alcohols, and simple sugars, as well as oligosaccharides, amino acids and peptides, vitamins, pesticides, herbicides, food additives, antioxidants, nitrosamines, polychlorinated biphenyls (PCBs), drugs, flavor compounds, and many more. The fact that GC has been used for these various applications does not necessarily mean that it is the best method – often better choices exist. GC is ideally suited to the analysis of thermally stable volatile substances. Substances that do not meet these requirements (e.g., sugars, oligosaccharides, amino acids, peptides, and vitamins) are more suited to analysis by a technique such as high-performance liquid chromatography (HPLC) or supercritical fluid chromatography (SFC). Yet gas chromatographic methods appear in the literature for these substances.

This chapter will discuss sample preparation for GC, gas chromatographic hardware, columns, and chromatographic theory as it uniquely applies to GC. Texts devoted to GC in general (2–4) and food applications in particular (5, 6) should be consulted for more detail.

29.2 SAMPLE PREPARATION FOR GAS CHROMATOGRAPHY

29.2.1 Introduction

One cannot generally directly inject a food product into a GC without some sample preparation. The high temperatures of the injection port will result in the degradation of nonvolatile constituents and create a number of false GC peaks corresponding to the volatile degradation products formed. In addition, very often the constituent of interest must be isolated from the food matrix simply to permit concentration such that it is at detectable limits for the GC or to isolate it from the bulk of the food. Thus, one must generally do some type of sample preparation, component isolation, and concentration prior to GC analysis.

Sample preparation often involves grinding, homogenization, or otherwise reducing particle size. There is substantial documentation in the literature showing that foods may undergo changes during sample storage and preparation. Many foods contain active enzyme systems that will alter the composition of the food product. This is very evident in the area of flavor work (7–9). Inactivation of enzyme systems via high-temperature-short-time thermal processing, sample storage under frozen conditions, drying the sample, or homogenization with alcohol may be necessary (see Chap. 5).

Microbial growth or chemical reactions also may occur in the food during sample preparation. Chemical reactions often will result in the formation of volatiles that will again give false peaks on the GC. Thus, the sample must be maintained under conditions such that degradation does not occur. Microorganisms often are inhibited by chemical means (e.g., sodium fluoride), thermal processing, drying, or frozen storage.

29.2.2 Isolation of Solutes from Foods

29.2.2.1 Introduction

The isolation procedure may be quite complicated depending upon the constituent to be analyzed. For example, if one were to analyze the triglyceride bound fatty acids in a food, one would first have to extract the lipids (free fatty acids; mono-, di-, and triglycerides; sterols; fat-soluble vitamins, etc.) from the food (e.g., by solvent extraction) and then isolate only the triglyceride fraction (e.g., by adsorption chromatography on silica). The isolated triglycerides then would have to be treated to first hydrolyze the fatty acids from the triglycerides and subsequently to form esters to improve gas chromatographic properties. The two latter steps might be accomplished in one reaction by transesterification (e.g., borontrifluoride in methanol) as described in Chap. 8, Sect. 8.3.1.6, and Chap. 14, Sect. 14.6.2. Thus many steps involving several types of chromatography may be used in sample preparation for GC analysis.

The analysis of volatiles in foods (e.g., packaging or environmental contaminants, alcohols, and flavors or off-flavors) may be a simpler task. These materials for GC analysis may be isolated by headspace analysis (static or dynamic), distillation, preparative chromatography (e.g., solid-phase extraction, column chromatography on silica gel), simple solvent extraction, or some combination of these basic methods. The procedure used will depend on the food matrix as well as the compounds to be analyzed. The primary considerations are to isolate the compounds of interest from nonvolatile food constituents (e.g., carbohydrates, proteins, vitamins) or those that would interfere with GC (e.g., lipids). Some of the chromatographic methods that might be applied to this task have been discussed in the basic chromatography chapter (Chap. 27). Methods for the isolation of volatile substances will be covered briefly as they pertain to the isolation of components for gas chromatographic analysis.

It should be emphasized that the isolation procedure used is critical in determining the results obtained. An improper choice of method or poor technique at this step negates the best gas chromatographic analysis of the isolated solutes. The influence of isolation technique on gas chromatographic analysis of aroma compounds has been demonstrated (10). These biases are discussed in the sections that follow and in more detail in books edited by Marsili (11) and Mussinan and Morello (12). While these books relate to the analysis of aroma compounds in foods, the techniques for the isolation of these volatiles are the same as used in the analysis of other volatiles in foods.

29.2.2.2 Headspace Methods

One of the simplest methods of isolating volatile compounds from foods is by direct injection of the headspace vapors above a food product. There are two types of headspace sampling: direct (or static) headspace sampling and dynamic headspace sampling.

Direct headspace sampling has been used extensively when rapid analysis is necessary and major component analysis is satisfactory. At equilibrium, the headspace of the sample is taken using a gas-tight syringe and then injected directly into the GC. Examples of method applications include measurement of hexanal as an indicator of oxidation (13, 14) and 2-methylpropanal, 2-methylbutanal, and 3-methylbutanal as indicators of nonenzymatic browning (15). The determination of residual solvents in packaging materials also may be approached by this method. Unfortunately, this method does not provide the sensitivity needed for trace analysis. Instrumental constraints typically limit headspace injection volumes to 5 ml or less. Therefore, only volatiles present in the headspace at concentrations greater than 10^{-7} g/l headspace would be at detectable levels [using a flame ionization detector (FID)].

Dynamic headspace sampling or **purge and trap** has found wide usage in recent years. This concentration method may involve simply passing large volumes of headspace vapors through a cryogenic trap or, alternatively, a more complicated extraction and/or adsorption trap. A simple **cryogenic trap** offers some advantages and disadvantages. A cryo trap (if properly designed and operated) will collect headspace vapors irrespective of compound polarity and boiling point. However, water is typically the most abundant volatile in a food product, and, therefore, one collects an aqueous distillate of the product aroma. This distillate must be extracted with an organic solvent, dried, and then concentrated for analysis. These additional steps add analysis time and provide opportunity for sample contamination. A more commonly used technique is adsorbent traps.

Adsorbent traps offer the advantages of providing a water-free volatile isolate (trap material typically has little affinity for water) and are readily automated. The adsorbent initially used for headspace trapping was charcoal. The charcoal was either solvent extracted (CS) or thermally desorbed with backflushing (inert gas) to recover the adsorbed volatiles. The use of synthetic porous polymers as headspace trap material now dominates. Initially, Tenax (a porous polymer very similar to the skeleton of ion-exchange resins) was most commonly used; however, combinations of Tenax and other polymers are now seeing greater application. These polymers exhibit good thermal stability and reasonable capacity. Adsorbent traps are generally placed in a closed system and loaded, desorbed, and so on via the use of automated multiport valving systems. The automated closed system approach provides reproducible GC retention times and quantitative precision necessary for some studies. The primary disadvantage of adsorbent traps is their differential adsorption affinity and limited capacity. Buckholz et al. (16) have shown that the most volatile peanut aroma constituents will break through two Tenax traps in series after purging at 40 ml/min for only 15 min. Therefore, the GC profile may only poorly represent the actual food composition due to biases introduced by the purging and trapping steps.

29.2.2.3 Distillation Methods

Distillation processes are quite effective at isolating volatile compounds from foods for GC analysis. Product moisture or outside steam is used to heat and codistill the volatiles from a food product. This means that a very dilute aqueous solution of volatiles results, and a solvent extraction must be performed on the distillate to permit concentration for analysis. The distillation method most commonly used today is some modification of the original Nickerson–Likens distillation head. In this apparatus, a sample is boiled in one side flask and an extracting solvent in another. The product steam and solvent vapors are intermixed and condensed; the solvent extracts the organic volatiles from the condensed steam. The solvent and extracted distillate return to their respective flasks and are distilled to again extract the volatiles from the food. While this method is convenient and efficient, artifacts from solvents used in extraction, antifoam agents, steam supply (contaminated water), thermally induced chemical changes, and leakage of contaminated laboratory air into the system may contaminate the volatile isolate.

29.2.2.4 Solvent Extraction

Solvent extraction is often the preferred method for the recovery of volatiles from foods. Recovery of volatiles will depend upon solvent choice and the solubility of the solutes being extracted. Solvent extraction typically involves the use of an organic solvent (unless sugars, amino acids, or some other water-soluble components are of interest). Extraction with organic solvents limits the method to the isolation of volatiles from fat-free foods (e.g., wines, some breads, fruit and berries, some vegetables, and alcoholic beverages), or an additional procedure must be employed to separate the extracted fat from the isolated volatiles (e.g., a chromatographic method). Fat will otherwise interfere with subsequent concentration and GC analysis.

Solvent extractions may be carried out in quite elaborate equipment, such as supercritical CO_2 extractors, or can be as simple as a batch process in a separatory funnel. Batch extractions can be quite efficient if multiple extractions and extensive shaking are used (17). The continuous extractors (liquid–liquid) are more efficient but require more costly and elaborate equipment.

29.2.2.5 Solid-Phase Extraction

The extractions discussed above involved the use of two immiscible phases (water and an organic solvent). However, a newer and a very rapidly growing alternative to such extractions is **solid-phase extraction** (18, 19). In one version of this technique, a liquid sample (most often aqueous based) is passed through a column (2–10 ml vol) filled with chromatographic packing or a Teflon^R filter disk (25–90 mm in diameter) that has the chromatographic packing embedded in it. The chromatographic packing may be any of a number of different materials (e.g., ion-exchange resins or a host of different reversed- or normal-phase HPLC column packings). When a sample is passed through the cartridge or filter, solutes that have an affinity for the chromatographic phase will be retained on the phase while those with little or no affinity will pass through. The phase is next rinsed with water, perhaps a weak solvent (e.g., pentane), and then a stronger solvent (e.g., dichloromethane). The strong eluent is chosen such that it will remove the solutes of interest.

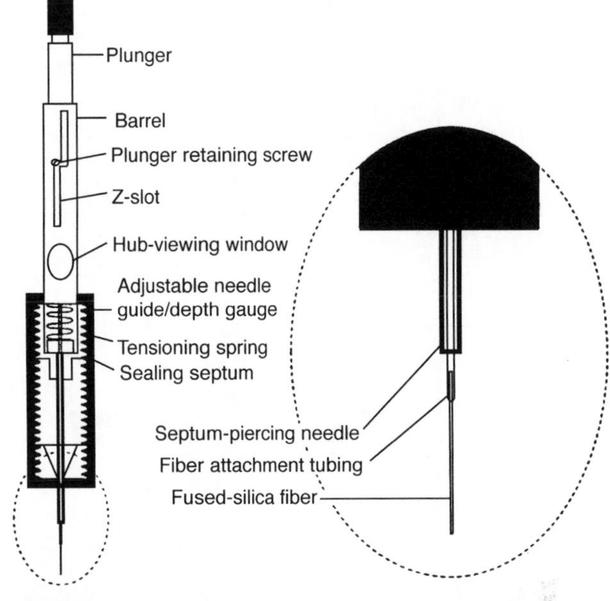

29-1 figure Schematic of a solid-phase microextraction (SPME) device (21). (Courtesy of Dr. Janusz Pawliszyn, Department of Chemistry, University of Waterloo, Waterloo, Ontario, Canada.)

Overall, solid-phase extraction has numerous *advantages* over traditional liquid–liquid extractions including: (1) less solvent is required; (2) speed; (3) less glassware is needed (less cost and potential for contamination); (4) better precision and accuracy; (5) minimal solvent evaporation for further analysis (e.g., GC); and (6) it is readily automated. Solid-phase extraction has limitations, but new variations of the technique seek to overcome some of these.

The most recent version of solid-phase extraction is called **solid-phase microextraction** (SPME). This method was developed originally for environmental work (20, 21). In this adaptation, the phase is bound onto a fine fused silica filament (ca. the size of a 10-μl syringe needle, Fig. 29-1). The filament is immersed in a sample or in the headspace above a sample. After the desired extraction time, the filament is pulled into a protective metal sheath, removed from the sample, and forced through the septum of a gas chromatograph where the adsorbed volatiles are thermally desorbed from the filament (Fig. 29-2).

SPME is an equilibrium technique and, therefore, the volatile profile (i.e., volatile recovery) that one obtains is strongly dependent upon sample composition and careful control of all sampling parameters. This includes the specific phase coating and thickness on the filament/fiber used. Several different phases of fiber are commercially available, and compounds with a wide range of polarity or volatility can be analyzed. PDMS (polydimethylsiloxane) is a nonpolar phase coating and can be used to extract

Extraction Procedure

Desorption Procedure

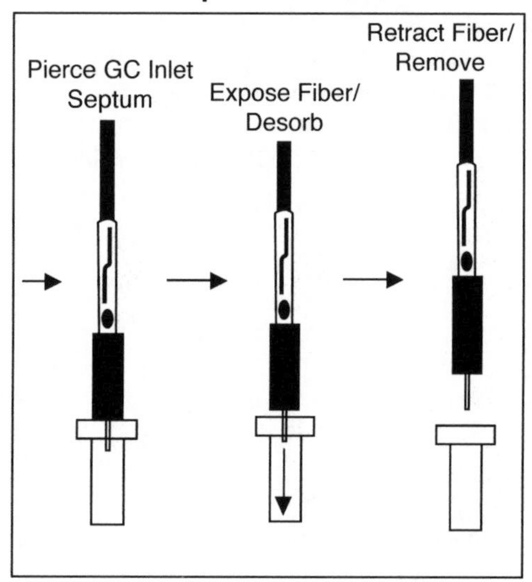

29-2 figure Schematic showing the steps involved in the use of a solid-phase microextraction (SPME) device. (Reprinted with permission of Supelco, Bellefonte, PA 16823, USA.)

nonpolar compounds. Polar analytes can be extracted with polar phases (e.g., polyacrylate and Carbowax coatings). Porous fibers such as Carboxen or divinylbenzene (DVB) coating are good for highly volatile compounds. The coating has various film thicknesses. Thicker film fibers (100 μm) are better for volatiles, whereas thinner film fibers (7 μm and 30 μm) are better for larger molecules. Multiphase fibers (such as Carboxen/PDMS, Carboxen/DVB/PDMS) are also available to extract both polar and nonpolar compounds.

Due to its simplicity, SPME is very popular for volatile aroma analysis of food and beverages. While Harmon (22) notes that the method can give excellent results, Coleman (23) cautions that the fibers have a definite linear range and competition between volatiles for binding sites can introduce errors. Other concerns are for sensitivity limitations, precision, and life of the filament. If the filament must be replaced (breakage), there is the issue of reproducibility of the new vs. the old filament.

Solid-phase dynamic extraction (SPDE) is another technique for volatile extraction. SPDE is similar to SPME, except the polymer is coated inside a special needle. A gas-tight syringe is used for SPDE to draw the headspace of food, and the volatiles are absorbed by the phase. The process can be repeated many times by moving the plunger up and down to achieve maximum absorption. The needle can then be injected into the GC for analysis. Different phases are available and the volume of the phase is about 4.5 μl compared with only 0.6 μl for SPME, so the SPDE has less issue with analyte saturation and competition.

Stir bar sorptive extraction (SBSE) is a new technique for volatile extraction. In SBSE (Fig. 29-3), a magnet stir bar is jacketed with glass, and the glass is coated with a layer of absorbent (PDMS). The bar spins in the sample solution and absorbs the analytes from the sample solution. The stir bar can also just hang on the headspace for volatile extraction the same way as the SPME. After the extraction, the volatiles are then thermally desorbed and introduced into a GC. Stir bar has almost 50 times more volume of absorbent than SPME. For SPME, the PDMS volume is about 0.5 μl; with SBSE, it is 24–126 μl. Due to the increased volume of absorbent phase, SBSE has much higher sensitivity than SPME and has minimum competition and saturation effects (24,25). The high sensitivity (ppt to ppg)

29-3 figure Diagram of stir bar sorptive extraction (SBSE) device. (Courtesy of Gerstel, Inc., Linthicum, MD.)

and flexibility of SBSE for nonpolar and medium polar compounds makes it an effective and time saving method for extracting trace volatile compounds from complex matrices (25). The PDMS phase is robust; does not absorb water, alcohol, or pigment; and is very good for flavor extraction in alcoholic beverages. However, the PDMS phase is not selective for shorter-chain acids and polar compounds. Other phases of SBSE need to be used to analyze polar compounds.

29.2.2.6 Direct Injection

It is theoretically possible to analyze some foods by direct injection of the food into a gas chromatograph. Assuming one can inject a 2- to 3-μl sample into a GC and the GC has a detection limit of 0.1 ng (0.1 ng/2 μl), one could detect volatiles in the sample at concentrations greater than 50 ppb. Problems with direct injection arise due to thermal degradation of any nonvolatile food constituents, damage to the GC column, decreased separation efficiency due to water in the food sample, contamination of the column and injection port by nonvolatile materials, and reduced column efficiency due to slow vaporization of volatiles from the food (injection port temperatures are reduced to minimize thermal degradation of the nonvolatile food constituents). Despite these concerns, direct injection is commonly used to determine oxidation in vegetable oils (26,27). A relatively large volume of oil (50–100 μl) can be directly injected into an injection port of a GC that has been packed with glass wool. Since vegetable oils are reasonably thermally stable and free of water, this method is particularly well suited to oil analysis.

There are numerous other approaches for the isolation of volatiles from foods. Some are simple variations of these methods, while others are unique. Several review articles are available that provide a more complete view of methodology (11,12,28).

29.2.3 Sample Derivatization

The compounds one wishes to determine by GC must be thermally stable under the GC conditions employed. Thus, for some compounds (e.g., pesticides, aroma compounds, PCBs, and volatile contaminants), the analyst can simply isolate the components of interest from a food as discussed above and directly inject them into the GC. For compounds that are thermally unstable, are too low in volatility (e.g., sugars and amino acids), or yield poor chromatographic separation due to polarity (e.g., phenols or acids), a derivatization step must be included prior to GC analysis (see also Chaps. 10 and 14). A listing of some of the reagents used in preparing volatile derivatives for GC is given in Table 29-1. The conditions of use for these reagents are often specified by the supplier or can be found in the literature.

Reagents Used for Making Volatile Derivatives of Food Components for GC Analysis

Reagent	*Chemical Group*	*Food Constituent*
Silyl reagents	Hydroxy, amino carboxylic acids	Sugars, sterols, amino acids
Trimethylchlorosilane/ hexamethyldisilazane		
BSA [*N*, *O*-bis(trimethylsilyl) acetamide		
BSTFA [*N*, *O*-bis (trimethylsilyl) trifluoroacetamide]		
t-BuDMCS (*t*-butyldimethylchlorosilyl/imidazole)		
Esterifying reagents	Carboxylic acids	Fatty acids, amines, amino acids, triglycerides, wax esters, phospholipids, cholesteryl esters
Methanolic HCl		
Methanolic sodium methoxide		
N, *N*-Dimethylformamide dimethyl acetal		
Boron trifluoride (or trichloride)/methanol		
Miscellaneous		
Acetic anhydride/pyridine	Alcoholic and phenolic	Phenols, aromatic hydroxyl groups, alcohols
N-trifluoroacetylimidizole/ *N*-heptafluorobutyrlimidizole	Hydroxy and amines	Same as above
Alkylboronic acids	Polar groups on neighboring atoms	
O-alkylhydroxylamine	Compounds containing both hydroxyl and carbonyl groups	Ketosteroids, prostaglandins

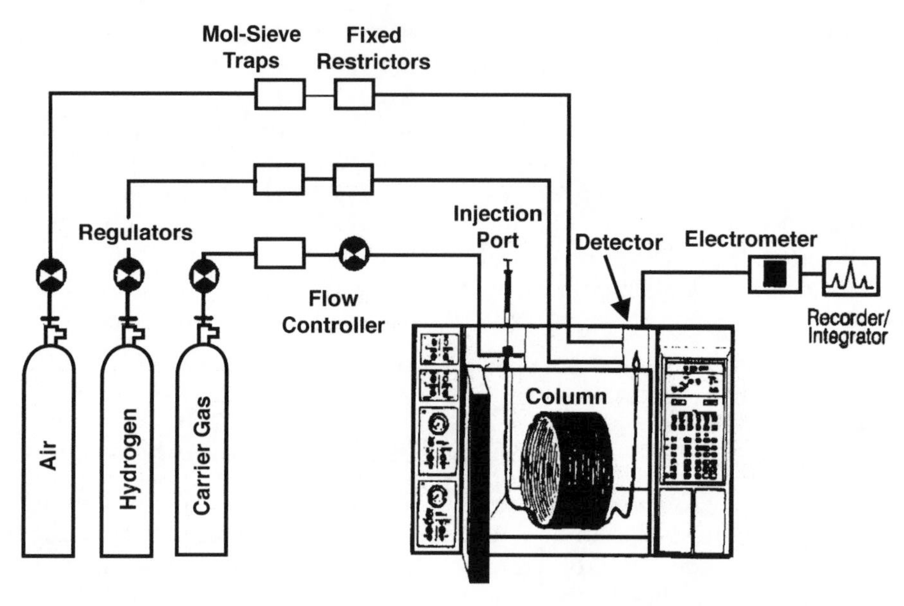

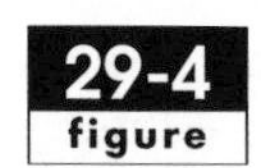

Diagram of a gas chromatographic system. (Courtesy of Hewlett Packard Co., Analytical Customer Training, Atlanta, GA.)

Gas Chromatographic Hardware and Operating Conditions to be Recorded for All GC Separations

Parameter	*Description*
Sample	Name and injection volume
Injection	Type of injection [e.g., split vs. splitless and conditions (injection port flow rates)]
Column	Length, diameter (material-packed columns), and manufacturer
Packing/phase	Packed columns – solid support; size mesh; coating; loading (%)
	Capillary columns – phase material and thickness
Temperatures	Injector; detector; oven and any programming information
Carrier gas	Flow rate (velocity) and type
Detector	Type
Data output	Attenuation and chart speed

29.3 GAS CHROMATOGRAPHIC HARDWARE AND COLUMNS

The major parts of a GC are the **gas supply** system, **injection port, oven, column, detector, electronics,** and **recorder/data handling system** (Fig. 29-4). The hardware as well as operating parameters used in any GC analysis must be accurately and completely recorded. The information that must be included is presented in Table 29-2.

29.3.1 Gas Supply System

The gas chromatograph will require at least a supply of carrier gas, and, most likely, gases for the detector (e.g., hydrogen and air for a FID). The gases used must be of high purity and all regulators, gas lines, and fittings must be of good quality. High-quality pressure regulators must be used to provide a stable and continuous gas supply. The regulators should have stainless steel rather than polymer diaphragms since polymers will give off volatiles that may contribute peaks to the analytical run. All gas lines must be clean and contain no residual drawing oil. Nitrogen, helium, and hydrogen gases are typically used as the carrier gas to transport the analytes in the GC column. The carrier gas line should have traps (moisture trap, oxygen trap, and hydrocarbon trap) in line to remove any moisture and contaminants from the incoming gas. These traps must be periodically replaced to maintain effectiveness.

29.3.2 Injection Port

29.3.2.1 Hardware

The injection port serves the purpose of providing a place for sample introduction, its vaporization, and possibly some dilution and splitting. Liquid samples make up the bulk of materials analyzed by GC, and they are always done by syringe injection (manual or automated). The injection port contains a soft septum that provides a gas-tight seal but can be penetrated by a syringe needle for sample introduction.

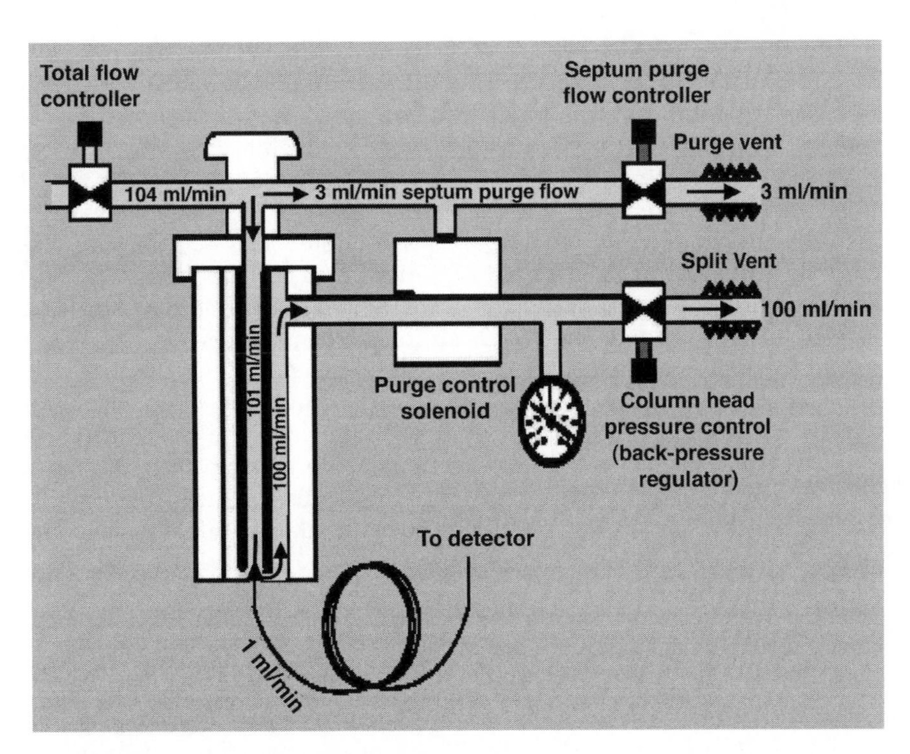

29-5 figure Schematic of a GC injection port. (Courtesy of Hewlett-Packard Co., Analytical Customer Training, Atlanta, GA.)

Samples may be introduced into the injection port using a **manual syringe technique** or an **automated sampling system**. Manual sample injection is generally the largest single source of poor precision in GC analysis. Ten-microliter syringes are usually chosen since they are more durable than the microsyringes, and sample injection volumes typically range from 1 to 3 μl. These syringes will hold about 0.6 μl in the needle and barrel (this is in addition to that measured on the barrel). Thus the amount of sample that is injected into the GC depends upon the proportion of this 0.6 μl that is included in the injection and the ability of the analyst to accurately read the desired sample volume on the syringe barrel. This can be quite variable for the same analyst and be grossly different between analysts. This variability between injections and the small sample volumes injected is the reason why internal (vs. external) standards are common for GC.

29.3.2.2 Sample Injection Techniques

The sample must be vaporized in the injection port in order to pass through the column for separation. This vaporization can occur quickly by flash evaporation (standard injection ports) or slowly in a gentler manner (temperature-programmed injection port or on-column injection). The choice depends upon the thermal stability of the analytes. Due to the various sample as well as instrumental requirements, there are several different designs of injection ports available.

29.3.2.2.1 Split Injection Capillary columns have limited capacity, and the injection volume may have to be reduced to permit efficient chromatography. The injection port may serve the additional function of splitting the injection so that only a portion of the analyte goes on the column (i.e., **split injection**) (Fig. 29-5). The injection port is operated about 20°C warmer than the maximum column oven temperature. The sample may be diluted with carrier gas to accomplish a split (1:50 to 1:100 preferred), whereby only a small portion (1 part) of the analyte (more exactly, 1 part of gas flow) goes on the column, and the majority (49/50–99/100 parts) of the analytes are vented to the split vent. High split ratio typically gives a sharp, narrow peak.

29.3.2.2.2 Splitless Injection To increase the sensitivity, a **splitless injection** mode can be used. In splitless injection, the split vent valve is closed and all of the analyte goes on the column (Fig. 29-5). Similar to the split injection, the temperature of the injector is operated at 20°C higher than the maximum column oven temperature. Splitless injection requires to set up the initial column temperature at least 20°C lower than the boiling point of the sample solvent, so the solvent can recondense in the column for acceptable chromatography of early eluting compounds.

29.3.2.2.3 Temperature Programmed Injection For **temperature-programmed injection** ports, the sample is introduced into an ambient temperature port and then it is temperature programmed to some desired temperature. Since the sample is not introduced to the hot injector, the technique is desired for temperature-sensitive analytes. In addition, this technique is very useful to inject a large amount sample when it is used together with a split/splitless injection mode to increase the sensitivity. For example, 10 μl of liquid sample can be injected at low temperature using a high split ratio to let the solvent vent out and then the injection mode can be changed to "splitless" as the injector is heated up to transfer all analytes onto the column.

29.3.2.2.4 On-Column Injections **On-column injection** is a technique whereby the sample is directly introduced into the column whose temperature is at that of the GC oven or that of the room. The sample is then slowly volatized as the oven heats up. The initial oven temperature needs to be below the boiling point of the solvent. This technique is good for thermally labile analytes.

29.3.2.2.5 Thermal Desorption Injection The volatiles can be introduced onto the head of GC column for chromatographic separation directly from food samples through **thermal desorption**. The sample is heated in a thermal desorption unit, and the volatiles are carried through a purge gas to a split/splitless injector. Cryofocusing with liquid nitrogen either in the injector or column is needed to attain sharp peaks. Alternatively, the volatiles can be retained using a trap such as Tenax™ during the purge stage and then thermally desorbed onto the column. The samples can be extracted with SPME or SBSE described previously (Sect. 29.2.2.5) and then thermally desorbed into the column for analysis. This technique has gained popularity to analyze volatile aroma compounds in foods including fruits (29,30) and wine (31).

29.3.3 Oven

The oven controls the temperature of the column. In GC, one takes advantage of both an interaction of the analyte with the stationary phase and the boiling point for separation of compounds. Thus, the injection is often made at a lower oven temperature and is then temperature programmed to some elevated temperature. While analyses may be done isothermally, compound elution time and resolution are extremely dependent upon temperature, so temperature-programmed runs are most common. It should be obvious that higher temperatures will cause the sample to elute faster and, therefore, be at a cost of resolution. Oven temperature program rates can range from as little as 0.1°C/min to the maximum temperature heating rate that the GC can provide. A rate of 2–10°C/min is most common.

The capillary column also can be directly heated with an insulated heating wire based on low thermal mass (LTM) technology. A temperature sensor is mounted on the column. The column, the heating wire, and the sensor are all coiled together and wrapped with aluminum foil. The column can be uniformly heated very rapidly to improve the separation and efficiency. Since the system does not have much void volume and other insulation materials, it cools very quickly. The total heating and cooling cycle is much shorter than the traditional standard GC oven, which makes it ideal for fast GC analysis. The module is available with almost any standard capillary GC column.

29.3.4 Column and Stationary Phases

The GC column may be either **packed** or **capillary**. Early chromatography was done on packed columns, but the advantages of capillary chromatography so greatly outweigh those of packed column chromatography that few packed column instruments are sold any longer (Fig. 29-6). While some use high resolution gas chromatography (HRGC) to designate capillary GC, GC today means capillary chromatography to most individuals.

29.3.4.1 Packed Columns

The packed column is most commonly made of stainless steel or glass and may range from 1.6 to 12.7 mm in outer diameter and be 0.5–5.0 m long (generally 2–3 m). It is packed with a granular material consisting of a "liquid" coated on an allegedly inert solid support. The **solid support** is most often diatomaceous earth (skeletons of algae) that has been purified, possibly chemically modified (e.g., silane treated), and then sieved to provide a definite mesh size (60/80, 80/100, or 100/120).

The liquid loading is usually applied to the solid support at 1–10% by weight of the solid support. While the liquid coating can be any one of the approximately 200 available, the most common are silicone-based phases (methyl, phenyl, or cyano substituted) and Carbowax (ester based).

The liquid phase and the percent loading are determined by the analysis desired. The choice of liquid is typically such that it is of similar polarity to the analytes to be separated. Loading influences time of analysis (retention time is proportional to loading), resolution (generally improved by increasing phase

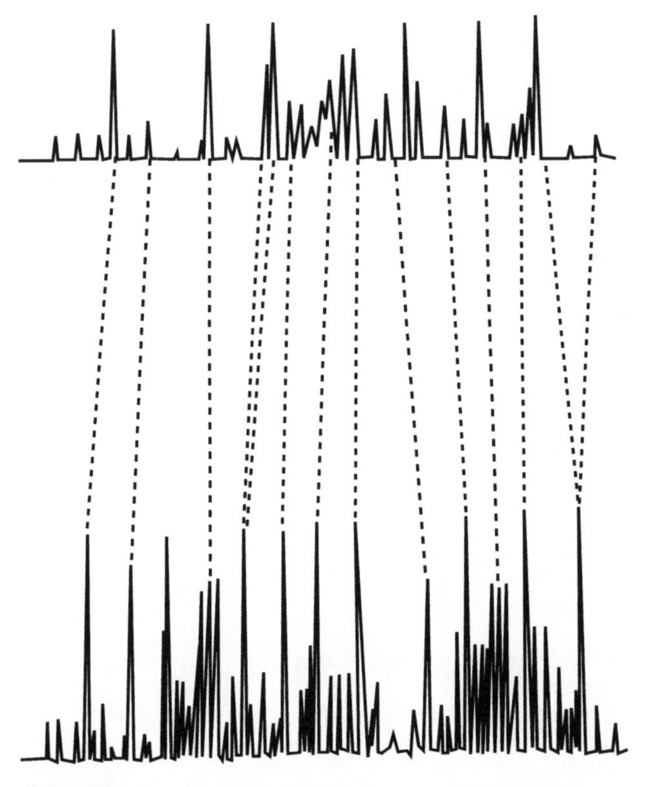

29-6 figure Comparison of gas chromatographic separation of perfume base using packed (*top*) and capillary columns (*bottom*). (Courtesy of Hewlett-Packard Co., Analytical Customer Training, Atlanta, GA.)

loading, within limits), and bleed. The liquid coatings are somewhat volatile and will be lost from the column at high temperatures (this is dependent upon the phase itself). This results in an increasing baseline (**column bleeding**) during temperature programming.

As many as 200 different liquid phases have been developed for GC. As GC has changed from packed to capillary columns, fewer stationary phases are now in use since column efficiency has substituted for phase selectivity (i.e., high efficiency has resulted in better separations even though the stationary phase is less suited for the separation). Now we find fewer than a dozen phases in common use (Table 29-3). The most durable and efficient phases are those based on polysiloxane (−Si−O−Si−).

Stationary phase selection involves some intuition, knowledge of chemistry, and help from the column manufacturer and the literature. There are general rules, such as choosing polar phases to separate polar compounds and the converse or phenyl-based column phase to separate aromatic compounds. However, the high efficiency of capillary columns often results in separation even though the phase is not optimal. For example, a 5% phenyl-substituted methyl silicone phase applied to a capillary column will separate polar as well as nonpolar compounds and is a commonly used phase coating.

29.3.4.2 Capillary Columns

The capillary column is a hollow fused silica glass (<100 ppm impurities) tube ranging in length from 5 to 100 m. The walls are so thin, ca. 25 μm, that they are flexible. The column outer walls are coated with a polyamide material to enhance strength and reduce breakage. Column inner diameters are typically 0.1 mm (**microbore**), 0.2–0.32 mm (**normal capillary**), or 0.53 mm (**megabore**).

Megabore columns (0.53 mm i.d.) were initially designed to replace packed columns without modification of instrumentation hardware. The most commonly used capillary columns are now 0.32 mm and 0.25 mm i.d. columns. Smaller diameter columns (0.10 mm and 0.18 mm i.d.) are used for fast GC analysis. The most common lengths of the GC column are 15, 30, and 60 m, although special columns can be over 100 m. Longer columns require longer analysis time. Although a longer column gives improved resolution, this benefit of better separation is not particularly obvious due to already high resolution power of capillary GC column.

Liquid coating is chemically bonded to the glass walls of capillary columns and internally crosslinked to give phase thicknesses ranging from 0.1 to 5 μm. Film thickness directly affects separation. Thicker films retain compounds longer in the stationary phase, thus the analytes will have longer interaction with the stationary phase to achieve separation. Generally, thick filmed column should be used to separate very volatile compounds. For example, a FFAP column with 1-μm film thickness can effectively hold and separate dimethyl sulfide (H_2S) and other highly volatile sulfur compounds (32). However, a thick film also will give a higher baseline due to bleeding. A thin film (0.25 μm) column is usually used to separate high molecular weight compounds; the analytes will stay in the stationary phase less time. Thin film columns also have less breeding at high temperature, and they are frequently used for GC-MS.

Most compounds can be separated using nonpolar 5% phenyl 95% dimethylpolysiloxane-based columns (e.g., DB-5, HP-5, RTX-5). This type of column has a very wide temperature range (−60°C to 325°C) and is very stable. However, to separate very polar compounds, such as alcohols and free fatty acids, a polar column is needed such as XX-WAX (polyethylene glycol) or XX-FFAP (polyethylene glycol treated with nitroterephthalic acid). A wax-type column has superior separation power; however, it has a narrow

Common Stationary Phases

Composition	Polarity	Applications[a]	Phases with Similar McReynolds Constants[b]	Temperature Limits[c]
100% Dimethyl polysiloxane (gum)	Nonpolar	Phenols, hydrocarbons, amines, sulfur compounds, pesticides, PCBs	OV-1, SE-30	−60°C to 325°C
100% Dimethyl polysiloxane (fluid)	Nonpolar	Amino acid derivatives, essential oils	OV-101, SP-2100	0–280°C
5% Phenyl 95% dimethyl polysiloxane	Nonpolar	Fatty acids, methyl esters alkaloids, drugs, halogenated compounds	SE-52, OV-23, SE-54	−60°C to 325°C
14% Cyanopropylphenyl methyl polysiloxane	Intermediate	Drugs, steroids, pesticides	OV-1701	−200°C to 280°C
50% Phenyl, 50% methyl methyl polysiloxane	Intermediate	Drugs, steroids, pesticides, glycols	OV-17	60–240°C
50% Cyanopropylmethyl, 50% phenyl methyl polysiloxane	Intermediate	Fatty acids, methyl esters, alditol acetates	OV-225	60–240°C
50% Trifluoropropyl polysiloxane	Intermediate	Halogenated compounds, aromatics	OV-210	45–240°C
Polyethylene glycol-TPA modified	Polar	Acids, alcohols, aldehydes, acrylates, nitrites, ketones	OV-351, SP-1000	60–240°C
Polyethylene glycol	Polar	Free acids, alcohols, esters, essential oils, glycols, solvents	Carbowax 20M	60–220°C

[a]Specific application notes from column suppliers provide information for choosing a specific column.
[b]McReynolds constants are used to group stationary phases together on the basis of separation properties.
[c]Stationary phases have both upper and lower temperature limits. Lower temperature limit is often due to a phase change (liquid to solid) and upper temperature limit to a volatilization of phase.

usable temperature range (40–240°C). It bleeds highly at high temperature and becomes solid (lost separation power) at low temperature. It is also sensitive to residue oxygen in the carrier gas, and it deteriorates quickly if oxygen is not removed in the carrier gas. Other specialty phase columns have been developed to improve specific resolution. Cyanopropyl-based columns (SP-2560, CP-Sil 88) are good for trans fatty acid esters. A cyclodex-based column is useful to separate stereoisomers of many flavor compounds.

29.3.4.3 Gas–Solid (PLOT) Chromatography

Gas–solid chromatography is a very specialized area of chromatography accomplished without using a liquid phase – the analyte interaction is with a porous material. This material has been applied both to packed and capillary columns. For the capillary column, the porous material is chemically or physically (by deposition) coated on the inner wall of the capillary and the column is called **porous-layer open-tubular** (PLOT) column. The most popular porous materials are alumina oxide, carbon, molecular sieve, and synthetic polymers such as Poropak or Chromosorb (trade names of polymers based on vinyl benzene). Separations usually involve water or other very volatile compounds such as headspace gas composition (N_2, O_2, CO_2, CO) in packaged food and ethylene during fruit ripening and storage.

29.3.5 Detectors

There are numerous detectors available for GC, each offering certain advantages in either sensitivity (e.g., electron capture) or selectivity (e.g., atomic emission detector). The most common detectors are the **FID, thermal conductivity** (TCD), **electron capture** (ECD), **flame photometric** (FPD), **pulsed flame photometric** (PFPD), and **photoionization** (PID) detectors. The operating principles and food applications of these detectors are discussed below. The characteristics of these detectors are summarized in Table 29-4.

29.3.5.1 Thermal Conductivity Detector

29.3.5.1.1 Operating Principles As the carrier gas passes over a hot filament (tungsten), it cools the filament at a certain rate depending on carrier gas velocity and composition. The temperature of the filament determines its resistance to electrical current.

29-4 table Characteristics of Most Common Detectors for Gas Chromatography

Characteristic	*Thermal Conductivity Detector*	*Flame Ionization Detector*	*Electron Capture Detector*	*Flame Photometric Detector*	*Photoionization Detector*
Specificity	Very little; detects almost anything, including H_2O; called the "universal detector"	Most organics	Halogenated compounds and those with nitro or conjugated double bonds	Organic compounds with S or P (determined by which filter is used)	Depends on ionization energy of lamp relative to bond energy of solutes
Sensitivity limits	ca. 400 pg; relatively poor; varies with thermal properties of compound	10–100 pg for most organics; very good	0.05–1 pg; excellent	2 pg for S and 0.9 pg for P compounds; excellent	1–10 pg depending on compound and lamp energy; excellent
Linear range	10^4 – poor; response easily becomes nonlinear	10^6–10^7 – excellent	10^4 – poor	10^4 for P; 10^3 for S	10^7 – excellent

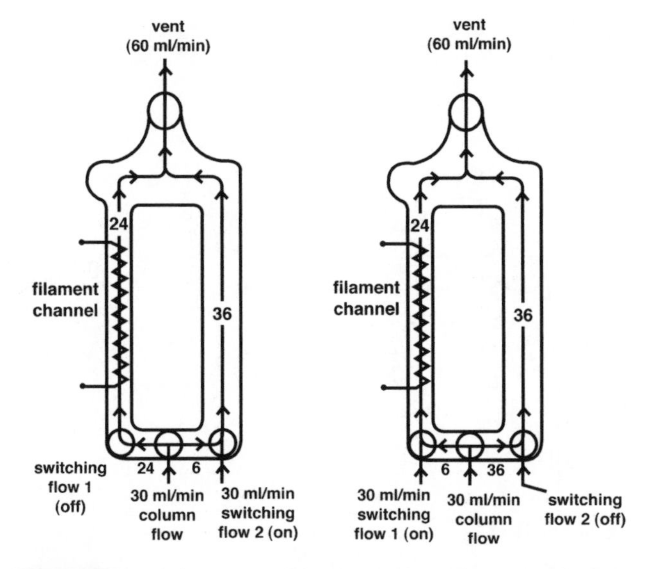

29-7 figure Schematic of the thermal conductivity detector. (Courtesy of Hewlett-Packard Co., Analytical Customer Training, Atlanta, GA.)

As a compound elutes with the carrier gas, the cooling effect on the filament is typically less, resulting in a temperature increase in the filament and an increase in resistance that is monitored by the GC electronics. Older style TCDs used two detectors and two matching columns; one system served as a reference and the other as the analytical system. Newer designs use only one detector (and column), which employs a carrier gas switching value to pass alternately carrier gas or column effluent through the detector (Fig. 29-7). The signal is then a change in cooling of the detector as a function of which gas is passing through the detector from the analytical column or carrier gas supply (reference gas flow).

The choice of carrier gas is important since differences between its thermal properties and the analytes determine response. While hydrogen is the best choice, helium is most commonly used since hydrogen is flammable.

29.3.5.1.2 Applications The most valuable properties of this detector are that it is *universal* in response and nondestructive to the sample. Thus, it is used in food applications for which there is no other detector that will adequately respond to the analytes (e.g., water, permanent gases, CO) or when the analyst wishes to recover the separated compounds for further analysis (e.g., trap the column effluent for infrared, nuclear magnetic resonance (NMR), or sensory analysis). It does not find broad use because it is relatively insensitive, and often the analyst desires specificity in detector response to remove interfering compounds from the chromatogram.

29.3.5.2 Flame Ionization Detector

29.3.5.2.1 Operating Principles As compounds elute from the analytical column, they are burned in a hydrogen flame (Fig. 29-8). A potential (often 300 V) is applied across the flame. The flame will carry a current across the potential which is proportional to the organic ions present in the flame from the burning of an organic compound. The current flowing across the flame is amplified and recorded. The FID responds to organics on a weight basis. It gives virtually no response to H_2O, NO_2, CO_2, H_2S and limited response to many other compounds. Response is best with compounds containing C–C or C–H bonds.

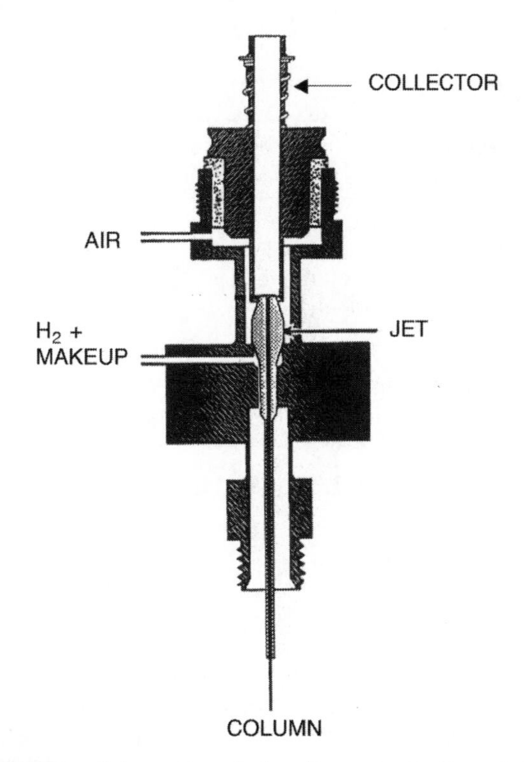

Schematic of the flame ionization detector designed for use with capillary columns. (Courtesy of Hewlett-Packard Co., Analytical Customer Training, Atlanta, GA.)

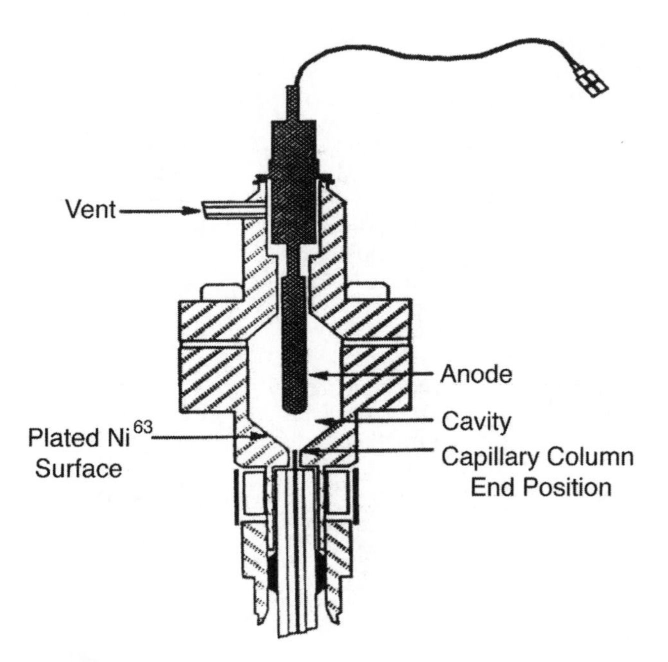

29-9 figure Schematic of the electron capture detector. (Courtesy of Hewlett-Packard Co., Analytical Customer Training, Atlanta, GA.)

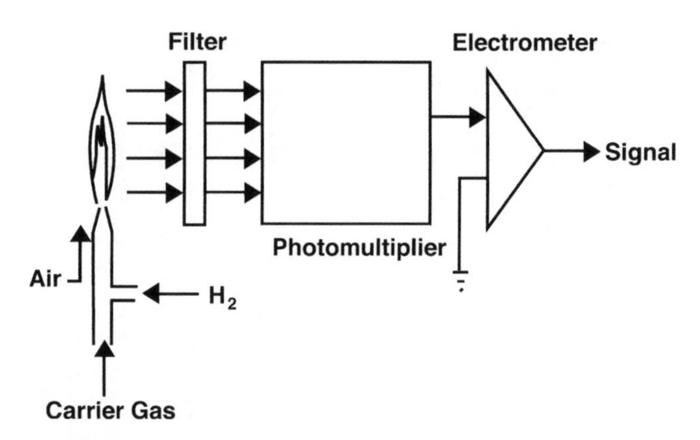

29-10 figure Schematic of the flame photometric detector. (Courtesy of Hewlett-Packard Co., Analytical Customer Training, Atlanta, GA.)

29.3.5.2.2 Applications The food analyst is most often working with organic compounds, to which this detector responds well. Its very good sensitivity, wide linear range in response (necessary in quantitation), and dependability make this detector the choice for most food work. Thus, this detector is used for virtually all food analyses for which a specific detector is not desired or sample destruction is acceptable (column eluant is burned in flame). This includes, for example, flavor studies, fatty acid analysis, carbohydrate analysis, sterols, contaminants in foods, and antioxidants.

29.3.5.3 Electron Capture Detector

29.3.5.3.1 Operating Principles The ECD contains a radioactive foil coating that emits electrons as it undergoes decay (Fig. 29-9). The electrons are collected on an anode, and the standing current is monitored by instrument electronics. As an analyte elutes from the GC column, it passes between the radioactive foil and the anode. Compounds that capture electrons reduce the standing current and thereby give a measurable response. Halogenated compounds or those with conjugated double bonds give the greatest detector response. Unfortunately this detector becomes saturated quite easily and thus has a very limited linear response range.

29.3.5.3.2 Applications In food applications, the ECD has found its greatest use in determining PCBs and pesticide residues (see Chap. 18). The specificity and sensitivity of this detector make it ideal for this application.

29.3.5.4 Flame Photometric Detector and Pulsed Flame Photometric Detector

29.3.5.4.1 Operating Principles The FPD detector works by burning all analytes eluting from the analytical column and then measuring specific wavelengths of light that are emitted from the flame using a filter and photometer (Fig. 29-10). The wavelengths

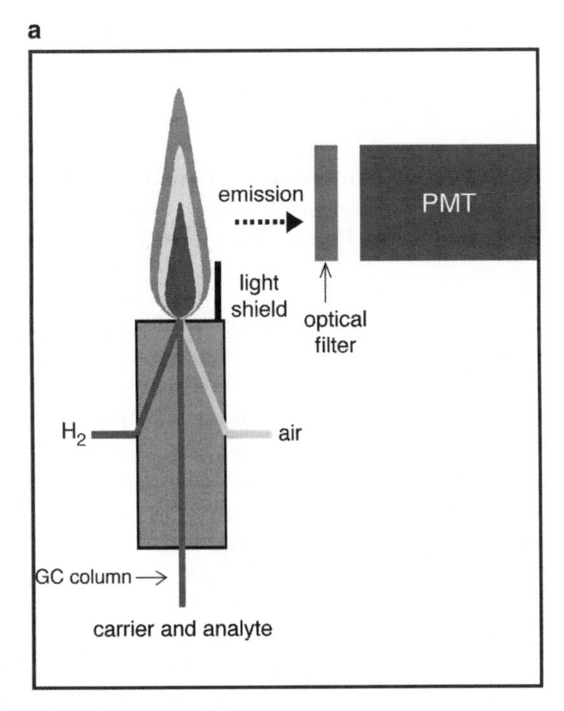

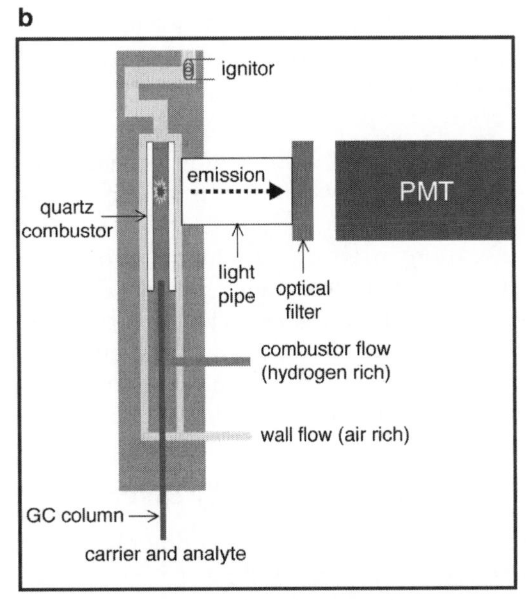

29-11 figure Comparison of flame photometric detector (**a**) and pulsed flame photometric detector (**b**). (Courtesy of Varian Inc., Palo Alto, CA.)

of light that are suitable in terms of intensity and uniqueness are characteristic of sulfur (S) and phosphorus (P). Thus this detector gives a greatly enhanced signal for these two elements (several thousandfold for S- or P-containing organic molecules vs. non-S or P-containing organic molecules). Detector response to S-containing molecules is nonlinear and thus quantification must be done with care.

The PFPD is very similar to FPD. Unlike traditional flame photometric detection (FPD), which uses a continuous flame, the PFPD ignites, propagates, and self-terminates 2–4 times per second (Fig. 29-11). Specific elements have their own emission profile: hydrocarbons will complete emission early while sulfur emissions begin at a relatively later time after combustion. Therefore, a timed "gate delay" can selectively allow for only emissions due to sulfur to be integrated, producing a clean chromatogram. This timed "gate delay" greatly improves the sensitivity. The PFPD can detect sulfur-containing compounds at a much lower detection limit than nearly all other methods of detection (33).

29.3.5.4.2 Applications Both the FPD and the PFPD have found major food applications in the determination of organophosphorus pesticides and volatile sulfur compounds in general. The determination of sulfur compounds has typically been in relation to flavor studies.

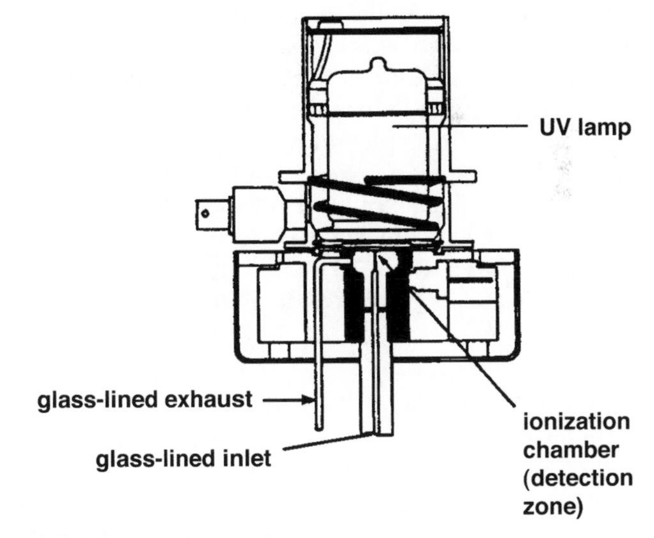

29-12 figure Schematic of the photoionization detector. (Courtesy of Hewlett-Packard Co., Analytical Customer Training, Atlanta, GA.)

29.3.5.5 Photoionization Detector

29.3.5.5.1 Operating Principles The **photoionization detector** (PID) uses ultraviolet (UV) irradiation (usually 10.2 eV) to ionize analytes eluting from the analytical column (Fig. 29-12). The ions are accelerated by a polarizing electrode to a collecting electrode. The small current formed is magnified by the electrometer of the GC to provide a measurable signal.

This detector offers the advantages of being quite sensitive and nondestructive and may be operated in a selective response mode. The selectivity comes from being able to control the energy of ionization, which will determine the classes of compounds that are ionized and thus detected.

29.3.5.5.2 Applications The PID finds primary use in analyses for which excellent sensitivity is required from a nondestructive detector. This is most often a flavor application in which the analyst wishes to smell the GC effluent to determine the sensory character of the individual GC peaks. While this detector might find broader use, the widespread availability of the FID (which is suitable for most of the same applications) meets most of these needs.

29.3.5.6 Electrolytic Conductivity Detector

29.3.5.6.1 Operating Principles Compounds entering the **electrolytic conductivity detector** (ELCD) are mixed with a reagent gas (oxidizing or reducing depending on the analysis) in a nickel reaction tube producing ionic species. These products are mixed with a deionized solvent, interfering ions are scrubbed from the effluent, and the ionic analyte-transformation product is detected within the electrolyte conductivity cell. This detector can be used for the specific detection of sulfur-, nitrogen-, or halogen-containing molecules. For example, when operated in the nitrogen mode, analyte is mixed with H_2 gas and hydrogenated over a nickel catalyst at 850°C. Acidic hydrogenation products are removed from the effluent by passage through a $Sr(OH)_2$ trap and the NH_3 from the analyte passes to the conductivity cell where it is measured (34).

29.3.5.6.2 Applications This detector can be used in many applications for which element specificity is desired. Examples would be pesticide, herbicide, nitrosamine, or flavor analysis. The ELCD is very selective and quite sensitive having detection limits of 0.1–1 pg of chlorinated compounds, 2 pg for sulfur, and 4 pg for nitrogen.

29.3.5.7 Thermionic Detector

29.3.5.7.1 Operating Principles The thermionic detector (also called the nitrogen phosphorus detector, NPD) is a modified FID in which a nonvolatile ceramic bead is used to suppress the ionization of hydrocarbons as they pass through a low-temperature fuel-poor hydrogen plasma. The ceramic bead is typically composed of rubidium which is heated to 600–800°C. Most commonly this detector is used for the selective detection of nitrogen- or phosphorus-containing compounds. It does not detect inorganic nitrogen or ammonia.

29.3.5.7.2 Applications This detector is primarily used for the measurement of specific classes of flavor compounds, nitrosamines, amines, and pesticides.

29.3.5.8 Hyphenated Gas Chromatographic Techniques

Hyphenated gas chromatographic techniques are those that combine GC with another major technique. Examples are **GC-AED** (atomic emission detector), **GC-FTIR** (Fourier transform infrared), and **GC-MS** (mass spectrometry). While all of the techniques are established methods of analysis in themselves, they become powerful tools when combined with a technique such as GC. GC provides the separation and the hyphenated technique the detector. GC-MS has long been known to be a most valuable tool for the identification of volatile compounds (see Chap. 26). The MS, however, may perform the task of serving as a specific detector for the GC by selectively focusing on ion fragments unique to the analytes of interest. The analyst can detect and quantify components without their gas chromatographic resolution in this manner. The same statements can be made about GC-FTIR (see Chap. 24). The FTIR can readily serve as a GC detector.

A relatively new combination is GC-AED. In this technique, the GC column effluent enters a microwave-generated helium plasma that excites the atoms present in the analytes. The atoms emit light at their characteristic wavelengths, and this emission is monitored using a diode ray detector similar to that used in HPLC. This results in a very sensitive and specific elemental detector.

29.3.5.9 Multidimensional Gas Chromatography

Multidimensional gas chromatography (MDGC) greatly increases the separation ability of gas chromatography (35). By simply coupling two GC columns, each of opposite polarity, an overall improvement in separation can be accomplished (36). However, this tandem operation of GC columns does not actually represent multidimensionality, but rather resembles the use of a mixed-stationary phase column (35). True MDGC involves a process known as orthogonal separation in which a sample is first dispersed by one column, and the simplified subsamples are then applied onto another column for further separation. MDGC techniques can be generally divided into two classes: (1) conventional, or "heart-cut," MDGC and (2) comprehensive two-dimensional gas chromatography (GC × GC).

29.3.5.9.1 Conventional Two-Dimensional GC **Conventional two-dimensional GC** is achieved by using coupled capillary columns for which a small portion, or heart-cut, of the effluent from the first ("preseparation") column is transferred to the second ("analytical") column. The concept of conventional MDGC is almost identical to that of preparative GC operations, for which one column is used to obtain a partially separated fraction of a complex aroma mixture, which is then reinjected onto another GC column, usually with an opposite stationary phase, for further separation. The only difference is that with MDGC there are no requirements for manual collection of the effluent obtained from the preseparation column since the two columns are directly connected.

Because the second column in the MDGC system is only injected with a small portion of the total sample at one time, a large quantity of the sample can be injected onto the first column without the worry of chromatographic band smearing during analytical separations (37). Therefore, trace compounds can be easily enriched for more successful detection and identification.

The MDGC technique is particularly useful to study enantiomers of flavor compounds. The interested compound can be "heart-cut" and transferred to an analytical column with an enantioselective stationary phase for good separation of targeted chiral compounds.

29.3.5.9.2 Comprehensive Two-Dimensional GC **Comprehensive two-dimensional MDGC** is among the most powerful two-dimensional gas chromatographic techniques that have been developed today (Fig. 29-13). Unlike conventional MDGC in which only particular segments are transferred from the preseparation column onto the analytical column, comprehensive MDGC, or GC × GC, involves the transfer of the entire effluent from the first column onto a second column by way of a modulation

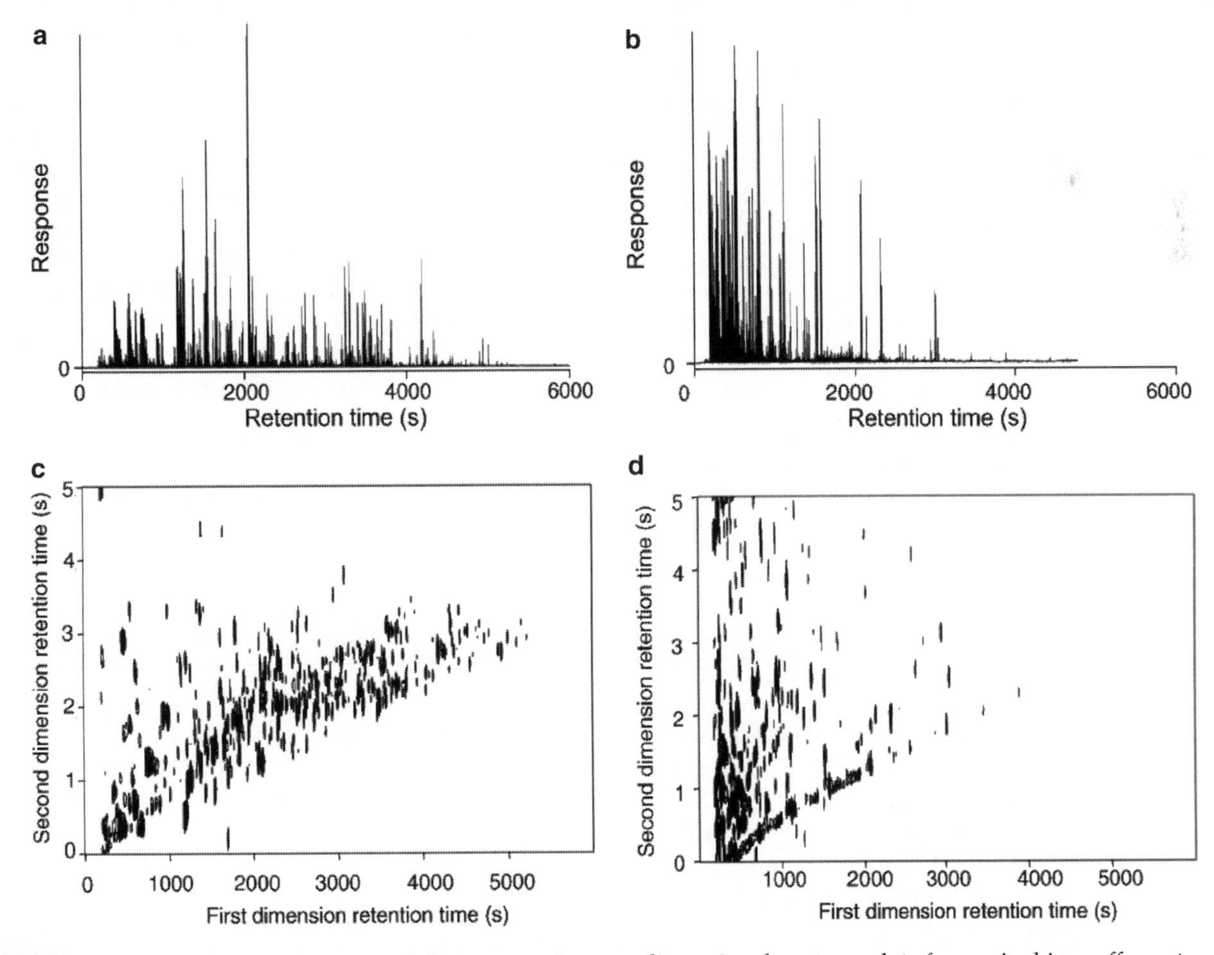

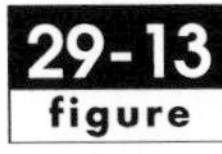

Total ion chromatograms and their respective two-dimensional contour plots for an Arabica coffee extract separated by GC × GC using two different column sets: polar × nonpolar (**a** and **c**) along with nonpolar × polar (**b** and **d**). [Reprinted from reference (45), used with permission.]

interface so that complete two-dimensional data can be obtained for the entire run of the first column. The operation of the modulator involves the generation of narrow injection bands from the first column, which are continuously, but individually, sent to the secondary column for final separation. GC × GC requires that the second column can operate quickly enough to generate a complete set of data during the time that a single peak elutes from the first GC column, generally within 5 s (35, 38). The data from both time axes are combined to create a set of coordinates for each peak so that the resultant chromatogram is actually a two-dimensional (2D) plane rather than a straight line. Peak area information can be obtained by summing the integration over both dimensions.

In comprehensive GC × GC, the two columns perform independently of each other, therefore the overall peak capacity becomes the product of the capacities for each column. Because analytes elute from the second column so quickly, data acquisition must be adequately fast enough for proper detection. Time-of-flight mass spectrometry (TOF-MS) and rapid-scanning quadrupole mass spectrometry (qMS) have both been used as effective detection methods for GC × GC to obtain mass spectral information (39, 40).

Although the instrumentation can be quite expensive, the use of comprehensive GC × GC for volatile aroma analysis has exponentially increased over the past few years as methodologies have become more established and systems have become commercially available. Overall, the application of MDGC, both conventional and comprehensive, has allowed for advanced separations of complex aromas to occur by using state-of-the-art instrumentation.

29.4 CHROMATOGRAPHIC THEORY

29.4.1 Introduction

GC may depend on several types (or principles) of chromatography for separation. The principles of chromatographic separation are discussed in Chap. 27, Sect. 27.4. For example, size-exclusion chromatography is used in the separation of permanent gases such as N_2, O_2, and H_2. A variation of size exclusion is used to separate chiral compounds on cyclodextrin-based columns; one enantiomorphic form will fit better into the cavity of the cyclodextrin than will the other form, resulting in separation. Adsorption chromatography is used to separate very volatile polar compounds (e.g., alcohols, water, and aldehydes) on porous polymer columns (e.g., Tenax^R phase). Partition chromatography is the workhorse for gas chromatographic separations. There are over 200 different liquid phases that have been developed for gas chromatographic use over time. Fortunately, the vast majority of separations can be accomplished with only a few of these phases, and the other phases have fallen into disuse. GC depends not only upon adsorption, partition, and/or size exclusion for separation, but also upon solute boiling point for additional resolving powers. Thus, the separations accomplished are based on several properties of the solutes. This gives GC virtually unequaled resolution powers as compared with most other types of chromatography (e.g., HPLC, paper, or thin-layer chromatography).

A brief discussion of chromatographic theory will follow. The purpose of this additional discussion is to apply this theory to GC to optimize separation efficiency so that analyses can be done faster, less expensively, or with greater precision and accuracy. If one understands the factors influencing resolution in GC, one can optimize the process and gain in efficiency of operation.

29.4.2 Separation Efficiency

A good separation has narrow-based peaks and ideally, but not essential to quality of data, baseline separation of compounds. This is not always achieved. Peaks broaden as they pass through the column – the more they broaden, the poorer is the separation and efficiency. As discussed in Chap. 27, Sect. 27.5.2.2.2, a measure of this broadening is **height equivalent to a theoretical plate** (HETP). This term is derived from N, the number of plates in the column, and L, the length of the column. A good packed column might have $N = 5000$, while a good capillary column should have about 3000–4000 plates per meter for a total of 100,000–500,000 plates depending on column length. HETP will range from about 0.1 to 1 mm for good columns.

29.4.2.1 Carrier Gas Flow Rates and Column Parameters

Several factors influence column efficiency (peak broadening). As presented in Chap. 27, these are related by the **Van Deemter equation** [1]: (HETP values should be small.)

$$\text{HETP} = A + B/u + Cu \qquad [1]$$

where:

HETP = height equivalent to a theoretical plate
A = eddy diffusion
B = band broadening due to diffusion
u = velocity of the mobile phase
C = resistance to mass transfer

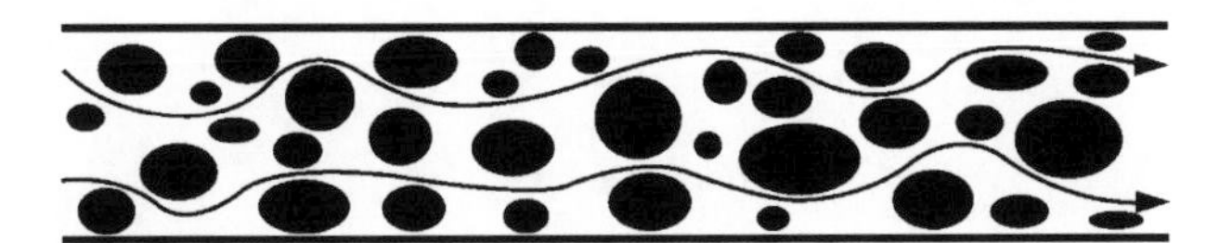

29-14 figure Illustration of flow properties that lead to large eddy diffusion (term A).

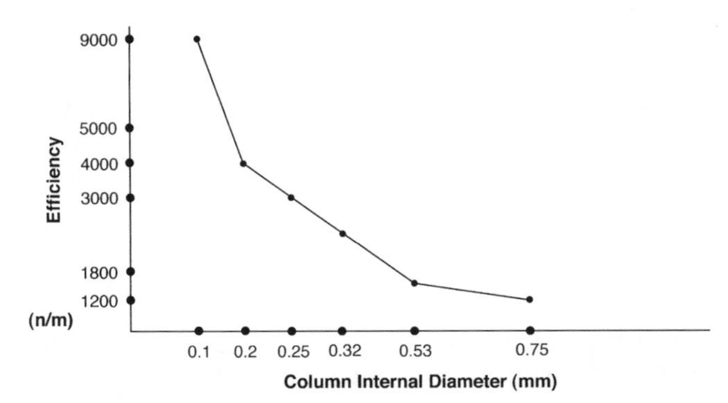

29-15 figure The influence of column diameter on column efficiency (plates/meter). (Courtesy of Hewlett-Packard, Analytical Customer Training, Atlanta, GA.)

A is eddy diffusion; this is a spreading of the analytes in the column due to the carrier gas having various pathways or nonuniform flow (Fig. 29-14). In packed column chromatography, poor uniformity in solid support size or poor packing results in channeling and multiple pathways for carrier flow, which results in spreading of the analyte in the column. Thus, improved efficiency is obtained by using the high performance solid supports and commercially packed columns.

In capillary chromatography, the *A* term is relatively very small. However, as the diameter of the capillary column increases, the flow properties deteriorate, and band spreading occurs. The most efficient capillary columns have small diameters (0.1 mm), and efficiency decreases rapidly as one goes to megabore columns (Fig. 29-15). Megabore columns are only slightly more efficient than packed columns. While column efficiency increases as we go to smaller columns, column capacity decreases rapidly. Microbore columns are easily overloaded (capacity may be 1–5 ng per analyte), resulting again in poor chromatography. Thus, column diameter is generally chosen as 0.2–0.32 mm to compromise efficiency with capacity.

B is band broadening due to diffusion; solutes will go from a high to a low concentration. The term *u* is velocity of the mobile phase. Thus, very slow flow rates result in large amounts of diffusion band broadening, and faster flow rates minimize this term. The term *u* is influenced by the carrier gas choice. Larger-molecular-weight carrier gases (e.g., nitrogen) are more viscous than the lighter-molecular-weight gases (e.g., helium or hydrogen) and thus peak spreading is less for nitrogen than for helium or hydrogen carrier gases. This results in nitrogen having the lowest HETP of the carrier gases and theoretically being the best choice for a carrier gas. However, other considerations that will be discussed in Sect. 29.4.2.2 make nitrogen a very poor choice for a carrier gas.

C is resistance to mass transfer. If the flow (u) is too fast, the equilibrium between the phases is not established, and poor efficiency results. This can be visualized in the following way: If one molecule of solute is dissolved in the stationary phase and another is not, the undissolved molecule continues to move through the column while the other is retained. This results in band spreading within the column. Other factors that influence this term are thickness of the stationary phase and uniformity of coating on the phase support. Thick films give greater capacity (ability to handle larger amounts of a solute) but at a cost in terms of band spreading (efficiency of separation) since thick films provide more variation in diffusion properties in and out of the stationary phase. Thus phase thickness is a compromise between maximizing separation efficiency and sample capacity (too much sample – overloading a column – destroys separation ability). Phase thicknesses of 0.25–1 μm are commonly used for most applications.

If the Van Deemter equation is plotted, giving the figure discussed in Fig. 27-13, we see an optimum in flow rate due to the opposing effects of the *B* and the *C* terms. It should be noted that the GC may not be operated at a carrier flow velocity yielding maximum efficiency (lowest HETP). Analysis time is directly proportional to carrier gas flow velocity. If the analysis time can be significantly shortened by operating above the optimum flow velocity and adequate resolution is still obtained, velocities well in excess of optimum should be used.

29.4.2.2 Carrier Gas Type

The relationship between HETP and carrier gas flow velocity is strongly influenced by carrier gas choice (Fig. 29-16). Nitrogen is the most efficient (lowest HETP) carrier gas, as discussed in Sect. 29.4.2.1, but its minimum HETP occurs at a very low flow velocity. This low mobile phase velocity results in unnecessarily long analysis times. Considering the data plotted in Fig. 29-16, nitrogen has an HETP of about 0.25 at an optimum flow velocity of 10 cm/s. The HETP of helium is only about 0.35 at 40 cm/s flow velocity. This is a small loss in resolution to reduce the analysis time fourfold (10 cm/s for nitrogen vs. 40 cm/s for helium).

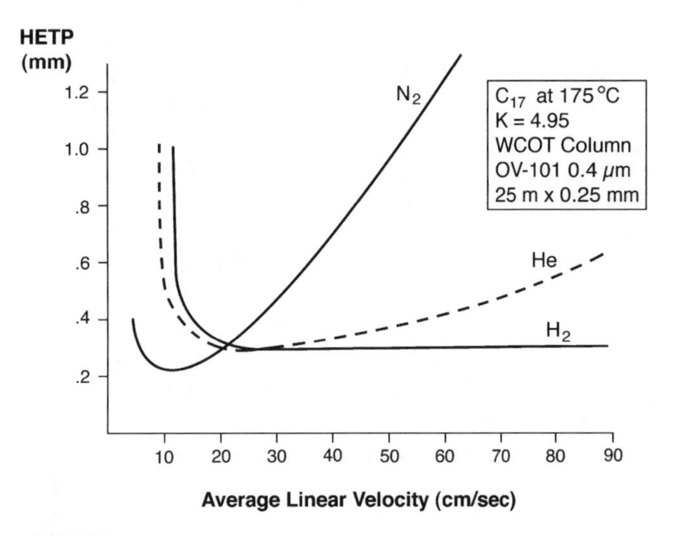

29-16 figure Influence of carrier gas type and flow rate on column efficiency. (Courtesy of Hewlett-Packard Co., Analytical Customer Training, Atlanta, GA.)

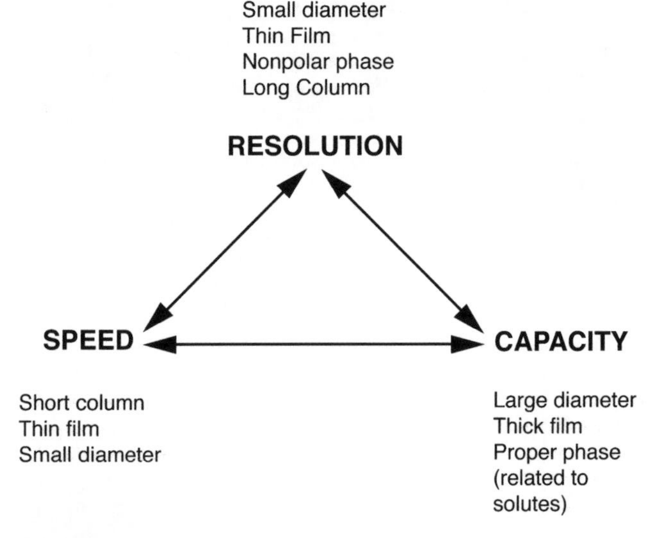

29-17 figure Relationships among column capacity, efficiency, resolution, and analysis speed.

One can potentially even push the flow velocity up to 60 or 70 cm/s and accomplish separation in even shorter times.

The plots in Fig. 29-16 suggest that hydrogen is an even better choice for a carrier gas than helium (i.e., has a flatter relationship between carrier gas flow velocity and HETP). However, there are some concerns about hydrogen being flammable and reports in the literature that some compounds may be hydrogenated in the GC system. Additionally, some detectors cannot use hydrogen as a carrier gas (e.g., a mass spectrometer) and, thus, one may be limited to helium as a good compromise.

29.4.2.3 Summary of Separation Efficiency

In summary, an important goal of analysis is to achieve the necessary separation in the minimum amount of time. The following factors should be considered:

1. In general, small diameter columns (packed or capillary) should be used since separation efficiency is strongly dependent on column diameter. While small diameter columns will limit column capacity, limited capacity often can be compensated for by increasing phase thickness. Increased phase thickness will also decrease column efficiency but to a lesser extent than increasing column diameter.
2. Lower column operating temperatures should be used – if elevated column temperatures are required for the compounds of interest to elute, use a shorter column if resolution is adequate.
3. One should keep columns as short as possible (analysis time is directly proportional to column length – resolution is proportional to the square root of length).
4. Use hydrogen as the carrier gas if the detector permits. Some detectors have specific carrier gas requirements.
5. Operate the GC at the maximum carrier gas velocity that provides resolution.

The pyramid shown in Fig. 29-17 summarizes the compromises that must be made in choosing the analytical column and gas chromatographic operating conditions. One cannot optimize any given operating conditions and column choices to get one of these properties without compromising another property. For example, optimizing chromatographic resolution (small bore capillary diameter, thin phase coating, long column lengths, and slow or optimum carrier gas flow rate) will be at the cost of capacity (large bore columns and thick phase coating) and speed (thin film coating, high carrier gas flow velocities, and short columns). Capacity will be at a cost of resolution and speed, etc. The choice of column and operating parameters must consider the needs of the analyst and the compromises involved in these choices.

29.5 APPLICATIONS OF GC

While some detail on the application of GC to food analyses has been presented in Chaps. 10, 14, and 18, a few additional examples will be presented below to illustrate separations and chromatographic conditions.

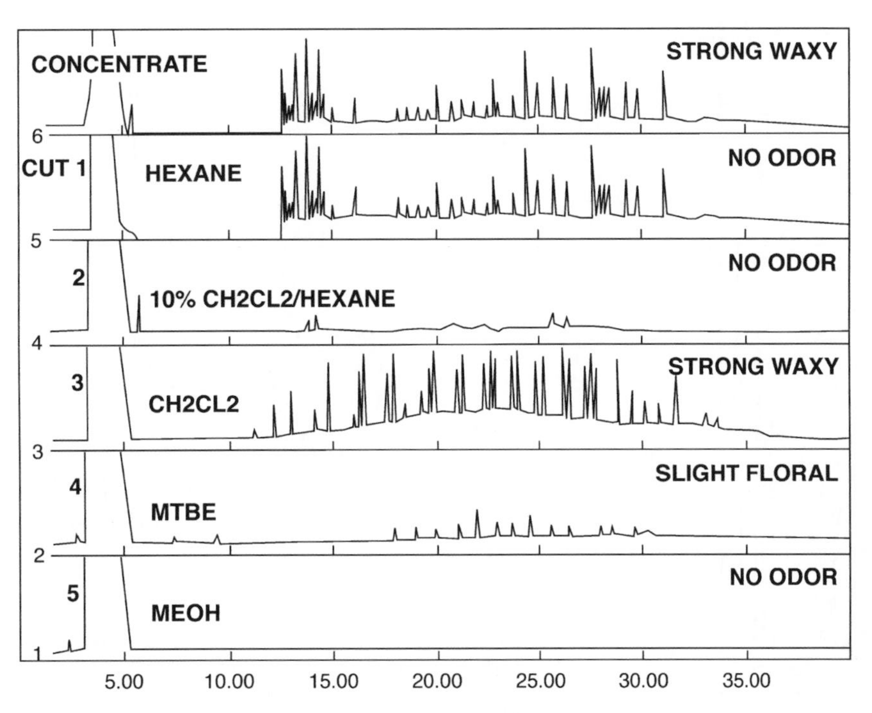

Typical capillary gas chromatographic separation of residual volatiles in a food packaging film. [From (41), used with permission.]

29.5.1 Residual Volatiles in Packaging Materials

Residual volatiles in packaging materials can be a problem both from health (if they are toxic) and quality standpoints (produce off flavors in the food). As the industry has turned from glass to polymeric materials, there have been more problems in this respect. GC is most commonly used to determine the residual volatiles in these materials.

The chromatograms presented in Fig. 29-18 were produced by steam distilling a food packaging film, extracting the volatiles from the distillate in an organic solvent, concentrating the solvent extract, and then chromatographing it on a capillary column (top chromatogram in Fig. 29-18) (41). The extreme complexity of the chromatogram required that the concentrate be further fractionated on silica gel and each fraction rechromatographed. The chromatograms labeled "cuts 1–5" are the chromatograms resulting from eluting the silica gel with: (1) hexane removing saturated hydrocarbons from the gel bed (cut 1); (2) 10% CH_2Cl_2/hexane removing the unsaturated and aromatic hydrocarbons (cut 2); (3) CH_2Cl_2 removing the ketones and aldehydes (cut 3); (4) methyl-*t*-butylether removing the acids, unsaturated ketones, and aldehydes (cut 4); and (5) alcohol removing the remaining polar volatiles (cut 5). One can see that the prefractionation of the extracted packaging volatiles greatly simplified the chromatography and permitted the researcher to focus on the volatiles responsible for the off odor in the packaging material.

29.5.2 Separation of Stereoisomers

GC has found extensive application in the separation of chiral volatile compounds in foods (e.g., D and L-carvone). Chiral separations are most commonly accomplished using cyclodextrin-based gas chromatographic columns. Cyclodextrins are molecules (6-, 7-, or 8-membered rings of glucose) that have an internal cavity of suitable dimensions to permit the inclusion of many small organic molecules. While optical isomers of molecules have virtually identical physical properties and thus they are difficult to separate by most chromatographic methods, they differ in spatial configuration. Stereoisomers of a given compound will be included in the cyclodextrin cavity of the gas chromatographic column to a lesser or greater extent as they flow through a cyclodextrin capillary column and become separated.

The chromatogram presented in Fig. 29-19 shows the separation of six stereoisomers of α and β-irone (42). This separation was accomplished using an octakis (6-*O*-methyl-2,3-di-*O*-pentyl)-γ-cyclodextrin/OV-17 capillary column.

29.5.3 Headspace Analysis of Ethylene Oxide in Spices

Ethylene oxide (ETO) is a highly volatile compound that has found use in the food industry as a fumigant for spices (43). It has been classified as a suspect human carcinogen and thus its residual concentration

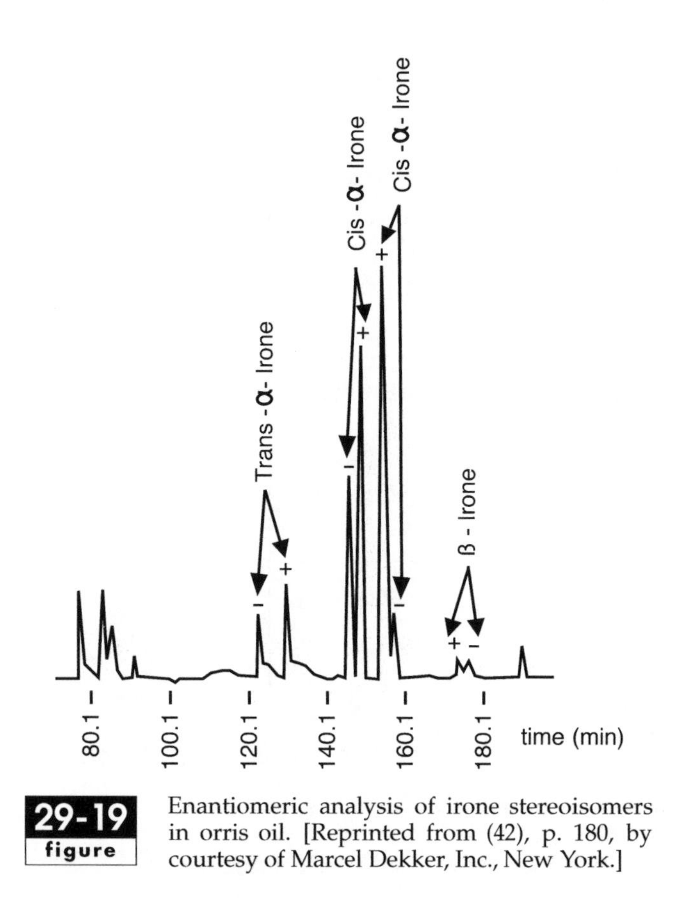

29-19 figure Enantiomeric analysis of irone stereoisomers in orris oil. [Reprinted from (42), p. 180, by courtesy of Marcel Dekker, Inc., New York.]

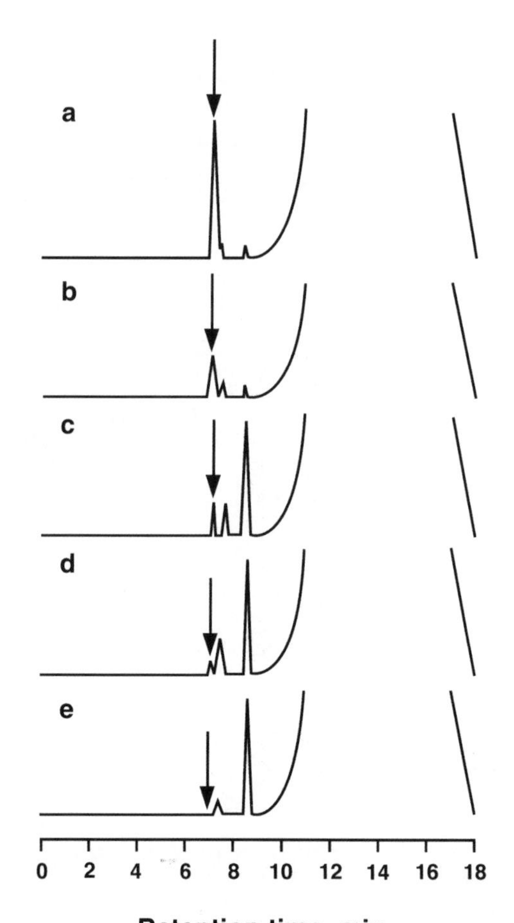

29-20 figure Headspace gas chromatographic analysis of ethylene oxide in spices. (**a**) 3 μg pure ETO, (**b**) 1 μg pure ETO, (**c**) 3 μg ETO in spice, (**d**) 1 μg in spice, (**e**) the pure spice. [From (43), used with permission.]

in spices is of concern. Because of its volatility, ETO is well suited to determination by GC.

Woodrow et al. (43) chose to use a headspace method for ETO determination. This is reasonable since ETO is very volatile, sensitivity is adequate, and headspace techniques are simple to perform. The method involved adding 1 g of ground spice to a 22-ml headspace vial (a vial that has a Teflon septum closure for sampling), adding internal standard (1-octanol), incubating the vial at 60°C for 20 min, and then removing and injecting ca. 1 ml of the headspace into the gas chromatograph. ETO was separated from other volatiles in the sample using a porous polymer capillary column (divinylbenzene homopolymer). Typical chromatograms of pure ETO, spice, and spice spiked with ETO are shown in Fig. 29-20.

29.5.4 Aroma Analysis of Heated Butter

Volatile aroma compounds are important contributors to the quality of foods. The composition and the concentration of volatile aroma compounds impact the flavor perceived. GC has been widely applied to define a volatile chemical fingerprint to characterize the flavor quality of food products. A chromatogram of the volatile compounds in heated butter isolated by a static headspace-Tenax absorbent technique and subsequently analyzed by GC on a wax capillary column is illustrated in Fig. 29-21. Seven select aroma compounds reported to contribute to the flavor of heated butter are displayed. Changes in the concentrations of the volatile flavor compounds can be related to changes in the flavor properties of foods and provide insights into the role of processing, storage, ingredients, packaging, etc. on food flavor. Volatile flavor compounds that originally contributed desirable flavor properties can also become undesirable at elevated levels and result in off-flavor development. For example, an off-flavor defect in butter developed during storage has been related to an increase in lactone concentration, such as δ-decalactone (44). The prediction of product shelf-life based on off-flavor development typically involves very volatile compounds, such as hexanal (a common indicator of lipid oxidation) (45).

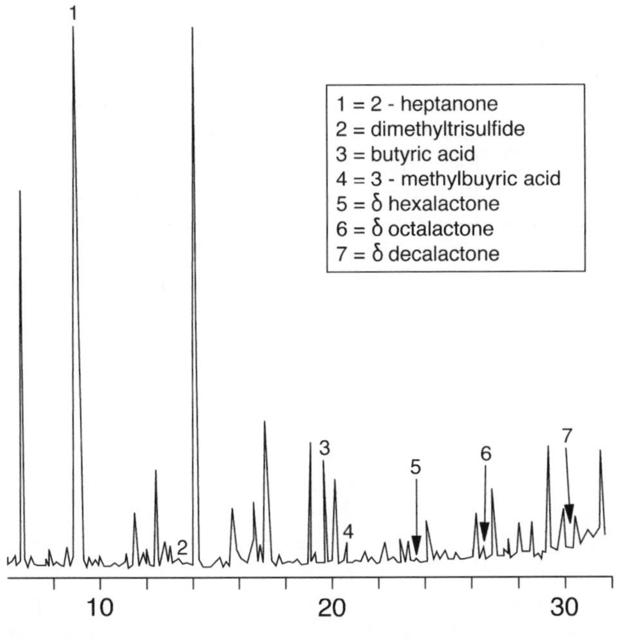

29-21 figure Heated butter static headspace GC chromatogram with select aroma compound displayed. [Adapted from (46), used with permission.]

29.6 SUMMARY

GC has found broad application in both the food industry and academia. It is exceptionally well suited to the analysis of volatile thermally stable compounds. This is due to the outstanding resolving properties of the method and the wide variety of detectors that can provide either sensitivity or selectivity in analysis.

Sample preparation generally involves the isolation of solutes from foods, which may be accomplished by headspace analysis, distillation, preparative chromatography (including solid-phase extraction), or extraction (liquid–liquid). Some solutes can then be directly analyzed, while others must be derivatized prior to analysis.

The gas chromatograph consists of a gas supply and regulators (pressure and flow control), injection port, column and column oven, detector, electronics, and a data recording and processing system. The analyst must be knowledgeable about each of these GC components: carrier and detector gases; injection port temperatures and operation in split, splitless, temperature-programmed, or on-column modes; column choices and optimization (gas flows and temperature profile during separation); and detectors (TCD, FID, NPD, ECD, FPD, PFPD, and PID). The characteristics of these GC components and an understanding of basic chromatographic theory are essential to balancing the properties of resolution, capacity, speed, and sensitivity.

Unlike most of the other chromatographic techniques, traditional GC has reached the theoretical limits in terms of both resolution and sensitivity. Thus, this method will not change significantly in the future other than for minor innovations in hardware or associated computer software. However, two-dimensional GCs, both heart-cut GC–GC and comprehensive GC × GC, are still developing quickly in both instrumentation and applications, especially in the field of flavor analysis.

GC as a separation technique has been combined with AED, FTIR, and MS as detection techniques to make GC an even more powerful tool. Such hyphenated techniques are likely to continue to be developed and refined.

29.7 STUDY QUESTIONS

1. For each of the following methods to isolate solutes from food prior to GC analysis, describe the procedure, the applications, and the cautions in use of the method:
   (a) Headspace methods
   (b) Distillation methods
   (c) Solvent extraction
2. What is solid-phase extraction and why is it advantageous over traditional liquid–liquid extractions?
3. Why must sugars and fatty acids be derivatized before GC analysis, while pesticides and aroma compounds need not be derivatized?
4. Why is the injection port of a GC at a higher temperature than the oven temperature?
5. Differentiate packed columns from capillary columns (microbore and megabore) with regard to physical characteristics and column efficiency.
6. You are doing GC with a packed column and notice that the baseline rises from the beginning to the end of each run. Explain a likely cause for this increase.
7. The most common detectors for GC are TCD, FID, ECD, FPD, and PID. Differentiate each of these with regard to the operating principles. Also, indicate below which detector(s) fits the description given.
   (a) Least sensitive
   (b) Most sensitive
   (c) Least specific
   (d) Greatest linear range
   (e) Nondestructive to sample
   (f) Commonly used for pesticides
   (g) Commonly used for volatile sulfur compounds
8. What types of chromatography does GC rely upon for separation of compounds?
9. In GC, explain why a balance has to be maintained between efficiency and capacity. Also, give an example situation in which you would sacrifice capacity for efficiency.
10. You plan to use GC to achieve good chromatographic separation of Compounds A, B, and C in your food

sample. You plan to use an internal standard to quantitate each compound. By answering the following questions, describe how using an internal standard works for this purpose (see also Chap. 27, Sect. 27.5.3).
   (a) How do you choose the internal standard for your application?
   (b) What do you do with the internal standard, relative to the standard solutions for Compounds A, B, and C and relative to the food sample? Be specific in your answer.
   (c) What do you measure?
   (d) If you were to prepare a standard curve, what would you plot?
   (e) Why are internal standards commonly used for GC?
11. A fellow lab worker is familiar with HPLC for food analysis but not with GC. As you consider each component of a typical chromatographic system (and specifically the components and conditions for GC and HPLC systems), explain GC to the fellow worker by comparing and contrasting it to HPLC. Following that, state in general terms the differences among the types of samples appropriate for analysis by GC vs. HPLC and give several examples of food constituents appropriate for analysis by each (see also Chap. 28).

29.8 REFERENCES

1. James AT, Martin AJP (1952) Gas–liquid chromatography: the separation and microestimation of volatile fatty acids from formic acid to dodecanoic acid. Biochem J 50:679
2. Niessen WMA (2001) Current practice of gas chromatography – mass spectrometry. Marcel Dekker, New York
3. Rood D (1999) A practical guide to the care, maintenance, and troubleshooting of capillary gas chromatographic systems, 3rd edn. Weinheim, New York
4. Schomburg G (1990) Gas chromatography: a practical course. Weinheim, New York
5. Gordon MH (1990) Principles and applications of gas chromatography in food analysis. E. Horwood, New York
6. O'Keeffe M (2000) Residue analysis in food: principles and applications. Harwood Academic, Amsterdam
7. Drawert F, Heimann W, Enberger R, Tressl R (1965) Enzymatische Verandrung des naturlichen Apfelaromass bei der Aurfarbeitung. Naturwissenschaften 52:304
8. Fleming HP, Fore SP, Goldblatt LA (1968) The formation of carbonyl compounds in cucumbers. J Food Sci 33:572
9. Kazeniak SJ, Hall RM (1970) Flavor chemistry of tomato volatiles. J Food Sci 35:519
10. Leahy MM, Reineccius GA (1984) Comparison of methods for the analysis of volatile compounds from aqueous model systems. In: Schreier P (ed) Analysis of volatiles: new methods and their application. DeGruyter, Berlin
11. Mresili R (1997) Techniques for analyzing food aroma. Marcel Dekker, New York
12. Mussinan CJ, Morello MJ (1998) Flavor analysis. American Chemical Society, Washington, DC
13. Sapers GM, Panasiuk O, and Talley FB (1973) Flavor quality and stability of potato flakes: effects of raw material and processing. J Food Sci 38:586
14. Seo EW, Joel DL (1980) Pentane production as an index of rancidity in freeze-dried pork. J Food Sci 45:26
15. Buttery RG, Teranishi R (1963) Measurement of fat oxidation and browning aldehydes in food vapors by direct injection gas–liquid chromatography. J Agric Food Chem 11:504
16. Buckholz LL, Withycombe DA, Daun H (1980) Application and characteristics of polymer adsorption method used to analyze flavor volatiles from peanuts. J Agric Food Chem 28:760
17. Reineccius GA, Keeney PA, Weiseberger W (1972) Factors affecting the concentration of pyrazines in cocoa beans. J Agric Food Chem 20:202
18. Majors RE (1986) Sample preparation for HPLC and GC using solid-phase extraction. LC-GC 4:972
19. Markel C, Hagen DF, Bunnelle VA (1991) New technologies in solid-phase extraction. LC–GC 9:332
20. Pawliszyn J (1997) Solid phase microextraction: theory and practice. VCH Publishers, New York
21. Zhang Z, Yang ML, Pawliszyn J (1994) Solid phase-microextraction: a solvent-free alternative for sample preparation. Anal Chem 66:844A–857A
22. Harmon AD (1997) Solid phase microextraction for the analysis of flavors. In: Marsili R (ed) Techniques for analyzing food aroma. Marcel Dekker, New York
23. Coleman WMI (1996) A study of the behavior of Maillard reaction products analyzed by solid-phase microextraction gas chromatography-mass selective detection. J Chromatogr Sci 34:213–218
24. Pfanncoch E, Whitecavage J (2002) Stir bar sorptive extraction capacity and competition effects. Gerstel Global, Baltimore, MD, pp 1–8
25. David F, Tienpont B, Sandra P (2003) Stir-bar sorptive extraction of trace organic compounds from aqueous matrices. LC-GC Europe 16:410
26. Dupuy HP, Fore SP, Goldbatt LA (1971) Elution and analysis of volatiles in vegetable oils by gas chromatography. J Am Oil Chem Soc 48:876
27. Legendre MG, Fisher GS, Fuller WH, Dupuy HP, Rayner ET (1979) Novel technique for the analysis of volatiles in aqueous and nonaqueous systems. J Am Oil Chem Soc 56:552
28. Widmer HM (1990) Recent developments in instrumental analysis. In: Bessiere Y, Thomas AF (eds) Flavor science and technology. Wiley, Chichester, p 181
29. Malowicki SMM, Martin R, Qian MC (2008) Volatile composition in raspberry cultivars grown in the Pacific Northwest determined by stir bar sorptive extraction-gas chromatography-mass spectrometry. J Agric Food Chem 56:4128–4133
30. Du X, Qian M (2008) Quantification of 2,5-dimethyl-4-hydroxy-3(2*H*)-furanone using solid-phase extraction and direct microvial insert thermal desorption gas chromatography-mass spectrometry. J Chromatogr A 1208:197–201

31. Fang Y, Qian MC (2006) Quantification of selected aroma-active compounds in Pinot noir wines from different grape maturities. J Agric Food Chem 54:8567–8573
32. Fang Y, Qian MC (2005) Sensitive quantification of sulfur compounds in wine by headspace solid-phase microextraction technique. J Chromatogr A 1080:177–185
33. Amirav A, Jing H (1995) Pulsed flame photometer detector for gas chromatography. Anal Chem 67:3305–3318
34. Buffington R, Wilson MK (1987) Detectors for gas chromatography. Hewlett-Packard Corp., Avondale, PA
35. Shellie R, Marriott P (2003) Opportunities for ultra-high resolution analysis of essential oils using comprehensive two-dimensional gas chromatography: a review. Flavour Fragrance J 18:179–191
36. Merritt C (1971) Application in flavor research. In: Zlatkis A, Pretorius V (Eds) Preparative gas chromatography. Wiley-Interscience, New York, pp 235–276
37. Kempfert KD (1989) Evaluation of apparent sensitivity enhancement in GC/FTIR using multidimensional GC techniques. J Chromatogr Sci 27:63–70
38. Phillips JB, Xu J (1995) Comprehensive multi-dimensional gas chromatography (Review). J Chromatogr A 703:327–334
39. Shellie R, Mondello L, Marriott P, Dugo G (2002) Characterization of lavender essential oils by using gas chromatography-mass spectrometry with correlation of linear retention indices and comparison with comprehensive two-dimensional gas chromatography. J Chromatogr 970(1/2):225–234
40. Adahchour M, Brandt M, Baier H-U, Vreuls RJJ, Batenburg AM, Brinkman UAT (2005) Comprehensive two-dimensional gas chromatography coupled to a rapid-scanning quadrupole mass spectrometer: principles and applications. J Chromatogr A 1067:245–254
41. Hodges K (1991) Sensory-directed analytical concentration techniques for aroma-flavor characterization and quantification. In: Risch SJ, Hotchkiss JH (eds) Food packaging interactions II. American Chemical Society, Washington, DC, p 174
42. Bernreuther A, Epperlein U, Koppenhoefer B (1997) In: Marsili R (ed) Techniques for analyzing food aroma. Marcel Dekker, New York, p 143
43. Woodrow JE, McChesney MM, Seiber JN (1995) Determination of ethylene oxide in spices using headspace gas chromatography. J Agric Food Chem 43:2126
44. Keeney PG, Patton S (1956) The coconut-like flavor defect of milk fat. I. Isolation of the flavor compounds from butter oil and its identification as delta-decalactone. J Dairy Sci 39:1104–1113
45. Ryan D, Shellie R, Tranchida P, Casilli A, Mondello L, Marriott P (2004) Analysis of roasted coffee bean volatiles by using comprehensive two-dimensional gas chromatography-time-of-flight mass spectrometry. J Chromatogr A 1054:57–65
46. Peterson DG, Reineccius GA (2002) Determination of the aroma impact compounds in heated sweet cream butter. Flavour Fragrance J 18:320–324

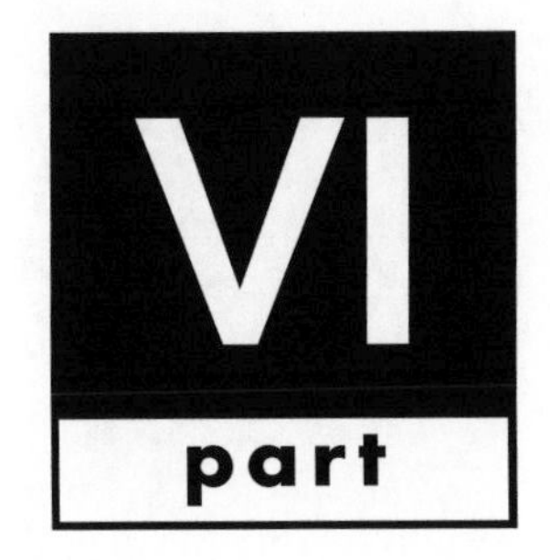

Physical Properties of Foods

Rheological Principles for Food Analysis

Christopher R. Daubert and E. Allen Foegeding*

Department of Food, Bioprocessing & Nutrition Sciences, NC State University, Raleigh, NC 27695-7624, USA
cdaubert@ncsu.edu
allen_foegeding@ncsu.edu

30.1 Introduction 543
30.1.1 Rheological and Textural Properties 543
30.1.2 Fundamental and Empirical Methods 543
30.1.3 Basic Assumptions for Fundamental Rheological Methods 543
30.2 Fundamentals of Rheology 543
30.2.1 Concepts of Stress 543
30.2.2 Concepts of Strain and Strain (Shear) Rate 544
30.2.3 Solids: Elastic and Shear Moduli 545
30.2.4 Fluid Viscosity 545
30.2.5 Fluid Rheograms 546
30.3 Rheological Fluid Models 546
30.3.1 Herschel-Bulkley Model 547
30.3.2 Newtonian Model 547
30.3.3 Power Law Model 547
30.3.4 Bingham Plastic Model 547

S.S. Nielsen, *Food Analysis*, Food Science Texts Series, DOI 10.1007/978-1-4419-1478-1_30,

30.4 Rheometry 548
30.4.1 Rotational Viscometry 548
30.4.1.1 Concentric Cylinders 548
30.4.1.2 Cone and Plate 549
30.4.1.3 Experimental Procedure for Steady Shear Rotational Viscometry 550
30.4.1.3.1 Test Fixture Selection 550
30.4.1.3.2 Speed (Shear Rate) Selection 550
30.4.1.3.3 Data Collection 550
30.4.1.3.4 Shear Calculations 550
30.4.1.3.5 Model Parameter Determination 550
30.4.2 Compression, Extension, and Torsion (Shear) Analysis 550
30.4.2.1 Large Strain Testing 550
30.4.2.1.1 Determining Stress, Strain, and Elastic Modulus (E) in Compression 551
30.4.2.1.2 Texture Profile Analysis 552
30.4.2.2 Small Strain Testing 552
30.5 Summary 552
30.6 Glossary 552
30.7 Nomenclature 553
30.8 Study Questions 553
30.9 References 554

30.1 INTRODUCTION

30.1.1 Rheological and Textural Properties

Food scientists are routinely confronted with the need to measure physical properties related to sensory texture and processing needs. These properties are determined by **rheological methods**, where ***rheology is a science devoted to the deformation and flow of all materials***. Rheological properties should be considered a subset of the textural properties of foods, because the sensory detection of texture encompasses factors beyond rheological properties. Specifically, rheological methods accurately measure "force," "deformation," and "flow," and food scientists and engineers must determine how best to apply this information. For example, the flow of salad dressing from a bottle, the snapping of a candy bar, or the pumping of cream through a homogenizer are each related to the rheological properties of these materials. In this chapter, we describe fundamental concepts pertinent to the understanding of the subject and discuss typical examples of rheological tests for common foods. A glossary is included as Sect. 30.6 to clarify and summarize rheological definitions throughout the chapter.

30.1.2 Fundamental and Empirical Methods

Rheological properties are determined by applying and measuring forces and deformations as a function of time. The difference between **fundamental** and **empirical** rheological methods is that, unlike the latter, the fundamental techniques account for the magnitude and direction of forces and deformations while placing restrictions on acceptable sample shapes and composition. Fundamental tests have the advantage of being based on known concepts and equations of physics. When sample composition or geometry is too complex to account for forces and deformations, empirical methods often are used. These applied methods are typically descriptive in nature and ideal for rapid analysis. Empirical tests are of value especially when they correlate with a property of interest, whereas fundamental tests determine true physical properties.

30.1.3 Basic Assumptions for Fundamental Rheological Methods

Two important assumptions for fundamental methodologies are that the material is **homogeneous** and **isotropic**. **Homogeneity** implies a well-mixed and compositionally similar material, an assumption generally valid for fluid foods provided they are not a suspension of large particles, such as vegetable soup. For example, milk, infant formula, and apple juice are each considered homogeneous and isotropic. Homogeneity is more problematic in solid foods. For example, frankfurters without the skins can be considered homogeneous. When particle size is significant, such as fat particles in some processed meats like salami, one must determine if homogeneity is a valid assumption. **Isotropic** materials display a consistent response to a load regardless of the applied direction. In foods such as a steak, muscle fibers make the material anisotropic (the response varies with the direction of the force or deformation).

30.2 FUNDAMENTALS OF RHEOLOGY

Rheology is concerned with how all materials respond to applied forces and deformations, and food rheology is the material science devoted to foods. Basic concepts of **stress** (force per area) and **strain** (relative deformation) are key to all rheological evaluations. Special constants of proportionality, called **moduli**, link stress with strain. With ideal solids, the material obeys **Hooke's Law** – the stress is related directly with strain. Whereas with ideal fluids, the material follows **Newtonian** principles, and the proportionality constant is commonly referred to as **viscosity**, ***defined as an internal resistance to flow***. These principles form the foundation for the entire chapter and are described in detail in Steffe and Daubert (9) and throughout this section.

30.2.1 Concepts of Stress

Stress (σ) is always a measurement of **force**. Defined as the force (F, **Newtons**) divided by the area (A, **meters**^{2}) over which the force is applied, stress is generally expressed with units of **Pascals** (Pa). To illustrate the notion of stress, conceptualize placing a water balloon on a table as opposed to placing it on the tip of a pair of scissors (Fig. 30-1). Obviously the scissors tip has a considerably smaller surface area, causing

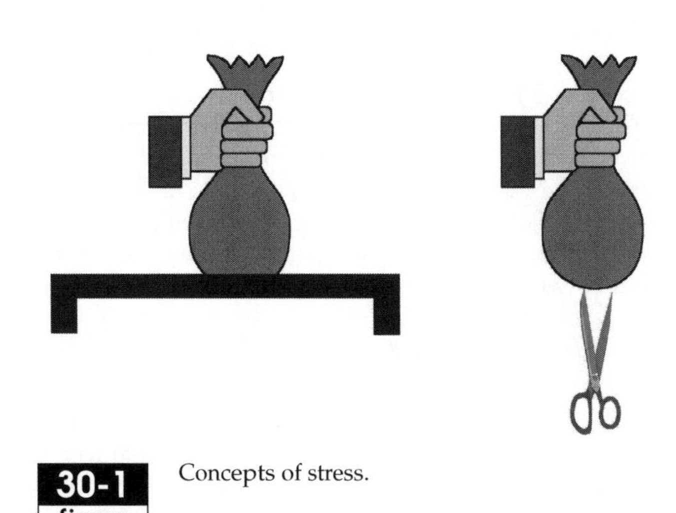

30-1 figure Concepts of stress.

the stress or weight/unit area of contact to be larger when compared with the stress of a balloon resting on a tabletop. Although the force magnitude – the weight of the water balloon – is a constant value for each case, the final outcomes for this demonstration will be very different.

The direction of the force, with respect to the surface area impacted, determines the type of stress. For example, if the force is directly perpendicular to a surface, a **normal stress** results and can be achieved under tension or compression. Should the force act in parallel to the sample surface, a **shear stress** is experienced. Examples of normal stresses include the everyday practice of chewing a piece of gum and the kneading of dough. During breadmaking, dough is continuously pressed and pulled until the proper consistency is achieved. Chewing a stick of gum involves the repetitive compression and extension of the material. Examples of shear forces, on the other hand, occur when spreading butter over a slice of toast, brushing barbecue sauce on chicken, or stirring milk into a hot cup of coffee.

30.2.2 Concepts of Strain and Strain (Shear) Rate

When a stress is applied to a food, the food deforms or flows. **Strain** is a dimensionless quantity representing the relative deformation of a material, and the direction of the applied stress with respect to the material surface will determine the type of strain. If the stress is normal (perpendicular) to a sample surface, the material will experience **normal strain** (ε). Foods show normal strains when they are compressed (compressive stress) or stretched (tensile stress).

Normal strain (ε) can be calculated as a true strain from an integration over the deformed length of the material (Fig. 30-2).

$$\varepsilon = \int_{L_i}^{L_i+\Delta_L} \frac{dL}{L} = \ln\left(1 + \frac{\Delta_L}{L_i}\right) \qquad [1]$$

According to Steffe (1), a true strain is more applicable to larger deformations such as may occur when texture testing. Strain calculations result in negative values for compression and positive values for extension (tensile). Rather than expressing a negative strain, many typically record the absolute value of the strain and denote the compression test mode.

$$\varepsilon = -0.05 \equiv 0.05_{\text{compression}} \qquad [2]$$

On the other hand, when a sample encounters a shear stress – such as the pumping of tomato paste through a pipe – a **shear strain** (γ) is observed. Figure 30-3

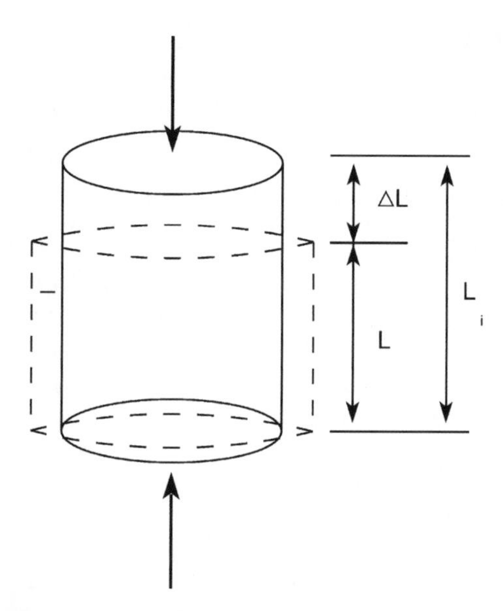

30-2 figure Normal strain in a cylinder.

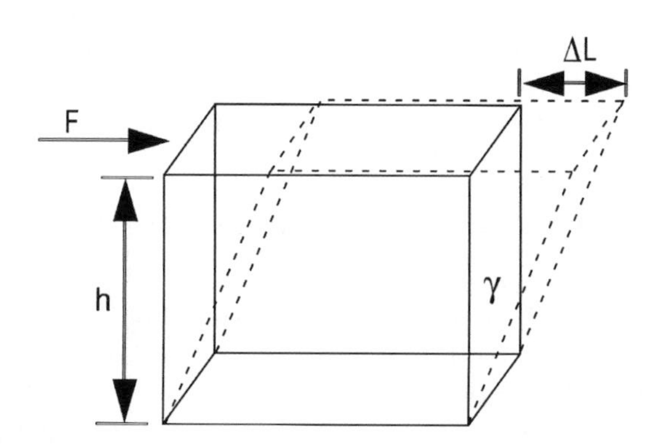

30-3 figure Shear strain in a cube.

shows how a sample deforms when a shear stress is applied. Shear strain is determined from applications of geometry as

$$\tan(\gamma) = \frac{\Delta L}{h} \qquad [3]$$

or

$$\gamma = \tan^{-1}\left(\frac{\Delta L}{h}\right) \qquad [4]$$

where h is the specimen height. For simplification, during exposure to small strains, the angle of shear may be considered equal to the shear strain.

$$\tan(\gamma) \approx \gamma \qquad [5]$$

When the material is a liquid, this approach for strain quantification is a bit more challenging. As coffee is stirred, water is pumped, or milk is pasteurized, these fluids all are exposed to shear and display irrecoverable deformation. Therefore, a **shear (strain)**

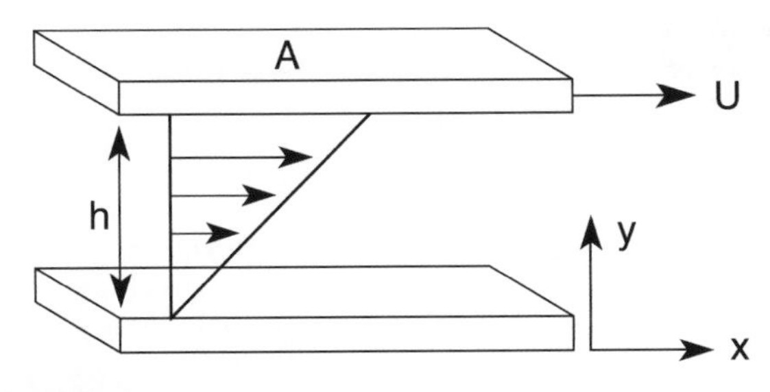

30-4 figure Shear flow between parallel plates.

rate ($\dot{\gamma}$) is typically used to quantify strain during fluid flow. Shear rate is the degree of deformation with respect to time with units of s^{-1}.

To further the shear rate concept, consider a fluid filling the gap between two moveable, parallel plates separated by a known distance, h, as illustrated in Fig. 30-4. Now, set one plate in motion with respect to the other at a constant horizontal velocity, U.

$$\gamma = \frac{\Delta L}{h} \quad [6]$$

$$\frac{d(\gamma)}{dt} = \frac{d\left(\frac{\Delta L}{h}\right)}{dt} = \dot{\gamma} \quad [7]$$

$$\dot{\gamma} = \frac{1}{h}\frac{d}{dt}(\Delta L) \quad [8]$$

$$\dot{\gamma} = \frac{U}{h} \quad [9]$$

The shear strain rate, or shear rate, for this system can be approximated as the quotient of the plate velocity and the fluid gap height, producing units of s^{-1}. This shear rate may be more easily understood through the deck of cards analogy. Imagine a stack of playing cards, with each card representing an infinitely thin layer of fluid. When the top card is stroked with some force, the entire deck deforms to some degree proportional with the magnitude of the force. This type of straining is commonly called **simple shear** and may be defined as a laminar deformation along a plane parallel to the applied force.

30.2.3 Solids: Elastic and Shear Moduli

Hooke's Law states that when a solid material is exposed to a stress, it experiences an amount of deformation or strain proportional with the magnitude of the stress. The constants of proportionality, used to equate stress with strain, are called **moduli**.

$$\text{Stress } (\sigma) \propto \text{Strain } (\varepsilon \text{ or } \gamma) \quad [10]$$

$$\text{Stress} = \text{Modulus} \times \text{Strain} \quad [11]$$

30-1 table Elastic and Shear Moduli for Common Materials

Material	*E, Elastic Moduli (Pa)*	*G, Shear Moduli (Pa)*
Apple	1.0×10^7	0.38×10^7
Potato	1.0×10^7	0.33×10^7
Spaghetti, dry	0.27×10^{10}	0.11×10^{10}
Glass	7.0×10^{10}	2.0×10^{10}
Steel	25.0×10^{10}	8.0×10^{10}

If a normal stress is applied to a sample, the proportionality constant is known as **elastic modulus** (E), sometimes called Young's modulus.

$$\sigma = \frac{F}{A} = E\varepsilon \quad [12]$$

Likewise, if the stress is shear, the constant is the **shear modulus** (G).

$$\sigma = G_{\gamma} \quad [13]$$

The moduli are inherent properties of the material and have been used as indicators of quality. Moduli of select foods and materials are seen in Table 30-1.

30.2.4 Fluid Viscosity

For the case of the simplest kind of fluid, the viscosity is constant and independent of shear rate and time. In other words, **Newton's postulate** is obeyed. This postulate stipulates that if the shear stress is doubled, the velocity gradient (shear rate) within the fluid is doubled. Typically for fluids, the shear stress is expressed as some function of shear rate and a viscosity term that provides an indication of the internal resistance of a fluid to flow. For **Newtonian fluids**, the viscosity function is constant and called the **coefficient of viscosity** or **Newtonian viscosity** (μ).

$$\sigma = \mu\dot{\gamma} \quad [14]$$

However, for most liquids the viscosity term is not constant, but rather changes as a function of the shear rate, and the material is considered **non-Newtonian**. A function called the **apparent viscosity** (η) is defined as the shear-dependent viscosity. Mathematically, the apparent viscosity function is the result of the shear stress divided by the shear rate.

$$\eta = f(\dot{\gamma}) = \frac{\sigma}{\dot{\gamma}} \quad [15]$$

Table 30-2 displays Newtonian viscosities for some common items at 20°C (1, 2). Temperature is a very important parameter when describing rheological properties. Typically viscosity decreases as temperature increases.

table 30-2 Newtonian Viscosities for Common Materials at 20°C

Material	*Viscosity, μ (Pa s)*
Honey	11.0
Rapeseed oil	0.163
Olive oil	0.084
Cottonseed oil	0.070
Raw milk	0.002
Water	0.001
Air	0.0000181

Adapted from (1, 2).

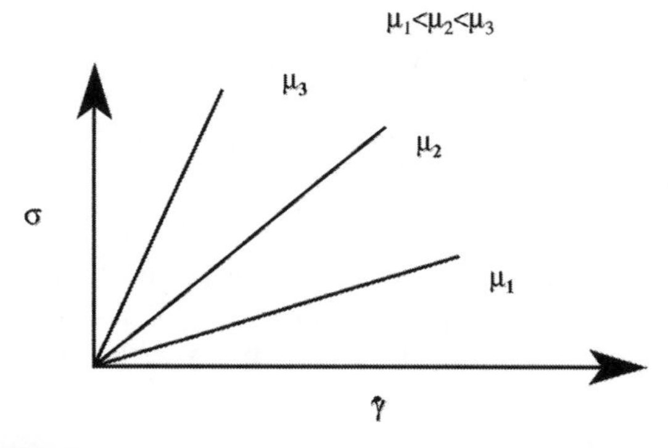

figure 30-5 Rheogram for three Newtonian fluids.

30.2.5 Fluid Rheograms

The flow behavior of a food can be described with appropriate rheological methods. A **rheogram**, for instance, is a graphical representation of the flow behavior, showing the relationship between stress and strain or shear rate. Much can be learned from inspection of rheograms. For example, if a plot of shear stress vs. shear rate results in a straight line passing through the origin, the material is a Newtonian fluid, with the slope of the line equaling the Newtonian viscosity (μ), shown in Fig. 30-5. Many common foods exhibit this ideal response, including water, milk, vegetable oils, and honey.

As established earlier, the majority of fluid foods do not show Newtonian flow behavior. The flow changes with shear rate (i.e., mixing speed) or with time at a constant shear rate. Time-independent deviation from ideal Newtonian behavior will cause the relationship between shear stress and shear rate to be nonlinear. Should the viscosity diminish as shear rate increases, the material is referred to as **shear thinning** or **pseudoplastic**. Examples of pseudoplastic food items are applesauce and pie fillings. On the other hand, if the material responds in a thickening fashion (i.e., the viscosity increases with shear rate) the sample is called **shear thickening** or **dilatent**. Corn starch slurries are well known for dilatent behavior.

table 30-3 Summary of Shear-Dependent Terminology

	Time Independent	*Time Dependent*
Thinning	Pseudoplastic	Thixotropic
Thickening	Dilatent	Antithixotropic (Rheopectic)

Pseudoplasticity and dilatency are **time-independent properties**. Materials that thin and thicken with time are known as **thixotropic** and **antithixotropic** liquids, respectively. These fluids are easily detected by monitoring the viscosity at a constant shear rate with respect to time. If pumpkin pie filling is mixed at a constant speed, the material thins (thixotropy) with time due to the destruction of weak bonds linking the molecules. Table 30-3 summarizes the terminology for time-dependent and independent responses.

Many fluids do not flow at low magnitudes of stresses. In fact, a certain catsup brand once staked numerous marketing claims on the "anticipation" of flow from the bottle. Often, additional force was applied to the catsup container to expedite the pouring. The minimum force, or stress, required to initiate flow is known as a **yield stress** (σ_0). Because Newtonian fluids require the stress–shear rate relationship to be a continuous straight line passing through the origin, any material with a yield stress is automatically non-Newtonian. A few common foods possessing yield stresses are catsup, yogurt, mayonnaise, and salad dressing.

Many foods are explicitly designed to include a certain yield stress. For example, if melted cheese did not have a yield stress, the cheese would flow off a cheeseburger or pizza. If salad dressings flowed at the lowest of applied stresses, the force of gravity would cause the dressing to run off the salad leaves. There are many fascinating rheological features of foods that consumers may never consider!

30.3 RHEOLOGICAL FLUID MODELS

Once you have the data for shear stress and shear rate, you can use **rheological models** to gain a greater understanding of the flow response. Rheological models are mathematical expressions relating shear stress to shear rate, providing what we like to call a "flow fingerprint" for a particular food. In addition, the models permit prediction of rheological behavior across a wide range of processing conditions.

30.3.1 Herschel-Bulkley Model

For most practical purposes, the **Herschel-Bulkley model** can account for the steady-state rheological performance of many fluid foods.

$$\sigma = \sigma_o + K\dot{\gamma}^n \quad [16]$$

K and n represent material constants called the **consistency coefficient** and **flow behavior index**, respectively. The flow behavior index provides an indication of Newtonian or non-Newtonian flow, provided the material has no yield stress. Table 30-4 illustrates how the Herschel-Bulkley model is used to identify specific flow characteristics.

Table 30-5 displays Herschel-Bulkley model data obtained for a variety of food products. Take, for example, the data for peanut oil. Table 30-5 reports no yield stress and a flow behavior index of 1.00 – the scenario for a Newtonian fluid. Accordingly, the Herschel-Bulkley model does in fact collapse to the special Newtonian case. The following models are considered as simple modifications to the Herschel-Bulkley model (1).

30-4 table Manipulation of Herschel-Bulkley Model to Describe Flow Behavior

Fluid Type	σ_o	*n*
Newtonian	0	1.0
Non-Newtonian		
Pseudoplastic	0	<1.0
Dilatent	0	>1.0
Yield stress	>0	Any

σ_o, yield stress; n, flow behavior index.

30.3.2 Newtonian Model ($n = 1$; $K = \mu$; $\sigma_o = 0$)

For Newtonian fluids, Equation [17] is manipulated with the flow behavior index (n) equaling 1.0 and the consistency coefficient (K) equaling the Newtonian viscosity (μ).

$$\sigma = 0 + \mu\dot{\gamma}^1 \quad [17]$$

or

$$\sigma = \mu\dot{\gamma} \quad [18]$$

30.3.3 Power Law Model ($\sigma_o = 0$)

Power law fluids show no yield stress (σ_o) and a nonlinear relationship between shear stress and shear rate. Pseudoplastic and dilatent fluids may be considered power law fluids, each with different ranges for flow behavior index values; refer to Table 30-4.

$$\sigma = 0 + K\dot{\gamma}^n \quad [19]$$

or

$$\sigma = K\dot{\gamma}^n \quad [20]$$

30.3.4 Bingham Plastic Model ($n = 1$; $K = \mu_{pl}$)

Bingham plastic materials have a distinguishing feature – a yield stress is present. Once flow is established, the relationship between shear stress and shear rate is linear, explaining why $n = 1.0$ and K is a constant value known as the **plastic viscosity**, μ_{pl}. Caution: The plastic value is ***not*** the same as the apparent viscosity (η) or the Newtonian viscosity (μ)!

$$\sigma = \sigma_o + \mu_{pl}\dot{\gamma}^1 \quad [21]$$

or

$$\sigma = \sigma_o + \mu_{pl}\dot{\gamma} \quad [22]$$

30-5 table Herschel-Bulkley Model Data for Common Foods

Product	*Temp. (°C)*	*Shear Rate (s^{-1})*	*K (Pa s^n)*	*n (–)*	*σ_o (Pa s)*
Orange juice (13°Brix)	30	100–600	0.06	0.86	–
Orange juice (22°Brix)	30	100–600	0.12	0.79	–
Orange juice (33°Brix)	30	100–600	0.14	0.78	–
Apple juice (35°Brix)	25	3–2000	0.001	1.00	–
Ketchup	25	50–2000	6.1	0.40	–
Apple sauce	32	–	200	0.42	240
Mustard	25	–	3.4	0.56	20
Melted chocolate	46	–	0.57	1.16	1.16
Peanut oil	21.1	0.32–64	0.065	1.00	–

Sources: Steffe (1); Rao (3)

30.4 RHEOMETRY

Rheometers are devices used to determine viscosity and other rheological properties of materials. Relationships between shear stress and shear rate are derived from physical values of system configurations, pressures, flow rates, and other applied conditions.

30.4.1 Rotational Viscometry

For fluids, the primary mode for rheological measurement in the food industry is rotational viscometry, providing rapid and fundamental information. **Rotational viscometry** involves a known test fixture (geometrical shape) in contact with a sample, and through some mechanical, rotational means, the fluid is sheared by the fixture. Primary assumptions are made for the development of **constitutive equations** (relationships between shear stress and rate), and a few include the following:

- ***Laminar flow.*** Laminar flow is synonymous with streamline flow. In other words, if we were to track velocity and position of a fluid particle through a horizontal pipe, the path line would only be in the horizontal direction without a shift toward the pipe wall.
- ***Steady state.*** Steady state is synonymous with time-independent effects.
- ***No-slip boundary condition.*** When the test fixture is immersed in the fluid sample, the wall of the fixture and the sample container serve as boundaries for the fluid. This condition assumes that at whatever speed either boundary is moving, an infinitely thin layer of fluid immediately adjacent to the boundary is moving at precisely the same velocity.

Rotational rheometers may operate in two modes: **steady shear** or **oscillatory**. At this point we consider steady shear rotational viscometry. **Steady shear** is a condition in which the sheared fluid velocity, contained between the boundaries, remains constant at any single position. Furthermore, the velocity gradient across the fluid is a constant. Two test fixtures most often used in steady shear rotational viscometry are **concentric cylinder** and the **cone and plate**.

30.4.1.1 Concentric Cylinders

This rheological attachment consists of a cylindrical fixture shape, commonly called a **bob** with radius R_b, suspended from a **torque** (M) measuring device that is immersed in a sample fluid contained in a slightly larger cylinder, referred to as the **cup** with radius R_c (see Figs. 30-6 and 30-7). Torque is an action that generates rotation about an axis and is the product of a force and the perpendicular distance (r), called the **moment arm**, to the **axis of rotation**. The principles involved can be described relative to changing tires on a car. To loosen the lugnuts, a larger tire iron is often required. Essentially, this longer tool increases the moment arm, resulting in a greater torque about the lugnut. Even though you are still applying the same force on the iron, the longer device provides greater torque!

30-6 figure Photo of cup and bob test fixtures for rheological measurements.

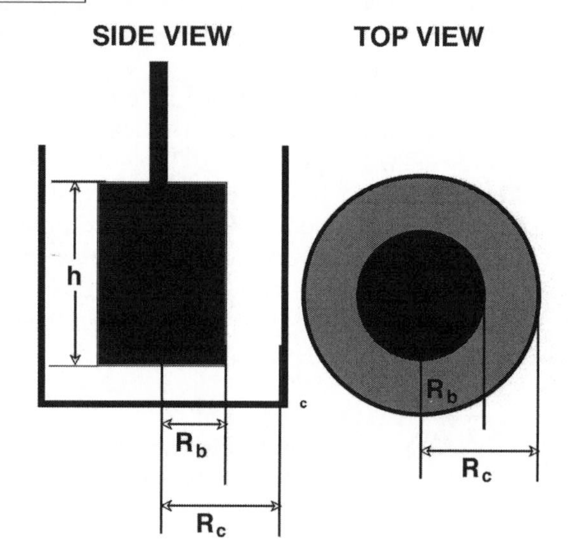

30-7 figure Concentric cylinder geometry.

To derive rheological data from experiments, expressions for shear stress and shear rate are developed. Shear stress at the surface of the bob (σ_b) may be calculated from a force balance as

$$\sigma_b = \frac{M}{2\pi h R_b^2} \qquad [23]$$

Therefore, to determine shear stress, all we need to know is the bob geometry (h and R_b) and the torque response (M) of the fluid on the measuring sensor.

A **simple shear approximation** commonly calculates a shear rate at the bob surface and assumes a constant shear rate across the fluid gap. This approximation is valid for small gap widths where $R_c/R_b \leq 1.1$.

$$\dot{\gamma}_b = \frac{\Omega R_b}{R_c - R_b} \qquad [24]$$

This calculation requires the **rotational speed**, or **angular velocity** (Ω), of the bob, typically expressed in radians per second. Converting units of revolutions per minute (rpm) to radians per second is simply achieved by multiplying by $2\pi/60$, and the following example converts rpm (10) to radians per second.

$$\frac{10 \text{ revolutions}}{1 \text{ min}} \times \frac{2\pi \text{ radians}}{1 \text{ revolution}} \times \frac{1 \text{ min}}{60 \text{ s}} = 1.047 \frac{\text{radians}}{\text{s}} \qquad [25]$$

One of the most common rheological devices found in the food industry is the **Brookfield viscometer** (Fig. 30-8). This affordable apparatus uses a spring as a torque sensor. The operator selects a rotational speed (rpm) of the bob, attached to the spring. Once the rotational speed is converted to an angular velocity, the simple shear approximation Equation [24] can be applied to calculate a shear rate. As the bob moves through the sample fluid, the viscosity impedes free rotation, causing the spring to wind. The degree of spring windup is a direct reflection of the torque magnitude (M), used to determine a shear stress at the bob surface.

Following progression through a series of rotational speeds, a rheogram can be created showing shear stress (σ) vs. shear rate ($\dot{\gamma}$). The importance of rheograms has been discussed, with a primary significance being apparent viscosity determination Equation [15]. The following tomato catsup data in Table 30-6 were collected with a standard cup and bob system ($R_c = 21$ mm, $R_b = 20$ mm, and $h = 60$ mm). Using Equations [15], [23], and [24] you should verify the results.

30-6 table **Rheological Data of Tomato Catsup Collected Using a Concentric Cylinder Test Fixture**

RPM	*Torque (N m)*	*Shear Rate (s^{-1})*	*Shear Stress (Pa)*	*Apparent Viscosity (Pa s)*
1.0	0.00346	2.09	22.94	10.98
2.0	0.00398	4.19	26.39	6.30
4.0	0.00484	8.38	32.10	3.83
8.0	0.00606	16.76	40.18	2.40
16.0	0.00709	33.51	47.02	1.40
32.0	0.00848	67.02	56.23	0.84
64.0	0.01060	134.04	70.29	0.52
128.0	0.01460	268.08	96.82	0.36
256.0	0.01970	536.16	130.63	0.24

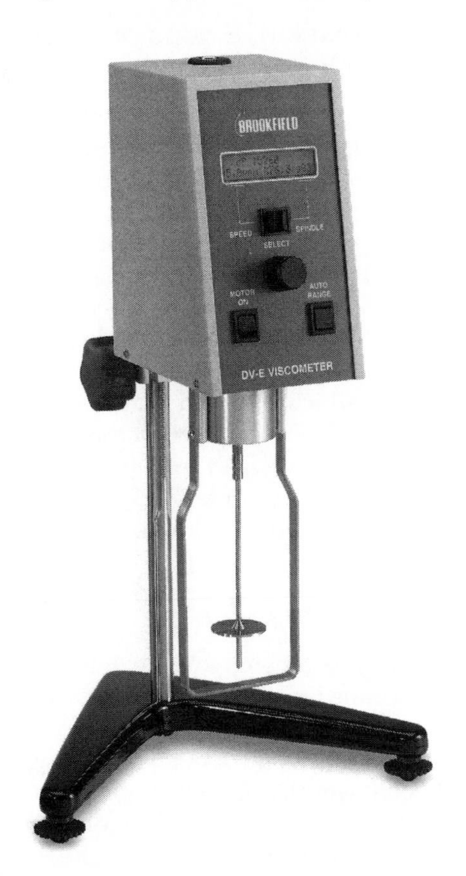

Photo of Brookfield viscometer. (Courtesy of Brookfield Engineering Laboratories, Inc., Middleboro, MA.)

30.4.1.2 Cone and Plate

Another popular system for rotational measurement is the cone and plate configuration (Figs. 30-9 and 30-10). Its special design permits the shear stress and shear rate to remain constant for any location of sample in the fluid gap. Test quality is best when the cone angle (θ) is small, and large errors may be encountered when the gap is improperly set or not well maintained.

The shear stress may be determined for a cone and plate configuration as:

$$\sigma = \frac{3M}{2\pi R^3} \qquad [26]$$

while the shear rate is calculated as:

$$\dot{\gamma} = \frac{r\Omega}{r \tan\theta} = \frac{\Omega}{\tan\theta} \qquad [27]$$

A primary advantage of the cone and plate test fixture is that shear stress and rate are independent of position – constant throughout the sample.

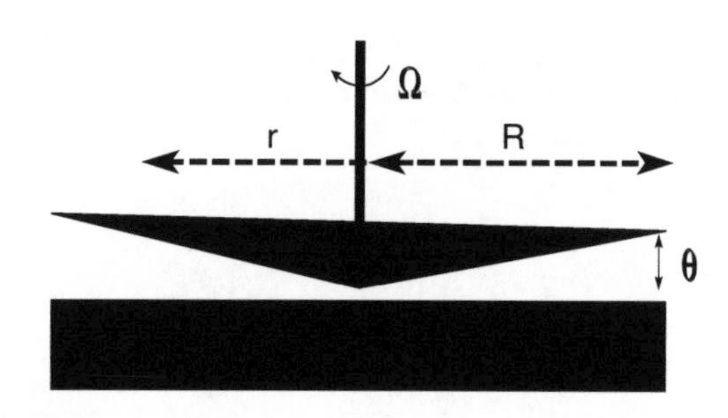

30-9 figure Cone and plate configuration.

30-10 figure Photo of cone and plate test fixtures for rheological measurements.

30.4.1.3 Experimental Procedure for Steady Shear Rotational Viscometry

30.4.1.3.1 Test Fixture Selection Many considerations go into the decision of selecting a fixture for a rheological test. To simplify the process, the information in Table 30-7 should be considered (4).

30.4.1.3.2 Speed (Shear Rate) Selection When performing a rheological test, it is necessary to have an understanding of the process for which the measurement is being performed. From the earlier example of tomato catsup, the apparent viscosity continuously decreased, exhibiting shear thinning behavior, as the shear rate increased. How would one report a viscosity? To answer that question, the process must be considered. For example, if a viscosity for molten milk chocolate is required for pipeline design and pump specification, a shear rate for this process should be known. All fluid processes administer a certain degree of shear on the fluid, and a good food scientist will consider the processing shear rate for proper rheological property determination. Barnes et al. (5) have prepared a list of common shear rates for typical processes, many of which are shown in Table 30-8.

30.4.1.3.3 Data Collection Once the test fixture and shear rate ranges have been selected, the experiment can begin. Record values of torque for each viscometer speed.

30.4.1.3.4 Shear Calculations Values for shear stress and shear rate are solved based on test fixture, fixture geometry, and angular velocity.

30.4.1.3.5 Model Parameter Determination Shear stress and shear rate can now be inserted into various rheological models previously described in Sect. 30.3. Rheological model parameters such as viscosity (μ, η, μ_{pl}), yield stress (σ_o), consistency coefficient (K), and the flow behavior index (n) may be analyzed for an even greater understanding of the flow of the material. For example, one may want to know: Does the material have a yield stress? Is the material shear thinning or shear thickening? What is the viscosity at a specific processing rate? Answering these and similar questions gives the food scientist a greater command of the behavior of the material for process design or quality determination.

30.4.2 Compression, Extension, and Torsion (Shear) Analysis

The rheological properties of solid foods are measured by compressing, extending, or twisting the material, and can be accomplished by two general approaches called **small** or **large strain** testing. In small strain tests the goal is to apply the minimal amount of strain or stress required to measure the rheological behavior, while at the same time preventing (or at least minimizing) damage to the sample. The goal in large strain or **fracture** tests is the opposite. Samples are deformed to an extent at which the food matrix is significantly strained, damaged, or possibly fractured. Small strain tests are used to understand properties of a food network, whereas large strain tests give an indication of sensory texture or product durability.

30.4.2.1 Large Strain Testing

Compression and **tension** (i.e., extension) tests are used to determine large strain and fracture food properties. Compression tests are generally selected for solid or viscoelastic solid foods when a tight attachment between sample and the testing fixture is not required, thereby simplifying sample preparation. Tension and **torsion** tests are well suited for highly deformable foods, and a high level of strain is needed to fracture the sample. The main disadvantage to tension and torsion tests is that the sample must be

Table 30-7 Advantages and Disadvantages of Rotational Viscometry Attachments

Rotational Geometry	*Advantages*	*Disadvantages*
Concentric cylinder	Good for low-viscosity fluids Good for suspensions Large surface area increases sensitivity at low shear rates	Potential end effects Large sample required
Cone and plate	Constant shear stress and shear rate in gap Good for high shear rates Good for medium- and high-viscosity samples Small sample required Quick-and-easy cleanup	Large particles interfere with sensitivity Potential edge effects Must maintain constant gap height

Table 30-8 Predicted Shear Rates for Typical Food Processes

Process	*Shear Rate (s^{-1})*
Sedimentation of powders in a liquid	10^{-6} to 10^{-4}
Draining under gravity	10^{-1} to 10^{1}
Extruding	10^{0} to 10^{2}
Chewing and swallowing	10^{1} to 10^{2}
Coating	10^{1} to 10^{2}
Mixing	10^{1} to 10^{3}
Pipe flow	10^{0} to 10^{3}
Spraying and brushing	10^{3} to 10^{4}

attached to the test fixture (6). Hamann (8) provides a detailed comparison and analysis of large deformation rheological testing.

30.4.2.1.1 Determining Stress, Strain, and Elastic Modulus (E) in Compression There are several assumptions to consider when doing compression testing. Along with the previously mentioned considerations of homogeneous and isotropic, the assumption that the food is an **incompressible material** greatly simplifies matters. An incompressible material is one that changes in shape but not volume when compressed. Foods such as frankfurters, cheese, cooked egg white, and other high-moisture gel-like foods generally are considered incompressible. The calculations for strain are as discussed previously Equation [1].

During compression, the initial cross sectional area (A_i) increases as the length decreases. To account for this change, a correction term incorporating a ratio of the cylinder lengths (L/L_i) is applied to the stress calculation.

$$\sigma = \frac{F}{A_i} \times \left(\frac{L}{L_i}\right) \quad [28]$$

In compression testing one should use a cylindrical shaped sample with a length (L) to diameter ratio of >1.0. The sample should be compressed between two flat plates with diameters exceeding the lateral expansion of the compressed sample (i.e., the sample should be in contact with the plates during testing). The equations are based on the sample maintaining a cylindrical shape when compressed. If this is not the case, the contact surface between the plate and the sample may need lubrication. Water or oil can be used, and one should pick the fluid that provides the desired lubrication without causing any deleterious effects to the sample.

A cylinder of Cheddar cheese 3 cm in length (L_i), with an initial radius (R_i) of 1 cm, was compressed at a constant rate to 1.8 cm (L) and recorded a force of 15 N. Then:

$$\varepsilon = \ln(1 + (-0.4)) = -0.5 = 0.5_{\text{compression}} \quad [29]$$

$$A_i = \pi R_i^2 = 0.000314\ \text{m}^2 \quad [30]$$

$$\sigma = \frac{F}{A_i} \times \left(\frac{L}{L_i}\right) = \frac{15\,N}{0.000314\ \text{m}^2} \times \left(\frac{1.8}{3.0}\right) = 28{,}700\ \text{Pa}_{\text{compression}} = 28.7\ kPa_{\text{compression}} \quad [31]$$

$$E = \frac{\sigma}{\varepsilon} = \frac{28.7\ kPa}{0.5} = 57.4\ kPa \quad [32]$$

If the material compressed is a pure elastic solid, the compression rate does not matter. However, if the material is viscoelastic (as is the case with most foods), then the values for stress, strain, and elastic modulus may change with the speed of compression.

30-11 figure Photo of universal testing machine. (Courtesy of Instron®, Norwood, MA.)

A complete characterization of a viscoelastic material requires determining these values at a variety of compression rates. Another factor to consider is the level of compression. The sample can be compressed to fracture or some level below fracture. The goal should be compression to fracture if correlating rheological with sensory properties.

30.4.2.1.2 Texture Profile Analysis **Texture profile analysis** (TPA) is an empirical technique using a two-cycle compression test, typically with a **universal testing machine** (Fig. 30-11). This test was developed by a group of food scientists from the General Foods Corporation and is compiled as force during compression and time. Data analyses correlated numerous sensory parameters, including hardness, cohesiveness, and springiness, with texture terms determined from the TPA test curve. For example, the peak force required to fracture a specimen has been strongly related to sample hardness. Bourne (7) provides a more detailed description of TPA.

30.4.2.2 Small Strain Testing

The goal in **small strain testing** is to characterize the rheological properties of a material without damaging the material. This procedure requires small forces and deformations. In addition, since most foods are viscoelastic, viscous and elastic properties must be distinguished. These goals are accomplished by applying: (1) a stress or strain in **oscillation** and measuring the respective strain or stress and phase angle between stress and strain, (2) a **constant strain** and measuring the decrease (relaxation) in stress, or (3) a **constant stress** and measuring the rate of deformation (creep). A more detailed description of these techniques may be found in Steffe (1) and Rao (3).

30.5 SUMMARY

Rheological testing is simple in that it only requires the measurement of force, deformation, and time. To convert these measurements into fundamental physics-based rheological properties requires an understanding of the material and testing method. Materials should be homogeneous and isotropic – true for most fluid foods and many solid foods. Fundamental rheological properties are determined based on knowledge of the stress or strain applied to the sample and the geometry of the testing fixture. Once rheological properties are determined they can be described by physical or mathematical models to gain a more complete understanding of the rheological properties. The advantage of determining fundamental, rather than empirical, rheological properties is the use of common units, independent of the specific instrument, to determine the rheological property. This approach not only allows for comparison among values determined on different instruments, but it also permits a comparison of the flow of honey with the flow of paint. Through rheological methods, food scientists have the ability to relate theoretical and experimental information from a range of disciplines, including polymer chemistry and materials sciences, to gain a greater understanding of the quality and behavior of food materials.

30.6 GLOSSARY

Compression A force acting in a perpendicular (normal) direction toward the body

Concentric cylinder A test fixture for rotational viscometry frequently called a cup and bob

Cone and plate A test fixture for rotational viscometry

Constitutive equation An equation relating stress with strain and sometimes other variables including time, temperature, and concentration

Dilatent Shear-dependent thickening

Empirical test Simple tests measuring poorly defined parameters but typically found to correlate with textural or other characteristics

Fundamental test A measurement of well-defined, physically based rheological properties

Homogeneous Well mixed and compositionally similar regardless of location
Incompressible No change in material density
Isotropic The material response is not a function of location or direction
Kinematic viscosity The viscosity divided by the density of the material
Laminar flow Nonturbulent flow
Modulus A ratio of stress to strain
Newtonian fluid A fluid with a linear relationship between shear stress and shear rate without a yield stress
Non-Newtonian fluid Any fluid deviating from Newtonian behavior
No-slip The fluid velocity adjacent to a moving boundary has the same velocity as the boundary
Oscillatory test Dynamic test using a controlled sinusoidally varying input function of stress or strain
Pseudoplastic Shear-thinning
Rheology A science studying how all materials respond to applied stresses or strains
Rheogram A graph showing rheological relationships
Rheometer An instrument measuring rheological properties
Rheopectic Time-dependent shear thickening
Shear (strain) rate Change in strain with respect to time
Simple shear The relative motion of a surface with respect to another parallel surface creating a shear field within the fluid contained between the surfaces
Simple shear approximation A prediction technique for shear rate estimation of fluids within a narrow gap
Steady shear A flow field in which the velocity is constant at each location with time
Steady state Independent of time
Strain Relative deformation
Stress Force per unit area
Tension A force acting in a perpendicular direction away from the body
Test fixture A rheological attachment, sometimes called a geometry, which shears the sample material
Thixotropic Time-dependent shear-thinning
Torque A force generating rotation about an axis, which is the product of the force and the perpendicular distance to the rotation axis
Torsion A twisting force applied to a specimen
Viscometer An instrument measuring viscosity
Viscosity An internal resistance to flow
Yield stress A minimum stress required for flow to occur

30.7 NOMENCLATURE

Symbol	Name	Units
A	Area	m^2
A_i	Initial sample area	m^2
E	Modulus of elasticity	Pa
F	Force	N
G	Shear modulus	Pa
H	Height	m
K	Consistency coefficient	Pa s^n
L	Length	m
L_i	Initial length	m
ΔL	Change in length	m
M	Torque	N m
N	Flow behavior index	unitless
R	Radial distance	m
R	Radius	m
R_i	Initial radius	m
R_b	Bob radius	m
R_c	Cup radius	m
T	Time	s
U	Velocity	m s^{-1}
ε	Normal strain	unitless
γ	Shear strain	unitless
γ	Angle of shear	radians or degrees
$\dot{\gamma}$	Shear (strain) rate	s^{-1}
η	Apparent viscosity	Pa s
θ	Cone angle	radians or degrees
μ	Newtonian viscosity	Pa s
μ_{pl}	Plastic viscosity	Pa s
σ	Stress	Pa
σ_b	Shear stress at the bob	Pa
σ_o	Yield stress	Pa
Ω	Angular velocity	radians s^{-1}

30.8 STUDY QUESTIONS

1. What is the difference between a shear and normal stress?
2. How is stress different from force?
3. What is the definition of apparent viscosity? When does the apparent viscosity equal the Newtonian viscosity?
4. Vegetable oil is Newtonian fluid and catsup is a Bingham Plastic fluid. What are the differences in flow behavior and how does it alter the food applications of these fluids?
5. Apple sauce at 26°C may be described by the following mathematical expression:

$$\sigma = 5.6\dot{\gamma}^{0.45}$$

where:

$$\sigma \equiv \text{Pa}$$
$$\dot{\gamma} \equiv {}^1/_{\text{s}}$$

a. Which rheological model is it? Identify the constants, K and n, with proper units.
   Honey at 26°C obeys a Newtonian model:

$$\sigma = 8.9\dot{\gamma}$$

   Evaluate the apparent viscosity for apple sauce and honey at a shear rate of 0.43, 1.00, 2.80, and 5.60 s^{-1}.

b. How do the viscosities compare for each material? How do the viscosities compare at each shear rate?

c. Describe the importance of multipoint testing.

6. If you were designing a new "chip dip," what type of rheological considerations might you suggest?
7. What are the differences between empirical and fundamental rheological tests? Be creative and develop two new empirical tests, like recording the amount of time required for spaghetti sauce to drain from a colander. Then, identify fundamental rheological tests that could be used to determine similar properties from your empirical tests. Explain advantages of using fundamental rheological tests.

30.9 REFERENCES

1. Steffe JF (1996) Rheological methods in food process engineering, 2nd edn. Freeman, East Lansing, MI
2. Muller HG (1973) An introduction to food rheology. Crane, Russak, Inc., New York
3. Rao MA (1999) Rheology of fluid and semisolid foods: principles and applications. Aspen, Gaithersburg, MD
4. Macosko CW (1994) Rheology: principles, measurements, and applications. VCH, New York
5. Barnes HA, Hutton JF, Walters K (1989) An introduction to rheology. Elsevier Science, New York
6. Diehl KC, Hamann DD, Whitfield JK (1979) Structural failure in selected raw fruits and vegetables. J Texture Stud 10:371–400
7. Bourne MC (1982) Food texture and viscosity: concept and measurement. Academic, New York
8. Hamann D, Zhang J, Daubert CR, Foegeding EA, Diehl KC (2006) Analysis of compression, tension and torsion for testing food gel fracture properties. J Texture Stud 37:620–639
9. Steffe JF, Daubert CR (2006) Bioprocessing pipelines: rheology and analysis. Freeman, East Lansing, MI

Thermal Analysis

Leonard C. Thomas*
DSC Solutions LLC, 27 E. Braeburn Drive, Smyrna, DE 19977, USA
LThomas@TAinstruments.com

and

Shelly J. Schmidt
University of Illinois at Urbana-Champaign, Department of Food Science and Human Nutrition, Urbana, IL 61801, USA
sjs@uiuc.edu

31.1 Introduction 557
31.2 Material Science 557
31.2.1 Amorphous Structure 557
31.2.2 Crystalline Structure 558
31.2.3 Semi-Crystalline Structure 559
31.2.4 Thermodynamic and Kinetic Properties 560
31.3 Principles and Methods 561
31.3.1 Thermogravimetric Analysis 561
31.3.1.1 Overview 561

S.S. Nielsen, *Food Analysis*, Food Science Texts Series, DOI 10.1007/978-1-4419-1478-1_31,

31.3.1.2 Experimental Conditions 562
31.3.1.3 Common Measurements 562
31.3.2 Differential Scanning Calorimetry 563
31.3.2.1 Overview 563
31.3.2.2 Experimental Conditions 564
31.3.2.3 Common Measurements 564
31.3.3 Modulated DSC® 565
31.3.3.1 Overview 565
31.3.3.2 Experimental Conditions 566
31.3.3.3 Common Measurements 566
31.4 Applications 567
31.5 Summary 569
31.6 Study Questions 570
31.7 Acknowledgment 570
31.8 References 571

31.1 INTRODUCTION

Thermal analysis is a term used to describe a broad range of analytical techniques that measure physical and chemical properties as a function of temperature, time, and atmosphere (inert or oxidizing gas, pressure, and relative humidity). Depending on the technique, test temperatures can range from −180 to 1000°C or more, allowing investigation into a range of applications, including low temperature stability and processing (e.g., freezing and freeze-drying) to high temperature processing and cooking (e.g., extrusion, spray drying, and frying).

Thermal analysis results provide insight into the structure and quality of starting materials, as well as finished products. The physical structure (amorphous, crystalline, semi-crystalline) of a material creates a set of physical properties, which in turn define end-use properties, such as texture and storage stability. Areas of application include quality assurance, product development, and research into new materials, formulations, and processing conditions (1–4).

Specific instruments are typically used for characterization of a particular property. Instrumentation includes a transducer, used to measure the property of interest, and a temperature-measuring device, such as a thermocouple, thermopile, or platinum resistance thermometer, used to record the sample's temperature. Experiments are performed while heating, cooling, or at a constant temperature (isothermal), and measured signals are stored for analysis.

Thermal analysis techniques of major interest to the food researcher and the properties they measure are listed in Table 31-1, with the first three widely used techniques discussed in detail in Sect. 31.2.

31-1 table **Techniques of Major Interest to the Food Researcher and the Properties They Measure**

Techniques	*Abbreviation*	*Property Measured*
Thermogravimetric analysis	TGA	Weight change
Differential scanning calorimetry	DSC	Heat flow
Modulated temperature DSC	MDSC®	Heat flow and heat capacity
Thermomechanical analysis	TMA	Dimensional change
Dynamic mechanical analysis	DMA	Stiffness and energy dissipation
Rheology	Rheometer	Viscosity/flow behavior
Moisture sorption analysis	MSA	Moisture sorption

Thermogravimetric analysis (TGA) is typically the first thermal analysis measurement done when characterizing a new material. TGA data can detect and quantify the presence of bulk water and/or associated water, and identify the temperature where molecular decomposition (chemical change) begins. The change in weight measured by TGA is quantitative; however, no information on the chemistry of evolved gases is obtained. If chemical knowledge of evolved gases is desired, TGA can be coupled to a mass spectrometer (MS) and/or Fourier Transform Infrared (FTIR) spectrometer (5).

Once composition and thermal stability are obtained from TGA results, the physical structure or "form" of the material is typically determined using **differential scanning calorimetry** (DSC) and/or **modulated temperature DSC** (MDSC®). Structure and the temperature(s) at which the structure changes [transition(s)] significantly influence physical and chemical properties of a material. By understanding structure and related physical properties, formulations can be developed that provide desired end-use properties, such as crispness, dissolution rate, and storage stability.

31.2 MATERIAL SCIENCE

Since the pioneering work of Slade and Levine [e.g., (6, 7)], food scientists have been actively applying the principles of material science to the study of food materials [e.g., (8, 9)]. One of the main driving forces underlying this application is that the end-use properties (functionality) of a material at a specific temperature are dependent on the structure of the components at that temperature. Therefore, it is necessary to measure structure as a function of temperature. The primary use of thermal analysis is to determine structure by measuring the physical properties (e.g., heat capacity, flow, expansion, rigidity) associated with that structure.

31.2.1 Amorphous Structure

Many food products are amorphous or have a high amorphous content (semi-crystalline). Examples include extruded snacks and breakfast cereals, low-moisture cookies and crackers, hard sugar-based candies, and powdered drink mixes. **Amorphous structure** has no regular or systematic molecular order, which means that it has the highest energy content, the highest molecular mobility, and the fastest rate of dissolution. A potential problem with amorphous structure is that physical properties can change by orders of magnitude at a specific temperature, termed the **glass transition temperature** (Tg).

At temperatures below *Tg*, an amorphous material acts like a glass. It is rigid, has low molecular mobility, and very high viscosity (low ability to flow). Above *Tg*, these materials act like a rubbery material, viscous liquid, or gel with greater free volume and much higher molecular mobility. Some amorphous materials (e.g., fats, oils, and water) have the ability to crystallize when cooled to lower temperatures. However, these materials may associate (e.g., hydrogen bond) with another material in the formulation and not crystallize, even at temperatures well below their freezing point. In the case of water, it is well known that not all the water freezes in a food material, even at very low storage temperatures (10).

Since properties change so significantly at *Tg*, it is important to be able to measure and control *Tg* by selecting appropriate ingredients in the correct weight ratio to other ingredients in a formulation or recipe. In foods, *Tg* can change by 50°C or more via changes in moisture content. Crisp snack foods, such as crackers and potato chips, are typical of foods for which water changes the *Tg* and the resulting physical properties. For example, the *Tg* of freshly processed potato chips is much higher than room temperature; thus, the chips are crisp and have a pleasant texture. If the chips are exposed to ambient temperature and high humidity conditions for several hours, they start to become soft and pliable, typical of being at a temperature above *Tg*. Once the package of chips is opened, the low-moisture chips begin to absorb moisture from the air, which lowers *Tg* and creates a different set of physical properties at the consumption temperature.

Tg can be measured with most thermal analysis techniques due to significant changes in physical properties, such as heat capacity (for DSC), coefficient of thermal expansion (for **thermomechanical analysis**, TMA), and stiffness (for **dynamic mechanical analysis**, DMA), that accompany *Tg* (8). However, DSC is usually the technique of choice because of easy sample preparation, short test times, ease of data interpretation, and the ability to use sealed pans (hermetic), which prevent loss of moisture as the sample is heated. As a result of the significant increase in molecular mobility and heat capacity that occurs as the sample is heated to a temperature above *Tg*, there is a corresponding increase in the heat flow rate measured by DSC. Data analysis software measures the temperature and magnitude of the change in heat flow that occurs, which is proportional to the amount of amorphous material in the sample (Fig. 31-1).

31.2.2 Crystalline Structure

Crystalline structure is different from amorphous structure in many ways. Molecules have long-range order, lower energy (heat content), higher density, and a different set of physical properties. Molecular mobility is low, which means that heat capacity is low. Melting of the crystalline material creates an amorphous liquid. Upon cooling of the amorphous liquid, some

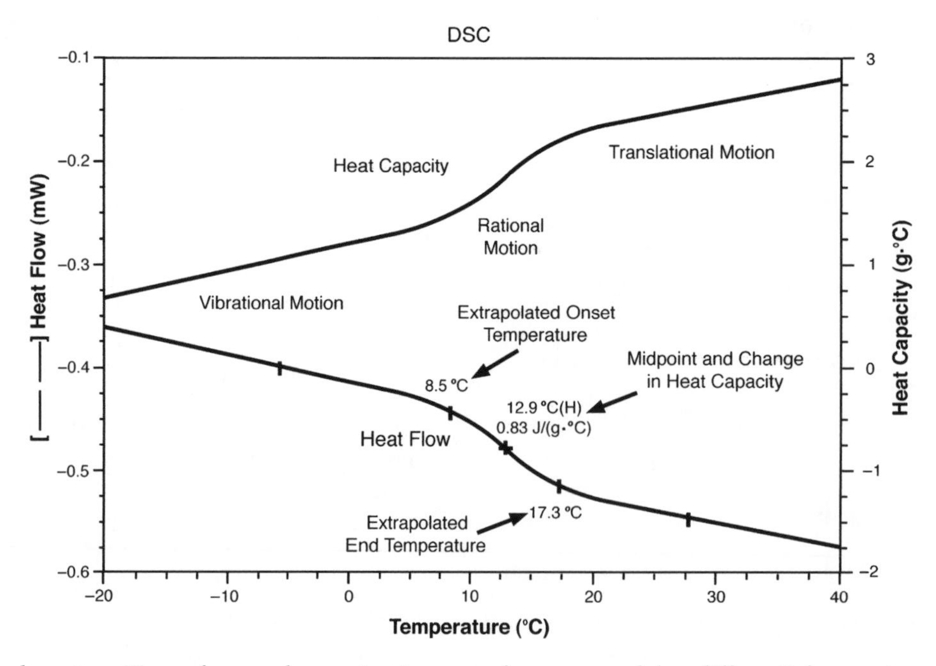

The glass transition of amorphous structure can be measured by differential scanning calorimetry (DSC) due to the significant increase in heat capacity that occurs as the material is heated to a temperature above the glass transition temperature (*Tg*). Typical *Tg* analysis includes the extrapolated onset, mid-point (temperature of one-half of the heat capacity change), and end point temperatures, as well as the difference in heat capacity (ΔCp, J/g°C).

materials (e.g., fats) will crystallize, while others (e.g., sucrose and fructose) remain amorphous (glassy).

Because of the increased density and molecular order within the crystal, there is a reduction in the ability of crystalline material to form hydrogen bonds, and thus a reduced tendency to absorb moisture from the atmosphere. In addition, melting occurs at a higher temperature than the glass transition, which makes crystalline material more stable and much more rigid (often gritty like table sugar and salt) than amorphous material over a wide temperature range. Because the crystalline structure is more stable, physical properties change less with time.

Since crystalline material has lower heat content than amorphous material, crystalline material must absorb heat (endothermic process) to become amorphous. The absorption of heat during a DSC experiment is seen as an endothermic peak, termed the melting peak. Data analysis software can measure the temperature of the peak and calculate the heat (J/g) required to melt the sample. The area of the melting peak (J/g) increases as the crystallinity of the material increases. Figure 31-2 shows the DSC data for melting of the crystalline sugar alcohol (polyol) mannitol, which is commonly used in confectioneries, such as "breath-freshening" mints and gums.

Normally, crystalline structure is converted to amorphous structure by heating the sample to a temperature (an energy level) that is high enough to overcome the energy associated with the crystalline lattice (termed thermodynamic melting). However, crystalline structure also can be lost due to processes such as dissolving the crystals in a solvent (dissolution), dehydration of a hydrated crystalline form, chemical interaction of functional groups between two materials in a mixture, and breaking of chemical bonds (decomposition) at temperatures below the true melting point of the material. Since these are time-dependent (kinetic) processes, the endothermic peak observed in DSC data shifts to higher temperature as heating rate increases. A material that illustrates this behavior is the monosaccharide sugar fructose (Fig. 31-3).

31.2.3 Semi-Crystalline Structure

Many foods contain both amorphous and crystalline structures. In some cases, such as a lipid that melts over a temperature range, one ingredient can exist in both phases. A term that is sometimes used to describe a mixture of phases in a lipid is "**solid-fat index**." At a particular temperature (usually room temperature, 72°F or 22°C), a certain fraction or percentage of the lipid material is solid (crystalline) and the remainder is liquid (amorphous). One lipid that illustrates this property at room temperature is cocoa butter, a common ingredient in chocolate. Since cocoa butter can exist in different crystal forms (termed **polymorphs**)

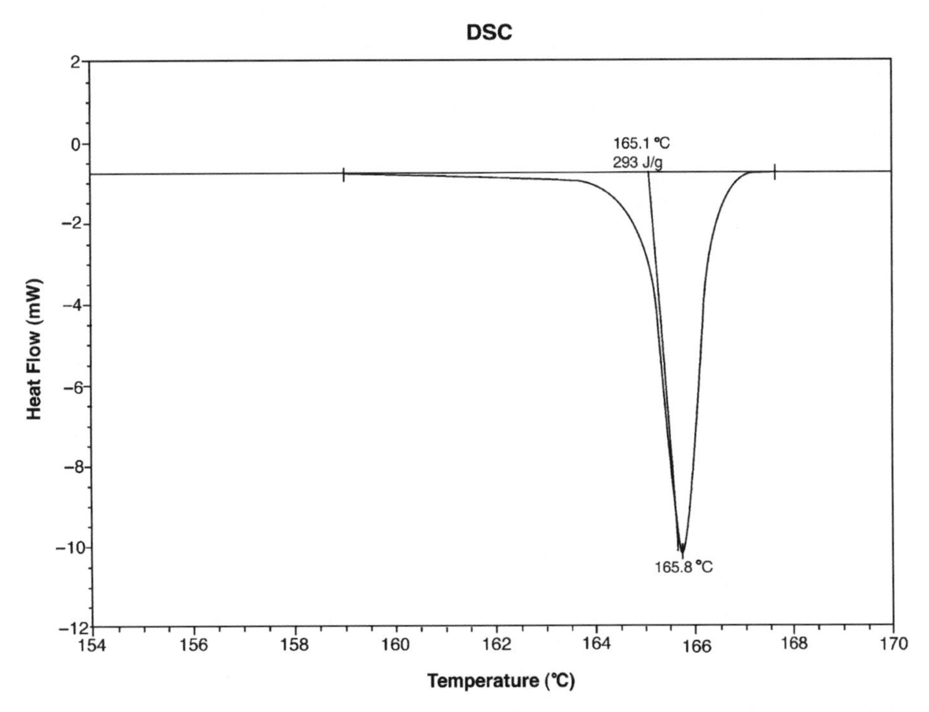

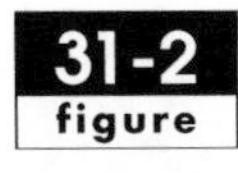

DSC data showing melting of the crystalline sugar alcohol (polyol) mannitol. Analysis of the data shows the extrapolated onset and peak melting temperatures, and heat required to melt the crystalline structure (heat of fusion, Joules/gram, J/g).

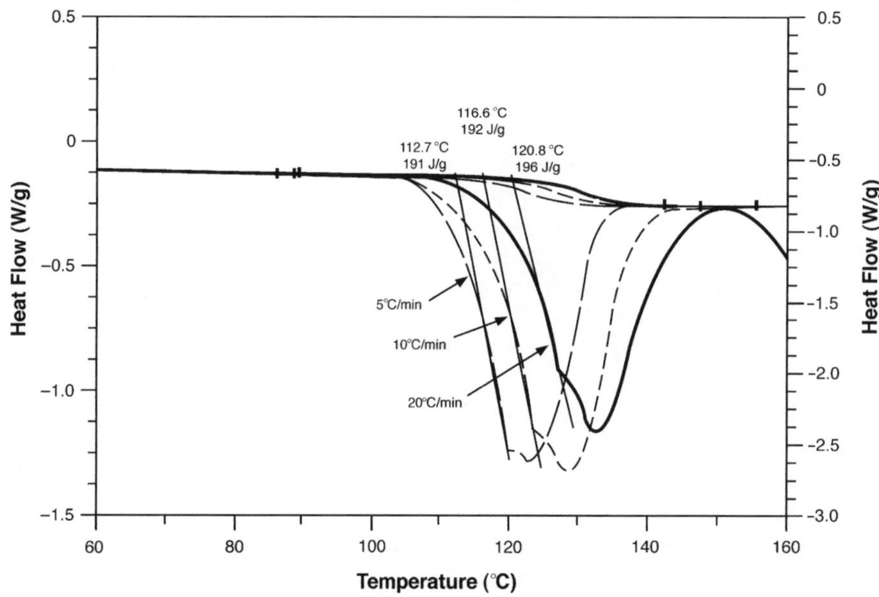

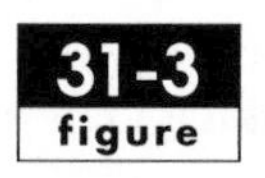
31-3 figure

The endothermic peak associated with loss of crystalline structure in fructose shifts to higher temperatures at higher heating rates, indicating the influence of a time-dependent (kinetic) process.

and these different forms have different melting temperatures, the percentage of liquid lipid in the chocolate at room temperature can be controlled by varying the ratio of the different crystal forms.

Figure 31-4 shows the broad melt of a sugar-coated chocolate candy. There are several small overlapping melting peaks below 22°C that are due to melting of the less stable crystal forms in the sample. By measuring the percentage of melting below and above 22°C, the ratio of liquid to solid phases can be determined for room temperature. A feature of the DSC data analysis software, termed the "running integral," can plot percent melted vs. temperature. Results show that 22.9% of the cocoa butter is liquid (melted) at room temperature in this particular sample. The presence of the liquid lipid provides a creamy texture to the chocolate.

31.2.4 Thermodynamic and Kinetic Properties

As discussed in the next section, a natural limitation of DSC, and other thermal analysis techniques used to measure structure, is that heating of the sample is typically required. As temperature increases, molecular mobility increases, which permits the structure to change in ways that are not always obvious in the data. Since the purpose of the experiment is typically to measure the existing structure in the material, users of thermal analysis instrumentation must be able to recognize when structure or composition (e.g., solvent evaporation) is changing.

Thermodynamic properties (e.g., heat capacity, enthalpy, and density) have absolute values as a function of temperature, while kinetic properties are always a function of time and temperature. Some common examples of kinetic processes observed in foods are freezing of water during cold storage, crystallization of fats, adsorption/desorption of water, staling of bread, and decomposition/oxidation during processing, such as deep-frying.

The easiest way to determine if an observed event in the data is due to thermodynamics or kinetics is to change the heating rate. Since heating rate has the units of °C/min, the reciprocal is min/°C. The higher the heating rate, the less time the sample experiences at each temperature. Therefore, high heating rates reduce the ability of the structure to change, while low heating rates increase the probability of structural change. If the onset temperature of the observed event remains relatively constant (changes <1°C) with a tenfold change in heating rate, it is typically a thermodynamic event, while an increase in event temperature with heating rate indicates influence of a kinetic process.

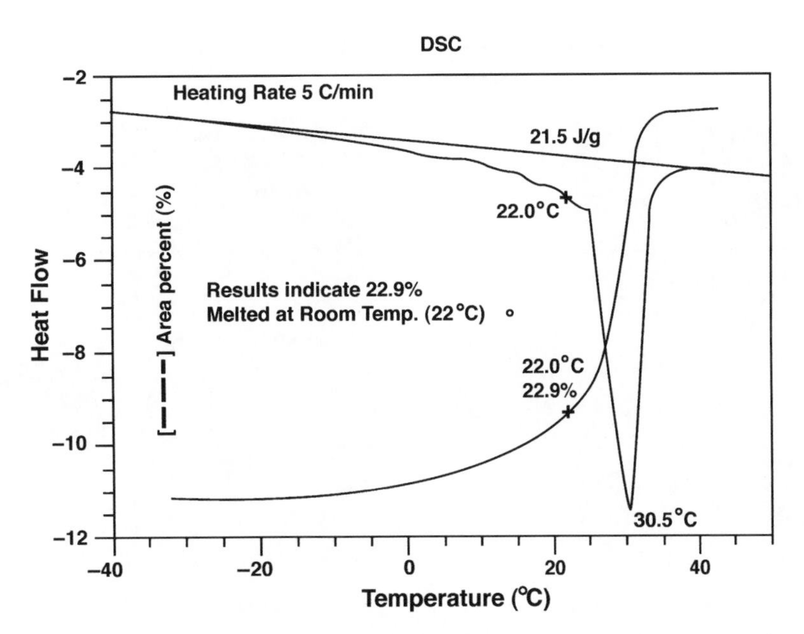

DSC data for sugar-coated chocolate candy, containing cocoa butter, shows several small overlapping melting peaks below 22°C, due to melting of the less stable crystal lipid forms in the sample. In this particular sample, results show that 22.9% of the cocoa butter is liquid (melted) at 22°C; the remainder is crystalline (solid).

31.3 PRINCIPLES AND METHODS

This section describes the working principles of the most frequently used techniques and makes recommendations on optimum experimental conditions for characterizing common materials.

31.3.1 Thermogravimetric Analysis

31.3.1.1 Overview

Thermogravimetric analysis (TGA) should be the first thermal analysis technique used to characterize a new material. TGA provides information about the composition (number of components) of the material and its thermal or oxidative stability (decomposition in inert and oxidizing atmospheres, respectively). TGA instruments use a specially designed and very sensitive analytical balance to measure weight changes as the sample is typically heated from room temperature to a 1000°C or more. A thermocouple is located close to the sample to continuously record the temperature as weight changes occur. The heated sample chamber is typically purged with an inert gas, such as nitrogen or helium; however, air or oxygen can be used when measuring oxidative stability. Most weight changes are weight losses due to volatilization or decomposition, but weight gain is observed during early stages of oxidation. A specialized version of TGA is designed with humidity control so that the rate of moisture sorption (both absorption and desorption) can be measured as a function of time, temperature, and relative humidity.

Figure 31-5 shows a schematic of possible designs for both conventional TGA and a humidity-controlled sorption analysis TGA. Both contain sample and reference pans attached to the balance. The reference pan is empty, since its purpose is to offset the weight of the sample pan. With conventional TGA, the reference pan is typically not heated. The reference pan for sorption analysis is exposed to the same temperature and relative humidity as the sample. This greatly improves the stability and baseline performance of the measurement. Most modern instruments have auto samplers so that many samples can be run sequentially without need of operator presence.

Most TGA instruments have several natural limitations. Although TGA provides a quantitative measurement of weight change, it is often difficult to quantify the weight of a specific component because weight losses typically overlap in temperature. Improved temperature resolution of these weight losses can usually be obtained by slowing the heating rate and reducing sample weight. However, the lower heating rate increases test time (reduced productivity) and the smaller sample size reduces accuracy of small weight changes. Some manufacturers of TGA instrumentation have specialized software that automatically slows the heating rate during weight losses to improve resolution in an automated way (HiRes TGA™, TA Instruments, New Castle, DE).

Another limitation of TGA is that it cannot identify the chemistry of gases evolved from the sample. Knowledge of the gas composition helps to distinguish between water loss and loss of low molecular weight

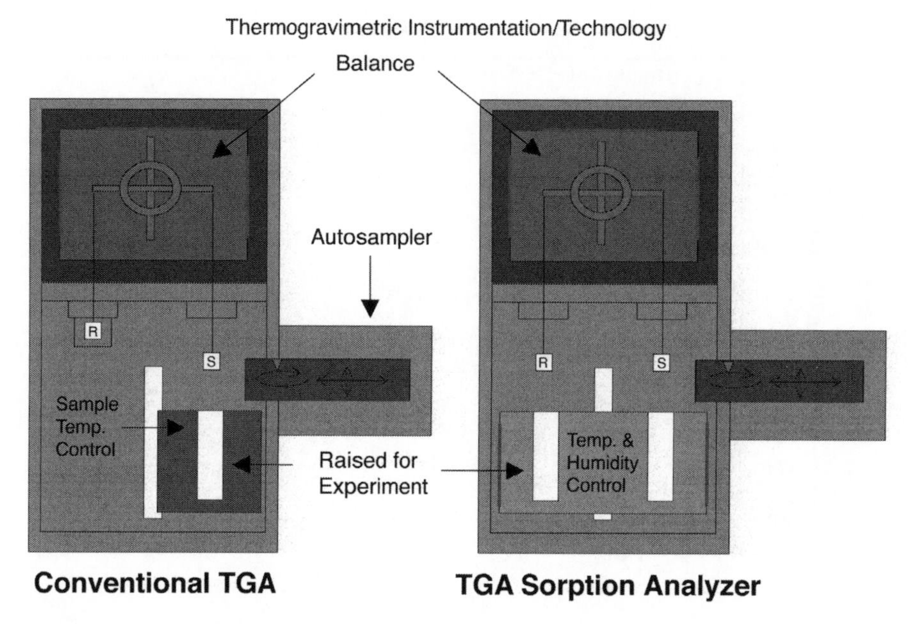

31-5 figure Thermogravimetric instrumentation measures weight change using a sensitive analytical balance. The sample container is typically suspended into a temperature and atmosphere controlled chamber. Adapted from figure by TA instruments, New Castle, DE.

additives, such as flavors and fragrances, and helps to determine the chemical mechanisms involved in cooking and decomposition. Most manufacturers of MSs and FTIR instruments offer interfaces between their products and the TGA instrument so that off-gases can be chemically identified.

31.3.1.2 Experimental Conditions

Although TGA experiments can be performed over a wide range of conditions, a good starting point for most materials is as follows:

- Sample weight: 10–15 mg
- Pan type: Platinum
- Purge gas: Nitrogen
- Start temperature: Room temperature, typically 20°C
- Heating rate: 10°C/min
- Final temperature: 300°C

31.3.1.3 Common Measurements

TGA experiments are primarily heating experiments; however, isothermal (constant temperature) conditions can be used to determine drying rates or follow weight changes at processing/cooking temperatures. The most common measurements include the following:

- Temperature of weight change
- Free (or bulk) moisture content
- "Bound" or associated water content (part of the structure)
- Composition (multiple components)
- Decomposition temperature; in reality, there is no such thing as a decomposition temperature. Decomposition is a kinetic process, which means that it is a function of both time and temperature. Therefore, the temperature of weight loss due to decomposition increases if the heating rate is increased.

Figure 31-6 shows TGA data from the center of a loaf of bread about 24 h after baking. The solid curve is the remaining weight %, where a value of 99% means that the sample has lost 1% of its original weight. The dashed curve is the time-based derivative of the weight curve and has the units of %/min, which is the rate of weight loss. The derivative signal is extremely useful in detecting weight losses that overlap in temperature and shows that two components are lost over the temperature range from room temperature to 200°C. The first weight loss begins immediately, which is typical of moisture evaporation. The second weight loss at higher temperature is likely the breakdown of some component (starch or protein) in the flour. Even though the baking temperature was 375°F (190°C), the center of the bread probably did not reach much above 212°F (100°C) during baking due to the cooling effect of moisture evaporation. Therefore, component 2 remained in the bread because it does not break down until temperatures above 100°C are reached. Peak 3 in the derivative signal is due to charring of remaining materials in the flour.

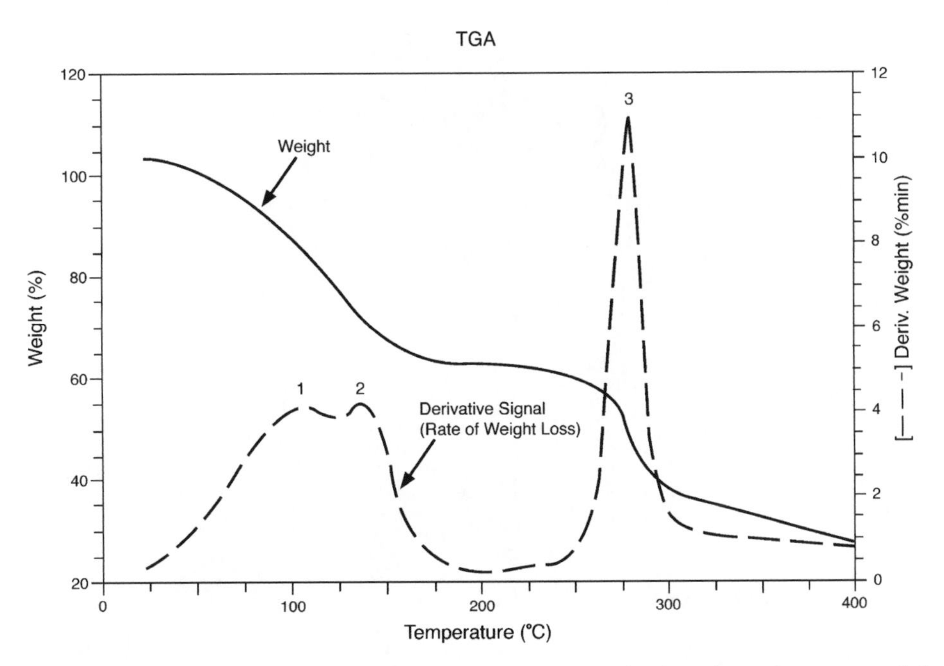

Thermogravimetric analysis (TGA) data for a sample of bread taken from the center of the loaf shows three components. The first is moisture, while the second and third are decomposition of components contained in the flour. The solid curve is the remaining weight %, and the dashed curve is the time-based derivative of the weight curve with units of %/min.

31.3.2 Differential Scanning Calorimetry

31.3.2.1 Overview

Differential scanning calorimetry (DSC) is the most frequently used thermal analysis technique and probably accounts for 70% of all thermal analysis measurements. Since *every* change in structure (transition) either absorbs or releases heat, DSC is the universal detector for measuring structure. The only limitation is the sensitivity of the instrument, which is its ability to detect small transitions or very slow kinetic processes where the rate of heat flow is similar to or less than the signal noise of the instrument.

The ability of DSC to measure very small rates (microwatts, μJ/s) of heat flow is greatly enhanced because it uses a differential signal. An empty reference pan is subjected to the same thermal environment as the sample pan, and the measured signal from the DSC is the difference between the sample and reference. This effectively eliminates signal noise and drift caused by heat exchanges with the environment or atmosphere around the pans.

There are two general approaches to making the differential measurement. One approach uses a common furnace for both the sample and reference (heat flux design), while the other uses individual furnaces (power compensation design). Each design has theoretical advantages and limitations, but performance is mostly based on the manufacturer's ability to build a completely symmetric system so that instrumental effects on the measurement are minimized. This provides the highest sensitivity (ability to detect weak transitions), resolution (ability to separate transitions close in temperature), accuracy, and precision. Heat flux is the most common approach to making DSC measurements and is discussed below.

Figure 31-7 is a cross-sectional view of a heat flux DSC. A common furnace is used for the sample and reference pans, and is typically purged with high purity, ultra-dry nitrogen. The furnace and cooling accessory provide a wide temperature range from −180 to 725°C. The sample and reference sensors are temperature sensors, such as thermocouples, thermopiles, or platinum resistance thermometers. They provide the basis of the differential heat flow measurement and directly measure temperature.

Even though DSC is the most useful of the thermal techniques for measuring structure and changes in structure, it has a number of natural limitations, including the following:

- DSC measures the *sum* of all heat flows within the calorimeter. It is sometimes difficult to interpret data because of overlapping events (multiple transitions occurring at the same time and temperature).
- Most measurements involve heating the sample to higher temperatures. As temperature increases, mobility increases and this permits the structure to change in ways that are not

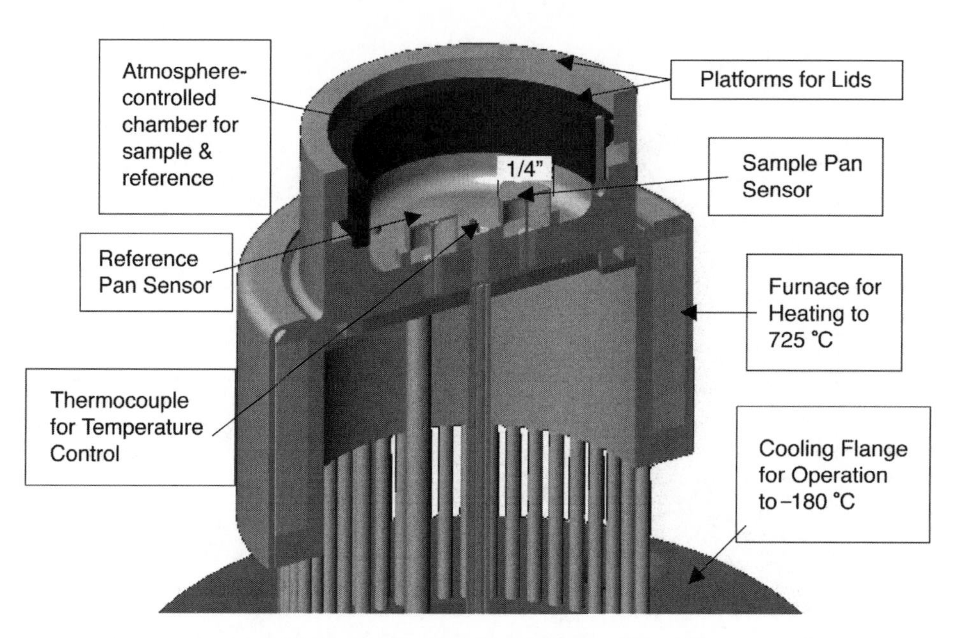

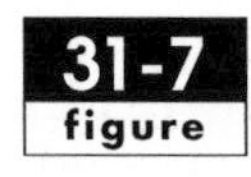

Cross-sectional view of a heat flux DSC. The sample and reference pans are located in a common chamber that is temperature controlled over the range of −180 to 725°C. The chamber is typically purged with high purity, ultra-dry nitrogen gas. (Courtesy of TA Instruments, New Castle, DE.)

always obvious. The measured structure may not be the original structure at the start of the experiment.

- DSC uses a single heating rate. However, higher heating rates provide better sensitivity, while lower heating rates provide better resolution. Therefore, it is not possible to optimize both sensitivity and resolution in a single DSC experiment.
- DSC cannot measure heat capacity under isothermal conditions. Therefore, DSC cannot use heat capacity as a way to follow changes in structure at constant temperature.

31.3.2.2 Experimental Conditions

Significantly different conditions are used depending on the measurement. However, larger sample weights and heating rates always improve sensitivity, while smaller sample weights and heating rates always improve resolution. The conditions listed below provide a good starting point, but must be optimized for the best results. The first decision in selecting conditions is the type of pan that will be used. If the sample is dry (less than 0.5% volatile components at 100°C in TGA) then standard aluminum crimped pans will provide the best results. Crimped pans have lower mass and better contact with the sensor; however, they are not sealed, which permits evaporation of volatile components. This creates a broad endothermic peak that may hide other transitions of interest. Hermetic pans are sealed and recommended for samples with volatile components. They are heavier than crimped pans and generally provide poorer contact between the sample and sensor, which slightly lowers sensitivity and resolution. Recommended starting conditions are as follows:

- Sample weight: 10–15 mg
- Start temperature: At least 25°C below the first transition of interest. This gives the baseline time to stabilize before the temperature of the event is reached, and permits better quantification of the change in heat content (**enthalpy**) resulting from the change in structure.
- Purge gas: Dry nitrogen
- Heating rate: 10°C/min
- Final temperature: Temperature of 5% weight loss due to decomposition in TGA data. In general, it is bad to decompose samples in the DSC cell because decomposition products can condense and affect the quality of future data.

31.3.2.3 Common Measurements

DSC can be used to measure the properties and structure of most ingredients used in the food industry. Experiments are typically heating experiments; however, measurements also are made while cooling or under constant temperature (isothermal). Examples of common measurements include the following:

- Glass transition temperature of amorphous structure
- Melting temperature of crystalline structure

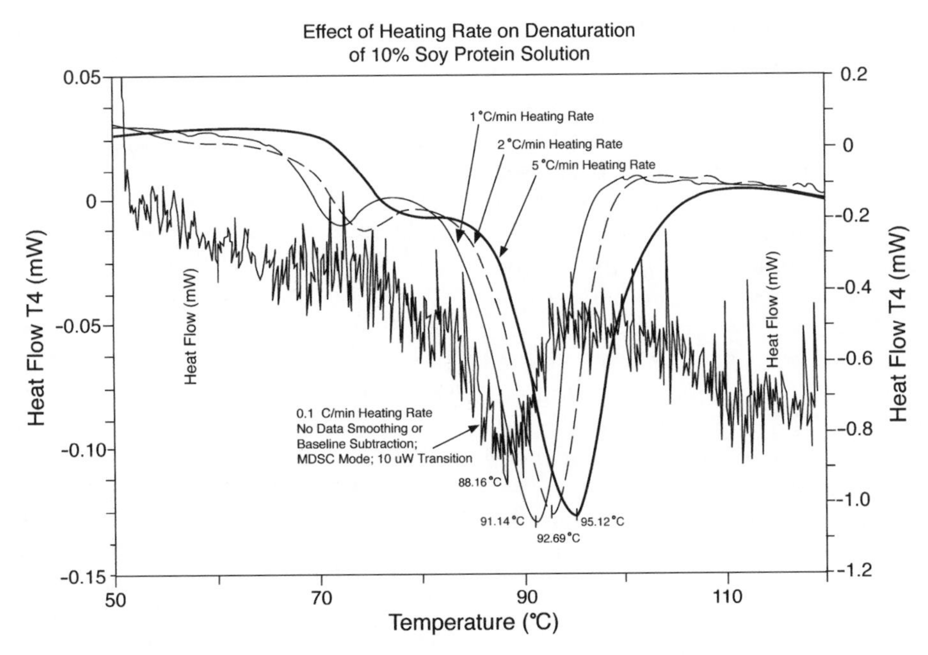

31-8 figure DSC data from a series of heating rates applied to a 10% (w/w) solution of soy protein. The endothermic peaks show that the two-stage denaturation process shifts to higher temperatures at higher heating rates. The amount of shift provides information on the kinetics of the denaturation process. (Courtesy of TA Instruments, New Castle, DE.)

- Percent crystallinity of semi-crystalline material
- Crystallization of amorphous material
- Denaturation of proteins
- Gelatinization of starch
- Analysis of frozen solutions used for freeze drying
- Oxidative stability of fats and oils

Changes in structure (transitions) are either **endothermic**, where the sample absorbs additional energy (heat), or **exothermic**, where heat is released. Figure 31-8 shows endothermic transitions associated with denaturation of soy protein in a 10% (w/w) solution with water. Unfolding of the protein results in an increase in free volume and molecular mobility, and this requires heat to occur. Transitions that involve a process (denaturation and gelatinization) shift to higher temperatures at higher heating rates. This is due to the fact that these processes take time as well as temperature (kinetic), and the amount of time at a given temperature decreases as heating rate increases.

31.3.3 Modulated DSC®

31.3.3.1 Overview

Modulated DSC® (MDSC®) is a special type of DSC that applies two simultaneous heating rates to the sample (11). A linear rate is used to obtain the same information as provided by conventional DSC, while a sinusoidal rate superimposed on the linear rate permits measurement of the **heat capacity** component of the total heat flow signal. As described above, one of the natural limitations of the DSC technique is that it can only measure the sum of all heat flows and this often makes interpretation of the data difficult. This can be seen by a brief review of Equation [1] often used to describe the heat flow signal from DSC.

$$\frac{dH}{dt} = Cp\frac{dT}{dt} + f(T,t) \qquad [1]$$

where:

dH/dt = measured heat flow rate (mW = mJ/s)
Cp = heat capacity (J/°C); product of specific heat (J/g°C) × sample weight (g)
dT/dt = heating rate (°C/min)
$f(T,t)$ = heat flow rate due to time-dependent, kinetic processes (mW)

As seen from Equation [1], the heat flow signal measured by conventional DSC has two components, one associated with heat capacity and the other with kinetic processes that are a function of both time and temperature. DSC only measures the sum of the two components. By applying two simultaneous heating rates, MDSC® can separate the total signal into its individual components. Figure 31-9 shows temperature vs. time and heating rate vs. time for an MDSC® experiment. The MDSC® average temperature and rate would be typical of a DSC experiment, while the modulated temperature and rate only occur with MDSC®.

As with every analytical technique, MDSC® also has limitations, including:

- Slow average heating rates (typically 1–5°C/min) must be used to obtain good separation of overlapping events. This decreases the productivity (number of samples per day) of MDSC® as compared with DSC.
- MDSC® is more complex because it requires additional experimental parameters and creates more signals than DSC.
- Separation of overlapping events requires the ability to modulate the sample's temperature during the events. This is not possible during melting of relatively pure materials that melt over just a few degrees.
- The probability of structural change is increased in the sample while heating due to the slower heating rates of MDSC®.

31.3.3.2 Experimental Conditions

As seen in Fig. 31-9, the applied temperature of MDSC® has both linear and sinusoidal components. Therefore, it is necessary to specify conditions for both. Recommended starting conditions that will work for most samples include the following:

- Average linear heating rate: 2°C/min
- Temperature modulation period: 60 s
- Temperature modulation amplitude: ±0.5°C
- Other conditions: same as DSC

31.3.3.3 Common Measurements

MDSC® is used to make the same measurements as DSC, but has the significant advantage of being able to separate the heat flow signal into the heat capacity and kinetic components. The benefit of this can be seen in Fig. 31-10, which is an analysis of the structure of amorphous polydextrose, a bulking agent commonly used in the food industry as a replacement for sucrose, starch, and lipids in a number of commercial products, such as baked products, candies, and salad dressings. The **total heat flow signal** (equivalent to conventional DSC) contains an enthalpic peak (termed physical aging, a common observation in glassy amorphous materials held below their *Tg*) superimposed on the glass transition, making it very difficult to accurately analyze the *Tg*. However, MDSC® separates the contribution of the enthalpic peak (kinetic component) from the change in heat capacity, allowing for straightforward analysis of the *Tg*.

Thus, one of the major advantages of MDSC® is that it separates the total heat flow signal into the heat capacity ("**reversing signal**") and kinetic ("**nonreversing signal**") components and permits their individual analyses. In general, the reversing signal includes heat capacity, changes in heat capacity,

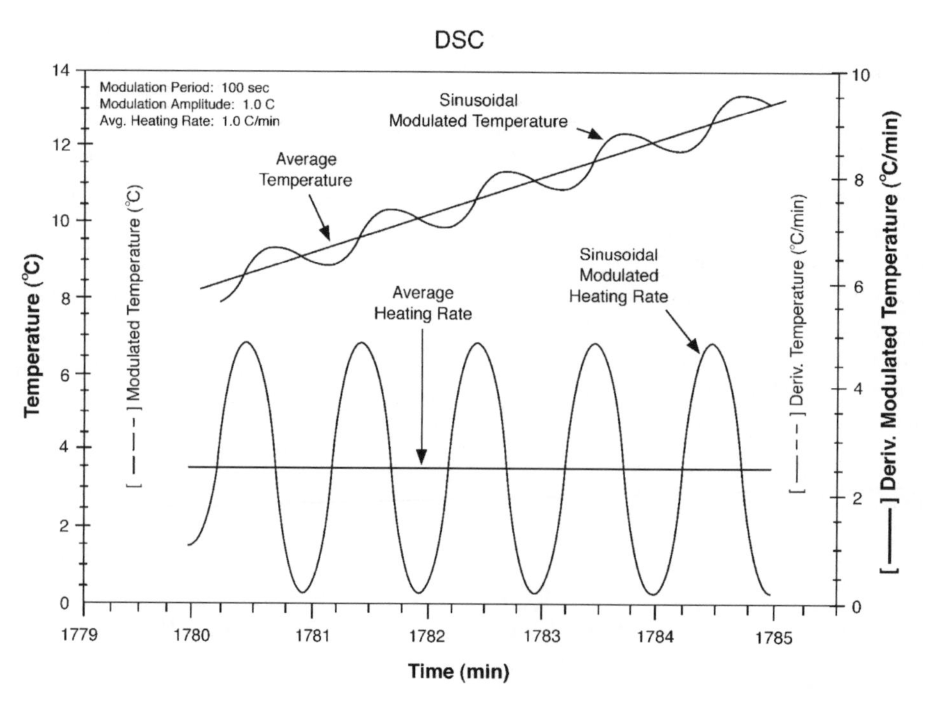

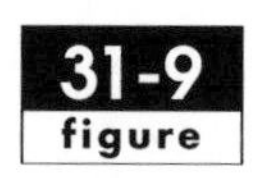
31-9 figure Modulated temperature DSC (MDSC®) applies two simultaneous heating rates (linear and sinusoidal) to separate the total heat flow (equivalent to conventional DSC) signal into the heat capacity and kinetic components.

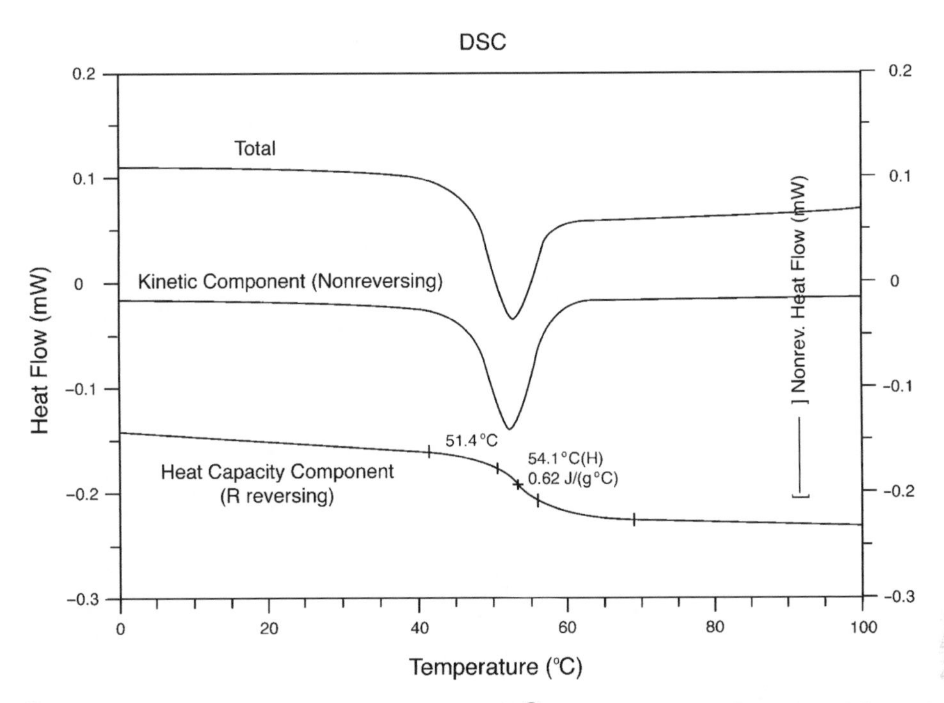

figure 31-10 MDSC® analysis of amorphous polydextrose. MDSC® separates the contribution of the enthalpic peak (kinetic component) from the glass transition (heat capacity component), allowing for straightforward analysis of the glass transition temperature.

and most melting. All kinetic events such as crystallization, decomposition, evaporation, and physical aging are seen in the nonreversing signal.

31.4 APPLICATIONS

The most common applications of each technique were summarized in the previous sections. This section will illustrate some practical uses of the techniques for determining composition and structure in a variety of food products.

Figure 31-6 showed the TGA weight loss profile for bread taken from the center of a loaf. The first two components comprise almost 40% of the total weight and most of that is likely water. Interpretation of the data was that the center of the loaf probably never exceeded 100°C due to the cooling effect of water evaporation. Since component number 2 (most likely a component of the flour) is not lost until the temperature exceeds about 125°C, it remains in the center of the loaf. The crust of the loaf should reach a much higher temperature, but probably still below oven temperature (190°C) due to evaporation of the water from the surface of the loaf. If this assumption is correct, TGA data from the crust of the loaf would not likely show component number 2.

Figure 31-11 is a comparison of the TGA results from the center and crust of the loaf of bread. As expected, there is no indication of component number 2 in the crust, indicating that the crust exceeded the decomposition temperature of that material during baking. Since the loaf was allowed to sit for about 24 h prior to analysis, moisture had diffused from the center of the loaf to the surface. Data indicates that the moisture content of the crust was almost 20% (wet basis) after that period of time. The physical properties of the crust change continuously during this time.

As seen in the loaf of bread example, TGA is an excellent tool for determining the composition of materials, as well as the temperature at which those materials decompose due to either thermal or oxidative degradation.

Figure 31-12 is a comparison of DSC results from a sugar-coated chocolate candy and the sugar-coating alone. Note that the melting of the sugar is different in the presence of the melted chocolate. This is due to the sugar starting to dissolve in the liquid chocolate.

Many materials can exist in either an amorphous or crystalline form depending on how the material was processed. Since the structure or form of the material is a function of its previous thermal history (e.g., time, temperature, relative humidity, pressure), the structure can change as it is heated during the DSC experiment. This can be seen in Fig. 31-13, a DSC experiment on amorphous sucrose that was prepared by freeze-drying. The first observed transition is the glass transition near 40°C. The Tg of dry (<0.1% moisture content) sucrose is approximately 68°C, which decreases as moisture content increases. A Tg of 40°C

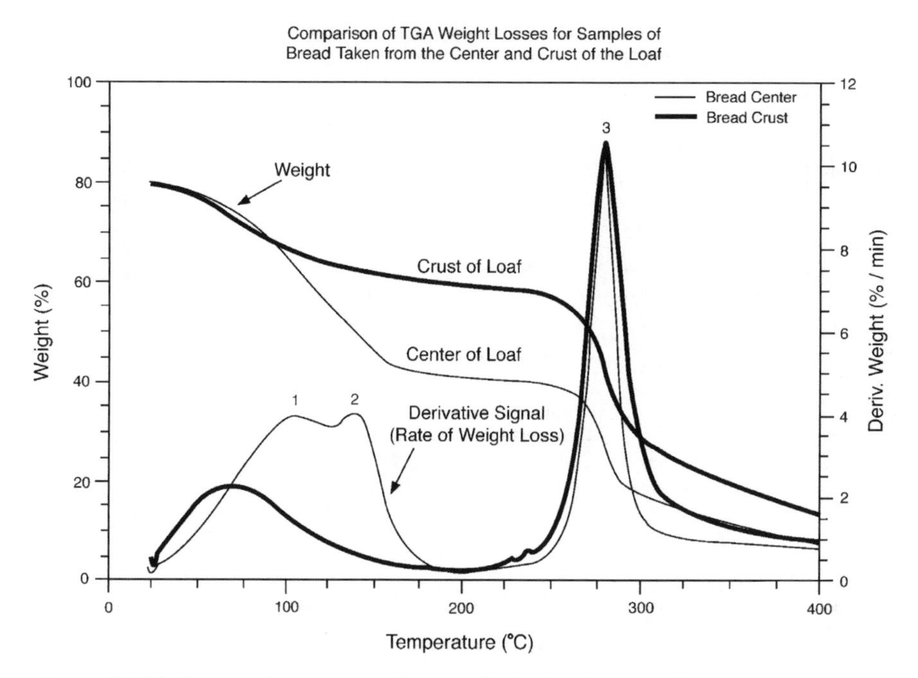

31-11 figure A comparison of TGA data on the center and crust of a loaf of bread baked at 190°C shows substantial differences at temperatures below 200°C. This is because the center of the loaf remains much cooler than the crust during baking due to the evaporation of water.

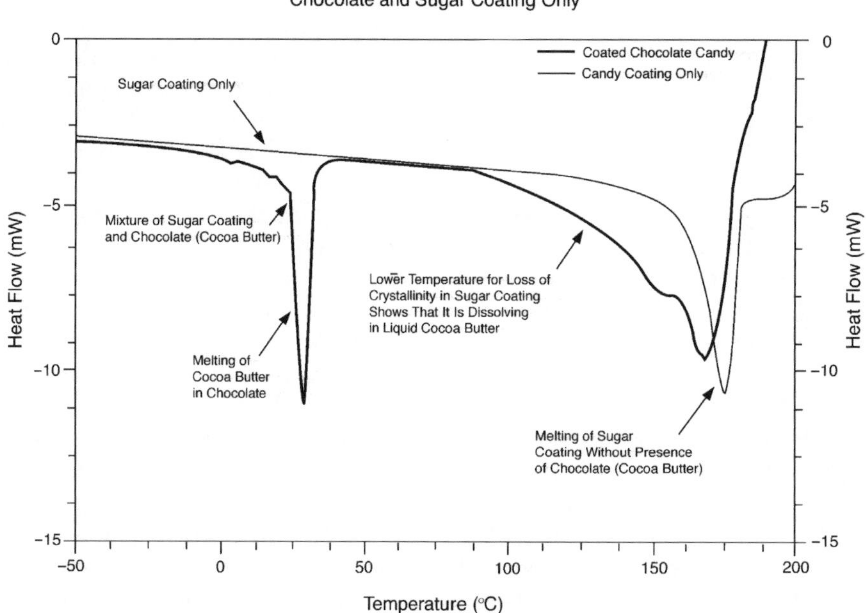

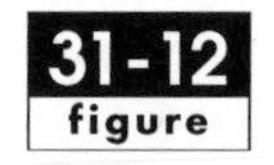
31-12 figure Comparison of DSC data on sugar-coated chocolate candy and the sugar-coating alone. Melting of the sugar changes in the presence of the liquid chocolate due to its solubility in the chocolate.

indicates that this sample has moisture content of 3.0–4.0% (dry basis) (12). The second useful piece of information is the size of the glass transition, which is the magnitude of the step change in heat capacity and is typically expressed in heat capacity units of J/g°C. A value of 0.91 J/g°C for this sample indicates that it is very close to 100% amorphous.

The second transition near 100°C is an exothermic peak caused by crystallization (commonly termed thermally induced crystallization or cold crystallization) of the amorphous material. The size of the peak (J/g) provides information on the amount of amorphous material that crystallizes. By dividing the measured area of the peak (71.8 J/g) by the heat of

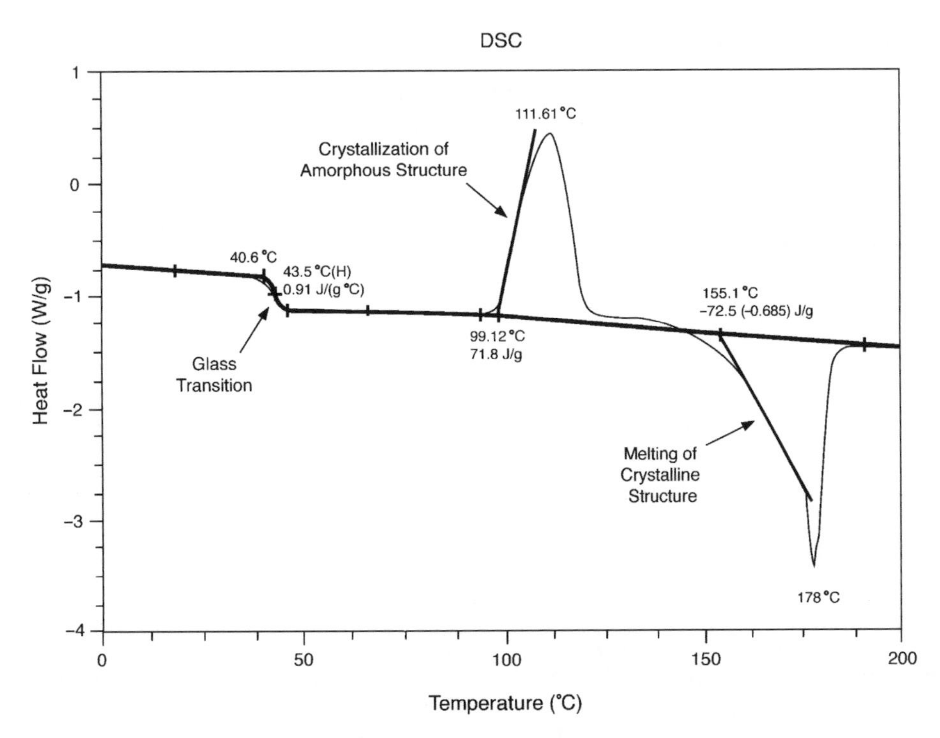

31-13 figure DSC data shows that a sample of freeze-dried sucrose is 100% amorphous based on the size of the glass transition and a comparison of the energies (J/g) of crystallization and melting.

crystallization for 100% conversion (approximately 115 J/g obtained from a separate DSC experiment on 100% crystalline sucrose), it appears that only about 62% crystallized during heating.

The last observed transition is an endothermic peak between approximately 150 and 180°C that is caused by conversion of the crystalline structure to an amorphous form. The shape of the peak is not very symmetric, because sucrose begins to decompose (chemical transformation) at the same temperature. The overlapping of the two processes creates the nonsymmetric peak. As with thermally induced crystallization, the size of the peak (J/g) is a quantitative measure of the amount of structural change. In this case, the size of the melting peak (72.5 J/g) is within experimental error (typically ±2%) of the size of the crystallization peak (71.8 J/g), indicating that all observed melting is the result of the crystallization that occurred during heating and that the original structure was 100% amorphous.

The last example is MDSC® analysis of a soft chocolate chip cookie. The experiment was started just above room temperature and involved cooling to −30°C at 1°C/min. The glass transition is observed in the reversing signal of Fig. 31-14 as a step change over the temperature range of 7 to −10°C. Since room temperature is above the glass transition temperature of the cookie, the cookie would be soft and chewy at the typical consumption temperature. The nonreversing signal, which shows kinetic/time-dependent processes, shows two broad exothermic peaks that are most likely crystallization of fats used in the recipe. The **total signal** (only signal available with conventional DSC) is the sum of the glass transition and fat crystallization events and is difficult to interpret compared with the separated reversing and nonreversing signals of MDSC®.

31.5 SUMMARY

Thermal analysis is a series of laboratory techniques that measures physical and chemical properties of materials as a function of temperature and time. In a thermal analysis experiment, temperature is typically either held constant (isothermal) or programmed to increase or decrease at a linear rate. Since temperature and time are controlled in all food preparation processes, thermal analysis instruments can simulate these processes on a very small scale (milligrams) and measure the response of the material. The most frequently used techniques include DSC and TGA.

The utility of thermal analysis to the food scientist is due to the fact that end-use properties (functionality) at a specific temperature are dependent on the structure of the components at that temperature. Therefore,

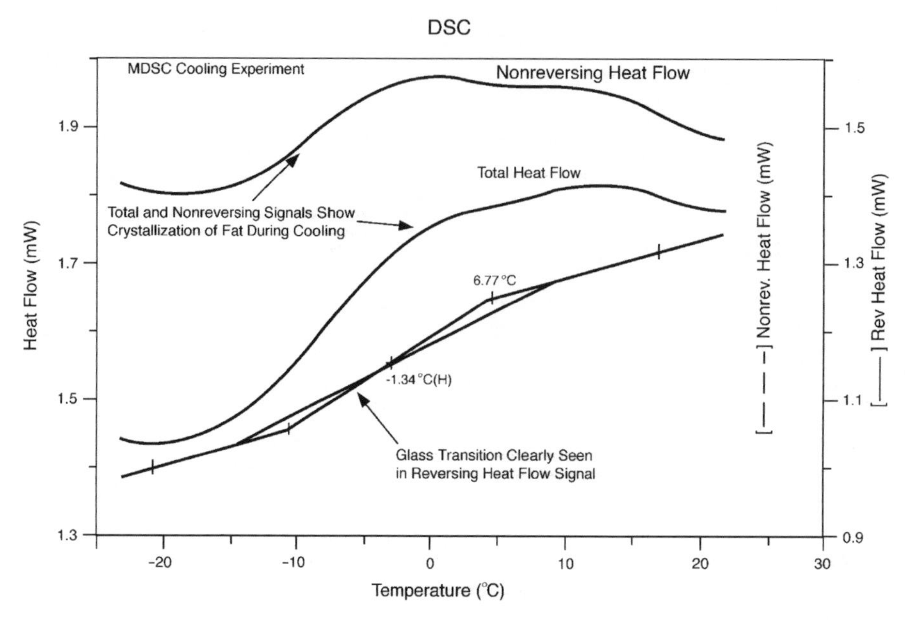

31-14 figure

MDSC® analysis of a soft chocolate chip cookie shows a glass transition below room temperature (22°C). Since consumption temperature is above the glass transition temperature, the cookie will be soft and chewy.

it is necessary to measure structure as a function of temperature. Structure can be defined as amorphous (no molecular order), crystalline, or semi-crystalline, with most food products having both amorphous and crystalline components.

Amorphous structure is characterized by analysis of the glass transition, which involves a significant change in the material's heat capacity and molecular mobility. Crystalline structure melts as temperature is increased, creating an endothermic peak that provides information on the quantity of crystalline structure and its melting temperature. This chapter has illustrated application of thermal analysis to a variety of food materials.

31.6 STUDY QUESTIONS

1. How would DSC data be useful in understanding differences in the texture of two batches of crackers?
2. How would thermal analysis data be useful in helping to understand the effect of a change in the processing conditions of a food product?
3. Why is it important to determine the glass transition temperature (Tg) of food products?
4. Assume that you have been asked to develop a recipe for a crisp cookie based on your knowledge of material properties. To develop the recipe, you need to answer the following questions:
   (a) Should most of the recipe be in the glassy or rubbery state at room temperature? Explain your choice.
   (b) In selecting TGA experimental conditions, would you use a sealed or unsealed sample pan and why?
   (c) In selecting DSC experimental conditions, would you typically use a sealed or unsealed pan and why?
   (d) Which MDSC® signal would contain the glass transition and why?
   (e) Assuming that a lipid is used that melts over the same temperature range as the glass transition of the crisp cookie, show what the MDSC reversing heat capacity signals would look like in the cooling and heating modes. Make a plot of reversing heat capacity vs. temperature.
   (f) You find out that high fructose corn syrup (HFCS) is less expensive than sucrose. Can you substitute the sucrose in your crisp cookie formula for HFCS? Be sure to give reasons for your response.
   (g) While performing a DSC or MDSC® experiment, an amorphous component in the sample begins to crystallize. Would this cause an increase or decrease in the measured heat capacity?
   (h) As stated in Sect. 31.3.2, a limitation of DSC is that it cannot measure heat capacity under isothermal conditions. However, MDSC® can measure heat capacity during an isothermal experiment. Explain this difference between DSC and MDSC® using Equation [1] in Sect. 31.3.3.

31.7 ACKNOWLEDGMENT

The authors of this chapter wish to acknowledge Dr. Thomas Schenz, who authored the thermal analysis chapter for the third edition of this book, and Dr. Eugenia Davis, who wrote the original version of this chapter for the first edition. Some ideas for the content and organization of the current chapter came from these previous versions of the chapter.

31.8 REFERENCES

1. Farkas J, Mohácsi-Farkas C (1996) Application of differential scanning calorimetry in food research and food quality assurance. J Therm Anal 47:1787–1803
2. Eliasson A-C (2003) Utilization of thermal properties for understanding baking and staling processes, Ch. 3. In: Kaletunc G, Breslauer KJ (eds) Characterization of cereals and flours: properties, analysis, and applications. CRC Press, Raton, FL, pp 64–115
3. Ievolella J, Wang M, Slade L, Levine H (2003) Application of thermal analysis to cookie, cracker, and pretzel manufacturing, Ch. 2. In: Kaletunc G, Breslauer KJ (eds) Characterization of cereals and flours: properties, analysis, and applications. CRC Press, Boca Raton, FL, pp 37–63
4. Sahin S, Sumnu SG (2006) Physical properties of foods. Springer, New York, p 257
5. Kamruddin M, Ajikumar PK, Dash S, Tyagi K, Baldev RAJ (2003) Thermogravimetry-evolved gas analysis–mass spectrometry system for materials research. Bull Mater Sci 26(4):449–460
6. Schmidt SJ (2004) Water and solids mobility in foods. *Advances in Food and Nutrition Research*, vol 48. Academic Press, London, UK, pp 1–101
7. Slade L, Levine H (1988) Non-equilibrium behavior of small carbohydrate-water systems. Pure Appl Chem 60(12):1841–1864
8. Slade L, Levine H (1991) Beyond water activity: recent advances based on an alternative approach to the assessment of food quality and safety. Crit Rev Food Sci Nutr 30(2–3):115–360
9. Aguilera JM, Lillford PJ (2007) Food materials science: principles and practice. Springer, New York, NY, p 622
10. Sun D-W (2005) Handbook of frozen food processing and packaging. CRC Press, Raton, FL, p 760
11. Thomas L (2006) Modulated DSC technology manual. TA Instruments, New Castle, DE
12. Yu X, Kappes SM, Bello-Perez LA, Schmidt SJ (2008) Investigating the moisture sorption behavior of amorphous sucrose using a dynamic humidity generating instrument. J Food Sci 73(1):E25–E35

Color Analysis

Ronald E. Wrolstad* and Daniel E. Smith
Department of Food Science and Technology, Oregon State University,
Corvallis, OR 97331–6602, USA
ron.wrolstad@oregonstate.edu
dan.smith@oregonstate.edu

32.1 Introduction 575
32.2 Physiological Basis of Color 575
32.3 Color Specification Systems 577
 32.3.1 Visual Systems 577
 32.3.2 Instrumental Measurement of Color 578
  32.3.2.1 Historical Development 578
  32.3.2.2 The CIE Tristimulus System 579
 32.3.3 Tristimulus Colorimeters and Color Spaces 580
32.4 Practical Considerations in Color Measurement 582
 32.4.1 Interaction of Light with Sample 582
 32.4.2 Instrument Choice 583
 32.4.3 Color Difference Equations and Color Tolerances 583
 32.4.4 Sample Preparation and Presentation 583
32.5 Summary 585
32.6 Study Questions 585
32.7 Acknowledgments 586
32.8 References 586

S.S. Nielsen, *Food Analysis*, Food Science Texts Series, DOI 10.1007/978-1-4419-1478-1_32,

32.1 INTRODUCTION

Color, flavor, and texture are the three principal quality attributes that determine food acceptance, and color has a far greater influence on our judgment than most of us appreciate. We use color to determine if a banana is at our preferred ripeness level, and a discolored meat product can warn us that the product may be spoiled. The marketing departments of our food corporations know that, for their customers, the color must be "right." The University of California Davis scorecard for wine quality designates four points out of 20, or 20% of the total score, for color and appearance (1). Food scientists who establish quality control specifications for their product are very aware of the importance of color and appearance. While subjective visual assessment and use of visual color standards are still used in the food industry, instrumental color measurements are extensively employed. Objective measurement of color is desirable for both research and industrial applications, and the ruggedness, stability, and ease of use of today's color measurement instruments have resulted in their widespread adoption.

Color can be defined as the sensation that is experienced by an individual when radiant energy within the visible spectrum (380–770 nm) falls upon the retina of the eye (2), and a **colorant** is a pigment that is used to color a product. For the phenomenon of color to occur there must be (1) a colored object, (2) light in the visible region of the spectrum, and (3) an observer. All three of these factors must be taken into account when assessing and measuring color. When white light strikes an object it can be absorbed, reflected, and/or scattered. Selective absorption of certain wavelengths of light is the primary basis for the color of an object. Color, as seen by the eye, is an interpretation by the brain of the character of light coming from an object. **Colorimetry** is the science of color measurement (3). It is possible to define color in mathematical units; however, those numbers do not easily relate to the observed color. A number of color-ordering systems and color spaces have been developed that better agree with visual assessment. In food research and quality control, instruments are needed which provide repeatable data that correspond to how the eye sees color. This chapter will provide a brief description of human physiology of vision, and an overview of the different color-ordering and color-measuring systems. The chapter is limited to presenting the basic underlying principles that will hopefully allow for an understanding of how color of food products should be measured. Color measurement is a very complex subject, and for more detailed exploration of the subject the following references are recommended (2–7).

32.2 PHYSIOLOGICAL BASIS OF COLOR

Humans have excellent color perception and they can detect up to 10,000,000 different colors (8). They have very poor color memory, however, and cannot accurately recall colors of objects previously observed (5,9) (hence the need for objective measurement of color). While color perception varies somewhat with humans, it is much less variable than that for the senses of taste and smell. Color perception is comparatively uniform for people with normal color vision; however, 8% of males and 0.5% of females have physiological defects and perceive colors in a markedly different way (2,5).

Figure 32-1 is a simplified diagram of the human eye. Light enters the eye through the cornea, passes through the aqueous and vitreous humor and is focused on the **retina**, which contains the receptor system (10). The **macula** is a small (approximately 5 mm in diameter) and highly sensitive part of the retina that is responsible for detailed central vision. It is located roughly in the center of the retina. It is yellow-orange colored and contains a high concentration of the carotenoid pigments, lutein and zeaxanthin. It is believed that these dietary antioxidants may protect the retina from photo damage (11). Age-related macular degeneration results in loss of central vision, and is a major health issue in our aging population. The **fovea**, the very center of the macula, is about 2 mm in diameter and contains a high concentration of **cones**, which are responsible for daylight and color vision, known as "**photopic**" vision. The cones contain receptors that are sensitive to red, green, and blue light. Figure 32-2 shows the spectral sensitivity curves for the three respective cones. **Rods** are more widely distributed in the retina and are sensitive to low-intensity light. They have no color discrimination and are responsible for night or "**scotopic**" vision. Figure 32-3 shows the spectral sensitivity curves for scotopic (rod) and photopic (cone) vision, the latter being an integration of the curves shown in Fig. 32-2. Note the sensitivity maximum is at 510 nm for scotopic vision and 580 nm for photopic vision. This accounts for blues appearing to be brighter and reds darker at twilight when both scotopic vision and photopic vision are functioning.

Signals are sent via the optic nerve to the brain, where "vision" occurs. According to the "Color Opponent Theory" (4), the signals from the red, green, and blue receptors are transformed to one brightness signal indicating darkness and lightness and two hue signals, red vs. green and blue vs. yellow. Figure 32-4 shows a diagram of the opponent color model. The brain's interpretation of signals is a complex phenomenon and is influenced by a variety of psychological aspects. One such aspect is **color constancy**. The same sheet of white paper will appear white when seen in bright

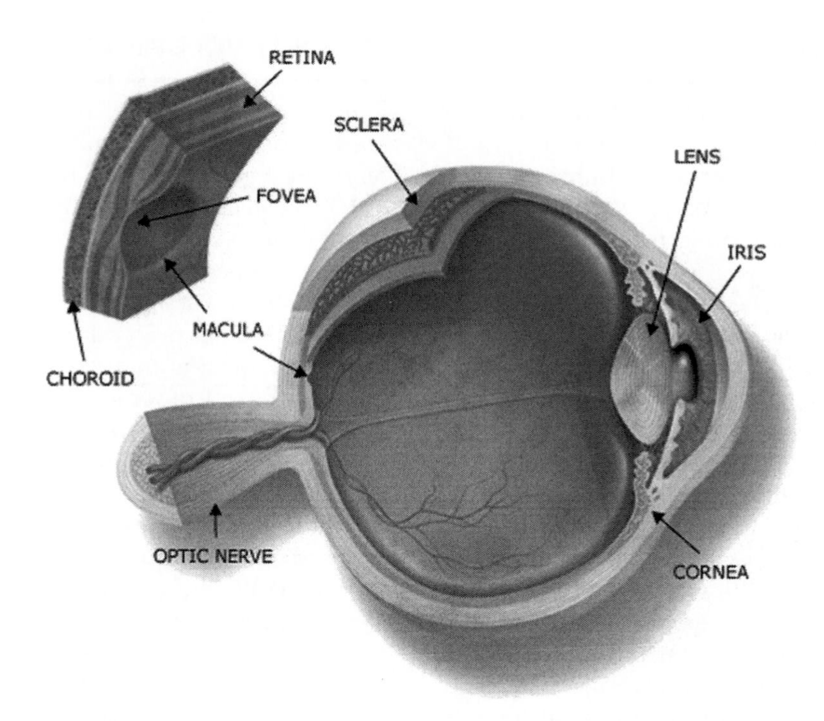

32-1 figure Diagram of the human eye. http://www.amdcanada.com/template.php?lang = eng& section = 4& subSec = 2d& content = 4_2.

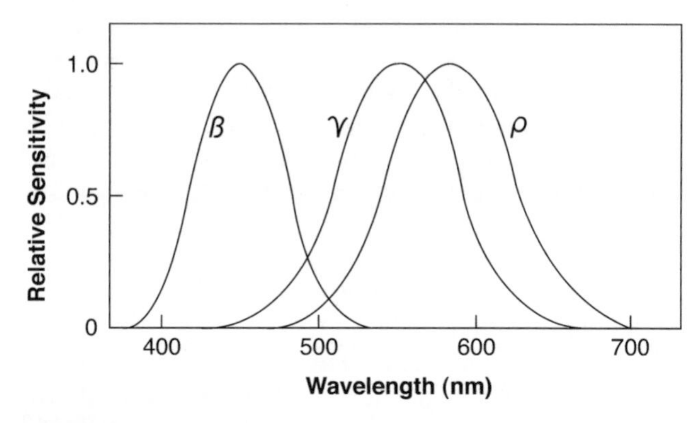

32-2 figure Spectral sensitivity curves of the three types of cones comprising photopic vision. [From (5), with kind permission of Springer + Business Media.]

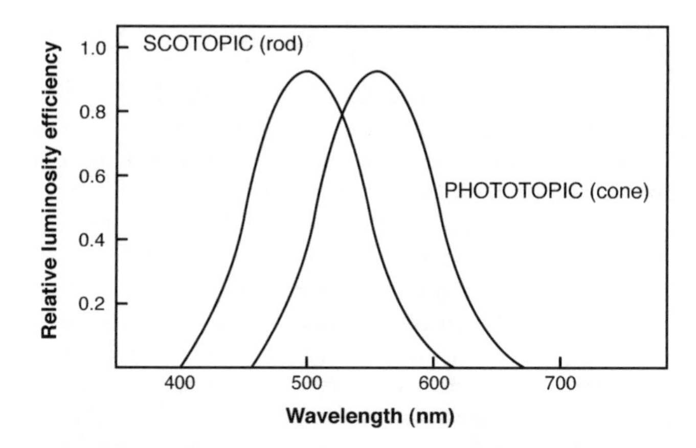

32-3 figure Spectral sensitivity curves for scotopic (rods) and photopic (cones) vision. [From (5), with kind permission of Springer + Business Media.]

sunlight, and also when it is viewed indoors under dim light. The physical stimuli in each case are obviously quite different, but the brain knows that the paper should be white and draws on its experience. A second aspect occurs when a large expanse of color appears brighter than the same color in a small area. One only needs the experience of painting a whole wall of a room, and then seeing how different it appears from the small color chip obtained from the paint store.

Some trivia (10): The color of the iris, which regulates the amount of light entering the eye, appears brown, blue, or green depending on the amount of melanin pigment. Blue eyes result from the absence of light-absorbing melanin pigment, which results in greater reflection of highly scattered short wavelength blue light. Birds, fish, reptiles, and primates have strong color vision, while nocturnal mammals such as cats have keen night vision, but very limited color vision, probably seeing a pastel world during daylight (10). Hawks have eight times the visual acuity of humans, having a concentration of 1,000,000 cones per mm of fovea compared with 160,000 for man. Dogs have poorer visual acuity than humans, but are very good at sensing motion, which facilitates their pursuit of a moving object (12).

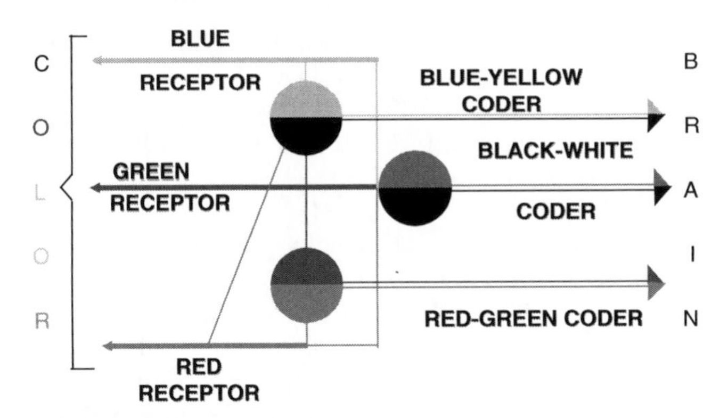

32-4 figure The color opponent model. (Courtesy of HunterLab, Reston, VA.)

32.3 COLOR SPECIFICATION SYSTEMS

There are verbal, visual matching, and instrumental methods for describing and specifying color. Color is three dimensional, and any color-order system will need to address **hue**, what we instinctively think of as color (e.g., red, blue, green), **value**, which represents lightness and darkness, and **chroma** or **saturation** which indicates intensity. When attempting to verbally describe a color defect or problem, one should attempt to use these three qualities in formulating a color description.

32.3.1 Visual Systems

The **Munsell** system is probably the best known and most widely used visual color-ordering system. It was developed by A.H. Munsell, a Boston art teacher, in 1905. In this system, Red, Yellow, Green, Blue, and Purple plus five adjacent pairs, Green-Yellow, Yellow-Red, Red-Purple, Purple-Blue, and Blue-Green describe **hue**. **Value** is that quality of color described by lightness and darkness, from white to gray to black. Value is designated from zero (absolute black) to ten (absolute white). **Chroma** is that quality that describes the extent a color differs from a gray of the same value. It is designated in increasing numbers starting with 0 (neutral gray) and extending to/16 or even higher. A change from pink to red is an example of an increase in chroma. In Munsell notation, hue is listed first and designated by a number and letter combination. Numbers run from 1 to 100, and the letters are taken from the ten major hue names, e.g., 10 GY. Value follows with a number from 0 to 10 followed by a slash mark, which is followed by a number for chroma (e.g., 5R 5/10).

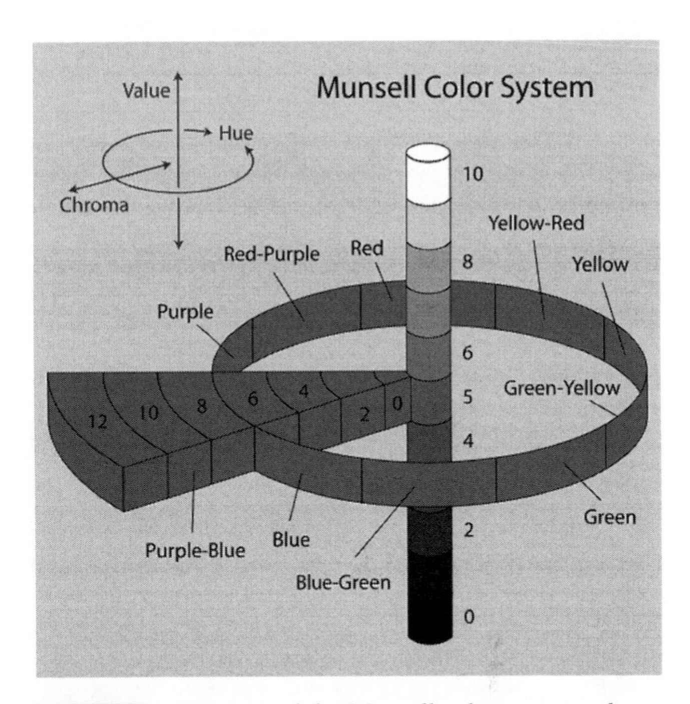

32-5 figure Diagram of the Munsell color system, showing: a circle of hues at value 5 chroma 6; the neutral values from 0 to 10; and the chromas of purple-blue (5PB) at value 5. [Source, Wikipedia: Jacobolus (http://en.wikipedia.org/wiki/User:Jacobolus).]

One of Munsell's objectives was to develop a system based on equal visual perception, with equal steps of perception for each of the coordinates. For example, the difference in value between 2 and 3 is visually equivalent to the difference between 5 and 6. This visual linearity applies to the other coordinates as well. The Munsell systems' visual linearity undoubtedly contributes to its success and wide popularity in many different fields. Figure 32-5 illustrates the Munsell color system, showing a circle of hues at value 5 chroma 6; the neutral values from 0 to 10; and the chroma of purple-blue (5PB) at value 5. The ten named hues are shown with additional intermediate hues interspersed. The distance from the core to the edge shows increasing chroma, the maximum chroma differing considerably for different hues (e.g., R5 has a maximum of 12 and yellow has a maximum of 6). Interactive kits that demonstrate the relationships between Munsell hue, value, and chroma are available for purchase (13). Also available is the Munsell Book of Color with 1605 colored chips, each with a numerical designation.

Assessing color of foods by visual comparison with color standards is an option for a number of food products. USDA color standards are available for honey, frozen French fried potatoes, peanut butter, and canned ripe olives, for example (13). This method is simple, convenient, and easy to understand; however, it is subjective.

32.3.2 Instrumental Measurement of Color

32.3.2.1 Historical Development

Early investigators believed that it was possible to measure color since the basic mechanisms for color perception were well understood. It was realized that to objectively measure color there needed to be standardization of light sources. The **CIE (Commission Internationale de l'Eclairage** or the **International Commission on Illumination)** is the main international organization concerned with color and color measurement (3). Standard illuminants for color measurement were first established in 1931 by the CIE. Figure 32-6 shows the spectral power distribution curves of three standard CIE illuminants, A, C, and D_{65}. **Illuminant C** was adopted in 1931 and represents overcast daylight, while **illuminant D_{65}**, which was adopted in 1965, also represents average daylight, but includes the ultraviolet wavelength region. **Illuminant A**, adopted in 1931, represents an incandescent light bulb. Objects will appear to have different colors when viewed under illuminants A and C. Because of the predominance of long wavelength light and lesser amounts of shorter wavelength light of illuminant A, one can predict that objects will appear to have a "warmer" color under illuminant A than under other illuminants. **Metamerism** occurs when two objects appear to have the same color under one light source, but exhibit different colors under another source. In 1965, the CIE adopted **illuminant D_{65}**, which contains more UV than illuminant C and is believed to better represent average daylight. It is today's most widely used standard illuminant for color measurement. When evaluating food samples for color and appearance, one should make an effort to control and standardize lighting. Commercial light booths that are equipped with standard illuminants are available for standardizing viewing conditions for controlled visual assessment (4).

Scientists knew that a color sensation could be matched by mixing three colored lights (3). W.D. Wright in 1928 and J. Guild in 1931 conducted independent experiments in which people with normal color vision visually matched spectral (single wavelength) light by mixing different amounts of three primary lights (red, green, and blue) using rheostats (Fig. 32-7). The process was repeated for test colors covering the entire visible spectrum. The **field of view** for these experiments is described as **2°**, which is similar to viewing a dime at an arm's length. The purpose of these viewing conditions was to have primary involvement of the fovea, the retinal area of greatest visual acuity. The red, green, and blue response factors were averaged and mathematically converted to and x, y, and z functions that quantify the red, green, and blue cone sensitivity of the average human observer. The observer functions were standardized and adopted by CIE in 1931 as the **CIE 2° Standard Color Observer**. The Standard Observer curves provide human sensory response factors that are used in color measurement worldwide (Fig. 32-8). Later it was realized that the cones spread beyond the fovea, and more realistic data could be obtained from a larger field of view. The experiment was repeated using a 10° field of view and adopted by CIE in 1964 as the **10° Standard Observer**. Both sets of data are used today, but the 10° Standard Observer is preferable because it better correlates to the visual assessments typically made with a larger field of view.

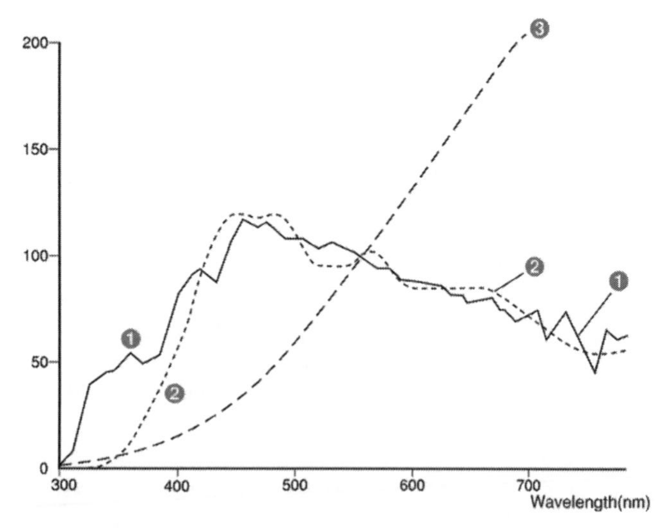

32-6 figure The spectral power distribution curves of three standard CIE illuminants. Standard illuminant D_{65}: Average daylight including ultraviolet wavelength (1); Standard illuminant C: Average daylight (not including ultraviolet wavelength region) (2), and Standard illuminant A: Incandescent light (3). (Courtesy of Konica Minolta Sensing Americas, Inc., Ramsey, NJ.)

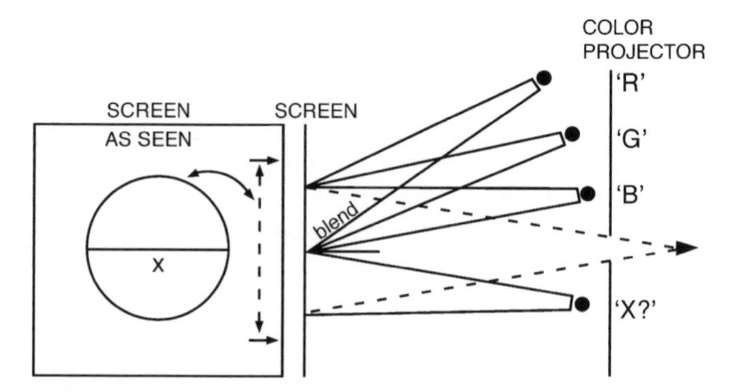

32-7 figure Diagram Showing three projectors focused on the upper half of a circle on the screen the color to be measured is projected on the lower half and the eye can see both halves simultaneously.

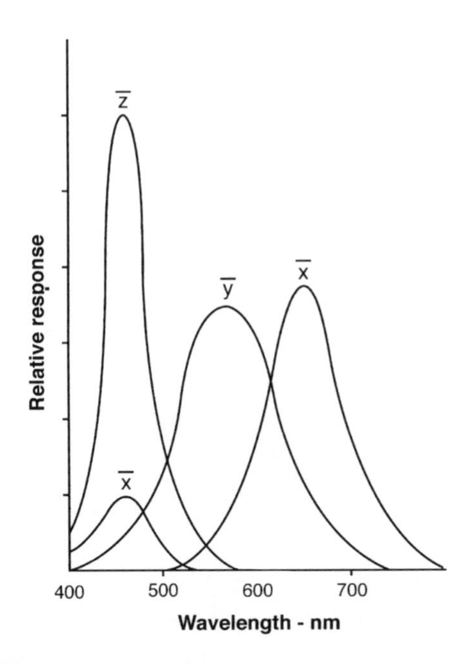

32-8 figure Standard observer curves showing the relationship between the red (x), blue (z), and green (y) cone sensitivity and the visible spectrum.

32.3.2.2 The CIE Tristimulus System

With the adoption of standard observer functions and standard illuminants, it became possible to convert the spectral transmission or reflectance curve of any object to three numerical values. These numbers are known as the **CIE tristimulus values**, X, Y, and Z, the amounts of red, green, and blue primaries required to give a color match. The data values for a standard illuminant and the standard observer functions are multiplied by the % reflectance or % transmission values for the object at selected wavelengths. Summation of the products for the wavelengths in the visible spectrum (essentially integrating the areas under the three curves) gives the resulting X, Y, and Z tristimulus values. This can mathematically be represented as follows:

$$X = \int_{380}^{750} RE\bar{x}\,dx \qquad [1]$$

$$Y = \int_{380}^{750} RE\bar{y}\,dy \qquad [2]$$

$$Z = \int_{380}^{750} RE\bar{z}\,dz \qquad [3]$$

where:

R = sample spectrum
E = source light spectrum
$\bar{x}, \bar{y}, \bar{z}$ = standard observer curves

With the objective of plotting the three coordinates in two dimensions, the CIE converted the X, Y, and Z tristimulus values to x, y, and z coordinates by the following mathematical operation:

$$x = \frac{X}{X+Y+Z} \qquad [4]$$

$$y = \frac{Y}{X+Y+Z} \qquad [5]$$

$$z = \frac{Z}{X+Y+Z} \qquad [6]$$

Since $x + y + z = 1$, only two coordinates are needed to describe color as $z = 1 - (x + y)$.

Figure 32-9 shows the 1931 **chromaticity diagram** where x vs. y are plotted to give the horseshoe-shaped locus. Spectral colors lie around the perimeter and white light (illuminant D_{65}) has the coordinates $x = 0.314$, $y = 0.331$. Copies of the chromaticity diagram are available for plotting the calculated coordinates x and y for an object (14). With the aid of a ruler, a line can be drawn from the coordinates for white light through the object coordinates to the edge, which gives the **dominant wavelength,** λ_d. Dominant wavelength is analogous to hue in the Munsell system. The distance from the white light coordinates to the object coordinates, relative to the distance from the white light coordinates to λ_d, is described as **% purity** and is analogous to chroma in the Munsell system. The standard observer curve for y (green) shown in Fig. 32-8 is very similar to the sensitivity curve for human photopic vision shown in Fig. 32-3. Because of this, tristimulus value **Y** is known as **luminosity**, and is analogous to value in the Munsell system.

Manual calculation of XYZ tristimulus values from reflectance/transmission spectra is a tedious operation. Modern colorimetric spectrophotometers measure the light reflected or transmitted from an object, and the data are sent to a processor where it is multiplied by standard illuminant and standard observer functions to give the XYZ tristimulus values. Since objects with identical XYZ tristimulus values will provide a color match, they find application in the paper, paint, and textile industries. Unfortunately, the XYZ numbers do not easily relate to observed color and they have the limitation of not having equivalent visual spacing. [Referral to Fig. 32-9 reveals that the wavelength spacing in the green region (500–540 nm) is much larger than that in the red (600–700 nm) or blue (380–480 nm) regions.] The same numerical color differences between colors will not equate to the same visual difference for all colors. This is a severe limitation in measurement of color of food products, as major interest is in how food product color deviates from a standard or changes during processing and storage. Statistical analysis of color data for which numerical units were nonequivalent would be problematic.

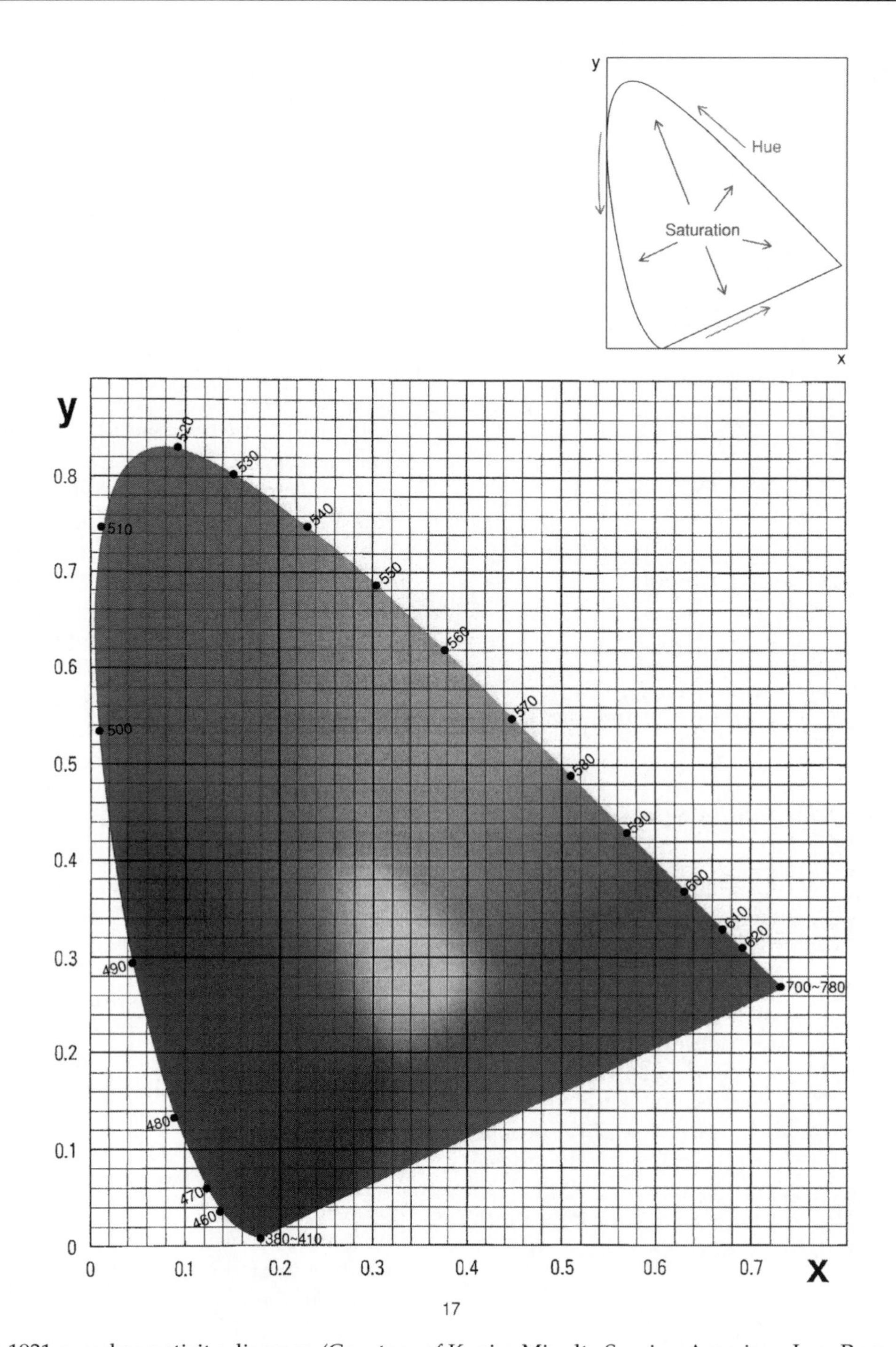

32-9 figure The 1931 *x*, *y* chromaticity diagram. (Courtesy of Konica Minolta Sensing Americas, Inc., Ramsey, NJ.)

32.3.3 Tristimulus Colorimeters and Color Spaces

Richard S. Hunter, Deane B. Judd, and Henry A. Gardner were among the pioneering scientists who in the 1940s were working to develop color-measuring instruments that would overcome the disadvantages of the CIE Spectrophotometric tristiumlus system (2,5,6). Light sources that were similar to Illuminant C were used, along with filter systems that approximated the sensitivity of the cones in the human eye. Empirical approaches were taken to get more equivalent visual spacing. In an effort to get numerical values that better related to observed color, a system that applied the color opponent theory of color perception was developed (3).

The **Hunter color solid** (Fig. 32-10) was first published in 1942 where **L** indicated **lightness**, ***a***, the **red**

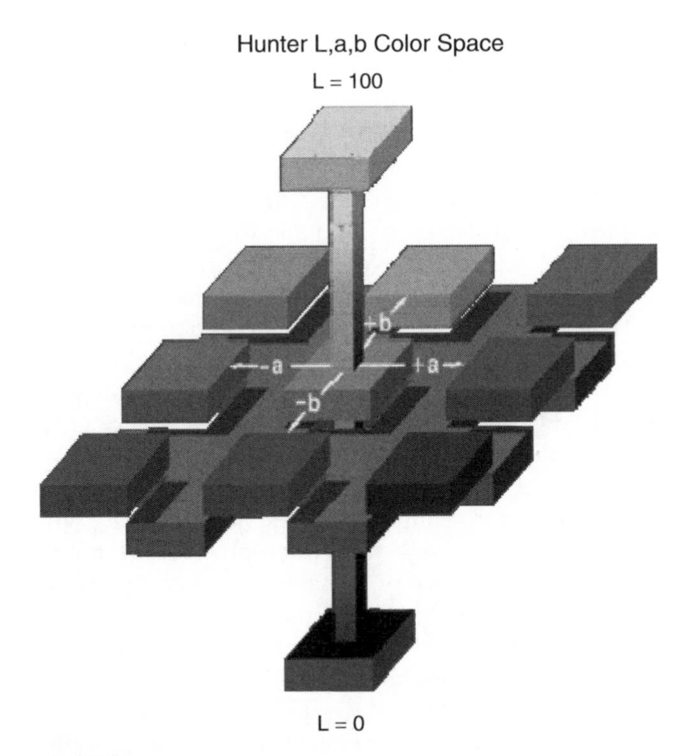

32-10 figure The Hunter *L, a, b* Color Solid. (Courtesy of HunterLab, Reston, VA.)

(+) or **green** (−) coordinate, and **b**, the **yellow** (+) or **blue** (−) coordinate. The **Hunter L a b color space** has been widely adopted by the food industry. It is very effective for measuring color differences. The *Lab* system was subsequently improved to give more uniform color spacing. In 1976, the CIE officially adopted the modified system as **CIELAB** with the parameters $L^*a^*b^*$. L^* indicates **lightness** (0 to 100) with 0 being black and 100 being white. The coordinate $\mathbf{a}^*$ is for **red** (+) and **green** (−), and $\mathbf{b}^*$ is for **yellow** (+) and **blue** (−). The limits for a^* and b^* are approximately + or −80. Figure 32-11 shows a portion of the a^*, b^* chromaticity diagram where a^* and b^* are both positive, representing a color range from red to yellow. Point A is the plot of a^* and b^* for a red apple. The angle from the start of the $+a^*$ axis to point A can be calculated as $\mathbf{arctan}^{b^*/a^*}$, and is known as **hue angle, h** or $\mathbf{H}^*$. The distance from the center to point A is **chroma**, which is calculated as the hypotenuse of the right triangle formed by the origin and the values of coordinates *a* and *b*. $(\mathbf{a}^{*2} + \mathbf{b}^{*2})^{1/2} = \mathbf{C}^*$. The CIE has also recommended adoption of this color scale known as **CIELCH** or $L^*C^*H^*$. This color space (which is illustrated in Fig. 32-12) designates hue (H^*) as one of the three dimensions, the other two being lightness (L^*)

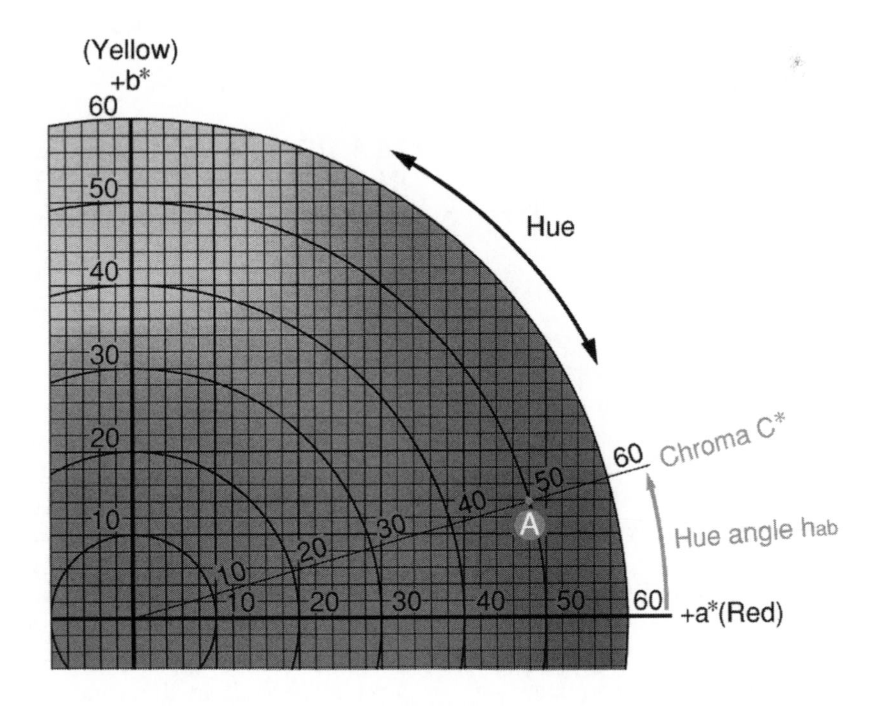

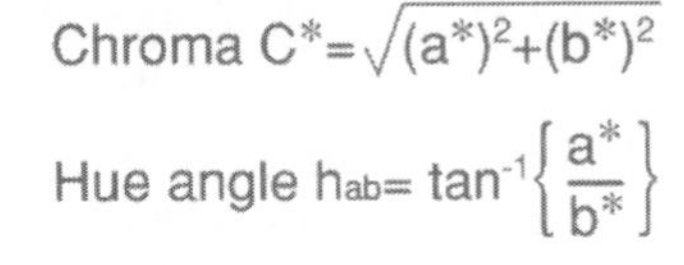

A portion of an a^*, b^* chromaticity diagram showing the position A for a red apple. (Courtesy of Konica Minolta Sensing Americas, Inc., Ramsey NJ.)

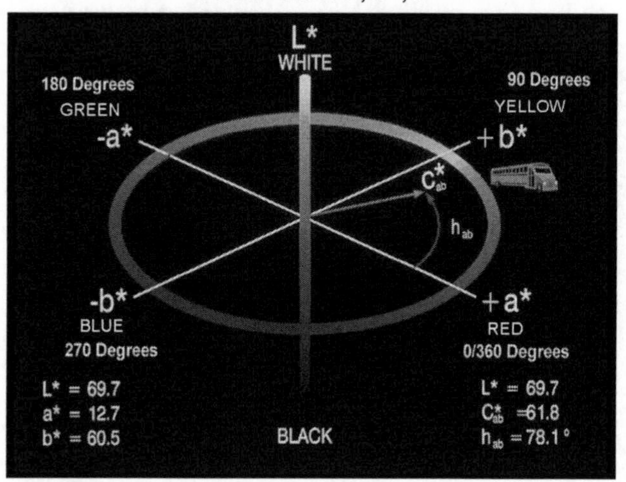

32-12 figure The CIE $L^*C^*H^*$ Color Space showing the location of "School Bus Yellow". (Courtesy of HunterLab, Reston, VA.)

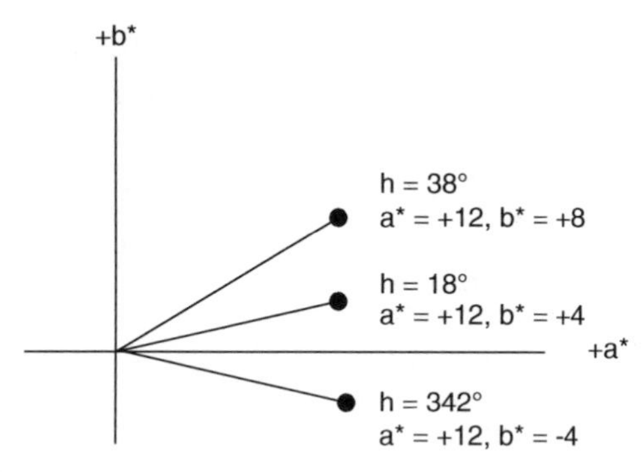

32-13 figure Plots of a^* and b^* for three hypothetical samples.

and chroma (C^*), which have an obvious parallel to Munsell hue, value, and chroma. This color space is advantageous as hue is most critical to humans with normal color vision for perception and acceptability. In this system, 0° represents red, 90° – yellow, 180° – green, and 270° – blue. Figure 32-13 shows plots of a^* and b^* for three hypothetical objects having the following a^*b^* coordinates: $a^* = +12$ and $b^* = +8$; $a^* = +12$ and $b^* = +4$; $a^* = +12$ and $b^* = -4$. While all objects have identical a* values, their colors range from purplish red ($H^* = 342°$) to red ($H^* = 18°$) to orange ($H^* = 34°$). A common error in interpretation of color measurements is to use only the coordinate a* as a measure of "redness." Monitoring color change is more understandable if one measures lightness (L^*), hue angle (H^* from 0–360°), and chroma. Chroma will increase with increasing pigment concentration, and then decrease as the sample becomes darker. Thus, it is possible for two light and dark samples to have the same hue angle and the same chroma. They will readily be distinguished, however, because of their different lightness values.

The colorimeters that are available in the market today have vastly improved from earlier models with respect to stability, ruggedness, and ease of use. There are instruments that are portable for use in the field, on-line instruments for process control, and specialized colorimeters for specific commodities. They vary with respect to operating in transmission or reflectance mode and size of sample viewing area. Colorimeters have a high degree of precision, but do not have a high degree of accuracy with respect to identifying or matching colors. Most colorimeters used in research are color spectrophotometers with a diffraction grating for scanning the visible spectrum, with these data being sent to a microprocessor for conversion of reflectance or transmission data to tristimulus numbers. In operating the instrument, choices must be made as to **illuminant**, **2° or 10° viewing angle**, and data presentation as **XYZ**, ***Lab***, **CIEL*a*b***, or **L*C*H***. Illuminant D_{65}, 10° viewing angle, and $L^*C^*H^*$ are appropriate for most food applications. It should be obvious that different numbers will be obtained with different illuminants, viewing angles, and color scales. It is critical that the illuminant, viewing angle, and color scale used in color measurement be reported in technical reports and research publications.

32.4 PRACTICAL CONSIDERATIONS IN COLOR MEASUREMENT

Choice of an appropriate instrument, sample preparation, sample presentation, and handling of data are issues that must be dealt with in color measurement.

32.4.1 Interaction of Light with Sample

When a sample is illuminated with light a number of things occur that are illustrated in Fig. 32-14. Light for which the angle of reflection is equal to the angle of incidence is described as **specular light**. Smooth polished surfaces will appear **glossy** because of the high degree of **specular reflection**. Rough surfaces will have a great deal of **diffuse reflection** and will have a dull or **matte** appearance. Selective absorption of light will result in the appearance of color. **Opaque** samples will **reflect** light. **Transparent** samples will primarily **transmit** light, and **translucent** samples will both **reflect** and **transmit** light. Ideal samples for color measurement will be flat, smooth, uniform, matte, and either opaque or transparent. A brick of colored Cheddar cheese is one of the few food examples that come close to having those characteristics.

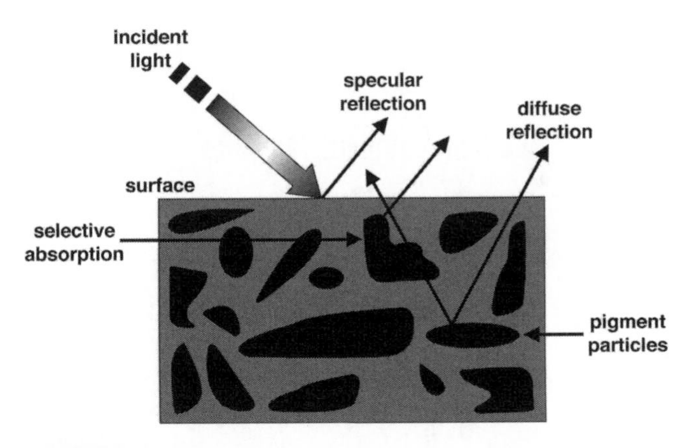

32-14 figure Interaction of light with an object. [From (4), Used with permission.]

32.4.2 Instrument Choice

Instrument geometry refers to the **arrangement of light source, sample placement**, and **detector**. The CIE recognizes the following instrument geometries: **45°/0°** where the specimen is illuminated at 45° and measured at 0°, and the inverse, **0°/45°** where the specimen is illuminated at 0° and measured at 45°. Diffuse reflectance is measured since specular light is excluded. These are illustrated in Fig. 32-15. **Diffuse sphere geometry** (Fig. 32-16) is the third type where a white-coated sphere is used to illuminate a sample. With some sphere geometry instruments, measurements can either include or exclude specular reflectance. These instruments are versatile in that they can measure in transmission for transparent samples and in reflectance for opaque samples. Some can also measure the amount of light scattering, turbidity or haze in liquid samples, and the amount of gloss in solid samples. Instruments with 45°/0° and 0°/45° geometries can only measure reflectance.

32.4.3 Color Difference Equations and Color Tolerances

When colorimeter measurements are conducted under carefully controlled conditions, data with a high degree of precision can be obtained. In both industrial and research applications, the interest is primarily in how color dimensions deviate from a standard, or how they change from batch to batch, year to year, or during processing and storage. Color differences are calculated by subtracting $L^*a^*b^*$ and $L^*C^*H^*$ values for the sample from the standard, e.g.,

Delta $L^* = L^*_{\text{sample}} - L^*_{\text{standard}}$. Positive dL^* numbers will be lighter than the standard, and negative dL^* numbers will be darker.

Delta $a^* = a^*_{\text{sample}} - a^*_{\text{standard}}$. Positive da^* numbers will be more "red" (or less "green") than the standard, and negative da^* numbers will be more "green" (or less "red").

Delta $b^* = b^*_{\text{sample}} - b^*_{\text{standard}}$. Positive db^* numbers will be more "yellow" (or less "blue"), and negative db^* numbers will be more "blue" (or less "yellow").

Delta $C^* = C^*_{\text{sample}} - C^*_{\text{standard}}$. Positive dC^* numbers mean the sample has greater intensity or is more saturated, and negative dC^* numbers mean that the sample is less saturated.

Delta $H^* = H^*_{\text{sample}} - H^*_{\text{standard}}$. Positive H^* numbers indicate the hue angle is in the counterclockwise direction from the standard, and negative numbers are in the clockwise direction. If the standard has a hue angle of 90°, a positive dH* is a shift in the green direction, and a negative dH* number is a shift in the red direction.

A single number is often desired in industry for establishing pass/fail acceptability limits. **Total Color Difference (dE*)** is calculated by the following equation:

$$dE^* = (dL^{*2} + da^{*2} + db^{*2})^{1/2} \quad [7]$$

A limitation of dE^* is that the single number will only indicate the magnitude of color difference, not the direction. Samples with identical dE^* numbers will not necessarily have the same visual appearance.

In establishing color tolerances, dL^*, dC^*, and dH^* numbers are preferred since they correlate well with visual appearance. A diagram showing acceptable tolerances based on dL^*, dC^*, and dH^* numbers is shown in Fig. 32-17. The elliptical shape of the solid arises since tolerances for dH^* are considerably narrower than for dC^* and dL^*. In 1984, CIE recommended a new DE_{CMC} formula for industrial pass/fail decisions that utilizes dL^*, dC^*, and dH^* numbers.

32.4.4 Sample Preparation and Presentation

For color measurement data to be at all useful, the numbers must be consistent and repeatable. Sampling of product must be done so that it is representative of the product, and prepared so that it represents the product's color characteristics. Many food samples are far from ideal in that they may be partially transmitting and partially reflecting. Rather than being uniform they may be mottled or highly variable in color. The number of readings that need to be taken for acceptable repeatability is dependent on the nature of the sample. Another problem is that often the only instrument available is one that is less than ideal for the sample. Gordon Leggett (15) provides some practical tips and a systematic protocol for consistent color measurement of different food categories. Transparent

45°/0° and 0°/45° Specular Excluded Geometry

45° Illumination/0° Measure

Spectrometer
Source
Specular
0°
45°
Diffuse
Diffuse
Specimen

0° Illumination/45° Measure

Source
Spectrometer
Specular
0°
45°
Diffuse
Diffuse
Specimen

32-15 figure CIE standardized geometries for 45°/0° and 0°/45° instruments. (Courtesy of HunterLab, Reston, VA.)

Sphere Geometry d/8°

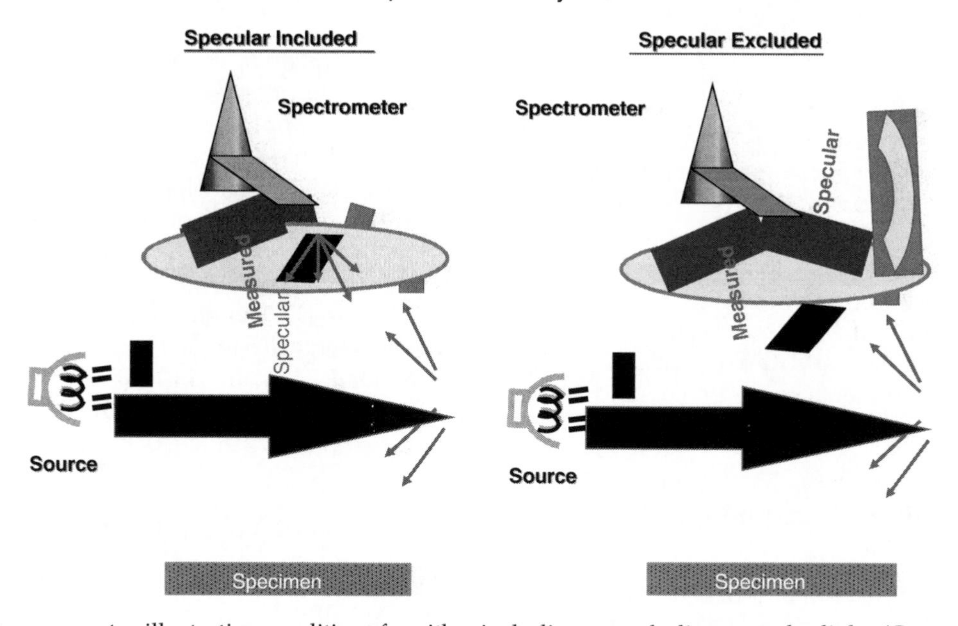

32-16 figure Sphere geometry illustration conditions for either including or excluding specular light. (Courtesy of HunterLab, Reston, VA.)

liquids should be measured with a sphere instrument, using a clear glass or plastic cell. A cell filled with distilled water can be used as a blank to negate the effects of cell and solvent. Cell path length is selected based on color intensity. A 20 mm cell is used for most colored liquids, with 10 mm cells for highly absorbing liquids. A very thin 2 mm cell may be appropriate for highly absorbent transparent liquids such as soy sauce. For nearly colorless liquids, a 50 mm cell may be necessary. For clear transparent liquids, a single measurement using a viewing area of 15 mm diameter or greater may be sufficient for good repeatability. For hazy transparent liquids, two to four readings with replacement of the liquid between readings were necessary to get acceptable repeatability when using a 10 mm path length cell and a sphere instrument.

Liquid samples with high solids are translucent rather than transparent. They can be measured by transmission using a very thin 2 mm path length cell, or measured in reflectance. Here it is necessary to

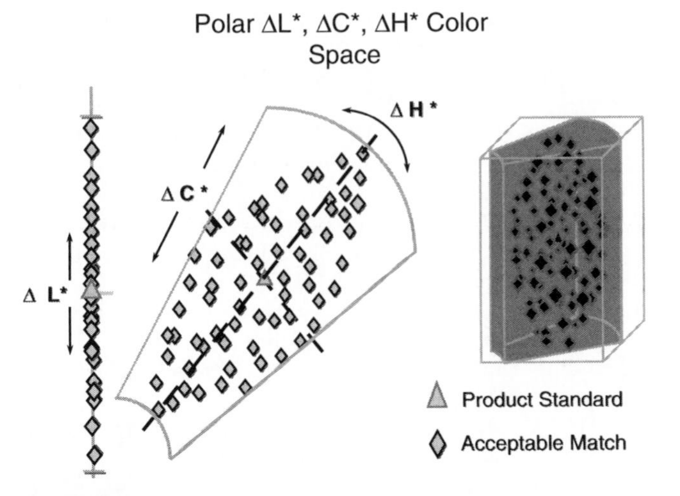

32-17 figure Diagram showing acceptable ΔL^*, ΔC^*, ΔH^* tolerance limits for a product. (Courtesy of HunterLab, Reston VA.)

control the thickness of the sample so that it is effectively opaque. Solid foods vary with respect to size, geometry, and uniformity. With some colorimeters, reflectance measurements can be taken directly on the sample. Ideally, the surface should be flat. Readings of an apple or orange may be distorted because of the "pillowing" effect, which is a result of the distorted reflectance values from the uneven surface. Pureeing nonuniform materials such as strawberries will give a uniform sample; however, the incorporation of air renders a color extremely different from the sample of interest. For opaque foods, instruments with 45*/0° and 0°/45° geometries are recommended as the measurements correlate better with visual assessment than those obtained with sphere instruments. Instruments with a large area of view, e.g., 25–50 mm are helpful for area-averaging nonuniform color. For powders, two readings with replacement of the powder between readings may be sufficient, but for flakes, chunks, and large particulates, a large field of view (40 mm or larger) with three to six readings and sample replacement between readings is recommended.

Different commodities present their own peculiarities when it comes to measuring color and appearance. The proceedings of a recent American Chemical Society symposium are recommended; different authors discuss methodology for color measurement of meat, fish, wine, beer, and several fruits and vegetables (16).

32.5 SUMMARY

Color is three dimensional, and any color-ordering or color-measuring system needs to address that fact. The Munsell system is a visual system that designates color in terms of hue, value, and chroma. Each of these dimensions has equivalent visual spacing, which is advantageous. The physiology of color vision has been long understood, and it provided the necessary background information for development of the CIE tristimulus system. Standardization of illuminants and experiments using humans with normal color version was necessary to develop color-matching functions that corresponded to the color sensitivity of the human eye. The system permits calculation of numerical XYZ tristimulus values that can accurately represent a color and are useful in color matching. The system does not have equivalent visual spacing, which is a disadvantage when measuring how a sample differs from a standard or changes during processing and storage. Color-order systems have been developed that are more suitable for measuring color differences. These include the Hunter *Lab* system, the *CIEL*a*b** system, and the $L^*C^*H^*$ system. The latter two systems are recommended by the CIE, the International Association with responsibility for standardization and measurement of light. They have been widely adopted by the food industry for color measurement. A wide range of colorimeters is available for industrial and research application. Today's instruments are rugged, easy to standardize, and user friendly. They vary with respect to presentation of sample, size of viewing area, portability, and the ability to measure by transmittance or reflectance. Many food samples are less than ideal for color measurements because they may be partial transmitting and partial reflecting, nonuniform, and of varying size and shape. A number of factors need to be considered with respect to sample preparation and presentation to get measurements that are repeatable and that correspond to visual appearance.

There are a number of excellent illustrative tutorials dealing with color measurement that are available on various Web sites that have been developed by organizations and commercial companies. The following are recommended: HunterLab (17), Konica-Minolta (18), Color Eng. Inc. (19), CIE (20), Munsell (21), Color Models Technical Guides (22), A Review of RGB Color Spaces (23), and Beer Color Laboratories (24).

32.6 STUDY QUESTIONS

1. Dominant wavelength (λ_d), % purity and luminosity (Y) in the CIE *XYZ* system correspond to what indices in the Munsell system? In the CIE $L^*C^*H^*$ system?
2. Using a calculator, determine hue angle and chroma for the following sets of a^*, b^* data: $a^* = +12$ and $b^* = +8$; $a^* = +12$ and $b^* = +4$; $a^* = 12$ and $b^* = -4$.
3. If one wants to use a colorimeter to measure of the amount of browning in maple syrup, what indices would you expect to correspond well with visual assessment?

4. How variable is human color perception when compared with that of taste and smell? What are the human capabilities for color perception and color memory?
5. Why is CIE Tristimulus Y used as a measure of luminosity?
6. Give examples where it is appropriate to use a colorimeter with diffuse sphere geometry, and conversely, a colorimeter with $0°/45°$ reflectance geometry.
7. How can you determine how many readings should be taken for a given sample?

32.7 ACKNOWLEDGMENTS

The authors of this chapter wish to acknowledge Dr. Jack Francis, a legend in the area of color analysis and the person who wrote the chapter on this topic in the previous two editions of this book. Ideas for the content or organization, along with some of the text, came from his chapter. Dr. Francis offered the use of his chapter contents.

32.8 REFERENCES

1. Amerine MA, Roessler EB (1976) Wines: their sensory evaluation. W. H. Freeman & Co., San Francisco, CA
2. Berns RS (2000) Billmeyer and Saltzman's principles of color technology, 3rd edn. John Wiley and Sons, Inc., NY
3. Loughrey K (2005) Overview of color analysis. Unit F5.1. In: Wrolstad RE, Acree TE, Decker EA, Penner MH, Reid DS, Schwartz SJ, Shoemaker CF, Smith D, Sporns P. (eds) Handbook of food analytical chemistry – pigments, colorants, flavors, texture, and bioactive food components. Wiley, NY
4. Loughry K (2000) The measurement of color, Chap 13. In: Francis FJ, Lauro GJ (eds) Natural food colorants. Marcel Dekker, NY
5. Hutchings JB (1999) Food color and appearance, 2nd edn. Aspen Publishers, Gaithersburg, MD
6. Hunter RS, Harold RW (1987) The measurement of appearance, 2nd edn. Wiley, NY
7. Wright WD (1971) The measurement of color. Van Nostrand Reinhold, NY
8. Francis FJ (1999) Colorants – Eagen Press Handbook Series. Eagan Press, St. Paul, MN
9. Bartleson CJ (1960) Memory colors of familiar objects. J Opt Soc Am 50:73–77
10. Campbell NA, Reece JB, Mitchell LG (1999) Biology, 5th edn. Benjamin/Cummings, Addison Wesley Longman, Menlo Park, CA
11. Krinsky NI, Landrum JT, Bone RA (2003) Biologic mechanisms of the protective role of lutein and zeaxanthin in the eye. Ann Rev Nutr 23:171–201
12. Coren S (2004) How dogs think: understanding the canine mind. Free Press Publishing Co., Tampa, FL
13. X-Rite Home Page (2009) http://gretagmacbethstore.com. Accessed 7 Jan 2009
14. The xy Chromaticity Diagram (2009) http://www.digitalcolour.org/understanding/Chromaticity.htm. Accessed 1 Mar 2009
15. Leggett GJ (2008) Color measurement techniques for food products, chap 2. In: Culver CA, Wrolstad RE (eds) Color quality of fresh and processed foods. ACS Symposium Series No. 983, American Chemical Society, Washington DC
16. Culver CA, Wrolstad RE (2008) Color quality of fresh and processed foods. ACS Symposium Series No. 983, American Chemical Society, Washington, DC
17. Hunter Lab (2009) http://www.hunterlab.com/ColorEducation/ColorTheory. Accessed 21 Jan 2009
18. Konica Minolta (2009) http://www.konicaminolta.us. Color Measurement Tutorial, Precise Color Communication. Accessed 21 Jan 2009
19. Color Eng Inc (2009) http://www.colorpro.com/info/tools/labcalc.htm. Accessed 21 Jan 2009
20. CIE International Commission on Illumination (2009) http://www.cie.co.at/index_ie.html. Accessed 21 Jan 2009
21. Munsell (2009) http://www.xrite.com/custom_page.aspx?PageID=46. Accessed 28 Feb 2009
22. Color Models Technical Guides (2009) http://dba.med.sc.edu/price/irf/Adobe_tg/models/main.html. Accessed 27 Feb 2009
23. A Review of RGB Color Spaces (2009) http://www.babelcolor.com/download/A%20review%20of%20RGB%20color%20spaces.pdf. Accessed 2 Mar 2009
24. Beer Color Laboratories (2009) http://www.beercolor.com/glossary_of_selected_light_and_c.htm. Accessed 27 Feb 2009

Index

A

AACC International. *See* Standard methods
Abbe refractometer, 99
Absolute error, 58
Absorption of radiation, 382, 383, 389, 390, 392, 400, 418, 427
Absorption spectrum, 382, 383, 392, 397, 403, 404, 424, 425, 505
Accelerated solvent extraction (ASE), 323, 330, 475
Accuracy, analyses, 55
Acid value, oils, 248
Acid-base equilibria, 223, 224
Acid-base titrations, 228
Acidity. *See also* pH; Titratable acidity
- standard acid, 231
- standard alkali, 230, 231
- titration
  - endpoint, 229
  - equivalence point, 229
  - indicators, 229, 230

Acids, organic
- in foods, 222, 227–229, 232, 233
- ion-chromatographic analysis, 509, 510
- malic, enzymatic assay, 295, 296

Acrylamide, 268, 270, 343, 467
Activation energy (E_a), 290
Active oxygen method (AOM), 254
Adsorption chromatography, 264, 265, 481–484, 488, 530
Adulteration, 19, 72, 76, 353, 358, 361, 363, 453, 454
Affinity chromatography
- applications, 265, 489
- elution methods, 489
- ligand, 488
- principles, 488, 489, 511
- spacers, 489

Aflatoxins. *See* Mycotoxin residues
Agricultural biotechnology, 335, 336
Alcoholometers, 97
Alkaline phosphatase, assay, 296, 297
Allergens, 339–341
- DNA methods, 341
- protein methods, 340, 341

Amici prism, 98
Amino acid
- analysis, 271, 272, 274, 510
- classification, 135, 272
- scoring patterns, 273 (Found in table only)

Ammonium sulfate fractionation of proteins, 263
Amylase, used in assay, 160–162, 295, 297
Amylopectin, 162, 166, 510, 511
Amylose, 160, 162, 166, 418, 511
Analytical microbiology. *See* Microbiological assays
Anionic dye-binding. *See* Dye-binding methods
Anisidine value, 250–252, 254
Anthocyanins, 227, 509
Antibiotic residue, assay, 332–335
Antibodies, 265, 269, 303–305, 308–311, 313, 314, 326, 331, 336, 337, 340, 361, 489
Antigens, 303–306, 308, 311, 313, 361
Antinutritional factors, 272
Antioxidants, 80, 184, 188, 189, 243, 253, 254, 258, 275, 469, 509, 515, 526
AOAC International. *See* Standard methods
Archimedes principle, 97
Arrhenius
- equation, 290
- plot, 290

Ascarite trap, 230
Ash. *See also* Minerals
- alkalinity, 112
- comparison of methods, 112, 113
- contamination during assay, 108, 112
- content of foods, 107, 108
- definitions, 107
- importance of analysis, 107
- insoluble, 109, 112
- methods of determination
  - dry ashing
    - crucibles, 108
    - for elemental analyses, 436
    - furnaces, 107, 108
    - losses during, 109
    - modified procedures, 110
    - preparation of sample, 108
    - principles, 108
    - procedures, 109
    - temperature, 108
  - microwave ashing
    - dry ashing, 110, 112
    - wet ashing
      - closed-vessel system, 111
      - open-vessel system, 111, 112
  - wet ashing
    - acids used, 110
    - for elemental analysis, 107, 109, 113, 436
    - precautions, 110
    - principles, 109, 110
    - procedures, 110
- sample preparation, 108
- soluble, 112
- sulfated, 112

Aspartame, 509
Assay methods, general
- selection, 7–10
- standard methods (*see* Standard methods)
- steps in analysis, 6, 7
- validity, 9, 10

Assessing the nutritional value of proteins. *See* Protein, nutritional quality
Atomic absorption spectroscopy
- applications, 436, 437
- atomization, 424
- calibration
  - sensitivity check, 429
  - standards, 429

S.S. Nielsen, *Food Analysis*, Food Science Texts Series, DOI 10.1007/978-1-4419-1478-1,

comparison to inductively coupled plasma-atomic emission spectroscopy, 436
energy transitions in atoms, 423, 424
flame, 425
graphite furnace, 426
instrument components
atomizer
burner, 427
cold vapor technique, 428
electrothermal, 428
flame, 427
graphite furnace, 426
hydride generation, 428
nebulizer, 427
chopper, 427
detector, 428
hollow cathode lamp (HCL), 426, 427
monochromator, 428
interferences
nonspectral
ionization, 430
solute volatilization, 430
transport, 430
spectral
absorption of source radiation, 429
background absorption of source radiation, 430
labware, 437
operation, 428, 429
principles, 425, 426
procedure, 428, 429
reagents, 437
standards, 437
Atomic emission spectroscopy
applications, 436, 437
flame
instrument components, 431
principles, 431
inductively coupled plasma
comparison to atomic absorption spectroscopy, 436
instrument components
argon plasma torch, 432, 433
detectors
charge coupled device, 431, 434
charge injection device, 431, 434
photomultiplier tube, 431, 434
echelle optical system, 431, 434
monochromator, 431
optical system, 434, 435
polychromator, 431
interferences, 435, 436
labware, 437
principles, 431
procedures, 435
reagents, 437
standards, 437
Atomization, 424–426, 428, 430–433, 437
Automation, 131, 137, 212, 453, 485
Autosamplers, 7, 503

B

Babcock method, fat analysis, 127, 130
Bacteria, in microbiological assays
drug residues, 333, 334
vitamins, 185–187, 194
Baumé hydrometer, 97, 98
Beer's law, 390–394, 402, 412, 425, 429
Benzene, 257, 343, 344, 382, 399
Beta-glucan, 10, 149, 159, 162, 166, 169, 450–452
Bicinchoninic acid method, proteins, 141–144, 495
Bioassays
protein quality, 272–274
vitamins, 184–186
Biochemical oxygen demand (BOD), 24, 369–371
Biosensors, 297, 298, 331, 332, 340
Bisphenol A (BPA), 343
Biuret method, proteins, 139, 143
Bradford method, proteins, 141, 143
Brix, 23, 87, 97–99, 172, 233
Brix/acid ratio, 233
Bronsted-Lowry theory, 223
Buffering, 221, 228–231, 289, 291

C

Caffeine, 61–64, 467, 468, 506, 509
Calculations
ash, 109
chromatographic parameters, 491–494
fat, 122–124, 127
fiber, 170, 172
Mohr titration, 207
moisture, 91, 96
neutralization reaction, 221–223
protein, 137
titratable acidity, 232
vitamins, 187–194
Vohlard titration, 208
Calibration curves. *See* Data evaluation, standard curves
Calorie content, 40–42
Calorimeters, 563
Calorimetry, 41, 42, 563. *See also* Thermal analysis
Capillary electrophoresis, 160, 270, 271, 278, 332
Capillary isoelectric focusing, 271
Carbazole assay, 165
Carbohydrates. *See also* individual carbohydrates by name
beta-glucan, 10, 149, 159, 166, 450–452
calculation by difference, 151
Carrez treatment, 159
cellulose, 166, 167
content in foods, 151
definitions, 149, 151
extraction, 152
gums, 162–167
hemicellulose, 163, 166–168
hydrocolloids, 162–167
methods of analysis
chemical, for monosaccharides and oligosaccharides
dinitrosalicylic acid, 154
enzymatic, 159
Lane-Eynon, 154
Munson–Walker, 154
phenol-sulfuric acid, 152, 153
reducing sugars, 153, 154
Somogyi-Nelson, 153, 154
chromatographic
gas chromatography
derivatization, 157, 158
hydrolyzates of polysaccharides, 158
neutral sugars, 157
high-performance liquid chromatography
detection, 156, 157
stationary phases, 155, 156

dietary fiber, 168–171
extraction, 152
fiber, 168–171
glucose, 159, 160, 295
gums, 163–165
hydrocolloids, 163–167
microscopy, 171
near-infrared spectroscopy, 171
pectin
degree of esterification, 165
determination, 165
physical
microscopy, 171
refractive index, 172
specific gravity
hydrometer, 172
pycnometer, 97
spectrometry, 171
resistant starch, 167
sample preparation, 151, 152, 159, 168
starch
degree of gelatization, 162
degree of retrogradation, 162
determination, 160, 161
resistant, 167
total carbohydrate, 152, 153
monosaccharides, 152–159, 163, 172, 173
occurrence, 149, 150
oligosaccharides, 149, 152–160, 162, 167–169, 172
pectin
component of dietary fiber, 167
degree of esterification, 165
determination, 165
nature, 165, 167
polysaccharides, 160–165
reducing sugars, 153, 154
starch
degree of gelatinization, 162
degree of retrogradation, 162
determination, 160–162
gelatinization, 162
in foods, 160
methods of analysis, 160–163, 295
problematic in fiber determination, 168
resistant, 160–162, 166–169, 171
retrogradation, 162
sugar alcohols, 150–152, 155
sugars, 151–160, 162–165, 167, 171, 172
Carotenoids, assay, 181, 481, 508, 509
Carrez reagent, 159
Cellulose, 166, 167, 323, 370, 477, 478, 483, 484, 489, 511
Charge coupled device (CCD), 359, 431, 434
Charge injection device (CID), 431, 434
CHARM II® test, 334
Check sample, 9, 10
Chemical constituents of concern. *See* Contaminants
Chemical ionization (CI), 463
Chemical oxygen demand (COD), 370, 371
Chlorophylls, 481, 509
Choices of method, general
characteristics, 7
food matrix, 8, 9
objective, 7, 8
validity, 9, 10
Cholesterol, 5, 10, 38, 39, 46, 48, 149, 167, 241, 243, 257, 258, 515
Chromatography, 475–497, 501–512, 515–535. *See also* individual types
Chromatography, applications
amino acids, 271, 272, 478, 509
antioxidants, 509, 526
aspartame, 509
carbohydrates, 155–160, 479, 483, 485, 497, 505, 508, 509, 516, 526
cholesterol, 257, 258, 515
drug residues, 319–321, 333, 485, 515
flavors, 515, 519, 526–528
lipids, 125–127, 255–257, 481, 483, 509, 515, 516
mycotoxin residues, 320, 321, 327–332, 478
organic acids, 483, 485, 509
pesticide residues, 319–327, 329, 341, 481, 524, 526, 528
pigments, 478, 483, 508, 509, 519
polyphenols, 483, 508, 509
protein separation, 264–267, 484, 487–489, 511
sulfite, 341, 342, 510
vitamins, 182–195, 478, 479, 481, 508, 515
Chromatography, principles
adsorption, 481
affinity, 488, 489
developing a separation, 490, 491
elution
gradient, 490, 491
isocratic, 490
extraction, 475, 517–519
gradient elution, 490, 491
historical perspective, 475, 476
ion exchange
anion exchangers, 484
cation exchangers, 484
liquid-liquid, 482, 483
liquid-solid, 481
paper, 477, 478
partition
bonded supports, 483
coated supports, 483
normal-phase, 482
reversed-phase, 482
qualitative analysis, 495
quantitative analysis
external standards, 495
internal standards, 495, 496
peak area, 495
peak height, 491, 495, 496
resolution
capacity factor, 494, 495
efficiency
height equivalent to a theoretical plate, 493
theoretical plates, 492, 493, 530
Van Deemter equation, 493, 530
selectivity, 494
separation, 481–489
size exclusion, 485–488
supercritical fluid, 480, 481
terminology, 476
thin layer
aflatoxins, 331, 479
amino acids, 479
carbohydrates, 479
lipids, 258, 479
mycotoxin residues, 331, 479
pesticide residues, 326
principles, 478
vitamins, 479

Code of Federal Regulations (CFR), 17, 18, 21–25, 27, 37–43, 46, 47, 49, 60, 74, 241, 243, 273, 333, 341, 353
Codex Alimentarius Commission, 29, 30
Coefficient of determination, 62
Coefficient of variation (CV), 57
Collision reaction cells (CRC), 438, 439
Color
  colorimeters, 246, 575, 580–582
  equations, 583
  instrument choice, 583
  interaction of light with sample, 582
  measurement, 582–585
  physiological basis, 575, 576
  sample preparation, 583–585
  sample presentation, 583–585
  space, 580–582
  systems, to describe color
    instrumental
      CIE system, 579, 580
      standard observer, 578, 579
    visual
      Munsell System, 577
  tolerances, 583
  tristimulus colorimeters, 580–582
Colorimeter, 246, 580–583
Colorimetric methods
  carbohydrates, 152–154
  minerals, 208, 209
  proteins, 139–142
Column chromatography, 322, 476, 479, 481, 503, 531
Columns
  analytical HPLC
    capillary, 503
    microbore, 503
    microcolumn, 503
    ultra-HPLC, 503
  capillary
    megabore, 523
    microbore, 523
    normal, 523
  efficiency, 492, 493
  guard, 503
  packed, 522, 523
  packing material, 503, 520
  porous-layer open-tubular (PLOT), 524
  precolumn, 503
Concentration units, 221–223, 397
Conductivity
  detector for GC, 138
  detector for HPLC, 509
Confidence interval, 57, 58, 63, 76
Conjugated acid, 228
Conjugated base, 228
Constituents of concern. *See* Contaminants
Contaminants
  3-monochloropropane 1,2-diol, 344
  4-methylbenzophenone, 343
  acrylamide, 268, 270, 343
  allergens, 339–341
  benzene, 343, 344
  bisphenol A, 343
  choice of methods, 320, 321, 341
  drug residues, 332–335
  furans, 343
  heavy metals, 319
  melamine, 342
  mycotoxins, 327–331
  perchlorate, 344
  pesticide residues, 324–327
  sample preparation
    derivatization, 323
    extraction, 322
  sulfites, 294, 295, 341, 342
Correlation coefficient, 61–63

D

Daily Value (DV), 38–42, 49, 50, 181, 182, 273, 274
Data evaluation
  errors, 59, 62, 63
  measures of central tendency, 55
  reliability of analysis
    absolute error, 58
    accuracy, 55–59
    coefficient of variation, 57
    confidence interval, 57
    limit of detection, 59, 60
    limit of quantitation, 60
    precision, 55–59
    Q-test, 65
    Q-value, 65
    range, 56
    relative error, 58
    sensitivity, 59
    sources of error, 59
    specificity, 59
    standard deviation, 56
    standard error of the mean, 58
    t-score, 77
    t-test, 77
    t-value, 7, 58
    Z value, 57
  reporting results
    rejecting data, 65
    significant figures, 64
  standard curves
    applications, 208, 209, 393, 394, 401, 402, 429
    coefficient of determination, 62
    confidence intervals, 63
    correlation coefficient, 61, 62
    errors, 62, 63
    linear regression, 60, 61
    plotting curve, 61–63
Databases
  mass spectroscopy, 462
  nutrient content, 43
Defect action level (DAL), 353
Denaturation of proteins, 264, 311, 565
Density
  fat determination, 130, 131
  hydrometers, 97, 172
  pycnometers, 97, 98
Derivatization
  carbohydrate analysis, 155, 160
  contaminants, 321, 323
  GC analysis, 519
  HPLC analysis, 157
  lipid analysis, 125, 126, 255
  pesticide residues, 323
Detection limit, 60, 340, 423, 433, 435, 436, 439, 460, 467, 505, 519, 527, 528
Detectors
  atomic absorption spectroscopy, 426, 428
  atomic emission spectroscopy, 431, 434, 435

gas chromatography, 524–528
high-performance liquid chromatography, 505, 506
infrared-spectroscopy, 410, 415
UV-visible spectroscopy, 396, 397
Dextinizing activity, 297
Dextrins, 295
Dialysis, 144, 164, 265–267, 488
Dielectric method for moisture analysis, 96
Dietary fiber. *See* Fiber, dietary
Dietary supplements, 17, 40, 43, 48
Differential scanning calorimetry. *See* Thermal analysis
Digestibility, proteins, 42, 144, 272–275. *See also* Protein, nutritional quality
Dilatometry, 248
Distillation methods. *See* Moisture, distillation methods
Dough formation. *See* Protein, functional properties
Drug residues
confirmatory methods, 335
determinative methods, 335
regulations, 332, 333
screening methods, 333–335
Dry ashing, 107–110, 112, 113. *See also* Ash, methods of determination
Drying methods. *See* Moisture
Dye-binding methods, proteins
anionic dye-binding, 140, 141
Bradford, 141
Dynamic mechanical thermal analysis, 558. *See also* Thermal analysis

E
Echelle optical system, 431, 434
EDTA complexometric titration, mineral determination, 205, 206
Effluent composition, 24, 25
Electrical methods
extraneous matter, 360
moisture determination, 96
Electrode potential, 211, 225
Electrodes
combination, 227
enzyme, 297
glass, 226
indicator, 226, 227
ion selective, 209–212
oxygen-sensitive membrane, 369
reference, 210, 226
saturated calomel, 226
silver-silver chloride, 226
Electromagnetic radiation, 172, 323, 377, 389, 400, 459, 478
Electron impact ionization, 327, 460, 462, 464
Electrophoresis
applications, 267, 311
capillary, 160, 270, 271
isoelectric focusing, 269, 270
mobility calculations, 267
molecular weight estimation, 269
native, 268
polyacrylamide gel, 267–269
sodium dodecyl sulfate (SDS), 268
two-dimensional, 270
Elemental analysis, 107, 113, 203–212, 423–439. *See also* Minerals, determination
ELISA. *See* Enzyme-linked immunosorbent assay
Emission of radiation, 377, 383, 384, 423, 437
Emission spectrum, 424, 425
Emulsification. *See* Protein, functional properties
Emulsions. *See* Protein, functional properties
Endpoint
colorimetric, 229
equivalence point, 229
phenolpthalein, 229
Energy level transitions. *See* Spectroscopy
Energy states of matter. *See* Spectroscopy
Enthalpy, 560, 564
Environmental Protection Agency (EPA), 17, 23–25, 319, 324
Enzymatic methods
as analytical aids
fiber analysis, 168–171, 285
protein hydrolysis, 272, 274, 285
vitamin assays, 184, 193, 285
biosensors, 297, 298
electrodes, 298
enzyme activity assays
alkaline phosphatase, 296, 297
alpha-amylase, 297
lipoxygenase, 296
peroxidase, 296
rennet, 297
immobilized enzymes, 297, 298
substrate determination
carbohydrate analysis, 159–162, 295
endpoint method, 294
glucose, 295
malic acid, 295
monosaccharides, 149, 159, 295
oligosaccharides, 159
polysaccharides, 159–162, 295
sample preparation, 159, 294
starch, 160–162, 295
sulfite, 294, 295
total change, 294
Enzymic methods. *See* Enzymatic methods
Enzyme
activity
affected by
activators, 291, 292
enzyme concentration, 288, 289
inhibitors
irreversible, 292
reversible, 292
pH, 290–291
substrate concentration, 289
temperature, 289, 290
assays (*see* Enzymatic methods)
coupled reactions, 293
methods of measurement, 292, 293
order of reactions, 287
reaction rate, 286–292
electrodes, 298
immunoassays, 306–311
kinetics, 285–288
Enzyme immunoassays. *See* Immunoassays
Enzyme-linked immunosorbent assay (ELISA), 306–311, 321, 325, 326, 331, 332, 334–341
Equivalent weight, 221, 222, 232, 236
Error, data, 59
Ethylene oxide, 533, 534
External standards, 495, 521

Extraction
accelerated solvent, 323
batch, 475
continuous, 122, 475
countercurrent, 475
lipids, 120–125
microwave-assisted solvent, 323
monosaccharides, 152
nonsolvent, 127, 130
oligosaccharides, 152
partition coefficient, 475
polysaccharides, 152
pressurized liquid, 323
QuEChERS, 323
solid phase
dynamic extraction, 518, 519
microextraction, 252, 322, 323, 517, 518
solvent, 120–125, 517
stir bar sorptive extraction, 518, 519
Extractor types, 122–124, 517, 518
Extraneous matter
definition, 353, 354
diagnostic characteristics, 354
filth, 354
isolation principles, 354–358, 363–364
methods of analysis
comparison of methods, 361, 362
electrical conductance, 360
flotation, 356–358
immunoassays, 361
impact-acoustic emission, 360
microscopy, 360, 361
near-infrared spectroscopy, 361
objectivity/subjectivity, 358
x-ray microtomogtraphy, 359
x-ray radiography, 359
regulations
defect action levels (DALs), 353
Federal Food, Drug, and Cosmetic Act, 353
good manufacturing practices, 353

F
Falling number method, 297
Farinograph®, 278
Fat substitutes, 5, 254, 255, 481
Fats. *See* Lipids
Fatty acids, 12, 48, 59, 89, 119, 125, 127, 186, 232, 241, 243, 244, 247, 248, 250, 254–258, 296, 453, 464, 467, 481, 515, 523
Fiber
dietary
components, 167, 168
definition, 166, 167
importance in diet, 166
insoluble fiber, 166
methods of analysis, 168–171
principles, 169–171
procedures, 170, 171
resistant starch, 167
sample preparation, 168
soluble fiber, 166
total, 170
Filth. *See* Extraneous matter
Fixed acidity, 233
Flavor analysis, 515, 519, 526–528
Fluorescence microscopy, 250
Fluorescence spectroscopy
calibration curve, 401, 402
detector, 400, 401
emission beam, 400
excitation beam, 400
quantum efficiency, 401
vitamin analysis, 191–194
Foaming. *See* Protein, functional properties
Folate, 184–187, 195, 511
Folin-Ciocalteau phenol reagent, 139
Food additives, 12, 17, 19, 23, 29, 30, 319, 339, 515
Food analysis, general
necessity, 5–7
standard methods (*see* Standard methods)
steps in analysis, 6, 7
types of samples, 6
Food and Drug Administration (FDA), 17, 37, 38, 41, 43–48, 71, 319
Food Chemicals Codex (FCC), 12, 30
Food composition. *See* Specifications
Food dyes, assay, 509
Food Safety Inspection Service (FSIS), 21
Food, Drug, and Cosmetic Act, 17, 37, 353
Foreign matter. *See* Extraneous matter
Foss-Let fat determination method, 130, 131
Fourier transform
gas chromatography, 528
infrared, 241, 256, 257, 332, 410–412, 415
ion cyclotrons, 460
mass spectrometry, 460, 528
Free fatty acids (FFAs), 59, 241, 244, 248, 523
Freezing point, 100, 101, 558
Functional properties. *See* Protein, functional properties
Furans, 343

G
Gas chromatography
applications
3-monochloropropane 1,2-diol, 344
aroma analysis, 534, 535
benzene, 343, 344
carbohydrates, 157, 158, 519
cholesterol, 257, 258, 515
drug residues, 320, 515
flavors, 515, 519, 526–528
furans, 343
lipids, 125–127, 255–258, 516
melamine, 342
mycotoxin residues, 328, 331
packaging materials, 343, 533
pesticides residues, 325–329, 526
spices, 533, 534
stereoisomer separation, 533
columns
capillary, 523, 524
megabore, 523
microbore, 523
packed, 522–523
solid support, 522
stationary phases, 522
detectors
applications, 525–528
electrolytic conductivity, 528
electron capture, 526
flame ionization, 525, 526
flame photometric, 526, 527

photoionization, 527, 528
pulsed flame photometric, 526, 527
thermal conductivity, 524, 525
thermionic, 528
gas supply, 520
headspace methods
direct headspace sampling, 516
dynamic headspace sampling, 516
hyphenated techniques
atomic emission detector (AED), 528
Fourier transform infrared (FTIR), 528
mass spectrometry (MS), 459, 460, 464, 465, 528
injection port
hardware, 520, 521
sample injection, 521, 522
isolation of solutes, methods
direct injection, 519
distillation, 516, 517
headspace
adsorbent trap, 516
concentration, 516
cryogenic trap, 516
direct sampling, 516
solid-phase extraction, 517–519
solvent extraction, 517
multidimensional GC
comprehensive two-dimensional GC, 529, 530
conventional two-dimensional GC, 529
oven, 522
principles, 476, 477
sample derivatization, 519, 520
sample preparation, 515–519
separation efficiency
carrier gas
flow rate, 530, 531
type, 531, 532
column parameters, 530, 531
temperature programming, 521–523
Gel filtration. *See* Size-exclusion chromatography
Gel permeation. *See* Size-exclusion chromatography
Gelatinization, starch, 162
Gelation, protein, 275, 277
General Agreement on Tariffs and Trade (GATT), 29
Genetic engineering. *See* Agricultural biotechnology
Genetically modified organisms (GMOs)
DNA methods, 337, 338
labeling, 336
method comparison, 338
polymerase chain reaction (PCR) (DNA method), 337, 338
protein methods
enzyme-linked immunosorbent assays, 336, 337
lateral flow strips, 337
Western blots, 336
uses, 335
Gerber method, fat analysis, 130
Glass electrodes, 226
Glass transition temperature, 454, 557, 558, 564, 569, 570
Glucose, assay, 159, 160, 295
Goldfish method, fat analysis, 122
Good manufacturing practices (GMP), 6, 18, 43, 353
Government
acts
Agricultural Marketing Act, 27
Dietary Supplement Health and Education Act, 17
Egg Products Inspection Act, 22
Fair Packaging and Labeling Act, 26
Federal Alcohol Administration Act, 22
Federal Insecticide, Fungicide, and Rodenticide Act, 18, 23
Federal Trade Commission Act, 25
Federal Water Pollution and Control Act, 24
Food and Drug Administration Modernization Act, 37
Food Quality Protection Act, 18, 23
Food, Drug, and Cosmetic Act, 17, 37, 353
Nutrition Labeling and Education Act, 17, 37, 38
Poultry Products Inspection Act, 22
Safe Drinking Water Act, 24
U.S. Grain Standards Act, 22
agencies, bureaus, departments
Bureau of Alcohol, Tobacco, Firearms and Explosives, 22
Bureau of Consumer Protection, 25
Department of Health and Human Services, 17, 23–25, 319, 324
Environmental Protection Agency (EPA), 17
Federal Trade Commission (FTC), 25
Food and Drug Administration (FDA), 17, 37, 38, 41, 43–48, 71, 319
National Bureau of Standards (*see* National Institute of Standards and Technology)
National Conference on Weights and Measures (NCWM), 28
National Institute of Standards and Technology (NIST), 28
National Marine Fisheries Service (NMFS), 22
National Oceanic and Atmospheric Administration (NOAA), 22
U.S. Customs Service, 25
U.S. Department of Agriculture (USDA), 21, 22, 26, 27, 37, 40, 42, 43
U.S. Department of Commerce, 22
U.S. Department of Justice, 22
amendments
Color Additives Amendment, 17
Delaney Clause, 17
Food Additives Amendment, 17
inspection programs and services
Dairy Quality Program, 27
Food Safety and Inspection Service (FSIS), 21
Grain Inspection, Packers and Stockyard Administration (GIPSA), 22
Interstate Milk Shippers Program (IMS), 27
National Marine Fisheries Service (NMFS), 22
National Shellfish Sanitation Program (NSSP), 27
regulations, 5, 6, 17 (*see also* Government, acts; Government, amendments)
advertising, 26
alcoholic beverages, 22, 23
Code of Federal Regulations (CFR), 17, 37
drinking water, 24
effluent composition, 24, 25
extraneous matter, 353
fishery products, 22, 27, 28
Good manufacturing practice (GMP) regulations, 6, 18, 43, 353
Harmonized Tariff Schedules of the U.S., 25
imported goods, 25
labeling
alcoholic beverages, 22

dairy products, 26, 27
ingredient, 49
milk, 26, 27
nutrition
caloric content, 40–42
compliance, 42, 43
Daily Value (DV), 39, 40
databases, 37
designation of ingredients, 49
exemptions, 40
format, 38–40
health claims, 43, 46, 48, 49
methods of analysis, 42
national uniformity and preemptions, 49
nutrient content claims, 43–48
Nutrition Labeling and Education Act (NLEA), 17, 37, 49
protein quality, 42
rounding rules, 40
sample collection, 42
serving size, 39, 40
structure/function claims, 48
total carbohydrate, 151
meat and poultry, 21, 22, 37
milk
Grade A, 26
Grade B, 27
manufacturing grade, 27
Pasteurized Milk Ordinance (PMO), 26
pesticide residues, 23, 24, 28 (*see also* Pesticide residues)
shellfish, 27–28
standards
fill, 19
grades, 21
identity, 18
quality, 19
Gratings, monochromator, 396
Gravimetric analysis
fiber determination, 168
Grinding
applications, 78
equipment, 78
particle size determination, 78, 79
samples for fat analysis, 121
samples for moisture analysis, 87, 88
Gums, 162–164, 166, 167

H
Handbooks
National Institute of Standards and Technology (NIST) handbook, 9, 28
Pesticide Analytical Manual (PAM), 324, 325
Hazard Analysis Critical Control Point (HACCP), 6, 18, 19, 22, 29
Headspace methods
direct headspace sampling, 516
dynamic headspace sampling, 516
Health claims, 5, 18, 38, 43, 46, 48, 49
Heat capacity, 558, 560, 564–568, 570
Heavy metals, 319, 423
Hemicellulose, 163, 166–168
Henderson-Hasselbach equation, 228, 229
Hexanal, 250, 252, 516, 534
High-performance liquid chromatography (HPLC)
applications
4-methylbenzophenone, 343
acrylamide, 343, 467
amino acids, 272, 507, 510
antioxidants, 509
aspartame, 509
bisphenol A, 343
caffeine, 468, 506, 509
carbohydrates, 155–157, 507–511
drug residues, 335, 507
dyes, 511
flavor compounds, 509
folates, 511
inorganic ions, 506, 509, 510
lectins, 511
lipids, 501, 506, 509
melamine, 342, 467–469, 507
molecular weight estimation, 510
mycotoxin residues, 332, 507
organic acids, 509, 510
perchlorate, 344
pesticide residues, 326, 327, 507
phenolic compounds, 506
pigments, 508, 509
proteins, 265, 510
sugars, 501, 507, 510, 511
sulfites, 341, 342, 507
surfactants, 508
vitamins, 183, 188–190, 505, 507, 508
column hardware
analytical, 503
guard column, 503
precolumns, 503
column packing materials
carbohydrate analysis, 510
polymeric
macroporous, 504
microporous, 504
silica-based
bonded phases, 504
pellicular packing, 504
porous silica, 504
coupled techniques, 195, 465–467, 503, 506
data stations, 507
derivatization
postcolumn, 157, 271, 326, 327, 332, 506
precolumn, 157, 272, 506
detectors
amperometric, 155, 505, 506, 510
chemiluminescent nitrogen (CLND), 506
conductivity, 509, 510
diode array, 505
electrochemical, 157, 505, 506
fluorescence, 505
light scattering, 506
mass, 505
pulsed amperometric (PAD), 157, 510
radioactive, 506
refractive index (RI), 156, 467, 505, 511
ultraviolet-visible (UV-Vis) absorption
diode-array, 505
fixed wavelength, 505
variable wavelength, 505
gradient elution, 501, 502
hyphenated techniques mass spectrometry, 195, 459, 465–467, 503, 506
injector
autosamplers, 503
fixed-volume loop, 502
valve injectors, 502

postcolumn derivatization, 157, 271, 326, 327, 332, 506
precolumn derivatization, 157, 272, 506
pump, 501, 502
separation modes
affinity, 511
ion exchange
ion chromatography, 509, 510
normal phase, 507, 508
reversed phase, 508, 509
size exclusion, 510, 511
software packages, 507
with mass spectrometry, 195, 321, 465–467, 503, 506
Hollow cathode lamp, 426, 427
Hydrocolloids, 162–167
Hydrogenation, oils, 247
Hydrolysis
methods of starch determination, 160–162, 295
of lipids, 121, 122, 125
of pectins, 165
of proteins to amino acids, 135, 141, 271, 272
Hydrolytic rancidity, 248, 250
Hydrometry, 96–98, 172. *See also* Moisture, physical methods
Hyphenated techniques, 195, 321, 464–467, 503, 506, 528

I

Immunoassays
applications
allergens, 313, 339–341
bacterial toxins, 313
drug residues, 313, 334, 335
extraneous matter, 361
fish species, 313
genetically modified organisms, 313, 314, 336–338
meat species identification, 313
mycotoxin residues, 331, 332
pathogens, 313
pesticide residues, 313, 325, 326
definitions, 303, 304
enzyme
competitive
bound antibody format, 309
bound hapten format, 309
standard curve, 309, 310
direct, 307–310
indirect, 307, 310, 311
noncompetitive, 307, 308
sandwich, 308, 336, 337
immunoaffinity purification, 313
lateral flow strip test, 312, 313, 337
radioimmunoassay (RIA), 305, 306, 331
theory, 305, 306
Western blot, 311, 336
Immunoaffinity purification, 313
Impact-acoustic emission, extraneous matter, 360
Inductively coupled plasma (ICP)
atomic emission spectroscopy (AES) (*see* Atomic emission spectroscopy, inductively coupled plasma)
mass spectroscopy, 423, 437–439
Infrared spectroscopy
applications
carbohydrates, 414
extraneous matter, 361
fat, 130, 412
mid-infrared, 100, 130, 138, 342, 411, 414
moisture, 99, 100, 409, 414, 417, 418
mycotoxins, 332
near-infrared, 99, 100, 342, 413–416
protein, 138, 412, 414, 417, 418
Raman, 412, 413
sugar, 171, 418
trans fatty acids, 256, 257, 413
wheat hardness, 417
near-infrared
calibration, 417
Infrared spectroscopy mid-infrared
applications
absorption bands, 411
presentation of spectra, 411
qualitative, 412
quantitative, 412
instrumentation
dispersive, 410
fourier transforms, 410, 411
sample handling, 411
Infrared spectroscopy near-infrared
absorption bands, 414, 415
applications, 417, 418
calibration, 417
diffuse reflectance measurements, 413, 414
instrumentation, 415, 416
principles, 413–415
qualitative analysis, 417
quantitative methods, 416, 417
Infrared spectroscopy principles
frequency of vibration, 409, 410
infrared radiation, 409
molecular vibrations, 409
Infrared spectroscopy Raman, 412, 413
Infrared spectroscopy sample handling, 411
Insoluble fiber. *See* Fiber, insoluble
Internal standards, 495, 496
International Organization for Standardization, 30
International standards and policies
Codex Alimentarius Commission, 29
International Organization for Standardization, 30
others, 30
Iodine value, oils, 246, 247
Ion chromatography, 344, 506, 509, 510
Ion selective electrodes
activity, 210
applications, 211, 212
calibration curves, 211
electrodes, 209–211
end point of titration, 211
principle, 209, 210
standard addition, 211
Ion-exchange chromatography
anion exchangers, 264, 484
applications, 265, 509, 510
cation exchangers, 264, 484
HPLC, 265, 483–485, 509, 510
principles, 264, 483, 484
Iron analysis, 208, 209
Isoelectric focusing, 269–271
Isoelectric point, 141, 263, 270
Isoelectric precipitation, 263, 264

J

Joint FAO/WHO Expert Committee on Food Additives (JECFA), 30, 330

K

Karl Fischer titration, 94–96, 101, 102. *See also* Moisture
Kjeldahl method, proteins, 136–139, 141, 143, 170, 275

L

Labeling. *See* Government, regulations
Laboratory information management system (LIMS), 7
Lactometers, 97
Lane-Eynon method, 154
Lateral flow strips, 312, 313, 337
Ligand, affinity chromatography, 488, 489
Light. *See* Spectroscopy
Lignin, 149, 166–168, 371
Limit dextrin, 297
Limit of detection, 59, 60, 335, 341, 436, 519
Limit of quantitation, 60
Linear regression, 60–62, 412, 417
Lineweaver-Burk
  formula, 297
  plot, 287, 288
Lipids
  analyses of lipid fractions
    cholesterol content, 257, 258
    gas chromatography, 125–127, 255–258
    thin-layer chromatography, 258
  antioxidants, 243, 252–254, 258
  characterization, methods
    acid value, 248
    cloud point, 246
    cold test, 246
    color, Lovibond, 246
    consistency, 249
    fat substitutes, 5, 254, 255, 481
    fatty acid composition, 125–127, 254–257
    fire point, 245
    flash point, 245
    free fatty acids, 248
    hexanal, 252
    iodine value, 243, 246, 247
    melting point, 243, 245
    polar components, 249, 250
    refractive index, 243–245
    sample preparation, 244, 251
    saponification number, 243, 247, 248
    smoke point, 245
    solid fat content, 248, 249
    solid fat index, 248, 249
    trans isomer fatty acids, 256, 257
  choice of methods, 212
  cholesterol, 257, 258
  classification
    compound, 241
    derived, 241
    monounsaturated, 241
    polyunsaturated, 241, 256
    saturated, 241
    simple, 241
  color, Lovibond, 246
  comparison of methods, 131
  consistency, 249
  content in foods, 120, 243
  definitions, 119, 120, 241–243
  dilatometry, 248
  extraction, 120–130, 244, 255
  fat substitutes, 5, 254, 255, 481
  fractionation, 125–127, 255–257
  gas chromatography analysis, 125–128, 255–257
  high-performance liquid chromatography analysis, 233
  importance of analysis, 120, 243
  in dairy products, 124, 125, 127, 130
  in flour, 121, 124
  in pet food, 124
  monounsaturated fatty acids, 125, 241, 256
  nuclear magnetic resonance analysis, 131, 453
  oxidation
    evaluating present status
      anisidine value, 251, 252
      fluorescence microscopy, 250
      hexanal, 252
      peroxide value, 251
      sample preparation, 251
      thiobarbituric acid test, 252
      totox value, 251, 252
    evaluating stability
      active oxygen method
        oil stability index, 254
        oxygen bomb, 254
      Schaal oven test, 253
    sample protection, 251
  polar components, 249, 250
  polyunsaturated fatty acids, 241, 254, 256, 259
  rancidity
    hydrolytic, 248, 250
    oxidative, 250–254
  saturated fat, 241
  solid fat content, 248, 249
  solid fat index, 248, 249
  solvent extraction, 120–127, 244, 255
  total content, methods
    instrumental methods
      Foss-Let, 130
      infrared, 130
      NMR, 131
      specific gravity, 130
    nonsolvent wet extraction
      Babcock, 127, 130
      Gerber, 130
    solvent extraction
      accelerated, 253
      chloroform-methanol, 124, 125
      gas chromatography, 125–128, 255–257
      Goldfish, 122
      Mojonnier, 123, 124
      sample preparation, 121, 125
      solid-phase, 244
      solvent selection, 122
      Soxhlet, 123
  trans isomer fatty acids, 256, 257
Lipoxygenase, assay, 296
Liquid-liquid chromatography, 482, 483. *See also* Column chromatography
Lovibond method, 246
Lowry method, proteins, 139–144, 389
Lysine, availability, 140, 141, 275

M

Magnetic resonance. *See* Nuclear magnetic resonance
Malic acid, assay, 295, 296
Mass analyzers, 329, 459–463
Mass spectrometry (MS)
  applications, 159, 160, 195, 327, 332, 335, 342–344
  gas chromatography-MS, 459, 460, 464, 465, 530

inductively coupled plasma, 423, 436–439, 460
instrumentation, 459–462, 467
interfaces, 459, 460, 463–468, 470
interpretation of mass spectra, 462, 463
ionization, 327, 342, 425, 430, 431, 459, 460, 462–467, 469, 470
liquid chromatography-MS, 195, 459, 465–467, 503, 506
mass analyzers, 460–462
sample introduction, 459
Matrix, food, 8, 9, 12, 160, 204, 320, 324, 337, 342, 453, 515, 550
Measures of central tendency. *See* Data evaluation
Melamine, 143, 342, 467–469, 507
Melting point of lipids, 245, 559
Mesh size, 78, 79, 354, 356, 522
Methods of analysis. *See* Assay methods, general; Standard methods
4-Methylbenzopheone, 343
Michaelis constant, 286–288
Michaelis-Menten equation, 286, 287, 289
Micro Kjeldahl, proteins, 137, 138
Microbiological assays
drug residues, 333, 334
vitamins, 185–187, 194
Microfiltration, 266, 267
Microscopy
carbohydrate analysis, 171
extraneous matter, 355–361
lipid oxidation analysis, 250, 251
Microwave-assisted methods
ashing, 110–112
moisture analysis, 92, 93, 102
Mid-infrared spectroscopy (Mid-IR), 100, 130, 138, 342, 409–413, 416, 418. *See also* Infrared spectroscopy
Minerals. *See also* Ash
content in foods, 39, 48, 107, 109, 203–206, 208, 209
determination
atomic absorption spectroscopy, 28, 110, 203, 204, 423, 425–430, 436, 437
atomic emission spectroscopy, 203, 204, 423, 430–437
atomic emission spectroscopy-inductively couple plasma, 203, 204, 423, 431–437
colorimetric methods, 208, 209
comparison of methods, 212, 436
dry ashing, 113, 436
EDTA complexometric titration, 205, 206
inductively coupled plasma-MS, 423, 437–439
interferences, 203, 205, 429, 430, 435
ion selective electrodes, 209–212
mass spectrometry-inductively coupled plasma, 423, 437–439
microwave ashing, 110–112
precipitation titration Mohr titration, 206, 207
precipitation titration Volhard titration, 206–208
sample preparation, 204, 205
water hardness, 205, 206
wet ashing, 107, 109, 436
importance in diet, 203
Mixograph, 277, 278
Modulated temperature DSC®, 557, 565–567, 569, 570
Moisture
basis for reporting results, 87
comparison of methods
intended purpose, 102
nature of sample, 101, 102
principles, 101
content of foods, 87
distillation methods
elimination of errors, 94
receiver tubes, 93, 94
solvents, 93, 95
types, 93
drying methods
calculations, 91
decomposition, 89
ovens
forced draft, 91
infrared, 93
microwave, 92, 93, 102
rapid moisture analyzer, 93
vacuum, 91
pans
handling, 90
types, 90
removal of moisture, 88, 89
surface crust formation, 90, 91
temperature control, 89, 90
forms in foods
adsorbed, 87
free, 87
water of hydration, 87
importance of assay, 87
Karl Fischer titration method
applications, 96
endpoints, 95, 96
reactions, 94, 95
reagents, 95, 96
sources of error, 96
physical methods
electrical, 96
freezing point, 100, 101
hydrometry
hydrometer
alcoholometer, 97
Baumé, 97
Brix, 97
lactometer, 97
Twaddell, 97
pycnometer, 97, 98
infrared, 99, 100
refractometry
refractive index, 98
refractometer, 98
sample collection, 87, 88
total solids, 23, 73, 87, 91, 92, 100
Mojonnier method, fat analysis, 123, 124
Molarity, 221, 222
Molecular energy levels, 380, 381, 383, 384
Molecular exclusion. *See* Size-exclusion chromatography
3-Monochloropropane 1,2-diol, 344
Monochromator, 394–396, 398–400, 410, 412, 415, 416, 425–429, 431, 438, 505
Monosaccharides, 140, 149, 150, 155–157, 159, 163, 164, 173, 285, 559. *See also* Carbohydrates, methods of analysis
Multidimensional GC. *See* Gas chromatography
Munson-Walker method, 154
Mycotoxin residues
methods of analysis
capillary electrophoresis, 320, 332
gas chromatography, 320, 332
high performance liquid chromatography, 320, 332

immunoassays, 320, 331, 332
thin-layer chromatography, 320, 331
occurrence, 329
sampling, 330

N
Nanofiltration, 266, 267
National Institute of Standards and Technology (NIST), 28
Near-infrared spectroscopy. *See also* Infrared spectroscopy
carbohydrate, 171, 414
extraneous matter, 361
fat, 130, 418
moisture, 99, 100, 409, 414, 417, 418
mycotoxins, 332
protein, 138, 414, 415, 417, 418
Nernst equation, 210, 211, 225, 226, 234
Nessler colorimetric method, proteins, 136, 137, 140
Neutralization reactions, 221–223
Newtonian fluids, 545–547
Niacin, 37, 181, 184–186, 194
Ninhydrin method, proteins, 272, 362
Non-Newtonian fluids, 545–547
Nonprotein-N compounds
amino acids, 135, 141, 144
separation from protein, 144
Normal-phase chromatography, 155, 342, 482, 507, 508
Normality, 221, 222, 228, 230, 232
Nuclear magnetic resonance (NMR)
applications
fats, 131, 249, 453, 454
ingredients, 454
liquids, 450–452
magnetic resonance imaging, 452, 453
relaxometry, 453
solids, 452
water, 454
instrument, 449, 450
principles, 445–449
Nutrient content claims, 37, 38, 40, 43–48, 50, 131
Nutrient content databases, 37, 38
Nutrition labeling. *See* Government, regulations, labeling
Nutritional value of proteins. *See* Protein, nutritional quality

O
Oil stability index, 254
Oils. *See* Lipids
Oligosaccharides. *See* Carbohydrates
On-line analyses, 99, 130, 418
Organic acids, 222, 227–229, 232, 233
Ovens. *See* Moisture, drying methods
Oxidation, lipid, 80, 91, 121, 125, 241, 243, 244, 247, 250–254
Oxygen bomb, 254
Oxygen demand
biochemical oxygen demand (BOD), 24, 369–371
chemical oxygen demand (COD), 369–371
comparison of methods, 370, 371
sampling and handling, 371

P
Packaging material volatiles
residues, 343, 533
volatiles, 533
Paper chromatography, 149, 477, 478
Partition chromatography. *See* Chromatography, principles
Pathogens, 313
Pectin. *See* Carbohydrates
Perchlorate, 110, 319, 344
Peroxidase, assay, 296
Peroxide value, 28, 242, 250–252, 254, 258, 259, 418
Pesticide residues
methods of analysis
chromatographic separation and analysis
gas chromatography, 326, 327
high-performance liquid chromatography, 326, 327
thin-layer chromatography, 326
extraction, 324, 325
immunoassays, 320, 325, 326
mass spectrometry detection, 327
multiple-residue methods, 324
screening methods, 326
single-residue methods, 324, 326
regulations
other raw agricultural commodities, 20, 324
shellfish, 27, 28
tolerance levels, 23, 24, 324
safety, 324, 329
types, 324, 325
pH
acid-base equilibria, 223–224
activity coefficient, 224
meter
calibration, 227
electrodes, 226–227
operation, 227
principles, 224–226
Phenol-sulfuric acid method, 153, 389
Photomultiplier tube (PMT), 396, 397, 426, 428, 431, 506
Phototube, 396, 397
Phthalaldehyde method, proteins, 272, 506
Physical methods
color, 575–586
rheology, 543–554
thermal analysis, 557–570
Pigments
carotenoids, 181, 509
chlorophylls, 481, 509
synthetic food dyes, 509
Planck's constant, 379, 409, 410, 424
Polarimetry, 293
Polyacrylamide gel electrophoresis, 267–269, 271, 311
Polymerase chain reaction (PCR), 336–339, 341
Polysaccharides, 149, 150, 152–155, 157–168, 170–173, 451, 484, 486–489, 497, 504, 511, 512. *See also* Carbohydrates, methods of analysis
Polyunsaturated fatty acids, 241, 254, 256, 259
Potentiometer, 95, 224, 227
Potentiometry, 224, 225, 229–231, 234. *See also* Ion selective electrodes; pH
Precipitation titration, mineral determination
Mohr, 206–208
Volhard, 206–208
Precision, analyses, 8, 55–59, 76
Preparation of samples. *See* Sample, preparation
Protein. *See* Proteins
Protein digestibility-corrected amino acid score (PDCAAS), 42, 273, 274
Protein efficiency ratio (PER), 42, 144, 273–275
Protein needs, 50, 135, 272, 273
Protein, functional properties
dough formation
Farinograph®, 277, 278
mixograph, 277, 278
RapidViscoAnalyser (RVA), 277, 278

emulsification
applications, 277
emulsion, 275
emulsion stability, 276
foaming
applications, 277
foam stability, 276, 277
foam volume, 276, 277
gelation, 277–278
solubility
applications, 277
Protein, nutritional quality
amino acid composition, 271–274
amino acid score, 274
digestibility
corrected amino acid score, 273
pH-shift, 274
true, 273
essential amino acid index (EAAI), 274
estimates of protein needs, 273, 274
lysine, availability, 275
protein digestibility-corrected amino acid score, 273, 274
protein efficiency ratio, 42, 144, 273–275
regulations, 42
Protein, separation
adsorption, chromatography
affinity, 265
ion-exchange, 264, 265
differential solubility characteristics
denaturation, 264
isoelectric precipitation, 263, 264
salting out, 263
solvent fractionation, 264
electrophoresis
capillary electrophoresis, 270, 271
isoelectric focusing, 269–270
polyacrylamide gel electrophoresis, 267–269
high-performance liquid chromatography (HPLC), 265
reversed-phase HPLC, 271, 272
size
applications, 267
dialysis, 265, 266
membrane processes
microfilitration, 266
nanofiltration, 266
reverse osmosis, 266
ultrafiltration, 266
size-exclusion chromatography, 266, 267
Proteins
amino acid analysis, 271, 272
amino acids in, 278, 279
classification, 135
content in foods, 135
importance of analysis, 135
methods of analysis
bicinchoninic acid, 141–143
biuret, 139, 143
Bradford, 141, 143
comparison of, 143
Dumas, 138, 143
dye binding
anionic dye, 140, 141
Bradford method, 141, 143
infrared, 135, 138, 143, 144
Kjeldahl, 136, 139, 141, 143
Lowry, 139–144
Nessler colorimetric, 136
ninhydrin, 272
nitrogen combustion, 138, 143
nonprotein nitrogen, 135, 141, 144
o-phthaldehyde, 272
phenol reagent, 137
selection of, 135
ultraviolet 280 nm absorption, 142, 143
quality (*see* Protein, nutritional quality)
separation (*see* Protein, separation)
Proximate analysis, 8, 151
Pycnometers, 97, 98

Q
Q-value, 65
Quality assurance, 6, 7, 72, 149, 212, 250, 258, 451, 455, 557

R
Radioimmunoassay (RIA), 305, 306, 331
Raman spectroscopy, 412, 413
Rancidity
hydrolytic, 248, 250
oxidative, 250, 254
Rapid visco analyser (RVA), 277, 278
Redox reactions, 203
Reducing sugars, 140, 142, 149, 152–155, 172, 275, 295
Reference samples, 10, 389, 398
Refractive index
carbohydrate analysis, 156, 157
fat characterization, 244
total solids analysis, 98, 99
Refractometers
Abbe, 99
refractive index, 98, 99, 156, 157, 244
Relative error, 58, 59, 394
Relaxometry. *See* Nuclear magnetic resonance
Reliability of analysis. *See* Data evaluation
Rennet, assay, 297
Replication of analyses, 55, 65
Reporting results. *See* Data evaluation
Representative samples, 75, 76, 80, 204
Residues. *See* Contaminants
Resistant starch (RS), 160–162, 166–169, 171
Reverse osmosis (RO), 266, 267
Reversed-phase chromatography, 156, 482, 483, 504, 508
Rheology
Brookfield viscometer, 549
empirical methods, 543
fluid models
Bingham plastic, 547
Herschel-Bulkley, 547
Newtonian, 547
Power law, 547
fundamental methods, 543
glossary, 552–553
Newtonian fluids, 545–547
Newtonian viscosity, 545–547
nomenclature, 553
non-Newtonian fluids, 545–547
principles
elastic modulus, 545
fluid rheograms, 546
fluid viscosity, 545
shear modulus, 545, 553
strain, 544–545
stress, 545–551

rheometers
rotational viscometers, 548–550
concentric cylinder, 548, 549
Brookfield, 549
cone and plate, 549, 550
procedure, 550
solids
large strain, 550–552
small strain, 552
shear, 544, 545
strain, 544, 545
stress, 543, 544
texture profile analysis, 552
universal testing machine, 552
viscoelasticity, 550–552
viscometry, 548, 550, 551
rotational, 548
concentric cylinders, 548, 549
cone and plate, 549, 550
procedures, 550
viscosity, 545, 546
yield stress, 546, 553
Rheometers. *See* Rheology
Riboflavin, 37, 184, 193, 194
Roese-Gottlieb method, fat determination, 123, 124, 131
Rotary evaporator, 125, 153, 243, 257
Ro-Tap®, 79
Rounding rules, nutrition labeling, 40, 41

S

Salt analysis, 206–208
Salting out proteins, 263
Sampler types, 74, 75
Samples. *See also* individual food constituents
preparation
contamination, 80, 108, 204, 516, 519
derivatization, 125, 126, 155, 157, 160, 255, 323, 519
enzymatic inactivation, 79
extraction, 120–125, 322, 323, 337
grinding, 78, 79
homogenization, 322
lipid oxidation protection, 80, 125
microbial growth, 80
particle size, 78, 79, 121
size reduction, 77, 78, 80, 121
types, 7
Sampling
mycotoxins, 330
nutrition labeling, 42
plans, 71–76, 80, 330
problems, 77, 330
procedures, 74–77
selection of procedure, 71–74
factors affecting choice, 72
risks, 74
size, 76, 77
statistical considerations, 75–77
Saponification number, oils, 242, 247–250, 255
Scanning electron microscopy (SEM), 171, 359, 360
Schaal oven test, oils, 253
Screening tests in food analysis, 320, 333–335
Selection, assay methods, 7, 8
Sensitivity, analyses, 60
Shear. *See* Rheology
Significant figures, 64, 65
Size-exclusion chromatography
applications, 267, 485
fractionation range, 487
hydrophilic gels, 488
hydrophobia media, 488
molecular weight estimation, 487
Slip melting point, 245
Solid fat content, 248, 249, 454
Solid fat index, 248, 249, 559
Solid-phase extraction (SPE), 244, 323, 330, 333, 344, 475, 515–519
Solids, 23, 73, 87, 91, 92, 100, 172
Solubility criteria for lipids, 119
Soluble fiber. *See* Fiber, soluble
Solvent extraction
GC analysis, 121–125, 322, 517
lipids, 120–127
pesticides, 323
Solvent fractionation, proteins, 264
Somogyi-Nelson method, 153–154
Soxhlet method, fat analysis, 123, 125, 126
Specific gravity, 97, 101, 130, 131, 172, 356
Specifications
commercial item descriptions (CIDs), 28
commodity specifications, 28
Department of Defense, 28
federal specifications, 28
purchase product description (PPD), 28
USDA specifications, 28
Specificity, analyses, 59
Spectroscopy
absorption, basis of quantitation, 389–391
absorbance, 382, 389, 390, 392
Beer's law, 390–393, 402, 412, 425
molar absorptivity, 390, 392, 399, 400, 425
reference cell, 390, 392, 398, 399
transmittance, 171, 389–394, 396, 397, 411
calibration curves, 393–394
characteristics of UV-Vis absorbing species, 399, 400
energy level transitions
absorption, 382, 383
emission, 383, 384
resonance line, 424
resonance transition, 424
energy states of matter
levels
electronic, 380, 381
nuclear, 382
rotational, 380, 381
transitions
absorption of radiation, 382, 383
emission of radiation, 383, 384
vibrational, 380, 381
quantum nature, 379, 380
errors, 394
instrument design
double-beam, 398
single-beam, 398
instrumentation, UV-visible
cells, 392
detectors, 396, 397
light sources, 394, 395
monochromators, 395, 396
light
interferences, 378, 379
properties, 377, 378
terminology, 378
sample handling and preparation, 391, 392, 411

types
atomic absorption, 28, 110, 203, 204, 423, 425–430
atomic emission, 203, 204, 423, 430–437
fluorescence, 400–402
infrared
mid-, 100, 130, 138, 342, 411, 414
near-, 99, 100, 138, 361, 409, 413–418
mass, 151, 159–161, 171, 195, 255, 256, 292, 321, 327, 328, 342, 344, 421–441, 457–470, 481, 495, 503, 505, 506, 511, 528, 530, 557
nuclear magnetic resonance, 131, 249, 445, 454
Raman, 412, 413
ultraviolet, 389, 400
visible, 389, 400
UV-visible, 389, 400
wavelength selection, 392
Spices, moisture analysis, 93
Stability test of lipids, 253, 254
Standard acid, 231
Standard addition, 211, 294, 343
Standard curves. *See also* Data evaluation
atomic absorption spectroscopy, 429
colorimetric method, iron, 208, 209
fluorescence spectroscopy, 401, 402
ion selective electrodes, 210
UV-Vis spectroscopy, 393
Standard deviation, 56, 57, 60, 63–65, 76, 77, 90, 394, 492
Standard error of the mean, 58
Standard methods
AACC International, 9, 11, 12, 166, 169, 171, 195, 275, 330, 354, 355, 417
American Oil Chemists' Society (AOCS), 10–12, 28, 125, 242–247, 251–253, 255–257, 275, 330
American Society for Testing Materials (ASTM), 24
AOAC International, 8–11, 19, 20, 26, 42, 74, 77, 87, 88, 92, 93, 109, 121, 125, 149, 152, 163, 169, 170, 172, 182, 186, 187, 195, 203, 241, 242, 330–332, 353–355, 363, 441
extraneous matter, 354, 355
food additives, 12
Food Chemicals Codex (FCC), 12, 30
International Union of Pure and Applied Chemistry (IUPAC), 72, 242, 330
pesticides, 324–327
Standard Methods for the Examination of Dairy Products (SMEDP), 12, 26
Standard Methods for the Examination of Water and Wastewater, 24, 206, 369
Standards. *See* Government, standards
Starch. *See* Carbohydrates
Statistics
data analysis, 55, 63, 65
sampling, 72, 74–77, 80
Sterioisomer separation, 533
Strain. *See* Rheology
Stress. *See* Rheology
Structure/function claims, 48
Sulfite, 294, 295, 341, 342, 510
Supercritical fluid chromatography (SFC), 476, 480–481, 515
Synthetic food dyes, 508, 509

T

t-test, 77
t-value, 57, 58
Texture. *See* Rheology
Texture profile analysis (TPA), 552
Thermal analysis
amorphous structure, 557, 558, 570
applications, 567–569
calorimetry
calorimeter, 557
reactions
endothermic, 565
exothermic, 565
crystalline structure, 558–560, 564, 569
differential scanning calorimetry (DSC)
applications, 567–569
experimental conditions, 564
measurements, 564, 565
dynamic mechanical analysis (DMA), 557
enthalpy, 564
glass transition temperature, 454, 557, 558, 564, 569, 570
heat capacity, 560
modulated temperature DSC
applications, 569
experimental conditions, 566
measurements, 566
thermodynamic properties, 560
thermogravimetric analysis (TGA), 557, 561–563
thermomechanical analysis (TMA), 558
Thermogravimetric analysis. *See* Thermal analysis
Thiamin, 20, 37, 39, 183, 184, 191–193, 389
Thin-layer chromatography (TLC)
applications
lipids, 258, 478, 479
mycotoxin residues, 331, 479
pesticide residues, 326
high performance (HPTLC), 478
principles, 478
procedure, 478
Thiobarbituric acid (TBA) test, 242, 252, 389
Titratable acidity
analysis
high-performance liquid chromatography (HPLC), 233
titration
endpoint, 230
indicators, 229–230
reagents, 230–231
sample preparation and analysis, 231, 232
Brix/acid ratio, 233
calculations, 232
content in food, 232, 233
interference, 230
predominant acid, 232
principle, 227, 228
Total carbohydrate. *See* Carbohydrate, methods of analysis
Total dietary fiber. *See* Fiber, dietary
Total fiber. *See* Fiber, total
Total solids (TS), 23, 73, 87–91, 172
Totox value, 251
Trace metals determination. *See* Minerals, determination
Trans isomer fatty acids, 242, 243, 256, 257
Trends
consumer, 5
food industry, 5–6
government regulation, 6
international standards, 6
Turbidimetry, vitamins, 185

U

U.S.Department of Agriculture (USDA), 10, 18, 21–23, 25–30, 37, 40, 42, 43, 49, 71, 73, 155, 319, 325, 330, 423, 577
Ultrafiltration, 144, 266, 267, 278
Ultraviolet 280 nm absorption, proteins, 142, 143
Ultraviolet absorption spectroscopy, 389–400. *See also* Spectroscopy, types
Unsaturated fatty acids, 89, 243, 247, 453
Uronic acid, 157, 158, 165, 167

V

Validity of method, 7–10
Van Deemter equation, 493, 530, 531
Viscometers. *See* Rheology, rheometers
Viscometry. *See* Rheology
Viscosity. *See* Rheology
Visible absorption spectroscopy, 389–400. *See also* Spectroscopy, types
Vitamins
  bioassay methods, 184, 185
    vitamin D line test, 184–186
  chemical methods
    riboflavin, 193, 194
    thiamin, 191–193
    vitamin A, 188, 189
    vitamin C
      2,6-dichloroindophenol, 190, 191
      fluorometric, 191, 192
    vitamin D, 184–186
    vitamin E, 189, 190
  comparison of methods, 194, 195
  definition, 181
  extraction methods, 184
  high-performance liquid chromatography (HPLC) analysis, 183, 188, 189, 194, 195
  importance
    in diet, 181, 182
    of analysis, 181
  microbiological assays
    folate, 185, 187
    niacin, 185, 186
    principles, 185, 187
  units, 181
Volatile acidity, 23, 233

W

Water activity (a_w), 87, 101
Water hardness, 205, 206, 212
Water, drinking, regulations, 24
Western blots, 311, 336
Wet ashing, 109–111, 113, 436. *See also* Ash, methods of determination

X

X-ray microtomography (XMT), extraneous matter, 359, 360
X-ray radiography, extraneous matter, 359

Z

Z value, 57